FINITE ELEMENT ANALYSIS

Theory and Application with ANSYS

Third Edition

Saeed Moaveni
Minnesota State University, Mankato

PEARSON
Prentice Hall
Pearson Education International

Vice President and Editorial Director, ECS: *Marcia J. Horton*
Acquisitions Editor: *Laura Fischer*
Editorial Assistant: *Erin Katchmar*
Vice President and Director of Production and Manufacturing, ESM: *David W. Riccardi*
Executive Managing Editor: *Vince O'Brien*
Managing Editor: *David A. George*
Production Editor: *Winifred S. Sanchez*
Director of Creative Services: *Paul Belfanti*
Creative Director: *Carole Anson*
Art and Cover Director: *Jayne Conte*
Art Editor: *Greg Dulles*
Cover Designer: *Saeed Moaveni*
Manufacturing Manager: *Trudy Pisciotti*
Manufacturing Buyer: *Lynda Castillo*
Marketing Manager: *Holly Stark*

Pearson Prentice Hall
Pearson Education, Inc.
Upper Saddle River, NJ 07458

Printed in the United States of America

10 9 8 7 6 5 4 3 2
ISBN 13: 978-0-13-241651-1
ISBN 10: 0-13-241651-4

Pearson Education LTD., London
Pearson Education Australia PTY, Limited
Pearson Education Singapore, Pte. Ltd
Pearson Education North Asia Ltd
Pearson Education Canada, Inc.
Pearson Educación de Mexico, S.A. de C.V.
Pearson Education -- Japan
Pearson Education Malaysia, Pte. Ltd
Pearson Education, Upper Saddle River, New Jersey

To my mother and father

Contents

Preface

CHANGES IN THE THIRD EDITION

The third edition, consisting of 15 chapters, includes a number of new additions and changes that were incorporated in response to suggestions and requests made by professors, students, and professionals using the second edition of the book. The major changes include:

- A new section on using Excel to manipulate matrices
- A new comprehensive appendix on MATLAB
- More detail on theory and expanded derivations
- Reorganization of Chapter 15 to include examples of batch files
- Additional problems

ORGANIZATION

There are many good textbooks already in existence that cover the theory of finite element methods for advanced students. However, none of these books incorporate ANSYS as an integral part of their materials to introduce finite element modeling to undergraduate students and newcomers. In recent years, the use of finite element analysis as a design tool has grown rapidly. Easy-to-use, comprehensive packages such as ANSYS, a general-purpose finite element computer program, have become common tools in the hands of design engineers. Unfortunately, many engineers who lack the proper training or understanding of the underlying concepts have been using these tools. This introductory book is written to assist engineering students and practicing engineers new to the field of finite element modeling to gain a clear understanding of the basic concepts. The text offers insight into the theoretical aspects of finite element analysis and also covers some practical aspects of modeling. Great care has been exercised to avoid overwhelming students with theory, yet enough theoretical background is offered to allow individuals to use ANSYS intelligently and effectively. ANSYS is an integral part of this text. In each chapter, the relevant basic theory is discussed first and demonstrated using simple problems with hand calculations. These problems are followed by examples that are solved using ANSYS. Exercises in the text are also presented in this manner. Some exercises require manual calculations, while others, more complex in nature, require the use of ANSYS. The simpler hand-calculation problems will enhance students' understanding of the concepts by encouraging them to go through the necessary steps in a finite element analysis. Design problems are also included at the end of Chapters 3, 4, 6, and 9 through 14.

Various sources of error that can contribute to incorrect results are discussed. A good engineer must always find ways to check the results. While experimental testing of models may be the best way, such testing may be expensive or time consuming. Therefore, whenever possible, throughout this text emphasis is placed on doing a "sanity check" to verify one's finite element analysis (FEA). A section at the end of each appropriate chapter is devoted to possible approaches for verifying ANSYS results.

Another unique feature of this book is that the last two chapters are devoted to the introduction of design, material selection, optimization, and parametric programming with ANSYS.

The book is organized into 15 chapters. Chapter 1 reviews basic ideas in finite element analysis. Common formulations, such as direct, potential energy, and weighted residual methods are discussed. Chapter 2 provides a comprehensive review of matrix algebra. Chapter 3 deals with the analysis of trusses, because trusses offer economical solutions to many engineering structural problems. An overview of the ANSYS program is given in Chapter 3 so that students can begin to use ANSYS right away. Finite element formulation of members under axial loading, beams, and frames are introduced in Chapter 4. Chapter 5 lays the foundation for analysis of one-dimensional problems by introducing one-dimensional linear, quadratic, and cubic elements. Global, local, and natural coordinate systems are also discussed in detail in Chapter 5. An introduction to isoparametric formulation and numerical integration by Gauss–Legendre formulae are also presented in Chapter 5. Chapter 6 considers Galerkin formulation of one-dimensional heat transfer and fluid problems. Two-dimensional linear and higher order elements are introduced in Chapter 7. Gauss–Legendre formulae for two-dimensional integrals are also presented in Chapter 7. In Chapter 8 the essential capabilities and the organization of the ANSYS program are covered. The basic steps in creating and analyzing a model with ANSYS is discussed in detail. Chapter 9 includes the analysis of two-dimensional heat transfer problems with a section devoted to unsteady situations. Chapter 10 provides analysis of torsion of noncircular shafts and plane stress problems. Dynamic problems are explored in Chapter 11. Review of dynamics and vibrations of mechanical and structural systems are also given in this chapter. In Chapter 12, two-dimensional, ideal fluid-mechanics problems are analyzed. Direct formulation of the piping network problems and underground seepage flow are also discussed. Chapter 13 provides a discussion of three-dimensional elements and formulations. This chapter also presents basic ideas regarding top-down and bottom-up solid modeling methods. The last two chapters of the book are devoted to design and optimization ideas. Design process and material selection are explained in Chapter 14. Design optimization ideas and parametric programming are discussed in Chapter 15. Examples of ANSYS batch files are also given in Chapter 15. Each chapter begins by stating the objectives and concludes by summarizing what the reader should have gained from studying that chapter.

The examples that are solved using ANSYS show in great detail how to use ANSYS to model and analyze a variety of engineering problems. Chapter 8 is also written such that it can be taught right away if the instructor sees the need to start with ANSYS.

A brief review of appropriate fundamental principles in solid mechanics, heat transfer, dynamics, and fluid mechanics is also provided throughout the book. Additionally, when appropriate, students are warned about becoming too quick to generate finite element models for problems for which there exist simple analytical solutions. Mechanical and thermophysical properties of some common materials used in engineering are given in appendices A and B. Appendices C and D give properties of common area shapes and properties of structural steel shapes, respectively. A comprehensive introduction to MATLAB is given in Appendix F.

Finally, a Web site at *http://www.prenhall.com/Moaveni* will be maintained for the following purposes: (1) to share any changes in the upcoming versions of ANSYS; (2) to share additional information on upcoming text revisions; (3) to provide additional homework problems and design problems; and (4) although I have done my best to eliminate errors and mistakes, as is with most books, some errors may still exist. I will post the corrections that are brought to my attention at the site. The Web site will be accessible to all instructors and students.

Thank you for considering this book and I hope you enjoy the third edition.

SAEED MOAVENI

Acknowledgments

I would like to express my sincere gratitude to Mr. Raymond Browell of ANSYS, Inc. for providing the photographs for the cover of this book. Descriptions for the cover photoghraphs are given on page 7. I would also like to thank ANSYS, Inc. for giving me permission to adapt material from various ANSYS documents, related to capabilities and the organization of ANSYS. The essential capabilities and organizations of ANSYS are covered in Chapters 3, 8, 13, and 15.

As I have mentioned in the Preface, there are many good published books in finite element analysis. When writing this book, several of these books were consulted. They are cited at the end of each appropriate chapter. The reader can benefit from referring to these books and articles.

I am thankful to all reviewers who offered general and specific comments. I am also grateful to Ms Tacy Quinn of Prentice Hall and Ms Winifred S. Sanchez of ICC Macmillan Inc. for their assistance in the preparation of this book.

CHAPTER 1

Introduction

The finite element method is a numerical procedure that can be used to obtain solutions to a large class of engineering problems involving stress analysis, heat transfer, electromagnetism, and fluid flow. This book was written to help you gain a clear understanding of the fundamental concepts of finite element modeling. Having a clear understanding of the basic concepts will enable you to use a general-purpose finite element software, such as ANSYS, effectively. ANSYS is an integral part of this text. In each chapter, the relevant basic theory behind each respective concept is discussed first. This discussion is followed by examples that are solved using ANSYS. Throughout this text, emphasis is placed on methods by which you may verify your findings from finite element analysis (FEA). In addition, at the end of particular chapters, a section is devoted to the approaches you should consider to verify results generated by using ANSYS.

Some of the exercises provided in this text require manual calculations. The purpose of these exercises is to enhance your understanding of the concepts by encouraging you to go through the necessary steps of FEA. This book can also serve as a reference text for readers who may already be design engineers who are beginning to get involved in finite element modeling and need to know the underlying concepts of FEA.

The objective of this chapter is to introduce you to basic concepts in finite element formulation, including direct formulation, the minimum potential energy theorem, and the weighted residual methods. The main topics of Chapter 1 include the following:

1.1 Engineering Problems
1.2 Numerical Methods
1.3 A Brief History of the Finite Element Method and ANSYS
1.4 Basic Steps in the Finite Element Method
1.5 Direct Formulation
1.6 Minimum Total Potential Energy Formulation
1.7 Weighted Residual Formulations
1.8 Verification of Results
1.9 Understanding the Problem

1.1 ENGINEERING PROBLEMS

In general, engineering problems are mathematical models of physical situations. Mathematical models of many engineering problems are differential equations with a set of corresponding boundary and/or initial conditions. The differential equations are derived by applying the fundamental laws and principles of nature to a system or a control volume. These governing equations represent balance of mass, force, or energy. When possible, the exact solution of these equations renders detailed behavior of a system under a given set of conditions, as shown by some examples in Table 1.1. The analytical solutions are composed of two parts: (1) a homogenous part and (2) a particular part. In any given engineering problem, there are two sets of design parameters that influence the way in which a system behaves. First, there are those parameters that

TABLE 1.1 Examples of governing differential equations, boundary conditions, initial conditions, and exact solutions for some engineering problems

Problem Type	Governing Equation, Boundary Conditions, or Initial Conditions	Solution
A beam: 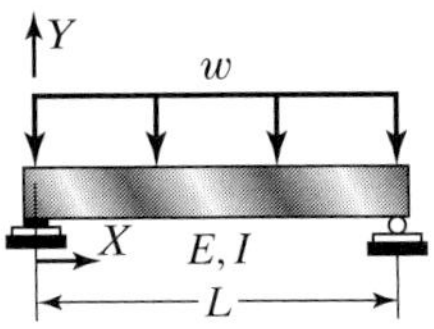	$EI\frac{d^2Y}{dX^2} = \frac{wX(L - X)}{2}$ Boundary conditions: at $X = 0, Y = 0$ and at $X = L, Y = 0$	Deflection of the beam Y as the function of distance X: $Y = \frac{w}{24EI}(-X^4 + 2LX^3 - L^3X)$
An elastic system: 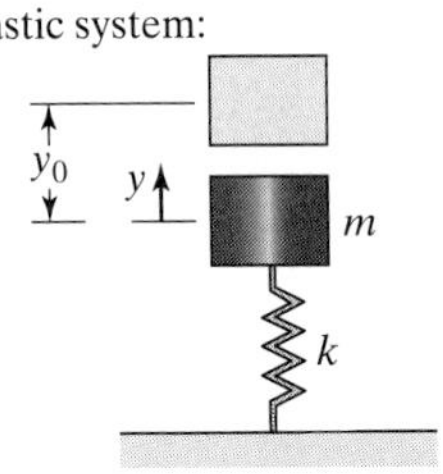	$\frac{d^2y}{dt^2} + \omega_n^2 y = 0$ where $\omega_n^2 = \frac{k}{m}$ Initial conditions: at time $t = 0, y = y_0$ and at time $t = 0, \frac{dy}{dt} = 0$	The position of the mass y as the function of time: $y(t) = y_0 \cos \omega_n t$
A fin: 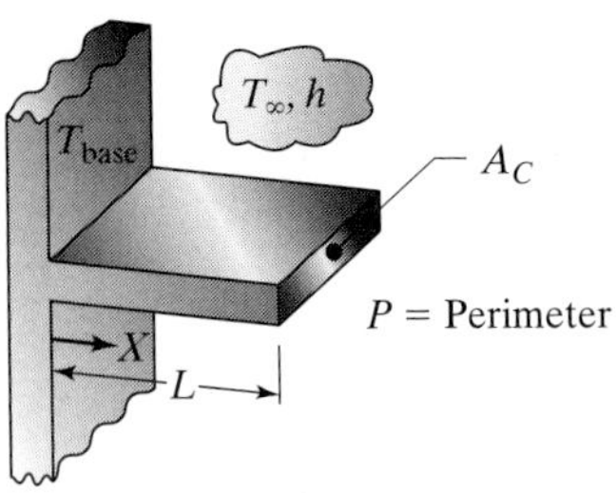	$\frac{d^2T}{dX^2} - \frac{hp}{kA_c}(T - T_\infty) = 0$ Boundary conditions: at $X = 0, T = T_{base}$ as $L \rightarrow \infty, T = T_\infty$	Temperature distribution along the fin as the function of X: $T = T_\infty + (T_{base} - T_\infty)e^{-\sqrt{\frac{hp}{kA_c}}X}$

provide information regarding the *natural behavior* of a given system. These parameters include material and geometric properties such as modulus of elasticity, thermal conductivity, viscosity, and area, and second moment of area. Table 1.2 summarizes the physical properties that define the natural characteristics of various problems.

TABLE 1.2 Physical properties characterizing various engineering systems

Problem Type	Examples of Parameters That Characterize a System
Solid Mechanics Examples	
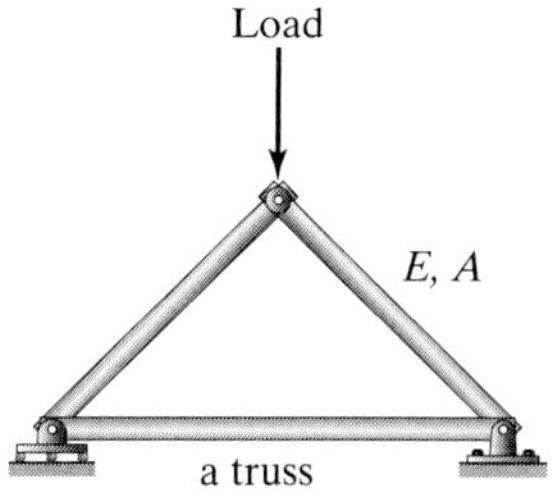 a truss	modulus of elasticity, E; cross-sectional area, A
 an elastic plate	modulus of elasticity, E
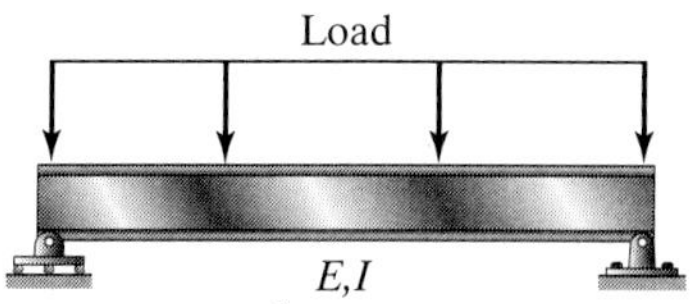 a beam	modulus of elasticity, E; second moment of area, I
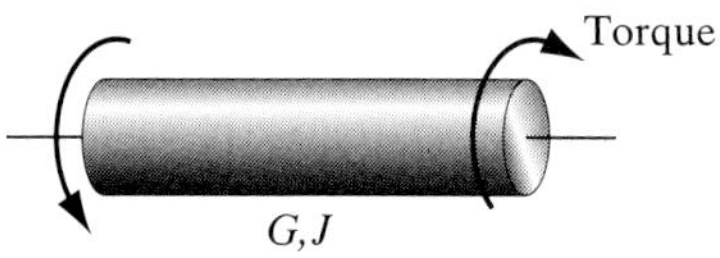 a shaft	modulus of rigidity, G; polar moment of inertia of the area, J
Heat Transfer Examples	
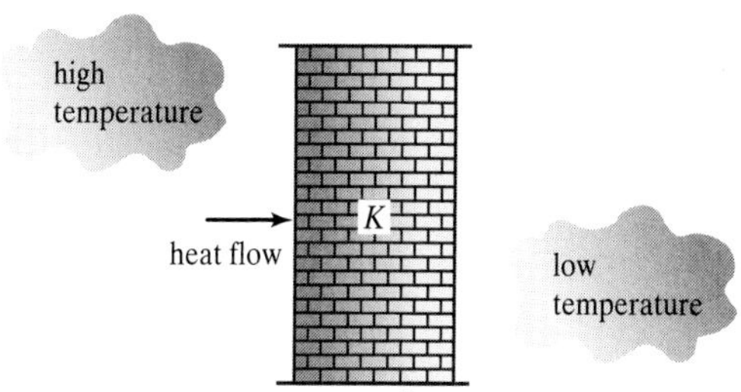 a wall	thermal conductivity, K; thickness, L; Area, A

continued

TABLE 1.2 *Continued*

Problem Type	Examples of Parameters That Characterize a System
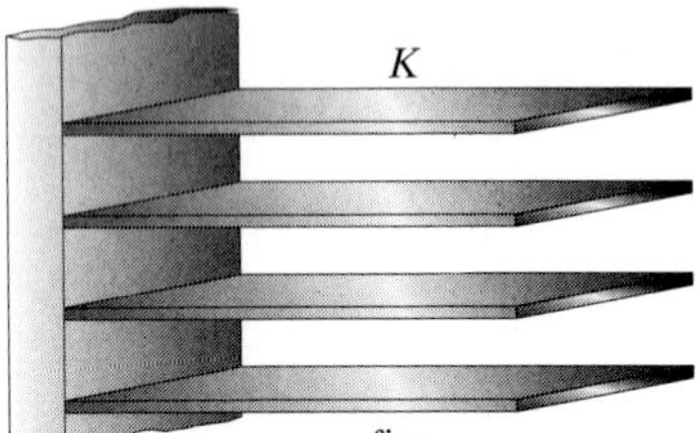 fins	thermal conductivity, K; Cross-Sectional Area, A
Fluid Flow Examples	
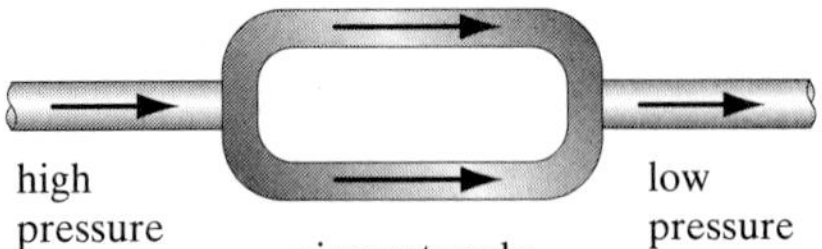 pipe networks	viscosity, μ; pipe roughness, e; pipe diameter, D
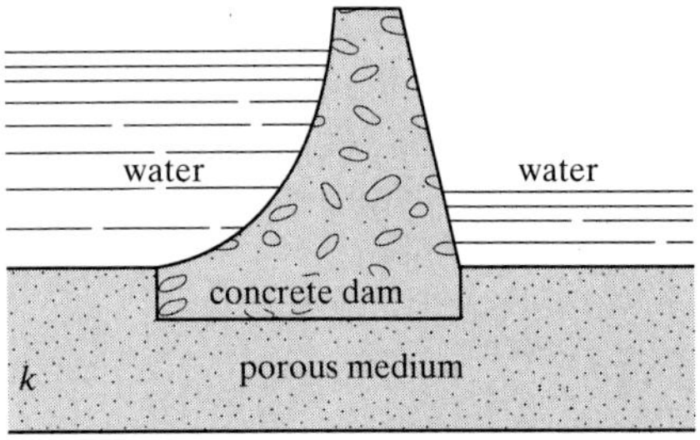 a concrete dam	soil permeability, k
Electrical and Magnetism Problems	
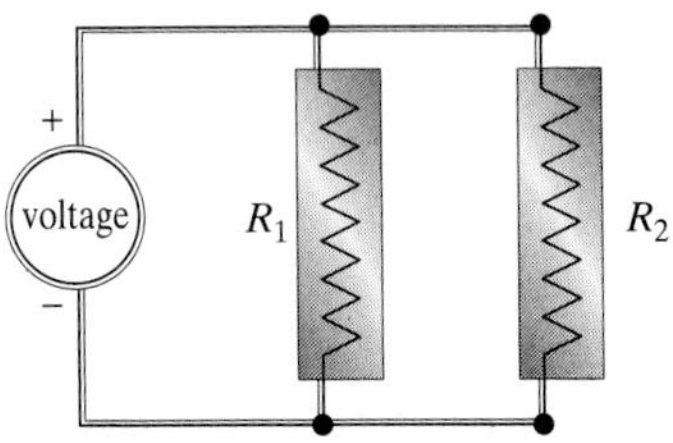 electrical network	resistance, R
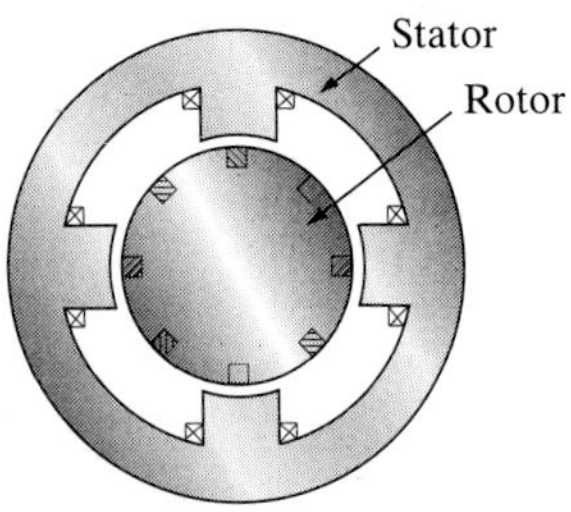 magnetic field of an electric motor	permeability, μ

TABLE 1.3 Parameters causing disturbances in various engineering systems

Problem Type	Examples of Parameters that Produce Disturbances in a System
Solid Mechanics	external forces and moments; support excitation
Heat Transfer	temperature difference; heat input
Fluid Flow and Pipe Networks	pressure difference; rate of flow
Electrical Network	voltage difference

On the other hand, there are parameters that produce *disturbances* in a system. These types of parameters are summarized in Table 1.3. Examples of these parameters include external forces, moments, temperature difference across a medium, and pressure difference in fluid flow.

The system characteristics as shown in Table 1.2 dictate the natural behavior of a system, and they always appear in the *homogenous part of the solution* of a governing differential equation. In contrast, the parameters that cause the disturbances appear in the *particular solution*. It is important to understand the role of these parameters in finite element modeling in terms of their respective appearances in stiffness or conductance matrices and load or forcing matrices. The system characteristics will always show up in the stiffness matrix, conductance matrix, or resistance matrix, whereas the disturbance parameters will always appear in the load matrix. We will explain the concepts of stiffness, conductance, and load matrices in Section 1.5.

1.2 NUMERICAL METHODS

There are many practical engineering problems for which we cannot obtain exact solutions. This inability to obtain an exact solution may be attributed to either the complex nature of governing differential equations or the difficulties that arise from dealing with the boundary and initial conditions. To deal with such problems, we resort to numerical approximations. In contrast to analytical solutions, which show the exact behavior of a system at any point within the system, numerical solutions approximate exact solutions only at discrete points, called nodes. The first step of any numerical procedure is discretization. This process divides the medium of interest into a number of small subregions (elements) and nodes. There are two common classes of numerical methods: (1) *finite difference methods* and (2) *finite element methods*. With finite difference methods, the differential equation is written for each node, and the derivatives are replaced by *difference equations*. This approach results in a set of simultaneous linear equations. Although finite difference methods are easy to understand and employ in simple problems, they become difficult to apply to problems with complex geometries or complex boundary conditions. This situation is also true for problems with nonisotropic material properties.

In contrast, the finite element method uses *integral formulations* rather than difference equations to create a system of algebraic equations. Moreover, a continuous function is assumed to represent the approximate solution for each element. The complete solution is then generated by connecting or assembling the individual solutions, allowing for continuity at the interelemental boundaries.

1.3 A BRIEF HISTORY* OF THE FINITE ELEMENT METHOD AND ANSYS

The finite element method is a numerical procedure that can be applied to obtain solutions to a variety of problems in engineering. Steady, transient, linear, or nonlinear problems in stress analysis, heat transfer, fluid flow, and electromagnetism problems may be analyzed with finite element methods. The origin of the modern finite element method may be traced back to the early 1900s when some investigators approximated and modeled elastic continua using discrete equivalent elastic bars. However, Courant (1943) has been credited with being the first person to develop the finite element method. In a paper published in the early 1940s, Courant used piecewise polynomial interpolation over triangular subregions to investigate torsion problems.

The next significant step in the utilization of finite element methods was taken by Boeing in the 1950s when Boeing, followed by others, used triangular stress elements to model airplane wings. Yet, it was not until 1960 that Clough made the term *finite element* popular. During the 1960s, investigators began to apply the finite element method to other areas of engineering, such as heat transfer and seepage flow problems. Zienkiewicz and Cheung (1967) wrote the first book entirely devoted to the finite element method in 1967. In 1971, ANSYS was released for the first time.

ANSYS is a comprehensive general-purpose finite element computer program that contains more than 100,000 lines of code. ANSYS is capable of performing static, dynamic, heat transfer, fluid flow, and electromagnetism analyses. ANSYS has been a leading FEA program for well over 30 years. The current version of ANSYS has a completely new look, with multiple windows incorporating a graphical user interface (GUI), pull-down menus, dialog boxes, and a tool bar. Today, you will find ANSYS in use in many engineering fields, including aerospace, automotive, electronics, and nuclear. In order to use ANSYS or any other "canned" FEA computer program intelligently, it is imperative that one first fully understands the underlying basic concepts and limitations of the finite element methods.

ANSYS is a very powerful and impressive engineering tool that may be used to solve a variety of problems (see Table 1.4). However, a user without a basic understanding of the finite element methods will find himself or herself in the same predicament as a computer technician with access to many impressive instruments and tools, but who cannot fix a computer because he or she does not understand the inner workings of a computer!

1.4 BASIC STEPS IN THE FINITE ELEMENT METHOD

The basic steps involved in any finite element analysis consist of the following:

Preprocessing Phase

1. Create and discretize the solution domain into finite elements; that is, subdivide the problem into nodes and elements.

---

*See Cook et al. (1989) for more detail.

TABLE 1.4 Examples of the capabilities of ANSYS*

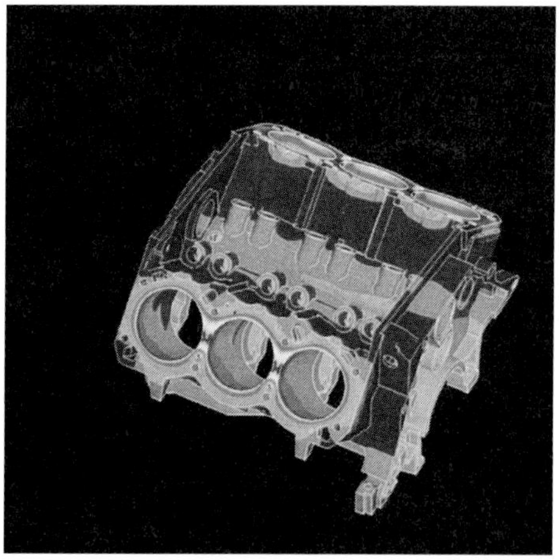

A V6 engine used in front-wheel-drive automobiles was analyzed using ANSYS heat transfer capabilities. The analyses were conducted by Analysis & Design Appl. Co. Ltd. (ADAPCO) on behalf of a major U.S. automobile manufacturer to improve product performance. Contours of thermal stress in the engine block are shown in the figure above.

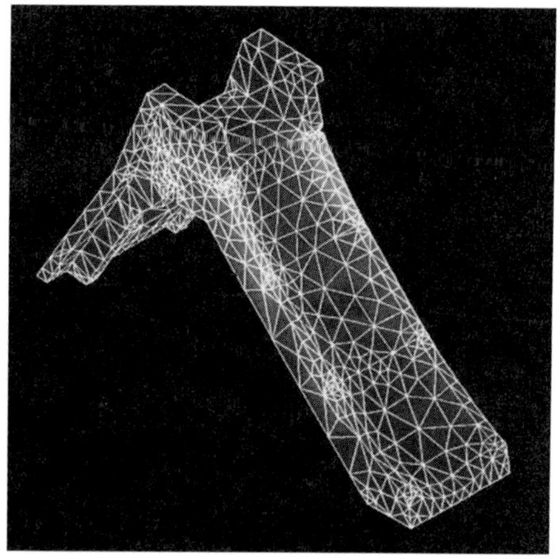

Large deflection capabilities of ANSYS were utilized by engineers at Today's Kids, a toy manufacturer, to confirm failure locations on the company's play slide, shown in the figure above, when the slide is subjected to overload. This nonlinear analysis capability is required to detect these stresses because of the product's structural behavior.

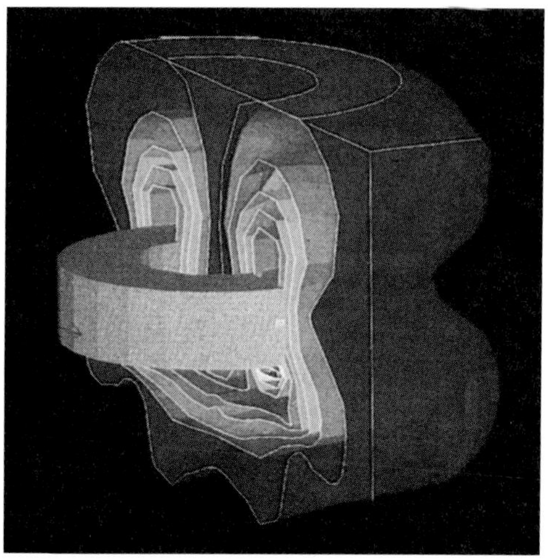

Electromagnetic capabilities of ANSYS, which include the use of both vector and scalar potentials interfaced through a specialized element, as well as a three-dimensional graphics representation of far-field decay through infinite boundary elements, are depicted in this analysis of a bath plate, shown in the figure above. Isocontours are used to depict the intensity of the H-field.

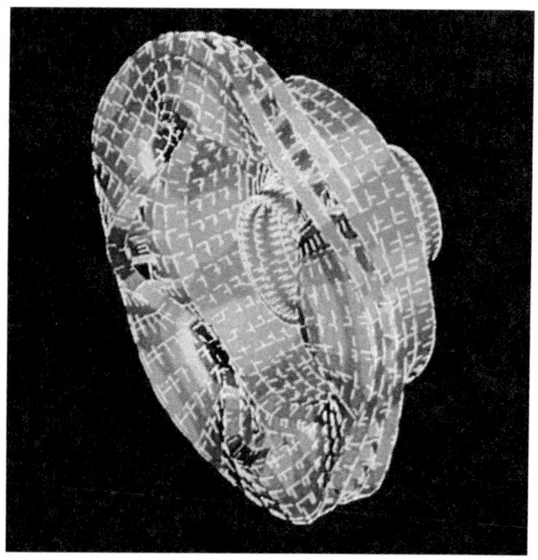

Structural Analysis Engineering Corporation used ANSYS to determine the natural frequency of a rotor in a disk-brake assembly. In this analysis, 50 modes of vibration, which are considered to contribute to brake squeal, were found to exist in the light-truck brake rotor.

*Photographs courtesy of ANSYS, Inc., Canonsburg, PA.

2. Assume a shape function to represent the physical behavior of an element; that is, a continuous function is assumed to represent the approximate behavior (solution) of an element.
3. Develop equations for an element.
4. Assemble the elements to present the entire problem. Construct the global stiffness matrix.
5. Apply boundary conditions, initial conditions, and loading.

Solution Phase

6. Solve a set of linear or nonlinear algebraic equations simultaneously to obtain nodal results, such as displacement values at different nodes or temperature values at different nodes in a heat transfer problem.

Postprocessing Phase

7. Obtain other important information. At this point, you may be interested in values of principal stresses, heat fluxes, and so on.

In general, there are several approaches to formulating finite element problems: (1) *direct formulation*, (2) *the minimum total potential energy formulation*, and (3) *weighted residual formulations*. Again, it is important to note that the basic steps involved in any finite element analysis, regardless of how we generate the finite element model, will be the same as those listed above.

1.5 DIRECT FORMULATION

The following problem illustrates the steps and the procedure involved in direct formulation.

EXAMPLE 1.1

Consider a bar with a variable cross section supporting a load P, as shown in Figure 1.1. The bar is fixed at one end and carries the load P at the other end. Let us designate the width of the bar at the top by w_1, at the bottom by w_2, its thickness by t, and its length by L. The bar's modulus of elasticity will be denoted by E. We are interested in approximating how much the bar will deflect at various points along its length when it is subjected to the load P. We will neglect the weight of the bar in the following analysis, assuming that the applied load is considerably larger than the weight of the bar:

Preprocessing Phase

1. *Discretize the solution domain into finite elements.*
We begin by subdividing the problem into nodes and elements. In order to highlight the basic steps in a finite element analysis, we will keep this problem simple and thus represent it by a model that has five nodes and four elements, as shown in Figure 1.2. However, note that we can increase the accuracy of our results by generating a model with additional nodes and elements. This task is left as an exercise for you to complete. (See Problem 1 at the end of this chapter.) The given bar

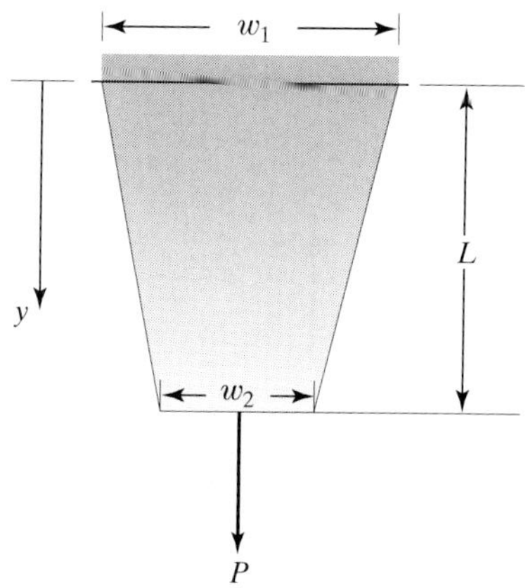

FIGURE 1.1 A bar under axial loading.

is modeled using four individual segments (elements), with each segment having a uniform cross section. The cross-sectional area of each element is represented by an average area of the cross sections at the nodes that make up (define) the element. This model is shown in Figure 1.2.

2. *Assume a solution that approximates the behavior of an element.*
In order to study the behavior of a typical element, let's consider the deflection of a solid member with a uniform cross section A that has a length ℓ when subjected to a force F, as shown in Figure 1.3.

The average stress σ in the member is given by

$$\sigma = \frac{F}{A} \tag{1.1}$$

The average normal strain ε of the member is defined as the change in length $\Delta\ell$ per unit original length ℓ of the member:

$$\varepsilon = \frac{\Delta\ell}{\ell} \tag{1.2}$$

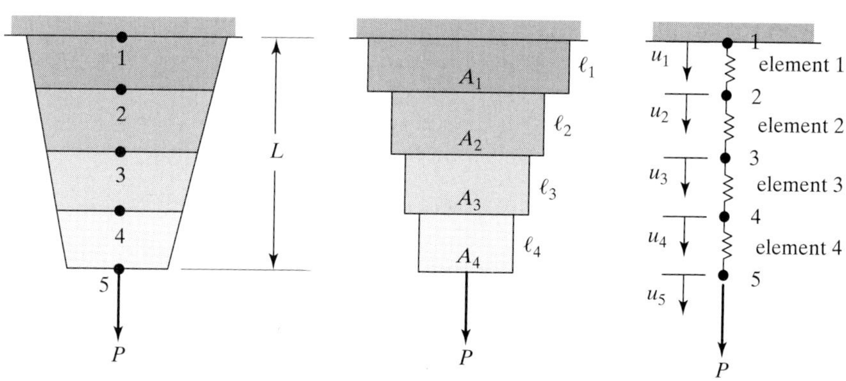

FIGURE 1.2 Subdividing the bar into elements and nodes.

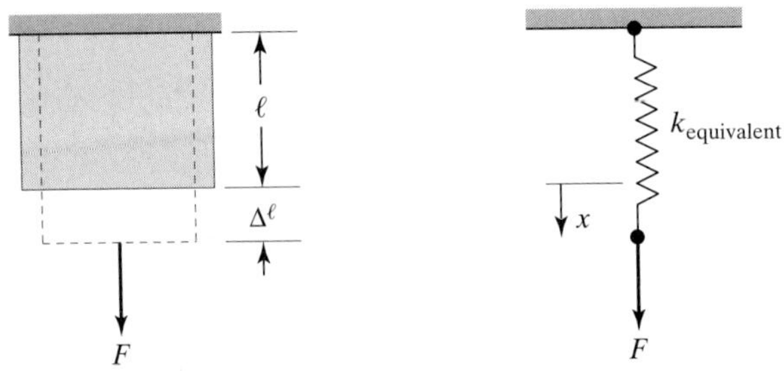

FIGURE 1.3 A solid member of uniform cross section subjected to a force F.

Over the elastic region, the stress and strain are related by Hooke's law, according to the equation

$$\sigma = E\varepsilon \tag{1.3}$$

where E is the modulus of elasticity of the material. Combining Eqs. (1.1), (1.2), and (1.3) and simplifying, we have

$$F = \left(\frac{AE}{\ell}\right)\Delta\ell \tag{1.4}$$

Note that Eq. (1.4) is similar to the equation for a linear spring, $F = kx$. Therefore, a centrally loaded member of uniform cross section may be modeled as a spring with an equivalent stiffness of

$$k_{eq} = \frac{AE}{\ell} \tag{1.5}$$

Turning our attention to Example 1.1, we note once again that the bar's cross section varies in the y-direction. As a first approximation, we model the bar as a series of centrally loaded members with different cross sections, as shown in Figure 1.2. Thus, the bar is represented by a model consisting of four elastic springs (elements) in series, and the elastic behavior of an element with nodes i and $i+1$ is modeled by an equivalent linear spring according to the equation

$$f = k_{eq}(u_{i+1} - u_i) = \frac{A_{avg}E}{\ell}(u_{i+1} - u_i) = \frac{(A_{i+1} + A_i)E}{2\ell}(u_{i+1} - u_i) \tag{1.6}$$

where the equivalent element stiffness is given by

$$k_{eq} = \frac{(A_{i+1} + A_i)E}{2\ell} \tag{1.7}$$

A_i and A_{i+1} are the cross-sectional areas of the member at nodes i and $i + 1$ respectively, and ℓ is the length of the element. Employing the above model, let us consider the forces acting on each node. The free-body diagram of nodes, which shows the forces acting on nodes 1 through 5 of this model, is depicted in Figure 1.4.

Static equilibrium requires that the sum of the forces acting on each node be zero. This requirement creates the following five equations:

$$
\begin{aligned}
&\text{node 1:} \quad R_1 - k_1(u_2 - u_1) = 0 \\
&\text{node 2:} \quad k_1(u_2 - u_1) - k_2(u_3 - u_2) = 0 \\
&\text{node 3:} \quad k_2(u_3 - u_2) - k_3(u_4 - u_3) = 0 \\
&\text{node 4:} \quad k_3(u_4 - u_3) - k_4(u_5 - u_4) = 0 \\
&\text{node 5:} \quad k_4(u_5 - u_4) - P = 0
\end{aligned}
\tag{1.8}
$$

Rearranging the equilibrium equations given by Eq. (1.8) by separating the reaction force R_1 and the applied external force P from the internal forces, we have

$$
\begin{array}{lllllllll}
k_1u_1 & -k_1u_2 & & & & & & & = -R_1 \\
-k_1u_1 & +k_1u_2 & +k_2u_2 & -k_2u_3 & & & & & = 0 \\
 & & -k_2u_2 & +k_2u_3 & +k_3u_3 & -k_3u_4 & & & = 0 \\
 & & & & -k_3u_3 & +k_3u_4 & +k_4u_4 & -k_4u_5 & = 0 \\
 & & & & & & -k_4u_4 & +k_4u_5 & = P
\end{array}
\tag{1.9}
$$

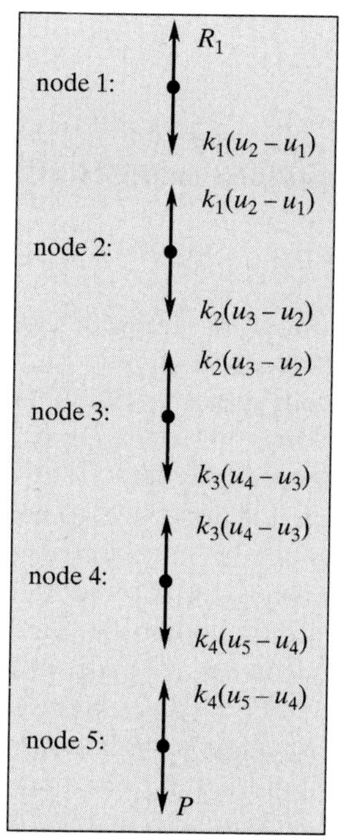

FIGURE 1.4 Free body diagram of the nodes in Example 1.1.

Presenting the equilibrium equations of Eq. (1.9) in a matrix form, we have

$$\begin{bmatrix} k_1 & -k_1 & 0 & 0 & 0 \\ -k_1 & k_1 + k_2 & -k_2 & 0 & 0 \\ 0 & -k_2 & k_2 + k_3 & -k_3 & 0 \\ 0 & 0 & -k_3 & k_3 + k_4 & -k_4 \\ 0 & 0 & 0 & -k_4 & k_4 \end{bmatrix} \begin{Bmatrix} u_1 \\ u_2 \\ u_3 \\ u_4 \\ u_5 \end{Bmatrix} = \begin{Bmatrix} -R_1 \\ 0 \\ 0 \\ 0 \\ P \end{Bmatrix} \quad (1.10)$$

It is also important to distinguish between the reaction forces and the applied loads in the load matrix. Therefore, the matrix relation of Eq. (1.10) can be written as

$$\begin{Bmatrix} -R_1 \\ 0 \\ 0 \\ 0 \\ 0 \end{Bmatrix} = \begin{bmatrix} k_1 & -k_1 & 0 & 0 & 0 \\ -k_1 & k_1 + k_2 & -k_2 & 0 & 0 \\ 0 & -k_2 & k_2 + k_3 & -k_3 & 0 \\ 0 & 0 & -k_3 & k_3 + k_4 & -k_4 \\ 0 & 0 & 0 & -k_4 & k_4 \end{bmatrix} \begin{Bmatrix} u_1 \\ u_2 \\ u_3 \\ u_4 \\ u_5 \end{Bmatrix} - \begin{Bmatrix} 0 \\ 0 \\ 0 \\ 0 \\ P \end{Bmatrix} \quad (1.11)$$

We can readily show that under additional nodal loads and other fixed boundary conditions, the relationship given by Eq. (1.11) can be put into the general form

$$\{\mathbf{R}\} = [\mathbf{K}]\{\mathbf{u}\} - \{\mathbf{F}\} \quad (1.12)$$

which stands for

$$\{\textbf{reaction matrix}\} = [\textbf{stiffness matrix}]\{\textbf{displacement matrix}\} - \{\textbf{load matrix}\}$$

Note the difference between applied load matrix $\{\mathbf{F}\}$ and the reaction force matrix $\{\mathbf{R}\}$.

Turning our attention to Example 1.1 again, we find that because the bar is fixed at the top, the displacement of node 1 is zero. Hence, there are only four unknown nodal displacement values, u_2, u_3, u_4, and u_5. The reaction force at node 1, R_1, is also unknown—all together, there are five unknowns. Because there are five equilibrium equations, as given by Eq. (1.11), we should be able to solve for all of the unknowns. However, it is important to note that even though the number of unknowns match the number of equations, the system of equations contains two different types of unknowns—displacement and reaction force. In order to eliminate the need to consider the unknown reaction force simultaneously and focus first on unknown displacements, we make use of the known boundary condition and replace the first row of Eq. (1.10) with a row that reads $u_1 = 0$. The application of the boundary condition $u_1 = 0$ eliminates the need to consider the unknown reaction force in our system of equations and creates a set of equations with the displacements being the only unknowns. Thus, application of the boundary condition leads to the following matrix equation:

$$\begin{bmatrix} 1 & 0 & 0 & 0 & 0 \\ -k_1 & k_1 + k_2 & -k_2 & 0 & 0 \\ 0 & -k_2 & k_2 + k_3 & -k_3 & 0 \\ 0 & 0 & -k_3 & k_3 + k_4 & -k_4 \\ 0 & 0 & 0 & -k_4 & k_4 \end{bmatrix} \begin{Bmatrix} u_1 \\ u_2 \\ u_3 \\ u_4 \\ u_5 \end{Bmatrix} = \begin{Bmatrix} 0 \\ 0 \\ 0 \\ 0 \\ P \end{Bmatrix} \quad (1.13)$$

The solution of the above matrix yields the nodal displacement values. It should be clear from the above explanation and examining Eq. (1.13) that for solid mechanics problems, the application of boundary conditions to the finite element formulations transforms the system of equations as given by Eq. (1.11) to a new general form that is made up of only the stiffness matrix, the displacement matrix, and the load matrix:

$$[\textbf{stiffness matrix}]\{\textbf{displacement matrix}\} = \{\textbf{load matrix}\}$$

After we solve for the nodal displacement values, from the above relationship, we use Eq. (1.12) to solve for the reaction force(s). In the next section, we will develop the general elemental stiffness matrix and discuss the construction of the global stiffness matrix by inspection.

3. *Develop equations for an element.*
Because each of the elements in Example 1.1 has two nodes, and with each node we have associated a displacement, we need to create two equations for each element. These equations must involve nodal displacements and the element's stiffness. Consider the internally transmitted forces f_i and f_{i+1} and the end displacements u_i and u_{i+1} of an element, which are shown in Figure 1.5.

Static equilibrium conditions require that the sum of f_i and f_{i+1} be zero. Note that the sum of f_i and f_{i+1} is zero regardless of which representation of Figure 1.5 is selected. However, for the sake of consistency in the forthcoming derivation, we will use the representation given by Figure 1.5(b), so that f_i and f_{i+1} are given in the positive y-direction. Thus, we write the transmitted forces at

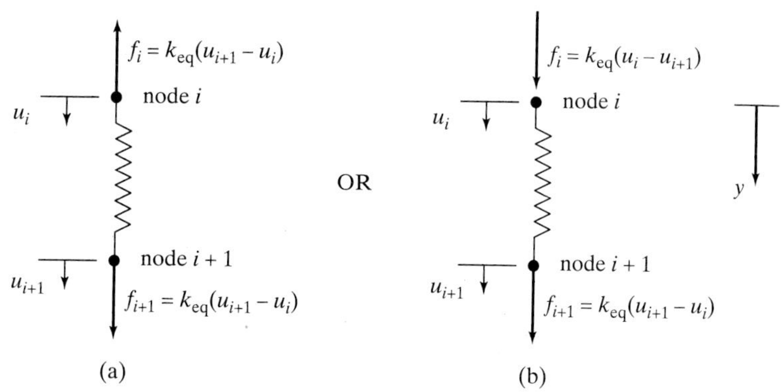

FIGURE 1.5 Internally transmitted forces through an arbitrary element.

nodes i and $i + 1$ according to the following equations:

$$\begin{aligned} f_i &= k_{eq}(u_i - u_{i+1}) \\ f_{i+1} &= k_{eq}(u_{i+1} - u_i) \end{aligned} \tag{1.14}$$

Equation (1.14) can be expressed in a matrix form by

$$\begin{Bmatrix} f_i \\ f_{i+1} \end{Bmatrix} = \begin{bmatrix} k_{eq} & -k_{eq} \\ -k_{eq} & k_{eq} \end{bmatrix} \begin{Bmatrix} u_i \\ u_{i+1} \end{Bmatrix} \tag{1.15}$$

4. *Assemble the elements to present the entire problem.*
Applying the elemental description given by Eq. (1.15) to all elements and assembling them (putting them together) will lead to the formation of the global stiffness matrix. The stiffness matrix for element (1) is given by

$$[\mathbf{K}]^{(1)} = \begin{bmatrix} k_1 & -k_1 \\ -k_1 & k_1 \end{bmatrix}$$

and its position in the global stiffness matrix is given by

$$[\mathbf{K}]^{(1G)} = \begin{bmatrix} k_1 & -k_1 & 0 & 0 & 0 \\ -k_1 & k_1 & 0 & 0 & 0 \\ 0 & 0 & 0 & 0 & 0 \\ 0 & 0 & 0 & 0 & 0 \\ 0 & 0 & 0 & 0 & 0 \end{bmatrix} \begin{matrix} u_1 \\ u_2 \\ u_3 \\ u_4 \\ u_5 \end{matrix}$$

The nodal displacement matrix is shown alongside the position of element 1 in the global stiffness matrix to aid us to observe the contribution of a node to its neighboring elements. Similarly, for elements (2), (3), and (4), we have

$$[\mathbf{K}]^{(2)} = \begin{bmatrix} k_2 & -k_2 \\ -k_2 & k_2 \end{bmatrix}$$

and its position in the global matrix

$$[\mathbf{K}]^{(2G)} = \begin{bmatrix} 0 & 0 & 0 & 0 & 0 \\ 0 & k_2 & -k_2 & 0 & 0 \\ 0 & -k_2 & k_2 & 0 & 0 \\ 0 & 0 & 0 & 0 & 0 \\ 0 & 0 & 0 & 0 & 0 \end{bmatrix} \begin{matrix} u_1 \\ u_2 \\ u_3 \\ u_4 \\ u_5 \end{matrix}$$

$$[\mathbf{K}]^{(3)} = \begin{bmatrix} k_3 & -k_3 \\ -k_3 & k_3 \end{bmatrix}$$

and its position in the global matrix

$$[\mathbf{K}]^{(3G)} = \begin{bmatrix} 0 & 0 & 0 & 0 & 0 \\ 0 & 0 & 0 & 0 & 0 \\ 0 & 0 & k_3 & -k_3 & 0 \\ 0 & 0 & -k_3 & k_3 & 0 \\ 0 & 0 & 0 & 0 & 0 \end{bmatrix} \begin{matrix} u_1 \\ u_2 \\ u_3 \\ u_4 \\ u_5 \end{matrix}$$

and

$$[\mathbf{K}]^{(4)} = \begin{bmatrix} k_4 & -k_4 \\ -k_4 & k_4 \end{bmatrix}$$

and its position in the global matrix

$$[\mathbf{K}]^{(4G)} = \begin{bmatrix} 0 & 0 & 0 & 0 & 0 \\ 0 & 0 & 0 & 0 & 0 \\ 0 & 0 & 0 & 0 & 0 \\ 0 & 0 & 0 & k_4 & -k_4 \\ 0 & 0 & 0 & -k_4 & k_4 \end{bmatrix} \begin{matrix} u_1 \\ u_2 \\ u_3 \\ u_4 \\ u_5 \end{matrix}$$

The final global stiffness matrix is obtained by assembling, or adding together, each element's position in the global stiffness matrix:

$$[\mathbf{K}]^{(G)} = [\mathbf{K}]^{(1G)} + [\mathbf{K}]^{(2G)} + [\mathbf{K}]^{(3G)} + [\mathbf{K}]^{(4G)}$$

$$[\mathbf{K}]^{(G)} = \begin{bmatrix} k_1 & -k_1 & 0 & 0 & 0 \\ -k_1 & k_1 + k_2 & -k_2 & 0 & 0 \\ 0 & -k_2 & k_2 + k_3 & -k_3 & 0 \\ 0 & 0 & -k_3 & k_3 + k_4 & -k_4 \\ 0 & 0 & 0 & -k_4 & k_4 \end{bmatrix} \quad (1.16)$$

Note that the global stiffness matrix obtained using elemental description, as given by Eq. (1.16), is identical to the global stiffness matrix we obtained earlier from the analysis of the free-body diagrams of the nodes, as given by the left-hand side of Eq. (1.10).

5. *Apply boundary conditions and loads.*
The bar is fixed at the top, which leads to the boundary condition $u_1 = 0$. The external load P is applied at node 5. Applying these conditions results in the following set of linear equations.

$$\begin{bmatrix} 1 & 0 & 0 & 0 & 0 \\ -k_1 & k_1 + k_2 & -k_2 & 0 & 0 \\ 0 & -k_2 & k_2 + k_3 & -k_3 & 0 \\ 0 & 0 & -k_3 & k_3 + k_4 & -k_4 \\ 0 & 0 & 0 & -k_4 & k_4 \end{bmatrix} \begin{Bmatrix} u_1 \\ u_2 \\ u_3 \\ u_4 \\ u_5 \end{Bmatrix} = \begin{Bmatrix} 0 \\ 0 \\ 0 \\ 0 \\ P \end{Bmatrix} \quad (1.17)$$

Again, note that the first row of the matrix in Eq. (1.17) must contain a 1 followed by four 0s to read $u_1 = 0$, the given boundary condition. As explained earlier, also note that in solid mechanics problems, the finite element formulation will always lead to the following general form:

$$[\textbf{stiffness matrix}]\{\textbf{displacement matrix}\} = \{\textbf{load matrix}\}$$

Solution Phase

6. *Solve a system of algebraic equations simultaneously.*
In order to obtain numerical values of thc nodal displacements, let us assume that $E = 10.4 \times 10^6$ lb/in^2 (aluminum), $w_1 = 2$ in, $w_2 = 1$ in, $t = 0.125$ in, $L = 10$ in, and $P = 1000$ lb. You may consult Table 1.5 while working toward the solution.

TABLE 1.5 Properties of the elements in Example 1.1

Element	Nodes	Average Cross-Sectional Area (in^2)	Length (in)	Modulus of Elasticity (lb/in^2)	Element's Stiffness Coefficient (lb/in)
1	1 2	0.234375	2.5	10.4×10^6	975×10^3
2	2 3	0.203125	2.5	10.4×10^6	845×10^3
3	3 4	0.171875	2.5	10.4×10^6	715×10^3
4	4 5	0.140625	2.5	10.4×10^6	585×10^3

The variation of the cross-sectional area of the bar in the y-direction can be expressed by:

$$A(y) = \left(w_1 + \left(\frac{w_2 - w_1}{L}\right)y\right)t = \left(2 + \frac{(1-2)}{10}y\right)(0.125) = 0.25 - 0.0125y \quad (1.18)$$

Using Eq. (1.18), we can compute the cross-sectional areas at each node:

$A_1 = 0.25 \text{ in}^2$ $\qquad A_2 = 0.25 - 0.0125(2.5) = 0.21875 \text{ in}^2$

$A_3 = 0.25 - 0.0125(5.0) = 0.1875 \text{ in}^2$ $\qquad A_4 = 0.25 - 0.0125(7.5) = 0.15625 \text{ in}^2$

$A_5 = 0.125 \text{ in}^2$

Next, the equivalent stiffness coefficient for each element is computed from the equations

$$k_{eq} = \frac{(A_{i+1} + A_i)E}{2\ell}$$

$$k_1 = \frac{(0.21875 + 0.25)(10.4 \times 10^6)}{2(2.5)} = 975 \times 10^3 \frac{\text{lb}}{\text{in}}$$

$$k_2 = \frac{(0.1875 + 0.21875)(10.4 \times 10^6)}{2(2.5)} = 845 \times 10^3 \frac{\text{lb}}{\text{in}}$$

$$k_3 = \frac{(0.15625 + 0.1875)(10.4 \times 10^6)}{2(2.5)} = 715 \times 10^3 \frac{\text{lb}}{\text{in}}$$

$$k_4 = \frac{(0.125 + 0.15625)(10.4 \times 10^6)}{2(2.5)} = 585 \times 10^3 \frac{\text{lb}}{\text{in}}$$

and the elemental matrices are

$$[\mathbf{K}]^{(1)} = \begin{bmatrix} k_1 & -k_1 \\ -k_1 & k_1 \end{bmatrix} = 10^3\begin{bmatrix} 975 & -975 \\ -975 & 975 \end{bmatrix}$$

$$[\mathbf{K}]^{(2)} = \begin{bmatrix} k_2 & -k_2 \\ -k_2 & k_2 \end{bmatrix} = 10^3\begin{bmatrix} 845 & -845 \\ -845 & 845 \end{bmatrix}$$

$$[\mathbf{K}]^{(3)} = \begin{bmatrix} k_3 & -k_3 \\ -k_3 & k_3 \end{bmatrix} = 10^3\begin{bmatrix} 715 & -715 \\ -715 & 715 \end{bmatrix}$$

$$[\mathbf{K}]^{(4)} = \begin{bmatrix} k_4 & -k_4 \\ -k_4 & k_4 \end{bmatrix} = 10^3\begin{bmatrix} 585 & -585 \\ -585 & 585 \end{bmatrix}$$

Assembling the elemental matrices leads to the generation of the global stiffness matrix:

$$[\mathbf{K}]^{(G)} = 10^3\begin{bmatrix} 975 & -975 & 0 & 0 & 0 \\ -975 & 975+845 & -845 & 0 & 0 \\ 0 & -845 & 845+715 & -715 & 0 \\ 0 & 0 & -715 & 715+585 & -585 \\ 0 & 0 & 0 & -585 & 585 \end{bmatrix}$$

Applying the boundary condition $u_1 = 0$ and the load $P = 1000$ lb, we get

$$10^3\begin{bmatrix} 1 & 0 & 0 & 0 & 0 \\ -975 & 1820 & -845 & 0 & 0 \\ 0 & -845 & 1560 & -715 & 0 \\ 0 & 0 & -715 & 1300 & -585 \\ 0 & 0 & 0 & -585 & 585 \end{bmatrix}\begin{Bmatrix} u_1 \\ u_2 \\ u_3 \\ u_4 \\ u_5 \end{Bmatrix} = \begin{Bmatrix} 0 \\ 0 \\ 0 \\ 0 \\ 10^3 \end{Bmatrix}$$

Because in the second row, the -975 coefficient gets multiplied by $u_1 = 0$, we need only to solve the following 4×4 matrix:

$$10^3\begin{bmatrix} 1820 & -845 & 0 & 0 \\ -845 & 1560 & -715 & 0 \\ 0 & -715 & 1300 & -585 \\ 0 & 0 & -585 & 585 \end{bmatrix}\begin{Bmatrix} u_2 \\ u_3 \\ u_4 \\ u_5 \end{Bmatrix} = \begin{Bmatrix} 0 \\ 0 \\ 0 \\ 10^3 \end{Bmatrix}$$

The displacement solution is $u_1 = 0$, $u_2 = 0.001026$ in, $u_3 = 0.002210$ in, $u_4 = 0.003608$ in, and $u_5 = 0.005317$ in.

Postprocessing Phase

7. *Obtain other information.*
For Example 1.1, we may be interested in obtaining other information, such as the average normal stresses in each element. These values can be determined

from the equation

$$\sigma = \frac{f}{A_{avg}} = \frac{k_{eq}(u_{i+1} - u_i)}{A_{avg}} = \frac{\frac{A_{avg}E}{\ell}(u_{i+1} - u_i)}{A_{avg}} = E\left(\frac{u_{i+1} - u_i}{\ell}\right) \tag{1.19}$$

Since the displacements of different nodes are known, Eq. (1.19) could have been obtained directly from the relationship between the stresses and strains,

$$\sigma = E\varepsilon = E\left(\frac{u_{i+1} - u_i}{\ell}\right) \tag{1.20}$$

Employing Eq. (1.20) in Example 1.1, we compute the average normal stress for each element as

$$\sigma^{(1)} = E\left(\frac{u_2 - u_1}{\ell}\right) = \frac{(10.4 \times 10^6)(0.001026 - 0)}{2.5} = 4268\ \frac{\text{lb}}{\text{in}^2}$$

$$\sigma^{(2)} = E\left(\frac{u_3 - u_2}{\ell}\right) = \frac{(10.4 \times 10^6)(0.002210 - 0.001026)}{2.5} = 4925\ \frac{\text{lb}}{\text{in}^2}$$

$$\sigma^{(3)} = E\left(\frac{u_4 - u_3}{\ell}\right) = \frac{(10.4 \times 10^6)(0.003608 - 0.002210)}{2.5} = 5816\ \frac{\text{lb}}{\text{in}^2}$$

$$\sigma^{(4)} = E\left(\frac{u_5 - u_4}{\ell}\right) = \frac{(10.4 \times 10^6)(0.005317 - 0.003608)}{2.5} = 7109\ \frac{\text{lb}}{\text{in}^2}$$

In Figure 1.6, we note that for the given problem, regardless of where we cut a section through the bar, the internal force at the section is equal to 1000 lb. So,

$$\sigma^{(1)} = \frac{f}{A_{avg}} = \frac{1000}{0.234375} = 4267\ \frac{\text{lb}}{\text{in}^2}$$

$$\sigma^{(2)} = \frac{f}{A_{avg}} = \frac{1000}{0.203125} = 4923\ \frac{\text{lb}}{\text{in}^2}$$

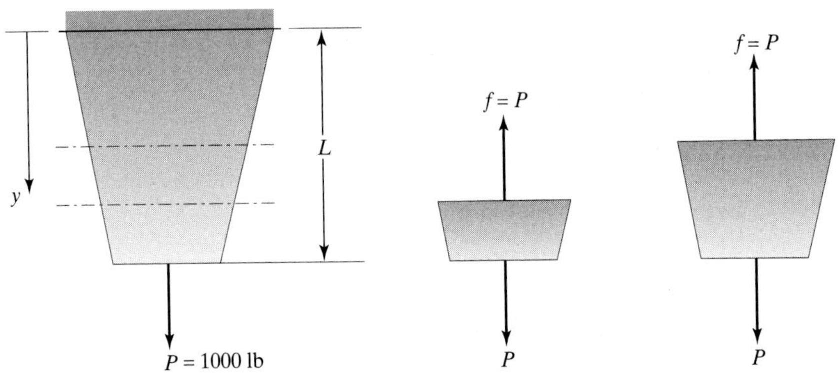

FIGURE 1.6 The internal forces in Example 1.1.

$$\sigma^{(3)} = \frac{f}{A_{avg}} = \frac{1000}{0.171875} = 5818\frac{\text{lb}}{\text{in}^2}$$

$$\sigma^{(4)} = \frac{f}{A_{avg}} = \frac{1000}{0.140625} = 7111\frac{\text{lb}}{\text{in}^2}$$

Ignoring the errors we get from rounding off our answers, we find that these results are identical to the element stresses computed from the displacement information. This comparison tells us that our displacement calculations are good for this problem.

Reaction Forces For Example 1.1, the reaction force may be computed in a number of ways. First, referring to Figure 1.4, we note that the statics equilibrium at node 1 requires

$$R_1 = k_1(u_2 - u_1) = 975 \times 10^3(0.001026 - 0) = 1000\text{ lb}$$

The statics equilibrium for the entire bar also requires that

$$R_1 = P = 1000\text{ lb}$$

As you may recall, we can also compute the reaction forces from the general reaction equation

$$\{\mathbf{R}\} = [\mathbf{K}]\{\mathbf{u}\} - \{\mathbf{F}\}$$

or

$$\{\textbf{reaction matrix}\} = [\textbf{stiffness matrix}]\{\textbf{displacement matrix}\} - \{\textbf{load matrix}\}$$

Because Example 1.1 is a simple problem, we do not actually need to go through the matrix operations in the aforementioned general equation to compute the reaction forces. However, as a demonstration, the procedure is shown here. From the general equation, we get

$$\begin{Bmatrix} R_1 \\ R_2 \\ R_3 \\ R_4 \\ R_5 \end{Bmatrix} = 10^3 \begin{bmatrix} 975 & -975 & 0 & 0 & 0 \\ -975 & 1820 & -845 & 0 & 0 \\ 0 & -845 & 1560 & -715 & 0 \\ 0 & 0 & -715 & 1300 & -585 \\ 0 & 0 & 0 & -585 & 585 \end{bmatrix} \begin{Bmatrix} 0 \\ 0.001026 \\ 0.002210 \\ 0.003608 \\ 0.005317 \end{Bmatrix} - \begin{Bmatrix} 0 \\ 0 \\ 0 \\ 0 \\ 10^3 \end{Bmatrix}$$

where R_1, R_2, R_3, R_4, and R_5 represent the reactions forces at nodes 1 through 5 respectively. Performing the matrix operation, we have

$$\begin{Bmatrix} R_1 \\ R_2 \\ R_3 \\ R_4 \\ R_5 \end{Bmatrix} = \begin{Bmatrix} -1000 \\ 0 \\ 0 \\ 0 \\ 0 \end{Bmatrix}$$

The negative value of R_1 simply means that the direction of the reaction force is up (because we assumed that the positive y-direction points down). Of course, as expected, the outcome is the samc as in our earlier calculations because the rows of the above matrix represent the static equilibrium conditions at each node. When solving for reaction forces, note that you must use the complete stiffness matrix—without the influence of boundary conditions—in Equation (1.12). Next, we will consider finite element formulation of a heat transfer problem.

EXAMPLE 1.2

A typical exterior frame wall (made up of 2 × 4 studs) of a house contains the materials shown in the table below. Let us assume an inside room temperature of 70°F and an outside air temperature of 20°F, with an exposed area of 150 ft². We are interested in determining the temperature distribution through the wall.

Items	Resistance hr · ft² · °F/Btu	*U*-factor Btu/hr · ft² · °F
1. Outside film resistance (winter, 15-mph wind)	0.17	5.88
2. Siding, wood (1/2 × 8 lapped)	0.81	1.23
3. Sheathing (1/2 in regular)	1.32	0.76
4. Insulation batt (3 – 3½ in)	11.0	0.091
5. Gypsum wall board (1/2 in)	0.45	2.22
6. Inside film resistance (winter)	0.68	1.47

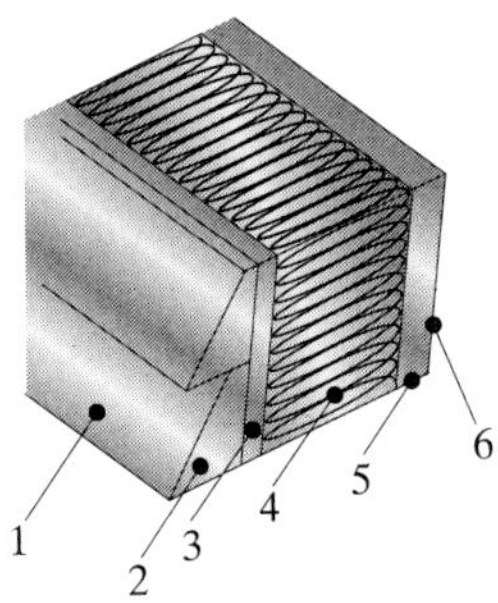

Preprocessing Phase

1. *Discretize the solution domain into finite elements.*
   We will represent this problem by a model that has seven nodes and six elements, as shown in Figure 1.7.
2. *Assume a solution that approximates the behavior of an element.*
   For Example 1.2, there are two modes of heat transfer (conduction and convection) that we must first understand before we can proceed with formulating the conductance matrix and the thermal load matrix. The steady-state thermal behavior

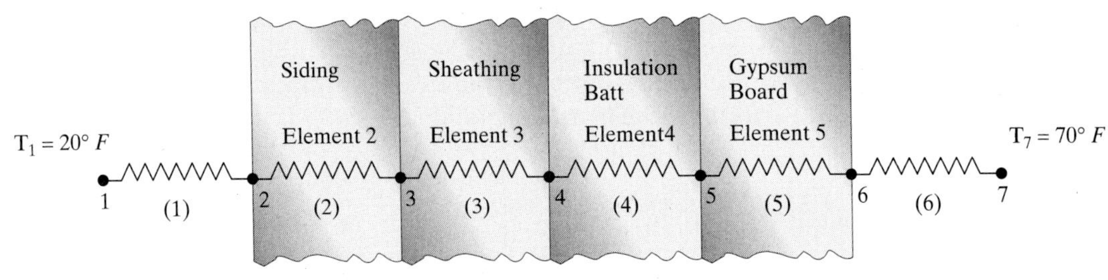

FIGURE 1.7 Finite element model of Example 1.2.

of the elements (2), (3), (4), and (5) may be modeled using Fourier's law. When there exists a temperature gradient in a medium, conduction heat transfer occurs, as shown in Figure 1.8. The energy is transported from the high-temperature region to the low-temperature region by molecular activities. The heat transfer rate is given by Fourier's law:

$$q_X = -kA\frac{\partial T}{\partial X} \tag{1.21}$$

q_X is the X-component of the heat transfer rate, k is the thermal conductivity of the medium, A is the area normal to heat flow, and $\frac{\partial T}{\partial X}$ is the temperature gradient. The minus sign in Eq. (1.21) is due to the fact that heat flows in the direction of decreasing temperature. Equation (1.21) can be written in a difference form in terms of the spacing between the nodes (length of the element) ℓ and the respective temperatures of the nodes i and $i + 1$, T_i and T_{i+1}, according to the equation

$$q = \frac{kA(T_{i+1} - T_i)}{\ell} \tag{1.22}$$

In the field of heat transfer, it is also common to write Eq. (1.22) in terms of the thermal transmittance coefficient U, or, as it is often called, the U-factor $(U = \frac{k}{\ell})$. The U-factor represents thermal transmission through a unit area and has the units of Btu/hr $\cdot$ ft^2 $\cdot$ °F. It is the reciprocal of thermal resistance. So,

$$q = UA(T_{i+1} - T_i) \tag{1.23}$$

The steady-state thermal behavior of elements (1) and (6) may be modeled using Newton's law of cooling. Convection heat transfer occurs when a fluid in motion comes into contact with a surface whose temperature differs from the moving fluid. The overall heat transfer rate between the fluid and the surface is

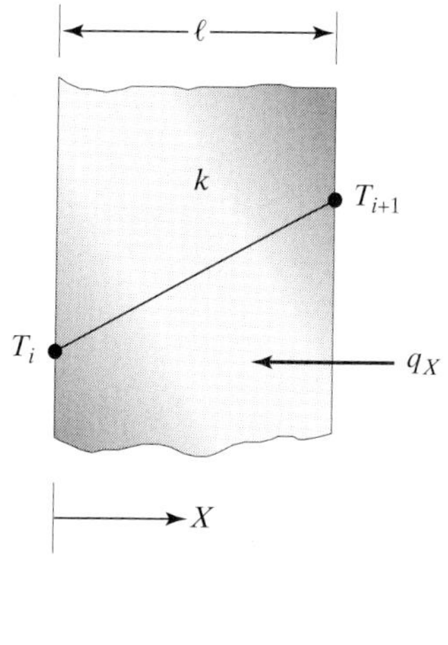

FIGURE 1.8 Heat transfer in a medium by conduction.

governed by Newton's law of cooling, according to the equation

$$q = hA(T_s - T_f) \tag{1.24}$$

where h is the heat transfer coefficient, T_s is the surface temperature, and T_f represents the temperature of the moving fluid. Newton's law of cooling can also be written in terms of the U-factor, such that

$$q = UA(T_s - T_f) \tag{1.25}$$

where $U = h$, and it represents the reciprocal of thermal resistance due to convection boundary conditions. Under steady-state conduction, the application of energy balance to a surface, with a convective heat transfer, requires that the energy transferred to this surface via conduction must be equal to the energy transfer by convection. This principle,

$$-kA\frac{\partial T}{\partial X} = hA[T_s - T_f] \tag{1.26}$$

is depicted in Figure 1.9.

Now that we understand the two modes of heat transfer involved in this problem, we can apply the energy balance to the various surfaces of the wall, starting with the wall's exterior surface located at node 2. The heat loss through the wall due to conduction must equal the heat loss to the surrounding cold air by convection. That is,

$$U_2A(T_3 - T_2) = U_1A(T_2 - T_1)$$

The application of energy balance to surfaces located at nodes 3, 4, and 5 yields the equations

$$\begin{aligned} U_3A(T_4 - T_3) &= U_2A(T_3 - T_2) \\ U_4A(T_5 - T_4) &= U_3A(T_4 - T_3) \\ U_5A(T_6 - T_5) &= U_4A(T_5 - T_4) \end{aligned}$$

For the interior surface of the wall, located at node 6, the heat loss by convection of warm air is equal to the heat transfer by conduction through the gypsum board,

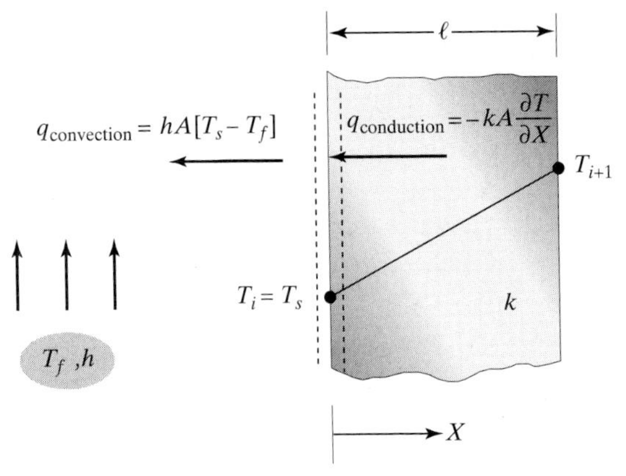

FIGURE 1.9 Energy balance at a surface with a convective heat transfer.

according to the equation

$$U_6A(T_7 - T_6) = U_5A(T_6 - T_5)$$

Separating the known temperatures from the unknown temperatures, we have

$$\begin{aligned}
+(U_1 + U_2)AT_2 \quad -U_2AT_3 \quad &= U_1AT_1 \\
-U_2AT_2 \quad +(U_2 + U_3)AT_3 \quad -U_3AT_4 \quad &= 0 \\
-U_3AT_3 \quad +(U_3 + U_4)AT_4 \quad -U_4AT_5 \quad &= 0 \\
-U_4AT_4 \quad +(U_4 + U_5)AT_5 \quad -U_5AT_6 \quad &= 0 \\
-U_5AT_5 \quad +(U_5 + U_6)AT_6 &= U_6AT_7
\end{aligned}$$

The above relationships can be represented in matrix form as

$$A\begin{bmatrix} U_1 + U_2 & -U_2 & 0 & 0 & 0 \\ -U_2 & U_2 + U_3 & -U_3 & 0 & 0 \\ 0 & -U_3 & U_3 + U_4 & -U_4 & 0 \\ 0 & 0 & -U_4 & U_4 + U_5 & -U_5 \\ 0 & 0 & 0 & -U_5 & U_5 + U_6 \end{bmatrix} \begin{Bmatrix} T_2 \\ T_3 \\ T_4 \\ T_5 \\ T_6 \end{Bmatrix} = \begin{Bmatrix} U_1AT_1 \\ 0 \\ 0 \\ 0 \\ U_6AT_7 \end{Bmatrix} \tag{1.27}$$

Note that the relationship given by Eq. (1.27) was developed by applying the conservation of energy to the surfaces located at nodes 2, 3, 4, 5, and 6. Next, we will consider the elemental formulation of this problem, which will lead to the same results.

3. *Develop equations for an element.*
In general, for conduction problems, the heat transfer rates q_i and q_{i+1} and the nodal temperatures T_i and T_{i+1} for an element are related according to the equations

$$q_i = \frac{kA}{\ell}(T_i - T_{i+1})$$

$$q_{i+1} = \frac{kA}{\ell}(T_{i+1} - T_i) \tag{1.28}$$

The heat flow through nodes i and $i + 1$ is depicted in Figure 1.10.

Because each of the elements in Example 1.2 has two nodes, and we have associated a temperature with each node, we want to create two equations for each element. These equations must involve nodal temperatures and the element's thermal conductivity or U-factor, based on Fourier's law. Under steady-state conditions, the application of the conservation of energy requires that the sum of q_i and q_{i+1} into an element be zero; that is, the energy flowing into node $i + 1$ must be equal to the energy flowing out of node i. Note that the sum of q_i and q_{i+1} is zero regardless of which representation of Figure 1.10 is selected. However, for the sake of consistency in the forthcoming derivation, we will use the representation given by Figure 1.10(b). Elemental description given by Eq. (1.28) can be expressed in matrix form by

$$\begin{Bmatrix} q_i \\ q_{i+1} \end{Bmatrix} = \frac{kA}{\ell}\begin{bmatrix} 1 & -1 \\ -1 & 1 \end{bmatrix}\begin{Bmatrix} T_i \\ T_{i+1} \end{Bmatrix} \tag{1.29}$$

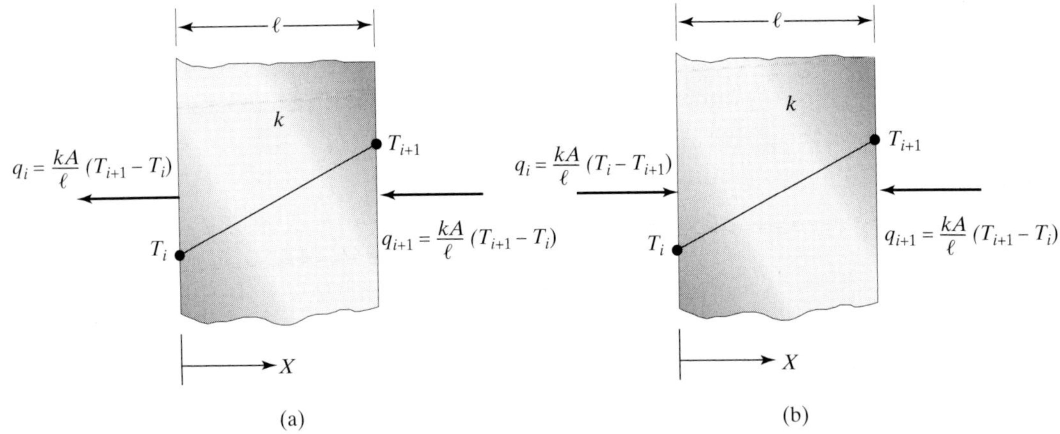

FIGURE 1.10 Heat flow through nodes i and $i + 1$.

The thermal conductance matrix for an element is

$$[\mathbf{K}]^{(e)} = \frac{kA}{\ell}\begin{bmatrix} 1 & -1 \\ -1 & 1 \end{bmatrix} \tag{1.30}$$

The conductance matrix can also be written in terms of the U-factor $\left(U = \frac{k}{\ell}\right)$:

$$[\mathbf{K}]^{(e)} = UA\begin{bmatrix} 1 & -1 \\ -1 & 1 \end{bmatrix} \tag{1.31}$$

Similarly, under steady-state conditions, the application of the conservation of energy to the nodes of a convective element gives

$$\begin{aligned} q_i &= hA(T_i - T_{i+1}) \\ q_{i+1} &= hA(T_{i+1} - T_i) \end{aligned} \tag{1.32}$$

Equation (1.32) expressed in a matrix form is

$$\begin{Bmatrix} q_i \\ q_{i+1} \end{Bmatrix} = hA\begin{bmatrix} 1 & -1 \\ -1 & 1 \end{bmatrix}\begin{Bmatrix} T_i \\ T_{i+1} \end{Bmatrix}$$

The thermal conductance matrix for a convective element then becomes

$$[\mathbf{K}]^{(e)} = hA\begin{bmatrix} 1 & -1 \\ -1 & 1 \end{bmatrix} \tag{1.33}$$

Equation (1.33) can also be written in terms of the U-factor ($U = h$):

$$[\mathbf{K}]^{(e)} = UA\begin{bmatrix} 1 & -1 \\ -1 & 1 \end{bmatrix} \tag{1.34}$$

4. *Assemble the elements to present the entire problem.*
Applying the elemental description given by Eqs. (1.31) and (1.34) to all of the elements in Example 1.2 and assembling leads to the formation of the global stiffness matrix. So,

$$[\mathbf{K}]^{(1)} = A\begin{bmatrix} U_1 & -U_1 \\ -U_1 & U_1 \end{bmatrix}$$

and its position in the global matrix is

$$[\mathbf{K}]^{(1G)} = A\begin{bmatrix} U_1 & -U_1 & 0 & 0 & 0 & 0 & 0 \\ -U_1 & U_1 & 0 & 0 & 0 & 0 & 0 \\ 0 & 0 & 0 & 0 & 0 & 0 & 0 \\ 0 & 0 & 0 & 0 & 0 & 0 & 0 \\ 0 & 0 & 0 & 0 & 0 & 0 & 0 \\ 0 & 0 & 0 & 0 & 0 & 0 & 0 \\ 0 & 0 & 0 & 0 & 0 & 0 & 0 \end{bmatrix}\begin{matrix} T_1 \\ T_2 \\ T_3 \\ T_4 \\ T_5 \\ T_6 \\ T_7 \end{matrix}$$

The nodal temperature matrix is shown along with the global thermal conductance matrix to help you observe the contribution of a node to its neighboring elements:

$$[\mathbf{K}]^{(2)} = A\begin{bmatrix} U_2 & -U_2 \\ -U_2 & U_2 \end{bmatrix} \text{ and } [\mathbf{K}]^{(2G)} = A\begin{bmatrix} 0 & 0 & 0 & 0 & 0 & 0 & 0 \\ 0 & U_2 & -U_2 & 0 & 0 & 0 & 0 \\ 0 & -U_2 & U_2 & 0 & 0 & 0 & 0 \\ 0 & 0 & 0 & 0 & 0 & 0 & 0 \\ 0 & 0 & 0 & 0 & 0 & 0 & 0 \\ 0 & 0 & 0 & 0 & 0 & 0 & 0 \\ 0 & 0 & 0 & 0 & 0 & 0 & 0 \end{bmatrix}\begin{matrix} T_1 \\ T_2 \\ T_3 \\ T_4 \\ T_5 \\ T_6 \\ T_7 \end{matrix}$$

$$[\mathbf{K}]^{(3)} = A\begin{bmatrix} U_3 & -U_3 \\ -U_3 & U_3 \end{bmatrix} \text{ and } [\mathbf{K}]^{(3G)} = A\begin{bmatrix} 0 & 0 & 0 & 0 & 0 & 0 & 0 \\ 0 & 0 & 0 & 0 & 0 & 0 & 0 \\ 0 & 0 & U_3 & -U_3 & 0 & 0 & 0 \\ 0 & 0 & -U_3 & U_3 & 0 & 0 & 0 \\ 0 & 0 & 0 & 0 & 0 & 0 & 0 \\ 0 & 0 & 0 & 0 & 0 & 0 & 0 \\ 0 & 0 & 0 & 0 & 0 & 0 & 0 \end{bmatrix}\begin{matrix} T_1 \\ T_2 \\ T_3 \\ T_4 \\ T_5 \\ T_6 \\ T_7 \end{matrix}$$

$$[\mathbf{K}]^{(4)} = A\begin{bmatrix} U_4 & -U_4 \\ -U_4 & U_4 \end{bmatrix} \text{ and } [\mathbf{K}]^{(4G)} = A\begin{bmatrix} 0 & 0 & 0 & 0 & 0 & 0 & 0 \\ 0 & 0 & 0 & 0 & 0 & 0 & 0 \\ 0 & 0 & 0 & 0 & 0 & 0 & 0 \\ 0 & 0 & 0 & U_4 & -U_4 & 0 & 0 \\ 0 & 0 & 0 & -U_4 & U_4 & 0 & 0 \\ 0 & 0 & 0 & 0 & 0 & 0 & 0 \\ 0 & 0 & 0 & 0 & 0 & 0 & 0 \end{bmatrix}\begin{matrix} T_1 \\ T_2 \\ T_3 \\ T_4 \\ T_5 \\ T_6 \\ T_7 \end{matrix}$$

$$[\mathbf{K}]^{(5)} = A\begin{bmatrix} U_5 & -U_5 \\ -U_5 & U_5 \end{bmatrix} \text{ and } [\mathbf{K}]^{(5G)} = A\begin{bmatrix} 0 & 0 & 0 & 0 & 0 & 0 & 0 \\ 0 & 0 & 0 & 0 & 0 & 0 & 0 \\ 0 & 0 & 0 & 0 & 0 & 0 & 0 \\ 0 & 0 & 0 & 0 & 0 & 0 & 0 \\ 0 & 0 & 0 & 0 & U_5 & -U_5 & 0 \\ 0 & 0 & 0 & 0 & -U_5 & U_5 & 0 \\ 0 & 0 & 0 & 0 & 0 & 0 & 0 \end{bmatrix} \begin{matrix} T_1 \\ T_2 \\ T_3 \\ T_4 \\ T_5 \\ T_6 \\ T_7 \end{matrix}$$

$$[\mathbf{K}]^{(6)} = A\begin{bmatrix} U_6 & -U_6 \\ -U_6 & U_6 \end{bmatrix} \text{ and } [\mathbf{K}]^{(6G)} = A\begin{bmatrix} 0 & 0 & 0 & 0 & 0 & 0 & 0 \\ 0 & 0 & 0 & 0 & 0 & 0 & 0 \\ 0 & 0 & 0 & 0 & 0 & 0 & 0 \\ 0 & 0 & 0 & 0 & 0 & 0 & 0 \\ 0 & 0 & 0 & 0 & 0 & 0 & 0 \\ 0 & 0 & 0 & 0 & 0 & U_6 & -U_6 \\ 0 & 0 & 0 & 0 & 0 & -U_6 & U_6 \end{bmatrix} \begin{matrix} T_1 \\ T_2 \\ T_3 \\ T_4 \\ T_5 \\ T_6 \\ T_7 \end{matrix}$$

The global conductance matrix is

$$[\mathbf{K}]^{(G)} = [\mathbf{K}]^{(1G)} + [\mathbf{K}]^{(2G)} + [\mathbf{K}]^{(3G)} + [\mathbf{K}]^{(4G)} + [\mathbf{K}]^{(5G)} + [\mathbf{K}]^{(6G)}$$

$$[\mathbf{K}]^{(G)} = A\begin{bmatrix} U_1 & -U_1 & 0 & 0 & 0 & 0 & 0 \\ -U_1 & U_1 + U_2 & -U_2 & 0 & 0 & 0 & 0 \\ 0 & -U_2 & U_2 + U_3 & -U_3 & 0 & 0 & 0 \\ 0 & 0 & -U_3 & U_3 + U_4 & -U_4 & 0 & 0 \\ 0 & 0 & 0 & -U_4 & U_4 + U_5 & -U_5 & 0 \\ 0 & 0 & 0 & 0 & -U_5 & U_5 + U_6 & -U_6 \\ 0 & 0 & 0 & 0 & 0 & -U_6 & U_6 \end{bmatrix} \quad (1.35)$$

5. *Apply boundary conditions and thermal loads.*
For the given problem, the exterior of the wall is exposed to a known air temperature T_1, and the room temperature, T_7, is also known. Thus, we want the first row to read $T_1 = 20°\text{F}$ and the last row to read $T_7 = 70°\text{F}$. So, we have

$$A\begin{bmatrix} 1/A & 0 & 0 & 0 & 0 & 0 & 0 \\ -U_1 & U_1 + U_2 & -U_2 & 0 & 0 & 0 & 0 \\ 0 & -U_2 & U_2 + U_3 & -U_3 & 0 & 0 & 0 \\ 0 & 0 & -U_3 & U_3 + U_4 & -U_4 & 0 & 0 \\ 0 & 0 & 0 & -U_4 & U_4 + U_5 & -U_5 & 0 \\ 0 & 0 & 0 & 0 & -U_5 & U_5 + U_6 & -U_6 \\ 0 & 0 & 0 & 0 & 0 & 0 & 1/A \end{bmatrix} \begin{Bmatrix} T_1 \\ T_2 \\ T_3 \\ T_4 \\ T_5 \\ T_6 \\ T_7 \end{Bmatrix} = \begin{Bmatrix} 20°\text{F} \\ 0 \\ 0 \\ 0 \\ 0 \\ 0 \\ 70°\text{F} \end{Bmatrix} \quad (1.36)$$

Note that the finite element formulation of heat transfer problems will always lead to an equation of the form

$$[\mathbf{K}]\{\mathbf{T}\} = \{\mathbf{q}\}$$

$$[\textbf{conductance matrix}]\{\textbf{temperature matrix}\} = \{\textbf{heat flow matrix}\}$$

Also note that for Example 1.2, the heat transfer rate through each element was caused by temperature differences across the nodes of a given element. Thus, the external nodal heat flow values are zero in the heat flow matrix. An example of a situation in which external nodal heat values are not zero is a heating strip attached to a solid surface (e.g., the base of a pressing iron); for such a situation, the external nodal heat value is equal to the amount of heat being generated by the heating strip over the surface. Turning our attention to the matrices given by Eq. (1.36) and incorporating the known boundary conditions into rows 2 and 6 of the conductance matrix, we can reduce Eq. (1.36) to

$$A\begin{bmatrix} U_1 + U_2 & -U_2 & 0 & 0 & 0 \\ -U_2 & U_2 + U_3 & -U_3 & 0 & 0 \\ 0 & -U_3 & U_3 + U_4 & -U_4 & 0 \\ 0 & 0 & -U_4 & U_4 + U_5 & -U_5 \\ 0 & 0 & 0 & -U_5 & U_5 + U_6 \end{bmatrix}\begin{Bmatrix} T_2 \\ T_3 \\ T_4 \\ T_5 \\ T_6 \end{Bmatrix} = \begin{Bmatrix} U_1 A T_1 \\ 0 \\ 0 \\ 0 \\ U_6 A T_7 \end{Bmatrix}$$

Keep in mind that the above matrix was obtained by assembling the elemental description and applying the boundary conditions. Moreover, the results of this approach are identical to the relations we obtained earlier by balancing the heat flows at the nodes, as given by Eq. (1.27). This equality in the outcome is expected because the elemental formulations are based on the application of energy balance as well.

Referring to the original global matrix, substituting for the U-values and employing the given boundary conditions, we have

$$150\begin{bmatrix} \frac{1}{150} & 0 & 0 & 0 & 0 & 0 & 0 \\ -5.88 & 5.88 + 1.23 & -1.23 & 0 & 0 & 0 & 0 \\ 0 & -1.23 & 1.23 + 0.76 & -0.76 & 0 & 0 & 0 \\ 0 & 0 & -0.76 & 0.76 + 0.091 & -0.091 & 0 & 0 \\ 0 & 0 & 0 & -0.091 & 0.091 + 2.22 & -2.22 & 0 \\ 0 & 0 & 0 & 0 & -2.22 & 2.22 + 1.47 & -1.47 \\ 0 & 0 & 0 & 0 & 0 & 0 & \frac{1}{150} \end{bmatrix}\begin{Bmatrix} T_1 \\ T_2 \\ T_3 \\ T_4 \\ T_5 \\ T_6 \\ T_7 \end{Bmatrix} = \begin{Bmatrix} 20^\circ\mathrm{F} \\ 0 \\ 0 \\ 0 \\ 0 \\ 0 \\ 70^\circ\mathrm{F} \end{Bmatrix}$$

Simplifying, we obtain

$$\begin{bmatrix} 7.11 & -1.23 & 0 & 0 & 0 \\ -1.23 & 1.99 & -0.76 & 0 & 0 \\ 0 & -0.76 & 0.851 & -0.091 & 0 \\ 0 & 0 & -0.091 & 2.311 & -2.22 \\ 0 & 0 & 0 & -2.22 & 3.69 \end{bmatrix} \begin{Bmatrix} T_2 \\ T_3 \\ T_4 \\ T_5 \\ T_6 \end{Bmatrix} = \begin{Bmatrix} (5.88)(20) \\ 0 \\ 0 \\ 0 \\ (1.47)(70) \end{Bmatrix}$$

Solution Phase

6. *Solve a system of algebraic equations simultaneously.*
Solving the previous matrix yields the temperature distribution along the wall:

$$\begin{Bmatrix} T_1 \\ T_2 \\ T_3 \\ T_4 \\ T_5 \\ T_6 \\ T_7 \end{Bmatrix} = \begin{Bmatrix} 20.00 \\ 20.59 \\ 23.41 \\ 27.97 \\ 66.08 \\ 67.64 \\ 70.00 \end{Bmatrix} {}^{\circ}\mathrm{C}$$

For problems similar to the type discussed here, the knowledge of temperature distribution within the wall is important in determining where condensation may occur in the wall and thus where one should place a vapor barrier to avoid moisture condensation. To demonstrate this concept, let us assume that moisture can diffuse through the gypsum board and that the inside air has a relative humidity of 40%. With the help of a psychometric chart, using a dry bulb temperature of 70°F and the value $\phi = 40\%$, we identify the condensation temperature to be 44°F. Therefore, the water vapor in the air at any surface whose temperature is 44°F or below will condense. In the absence of a vapor barrier, the water vapor in the air will condense somewhere between surface 5 and 4 for the assumed conditions in this problem.

Postprocessing Phase

7. *Obtain other information.*
For this example, we may be interested in obtaining other information, such as heat loss through the wall. Such information is important in computing the heat load for a building. Because we have assumed steady-state conditions, the heat loss through the wall should be equal to the heat transfer through each element. This value can be determined from the equation

$$q = UA(T_{i+1} - T_i) \tag{1.37}$$

The heat transfer through each element is

$$q = UA(T_{i+1} - T_i) = (1.47)(150)(70 - 67.64) = (2.22)(150)(67.64 - 66.08) = \cdots$$
$$= (5.88)(150)(20.59 - 20) = 520\frac{\text{Btu}}{\text{hr}}$$

We also could have calculated the heat loss through the wall using the overall U-factor in the following manner:

$$q = U_{\text{overall}}A(T_{\text{inside}} - T_{\text{outside}}) = \frac{1}{\Sigma\ \text{Resistance}}A(T_{\text{inside}} - T_{\text{outside}})$$
$$= (0.0693)(150)(70 - 20) = 520\frac{\text{Btu}}{\text{hr}}$$

This problem is just another example of how we can generate finite element models using the direct method.

A Torsional Problem: Direct Formulation

EXAMPLE 1.3

Consider the torsion of a circular shaft, shown in Figure 1.11. Recall from your previous study of the mechanics of materials that the angle of twist θ for a shaft with a constant cross-sectional area with a polar moment of inertia J and length ℓ, made of homogenous material with a shear modulus of elasticity G, subject to a torque T is given by

$$\theta = \frac{T\ell}{JG}$$

Using direct formulation, equilibrium conditions, and

$$\theta = \frac{T\ell}{JG}$$

we can show that for an element comprising two nodes, the stiffness matrix, the angle of twists, and the torques are related according to the equation

$$\frac{JG}{\ell}\begin{bmatrix} 1 & -1 \\ -1 & 1 \end{bmatrix}\begin{Bmatrix} \theta_1 \\ \theta_2 \end{Bmatrix} = \begin{Bmatrix} T_1 \\ T_2 \end{Bmatrix} \tag{1.38}$$

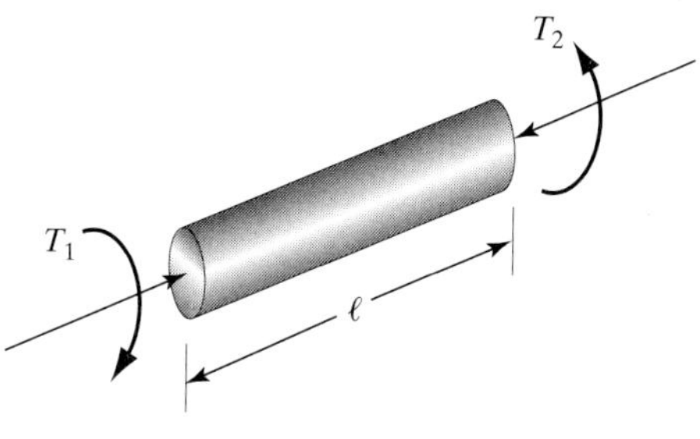

FIGURE 1.11 A torsion of circular shaft.

We will discuss torsional problems in much more detail in Chapter 10. For now, let us consider a shaft that is made of two parts, as shown in Figure 1.12. Part AB is made of material with a shear modulus of elasticity of $G_{AB} = 3.9 \times 10^6\ \text{lb/in}^2$ and has a diameter of 1.5 in. Segment BC is made of slightly different material with a shear modulus of elasticity of $G_{BC} = 4.0 \times 10^6\ \text{lb/in}^2$ and with a diameter of 1 in. The shaft is fixed at both ends. A torque of 200 lb · ft is applied at D. Using three elements, let us determine the angle of twist at D and B, and the torsional reactions at the boundaries.

We will represent this problem by a model that has four nodes at A, B, C, and D, respectively, and three elements (AD, DB, BC).

The polar moment of inertia for each element is given by

$$J_1 = J_2 = \frac{1}{2}\pi r^4 = \frac{1}{2}\pi\left(\frac{1.5}{2}\text{ in}\right)^4 = 0.497\text{ in}^4$$

$$J_3 = \frac{1}{2}\pi r^4 = \frac{1}{2}\pi\left(\frac{1.0}{2}\text{ in}\right)^4 = 0.0982\text{ in}^4$$

The stiffness matrix for each element is computed from Eq. (1.38) as

$$[\mathbf{K}]^{(e)} = \frac{JG}{\ell}\begin{bmatrix} 1 & -1 \\ -1 & 1 \end{bmatrix}$$

So, for element (1), the stiffness matrix is

$$[\mathbf{K}]^{(1)} = \frac{(0.497\text{ in}^4)(3.9 \times 10^6\text{ lb/in}^2)}{(12 \times 2.5)\text{ in}}\begin{bmatrix} 1 & -1 \\ -1 & 1 \end{bmatrix} = \begin{bmatrix} 64610 & -64610 \\ -64610 & 64610 \end{bmatrix}\text{lb} \cdot \text{in}$$

and its position in the global stiffness matrix is

$$[\mathbf{K}]^{(1G)} = \begin{bmatrix} 64610 & -64610 & 0 & 0 \\ -64610 & 64610 & 0 & 0 \\ 0 & 0 & 0 & 0 \\ 0 & 0 & 0 & 0 \end{bmatrix}\begin{matrix} \theta_1 \\ \theta_2 \\ \theta_3 \\ \theta_4 \end{matrix}$$

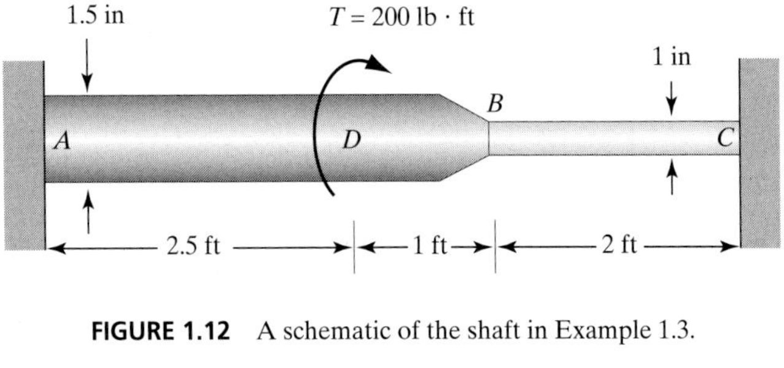

FIGURE 1.12 A schematic of the shaft in Example 1.3.

Similarly, for elements (2) and (3), their respective stiffness matrices and positions in the global stiffness matrix are as follows:

$$[\mathbf{K}]^{(2)} = \frac{(3.9 \times 10^6\,\text{lb/in}^2)(0.497\,\text{in}^4)}{(12 \times 1.0)\,\text{in}}\begin{bmatrix} 1 & -1 \\ -1 & 1 \end{bmatrix} = \begin{bmatrix} 161525 & -161525 \\ -161525 & 161525 \end{bmatrix}\text{lb}\cdot\text{in}$$

$$[\mathbf{K}]^{(2G)} = \begin{bmatrix} 0 & 0 & 0 & 0 \\ 0 & 161525 & -161525 & 0 \\ 0 & -161525 & 161525 & 0 \\ 0 & 0 & 0 & 0 \end{bmatrix}\begin{matrix} \theta_1 \\ \theta_2 \\ \theta_3 \\ \theta_4 \end{matrix}$$

$$[\mathbf{K}]^{(3)} = \frac{(4.0 \times 10^6\,\text{lb/in}^2)(0.0982\,\text{in}^4)}{(12 \times 2.0)\,\text{in}}\begin{bmatrix} 1 & -1 \\ -1 & 1 \end{bmatrix} = \begin{bmatrix} 16367 & -16367 \\ -16367 & 16367 \end{bmatrix}\text{lb}\cdot\text{in}$$

$$[\mathbf{K}]^{(3G)} = \begin{bmatrix} 0 & 0 & 0 & 0 \\ 0 & 0 & 0 & 0 \\ 0 & 0 & 16367 & -16367 \\ 0 & 0 & -16367 & 16367 \end{bmatrix}\begin{matrix} \theta_1 \\ \theta_2 \\ \theta_3 \\ \theta_4 \end{matrix}$$

The final global matrix is obtained simply by assembling, or adding, elemental descriptions:

$$[\mathbf{K}]^{(G)} = [\mathbf{K}]^{(1G)} + [\mathbf{K}]^{(2G)} + [\mathbf{K}]^{(3G)}$$

$$[\mathbf{K}]^{(G)} = \begin{bmatrix} 64610 & -64610 & 0 & 0 \\ -64610 & 64610 + 161525 & -161525 & 0 \\ 0 & -161525 & 161525 + 16367 & -16367 \\ 0 & 0 & -16367 & 16367 \end{bmatrix}$$

Applying the fixed boundary conditions at points A and C and applying the external torque, we have

$$\begin{bmatrix} 1 & 0 & 0 & 0 \\ -64610 & 226135 & -161525 & 0 \\ 0 & -161525 & 177892 & -16367 \\ 0 & 0 & 0 & 1 \end{bmatrix}\begin{Bmatrix} \theta_1 \\ \theta_2 \\ \theta_3 \\ \theta_4 \end{Bmatrix} = \begin{Bmatrix} 0 \\ -(200 \times 12)\,\text{lb}\cdot\text{in} \\ 0 \\ 0 \end{Bmatrix}$$

Solving the above set of equations, we obtain

$$\begin{Bmatrix} \theta_1 \\ \theta_2 \\ \theta_3 \\ \theta_4 \end{Bmatrix} = \begin{Bmatrix} 0 \\ -0.03020\,\text{rad} \\ -0.02742\,\text{rad} \\ 0 \end{Bmatrix}$$

The reaction moments at boundaries A and C can be determined as follows:

$$\{\mathbf{R}\} = [\mathbf{K}]\{\boldsymbol{\theta}\} - \{\mathbf{T}\}$$

$$\begin{Bmatrix} R_A \\ R_D \\ R_B \\ R_C \end{Bmatrix} = \begin{bmatrix} 64610 & -64610 & 0 & 0 \\ -64610 & 226135 & -161525 & 0 \\ 0 & -161525 & 177892 & -16367 \\ 0 & 0 & -16367 & 16367 \end{bmatrix} \begin{Bmatrix} 0 \\ -0.03020 \text{ rad} \\ -0.02742 \text{ rad} \\ 0 \end{Bmatrix} - \begin{Bmatrix} 0 \\ -(200 \times 12) \text{ lb} \cdot \text{in} \\ 0 \\ 0 \end{Bmatrix}$$

$$\begin{Bmatrix} R_A \\ R_D \\ R_B \\ R_C \end{Bmatrix} = \begin{Bmatrix} 1951 \text{ lb} \cdot \text{in} \\ 0 \\ 0 \\ 449 \text{ lb} \cdot \text{in} \end{Bmatrix}$$

Note that the sum of R_A and R_C is equal to the applied torque of 2400 lb·in. Also note that the change in the diameter of the shafts will give rise to stress concentrations that are not accounted for by the model we used here.

EXAMPLE 1.4

A steel plate is subjected to an axial load, as shown in Figure 1.13. Approximate the deflections and average stresses along the plate. The plate is 1/16 in thick and has a modulus of elasticity $E = 29 \times 10^6 \text{ lb/in}^2$.

We may model this problem using four nodes and four elements, as shown in Figure 1.13. Next, we compute the equivalent stiffness coefficient for each element:

$$k_1 = \frac{A_1 E}{\ell_1} = \frac{(5)(0.0625)(29 \times 10^6)}{1} = 9{,}062{,}500 \text{ lb/in}$$

$$k_2 = k_3 = \frac{A_2 E}{\ell_2} = \frac{(2)(0.0625)(29 \times 10^6)}{4} = 906{,}250 \text{ lb/in}$$

$$k_4 = \frac{A_4 E}{\ell_4} = \frac{(5)(0.0625)(29 \times 10^6)}{2} = 4{,}531{,}250 \text{ lb/in}$$

The stiffness matrix for element (1) is

$$[\mathbf{K}]^{(1)} = \begin{bmatrix} k_1 & -k_1 \\ -k_1 & k_1 \end{bmatrix}$$

and its position in the global stiffness matrix is

$$[\mathbf{K}]^{(1G)} = \begin{bmatrix} k_1 & -k_1 & 0 & 0 \\ -k_1 & k_1 & 0 & 0 \\ 0 & 0 & 0 & 0 \\ 0 & 0 & 0 & 0 \end{bmatrix} \begin{matrix} u_1 \\ u_2 \\ u_3 \\ u_4 \end{matrix}$$

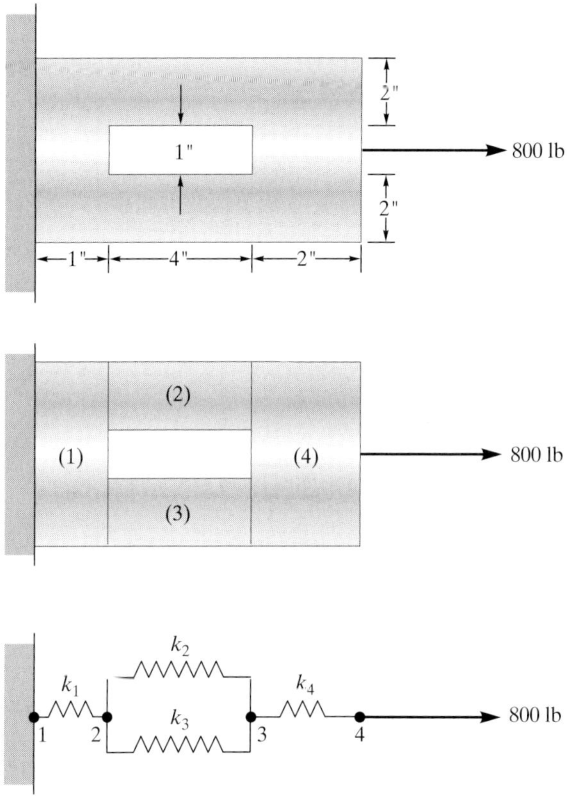

FIGURE 1.13 A schematic of the steel plate in Example 1.4.

Similarly, the respective stiffness matrices and positions in the global stiffness matrix for elements (2), (3), and (4) are

$$[\mathbf{K}]^{(2)} = \begin{bmatrix} k_2 & -k_2 \\ -k_2 & k_2 \end{bmatrix} \qquad [\mathbf{K}]^{(2G)} = \begin{bmatrix} 0 & 0 & 0 & 0 \\ 0 & k_2 & -k_2 & 0 \\ 0 & -k_2 & k_2 & 0 \\ 0 & 0 & 0 & 0 \end{bmatrix} \begin{matrix} u_1 \\ u_2 \\ u_3 \\ u_4 \end{matrix}$$

$$[\mathbf{K}]^{(3)} = \begin{bmatrix} k_3 & -k_3 \\ -k_3 & k_3 \end{bmatrix} \qquad [\mathbf{K}]^{(3G)} = \begin{bmatrix} 0 & 0 & 0 & 0 \\ 0 & k_3 & -k_3 & 0 \\ 0 & -k_3 & k_3 & 0 \\ 0 & 0 & 0 & 0 \end{bmatrix} \begin{matrix} u_1 \\ u_2 \\ u_3 \\ u_4 \end{matrix}$$

$$[\mathbf{K}]^{(4)} = \begin{bmatrix} k_4 & -k_4 \\ -k_4 & k_4 \end{bmatrix} \qquad [\mathbf{K}]^{(4G)} = \begin{bmatrix} 0 & 0 & 0 & 0 \\ 0 & 0 & 0 & 0 \\ 0 & 0 & k_4 & -k_4 \\ 0 & 0 & -k_4 & k_4 \end{bmatrix} \begin{matrix} u_1 \\ u_2 \\ u_3 \\ u_4 \end{matrix}$$

The final global matrix is obtained simply by assembling, or adding, the individual elemental matrices:

$$[\mathbf{K}]^{(G)} = [\mathbf{K}]^{(1G)} + [\mathbf{K}]^{(2G)} + [\mathbf{K}]^{(3G)} + [\mathbf{K}]^{(4G)}$$

$$[\mathbf{K}]^{(G)} = \begin{bmatrix} k_1 & -k_1 & 0 & 0 \\ -k_1 & k_1 + k_2 + k_3 & -k_2 - k_3 & 0 \\ 0 & -k_2 - k_3 & k_2 + k_3 + k_4 & -k_4 \\ 0 & 0 & -k_4 & k_4 \end{bmatrix}$$

Substituting for the elements' respective stiffness coefficients, the global stiffness matrix becomes

$$[\mathbf{K}]^{(G)} = \begin{bmatrix} 9{,}062{,}500 & -9{,}062{,}500 & 0 & 0 \\ -9{,}062{,}500 & 10{,}875{,}000 & -1{,}812{,}500 & 0 \\ 0 & -1{,}812{,}500 & 6{,}343{,}750 & -4{,}531{,}250 \\ 0 & 0 & -4{,}531{,}250 & 4{,}531{,}250 \end{bmatrix}$$

Applying the boundary condition $u_1 = 0$ and the load to node 4, we obtain

$$\begin{bmatrix} 1 & 0 & 0 & 0 \\ -9{,}062{,}500 & 10{,}875{,}000 & -1{,}812{,}500 & 0 \\ 0 & -1{,}812{,}500 & 6{,}343{,}750 & -4{,}531{,}250 \\ 0 & 0 & -4{,}531{,}250 & 4{,}531{,}250 \end{bmatrix} \begin{Bmatrix} u_1 \\ u_2 \\ u_3 \\ u_4 \end{Bmatrix} = \begin{Bmatrix} 0 \\ 0 \\ 0 \\ 800 \end{Bmatrix}$$

Solving the system of equations yields the displacement solution as

$$\begin{Bmatrix} u_1 \\ u_2 \\ u_3 \\ u_4 \end{Bmatrix} = \begin{Bmatrix} 0 \\ 8.827 \times 10^{-5} \\ 5.296 \times 10^{-4} \\ 7.062 \times 10^{-4} \end{Bmatrix} \text{in}$$

and the stresses in each element are

$$\sigma^{(1)} = E\left(\frac{u_2 - u_1}{\ell}\right) = \frac{(29 \times 10^6)(8.827 \times 10^{-5} - 0)}{1} = 2560 \frac{\text{lb}}{\text{in}^2}$$

$$\sigma^{(2)} = \sigma^{(3)} = E\left(\frac{u_3 - u_2}{\ell}\right) = \frac{(29 \times 10^6)(5.296 \times 10^{-4} - 8{,}827 \times 10^{-5})}{4} = 3200 \frac{\text{lb}}{\text{in}^2}$$

$$\sigma^{(4)} = E\left(\frac{u_4 - u_3}{\ell}\right) = \frac{(29 \times 10^6)(7.062 \times 10^{-4} - 5.296 \times 10^{-4})}{2} = 2560 \frac{\text{lb}}{\text{in}^2}$$

Note that the model used to analyze this problem consisted of springs in parallel as well as in series. The two springs in parallel could have been combined and represented by a single spring having a stiffness equal to $k_2 + k_3$ (see Problem 25). Also note that because of the hole, the abrupt changes in the cross section of the strip will give rise to stress concentrations with values exceeding those average values we computed

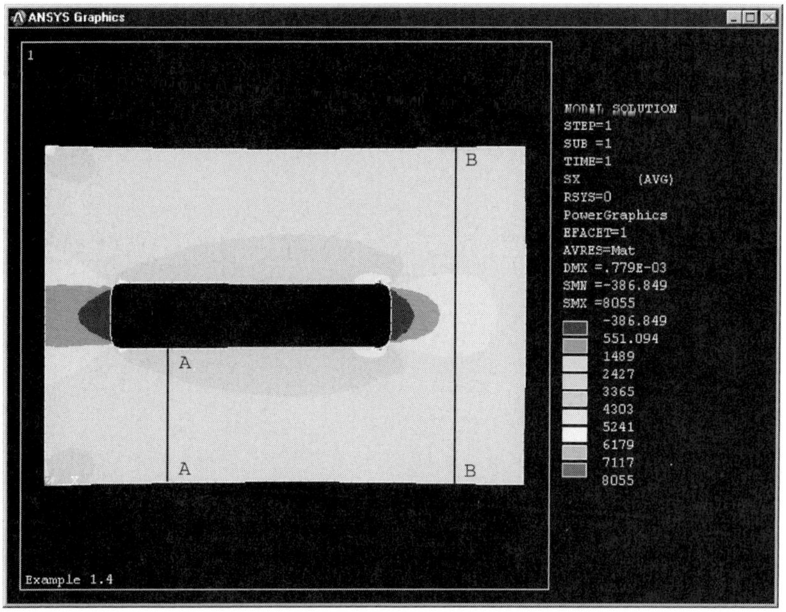

FIGURE 1.14 The x-component of stress distribution for the plate in Example 1.4, as computed by ANSYS.

here. After you study plane-stress finite element formulation (discussed in Chapter 10), you will revisit this problem (see Problem 10.13) and be asked to solve it using ANSYS. Furthermore, you will be asked to plot the components of the stress distributions in the plate and thus identify the location and magnitude of maximum stresses.

To give you just a taste of what is to come in Chapter 10 and also to shed more light on our discussion about the stress concentration regions, we have solved Example 1.4 using ANSYS and have determined the x-component of the stress distribution in the plate, as shown in Figure 1.14. In the results shown in Figure 1.14, the load was applied as a pressure over the entire right surface of the bar. Note the variation of the stresses at section A–A from approximately 3000 psi to 3500 psi. At section B–B, the x-component of the stresses varies from approximately 2300 psi to 2600 psi. These values are not that far off from the average stress values obtained using the direct model. Also note that the maximum and minimum stress values given by ANSYS could change, depending upon how we apply the load to the bar, especially in the regions near the point of load application and the regions near the hole. Keeping in mind Example 1.4 and Figure 1.13, remember that in a real situation, the load would be applied over an area, not at a single point. Thus, remember that how you apply the external load to your finite element model will influence the stress distribution results, particularly in the region near where the load is applied. This principle is especially true in Example 1.4 because it deals with a short plate with a hole.

For the sake of convenience, the results of Section 1.5 is summarized in Table 1.6. In the table, carefully examine what constitutes an element, its degrees of freedom, and the physical balance requirements.

TABLE 1.6 Examples of elements and nodes

Element	Degrees of Freedom	Physical Balance Requirement
Linear Elastic Element (linear spring) 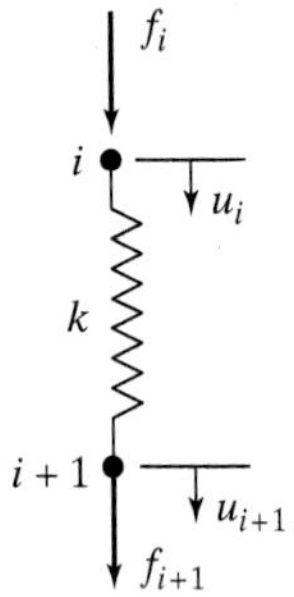 $f_i = k(u_i - u_{i+1})$ $f_{i+1} = k(u_{i+1} - u_i)$	nodal displacements: u_i, u_{i+1}	force balance: $f_i + f_{i+1} = 0$
Torsional Elastic Element (torsional spring) 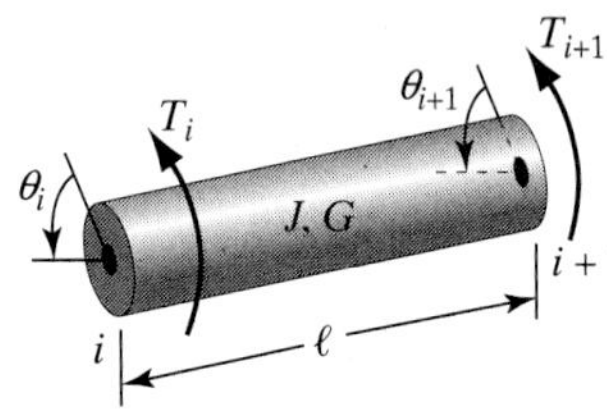 $T_i = \frac{JG}{\ell}(\theta_i - \theta_{i+1})$ $T_{i+1} = \frac{JG}{\ell}(\theta_{i+1} - \theta_i)$	nodal angle of twist: θ_i, θ_{i+1}	torque balance: $T_i + T_{i+1} = 0$
Conduction Element 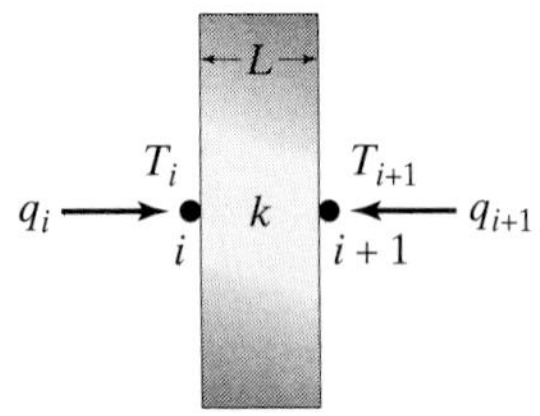 $q_i = \frac{kA}{L}(T_i - T_{i+1})$ $q_{i+1} = \frac{kA}{L}(T_{i+1} - T_i)$	nodal temperatures: T_i, T_{i+1}	energy balance: $q_i + q_{i+1} = 0$
Laminar Pipe Flow Element (See Section 12.1) 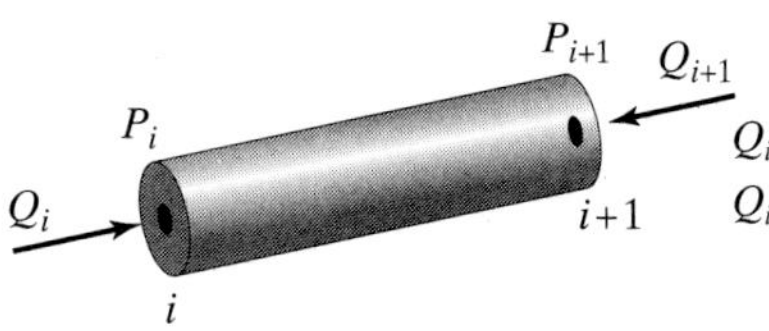 $Q_i = C(P_i - P_{i+1})$ $Q_{i+1} = C(P_{i+1} - P_i)$	nodal pressures: P_i, P_{i+1}	flow balance: $Q_i + Q_{i+1} = 0$
Electrical Resistance Element 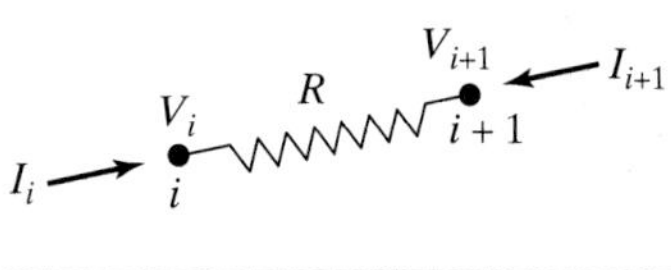 $I_i = \frac{1}{R}(V_i - V_{i+1})$ $I_{i+1} = \frac{1}{R}(V_{i+1} - V_i)$	nodal voltages: V_i, V_{i+1}	electric current balance: $I_i + I_{i+1} = 0$

1.6 MINIMUM TOTAL POTENTIAL ENERGY FORMULATION

The minimum total potential energy formulation is a common approach in generating finite element models in solid mechanics. External loads applied to a body will cause the body to deform. During the deformation, the work done by the external forces is stored in the material in the form of elastic energy, called strain energy. Let us consider the strain energy in a solid member when it is subjected to a central force F, as shown in Figure 1.15.

Also shown in Figure 1.15 is a piece of material from the member in the form of differential volume and the normal stresses acting on the surfaces of this volume. Earlier, it was shown that the elastic behavior of the member may be modeled as a linear spring. In Figure 1.15 note that y' is a variable measuring deformation of the member and its value varies from 0 to $\Delta\ell$. When the member is stretched by a differential amount dy', the stored energy in the material is

$$\Lambda = \int_0^{y'} F dy' = \int_0^{y'} ky'dy' = \frac{1}{2}ky'^2 = \left(\frac{1}{2}ky'\right)y' \tag{1.39}$$

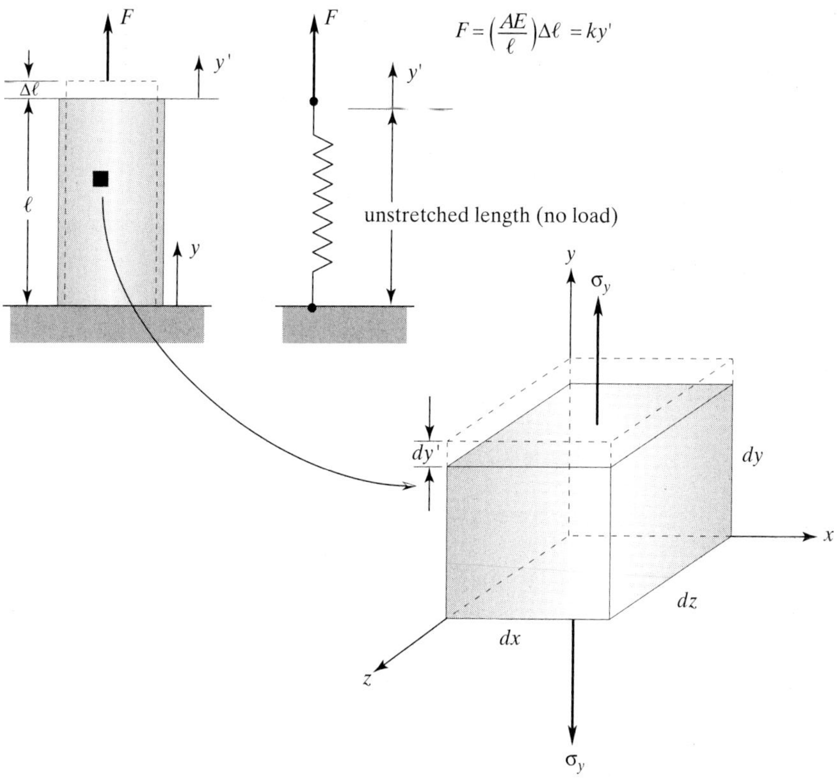

FIGURE 1.15 The elastic behavior of a member subjected to a central load.

We can write Eq. (1.39)—for a piece of material from the member in the form of differential volume—in terms of the normal stress (σ) and strain (ε):

$$d\Lambda = \frac{1}{2}\overbrace{(ky')}^{\text{elastic force}}\, dy' = \frac{1}{2}\overbrace{(\sigma_y dxdz)}^{\text{elastic force}}\overbrace{\varepsilon dy}^{dy'} = \frac{1}{2}\sigma\varepsilon\, dV$$

Therefore, for a member or an element under axial loading, the strain energy $\Lambda^{(e)}$ is obtained by adding up the stored energy in all pieces (differential volumes) making up the member:

$$\Lambda^{(e)} = \int d\Lambda = \int_V \frac{\sigma\varepsilon}{2} dV = \int_V \frac{E\varepsilon^2}{2} dV \qquad (1.40)$$

where V is the volume of the member. The total potential energy Π for a body consisting of n elements and m nodes is the difference between the total strain energy and the work done by the external forces:

$$\Pi = \sum_{e=1}^{n} \Lambda^{(e)} - \sum_{i=1}^{m} F_i u_i \qquad (1.41)$$

The minimum total potential energy principle simply states that for a stable system, the displacement at the equilibrium position occurs such that the value of the system's total potential energy is a minimum.

$$\frac{\partial \Pi}{\partial u_i} = \frac{\partial}{\partial u_i}\sum_{e=1}^{n} \Lambda^{(e)} - \frac{\partial}{\partial u_i}\sum_{i=1}^{m} F_i u_i = 0 \quad \text{for } i = 1, 2, 3, \ldots, n \qquad (1.42)$$

The following examples offer insight into the physical meaning of Eq. (1.42).

EXAMPLE 1.5

Consider the following situations: (a) We have applied a force F to a linear spring as shown in Figure 1.16. Depending on the stiffness value of the spring, the spring stretches by a certain amount x. The static equilibrium requires that the applied force F be equal to the internal force in the spring kx.

$$F = kx \quad \text{or} \quad x = \frac{F}{k}$$

Now, let us consider the total potential energy of the system as defined by Eq. (1.41). The stored elastic energy in the spring is $\Lambda = \frac{1}{2}kx^2$ and the work done by the external force F is Fx (force times displacement). Thus, the total potential energy of the system is

$$\Pi = \frac{1}{2}kx^2 - Fx$$

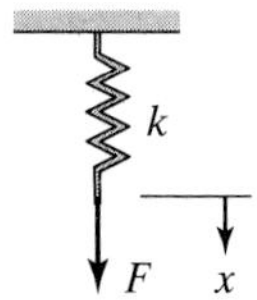

FIGURE 1.16 A linear spring subjected to a force F.

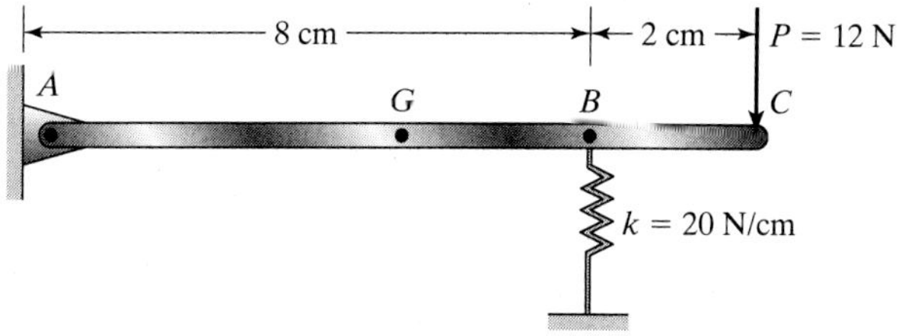

FIGURE 1.17 The rod of Example 1.5.

Minimizing Π with respect to x, we have

$$\frac{d\Pi}{dx} = \frac{d}{dx}\left(\frac{1}{2}kx^2 - Fx\right) = kx - F = 0$$

which results in $x = \dfrac{F}{k}$.

(b) The slender rod shown in Figure 1.17 weighs 8 N and is supported by a spring with a stiffness $k = 20\ N/\text{cm}$. A force $P = 12\ N$ is applied to the end of the rod at point C. We are interested in determining the deflection of the spring.

First, we solve this problem by applying the static equilibrium conditions and then apply the minimum total potential energy concept. Static equilibrium requires that sum of the moments of the forces acting on the rod about point A be zero. Considering the free-body diagram of the rod shown in Figure 1.18, we find

$$\circlearrowleft\!+ \sum M_A = 0 \qquad -(8N)(5\text{ cm}) + F_s(8\text{ cm}) - (12\ N)(10\text{ cm}) = 0$$

$$F_s = 20\ N \quad \text{and} \quad kx = (20\ N/\text{cm})(x) = 20\ N$$

$$x = 1\text{ cm}$$

Now, we solve the problem using the minimum total potential energy approach. We note that elastic energy stored in the system is predominantly due to elastic energy of the spring and is given by

$$\Lambda = \frac{1}{2}kx^2 = \frac{1}{2}(20\ N/\text{cm})(x^2) = 10x^2$$

The work done by the external forces is calculated by multiplying the weight of the rod by the displacement of point G, and force P by the displacement of endpoint C. Through

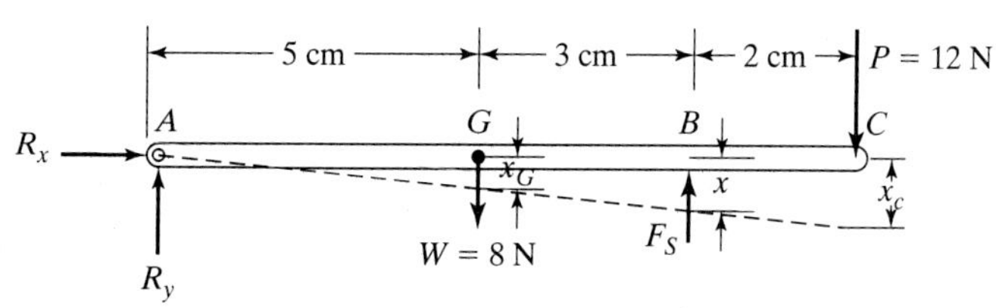

FIGURE 1.18 The free-body diagram of the rod in Example 1.5.

similar triangles, we can relate the displacements of points G and C to the displacement of the spring (point B) according to

$$\frac{x}{8} = \frac{x_G}{5} \quad \text{or} \quad x_G = \frac{5}{8}x$$

$$\frac{x}{8} = \frac{x_C}{10} \quad \text{or} \quad x_C = \frac{5}{4}x$$

Thus, the work done by the external forces is given by

$$\sum F_i u_i = (8\,N)\left(\frac{5}{8}x\right) + (12\,N)\left(\frac{5}{4}x\right) = 5x + 15x = 20x$$

The total potential energy of the system is

$$\Pi = \sum \Lambda - \sum F_i u_i = 10x^2 - 20x$$

and

$$\frac{d\Pi}{dx} = \frac{d}{dx}\left(10x^2 - 20x\right) = 20x - 20 = 0$$

Solving the above equation for x, we find $x = 1$ cm. Because there is only one unknown displacement, note that when we employed Eqs. (1.41) and (1.42), we replaced the displacement u_i with x and the partial derivative symbol with the ordinary symbol. We have plotted the total potential energy $\Pi = 10x^2 - 20x$ as a function of displacement x in Figure 1.19. It is clear from examining Figure 1.19 that the minimum total potential energy occurs at $x = 1$ cm.

Now, let us turn our attention back to Example 1.1. The strain energy for an arbitrary element (e) can be determined from Eq. (1.40) as

$$\Lambda^{(e)} = \int_V \frac{E\varepsilon^2}{2}\,dV = \frac{A_{\text{avg}}E}{2\ell}(u_{i+1}^2 + u_i^2 - 2u_{i+1}u_i) \tag{1.43}$$

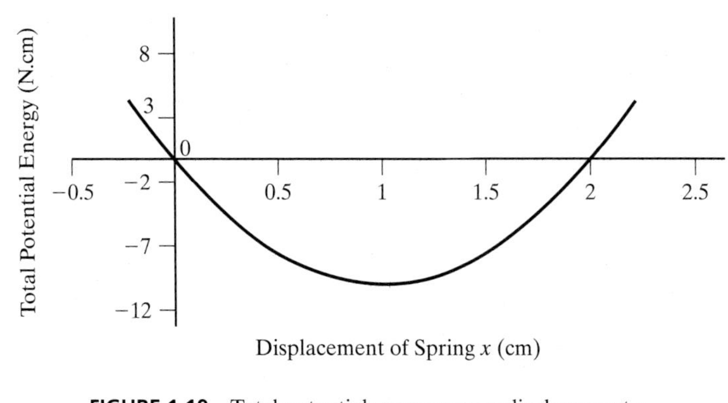

FIGURE 1.19 Total potential energy versus displacement x.

where $\varepsilon = (u_{i+1} - u_i)/\ell$ was substituted for the axial strain. Minimizing the strain energy with respect to u_i and u_{i+1} leads to

$$\frac{\partial \Lambda^{(e)}}{\partial u_i} = \frac{A_{\text{avg}}E}{\ell}(u_i - u_{i+1})$$
$$\frac{\partial \Lambda^{(e)}}{\partial u_{i+1}} = \frac{A_{\text{avg}}E}{\ell}(u_{i+1} - u_i) \tag{1.44}$$

and, in matrix form,

$$\begin{Bmatrix} \dfrac{\partial \Lambda^{(e)}}{\partial u_i} \\ \dfrac{\partial \Lambda^{(e)}}{\partial u_{i+1}} \end{Bmatrix} = \begin{bmatrix} k_{\text{eq}} & -k_{\text{eq}} \\ -k_{\text{eq}} & k_{\text{eq}} \end{bmatrix} \begin{Bmatrix} u_i \\ u_{i+1} \end{Bmatrix} \tag{1.45}$$

where $k_{\text{eq}} = (A_{\text{avg}}E)/\ell$. Minimizing the work done by the external forces at nodes i and $i + 1$ of an arbitrary element (e), we get

$$\frac{\partial}{\partial u_i}(F_i u_i) = F_i$$
$$\frac{\partial}{\partial u_{i+1}}(F_{i+1} u_{i+1}) = F_{i+1} \tag{1.46}$$

For Example 1.1, the minimum total potential energy formulation leads to a global stiffness matrix that is identical to the one obtained from direct formulation:

$$[\mathbf{K}]^{(G)} = \begin{bmatrix} k_1 & -k_1 & 0 & 0 & 0 \\ -k_1 & k_1 + k_2 & -k_2 & 0 & 0 \\ 0 & -k_2 & k_2 + k_3 & -k_3 & 0 \\ 0 & 0 & -k_3 & k_3 + k_4 & -k_4 \\ 0 & 0 & 0 & -k_4 & k_4 \end{bmatrix}$$

Furthermore, application of the boundary condition and the load results in

$$\begin{bmatrix} 1 & 0 & 0 & 0 & 0 \\ -k_1 & k_1 + k_2 & -k_2 & 0 & 0 \\ 0 & -k_2 & k_2 + k_3 & -k_3 & 0 \\ 0 & 0 & -k_3 & k_3 + k_4 & -k_4 \\ 0 & 0 & 0 & -k_4 & k_4 \end{bmatrix} \begin{Bmatrix} u_1 \\ u_2 \\ u_3 \\ u_4 \\ u_5 \end{Bmatrix} = \begin{Bmatrix} 0 \\ 0 \\ 0 \\ 0 \\ P \end{Bmatrix} \tag{1.47}$$

The displacement results will be identical to the ones obtained earlier from the direct method, as given by Eq. (1.17). The concepts of strain energy and minimum total potential energy will be used to formulate solid mechanics problems in Chapters 4, 10, and 13. Therefore, spending a little extra time now to understand the basic ideas will benefit you enormously later.

Example 1.1: Exact Solution*

In this section, we will derive the exact solution to Example 1.1 and compare the finite element formulation displacement results for this problem to the exact displacement solutions. As shown in Figure 1.20, the statics equilibrium requires the sum of the forces in the y-direction to be zero. This requirement leads to the relation

$$P - (\sigma_{avg})A(y) = 0 \tag{1.48}$$

Once again, using Hooke's law ($\sigma = E\varepsilon$) and substituting for the average stress in terms of the strain, we have

$$P - E\varepsilon A(y) = 0 \tag{1.49}$$

Recall that the average normal strain is the change in length du per unit original length of the differential segment dy. So,

$$\varepsilon = \frac{du}{dy}$$

If we substitute this relationship into Eq. (1.49), we now have

$$P - EA(y)\frac{du}{dy} = 0 \tag{1.50}$$

Rearranging Eq. (1.50), we get

$$du = \frac{Pdy}{EA(y)} \tag{1.51}$$

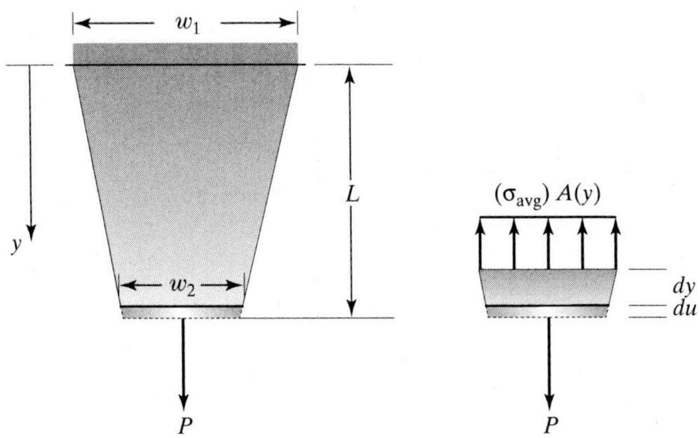

FIGURE 1.20 The relationship between the external force P and the average stresses for the bar in Example 1.1.

*The contribution of shear stresses is neglected.

TABLE 1.7 Comparison of displacement results

Location of a Point Along the Bar (in)	Results from the Exact Displacement Method (in) Eq. (1.53)	Results from the Direct Method (in)	Results from the Energy Method (in)
$y = 0$	0	0	0
$y = 2.5$	0.001027	0.001026	0.001026
$y = 5.0$	0.002213	0.002210	0.002210
$y = 7.5$	0.003615	0.003608	0.003608
$y = 10$	0.005333	0.005317	0.005317

The exact solution is then obtained by integrating Eq. (1.51) over the length of the bar

$$\int_0^u du = \int_0^L \frac{Pdy}{EA(y)}$$

$$u(y) = \int_0^y \frac{Pdy}{EA(y)} = \int_0^y \frac{Pdy}{E\left(w_1 + \left(\frac{w_2 - w_1}{L}\right)y\right)t} \tag{1.52}$$

where the area is

$$A(y) = \left(w_1 + \left(\frac{w_2 - w_1}{L}\right)y\right)t$$

The deflection profile along the bar is obtained by integrating Eq. (1.52), resulting in

$$u(y) = \frac{PL}{Et(w_2 - w_1)}\left[\ln\left(w_1 + \left(\frac{w_2 - w_1}{L}\right)y\right) - \ln w_1\right] \tag{1.53}$$

Equation (1.53) can be used to generate displacement values at various points along the bar. It is now appropriate to examine the accuracy of the direct and potential energy methods by comparing their displacement results with the values. Table 1.7 shows nodal displacements computed using direct and energy methods.

It is clear from examination of Table 1.7 that all of the results are in agreement with each other.

1.7 WEIGHTED RESIDUAL FORMULATIONS

The *weighted residual methods* are based on assuming an approximate solution for the governing differential equation. The assumed solution must satisfy the initial and boundary conditions of the given problem. Because the assumed solution is not exact, substitution of the solution into the differential equation will lead to some *residuals* or *errors*. Simply stated, each residual method requires the error to vanish over some selected intervals or at some points. To demonstrate this concept, let's turn our attention to

Example 1.1. The governing differential equation and the corresponding boundary condition for this problem are as follows:

$$A(y)E\frac{du}{dy} - P = 0 \quad \text{subject to the boundary condition } u(0) = 0 \tag{1.54}$$

Next, we need to assume an approximate solution. Again, keep in mind that the assumed solution must satisfy the boundary condition. We choose

$$u(y) = c_1 y + c_2 y^2 + c_3 y^3 \tag{1.55}$$

where c_1, c_2, and c_3 are unknown coefficients. Equation (1.55) certainly satisfies the fixed boundary condition represented by $u(0) = 0$. Substitution of the assumed solution, Eq. (1.55), into the governing differential equation, Eq. (1.54), yields the error function $\mathcal{R}$:

$$\overbrace{\left(w_1 + \left(\frac{w_2 - w_1}{L}\right)y\right)}^{A(y)} tE\overbrace{(c_1 + 2c_2 y + 3c_3 y^2)}^{\frac{du}{dy}} - P = \mathcal{R} \tag{1.56}$$

Substituting for values of w_1, w_2, L, t, and E in Example 1.1 and simplifying, we get

$$\mathcal{R}/E = (0.25 - 0.0125y)(c_1 + 2c_2 y + 3c_3\, y^2) - 96.154 \times 10^{-6}$$

Collocation Method

In the *collocation method* the error, or residual, function $\mathcal{R}$ is forced to be zero at as many points as there are unknown coefficients. Because the assumed solution in this example has three unknown coefficients, we will force the error function to equal zero at three points. We choose the error function to vanish at $y = L/3$, $y = 2L/3$, and $y = L$:

$$\mathcal{R}(c, y)\Big|_{y=\frac{L}{3}} = 0$$

$$\mathcal{R} = \left(0.25 - 0.0125\left(\frac{10}{3}\right)\right)\left(c_1 + 2c_2\left(\frac{10}{3}\right) + 3c_3\left(\frac{10}{3}\right)^2\right) - 96.154 \times 10^{-6} = 0$$

$$\mathcal{R}(c, y)\Big|_{y=\frac{2L}{3}} = 0$$

$$\mathcal{R} = \left(0.25 - 0.0125\left(\frac{20}{3}\right)\right)\left(c_1 + 2c_2\left(\frac{20}{3}\right) + 3c_3\left(\frac{20}{3}\right)^2\right) - 96.154 \times 10^{-6} = 0$$

$$\mathcal{R}(c, y)\Big|_{y=L} = 0$$

$$\mathcal{R} = (0.25 - 0.0125(10))(c_1 + 2c_2(10) + 3c_3(10)^2) - 96.154 \times 10^{-6} = 0$$

This procedure creates three linear equations that we can solve to obtain the unknown coefficients c_1, c_2, and c_3:

$$\begin{aligned} c_1 + \frac{20}{3}c_2 + \frac{100}{3}c_3 &= 461.539 \times 10^{-6} \\ c_1 + \frac{40}{3}c_2 + \frac{400}{3}c_3 &= 576.924 \times 10^{-6} \\ c_1 + 20c_2 + 300c_3 &= 769.232 \times 10^{-6} \end{aligned}$$

Solving the above equations yields $c_1 = 423.0776 \times 10^{-6}$, $c_2 = 21.65 \times 10^{-15}$, and $c_3 = 1.153848 \times 10^{-6}$. Substitution of the c-coefficients into Eq. (1.55) yields the approximate displacement profile:

$$u(y) = 423.0776 \times 10^{-6}y + 21.65 \times 10^{-15}y^2 + 1.153848 \times 10^{-6}y^3 \tag{1.57}$$

In order to get an idea of how accurate the collocation approximate results are, we will compare them to the exact results later in this chapter.

Subdomain Method

In the *subdomain method*, the integral of the error function over some selected subintervals is forced to be zero. The number of subintervals chosen must equal the number of unknown coefficients. Thus, for our assumed solution, we will have three integrals:

$$\int_0^{\frac{L}{3}} \mathcal{R}\,dy = 0 \tag{1.58}$$

$$\int_0^{\frac{L}{3}} [(0.25 - 0.0125y)(c_1 + 2c_2y + 3c_3y^2) - 96.154 \times 10^{-6}]dy = 0$$

$$\int_{\frac{L}{3}}^{\frac{2L}{3}} \mathcal{R}\,dy = 0$$

$$\int_{\frac{L}{3}}^{\frac{2L}{3}} [(0.25 - 0.0125y)(c_1 + 2c_2y + 3c_3y^2) - 96.154 \times 10^{-6}]dy = 0$$

$$\int_{\frac{2L}{3}}^{L} \mathcal{R}\,dy = 0$$

$$\int_{\frac{2L}{3}}^{L} [(0.25 - 0.0125y)(c_1 + 2c_2y + 3c_3y^2) - 96.154 \times 10^{-6}]dy = 0$$

Integration of equations given by Eq. (1.58) results in three linear equations that we can solve to obtain the unknown coefficients c_1, c_2, and c_3:

$$\begin{aligned} 763.88889 \times 10^{-3}c_1 + 2.4691358c_2 + 8.1018519c_3 &= 320.513333 \times 10^{-6} \\ 0.625c_1 + 6.1728395c_2 + 47.4537041c_3 &= 3.2051333 \times 10^{-4} \\ 0.4861111c_1 + 8.0246917c_2 + 100.694444c_3 &= 3.2051333 \times 10^{-4} \end{aligned}$$

Solving the above equations yields $c_1 = 391.35088 \times 10^{-6}$, $c_2 = 6.075 \times 10^{-6}$, and $c_3 = 809.61092 \times 10^{-9}$. Substitution of the c-coefficients into Eq. (1.55) yields the approximate displacement profile:

$$u(y) = 391.35088 \times 10^{-6}y + 6.075 \times 10^{-6}y^2 + 809.61092 \times 10^{-9}y^3 \qquad (1.59)$$

We will compare the displacement results obtained from the subdomain method to the exact results later in this chapter.

Galerkin Method

The *Galerkin method* requires the error to be orthogonal to some weighting functions Φ_i, according to the integral

$$\int_a^b \Phi_i \mathcal{R}\, dy = 0 \qquad i = 1, 2, \ldots, N \qquad (1.60)$$

The weighting functions are chosen to be members of the approximate solution. Because there are three unknowns in the assumed approximate solution for Example 1.1, we need to generate three equations. Recall that the assumed solution is $u(y) = c_1y + c_2y^2 + c_3y^3$; thus, the weighting functions are selected to be $\Phi_1 = y$, $\Phi_2 = y^2$, and $\Phi_3 = y^3$. This selection leads to the following equations:

$$\int_0^L y[(0.25 - 0.0125y)(c_1 + 2c_2y + 3c_3y^2) - 96.154 \times 10^{-6}]dy = 0 \qquad (1.61)$$

$$\int_0^L y^2[(0.25 - 0.0125y)(c_1 + 2c_2y + 3c_3y^2) - 96.154 \times 10^{-6}]dy = 0$$

$$\int_0^L y^3[0.25 - 0.0125y)(c_1 + 2c_2y + 3c_3y^2) - 96.154 \times 10^{-6}]dy = 0$$

Integration of Eq. (1.61) results in three linear equations that we can solve to obtain the unknown coefficients c_1, c_2, and c_3:

$$8.333333c_1 + 104.1666667c_2 + 1125c_3 = 0.0048077$$

$$52.083333c_1 + 750c_2 + 8750c_3 = 0.0320513333$$

$$375c_1 + 5833.3333c_2 + 71428.57143c_3 = 0.240385$$

Solving the above equations yields $c_1 = 400.642 \times 10^{-6}$, $c_2 = 4.006 \times 10^{-6}$, and $c_3 = 0.935 \times 10^{-6}$. Substitution of the c-coefficients into Eq. (1.55) yields the approximate displacement profile:

$$u(y) = 400.642 \times 10^{-6}y + 4.006 \times 10^{-6}y^2 + 0.935 \times 10^{-6}y^3 \qquad (1.62)$$

We will compare the displacement results obtained from the Galerkin method to the exact results later in this chapter.

Least-Squares Method

The *least-squares method* requires the error to be minimized with respect to the unknown coefficients in the assumed solution, according to the relationship

$$\text{Minimize}\left(\int_a^b \mathcal{R}^2 dy\right)$$

which leads to

$$\int_a^b \mathcal{R}\frac{\partial \mathcal{R}}{\partial c_i}dy = 0 \qquad i = 1, 2, \ldots, N \tag{1.63}$$

Because there are three unknowns in the approximate solution of Example 1.1, Eq. (1.63) generates three equations. Recall that the error function is

$$\mathcal{R}/E = (0.25 - 0.0125y)(c_1 + 2c_2y + 3c_3y^2) - 96.154 \times 10^{-6}$$

Differentiating the error function with respect to c_1, c_2, and c_3 and substituting into Eq. (1.63), we have:

$$\int_0^{10} \overbrace{[(0.25 - 0.0125y)(c_1 + 2c_2y + 3c_3y^2) - 96.154\times 10^{-6}]}^{\mathcal{R}}\overbrace{(0.25 - 0.0125y)}^{\frac{\partial \mathcal{R}}{\partial c_1}}dy = 0$$

$$\int_0^{10} \overbrace{[(0.25 - 0.0125y)(c_1 + 2c_2y + 3c_3\,y^2) - 96.154\times 10^{-6}]}^{\mathcal{R}}\overbrace{(0.25 - 0.0125y)2y}^{\frac{\partial \mathcal{R}}{\partial c_2}}dy = 0$$

$$\int_0^{10} \overbrace{[(0.25 - 0.0125y)(c_1 + 2c_2y + 3c_3y^2) - 96.154 \times 10^{-6}]}^{\mathcal{R}}\overbrace{(0.25 - 0.0125y)3y^2}^{\frac{\partial \mathcal{R}}{\partial c_3}}dy = 0$$

Integration of the above equations results in three linear equations that we can solve to obtain the unknown coefficients c_1, c_2, and c_3:

$$0.364583333c_1 + 2.864583333c_2 + 25c_3 = 0.000180289$$
$$2.864583333c_1 + 33.333333c_2 + 343.75c_3 = 0.001602567$$
$$25c_1 + 343.75c_2 + 3883.928571c_3 = 0.015024063$$

Solving the set of equations simultaneously yields $c_1 = 389.773 \times 10^{-6}$, $c_2 = 6.442 \times 10^{-6}$, and $c_3 = 0.789 \times 10^{-6}$. Substitution of the c-coefficients into Eq. (1.55) yields the approximate displacement profile:

$$u(y) = 389.733 \times 10^{-6}y + 6.442 \times 10^{-6}y^2 + 0.789 \times 10^{-6}y^3 \tag{1.64}$$

Next, we will compare the displacement results obtained from the least-squares method and the other weighted residual methods to the exact results.

TABLE 1.8 Comparison of weighted residual results

Location of a Point Along the Bar (in)	Displacement Results from the Exact Solution Eq. (1.53) (in)	Displacement Results from the Collocation Method Eq. (1.57) (in)	Displacement Results from the Subdomain Method Eq. (1.59) (in)	Displacement Results from the Galerkin Method Eq. (1.62) (in)	Displacement Results from the Least-Squares Method Eq. (1.64) (in)
$y = 0$	0	0	0	0	0
$y = 2.5$	0.001027	0.001076	0.001029	0.001041	0.001027
$y = 5.0$	0.002213	0.002259	0.002209	0.002220	0.002208
$y = 7.5$	0.003615	0.003660	0.003618	0.003624	0.003618
$y = 10$	0.005333	0.005384	0.005330	0.005342	0.005331

Comparison of Weighted Residual Solutions

Now we will examine the accuracy of weighted residual methods by comparing their displacement results with the exact values. Table 1.8 shows nodal displacements computed using the exact, collocation, subdomain, Galerkin, and least-squares methods.

It is clear from an examination of Table 1.8 that the results are in agreement with each other. It is also important to note here that the primary purpose of Section 1.7 was to introduce you to the general concepts of weighted residual methods and the basic procedures in the simplest possible way. Because the Galerkin method is one of the most commonly used procedures in finite element formulations, more detail and an in-depth view of the Galerkin method will be offered later in Chapters 6 and 9. We will employ the Galerkin method to formulate one- and two-dimensional problems once you have become familiar with the ideas of one- and two-dimensional elements. Also note that in the above examples of the use of weighted residual methods, we assumed a solution that was to provide an approximate solution over the entire domain of the given problem. As you will see later, we will use piecewise solutions with the Galerkin method. That is to say, we will assume linear or nonlinear solutions that are valid only over each element and then combine, or assemble, the elemental solutions.

1.8 VERIFICATION OF RESULTS

In recent years, the use of finite element analysis as a design tool has grown rapidly. Easy-to-use, comprehensive packages such as ANSYS have become a common tool in the hands of design engineers. Unfortunately, many engineers without the proper training or a solid understanding of the underlying concepts have been using finite element analysis. Engineers who use finite element analysis must understand the limitations of the finite element procedures. There are various sources of error that can contribute to incorrect results. They include

1. *Wrong input data, such as physical properties and dimensions*
   This mistake can be corrected by simply listing and verifying physical properties and coordinates of nodes or keypoints (points defining the vertices of an object;

they are covered in more detail in Chapters 8 and 13) before proceeding any further with the analysis.

2. *Selecting inappropriate types of elements*
Understanding the underlying theory will benefit you the most in this respect. You need to fully grasp the limitations of a given type of element and understand to which type of problems it applies.
3. *Poor element shape and size after meshing*
This area is a very important part of any finite element analysis. Inappropriate element shape and size will influence the accuracy of your results. It is important that the user understands the difference between free meshing (using mixed-area element shapes) and mapped meshing (using all quadrilateral area elements or all hexahedral volume elements) and the limitations associated with them. These concepts will be explained in more detail in Chapters 8 and 13.
4. *Applying wrong boundary conditions and loads*
This step is usually the most difficult aspect of modeling. It involves taking an actual problem and estimating the loading and the appropriate boundary conditions for a finite element model. This step requires good judgment and some experience.

You must always find ways to check your results. While experimental testing of your model may be the best way to do so, it may be expensive or time consuming. You should always start by applying equilibrium conditions and energy balance to different portions of a model to ensure that the physical laws are not violated. For example, for static models, the sum of the forces acting on a free-body diagram of your model must be zero. This concept will allow you to check for the accuracy of computed reaction forces. You may want to consider defining and mapping stresses along an arbitrary cross section and integrating this information. The resultant internal forces computed in this manner must balance against external forces. In a heat transfer problem under steady-state conditions, apply conservation of energy to a control volume surrounding an arbitrary node. Are the energies flowing into and out of a node balanced? At the end of particular chapters in this text, a section is devoted to verifying the results of your models. In these sections, problems will be solved using ANSYS, and the steps for verifying results will be shown.

1.9 UNDERSTANDING THE PROBLEM

You can save lots of time and money if you first spend a little time with a piece of paper and a pencil to try to understand the problem you are planning to analyze. Before initiating numerical modeling on the computer and generating a finite element model, it is imperative that you develop a sense of or a feel for the problem. There are many questions that a good engineer will ask before proceeding with the modeling process: Is the material under axial loading? Is the body under bending moments or twisting moments or a combination of the two? Do we need to worry about buckling? Can we approximate the behavior of the material with a two-dimensional model? Does heat transfer play a significant role in the problem? Which modes of heat transfer are influential? If you choose to employ FEA, "back-of-the-envelope" calculations will greatly enhance your understanding of the problem, in turn helping you to develop a good, reasonable finite element model, particularly in terms of your selection of element types.

Some practicing engineers still use finite element analysis to solve a problem that could have been solved more easily by hand by someone with a good grasp of the fundamental concepts of the mechanics of materials and heat transfer. To shed more light on this very important point, consider the following examples.

EXAMPLE 1.6

Imagine that by mistake, an empty coffee pot has been left on a heating element. Assuming that the heater puts approximately 20 Watts (typically, a heater creates a lower wattage) into the bottom of the pot, determine the temperature distribution within the glass if the surrounding air is at 25°C, with a corresponding heat transfer coefficient $h = 15\ \mathrm{W/m^2 \cdot K}$. The pot is cylindrical in shape, with a diameter of 14 cm and height of 14 cm, and the glass is 3 mm thick.

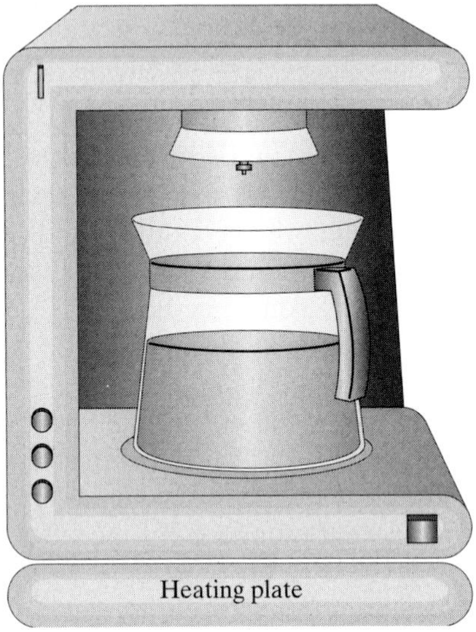

This problem is first analyzed using a finite element model. After you study three-dimensional thermal-solid elements (discussed in Chapter 13), you will revisit this problem (see problem 13.11) and be asked to solve it using ANSYS. As you will learn later, a solid model of the pot is created and meshed and the appropriate boundary conditions are applied and the temperature solutions is then obtained. The results of this analysis is shown in Figure 1.21.

From the results of finite element analysis we find that the maximum temperature of 113.18°C occurs at the bottom of the pot in the center location, as shown in Figure 1.21. This is a good example of a problem that could have been solved more easily by hand by someone with a good grasp of the fundamental concepts of heat transfer. We can approximate the temperature of the glass by applying the energy balance to the bottom of the pot and assuming a one-dimensional model. Because the pot is made of thin glass, we can neglect the spatial temperature variation within the glass. Under steady-state conditions, the heat flux added into the bottom of the glass is approximately equal to the rate

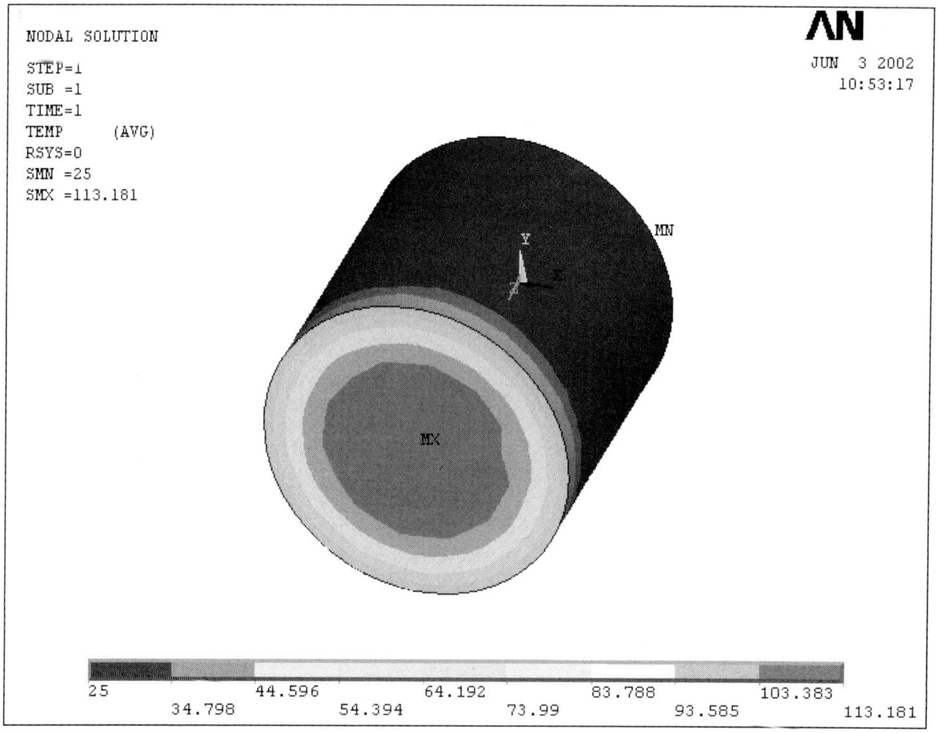

FIGURE 1.21 The temperature distribution in the pot of Example 1.6.

of energy convected away by air. Thus, we employ Newton's law of cooling

$$q'' = h(T_s - T_f) \tag{1.65}$$

where

$$\begin{aligned} q'' &= \text{heat flux, W/m}^2 \\ h &= \text{heat transfer coefficient, W/m}^2\cdot{}^\circ\text{C (W/m}^2\cdot\text{K)} \\ T_s &= \text{surface temperature of the coffee pot,}{}^\circ\text{C} \\ T_f &= \text{surrounding air temperature,}{}^\circ\text{C} \end{aligned}$$

We can estimate the heat flux into the bottom of the pot:

$$q'' = \frac{20 \text{ W}}{\frac{\pi}{4}(0.14 \text{ m})^2} = 1299 \text{ W/m}^2$$

and substituting for heat flux, h, and T_f into Eq. (1.65), and solving for T_s,

$$1299 \text{ W/m}^2 = (15 \text{ W/m}\cdot{}^\circ\text{C})(T_s - 25) \quad \rightarrow \quad T_s = 111.6^\circ\text{C}$$

As you can see, the temperature result obtained by hand calculation ($T_s = 111.6^\circ$C) is very close to the result of our finite element model ($T_{max} = 113.18^\circ$C). Thus, there was no need to resort to finite element formulation to solve the above problem.

EXAMPLE 1.7

Consider the torsion of a steel bar ($G = 11 \times 10^3$ ksi) having a rectangular cross section, as shown in the accompanying figure. Under the loading shown, the angle of twist is measured to be $\theta = 0.0005$ rad/in. We are interested in determining the location(s) and magnitude of the maximum shear stress.

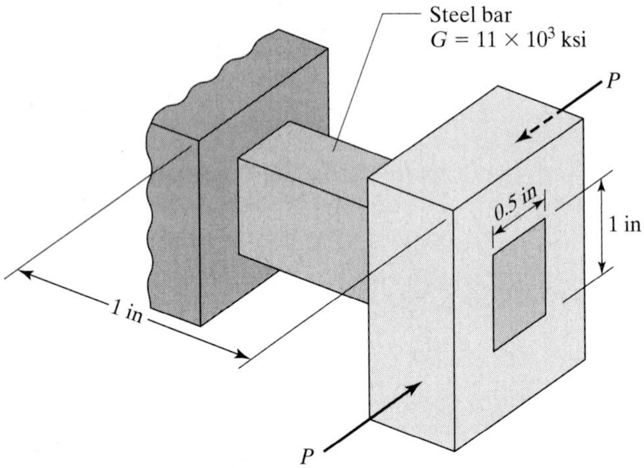

Again, we have analyzed this problem using a finite element model. All of the steps leading to the ANSYS solution are given in Example 10.1 (revisited). The results of this analysis are shown in Figure 1.22.

The results of finite element analysis show that the maximum shear stress of 2558 lb/in^2 occurs at the midsection of the rectangle. This is another example of a problem that could have been solved more easily by hand by someone with a good grasp of the fundamental concepts of mechanics of materials.

As you will learn in Chapter 10, Section 10.1, there are analytical solutions that we could employ to solve problems dealing with torsion of members with rectangular cross-sectional area. When a torque is applied to a straight bar with a rectangular cross-sectional area, within the elastic region of the material, the maximum shearing stress and an angle of twist caused by the torque are given by

$$\tau_{max} = \frac{T}{c_1 w h^2}$$

where

τ_{max} = maximum shear stress, lb/in^2

T = applied torque, lb · in

w = width of the rectangular cross-section, in

h = height of the rectangular cross-section, in

c_1 = a constant coefficient that depends on aspect ratio of the cross section, 0.246; see Table 10.1

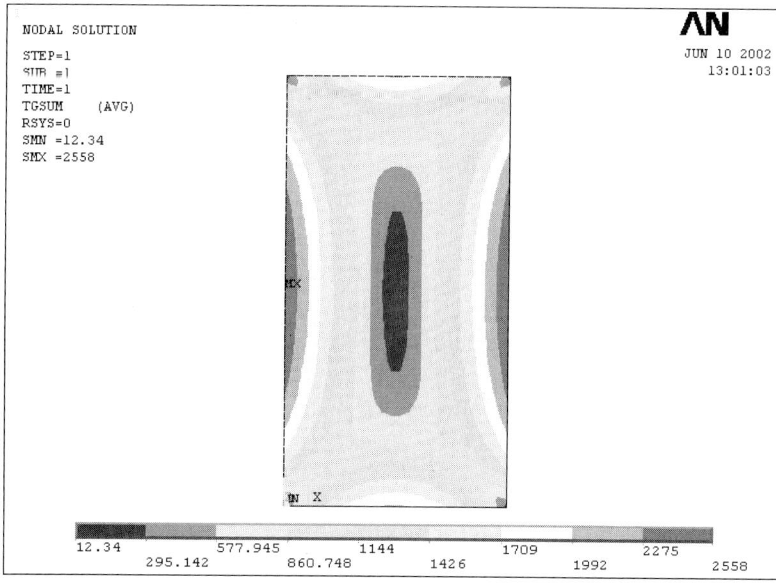

FIGURE 1.22 The shear stress distribution for the steel bar of Example 1.7.

and

$$\theta = \frac{TL}{c_2 Gwh^3}$$

L = length of the bar, in

G = shear modulus or modulus of rigidity of material, lb/in^2

c_2 = a constant coefficient that depends on aspect ratio of the cross section, 0.229; see Table 10.1

Substituting into the above equations appropriate values, we get

$$\theta = \frac{TL}{c_2 Gwh^3} = 0.0005 \text{ rad/in} = \frac{T(1 \text{ in})}{0.229(11 \times 10^6 \text{ lb/in}^2)(1 \text{ in})(0.5 \text{ in})^3} \Rightarrow T = 157.5 \text{ lb} \cdot \text{in}$$

$$\tau_{max} = \frac{T}{c_1 wh^2} = \frac{157.5 \text{ lb} \cdot \text{in}}{0.246(1 \text{ in})(0.5 \text{ in})^2} = 2560 \text{ lb/in}^2$$

When comparing 2560 lb/in^2 to the FEA results of 2558 lb/in^2, you see that we could have saved lots of time by calculating the maximum shear stress using the analytical solution and avoided generating a finite element model.

SUMMARY

At this point you should

1. have a good understanding of the physical properties and the parameters that characterize the behavior of an engineering system. Examples of these properties and design parameters are given in Tables 1.2 and 1.3.

2. realize that a good understanding of the fundamental concepts of the finite element method will benefit you by enabling you to use ANSYS more effectively.
3. know the seven basic steps involved in any finite element analysis, as discussed in Section 1.4.
4. understand the differences among direct formulation, minimum total potential energy formulation, and the weighted residual methods (particularly the Galerkin formulation).
5. know that it is wise to spend some time to gain a full understanding of a problem before initiating a finite element model of the problem. There may even exist a reasonable closed-form solution to the problem, and thus you can save lots of time and money.
6. realize that you must always find a way to verify your FEA results.

REFERENCES

ASHRAE Handbook, *Fundamental Volume*, American Society of Heating, Refrigerating, and Air-Conditioning Engineers, Atlanta, 1993.

Bickford, B. W., *A First Course in the Finite Element Method*, Richard D. Irwin, Burr Ridge, 1989.

Clough, R. W., "The Finite Element Method in Plane Stress Analysis, Proceedings of American Society of Civil Engineers, 2nd Conference on Electronic Computations," Vol. 23, 1960, pp. 345–378.

Cook, R. D., Malkus, D. S., and Plesha, M. E., *Concepts and Applications of Finite Element Analysis*, 3rd. ed., New York, John Wiley and Sons, 1989.

Courant, R., "Variational Methods for the Solution of Problems of Equilibrium and Vibrations," Bulletin of the American Mathematical Society, Vol. 49, 1943, pp. 1–23.

Hrennikoff, A., "Solution of Problems in Elasticity by the Framework Method," *J. Appl. Mech.*, Vol. 8, No. 4, 1941, pp. A169–A175.

Levy, S., "Structural Analysis and Influence Coefficients for Delta Wings," *Journal of the Aeronautical Sciences*, Vol. 20, No. 7, 1953, pp. 449–454.

Patankar, S. V., *Numerical Heat Transfer and Fluid Flow*, New York, McGraw-Hill, 1991.

Zienkiewicz, O. C., and Cheung, Y. K. K., *The Finite Element Method in Structural and Continuum Mechanics*, London, McGraw-Hill, 1967.

Zienkiewicz, O. C., *The Finite Element Method*, 3d. ed., London, McGraw-Hill, 1979.

PROBLEMS

1. Solve Example 1.1 using (1) two elements and (2) eight elements. Compare your results to the exact values.
2. A concrete table column-support with the profile shown in the accompanying figure is to carry a load of approximately 500 lb. Using the direct method discussed in Section 1.5, determine the deflection and average normal stresses along the column. Divide the column into five elements. ($E = 3.27 \times 10^3$ ksi)

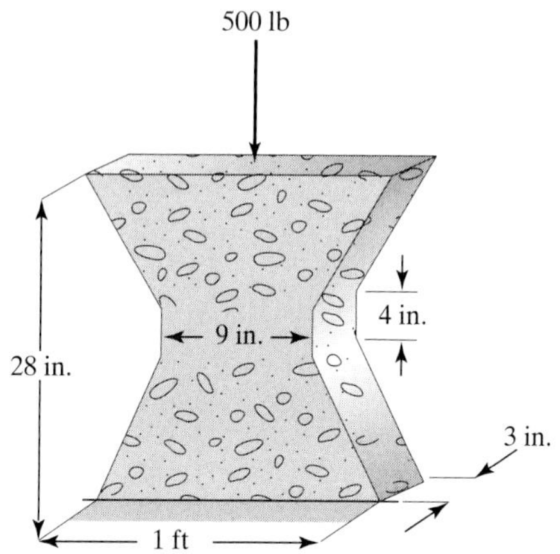

3. An aluminum strap with a thickness of 6 mm and the profile shown in the accompanying figure is to carry a load of 1800 N. Using the direct method discussed in Section 1.5, determine the deflection and the average normal stress along the strap. Divide the strap into three elements. This problem may be revisited again in Chapter 10, where a more in-depth analysis may be sought. ($E = 68.9$ GPa)

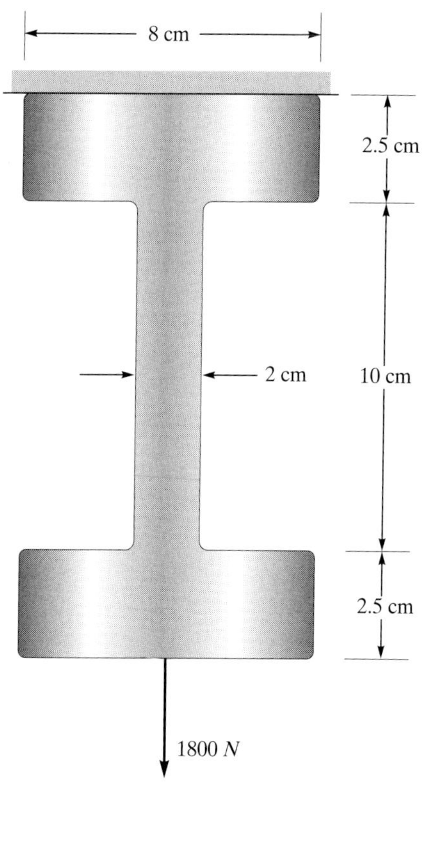

4. A thin steel plate with the profile shown in the accompanying figure is subjected to an axial load. Approximate the deflection and the average normal stresses along the plate using the model shown in the figure. The plate has a thickness of 0.125 in and a modulus of elasticity $E = 28 \times 10^3$ ksi. You will be asked to use ANSYS to analyze this problem again in Chapter 10.

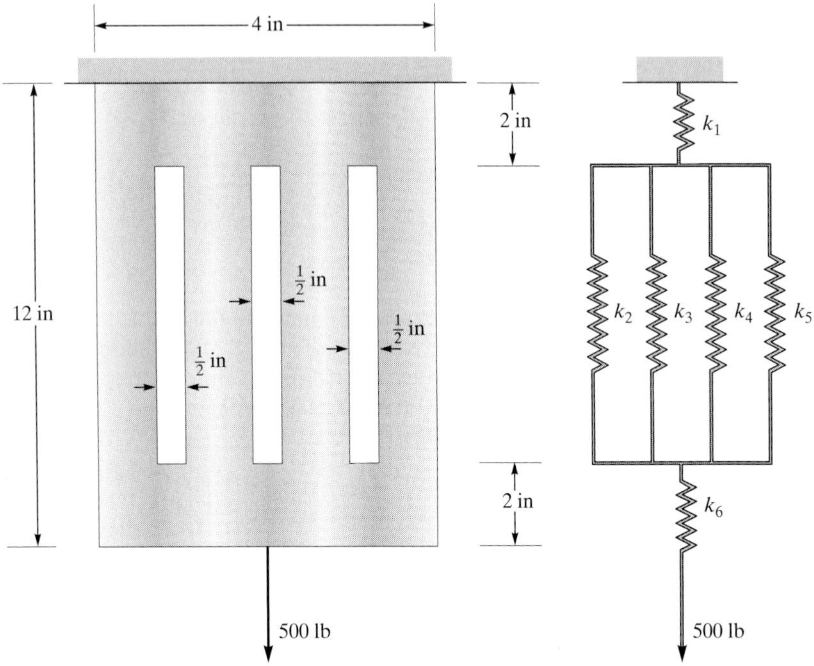

5. Apply the statics equilibrium conditions directly to each node of the thin steel plate (using a finite element model) in Problem 4.

6. For the spring system shown in the accompanying figure, determine the displacement of each node. Start by identifying the size of the global matrix. Write down elemental stiffness matrices, and show the position of each elemental matrix in the global matrix. Apply the boundary conditions and loads. Solve the set of linear equations. Also compute the reaction forces.

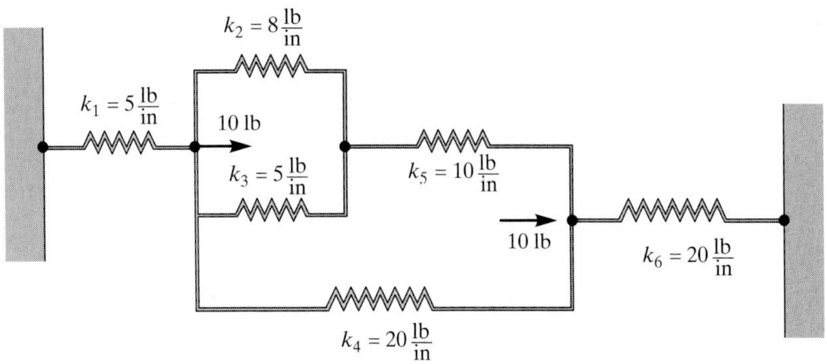

7. A typical exterior masonry wall of a house, shown in the accompanying figure, consists of the items in the accompanying table. Assume an inside room temperature of 68°F and an outside air temperature of 10°F, with an exposed area of 150 ft^2. Determine the temperature distribution through the wall. Also calculate the heat loss through the wall

Items	Resistance $hr \cdot ft^2 \cdot {}^\circ F/Btu$	U-factor $Btu/hr \cdot ft^2 \cdot {}^\circ F$
1. Outside film resistance (winter, 15-mph wind)	0.17	5.88
2. Face brick (4 in)	0.44	2.27
3. Cement mortar (1/2 in)	0.1	10.0
4. Cinder block (8 in)	1.72	0.581
5. Air space (3/4 in)	1.28	0.781
6. Gypsum board (1/2 in)	0.45	2.22
7. Inside film resistance (winter)	0.68	1.47

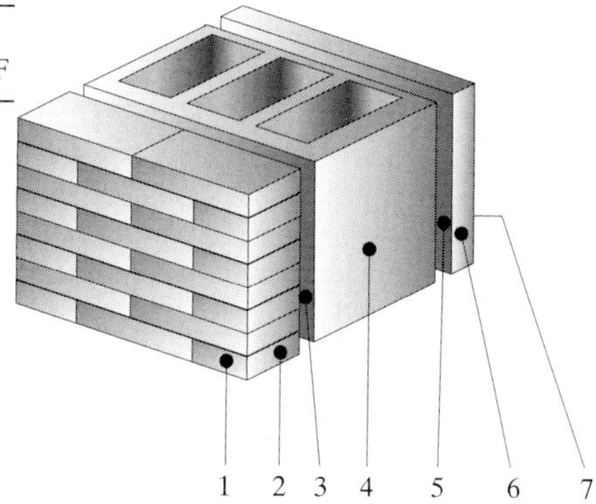

8. In order to increase the thermal resistance of a typical exterior frame wall, such as the one in Example 1.2, it is customary to use 2 × 6 studs instead of 2 × 4 studs to allow for placement of more insulation within the wall cavity. A typical exterior (2 × 6) frame wall of a house consists of the materials shown in the accompanying figure. Assume an inside room temperature of 68°F and an outside air temperature of 20°F, with an exposed area of 150 ft^2. Determine the temperature distribution through the wall.

Items	Resistance $hr \cdot ft^2 \cdot {}^\circ F/Btu$	U-factor $Btu/hr \cdot ft^2 \cdot {}^\circ F$
1. Outside film resistance (winter, 15-mph wind)	0.17	5.88
2. Siding, wood (1/2 × 8 lapped)	0.81	1.23
3. Sheathing (1/2 in regular)	1.32	0.76
4. Insulation batt ($5\frac{1}{2}$ in)	19.0	0.053
5. Gypsum wall board (1/2 in)	0.45	2.22
6. Inside film resistance (winter)	0.68	1.47

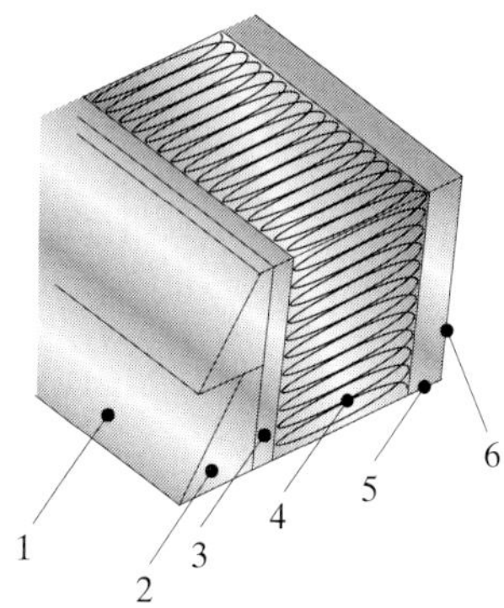

9. Assuming the moisture can diffuse through the gypsum board in Problem 8, where should you place a vapor barrier to avoid moisture condensation? Assume an indoor air temperature of 68°F with relative humidity of 40%.

10. A typical ceiling of a house consists of the items in the accompanying table. Assume an inside room temperature of 70°F and an attic air temperature of 15°F, with an exposed area of 1000 ft^2. Determine the temperature distribution through the ceiling. Also calculate heat loss through the ceiling.

Items	Resistance hr · ft² · °F/Btu	*U*-factor Btu/hr · ft² · °F
1. Inside attic film resistance	0.68	1.47
2. Insulation batt (6 in)	19.0	0.053
3. Gypsum board (1/2 in)	0.45	2.22
4. Inside film resistance (winter)	0.68	1.47

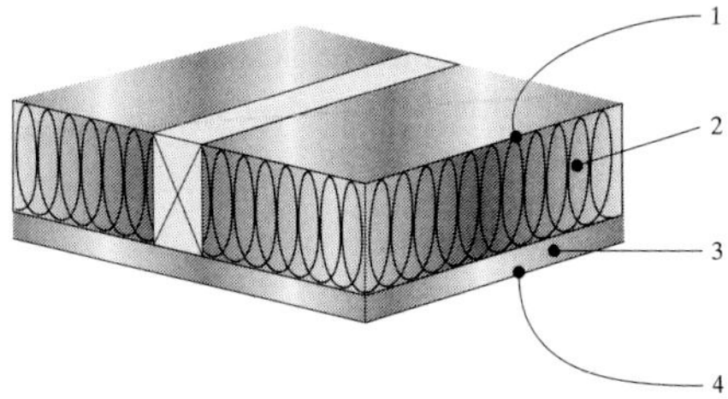

11. A typical $1\frac{3}{8}$-in solid wood core door exposed to winter conditions has the characteristics shown in the accompanying table. Assume an inside room temperature of 70°F and an outside air temperature of 20°F, with an exposed area of 22.5 ft². (a) Determine the inside and outside temperatures of the door's surface. (b) Determine heat loss through the door.

Items	Resistance hr · ft² · °F/Btu	*U*-factor Btu/hr · ft² · °F
1. Outside film resistance (winter, 15-mph wind)	0.17	5.88
2. $1\frac{3}{8}$-in solid wood core	0.39	2.56
3. Inside film resistance (winter)	0.68	1.47

12. The concrete table column-support in Problem 2 is reinforced with three $\frac{1}{2}$-in steel rods, as shown in the accompanying figure. Determine the deflection and average normal stresses along the column under a load of 1000 lb. Divide the column into five elements. ($E_C = 3.27 \times 10^3$ ksi; $E_s = 29 \times 10^3$ ksi)

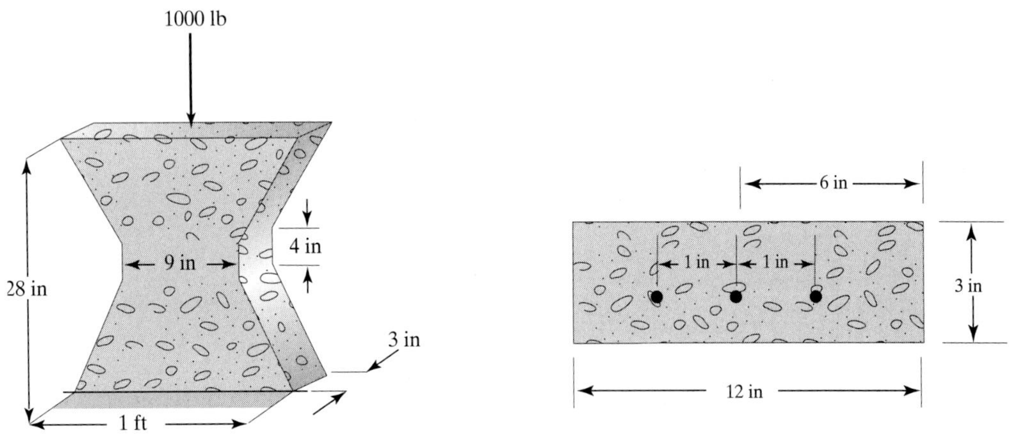

13. Compute the total strain energy for the concrete table column-support in Problem 12.

14. A 10-in slender rod weighing 6 lb is supported by a spring with a stiffness $k = 60$ lb/in. A force $P = 35$ lb is applied to the rod at the location shown in the accompanying figure. Determine the deflection of the spring (a) by drawing a free-body diagram of the rod and applying the statics equilibrium conditions, and (b) by applying the minimum total potential energy concept.

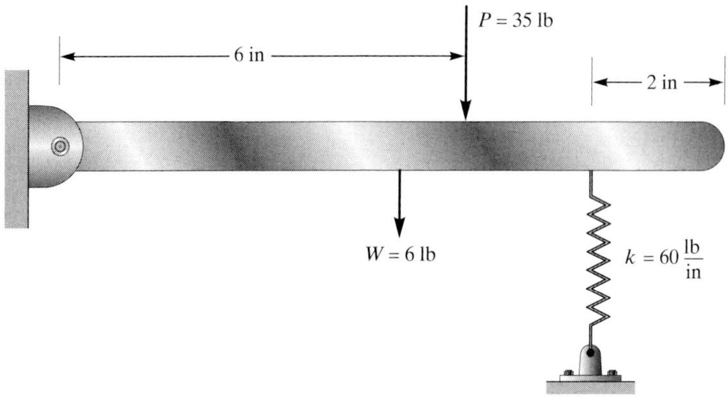

15. In a DC electrical circuit, Ohm's law relates the voltage drop $V_2 - V_1$ across a resistor to a current I flowing through the element and the resistance R according to the equation $V_2 - V_1 = RI$.

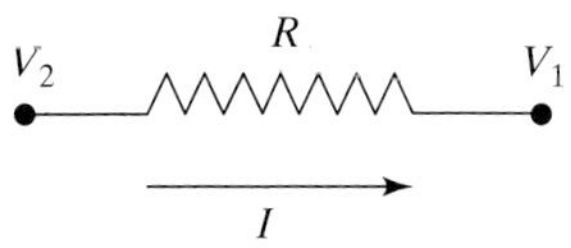

Using direct formulation, show that for a resistance element comprising two nodes, the conductance matrix, the voltage drop, and the currents are related according to the equation

$$\frac{1}{R}\begin{bmatrix} 1 & -1 \\ -1 & 1 \end{bmatrix}\begin{Bmatrix} V_1 \\ V_2 \end{Bmatrix} = \begin{Bmatrix} I_1 \\ I_2 \end{Bmatrix}$$

16. Use the results of Problem 15 to set up and solve for the voltage drop in each branch of the circuit shown in the accompanying figure.

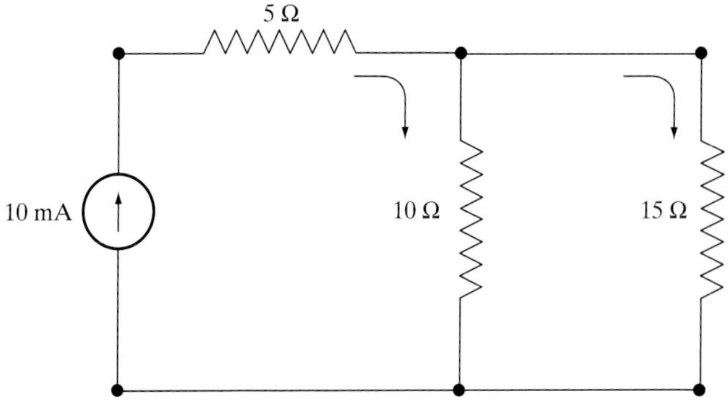

17. The deformation of a simply supported beam under a distributed load, shown in the accompanying figure, is governed by the relationship

$$\frac{d^2Y}{dX^2} = \frac{M(X)}{EI}$$

where $M(X)$ is the internal bending moment and is given by

$$M(X) = \frac{wX(L - X)}{2}$$

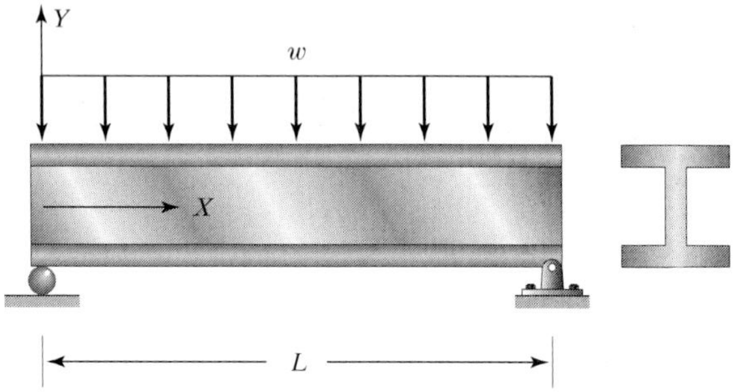

Derive the equation for the exact deflection. Assume an approximate deflection solution of the form

$$Y(X) = c_1\left[\left(\frac{X}{L}\right)^2 - \left(\frac{X}{L}\right)\right]$$

Use the following methods to evaluate c_1: (a) the collocation method and (b) the subdomain method. Also, using the approximate solutions, determine the maximum deflection of the beam if a W24 × 104 (wide flange shape) with a span of $L = 20$ ft supports a distributed load of $w = 5$ kips/ft.

18. For the example problem used throughout Section 1.7, assume an approximate solution of the form $u(y) = c_1y + c_2y^2 + c_3y^3 + c_4y^4$. Using the collocation, subdomain, Galerkin, and least-squares methods, determine the unknown coefficients c_1, c_2, c_3, and c_4. Compare your results to those obtained in Section 1.7.

19. The leakage flow of hydraulic fluid through the gap between a piston–cylinder arrangement may be modeled as laminar flow of fluid between infinite parallel plates, as shown in the accompanying figure. This model offers reasonable results for relatively small gaps. The differential equation governing the flow is

$$\mu\frac{d^2u}{dy^2} = \frac{dp}{dx}$$

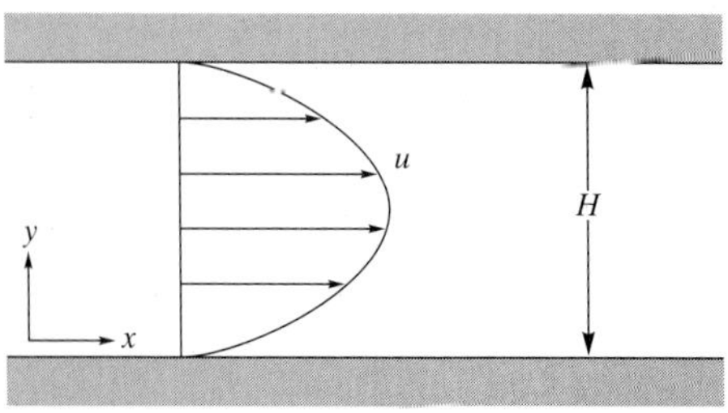

where μ is the dynamic viscosity of the hydraulic fluid, u is the fluid velocity, and $\frac{dp}{dx}$ is the pressure drop and is constant. Derive the equation for the exact fluid velocities. Assume an approximate fluid velocity solution of the form $u(y) = c_1\left[\sin\left(\frac{\pi y}{H}\right)\right]$. Use the following methods to evaluate c_1: (a) the collocation method and (b) the subdomain method. Compare the approximate results to the exact solution.

20. Use the Galerkin and least-squares methods to solve Problem 19. Compare the approximate results to the exact solution.

21. For the cantilever beam shown in the accompanying figure, the deformation of the beam under a load P is governed by the relationship

$$\frac{d^2Y}{dX^2} = \frac{M(X)}{EI}$$

where $M(X)$ is the internal bending moment and is

$$M(X) = -PX$$

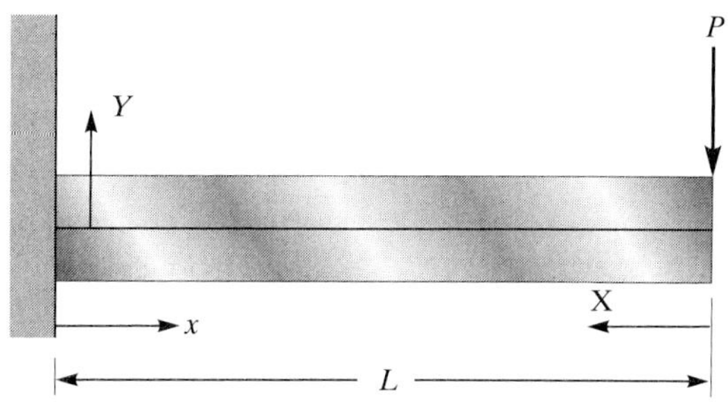

Derive the equation for the exact deflection. Assume an appropriate form of a polynomial function. Keep in mind that the assumed solution must satisfy the given boundary conditions. Use the subdomain method and the Galerkin method to solve for the unknown coefficients of the assumed solution.

22. A shaft is made of three parts, as shown in the accompanying figure. Parts AB and CD are made of the same material with a modulus of rigidity of $G = 9.8 \times 10^3$ ksi, and each has a diameter of 1.5 in. Segment BC is made of a material with a modulus of rigidity of $G = 11.2 \times 10^3$ ksi and has a diameter of 1 in. The shaft is fixed at both ends. A torque of 2400 lb · in is applied at C. Using three elements, determine the angle of twist at B and C and the torsional reactions at the boundaries.

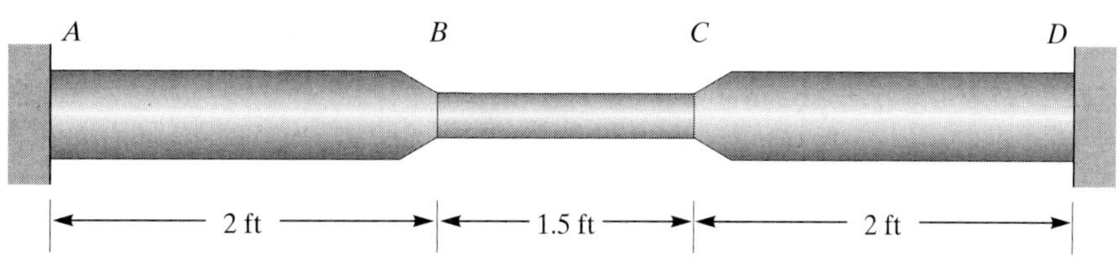

23. For the shaft in Problem 22, replace the torque at C by two equal torques of 1200 lb · in at B and C. Compute the angle of twist at B and C and the torsional reactions at the boundaries.

24. Consider a plate with a variable cross section supporting a load of 1500 lb, as shown in the accompanying figure. Using direct formulation, determine the deflection of the bar at locations $y = 2.5$ in, $y = 7.5$ in, and $y = 10$ in. The plate is made of a material with a modulus of elasticity $E = 10.6 \times 10^3$ ksi.

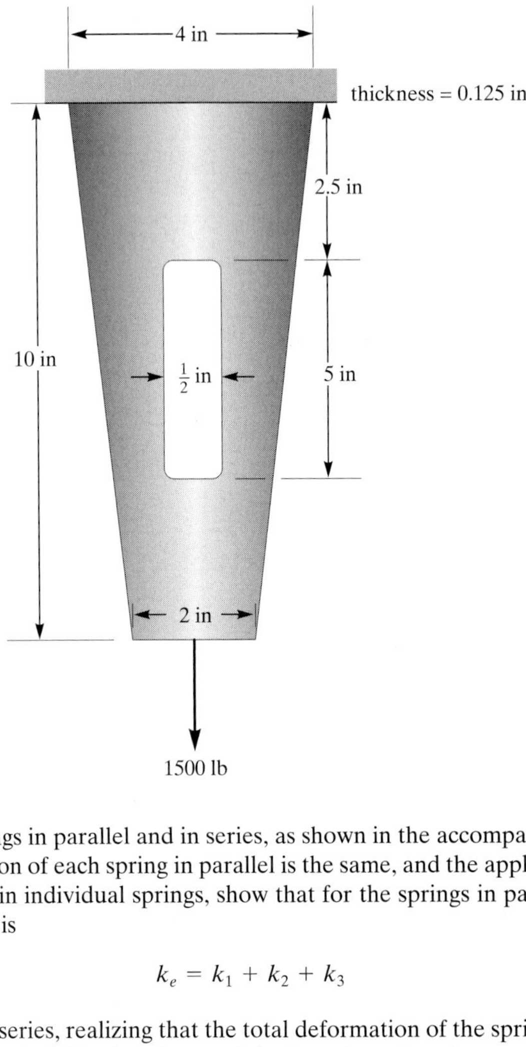

25. Consider the springs in parallel and in series, as shown in the accompanying figure. Realizing that deformation of each spring in parallel is the same, and the applied force must equal the sum of forces in individual springs, show that for the springs in parallel the equivalent spring constant k_e is

$$k_e = k_1 + k_2 + k_3$$

For the springs in series, realizing that the total deformation of the springs is the sum of the deformations of the individual springs, and the force in each spring equals the applied force, show that for the springs in series, the equivalent spring constant is

$$k_e = \frac{1}{\dfrac{1}{k_1} + \dfrac{1}{k_2} + \dfrac{1}{k_3}}$$

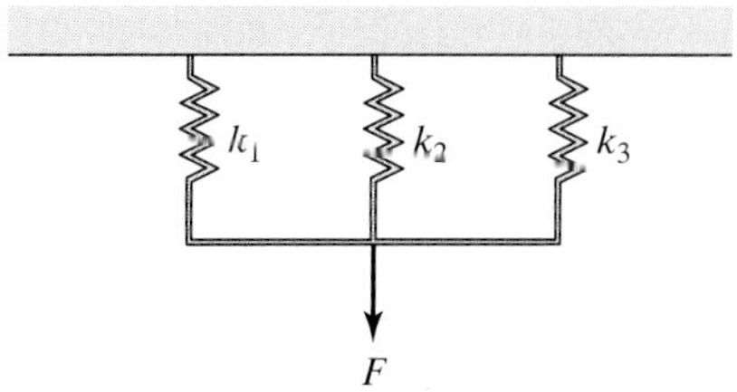

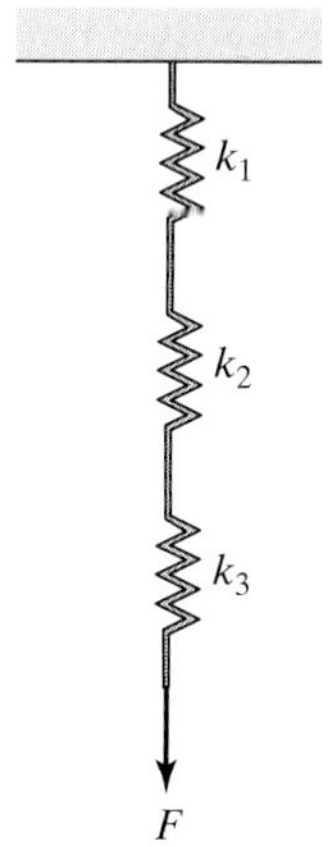

26. Use the results of problem 25 and determine the equivalent spring constant for the system of the springs shown in the accompanying figure.

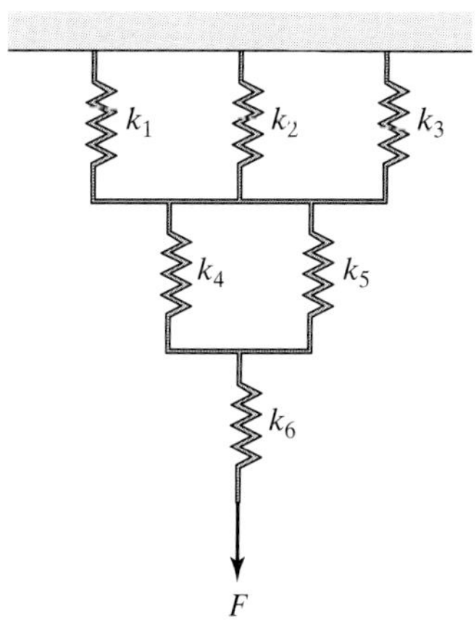

27. Determine the equivalent spring constant for the cantilever beam shown in the accompanying figure.

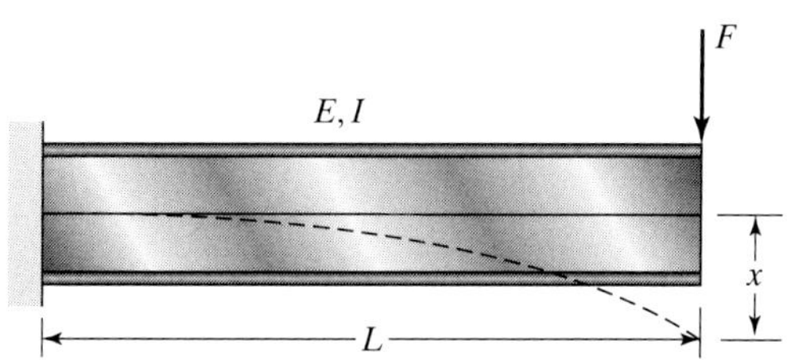

28. Use the results of Problem 27 and Eq. (1.5) to determine the equivalent spring constant for the system shown in the accompanying figure.

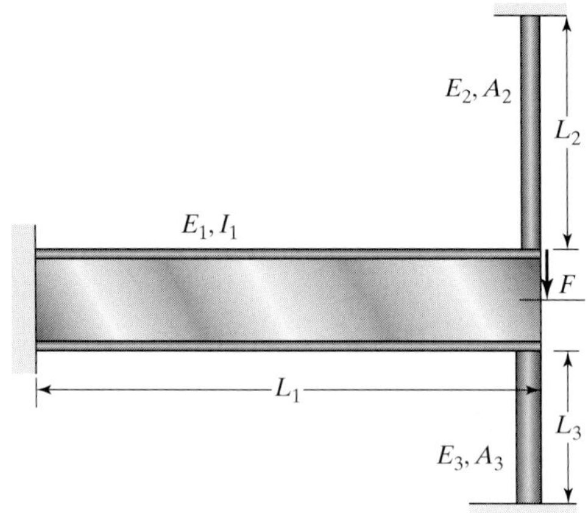

29. Determine the equivalent spring constant for the system shown. Determine the deflection of point A, using the minimum total potential energy concept.

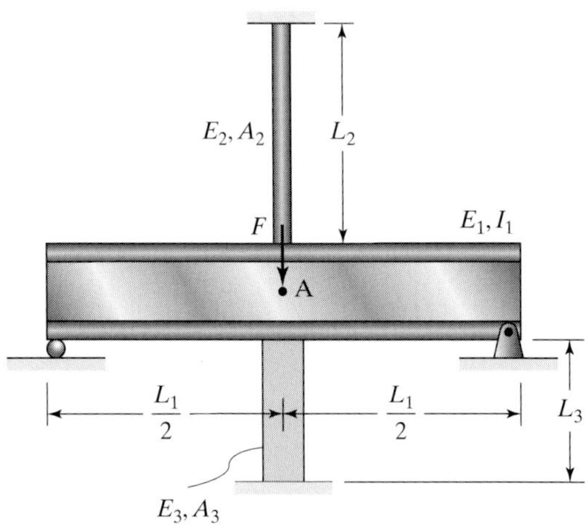

30. Neglect the mass of the connecting rod and determine the deflection of each spring for the system shown in the accompanying figure (a) by applying the statics equilibrium conditions, and (b) by applying the minimum total potential energy concept.

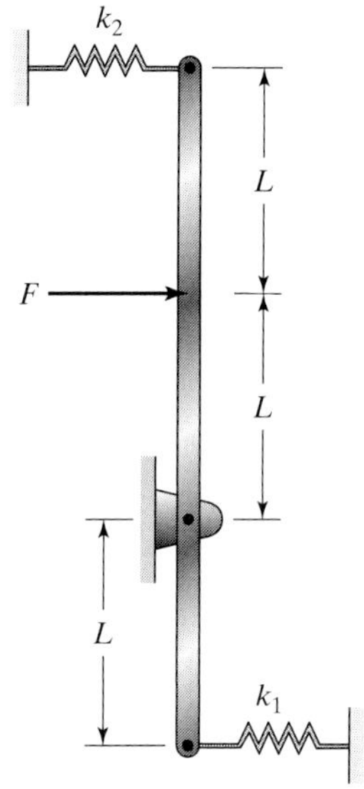

C H A P T E R 2

Matrix Algebra

In Chapter 1 we discussed the basic steps involved in any finite element analysis. These steps include discretizing the problem into elements and nodes, assuming a function that represents behavior of an element, developing a set of equations for an element, assembling the elemental formulations to present the entire problem, and applying the boundary conditions and loading. These steps lead to a set of linear (nonlinear for some problems) algebraic equations that must be solved simultaneously. A good understanding of matrix algebra is essential in formulation and solution of finite element models. As is the case with any topic, matrix algebra has its own terminology and follows a set of rules. We provide an overview of matrix terminology and matrix algebra in this chapter. The main topics discussed in Chapter 2 include

2.1 Basic Definitions
2.2 Matrix Addition or Subtraction
2.3 Matrix Multiplication
2.4 Partitioning of a Matrix
2.5 Transpose of a Matrix
2.6 Determinant of a Matrix
2.7 Solutions of Simultaneous Linear Equations
2.8 Inverse of a Matrix
2.9 Eigenvalues and Eigenvectors
2.10 Using MATLAB® to Manipulate Matrices
2.11 Using Excel® to Manipulate Matrices

2.1 BASIC DEFINITIONS

A matrix is an array of numbers or mathematical terms. The numbers or the mathematical terms that make up the matrix are called the *elements of matrix*. The *size* of a matrix is defined by its number of rows and columns. A matrix may consist of m rows and n columns. For example,

$$[\boldsymbol{N}] = \begin{bmatrix} 6 & 5 & 9 \\ 1 & 26 & 14 \\ -5 & 8 & 0 \end{bmatrix} \quad [\boldsymbol{T}] = \begin{bmatrix} \cos\theta & -\sin\theta & 0 & 0 \\ \sin\theta & \cos\theta & 0 & 0 \\ 0 & 0 & \cos\theta & -\sin\theta \\ 0 & 0 & \sin\theta & \cos\theta \end{bmatrix}$$

$$\{\boldsymbol{L}\} = \begin{Bmatrix} \dfrac{\partial f(x, y, z)}{\partial x} \\ \dfrac{\partial f(x, y, z)}{\partial y} \\ \dfrac{\partial f(x, y, z)}{\partial z} \end{Bmatrix} \quad [\boldsymbol{I}] = \begin{bmatrix} \displaystyle\int_0^L x\,dx & \displaystyle\int_0^W y\,dy \\ \displaystyle\int_0^L \frac{x^2}{L}\,dx & \displaystyle\int_0^W \frac{y^2}{L}\,dy \end{bmatrix}$$

Matrix $[\boldsymbol{N}]$ is a 3 by 3 (or 3×3) matrix whose elements are numbers, $[\boldsymbol{T}]$ is a 4×4 that has *sine* and *cosine* terms as its elements, $\{\boldsymbol{L}\}$ is a 3×1 matrix with its elements representing partial derivatives, and $[\boldsymbol{I}]$ is a 2×2 matrix with integrals for its elements. The $[\boldsymbol{N}]$, $[\boldsymbol{T}]$, and $[\boldsymbol{I}]$ are square matrices. A *square* matrix has the same number of rows and columns. The element of a matrix is denoted by its location. For example, the element in the first row and the third column of a matrix $[\boldsymbol{B}]$ is denoted by b_{13}, and an element occurring in matrix $[\boldsymbol{A}]$ in row 2 and column 3 is denoted by the term a_{23}. In this book, we denote the matrix by a **bold-face letter** in brackets [] and {}, for example: [**K**], [**T**], {**F**}, and the elements of matrices are represented by regular lowercase letters. The {} is used to distinguish a column matrix.

Column Matrix and Row Matrix

A column matrix is defined as a matrix that has one column but could have many rows. On the other hand, a row matrix is a matrix that has one row but could have many columns. Examples of column and row matrices follow.

$$\{\boldsymbol{A}\} = \begin{Bmatrix} 1 \\ 5 \\ -2 \\ 3 \end{Bmatrix}, \{\boldsymbol{X}\} = \begin{Bmatrix} x_1 \\ x_2 \\ x_3 \end{Bmatrix}, \text{ and } \{\boldsymbol{L}\} = \begin{Bmatrix} \dfrac{\partial f(x, y, z)}{\partial x} \\ \dfrac{\partial f(x, y, z)}{\partial y} \\ \dfrac{\partial f(x, y, z)}{\partial z} \end{Bmatrix} \text{ are examples of column matrices.}$$

whereas $[\boldsymbol{C}] = [5 \quad 0 \quad 2 \quad -3]$ and $[\boldsymbol{Y}] = [y_1 \quad y_2 \quad y_3]$ are examples of row matrices.

Diagonal, Unit, and Band (Banded) Matrix

A diagonal matrix is one that has elements only along its principal diagonal; the elements are zero everywhere else ($a_{ij} = 0$ when $i \neq j$). An example of a

4 × 4 diagonal matrix follows.

$$[\boldsymbol{A}] = \begin{bmatrix} a_1 & 0 & 0 & 0 \\ 0 & a_2 & 0 & 0 \\ 0 & 0 & a_3 & 0 \\ 0 & 0 & 0 & a_4 \end{bmatrix}$$

The diagonal along which a_1, a_2, a_3, and a_4 lies is called the *principal diagonal*. An *identity* or *unit matrix* is a diagonal matrix whose elements consist of a value 1. An example of an identity matrix follows.

$$[\boldsymbol{I}] = \begin{bmatrix} 1 & 0 & 0 & . & . & 0 & 0 \\ 0 & 1 & 0 & . & . & 0 & 0 \\ 0 & 0 & 1 & . & . & 0 & 0 \\ . & . & . & . & . & . & . \\ . & . & . & . & . & . & . \\ 0 & 0 & 0 & . & . & 1 & 0 \\ 0 & 0 & 0 & . & . & 0 & 1 \end{bmatrix}$$

A *banded matrix* is a matrix that has a band of nonzero elements parallel to its principal diagonal. As shown in the example that follows, all other elements outside the band are zero.

$$[\boldsymbol{B}] = \begin{bmatrix} b_{11} & b_{12} & 0 & 0 & 0 & 0 & 0 \\ b_{21} & b_{22} & b_{23} & 0 & 0 & 0 & 0 \\ 0 & b_{32} & b_{33} & b_{34} & 0 & 0 & 0 \\ 0 & 0 & b_{43} & b_{44} & b_{45} & 0 & 0 \\ 0 & 0 & 0 & b_{54} & b_{55} & b_{56} & 0 \\ 0 & 0 & 0 & 0 & b_{65} & b_{66} & b_{67} \\ 0 & 0 & 0 & 0 & 0 & b_{76} & b_{77} \end{bmatrix}$$

Upper and Lower Triangular Matrix

An *upper triangular matrix* is one that has zero elements below the principal diagonal ($u_{ij} = 0$ when $i > j$), and the *lower triangular matrix* is one that has zero elements above the principal diagonal ($l_{ij} = 0$ when $i < j$). Examples of upper triangular and lower triangular matrices are shown below.

$$[\boldsymbol{U}] = \begin{bmatrix} u_{11} & u_{12} & u_{13} & u_{14} \\ 0 & u_{22} & u_{23} & u_{24} \\ 0 & 0 & u_{33} & u_{34} \\ 0 & 0 & 0 & u_{44} \end{bmatrix} \qquad [\boldsymbol{L}] = \begin{bmatrix} l_{11} & 0 & 0 & 0 \\ l_{21} & l_{22} & 0 & 0 \\ l_{31} & l_{32} & l_{33} & 0 \\ l_{41} & l_{42} & l_{43} & l_{44} \end{bmatrix}$$

2.2 MATRIX ADDITION OR SUBTRACTION

Two matrices can be added together or subtracted from each other provided that they are of the same size—each matrix must have the same number of rows and columns. We can add matrix $[\boldsymbol{A}]_{m \times n}$ of dimension m by n to matrix $[\boldsymbol{B}]_{m \times n}$ of the same dimension by adding the like elements. Matrix subtraction follows a similar rule, as shown.

$$[\boldsymbol{A}] \pm [\boldsymbol{B}] = \begin{bmatrix} a_{11} & a_{12} & . & . & a_{1n} \\ a_{21} & a_{22} & . & . & a_{2n} \\ . & . & . & . & . \\ . & . & . & . & . \\ a_{m1} & a_{m2} & . & . & a_{mn} \end{bmatrix} \pm \begin{bmatrix} b_{11} & b_{12} & . & . & b_{1n} \\ b_{21} & b_{22} & . & . & b_{2n} \\ . & . & . & . & . \\ . & . & . & . & . \\ b_{m1} & b_{m2} & . & . & b_{mn} \end{bmatrix}$$

$$= \begin{bmatrix} (a_{11} \pm b_{11}) & (a_{12} \pm b_{12}) & . & . & (a_{1n} \pm b_{1n}) \\ (a_{21} \pm b_{21}) & (a_{22} \pm b_{22}) & . & . & (a_{2n} \pm b_{2n}) \\ . & . & . & . & . \\ . & . & . & . & . \\ (a_{m1} \pm b_{m1}) & (a_{m2} \pm b_{m2}) & . & . & (a_{mn} \pm b_{mn}) \end{bmatrix}$$

The rule for matrix addition or subtraction can be generalized in the following manner. Let us denote the elements of matrix $[\boldsymbol{A}]$ by a_{ij} and the elements of matrix $[\boldsymbol{B}]$ by b_{ij}, where the number of rows i varies from 1 to m and the number of columns j varies from 1 to n. If we were to add matrix $[\boldsymbol{A}]$ to matrix $[\boldsymbol{B}]$ and denote the resulting matrix by $[\boldsymbol{C}]$, it follows that

$$[\boldsymbol{A}] + [\boldsymbol{B}] = [\boldsymbol{C}]$$

and

$$c_{ij} = a_{ij} + b_{ij} \text{ for } i = 1, 2, \ldots, m \text{ and } j = 1, 2, \ldots, n \tag{2.1}$$

2.3 MATRIX MULTIPLICATION

In this section we discuss the rules for multiplying a matrix by a scalar quantity and by another matrix.

Multiplying a Matrix by a Scalar Quantity

When a matrix $[\boldsymbol{A}]$ of size $m \times n$ is multiplied by a scalar quantity such as β, the operation results in a matrix of the same size $m \times n$, whose elements are the product of elements in the original matrix and the scalar quantity. For example, when we multiply matrix $[\boldsymbol{A}]$ of size $m \times n$ by a scalar quantity β, this operation results in another matrix of size $m \times n$, whose elements are computed by multiplying each element of matrix $[\boldsymbol{A}]$ by β, as shown below.

$$\beta[\boldsymbol{A}] = \beta \begin{bmatrix} a_{11} & a_{12} & . & . & a_{1n} \\ a_{21} & a_{22} & . & . & a_{2n} \\ . & . & . & . & . \\ . & . & . & . & . \\ a_{m1} & a_{m2} & . & . & a_{mn} \end{bmatrix} = \begin{bmatrix} \beta a_{11} & \beta a_{12} & . & . & \beta a_{1n} \\ \beta a_{21} & \beta a_{22} & . & . & \beta a_{2n} \\ . & . & . & . & . \\ . & . & . & . & . \\ \beta a_{m1} & \beta a_{m2} & . & . & \beta a_{mn} \end{bmatrix} \qquad (2.2)$$

Multiplying a Matrix by Another Matrix

Whereas any size matrix can be multiplied by a scalar quantity, matrix multiplication can be performed only when the number of columns in the *premultiplier* matrix is equal to the number of rows in the *postmultiplier* matrix. For example, matrix $[\boldsymbol{A}]$ of size $m \times n$ can be premultiplied by matrix $[\boldsymbol{B}]$ of size $n \times p$ because the number of columns n in matrix $[\boldsymbol{A}]$ is equal to number of rows n in matrix $[\boldsymbol{B}]$. Moreover, the multiplication results in another matrix, say $[\boldsymbol{C}]$, of size $m \times p$. Matrix multiplication is carried out according to the following rule.

$$[\boldsymbol{A}]_{m\times \underset{\uparrow}{n}}[\boldsymbol{B}]_{\underset{\uparrow}{n}\times p} = [\boldsymbol{C}]_{m\times p}$$

must match

$$[\boldsymbol{A}][\boldsymbol{B}] = \begin{bmatrix} a_{11} & a_{12} & . & . & a_{1n} \\ a_{21} & a_{22} & . & . & a_{2n} \\ . & . & . & . & . \\ . & . & . & . & . \\ a_{m1} & a_{m2} & . & . & a_{mn} \end{bmatrix} \begin{bmatrix} b_{11} & b_{12} & . & . & b_{1p} \\ b_{21} & b_{22} & . & . & b_{2p} \\ . & . & . & . & . \\ . & . & . & . & . \\ b_{n1} & b_{n2} & . & . & b_{np} \end{bmatrix} = \begin{bmatrix} c_{11} & c_{12} & . & . & c_{1p} \\ c_{21} & c_{22} & . & . & c_{2p} \\ . & . & . & . & . \\ . & . & . & . & . \\ c_{m1} & c_{m2} & . & . & c_{mp} \end{bmatrix}$$

where the elements in the first column of the $[\boldsymbol{C}]$ matrix are computed from

$$\begin{aligned} c_{11} &= a_{11}b_{11} + a_{12}b_{21} + \cdots + a_{1n}b_{n1} \\ c_{21} &= a_{21}b_{11} + a_{22}b_{21} + \cdots + a_{2n}b_{n1} \\ &\cdots\cdots\cdots\cdots\cdots\cdots\cdots\cdots \\ c_{m1} &= a_{m1}b_{11} + a_{m2}b_{21} + \cdots + a_{mn}b_{n1} \end{aligned}$$

and the elements in the second column of the $[\boldsymbol{C}]$ matrix are

$$\begin{aligned} c_{12} &= a_{11}b_{12} + a_{12}b_{22} + \cdots + a_{1n}b_{n2} \\ c_{22} &= a_{21}b_{12} + a_{22}b_{22} + \cdots + a_{2n}b_{n2} \\ &\cdots\cdots\cdots\cdots\cdots\cdots\cdots\cdots \\ c_{m2} &= a_{m1}b_{12} + a_{m2}b_{22} + \cdots + a_{mn}b_{n2} \end{aligned}$$

and similarly, the elements in the other columns are computed, leading to the last column of the $[\boldsymbol{C}]$ matrix

$$\begin{aligned} c_{1p} &= a_{11}b_{1p} + a_{12}b_{2p} + \cdots + a_{1n}b_{np} \\ c_{2p} &= a_{21}b_{1p} + a_{22}b_{2p} + \cdots + a_{2n}b_{np} \\ &\cdots\cdots\cdots\cdots\cdots\cdots\cdots\cdots \\ c_{mp} &= a_{m1}b_{1p} + a_{m2}b_{2p} + \cdots + a_{mn}b_{np} \end{aligned}$$

The multiplication procedure that leads to the values of the elements in the $[\boldsymbol{C}]$ matrix may be represented in a compact summation form by

$$c_{mp} = \sum_{k=1}^{n} a_{mk} b_{kp} \tag{2.3}$$

When multiplying matrices, keep in mind the following rules. Matrix multiplication is not commutative except for very special cases.

$$[\boldsymbol{A}][\boldsymbol{B}] \neq [\boldsymbol{B}][\boldsymbol{A}] \tag{2.4}$$

Matrix multiplication is associative; that is

$$[\boldsymbol{A}]([\boldsymbol{B}][\boldsymbol{C}]) = ([\boldsymbol{A}][\boldsymbol{B}])[\boldsymbol{C}] \tag{2.5}$$

The distributive law holds true for matrix multiplication; that is

$$([\boldsymbol{A}] + [\boldsymbol{B}])[\boldsymbol{C}] = [\boldsymbol{A}][\boldsymbol{C}] + [\boldsymbol{B}][\boldsymbol{C}] \tag{2.6}$$

or

$$[\boldsymbol{A}]([\boldsymbol{B}] + [\boldsymbol{C}]) = [\boldsymbol{A}][\boldsymbol{B}] + [\boldsymbol{A}][\boldsymbol{C}] \tag{2.7}$$

For a square matrix, the matrix may be raised to an integer power n in the following manner:

$$[\boldsymbol{A}]^n = \overbrace{[\boldsymbol{A}][\boldsymbol{A}]\ldots[\boldsymbol{A}]}^{n \text{ times}} \tag{2.8}$$

This may be a good place to point out that if $[\boldsymbol{I}]$ is an identity matrix and $[\boldsymbol{A}]$ is a square matrix of matching size, then it can be readily shown that the product of $[\boldsymbol{I}][\boldsymbol{A}] = [\boldsymbol{A}][\boldsymbol{I}] = [\boldsymbol{A}]$. See Example 2.1 for the proof.

EXAMPLE 2.1

Given matrices

$$[\boldsymbol{A}] = \begin{bmatrix} 0 & 5 & 0 \\ 8 & 3 & 7 \\ 9 & -2 & 9 \end{bmatrix}, [\boldsymbol{B}] = \begin{bmatrix} 4 & 6 & -2 \\ 7 & 2 & 3 \\ 1 & 3 & -4 \end{bmatrix}, \text{ and } \{\boldsymbol{C}\} = \begin{Bmatrix} -1 \\ 2 \\ 5 \end{Bmatrix}$$

perform the following operations:

a. $[\boldsymbol{A}] + [\boldsymbol{B}] = ?$

b. $[\boldsymbol{A}] - [\boldsymbol{B}] = ?$

c. $3[\boldsymbol{A}] = ?$

d. $[\boldsymbol{A}][\boldsymbol{B}] = ?$

e. $[\boldsymbol{A}]\{\boldsymbol{C}\} = ?$

f. $[\boldsymbol{A}]^2 = ?$

g. Show that $[\boldsymbol{I}][\boldsymbol{A}] = [\boldsymbol{A}][\boldsymbol{I}] = [\boldsymbol{A}]$

We will use the operation rules discussed in the preceding sections to answer these questions.

a. $[\mathbf{A}] + [\mathbf{B}] = ?$

$$[\mathbf{A}] + [\mathbf{B}] = \begin{bmatrix} 0 & 5 & 0 \\ 8 & 3 & 7 \\ 9 & -2 & 9 \end{bmatrix} + \begin{bmatrix} 4 & 6 & -2 \\ 7 & 2 & 3 \\ 1 & 3 & -4 \end{bmatrix}$$

$$= \begin{bmatrix} (0+4) & (5+6) & (0+(-2)) \\ (8+7) & (3+2) & (7+3) \\ (9+1) & (-2+3) & (9+(-4)) \end{bmatrix} = \begin{bmatrix} 4 & 11 & -2 \\ 15 & 5 & 10 \\ 10 & 1 & 5 \end{bmatrix}$$

b. $[\mathbf{A}] - [\mathbf{B}] = ?$

$$[\mathbf{A}] - [\mathbf{B}] = \begin{bmatrix} 0 & 5 & 0 \\ 8 & 3 & 7 \\ 9 & -2 & 9 \end{bmatrix} - \begin{bmatrix} 4 & 6 & -2 \\ 7 & 2 & 3 \\ 1 & 3 & -4 \end{bmatrix}$$

$$= \begin{bmatrix} (0-4) & (5-6) & (0-(-2)) \\ (8-7) & (3-2) & (7-3) \\ (9-1) & (-2-3) & (9-(-4)) \end{bmatrix} = \begin{bmatrix} -4 & -1 & 2 \\ 1 & 1 & 4 \\ 8 & -5 & 13 \end{bmatrix}$$

c. $3[\mathbf{A}] = ?$

$$3[\mathbf{A}] = 3\begin{bmatrix} 0 & 5 & 0 \\ 8 & 3 & 7 \\ 9 & -2 & 9 \end{bmatrix} = \begin{bmatrix} 0 & (3)(5) & 0 \\ (3)(8) & (3)(3) & (3)(7) \\ (3)(9) & (3)(-2) & (3)(9) \end{bmatrix} = \begin{bmatrix} 0 & 15 & 0 \\ 24 & 9 & 21 \\ 27 & -6 & 27 \end{bmatrix}$$

d. $[\mathbf{A}][\mathbf{B}] = ?$

$$[\mathbf{A}][\mathbf{B}] = \begin{bmatrix} 0 & 5 & 0 \\ 8 & 3 & 7 \\ 9 & -2 & 9 \end{bmatrix}\begin{bmatrix} 4 & 6 & -2 \\ 7 & 2 & 3 \\ 1 & 3 & -4 \end{bmatrix} =$$

$$\begin{bmatrix} (0)(4)+(5)(7)+(0)(1) & (0)(6)+(5)(2)+(0)(3) & (0)(-2)+(5)(3)+(0)(-4) \\ (8)(4)+(3)(7)+(7)(1) & (8)(6)+(3)(2)+(7)(3) & (8)(-2)+(3)(3)+(7)(-4) \\ (9)(4)+(-2)(7)+(9)(1) & (9)(6)+(-2)(2)+(9)(3) & (9)(-2)+(-2)(3)+(9)(-4) \end{bmatrix}$$

$$= \begin{bmatrix} 35 & 10 & 15 \\ 60 & 75 & -35 \\ 31 & 77 & -60 \end{bmatrix}$$

e. $[\mathbf{A}]\{\mathbf{C}\} = ?$

$$[\mathbf{A}]\{\mathbf{C}\} = \begin{bmatrix} 0 & 5 & 0 \\ 8 & 3 & 7 \\ 9 & -2 & 9 \end{bmatrix}\begin{Bmatrix} -1 \\ 2 \\ 5 \end{Bmatrix} = \begin{Bmatrix} (0)(-1)+(5)(2)+(0)(5) \\ (8)(-1)+(3)(2)+(7)(5) \\ (9)(-1)+(-2)(2)+(9)(5) \end{Bmatrix} = \begin{Bmatrix} 10 \\ 33 \\ 32 \end{Bmatrix}$$

f. $[\mathbf{A}]^2 = ?$

$$[\mathbf{A}]^2 = [\mathbf{A}][\mathbf{A}] = \begin{bmatrix} 0 & 5 & 0 \\ 8 & 3 & 7 \\ 9 & -2 & 9 \end{bmatrix}\begin{bmatrix} 0 & 5 & 0 \\ 8 & 3 & 7 \\ 9 & -2 & 9 \end{bmatrix} = \begin{bmatrix} 40 & 15 & 35 \\ 87 & 35 & 84 \\ 65 & 21 & 67 \end{bmatrix}$$

g. Show that $[\boldsymbol{I}][\mathbf{A}] = [\mathbf{A}][\boldsymbol{I}] = [\mathbf{A}]$

$$[\boldsymbol{I}][\mathbf{A}] = \begin{bmatrix} 1 & 0 & 0 \\ 0 & 1 & 0 \\ 0 & 0 & 1 \end{bmatrix}\begin{bmatrix} 0 & 5 & 0 \\ 8 & 3 & 7 \\ 9 & -2 & 9 \end{bmatrix} = \begin{bmatrix} 0 & 5 & 0 \\ 8 & 3 & 7 \\ 9 & -2 & 9 \end{bmatrix} \text{ and}$$

$$[\mathbf{A}][\boldsymbol{I}] = \begin{bmatrix} 0 & 5 & 0 \\ 8 & 3 & 7 \\ 9 & -2 & 9 \end{bmatrix}\begin{bmatrix} 1 & 0 & 0 \\ 0 & 1 & 0 \\ 0 & 0 & 1 \end{bmatrix} = \begin{bmatrix} 0 & 5 & 0 \\ 8 & 3 & 7 \\ 9 & -2 & 9 \end{bmatrix}$$

2.4 PARTITIONING OF A MATRIX

Finite element formulation of complex problems typically involves relatively large sized matrices. For these situations, when performing numerical analysis dealing with matrix operations, it may be advantageous to partition the matrix and deal with a subset of elements. The partitioned matrices require less computer memory to perform the operations. Traditionally, dashed horizontal and vertical lines are used to show how a matrix is partitioned. For example, we may partition matrix $[\mathbf{A}]$ into four smaller matrices in the following manner:

$$[\mathbf{A}] = \left[\begin{array}{ccc:ccc} a_{11} & a_{12} & a_{13} & a_{14} & a_{15} & a_{16} \\ a_{21} & a_{22} & a_{23} & a_{24} & a_{25} & a_{26} \\ \hdashline a_{31} & a_{32} & a_{33} & a_{34} & a_{35} & a_{36} \\ a_{41} & a_{42} & a_{43} & a_{44} & a_{45} & a_{46} \\ a_{51} & a_{52} & a_{53} & a_{54} & a_{55} & a_{56} \end{array}\right]$$

and in terms of submatrices $[\mathbf{A}] = \begin{bmatrix} \mathbf{A}_{11} & \mathbf{A}_{12} \\ \mathbf{A}_{21} & \mathbf{A}_{22} \end{bmatrix}$

where

$$[\mathbf{A}_{11}] = \begin{bmatrix} a_{11} & a_{12} & a_{13} \\ a_{21} & a_{22} & a_{23} \end{bmatrix} \qquad [\mathbf{A}_{12}] = \begin{bmatrix} a_{14} & a_{15} & a_{16} \\ a_{24} & a_{25} & a_{26} \end{bmatrix}$$

$$[\mathbf{A}_{21}] = \begin{bmatrix} a_{31} & a_{32} & a_{33} \\ a_{41} & a_{42} & a_{43} \\ a_{51} & a_{52} & a_{53} \end{bmatrix} \qquad [\mathbf{A}_{22}] = \begin{bmatrix} a_{34} & a_{35} & a_{36} \\ a_{44} & a_{45} & a_{46} \\ a_{54} & a_{55} & a_{56} \end{bmatrix}$$

It is important to note that matrix $[\mathbf{A}]$ could have been partitioned in a number of other ways, and the way a matrix is partitioned would define the size of submatrices.

Addition and Subtraction Operations Using Partitioned Matrices

Now let us turn our attention to matrix operations dealing with addition, subtraction, or multiplication of two matrices that are partitioned. Consider matrix $[\boldsymbol{B}]$ having the same size (5×6) as matrix $[\boldsymbol{A}]$. If we partition matrix $[\boldsymbol{B}]$ in exactly the same way we partitioned $[\boldsymbol{A}]$ previously, then we can add the submatrices in the following manner:

$$[\boldsymbol{B}] = \left[\begin{array}{ccc|ccc} b_{11} & b_{12} & b_{13} & b_{14} & b_{15} & b_{16} \\ b_{21} & b_{22} & b_{23} & b_{24} & b_{25} & b_{26} \\ \hline b_{31} & b_{32} & b_{33} & b_{34} & b_{35} & b_{36} \\ b_{41} & b_{42} & b_{43} & b_{44} & b_{45} & b_{46} \\ b_{51} & b_{52} & b_{53} & b_{54} & b_{55} & b_{56} \end{array}\right]$$

where

$$[\boldsymbol{B}_{11}] = \begin{bmatrix} b_{11} & b_{12} & b_{13} \\ b_{21} & b_{22} & b_{23} \end{bmatrix} \qquad [\boldsymbol{B}_{12}] = \begin{bmatrix} b_{14} & b_{15} & b_{16} \\ b_{24} & b_{25} & b_{26} \end{bmatrix}$$

$$[\boldsymbol{B}_{21}] = \begin{bmatrix} b_{31} & b_{32} & b_{33} \\ b_{41} & b_{42} & b_{43} \\ b_{51} & b_{52} & b_{53} \end{bmatrix} \qquad [\boldsymbol{B}_{22}] = \begin{bmatrix} b_{34} & b_{35} & b_{36} \\ b_{44} & b_{45} & b_{46} \\ b_{54} & b_{55} & b_{56} \end{bmatrix}$$

Then, using submatrices we can write

$$[\boldsymbol{A}] + [\boldsymbol{B}] = \begin{bmatrix} \boldsymbol{A}_{11} + \boldsymbol{B}_{11} & \boldsymbol{A}_{12} + \boldsymbol{B}_{12} \\ \boldsymbol{A}_{21} + \boldsymbol{B}_{21} & \boldsymbol{A}_{22} + \boldsymbol{B}_{22} \end{bmatrix}$$

Matrix Multiplication Using Partitioned Matrices

As mentioned earlier, matrix multiplication can be performed only when the number of columns in the *premultiplier* matrix is equal to the number of rows in the *postmultiplier* matrix. Referring to $[\boldsymbol{A}]$ and $[\boldsymbol{B}]$ matrices of the preceding section, because the number of columns in matrix $[\boldsymbol{A}]$ does not match the number of rows of matrix $[\boldsymbol{B}]$, then matrix $[\boldsymbol{B}]$ cannot be premultiplied by matrix $[\boldsymbol{A}]$. To demonstrate matrix multiplication using submatrices, consider matrix $[\boldsymbol{C}]$ of size 6×3, which is partitioned in the manner shown below.

$$[\boldsymbol{C}] = \left[\begin{array}{cc|c} c_{11} & c_{12} & c_{13} \\ c_{21} & c_{22} & c_{23} \\ c_{31} & c_{32} & c_{33} \\ \hline c_{41} & c_{42} & c_{43} \\ c_{51} & c_{52} & c_{53} \\ c_{61} & c_{62} & c_{63} \end{array}\right]$$

$$[\boldsymbol{C}] = \begin{bmatrix} \boldsymbol{C}_{11} & \boldsymbol{C}_{12} \\ \boldsymbol{C}_{21} & \boldsymbol{C}_{22} \end{bmatrix}$$

where

$$[\boldsymbol{C}_{11}] = \begin{bmatrix} c_{11} & c_{12} \\ c_{21} & c_{22} \\ c_{31} & c_{32} \end{bmatrix} \quad \text{and } \{\boldsymbol{C}_{12}\} = \begin{Bmatrix} c_{13} \\ c_{23} \\ c_{33} \end{Bmatrix}$$

$$[\boldsymbol{C}_{21}] = \begin{bmatrix} c_{41} & c_{42} \\ c_{51} & c_{52} \\ c_{61} & c_{62} \end{bmatrix} \quad \text{and } \{\boldsymbol{C}_{22}\} = \begin{Bmatrix} c_{43} \\ c_{53} \\ c_{63} \end{Bmatrix}$$

Next, consider premultiplying matrix $[\boldsymbol{C}]$ by matrix $[\boldsymbol{A}]$. Let us refer to the results of this multiplication by matrix $[\boldsymbol{D}]$ of size 5×3. In addition to paying attention to the size requirement for matrix multiplication, to carry out the multiplication using partitioned matrices, the premultiplying and postmultiplying matrices must be partitioned in such a way that the resulting submatrices conform to the multiplication rule. That is, if we partition matrix $[\boldsymbol{A}]$ between the third and the fourth columns, then matrix $[\boldsymbol{C}]$ must be partitioned between the third and the fourth *rows*. However, the *column* partitioning of matrix $[\boldsymbol{C}]$ may be done arbitrarily, because regardless of how the columns are partitioned, the resulting submatrices will still conform to the multiplication rule. In other words, instead of partitioning matrix $[\boldsymbol{C}]$ between columns two and three, we could have partitioned the matrix between columns one and two and still carried out the multiplication using the resulting submatrices.

$$[\boldsymbol{A}][\boldsymbol{C}] = [\boldsymbol{D}] = \begin{bmatrix} \boldsymbol{A}_{11} & \boldsymbol{A}_{12} \\ \boldsymbol{A}_{21} & \boldsymbol{A}_{22} \end{bmatrix} \begin{bmatrix} \boldsymbol{C}_{11} & \boldsymbol{C}_{12} \\ \boldsymbol{C}_{21} & \boldsymbol{C}_{22} \end{bmatrix} = \begin{bmatrix} \boldsymbol{A}_{11}\boldsymbol{C}_{11} + \boldsymbol{A}_{12}\boldsymbol{C}_{21} & \boldsymbol{A}_{11}\boldsymbol{C}_{12} + \boldsymbol{A}_{12}\boldsymbol{C}_{22} \\ \boldsymbol{A}_{21}\boldsymbol{C}_{11} + \boldsymbol{A}_{22}\boldsymbol{C}_{21} & \boldsymbol{A}_{21}\boldsymbol{C}_{12} + \boldsymbol{A}_{22}\boldsymbol{C}_{22} \end{bmatrix}$$

$$= \begin{bmatrix} \boldsymbol{D}_{11} & \boldsymbol{D}_{12} \\ \boldsymbol{D}_{21} & \boldsymbol{D}_{22} \end{bmatrix}$$

where

$$[\boldsymbol{D}_{11}] = [\boldsymbol{A}_{11}][\boldsymbol{C}_{11}] + [\boldsymbol{A}_{12}][\boldsymbol{C}_{21}] \quad \text{and} \quad [\boldsymbol{D}_{12}] = [\boldsymbol{A}_{11}]\{\boldsymbol{C}_{12}\} + [\boldsymbol{A}_{12}]\{\boldsymbol{C}_{22}\}$$

$$[\boldsymbol{D}_{21}] = [\boldsymbol{A}_{21}][\boldsymbol{C}_{11}] + [\boldsymbol{A}_{22}][\boldsymbol{C}_{21}] \qquad [\boldsymbol{D}_{22}] = [\boldsymbol{A}_{21}]\{\boldsymbol{C}_{12}\} + [\boldsymbol{A}_{22}]\{\boldsymbol{C}_{22}\}$$

EXAMPLE 2.2

Given the following partitioned matrices, calculate the product of $[\boldsymbol{A}][\boldsymbol{B}] = [\boldsymbol{C}]$ using the submatrices shown.

$$[\boldsymbol{A}] = \left[\begin{array}{ccc|ccc} 5 & 7 & 2 & 0 & 3 & 5 \\ 3 & 8 & -3 & -5 & 0 & 8 \\ \hline 1 & 4 & 0 & 7 & 15 & 9 \\ 0 & 10 & 5 & 12 & 3 & -1 \\ 2 & -5 & 9 & 2 & 18 & -10 \end{array}\right] \quad \text{and} \quad [\boldsymbol{B}] = \left[\begin{array}{cc|c} 2 & 10 & 0 \\ 8 & 7 & 5 \\ -5 & 2 & -4 \\ \hline 4 & 8 & 13 \\ 3 & 12 & 0 \\ 1 & 5 & 7 \end{array}\right]$$

$$[\boldsymbol{A}] = \begin{bmatrix} \boldsymbol{A}_{11} & \boldsymbol{A}_{12} \\ \boldsymbol{A}_{21} & \boldsymbol{A}_{22} \end{bmatrix}$$

where

$$[\boldsymbol{A}_{11}] = \begin{bmatrix} 5 & 7 & 2 \\ 3 & 8 & -3 \end{bmatrix} \quad [\boldsymbol{A}_{12}] = \begin{bmatrix} 0 & 3 & 5 \\ -5 & 0 & 8 \end{bmatrix}$$

$$[\boldsymbol{A}_{21}] = \begin{bmatrix} 1 & 4 & 0 \\ 0 & 10 & 5 \\ 2 & -5 & 9 \end{bmatrix} \quad [\boldsymbol{A}_{22}] = \begin{bmatrix} 7 & 15 & 9 \\ 12 & 3 & -1 \\ 2 & 18 & -10 \end{bmatrix}$$

and

$$[\boldsymbol{B}] = \begin{bmatrix} \boldsymbol{B}_{11} & \boldsymbol{B}_{12} \\ \boldsymbol{B}_{21} & \boldsymbol{B}_{22} \end{bmatrix}$$

where

$$[\boldsymbol{B}_{11}] = \begin{bmatrix} 2 & 10 \\ 8 & 7 \\ -5 & 2 \end{bmatrix} \quad \text{and } \{\boldsymbol{B}_{12}\} = \begin{Bmatrix} 0 \\ 5 \\ -4 \end{Bmatrix}$$

$$[\boldsymbol{B}_{21}] = \begin{bmatrix} 4 & 8 \\ 3 & 12 \\ 1 & 5 \end{bmatrix} \quad \text{and } \{\boldsymbol{B}_{22}\} = \begin{Bmatrix} 13 \\ 0 \\ 7 \end{Bmatrix}$$

$$[\boldsymbol{A}][\boldsymbol{B}] = [\boldsymbol{C}] = \begin{bmatrix} \boldsymbol{A}_{11} & \boldsymbol{A}_{12} \\ \boldsymbol{A}_{21} & \boldsymbol{A}_{22} \end{bmatrix}\begin{bmatrix} \boldsymbol{B}_{11} & \boldsymbol{B}_{12} \\ \boldsymbol{B}_{21} & \boldsymbol{B}_{22} \end{bmatrix} = \begin{bmatrix} \boldsymbol{A}_{11}\boldsymbol{B}_{11} + \boldsymbol{A}_{12}\boldsymbol{B}_{21} & \boldsymbol{A}_{11}\boldsymbol{B}_{12} + \boldsymbol{A}_{12}\boldsymbol{B}_{22} \\ \boldsymbol{A}_{21}\boldsymbol{B}_{11} + \boldsymbol{A}_{22}\boldsymbol{B}_{21} & \boldsymbol{A}_{21}\boldsymbol{B}_{12} + \boldsymbol{A}_{22}\boldsymbol{B}_{22} \end{bmatrix} = \begin{bmatrix} \boldsymbol{C}_{11} & \boldsymbol{C}_{12} \\ \boldsymbol{C}_{21} & \boldsymbol{C}_{22} \end{bmatrix}$$

where

$$[\boldsymbol{C}_{11}] = [\boldsymbol{A}_{11}][\boldsymbol{B}_{11}] + [\boldsymbol{A}_{12}][\boldsymbol{B}_{21}] = \begin{bmatrix} 5 & 7 & 2 \\ 3 & 8 & -3 \end{bmatrix}\begin{bmatrix} 2 & 10 \\ 8 & 7 \\ -5 & 2 \end{bmatrix}$$

$$+ \begin{bmatrix} 0 & 3 & 5 \\ -5 & 0 & 8 \end{bmatrix}\begin{bmatrix} 4 & 8 \\ 3 & 12 \\ 1 & 5 \end{bmatrix} = \begin{bmatrix} 70 & 164 \\ 73 & 80 \end{bmatrix}$$

$$[\boldsymbol{C}_{12}] = [\boldsymbol{A}_{11}]\{\boldsymbol{B}_{12}\} + [\boldsymbol{A}_{12}]\{\boldsymbol{B}_{22}\} = \begin{bmatrix} 5 & 7 & 2 \\ 3 & 8 & -3 \end{bmatrix}\begin{Bmatrix} 0 \\ 5 \\ -4 \end{Bmatrix}$$

$$+ \begin{bmatrix} 0 & 3 & 5 \\ -5 & 0 & 8 \end{bmatrix}\begin{Bmatrix} 13 \\ 0 \\ 7 \end{Bmatrix} = \begin{Bmatrix} 62 \\ 43 \end{Bmatrix}$$

$$[\boldsymbol{C}_{21}] = [\boldsymbol{A}_{21}][\boldsymbol{B}_{11}] + [\boldsymbol{A}_{22}][\boldsymbol{B}_{21}] = \begin{bmatrix} 1 & 4 & 0 \\ 0 & 10 & 5 \\ 2 & -5 & 9 \end{bmatrix}\begin{bmatrix} 2 & 10 \\ 8 & 7 \\ -5 & 2 \end{bmatrix}$$

$$+ \begin{bmatrix} 7 & 15 & 9 \\ 12 & 3 & -1 \\ 2 & 18 & -10 \end{bmatrix}\begin{bmatrix} 4 & 8 \\ 3 & 12 \\ 1 & 5 \end{bmatrix} = \begin{bmatrix} 116 & 319 \\ 111 & 207 \\ -29 & 185 \end{bmatrix}$$

$$[C_{22}] = [A_{21}]\{B_{12}\} + [A_{22}]\{B_{22}\} = \begin{bmatrix} 1 & 4 & 0 \\ 0 & 10 & 5 \\ 2 & -5 & 9 \end{bmatrix} \begin{Bmatrix} 0 \\ 5 \\ -4 \end{Bmatrix}$$

$$+ \begin{bmatrix} 7 & 15 & 9 \\ 12 & 3 & -1 \\ 2 & 18 & -10 \end{bmatrix} \begin{Bmatrix} 13 \\ 0 \\ 7 \end{Bmatrix} = \begin{Bmatrix} 174 \\ 179 \\ -105 \end{Bmatrix}$$

The final result is given by

$$[\boldsymbol{A}][\boldsymbol{B}] = [\boldsymbol{C}] = \begin{bmatrix} \boldsymbol{A}_{11} & \boldsymbol{A}_{12} \\ \boldsymbol{A}_{21} & \boldsymbol{A}_{22} \end{bmatrix} \begin{bmatrix} \boldsymbol{B}_{11} & \boldsymbol{B}_{12} \\ \boldsymbol{B}_{21} & \boldsymbol{B}_{22} \end{bmatrix} = \begin{bmatrix} \boldsymbol{C}_{11} & \boldsymbol{C}_{12} \\ \boldsymbol{C}_{21} & \boldsymbol{C}_{22} \end{bmatrix} = \left[\begin{array}{cc:c} 70 & 164 & 62 \\ 73 & 80 & 43 \\ \hdashline 116 & 319 & 174 \\ 111 & 207 & 179 \\ -29 & 185 & -105 \end{array}\right]$$

As explained earlier, the column partitioning of matrix $[\boldsymbol{B}]$ may be done arbitrarily because the resulting submatrices still conform to the multiplication rule. It is left as an exercise for you (see problem 3 at the end of this chapter) to show that we could have partitioned matrix $[\boldsymbol{B}]$ between columns one and two and used the resulting submatrices to compute $[\boldsymbol{A}][\boldsymbol{B}] = [\boldsymbol{C}]$.

2.5 TRANSPOSE OF A MATRIX

As you will see in the following chapters, the finite element formulation lends itself to situations wherein it is desirable to rearrange the rows of a matrix into the columns of another matrix. To demonstrate this idea, let us go back and consider step 4 in Example 1.1. In step 4 we assembled the elemental stiffness matrices to obtain the global stiffness matrix. You will recall that we constructed the stiffness matrix for each element with its position in the global stiffness matrix by inspection. Let's recall the stiffness matrix for element (1), which is shown here again for the sake of continuity and convenience.

$$[\boldsymbol{K}]^{(1)} = \begin{bmatrix} k_1 & -k_1 \\ -k_1 & k_1 \end{bmatrix}$$ and its position in the global matrix

$$[\boldsymbol{K}]^{(1G)} = \begin{bmatrix} k_1 & -k_1 & 0 & 0 & 0 \\ -k_1 & k_1 & 0 & 0 & 0 \\ 0 & 0 & 0 & 0 & 0 \\ 0 & 0 & 0 & 0 & 0 \\ 0 & 0 & 0 & 0 & 0 \end{bmatrix}$$

Instead of putting together $[\boldsymbol{K}]^{(1G)}$ by inspection as we did, we could obtain $[\boldsymbol{K}]^{(1G)}$ using the following procedure:

$$[\boldsymbol{K}]^{(1G)} = [\boldsymbol{A}_1]^T[\boldsymbol{K}]^{(1)}[\boldsymbol{A}_1] \tag{2.9}$$

where

$$[\mathbf{A}_1] = \begin{bmatrix} 1 & 0 & 0 & 0 & 0 \\ 0 & 1 & 0 & 0 & 0 \end{bmatrix}$$

and

$$[\mathbf{A}_1]^T = \begin{bmatrix} 1 & 0 \\ 0 & 1 \\ 0 & 0 \\ 0 & 0 \\ 0 & 0 \end{bmatrix}$$

$[\mathbf{A}_1]^T$, called the transpose of $[\mathbf{A}_1]$, is obtained by taking the first and the second rows of $[\mathbf{A}_1]$ and making them into the first and the second columns of the transpose matrix. It is easily verified that by carrying out the multiplication given by Eq. (2.9), we will arrive at the same result that was obtained by inspection.

$$[\boldsymbol{K}]^{(1G)} = \begin{bmatrix} 1 & 0 \\ 0 & 1 \\ 0 & 0 \\ 0 & 0 \\ 0 & 0 \end{bmatrix} \begin{bmatrix} k_1 & -k_1 \\ -k_1 & k_1 \end{bmatrix} \begin{bmatrix} 1 & 0 & 0 & 0 & 0 \\ 0 & 1 & 0 & 0 & 0 \end{bmatrix} = \begin{bmatrix} k_1 & -k_1 & 0 & 0 & 0 \\ -k_1 & k_1 & 0 & 0 & 0 \\ 0 & 0 & 0 & 0 & 0 \\ 0 & 0 & 0 & 0 & 0 \\ 0 & 0 & 0 & 0 & 0 \end{bmatrix}$$

Similarly, we could have performed the following operation to obtain $[\boldsymbol{K}]^{(2G)}$:

$$[\boldsymbol{K}]^{(2G)} = [\mathbf{A}_2]^T[\boldsymbol{K}]^{(2)}[\mathbf{A}_2]$$

where

$$[\mathbf{A}_2] = \begin{bmatrix} 0 & 1 & 0 & 0 & 0 \\ 0 & 0 & 1 & 0 & 0 \end{bmatrix}$$

and

$$[\mathbf{A}_2]^T = \begin{bmatrix} 0 & 0 \\ 1 & 0 \\ 0 & 1 \\ 0 & 0 \\ 0 & 0 \end{bmatrix}$$

and

$$[\boldsymbol{K}]^{(2G)} = \begin{bmatrix} 0 & 0 \\ 1 & 0 \\ 0 & 1 \\ 0 & 0 \\ 0 & 0 \end{bmatrix} \begin{bmatrix} k_2 & -k_2 \\ -k_2 & k_2 \end{bmatrix} \begin{bmatrix} 0 & 1 & 0 & 0 & 0 \\ 0 & 0 & 1 & 0 & 0 \end{bmatrix} = \begin{bmatrix} 0 & 0 & 0 & 0 & 0 \\ 0 & k_2 & -k_2 & 0 & 0 \\ 0 & -k_2 & k_2 & 0 & 0 \\ 0 & 0 & 0 & 0 & 0 \\ 0 & 0 & 0 & 0 & 0 \end{bmatrix}$$

As you have seen from the previous examples, we can use a positioning matrix, such as $[\boldsymbol{A}]$ and its transpose, and methodically create the global stiffness matrix for finite element models.

In general, to obtain the transpose of a matrix $[\boldsymbol{B}]$ of size $m \times n$, the first row of the given matrix becomes the first column of the $[\boldsymbol{B}]^T$, the second row of $[\boldsymbol{B}]$ becomes the second column of $[\boldsymbol{B}]^T$, and so on, leading to the mth row of $[\boldsymbol{B}]$ becoming the mth column of the $[\boldsymbol{B}]^T$, resulting in a matrix with the size of $n \times m$. Clearly, if you take the transpose of $[\boldsymbol{B}]^T$, you will end up with $[\boldsymbol{B}]$. That is,

$$([\boldsymbol{B}]^T)^T = [\boldsymbol{B}] \tag{2.10}$$

As you will see in succeeding chapters, in order to save space, we write the solution matrices, which are column matrices, as row matrices using the transpose of the solution—another use for transpose of a matrix. For example, we represent the displacement solution

$$\{\boldsymbol{U}\} = \begin{Bmatrix} U_1 \\ U_2 \\ U_3 \\ \cdot \\ U_n \end{Bmatrix} \text{ by } [\boldsymbol{U}]^T = [U_1 \quad U_2 \quad U_3 \quad \cdot \quad U_n]$$

When performing matrix operations dealing with transpose of matrices, the following identities are true:

$$([\boldsymbol{A}] + [\boldsymbol{B}] + \cdots + [\boldsymbol{N}])^T = [\boldsymbol{A}]^T + [\boldsymbol{B}]^T + \cdots + [\boldsymbol{N}]^T \tag{2.11}$$

$$([\boldsymbol{A}][\boldsymbol{B}]\ldots[\boldsymbol{N}])^T = [\boldsymbol{N}]^T\ldots[\boldsymbol{B}]^T[\boldsymbol{A}]^T \tag{2.12}$$

In Eq. (2.12), note the change in the order of multiplication.

This is a good place to define a symmetric matrix. A *symmetric matrix* is a square matrix whose elements are symmetrical with respect to its principal diagonal. An example of a symmetric matrix follows.

$$[\boldsymbol{A}] = \begin{bmatrix} 1 & 4 & 2 & -5 \\ 4 & 5 & 15 & 20 \\ 2 & 15 & -3 & 8 \\ -5 & 20 & 8 & 0 \end{bmatrix}$$

Note that for a symmetric matrix, element a_{mn} is equal to a_{nm}. That is, $a_{mn} = a_{nm}$ for all values of n and m. Therefore, for a symmetric matrix, $[\boldsymbol{A}] = [\boldsymbol{A}]^T$.

EXAMPLE 2.3

Given the following matrices:

$$[\boldsymbol{A}] = \begin{bmatrix} 0 & 5 & 0 \\ 8 & 3 & 7 \\ 9 & -2 & 9 \end{bmatrix} \text{ and } [\boldsymbol{B}] = \begin{bmatrix} 4 & 6 & -2 \\ 7 & 2 & 3 \\ 1 & 3 & -4 \end{bmatrix}$$

perform the following operations:

a. $[\mathbf{A}]^T = ?$ and $[\mathbf{B}]^T = ?$

b. verify that $([\mathbf{A}] + [\mathbf{B}])^T = [\mathbf{A}]^T + [\mathbf{B}]^T$

c. verify that $([\mathbf{A}][\mathbf{B}])^T = [\mathbf{B}]^T[\mathbf{A}]^T$

a. $[\mathbf{A}]^T = ?$ and $[\mathbf{B}]^T = ?$

As explained earlier, the first, second, third, . . . , and the mth rows of a matrix become the first, second, third, . . . , and the mth columns of the transpose matrix respectively.

$$[\mathbf{A}]^T = \begin{bmatrix} 0 & 8 & 9 \\ 5 & 3 & -2 \\ 0 & 7 & 9 \end{bmatrix}$$

Similarly,

$$[\mathbf{B}]^T = \begin{bmatrix} 4 & 7 & 1 \\ 6 & 2 & 3 \\ -2 & 3 & -4 \end{bmatrix}$$

b. Verify that $([\mathbf{A}] + [\mathbf{B}])^T = [\mathbf{A}]^T + [\mathbf{B}]^T$.

$$[\mathbf{A}] + [\mathbf{B}] = \begin{bmatrix} 4 & 11 & -2 \\ 15 & 5 & 10 \\ 10 & 1 & 5 \end{bmatrix} \text{ and } ([\mathbf{A}] + [\mathbf{B}])^T = \begin{bmatrix} 4 & 15 & 10 \\ 11 & 5 & 1 \\ -2 & 10 & 5 \end{bmatrix}$$

$$[\mathbf{A}]^T + [\mathbf{B}]^T = \begin{bmatrix} 0 & 8 & 9 \\ 5 & 3 & -2 \\ 0 & 7 & 9 \end{bmatrix} + \begin{bmatrix} 4 & 7 & 1 \\ 6 & 2 & 3 \\ -2 & 3 & -4 \end{bmatrix} = \begin{bmatrix} 4 & 15 & 10 \\ 11 & 5 & 1 \\ -2 & 10 & 5 \end{bmatrix}$$

Comparing results, it is clear that the given identity is true.

c. Verify that $([\mathbf{A}][\mathbf{B}])^T = [\mathbf{B}]^T[\mathbf{A}]^T$.

In Example 2.2 we computed the product of $[\mathbf{A}][\mathbf{B}]$:

$$[\mathbf{A}][\mathbf{B}] = \begin{bmatrix} 0 & 5 & 0 \\ 8 & 3 & 7 \\ 9 & -2 & 9 \end{bmatrix}\begin{bmatrix} 4 & 6 & -2 \\ 7 & 2 & 3 \\ 1 & 3 & -4 \end{bmatrix} = \begin{bmatrix} 35 & 10 & 15 \\ 60 & 75 & -35 \\ 31 & 77 & -60 \end{bmatrix}$$

and using the results we get

$$([\mathbf{A}][\mathbf{B}])^T = \begin{bmatrix} 35 & 60 & 31 \\ 10 & 75 & 77 \\ 15 & -35 & -60 \end{bmatrix}$$

Alternatively,

$$[\mathbf{B}]^T[\mathbf{A}]^T = \begin{bmatrix} 4 & 7 & 1 \\ 6 & 2 & 3 \\ -2 & 3 & -4 \end{bmatrix}\begin{bmatrix} 0 & 8 & 9 \\ 5 & 3 & -2 \\ 0 & 7 & 9 \end{bmatrix} = \begin{bmatrix} 35 & 60 & 31 \\ 10 & 75 & 77 \\ 15 & -35 & -60 \end{bmatrix}$$

Again, by comparing results we can see that the given identity is true.

2.6 DETERMINANT OF A MATRIX

Up to this point we have defined essential matrix terminology and discussed basic matrix operations. In this section we define what is meant by a *determinant of a matrix*. As you will see in the succeeding sections, determinant of a matrix is used in solving a set of simultaneous equations, obtaining the inverse of a matrix, and forming the characteristic equations for a dynamic problem (eigenvalue problem).

Let us consider the solution to the following set of simultaneous equations:

$$a_{11}x_1 + a_{12}x_2 = b_1 \tag{2.13a}$$

$$a_{21}x_1 + a_{22}x_2 = b_2 \tag{2.13b}$$

We can represent Eqs. (2.13a) and (2.13b) in a matrix form by

$$\begin{bmatrix} a_{11} & a_{12} \\ a_{21} & a_{22} \end{bmatrix} \begin{Bmatrix} x_1 \\ x_2 \end{Bmatrix} = \begin{Bmatrix} b_1 \\ b_2 \end{Bmatrix} \tag{2.14}$$

or in a compact form by

$$[\mathbf{A}]\{\mathbf{X}\} = \{\mathbf{B}\}$$

To solve for the unknowns x_1 and x_2, we may first solve for x_2 in terms of x_1, using Eq. (2.13b), and then substitute that relationship into Eq. (2.13a). These steps are shown next.

$$x_2 = \frac{b_2 - a_{21}x_1}{a_{22}} \quad \Rightarrow \quad a_{11}x_1 + a_{12}\left(\frac{b_2 - a_{21}x_1}{a_{22}}\right) = b_1$$

Solving for x_1

$$x_1 = \frac{b_1a_{22} - a_{12}b_2}{a_{11}a_{22} - a_{12}a_{21}} \tag{2.15a}$$

After we substitute for x_1 in either Eq. (2.13a) or (2.13b), we get

$$x_2 = \frac{a_{11}b_2 - b_1a_{21}}{a_{11}a_{22} - a_{12}a_{21}} \tag{2.15b}$$

Referring to the solutions given by Eqs. (2.15a) and (2.15b), we see that the denominators in these equations represent the product of coefficients in the main diagonal minus the product of the coefficient in the other diagonal of the $[\mathbf{A}]$ matrix.

The $a_{11}a_{22} - a_{12}a_{21}$ is the *determinant* of the 2×2 $[\mathbf{A}]$ matrix and is represented in one of following ways:

$$\mathbf{Det}[\mathbf{A}] \text{ or } \mathbf{det}[\mathbf{A}] \text{ or } \begin{vmatrix} a_{11} & a_{12} \\ a_{21} & a_{22} \end{vmatrix} = a_{11}a_{22} - a_{12}a_{21} \tag{2.16}$$

Only the determinant of a square matrix is defined. Moreover, keep in mind that the determinant of the $[\mathbf{A}]$ matrix is a single number. That is, after we substitute for the values of a_{11}, a_{22}, a_{12}, and a_{21} into $a_{11}a_{22} - a_{12}a_{21}$, we get a single number. In general, the determinant of a matrix is a single value. However, as you will see later, for dynamic problems the determinant of a matrix resulting from equations of motions is a polynomial expression.

Cramer's rule is a numerical technique that can be used to obtain solutions to a relatively small set of equations similar to the previous example. Using Cramer's rule we can represent the solutions to the set of simultaneous equations given by Eqs. (2.13a) and (2.13b) with the following determinants:

$$x_1 = \frac{\begin{vmatrix} b_1 & a_{12} \\ b_2 & a_{22} \end{vmatrix}}{\begin{vmatrix} a_{11} & a_{12} \\ a_{21} & a_{22} \end{vmatrix}} \quad \text{and} \quad x_2 = \frac{\begin{vmatrix} a_{11} & b_1 \\ a_{21} & b_2 \end{vmatrix}}{\begin{vmatrix} a_{11} & a_{12} \\ a_{21} & a_{22} \end{vmatrix}} \tag{2.17}$$

Let us now consider the determinant of a 3×3 matrix such as

$$[\boldsymbol{C}] = \begin{bmatrix} c_{11} & c_{12} & c_{13} \\ c_{21} & c_{22} & c_{23} \\ c_{31} & c_{32} & c_{33} \end{bmatrix}$$

which is computed in the following manner:

$$\begin{vmatrix} c_{11} & c_{12} & c_{13} \\ c_{21} & c_{22} & c_{23} \\ c_{31} & c_{32} & c_{33} \end{vmatrix} = \begin{matrix} c_{11}c_{22}c_{33} + c_{12}c_{23}c_{31} + c_{13}c_{21}c_{32} - c_{13}c_{22}c_{31} \\ - c_{11}c_{23}c_{32} - c_{12}c_{21}c_{33} \end{matrix} \tag{2.18}$$

There is a simple procedure called *direct expansion*, which you can use to obtain the results given by Eq. (2.18). Direct expansion proceeds in the following manner. First, we repeat and place the first and the second columns of the matrix $[\boldsymbol{C}]$ next to the third column, as shown in Figure 2.1. Then we add the products of the diagonal elements lying on the solid arrows and subtract them from the products of the diagonal elements lying on the dashed arrows. This procedure, shown in Figure 2.1, results in the determinant value given by Eq. (2.18).

The direct expansion procedure cannot be used to obtain higher order determinants. Instead, we resort to a method that first reduces the order of the determinant—to what is called a *minor*—and then evaluates the lower order determinants. To demonstrate this method, let's consider the right-hand side of Eq. (2.18) and factor out c_{11}, $-c_{12}$, and c_{13} from it. This operation is shown below.

$$\begin{aligned} &c_{11}c_{22}c_{33} + c_{12}c_{23}c_{31} + c_{13}c_{21}c_{32} - c_{13}c_{22}c_{31} - c_{11}c_{23}c_{32} - c_{12}c_{21}c_{33} \\ &= c_{11}(c_{22}c_{33} - c_{23}c_{32}) - c_{12}(c_{21}c_{33} - c_{23}c_{31}) + c_{13}(c_{21}c_{32} - c_{22}c_{31}) \end{aligned}$$

As you can see, the expressions in the parentheses represent the determinants of reduced 2×2 matrices. Thus, we can express the determinant of the given 3×3 matrix

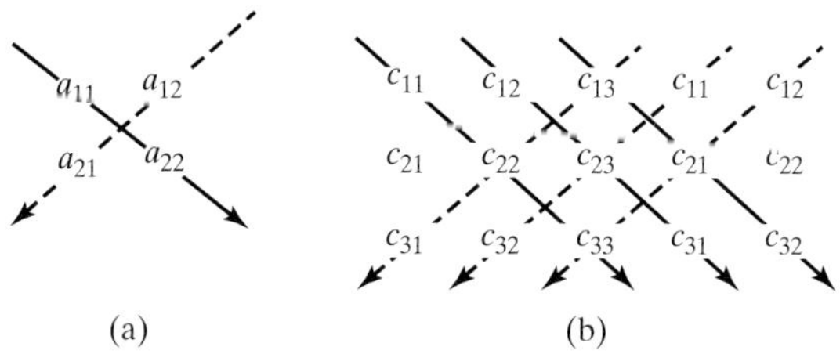

FIGURE 2.1 Direct expansion procedure for computing the determinant of (a) 2×2 matrix, and (b) 3×3 matrix.

in terms of the determinants of the reduced 2×2 matrices (*minors*) in the following manner:

$$\begin{vmatrix} c_{11} & c_{12} & c_{13} \\ c_{21} & c_{22} & c_{23} \\ c_{31} & c_{32} & c_{33} \end{vmatrix} = c_{11}\begin{vmatrix} c_{22} & c_{23} \\ c_{32} & c_{33} \end{vmatrix} - c_{12}\begin{vmatrix} c_{21} & c_{23} \\ c_{31} & c_{33} \end{vmatrix} + c_{13}\begin{vmatrix} c_{21} & c_{22} \\ c_{31} & c_{32} \end{vmatrix}$$

A simple way to visualize this reduction of a third-order determinant into three second-order minors is to draw a line through the first row, first column, second column, and third column. Note that the elements shown in the box—the common element contained in elimination rows and columns—are factors that get multiplied by the lower order minors. The plus and the minus signs before these factors are assigned based on the following procedure. We add the row and the column number of the factor, and if the sum is an even number, we assign a positive sign, and if the sum is an odd number we assign a negative sign to the factor. For example, when deleting the *first* row and the *second* column $(1 + 2)$, a negative sign should then appear before the c_{12} factor.

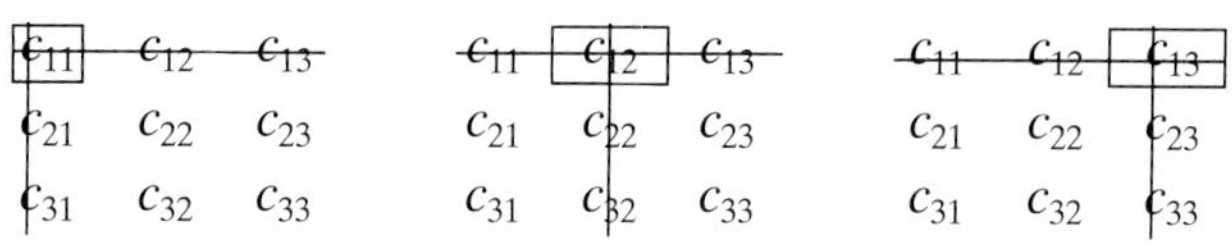

It is important to note that alternatively to compute the determinant of $[\mathbf{A}]$ we could have eliminated the second or the third row—instead of the first row—to reduce the given determinant into three other second-order minors. This point is demonstrated in Example 2.4.

Our previous discussion demonstrates that the order of a determinant may be reduced into other lower order minors, and the lower order determinants may be used to evaluate the value of the higher order determinant.

Here are two useful properties of determinants: (1) The determinant of a matrix $[\mathbf{A}]$ is equal to the determinant of its transpose $[\mathbf{A}]^T$. This property may be readily verified (see Example 2.4). (2) If you multiply the elements of a row or a column of a matrix by a scalar quantity, then its determinant gets multiplied by that quantity as well.

EXAMPLE 2.4

Given the following matrix:

$$[\mathbf{A}] = \begin{bmatrix} 1 & 5 & 0 \\ 8 & 3 & 7 \\ 6 & -2 & 9 \end{bmatrix}$$

calculate

a. determinant of $[\mathbf{A}]$
b. determinant of $[\mathbf{A}]^T$

a. For this example, we use both the direct expansion and the minor methods to compute the determinant of $[\mathbf{A}]$. As explained earlier, using the direct expansion method, we repeat and place the first and the second columns of the matrix next to the third column as shown, and compute the products of the elements along the solid arrows, then subtract them from the products of elements along the dashed arrow.

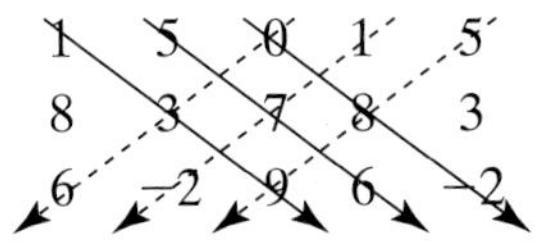

$$\begin{vmatrix} 1 & 5 & 0 \\ 8 & 3 & 7 \\ 6 & -2 & 9 \end{vmatrix} = \begin{matrix} (1)(3)(9) + (5)(7)(6) + (0)(8)(-2) - (5)(8)(9) \\ - (1)(7)(-2) - (0)(3)(6) = -109 \end{matrix}$$

Next, we use the minor to compute the determinant of $[\mathbf{A}]$. For this example, we eliminate the elements in the first row and in first, second, and third columns, as shown.

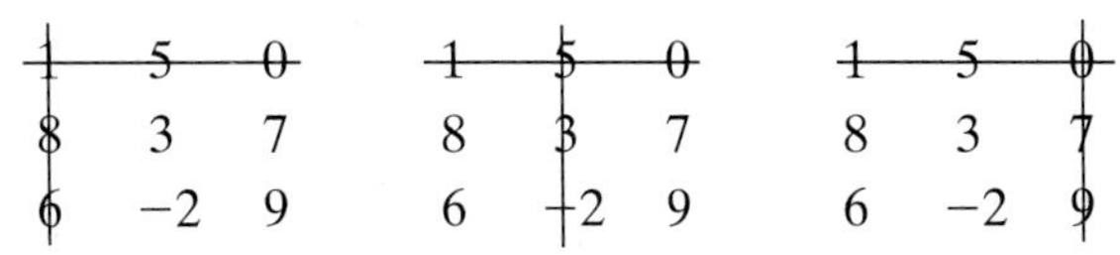

$$\begin{vmatrix} 1 & 5 & 0 \\ 8 & 3 & 7 \\ 6 & -2 & 9 \end{vmatrix} = 1\begin{vmatrix} 3 & 7 \\ -2 & 9 \end{vmatrix} - 5\begin{vmatrix} 8 & 7 \\ 6 & 9 \end{vmatrix} + 0\begin{vmatrix} 8 & 3 \\ 6 & -2 \end{vmatrix}$$

$$= (1)[(3)(9)-(7)(-2)]-(5)[(8)(9)-(7)(6)] = -109$$

Alternatively, to compute the determinant of the $[\mathbf{A}]$ matrix, we can eliminate the elements in the second row, and first, second, and third column as shown:

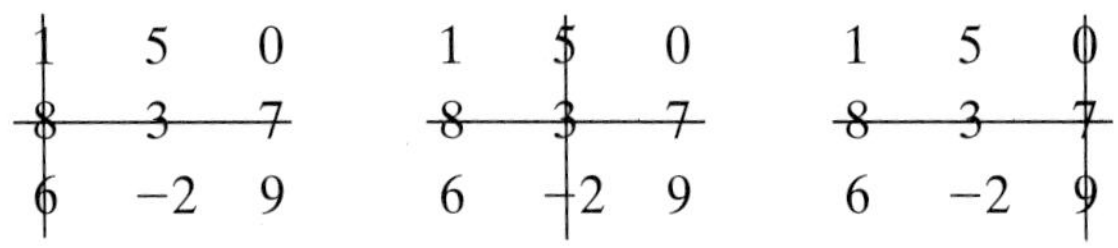

$$\begin{vmatrix} 1 & 5 & 0 \\ 8 & 3 & 7 \\ 6 & -2 & 9 \end{vmatrix} = -8\begin{vmatrix} 5 & 0 \\ -2 & 9 \end{vmatrix} + 3\begin{vmatrix} 1 & 0 \\ 6 & 9 \end{vmatrix} - 7\begin{vmatrix} 1 & 5 \\ 6 & -2 \end{vmatrix}$$

$$= -(8)[(5)(9)-(0)(-2)]+(3)[(1)(9)-(0)(6)]-(7)[(1)(-2)-(5)(6)] = -109$$

b. As already mentioned, the determinant of $[\mathbf{A}]^T$ is equal to the determinant of $[\mathbf{A}]$. Therefore, there is no need to perform any additional calculations. However, as a means of verifying this identity, we will compute and compare determinant of $[\mathbf{A}]^T$ to the determinant of $[\mathbf{A}]$. Recall that $[\mathbf{A}]^T$ is obtained by interchanging the first, second, and third rows of $[\mathbf{A}]$ into the first, second, and third columns of the $[\mathbf{A}]^T$, leading to $[\mathbf{A}]^T = \begin{bmatrix} 1 & 8 & 6 \\ 5 & 3 & -2 \\ 0 & 7 & 9 \end{bmatrix}$. Using minors, we get

$$\begin{matrix} \not{1} & \not{8} & \not{6} \\ \not{5} & 3 & -2 \\ \not{0} & 7 & 9 \end{matrix} \qquad \begin{matrix} \not{1} & \not{8} & \not{6} \\ 5 & \not{3} & -2 \\ 0 & \not{7} & 9 \end{matrix} \qquad \begin{matrix} \not{1} & \not{8} & \not{6} \\ 5 & 3 & \not{-2} \\ 0 & 7 & \not{9} \end{matrix}$$

$$\begin{vmatrix} 1 & 8 & 6 \\ 5 & 3 & -2 \\ 0 & 7 & 9 \end{vmatrix} = 1\begin{vmatrix} 3 & -2 \\ 7 & 9 \end{vmatrix} - 8\begin{vmatrix} 5 & -2 \\ 0 & 9 \end{vmatrix} + 6\begin{vmatrix} 5 & 3 \\ 0 & 7 \end{vmatrix}$$

$$= (1)[(3)(9)-(-2)(7)]-(8)[(5)(9)-(-2)(0)] + (6)[(5)(7)-(3)(0)] = -109$$

When the determinant of a matrix is zero, the matrix is called *a singular*. A singular matrix results when the elements in two or more rows of a given matrix are identical. For example, consider the following matrix:

$$[\mathbf{A}] = \begin{bmatrix} 2 & 1 & 4 \\ 2 & 1 & 4 \\ 1 & 3 & 5 \end{bmatrix}$$

whose rows one and two are identical. As shown below, the determinant of $[\mathbf{A}]$ is zero.

$$\begin{vmatrix} 2 & 1 & 4 \\ 2 & 1 & 4 \\ 1 & 3 & 5 \end{vmatrix} = \begin{matrix} (2)(1)(5) + (1)(4)(1) + (4)(2)(3) - (1)(2)(5) \\ - (2)(4)(3) - (4)(1)(1) = 0 \end{matrix}$$

Matrix singularity can also occur when the elements in two or more rows of a matrix are linearly dependent. For example, if we multiply the elements of the second row of matrix $[\mathbf{A}]$ by a scalar factor such as 7, then the resulting matrix,

$$[\mathbf{A}] = \begin{bmatrix} 2 & 1 & 4 \\ 14 & 7 & 28 \\ 1 & 3 & 5 \end{bmatrix}$$

is singular because rows one and two are now linearly dependent. As shown below, the determinant of the new [**A**] matrix is zero.

$$\begin{vmatrix} 2 & 1 & 4 \\ 14 & 7 & 28 \\ 1 & 3 & 5 \end{vmatrix} = \begin{matrix} (2)(7)(5) + (1)(28)(1) + (4)(14)(3) - (1)(14)(5) \\ - (2)(28)(3) - (4)(7)(1) = 0 \end{matrix}$$

2.7 SOLUTIONS OF SIMULTANEOUS LINEAR EQUATIONS

As you saw in Chapter 1, the finite element formulation leads to a system of algebraic equations. Recall that for Example 1.1, the bar with a variable cross section supporting a load, the finite element approximation and the application of the boundary condition and the load resulted in a set of four linear equations:

$$10^3\begin{bmatrix} 1820 & -845 & 0 & 0 \\ -845 & 1560 & -715 & 0 \\ 0 & -715 & 1300 & -585 \\ 0 & 0 & -585 & 585 \end{bmatrix}\begin{Bmatrix} u_2 \\ u_3 \\ u_4 \\ u_5 \end{Bmatrix} = \begin{Bmatrix} 0 \\ 0 \\ 0 \\ 10^3 \end{Bmatrix}$$

In the sections that follow we discuss two methods that you can use to obtain solutions to a set of linear equations.

Gauss Elimination Method

We begin our discussion by demonstrating the Gauss elimination method using an example. Consider the following three linear equations with three unknowns, x_1, x_2, and x_3.

$$2x_1 + x_2 + x_3 = 13 \quad (2.19a)$$

$$3x_1 + 2x_2 + 4x_3 = 32 \quad (2.19b)$$

$$5x_1 - x_2 + 3x_3 = 17 \quad (2.19c)$$

1. We begin by dividing the first equation, Eq. (2.19a), by 2, the coefficient of x_1 term. This operation leads to

$$x_1 + \frac{1}{2}x_2 + \frac{1}{2}x_3 = \frac{13}{2} \quad (2.20)$$

2. We multiply Eq. (2.20) by 3, the coefficient of x_1 in Eq. (2.19b).

$$3x_1 + \frac{3}{2}x_2 + \frac{3}{2}x_3 = \frac{39}{2} \quad (2.21)$$

We then subtract Eq. (2.21) from Eq. (2.19b). This step eliminates x_1 from Eq. (2.19b). This operation leads to

$$\begin{array}{r} 3x_1 + 2x_2 + 4x_3 = 32 \\ -\left(3x_1 + \frac{3}{2}x_2 + \frac{3}{2}x_3 = \frac{39}{2}\right) \\ \hline \frac{1}{2}x_2 + \frac{5}{2}x_3 = \frac{25}{2} \end{array} \quad (2.22)$$

3. Similarly, to eliminate x_1 from Eq. (2.19c), we multiply Eq. (2.20) by 5, the coefficient of x_1 in Eq. (2.19c)

$$5x_1 + \frac{5}{2}x_2 + \frac{5}{2}x_3 = \frac{65}{2} \quad (2.23)$$

We then subtract the above equation from Eq. (2.19c), which eliminates x_1 from Eq. (2.19c). This operation leads to

$$\begin{array}{r} 5x_1 - x_2 + 3x_3 = 17 \\ -\left(5x_1 + \frac{5}{2}x_2 + \frac{5}{2}x_3 = \frac{65}{2}\right) \\ \hline -\frac{7}{2}x_2 + \frac{1}{2}x_3 = -\frac{31}{2} \end{array} \quad (2.24)$$

Let us summarize the results of the operations performed during steps 1 through 3. These operations eliminated the x_1 from Eqs. (2.19b) and (2.19c).

$$x_1 + \frac{1}{2}x_2 + \frac{1}{2}x_3 = \frac{13}{2} \quad (2.25a)$$

$$\frac{1}{2}x_2 + \frac{5}{2}x_3 = \frac{25}{2} \quad (2.25b)$$

$$-\frac{7}{2}x_2 + \frac{1}{2}x_3 = -\frac{31}{2} \quad (2.25c)$$

4. To eliminate x_2 from Eq. (2.25c), first we divide Eq. (2.25b) by 1/2, the coefficient of x_2.

$$x_2 + 5x_3 = 25 \quad (2.26)$$

Then we multiply Eq. (2.26) by $-7/2$, the coefficient of x_2 in Eq. (2.25c), and subtract that equation from Eq. (2.25c). These operations lead to

$$\begin{array}{r} -\frac{7}{2}x_2 + \frac{1}{2}x_3 = -\frac{31}{2} \\ -\left(-\frac{7}{2}x_2 - \frac{35}{2}x_3 = -\frac{175}{2}\right) \\ \hline 18x_3 = 72 \end{array} \quad (2.27)$$

Dividing both sides of Eq. (2.27) by 18, we get

$$x_3 = 4$$

Summarizing the results of the previous steps, we have

$$x_1 + \frac{1}{2}x_2 + \frac{1}{2}x_3 = \frac{13}{2} \tag{2.28}$$

$$x_2 + 5x_3 = 25 \tag{2.29}$$

$$x_3 = 4 \tag{2.30}$$

Note Eqs. (2.28) and (2.29) are the same as Eqs. (2.25a) and (2.26), which are renumbered for convenience. Now we can use back substitution to compute the values of x_2 and x_1. We substitute for x_3 in Eq. (2.29) and solve for x_2.

$$x_2 + 5(4) = 25 \quad \rightarrow \quad x_2 = 5$$

Next, we substitute for x_3 and x_2 in Eq. (2.28) and solve for x_1.

$$x_1 + \frac{1}{2}(5) + \frac{1}{2}(4) = \frac{13}{2} \quad \rightarrow \quad x_1 = 2$$

The Lower Triangular, Upper Triangular (LU) Decomposition Method

When designing structures, it is often necessary to change the load and consequently the load matrix to determine its effect on the resulting displacements and stresses. Some heat transfer analysis also requires experimenting with the heat load in reaching the desirable temperature distribution within the medium. The Gauss elimination method requires full implementation of the coefficient matrix (the stiffness or the conductance matrix) and the right-hand side matrix (load matrix) in order to solve for the unknown displacements (or temperatures). When using Gauss elimination, the entire process must be repeated each time a change in the load matrix is made. Whereas the Gauss elimination is not well suited for such situations, the LU method handles any changes in the load matrix much more efficiently. The LU method consists of two major parts: a decomposition part and a solution part. We explain the LU method using the following three equations

$$a_{11}x_1 + a_{12}x_2 + a_{13}x_3 = b_1 \tag{2.31}$$

$$a_{21}x_1 + a_{22}x_2 + a_{23}x_3 = b_2 \tag{2.32}$$

$$a_{31}x_1 + a_{32}x_2 + a_{33}x_3 = b_3 \tag{2.33}$$

or in a matrix form,

$$\begin{bmatrix} a_{11} & a_{12} & a_{13} \\ a_{21} & a_{22} & a_{23} \\ a_{31} & a_{32} & a_{33} \end{bmatrix} \begin{Bmatrix} x_1 \\ x_2 \\ x_3 \end{Bmatrix} = \begin{Bmatrix} b_1 \\ b_2 \\ b_3 \end{Bmatrix} \text{ or } [\mathbf{A}]\{\boldsymbol{x}\} = \{\boldsymbol{b}\} \tag{2.34}$$

Decomposition Part The main idea behind the LU method is first to decompose the coefficient matrix $[A]$ into lower $[L] = \begin{bmatrix} 1 & 0 & 0 \\ l_{21} & 1 & 0 \\ l_{31} & l_{32} & 1 \end{bmatrix}$ and upper triangular $[U] = \begin{bmatrix} u_{11} & u_{12} & u_{13} \\ 0 & u_{22} & u_{23} \\ 0 & 0 & u_{33} \end{bmatrix}$ matrices so that

$$\begin{bmatrix} a_{11} & a_{12} & a_{13} \\ a_{21} & a_{22} & a_{23} \\ a_{31} & a_{32} & a_{33} \end{bmatrix} = \begin{bmatrix} 1 & 0 & 0 \\ l_{21} & 1 & 0 \\ l_{31} & l_{32} & 1 \end{bmatrix} \begin{bmatrix} u_{11} & u_{12} & u_{13} \\ 0 & u_{22} & u_{23} \\ 0 & 0 & u_{33} \end{bmatrix} \tag{2.35}$$

Carrying out the multiplication operation, we get

$$\begin{aligned} \begin{bmatrix} a_{11} & a_{12} & a_{13} \\ a_{21} & a_{22} & a_{23} \\ a_{31} & a_{32} & a_{33} \end{bmatrix} &= \begin{bmatrix} 1 & 0 & 0 \\ l_{21} & 1 & 0 \\ l_{31} & l_{32} & 1 \end{bmatrix} \begin{bmatrix} u_{11} & u_{12} & u_{13} \\ 0 & u_{22} & u_{23} \\ 0 & 0 & u_{33} \end{bmatrix} \\ &= \begin{bmatrix} u_{11} & u_{12} & u_{13} \\ l_{21}u_{11} & l_{21}u_{12} + u_{22} & l_{21}u_{13} + u_{23} \\ l_{31}u_{11} & l_{31}u_{12} + l_{32}u_{22} & l_{31}u_{13} + l_{32}u_{23} + u_{33} \end{bmatrix} \end{aligned} \tag{2.36}$$

Now let us compare the elements in the first row of $[A]$ matrix in Eq. (2.36) to the elements in the first row of the $[L][U]$ multiplication results. From this comparison we can see the following relationships:

$$u_{11} = a_{11} \text{ and } u_{12} = a_{12} \text{ and } u_{13} = a_{13}$$

Now, by comparing the elements in the first column of $[A]$ matrix in Eq. (2.36) to the elements in the first column of the $[L][U]$ product, we can obtain the values of l_{21} and l_{31}:

$$l_{21}u_{11} = a_{21} \quad \rightarrow \quad l_{21} = \frac{a_{21}}{u_{11}} = \frac{a_{21}}{a_{11}} \tag{2.37}$$

$$l_{31}u_{11} = a_{31} \quad \rightarrow \quad l_{31} = \frac{a_{31}}{u_{11}} = \frac{a_{31}}{a_{11}} \tag{2.38}$$

Note the value of u_{11} was determined in the previous step. That is, $u_{11} = a_{11}$. We can obtain the values of u_{22} and u_{23} by comparing the *elements in the second rows of the matrices* in Eq. (2.36).

$$l_{21}u_{12} + u_{22} = a_{22} \quad \rightarrow \quad u_{22} = a_{22} - l_{21}u_{12} \tag{2.39}$$

$$l_{21}u_{13} + u_{23} = a_{23} \quad \rightarrow \quad u_{23} = a_{23} - l_{21}u_{13} \tag{2.40}$$

When examining Eqs. (2.39) and (2.40) remember that the values of l_{21}, u_{12}, and u_{13} are known from previous steps. Now we compare the *elements in the second columns of*

Eq. (2.36). This comparison leads to the value of l_{32}. Note, we already know values of u_{12}, l_{21}, u_{22}, and l_{31} from previous steps.

$$l_{31}u_{12} + l_{32}u_{22} = a_{32} \quad \rightarrow \quad l_{32} = \frac{a_{32} - l_{31}u_{12}}{u_{22}} \tag{2.41}$$

Finally, the comparison of *elements in the third rows* leads to the value of u_{33}.

$$l_{31}u_{13} + l_{32}u_{23} + u_{33} = a_{33} \quad \rightarrow \quad u_{33} = a_{33} - l_{31}u_{13} - l_{32}u_{23} \tag{2.42}$$

We used a simple 3×3 matrix to show how the LU decomposition is performed. We can now generalize the scheme for a square matrix of any size n in the following manner:

Step 1. The values of the elements in the first row of the $[U]$ matrix are obtained from

$$u_{1j} = a_{1j} \quad \text{for } j = 1 \text{ to } n \tag{2.43}$$

Step 2. The unknown values of the elements in the first column of the $[L]$ matrix are obtained from

$$l_{i1} = \frac{a_{i1}}{u_{11}} \quad \text{for } i = 2 \text{ to } n \tag{2.44}$$

Step 3. The unknown values of the elements in the second row of the $[U]$ matrix are computed from

$$u_{2j} = a_{2j} - l_{21}u_{1j} \quad \text{for } j = 2 \text{ to } n \tag{2.45}$$

Step 4. The values of the elements in the second column of $[L]$ matrix are calculated from

$$l_{i2} = \frac{a_{i2} - l_{i1}u_{12}}{u_{22}} \quad \text{for } i = 3 \text{ to } n \tag{2.46}$$

Next, we determine the unknown values of the elements in the third row of the $[U]$ matrix and the third column of $[L]$. By now you should see a clear pattern. We evaluate the values of the elements in a row first and then switch to a column. This procedure is repeated until all the unknown elements are computed. We can generalize the above steps in the following way. To obtain the values of the elements in the kth row of $[U]$ matrix, we use

$$u_{kj} = a_{kj} - \sum_{p=1}^{k-1} l_{kp}u_{pj} \quad \text{for } j = k \text{ to } n \tag{2.47}$$

We will then switch to the kth column of $[L]$ and determine the unknown values in that column.

$$l_{ik} = \frac{a_{ik} - \sum_{p=1}^{k-1} l_{ip}u_{pk}}{u_{kk}} \quad \text{for } i = k + 1 \text{ to } n \tag{2.48}$$

Solution Part So far you have seen how to decompose a square coefficient matrix $[A]$ into lower and upper triangular $[L]$ and $[U]$ matrices. Next, we use the $[L]$

and the $[\boldsymbol{U}]$ matrices to solve a set of linear equations. Let's turn our attention back to the three equations and three unknowns example and replace the coefficient matrix $[\boldsymbol{A}]$ with the $[\boldsymbol{L}]$ and $[\boldsymbol{U}]$ matrices:

$$[\boldsymbol{A}]\{\boldsymbol{x}\} = \{\boldsymbol{b}\} \tag{2.49}$$

$$[\boldsymbol{L}][\boldsymbol{U}]\{\boldsymbol{x}\} = \{\boldsymbol{b}\} \tag{2.50}$$

We now replace the product of $[\boldsymbol{U}]\{\boldsymbol{x}\}$ by a column matrix $\{\boldsymbol{z}\}$ such that

$$[\boldsymbol{U}]\{\boldsymbol{x}\} = \{\boldsymbol{z}\} \tag{2.51}$$

$$[\boldsymbol{L}]\overbrace{[\boldsymbol{U}]\{\boldsymbol{x}\}}^{\{z\}} = \{\boldsymbol{b}\} \rightarrow [\boldsymbol{L}]\{\boldsymbol{z}\} = \{\boldsymbol{b}\} \tag{2.52}$$

Because $[\boldsymbol{L}]$ is a lower triangular matrix, we can easily solve for the values of the elements in the $\{\boldsymbol{z}\}$ matrix, and then use the known values of the $\{\boldsymbol{z}\}$ matrix to solve for the unknowns in the $\{\boldsymbol{x}\}$ from the relationship $[\boldsymbol{U}]\{\boldsymbol{x}\} = \{\boldsymbol{z}\}$. These steps are demonstrated next.

$$\begin{bmatrix} 1 & 0 & 0 \\ l_{21} & 1 & 0 \\ l_{31} & l_{32} & 1 \end{bmatrix} \begin{Bmatrix} z_1 \\ z_2 \\ z_3 \end{Bmatrix} = \begin{Bmatrix} b_1 \\ b_2 \\ b_3 \end{Bmatrix} \tag{2.53}$$

From Eq. (2.53), it is clear that

$$z_1 = b_1 \tag{2.54}$$

$$z_2 = b_2 - l_{21}z_1 \tag{2.55}$$

$$z_3 = b_3 - l_{31}z_1 - l_{32}z_2 \tag{2.56}$$

Now that the values of the elements in the $\{\boldsymbol{z}\}$ matrix are known, we can solve for the unknown matrix $\{\boldsymbol{x}\}$ using

$$\begin{bmatrix} u_{11} & u_{12} & u_{13} \\ 0 & u_{22} & u_{23} \\ 0 & 0 & u_{33} \end{bmatrix} \begin{Bmatrix} x_1 \\ x_2 \\ x_3 \end{Bmatrix} = \begin{Bmatrix} z_1 \\ z_2 \\ z_3 \end{Bmatrix} \tag{2.57}$$

$$x_3 = \frac{z_3}{u_{33}} \tag{2.58}$$

$$x_2 = \frac{z_2 - u_{23}x_3}{u_{22}} \tag{2.59}$$

$$x_1 = \frac{z_1 - u_{12}x_2 - u_{13}x_3}{u_{11}} \tag{2.60}$$

Here we used a simple three equations and three unknowns to demonstrate how best to proceed to obtain solutions; we can now generalize the scheme to obtain the solutions for a set of n equations and n unknown.

$$z_1 = b_1 \quad \text{and} \quad z_i = b_i - \sum_{j=1}^{i-1} l_{ij} z_j \quad \text{for } i = 2, 3, 4, \ldots, n \tag{2.61}$$

$$x_n = \frac{z_n}{u_{nn}} \quad \text{and} \quad x_i = \frac{z_i - \sum_{j=i+1}^{n} u_{ij} x_j}{u_{ii}} \quad \text{for } i = n-1, n-2, n-3, \ldots, 3, 2, 1 \tag{2.62}$$

Next, we apply the LU method to the set of equations that we used to demonstrate the Gauss elimination method.

EXAMPLE 2.5

Apply the LU decomposition method to the following three equations and three unknowns set of equations:

$$2x_1 + x_2 + x_3 = 13$$
$$3x_1 + 2x_2 + 4x_3 = 32$$
$$5x_1 - x_2 + 3x_3 = 17$$

$$[\mathbf{A}] = \begin{bmatrix} 2 & 1 & 1 \\ 3 & 2 & 4 \\ 5 & -1 & 3 \end{bmatrix} \text{ and } \{\mathbf{b}\} = \begin{Bmatrix} 13 \\ 32 \\ 17 \end{Bmatrix}$$

Note that for the given problem, $n = 3$.

Decomposition Part

Step 1. The values of the elements in the first row of the $[\mathbf{U}]$ matrix are obtained from

$$u_{1j} = a_{1j} \qquad \text{for } j = 1 \text{ to } n$$

$$u_{11} = a_{11} = 2 \qquad u_{12} = a_{12} = 1 \qquad u_{13} = a_{13} = 1$$

Step 2. The unknown values of the elements in the first column of the $[\mathbf{L}]$ matrix are obtained from

$$l_{i1} = \frac{a_{i1}}{u_{11}} \qquad \text{for } i = 2 \text{ to } n$$

$$l_{21} = \frac{a_{21}}{u_{11}} = \frac{3}{2} \qquad l_{31} = \frac{a_{31}}{u_{11}} = \frac{5}{2}$$

Step 3. The unknown values of the elements in the second row of the $[\mathbf{U}]$ matrix are computed from

$$u_{2j} = a_{2j} - l_{21} u_{1j} \quad \text{for } j = 2 \text{ to } n$$

$$u_{22} = a_{22} - l_{21} u_{12} = 2 - \left(\frac{3}{2}\right)(1) = \frac{1}{2}$$

$$u_{23} = a_{23} - l_{21} u_{13} = 4 - \left(\frac{3}{2}\right)(1) = \frac{5}{2}$$

Step 4. The unknown values of the elements in the second column of the $[\boldsymbol{L}]$ matrix are determined from

$$l_{i2} = \frac{a_{i2} - l_{i1}u_{12}}{u_{22}} \quad \text{for } i = 3 \text{ to } n$$

$$l_{32} = \frac{a_{32} - l_{31}u_{12}}{u_{22}} = \frac{-1-\left(\dfrac{5}{2}\right)(1)}{\dfrac{1}{2}} = -7$$

Step 5. Compute the remaining unknown elements in the $[\boldsymbol{U}]$ and $[\boldsymbol{L}]$ matrices.

$$u_{kj} = a_{kj} - \sum_{p=1}^{k-1} l_{kp}u_{pj} \quad \text{for } j = k \text{ to } n$$

$$u_{33} = a_{33} - (l_{31}u_{13} + l_{32}u_{23}) = 3 - \left(\left(\frac{5}{2}\right)(1) + (-7)\left(\frac{5}{2}\right)\right) = 18$$

Because of the size of this problem ($n = 3$) and the fact that the elements along the main diagonal of the $[\boldsymbol{L}]$ matrix have values of 1, that is, $l_{33} = 1$, we do not need to proceed any further. Therefore, the application of the last step

$$l_{ik} = \frac{a_{ik} - \displaystyle\sum_{p=1}^{k-1} l_{ip}u_{pk}}{u_{kk}} \quad \text{for } i = k + 1 \text{ to } n$$

is omitted. We have now decomposed the coefficient matrix $[\boldsymbol{A}]$ into the following lower and upper triangular $[\boldsymbol{L}]$ and $[\boldsymbol{U}]$ matrices:

$$\begin{bmatrix} 2 & 1 & 1 \\ 3 & 2 & 4 \\ 5 & -1 & 3 \end{bmatrix} = \begin{bmatrix} 1 & 0 & 0 \\ \dfrac{3}{2} & 1 & 0 \\ \dfrac{5}{2} & -7 & 1 \end{bmatrix} \begin{bmatrix} 2 & 1 & 1 \\ 0 & \dfrac{1}{2} & \dfrac{5}{2} \\ 0 & 0 & 18 \end{bmatrix}$$

When performing this method by hand, here is a good place to check the decomposition results by premultiplying the $[\boldsymbol{L}]$ matrix by the $[\boldsymbol{U}]$ matrix to see if the $[\boldsymbol{A}]$ matrix is recovered.

We now proceed with the solution phase of the LU method, Eq. (2.61).

Solution Part

$$z_1 = b_1 \quad \text{and } z_i = b_i - \sum_{j=1}^{i-1} l_{ij}z_j \quad \text{for } i = 2 \text{ to } n$$

$$z_1 = 13 \qquad z_2 = b_2 - l_{21}z_1 = 32 - \left(\frac{3}{2}\right)(13) = \frac{25}{2}$$

$$z_3 = b_3 - (l_{31}z_1 + l_{32}z_2) = 17 - \left(\left(\frac{5}{2}\right)(13) + (-7)\left(\frac{25}{2}\right)\right) = 72$$

The solution is obtained from Eq. (2.62).

$$x_n = \frac{z_n}{u_{nn}} \quad \text{and} \quad x_i = \frac{z_i - \sum_{j=i+1}^{n} u_{ij}x_j}{u_{ii}} \qquad \text{for } i = n-1, n-2, n-3, \ldots, 3, 2, 1$$

Note for this Problem $n = 3$, therefore $i = 2, 1$.

$$x_3 = \frac{z_3}{u_{33}} = \frac{72}{18} = 4$$

$$x_2 = \frac{z_2 - u_{23}x_3}{u_{22}} = \frac{\frac{25}{2} - \left(\frac{5}{2}\right)(4)}{\frac{1}{2}} = 5$$

$$x_1 = \frac{z_1 - u_{12}x_2 - u_{13}x_3}{u_{11}} = \frac{13 - ((1)(5) + (1)(4))}{2} = 2$$

2.8 INVERSE OF A MATRIX

In the previous sections we discussed matrix addition, subtraction, and multiplication, but you may have noticed that we did not say anything about matrix division. That is because such an operation is not formally defined. Instead, we define an inverse of a matrix in such a way that when it is multiplied by the original matrix, the identity matrix is obtained.

$$[\mathbf{A}]^{-1}[\mathbf{A}] = [\mathbf{A}][\mathbf{A}]^{-1} = [\boldsymbol{I}] \tag{2.63}$$

In Eq. (2.63), $[\mathbf{A}]^{-1}$ is called the inverse of $[\mathbf{A}]$. Only a square and nonsingular matrix has an inverse. In Section 2.7 we explained the Gauss elimination and the LU methods that you can use to obtain solutions to a set of linear equations. Matrix inversion allows for yet another way of solving for the solutions of a set of linear equations. Once again, recall from our discussion in Chapter 1 that the finite element formulation of an engineering problem leads to a set of linear equations, and the solution of these equations render the nodal values. For instance, formulation of the problem in Example 1.1 led to the set of linear equations given by

$$[\boldsymbol{K}]\{\boldsymbol{u}\} = \{\boldsymbol{F}\} \tag{2.64}$$

To obtain the nodal displacement values $\{\boldsymbol{u}\}$, we premultiply Eq. (2.64) by $[\boldsymbol{K}]^{-1}$, which leads to

$$\overbrace{[\boldsymbol{K}]^{-1}[\boldsymbol{K}]}^{[\boldsymbol{I}]}\{\boldsymbol{u}\} = [\boldsymbol{K}]^{-1}\{\boldsymbol{F}\} \tag{2.65}$$

$$[\boldsymbol{I}]\{\boldsymbol{u}\} = [\boldsymbol{K}]^{-1}\{\boldsymbol{F}\} \tag{2.66}$$

and noting that $[\boldsymbol{I}]\{\boldsymbol{u}\} = \{\boldsymbol{u}\}$ and simplifying,

$$\{\boldsymbol{u}\} = [\boldsymbol{K}]^{-1}\{\boldsymbol{F}\} \tag{2.67}$$

From the matrix relationship given by Eq. (2.67), you can see that the nodal solutions can be easily obtained, provided the value of $[\boldsymbol{K}]^{-1}$ is known. This example shows the important role of the inverse of a matrix in obtaining the solution to a set of linear equations. Now that you see why the inverse of a matrix is important, the next question is, How do we compute the inverse of a square and nonsingular matrix? There are a number of established methods that we can use to determine the inverse of a matrix. Here we discuss a procedure based on the LU decomposition method. Let us refer back to the relationship given by Eq. (2.63), and decompose matrix $[\boldsymbol{A}]$ into lower and upper triangular $[\boldsymbol{L}]$ and $[\boldsymbol{U}]$ matrices.

$$\overbrace{[\boldsymbol{L}][\boldsymbol{U}]}^{[\boldsymbol{A}]}\ [\boldsymbol{A}]^{-1} = [\boldsymbol{I}] \tag{2.68}$$

Next, we represent the product of $[\boldsymbol{U}][\boldsymbol{A}]^{-1}$ by another matrix, say matrix $[\boldsymbol{Y}]$:

$$[\boldsymbol{U}][\boldsymbol{A}]^{-1} = [\boldsymbol{Y}] \tag{2.69}$$

and substitute for $[\boldsymbol{U}][\boldsymbol{A}]^{-1}$ in terms of $[\boldsymbol{Y}]$ in Eq. (2.68), which leads to

$$[\boldsymbol{L}][\boldsymbol{Y}] = [\boldsymbol{I}] \tag{2.70}$$

We then use the relationships given by Eq. (2.70) to solve for the unknown values of elements in matrix $[\boldsymbol{Y}]$, and then use Eq. (2.69) to solve for the values of the elements in matrix $[\boldsymbol{A}]^{-1}$. These steps are demonstrated using Example 2.6.

EXAMPLE 2.6

Given $[\boldsymbol{A}] = \begin{bmatrix} 2 & 1 & 1 \\ 3 & 2 & 4 \\ 5 & -1 & 3 \end{bmatrix}$, compute $[\boldsymbol{A}]^{-1}$.

Step 1. Decompose the given matrix into lower and upper triangular matrices. In Example 2.5 we showed the procedure for decomposing the $[\boldsymbol{A}]$ matrix into lower and upper triangular $[\boldsymbol{L}]$ and $[\boldsymbol{U}]$ matrices.

$$[\boldsymbol{L}] = \begin{bmatrix} 1 & 0 & 0 \\ \frac{3}{2} & 1 & 0 \\ \frac{5}{2} & -7 & 1 \end{bmatrix} \quad \text{and } [\boldsymbol{U}] = \begin{bmatrix} 2 & 1 & 1 \\ 0 & \frac{1}{2} & \frac{5}{2} \\ 0 & 0 & 18 \end{bmatrix}$$

Step 2. Use Eq. (2.70) to determine the unknown values of the elements in the $[\boldsymbol{Y}]$ matrix.

$$\overbrace{\begin{bmatrix} 1 & 0 & 0 \\ \frac{3}{2} & 1 & 0 \\ \frac{5}{2} & -7 & 1 \end{bmatrix}}^{[\boldsymbol{L}]} \overbrace{\begin{bmatrix} y_{11} & y_{12} & y_{13} \\ y_{21} & y_{22} & y_{23} \\ y_{31} & y_{32} & y_{33} \end{bmatrix}}^{[\boldsymbol{Y}]} = \begin{bmatrix} 1 & 0 & 0 \\ 0 & 1 & 0 \\ 0 & 0 & 1 \end{bmatrix}$$

First, let us consider the multiplication results pertaining to the first column of the $[\boldsymbol{Y}]$ matrix, as shown.

$$\begin{bmatrix} 1 & 0 & 0 \\ \frac{3}{2} & 1 & 0 \\ \frac{5}{2} & -7 & 1 \end{bmatrix} \begin{Bmatrix} y_{11} \\ y_{21} \\ y_{31} \end{Bmatrix} = \begin{Bmatrix} 1 \\ 0 \\ 0 \end{Bmatrix}$$

The solution of this system of equations leads to

$$y_{11} = 1 \qquad y_{21} = -\frac{3}{2} \qquad y_{31} = -13$$

Next, consider the multiplication results pertaining to the second column of $[\boldsymbol{Y}]$:

$$\begin{bmatrix} 1 & 0 & 0 \\ \frac{3}{2} & 1 & 0 \\ \frac{5}{2} & -7 & 1 \end{bmatrix} \begin{Bmatrix} y_{12} \\ y_{22} \\ y_{32} \end{Bmatrix} = \begin{Bmatrix} 0 \\ 1 \\ 0 \end{Bmatrix}$$

The solution of this system of equation yields

$$y_{12} = 0 \qquad y_{22} = 1 \qquad y_{32} = 7$$

Similarly, solve for the unknown values of the elements in the remaining column of the $[\boldsymbol{Y}]$ matrix:

$$\begin{bmatrix} 1 & 0 & 0 \\ \frac{3}{2} & 1 & 0 \\ \frac{5}{2} & -7 & 1 \end{bmatrix} \begin{Bmatrix} y_{13} \\ y_{23} \\ y_{33} \end{Bmatrix} = \begin{Bmatrix} 0 \\ 0 \\ 1 \end{Bmatrix}$$

$$y_{13} = 0 \qquad y_{23} = 0 \qquad y_{33} = 1$$

Now that the values of elements of the $[\boldsymbol{Y}]$ are known, we can proceed with calculation of the values of the elements comprising the $[\boldsymbol{A}]^{-1}$, denoted by $x_{11}, x_{12}, \ldots$, and so on, as shown. Using the relationship given by Eq. (2.69), we have

$$[\boldsymbol{U}][\boldsymbol{A}]^{-1} = [\boldsymbol{Y}]$$

$$\begin{bmatrix} 2 & 1 & 1 \\ 0 & \frac{1}{2} & \frac{5}{2} \\ 0 & 0 & 18 \end{bmatrix} \begin{bmatrix} x_{11} & x_{12} & x_{13} \\ x_{21} & x_{22} & x_{23} \\ x_{31} & x_{32} & x_{33} \end{bmatrix} = \begin{bmatrix} 1 & 0 & 0 \\ -\frac{3}{2} & 1 & 0 \\ -13 & 7 & 1 \end{bmatrix}$$

Again, we consider multiplication results pertaining to one column at a time. Considering the first column of the $[x]$ matrix,

$$\begin{bmatrix} 2 & 1 & 1 \\ 0 & \frac{1}{2} & \frac{5}{2} \\ 0 & 0 & 18 \end{bmatrix} \begin{Bmatrix} x_{11} \\ x_{21} \\ x_{31} \end{Bmatrix} = \begin{Bmatrix} 1 \\ -\frac{3}{2} \\ -13 \end{Bmatrix}$$

$$x_{31} = -\frac{13}{18} \qquad x_{21} = \frac{11}{18} \qquad x_{11} = \frac{10}{18}$$

Multiplication results for the second column render

$$\begin{bmatrix} 2 & 1 & 1 \\ 0 & \frac{1}{2} & \frac{5}{2} \\ 0 & 0 & 18 \end{bmatrix} \begin{Bmatrix} x_{12} \\ x_{22} \\ x_{32} \end{Bmatrix} = \begin{Bmatrix} 0 \\ 1 \\ 7 \end{Bmatrix}$$

$$x_{32} = \frac{7}{18} \qquad x_{22} = \frac{1}{18} \qquad x_{12} = -\frac{4}{18}$$

The multiplication results of the third column yield

$$\begin{bmatrix} 2 & 1 & 1 \\ 0 & \frac{1}{2} & \frac{5}{2} \\ 0 & 0 & 18 \end{bmatrix} \begin{Bmatrix} x_{13} \\ x_{23} \\ x_{33} \end{Bmatrix} = \begin{Bmatrix} 0 \\ 0 \\ 1 \end{Bmatrix}$$

$$x_{33} = \frac{1}{18} \qquad x_{23} = -\frac{5}{18} \qquad x_{13} = \frac{2}{18}$$

Therefore, the inverse of the $[\mathbf{A}]$ matrix is

$$[\mathbf{A}]^{-1} = \frac{1}{18}\begin{bmatrix} 10 & -4 & 2 \\ 11 & 1 & -5 \\ -13 & 7 & 1 \end{bmatrix}$$

We can check the result of our calculations by verifying that $[\mathbf{A}][\mathbf{A}]^{-1} = [\boldsymbol{I}]$.

$$\frac{1}{18}\begin{bmatrix} 2 & 1 & 1 \\ 3 & 2 & 4 \\ 5 & -1 & 3 \end{bmatrix}\begin{bmatrix} 10 & -4 & 2 \\ 11 & 1 & -5 \\ -13 & 7 & 1 \end{bmatrix} = \begin{bmatrix} 1 & 0 & 0 \\ 0 & 1 & 0 \\ 0 & 0 & 1 \end{bmatrix} \text{ Q.E.D.}$$

Finally, it is worth noting that the inversion of a diagonal matrix is computed simply by inversing its elements. That is, the inverse of a diagonal matrix is also a diagonal matrix with its elements being the inverse of the elements of the original matrix. For example, the inverse of the 4 × 4 diagonal matrix

$$[\mathbf{A}] = \begin{bmatrix} a_1 & 0 & 0 & 0 \\ 0 & a_2 & 0 & 0 \\ 0 & 0 & a_3 & 0 \\ 0 & 0 & 0 & a_4 \end{bmatrix} \text{ is } [\mathbf{A}]^{-1} = \begin{bmatrix} \frac{1}{a_1} & 0 & 0 & 0 \\ 0 & \frac{1}{a_2} & 0 & 0 \\ 0 & 0 & \frac{1}{a_3} & 0 \\ 0 & 0 & 0 & \frac{1}{a_4} \end{bmatrix}$$

This property of a diagonal matrix should be obvious because $[\mathbf{A}]^{-1}[\mathbf{A}] = [\mathbf{I}]$.

$$[\mathbf{A}]^{-1}[\mathbf{A}] = \begin{bmatrix} \frac{1}{a_1} & 0 & 0 & 0 \\ 0 & \frac{1}{a_2} & 0 & 0 \\ 0 & 0 & \frac{1}{a_3} & 0 \\ 0 & 0 & 0 & \frac{1}{a_4} \end{bmatrix} \begin{bmatrix} a_1 & 0 & 0 & 0 \\ 0 & a_2 & 0 & 0 \\ 0 & 0 & a_3 & 0 \\ 0 & 0 & 0 & a_4 \end{bmatrix} = \begin{bmatrix} 1 & 0 & 0 & 0 \\ 0 & 1 & 0 & 0 \\ 0 & 0 & 1 & 0 \\ 0 & 0 & 0 & 1 \end{bmatrix}$$

2.9 EIGENVALUES AND EIGENVECTORS

Up to this point we have discussed some of the methods that you can use to solve a set of linear equations of the form

$$[\mathbf{A}]\{\mathbf{x}\} = \{\mathbf{b}\} \tag{2.71}$$

For the set of linear equations that we have considered so far, the values of the elements of the $\{\mathbf{b}\}$ matrix were typically nonzero. This type of system of linear equations is commonly referred to as *nonhomogenous*. For a nonhomogenous system, unique solutions exist as long as the determinant of the coefficient matrix $[\mathbf{A}]$ is nonzero. We now discuss the type of problems that render a set of linear equations of the form

$$[\mathbf{A}]\{\mathbf{X}\} - \lambda\{\mathbf{X}\} = 0 \tag{2.72}$$

This type of problem, called an eigenvalue problem, occurs in analysis of buckling problems, vibration of elastic structures, and electrical systems. In general, this class of problems has nonunique solutions. That is, we can establish relationships among the unknowns, and many values can satisfy these relationships. It is common practice to write Eq. (2.72) as

$$[[\mathbf{A}] - \lambda[\mathbf{I}]]\{\mathbf{X}\} = 0 \tag{2.73}$$

where $[\mathbf{I}]$ is the identity matrix having the same dimension as the $[\mathbf{A}]$ matrix. In Eq. (2.73), the unknown matrix $\{\mathbf{X}\}$ is called the eigenvector. We demonstrate how to obtain the eigenvectors using the following vibration example.

EXAMPLE 2.7

Consider the two degrees of freedom system shown in Figure 2.2. We are interested in determining the natural frequencies of the system shown. We will discuss in detail the formulation and analysis of multiple degrees of freedom systems in Chapter 11. For the sake of presentation continuity, the derivation of the set of linear equations are shown here as well.

Using the free-body diagrams shown, the equations of motion

$$m_1\ddot{x_1} + 2kx_1 - kx_2 = 0 \tag{2.74}$$

$$m_2\ddot{x_2} - kx_1 + 2kx_2 = 0 \tag{2.75}$$

or in a matrix form,

$$\begin{bmatrix} m_1 & 0 \\ 0 & m_2 \end{bmatrix}\begin{Bmatrix} \ddot{x_1} \\ \ddot{x_2} \end{Bmatrix} + \begin{bmatrix} 2k & -k \\ -k & 2k \end{bmatrix}\begin{Bmatrix} x_1 \\ x_2 \end{Bmatrix} = \begin{Bmatrix} 0 \\ 0 \end{Bmatrix}$$

Note that Eqs. (2.74) and (2.75) are second-order homogenous differential equations. Also note that these equations are coupled, because both x_1 and x_2 appear in each equation. This type of system is called *elastically coupled* and may be represented in the general matrix form by

$$[\boldsymbol{M}]\{\ddot{\boldsymbol{x}}\} + [\boldsymbol{K}]\{\boldsymbol{x}\} = 0 \tag{2.76}$$

where $[\boldsymbol{M}]$ and $[\boldsymbol{K}]$ are the mass and the stiffness matrices respectively. We can simplify Eqs. (2.74) and (2.75) by dividing both sides of each equation by the values of the respective masses:

$$\ddot{x_1} + \frac{2k}{m_1}x_1 - \frac{k}{m_1}x_2 = 0 \tag{2.77}$$

$$\ddot{x_2} - \frac{k}{m_2}x_1 + \frac{2k}{m_2}x_2 = 0 \tag{2.78}$$

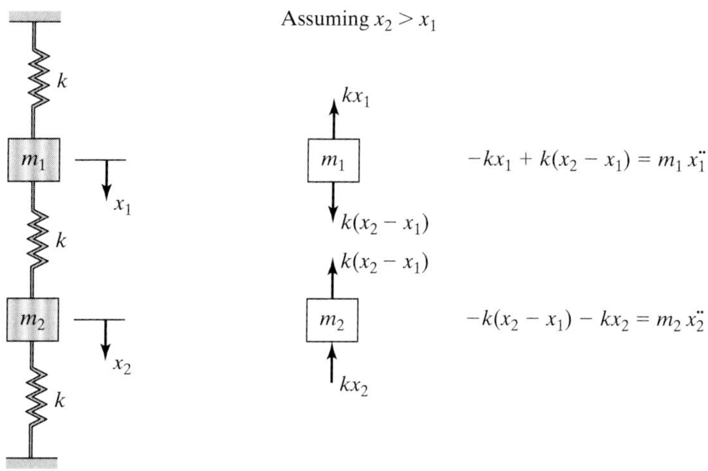

FIGURE 2.2 A schematic diagram of an elastic system with two degrees of freedom.

Using matrix notation, we premultiply the matrix form of the equations of motion by the inverse of the mass matrix $[\boldsymbol{M}]^{-1}$, which leads to

$$\{\ddot{\boldsymbol{x}}\} + [\boldsymbol{M}]^{-1}[\boldsymbol{K}]\{\boldsymbol{x}\} = 0$$

As a next step, we assume a harmonic solution of the form $x_1(t) = X_1 \sin(\omega t + \phi)$ and $x_2(t) = X_2 \sin(\omega t + \phi)$—or in matrix form, $\{\boldsymbol{x}\} = \{\boldsymbol{X}\}\sin(\omega t + \phi)$—and substitute the assumed solutions into the differential equations of motion, Eqs. (2.77) and (2.78), to create a set of linear algebraic equations.

This step leads to

$$-\omega^2 X_1 \sin(\omega t + \phi) + \frac{2k}{m_1} X_1 \sin(\omega t + \phi) - \frac{k}{m_1} X_2 \sin(\omega t + \phi) = 0$$

$$-\omega^2 X_2 \sin(\omega t + \phi) - \frac{k}{m_2} X_1 \sin(\omega t + \phi) + \frac{2k}{m_2} X_2 \sin(\omega t + \phi) = 0$$

After simplifying the $\sin(\omega t + \phi)$ terms,

$$-\omega^2 \begin{Bmatrix} X_1 \\ X_2 \end{Bmatrix} + \begin{bmatrix} \frac{2k}{m_1} & -\frac{k}{m_1} \\ -\frac{k}{m_2} & \frac{2k}{m_2} \end{bmatrix} \begin{Bmatrix} X_1 \\ X_2 \end{Bmatrix} = \begin{Bmatrix} 0 \\ 0 \end{Bmatrix} \tag{2.79}$$

or in a general matrix form,

$$-\omega^2\{\boldsymbol{X}\} + [\boldsymbol{M}]^{-1}[\boldsymbol{K}]\{\boldsymbol{X}\} = 0 \tag{2.80}$$

Note that $\{\boldsymbol{x}\} = \begin{Bmatrix} x_1(t) \\ x_2(t) \end{Bmatrix}$ represents the position of each mass as the function of time, the $\{\boldsymbol{X}\} = \begin{Bmatrix} X_1 \\ X_2 \end{Bmatrix}$ matrix denotes the amplitudes of each vibrating mass, and ϕ is the phase angle. Equation (2.79) may be written as

$$-\omega^2 \begin{bmatrix} 1 & 0 \\ 0 & 1 \end{bmatrix} \begin{Bmatrix} X_1 \\ X_2 \end{Bmatrix} + \begin{bmatrix} \frac{1}{m_1} & 0 \\ 0 & \frac{1}{m_2} \end{bmatrix} \begin{bmatrix} 2k & -k \\ -k & 2k \end{bmatrix} \begin{Bmatrix} X_1 \\ X_2 \end{Bmatrix} = 0 \tag{2.81}$$

or by

$$\left[\begin{bmatrix} \frac{2k}{m_1} & -\frac{k}{m_1} \\ -\frac{k}{m_2} & \frac{2k}{m_2} \end{bmatrix} - \omega^2 \begin{bmatrix} 1 & 0 \\ 0 & 1 \end{bmatrix} \right] \begin{Bmatrix} X_1 \\ X_2 \end{Bmatrix} = \begin{Bmatrix} 0 \\ 0 \end{Bmatrix} \tag{2.82}$$

Comparing Eq. (2.82) to Eq. (2.73), $[[A] - \lambda[I]]\{X\} = 0$ we note that $\omega^2 = \lambda$. Simplifying Eq. (2.82) further, we have

$$\begin{bmatrix} -\omega^2 + \dfrac{2k}{m_1} & -\dfrac{k}{m_1} \\ -\dfrac{k}{m_2} & -\omega^2 + \dfrac{2k}{m_2} \end{bmatrix} \begin{Bmatrix} X_1 \\ X_2 \end{Bmatrix} = 0 \tag{2.83}$$

Problems with governing equations of the type (2.73) or (2.83) have nontrivial solutions only when the determinant of the coefficient matrix is zero. Let's assign some numerical values to the above example problem and proceed with the solution. Let $m_1 = m_2 = 0.1$ kg and $k = 100$ N/m. Forming the determinant of the coefficient matrix and setting it equal to zero, we get

$$\begin{vmatrix} -\omega^2 + 2000 & -1000 \\ -1000 & -\omega^2 + 2000 \end{vmatrix} = 0 \tag{2.84}$$

$$(-\omega^2 + 2000)(-\omega^2 + 2000) - (-1000)(-1000) = 0 \tag{2.85}$$

Simplifying Eq. (2.85), we have

$$\omega^4 - 4000\omega^2 + 3{,}000{,}000 = 0 \tag{2.86}$$

Equation (2.86) is called the *characteristic equation*, and its roots are the natural frequencies of the system.

$$\omega_1^2 = \lambda_1 = 1000(\text{rad/s})^2 \quad \text{and} \quad \omega_1 = 31.62 \text{ rad/s}$$
$$\omega_2^2 = \lambda_2 = 3000(\text{rad/s})^2 \quad \text{and} \quad \omega_2 = 54.77 \text{ rad/s}$$

Once the ω^2 values are known, they can be substituted back into Eq. (2.83) to solve for the relationship between X_1 and X_2. The relationship between the amplitudes of mass oscillating at natural frequencies is called *natural modes*. We can use either relationship (rows) in Eq. (2.83).

$$(-\omega^2 + 2000)\mathbf{X}_1 - 1000\mathbf{X}_2 = 0 \quad \text{and substituting for } \omega_1^2 = 1000$$

$$(-1000 + 2000)\mathbf{X}_1 - 1000\mathbf{X}_2 = 0 \quad \rightarrow \quad \frac{\mathbf{X}_2}{\mathbf{X}_1} = 1$$

Or using the second row,

$$-1000\mathbf{X}_1 + (-\omega^2 + 2000)\mathbf{X}_2 = 0 \quad \text{and substituting for } \omega_1^2 = 1000$$

$$-1000\mathbf{X}_1 + (-1000 + 2000)\mathbf{X}_2 = 0 \quad \rightarrow \quad \frac{\mathbf{X}_2}{\mathbf{X}_1} = 1$$

As expected, the results are identical. The second mode is obtained in a similar manner by substituting for $\omega_2^2 = 3000$ in Eq. (2.83).

$$(-\omega^2 + 2000)\mathbf{X}_1 - 1000\mathbf{X}_2 = 0 \quad \text{and substituting for } \omega_1^2 = 3000$$

$$(-3000 + 2000)\mathbf{X}_1 - 1000\mathbf{X}_2 = 0 \quad \rightarrow \quad \frac{\mathbf{X}_2}{\mathbf{X}_1} = -1$$

It is important to note again that the solution of the eigenvalue problems leads to establishing a relationship among the unknowns, not specific values.

2.10 USING MATLAB TO MANIPULATE MATRICES

MATLAB is mathematical software available in most university computational labs today. MATLAB is a very powerful tool for manipulating matrices; in fact, it was originally designed for that purpose. There are many good textbooks that discuss the capabilities of MATLAB to solve a full range of problems. Here, we introduce only some basic ideas so that you can perform essential matrix operations. For more detail see Appendix F.

Once in a MATLAB environment, you can assign values to a variable or define elements of a matrix. For example, to assign a value 5 to the variable x, you simply type

$$x = 5$$

or to define the element of the matrix $[\mathbf{A}] = \begin{bmatrix} 1 & 5 & 0 \\ 8 & 3 & 7 \\ 6 & -2 & 9 \end{bmatrix}$, you type

$$A = [1\ \ 5\ \ 0;\ 8\ \ 3\ \ 7;\ 6\ \ -2\ \ 9]$$

Note that the elements of the matrix are enclosed in brackets and are separated by blank space, and the elements of each row are separated by a semicolon. The basic scalar operations are shown in Table 2.1.

MATLAB offers many tools for matrix operations and manipulations. Table 2.2 shows examples of these capabilities. Next, we demonstrate a few MATLAB commands with the aid of some examples.

TABLE 2.1 MATLAB's basic scalar operators

Operation	Symbol	Example: $x = 5$ and $y = 3$	Result
Addition	+	$x + y$	8
Subtraction	−	$x - y$	2
Multiplication	*	$x * y$	15
Division	/	$(x + y)/2$	4
Raised to a power	^	x^2	25

TABLE 2.2 Examples of MATLAB's matrix operations

Operation	Symbols or Commands	Example: A and B are matrices that you have defined
Addition	+	$A + B$
Subtraction	−	$A - B$
Multiplication	*	$A * B$
Transpose	*matrix name'*	A'
Inverse	inv(*matrix name*)	inv(A)
Determinant	det(*matrix name*)	det(A)
eigenvalues	eig(*matrix name*)	eig(A)
Matrix left division (uses Gauss elimination to solve a set of linear equations)	\	see Example 2.8

EXAMPLE 2.1 Revisited

Given the following matrices:

$$[A] = \begin{bmatrix} 0 & 5 & 0 \\ 8 & 3 & 7 \\ 9 & -2 & 9 \end{bmatrix}, [B] = \begin{bmatrix} 4 & 6 & -2 \\ 7 & 2 & 3 \\ 1 & 3 & -4 \end{bmatrix}, \text{ and } [C] = \begin{Bmatrix} -1 \\ 2 \\ 5 \end{Bmatrix}$$

using MATLAB, perform the following operations:

a. $[A] + [B] = ?$, **b.** $[A] - [B] = ?$, **c.** $3[A] = ?$, **d.** $[A][B] = ?$, **e.** $[A]\{C\} = ?$
f. $[A]^2 = ?$

Also compute $[A]^T$ and the determinant of $[A]$.

A screen capture of a MATLAB session is shown in Figure 2.3. When studying MATLAB examples (Figure 2.3), note that the response given by MATLAB is shown in boldface. Information that the user must type is shown in regular print. Hit the Enter key after you finish typing data. The MATLAB's prompt for input is >> .

```
        To get started, select "MATLAB Help" from the Help menu.
 >> A = [0 5 0;8 3 7;9 -2 9]
A =

  0   5   0
  8   3   7
  9  -2   9

 >> B = [4 6 -2;7 2 3;1 3 -4]
B =

  4   6  -2
  7   2   3
  1   3  -4

 >> C = [-1;2;5]
C =

 -1
  2
  5
```

FIGURE 2.3 Example of a MATLAB session.

```
>> A + B
ans =
   4   11   -2
  15    5   10
  10    1    5
>> A-B
ans =
-4   -1   °2
1     1   °4
8     5   13
>> 3*A
ans =
   0   15   °0
  24   °9   21
  27   °6   27
>> A*B
ans =
  35   10    15
  60   75   -35
  31   77   -60
>> A*C
ans =
  10
  33
  32
>> A^2
ans =
  40   15   35
  87   35   84
  65   21   67
>> A'
ans =
  0   8    9
  5   3   -2
  0   7    9
>> det(A)
ans =
−45
>>
```

FIGURE 2.3 *Continued*

EXAMPLE 2.8

Solve the following set of equations using the Gauss elimination and by inverting the $[A]$ matrix and multiplying it by the $\{b\}$ matrix.

$$2x_1 + x_2 + x_3 = 13$$
$$3x_1 + 2x_2 + 4x_3 = 32$$
$$5x_1 - x_2 + 3x_3 = 17$$

$$[A] = \begin{bmatrix} 2 & 1 & 1 \\ 3 & 2 & 4 \\ 5 & -1 & 3 \end{bmatrix} \text{ and } \{b\} = \begin{Bmatrix} 13 \\ 32 \\ 17 \end{Bmatrix}$$

We first use the MATLAB matrix left division operator \ to solve this problem. The \ operator solves the problem using Gauss elimination. We then solve the problem using the **inv** command.

```
To get started, select "MATLAB Help" from the Help menu.

>>A=[2 1 1;3 2 4;5 -1 3]
A =

  2  1  1
  3  2  4
  5 -1  3

>>b=[13;32;17]
b =

  13
  32
  17

>>x=A\b
x =

  2.0000
  5.0000
  4.0000

>>x=inv(A)*b
x =

  2.0000
  5.0000
  4.0000
```

2.11 USING EXCEL TO MANIPULATE MATRICES

Using the following two examples (Example 2.1 (revisited) and Example 2.8 (revisited)), we will show how to use Excel to perform matrix operations.

EXAMPLE 2.1 Revisited

Given matrices: $[A] = \begin{bmatrix} 0 & 5 & 0 \\ 8 & 3 & 7 \\ 9 & -2 & 9 \end{bmatrix}$, $[B] = \begin{bmatrix} 4 & 6 & -2 \\ 7 & 2 & 3 \\ 1 & 3 & -4 \end{bmatrix}$, and $\{C\} = \begin{Bmatrix} -1 \\ 2 \\ 5 \end{Bmatrix}$, Using Excel perform the following operations.

a. $[A] + [B] = ?$ **b.** $[A] - [B] = ?$ **c.** $[A][B] = ?$ **d.** $[A]\{C\} = ?$

1. In the cells shown in Figure 2.4 type the appropriate characters and values. Use the **Format Cells** and **Font** option to create the bold variables as shown. Note that you do not have to create the characters (**[A]=**, **[B]=**, and so on) and format them to carry out the matrix operations. This is done merely as a visual aide for presentation purposes.
2. In cell A10 type **[A]+[B]=**, and then using the left mouse button pick cells B9 through D11, as shown in Figure 2.5.
3. Next, in the formula bar type **=B3:D5+G3:I5** and while holding down the **Ctrl** and the **Shift** keys press the **Enter** key. Note that you could also pick the range B3:D5 or G3:I5, instead of typing them. This sequence of operations will create the result shown in Figure 2.6. You can follow similar procedure as outlined in step 2 to perform **[A]−[B]**, except in the formula bar type **=B3:D5−G3:I5**.

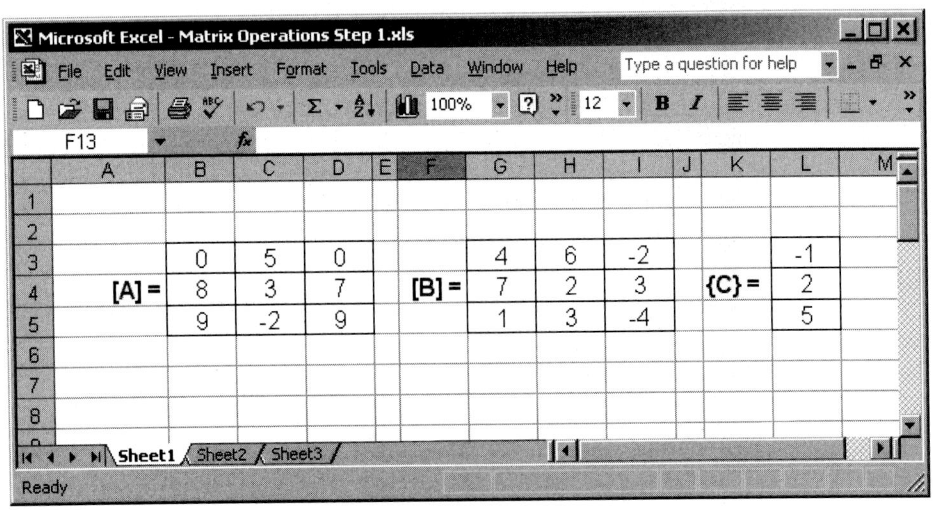

FIGURE 2.4 Example 2.1 Revisited with Excel **(step 1).**

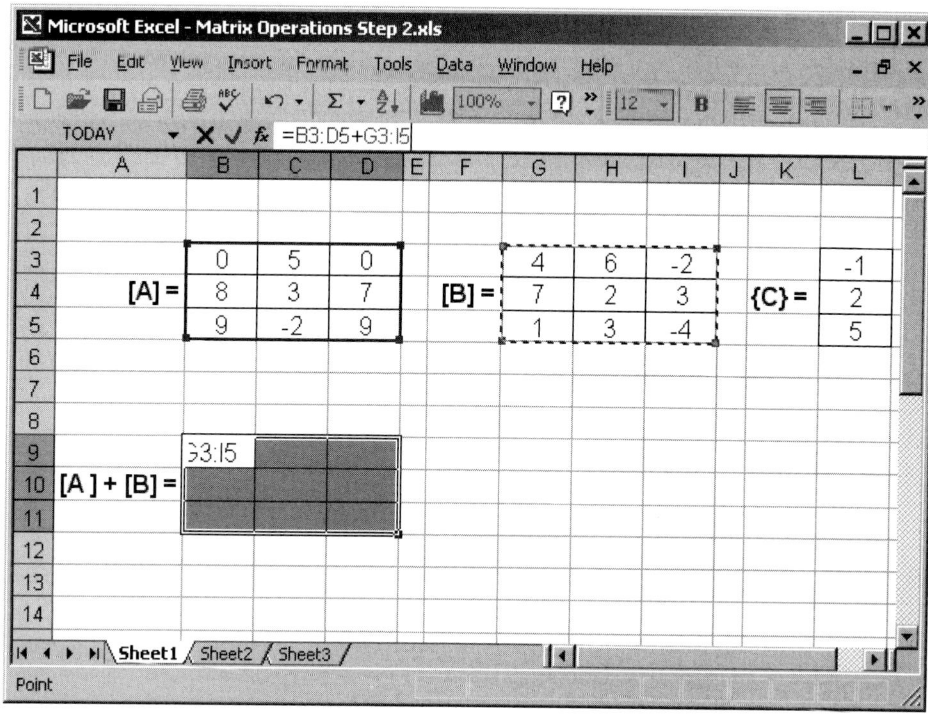

FIGURE 2.5 Example 2.1 Revisited with Excel **(step 2).**

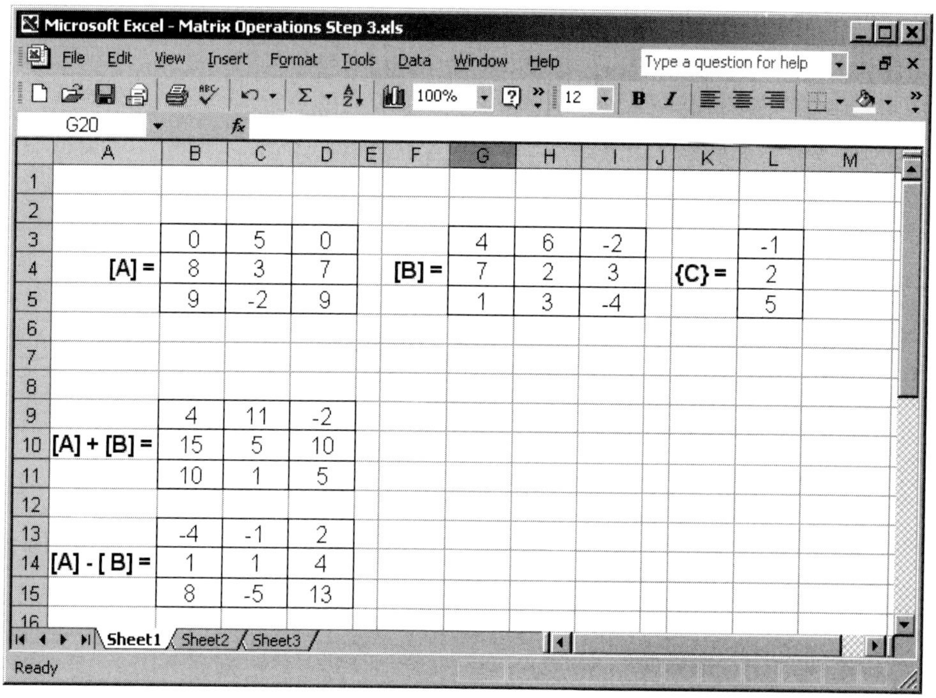

FIGURE 2.6 Example 2.1 Revisited with Excel **(step 3).**

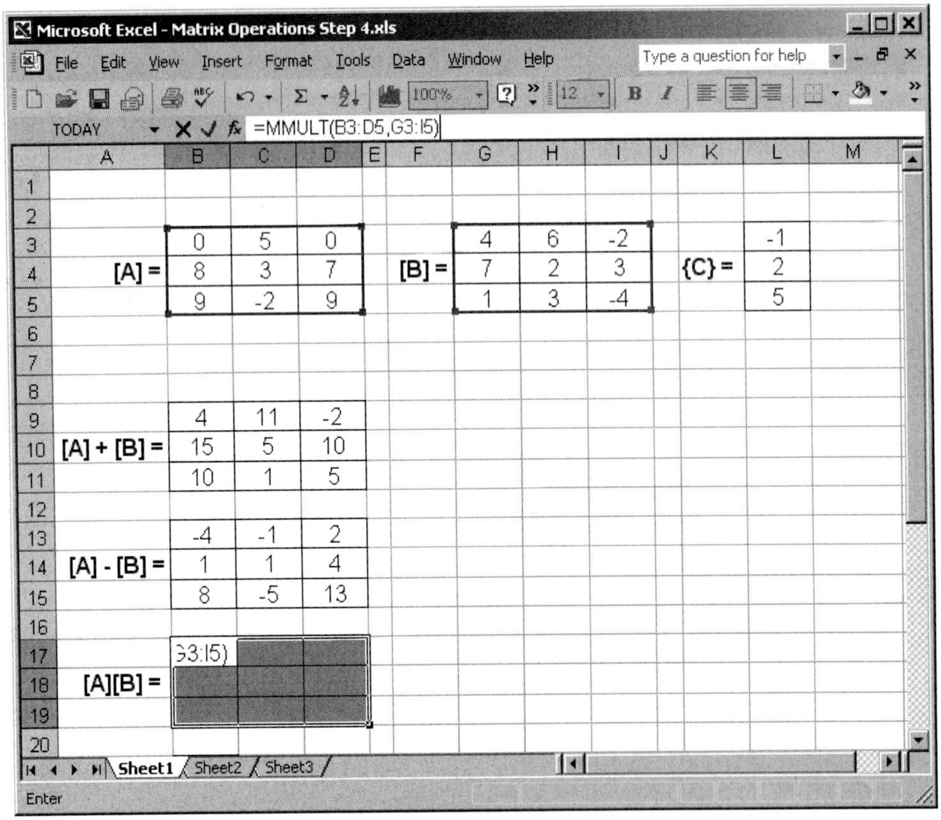

FIGURE 2.7 Example 2.1 Revisited with Excel **(step 4).**

4. To carry out the matrix multiplication, first type **[A][B]=** in cell A18, as shown in Figure 2.7. Then pick cells B17 through D19.
5. In the formula bar type **=MMULT(B3:D5,G3:I5)**, and while holding down the **Ctrl** and the **Shift** keys press the **Enter** key. Similarly you can perform the matrix operation $[A]\{C\}$. First you pick cells B21 through B23 and in the formula bar type **=MMULT(B3:D5, L3:L5)**, and while holding down the **Ctrl** and the **Shift** keys press the **Enter** key. This sequence of operations will create the result shown in Figure 2.8.

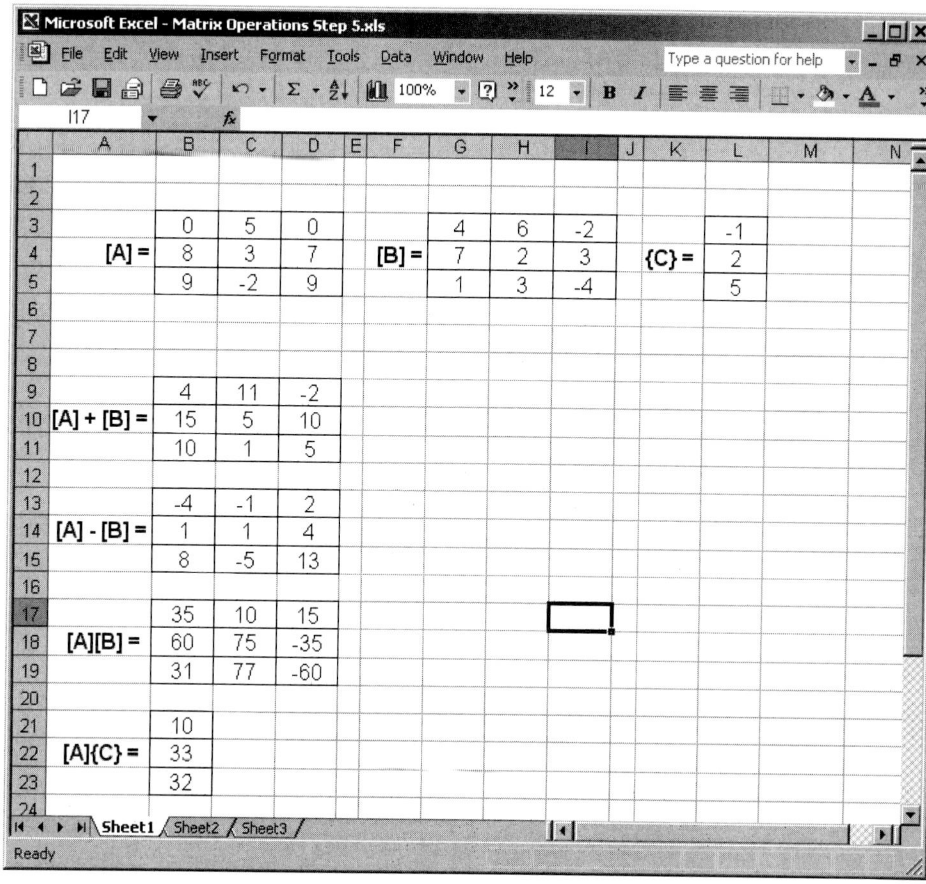

FIGURE 2.8 Example 2.1 Revisited with Excel **(step 5).**

EXAMPLE 2.8 Revisited

Consider the following 3 linear equations with three unknowns, x_1, x_2, and x_3. Our intent here is to show you how to use Excel to solve a set of linear equations.

$$2x_1 + x_2 + x_3 = 13$$

$$3x_1 + 2x_2 + 4x_3 = 32$$

$$5x_1 - x_2 + 3x_3 = 17$$

1. In the cells shown in Figure 2.9 type the appropriate characters and values. Use the **Format Cells** and **Font** option of create the bold and subscript variables.
2. In cell A10 type $[\mathbf{A}]^{-1} =$ and then using the left mouse button pick cells B9 through D11 as shown in Figure 2.10.
3. Next, in the formula bar type **=MINVERSE(B3:D5)** and while holding down the **Ctrl** and the **Shift** keys press the **Enter** key. This sequence of operations will create the result shown in Figure 2.11. The inverse of matrix [**A**] is now computed.

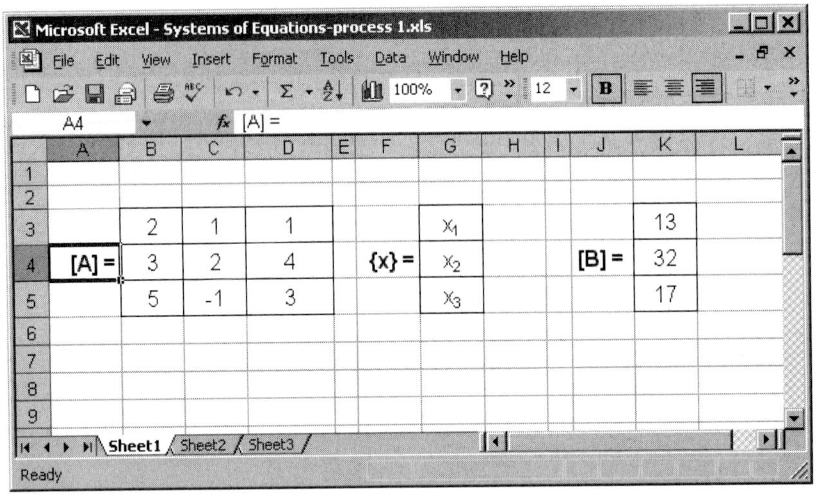

FIGURE 2.9 Example 2.8 Revisited with Excel **(step 1).**

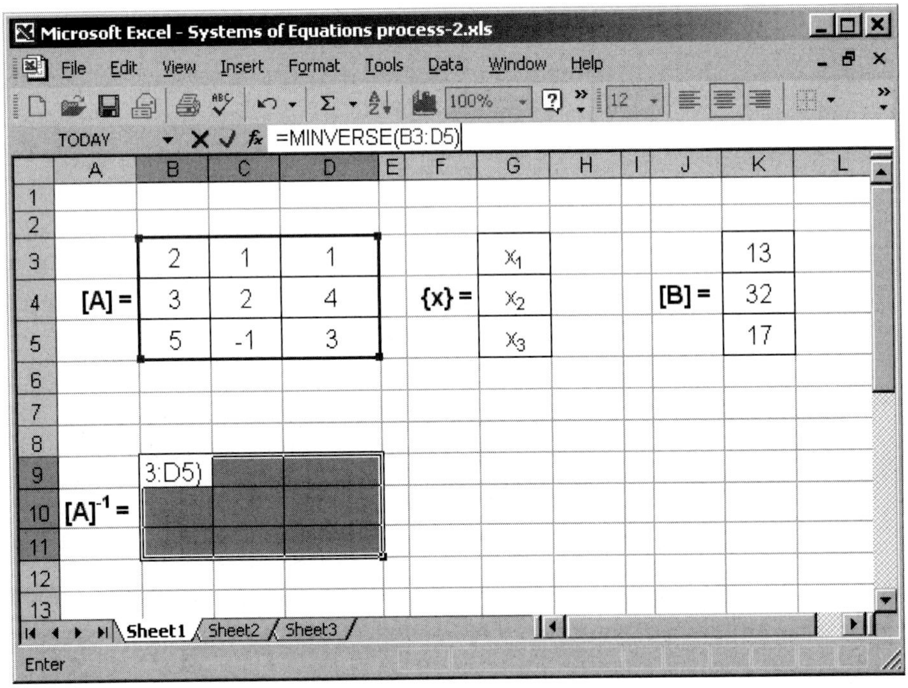

FIGURE 2.10 Example 2.8 Revisited with Excel **(step 2).**

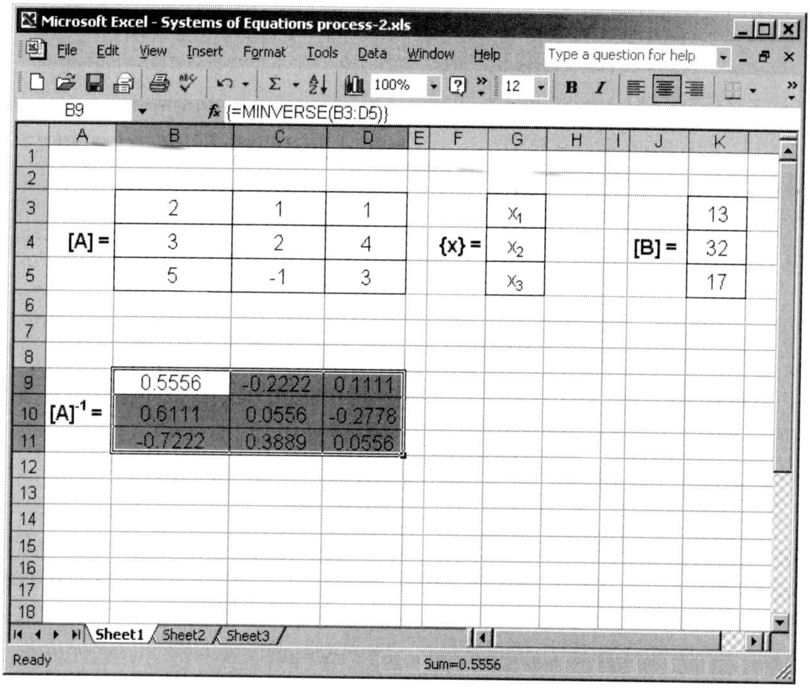

FIGURE 2.11 Example 2.8 Revisited with Excel **(step 3).**

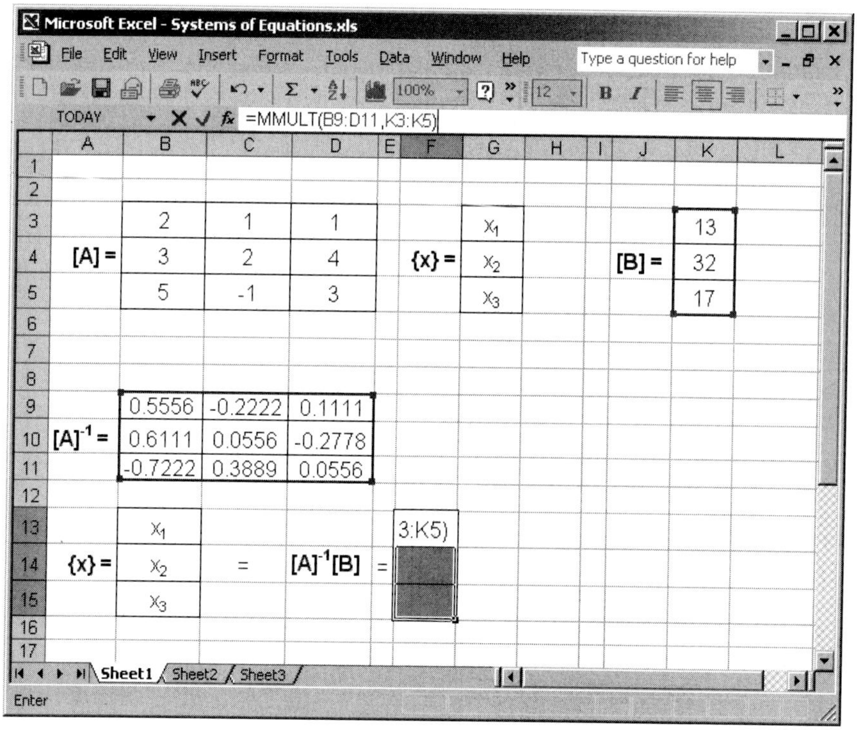

FIGURE 2.12 Example 2.8 Revisited with Excel **(step 4).**

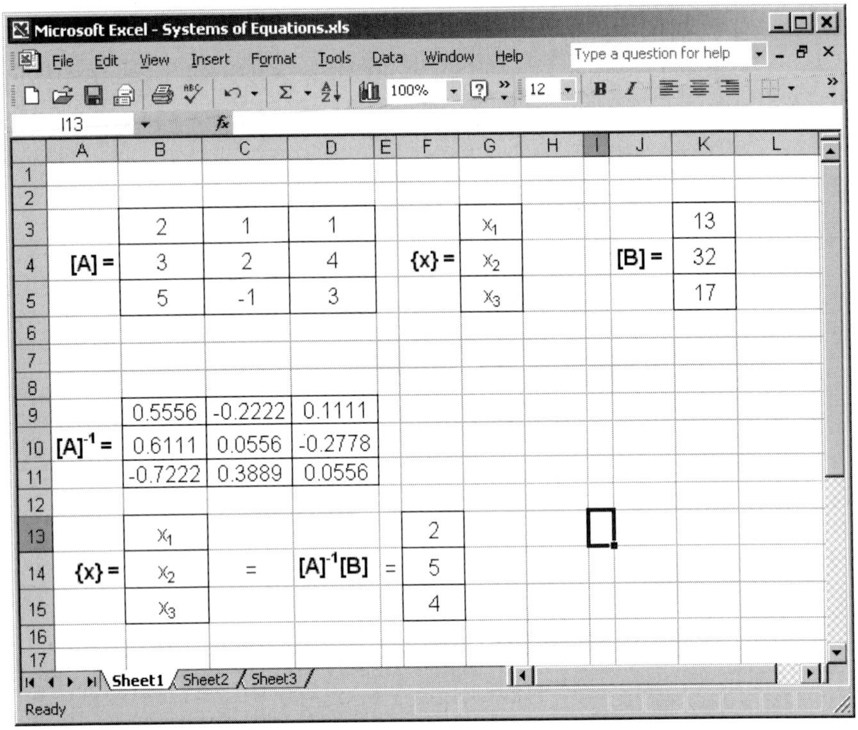

FIGURE 2.13 Example 2.8 Revisited with Excel **(step 5).**

4. Type the information shown in cells A14, B13 through B15, C14, D14, and E14 as shown in Figure 2.12.
5. Then pick cells F13 through F15 and in the formula bar type **=MMULT (B9:D11,K3:K5)**, and while holding down the **Ctrl** and the **Shift** keys press the **Enter** key. This sequence of operations will create the result shown in Figure 2.13. The values of x_1, x_2, and x_3 are now calculated.

SUMMARY

At this point you should

1. know the essential matrix terminology and the basic matrix operations.
2. know how to construct the transpose of a matrix.
3. know how to evaluate the determinant of a matrix.
4. know how to use Cramer's rule, Gauss elimination, and the LU decomposition methods to solve a set of linear equations.
5. know how to compute the inverse of a matrix.
6. be familiar with the solution of an eigenvalue problem.

REFERENCES

Gere, James M., and Weaver, William, *Matrix Algebra for Engineers*, New York, Van Nostrand, 1965.

James, M. L., Smith, G. M., and Wolford, J. C., *Applied Numerical Methods for Digital Computation*, 4th ed., New York, HarperCollins, 1993.

MATLAB Manual, *Learning MATLAB 6*, Release 12.

Nakamura, Shoichiro, *Applied Numerical Methods with Software*, Upper Saddle River, NJ, Prentice Hall, 1991.

PROBLEMS

1. Identify the size and the type of the given matrices. Denote whether the matrix is a square, column, diagonal, row, unit (identity), triangular, banded, or symmetric.

a. $\begin{bmatrix} 3 & 2 & 0 \\ 2 & 4 & 5 \\ 0 & 5 & 6 \end{bmatrix}$ **b.** $\begin{Bmatrix} x \\ x^2 \\ x^3 \\ x^4 \end{Bmatrix}$ **c.** $\begin{bmatrix} 4 & 0 \\ 0 & 8 \end{bmatrix}$ **d.** $[1 \quad y \quad y^2 \quad y^3]$

e. $\begin{bmatrix} 1 & 0 & 0 \\ 0 & 1 & 0 \\ 0 & 0 & 1 \end{bmatrix}$ **f.** $\begin{bmatrix} 3 & -1 & 0 & 0 & 0 \\ 2 & 0 & 6 & 0 & 0 \\ 0 & 4 & 1 & 4 & 0 \\ 0 & 0 & 5 & 4 & 2 \\ 0 & 0 & 0 & 7 & 8 \end{bmatrix}$ **g.** $\begin{bmatrix} 1 & 2 & 2 & 2 \\ 0 & 1 & 3 & 3 \\ 0 & 0 & 1 & 4 \\ 0 & 0 & 0 & 1 \end{bmatrix}$

h. $\begin{bmatrix} c_1 & 0 & 0 & 0 \\ 0 & c_2 & 0 & 0 \\ 0 & 0 & c_3 & 0 \\ 0 & 0 & 0 & c_4 \end{bmatrix}$

2. Given matrices

$$[\boldsymbol{A}] = \begin{bmatrix} 4 & 2 & 1 \\ 7 & 0 & -7 \\ 1 & -5 & 3 \end{bmatrix}, [\boldsymbol{B}] = \begin{bmatrix} 1 & 2 & -1 \\ 5 & 3 & 3 \\ 4 & 5 & -7 \end{bmatrix}, \text{and } \{\boldsymbol{C}\} = \begin{Bmatrix} 1 \\ -2 \\ 4 \end{Bmatrix}$$

perform the following operations:

a. $[\boldsymbol{A}] + [\boldsymbol{B}] = ?$

b. $[\boldsymbol{A}] - [\boldsymbol{B}] = ?$

c. $3[\boldsymbol{A}] = ?$

d. $[\boldsymbol{A}][\boldsymbol{B}] = ?$

e. $[\boldsymbol{A}]\{\boldsymbol{C}\} = ?$

f. $[\boldsymbol{A}]^2 = ?$

g. Show that $[\boldsymbol{I}][\boldsymbol{A}] = [\boldsymbol{A}][\boldsymbol{I}] = [\boldsymbol{A}]$

3. Given the following partitioned matrices, calculate the product of $[\mathbf{A}][\mathbf{B}] = [\mathbf{C}]$ using the submatrices given.

$$[\mathbf{A}] = \left[\begin{array}{ccc|ccc} 5 & 7 & 2 & 0 & 3 & 5 \\ 3 & 8 & -3 & -5 & 0 & 8 \\ \hline 1 & 4 & 0 & 7 & 15 & 9 \\ 0 & 10 & 5 & 12 & 3 & -1 \\ 2 & -5 & 9 & 2 & 18 & -10 \end{array}\right] \text{ and } [\mathbf{B}] = \left[\begin{array}{c|cc} 2 & 10 & 0 \\ 8 & 7 & 5 \\ -5 & 2 & -4 \\ \hline 4 & 8 & 13 \\ 3 & 12 & 0 \\ 1 & 5 & 7 \end{array}\right]$$

4. Given the matrices

$$[\mathbf{A}] = \begin{bmatrix} 1 & 4 & 2 \\ 8 & 3 & 6 \\ 7 & 1 & -2 \end{bmatrix} \text{ and } [\mathbf{B}] = \begin{bmatrix} 0 & 5 & -1 \\ -3 & 1 & 7 \\ 2 & 4 & -4 \end{bmatrix}$$

perform the following operations:

a. $[\mathbf{A}]^T = ?$ and $[\mathbf{B}]^T = ?$

b. verify that $([\mathbf{A}] + [\mathbf{B}])^T = [\mathbf{A}]^T + [\mathbf{B}]^T$

c. verify that $([\mathbf{A}][\mathbf{B}])^T = [\mathbf{B}]^T[\mathbf{A}]^T$

5. Given the following matrices

$$[\mathbf{A}] = \begin{bmatrix} 2 & 10 & 0 \\ 16 & 6 & 14 \\ 12 & -4 & 18 \end{bmatrix}, [\mathbf{B}] = \begin{bmatrix} 2 & 10 & 0 \\ 4 & 20 & 0 \\ 12 & -4 & 18 \end{bmatrix}$$

calculate

a. determinant of $[\mathbf{A}]$ and $[\mathbf{B}]$ by direct expansion and by minor lower order determinants methods.

b. determinant of $[\mathbf{A}]^T$

c. determinant of $5[\mathbf{A}]$

Which matrix is singular?

6. Given the following matrix:

$$[\mathbf{A}] = \begin{bmatrix} 0 & 5 & 0 \\ 8 & 3 & 7 \\ 9 & -2 & 9 \end{bmatrix}$$

calculate determinant of $[\mathbf{A}]$ and determinant of $[\mathbf{A}]^T$.

7. Solve the following set of equations using the Gauss elimination method, and compare your answer to the results of Example 1.4.

$$\begin{bmatrix} 10875000 & -1812500 & 0 \\ -1812500 & 6343750 & -4531250 \\ 0 & -4531250 & 4531250 \end{bmatrix} \begin{Bmatrix} u_2 \\ u_3 \\ u_4 \end{Bmatrix} = \begin{Bmatrix} 0 \\ 0 \\ 800 \end{Bmatrix}$$

8. Decompose the coefficient matrix in problem 6 into lower and upper triangular matrices.

9. Solve the following set of equations using the LU method, and compare your answer to the results of Example 1.4.

$$\begin{bmatrix} 10875000 & -1812500 & 0 \\ -1812500 & 6343750 & -4531250 \\ 0 & -4531250 & 4531250 \end{bmatrix} \begin{Bmatrix} u_2 \\ u_3 \\ u_4 \end{Bmatrix} = \begin{Bmatrix} 0 \\ 0 \\ 800 \end{Bmatrix}$$

10. Solve the following set of equations by finding the inverse of the coefficient matrix first. Compare your answer to the results of Example 1.4.

$$\begin{bmatrix} 10875000 & -1812500 & 0 \\ -1812500 & 6343750 & -4531250 \\ 0 & -4531250 & 4531250 \end{bmatrix} \begin{Bmatrix} u_2 \\ u_3 \\ u_4 \end{Bmatrix} = \begin{Bmatrix} 0 \\ 0 \\ 800 \end{Bmatrix}$$

11. Solve the following set of equations (a) using the Gaussian method, (b) using the LU decomposition method, and (c) by finding the inverse of the coefficient matrix.

$$\begin{bmatrix} 1 & 1 & 1 \\ 2 & 5 & 1 \\ -3 & 1 & 5 \end{bmatrix} \begin{Bmatrix} x_1 \\ x_2 \\ x_3 \end{Bmatrix} = \begin{Bmatrix} 6 \\ 15 \\ 14 \end{Bmatrix}$$

12. Determine the inverse of the following matrices:

$$[\mathbf{A}] = \begin{bmatrix} 5 & 0 & 0 & 0 \\ 0 & 2 & 0 & 0 \\ 0 & 0 & 8 & 0 \\ 0 & 0 & 0 & 4 \end{bmatrix} \quad [\mathbf{B}] = \begin{bmatrix} 1 & 1 & 1 \\ 2 & 5 & 1 \\ -3 & 1 & 5 \end{bmatrix} \quad [\mathbf{C}] = \begin{bmatrix} k_{11} & k_{12} \\ k_{21} & k_{22} \end{bmatrix}$$

13. Show that if we multiply the elements of a 4 × 4 matrix by a scalar quantity α, then the value of the determinant of the original matrix gets multiplied by α^4. That is:

$$\det\left(\alpha \begin{bmatrix} a_{11} & a_{12} & a_{13} & a_{14} \\ a_{21} & a_{22} & a_{23} & a_{24} \\ a_{31} & a_{32} & a_{33} & a_{34} \\ a_{41} & a_{42} & a_{43} & a_{44} \end{bmatrix}\right) = \alpha^4 \begin{vmatrix} a_{11} & a_{12} & a_{13} & a_{14} \\ a_{21} & a_{22} & a_{23} & a_{24} \\ a_{31} & a_{32} & a_{33} & a_{34} \\ a_{41} & a_{42} & a_{43} & a_{44} \end{vmatrix}$$

Also, show that if we multiply the elements of a 3 × 3 matrix by the same scalar quantity α, then the value of the determinant of the original matrix gets multiplied by α^3? How would you generalize the results of this example?

14. Determine the natural frequencies and the natural modes for the example problem in Section 2.9 for $m_1 = 0.1$ kg, $m_2 = 0.2$ kg, $k = 100$ N/m.

15. Solve problem 2 using MATLAB and Excel.

16. Solve problem 4 using MATLAB and Excel.

17. Solve problem 5 using MATLAB.

18. Solve problem 6 using MATLAB.

19. Solve problem 7 using MATLAB.

20. Solve problem 8 using MATLAB.

21. Solve problem 9 using MATLAB.

22. Solve problem 10 using MATLAB and Excel.

23. Solve problem 11 using MATLAB.

24. Solve problem 13 using MATLAB.

25. Solve the following set of equations, resulting from a model of composite wall, using MATLAB by employing the matrix left division command:

$$\begin{bmatrix} 7.11 & -1.23 & 0 & 0 & 0 \\ -1.23 & 1.99 & -0.76 & 0 & 0 \\ 0 & -0.76 & 0.851 & -0.091 & 0 \\ 0 & 0 & -0.091 & 2.31 & -2.22 \\ 0 & 0 & 0 & -2.22 & 3.69 \end{bmatrix} \begin{Bmatrix} T_2 \\ T_3 \\ T_4 \\ T_5 \\ T_6 \end{Bmatrix} = \begin{Bmatrix} (5.88)(20) \\ 0 \\ 0 \\ 0 \\ (1.47)(70) \end{Bmatrix}$$

and compare your solution to the results of Example 1.2.

26. Using MATLAB, solve problem 25 by first finding the inverse of the coefficient matrix and carrying out the necessary operations afterward.

27. Using MATLAB, solve problem 25 by first decomposing the coefficient matrix into lower and upper triangular matrices and carrying out the necessary operations afterward.

28. Solve the following set of equations, resulting from a model of a fin, using MATLAB.

$$\begin{bmatrix} 1 & 0 & 0 & 0 & 0 \\ -0.0408 & 0.0888 & -0.0408 & 0 & 0 \\ 0 & -0.0408 & 0.0888 & -0.0408 & 0 \\ 0 & 0 & -0.0408 & 0.0888 & -0.0408 \\ 0 & 0 & 0 & -0.0408 & 0.04455 \end{bmatrix} \begin{Bmatrix} T_1 \\ T_2 \\ T_3 \\ T_4 \\ T_5 \end{Bmatrix} = \begin{Bmatrix} 100 \\ 0.144 \\ 0.144 \\ 0.144 \\ 0.075 \end{Bmatrix}$$

29. Solve the following set of equations, resulting from a model of a truss, using MATLAB, and compare your solution to the results of Example 3.1:

$$10^5 \begin{bmatrix} 7.2 & 0 & 0 & 0 & -1.49 & -1.49 \\ 0 & 7.2 & 0 & -4.22 & -1.49 & -1.49 \\ 0 & 0 & 8.44 & 0 & -4.22 & 0 \\ 0 & -4.22 & 0 & 4.22 & 0 & 0 \\ -1.49 & -1.49 & -4.22 & 0 & 5.71 & 1.49 \\ -1.49 & -1.49 & 0 & 0 & 1.49 & 1.49 \end{bmatrix} \begin{Bmatrix} U_{2X} \\ U_{2Y} \\ U_{4X} \\ U_{4Y} \\ U_{5X} \\ U_{5Y} \end{Bmatrix} = \begin{Bmatrix} 0 \\ 0 \\ 0 \\ -500 \\ 0 \\ -500 \end{Bmatrix}$$

30. Solve problem 14 using MATLAB.

31. Solve problem 25 using Excel.

32. Solve problem 28 using Excel.

33. Solve problem 29 using Excel.

CHAPTER 3

Trusses

The objectives of this chapter are to introduce the basic concepts in finite element formulation of trusses and to provide an overview of the ANSYS program. A major section of this chapter is devoted to the Launcher, the Graphical User Interface, and the organization of the ANSYS program. The main topics discussed in Chapter 3 include the following:

3.1 Definition of a Truss
3.2 Finite Element Formulation
3.3 Space Trusses
3.4 Overview of the ANSYS Program
3.5 Examples Using ANSYS
3.6 Verification of Results

3.1 DEFINITION OF A TRUSS

A *truss* is an engineering structure consisting of straight members connected at their ends by means of bolts, rivets, pins, or welding. The members found in trusses may consist of steel or aluminum tubes, wooden struts, metal bars, angles, and channels. Trusses offer practical solutions to many structural problems in engineering, such as power transmission towers, bridges, and roofs of buildings. A plane truss is defined as a truss whose members lie in a single plane. The forces acting on such a truss must also lie in this plane. Members of a truss are generally considered to be *two-force members*. This term means that internal forces act in equal and opposite directions along the members, as shown in Figure 3.1.

In the analysis that follows, it is assumed that the members are connected by smooth pins and by a ball-and-socket joint in three-dimensional trusses. Moreover, it can be shown that as long as the center lines of the joining members intersect at a common point, trusses with bolted or welded joints may be treated as having smooth pins (no bending). Another important assumption deals with the way loads are applied. All loads must be applied at the joints. This assumption is true for most situations because

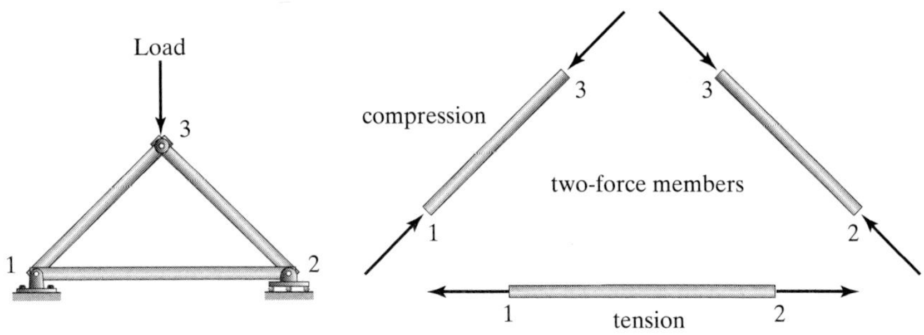

FIGURE 3.1 A simple truss subjected to a load.

trusses are designed so that the majority of the load is applied at the joints. Usually, the weights of members are negligible compared to those of the applied loads. However, if the weights of the members are to be considered, then half of the weight of each member is applied to the connecting joints. *Statically determinate* truss problems are covered in many elementary mechanics text. This class of problems is analyzed by the methods of joints or sections. These methods do not provide information on deflection of the joints because the truss members are treated as rigid bodies. Because the truss members are assumed to be rigid bodies, *statically indeterminate* problems are impossible to analyze. The finite element method allows us to remove the rigid body restriction and solve this class of problems. Figure 3.2 depicts examples of statically determinate and statically indeterminate problems.

3.2 FINITE ELEMENT FORMULATION

Let us consider the deflection of a single member when it is subjected to force F, as shown in Figure 3.3. The forthcoming derivation of the stiffness coefficient is identical to the analysis of a centrally loaded member that was presented in Chapter 1, Section 1.5. As a review and for the sake of continuity and convenience, the steps to derive the elements' equivalent stiffness coefficients are presented here again. Recall that the average stresses in any two-force member are given by

$$\sigma = \frac{F}{A} \tag{3.1}$$

The average strain of the member can be expressed by

$$\varepsilon = \frac{\Delta L}{L} \tag{3.2}$$

Over the elastic region, the stress and strain are related by Hooke's law,

$$\sigma = E\varepsilon \tag{3.3}$$

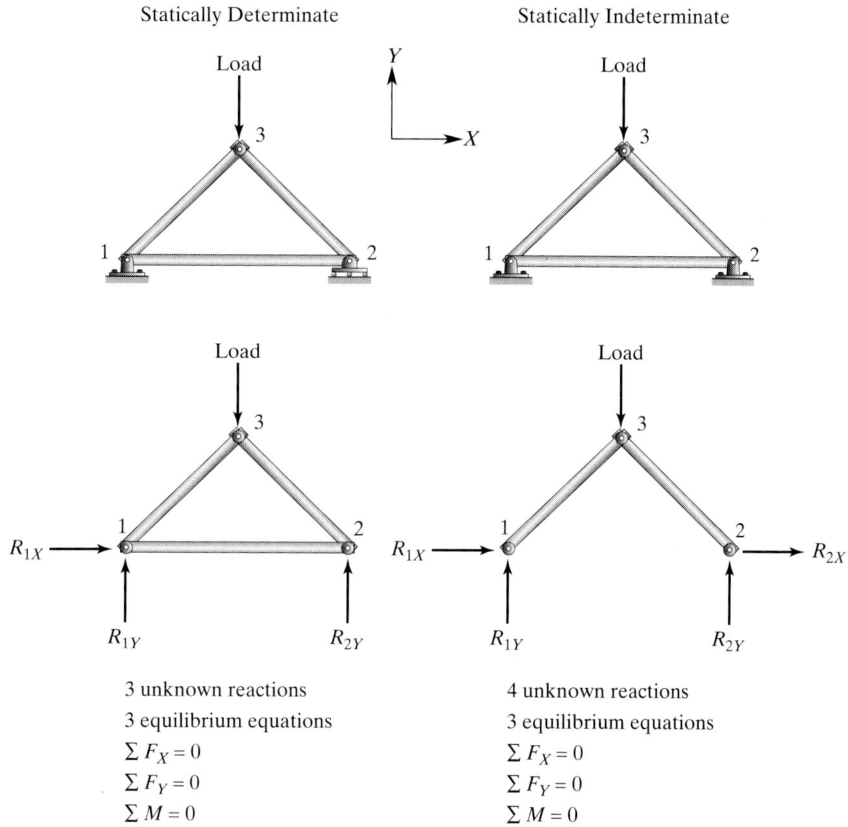

FIGURE 3.2 Examples of statically determinate and statically indeterminate problems.

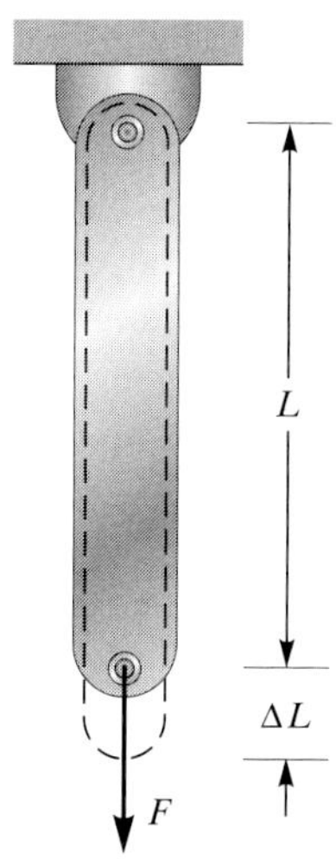

FIGURE 3.3 A two-force member subjected to a force F.

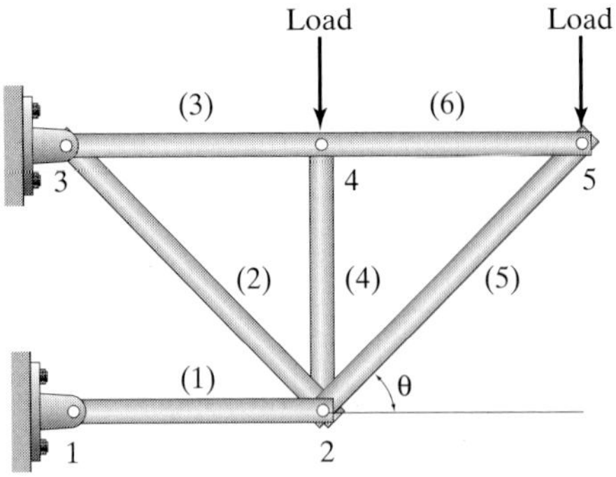

FIGURE 3.4 A balcony truss.

Combining Eqs. (3.1), (3.2), and (3.3) and simplifying, we have

$$F = \left(\frac{AE}{L}\right)\Delta L \tag{3.4}$$

Note that Eq. (3.4) is similar to the equation of a linear spring, $F = kx$. Therefore, a centrally loaded member of uniform cross section may be modeled as a spring with an equivalent stiffness of

$$k_{\text{eq}} = \frac{AE}{L} \tag{3.5}$$

A relatively small balcony truss with five nodes and six elements is shown in Figure 3.4. From this truss, consider isolating a member with an arbitrary orientation. Let us select element (5).

In general, two frames of reference are required to describe truss problems: a *global coordinate system* and a *local frame of reference*. We choose a fixed global coordinate system, XY (1) to represent the location of each joint (node) and to keep track of the orientation of each member (element), using angles such as θ; (2) to apply the constraints and the applied loads in terms of their respective global components; and (3) to represent the solution—that is, the displacement of each joint in global directions. We will also need a local, or an elemental, coordinate system xy, to describe the two-force member behavior of individual members (elements). The relationship between the local (element) descriptions and the global descriptions is shown in Figure 3.5.

The global displacements (U_{iX}, U_{iY} at node i and U_{jX}, U_{jY} at node j) are related to the local displacements (u_{ix}, u_{iy} at node i and u_{jx}, u_{jy} at node j) according to the equations

$$\begin{aligned}
U_{iX} &= u_{ix}\cos\theta - u_{iy}\sin\theta \\
U_{iY} &= u_{ix}\sin\theta + u_{iy}\cos\theta \\
U_{jX} &= u_{jx}\cos\theta - u_{jy}\sin\theta \\
U_{jY} &= u_{jx}\sin\theta + u_{jy}\cos\theta
\end{aligned} \tag{3.6}$$

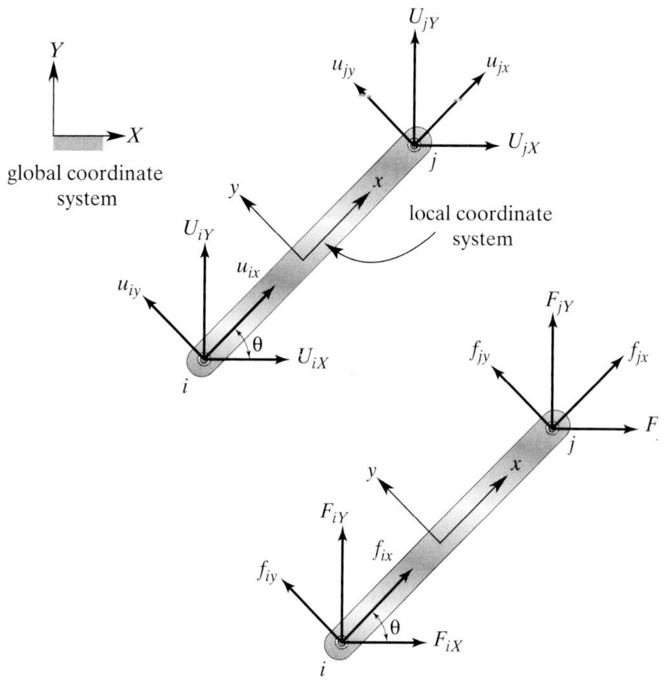

FIGURE 3.5 Relationship between local and global coordinates. Note that local coordinate x points from node i toward j.

If we write Eqs. (3.6) in matrix form, we have

$$\{\mathbf{U}\} = [\mathbf{T}]\{\mathbf{u}\} \tag{3.7}$$

where

$$\{\mathbf{U}\} = \begin{Bmatrix} U_{iX} \\ U_{iY} \\ U_{jX} \\ U_{jY} \end{Bmatrix}, [\mathbf{T}] = \begin{bmatrix} \cos\theta & -\sin\theta & 0 & 0 \\ \sin\theta & \cos\theta & 0 & 0 \\ 0 & 0 & \cos\theta & -\sin\theta \\ 0 & 0 & \sin\theta & \cos\theta \end{bmatrix}, \text{ and } \{\mathbf{u}\} = \begin{Bmatrix} u_{ix} \\ u_{iy} \\ u_{jx} \\ u_{jy} \end{Bmatrix}$$

$\{\mathbf{U}\}$ and $\{\mathbf{u}\}$ represent the displacements of nodes i and j with respect to the global XY and the local xy frame of references, respectively. $[\mathbf{T}]$ is the transformation matrix that allows for the transfer of local deformations to their respective global values. In a similar way, the local and global forces may be related according to the equations

$$\begin{aligned} F_{iX} &= f_{ix}\cos\theta - f_{iy}\sin\theta \\ F_{iY} &= f_{ix}\sin\theta + f_{iy}\cos\theta \\ F_{jX} &= f_{jx}\cos\theta - f_{jy}\sin\theta \\ F_{jY} &= f_{jx}\sin\theta + f_{jy}\cos\theta \end{aligned} \tag{3.8}$$

or, in matrix form,

$$\{\mathbf{F}\} = [\mathbf{T}]\{\mathbf{f}\} \tag{3.9}$$

where

$$\{\mathbf{F}\} = \begin{Bmatrix} F_{iX} \\ F_{iY} \\ F_{jX} \\ F_{jY} \end{Bmatrix}$$

are components of forces acting at nodes i and j with respect to global coordinates, and

$$\{\mathbf{f}\} = \begin{Bmatrix} f_{ix} \\ f_{iy} \\ f_{jx} \\ f_{jy} \end{Bmatrix}$$

represent the local components of the forces at nodes i and j.

A general relationship between the local and the global properties was derived in the preceding steps. However, we need to keep in mind that the displacements and the forces in the local y-direction are zero. This fact is simply because under the two-force assumption, the members can only be stretched or shortened along their longitudinal axis (local x-axis). Of course, this fact also holds true for the internal forces that act only in the local x-direction as shown in Figure 3.6. We do not initially set these terms equal to zero in order to maintain a general matrix description that will make the derivation of the element stiffness matrix easier. This process will become clear when we set the y-components of the displacements and forces equal to zero. The local internal forces and displacements are related through the stiffness matrix

$$\begin{Bmatrix} f_{ix} \\ f_{iy} \\ f_{jx} \\ f_{jy} \end{Bmatrix} = \begin{bmatrix} k & 0 & -k & 0 \\ 0 & 0 & 0 & 0 \\ -k & 0 & k & 0 \\ 0 & 0 & 0 & 0 \end{bmatrix} \begin{Bmatrix} u_{ix} \\ u_{iy} \\ u_{jx} \\ u_{jy} \end{Bmatrix} \tag{3.10}$$

where $k = k_{\text{eq}} = \dfrac{AE}{L}$, and using matrix form we can write

$$\{\mathbf{f}\} = [\mathbf{K}]\{\mathbf{u}\} \tag{3.11}$$

After substituting for $\{\mathbf{f}\}$ and $\{\mathbf{u}\}$ in terms of $\{\mathbf{F}\}$ and $\{\mathbf{U}\}$, we have

$$\overbrace{[\mathbf{T}]^{-1}\{\mathbf{F}\}}^{\{\mathbf{f}\}} = [\mathbf{K}]\overbrace{[\mathbf{T}]^{-1}\{\mathbf{U}\}}^{\{\mathbf{u}\}} \tag{3.12}$$

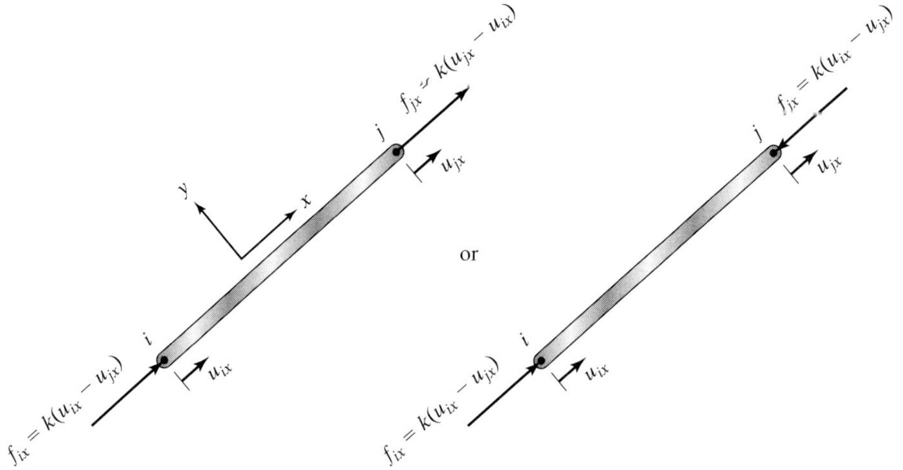

$f_{ix} = k(u_{ix} - u_{jx})$
$f_{iy} = 0$
$f_{jx} = k(u_{jx} - u_{ix}) = -k(u_{ix} - u_{jx})$ or in a matrix form $\begin{Bmatrix} f_{ix} \\ f_{iy} \\ f_{jx} \\ f_{jy} \end{Bmatrix} = \begin{bmatrix} k & 0 & -k & 0 \\ 0 & 0 & 0 & 0 \\ -k & 0 & k & 0 \\ 0 & 0 & 0 & 0 \end{bmatrix} \begin{Bmatrix} u_{ix} \\ u_{iy} \\ u_{jx} \\ u_{jy} \end{Bmatrix}$
$f_{jy} = 0$

FIGURE 3.6 Internal forces for an arbitrary truss element. Note that the static equilibrium conditions require that the sum of f_{ix} and f_{jx} be zero. Also note that the sum of f_{ix} and f_{jx} is zero regardless of which representation is selected.

where $[\mathbf{T}]^{-1}$ is the inverse of the transformation matrix $[\mathbf{T}]$ and is

$$[\mathbf{T}]^{-1} = \begin{bmatrix} \cos\theta & \sin\theta & 0 & 0 \\ -\sin\theta & \cos\theta & 0 & 0 \\ 0 & 0 & \cos\theta & \sin\theta \\ 0 & 0 & -\sin\theta & \cos\theta \end{bmatrix} \qquad (3.13)$$

Multiplying both sides of Eq. (3.12) by $[\mathbf{T}]$ and simplifying, we obtain

$$\{\mathbf{F}\} = [\mathbf{T}][\mathbf{K}][\mathbf{T}]^{-1}\{\mathbf{U}\} \qquad (3.14)$$

Substituting for values of the $[\mathbf{T}]$, $[\mathbf{K}]$, $[\mathbf{T}]^{-1}$, and $\{\mathbf{U}\}$ matrices in Eq. (3.14) and multiplying, we are left with

$$\begin{Bmatrix} F_{iX} \\ F_{iY} \\ F_{jX} \\ F_{jY} \end{Bmatrix} = k \begin{bmatrix} \cos^2\theta & \sin\theta\cos\theta & -\cos^2\theta & -\sin\theta\cos\theta \\ \sin\theta\cos\theta & \sin^2\theta & -\sin\theta\cos\theta & -\sin^2\theta \\ -\cos^2\theta & -\sin\theta\cos\theta & \cos^2\theta & \sin\theta\cos\theta \\ -\sin\theta\cos\theta & -\sin^2\theta & \sin\theta\cos\theta & \sin^2\theta \end{bmatrix} \begin{Bmatrix} U_{iX} \\ U_{iY} \\ U_{jX} \\ U_{jY} \end{Bmatrix} \qquad (3.15)$$

Equations (3.15) express the relationship between the applied forces, the element stiffness matrix $[\mathbf{K}]^{(e)}$, and the global deflection of the nodes of an arbitrary element. The stiffness matrix $[\mathbf{K}]^{(e)}$ for any member (element) of the truss is

$$[\mathbf{K}]^{(e)} = k\begin{bmatrix} \cos^2\theta & \sin\theta\cos\theta & -\cos^2\theta & -\sin\theta\cos\theta \\ \sin\theta\cos\theta & \sin^2\theta & -\sin\theta\cos\theta & -\sin^2\theta \\ -\cos^2\theta & -\sin\theta\cos\theta & \cos^2\theta & \sin\theta\cos\theta \\ -\sin\theta\cos\theta & -\sin^2\theta & \sin\theta\cos\theta & \sin^2\theta \end{bmatrix} \tag{3.16}$$

The next few steps involve assembling, or connecting, the elemental stiffness matrices, applying boundary conditions and loads, solving for displacements, and obtaining other information, such as normal stresses. These steps are best illustrated through an example problem.

EXAMPLE 3.1

Consider the balcony truss in Figure 3.4, shown here with dimensions. We are interested in determining the deflection of each joint under the loading shown in the figure. All members are made from Douglas-fir wood with a modulus of elasticity of $E = 1.90 \times 10^6$ lb/in^2 and a cross-sectional area of 8 in^2. We are also interested in calculating average stresses in each member. First, we will solve this problem manually. Later, once we learn how to use ANSYS, we will revisit this problem and solve it using ANSYS.

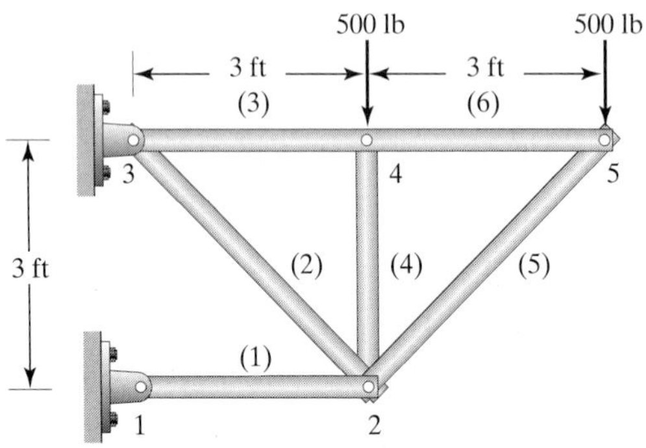

As discussed in Chapter 1, Section 1.4, there are seven steps involved in any finite element analysis. Here, these steps are discussed again to emphasize the three phases (preprocessing, solution, and postprocessing) associated with the analysis of truss problems.

Preprocessing Phase

1. *Discretize the problem into nodes and elements.*
   Each truss member is considered an element, and each joint connecting members is a node. Therefore, the given truss can be modeled with five nodes and six elements. Consult Table 3.1 while following the solution.

TABLE 3.1 The relationship between the elements and their corresponding nodes

Element	Node *i*	Node *j*	θ See Figures 3.7–3.10
(1)	1	2	0
(2)	2	3	135
(3)	3	4	0
(4)	2	4	90
(5)	2	5	45
(6)	4	5	0

2. *Assume a solution that approximates the behavior of an element.*
As discussed in Section 3.2, we will model the elastic behavior of each element as a spring with an equivalent stiffness of k as given by Eq. (3.5). Since elements (1), (3), (4), and (6) have the same length, cross-sectional area, and modulus of elasticity, the equivalent stiffness constant for these elements (members) is

$$k = \frac{AE}{L} = \frac{(8 \text{ in}^2)\left(1.90 \times 10^6 \frac{\text{lb}}{\text{in}^2}\right)}{36 \text{ in}} = 4.22 \times 10^5 \text{ lb/in.}$$

The stiffness constant for elements (2) and (5) is

$$k = \frac{AE}{L} = \frac{(8 \text{ in}^2)\left(1.90 \times 10^6 \frac{\text{lb}}{\text{in}^2}\right)}{50.9 \text{ in}} = 2.98 \times 10^5 \text{ lb/in.}$$

3. *Develop equations for elements.*
For elements (1), (3), and (6), the local and the global coordinate systems are aligned, which means that $\theta = 0$. This relationship is shown in Figure 3.7. Using Eq. (3.16), we find that the stiffness matrices are

$$[\mathbf{K}]^{(e)} = k\begin{bmatrix} \cos^2\theta & \sin\theta\cos\theta & -\cos^2\theta & -\sin\theta\cos\theta \\ \sin\theta\cos\theta & \sin^2\theta & -\sin\theta\cos\theta & -\sin^2\theta \\ -\cos^2\theta & -\sin\theta\cos\theta & \cos^2\theta & \sin\theta\cos\theta \\ -\sin\theta\cos\theta & -\sin^2\theta & \sin\theta\cos\theta & \sin^2\theta \end{bmatrix}$$

$$[\mathbf{K}]^{(1)} = 4.22 \times 10^5\begin{bmatrix} \cos^2(0) & \sin(0)\cos(0) & -\cos^2(0) & -\sin(0)\cos(0) \\ \sin(0)\cos(0) & \sin^2(0) & -\sin(0)\cos(0) & -\sin^2(0) \\ -\cos^2(0) & -\sin(0)\cos(0) & \cos^2(0) & \sin(0)\cos(0) \\ -\sin(0)\cos(0) & -\sin^2(0) & \sin(0)\cos(0) & \sin^2(0) \end{bmatrix}$$

$$[\mathbf{K}]^{(1)} = 4.22 \times 10^5\begin{bmatrix} 1 & 0 & -1 & 0 \\ 0 & 0 & 0 & 0 \\ -1 & 0 & 1 & 0 \\ 0 & 0 & 0 & 0 \end{bmatrix}\begin{matrix} U_{1X} \\ U_{1Y} \\ U_{2X} \\ U_{2Y} \end{matrix}$$

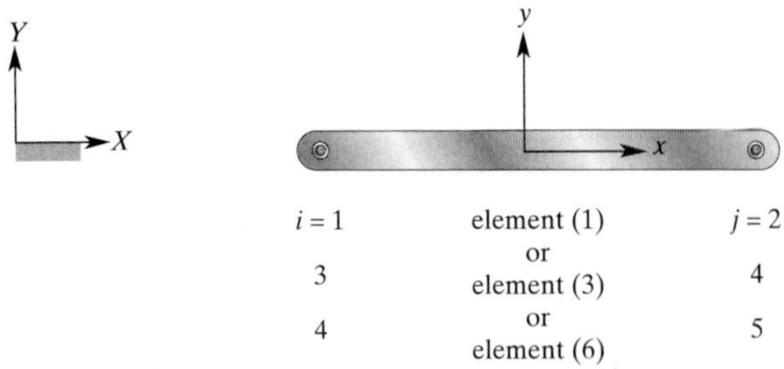

FIGURE 3.7 The orientation of the local coordinates with respect to the global coordinates for elements (1), (3), and (6).

and the position of element (1)'s stiffness matrix in the global matrix is

$$[\mathbf{K}]^{(1G)} = 10^5 \begin{bmatrix} 4.22 & 0 & -4.22 & 0 & 0 & 0 & 0 & 0 & 0 & 0 \\ 0 & 0 & 0 & 0 & 0 & 0 & 0 & 0 & 0 & 0 \\ -4.22 & 0 & 4.22 & 0 & 0 & 0 & 0 & 0 & 0 & 0 \\ 0 & 0 & 0 & 0 & 0 & 0 & 0 & 0 & 0 & 0 \\ 0 & 0 & 0 & 0 & 0 & 0 & 0 & 0 & 0 & 0 \\ 0 & 0 & 0 & 0 & 0 & 0 & 0 & 0 & 0 & 0 \\ 0 & 0 & 0 & 0 & 0 & 0 & 0 & 0 & 0 & 0 \\ 0 & 0 & 0 & 0 & 0 & 0 & 0 & 0 & 0 & 0 \\ 0 & 0 & 0 & 0 & 0 & 0 & 0 & 0 & 0 & 0 \\ 0 & 0 & 0 & 0 & 0 & 0 & 0 & 0 & 0 & 0 \end{bmatrix} \begin{matrix} U_{1X} \\ U_{1Y} \\ U_{2X} \\ U_{2Y} \\ U_{3X} \\ U_{3Y} \\ U_{4X} \\ U_{4Y} \\ U_{5X} \\ U_{5Y} \end{matrix}$$

Note that the nodal displacement matrix is shown alongside element (1)'s position in the global matrix to aid us in observing the location of element (1)'s stiffness matrix in the global matrix. Similarly, the stiffness matrix for element (3) is

$$[\mathbf{K}]^{(3)} = 4.22 \times 10^5 \begin{bmatrix} 1 & 0 & -1 & 0 \\ 0 & 0 & 0 & 0 \\ -1 & 0 & 1 & 0 \\ 0 & 0 & 0 & 0 \end{bmatrix} \begin{matrix} U_{3X} \\ U_{3Y} \\ U_{4X} \\ U_{4Y} \end{matrix}$$

and its position in the global matrix is

$$[\mathbf{K}]^{(3G)} = 10^5 \begin{bmatrix} 0 & 0 & 0 & 0 & 0 & 0 & 0 & 0 & 0 & 0 \\ 0 & 0 & 0 & 0 & 0 & 0 & 0 & 0 & 0 & 0 \\ 0 & 0 & 0 & 0 & 0 & 0 & 0 & 0 & 0 & 0 \\ 0 & 0 & 0 & 0 & 0 & 0 & 0 & 0 & 0 & 0 \\ 0 & 0 & 0 & 0 & 4.22 & 0 & -4.22 & 0 & 0 & 0 \\ 0 & 0 & 0 & 0 & 0 & 0 & 0 & 0 & 0 & 0 \\ 0 & 0 & 0 & 0 & -4.22 & 0 & 4.22 & 0 & 0 & 0 \\ 0 & 0 & 0 & 0 & 0 & 0 & 0 & 0 & 0 & 0 \\ 0 & 0 & 0 & 0 & 0 & 0 & 0 & 0 & 0 & 0 \\ 0 & 0 & 0 & 0 & 0 & 0 & 0 & 0 & 0 & 0 \end{bmatrix} \begin{matrix} U_{1X} \\ U_{1Y} \\ U_{2X} \\ U_{2Y} \\ U_{3X} \\ U_{3Y} \\ U_{4X} \\ U_{4Y} \\ U_{5X} \\ U_{5Y} \end{matrix}$$

The stiffness matrix for element (6) is

$$[\mathbf{K}]^{(6)} = 4.22 \times 10^5 \begin{bmatrix} 1 & 0 & -1 & 0 \\ 0 & 0 & 0 & 0 \\ -1 & 0 & 1 & 0 \\ 0 & 0 & 0 & 0 \end{bmatrix} \begin{matrix} U_{4X} \\ U_{4Y} \\ U_{5X} \\ U_{5Y} \end{matrix}$$

and its position in the global matrix is

$$[\mathbf{K}]^{(6G)} = 10^5 \begin{bmatrix} 0 & 0 & 0 & 0 & 0 & 0 & 0 & 0 & 0 & 0 \\ 0 & 0 & 0 & 0 & 0 & 0 & 0 & 0 & 0 & 0 \\ 0 & 0 & 0 & 0 & 0 & 0 & 0 & 0 & 0 & 0 \\ 0 & 0 & 0 & 0 & 0 & 0 & 0 & 0 & 0 & 0 \\ 0 & 0 & 0 & 0 & 0 & 0 & 0 & 0 & 0 & 0 \\ 0 & 0 & 0 & 0 & 0 & 0 & 0 & 0 & 0 & 0 \\ 0 & 0 & 0 & 0 & 0 & 0 & 4.22 & 0 & -4.22 & 0 \\ 0 & 0 & 0 & 0 & 0 & 0 & 0 & 0 & 0 & 0 \\ 0 & 0 & 0 & 0 & 0 & 0 & -4.22 & 0 & 4.22 & 0 \\ 0 & 0 & 0 & 0 & 0 & 0 & 0 & 0 & 0 & 0 \end{bmatrix} \begin{matrix} U_{1X} \\ U_{1Y} \\ U_{2X} \\ U_{2Y} \\ U_{3X} \\ U_{3Y} \\ U_{4X} \\ U_{4Y} \\ U_{5X} \\ U_{5Y} \end{matrix}$$

For element (4), the orientation of the local coordinate system with respect to the global coordinates is shown in Figure 3.8. Thus, for element (4), $\theta = 90$, which leads to the stiffness matrix

$$[\mathbf{K}]^{(4)} = 4.22 \times 10^5 \begin{bmatrix} \cos^2(90) & \sin(90)\cos(90) & -\cos^2(90) & -\sin(90)\cos(90) \\ \sin(90)\cos(90) & \sin^2(90) & -\sin(90)\cos(90) & -\sin^2(90) \\ -\cos^2(90) & -\sin(90)\cos(90) & \cos^2(90) & \sin(90)\cos(90) \\ -\sin(90)\cos(90) & -\sin^2(90) & \sin(90)\cos(90) & \sin^2(90) \end{bmatrix}$$

$$[\mathbf{K}]^{(4)} = 4.22 \times 10^5 \begin{bmatrix} 0 & 0 & 0 & 0 \\ 0 & 1 & 0 & -1 \\ 0 & 0 & 0 & 0 \\ 0 & -1 & 0 & 1 \end{bmatrix} \begin{matrix} U_{2X} \\ U_{2Y} \\ U_{4X} \\ U_{4Y} \end{matrix}$$

and its global position

$$[\mathbf{K}]^{(4G)} = 10^5 \begin{bmatrix} 0 & 0 & 0 & 0 & 0 & 0 & 0 & 0 & 0 & 0 \\ 0 & 0 & 0 & 0 & 0 & 0 & 0 & 0 & 0 & 0 \\ 0 & 0 & 0 & 0 & 0 & 0 & 0 & 0 & 0 & 0 \\ 0 & 0 & 0 & 4.22 & 0 & 0 & 0 & -4.22 & 0 & 0 \\ 0 & 0 & 0 & 0 & 0 & 0 & 0 & 0 & 0 & 0 \\ 0 & 0 & 0 & 0 & 0 & 0 & 0 & 0 & 0 & 0 \\ 0 & 0 & 0 & 0 & 0 & 0 & 0 & 0 & 0 & 0 \\ 0 & 0 & 0 & -4.22 & 0 & 0 & 0 & 4.22 & 0 & 0 \\ 0 & 0 & 0 & 0 & 0 & 0 & 0 & 0 & 0 & 0 \\ 0 & 0 & 0 & 0 & 0 & 0 & 0 & 0 & 0 & 0 \end{bmatrix} \begin{matrix} U_{1X} \\ U_{1Y} \\ U_{2X} \\ U_{2Y} \\ U_{3X} \\ U_{3Y} \\ U_{4X} \\ U_{4Y} \\ U_{5X} \\ U_{5Y} \end{matrix}$$

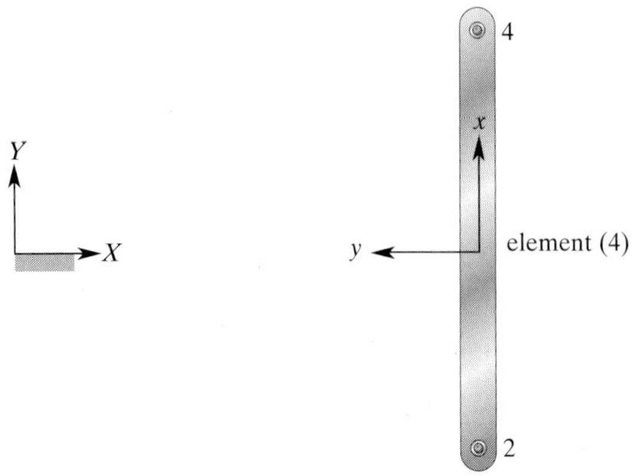

FIGURE 3.8 The orientation of the local coordinates with respect to the global coordinates for element (4).

For element (2), the orientation of the local coordinate system with respect to the global coordinates is shown in Figure 3.9. Thus, for element (2), yielding the stiffness matrix

$$[\mathbf{K}]^{(2)} = 2.98 \times 10^5 \begin{bmatrix} \cos^2(135) & \sin(135)\cos(135) & -\cos^2(135) & -\sin(135)\cos(135) \\ \sin(135)\cos(135) & \sin^2(135) & -\sin(135)\cos(135) & -\sin^2(135) \\ -\cos^2(135) & -\sin(135)\cos(135) & \cos^2(135) & \sin(135)\cos(135) \\ -\sin(135)\cos(135) & -\sin^2(135) & \sin(135)\cos(135) & \sin^2(135) \end{bmatrix}$$

$$[\mathbf{K}]^{(2)} = 2.98 \times 10^5 \begin{bmatrix} .5 & -.5 & -.5 & .5 \\ -.5 & .5 & .5 & -.5 \\ -.5 & .5 & .5 & -.5 \\ .5 & -.5 & -.5 & .5 \end{bmatrix} \begin{matrix} U_{2X} \\ U_{2Y} \\ U_{3X} \\ U_{3Y} \end{matrix}$$

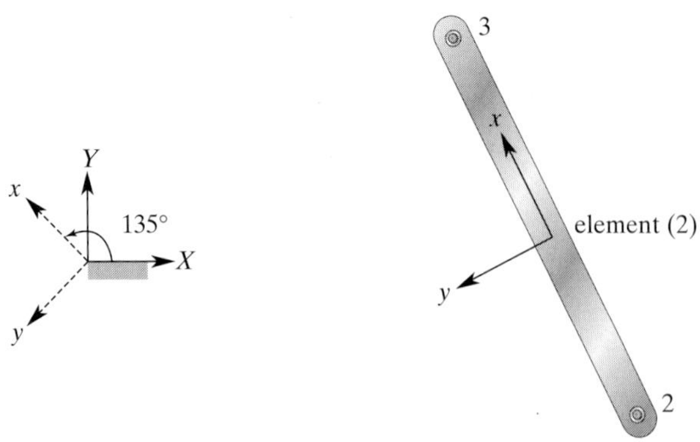

FIGURE 3.9 The orientation of the local coordinates with respect to the global coordinates for element (2).

Simplifying, we get

$$[\mathbf{K}]^{(2)} = 1.49 \times 10^5 \begin{bmatrix} 1 & -1 & -1 & 1 \\ -1 & 1 & 1 & 1 \\ -1 & 1 & 1 & -1 \\ 1 & -1 & -1 & 1 \end{bmatrix} \begin{matrix} U_{2X} \\ U_{2Y} \\ U_{3X} \\ U_{3Y} \end{matrix}$$

and its position in the global matrix is

$$[\mathbf{K}]^{(2G)} = 10^5 \begin{bmatrix} 0 & 0 & 0 & 0 & 0 & 0 & 0 & 0 & 0 & 0 \\ 0 & 0 & 0 & 0 & 0 & 0 & 0 & 0 & 0 & 0 \\ 0 & 0 & 1.49 & -1.49 & -1.49 & 1.49 & 0 & 0 & 0 & 0 \\ 0 & 0 & -1.49 & 1.49 & 1.49 & -1.49 & 0 & 0 & 0 & 0 \\ 0 & 0 & -1.49 & 1.49 & 1.49 & -1.49 & 0 & 0 & 0 & 0 \\ 0 & 0 & 1.49 & -1.49 & -1.49 & 1.49 & 0 & 0 & 0 & 0 \\ 0 & 0 & 0 & 0 & 0 & 0 & 0 & 0 & 0 & 0 \\ 0 & 0 & 0 & 0 & 0 & 0 & 0 & 0 & 0 & 0 \\ 0 & 0 & 0 & 0 & 0 & 0 & 0 & 0 & 0 & 0 \\ 0 & 0 & 0 & 0 & 0 & 0 & 0 & 0 & 0 & 0 \end{bmatrix} \begin{matrix} U_{1X} \\ U_{1Y} \\ U_{2X} \\ U_{2Y} \\ U_{3X} \\ U_{3Y} \\ U_{4X} \\ U_{4Y} \\ U_{5X} \\ U_{5Y} \end{matrix}$$

For element (5), the orientation of the local coordinate system with respect to the global coordinates is shown in Figure 3.10. Thus, for element (5), $\theta = 45$, yielding the stiffness matrix

$$[\mathbf{K}]^{(5)} = 2.98 \times 10^5 \begin{bmatrix} \cos^2(45) & \sin(45)\cos(45) & -\cos^2(45) & -\sin(45)\cos(45) \\ \sin(45)\cos(45) & \sin^2(45) & -\sin(45)\cos(45) & -\sin^2(45) \\ -\cos^2(45) & -\sin(45)\cos(45) & \cos^2(45) & \sin(45)\cos(45) \\ -\sin(45)\cos(45) & -\sin^2(45) & \sin(45)\cos(45) & \sin^2(45) \end{bmatrix}$$

$$[\mathbf{K}]^{(5)} = 2.98 \times 10^5 \begin{bmatrix} .5 & .5 & -.5 & -.5 \\ .5 & .5 & -.5 & -.5 \\ -.5 & -.5 & .5 & .5 \\ -.5 & -.5 & .5 & .5 \end{bmatrix} \begin{matrix} U_{2X} \\ U_{2Y} \\ U_{5X} \\ U_{5Y} \end{matrix}$$

and its position in the global stiffness matrix is

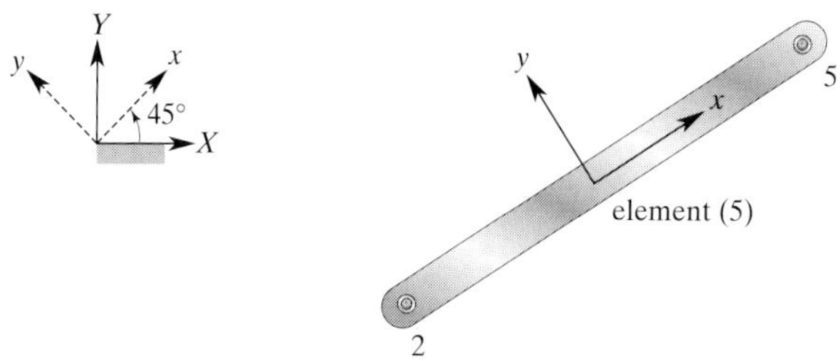

FIGURE 3.10 The orientation of the local coordinates with respect to the global coordinates for element (5).

$$[\mathbf{K}]^{(5G)} = 10^5 \begin{bmatrix} 0 & 0 & 0 & 0 & 0 & 0 & 0 & 0 & 0 & 0 \\ 0 & 0 & 0 & 0 & 0 & 0 & 0 & 0 & 0 & 0 \\ 0 & 0 & 1.49 & 1.49 & 0 & 0 & 0 & 0 & -1.49 & -1.49 \\ 0 & 0 & 1.49 & 1.49 & 0 & 0 & 0 & 0 & -1.49 & -1.49 \\ 0 & 0 & 0 & 0 & 0 & 0 & 0 & 0 & 0 & 0 \\ 0 & 0 & 0 & 0 & 0 & 0 & 0 & 0 & 0 & 0 \\ 0 & 0 & 0 & 0 & 0 & 0 & 0 & 0 & 0 & 0 \\ 0 & 0 & 0 & 0 & 0 & 0 & 0 & 0 & 0 & 0 \\ 0 & 0 & -1.49 & -1.49 & 0 & 0 & 0 & 0 & 1.49 & 1.49 \\ 0 & 0 & -1.49 & -1.49 & 0 & 0 & 0 & 0 & 1.49 & 1.49 \end{bmatrix} \begin{matrix} U_{1X} \\ U_{1Y} \\ U_{2X} \\ U_{2Y} \\ U_{3X} \\ U_{3Y} \\ U_{4X} \\ U_{4Y} \\ U_{5X} \\ U_{5Y} \end{matrix}$$

It is worth noting again that the nodal displacements associated with each element are shown next to each element's stiffness matrix. This practice makes it easier to connect (assemble) the individual stiffness matrices into the global stiffness matrix for the truss.

4. *Assemble elements.*
The global stiffness matrix is obtained by assembling, or adding together, the individual elements' matrices:

$$[\mathbf{K}]^{(G)} = [\mathbf{K}]^{(1G)} + [\mathbf{K}]^{(2G)} + [\mathbf{K}]^{(3G)} + [\mathbf{K}]^{(4G)} + [\mathbf{K}]^{(5G)} + [\mathbf{K}]^{(6G)}$$

$$[\mathbf{K}]^{(G)} = 10^5 \left[\begin{matrix} 4.22 & 0 & -4.22 & 0 & 0 \\ 0 & 0 & 0 & 0 & 0 \\ -4.22 & 0 & 4.22 + 1.49 + 1.49 & -1.49 + 1.49 & -1.49 \\ 0 & 0 & 1.49 - 1.49 & 4.22 + 1.49 + 1.49 & 1.49 \\ 0 & 0 & -1.49 & 1.49 & 4.22 + 1.49 \\ 0 & 0 & 1.49 & -1.49 & -1.49 \\ 0 & 0 & 0 & 0 & -4.22 \\ 0 & 0 & 0 & -4.22 & 0 \\ 0 & 0 & -1.49 & -1.49 & 0 \\ 0 & 0 & -1.49 & -1.49 & 0 \end{matrix}\right.$$

$$\left.\begin{matrix} 0 & 0 & 0 & 0 & 0 \\ 0 & 0 & 0 & 0 & 0 \\ 1.49 & 0 & 0 & -1.49 & -1.49 \\ -1.49 & 0 & -4.22 & -1.49 & -1.49 \\ -1.49 & -4.22 & 0 & 0 & 0 \\ 1.49 & 0 & 0 & 0 & 0 \\ 0 & 4.22 + 4.22 & 0 & -4.22 & 0 \\ 0 & 0 & 4.22 & 0 & 0 \\ 0 & -4.22 & 0 & 4.22 + 1.49 & 1.49 \\ 0 & 0 & 0 & 1.49 & 1.49 \end{matrix}\right] \begin{matrix} U_{1X} \\ U_{1Y} \\ U_{2X} \\ U_{2Y} \\ U_{3X} \\ U_{3Y} \\ U_{4X} \\ U_{4Y} \\ U_{5X} \\ U_{5Y} \end{matrix}$$

Simplifying, we get

$$[\mathbf{K}]^{(G)} = 10^5 \begin{bmatrix} 4.22 & 0 & -4.22 & 0 & 0 & 0 & 0 & 0 & 0 & 0 \\ 0 & 0 & 0 & 0 & 0 & 0 & 0 & 0 & 0 & 0 \\ -4.22 & 0 & 7.2 & 0 & -1.49 & 1.49 & 0 & 0 & -1.49 & -1.49 \\ 0 & 0 & 0 & 7.2 & 1.49 & -1.49 & 0 & -4.22 & -1.49 & -1.49 \\ 0 & 0 & -1.49 & 1.49 & 5.71 & -1.49 & -4.22 & 0 & 0 & 0 \\ 0 & 0 & 1.49 & -1.49 & -1.49 & 1.49 & 0 & 0 & 0 & 0 \\ 0 & 0 & 0 & 0 & -4.22 & 0 & 8.44 & 0 & -4.22 & 0 \\ 0 & 0 & 0 & -4.22 & 0 & 0 & 0 & 4.22 & 0 & 0 \\ 0 & 0 & -1.49 & -1.49 & 0 & 0 & -4.22 & 0 & 5.71 & 1.49 \\ 0 & 0 & -1.49 & -1.49 & 0 & 0 & 0 & 0 & 1.49 & 1.49 \end{bmatrix}$$

5. *Apply the boundary conditions and loads.*
The following boundary conditions apply to this problem: nodes 1 and 3 are fixed, which implies that $U_{1X} = 0, U_{1Y} = 0, U_{3X} = 0$, and $U_{3Y} = 0$. Incorporating these conditions into the global stiffness matrix and applying the external loads at nodes 4 and 5 such that $F_{4Y} = -500$ lb and $F_{5Y} = -500$ lb results in a set of linear equations that must be solved simultaneously:

$$10^5 \begin{bmatrix} 1 & 0 & 0 & 0 & 0 & 0 & 0 & 0 & 0 & 0 \\ 0 & 1 & 0 & 0 & 0 & 0 & 0 & 0 & 0 & 0 \\ -4.22 & 0 & 7.2 & 0 & -1.49 & 1.49 & 0 & 0 & -1.49 & -1.49 \\ 0 & 0 & 0 & 7.2 & 1.49 & -1.49 & 0 & -4.22 & -1.49 & -1.49 \\ 0 & 0 & 0 & 0 & 1 & 0 & 0 & 0 & 0 & 0 \\ 0 & 0 & 0 & 0 & 0 & 1 & 0 & 0 & 0 & 0 \\ 0 & 0 & 0 & 0 & -4.22 & 0 & 8.44 & 0 & -4.22 & 0 \\ 0 & 0 & 0 & -4.22 & 0 & 0 & 0 & 4.22 & 0 & 0 \\ 0 & 0 & -1.49 & -1.49 & 0 & 0 & -4.22 & 0 & 5.71 & 1.49 \\ 0 & 0 & -1.49 & -1.49 & 0 & 0 & 0 & 0 & 1.49 & 1.49 \end{bmatrix} \begin{Bmatrix} U_{1X} \\ U_{1Y} \\ U_{2X} \\ U_{2Y} \\ U_{3X} \\ U_{3Y} \\ U_{4X} \\ U_{4Y} \\ U_{5X} \\ U_{5Y} \end{Bmatrix} = \begin{Bmatrix} 0 \\ 0 \\ 0 \\ 0 \\ 0 \\ 0 \\ 0 \\ -500 \\ 0 \\ -500 \end{Bmatrix}$$

Because $U_{1X} = 0, U_{1Y} = 0, U_{3X} = 0$, and $U_{3Y} = 0$, we can eliminate the first, second, fifth, and sixth rows and columns from our calculation such that we need only solve a 6×6 matrix:

$$10^5\begin{bmatrix} 7.2 & 0 & 0 & 0 & -1.49 & -1.49 \\ 0 & 7.2 & 0 & -4.22 & -1.49 & -1.49 \\ 0 & 0 & 8.44 & 0 & -4.22 & 0 \\ 0 & -4.22 & 0 & 4.22 & 0 & 0 \\ -1.49 & -1.49 & -4.22 & 0 & 5.71 & 1.49 \\ -1.49 & -1.49 & 0 & 0 & 1.49 & 1.49 \end{bmatrix}\begin{Bmatrix} U_{2X} \\ U_{2Y} \\ U_{4X} \\ U_{4Y} \\ U_{5X} \\ U_{5Y} \end{Bmatrix} = \begin{Bmatrix} 0 \\ 0 \\ 0 \\ -500 \\ 0 \\ -500 \end{Bmatrix}$$

Solution Phase

6. *Solve a system of algebraic equations simultaneously.*
Solving the above matrix for the unknown displacements yields $U_{2X} = -0.00355$ in, $U_{2Y} = -0.01026$ in, $U_{4X} = 0.00118$ in, $U_{4Y} = -0.0114$ in, $U_{5X} = 0.00240$ in, and $U_{5Y} = -0.0195$ in. Thus, the global displacement matrix is

$$\begin{Bmatrix} U_{1X} \\ U_{1Y} \\ U_{2X} \\ U_{2Y} \\ U_{3X} \\ U_{3Y} \\ U_{4X} \\ U_{4Y} \\ U_{5X} \\ U_{5Y} \end{Bmatrix} = \begin{Bmatrix} 0 \\ 0 \\ -0.00355 \\ -0.01026 \\ 0 \\ 0 \\ 0.00118 \\ -0.0114 \\ 0.00240 \\ -0.0195 \end{Bmatrix} \text{in.}$$

Recognize that the displacements of the nodes are given with respect to the global coordinate system.

Postprocessing Phase

7. *Obtain other information.*

Reaction Forces As discussed in Chapter 1, the reaction forces can be computed from

$$\{\mathbf{R}\} = [\mathbf{K}]^{(G)}\{\mathbf{U}\} - \{\mathbf{F}\}$$

such that

$$\begin{Bmatrix} R_{1X} \\ R_{1Y} \\ R_{2X} \\ R_{2Y} \\ R_{3X} \\ R_{3Y} \\ R_{4X} \\ R_{4Y} \\ R_{5X} \\ R_{5Y} \end{Bmatrix} = 10^5 \begin{bmatrix} 4.22 & 0 & -4.22 & 0 & 0 & 0 & 0 & 0 & 0 & 0 \\ 0 & 0 & 0 & 0 & 0 & 0 & 0 & 0 & 0 & 0 \\ -4.22 & 0 & 7.2 & 0 & -1.49 & 1.49 & 0 & 0 & -1.49 & 1.49 \\ 0 & 0 & 0 & 7.2 & 1.49 & -1.49 & 0 & -4.22 & -1.49 & -1.49 \\ 0 & 0 & -1.49 & 1.49 & 5.71 & -1.49 & -4.22 & 0 & 0 & 0 \\ 0 & 0 & 1.49 & -1.49 & -1.49 & 1.49 & 0 & 0 & 0 & 0 \\ 0 & 0 & 0 & 0 & -4.22 & 0 & 8.44 & 0 & -4.22 & 0 \\ 0 & 0 & 0 & -4.22 & 0 & 0 & 0 & 4.22 & 0 & 0 \\ 0 & 0 & -1.49 & -1.49 & 0 & 0 & -4.22 & 0 & 5.71 & 1.49 \\ 0 & 0 & -1.49 & -1.49 & 0 & 0 & 0 & 0 & 1.49 & 1.49 \end{bmatrix}$$

$$\begin{Bmatrix} 0 \\ 0 \\ -0.00355 \\ -0.01026 \\ 0 \\ 0 \\ 0.00118 \\ -0.0114 \\ 0.00240 \\ -0.0195 \end{Bmatrix} - \begin{Bmatrix} 0 \\ 0 \\ 0 \\ 0 \\ 0 \\ 0 \\ 0 \\ -500 \\ 0 \\ -500 \end{Bmatrix}$$

Performing matrix operations yields the reaction results

$$\begin{Bmatrix} R_{1X} \\ R_{1Y} \\ R_{2X} \\ R_{2Y} \\ R_{3X} \\ R_{3Y} \\ R_{4X} \\ R_{4Y} \\ R_{5X} \\ R_{5Y} \end{Bmatrix} = \begin{Bmatrix} 1500 \\ 0 \\ 0 \\ 0 \\ -1500 \\ 1000 \\ 0 \\ 0 \\ 0 \\ 0 \end{Bmatrix} \text{lb}$$

Internal Forces and Normal Stresses Now let us compute internal forces and the average normal stresses in each member. The member internal forces f_{ix} and f_{jx}, which are equal and opposite in direction, are

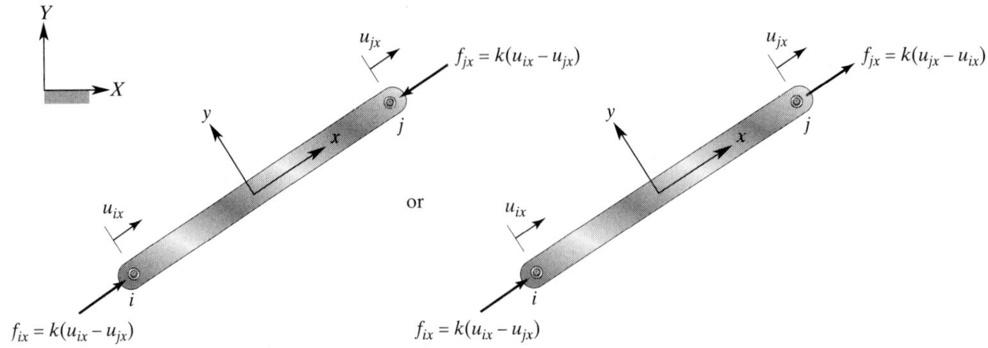

FIGURE 3.11 Internal forces in a truss member.

$$f_{ix} = k(u_{ix} - u_{jx}) \qquad (3.17)$$
$$f_{jx} = k(u_{jx} - u_{ix})$$

Note that the sum of f_{ix} and f_{jx} is zero regardless of which representation of Figure 3.11 we select. However, for the sake of consistency in the forthcoming derivation, we will use the second representation so that f_{ix} and f_{jx} are given in the positive local x-direction. In order to use Eq. (3.17) to compute the internal force in a given element, we must know the displacements of the element's end nodes, u_{ix} and u_{jx}, with respect to the local coordinate system, x, y. Recall that the global displacements are related to the local displacements through a transformation matrix, according to Eq. (3.7), repeated here for convenience,

$$\{\mathbf{U}\} = [\mathbf{T}]\{\mathbf{u}\}$$

and the local displacements in terms of the global displacements:

$$\{\mathbf{u}\} = [\mathbf{T}]^{-1}\{\mathbf{U}\}$$

$$\begin{Bmatrix} u_{ix} \\ u_{iy} \\ u_{jx} \\ u_{jy} \end{Bmatrix} = \begin{bmatrix} \cos\theta & \sin\theta & 0 & 0 \\ -\sin\theta & \cos\theta & 0 & 0 \\ 0 & 0 & \cos\theta & \sin\theta \\ 0 & 0 & -\sin\theta & \cos\theta \end{bmatrix} \begin{bmatrix} U_{iX} \\ U_{iY} \\ U_{jX} \\ U_{jY} \end{bmatrix}$$

Once the internal force in each member is computed, the normal stress in each member can be determined from the equation

$$\sigma = \frac{\text{internal force}}{\mathit{area}} = \frac{f}{A}$$

or alternatively, we can compute the normal stresses from

$$\sigma = \frac{f}{A} = \frac{k(u_{ix} - u_{jx})}{A} = \frac{\frac{AE}{L}(u_{ix} - u_{jx})}{A} = E\left(\frac{u_{ix} - u_{jx}}{L}\right) \qquad (3.18)$$

As an example, let us compute the internal force and the normal stress in element (5). For element (5), $\theta = 45$, $U_{2X} = -0.00355$ in, $U_{2Y} = -0.01026$ in, $U_{5X} = 0.0024$ in, and $U_{5Y} = -0.0195$ in. First, we solve for local displacements of nodes 2 and 5 from the relation

$$\begin{Bmatrix} u_{2x} \\ u_{2y} \\ u_{5x} \\ u_{5y} \end{Bmatrix} = \begin{bmatrix} \cos 45 & \sin 45 & 0 & 0 \\ -\sin 45 & \cos 45 & 0 & 0 \\ 0 & 0 & \cos 45 & \sin 45 \\ 0 & 0 & -\sin 45 & \cos 45 \end{bmatrix} \begin{Bmatrix} -0.00355 \\ -0.01026 \\ 0.00240 \\ -0.01950 \end{Bmatrix}$$

which reveals that $u_{2x} = -0.00976$ in and $u_{5x} = -0.01209$ in. Upon substitution of these values into Eqs. (3.17) and (3.18), the internal force and the normal stress in element (5) are 696 lb and 87 lb/in^2, respectively. Similarly, the internal forces and stresses can be obtained for other elements.

This problem will be revisited later in this chapter and solved using ANSYS. The verification of these results will also be discussed in detail later in Section 3.6.

3.3 SPACE TRUSSES

A three-dimensional truss is often called a space truss. A simple space truss has six members joined together at their ends to form a tetrahedron, as shown in Figure 3.12. We can create more complex structures by adding three new members to a simple truss. This addition should be done in a manner where one end of each new member is connected to a separate existing joint, attaching the other ends of the new members together to form a new joint. This structure is shown in Figure 3.13. As mentioned earlier, members of a truss are generally considered to be two-force members. In the analysis of space trusses, it is assumed that the members are connected together by

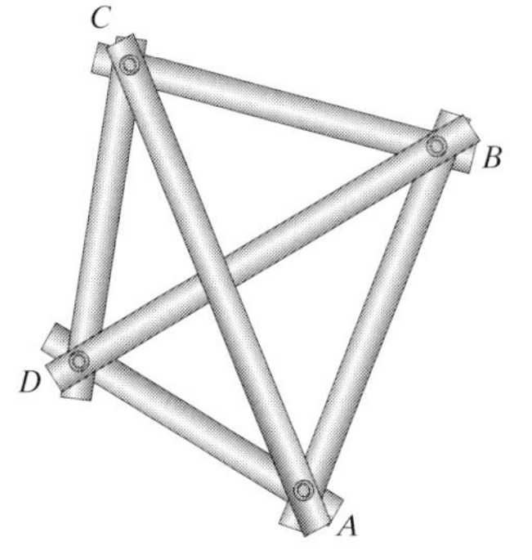

FIGURE 3.12 A simple truss.

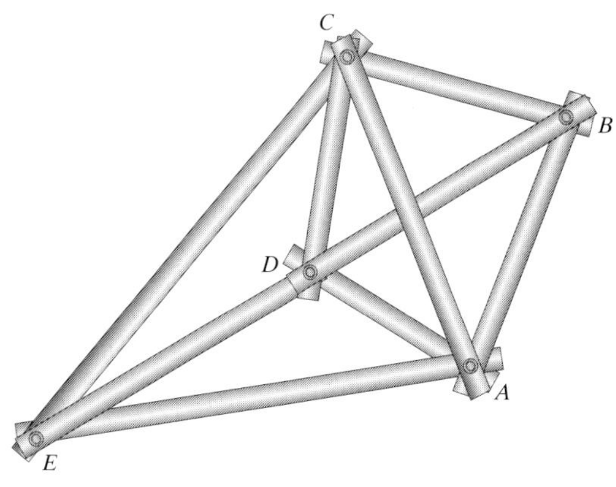

FIGURE 3.13 Addition of new elements to a simple truss to form complex structures.

ball-and-socket joints. It can be shown that as long as the center lines of the adjacent bolted members intersect at a common point, trusses with bolted or welded joints may also be treated under the ball-and-socket joints assumption (negligible bending moments at the joints). Another restriction deals with the assumption that all loads must be applied at the joints. This assumption is true for most situations. As stated earlier, the weights of members are usually negligible compared to the applied loads. However, if the weights of the members are to be considered, then half of the weight of each member is applied to the connecting joints.

Finite element formulation of space trusses is an extension of the analysis of plane trusses. In a space truss, the global displacement of an element is represented by six unknowns, U_{iX}, U_{iY}, U_{iZ}, U_{jX}, U_{jY}, and U_{jZ}, because each node (joint) can move in three directions. Moreover, the angles θ_X, θ_Y, and θ_Z define the orientation of a member with respect to the global coordinate system, as shown in Figure 3.14.

The directional cosines can be written in terms of the difference between the coordinates of nodes j and i of a member and the member's length according to the relationships

$$\cos\theta_X = \frac{X_j - X_i}{L} \tag{3.19}$$

$$\cos\theta_Y = \frac{Y_j - Y_i}{L} \tag{3.20}$$

$$\cos\theta_Z = \frac{Z_j - Z_i}{L} \tag{3.21}$$

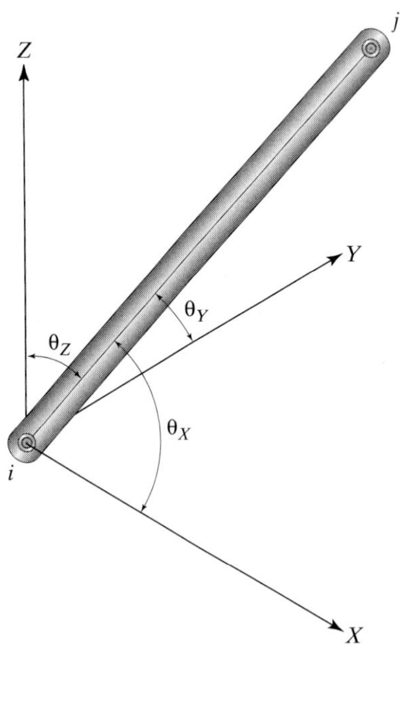

FIGURE 3.14 The angles formed by a member with the X-, Y-, and Z-axis.

where L is the length of the member and is given by

$$L = \sqrt{(X_j - X_i)^2 + (Y_j - Y_i)^2 + (Z_j - Z_i)^2} \tag{3.22}$$

The procedure for obtaining the element stiffness matrix for a space-truss member is identical to the one we followed to derive the two-dimensional truss element stiffness. We start the procedure by relating the global displacements and forces to local displacements and forces through a transformation matrix. We then make use of the two-force-member property of a member. We use a matrix relationship similar to the one given by Eq. (3.14). This relationship leads to the stiffness matrix $[\mathbf{K}]^{(e)}$ for an element. However, it is important to realize that the elemental stiffness matrix for a space-truss element is a 6×6 matrix rather than the 4×4 matrix that we obtained for the two-dimensional truss element. For a space-truss member, the elemental stiffness matrix is

$$[\mathbf{K}]^{(e)} = k \begin{bmatrix} \cos^2\theta_X & \cos\theta_X\cos\theta_Y & \cos\theta_X\cos\theta_Z & -\cos^2\theta_X & -\cos\theta_X\cos\theta_Y & -\cos\theta_X\cos\theta_Z \\ \cos\theta_X\cos\theta_Y & \cos^2\theta_Y & \cos\theta_Y\cos\theta_Z & -\cos\theta_X\cos\theta_Y & -\cos^2\theta_Y & -\cos\theta_Y\cos\theta_Z \\ \cos\theta_X\cos\theta_Z & \cos\theta_Y\cos\theta_Z & \cos^2\theta_Z & -\cos\theta_X\cos\theta_Z & -\cos\theta_Y\cos\theta_Z & -\cos^2\theta_Z \\ -\cos^2\theta_X & -\cos\theta_X\cos\theta_Y & -\cos\theta_X\cos\theta_Z & \cos^2\theta_X & \cos\theta_X\cos\theta_Y & \cos\theta_X\cos\theta_Z \\ -\cos\theta_X\cos\theta_Y & -\cos^2\theta_Y & -\cos\theta_Y\cos\theta_Z & \cos\theta_X\cos\theta_Y & \cos^2\theta_Y & \cos\theta_Y\cos\theta_Z \\ -\cos\theta_X\cos\theta_Z & -\cos\theta_Y\cos\theta_Z & -\cos^2\theta_Z & \cos\theta_X\cos\theta_Z & \cos\theta_Y\cos\theta_Z & \cos^2\theta_Z \end{bmatrix} \tag{3.23}$$

The procedure for the assembly of individual elemental matrices for a space-truss member—applying boundary conditions, loads, and solving for displacements—is exactly identical to the one we followed for a two-dimensional truss.

3.4 OVERVIEW OF THE ANSYS* PROGRAM

Entering ANSYS

This section provides a brief overview of the ANSYS program. More detailed information about how you should go about using ANSYS to model a physical problem is provided in Chapter 8. But for now, enough information will be provided to get you started. One way to enter the ANSYS program is through the ANSYS Launcher, shown in Figure 3.15. The Launcher has a menu that provides the choices you need to run the ANSYS program and other auxiliary programs.

---

*Materials were adapted with permission from ANSYS documents.

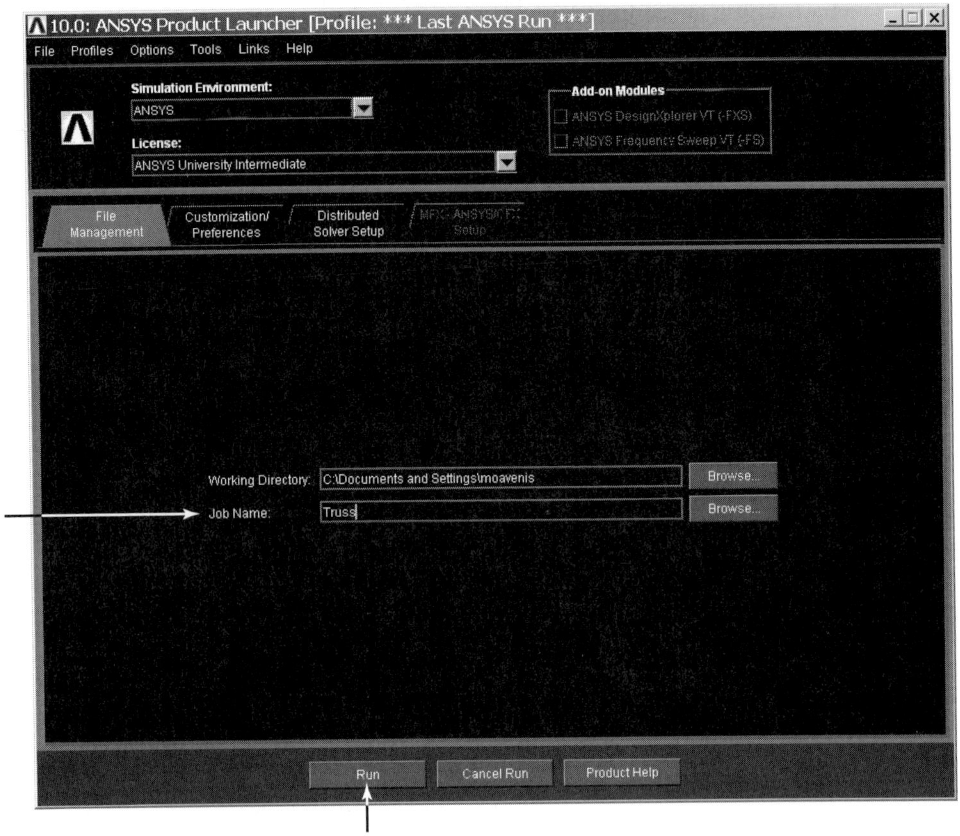

FIGURE 3.15 The ANSYS Product Launcher for a PC version.

When using the Launcher to enter ANSYS, follow these basic steps:

1. Activate the Launcher by issuing the proper command at the system prompt if you are running ANSYS on a UNIX Platform. On a PC platform, go to: **Start → Programs → ANSYS 10.0 → ANSYS Product Launcher®**.
2. Select the **ANSYS** option from the Launcher menu by positioning the cursor of the mouse over it and clicking the left mouse button. This command brings up a dialog box containing interactive entry options.
   a. **Working directory**: This directory is the one in which the ANSYS run will be executed. If the directory displayed is not the one you want to work in, pick the Browse button to the right of the directory name and specify the desired directory.
   b. **Jobname**: This jobname is the one that will be used as the prefix of the file name for all files generated by the ANSYS run. Type the desired jobname in this field of the dialog box.

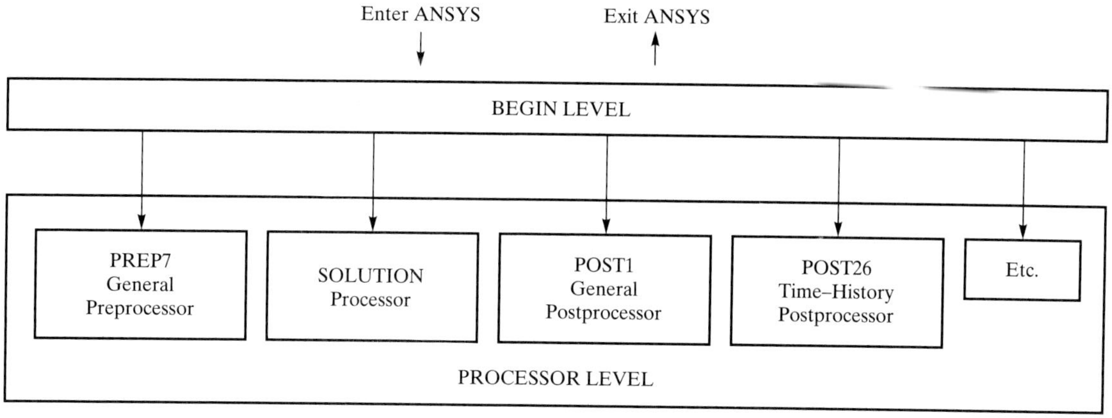

FIGURE 3.16 The organization of ANSYS.

3. Move the mouse cursor over the **Run** button at the bottom of the window and press it. The Graphical User Interface (**GUI**) will then be activated, and you are ready to begin.

Program Organization

Before introducing the GUI, we will discuss some basic concepts of the ANSYS program. The ANSYS program is organized into two levels: (1) the *Begin level* and (2) the *Processor level*. When you first enter the program, you are at the Begin level. From this level, you can enter the ANSYS processors, as shown in Figure 3.16.

You may have more or fewer processors available to you than the ones shown in Figure 3.16. The actual processors available depend on the particular ANSYS product you have. The Begin level acts as a gateway into and out of the ANSYS program. It is also used to access certain global program controls. At the Processor level, several routines (processors) are available; each accomplishes a specific task. Most of your analysis will be done at the Processor level. A typical analysis in ANSYS involves three distinct steps:

1. *Preprocessing*: Using the **PREP7** processor, you provide data such as the geometry, materials, and element type to the program.
2. *Solution*: Using the **Solution** processor, you define the type of analysis, set boundary conditions, apply loads, and initiate finite element solutions.
3. *Postprocessing*: Using **POST1** (for static or steady-state problems) or **POST26** (for transient problems), you review the results of your analysis through graphical displays and tabular listings.

You enter a processor by selecting it from the ANSYS main menu in the GUI. You can move from one processor to another by simply choosing the processor you want from the ANSYS main menu. The next section presents a brief overview of the GUI.

The Graphical User Interface (GUI)

The simplest way to communicate with ANSYS is by using the ANSYS menu system, called the Graphical User Interface. The GUI provides an interface between you and the ANSYS program. The program is internally driven by ANSYS commands. However, by using the GUI, you can perform an analysis with little or no knowledge of ANSYS commands. This process works because each GUI function ultimately produces one or more ANSYS commands that are automatically executed by the program.

Layout of the GUI The ANSYS GUI consists of six main regions, or windows, as shown in Figure 3.17.

A **Utility Menu:** Contains utility functions that are available throughout the ANSYS session, such as file controls, selecting, and graphics controls. You also exit the ANSYS program through this menu.

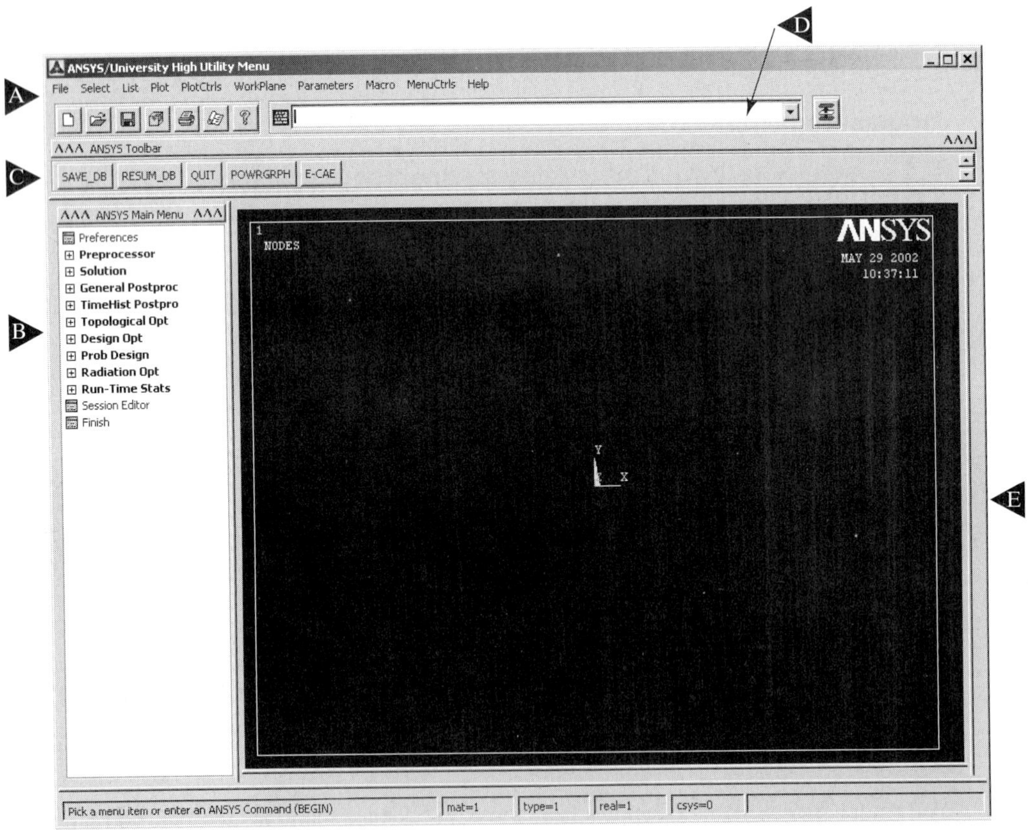

FIGURE 3.17 The ANSYS GUI.

B **Main Menu:** Contains the primary ANSYS functions, organized by processors. These functions include preprocessor, solution, general postprocessor, design optimizer, and so on.

C **Toolbar:** Contains push buttons that execute commonly used ANSYS commands and functions. You may add your own push buttons by defining abbreviations.

D **Input Window:** Allows you to type in commands directly. All previously typed-in commands also appear in this window for easy reference and access.

E **Graphics Window:** A window where graphics displays are drawn.

Output Window: Receives text output from the program. It is usually positioned behind the other windows and can be brought to the front when necessary.

The ANSYS main menu and the ANSYS utility menu, both of which you will use most often, are discussed next.

The Main Menu

The main menu, shown in Figure 3.18(a), contains main ANSYS functions such as preprocessing, solution, and postprocessing.

Each menu topic on the main menu either brings up a submenu or performs an action. The ANSYS main menu has a tree structure. Each menu topic can be expanded to reveal other menu options. The expansion of menu options is indicated by + . You click on the + or the topic name until you reach the desired action. As you reveal other subtopics, the + will turn into −, as shown in Figure 3.18(b). For example, to create a rectangle, you click on **Preprocessor,** then on **Modeling, Create, Areas,** and **Rectangle.** As you can see from Figure 3.18(b), you now have three options to create the rectangle: **By 2 Corners,** or **By Centr&Cornr,** or **By Dimensions.** Note that each time you revealed another subtopic, the + turned into −.

The left mouse button is used to select a topic from the main menu. The submenus in the main menu stay in place until you choose a different menu topic higher up in the hierarchy.

The Utility Menu

The utility menu, shown in Figure 3.19, contains ANSYS utility functions such as file controls, selecting, and graphic controls. Most of these functions are modeless; that is, they can be executed at any time during the ANSYS session. The modeless nature of the utility menu greatly enhances the productivity and user friendliness of the GUI.

Each menu topic on the utility menu activates a pull-down menu of subtopics, which in turn will either cascade to a submenu, indicated by a ▶ after the topic, or perform an action. The symbol to the right of the topic indicates the action:

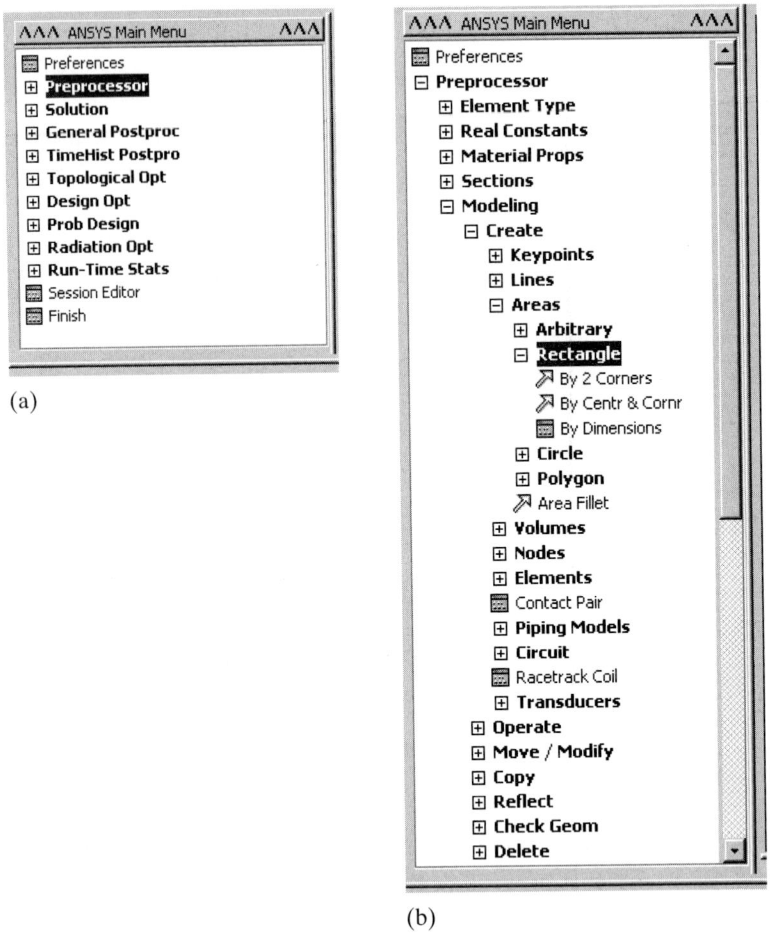

FIGURE 3.18 The main menu.

no symbol for immediate execution of the function

... for a dialog box

+ for a picking menu.

Clicking the left mouse button on a menu topic on the utility menu is used to "pull down" the menu topic. Dragging the cursor of the mouse allows you to move the cursor to the desired subtopic. The menus will disappear when you click on an action subtopic or elsewhere in the GUI.

FIGURE 3.19 The utility menu.

Graphical Picking

In order to use the GUI effectively, it is important to understand graphical picking. You can use the mouse to identify model entities and coordinate locations. There are two types of graphical-picking operations: *locational* picking, where you locate the coordinates of a new point, and *retrieval* picking, where you identify existing entities. For example, creating key points by picking their locations on the working plane is a locational-picking operation, whereas picking already-existing key points to apply a load on them is a retrieval-picking operation.

Whenever you use graphical picking, the GUI brings up a picking menu. Figure 3.20 shows the picking menus for locational and retrieval picking. The features of the picking menu that are used most frequently in upcoming examples are described in detail below.

B **Picking Mode:** Allows you to pick or unpick a location or entity. You can use either these toggle buttons or the right mouse button to switch between pick and unpick modes. The mouse pointer is an up arrow for picking and a down arrow for unpicking. For retrieval picking, you also have the option to choose from single pick, box, circle, and polygon mode.

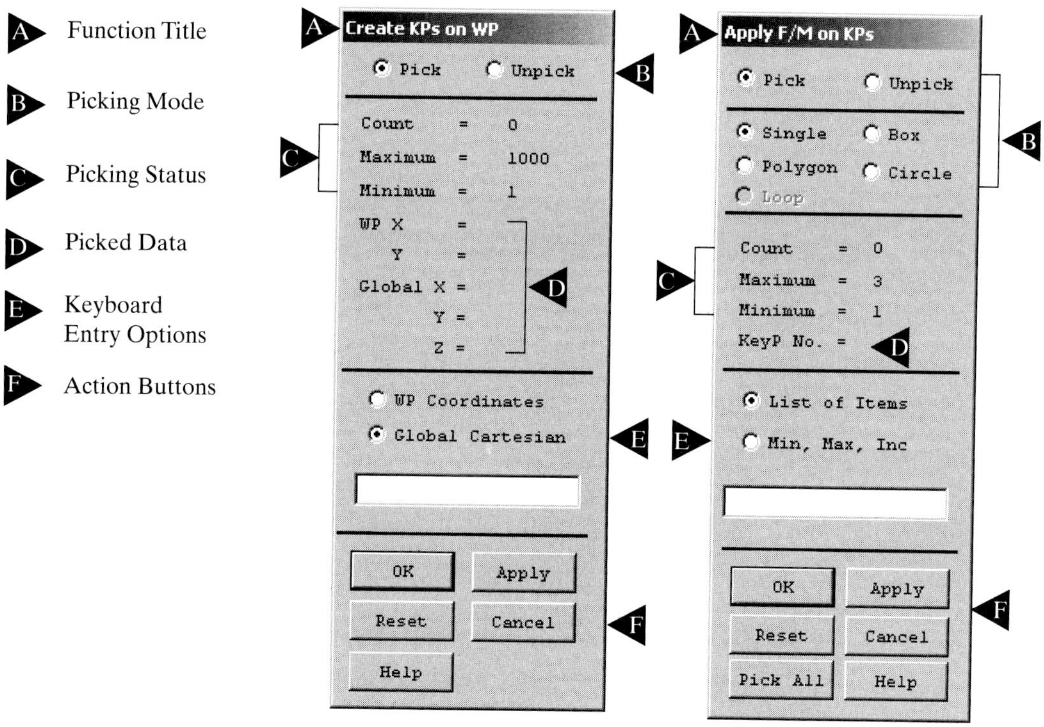

FIGURE 3.20 Picking menu for locational and retrieval picking.

Picked Data: Shows information about the item being picked. For locational picking, the working plane and global Cartesian coordinates of the point are shown. For retrieval picking, this area shows the entity number. You can see this data by pressing the mouse button and dragging the cursor of the mouse into the graphics area. This procedure allows you to preview the information before releasing the mouse button and picking the item.

Action Buttons: This area of the menu contains buttons that take certain actions on the picked entities, as follows:

OK: Applies the picked items to execute the function and closes the picking menu.

Apply: Applies the picked items to execute the function.

Reset: Unpicks all picked entities.

Cancel: Cancels the function and closes the picking menu.

Pick All: Picks all entities available for retrieval picking only.

Help: Brings up help information for the function being performed.

Mouse-Button Assignments for Picking A summary of the mouse-button assignments used during a picking operation is given below:

The left button picks or unpicks the entity or location closest to the cursor of the mouse. Pressing the left mouse button and dragging the cursor of the mouse allows you to preview the items being picked or unpicked.

The middle button applies the picked items to execute the function. Its function is the same as that of the **Apply** button on the picking menu.

The right button toggles between pick and unpick mode. Its function is the same as that of the toggle buttons on the picking menu.

The Help System

The ANSYS help system gives you information for virtually any component in the GUI and any ANSYS command or concept. It can be accessed within the GUI via the help topic on the utility menu or by pressing the help button from within a dialog box. You can access a help topic by choosing from a manual's table of contents or index. Other features of the help system include hypertext links, word search, and the ability to print out help topics. An in-depth explanation of the capabilities and the organization of the ANSYS program is offered in Chapter 8.

3.5 EXAMPLES USING ANSYS

In this section, ANSYS is used to solve truss problems. ANSYS offers two types of elements for the analysis of trusses: **LINK1** and **LINK8.** A two-dimensional spar, called

LINK1, with two nodes and two degrees of freedom (U_X, U_Y) at each node is commonly used to analyze plane truss problems. Input data must include node locations, crosssectional area of the member, and modulus of elasticity. If a member is prestressed, then the initial strain should be included in the input data as well. As we learned previously in our discussion on the theory of truss element, we cannot apply surface loads to this element; thus, all loads must be applied directly at the nodes. To analyze space-truss problems, ANSYS offers a three-dimensional spar element. This element, denoted by **LINK8,** offers three degrees of freedom (U_X, U_Y, U_Z) at each node. Required input data is similar to **LINK1** input information. To get additional information about these elements, run the ANSYS online help menu.

EXAMPLE 3.1 Revisited

Consider the balcony truss from Example 3.1, as shown in the accompanying figure. We are interested in determining the deflection of each joint under the loading shown in the figure. All members are made from Douglas-fir wood with a modulus of elasticity of $E = 1.90 \times 10^6$ lb/in^2 and a cross-sectional area of 8 in^2. We can now analyze this problem using ANSYS.

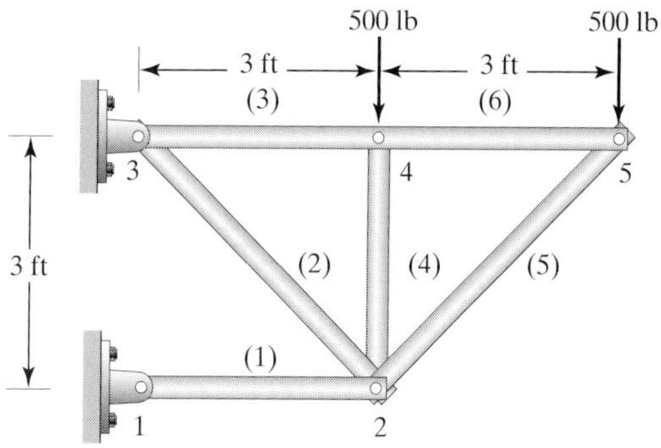

The following steps demonstrate how to create the truss geometry, choose the appropriate element type, apply boundary conditions and loads, and obtain results:

Enter the **ANSYS** program by using the Launcher.

Type **Truss** (or a file name of your choice) in the **Jobname** entry field of the dialog box.

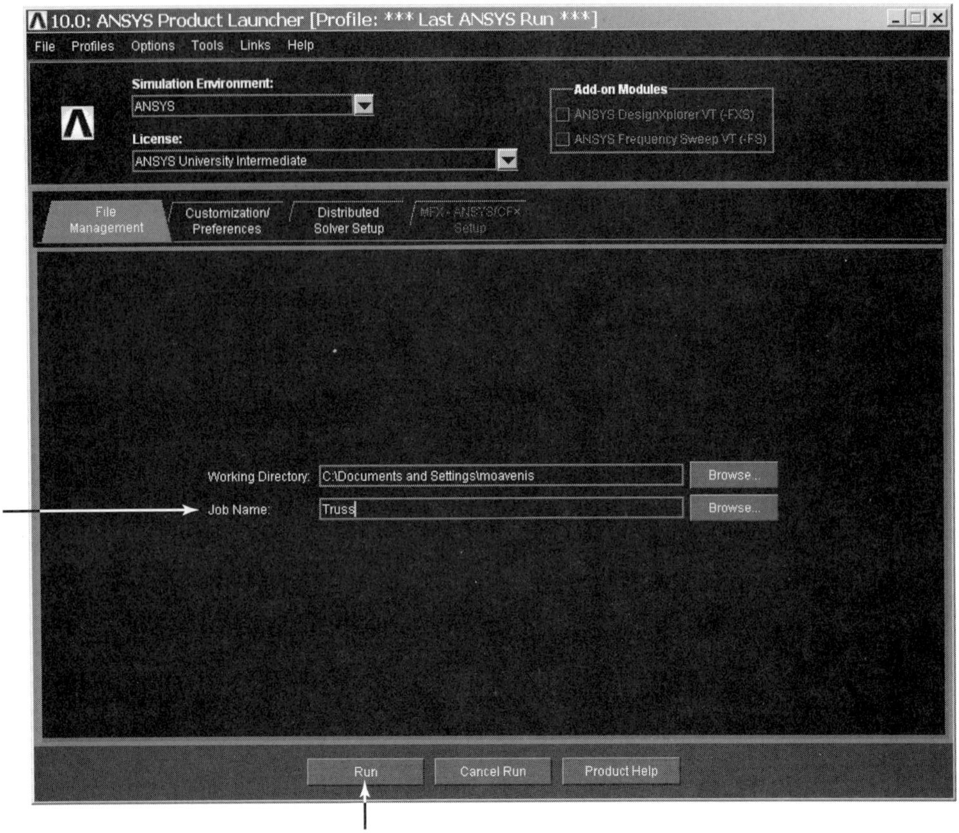

Pick **Run** to start the GUI. Create a title for the problem. This title will appear on ANSYS display windows to provide a simple way of identifying the displays. To create a title, issue the command

utility menu: **File → ChangeTitle** ...

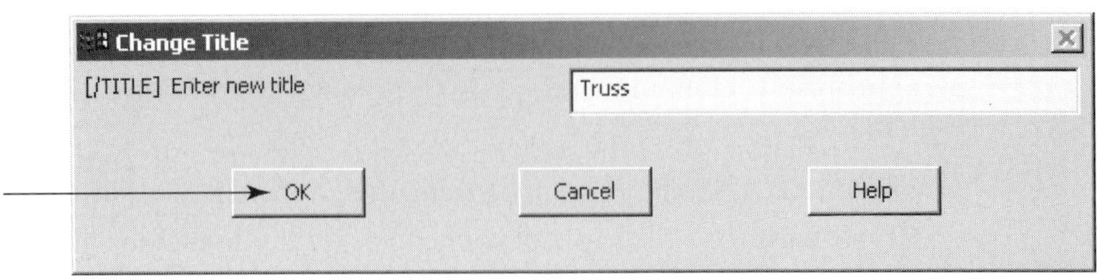

Define the element type and material properties:

main menu: **Preprocessor** → **Element type** → **Add/Edit/Delete**

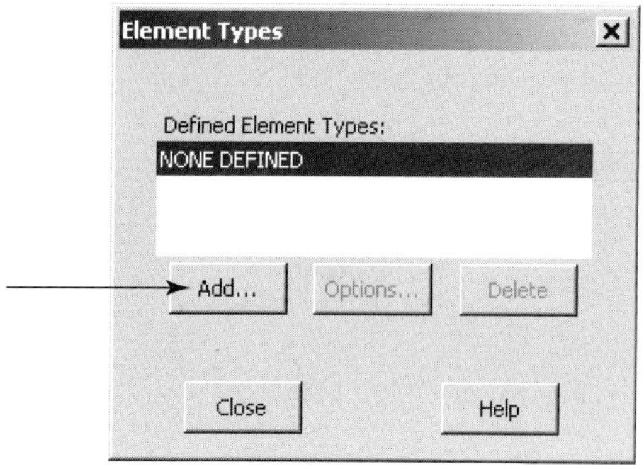

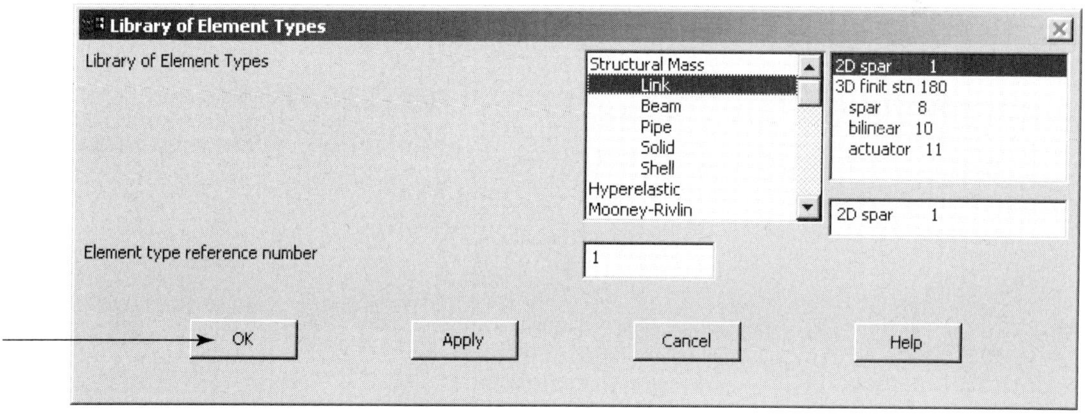

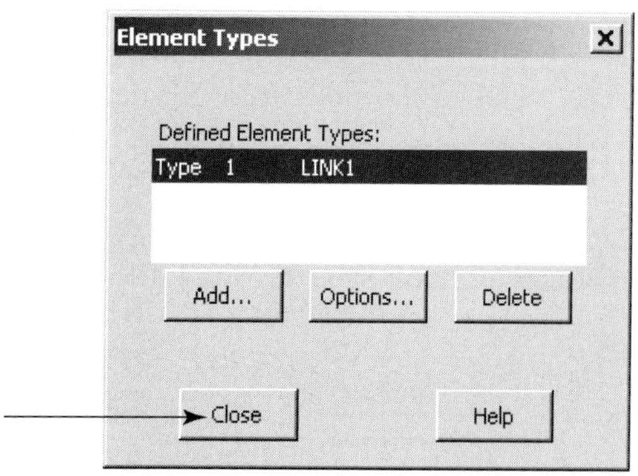

Assign the cross-sectional area of the truss members:

main menu: **Preprocessor → Real Constants → Add/Edit/Delete**

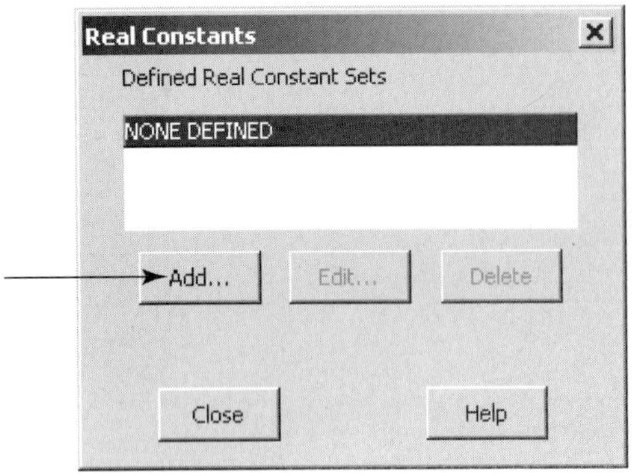

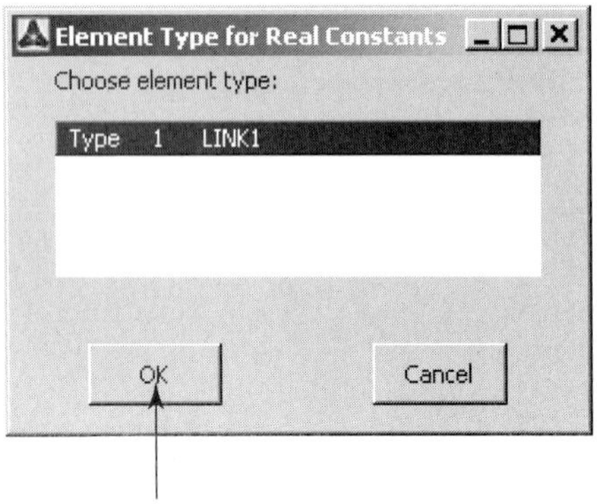

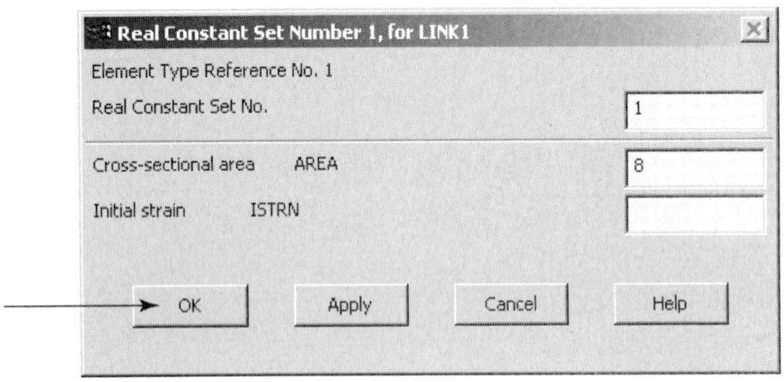

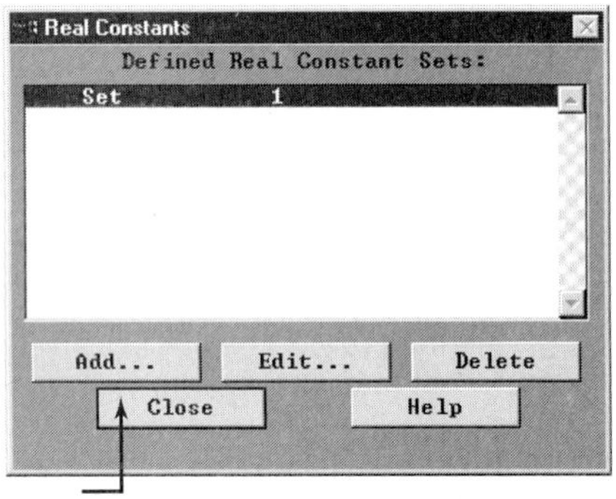

Assign the value of the modulus of elasticity:

main menu: **Preprocessor → Material Props → Material Models → Structural → Linear → Elastic → Isotropic**

Note: Double-click on Structural and then on Linear, Elastic, and Isotropic.

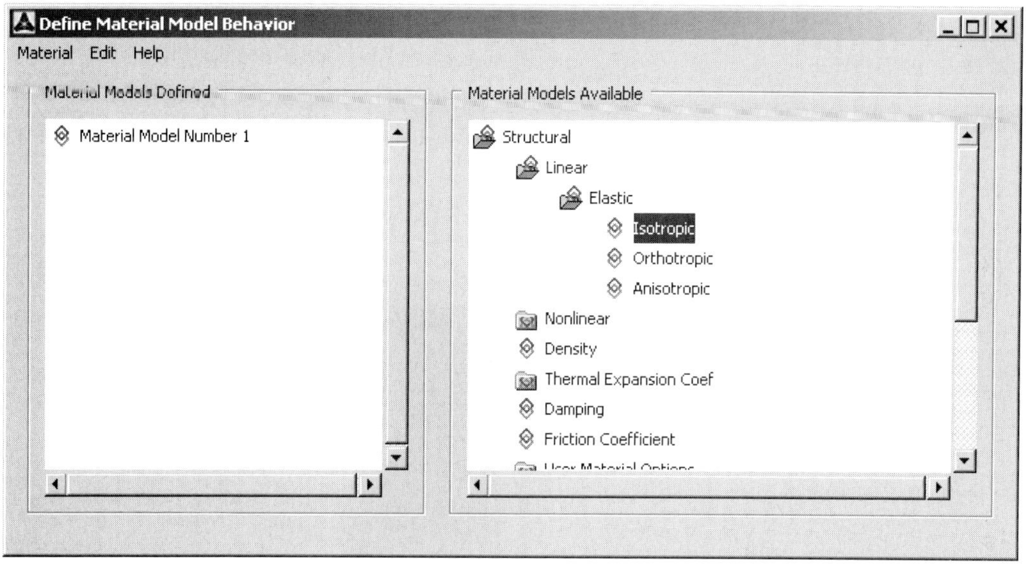

Note:
EX: Modulus of Elasticity
PRXY: Poisson's Ratio
Poisson's Ratio may be omitted for link elements.

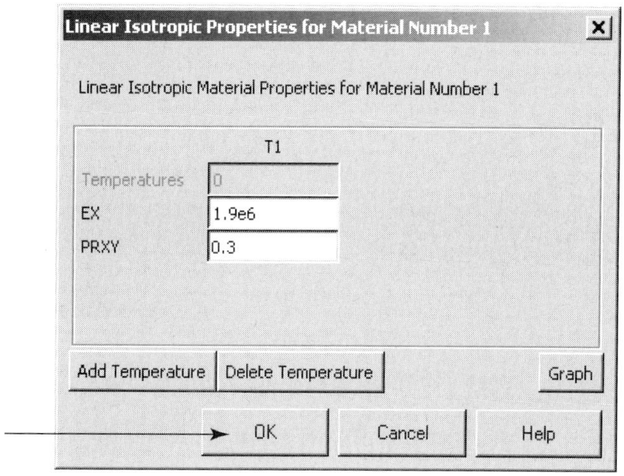

Close the Define Material Model Behavior window.

Save the input data:

ANSYS Toolbar: **SAVE_DB**

Set up the graphics area (i.e., workplane, zoom, etc.):

utility menu: **WorkPlane → WP Settings** . . .

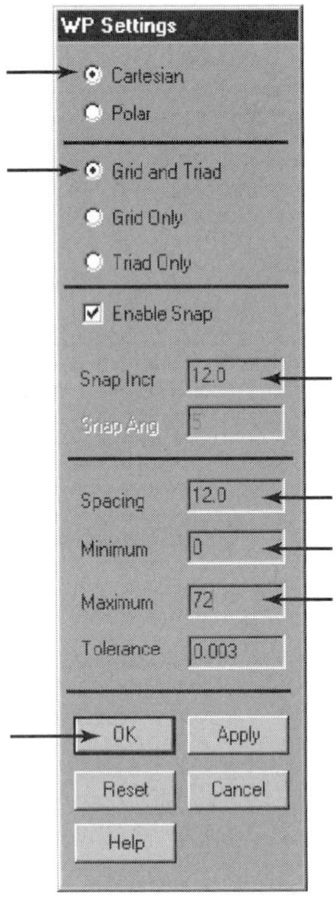

Toggle on the workplane by the following sequence:

utility menu: **Workplane → Display Working Plane**

Bring the workplane to view using the following sequence:

utility menu: **PlotCtrls → Pan, Zoom, Rotate** . . .

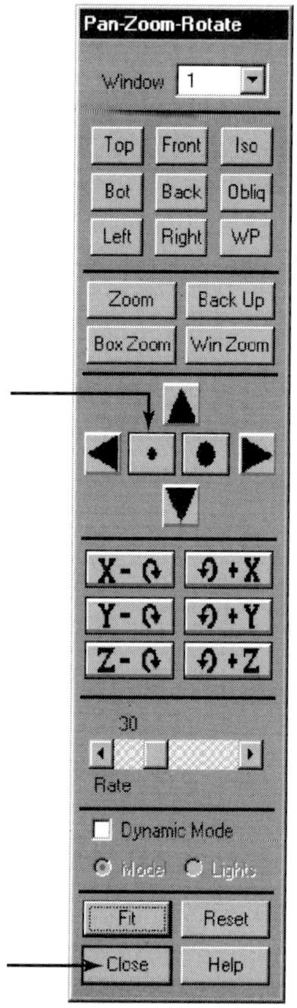

Click on the **small circle** until you bring the workplane to view. You can also use the **arrow** buttons to move the workplane in a desired direction. Then, create nodes by picking points on the workplane:

main menu: **Preprocessor** → **Modeling** → **Create** → **Nodes** → **On Working Plane**

On the workplane, pick the location of joints (nodes) and apply them:

[WP = 0,0]

[WP = 36,0]

[WP = 0,36]

[WP = 36,36]

[WP = 72,36]

OK

You may want to turn off the workplane now and turn on node numbering instead:

utility menu: **Workplane → Display Working Plane**

utility menu: **PlotCtrls → Numbering**...

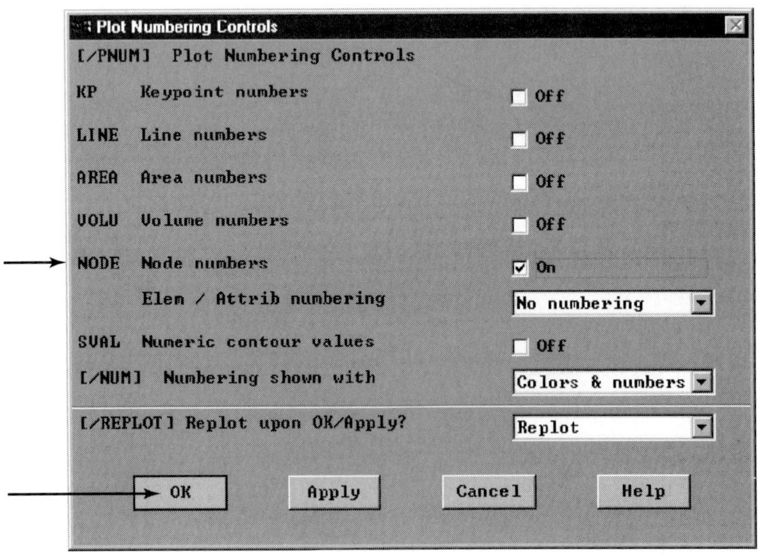

You may want to list nodes at this point in order to check your work:

utility menu: **List → Nodes** . . .

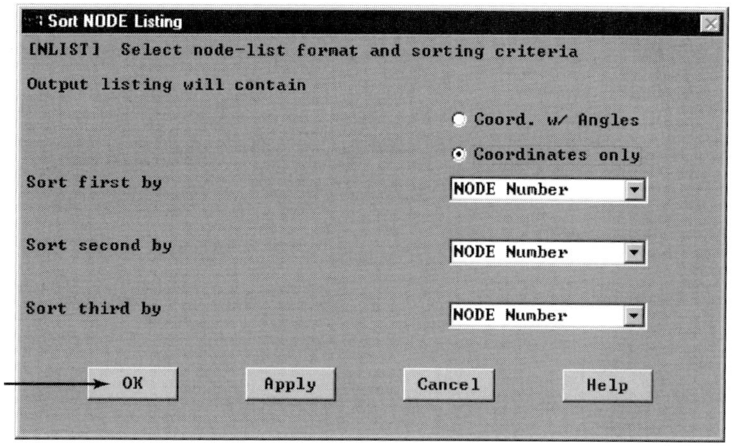

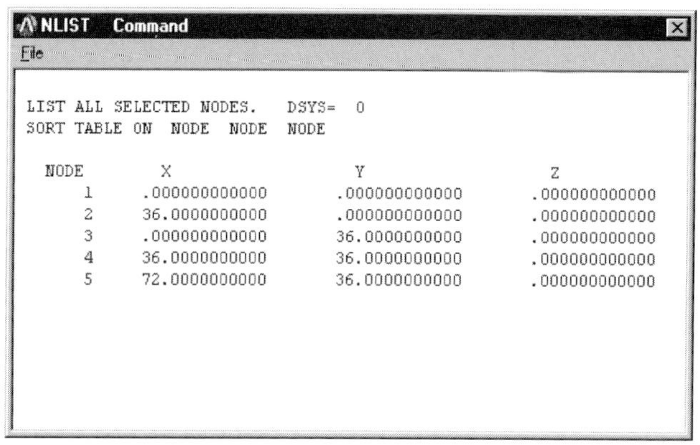

Close

ANSYS Toolbar: **SAVE_DB**

Define elements by picking nodes:

main menu: **Preprocessor → Modeling → Create → Elements → AutoNumbered → Thru Nodes**

[node 1 and then node 2]

[Use the middle button anywhere in the ANSYS graphics window to apply]

[node 2 and then node 3]

[anywhere in the ANSYS graphics window]

[node 3 and then node 4]

[anywhere in the ANSYS graphics window]

[node 2 and then node 4]

[anywhere in the ANSYS graphics window]

[node 2 and then node 5]

[anywhere in the ANSYS graphics window]

[node 4 and then node 5]

[anywhere in the ANSYS graphics window]

OK

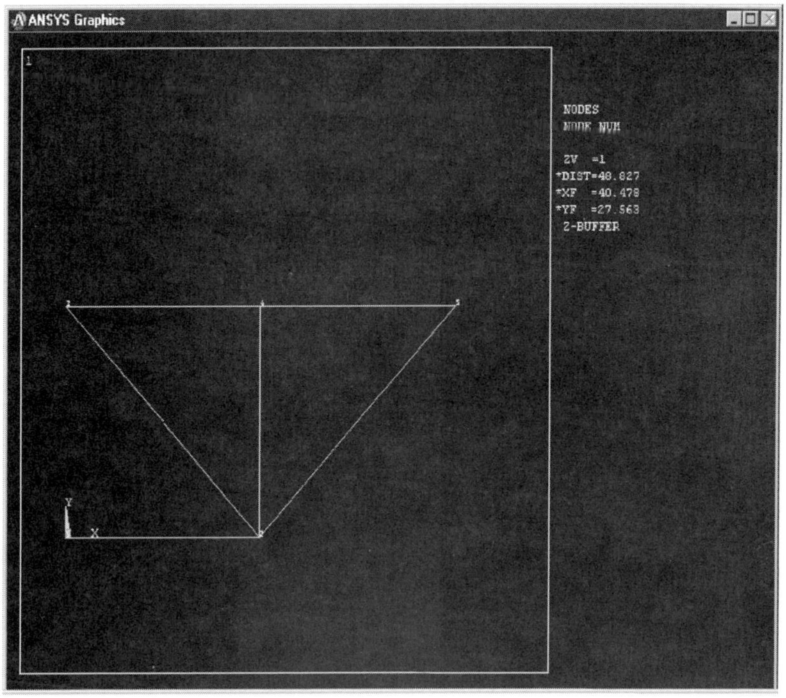

ANSYS Toolbar: **SAVE_DB**

Apply boundary conditions and loads:

main menu: **Solution → Define Loads → Apply → Structural → Displacement → On Nodes**

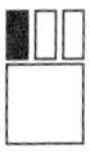 [node 1]

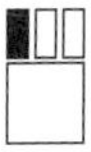 [node 3]

 [anywhere in the ANSYS graphics window]

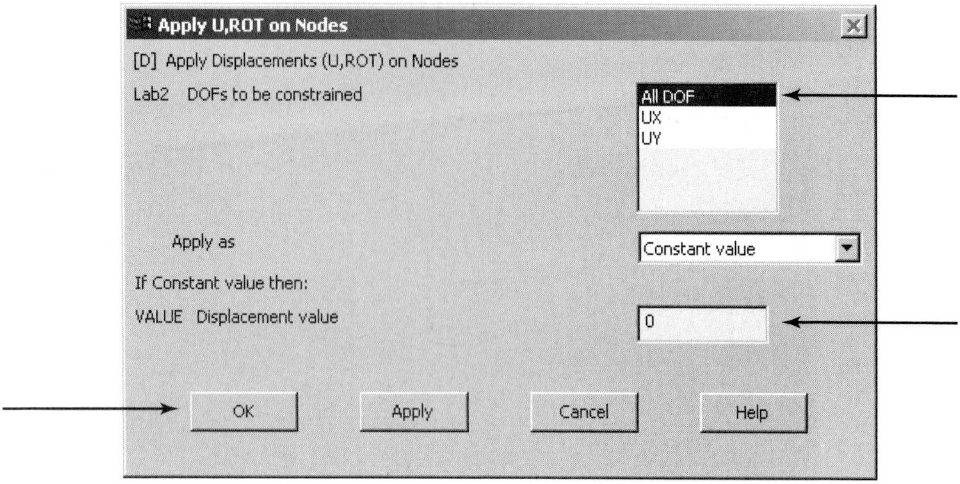

main menu: **Solution → Define Loads → Apply → Structural → Force/Moment → On Nodes**

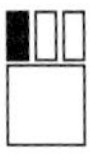 [node 4]

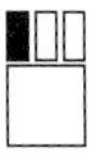 [node 5]

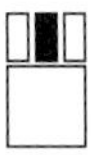 [anywhere in the ANSYS graphics window]

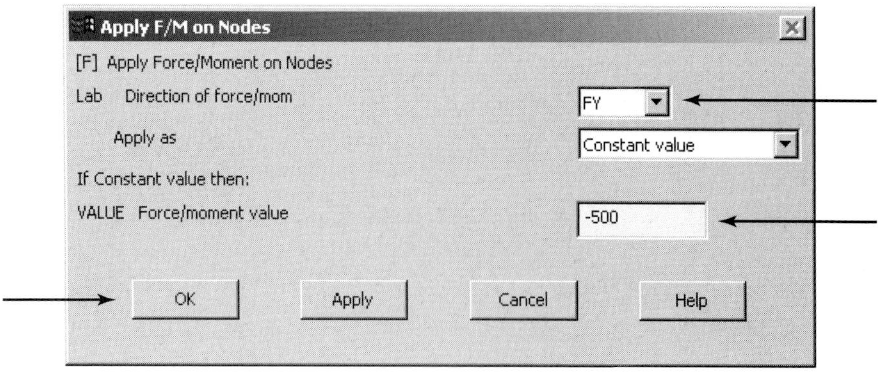

ANSYS Toolbar: **SAVE_DB**

Solve the problem:

main menu: **Solution → Solve → Current LS**

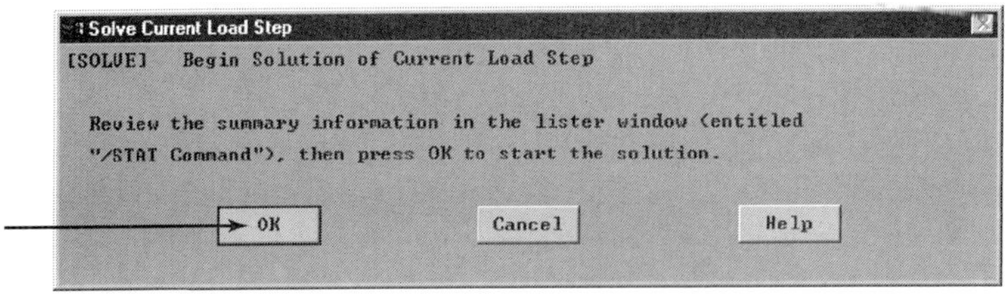

Close (the solution is done!) window.

Close (the /STAT Command) window.

For the postprocessing phase, first plot the deformed shape:

main menu: **General Postproc → Plot Results → Deformed Shape**

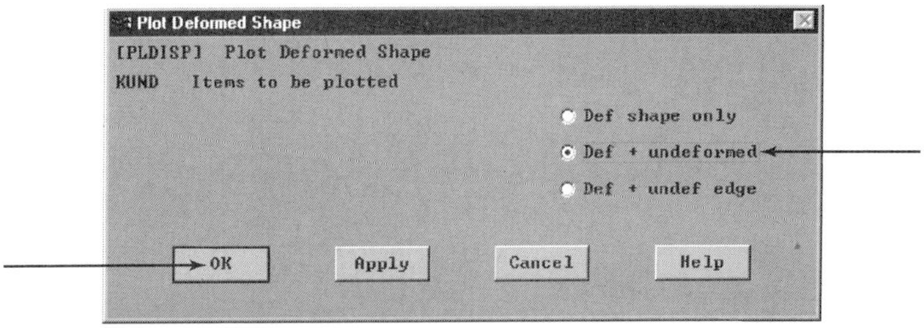

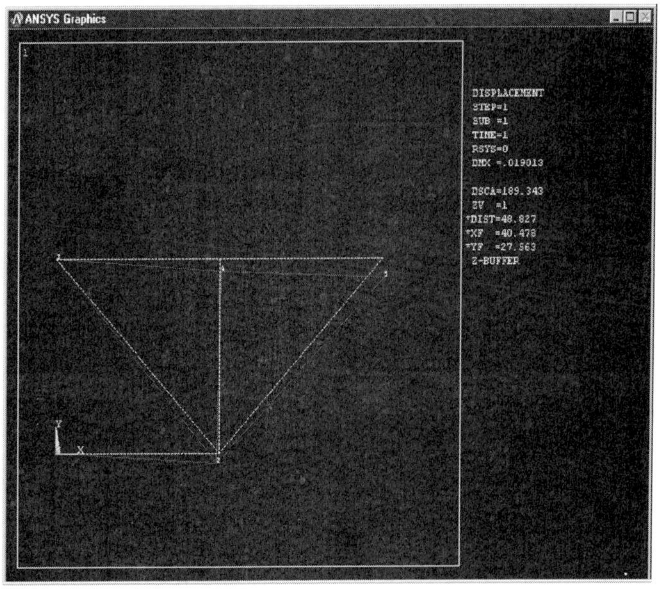

main menu: **General Postproc → List Results → Nodal Solution**

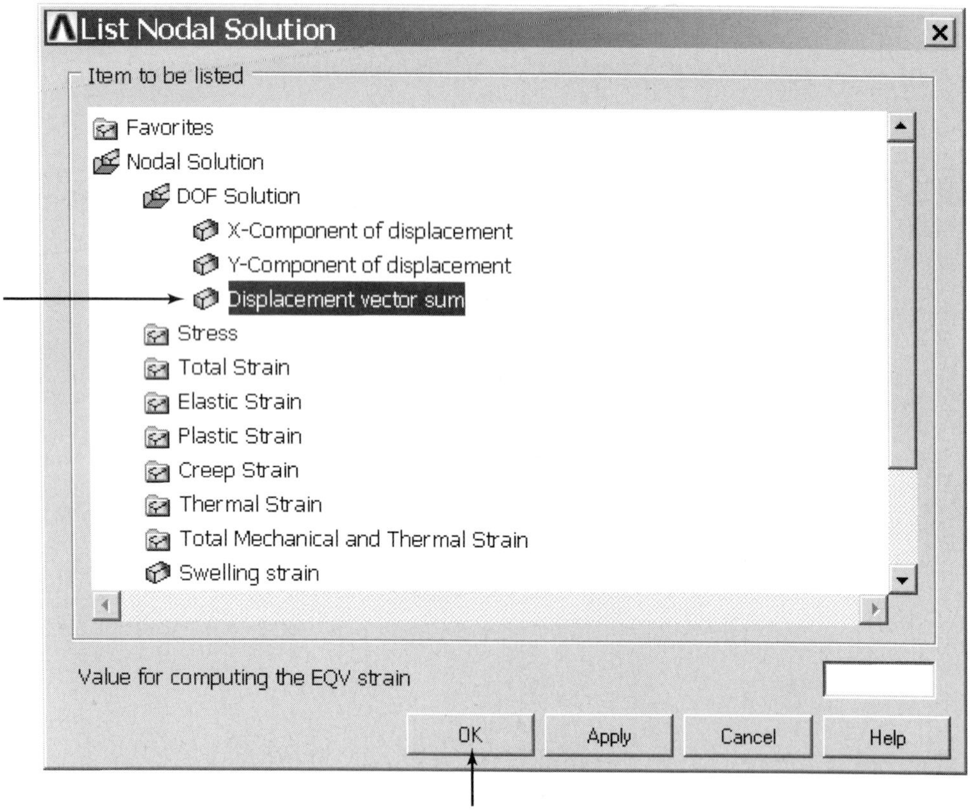

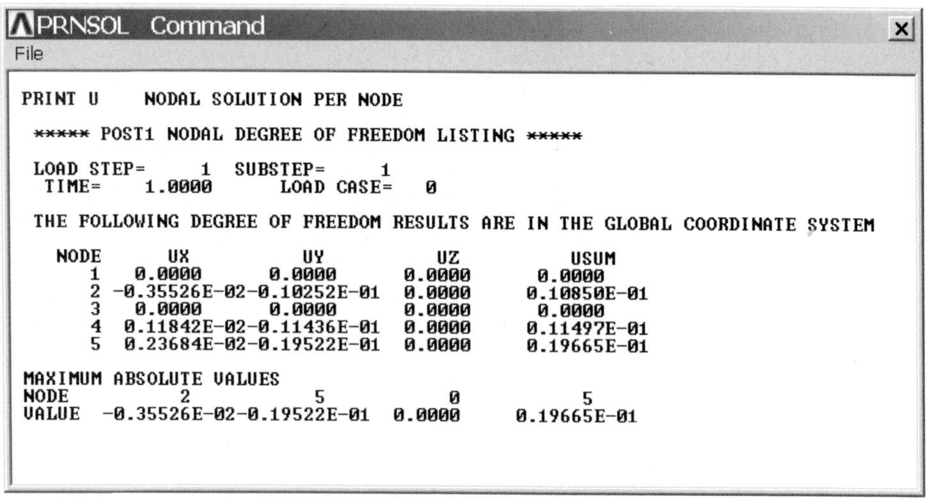
PRNSOL Command

File

```
PRINT U    NODAL SOLUTION PER NODE

 ***** POST1 NODAL DEGREE OF FREEDOM LISTING *****

 LOAD STEP=     1  SUBSTEP=     1
  TIME=    1.0000      LOAD CASE=   0

 THE FOLLOWING DEGREE OF FREEDOM RESULTS ARE IN THE GLOBAL COORDINATE SYSTEM

   NODE      UX          UY          UZ          USUM
      1    0.0000      0.0000      0.0000       0.0000
      2 -0.35526E-02-0.10252E-01   0.0000      0.10850E-01
      3    0.0000      0.0000      0.0000       0.0000
      4  0.11842E-02-0.11436E-01   0.0000      0.11497E-01
      5  0.23684E-02-0.19522E-01   0.0000      0.19665E-01

MAXIMUM ABSOLUTE VALUES
NODE          2           5           0           5
VALUE   -0.35526E-02-0.19522E-01   0.0000     0.19665E-01
```

Close

To review other results, such as axial forces and axial stresses, we must copy these results into element tables. These items are obtained using *item label* and *sequence numbers*, as given in the Table 4.1–4.3 section of the ANSYS elements manual. For truss elements, the values of internal forces and stresses, which ANSYS computes from the nodal displacement results, may be looked up and assigned to user-defined labels. For Example 3.1, we have assigned the internal force, as computed by ANSYS, in each member to a user defined label "Axforce." However, note that ANSYS allows up to eight characters to define such labels. Similarly, the axial stress result for each member is assigned to the label "Axstress." We now run the following sequence:

main menu: **General Postproc → Element Table → Define Table**

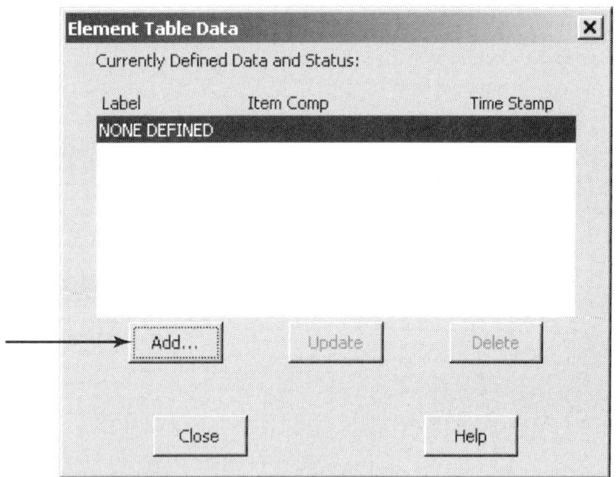

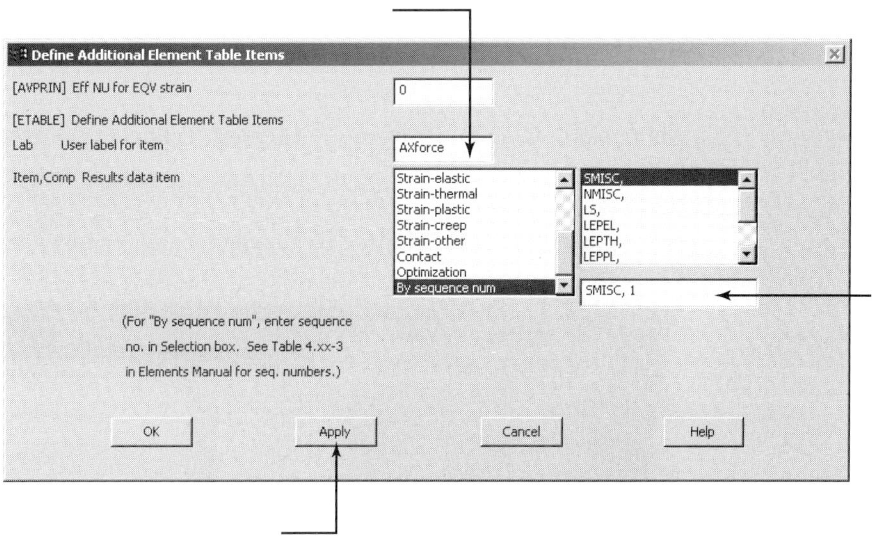

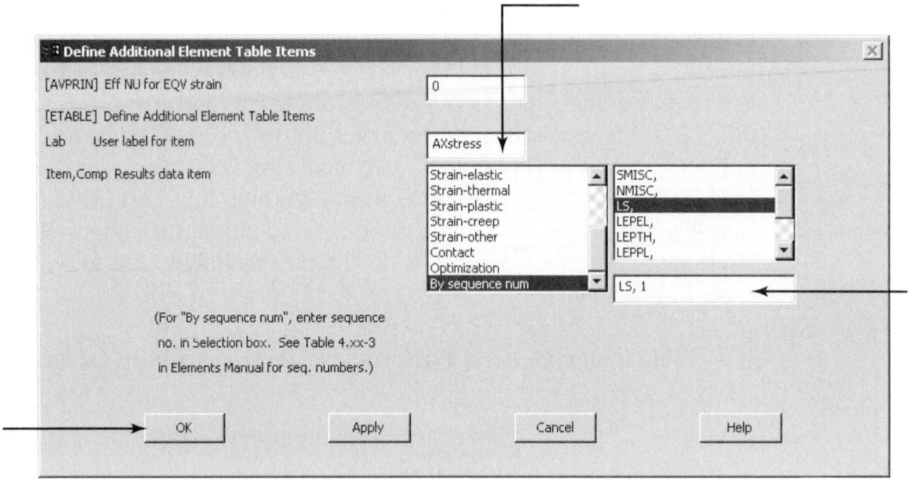

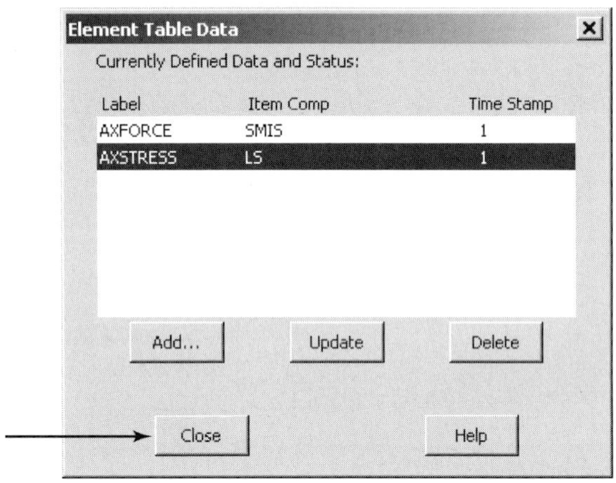

main menu: **General Postproc → Element Table → Plot Element Table**

or

main menu: **General Postproc → Element Table → List Element Table**

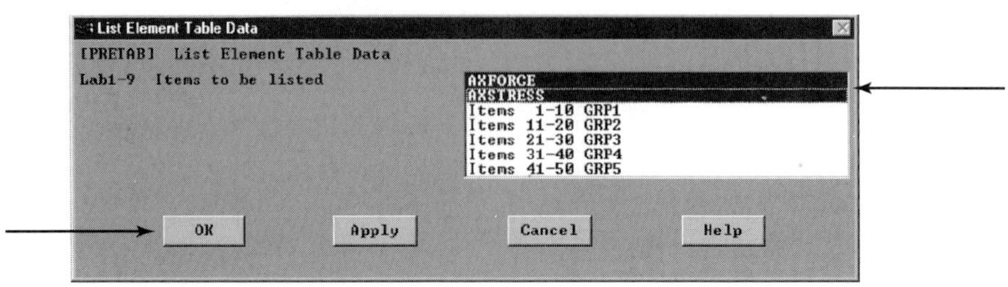

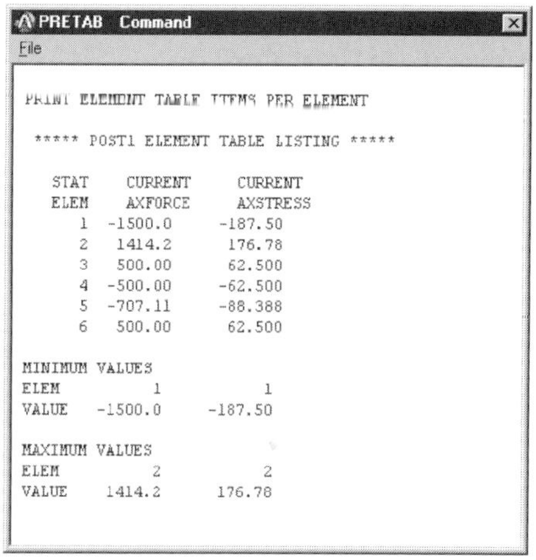

```
PRETAB Command
File

PRINT ELEMENT TABLE ITEMS PER ELEMENT

 ***** POST1 ELEMENT TABLE LISTING *****

   STAT     CURRENT      CURRENT
   ELEM     AXFORCE      AXSTRESS
      1  -1500.0       -187.50
      2   1414.2        176.78
      3   500.00        62.500
      4  -500.00       -62.500
      5  -707.11       -88.388
      6   500.00        62.500

MINIMUM VALUES
ELEM          1            1
VALUE   -1500.0      -187.50

MAXIMUM VALUES
ELEM          2            2
VALUE    1414.2       176.78
```

Close

List reaction solutions:

main menu: **General Postproc → List Results → Reaction Solu**

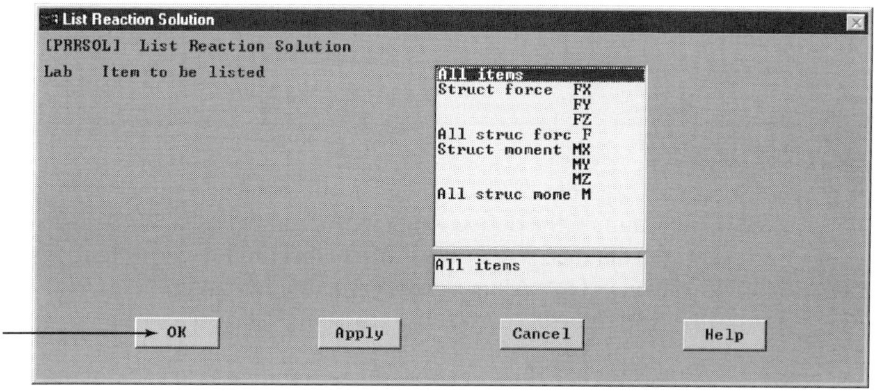

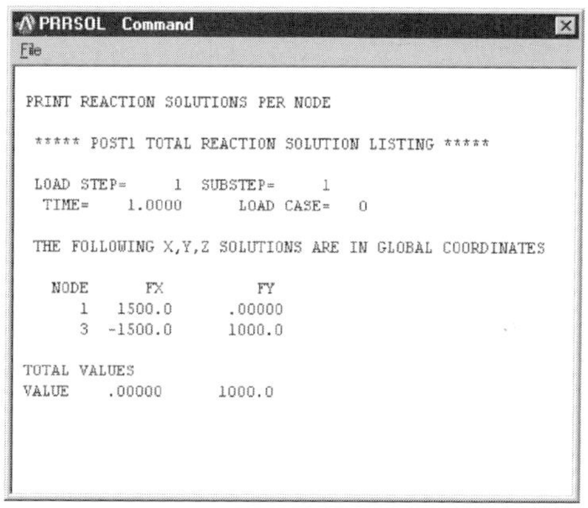

```
PRRSOL Command
File

PRINT REACTION SOLUTIONS PER NODE

 ***** POST1 TOTAL REACTION SOLUTION LISTING *****

 LOAD STEP=     1  SUBSTEP=     1
  TIME=    1.0000      LOAD CASE=   0

 THE FOLLOWING X,Y,Z SOLUTIONS ARE IN GLOBAL COORDINATES

   NODE       FX          FY
      1   1500.0      .00000
      3  -1500.0      1000.0

TOTAL VALUES
VALUE    .00000      1000.0
```

Exit ANSYS and save everything, including element tables and reaction forces:

ANSYS Toolbar: **QUIT**

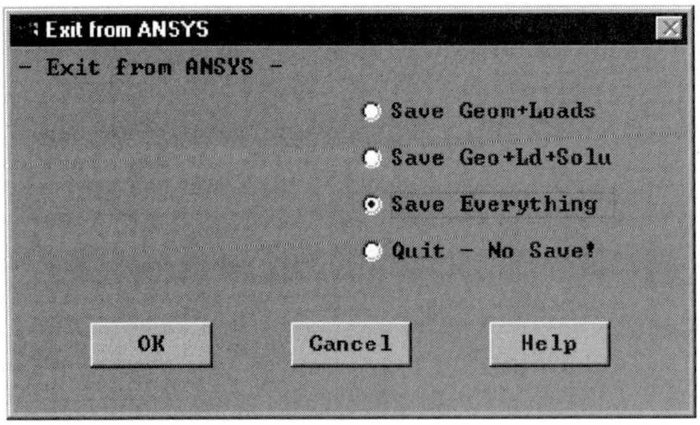

If, for any reason, you need to modify a model, first launch ANSYS and then type the **file name** of the model in the **Jobname** entry field of the **Launcher** dialog box. Then press **Run.** From the File menu, choose **Resume Jobname.DB.** Now you have complete access to your model. You can plot nodes, elements, and so on to make certain that you have chosen the right problem.

EXAMPLE 3.2

Consider the three-dimensional truss shown in the accompanying figure. We are interested in determining the deflection of joint 2 under the loading shown in the figure. The Cartesian coordinates of the joints with respect to the coordinate system shown in the figure are given in feet. All members are made from aluminum with a modulus of elasticity of $E = 10.6 \times 10^6$ lb/in^2 and a cross-sectional area of 1.56 in^2.

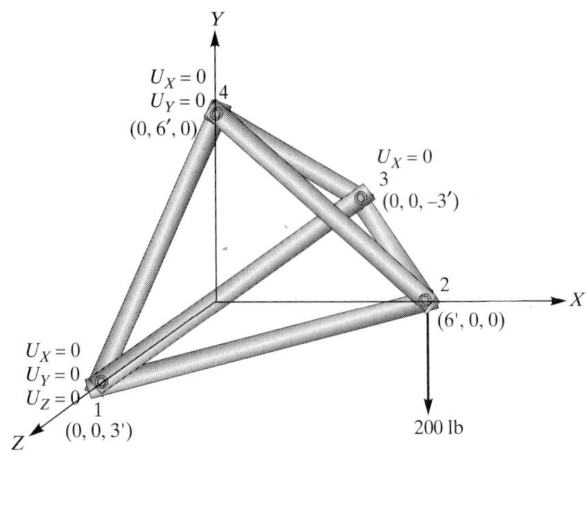

To solve this problem using ANSYS, we employ the following steps:

Enter the **ANSYS** program by using the Launcher.

Type **Truss3D** (or a file name of your choice) in the **Jobname** entry field of the dialog box.

Pick **Run** to start the GUI.

Create a title for the problem. This title will appear on ANSYS display windows to provide a simple way of identifying the displays:

utility menu: **File → Change Title** . . .

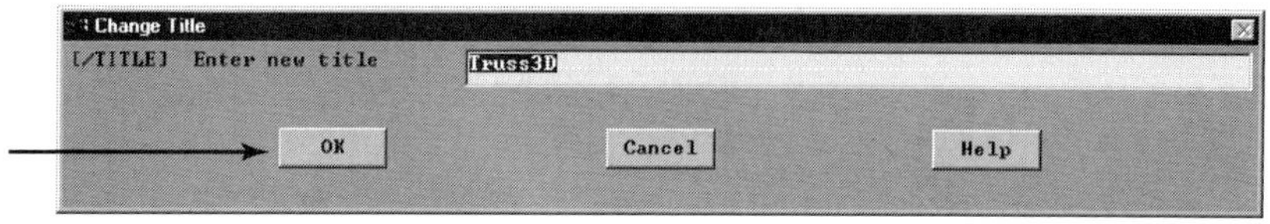

Define the element type and material properties:

main menu: **Preprocessor → Element Type → Add/Edit/Delete**

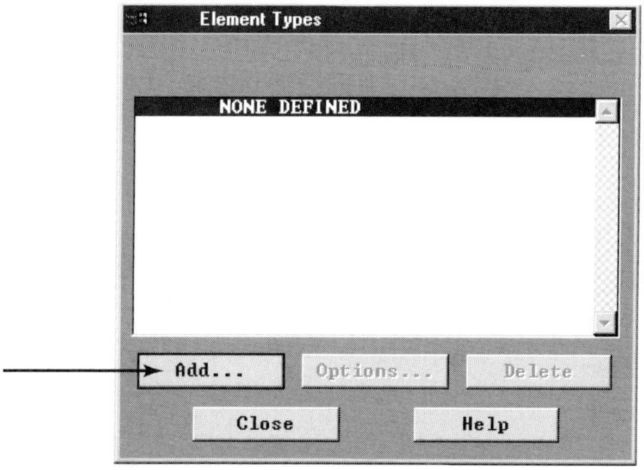

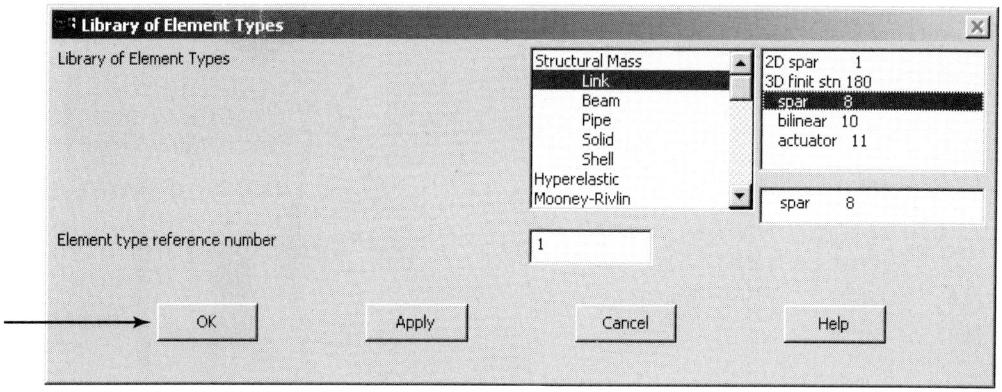

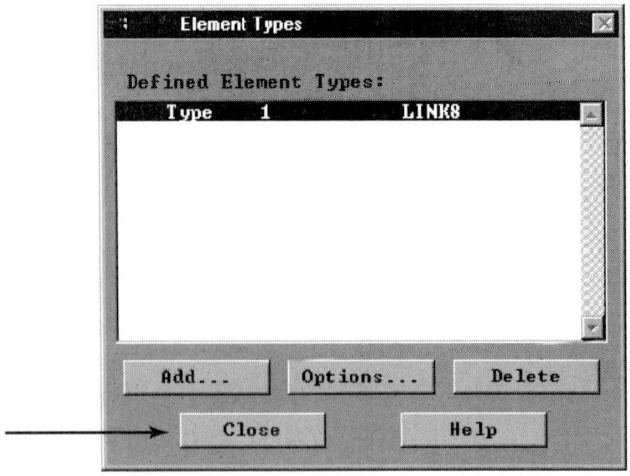

Assign the cross-sectional area of the truss members:

main menu: **Preprocessor → Real Constant → Add/Edit/Delete**

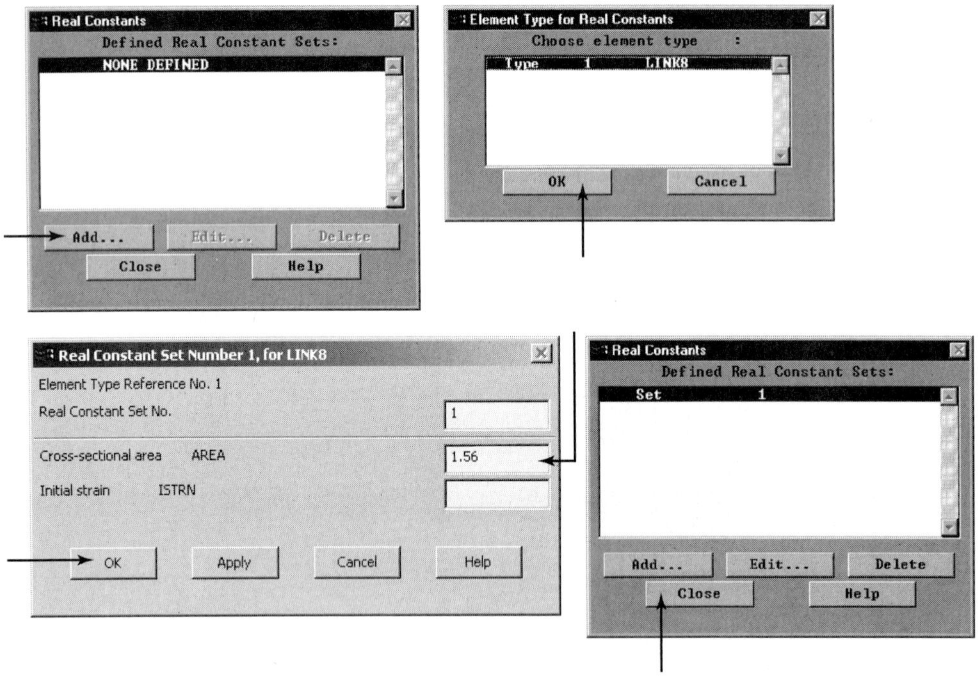

Assign the value of the modulus of elasticity:

main menu: **Preprocessor → Material Props → Material Models → Structural → Linear → Elastic → Isotropic**

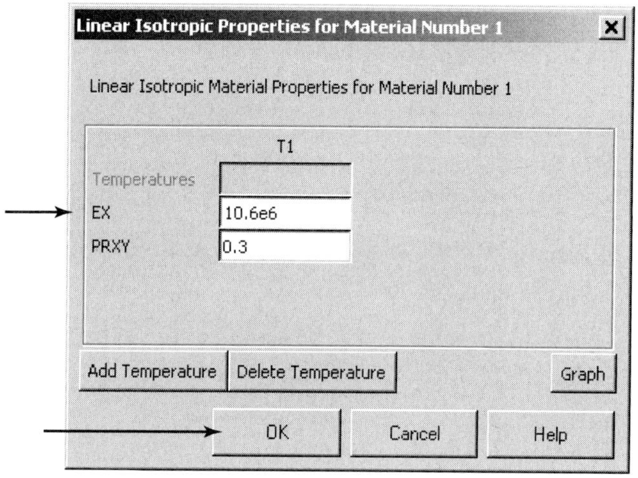

Close the Define Material Model Behavior window.

ANSYS Toolbar: **SAVE_DB**

Create nodes in active coordinate system:

main menu: **Preprocessor → Modeling → Create → Nodes → In Active CS**

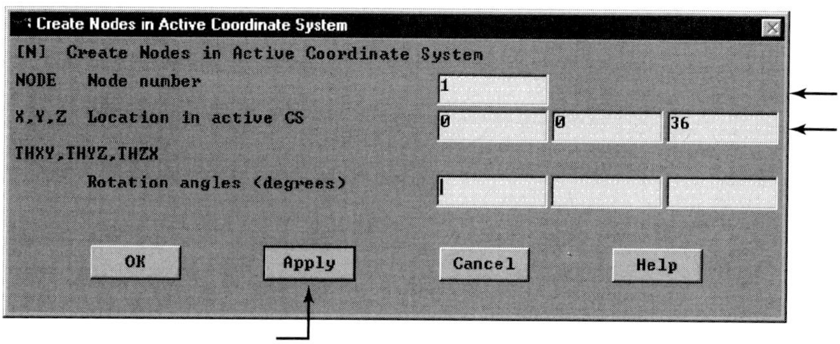

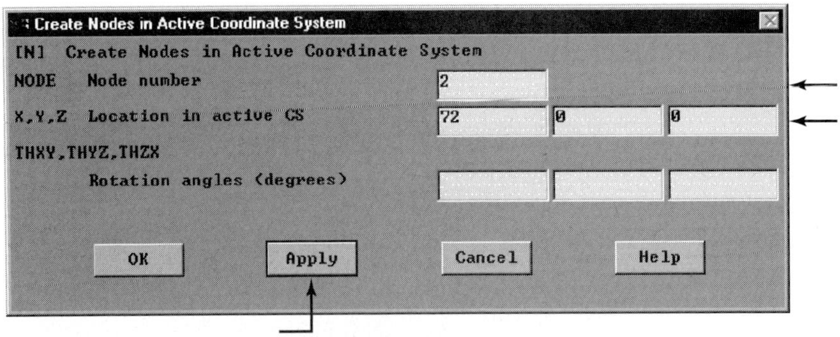

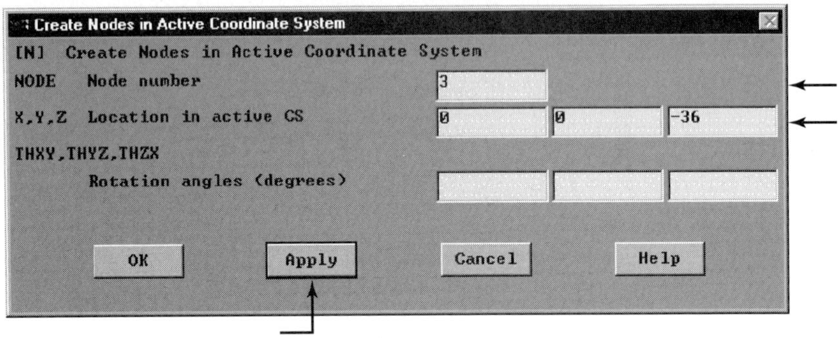

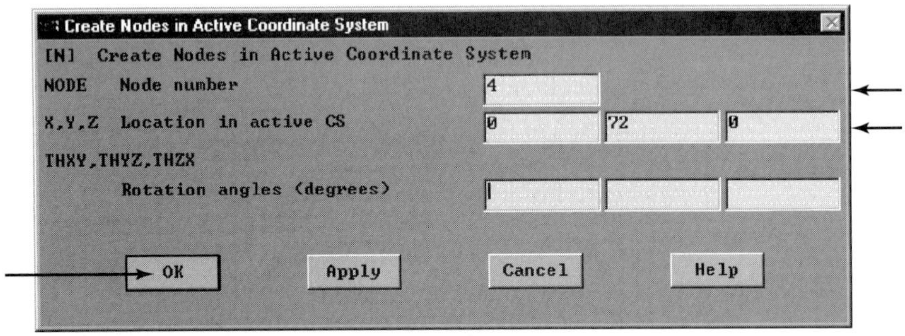

You may want to turn on node numbering:

utility menu: **PlotCtrls** → **Numbering**...

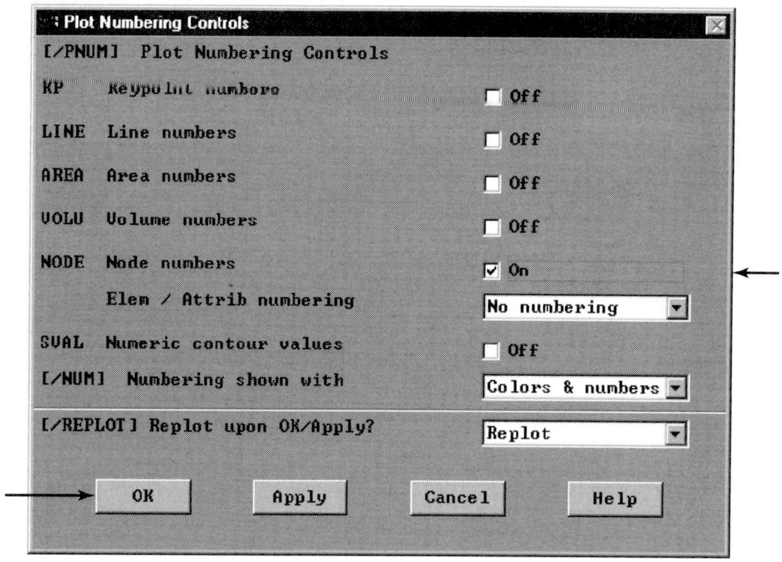

You may want to list nodes at this point in order to check your work:

utility menu: **List → Nodes** . . .

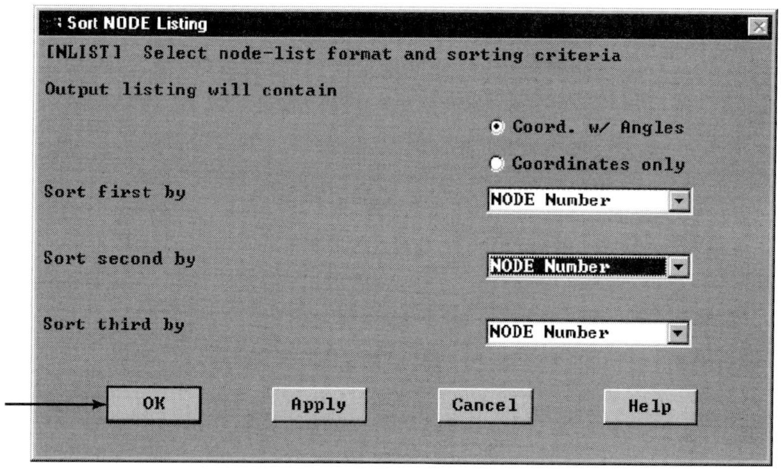

```
NLIST  Command
File

LIST ALL SELECTED NODES.   DSYS=  0
SORT TABLE ON  NODE  NODE  NODE

  NODE       X          Y          Z          THXY    THYZ    THZX
     1   .00000     .00000     36.000          .00     .00     .00
     2   72.000     .00000     .00000          .00     .00     .00
     3   .00000     .00000    -36.000          .00     .00     .00
     4   .00000     72.000     .00000          .00     .00     .00
```

Close

ANSYS Toolbar: **SAVE_DB**

Define elements by picking nodes. But first set the view angle:

utility menu: **PlotCtrls → Pan, Zoom, Rotate** . . .

Select the oblique (**Obliq**) or isometric (**Iso**) viewing.

main menu: **Preprocessor → Modeling → Create → Elements → Auto Numbered → Thru Nodes**

[node 1 and then node 2]

[Use the middle button anywhere in the ANSYS graphics window to apply]

[node 1 and then node 3]

[anywhere in the ANSYS graphics window]

[node 1 and then node 4]

[anywhere in the ANSYS graphics window]

[node 2 and then node 3]

[anywhere in the ANSYS graphics window]

[node 2 and then node 4]

[anywhere in the ANSYS graphics window]

[node 3 and then node 4]

[anywhere in the ANSYS graphics window]

OK

ANSYS Toolbar: **SAVE_DB**

Apply boundary conditions and loads:

main menu: **Solution → Define Loads → Apply → Structural → Displacement → On Nodes**

[node 1]

[node 3]

[node 4]

[anywhere in the ANSYS graphics window]

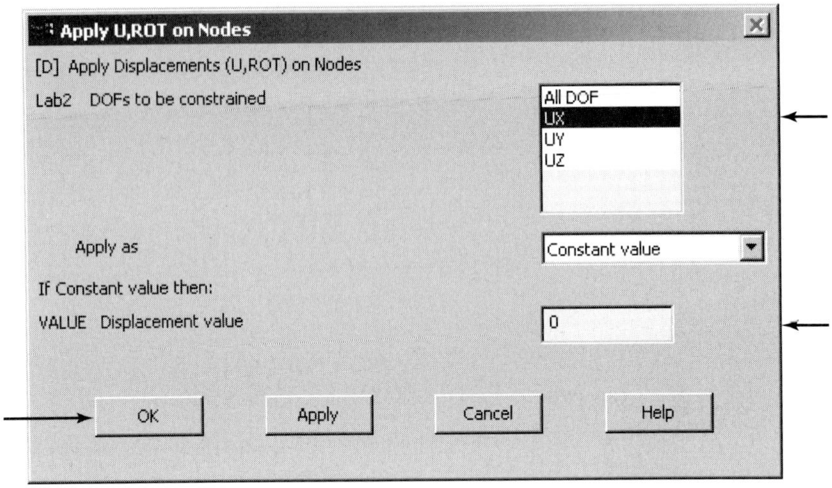

main menu: **Solution → Define Loads → Apply → Structural → Displacement → On Nodes**

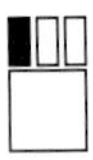

[node 1]

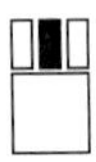

[anywhere in the ANSYS graphics window]

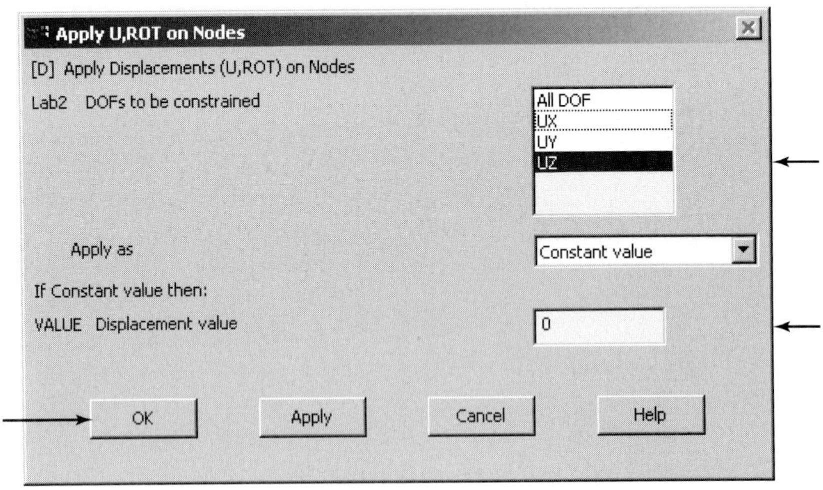

main menu: **Solution → Define Loads → Apply → Structural → Displacement → On Nodes**

[node 1]

[node 4]

[anywhere in the ANSYS graphics window]

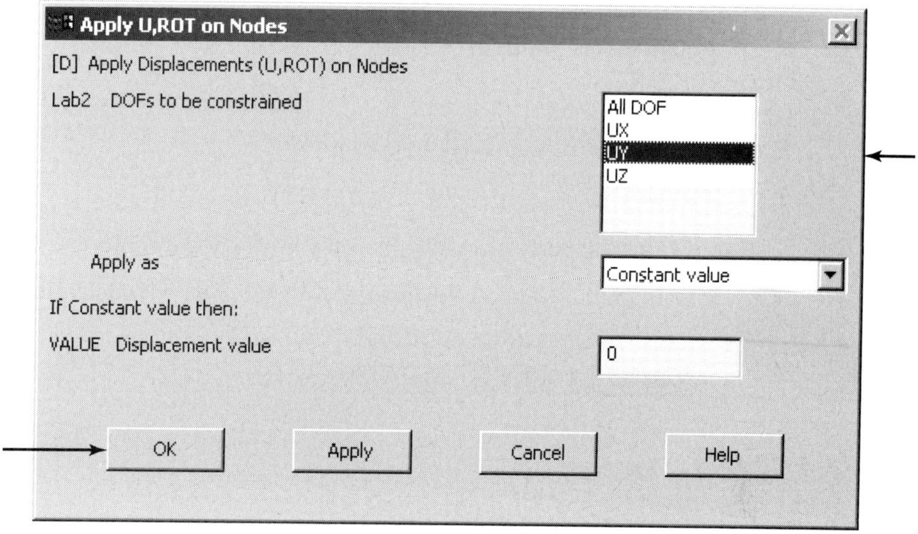

main menu: **Solution → Define Loads → Apply → Structural → Force/Moment → On Nodes**

[node 2]

[anywhere in the ANSYS graphics window]

ANSYS Toolbar: **SAVE_DB**

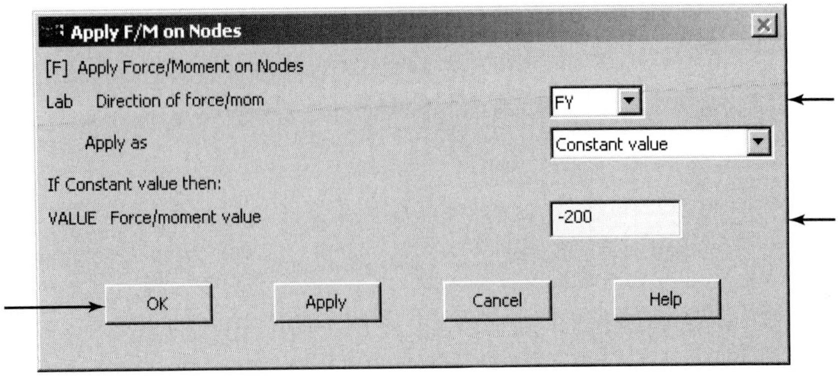

Solve the problem:

main menu: **Solution → Solve → Current LS**

OK

Close (the solution is done!) window.

Close (the /STAT Command) window.

Now we run the postprocessing phase by listing nodal solutions (displacements):

main menu: **General Postproc → List Results → Nodal Solution**

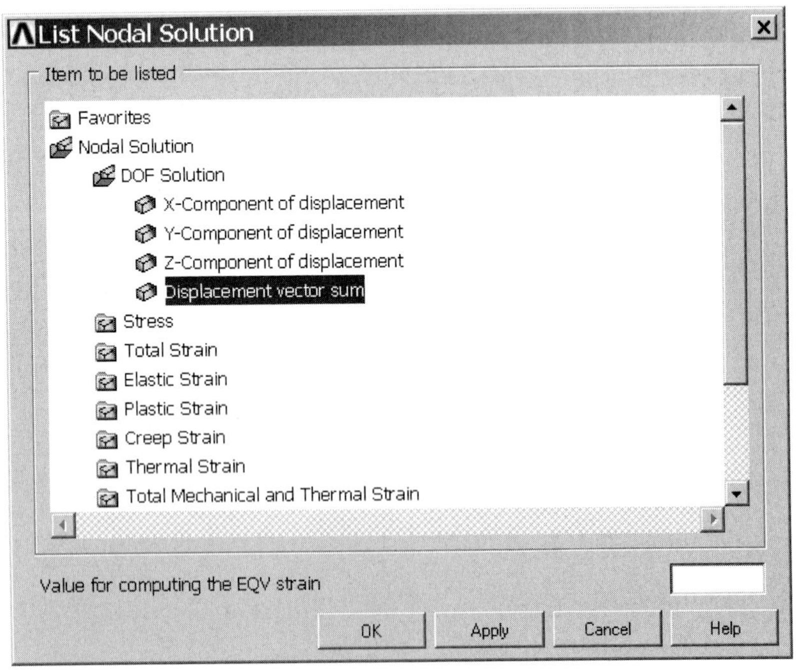

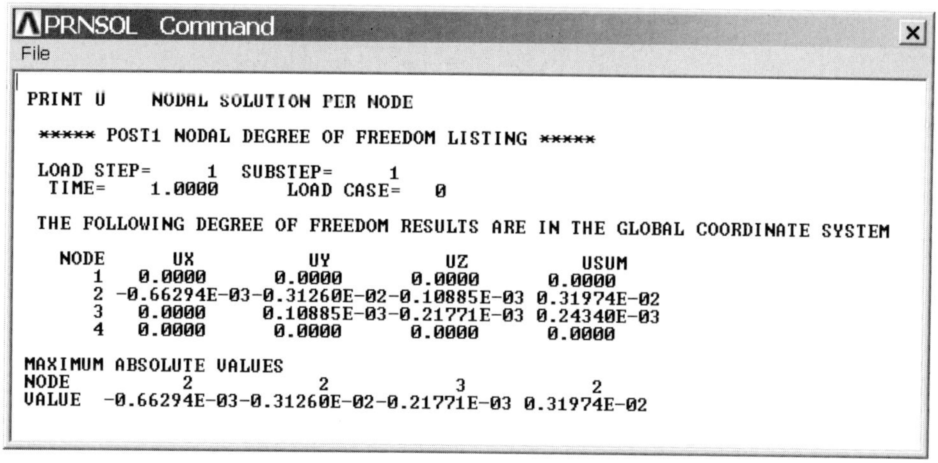

```
PRNSOL Command
File

PRINT U    NODAL SOLUTION PER NODE

 ***** POST1 NODAL DEGREE OF FREEDOM LISTING *****

 LOAD STEP=     1  SUBSTEP=     1
  TIME=    1.0000      LOAD CASE=   0

 THE FOLLOWING DEGREE OF FREEDOM RESULTS ARE IN THE GLOBAL COORDINATE SYSTEM

    NODE       UX          UY          UZ          USUM
       1   0.0000      0.0000      0.0000      0.0000
       2 -0.66294E-03-0.31260E-02-0.10885E-03 0.31974E-02
       3   0.0000     0.10885E-03-0.21771E-03 0.24340E-03
       4   0.0000      0.0000      0.0000      0.0000

MAXIMUM ABSOLUTE VALUES
NODE           2           2           3           2
VALUE   -0.66294E-03-0.31260E-02-0.21771E-03 0.31974E-02
```

To review other results, such as axial forces and axial stresses, we must copy these results into element tables. These items are obtained using *item label* and *sequence numbers*, as given in the Table 4.1–4.3 section of the ANSYS elements manual. So, we run the following sequence:

main menu: **General Postproc → Element Table → Defin**

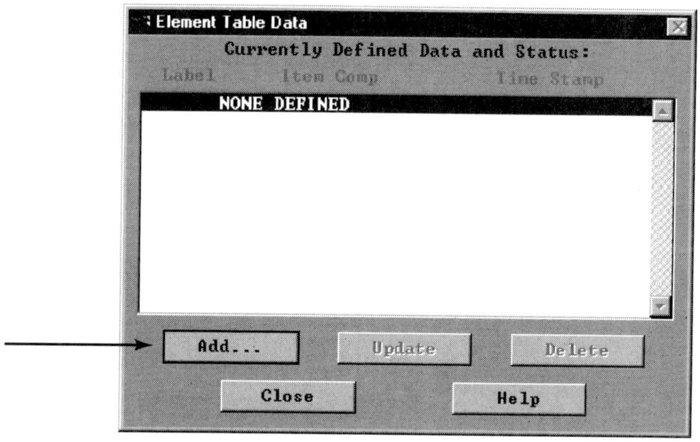

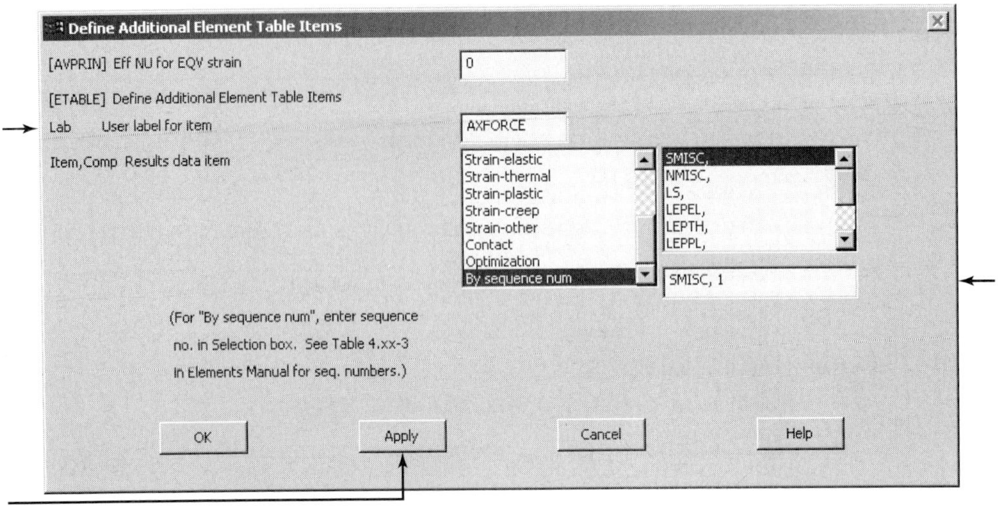
Define Additional Element Table Items
[AVPRIN] Eff NU for EQV strain
0
[ETABLE] Define Additional Element Table Items
Lab User label for item
AXFORCE
Item,Comp Results data item
Strain-elastic
Strain-thermal
Strain-plastic
Strain-creep
Strain-other
Contact
Optimization
By sequence num
SMISC,
NMISC,
LS,
LEPEL,
LEPTH,
LEPPL,
SMISC, 1
(For "By sequence num", enter sequence
no. in Selection box. See Table 4.xx-3
in Elements Manual for seq. numbers.)
OK
Apply
Cancel
Help

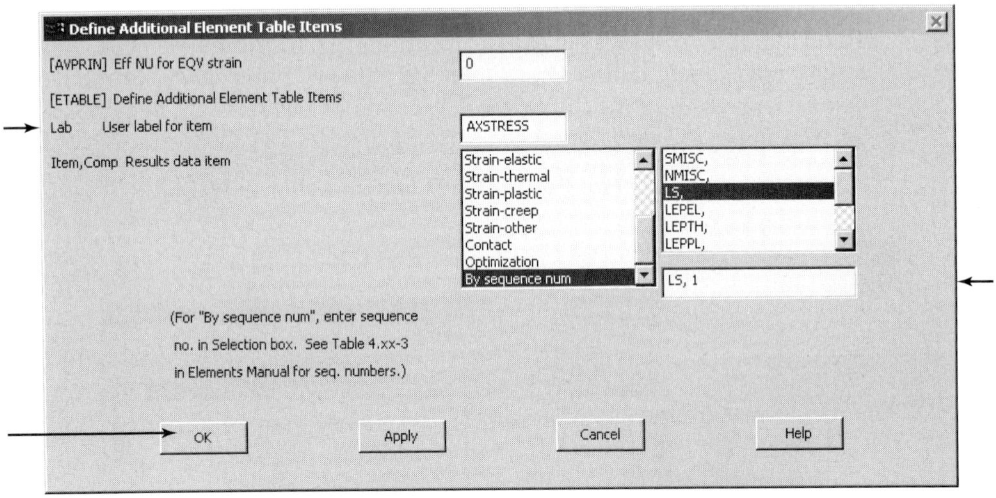
Define Additional Element Table Items
[AVPRIN] Eff NU for EQV strain
0
[ETABLE] Define Additional Element Table Items
Lab User label for item
AXSTRESS
Item,Comp Results data item
Strain-elastic
Strain-thermal
Strain-plastic
Strain-creep
Strain-other
Contact
Optimization
By sequence num
SMISC,
NMISC,
LS,
LEPEL,
LEPTH,
LEPPL,
LS, 1
(For "By sequence num", enter sequence
no. in Selection box. See Table 4.xx-3
in Elements Manual for seq. numbers.)
OK
Apply
Cancel
Help

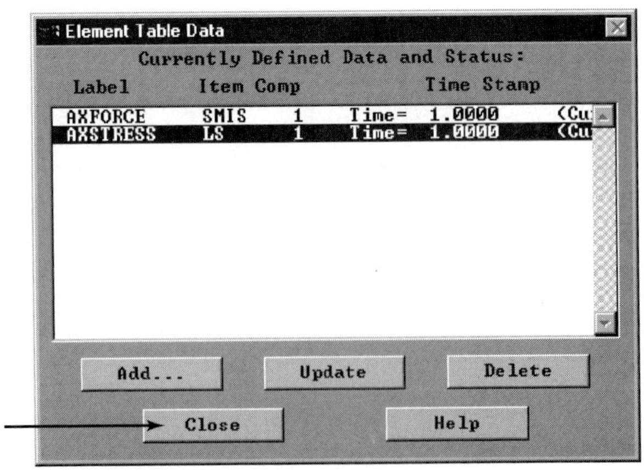
Element Table Data
Currently Defined Data and Status:
Label Item Comp Time Stamp
AXFORCE SMIS 1 Time= 1.0000
AXSTRESS LS 1 Time= 1.0000
Add...
Update
Delete
Close
Help

main menu: **General Postproc → Element Table → List Element Table**

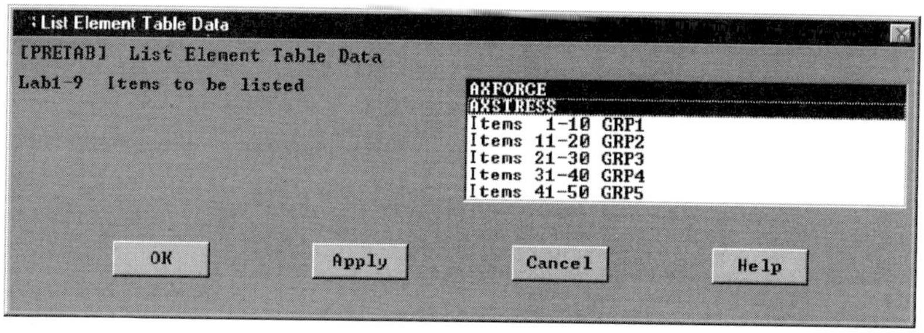

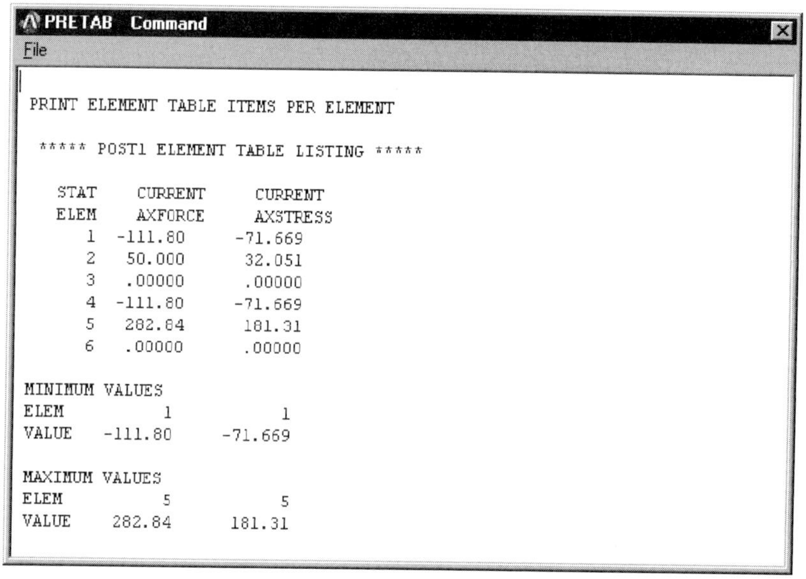


```
PRINT ELEMENT TABLE ITEMS PER ELEMENT

 ***** POST1 ELEMENT TABLE LISTING *****

   STAT    CURRENT     CURRENT
   ELEM    AXFORCE     AXSTRESS
      1  -111.80      -71.669
      2   50.000       32.051
      3   .00000       .00000
      4  -111.80      -71.669
      5   282.84       181.31
      6   .00000       .00000

MINIMUM VALUES
ELEM          1            1
VALUE   -111.80      -71.669

MAXIMUM VALUES
ELEM          5            5
VALUE    282.84       181.31
```

Close

List reaction solutions:

main menu: **General Postproc → List Results → Reaction Solu**

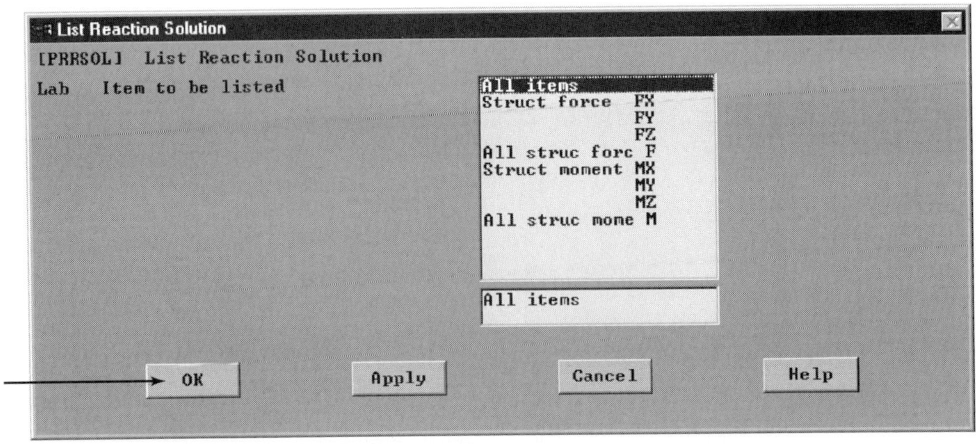

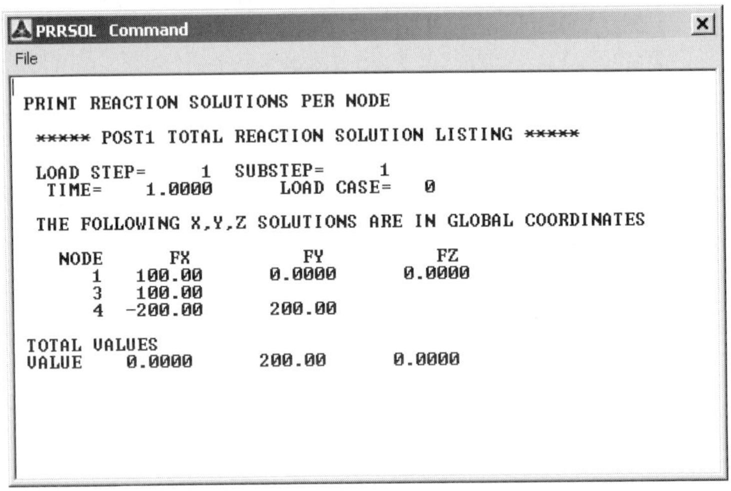

Close

Exit ANSYS and save everything, including element tables and reaction forces:

ANSYS Toolbar: **QUIT**

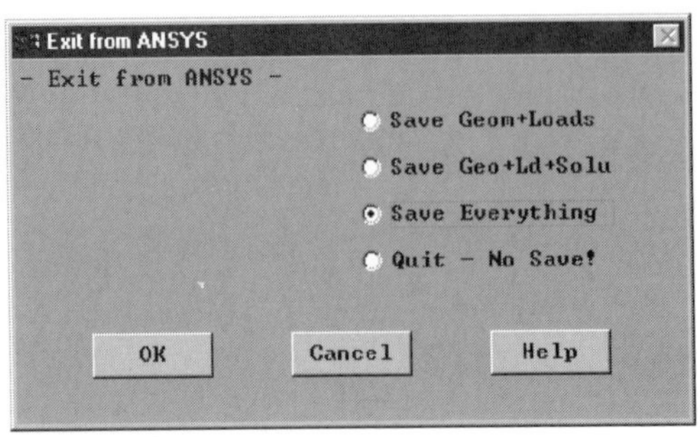

3.6 VERIFICATION OF RESULTS

There are various ways to verify your findings.

1. *Check the reaction forces.*

We can use the computed reaction forces and the external forces to check for statics equilibrium:

$$\Sigma F_X = 0$$
$$\Sigma F_Y = 0$$

and

$$\Sigma M_{\text{node}} = 0$$

The reaction forces computed by ANSYS are $F_{1X} = 1500$ lb; $F_{1Y} = 0$; $F_{3X} = -1500$ lb; and $F_{3Y} = 1000$ lb. Using the free-body diagram shown in the accompanying figure and applying the static equilibrium equations, we have:

$$\overset{+}{\rightarrow} \Sigma F_X = 0 \quad 1500-1500 = 0$$

$$+\uparrow \Sigma F_Y = 0 \quad 1000-500-500 = 0$$

$$\circlearrowleft\!+ \Sigma M_{\text{node1}} = 0 \quad (1500)(3) - (500)(3) - (500)(6) = 0$$

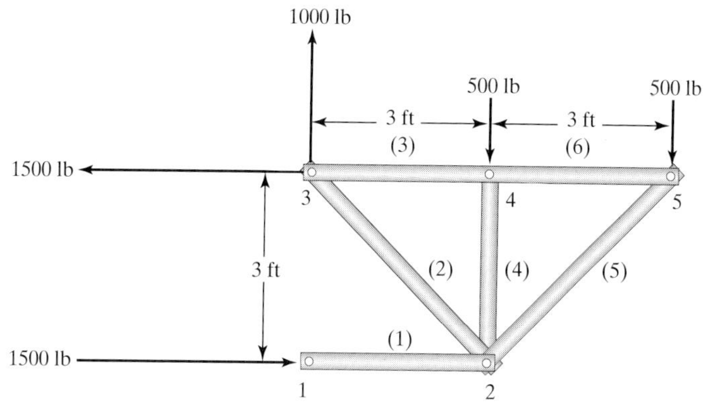

Now consider the internal forces of Example 3.1 as computed by ANSYS, shown in Table 3.2.

TABLE 3.2 Internal forces in each element as computed by ANSYS

Element Number	Internal Forces (lb)
1	−1500
2	1414
3	500
4	−500
5	−707
6	500

2. *The sum of the forces at each node should be zero.*
Choose an arbitrary node and apply the equilibrium conditions. As an example, let us choose node 5. Using the free-body diagram shown in the accompanying figure, we have

$$\overset{+}{\rightarrow} \Sigma F_X = 0 \quad -500 + 707 \cos 45 = 0$$

$$+\uparrow \Sigma F_Y = 0 \quad -500 + 707 \sin 45 = 0$$

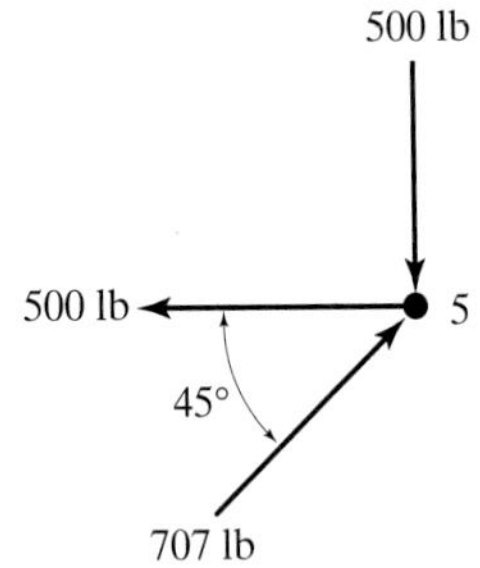

3. *Pass an arbitrary section through the truss.*
Another way of checking for the validity of your FEA findings is by arbitrarily cutting a section through the truss and applying the statics equilibrium conditions. For example, consider cutting a section through elements (1), (2), and (3), as shown in the accompanying figure.

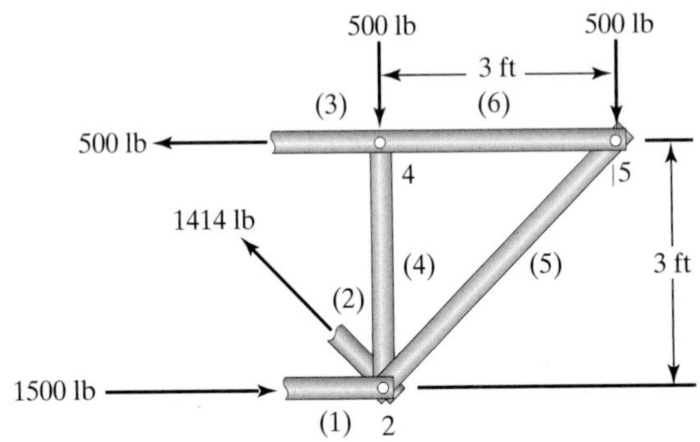

$$\overset{+}{\rightarrow} \Sigma F_X = 0 \qquad -500 + 1500 - 1414 \cos 45 = 0$$

$$+\uparrow \Sigma F_Y = 0 \qquad -500 - 500 + 1414 \cos 45 = 0$$

$$\circlearrowleft\!\!+ \Sigma M_{\text{node2}} = 0 \qquad -(500)(3) + (500)(3) = 0$$

Again, the validity of the computed internal forces is verified. Moreover, it is important to realize that when you analyze statics problems, statics equilibrium conditions must always be satisfied.

SUMMARY

At this point you should

1. have a good understanding of the underlying assumptions in truss analysis.
2. understand the significance of using global and local coordinate systems in describing a problem. You should also have a clear understanding of their role in describing nodal displacements and how information presented with respect to each frame of reference is related through the transformation matrix.
3. know the difference between the elemental stiffness matrix and the global stiffness matrix and know how to assemble elemental stiffness matrices to obtain a truss's global stiffness matrix.
4. know how to apply the boundary conditions and loads to a global matrix to obtain the nodal displacement solution.
5. know how to obtain internal forces and stresses in each member from displacement results.
6. have a good grasp of the basic concepts and commands of ANSYS. You should realize that a typical analysis using ANSYS involves the *preprocessing phase,* where you provide data such as geometry, materials, and element type to the program; the *solution phase,* where you apply boundary conditions, apply loads, and initiate a finite element solution; and the *postprocessing phase,* where you review the results of the analysis through graphics displays, tabular listings, or both.
7. know how to verify the results of your truss analysis.

REFERENCES

ANSYS User's Manual: Procedures, Vol. I, Swanson Analysis Systems, Inc.

ANSYS User's Manual: Commands, Vol. II, Swanson Analysis Systems, Inc.

ANSYS User's Manual: Elements, Vol. III, Swanson Analysis Systems, Inc.

Beer, F. P., and Johnston, E. R., *Vector Mechanics for Engineers: Statics*, 5th ed., New York, McGraw-Hill, 1988.

Segrlind, L., Applied Finite Element Analysis, 2nd ed., New York, John Wiley and Sons, 1984.

PROBLEMS

1. Starting with the transformation matrix, show that the inverse of the transformation matrix is its transpose. That is, show that

$$[\mathbf{T}]^{-1} = \begin{bmatrix} \cos\theta & \sin\theta & 0 & 0 \\ -\sin\theta & \cos\theta & 0 & 0 \\ 0 & 0 & \cos\theta & \sin\theta \\ 0 & 0 & -\sin\theta & \cos\theta \end{bmatrix}$$

2. Starting with Eq. (3.14), $\{\mathbf{F}\} = [\mathbf{T}][\mathbf{K}][\mathbf{T}]^{-1}\{\mathbf{U}\}$, and substituting for values of the $[\mathbf{T}]$, $[\mathbf{K}]$, $[\mathbf{T}]^{-1}$, and $\{\mathbf{U}\}$ matrices in Eq. (3.14), verify the elemental relationship

$$\begin{Bmatrix} F_{iX} \\ F_{iY} \\ F_{jX} \\ F_{jY} \end{Bmatrix} = k \begin{bmatrix} \cos^2\theta & \sin\theta\cos\theta & -\cos^2\theta & -\sin\theta\cos\theta \\ \sin\theta\cos\theta & \sin^2\theta & -\sin\theta\cos\theta & -\sin^2\theta \\ -\cos^2\theta & -\sin\theta\cos\theta & \cos^2\theta & \sin\theta\cos\theta \\ -\sin\theta\cos\theta & -\sin^2\theta & \sin\theta\cos\theta & \sin^2\theta \end{bmatrix} \begin{Bmatrix} U_{iX} \\ U_{iY} \\ U_{jX} \\ U_{jY} \end{Bmatrix}$$

3. The members of the truss shown in the accompanying figure have a cross-sectional area of 2.3 in^2 and are made of aluminum alloy ($E = 10.0 \times 10^6$ lb/in^2). Using hand calculations, determine the deflection of joint A, the stress in each member, and the reaction forces. Verify your results.

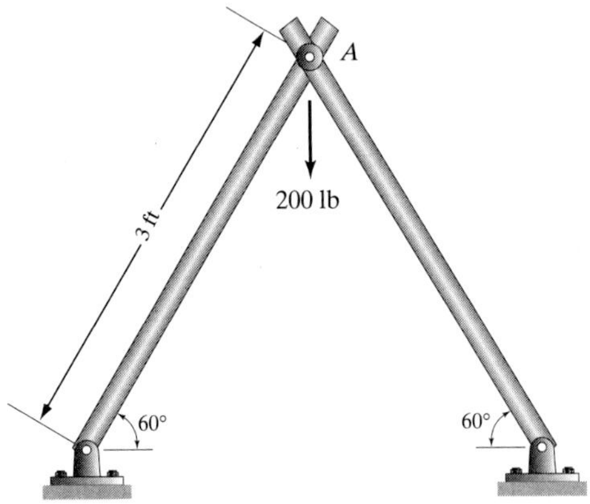

4. The members of the truss shown in the accompanying figure have a cross-sectional area of 8 cm^2 and are made of steel ($E = 200$ GPa). Using hand calculations, determine the deflection of each joint, the stress in each member, and the reaction forces. Verify your results.

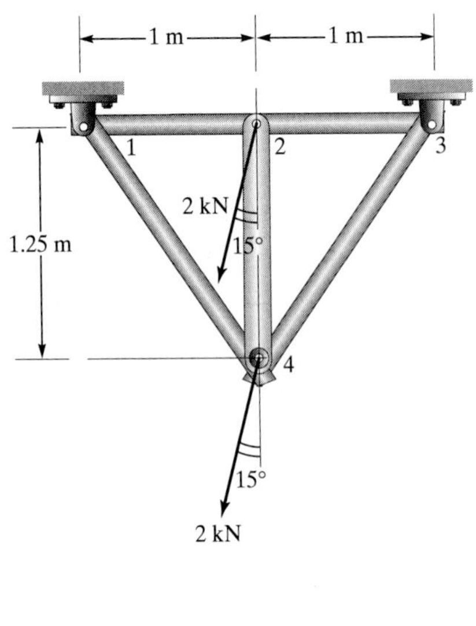

5. The members of the truss shown in the accompanying figure have a cross-sectional area of 15 cm^2 and are made of aluminum alloy ($E = 70$ GPa). Using hand calculations, determine the deflection of each joint, the stress in each member, and the reaction forces. Verify your results.

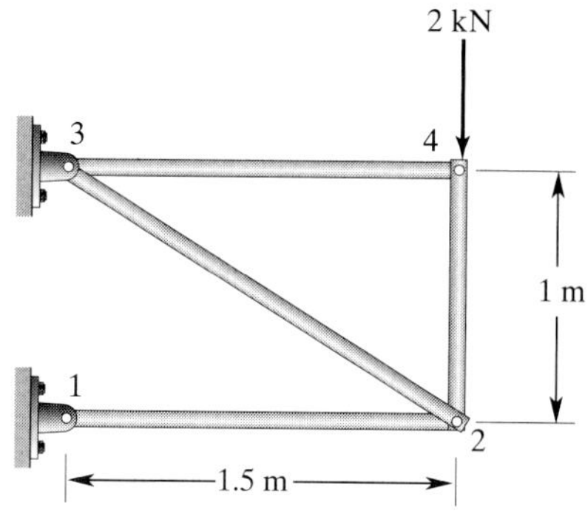

6. The members of the truss shown in the accompanying figure have a cross-sectional area of 2 in^2 and are made of structural steel ($E = 30.0 \times 10^6$ lb/in^2). Using hand calculations, determine the deflection of each joint, the stress in each member, and the reaction forces. Verify your results.

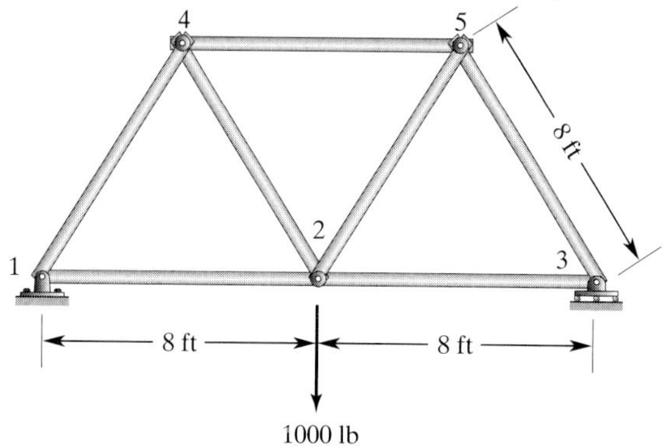

7. The members of the three-dimensional truss shown in the accompanying figure have a cross-sectional area of 2.5 in^2 and are made of aluminum alloy ($E = 10.0 \times 10^6$ lb/in^2). Using hand calculations, determine the deflection of joint A, the stress in each member, and the reaction forces. Verify your results.

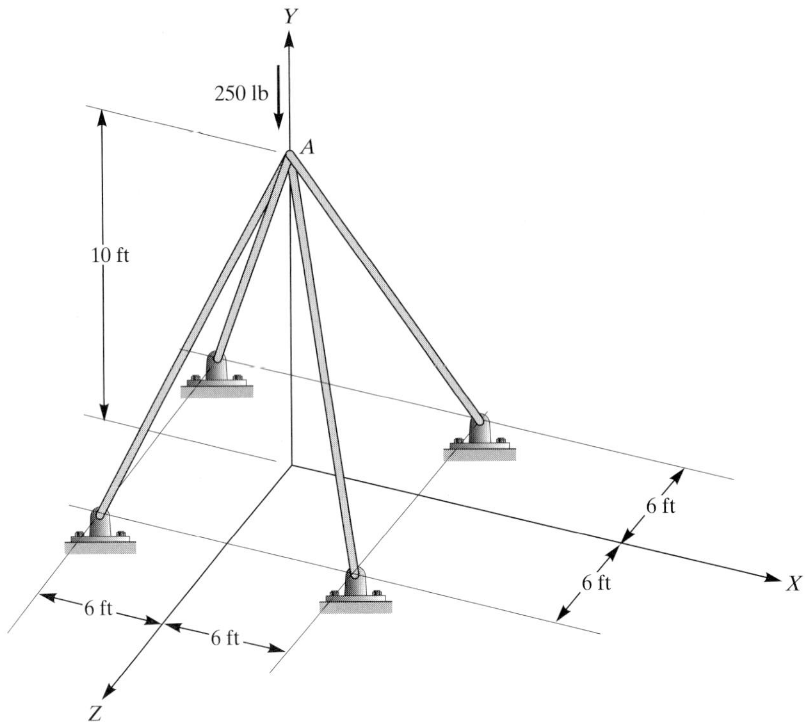

8. The members of the three-dimensional truss shown in the accompanying figure have a cross-sectional area of 15 cm^2 and are made of steel ($E = 200$ GPa). Using hand calculations, determine the deflection of joint A, the stress in each member, and the reaction forces. Verify your results.

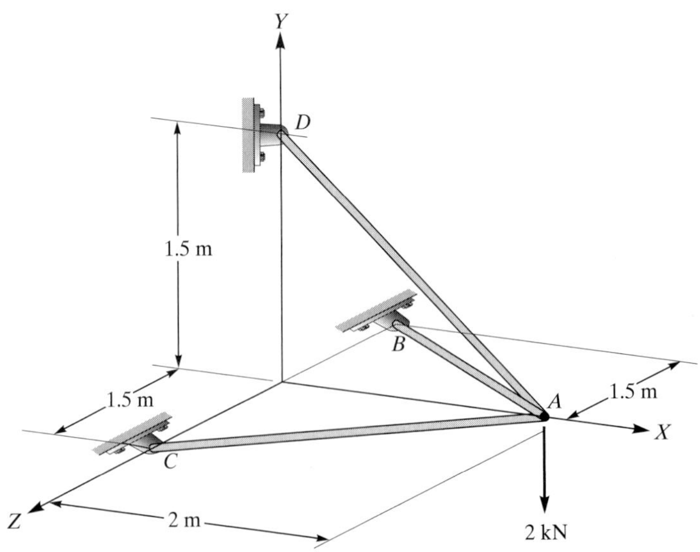

9. Consider the power transmission-line tower shown in the accompanying figure. The members have a cross-sectional area of 10 in^2 and a modulus of elasticity of $E = 29 \times 10^6$ lb/in^2. Using ANSYS, determine the deflection of each joint, the stress in each member, and the reaction forces at the base. Verify your results.

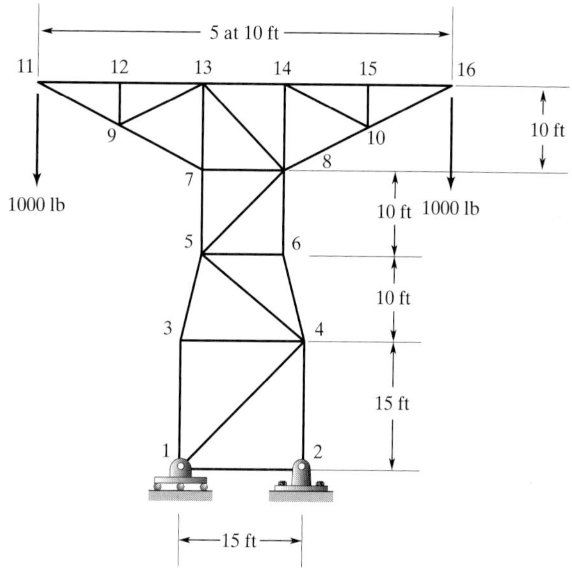

10. Consider the staircase truss shown in the accompanying figure. There are 14 steps, each with a rise of 8 in and a run of 12 in. The members have a cross-sectional area of 4 in^2 and are made of steel with a modulus of elasticity of $E = 29 \times 10^6$ lb/in^2. Using ANSYS, determine the deflection of each joint, the stress in each member, and the reaction forces. Verify your results.

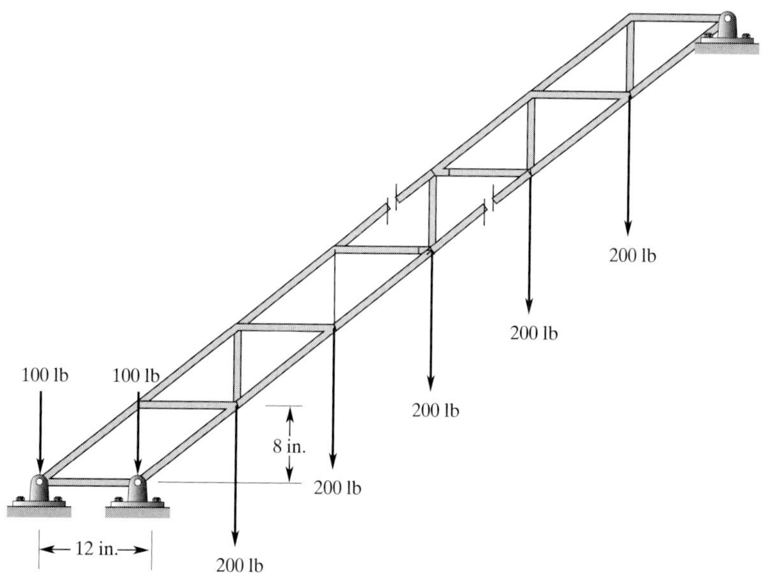

11. The members of the roof truss shown in the accompanying figure have a cross-sectional area of approximately 21.5 in^2 and are made of Douglas-fir wood with a modulus of elasticity of $E = 1.9 \times 10^6$ lb/in^2. Using ANSYS, determine the deflection of each joint, the stresses in each member, and the reaction forces. Verify your results. Also, replace one of the fixed boundary conditions with rollers and obtain the stresses in each member. Discuss the difference in results.

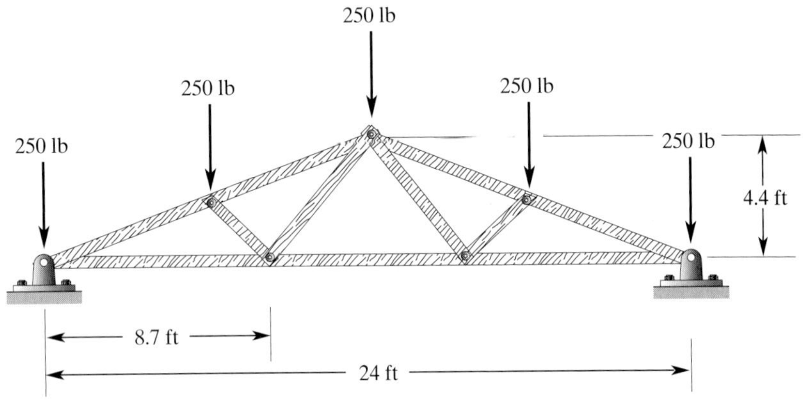

12. The members of the floor truss shown in the accompanying figure have a cross-sectional area of approximately 21.5 in^2 and are made of Douglas-fir wood with a modulus of elasticity of $E = 1.9 \times 10^6$ lb/in^2. Using ANSYS, determine the deflection of each joint, the stresses in each member, and the reaction forces. Verify your results. Also, replace one of the fixed boundary conditions with rollers and solve the problem again to obtain the stresses in each member. Discuss the difference in results.

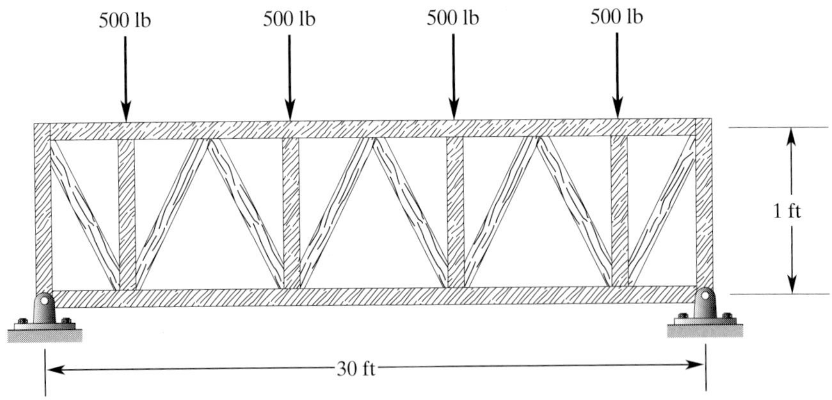

13. The three-dimensional truss shown in the accompanying figure is made of aluminum alloy ($E = 10.9 \times 10^6$ psi) and is to support a load of 500 lb. The Cartesian coordinates of the joints with respect to the coordinate system shown in the figure are given in feet. The cross-sectional area of each member is 2.246 in^2. Using ANSYS, determine the deflection of each joint, the stress in each member, and the reaction forces. Knowing that the second moment of area is 4.090 in^4, do you think that buckling is a concern for this truss? Verify your results.

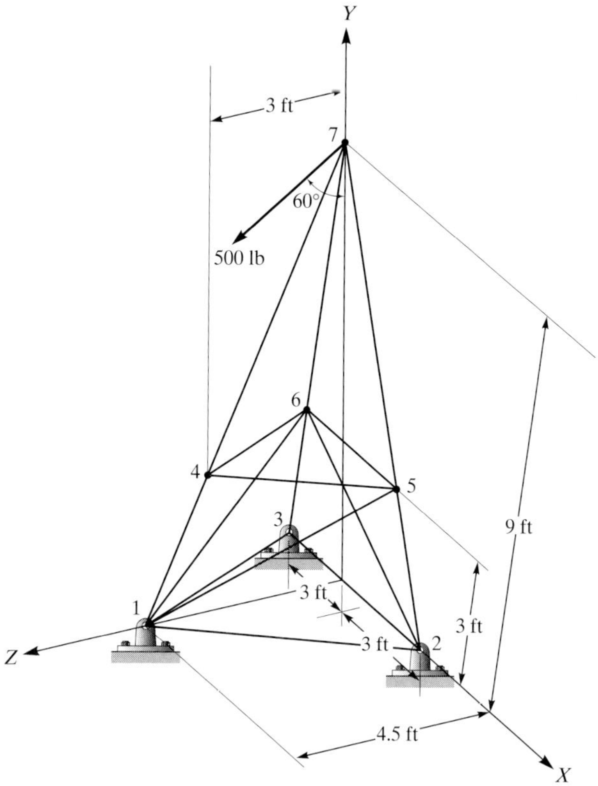

14. The three-dimensional truss shown in the accompanying figure is made of aluminum alloy ($E = 10.4 \times 10^6$ lb/in^2) and is to support a sign weighing 1000 lb. The Cartesian coordinates of the joints with respect to the coordinate system shown in the figure are given in feet. The cross-sectional area of each member is 3.14 in^2. Using ANSYS, determine the deflection of joint E, the stresses in each member, and the reaction forces. Verify your results.

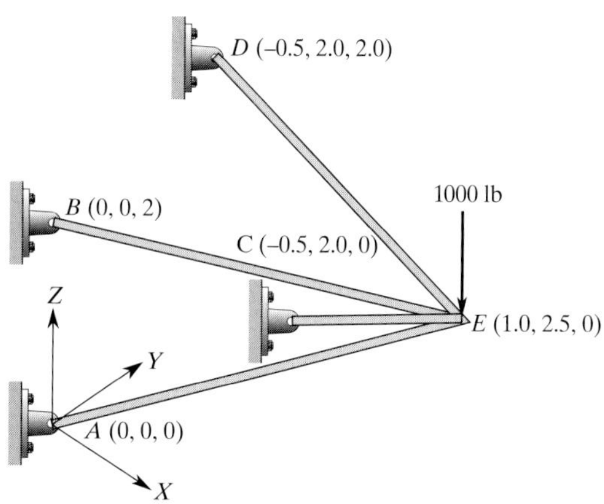

15. The three-dimensional truss shown in the accompanying figure is made of steel ($E = 29 \times 10^6$ psi) and is to support the load shown in the figure. The Cartesian coordinates of the joints with respect to the system shown in the figure are given in feet. The cross-sectional area of each member is 3.093 in^2. Using ANSYS, determine the deflection of each joint, the stresses in each member, and the reaction forces. Verify your results.

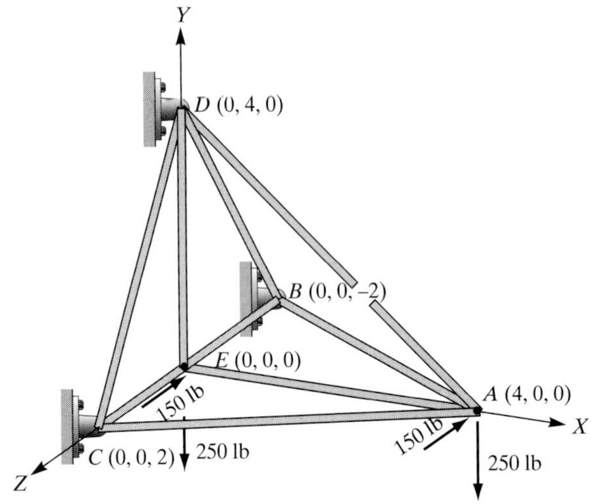

16. During a maintenance process on the three-dimensional truss in Problem 15, the AB member is replaced with a member with the following properties: $E = 28 \times 10^6$ psi and $A = 2.246$ in^2. Using ANSYS, determine the deflection of each joint and the stresses in each member. Hint: you may need to ask your instructor for some help with this problem or you may want to study Example 6.2 (revisited) on your own to learn about how to assign different attributes to an element in ANSYS.

17. During a maintenance process on the three-dimensional truss in Problem 13, members 4–5, 4–6, and 5–6 are replaced with steel members with the following properties: $E = 29 \times 10^6$ psi and $A = 1.25$ in^2. Member 1–5 is also replaced with a steel member with a cross-sectional area of 1.35 in^2. Using ANSYS, determine the deflection of each joint and the stresses in each member. See the hint given for Problem 16.

18. Derive the transformation matrix for an arbitrary member of a space truss, shown in the accompanying figure. The directional cosines, in terms of the difference between the coordinates of nodes j and i of a member and its length, are

$$\cos\theta_X = \frac{X_j - X_i}{L}; \quad \cos\theta_Y = \frac{Y_j - Y_i}{L}; \quad \cos\theta_Z = \frac{Z_j - Z_i}{L}$$

where L is the length of the member and is

$$L = \sqrt{(X_j - X_i)^2 + (Y_j - Y_i)^2 + (Z_j - Z_i)^2}$$

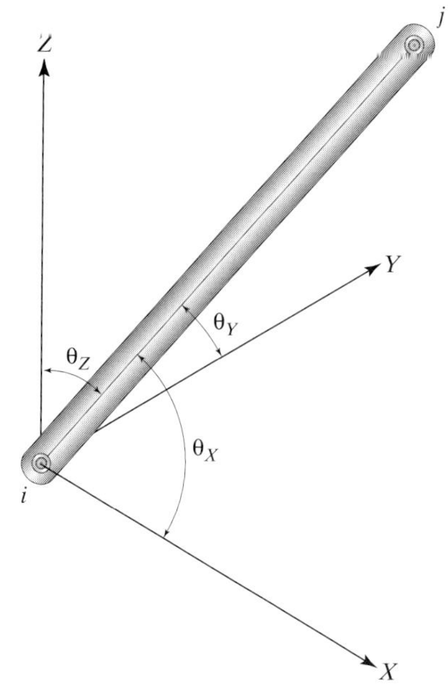

19. The three-dimensional truss shown in the accompanying figure is made of steel ($E = 29 \times 10^6$ psi) and is to support the load shown in the figure. Dimensions are given in feet. The cross-sectional area of each member is 3.25 in^2. Using ANSYS, determine the deflection of each joint, the stresses in each member, and the reaction forces. Verify your results.

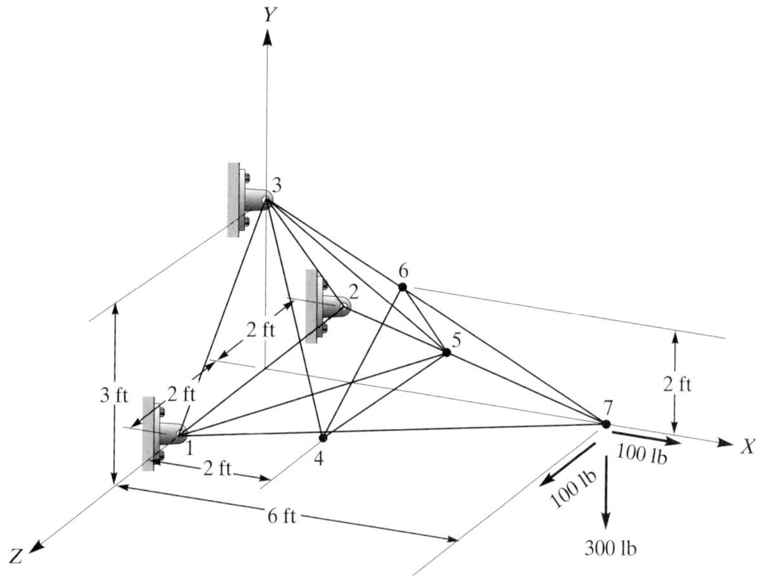

20. Design Problem Size the cross section of each member for the outdoor truss structure shown in the accompanying figure so that the end deflection of the truss is kept under 1 in. Select appropriate material and discuss how you arrived at your final design.

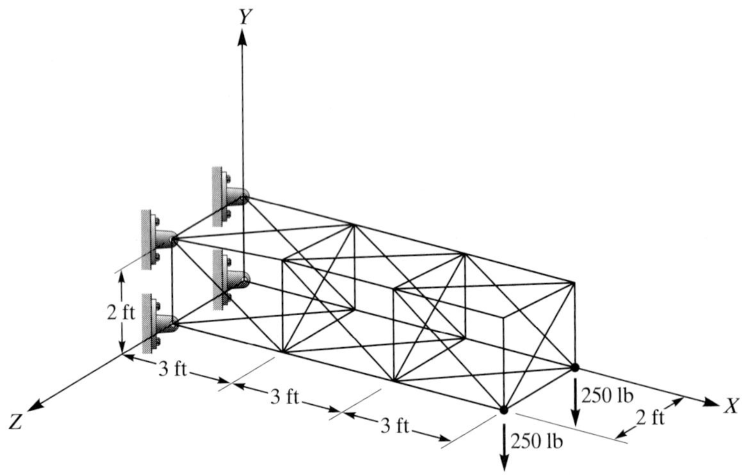

CHAPTER 4

Axial Members, Beams, and Frames

The objective of this chapter is to introduce you to the analysis of members under axial loading, beams, and frames. Structural members and machine components are generally subject to a push-pull, bending, or twisting type of loading. We will discuss twisting or torsion of structural members and plane stress formulation of machine components in Chapter 10. The main topics discussed in this chapter include the following:

4.1 Members Under Axial Loading
4.2 Beams
4.3 Finite Element Formulation of Beams
4.4 Finite Element Formulation of Frames
4.5 Three-Dimensional Beam Element
4.6 An Example Using ANSYS
4.7 Verification of Results

4.1 MEMBERS UNDER AXIAL LOADING

In this section, we use the minimum total potential energy formulation to generate finite element models for members under axial loading. However, before we proceed with finite element formulation of axial members, we should define what we mean by an axial element and corresponding shape functions and their properties.

A Linear Element

The structural example in this section is employed to introduce the basic ideas of one-dimensional element and shape functions. Steel columns are commonly used to support loads from various floors of multistory buildings, as shown in Figure 4.1. The column shown in the figure may be divided into four elements and five nodes to generate a

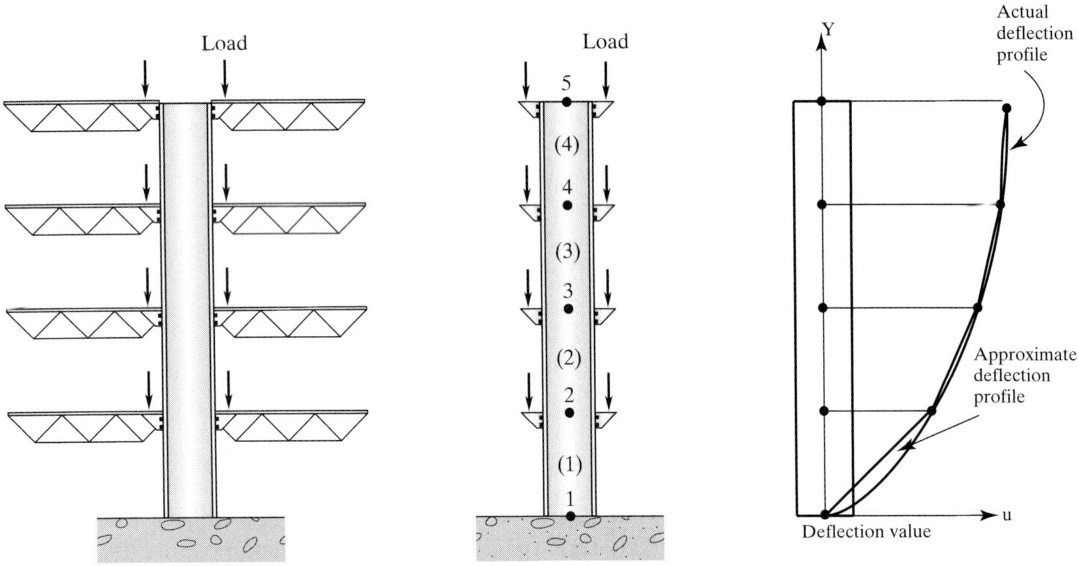

FIGURE 4.1 Deflection of a steel column subject to floor loading.

finite element model. The loading from the floors causes vertical displacements of various points along the column. Assuming axial central loading, we may approximate the actual *deflection* of the column by using a series of *linear functions*, describing the deflection over each element or each section of the column. Note that the deflection profile u represents the vertical (not the lateral) displacement of the column at various points along the column. The profile is merely plotted as a function of Y. We have modeled the example problem shown in Figure 4.1 by five nodes and four elements. Let us focus our attention on a typical element, as shown in Figure 4.2.

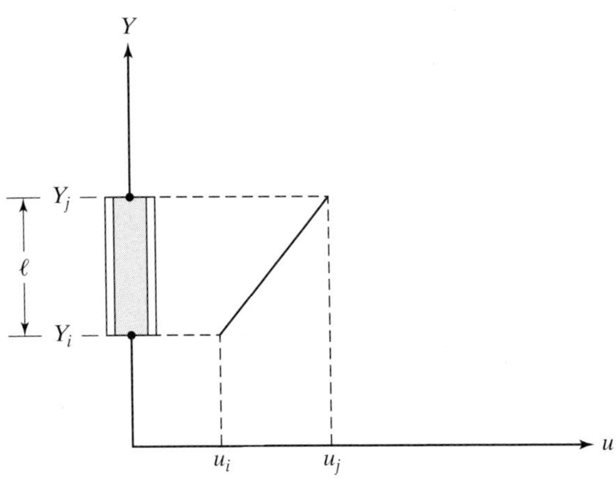

FIGURE 4.2 Linear approximation of deflection variation for an element.

The linear deflection distribution for a typical element may be expressed as

$$u^{(e)} = c_1 + c_2 Y \tag{4.1}$$

In order to solve for the unknown coefficients c_1 and c_2, we make use of the element's end deflection values which are given by the nodal deflections u_i and u_j, according to the conditions

$$\begin{aligned} u &= u_i \quad \text{at} \quad Y = Y_i \\ u &= u_j \quad \text{at} \quad Y = Y_j \end{aligned} \tag{4.2}$$

Substitution of nodal values into Eq. (4.1) results in two equations and two unknowns:

$$\begin{aligned} u_i &= c_1 + c_2 Y_i \\ u_j &= c_1 + c_2 Y_j \end{aligned} \tag{4.3}$$

Solving for the unknowns c_1 and c_2, we get

$$c_1 = \frac{u_i Y_j - u_j Y_i}{Y_j - Y_i} \tag{4.4}$$

$$c_2 = \frac{u_j - u_i}{Y_j - Y_i} \tag{4.5}$$

The element's deflection distribution in terms of its nodal values is

$$u^{(e)} = \frac{u_i Y_j - u_j Y_i}{Y_j - Y_i} + \frac{u_j - u_i}{Y_j - Y_i} Y \tag{4.6}$$

Grouping the u_i terms together and the u_j terms together, Eq. (4.6) becomes

$$u^{(e)} = \left(\frac{Y_j - Y}{Y_j - Y_i}\right) u_i + \left(\frac{Y - Y_i}{Y_j - Y_i}\right) u_j \tag{4.7}$$

We now define the *shape functions*, S_i and S_j using the terms in parentheses appearing before u_i and u_j, according to the equations

$$S_i = \frac{Y_j - Y}{Y_j - Y_i} = \frac{Y_j - Y}{\ell} \tag{4.8}$$

$$S_j = \frac{Y - Y_i}{Y_j - Y_i} = \frac{Y - Y_i}{\ell} \tag{4.9}$$

where ℓ is the length of the element. Thus, the deflection for an element in terms of the shape functions and the nodal deflection values can be written as

$$u^{(e)} = S_i u_i + S_j u_j \tag{4.10}$$

Equation (4.10) can also be expressed in matrix form as

$$u^{(e)} = [S_i \quad S_j] \begin{Bmatrix} u_i \\ u_j \end{Bmatrix} \tag{4.11}$$

It will become clear to you that we can use the same approach to approximate the spatial variation of any unknown variable, such as temperature or velocity, in the same manner. We will discuss the concept of one-dimensional elements and their properties in more detail in Chapter 5.

As we discussed in Chapter 3, in finite element modeling, it is convenient to use two frames of references: (1) a global coordinate system to represent the location of each node, orientation of each element, and to apply boundary conditions and loads. The nodal solutions of finite element models are generally expressed with respect to global coordinates as well. On the other hand, we employ (2) a local coordinate system to take advantage of the local characteristics of the system behavior.

For the one-dimensional element shown in Figure 4.2, the relationship between a global coordinate Y and a local coordinate y is given by $Y = Y_i + y$. This relationship is shown in Figure 4.3. Substituting for Y in terms of the local coordinate y in Eqs. (4.8) and (4.9), we get

$$S_i = \frac{Y_j - Y}{\ell} = \frac{Y_j - (Y_i + y)}{\ell} = 1 - \frac{y}{\ell} \tag{4.12}$$

$$S_j = \frac{Y - Y_i}{\ell} = \frac{(Y_i + y) - Y_i}{\ell} = \frac{y}{\ell} \tag{4.13}$$

where the local coordinate y varies from 0 to ℓ; that is, $0 \le y \le \ell$.

This is a good place to say a few words about the shape functions S_i and S_j. They possess unique properties that, once understood, can simplify the derivation of stiffness matrices. We now refer to Eqs. (4.12) and (4.13) and note that S_i and S_j each has a value of unity at its corresponding node and zero at the other adjacent node. For example, if we evaluate S_i at node i by substituting $y = 0$ in Eq. (4.12), we find that $S_i = 1$. Similarly, we can show that the value of S_j at node $j(y = \ell)$ is also 1. The value of the shape function S_i, Eq. (4.12), at the adjacent node $j(y = \ell)$ and S_j, Eq. (4.13), at its adjacent node $i(y = 0)$ are zero. We discuss the properties of shape functions in more detail in Chapter 5.

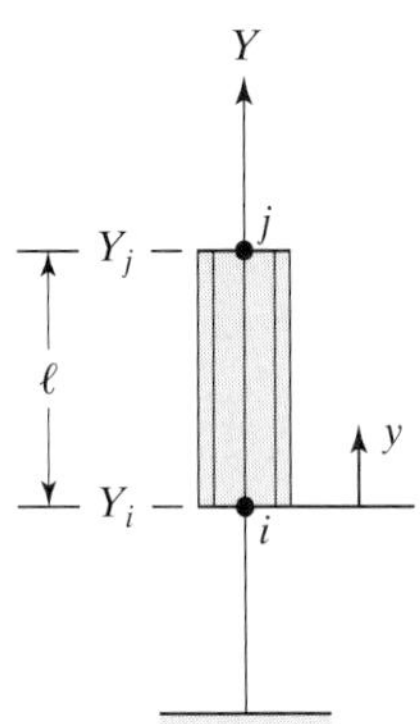

FIGURE 4.3 The relationship between a global coordinate Y and a local coordinate y.

EXAMPLE 4.1

Consider a four-story building with steel columns. One column is subjected to the loading shown in Figure 4.4. Under axial loading assumption and using linear elements, the vertical displacements of the column at various floor-column connection points were determined to be

$$\begin{Bmatrix} u_1 \\ u_2 \\ u_3 \\ u_4 \\ u_5 \end{Bmatrix} = -\begin{Bmatrix} 0 \\ 0.03283 \\ 0.05784 \\ 0.07504 \\ 0.08442 \end{Bmatrix} \text{in.}$$

The modulus of elasticity of $E = 29 \times 10^6$ lb/in^2 and area of $A = 39.7$ in^2 were used in the calculations. A detailed analysis of this problem is given in the next section. For now, given the nodal displacement values, we are interested in determining the deflections of points A and B.

a. Using the global coordinate Y, the displacement of point A is represented by element (1):

$$u^{(1)} = S_1^{(1)} u_1 + S_2^{(1)} u_2 = \frac{Y_2 - Y}{\ell} u_1 + \frac{Y - Y_1}{\ell} u_2$$

$$u = \frac{15 - 10}{15}(0) + \frac{10 - 0}{15}(-0.03283) = -0.02188 \text{ in}$$

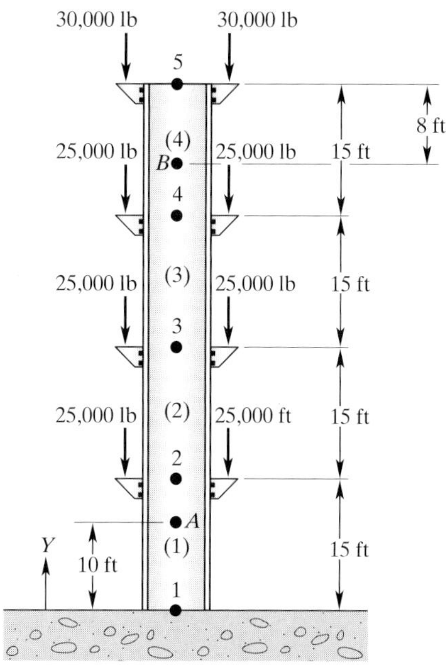

FIGURE 4.4 The column in Example 4.1.

b. The displacement of point B is represented by element (4):

$$u^{(4)} = S_4^{(4)}u_4 + S_5^{(4)}u_5 = \frac{Y_5 - Y}{\ell}u_4 + \frac{Y - Y_4}{\ell}u_5$$

$$u = \frac{60 - 52}{15}(-0.07504) + \frac{52 - 45}{15}(-0.08442) = -0.07941 \text{ in}$$

Stiffness and Load Matrices

In this section, we use the minimum total potential energy formulation to generate the stiffness and load matrices for members under axial loading. Previously, we showed that under axial loading, we can approximate the exact deflection of the column shown in Figure 4.1 by a series of linear functions. Moreover, as discussed in Section 1.6, applied external loads cause a body to deform. During the deformation, the work done by the external forces is stored in the material in the form of elastic energy, called strain energy. For a member (element) under axial loading, the strain energy $\Lambda^{(e)}$ is given by

$$\Lambda^{(e)} = \int_V \frac{\sigma\varepsilon}{2}\,dV = \int_V \frac{E\varepsilon^2}{2}\,dV \tag{4.14}$$

The total potential energy Π for a body consisting of n elements and m nodes is the difference between the total strain energy and the work done by the external forces:

$$\Pi = \sum_{e=1}^{n} \Lambda^{(e)} - \sum_{i=1}^{m} F_i u_i \tag{4.15}$$

The minimum total potential energy principle states that for a stable system, the displacement at the equilibrium position occurs such that the value of the system's total potential energy is a minimum. That is,

$$\frac{\partial \Pi}{\partial u_i} = \frac{\partial}{\partial u_i}\sum_{e=1}^{n} \Lambda^{(e)} - \frac{\partial}{\partial u_i}\sum_{i=1}^{m} F_i u_i = 0 \quad \text{for } i = 1, 2, 3, \ldots, m \tag{4.16}$$

where i takes on different values of node numbers. Recall that the deflection for an arbitrary element with nodes i and j in terms of local shape functions is given by

$$u^{(e)} = S_i u_i + S_j u_j \tag{4.17}$$

where $S_i = 1 - \frac{y}{\ell}$ and $S_j = \frac{y}{\ell}$ and y is the element's local coordinate, with its origin at node i. The strain in each member can be computed using the relation $\varepsilon = \frac{du}{dy}$ as

$$\varepsilon = \frac{du}{dy} = \frac{d}{dy}[S_i u_i + S_j u_j] = \frac{d}{dy}\left[\left(1 - \frac{y}{\ell}\right)u_i + \frac{y}{\ell}u_j\right] = \frac{-u_i + u_j}{\ell} \tag{4.18}$$

Incorporating Eq. (4.18) into Eq. (4.14) yields the strain energy for an arbitrary element (e):

$$\Lambda^{(e)} = \int_V \frac{E\varepsilon^2}{2} dV = \frac{AE}{2\ell}(u_j^2 + u_i^2 - 2u_j u_i) \tag{4.19}$$

Minimizing the strain energy with respect to u_i and u_j leads to

$$\frac{\partial \Lambda^{(e)}}{\partial u_i} = \frac{AE}{\ell}(u_i - u_j) \tag{4.20}$$

$$\frac{\partial \Lambda^{(e)}}{\partial u_j} = \frac{AE}{\ell}(u_j - u_i)$$

or, in matrix form,

$$\begin{Bmatrix} \dfrac{\partial \Lambda^{(e)}}{\partial u_i} \\ \dfrac{\partial \Lambda^{(e)}}{\partial u_j} \end{Bmatrix} = \frac{AE}{\ell}\begin{bmatrix} 1 & -1 \\ -1 & 1 \end{bmatrix}\begin{Bmatrix} u_i \\ u_j \end{Bmatrix} = \begin{bmatrix} k & -k \\ -k & k \end{bmatrix}\begin{Bmatrix} u_i \\ u_j \end{Bmatrix} \tag{4.21}$$

where $k = \dfrac{(AE)}{\ell}$. Minimizing the work done by external forces, the second term on the right-hand side of Eq. (4.16) results in the load matrix

$$\{\mathbf{F}\}^{(e)} = \begin{Bmatrix} F_i \\ F_j \end{Bmatrix} \tag{4.22}$$

Computing individual elemental stiffness and load matrices and connecting them leads to global stiffness and load matrices. This step is demonstrated by the next example.

EXAMPLE 4.2 A Column Problem

Consider a four-story building with steel columns. One column is subjected to the loading shown in Figure 4.5. Assuming axial loading, determine (a) vertical displacements of the column at various floor-column connection points and (b) the stresses in each portion of the column. $E = 29 \times 10^6$ lb/in^2, $A = 39.7$ in^2.

Because all elements have the same length, cross-sectional area, and physical properties, the elemental stiffness for elements (1),(2),(3), and (4) is given by

$$[\mathbf{K}]^{(e)} = \frac{AE}{\ell}\begin{bmatrix} 1 & -1 \\ -1 & 1 \end{bmatrix} = \frac{39.7 \times 29 \times 10^6}{15 \times 12}\begin{bmatrix} 1 & -1 \\ -1 & 1 \end{bmatrix} = 6.396 \times 10^6\begin{bmatrix} 1 & -1 \\ -1 & 1 \end{bmatrix}$$

$$[\mathbf{K}]^{(1)} = [\mathbf{K}]^{(2)} = [\mathbf{K}]^{(3)} = [\mathbf{K}]^{(4)} = 6.396 \times 10^6\begin{bmatrix} 1 & -1 \\ -1 & 1 \end{bmatrix}\frac{\text{lb}}{\text{in}}$$

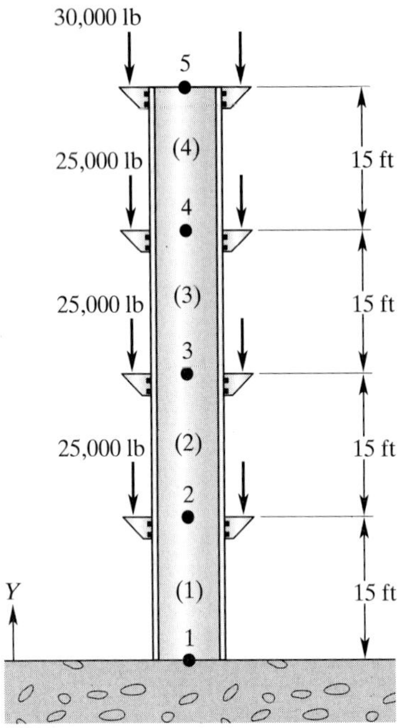

FIGURE 4.5 A schematic of the column in Example 4.2.

The global stiffness matrix is obtained by assembling the elemental matrices:

$$[\mathbf{K}]^{(G)} = 6.396 \times 10^6 \begin{bmatrix} 1 & -1 & 0 & 0 & 0 \\ -1 & 1+1 & -1 & 0 & 0 \\ 0 & -1 & 1+1 & -1 & 0 \\ 0 & 0 & -1 & 1+1 & -1 \\ 0 & 0 & 0 & -1 & 1 \end{bmatrix}$$

The global load matrix is obtained from

$$\{\mathbf{F}\}^{(G)} = \left\{\frac{\partial F_i u_i}{\partial u_i}\right\}_{i=1,5} = \begin{Bmatrix} F_1 \\ F_2 \\ F_3 \\ F_4 \\ F_5 \end{Bmatrix} = -\begin{Bmatrix} 0 \\ 50000 \\ 50000 \\ 50000 \\ 60000 \end{Bmatrix} \text{lb}$$

Note all applied forces act in the negative Y direction. Application of the boundary condition, $u_1 = 0$, and loads results in

$$6.396 \times 10^6 \begin{bmatrix} 1 & 0 & 0 & 0 & 0 \\ -1 & 2 & -1 & 0 & 0 \\ 0 & -1 & 2 & -1 & 0 \\ 0 & 0 & -1 & 2 & -1 \\ 0 & 0 & 0 & -1 & 1 \end{bmatrix} \begin{Bmatrix} u_1 \\ u_2 \\ u_3 \\ u_4 \\ u_5 \end{Bmatrix} = -\begin{Bmatrix} 0 \\ 50000 \\ 50000 \\ 50000 \\ 60000 \end{Bmatrix}$$

Solving for displacements, we have

$$\begin{Bmatrix} u_1 \\ u_2 \\ u_3 \\ u_4 \\ u_5 \end{Bmatrix} = -\begin{Bmatrix} 0 \\ 0.03283 \\ 0.05784 \\ 0.07504 \\ 0.08442 \end{Bmatrix} \text{in}$$

The axial stresses in each element are determined from

$$\sigma^{(1)} = \frac{E(u_j - u_i)}{\ell} = \frac{29 \times 10^6(-0.03283 - 0)}{15 \times 12} = -5289 \text{ lb/in}^2$$

$$\sigma^{(2)} = \frac{29 \times 10^6(-0.05784 - (-0.03283))}{15 \times 12} = -4029 \text{ lb/in}^2$$

$$\sigma^{(3)} = \frac{29 \times 10^6(-0.07504 - (-0.05784))}{15 \times 12} = -2771 \text{ lb/in}^2$$

$$\sigma^{(4)} = \frac{29 \times 10^6(-0.08442 - (-0.07504))}{15 \times 12} = -1511 \text{ lb/in}^2$$

4.2 BEAMS

Beams play significant roles in many engineering applications, including buildings, bridges, automobiles, and airplane structures. A beam is defined as a structural member whose cross-sectional dimensions are relatively smaller than its length. Beams are commonly subjected to transverse loading, which is a type of loading that creates bending in the beam. A beam subjected to a distributed load is shown in Figure 4.6.

In the previous chapter we defined trusses as structures consisting of two-force members. Moreover, recall that when using a truss model to analyze a physical problem, all loads are assumed to apply at the joints or the nodes of the truss. Therefore, no bending of the members are allowed. Note that for a structural member that is considered as a beam, loads may be applied anywhere along the beam and the loading will create bending in the beam. It is important to make these distinctions when modeling a physical problem.

The deflection of the neutral axis of a beam at any location x is represented by the variable v. For small deflections, the relationship between the normal stress σ at a

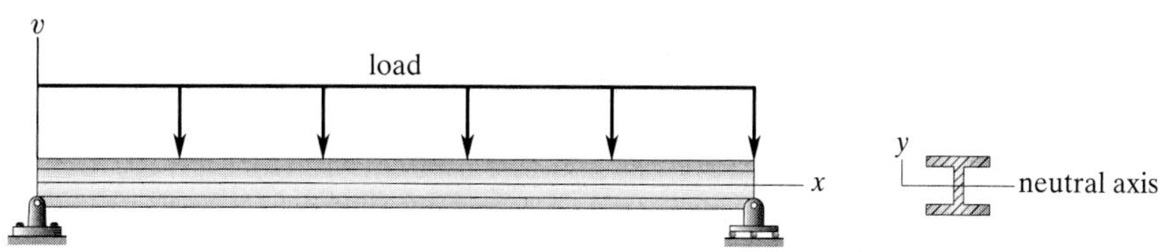

FIGURE 4.6 A beam subjected to a distributed load.

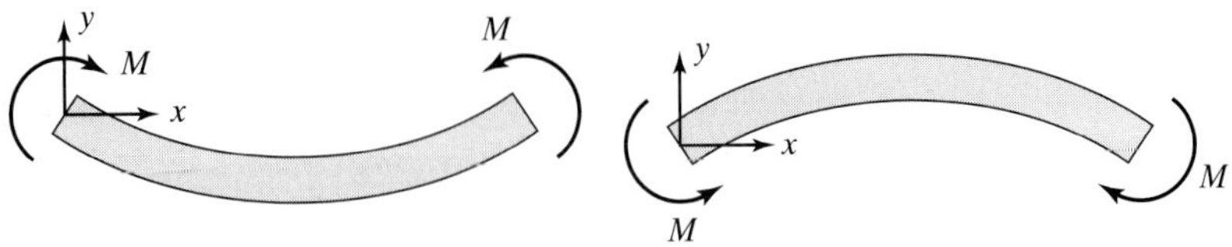

FIGURE 4.7 The positive and negative bending moments and curvature sign convention.

section, the bending moment at that section M, and the second moment of area I is given by the flexure formula. The flexure formula is the equation

$$\sigma = -\frac{My}{I} \tag{4.23}$$

where y locates a point in the cross section of the beam and represents the lateral distance from the neutral axis to that point. The deflection of the neutral axis v is also related to the internal bending moment $M(x)$, the transverse shear $V(x)$, and the load $w(x)$ according to the equations

$$EI\frac{d^2v}{dx^2} = M(x) \tag{4.24}$$

$$EI\frac{d^3v}{dx^3} = \frac{dM(x)}{dx} = V(x) \tag{4.25}$$

$$EI\frac{d^4v}{dx^4} = \frac{dV(x)}{dx} = w(x) \tag{4.26}$$

Note that the standard beam sign convention is assumed in the previous equations. The positive and negative bending moments and curvatures are shown in Figure 4.7. For your reference, the deflections and slopes of beams under some typical loads for simply supported and cantilevered supports are summarized in Table 4.1. If you come across problems that can be analyzed using equations (4.24), (4.25), and (4.26) and Table 4.1, solve them as such.

EXAMPLE 4.3

The cantilevered balcony beam shown in the accompanying figure is a wide-flange W18 $\times$ 35, with a cross-sectional area of 10.3 in^2 and a depth of 17.7 in. The second moment of area is 510 in^4. The beam is subjected to a uniformly distributed load of 1000 lb/ft. The modulus of elasticity of the beam is $E = 29 \times 10^6$ lb/in^2. Using the review

TABLE 4.1 Deflections and slopes of beams under some typical loads and supports

Beam Support and Load	Equation of Elastic Curve	Maximum Deflection	Slope
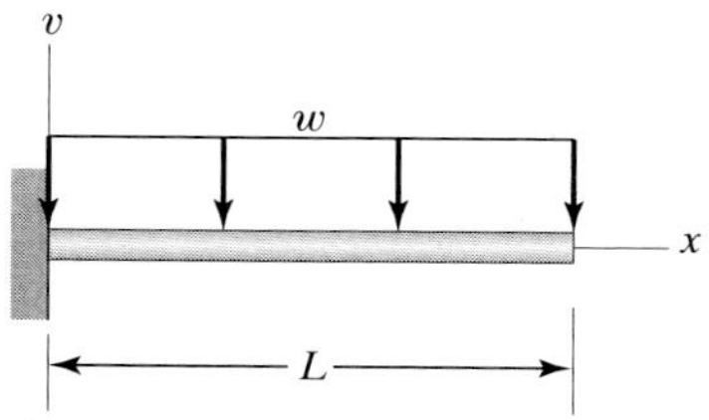	$v = \dfrac{-wx^2}{24EI}(x^2 - 4Lx + 6L^2)$	$v_{max} = \dfrac{-wL^4}{8EI}$	$\theta_{max} = \dfrac{-wL^3}{6EI}$
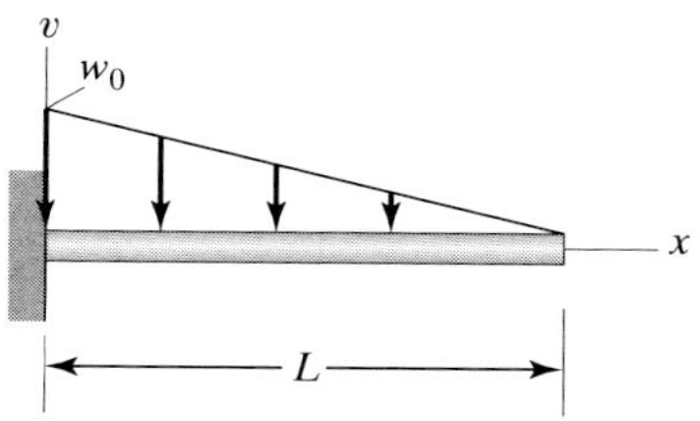	$v = \dfrac{-w_0x^2}{120LEI}(-x^3 + 5Lx^2 - 10L^2x + 10L^3)$	$v_{max} = \dfrac{-w_0L^4}{30EI}$	$\theta_{max} = \dfrac{-w_0L^3}{24EI}$
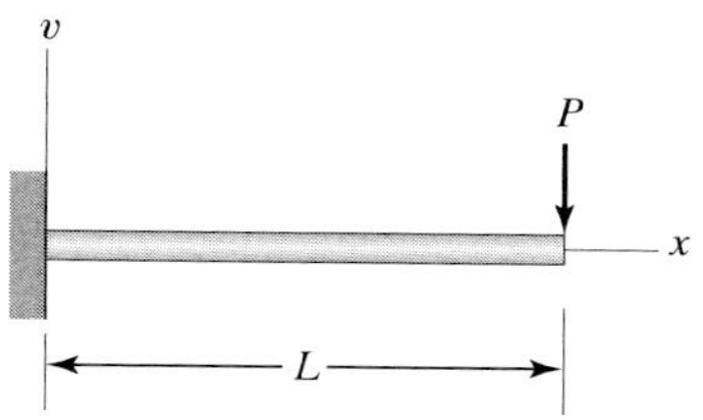	$v = \dfrac{-Px^2}{6EI}(3L - x)$	$v_{max} = \dfrac{-PL^3}{3EI}$	$\epsilon_{max} = \dfrac{-PL^2}{2EI}$

TABLE 4.1 *(continued)* Deflections and slopes of beams under some typical loads and supports

Beam Support and Load	Equation of Elastic Curve	Maximum Deflection	Slope
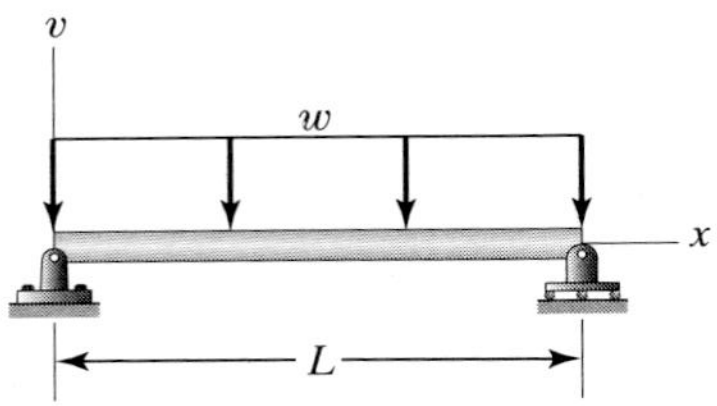	$v = \frac{-wx}{24EI}(x^3 - 2Lx^2 + L^3)$	$v_{max} = \frac{-5wL^4}{384EI}$	$\theta_{max} = \frac{-wL^3}{24EI}$
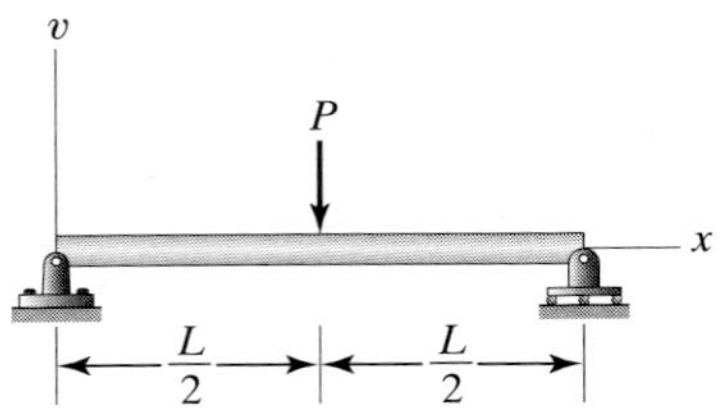	$v = \frac{-Px}{48EI}(3L^2 - 4x^2)$ for $\left(x \le \frac{L}{2}\right)$	$v_{max} = \frac{-PL^3}{48EI}$	$\theta_{max} = \frac{-PL^2}{16EI}$

materials presented in this section, we will determine the deflection of the beam at the midpoint B and the endpoint C. We will also compute the slope of the beam at point C.

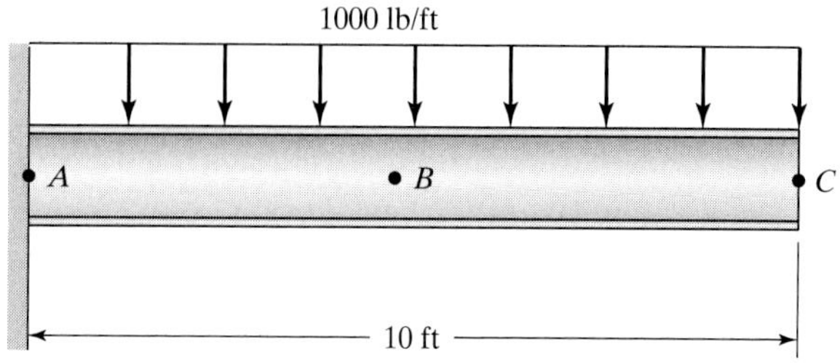

The deflection equation for a cantilever beam is given in Table 4.1.

$$v = \frac{-wx^2}{24EI}(x^2 - 4Lx + 6L^2)$$

The deflection of the beam at midpoint corresponding to $x = \frac{L}{2}$ is

$$v_B = \frac{-wx^2}{24EI}(x^2 - 4Lx + 6L^2)$$

$$= \frac{-(1000 \text{ lb/ft})(5 \text{ ft})^2}{24(29 \times 10^6 \text{ lb/in}^2)(510 \text{ in}^4)}((5^2 - 4(10)(5) + 6(10)^2)\text{ft}^2)\left(\frac{12 \text{ in}}{1 \text{ ft}}\right)^3 = -0.052 \text{ in.}$$

And the deflection of point C is

$$v_c = \frac{-wL^4}{8EI} = \frac{-(1000 \text{ lb/ft})(10 \text{ ft})^4\left(\frac{12 \text{ in}}{1 \text{ ft}}\right)^3}{8(29 \times 10^6 \text{ lb/in}^2)(510 \text{ in}^4)} = -0.146 \text{ in}$$

The maximum slope occurs at point C.

$$\theta_{max} = \frac{-wL^3}{6EI} = \frac{-(1000 \text{ lb/ft})(10 \text{ ft})^3}{6(29 \times 10^6 \text{ lb/in}^2)(510 \text{ in}^4)\left(\frac{1 \text{ ft}}{12 \text{ in}}\right)^2} = -0.00163 \text{ rad}$$

Let us also calculate the maximum bending stress in the beam. Because the maximum bending moment occurs at point A, the maximum bending stress in the beam will occur at point A. The resulting maximum bending stress at outer fiber of the beam at A is

$$\sigma = \frac{My}{I} = -\frac{\overbrace{(1000 \text{ lb/ft})(10 \text{ ft})(5 \text{ ft})}^{M}\left(\frac{12 \text{ in}}{1 \text{ ft}}\right)\overbrace{\left(\frac{17.7}{2} \text{ in}\right)}^{y}}{510 \text{ in}^4} = 10411 \text{ lb/in}^2$$

4.3 FINITE ELEMENT FORMULATION OF BEAMS

Before we proceed with finite element formulation of beams, we should define what we mean by a beam element. A simple beam element consists of two nodes. At each node, there are two degrees of freedom, a vertical displacement, and a rotation angle (slope), as shown in Figure 4.8.

There are four nodal values associated with a beam element. Therefore, we will use a third-order polynomial with four unknown coefficients to represent the displacement field. Moreover, we want the first derivatives of the shape functions to be continuous. The resulting shape functions are commonly referred to as Hermite shape functions. As you will see, they differ in some ways from the linear shape functions you have already studied. We start with the third-order polynomial

$$v = c_1 + c_2x + c_3x^2 + c_4x^3 \tag{4.27}$$

The element's end conditions are given by the following nodal values:

For node i: The vertical displacement at $x = 0 \quad v = c_1 = U_{i1}$

For node i: The slope at $x = 0 \quad \left.\dfrac{dv}{dx}\right|_{x=0} = c_2 = U_{i2}$

For node j: The vertical displacement at $x = L \quad v = c_1 + c_2L + c_3L^2 + c_4L^3 = U_{j1}$

For node j: The slope at $x = L \quad \left.\dfrac{dv}{dx}\right|_{x=L} = c_2 + 2c_3L + 3c_4L^2 = U_{j2}$

We now have four equations with four unknowns. Solving for c_1, c_2, c_3, and c_4; substituting into Eq. (4.27); and regrouping the $U_{i1}, U_{i2}, U_{j1}, U_{j2}$ terms results in the equation

$$v = S_{i1}U_{i1} + S_{i2}U_{i2} + S_{j1}U_{j1} + S_{j2}U_{j2} \tag{4.28}$$

where the shape functions are given by

$$S_{i1} = 1 - \frac{3x^2}{L^2} + \frac{2x^3}{L^3} \tag{4.29}$$

$$S_{i2} = x - \frac{2x^2}{L} + \frac{x^3}{L^2} \tag{4.30}$$

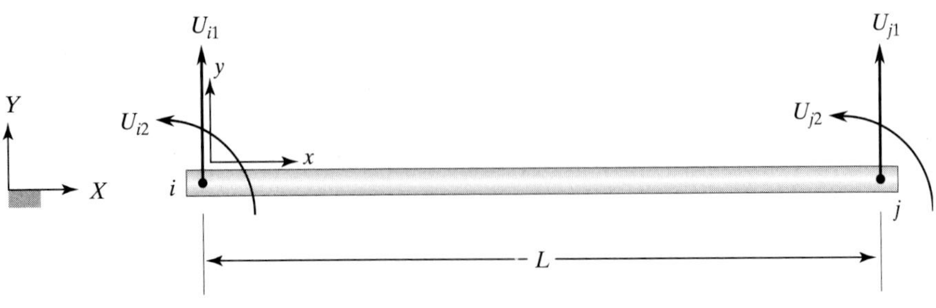

FIGURE 4.8 A beam element.

$$S_{j1} = \frac{3x^2}{L^2} - \frac{2x^3}{L^3} \tag{4.31}$$

$$S_{j2} = -\frac{x^2}{L} + \frac{x^3}{L^2} \tag{4.32}$$

It is clear that if we evaluate the shape functions, as given in Eqs. (4.29 through 4.32), at node i at $x = 0$, we find that $S_{i1} = 1$ and $S_{i2} = S_{j1} = S_{j2} = 0$. Also, if we evaluate the slopes of the shape functions at $x = 0$, we find that $\frac{dS_{i2}}{dx} = 1$ and $\frac{dS_{i1}}{dx} = \frac{dS_{j1}}{dx} = \frac{dS_{j2}}{dx} = 0$. If we evaluate the shape functions at node j at $x = L$, we find that $S_{j1} = 1$ and $S_{i1} = S_{i2} = S_{j2} = 0$, and if we evaluate the slopes of the shape functions at $x = L$, we determine that $\frac{dS_{j2}}{dx} = 1$ and $\frac{dS_{i1}}{dx} = \frac{dS_{i2}}{dx} = \frac{dS_{j1}}{dx} = 0$. These values are the properties of the Hermite third-order polynomials.

Now that you know what we mean by a beam element, we proceed with derivation of stiffness matrix. In the following derivation, we neglect the contribution of shear stresses to the strain energy. The strain energy for an arbitrary beam element (e) then becomes

$$\Lambda^{(e)} = \int_V \frac{\sigma\varepsilon}{2}\,dV = \int_V \frac{E\varepsilon^2}{2}\,dV = \frac{E}{2}\int_V \left(-y\frac{d^2v}{dx^2}\right)^2 dV \tag{4.33}$$

$$\Lambda^{(e)} = \frac{E}{2}\int_V \left(-y\frac{d^2v}{dx^2}\right)^2 dV = \frac{E}{2}\int_0^L \left(\frac{d^2v}{dx^2}\right)^2 dx \int_A y^2\,dA \tag{4.34}$$

Recognizing the integral $\int_A y^2 dA$ as the second moment of the area I, we have

$$\Lambda^{(e)} = \frac{EI}{2}\int_0^L \left(\frac{d^2v}{dx^2}\right)^2 dx \tag{4.35}$$

Next, we substitute for the displacement field v in terms of the shape functions and the nodal values. Let us begin by evaluating the equation

$$\frac{d^2v}{dx^2} = \frac{d^2}{dx^2}[S_{i1} \quad S_{i2} \quad S_{j1} \quad S_{j2}]\begin{Bmatrix} U_{i1} \\ U_{i2} \\ U_{j1} \\ U_{j2} \end{Bmatrix} \tag{4.36}$$

To simplify the next few steps of derivation and to avoid unnecessary mathematical operations, let us make use of matrix notations. First, let the second derivatives of the shape functions be defined in terms of the following relationships:

$$\begin{aligned} D_{i1} &= \frac{d^2 S_{i1}}{dx^2} = -\frac{6}{L^2} + \frac{12x}{L^3} \\ D_{i2} &= \frac{d^2 S_{i2}}{dx^2} = -\frac{4}{L} + \frac{6x}{L^2} \\ D_{j1} &= \frac{d^2 S_{j1}}{dx^2} = \frac{6}{L^2} - \frac{12x}{L^3} \\ D_{j2} &= \frac{d^2 S_{j2}}{dx^2} = -\frac{2}{L} + \frac{6x}{L^2} \end{aligned} \tag{4.36a}$$

Then, Eq. (4.36) takes on the compact-matrix form of

$$\frac{d^2 v}{dx^2} = [\mathbf{D}]\{\mathbf{U}\} \tag{4.37}$$

where $[\mathbf{D}] = [D_{i1} \quad D_{i2} \quad D_{j1} \quad D_{j2}]$ and $\{\mathbf{U}\} = \begin{Bmatrix} U_{i1} \\ U_{i2} \\ U_{j1} \\ U_{j2} \end{Bmatrix}$

The $\left(\frac{d^2 v}{dx^2}\right)^2$ term can be represented in terms of the $\{\mathbf{U}\}$ and $[\mathbf{D}]$ matrices as

$$\left(\frac{d^2 v}{dx^2}\right)^2 = ([\mathbf{D}]\{\mathbf{U}\})([\mathbf{D}]\{\mathbf{U}\}) = \{\mathbf{U}\}^T[\mathbf{D}]^T[\mathbf{D}]\{\mathbf{U}\} \tag{4.38}$$

In Eq. (4.38), note that $[\mathbf{D}]\{\mathbf{U}\} = \{\mathbf{U}\}^T[\mathbf{D}]^T$. The proof of this identity is left as an exercise for you to complete. See Problem 26 at the end of this chapter. Thus, using Eq. (4.38) the strain energy for an arbitrary beam element is

$$\Lambda^{(e)} = \frac{EI}{2}\int_0^L \{\mathbf{U}\}^T[\mathbf{D}]^T[\mathbf{D}]\{\mathbf{U}\}dx \tag{4.39}$$

Recall that the total potential energy Π for a body is the difference between the total strain energy and the work done by the external forces:

$$\Pi = \Sigma\Lambda^{(e)} - \Sigma FU \tag{4.40}$$

Also recall that the minimum total potential energy principle states that for a stable system, the displacement at the equilibrium position occurs such that the value of the system's total potential energy is a minimum. Thus, for a beam element, we have

$$\frac{\partial \Pi}{\partial U_k} = \frac{\partial}{\partial U_k}\Sigma\Lambda^{(e)} - \frac{\partial}{\partial U_k}\Sigma FU = 0 \quad \text{for } k = 1, 2, 3, 4 \tag{4.41}$$

where U_k takes on the values of the nodal degrees of freedom U_{i1}, U_{i2}, U_{j1}, and U_{j2}. Equation (4.40) has two main parts: the strain energy, and the work done by external forces. Differentiation of the strain energy with respect to the nodal degrees of freedom leads to the formulation of the beam's stiffness matrix and differentiation of the work done by external forces results in the load matrix. We begin minimizing the strain energy with respect to U_{i1}, U_{i2}, U_{j1}, and U_{j2} to obtain the stiffness matrix. Starting with the strain energy part of the total potential energy, we get

$$\frac{\partial \Lambda^{(e)}}{\partial U_k} = EI\int_0^L [\mathbf{D}]^T[\mathbf{D}]dx\{\mathbf{U}\} \tag{4.42}$$

Evaluating Eq. (4.42) leads to the expression

$$\frac{\partial \Lambda^{(e)}}{\partial U_k} = EI\int_0^L [\mathbf{D}]^T[\mathbf{D}]dx\{\mathbf{U}\} = \frac{EI}{L^3}\begin{bmatrix} 12 & 6L & -12 & 6L \\ 6L & 4L^2 & -6L & 2L^2 \\ -12 & -6L & 12 & -6L \\ 6L & 2L^2 & -6L & 4L^2 \end{bmatrix}\begin{Bmatrix} U_{i1} \\ U_{i2} \\ U_{j1} \\ U_{j2} \end{Bmatrix}$$

The stiffness matrix for a beam element with two degrees of freedom at each node—the vertical displacement and rotation—is

$$[\mathbf{K}]^{(e)} = \frac{EI}{L^3}\begin{bmatrix} 12 & 6L & -12 & 6L \\ 6L & 4L^2 & -6L & 2L^2 \\ -12 & -6L & 12 & -6L \\ 6L & 2L^2 & -6L & 4L^2 \end{bmatrix} \tag{4.43}$$

Starting with Eqs. (4.39) and (4.41), proof of steps leading to Eqs. (4.42) and (4.43) is left as an exercise for you to perform. See Problem 27.

Load Matrix

There are two ways in which we can formulate the nodal load matrices: (1) by minimizing the work done by the load as stated above, and (2) alternatively by computing the beam's reaction forces. Consider a uniformly distributed load acting on a beam of length L, as shown in Figure 4.9. The reaction forces and moments at the endpoints are also shown in the figure.

Using the first approach, we can compute the work done by this type of loading from $\int_L wv\,dx$. The next step involves substituting for the displacement function in terms of the shape functions and nodal values, and then integrating and differentiating the work term with respect to the nodal displacements. This approach will be

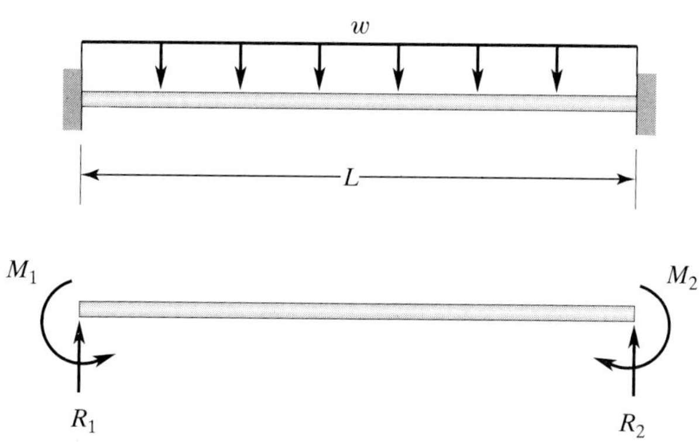

FIGURE 4.9 A beam element subjected to a uniform distributed load.

demonstrated in detail when we formulate the load matrix for a plane stress situation. In order to expose you to as many finite element formulations as possible, let us develop the load matrix using the alternate approach, starting with Eq. (4.26):

$$EI\frac{d^4v}{dx^4} = \frac{dV(x)}{dx} = w(x)$$

For a uniformly distributed load, $w(x)$ is constant. Integrating this equation, we get

$$EI\frac{d^3v}{dx^3} = -wx + c_1 \tag{4.44}$$

Applying the boundary condition (at $x = 0, V(x) = R_1$, and using Eq. (4.25)) $EI\left.\frac{d^3v}{dx^3}\right|_{x=0} = R_1$, we find that $c_1 = R_1$. Substituting for the value of c_1 and integrating Eq. (4.44) we obtain

$$EI\frac{d^2v}{dx^2} = -\frac{wx^2}{2} + R_1x + c_2 \tag{4.45}$$

Applying the boundary condition (at $x = 0, M(x) = -M_1$, and using Eq. (4.24)) $EI\left.\frac{d^2v}{dx^2}\right|_{x=0} = -M_1$, we find that $c_2 = -M_1$. Substituting for the value of c_2 and integrating, we obtain

$$EI\frac{dv}{dx} = -\frac{wx^3}{6} + \frac{R_1x^2}{2} - M_1x + c_3 \tag{4.46}$$

Applying the boundary condition (zero slope at $x = 0$) $\left.\frac{dv}{dx}\right|_{x=0} = 0$, we find that $c_3 = 0$. Integrating one last time, we have

$$EIv = -\frac{wx^4}{24} + \frac{R_1x^3}{6} - \frac{M_1x^2}{2} + c_4 \tag{4.47}$$

Applying the boundary condition (zero deflection at $x = 0$) $v(0) = 0$, we determine that $c_4 = 0$. To obtain the values of R_1 and M_1, we can apply two additional boundary conditions to this problem: $\left.\frac{dv}{dx}\right|_{x=L} = 0$ and $v(L) = 0$. Applying these conditions, we get

$$\left.\frac{dv}{dx}\right|_{x=L} = -\frac{wL^3}{6} + \frac{R_1L^2}{2} - M_1L = 0 \tag{4.48}$$

$$v(L) = -\frac{wL^4}{24} + \frac{R_1L^3}{6} - \frac{M_1L^2}{2} = 0 \tag{4.49}$$

Solving these equations simultaneously, we get $R_1 = \frac{wL}{2}$ and $M_1 = \frac{wL^2}{12}$. From the symmetry of the problem—that is, applying the statics equilibrium conditions—we find that the reactions at the other end of the beam are $R_2 = \frac{wL}{2}$ and $M_2 = \frac{wL^2}{12}$. All of the reactions are shown in Figure 4.10.

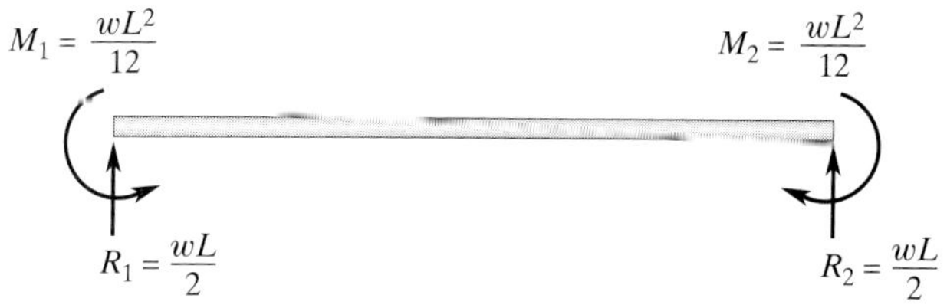

FIGURE 4.10 Reaction results for a beam subjected to a uniformly distributed load.

If we reverse the signs of the reactions at the endpoints, we can now represent the effect of a uniformly distributed load in terms of its equivalent nodal loads. Similarly, we can obtain the nodal load matrices for other loading situations. The relationships between the actual load and its equivalent nodal loads for some typical loading situations are summarized in Table 4.2.

TABLE 4.2 Equivalent nodal loading of beams

Loading	Equivalent Nodal Loading

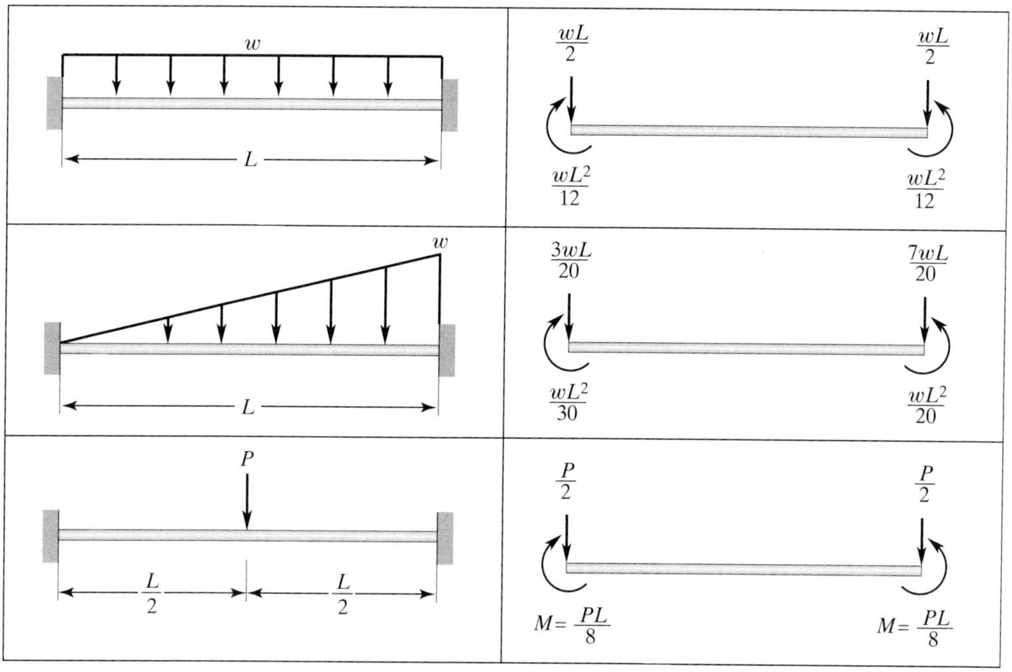

EXAMPLE 4.3 Revisited

Let us consider the cantilevered balcony beam of Example 4.3 again and solve it using a single beam element. Recall that the beam is a wide-flange W18 × 35, with a cross-sectional area of 10.3 in^2 and a depth of 17.7 in. The second moment of area is 510 in^4. The beam is subjected to a uniformly distributed load of 1000 lb/ft. The modulus of elasticity of the beam $E = 29 \times 10^6\,\text{lb/in}^2$. We are interested in determining the deflection of the beam at the midpoint B and the endpoint C. Also, we will compute the maximum slope that will occur at point C.

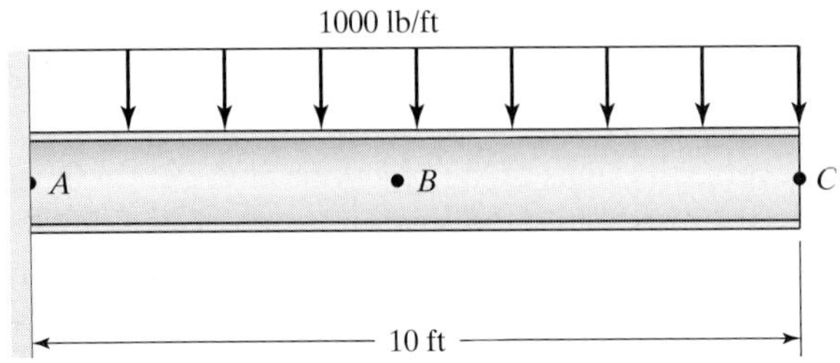

Because we are using a single element to model this problem, the elemental stiffness and load matrices are the same as the global matrices.

$$[\boldsymbol{K}]^{(e)} = [\boldsymbol{K}]^{(G)} = \frac{EI}{L^3}\begin{bmatrix} 12 & 6L & -12 & 6L \\ 6L & 4L^2 & -6L & 2L^2 \\ -12 & -6L & 12 & -6L \\ 6L & 2L^2 & -6L & 4L^2 \end{bmatrix} \quad \{\boldsymbol{F}\}^{(e)} = \{\boldsymbol{F}\}^{(G)} = \begin{Bmatrix} -\dfrac{wL}{2} \\ -\dfrac{wL^2}{12} \\ -\dfrac{wL}{2} \\ \dfrac{wL^2}{12} \end{Bmatrix}$$

$$\frac{EI}{L^3}\begin{bmatrix} 12 & 6L & -12 & 6L \\ 6L & 4L^2 & -6L & 2L^2 \\ -12 & -6L & 12 & -6L \\ 6L & 2L^2 & -6L & 4L^2 \end{bmatrix}\begin{Bmatrix} U_{11} \\ U_{12} \\ U_{21} \\ U_{22} \end{Bmatrix} = \begin{Bmatrix} -\dfrac{wL}{2} \\ -\dfrac{wL^2}{12} \\ -\dfrac{wL}{2} \\ \dfrac{wL^2}{12} \end{Bmatrix}$$

Applying the boundary conditions $U_{11} = 0$ and $U_{12} = 0$ at node 1, we have

$$\frac{EI}{L^3}\begin{bmatrix} 1 & 0 & 0 & 0 \\ 0 & 1 & 0 & 0 \\ -12 & -6L & 12 & -6L \\ 6L & 2L^2 & -6L & 4L^2 \end{bmatrix}\begin{Bmatrix} U_{11} \\ U_{12} \\ U_{21} \\ U_{22} \end{Bmatrix} = \begin{Bmatrix} 0 \\ 0 \\ -\dfrac{wL}{2} \\ \dfrac{wL^2}{12} \end{Bmatrix}$$

And simplifying, we get

$$\begin{bmatrix} 12 & -6L \\ -6L & 4L^2 \end{bmatrix} \begin{Bmatrix} U_{21} \\ U_{22} \end{Bmatrix} = \frac{L^3}{EI} \begin{Bmatrix} -\dfrac{wL}{2} \\ \dfrac{wL^2}{12} \end{Bmatrix}$$

$$\begin{bmatrix} 12 & -6(10\text{ ft}) \\ -6(10\text{ ft}) & 4(10\text{ ft})^2 \end{bmatrix} \begin{Bmatrix} U_{21} \\ U_{22} \end{Bmatrix} = \frac{(10\text{ ft})^3}{(29 \times 10^6 \text{ lb/in}^2)(510 \text{ in}^4)\left(\dfrac{1\text{ ft}}{12\text{ in.}}\right)^2} \begin{Bmatrix} -\dfrac{1000(10)}{2} \\ \dfrac{(1000)(10)^2}{12} \end{Bmatrix}$$

The deflection and the slope at endpoint C is

$$U_{21} = -0.01217 \text{ ft} = -0.146 \text{ in} \quad \text{and} \quad U_{22} = -0.00163 \text{ rad}$$

To determine the deflection at point B, we use the deflection equation for the beam element and evaluate the shape functions at $x = \dfrac{L}{2}$.

$$\begin{aligned} v &= S_{11}U_{11} + S_{12}U_{12} + S_{21}U_{21} + S_{22}U_{22} \\ &= S_{11}(0) + S_{12}(0) + S_{21}(-0.146) + S_{22}(-0.00163) \end{aligned}$$

Computing the values of the shape functions at point B, we have

$$S_{21} = \frac{3x^2}{L^2} - \frac{2x^3}{L^3} = \frac{3}{L^2}\left(\frac{L}{2}\right)^2 - \frac{2}{L^3}\left(\frac{L}{2}\right)^3 = \frac{1}{2}$$

$$S_{22} = -\frac{x^2}{L} + \frac{x^3}{L^2} = -\frac{\left(\frac{L}{2}\right)^2}{L} + \frac{\left(\frac{L}{2}\right)^3}{L^2} = -\frac{L}{8}$$

$$v_B = \left(\frac{1}{2}\right)(-0.146 \text{ in}) + \left(-\frac{120 \text{ in.}}{8}\right)(-0.00163 \text{ rad}) = -0.048 \text{ in}$$

Comparing results of our finite element model to the exact solutions given in Example 4.3, we note that they are in good agreement. We could improve our results for the midpoint deflection by using a model that uses two elements. We have left this as an exercise for you.

EXAMPLE 4.4

The beam shown in Figure 4.11 is a wide-flange W310 $\times$ 52 with a cross-sectional area of 6650 mm^2 and depth of 317 mm. The second moment of the area is 118.6×10^6 mm^4. The beam is subjected to a uniformly distributed load of 25,000 N/m. The modulus of elasticity of the beam is $E = 200$ GPa. Determine the vertical displacement at node 3 and the rotations at nodes 2 and 3. Also, compute the reaction forces and moment at nodes 1 and 2.

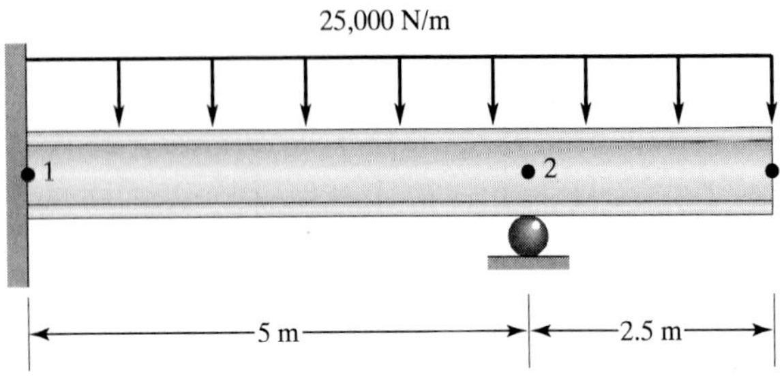

FIGURE 4.11 A schematic of the beam in Example 4.4

Note that this problem is statically indeterminate. We will use two elements to represent this problem. The stiffness matrices of the elements are computed from Eq. (4.43):

$$[\mathbf{K}]^{(e)} = \frac{EI}{L^3}\begin{bmatrix} 12 & 6L & -12 & 6L \\ 6L & 4L^2 & -6L & 2L^2 \\ -12 & -6L & 12 & -6L \\ 6L & 2L^2 & -6L & 4L^2 \end{bmatrix}$$

Substituting appropriate values for element (1), we have

$$[\mathbf{K}]^{(1)} = \frac{200 \times 10^9 \times 1.186 \times 10^{-4}}{5^3}\begin{bmatrix} 12 & 6(5) & -12 & 6(5) \\ 6(5) & 4(5)^2 & -6(5) & 2(5)^2 \\ -12 & -6(5) & 12 & -6(5) \\ 6(5) & 2(5)^2 & -6(5) & 4(5)^2 \end{bmatrix}$$

For convenience, the nodal degrees of freedom are shown alongside the stiffness matrices. For element (1), we have

$$[\mathbf{K}]^{(1)} = \begin{bmatrix} 2277120 & 5692800 & -2277120 & 5692800 \\ 5692800 & 18976000 & -5692800 & 9488000 \\ -2277120 & -5692800 & 2277120 & -5692800 \\ 5692800 & 9488000 & -5692800 & 18976000 \end{bmatrix}\begin{matrix} U_{11} \\ U_{12} \\ U_{21} \\ U_{22} \end{matrix}$$

Computing the stiffness matrix for element (2), we have

$$[\mathbf{K}]^{(2)} = \frac{200 \times 10^9 \times 1.186 \times 10^{-4}}{(2.5)^3}\begin{bmatrix} 12 & 6(2.5) & -12 & 6(2.5) \\ 6(2.5) & 4(2.5)^2 & -6(2.5) & 2(2.5)^2 \\ -12 & -6(2.5) & 12 & -6(2.5) \\ 6(2.5) & 2(2.5)^2 & -6(2.5) & 4(2.5)^2 \end{bmatrix}$$

Showing the nodal degrees of freedom alongside the stiffness matrix for element (2), we have

$$[\mathbf{K}]^{(2)} = \begin{bmatrix} 18216960 & 22771200 & -18216960 & 22771200 \\ 22771200 & 37952000 & -22771200 & 18976000 \\ -18216960 & -22771200 & 18216960 & -22771200 \\ 22771200 & 18976000 & -22771200 & 37952000 \end{bmatrix}\begin{matrix} U_{21} \\ U_{22} \\ U_{31} \\ U_{32} \end{matrix}$$

Assembling $[\mathbf{K}]^{(1)}$ and $[\mathbf{K}]^{(2)}$ to obtain the global stiffness matrix yields

$$[\mathbf{K}]^{(G)} = \begin{bmatrix} 2277120 & 5692800 & 2277120 & 5692800 & 0 & 0 \\ 5692800 & 18976000 & -5692800 & 9488000 & 0 & 0 \\ -2277120 & -5692800 & 20494080 & 17078400 & -18216960 & 22771200 \\ 5692800 & 9488000 & 17078400 & 56928000 & -22771200 & 18976000 \\ 0 & 0 & -18216960 & -22771200 & 18216960 & -22771200 \\ 0 & 0 & 22771200 & 18976000 & -22771200 & 37952000 \end{bmatrix}$$

Referring to Table 4.2, we can compute the load matrix for elements (1) and (2). The respective load matrices are

$$\{\mathbf{F}\}^{(1)} = \begin{Bmatrix} -\dfrac{wL}{2} \\ -\dfrac{wL^2}{12} \\ -\dfrac{wL}{2} \\ \dfrac{wL^2}{12} \end{Bmatrix} = \begin{Bmatrix} -\dfrac{25 \times 10^3 \times 5}{2} \\ -\dfrac{25 \times 10^3 \times 5^2}{12} \\ -\dfrac{25 \times 10^3 \times 5}{2} \\ \dfrac{25 \times 10^3 \times 5^2}{12} \end{Bmatrix} = \begin{Bmatrix} -62500 \\ -52083 \\ -62500 \\ 52083 \end{Bmatrix}$$

$$\{\mathbf{F}\}^{(2)} = \begin{Bmatrix} -\dfrac{wL}{2} \\ -\dfrac{wL^2}{12} \\ -\dfrac{wL}{2} \\ \dfrac{wL^2}{12} \end{Bmatrix} = \begin{Bmatrix} -\dfrac{25 \times 10^3 \times 2.5}{2} \\ -\dfrac{25 \times 10^3 \times 2.5^2}{12} \\ -\dfrac{25 \times 10^3 \times 2.5}{2} \\ \dfrac{25 \times 10^3 \times 2.5^2}{12} \end{Bmatrix} = \begin{Bmatrix} -31250 \\ -13021 \\ -31250 \\ 13021 \end{Bmatrix}$$

Combining the two load matrices to obtain the global load matrix, we obtain

$$\{\mathbf{F}\}^{(G)} = \begin{Bmatrix} -62500 \\ -52083 \\ -62500 - 31250 \\ 52083 - 13021 \\ -31250 \\ 13021 \end{Bmatrix} = \begin{Bmatrix} -62500 \\ -52083 \\ -93750 \\ 39062 \\ -31250 \\ 13021 \end{Bmatrix}$$

Applying the boundary conditions $U_{11} = U_{12} = 0$ at node 1 and the boundary condition $U_{21} = 0$ at node 2, we have

$$\begin{bmatrix} 1 & 0 & 0 & 0 & 0 & 0 \\ 0 & 1 & 0 & 0 & 0 & 0 \\ 0 & 0 & 1 & 0 & 0 & 0 \\ 5692800 & 9488000 & 17078400 & 56928000 & -22771200 & 18976000 \\ 0 & 0 & -18216960 & -22771200 & 18216960 & -22771200 \\ 0 & 0 & 22771200 & 18976000 & -22771200 & 37952000 \end{bmatrix} \begin{Bmatrix} U_{11} \\ U_{12} \\ U_{21} \\ U_{22} \\ U_{31} \\ U_{32} \end{Bmatrix} = \begin{Bmatrix} 0 \\ 0 \\ 0 \\ 39062 \\ -31250 \\ 13021 \end{Bmatrix}$$

Considering the applied boundary conditions, we reduce the global stiffness matrix and the load matrix to

$$\begin{bmatrix} 56928000 & -22771200 & 18976000 \\ -22771200 & 18216960 & -22771200 \\ 18976000 & -22771200 & 37952000 \end{bmatrix} \begin{Bmatrix} U_{22} \\ U_{31} \\ U_{32} \end{Bmatrix} = \begin{Bmatrix} 39062 \\ -31250 \\ 13021 \end{Bmatrix}$$

Solving the three equations simultaneously results in the unknown nodal values. The displacement result is

$$[\mathbf{U}]^T = [0 \quad 0 \quad 0 \quad -0.0013723(\text{rad}) \quad -0.0085772(\text{m}) \quad -0.004117(\text{rad})]$$

We can compute the nodal reaction forces and moments from the relationship

$$\{\mathbf{R}\} = [\mathbf{K}]\{\mathbf{U}\} - \{\mathbf{F}\} \tag{4.50}$$

where $\{\mathbf{R}\}$ is the reaction matrix. Substituting for the appropriate values in Eq. (4.50), we have

$$\begin{Bmatrix} R_1 \\ M_1 \\ R_2 \\ M_2 \\ R_3 \\ M_3 \end{Bmatrix} = \begin{bmatrix} 2277120 & 5692800 & -2277120 & 5692800 & 0 & 0 \\ 5692800 & 18976000 & -5692800 & 9488000 & 0 & 0 \\ -2277120 & -5692800 & 20494080 & 17078400 & -18216960 & 22771200 \\ 5692800 & 9488000 & 17078400 & 56928000 & -22771200 & 18976000 \\ 0 & 0 & -18216960 & -22771200 & 18216960 & -22771200 \\ 0 & 0 & 22771200 & 18976000 & -22771200 & 37952000 \end{bmatrix} \times$$

$$\begin{Bmatrix} 0 \\ 0 \\ 0 \\ -0.0013723 \\ -0.0085772 \\ -0.0041170 \end{Bmatrix} - \begin{Bmatrix} -62500 \\ -52083 \\ -93750 \\ 39062 \\ -31250 \\ 13021 \end{Bmatrix}$$

Performing the matrix operation results in the following reaction forces and moments at each node:

$$\begin{Bmatrix} R_1 \\ M_1 \\ R_2 \\ M_2 \\ R_3 \\ M_3 \end{Bmatrix} = \begin{Bmatrix} 54687(\text{N}) \\ 39062(\text{N} \cdot \text{m}) \\ 132814(\text{N}) \\ 0 \\ 0 \\ 0 \end{Bmatrix}$$

Note that by calculating the reaction matrix using the nodal displacement matrix, we can check the validity of our results. There is a reaction force and a reaction moment at node 1; there is a reaction force at node 2; there is no reaction moment at node 2, as expected; and there are no reaction forces or moments at node 3, as expected. The accuracy of the results is discussed further in Section 4.7.

4.4 FINITE ELEMENT FORMULATION OF FRAMES

Frames represent structural members that may be rigidly connected with welded joints or bolted joints. For such structures, in addition to rotation and lateral displacement, we also need to be concerned about axial deformations. Here, we focus on plane frames. The frame element, shown in Figure 4.12, consists of two nodes. At each node, there are three degrees of freedom: a longitudinal displacement, a lateral displacement, and a rotation.

Referring to Figure 4.12, note that u_{i1} represents the longitudinal displacement and u_{i2} and u_{i3} represent the lateral displacement and the rotation at node i, respectively. In the same manner, u_{j1}, u_{j2}, and u_{j3} represent the longitudinal displacement, the lateral displacement, and the rotation at node j, respectively. In general, two frames of reference are required to describe frame elements: a global coordinate system and a local frame of reference. We choose a fixed global coordinate system (X, Y) for several uses: (1) to represent the location of each joint (node) and to keep track of the orientation of each element using angles such as θ; (2) to apply the constraints and the applied

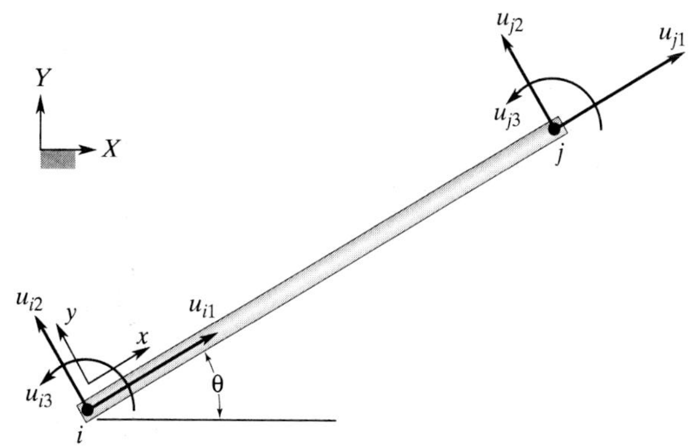

FIGURE 4.12 A frame element.

loads in terms of their respective global components; and (3) to represent the solution. We also need a local, or elemental, coordinate system to describe the axial-load behavior of an element. The relationship between the local coordinate system (x, y) and the global coordinate system (X, Y) is shown in Figure 4.12. Because there are three degrees of freedom associated with each node, the stiffness matrix for the frame element will be a 6×6 matrix. The local degrees of freedom are related to the global degrees of freedom through the transformation matrix, according to the relationship

$$[\mathbf{u}] = [\mathbf{T}][\mathbf{U}] \tag{4.51}$$

where the transformation matrix is

$$[\mathbf{T}] = \begin{bmatrix} \cos\theta & \sin\theta & 0 & 0 & 0 & 0 \\ -\sin\theta & \cos\theta & 0 & 0 & 0 & 0 \\ 0 & 0 & 1 & 0 & 0 & 0 \\ 0 & 0 & 0 & \cos\theta & \sin\theta & 0 \\ 0 & 0 & 0 & -\sin\theta & \cos\theta & 0 \\ 0 & 0 & 0 & 0 & 0 & 1 \end{bmatrix} \tag{4.52}$$

In the previous section, we developed the stiffness matrix attributed to bending for a beam element. This matrix accounts for lateral displacements and rotations at each node and is

$$[\mathbf{K}]_{xy}^{(e)} = \frac{EI}{L^3}\begin{array}{c} \begin{matrix} u_{i1} & u_{i2} & u_{i3} & u_{j1} & u_{j2} & u_{j3} \end{matrix} \\ \begin{bmatrix} 0 & 0 & 0 & 0 & 0 & 0 \\ 0 & 12 & 6L & 0 & -12 & 6L \\ 0 & 6L & 4L^2 & 0 & -6L & 2L^2 \\ 0 & 0 & 0 & 0 & 0 & 0 \\ 0 & -12 & -6L & 0 & 12 & -6L \\ 0 & 6L & 2L^2 & 0 & -6L & 4L^2 \end{bmatrix} \end{array} \begin{matrix} u_{i1} \\ u_{i2} \\ u_{i3} \\ u_{j1} \\ u_{j2} \\ u_{j3} \end{matrix} \tag{4.53}$$

To represent the contribution of each term to nodal degrees of freedom, the degrees of freedom are shown above and alongside the stiffness matrix in Eq. (4.53). In Section 4.1 we derived the stiffness matrix for members under axial loading as

$$[\mathbf{K}]_{\text{axial}}^{(e)} = \begin{array}{c} \begin{matrix} u_{i1} & u_{i2} & u_{i3} & u_{j1} & u_{j2} & u_{j3} \end{matrix} \\ \begin{bmatrix} \frac{AE}{L} & 0 & 0 & -\frac{AE}{L} & 0 & 0 \\ 0 & 0 & 0 & 0 & 0 & 0 \\ 0 & 0 & 0 & 0 & 0 & 0 \\ -\frac{AE}{L} & 0 & 0 & \frac{AE}{L} & 0 & 0 \\ 0 & 0 & 0 & 0 & 0 & 0 \\ 0 & 0 & 0 & 0 & 0 & 0 \end{bmatrix} \end{array} \begin{matrix} u_{i1} \\ u_{i2} \\ u_{i3} \\ u_{j1} \\ u_{j2} \\ u_{j3} \end{matrix} \tag{4.54}$$

Adding Eqs. (4.53) and (4.54) results in the stiffness matrix for a frame element with respect to local coordinate system x, y

$$[\mathbf{K}]_{xy}^{(e)} = \begin{bmatrix} \frac{AE}{L} & 0 & 0 & -\frac{AE}{L} & 0 & 0 \\ 0 & \frac{12EI}{L^3} & \frac{6EI}{L^2} & 0 & -\frac{12EI}{L^3} & \frac{6EI}{L^2} \\ 0 & \frac{6EI}{L^2} & \frac{4EI}{L} & 0 & -\frac{6EI}{L^2} & \frac{2EI}{L} \\ -\frac{AE}{L} & 0 & 0 & \frac{AE}{L} & 0 & 0 \\ 0 & -\frac{12EI}{L^3} & -\frac{6EI}{L^2} & 0 & \frac{12EI}{L^3} & -\frac{6EI}{L^2} \\ 0 & \frac{6EI}{L^2} & \frac{2EI}{L} & 0 & -\frac{6EI}{L^2} & \frac{4EI}{L} \end{bmatrix} \tag{4.55}$$

Note that we need to represent Eq. (4.55) with respect to the global coordinate system. To perform this task, we must substitute for the local displacements in terms of the global displacements in the strain energy equation, using the transformation matrix and performing the minimization. (See Problem 4.13.) These steps result in the relationship

$$[\mathbf{K}]^{(e)} = [\mathbf{T}]^T[\mathbf{K}]_{xy}^{(e)}[\mathbf{T}] \tag{4.56}$$

where $[\mathbf{K}]^{(e)}$ is the stiffness matrix for a frame element expressed in the global coordinate system X, Y. Next, we will demonstrate finite element modeling of frames with another example.

EXAMPLE 4.5

Consider the overhang frame shown in Figure 4.13. The frame is made of steel, with $E = 30 \times 10^6 \text{ lb/in}^2$. The cross-sectional areas and the second moment of areas for the two members are shown in Figure 4.13. The frame is fixed as shown in the figure, and we are interested in determining the deformation of the frame under the given distributed load.

We model the problem using two elements. For element (1), the relationship between the local and the global coordinate systems is shown in Figure 4.14.

Similarly, the relationship between the coordinate systems for element (2) is shown in Figure 4.15.

Note that for this problem, the boundary conditions are $U_{11} = U_{12} = U_{13} = U_{31} = U_{32} = U_{33} = 0$. For element (1), the local and the global frames of reference are aligned in the same direction; therefore, the stiffness matrix for element (1) can be computed from Eq. (4.55) resulting in

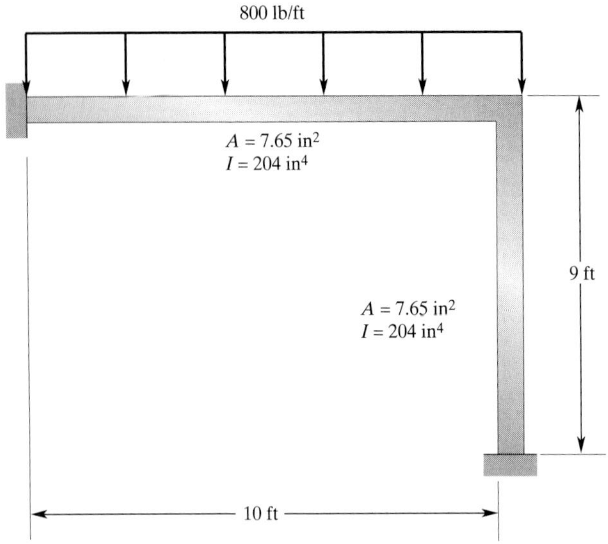

FIGURE 4.13 An overhang frame supporting a distributed load.

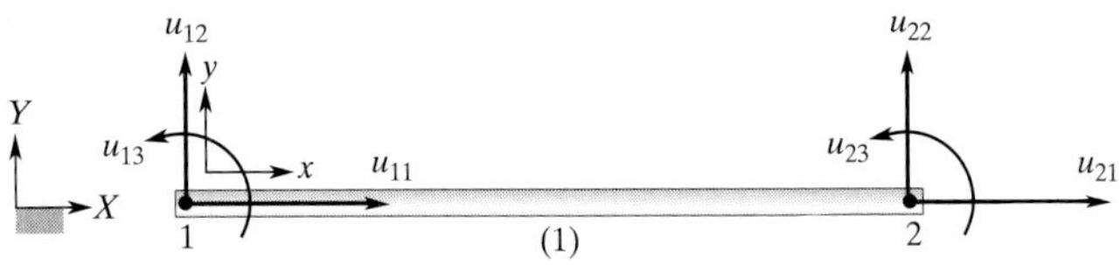

FIGURE 4.14 The configuration of element (1).

$$[\mathbf{K}]^{(1)} = 10^3 \begin{bmatrix} 1912.5 & 0 & 0 & -1912.5 & 0 & 0 \\ 0 & 42.5 & 2550 & 0 & -42.5 & 2550 \\ 0 & 2550 & 204000 & 0 & -2550 & 102000 \\ -1912.5 & 0 & 0 & 1912.5 & 0 & 0 \\ 0 & -42.5 & -2550 & 0 & 42.5 & -2550 \\ 0 & 2550 & 102000 & 0 & -2550 & 204000 \end{bmatrix}$$

For element (2), the stiffness matrix represented with respect to the local coordinate system is

$$[\mathbf{K}]^{(2)}_{xy} = 10^3 \begin{bmatrix} 2125 & 0 & 0 & -2125 & 0 & 0 \\ 0 & 58.299 & 3148.148 & 0 & -58.299 & 3148.148 \\ 0 & 3148.148 & 226666 & 0 & -3148.148 & 113333 \\ -2125 & 0 & 0 & 2125 & 0 & 0 \\ 0 & -58.299 & -3148.148 & 0 & 58.299 & -3148.148 \\ 0 & 3148.148 & 113333 & 0 & -3148.148 & 226666 \end{bmatrix}$$

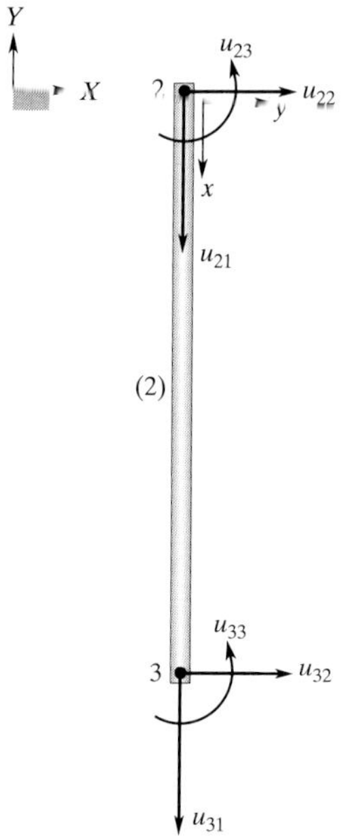

FIGURE 4.15 The configuration of element (2).

For element (2), the transformation matrix is

$$[\mathbf{T}] = \begin{bmatrix} \cos(270) & \sin(270) & 0 & 0 & 0 & 0 \\ -\sin(270) & \cos(270) & 0 & 0 & 0 & 0 \\ 0 & 0 & 1 & 0 & 0 & 0 \\ 0 & 0 & 0 & \cos(270) & \sin(270) & 0 \\ 0 & 0 & 0 & -\sin(270) & \cos(270) & 0 \\ 0 & 0 & 0 & 0 & 0 & 1 \end{bmatrix}$$

$$[\mathbf{T}] = \begin{bmatrix} 0 & -1 & 0 & 0 & 0 & 0 \\ 1 & 0 & 0 & 0 & 0 & 0 \\ 0 & 0 & 1 & 0 & 0 & 0 \\ 0 & 0 & 0 & 0 & -1 & 0 \\ 0 & 0 & 0 & 1 & 0 & 0 \\ 0 & 0 & 0 & 0 & 0 & 1 \end{bmatrix}$$

The transpose of the transformation matrix is

$$[\mathbf{T}]^T = \begin{bmatrix} 0 & 1 & 0 & 0 & 0 & 0 \\ -1 & 0 & 0 & 0 & 0 & 0 \\ 0 & 0 & 1 & 0 & 0 & 0 \\ 0 & 0 & 0 & 0 & 1 & 0 \\ 0 & 0 & 0 & -1 & 0 & 0 \\ 0 & 0 & 0 & 0 & 0 & 1 \end{bmatrix}$$

Substituting for $[\mathbf{T}]^T$, $[\mathbf{K}]^{(2)}_{xy}$, and $[\mathbf{T}]$ into Eq. (4.56), we have

$$[\mathbf{K}]^{(2)} = 10^3 \begin{bmatrix} 0 & 1 & 0 & 0 & 0 & 0 \\ -1 & 0 & 0 & 0 & 0 & 0 \\ 0 & 0 & 1 & 0 & 0 & 0 \\ 0 & 0 & 0 & 0 & 1 & 0 \\ 0 & 0 & 0 & -1 & 0 & 0 \\ 0 & 0 & 0 & 0 & 0 & 1 \end{bmatrix} \begin{bmatrix} 2125 & 0 & 0 & -2125 & 0 & 0 \\ 0 & 58.299 & 3148.148 & 0 & -58.299 & 3148.148 \\ 0 & 3148.148 & 226666 & 0 & -3148.148 & 113333 \\ -2125 & 0 & 0 & 2125 & 0 & 0 \\ 0 & -58.299 & -3148.148 & 0 & 58.299 & -3148.148 \\ 0 & 3148.148 & 113333 & 0 & -3148.148 & 226666 \end{bmatrix} \begin{bmatrix} 0 & -1 & 0 & 0 & 0 & 0 \\ 1 & 0 & 0 & 0 & 0 & 0 \\ 0 & 0 & 1 & 0 & 0 & 0 \\ 0 & 0 & 0 & 0 & -1 & 0 \\ 0 & 0 & 0 & 1 & 0 & 0 \\ 0 & 0 & 0 & 0 & 0 & 1 \end{bmatrix}$$

and performing the matrix operation, we obtain

$$[\mathbf{K}]^{(2)} = 10^3 \begin{bmatrix} 58.299 & 0 & 3148.148 & -58.299 & 0 & 3148.148 \\ 0 & 2125 & 0 & 0 & -2125 & 0 \\ 3148.148 & 0 & 226666 & -3148.148 & 0 & 113333 \\ -58.299 & 0 & -3148.148 & 58.299 & 0 & -3148.1480 \\ 0 & -2125 & 0 & 0 & 2125 & 0 \\ 3148.148 & 0 & 113333 & -3148.148 & 0 & 226666 \end{bmatrix}$$

Constructing the global stiffness matrix by assembling $[\mathbf{K}]^{(1)}$ and $[\mathbf{K}]^{(2)}$, we have

$$[\mathbf{K}]^{(G)} = 10^3 \begin{bmatrix} 1912.5 & 0 & 0 & -1912.5 & 0 & 0 & 0 & 0 & 0 \\ 0 & 42.5 & 2550 & 0 & -42.5 & 2550 & 0 & 0 & 0 \\ 0 & 2550 & 204000 & 0 & -2550 & 102000 & 0 & 0 & 0 \\ -1912.5 & 0 & 0 & 1912.5 + 58.299 & 0 & 0 + 3148.148 & -58.299 & 0 & 3148.148 \\ 0 & -42.5 & -2550 & 0 & 42.5 + 2125 & -2550 & 0 & -2125 & 0 \\ 0 & 2550 & 102000 & 0 + 3148.148 & -2550 & 204000 + 226666 & -3148.148 & 0 & 113333 \\ 0 & 0 & 0 & -58.299 & 0 & -3148.148 & 58.299 & 0 & -3148.1480 \\ 0 & 0 & 0 & 0 & -2125 & 0 & 0 & 2125 & 0 \\ 0 & 0 & 0 & 3148.148 & 0 & 113333 & -3148.148 & 0 & 226666 \end{bmatrix}$$

The load matrix is

$$\{\mathbf{F}\}^{(1)} = \begin{Bmatrix} 0 \\ -\dfrac{wL}{2} \\ -\dfrac{wL^2}{12} \\ 0 \\ -\dfrac{wL}{2} \\ \dfrac{wL^2}{12} \end{Bmatrix} = \begin{Bmatrix} 0 \\ -\dfrac{800 \times 10}{2} \\ -\dfrac{800 \times 10^2 \times 12}{12} \\ 0 \\ -\dfrac{800 \times 10}{2} \\ \dfrac{800 \times 10^2 \times 12}{12} \end{Bmatrix} = \begin{Bmatrix} 0 \\ -4000 \\ -80000 \\ 0 \\ -4000 \\ 80000 \end{Bmatrix}$$

In the load matrix, the force terms have the units of lb, whereas the moment terms have the units of lb·in. Application of the boundary conditions ($U_{11} = U_{12} = U_{13} = U_{31} = U_{32} = U_{33} = 0$) reduces the 9 × 9 global stiffness matrix to the following 3 × 3 matrix:

$$10^3 \begin{bmatrix} 1970.799 & 0 & 3148.148 \\ 0 & 2167.5 & -2550 \\ 3148.148 & -2550 & 430666 \end{bmatrix} \begin{Bmatrix} U_{21} \\ U_{22} \\ U_{23} \end{Bmatrix} = \begin{Bmatrix} 0 \\ -4000 \\ 80000 \end{Bmatrix}$$

Solving these equations simultaneously results in the following displacement matrix:

$$[\mathbf{U}]^T = [0 \quad 0 \quad 0 \quad -0.0002845(\text{in}) \quad -0.0016359(\text{in}) \quad 0.00017815(\text{rad}) \quad 0 \quad 0 \quad 0]$$

This problem will be revisited later in the chapter and solved with ANSYS.

4.5 THREE-DIMENSIONAL BEAM ELEMENT

ANSYS's three-dimensional beam element is suited for situations wherein the beam may be subjected to loads that can create tension, compression, bending about different axes, and twisting (torsion). At each node, there are six degrees of freedom, displacements in X-, Y-, and Z-directions, and rotation about X-, Y-, and Z-axes. Therefore, the elemental matrix for a three-dimensional beam element is a 12 × 12 matrix. ANSYS's three-dimensional elastic beam element is shown in Figure 4.16.

The element input data include node locations, the cross-sectional area, two moments of inertia (I_{ZZ} and I_{YY}), two thickness (TKY and TKZ), an angle of orientation (Θ) about the element x-axis, the torsional moment of inertia (I_{XX}), and the material properties. If I_{XX} is not specified or is equal to 0.0, ANSYS assumes that it is equal to the polar moment of inertia ($I_{YY} + I_{ZZ}$). Note that BEAM4 element is defined by two or three nodes. The element x-axis is oriented from node I toward node J. For the two-node option, the default ($\Theta = 0°$) orientation of the element y-axis, ANSYS automatically sets the orientation parallel to the global X–Y plane. For the case where the element is parallel to the global Z axis, the element y-axis is oriented parallel to the global Y-axis (as shown). For user control of the element orientation about the

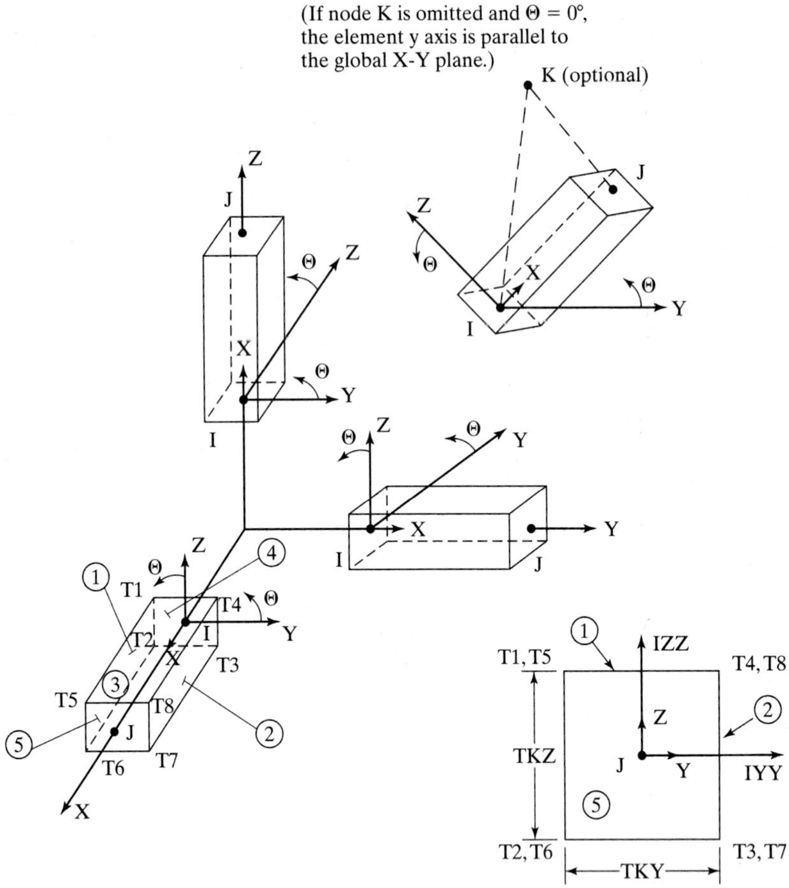

FIGURE 4.16 BEAM4 element, the three-dimensional elastic beam element used by ANSYS.

element x-axis, use the θ angle (THETA) or the third-node option. If both are defined, the third-node option takes precedence. The third node (K), if used, defines a plane (with I and J) containing the element x and z axes (as shown). The input data for BEAM4 is summarized below:

Nodes

I, J, K (K orientation node is optional)

Degrees of Freedom

UX, UY, UZ (displacements in X, Y, and Z-directions)

ROTX (rotation about X-axis), ROTY (rotation about Y-axis), ROTZ (rotation about Z-axis)

Real Constants

AREA, I_{ZZ}, I_{YY}, TKZ, TKY, THETA,
ISTRN (initial strain in the element), I_{XX}

Material Properties

EX (modulus of elasticity), ALPX (Poisson's ratio), DENS (density), GXY (shear modulus), DAMP (damping)

Surface Loads

Pressures
face 1 (I − J) (−Z normal direction)
face 2 (I − J) (−Y normal direction)
face 3 (I − J) (+X tangential direction)
face 4 (I) (+X axial direction)
face 5 (J) (−X axial direction)
(use negative value for opposite loading)

Temperatures

T1, T2, T3, T4, T5, T6, T7, T8

Stresses

As you will see in Example 4.5, to review stresses in beams, you must first copy these results into element tables, and then you can list them or plot them. These items are obtained using *item label* and *sequence numbers*. For a BEAM4 element, the following output information is available: the maximum stress, which is computed as the direct stress plus the absolute values of both bending stresses; the minimum stress, which is calculated as the direct stress minus the absolute value of both bending stresses. BEAM4 output includes additional stress values—examples of these stresses are given in Table 4.3.

TABLE 4.3 Examples of stresses computed by ANSYS

SDIR	Axial direct stress
SBYT	Bending stress on the element + Y side of the beam
SBYB	Bending stress on the element − Y side of the beam
SBZT	Bending stress on the element + Z side of the beam
SBZB	Bending stress on the element − Z side of the beam
SMAX	Maximum stress (direct stress + bending stress)
SMIN	Minimum stress (direct stress − bending stress)

TABLE 4.4 Item and sequence numbers for the BEAM4 element

Name	Item	E	I	J
SDIR	LS	-	1	6
SBYT	LS	-	2	7
SBYB	LS	-	3	8
SBZT	LS	-	4	9
SBZB	LS	-	5	10
SMAX	NMISC	-	1	3
SMIN	NMISC	-	2	4

Once you decide which stress values you want to look at, you can read them into a table using item labels and sequence numbers. Examples of the item labels and sequence numbers for BEAM4 are summarized in Table 4.4. See Example 4.5 for details on how to read stress values into a table for a beam element.

4.6 AN EXAMPLE USING ANSYS

ANSYS offers a number of beam elements that can be used to model structural problems.

BEAM3 is a uniaxial element with tension, compression, and bending capabilities. The element has three degrees of freedom at each node: translation in the x- and y-directions and rotation about the z-axis. The element input data include node locations, the cross-sectional area, the second moment of area, the height, and the material properties. Output data include nodal displacements and additional elemental output. Examples of elemental output include axial stress, bending stress at the top or bottom of the beam's cross section, maximum (axial + bending), and minimum (axial − bending). **BEAM4** is a three-dimensional version of BEAM3.

EXAMPLE 4.5 Revisited

Let us consider the overhang frame again in order to solve this problem using ANSYS. Recall that the frame is made of steel with $E = 30 \times 10^6$ lb/in^2. The respective cross-sectional areas and the second moments of areas for the two members are shown in Figure 4.13 (repeated in Figure 4.17 for your convenience). The members have a depth of 12.22 in. The frame is fixed as shown in the figure. We are interested in determining the deflections and the rotation of the frame under the given distributed load.

Enter the **ANSYS** program by using the Launcher.

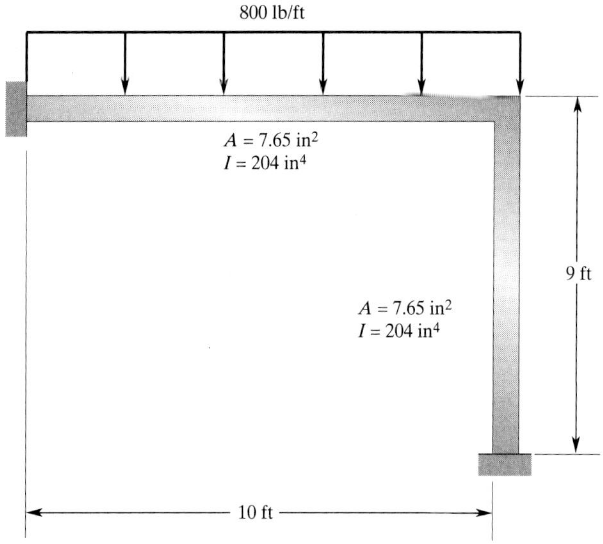

FIGURE 4.17 An overhang frame supporting a distributed load.

Type **Frame2D** (or a file name of your choice) in the **Jobname** entry field of the ANSYS Product Launcher and Pick **Run** to start GUI.

Create a title for the problem. This title will appear on ANSYS display windows to provide a simple way to identify the displays. Use the following command sequences:

utility menu: **File → Change Title** . . .

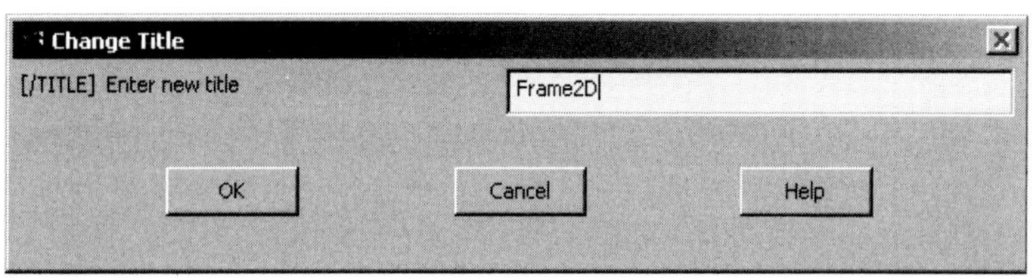

main menu: **Preprocessor → Element Type → Add/Edit/Delete**

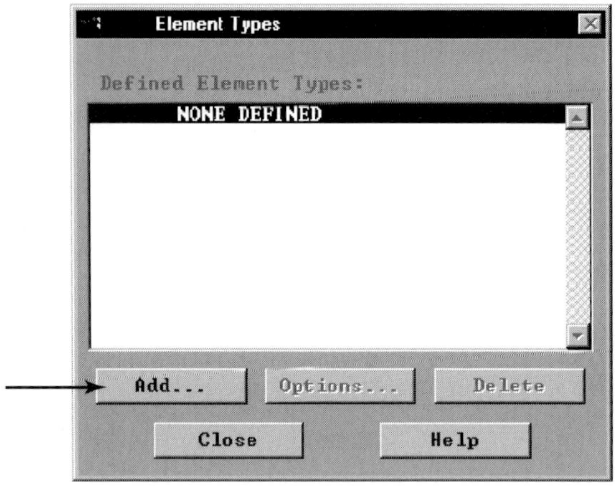
Element Types
Defined Element Types:
NONE DEFINED
Add...
Options...
Delete
Close
Help

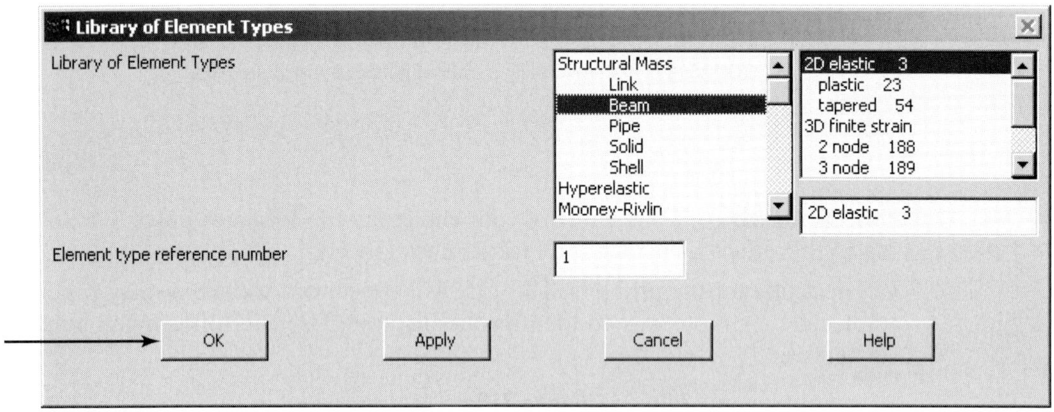
Library of Element Types
Library of Element Types
Structural Mass
Link
Beam
Pipe
Solid
Shell
Hyperelastic
Mooney-Rivlin
2D elastic 3
plastic 23
tapered 54
3D finite strain
2 node 188
3 node 189
2D elastic 3
Element type reference number
1
OK
Apply
Cancel
Help

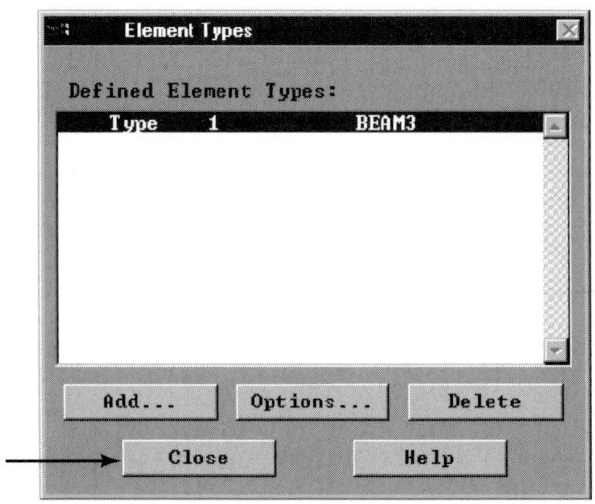
Element Types
Defined Element Types:
Type 1 BEAM3
Add...
Options...
Delete
Close
Help

Assign the modulus of elasticity by using the following commands:

main menu: **Preprocessor → Material Props → Material Models → Structural → Linear → Elastic → Isotropic**

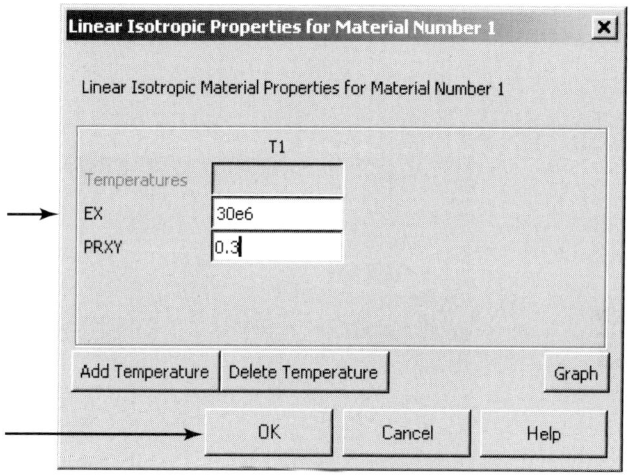

Close the "Define Material Model Behavior" window.

main menu: **Preprocessor → Real Constants → Add/Edit/Delete**

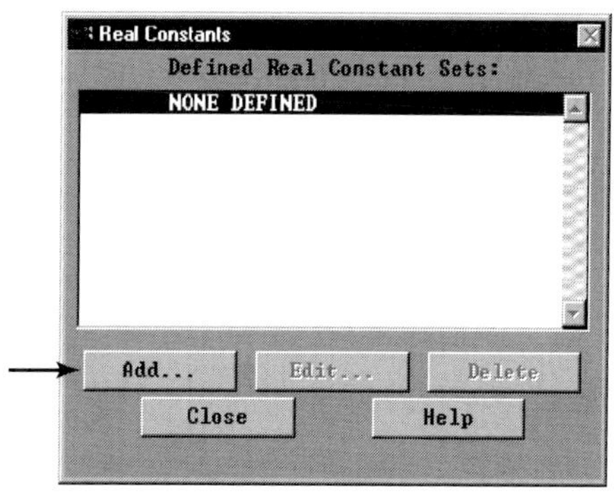

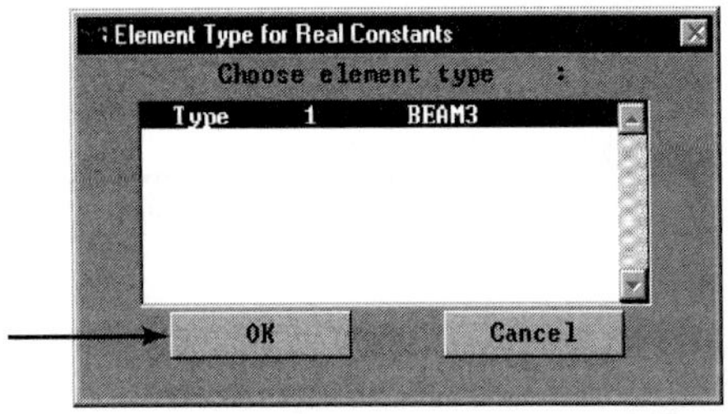
Element Type for Real Constants
Choose element type :
Type 1 BEAM3
OK
Cancel

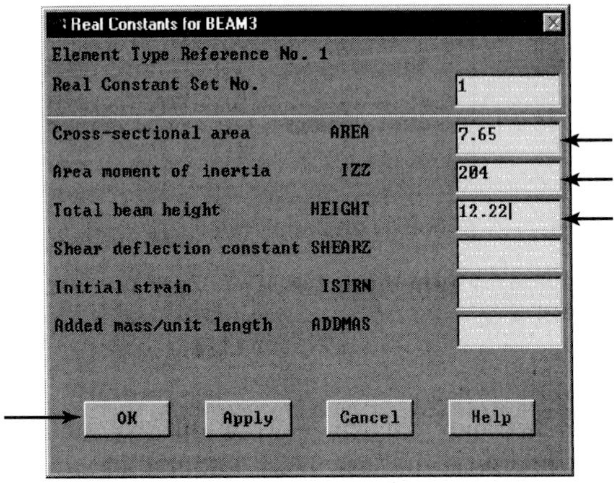
Real Constants for BEAM3
Element Type Reference No. 1
Real Constant Set No.
1
Cross-sectional area AREA
7.65
Area moment of inertia IZZ
204
Total beam height HEIGHT
12.22
Shear deflection constant SHEARZ
Initial strain ISTRN
Added mass/unit length ADDMAS
OK
Apply
Cancel
Help

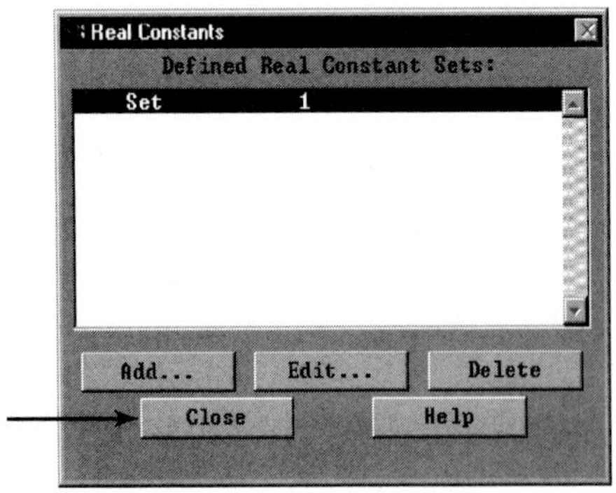
Real Constants
Defined Real Constant Sets:
Set 1
Add...
Edit...
Delete
Close
Help

ANSYS Toolbar: **SAVE_DB**

Set up the graphics area (i.e., work plane, zoom, etc.) with the following commands:

utility menu: **Workplane → WP Settings** . . .

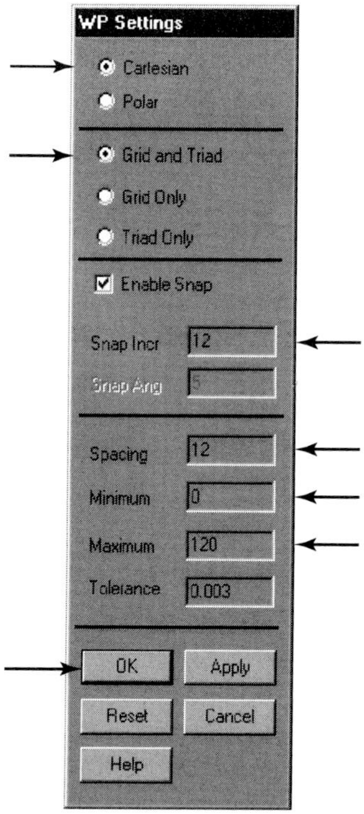

utility menu: **Workplane → Display Working Plane**

Bring the workplane to view by the command

utility menu: **PlotCtrls → Pan, Zoom, Rotate** . . .

Click on the small circle until you bring the workplane to view. Then create the nodes and elements:

main menu: **Preprocessor → Modeling → Create → Nodes → On Working Plane**

[WP = 0,108]

[WP = 120,108]

[WP = 120,0]

OK

main menu: **Preprocessor → Modeling → Create → Elements → Auto Numbered → Thru Nodes**

[pick node 1]

[pick node 2]

[apply anywhere in the ANSYS graphics window]

[pick node 2]

[pick node 3]

[anywhere in the ANSYS graphics window]

OK

utility menu: **Plot → Elements**

Toolbar: **SAVE_DB**

Apply boundary conditions with the following commands:

main menu: **Solution → Define Loads → Apply → Structural → Displacement → On Nodes**

[pick node 1]

[pick node 3]

[anywhere in the ANSYS graphics window]

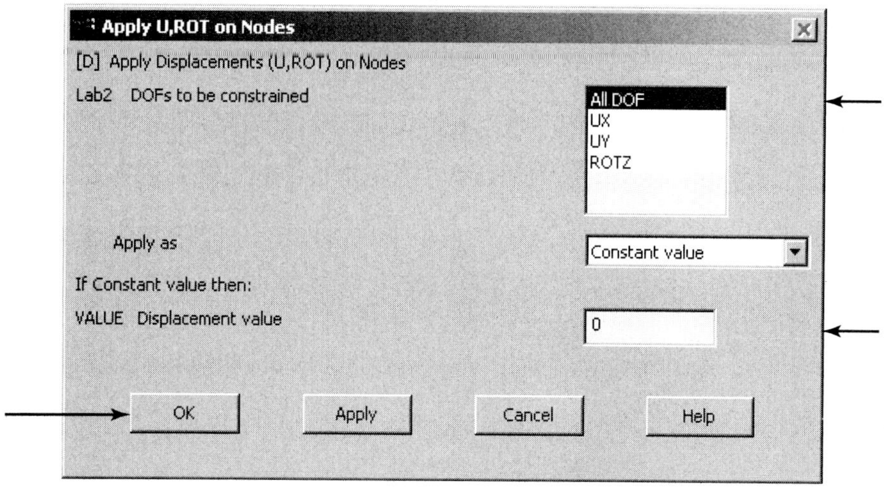

main menu: **Solution → Define Loads → Apply → Structural → Pressure → On Beams**

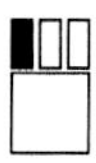

[pick element 1]

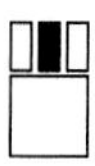

[anywhere in the ANSYS graphics window]

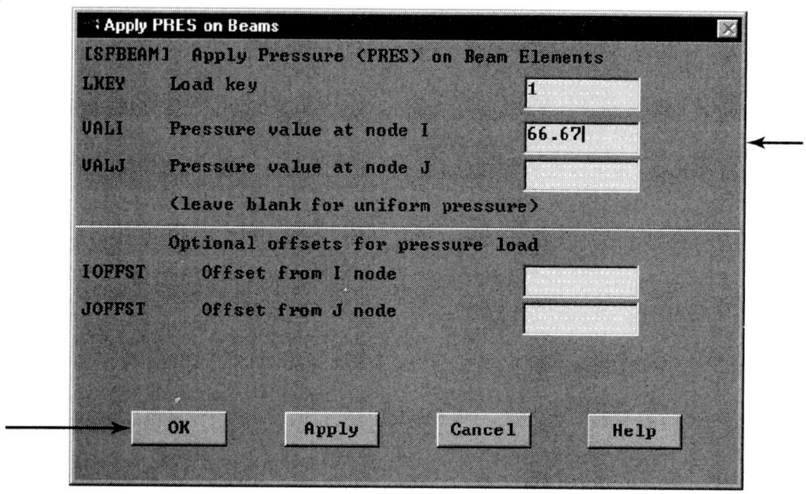

To see the applied distributed load and boundary conditions, use the following commands:

utility menu: **Plot Ctrls → Symbols** . . .

utility menu: **Plot → Elements**

ANSYS Toolbar: **SAVE_DB**

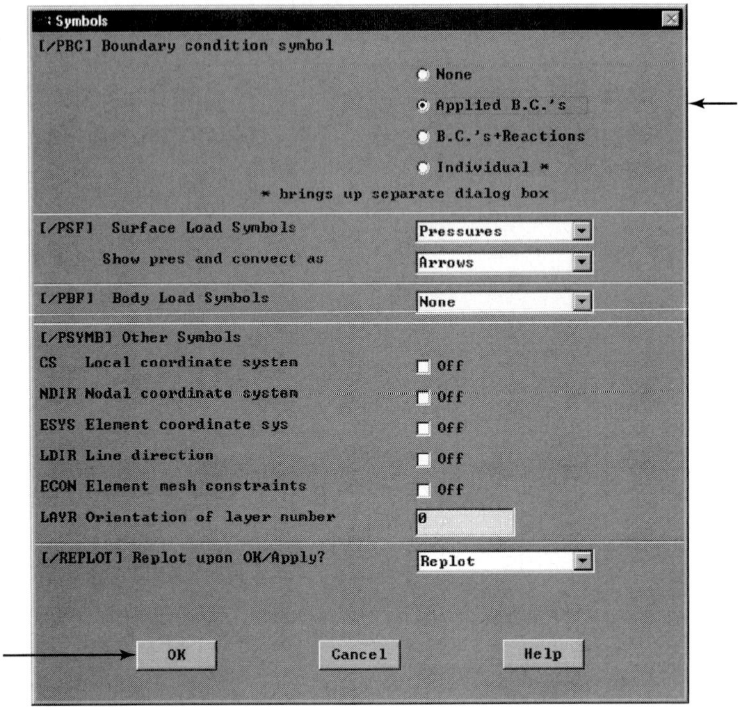

Solve the problem:

main menu: **Solution → Solve → Current LS**

OK

Close (the solution is done!) window.

Close (the/STAT Command) window.

Begin the postprocessing phase and plot the deformed shape with the following commands:

main menu: **General Postproc → Plot Results → Deformed Shape**

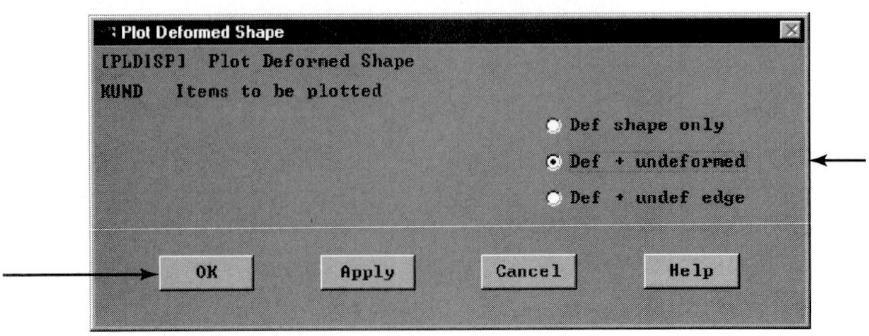

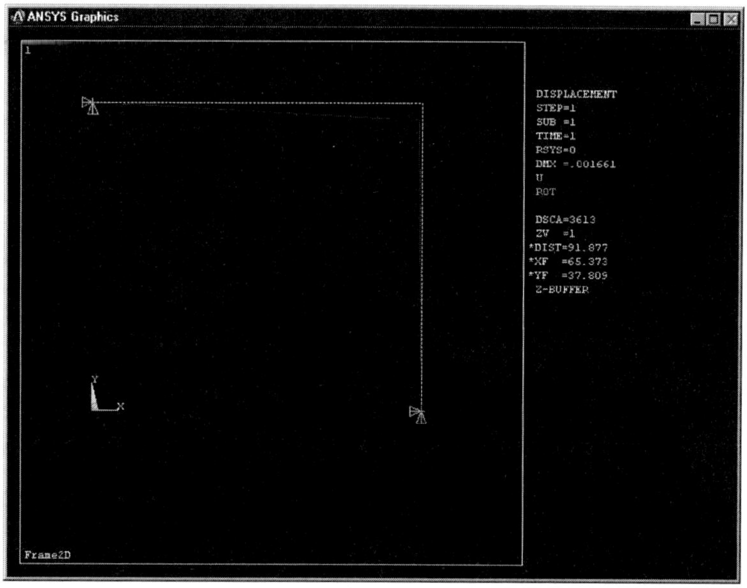

List the nodal displacements with the following commands:

main menu: **General Postproc → List Results → Nodal Solution**

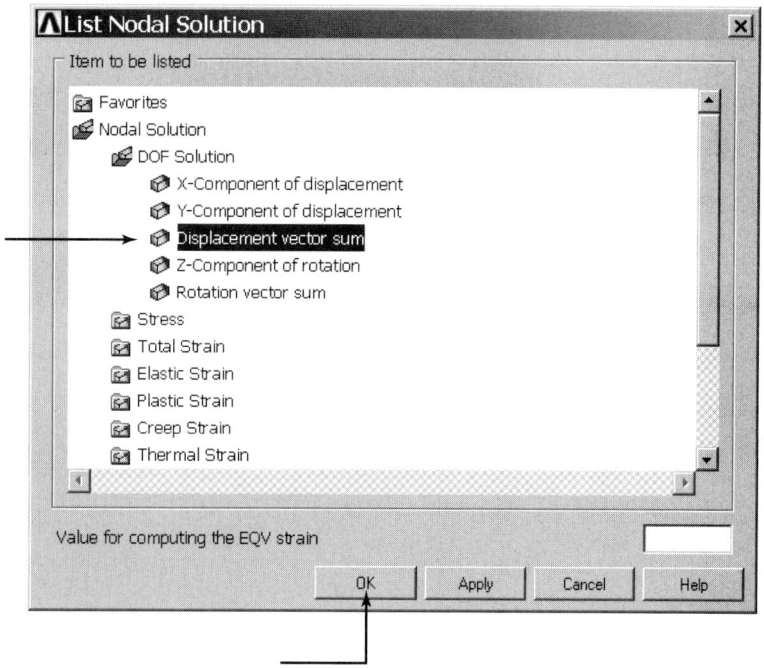

Also list Rotation vector sum.

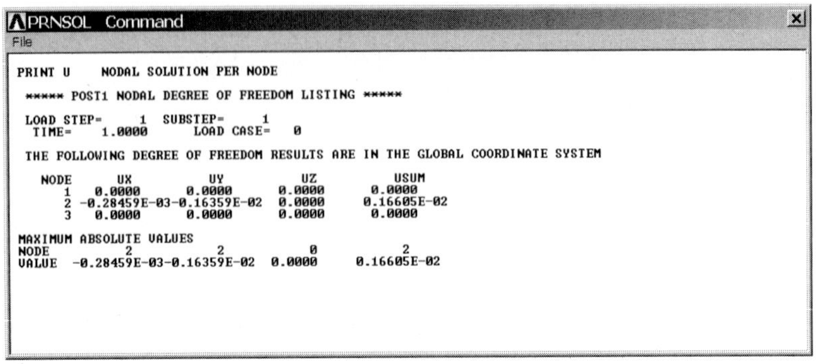

List the reactions with the following commands:

main menu: **General Postproc → List Results → Reaction Solution**

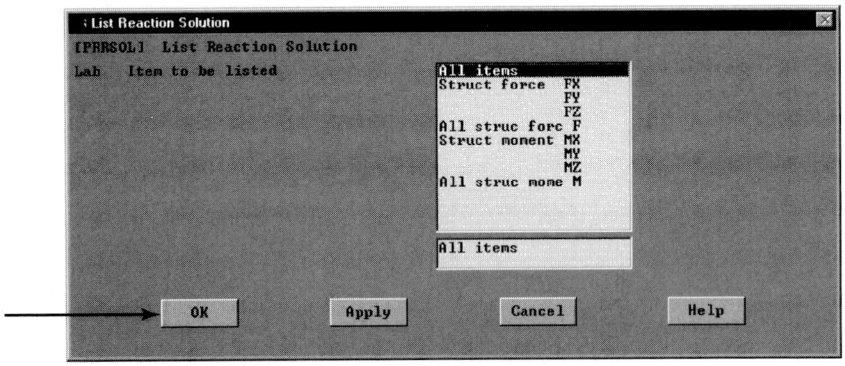

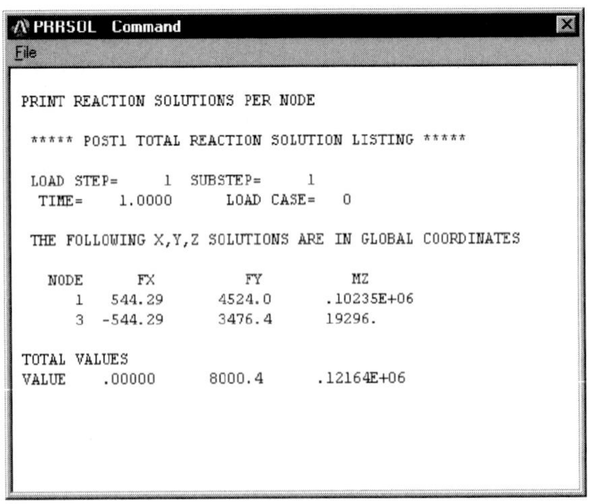

Stresses

To view stresses in beams and frames, we must first copy these results into element tables. These items are obtained using *item label* and *sequence numbers*. This approach is similar to the procedure we followed to obtain axial forces and axial stresses for truss elements. For a beam element, the following output information is available: direct axial stress, bending stress on +Y side, bending stress on −Y side, direct stress + bending stress, and direct stress – bending stress. These stresses are shown in Figure 4.18. The item labels and sequence numbers for stresses for a beam element are given in Table 4.5.

For Example 4.5, we have assigned the maximum and minimum stresses, as computed by ANSYS, in each element to user-defined labels (User label for item) MaxSts and MinSts. We now run the following commands:

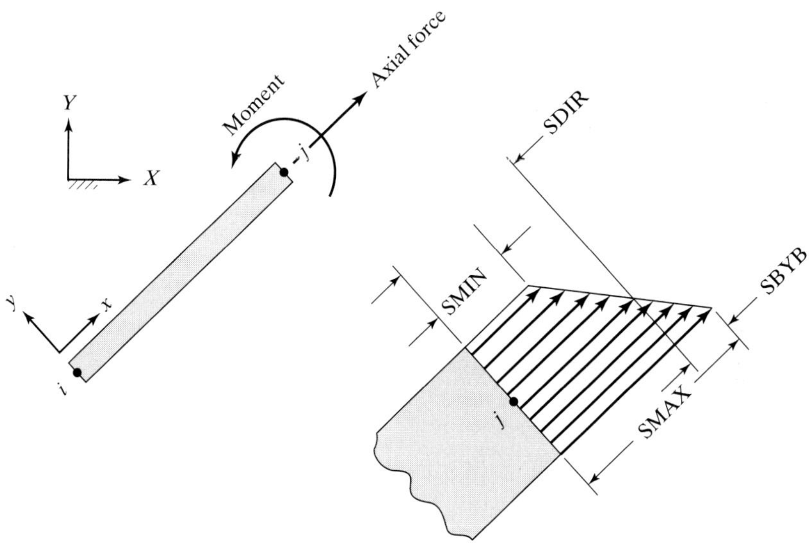

FIGURE 4.18 Stresses for a beam element.

TABLE 4.5 Item labels and sequence numbers for a beam element

Stresses	Data	Item label	Sequence number
axial direct stress	SDIR	LS	1,4
bending stress on +Y side	SBYT	LS	2,5
bending stress on −Y side	SBYB	LS	3,6
direct stress + bending stress	SMAX	NMISC	1,3
direct stress − bending stress	SMIN	NMISC	2,4

main menu: **General Postproc → Element Table → Define Table**

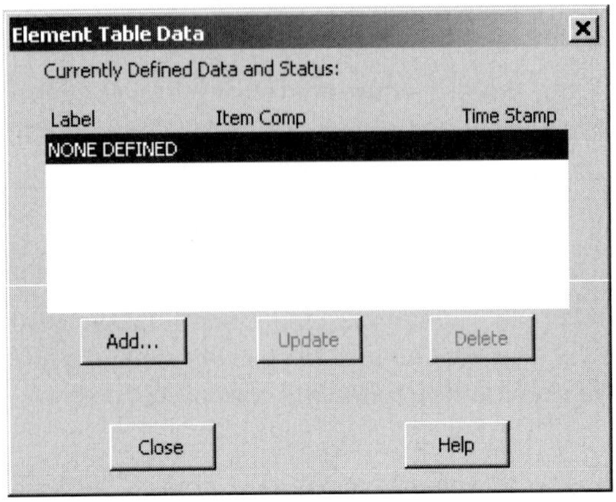

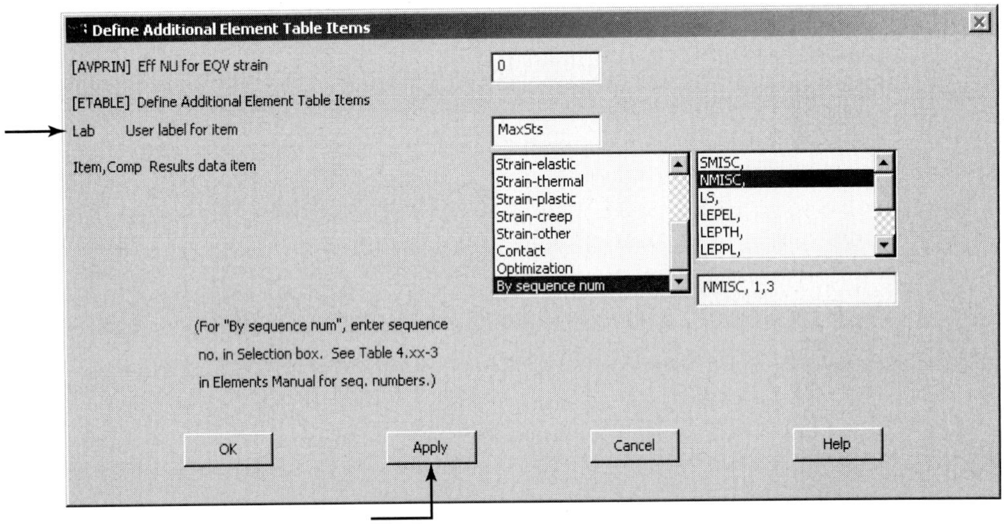

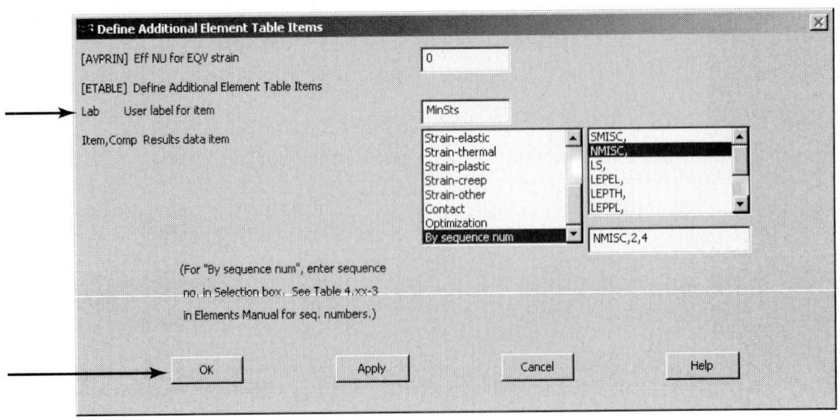

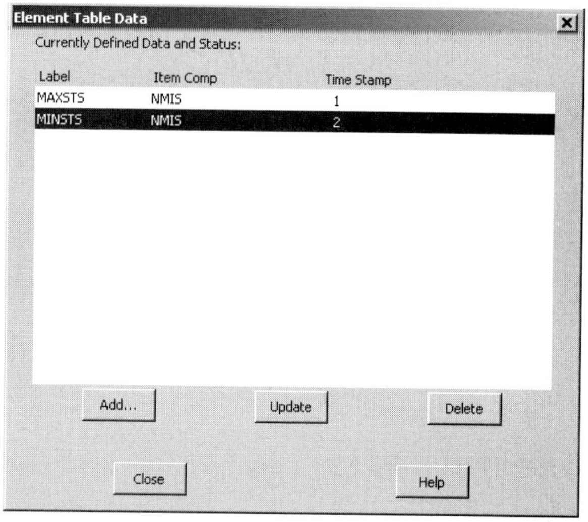

main: **General Postproc → Element Table → List Element Table**

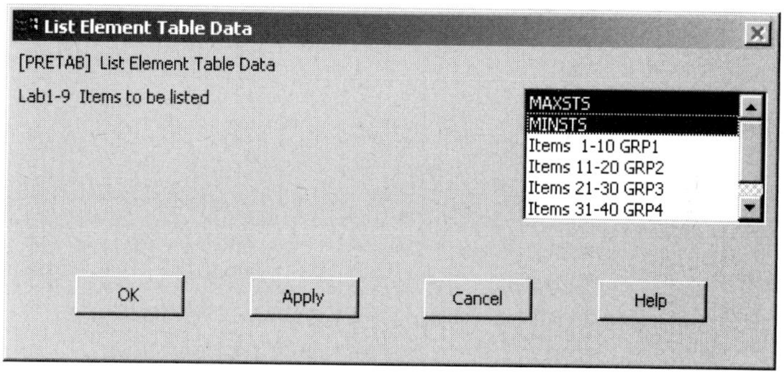

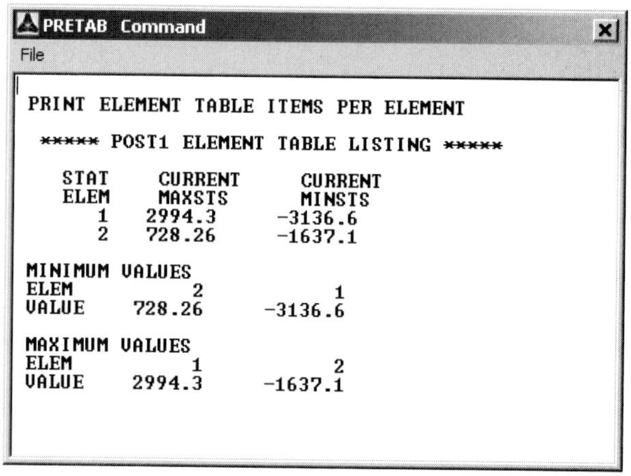

Exit ANSYS and save everything:

ANSYS Toolbar: **QUIT**

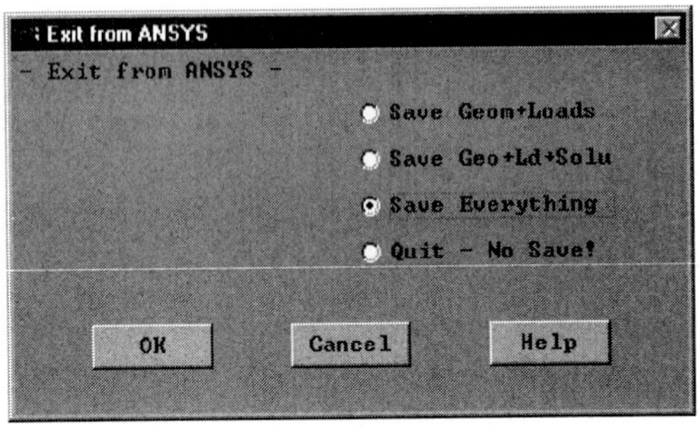

4.7 VERIFICATION OF RESULTS

Refer to Example 4.2. One way of checking for the validity of our FEA findings of Example 4.2 is to arbitrarily cut a section through the column and apply the static equilibrium conditions. As an example, consider cutting a section through the column containing element (2), as shown in the accompanying illustration.

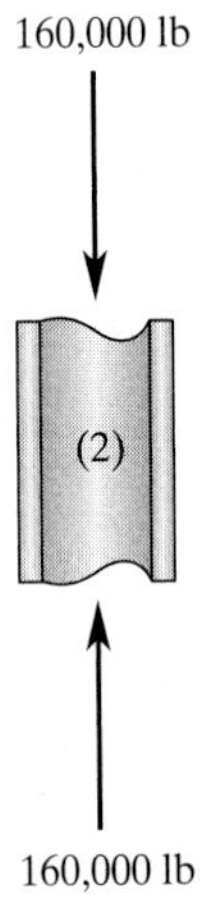

The average normal stress in that section of the column is

$$\sigma^{(2)} = \frac{f_{\text{internal}}}{A} = \frac{160{,}000}{39.7} = 4030 \text{ lb/in}^2$$

In a similar way, the average stress in element (4) can be checked by

$$\sigma^{(4)} = \frac{f_{\text{internal}}}{A} = \frac{60{,}000}{39.7} = 1511 \text{ lb/in}^2$$

The stresses computed in this manner are identical to the results obtained earlier using the energy method.

It is always necessary to compute the reaction forces and moments for beam and frame problems. The nodal reaction forces and moments can be computed from the relationship

$$\{\mathbf{R}\} = [\mathbf{K}]\{\mathbf{U}\} - \{\mathbf{F}\}$$

We computed the reaction matrix for Example 4.4, repeated here:

$$\begin{Bmatrix} R_1 \\ M_1 \\ R_2 \\ M_2 \\ R_3 \\ M_3 \end{Bmatrix} = \begin{Bmatrix} 54687(\text{N}) \\ 39062(\text{N}\cdot\text{m}) \\ 132814(\text{N}) \\ 0 \\ 0 \\ 0 \end{Bmatrix}$$

Earlier, we discussed how to check the validity of results qualitatively. It was mentioned that the results indicated that there is a reaction force and a reaction moment at node 1; there is a reaction force at node 2; there is no reaction moment at node 2, as expected; and there are no reaction forces or moments at node 3, as expected for the given problem. Let us also perform a quantitative check on the accuracy of the results. We can use the computed reaction forces and moments against the external loading to check for static equilibrium (see Figure 4.19):

$$+\uparrow\Sigma F_Y = 0 \qquad 13{,}2814 + 54{,}687 - (25{,}000)(7.5) = -1 \approx 0$$

and

$$\circlearrowleft\!+\Sigma M_{\text{node 2}} = 0 \qquad 39{,}062 - 54{,}687(5) + (25{,}000)(7.5)(1.25) = 2 \approx 0$$

Similarly, in reference to Example 4.5, we find that the reaction results generated using ANSYS are shown in Figure 4.20. Checking for static equilibrium, we find that

$$\overset{+}{\rightarrow}\Sigma F_X = 0 \qquad 544.26 - 544.26 = 0$$

$$+\uparrow\Sigma F_Y = 0 \qquad 4523.8 + 3476.2 - (800)(10) = 0$$

$$\circlearrowleft\!+\Sigma M_{\text{node 1}} = 0 \qquad 102{,}340 + 19{,}295 + 3476.2(10)(12) - (544.26)(9)(12) - (800)(10)(5)(12) = -1 \approx 0$$

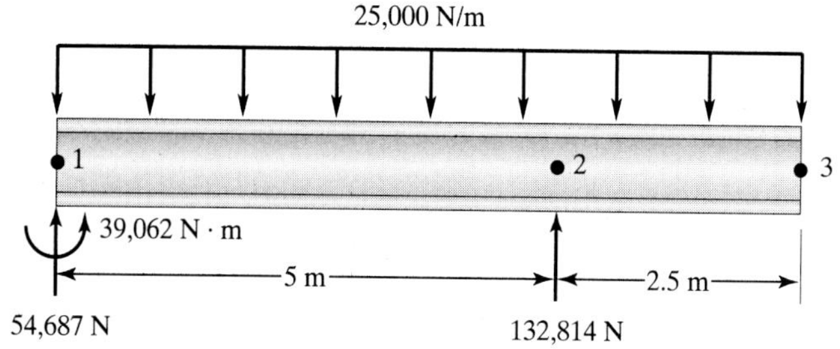

FIGURE 4.19 The free-body diagram for Example 4.4.

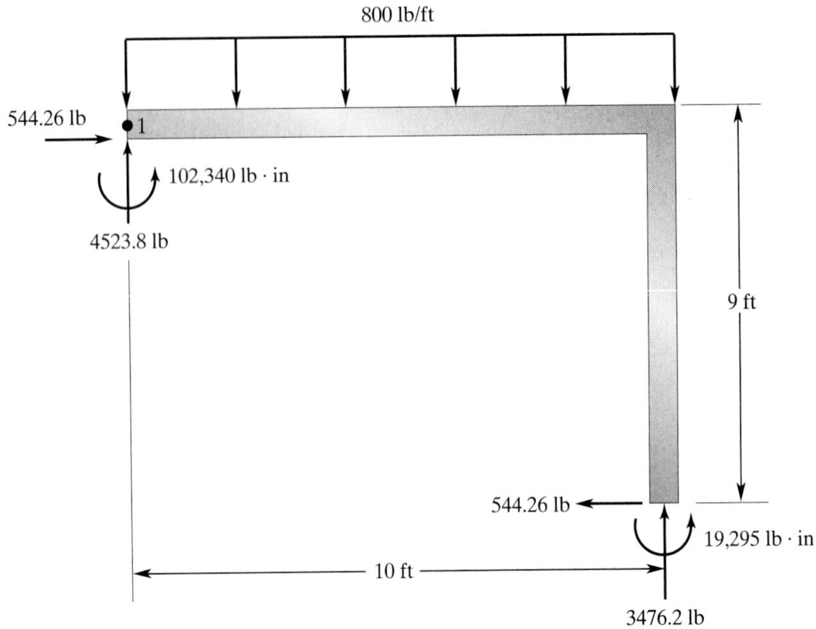

FIGURE 4.20 The free-body diagram for Example 4.5.

These simple problems illustrate the importance of checking for equilibrium conditions when verifying results.

SUMMARY

At this point you should

1. know how to formulate stiffness matrix for a member under axial loading.
2. know that it is wise to use simple analytical solutions rather than finite element modeling for a simple problem whenever appropriate. Use finite element modeling only when it is necessary to do so.
3. know that the stiffness matrix for a beam element with two degrees of freedom at each node (the vertical displacement and rotation) is

$$[\mathbf{K}]^{(e)} = \frac{EI}{L^3}\begin{bmatrix} 12 & 6L & -12 & 6L \\ 6L & 4L^2 & -6L & 2L^2 \\ -12 & -6L & 12 & -6L \\ 6L & 2L^2 & -6L & 4L^2 \end{bmatrix}$$

4. know how to compute the load matrix for a beam element by consulting Table 4.2 for equivalent nodal forces.

5. know that the stiffness matrix for a frame element (with local and global coordinate systems aligned) consisting of two nodes with three degrees of freedom at each node (axial displacement, lateral displacement and rotation) is

$$[\mathbf{K}]^{(e)} = \begin{bmatrix} \frac{AE}{L} & 0 & 0 & -\frac{AE}{L} & 0 & 0 \\ 0 & \frac{12EI}{L^3} & \frac{6EI}{L^2} & 0 & -\frac{12EI}{L^3} & \frac{6EI}{L^2} \\ 0 & \frac{6EI}{L^2} & \frac{4EI}{L} & 0 & -\frac{6EI}{L^2} & \frac{2EI}{L} \\ -\frac{AE}{L} & 0 & 0 & \frac{AE}{L} & 0 & 0 \\ 0 & -\frac{12EI}{L^3} & -\frac{6EI}{L^2} & 0 & \frac{12EI}{L^3} & -\frac{6EI}{L^2} \\ 0 & \frac{6EI}{L^2} & \frac{2EI}{L} & 0 & -\frac{6EI}{L^2} & \frac{4EI}{L} \end{bmatrix}$$

Note that for members that are not horizontal, the local degrees of freedom are related to the global degrees of freedom through the transformation matrix, according to the relationship

$$\{\mathbf{u}\} = [\mathbf{T}]\{\mathbf{U}\}$$

where the transformation matrix is

$$[\mathbf{T}] = \begin{bmatrix} \cos\theta & \sin\theta & 0 & 0 & 0 & 0 \\ -\sin\theta & \cos\theta & 0 & 0 & 0 & 0 \\ 0 & 0 & 1 & 0 & 0 & 0 \\ 0 & 0 & 0 & \cos\theta & \sin\theta & 0 \\ 0 & 0 & 0 & -\sin\theta & \cos\theta & 0 \\ 0 & 0 & 0 & 0 & 0 & 1 \end{bmatrix}$$

6. know how to compute the stiffness matrix for a frame element with an arbitrary orientation with respect to the global coordinate system using the relationship.

$$[\mathbf{K}]^{(e)} = [\mathbf{T}]^T[\mathbf{K}]^{(e)}_{xy}[\mathbf{T}]$$

7. know how to compute the load matrix for a frame element by consulting Table 4.2 for equivalent nodal forces.

REFERENCES

ANSYS User's Manual: Procedures, Vol. I, Swanson Analysis Systems, Inc.

ANSYS User's Manual: Commands, Vol. II, Swanson Analysis Systems, Inc.

ANSYS User's Manual: Elements, Vol. III, Swanson Analysis Systems, Inc.

Beer, P., and Johnston, E. R., *Mechanics of Materials*, 2nd ed., New York, McGraw-Hill, 1992.
Hibbleer, R. C., *Mechanics of Materials*, 2nd ed., New York, Macmillan, 1994.
Segrlind, L., *Applied Finite Element Analysis*, 2nd ed., New York, John Wiley and Sons, 1984.

PROBLEMS

1. Determine the deflections of point D and point F and the axial stress in each member of the system shown in the accompanying figure. ($E = 29 \times 10^3$ ksi.)

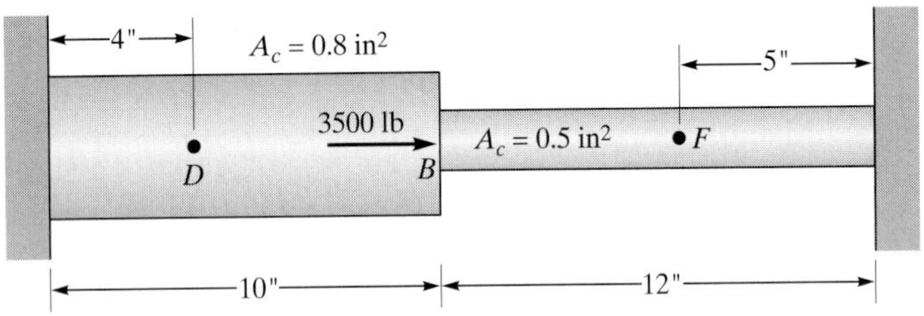

2. Consider a four-story building with steel columns similar to the one presented in Example 4.2. The column is subjected to the loading shown in the accompanying figure. Assuming axial loading, (a) determine vertical displacements of the column at various floor-column connection points and (b) determine the stresses in each portion of the column. ($E = 29 \times 10^6$ lb/in², $A = 59.1$ in².)

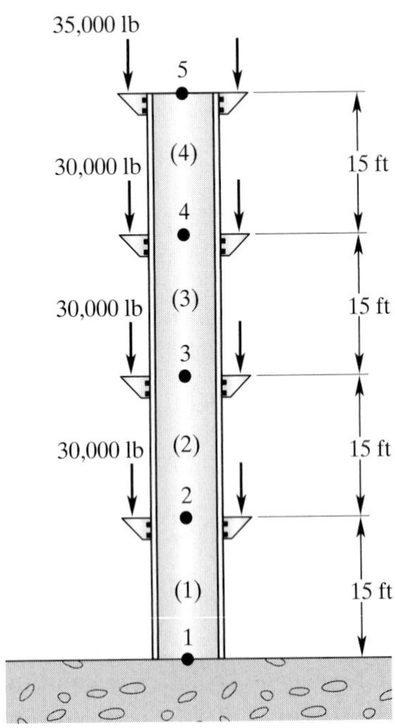

3. Determine the deflection of point D and the axial stress in each member in the system shown in the accompanying figure. ($E = 10.6 \times 10^3$ ksi.)

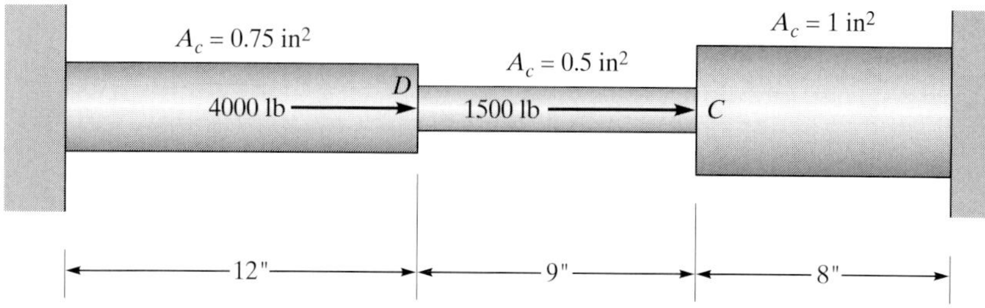

4. A 20-ft-tall post is used to support advertisement signs at various locations along its height, as shown in the accompanying figure. The post is made of structural steel with a modulus of elasticity of $E = 29 \times 10^6$ lb/in². Not considering wind loading on the signs, (a) determine displacements of the post at the points of load application and (b) determine stresses in the post.

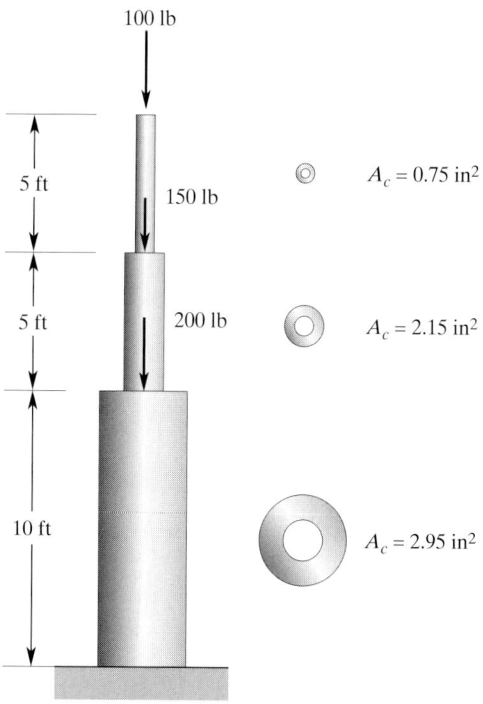

5. Determine the deflections of point D and point F in the system in the accompanying figure. Also compute the axial force and stress in each member. ($E = 29 \times 10^3$ ksi.)

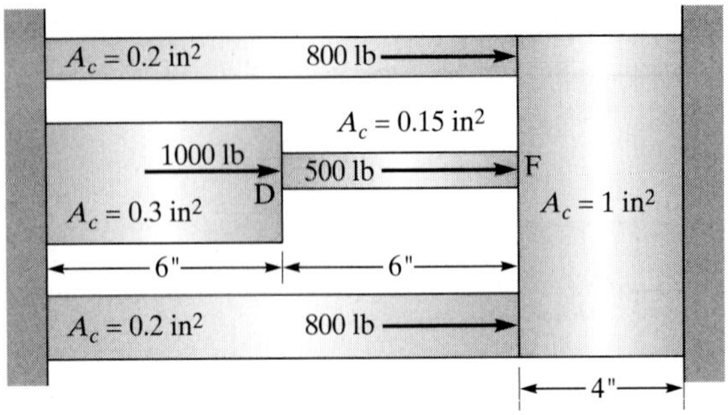

6. Determine the deflections of point D and point F in the system in the accompanying figure. Also compute the axial force and stress in each member.

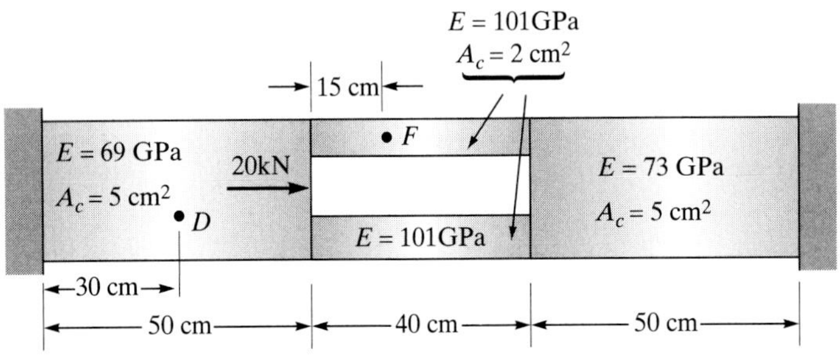

7. The beam shown in the accompanying figure is a wide-flange W18 $\times$ 35, with a cross-sectional area of 10.3 in^2 and a depth of 17.7 in. The second moment of area is 510 in^4. The beam is subjected to a uniformly distributed load of 2000 lb/ft. The modulus of elasticity of the beam is $E = 29 \times 10^6$ lb/in^2. Using manual calculations, determine the vertical displacement at node 3 and the rotations at nodes 2 and 3. Also, compute the reaction forces at nodes 1 and 2 and reaction moment at node 1.

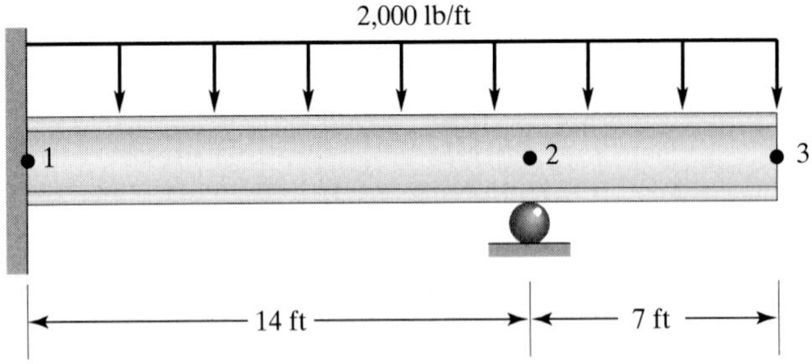

8. The beam shown in the accompanying figure is a wide-flange W16 $\times$ 31 with a cross-sectional area of 9.12 in^2 and a depth of 15.88 in. The second moment of area is 375 in^4. The beam is subjected to a uniformly distributed load of 1000 lb/ft and a point load of 500 lb. The modulus of elasticity of the beam is $E = 29 \times 10^6$ lb/in^2. Using manual calculations,

determine the vertical displacement at node 3 and the rotations at nodes 2 and 3. Also, compute the reaction forces at nodes 1 and 2 and reaction moment at node 1.

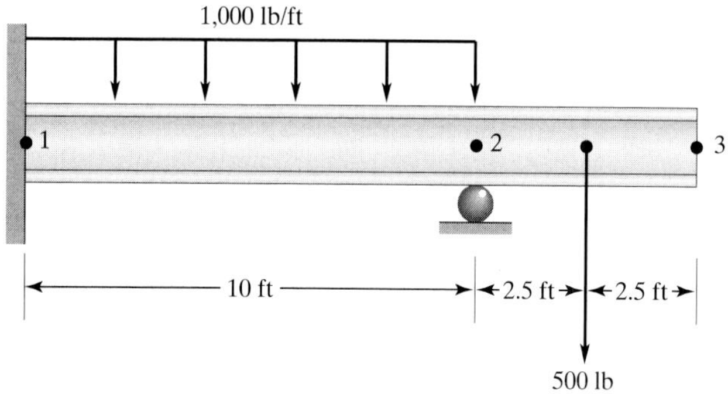

9. The lamp frame shown in the accompanying figure has hollow, square cross sections and is made of steel, with $E = 29 \times 10^6$ lb/in^2. Using hand calculations, determine the endpoint deflection of the cross member where the lamp is attached.

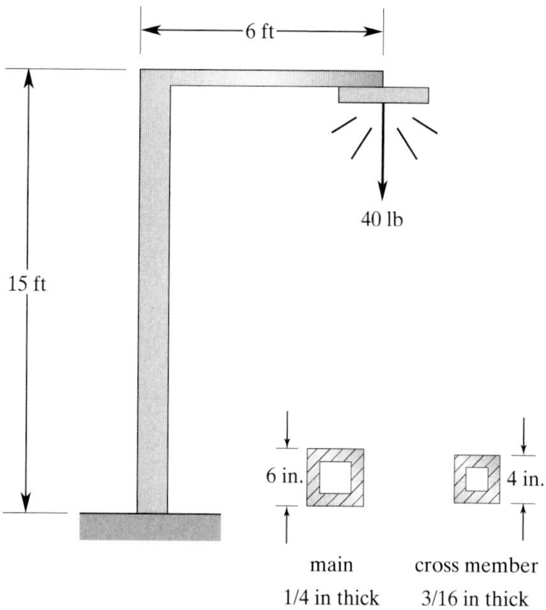

10. A park picnic-table top is supported by two identical metal frames; one such frame is shown in the accompanying figure. The frames are embedded in the ground and have hollow, circular cross-sectional areas. The tabletop is designed to support a distributed load of 250 lb/ft^2. Using ANSYS, size the cross section of the frame to support the load safely.

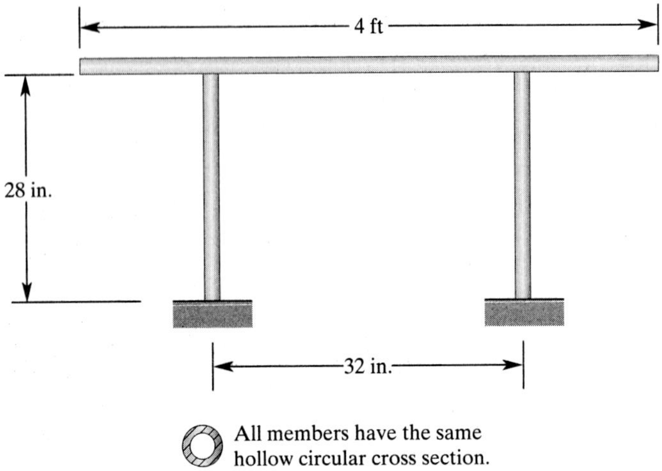

11. The frame shown in the accompanying figure is used to support a load of 2000 lb. The main vertical section of the frame has an annular cross section with an area of 8.63 in^2 and a polar radius of gyration of 2.75 in. The outer diameter of the main tubular section is 6 in. All other members also have annular cross sections with respective areas of 2.24 in^2 and polar radii of gyration of 1.91 in. The outer diameter of these members is 4 in. Using ANSYS, determine the deflections at the points where the load is applied. The frame is made of steel, with a modulus elasticity of $E = 29 \times 10^6$ lb/in^2.

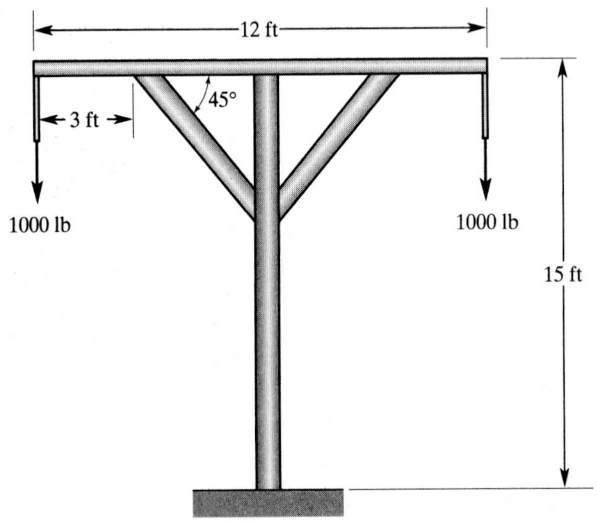

12. Verify the equivalent nodal loading for a beam element subjected to a triangular load, as shown in the accompanying figure.

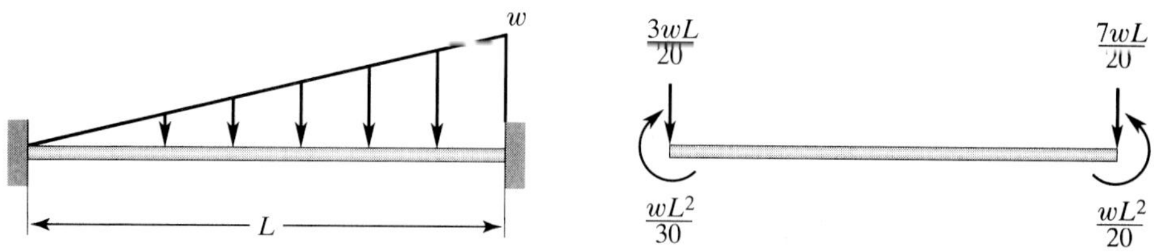

13. Referring to the section in this chapter discussing the frame elements, show that the stiffness matrix represented with respect to the global coordinate system is related to the stiffness matrix described with respect to the frame's local coordinate system, according to the relationship

$$[\mathbf{K}]^{(e)} = [\mathbf{T}]^T[\mathbf{K}]_{xy}^{(e)}[\mathbf{T}]$$

14. The frame shown in the accompanying figure is used to support a load of 500 lb/ft. Using ANSYS, size the cross sections of each member if standard-size steel square tubing is to be used. Use three different sizes. The deflection of the centerpoint is to be kept under 0.05 in.

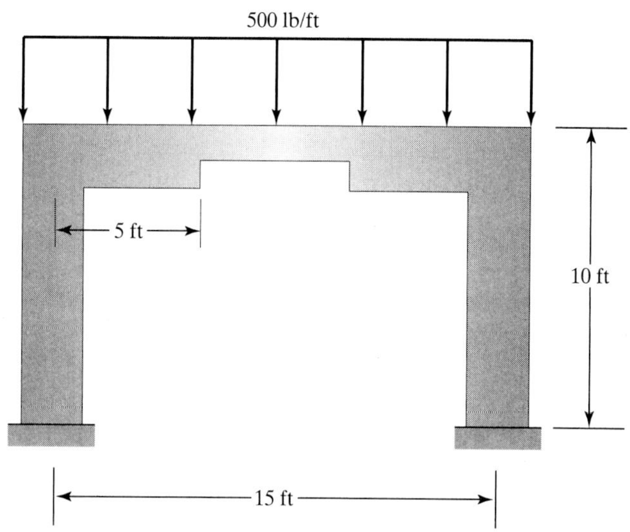

15. The frame shown in the accompanying figure is used to support the load given in the figure. Using ANSYS, size the members if standard sizes of steel I-beams are to be used.

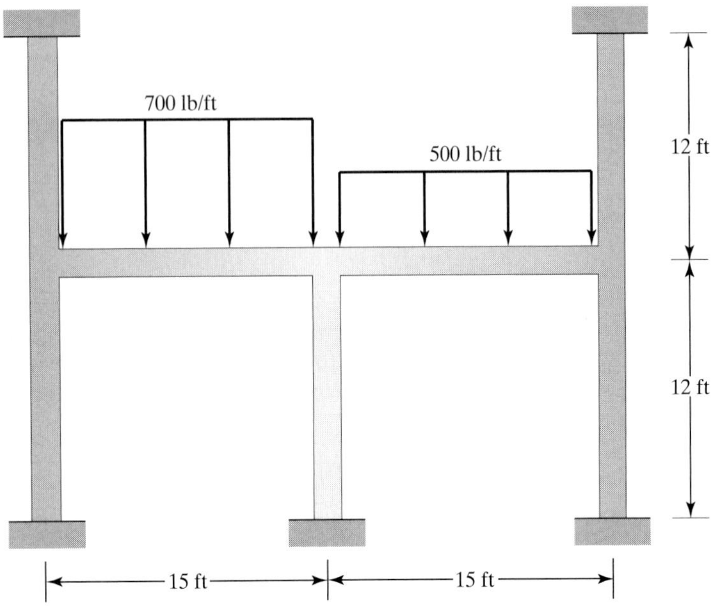

16. Verify the equivalent nodal loading for a beam element subjected to the load shown in the accompanying figure.

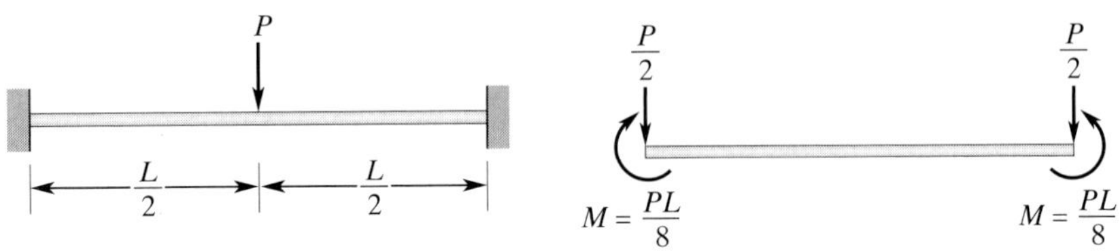

17. Use a one-element model and calculate the deflection and slope at the endpoint of the beam shown in the accompanying figure. Compare your results to deflection and slope values given in Table 4.1.

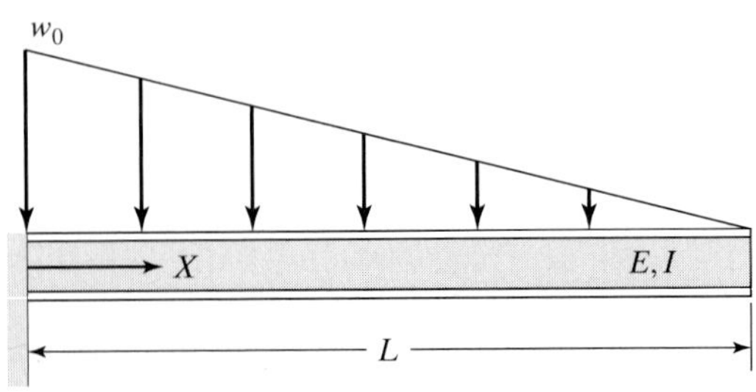

18. Use a two-element model and solve Example 4.3. Compare your results to the deflections and the end slope of the single element model.

19. Use a one-element model and calculate the deflection and slope at the endpoint of the beam shown in the accompanying figure. Compare your results to the deflection and slope values given in Table 4.1.

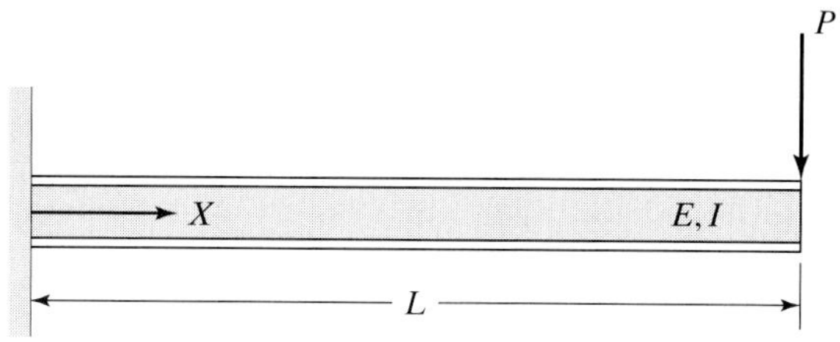

20. Use a one-element model and calculate the deflection and the slope at the midpoint of the beam shown in the accompanying figure. Compare your results to the deflection and slope values given in Table 4.1.

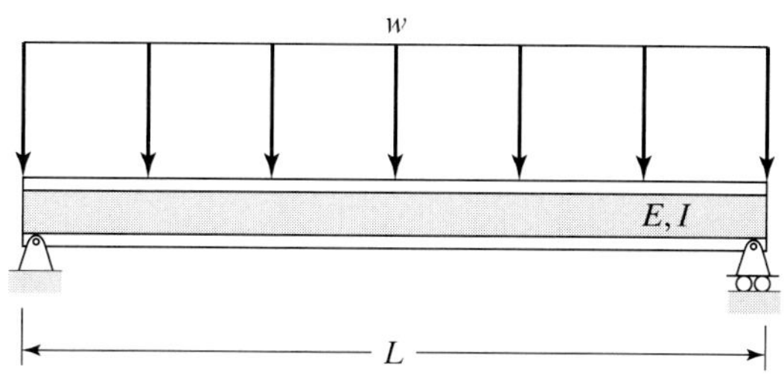

21. The beam shown in the accompanying figure is a wide-flange W18 × 35, with a cross-sectional area of 10.3 in^2 and a depth of 17.7 in. The second moment of area is 510 in^4. The beam is subjected to a point load of 2500 lb. The modulus of elasticity of the beam $E = 29 \times 10^6$ lb/in^2. Use a two-element model and calculate the deflection of the midpoint of the beam. Compare your results to exact values.

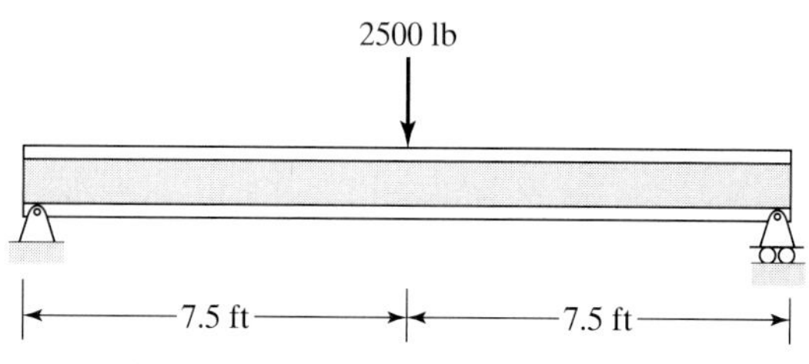

22. The frame shown in the accompanying figure is used to support a load of 1000 lb. Using ANSYS, size the cross sections of each member if standard sizes of steel I-beams are used.

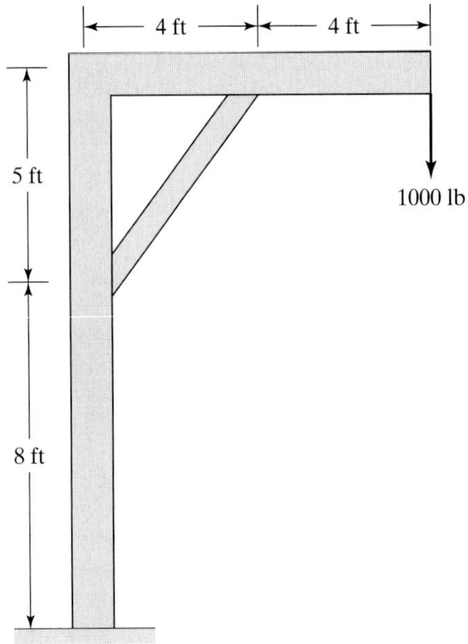

23. The frame shown in the accompanying figure is used to support the indicated load. Using ANSYS, size the cross sections of each member if standard sizes of steel I-beams are used.

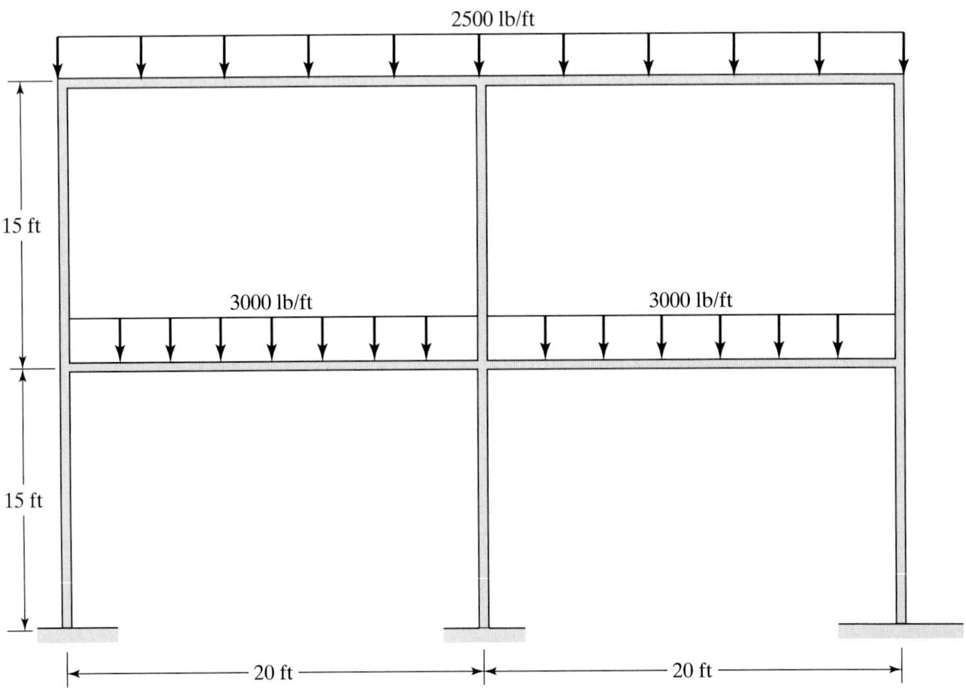

24. The frame shown in the accompanying figure is used to support the indicated load. Using ANSYS, size the cross sections of each member if standard sizes of steel I-beams are used.

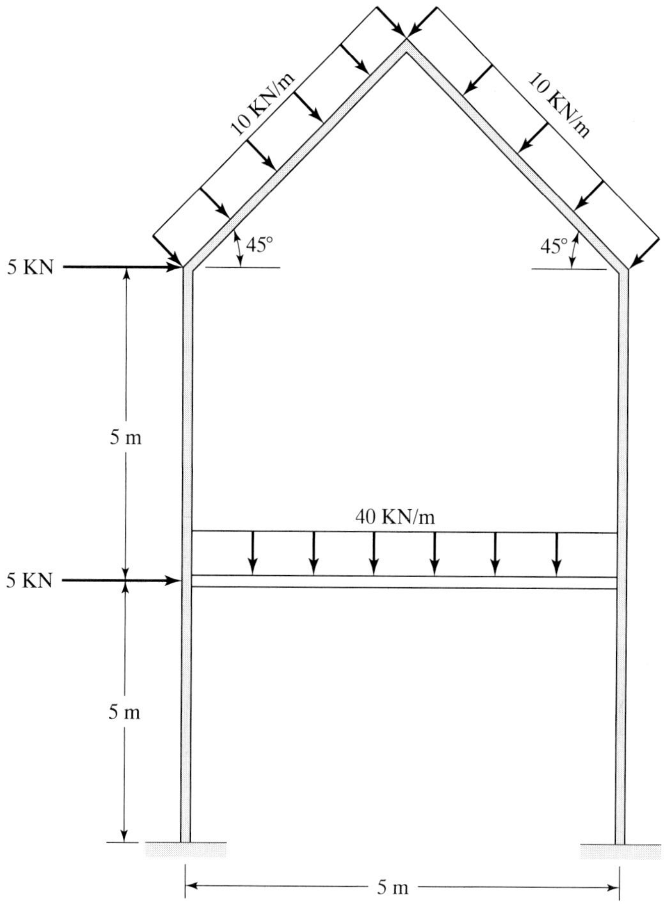

25. The beams shown in the accompanying figure are used to support the indicated load. Using ANSYS, size the cross sections of each beam if standard sizes of steel I-beams are used.

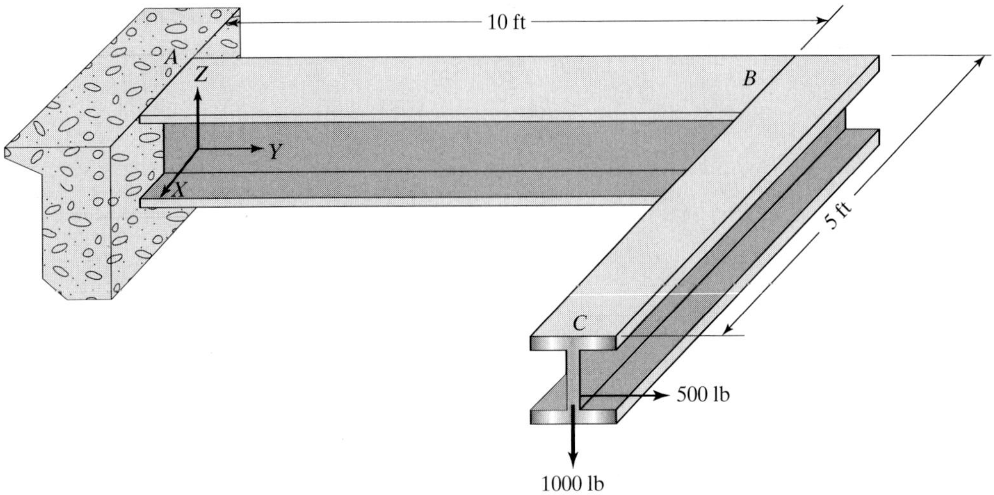

26. Show that $[\mathbf{D}]\{\mathbf{U}\} = \{\mathbf{U}\}^T[\mathbf{D}]^T$.

27. Starting with Equations (4.39) and (4.41), show that the relationship given by Eqs. (4.42) is true.

28. Starting with Equation (4.42) and using the results given by Eq. (4.36a) perform the integrations to obtain the results given in the first column of the stiffness matrix, Equation (4.43).

29. Starting with Equation (4.42) and using the results given by Eq. (4.36a) perform the integrations to obtain the results given in the fourth row of the stiffness matrix, Equation (4.43).

30. Design Project Size the members of the bridge shown in the accompanying figure for a case in which traffic is backed up with a total of four trucks equally spaced on the bridge. A typical truck has a payload weight of 64,000 lb and a cab weight of 8000 lb. As a starting point, you may use one cross section for all beam elements. You may also assume one cross section for all truss members. The roadbed weighs 1500 lb/ft and is supported by I-beams. Use standard steel I-beam sizes. Design your own truss configuration. In your analysis, you may assume that the concrete column does not deflect significantly. Write a brief report discussing how you came up with the final design.

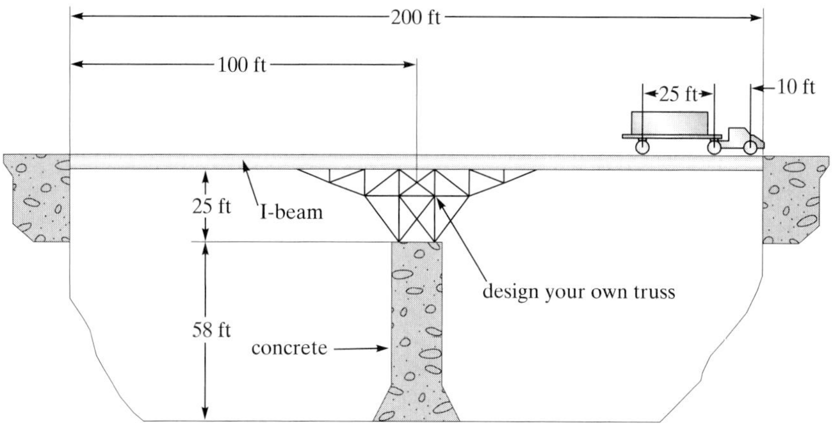

CHAPTER 5

One-Dimensional Elements

The objectives of this chapter are to introduce the concepts of one-dimensional elements and shape functions and their properties in more detail. The idea of local and natural coordinate systems is also presented here. In addition, one-dimensional elements used by ANSYS are discussed. These are the main topics discussed in Chapter 5:

5.1 Linear Elements
5.2 Quadratic Elements
5.3 Cubic Elements
5.4 Global, Local, and Natural Coordinates
5.5 Isoparametric Elements
5.6 Numerical Integration: Gauss–Legendre Quadrature
5.7 Examples of One-Dimensional Elements in ANSYS

5.1 LINEAR ELEMENTS

The heat transfer example in this section is employed to introduce the basic ideas of one-dimensional elements and shape functions. Fins are commonly used in a variety of engineering applications to enhance cooling. Common examples include a motorcycle engine head, a lawn mower engine head, extended surfaces (heat sinks) used in electronic equipment, and finned-tube heat exchangers. A straight fin of a uniform cross section is shown in Figure 5.1, along with a typical temperature distribution along the fin. As a first approximation, let us divide the fin into three elements and four nodes. The actual temperature distribution may be approximated by a combination of linear functions, as shown in Figure 5.1. To better approximate the actual temperature gradient near the base of the fin in our finite element model, we have placed the nodes closer to each other in that region. It should be clear that we can improve the accuracy of our approximation by increasing the number of elements as well. However, for now, let us be content with the three-element model and focus our attention on a typical element, as

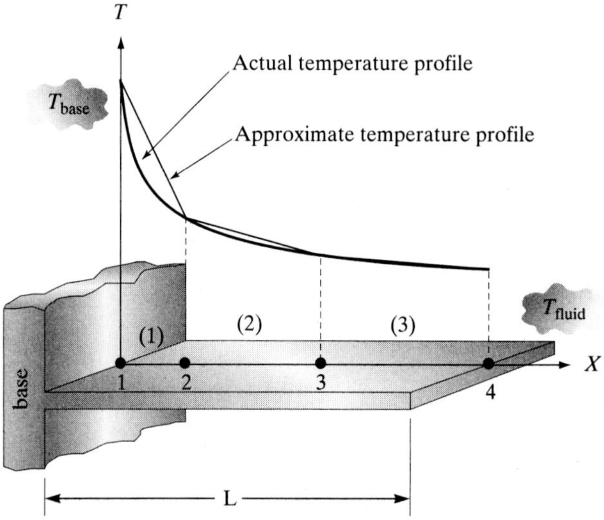

FIGURE 5.1 Temperature distribution for a fin of uniform cross section.

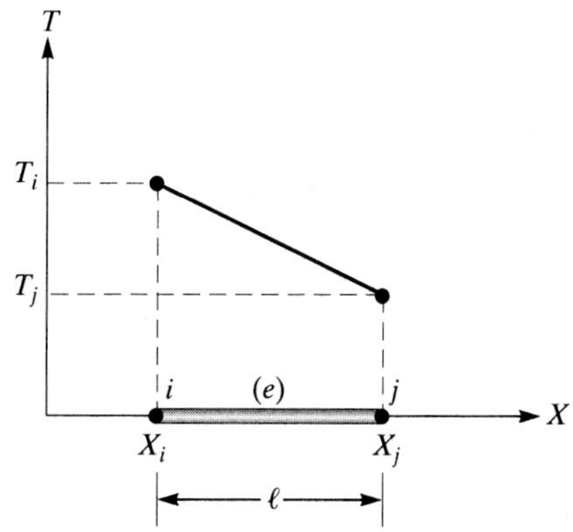

FIGURE 5.2 Linear approximation of temperature distribution for an element.

shown in Figure 5.2. The temperature distribution along the element may be interpolated (or approximated) using a linear function, as depicted in Figure 5.2.

The forthcoming derivation of the shape functions are similar to the one we showed in Chapter 4, Section 4.1. As a review and for the sake of continuity and convenience, the steps to derive the shape functions are presented here again.

The linear temperature distribution for a typical element may be expressed as

$$T^{(e)} = c_1 + c_2 X \tag{5.1}$$

The element's end conditions are given by the nodal temperatures T_i and T_j, according to the conditions

$$\begin{aligned} T &= T_i \quad \text{at} \quad X = X_i \\ T &= T_j \quad \text{at} \quad X = X_j \end{aligned} \tag{5.2}$$

Substitution of nodal values into Eq. (5.1) results in two equations and two unknowns:

$$T_i = c_1 + c_2X_i \tag{5.3}$$
$$T_j = c_1 + c_2X_j$$

Solving for the unknowns c_1 and c_2, we get

$$c_1 = \frac{T_iX_j - T_jX_i}{X_j - X_i} \tag{5.4}$$

$$c_2 = \frac{T_j - T_i}{X_j - X_i} \tag{5.5}$$

The element's temperature distribution in terms of its nodal values is

$$T^{(e)} = \frac{T_iX_j - T_jX_i}{X_j - X_i} + \frac{T_j - T_i}{X_j - X_i}X \tag{5.6}$$

Grouping the T_i terms together and the T_j terms together, we obtain

$$T^{(e)} = \left(\frac{X_j - X}{X_j - X_i}\right)T_i + \left(\frac{X - X_i}{X_j - X_i}\right)T_j \tag{5.7}$$

We now define the *shape functions*, S_i and S_j, according to the equations

$$S_i = \frac{X_j - X}{X_j - X_i} = \frac{X_j - X}{\ell} \tag{5.8}$$

$$S_j = \frac{X - X_i}{X_j - X_i} = \frac{X - X_i}{\ell} \tag{5.9}$$

where ℓ is the length of the element. Thus, the temperature distribution of an element in terms of the shape functions can be written as

$$T^{(e)} = S_iT_i + S_jT_j \tag{5.10}$$

Equation (5.10) can also be expressed in matrix form as

$$T^{(e)} = [S_i \quad S_j]\begin{Bmatrix} T_i \\ T_j \end{Bmatrix} \tag{5.11}$$

As you recall, for the structural example in Chapter 4, the deflection $u^{(e)}$ for a typical column element is represented by

$$u^{(e)} = [S_i \quad S_j]\begin{Bmatrix} u_i \\ u_j \end{Bmatrix} \tag{5.12}$$

where u_i and u_j represent the deflections of nodes i and j of an arbitrary element (e). It should be clear by now that we can represent the spatial variation of any unknown variable over a given element by using shape functions and the corresponding nodal values.

Thus, in general, we can write

$$\Psi^{(e)} = [S_i \quad S_j] \begin{Bmatrix} \Psi_i \\ \Psi_j \end{Bmatrix} \tag{5.13}$$

where Ψ_i and Ψ_j represent the nodal values of the unknown variable, such as temperature, deflection, or velocity.

Properties of Shape Functions

The shape functions possess unique properties that are important for us to understand because they simplify the evaluation of certain integrals when we are deriving the conductance or stiffness matrices. One of the inherent properties of a shape function is that it has a value of unity at its corresponding node and has a value of zero at the adjacent node. Let us demonstrate this property by evaluating the shape functions at $X = X_i$ and $X = X_j$. Evaluating S_i at $X = X_i$ and $X = X_j$, we get

$$S_i|_{X=X_i} = \left.\frac{X_j - X}{\ell}\right|_{X=X_i} = \frac{X_j - X_i}{\ell} = 1 \text{ and } S_i|_{X=X_j} = \left.\frac{X_j - X}{\ell}\right|_{X=X_j} = \frac{X_j - X_j}{\ell} = 0 \tag{5.14}$$

Evaluating S_j at $X = X_i$ and $X = X_j$, we obtain

$$S_j|_{X=X_i} = \left.\frac{X - X_i}{\ell}\right|_{X=X_i} = \frac{X_i - X_i}{\ell} = 0 \text{ and } S_j|_{X=X_j} = \left.\frac{X - X_i}{\ell}\right|_{X=X_j} = \frac{X_j - X_i}{\ell} = 1 \tag{5.15}$$

This property is also illustrated in Figure 5.3.

Another important property associated with shape functions is that the shape functions add up to a value of unity. That is,

$$S_i + S_j = \frac{X_j - X}{X_j - X_i} + \frac{X - X_i}{X_j - X_i} = 1 \tag{5.16}$$

It can also be readily shown that for linear shape functions, the sum of the derivatives with respect to X is zero. That is,

$$\frac{d}{dX}\left(\frac{X_j - X}{X_j - X_i}\right) + \frac{d}{dX}\left(\frac{X - X_i}{X_j - X_i}\right) = -\frac{1}{X_j - X_i} + \frac{1}{X_j - X_i} = 0 \tag{5.17}$$

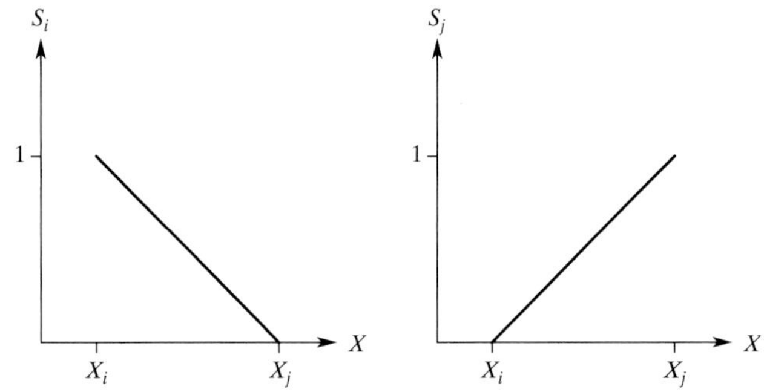

FIGURE 5.3 Linear shape functions.

EXAMPLE 5.1

We have used linear one-dimensional elements to approximate the temperature distribution along a fin. The nodal temperatures and their corresponding positions are shown in Figure 5.4. What is the temperature of the fin at (a) $X = 4$ cm and (b) $X = 8$ cm?

In Chapter 6, we will discuss in detail the analysis of one-dimensional fin problems, including the computation of nodal temperatures. However, for now, using the given nodal temperatures, we can proceed to answer both parts of the question:

a. The temperature of the fin at $X = 4$ cm is represented by element (2);

$$T^{(2)} = S_2^{(2)}T_2 + S_3^{(2)}T_3 = \frac{X_3 - X}{\ell}T_2 + \frac{X - X_2}{\ell}T_3$$

$$T = \frac{5 - 4}{3}(41) + \frac{4 - 2}{3}(34) = 36.3\,^\circ\text{C}$$

b. The temperature of the fin at $X = 8$ cm is represented by element (3);

$$T^{(3)} = S_3^{(3)}T_3 + S_4^{(3)}T_4 = \frac{X_4 - X}{\ell}T_3 + \frac{X - X_3}{\ell}T_4$$

$$T = \frac{10 - 8}{5}(34) + \frac{8 - 5}{5}(20) = 25.6^\circ\text{C}$$

For this example, note the difference between $S_3^{(2)}$ and $S_3^{(3)}$.

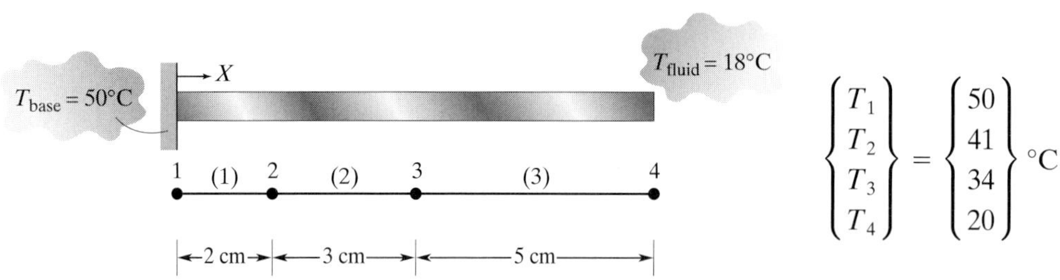

FIGURE 5.4 The nodal temperatures and their corresponding positions along the fin in Example 5.1.

5.2 QUADRATIC ELEMENTS

We can increase the accuracy of our finite element findings either by increasing the number of linear elements used in the analysis or by using higher order interpolation functions. For example, we can employ a quadratic function to represent the spatial variation of an unknown variable. Using a quadratic function instead of a linear function requires that we use three nodes to define an element. We need three nodes to define an element because in order to fit a quadratic function, we need three points. The third point can be created by placing a node, such as node k, in the middle of an element, as shown in Figure 5.5. Referring to the previous example of a fin, using quadratic approximation, the temperature distribution for a typical element can be represented by

$$T^{(e)} = c_1 + c_2X + c_3X^2 \tag{5.18}$$

and the nodal values are

$$\begin{aligned} T &= T_i \quad \text{at} \quad X = X_i \\ T &= T_k \quad \text{at} \quad X = X_k \\ T &= T_j \quad \text{at} \quad X = X_j \end{aligned} \tag{5.19}$$

Three equations and three unknowns are created upon substitution of the nodal values into Eq. (5.18):

$$\begin{aligned} T_i &= c_1 + c_2X_i + c_3X_i^2 \\ T_k &= c_1 + c_2X_k + c_3X_k^2 \\ T_j &= c_1 + c_2X_j + c_3X_j^2 \end{aligned} \tag{5.20}$$

Solving for c_1, c_2, and c_3 and rearranging terms leads to the element's temperature distribution in terms of the nodal values and the shape functions:

$$T^{(e)} = S_iT_i + S_jT_j + S_kT_k \tag{5.21}$$

In matrix form, the above expression is

$$T^{(e)} = [S_i \quad S_j \quad S_k] \begin{Bmatrix} T_i \\ T_j \\ T_k \end{Bmatrix} \tag{5.22}$$

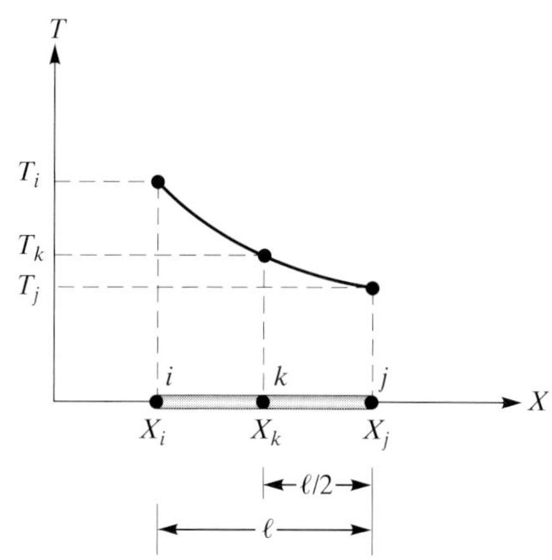

FIGURE 5.5 Quadratic approximation of the temperature distribution for an element.

where the shape functions are

$$S_i = \frac{2}{\ell^2}(X - X_j)(X - X_k) \tag{5.23}$$

$$S_j = \frac{2}{\ell^2}(X - X_i)(X - X_k)$$

$$S_k = \frac{-4}{\ell^2}(X - X_i)(X - X_j)$$

In general, for a given element the variation of any parameter Ψ in terms of its nodal values may be written as

$$\Psi^{(e)} = [S_i \quad S_j \quad S_k]\begin{Bmatrix} \Psi_i \\ \Psi_j \\ \Psi_k \end{Bmatrix} \tag{5.24}$$

It is important to note here that the quadratic shape functions possess properties similar to those of the linear shape functions; that is, (1) a shape function has a value of unity at its corresponding node and a value of zero at the other adjacent node, and (2) if we sum up the shape functions, we will again come up with a value of unity. The main difference between linear shape functions and quadratic shape functions is in their derivatives. The derivatives of the quadratic shape functions with respect to X are not constant.

5.3 CUBIC ELEMENTS

The quadratic interpolation functions offer good results in finite element formulations. However, if additional accuracy is needed, we can resort to even higher order interpolation functions, such as third-order polynomials. Thus, we can use cubic functions to represent the spatial variation of a given variable. Using a cubic function instead of a quadratic function requires that we use four nodes to define an element. We need four nodes to define an element because in order to fit a third-order polynomial, we need four points. The element is divided into three equal lengths. The placement of the four nodes is depicted in Figure 5.6. Referring to the previous example of a fin, using cubic approximation, the temperature distribution for a typical element can be represented by

$$T^{(e)} = c_1 + c_2 X + c_3 X^2 + c_4 X^3 \tag{5.25}$$

and the nodal values are

$$\begin{aligned} T &= T_i \quad \text{at} \quad X = X_i \\ T &= T_k \quad \text{at} \quad X = X_k \\ T &= T_m \quad \text{at} \quad X = X_m \\ T &= T_j \quad \text{at} \quad X = X_j \end{aligned} \tag{5.26}$$

Four equations and four unknowns are created upon substitution of the nodal values into Eq. (5.25). Solving for c_1, c_2, c_3, and c_4 and rearranging terms leads to the element's

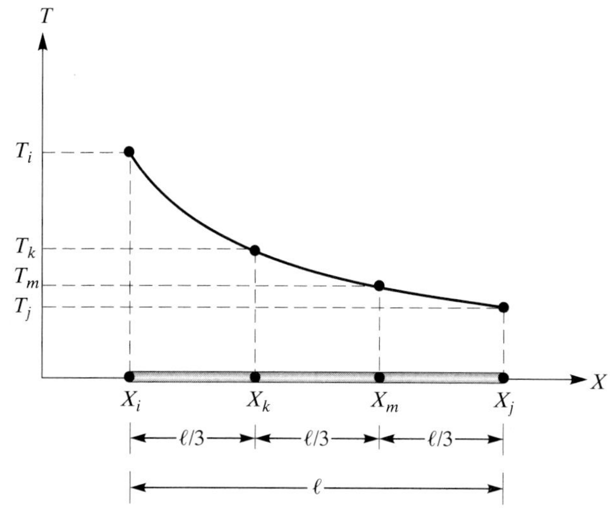

FIGURE 5.6 Cubic approximation of the temperature distribution for an element.

temperature distribution in terms of the nodal values and the shape functions:

$$T^{(e)} = S_iT_i + S_jT_j + S_kT_k + S_mT_m \tag{5.27}$$

In matrix form, the above expression is

$$T^{(e)} = [S_i \quad S_j \quad S_k \quad S_m]\begin{Bmatrix} T_i \\ T_j \\ T_k \\ T_m \end{Bmatrix} \tag{5.28}$$

where the shape functions are

$$S_i = -\frac{9}{2\ell^3}(X - X_j)(X - X_k)(X - X_m) \tag{5.29}$$

$$S_j = \frac{9}{2\ell^3}(X - X_i)(X - X_k)(X - X_m)$$

$$S_k = \frac{27}{2\ell^3}(X - X_i)(X - X_j)(X - X_m)$$

$$S_m = -\frac{27}{2\ell^3}(X - X_i)(X - X_j)(X - X_k)$$

It is worth noting that when the order of the interpolating function increases, it is necessary to employ **Lagrange interpolation functions** instead of taking the above approach to obtain the shape functions. The main advantage the Lagrange method offers is that using it, we do not have to solve a set of equations simultaneously to obtain the unknown coefficients of the interpolating function. Instead, we represent the shape functions in terms of the products of three linear functions. For cubic interpolating functions, the shape function associated with each node can be represented in terms of the product of three linear functions. For a given node—for example, *i*—we select the functions such that their product will produce a value of zero at other nodes—namely, j, k, and m—and a value of unity at the given node, i. Moreover, the product of the functions must produce linear and nonlinear terms similar to the ones given by a general third-order polynomial function.

To demonstrate this method, let us consider node i, with the global coordinate X_i. First, the functions must be selected such that when evaluated at nodes j, k, and m, the outcome is a value of zero. We select

$$S_i = a_1(X - X_j)(X - X_k)(X - X_m) \tag{5.30}$$

which satisfies the above condition. That is, if you substitute for $X = X_j$, or $X = X_k$, or $X = X_m$, the value of S_i is zero. We then evaluate a_1 such that when the shape function S_i is evaluated at node $i(X = X_i)$, it will produce a value of unity:

$$1 = a_1(X_i - X_j)(X_i - X_k)(X_i - X_m) = a_1(-\ell)\left(-\frac{\ell}{3}\right)\left(-\frac{2\ell}{3}\right)$$

Solving for a_1, we get

$$a_1 = -\frac{9}{2\ell^3}$$

and substituting into Eq. (5.30), we have

$$S_i = -\frac{9}{2\ell^3}(X - X_j)(X - X_k)(X - X_m)$$

The other shape functions are obtained in a similar fashion. Keeping in mind the explanation offered above, we can generate shape functions of an $(N - 1)$-order polynomial directly from the Lagrange polynomial formula:

$$S_K = \prod_{M=1}^{N} \frac{X - X_M \text{ omitting } (X - X_K)}{X_K - X_M \text{ omitting } (X_K - X_K)} = \frac{(X - X_1)(X - X_2)\cdots(X - X_N)}{(X_K - X_1)(X_K - X_2)\cdots(X_K - X_N)} \tag{5.31}$$

Note that in order to accommodate any order polynomial representation in Eq. (5.31) numeral values are assigned to the nodes and the subscripts of the shape functions.

In general, using a cubic interpolation function, the variation of any parameter Ψ in terms of its nodal values may be written as

$$\Psi^{(e)} = [S_i \quad S_j \quad S_k \quad S_m]\begin{Bmatrix} \Psi_i \\ \Psi_j \\ \Psi_k \\ \Psi_m \end{Bmatrix}$$

Once again, note that the cubic shape functions possess properties similar to those of the linear and the quadratic shape functions; that is, (1) a shape function has a value of unity at its corresponding node and a value of zero at the other adjacent node, and (2) if we sum up the shape functions, we will come up with a value of unity. However, note that taking the spatial derivative of cubic shape functions will produce quadratic results.

EXAMPLE 5.2

Using Lagrange interpolation functions, generate the quadratic shape functions directly from the Lagrange polynomial formula, Eq. (5.31):

$$S_K = \prod_{M=1}^{N} \frac{(X - X_M) \text{ omitting } (X - X_K)}{(X_K - X_M) \text{ omitting } (X_K - X_K)}$$

For quadratic shape functions, $N-1 = 2$ and $K = 1, 2, 3$. Refer to Figure 5.5 and note that subscripts 1, 2, and 3 correspond to nodes i, k, and j respectively. Also, it is important to distinguish the difference between lowercase k, denoting a specific node, and uppercase K, a variable subscript, denoting various nodes.

For node i or $K = 1$,

$$S_i = S_1 = \frac{(X - X_2)(X - X_3)}{(X_1 - X_2)(X_1 - X_3)} = \frac{(X - X_2)(X - X_3)}{\left(-\frac{\ell}{2}\right)(-\ell)} = \frac{2}{\ell^2}(X - X_2)(X - X_3)$$

For node k or $K = 2$,

$$S_k = S_2 = \frac{(X - X_1)(X - X_3)}{(X_2 - X_1)(X_2 - X_3)} = \frac{(X - X_1)(X - X_3)}{\left(\frac{\ell}{2}\right)\left(-\frac{\ell}{2}\right)} = \frac{-4}{\ell^2}(X - X_1)(X - X_3)$$

For node j or $K = 3$,

$$S_j = S_3 = \frac{(X - X_1)(X - X_2)}{(X_3 - X_1)(X_3 - X_2)} = \frac{(X - X_1)(X - X_2)}{(\ell)\left(\frac{\ell}{2}\right)} = \frac{2}{\ell^2}(X - X_1)(X - X_2)$$

The results are identical to shape functions given by Eq. (5.23).

5.4 GLOBAL, LOCAL, AND NATURAL COORDINATES

Most often, in finite element modeling, it is convenient to use several frames of reference, as we briefly discussed in Chapters 3 and 4. We need a global coordinate system to represent the location of each node, orientation of each element, and to apply boundary conditions and loads (in terms of their respective global components). Moreover, the solution, such as nodal displacements, is generally represented with respect to the global directions. On the other hand, we need to employ local and natural coordinates because they offer certain advantages when we construct the geometry or compute integrals. The advantage becomes apparent particularly when the integrals contain products of shape functions. For one-dimensional elements, the relationship between a global coordinate X and a local coordinate x is given by $X = X_i + x$, as shown in Figure 5.7.

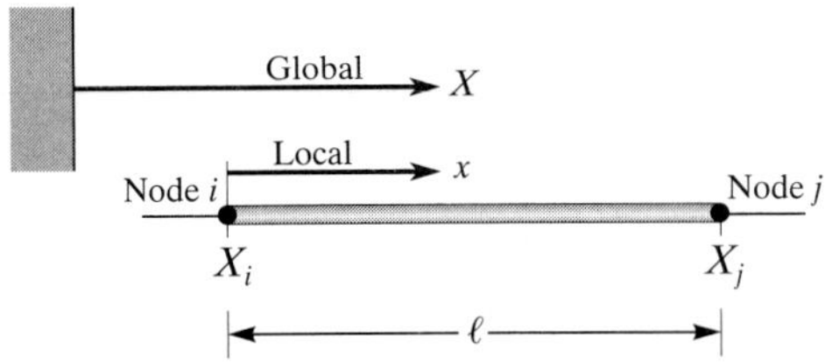

FIGURE 5.7 The relationship between a global coordinate X and a local coordinate x.

Substituting for X in terms of the local coordinate x in Eqs. (5.8) and (5.9), we get

$$S_i = \frac{X_j - X}{\ell} = \frac{X_j - (X_i + x)}{\ell} = 1 - \frac{x}{\ell} \tag{5.32}$$

$$S_j = \frac{X - X_i}{\ell} = \frac{(X_i + x) - X_i}{\ell} = \frac{x}{\ell} \tag{5.33}$$

where the local coordinate x varies from 0 to ℓ; that is $0 \leq x \leq \ell$.

One-Dimensional Linear Natural Coordinates

Natural coordinates are basically local coordinates in a dimensionless form. It is often necessary to use numerical methods to evaluate integrals for the purpose of calculating elemental stiffness or conductance matrices. Natural coordinates offer the convenience of having -1 and 1 for the limits of integration. For example, if we let

$$\xi = \frac{2x}{\ell} - 1$$

where x is the local coordinate, then we can specify the coordinates of node i as -1 and node j by 1. This relationship is shown in Figure 5.8.

We can obtain the natural linear shape functions by substituting for x in terms of ξ into Eqs. (5.32) and (5.33). This substitution yields

$$S_i = \frac{1}{2}(1 - \xi) \tag{5.34}$$

$$S_j = \frac{1}{2}(1 + \xi) \tag{5.35}$$

Natural linear shape functions possess the same properties as linear shape functions; that is, a shape function has a value of unity at its corresponding node and has a value

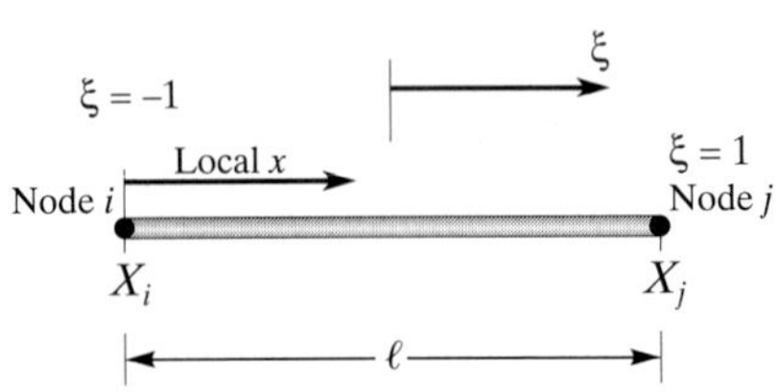

FIGURE 5.8 The relationship between the local coordinate x and the natural coordinate ξ.

of zero at the adjacent node in a given element. As an example, the temperature distribution over an element of a one-dimensional fin may expressed by

$$T^{(e)} = S_iT_i + S_jT_j = \frac{1}{2}(1 - \xi)T_i + \frac{1}{2}(1 + \xi)T_j \tag{5.36}$$

It is clear that at $\xi = -1, T = T_i$ and at $\xi = 1, T = T_j$.

5.5 ISOPARAMETRIC ELEMENTS

By now, it should be clear that we can represent other variables, such as the displacement u, in terms of the natural shape functions S_i and S_j according to the equation

$$u^{(e)} = S_iu_i + S_ju_j = \frac{1}{2}(1 - \xi)u_i + \frac{1}{2}(1 + \xi)u_j \tag{5.36a}$$

Also note that the transformation from the global coordinate $X(X_i \le X \le X_j)$ or the local coordinate $x(0 \le x \le \ell)$ to ξ can be made using the same shape functions S_i and S_j. That is,

$$X = S_iX_i + S_jX_j = \frac{1}{2}(1 - \xi)X_i + \frac{1}{2}(1 + \xi)X_j \tag{5.36b}$$

or

$$x = S_ix_i + S_jx_j = \frac{1}{2}(1 - \xi)x_i + \frac{1}{2}(1 + \xi)x_j$$

Comparing the relationships given by Eqs. (5.36), (5.36a), and (5.36b), we note that we have used a *single* set of parameters (such as S_i, S_j) to define the unknown variables u, T, and so on, and we used the same parameters (S_i, S_j) to express the geometry. Finite element formulation that makes use of this idea is commonly referred to as *isoparametric* (*iso* meaning the same or uniform) formulation, and an element expressed in such a manner is called an isoparametric element. We discuss isoparametric formulation further in Chapters 7 and 10.

EXAMPLE 5.3

Determine the temperature of the fin in Example 5.1 at the global location $X = 8$ cm using local coordinates. Also determine the temperature of the fin at the global location $X = 7.5$ cm using natural coordinates.

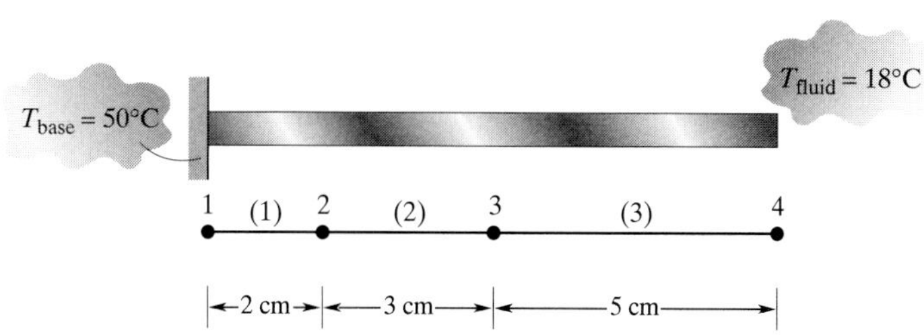

$$\begin{Bmatrix} T_1 \\ T_2 \\ T_3 \\ T_4 \end{Bmatrix} = \begin{Bmatrix} 50 \\ 41 \\ 34 \\ 20 \end{Bmatrix} \text{°C}$$

a. Using local coordinates, we find that the temperature of the fin at $X = 8$ cm is represented by element (3) according to the equation

$$T^{(3)} = S_3^{(3)}T_3 + S_4^{(3)}T_4 = \left(1 - \frac{x}{\ell}\right)T_3 + \frac{x}{\ell}T_4$$

Note that element (3) has a length of 5 cm, and the location of a point 8 cm from the base is represented by the local coordinate $x = 3$:

$$T = \left(1 - \frac{3}{5}\right)(34) + \frac{3}{5}(20) = 25.6\ ^\circ\text{C}$$

b. Using natural coordinates, we find that the temperature of the fin at $X = 7.5$ cm is represented by element (3) according to the equation

$$T^{(3)} = S_3^{(3)}T_3 + S_4^{(3)}T_4 = \frac{1}{2}(1 - \xi)T_3 + \frac{1}{2}(1 + \xi)T_4$$

Because the point with the global coordinate $X = 7.5$ cm is located in the middle of element (3), the natural coordinate of this point is given by $\xi = 0$:

$$T^{(3)} = \frac{1}{2}(1 - 0)(34) + \frac{1}{2}(1 + 0)(20) = 27\ ^\circ\text{C}$$

One-Dimensional Natural Quadratic and Cubic Shape Functions

The natural one-dimensional quadratic and cubic shape functions can be obtained in a way similar to the method discussed in the previous section. The quadratic natural shape functions are

$$S_i = -\frac{1}{2}\xi(1 - \xi) \tag{5.37}$$

$$S_j = \frac{1}{2}\xi(1 + \xi) \tag{5.38}$$

$$S_k = (1 + \xi)(1 - \xi) \tag{5.39}$$

The natural one-dimensional cubic shape functions are

$$S_i = \frac{1}{16}(1 - \xi)(3\xi + 1)(3\xi - 1) \tag{5.40}$$

$$S_j = \frac{1}{16}(1 + \xi)(3\xi + 1)(3\xi - 1) \tag{5.41}$$

$$S_k = \frac{9}{16}(1 + \xi)(\xi - 1)(3\xi - 1) \tag{5.42}$$

$$S_m = \frac{9}{16}(1 + \xi)(1 - \xi)(3\xi + 1) \tag{5.43}$$

For the sake of convenience, the results of Sections 5.1 to 5.4 are summarized in Table 5.1. Make sure to distinguish the differences among presentations of the shape functions using global, local, and natural coordinates.

EXAMPLE 5.4

Evaluate the integral $\int_{X_i}^{X_j} S_j^2 dX$ using (a) global coordinates and (b) local coordinates.

a. Using global coordinates, we obtain

$$\int_{X_i}^{X_j} S_j^2 dX = \int_{X_i}^{X_j}\left(\frac{X - X_i}{\ell}\right)^2 dX = \frac{1}{3\ell^2}(X - X_i)^3\bigg|_{X_i}^{X_j} = \frac{\ell}{3}$$

b. Using local coordinates, we obtain

$$\int_{X_i}^{X_j} S_j^2 dX = \int_0^{\ell}\left(\frac{x}{\ell}\right)^2 dx = \frac{x^3}{3\ell^2}\bigg|_0^{\ell} = \frac{\ell}{3}$$

This simple example demonstrates that local coordinates offer a simple way to evaluate integrals containing products of shape functions.

5.6 NUMERICAL INTEGRATION: GAUSS–LEGENDRE QUADRATURE

As we discussed earlier, natural coordinates are basically local coordinates in a dimensionless form. Moreover, most finite element programs perform element numerical integration by Gaussian quadratures, and as the limit of integration, they use an interval from -1 to 1. This approach is taken because when the function being integrated is known, the Gauss–Legendre formulae offer a more efficient way of evaluating an integral as compared to other numerical integration methods such as the trapezoidal method. Whereas the trapezoidal method or Simpson's method can be used to evaluate integrals dealing with discrete data (see Problem 24), the Gauss–Legendre method is based on the evaluation of a known function at nonuniformally spaced points to compute the integral. The two-point Gauss–Legendre formula is developed next in this section. The basic goal behind the Gauss–Legendre formulae is to represent an integral in terms of the sum of the product of certain weighting coefficients and the value of the function at some selected points. So, we begin with

$$I = \int_a^b f(x)dx = \sum_{i=1}^{n} w_i f(x_i) \tag{5.44}$$

Next, we must ask (1) How do we determine the value of the weighting coefficients, represented by the w_i's? (2) Where do we evaluate the function, or in other words, how do we select these points (x_i)? We begin by changing the limits of integration from $a - b$ to -1 to 1 with the introduction of the variable λ such that

$$x = c_0 + c_1\lambda$$

TABLE 5.1 One-dimensional shape functions

Interpolation function	In terms of global coordinate X $X_i \le X \le X_j$	In terms of local coordinate x $0 \le x \le \ell$	In terms of natural coordinate ξ $-1 \le \xi \le 1$
Linear	$S_i = \frac{X_j - X}{\ell}$	$S_i = 1 - \frac{x}{\ell}$	$S_i = \frac{1}{2}(1 - \xi)$
	$S_j = \frac{X - X_i}{\ell}$	$S_j = \frac{x}{\ell}$	$S_j = \frac{1}{2}(1 + \xi)$
Quadratic	$S_i = \frac{2}{\ell^2}(X - X_j)(X - X_k)$	$S_i = \left(\frac{x}{\ell} - 1\right)\left(2\left(\frac{x}{\ell}\right) - 1\right)$	$S_i = -\frac{1}{2}\xi(1 - \xi)$
	$S_j = \frac{2}{\ell^2}(X - X_i)(X - X_k)$	$S_j = \left(\frac{x}{\ell}\right)\left(2\left(\frac{x}{\ell}\right) - 1\right)$	$S_j = \frac{1}{2}\xi(1 + \xi)$
	$S_k = \frac{-4}{\ell^2}(X - X_i)(X - X_j)$	$S_k = 4\left(\frac{x}{\ell}\right)\left(1 - \left(\frac{x}{\ell}\right)\right)$	$S_k = (1 - \xi)(1 + \xi)$
Cubic	$S_i = -\frac{9}{2\ell^3}(X - X_j)(X - X_k)(X - X_m)$	$S_i = \frac{1}{2}\left(1 - \frac{x}{\ell}\right)\left(2 - 3\left(\frac{x}{\ell}\right)\right)\left(1 - 3\left(\frac{x}{\ell}\right)\right)$	$S_i = \frac{1}{16}(1 - \xi)(3\xi + 1)(3\xi - 1)$
	$S_j = \frac{9}{2\ell^3}(X - X_i)(X - X_k)(X - X_m)$	$S_j = \frac{1}{2}\left(\frac{x}{\ell}\right)\left(2 - 3\left(\frac{x}{\ell}\right)\right)\left(1 - 3\left(\frac{x}{\ell}\right)\right)$	$S_j = \frac{1}{16}(1 + \xi)(3\xi + 1)(3\xi - 1)$
	$S_k = \frac{27}{2\ell^3}(X - X_i)(X - X_j)(X - X_m)$	$S_k = \frac{9}{2}\left(\frac{x}{\ell}\right)\left(2 - 3\left(\frac{x}{\ell}\right)\right)\left(1 - \left(\frac{x}{\ell}\right)\right)$	$S_k = \frac{9}{16}(1 + \xi)(\xi - 1)(3\xi - 1)$
	$S_m = -\frac{27}{2\ell^3}(X - X_i)(X - X_j)(X - X_k)$	$S_m = \frac{9}{2}\left(\frac{x}{\ell}\right)\left(3\left(\frac{x}{\ell}\right) - 1\right)\left(1 - \left(\frac{x}{\ell}\right)\right)$	$S_m = \frac{9}{16}(1 + \xi)(1 - \xi)(3\xi + 1)$

Matching the limits, we get

$$a = c_0 + c_1(-1)$$
$$b = c_0 + c_1(1)$$

and solving for c_0 and c_1, we have

$$c_0 = \frac{(b + a)}{2}$$

and

$$c_1 = \frac{(b - a)}{2}$$

Therefore,

$$x = \frac{(b + a)}{2} + \frac{(b - a)}{2}\lambda \tag{5.45}$$

and

$$dx = \frac{(b - a)}{2}\, d\lambda \tag{5.46}$$

Thus, using Eqs. (5.45) and (5.46), we find that any integral in the form of Eq. (5.44) can be expressed in terms of an integral with its limits at -1 and 1:

$$I = \int_{-1}^{1} f(\lambda)\, d\lambda = \sum_{i=1}^{n} w_i f(\lambda_i) \tag{5.47}$$

The two-point Gauss–Legendre formulation requires the determination of two weighting factors w_1 and w_2 and two sampling points λ_1 and λ_2 to evaluate the function at these points. Because there are four unknowns, four equations are created using Legendre polynomials $(1, \lambda, \lambda^2, \lambda^3)$ as follows:

$$w_1 f(\lambda_1) + w_2 f(\lambda_2) = \int_{-1}^{1} 1\, d\lambda = 2$$

$$w_1 f(\lambda_1) + w_2 f(\lambda_2) = \int_{-1}^{1} \lambda\, d\lambda = 0$$

$$w_1 f(\lambda_1) + w_2 f(\lambda_2) = \int_{-1}^{1} \lambda^2\, d\lambda = \frac{2}{3}$$

$$w_1 f(\lambda_1) + w_2 f(\lambda_2) = \int_{-1}^{1} \lambda^3\, d\lambda = 0$$

The above equations lead to the equations

$$w_1(1) + w_2(1) = 2$$
$$w_1(\lambda_1) + w_2(\lambda_2) = 0$$

TABLE 5.2 Weighting factors and sampling points for Gauss–Legendre formulae

Points (n)	Weighting factors (w_i)	Sampling points (λ_i)
2	$w_1 = 1.00000000$ $w_2 = 1.00000000$	$\lambda_1 = -0.577350269$ $\lambda_2 = 0.577350269$
3	$w_1 = 0.55555556$ $w_2 = 0.88888889$ $w_3 = 0.55555556$	$\lambda_1 = -0.774596669$ $\lambda_2 = 0$ $\lambda_3 = 0.774596669$
4	$w_1 = 0.3478548$ $w_2 = 0.6521452$ $w_3 = 0.6521452$ $w_4 = 0.3478548$	$\lambda_1 = -0.861136312$ $\lambda_2 = -0.339981044$ $\lambda_3 = 0.339981044$ $\lambda_4 = 0.861136312$
5	$w_1 = 0.2369269$ $w_2 = 0.4786287$ $w_3 = 0.5688889$ $w_4 = 0.4786287$ $w_5 = 0.2369269$	$\lambda_1 = -0.906179846$ $\lambda_2 = -0.538469310$ $\lambda_3 = 0$ $\lambda_4 = 0.538469310$ $\lambda_5 = 0.906179846$

$$w_1(\lambda_1)^2 + w_2(\lambda_2)^2 = \frac{2}{3}$$

$$w_1(\lambda_1)^3 + w_2(\lambda_2)^3 = 0$$

Solving for w_1, w_2, λ_1, and λ_2, we have $w_1 = w_2 = 1, \lambda_1 = -0.577350269$, and $\lambda_2 = 0.577350269$. The weighting factors and the 2, 3, 4, and 5 sampling points for Gauss–Legendre formulae are given in Table 5.2. Note that as the number of sampling points increases, so does the accuracy of the calculations. As you will see in Chapter 7, we can readily extend the Gauss–Legendre quadrature formulation to two- or three-dimensional problems.

EXAMPLE 5.5

Evaluate the integral $I = \int_2^6 (x^2 + 5x + 3)\,dx$ using the Gauss–Legendre two-point sampling formula.

This integral is simple and can be evaluated analytically, leading to the solution $I = 161.333333333$. The purpose of this example is to demonstrate the Gauss–Legendre procedure. We begin by changing the variable x to λ by using Eq. (5.45). So, we obtain

$$x = \frac{(b + a)}{2} + \frac{(b - a)}{2}\lambda = \frac{(6 + 2)}{2} + \frac{(6 - 2)}{2}\lambda = 4 + 2\lambda$$

and

$$dx = \frac{(b - a)}{2}\,d\lambda = \frac{(6 - 2)}{2}\,d\lambda = 2\,d\lambda$$

Thus, the integral I can be expressed in terms of λ:

$$I = \int_2^6 \overbrace{(x^2 + 5x + 3)}^{f(x)}dx = \int_{-1}^{1} \overbrace{(2)[(4 + 2\lambda)^2 + 5(4 + 2\lambda) + 3)]}^{f(\lambda)}d\lambda$$

Using the Gauss–Legendre two-point formula and Table 5.2, we compute the value of the integral I from

$$I \cong w_1 f(\lambda_1) + w_2 f(\lambda_2)$$

From Table 5.2, we find that $w_1 = w_2 = 1$, and evaluating $f(\lambda)$ at $\lambda_1 = -0.577350269$ and $\lambda_2 = 0.577350269$, we obtain

$$f(\lambda_1) = (2)[[4 + 2(-0.577350269)]^2 + 5(4 + 2(-0.577350269) + 3)] = 50.6444526769$$

$$f(\lambda_2) = (2)[[4 + 2(0.577350269)]^2 + 5(4 + 2(0.577350269) + 3)] = 110.688880653$$

$$I = (1)(50.6444526769) + (1)110.688880653 = 161.33333333$$

EXAMPLE 5.6

Evaluate the integral $\int_{X_i}^{X_j} S_j^2 dX$ in Example 5.4 using the Gauss–Legendre two-point formula.

Recall from Eq. (5.35) that $S_j = \frac{1}{2}(1 + \xi)$ and by differentiating the relationship between the local coordinate x and the natural coordinate ξ (i.e., $\xi = \frac{2x}{\ell} - 1 \Rightarrow d\xi = \frac{2}{\ell}dx$) we find $dx = \frac{\ell}{2}d\xi$. Also note that for this problem, $\xi = \lambda$. So,

$$I = \int_{X_i}^{X_j} S_j^2\, dX = \int_{X_i}^{X_j}\left(\frac{X - X_i}{\ell}\right)^2 dX = \int_0^{\ell}\left(\frac{x}{\ell}\right)^2 dx = \frac{\ell}{2}\int_{-1}^{1}\left[\frac{1}{2}(1 + \xi)\right]^2 d\xi$$

Using the Gauss–Legendre two-point formula and Table 5.2, we compute the value of the integral I from

$$I \cong w_1 f(\lambda_1) + w_2 f(\lambda_2)$$

From Table 5.2, we find that $w_1 = w_2 = 1$, and evaluating $f(\lambda)$ at $\lambda_1 = -0.577350269$ and $\lambda_2 = 0.577350269$, we obtain

$$f(\xi_1) = \frac{\ell}{2}\left[\frac{1}{2}(1 + \xi_1)\right]^2 = \frac{\ell}{2}\left[\frac{1}{2}(1 - 0.577350269)\right]^2 = 0.022329099389\ell$$

$$f(\xi_2) = \frac{\ell}{2}\left[\frac{1}{2}(1 + \xi_2)\right]^2 = \frac{\ell}{2}\left[\frac{1}{2}(1 + 0.577350269)\right]^2 = 0.31100423389\ell$$

$$I = (1)(0.022329099389\ell) + (1)(0.31100423389\ell) = 0.333333333\ell$$

Note that the above result is identical to the results of Example 5.4.

5.7 EXAMPLES OF ONE-DIMENSIONAL ELEMENTS IN ANSYS

ANSYS offers uniaxial link elements that may be used to represent one-dimensional problems. These link elements include LINK31, LINK32, and LINK34. The LINK32 element is a uniaxial heat conduction element. It allows for the transfer of heat between its two nodes via conduction mode. The nodal degree of freedom associated with this element is temperature. The element is defined by its two nodes, cross-sectional area, and material properties such as thermal conductivity. The LINK34 element is a uniaxial convection link that allows for heat transfer between its nodes by convection. This element is defined by its two nodes, a convective surface area, and a convective heat transfer (film) coefficient. The LINK31 element can be used to model radiation heat transfer between two points in space. The element is defined by its two nodes, a radiation surface area, a geometric shape factor, emissivity, and the Stefan–Boltzman constant. In Chapter 6, we will use LINK32 and LINK34 to solve a one-dimensional heat-conduction problem.

SUMMARY

At this point you should

1. have a good understanding of the linear one-dimensional elements and shape functions, their properties, and their limitations.
2. have a good understanding of the quadratic and cubic one-dimensional elements and shape functions, their properties, and their advantages over linear elements.
3. know why it is important to use local and natural coordinate systems.
4. know what is meant by isoparametric element and formulation.
5. have a good understanding of Gauss–Legendre quadrature.
6. know examples of one-dimensional elements in ANSYS.

REFERENCES

ANSYS User's Manual: Elements, Vol. III, Swanson Analysis Systems, Inc.

Chandrupatla, T., and Belegundu, A., *Introduction to Finite Elements in Engineering*, Prentice Hall, 1991.

Incropera, F. P., and DeWitt, D. P., *Fundamentals of Heat and Mass Transfer*, 2nd ed., New York, John Wiley and Sons, 1985.

Segrlind, L., *Applied Finite Element Analysis*, 2nd ed., New York John Wiley and Sons, 1984.

PROBLEMS

1. We have used linear one-dimensional elements to approximate the temperature distribution along a fin. The nodal temperatures and their corresponding positions are shown in the accompanying figure. (a) What is the temperature of the fin at $X = 7$ cm? (b) Evaluate the heat loss from the fin using the relationship

$$Q = -kA\frac{dT}{dX}\Big|_{X=0}$$

where $k = 180$ W/m · K and A $= 10$ mm^2.

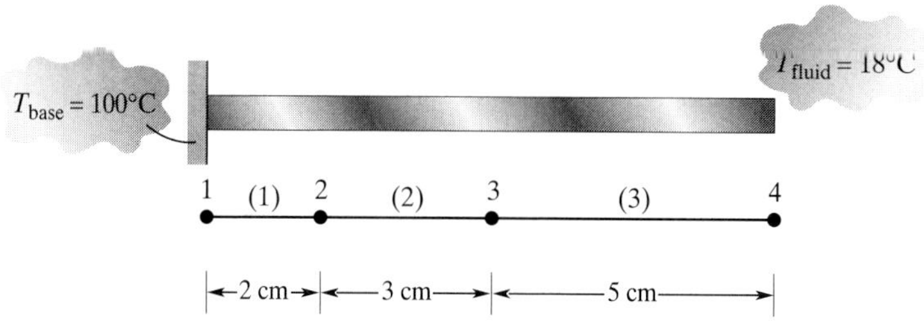

$$\begin{Bmatrix} T_1 \\ T_2 \\ T_3 \\ T_4 \end{Bmatrix} = \begin{Bmatrix} 100 \\ 800 \\ 64 \\ 41 \end{Bmatrix} °\text{C}$$

2. Evaluate the integral $\int_{X_i}^{X_j} S_i^2\,dX$ for a linear shape function using (a) global coordinates and (b) local coordinates.

3. Starting with the equations

$$T_i = c_1 + c_2X_i + c_3X_i^2$$

$$T_k = c_1 + c_2X_k + c_3X_k^2$$

$$T_j = c_1 + c_2X_j + c_3X_j^2$$

solve for c_1, c_2, and c_3, and rearrange terms to verify the shape functions given by

$$S_i = \frac{2}{\ell^2}(X - X_j)(X - X_k)$$

$$S_j = \frac{2}{\ell^2}(X - X_i)(X - X_k)$$

$$S_k = \frac{-4}{\ell^2}(X - X_i)(X - X_j)$$

4. For problem 3, use the Lagrange functions to derive the quadratic shape functions by the method discussed in Section 5.3.

5. Derive the expressions for quadratic shape functions in terms of the local coordinates and compare your results to the results given in Table 5.1.

6. Verify the results given for one-dimensional quadratic natural shape functions in Table 5.1 by showing that (1) a shape function has a value of unity at its corresponding node and a value of zero at the other nodes, and (2) if we sum up the shape functions, we will come up with a value of unity.

7. Verify the results given for the local cubic shape functions in Table 5.1 by showing that (1) a shape function has a value of unity at its corresponding node and a value of zero at the other nodes and (2) if we sum up the shape functions, we will come up with a value of unity.

8. Verify the results given for the natural cubic shape functions in Table 5.1 by showing that (1) a shape function has a value of unity at its corresponding node and a value of zero at the other nodes and (2) if we sum up the shape functions, we will come up with a value of unity.

9. Obtain expressions for the spatial derivatives of the quadratic and cubic shape functions.

10. As previously explained, we can increase the accuracy of our finite element findings either by increasing the number of elements used in the analysis to represent a problem or by using a higher order approximation. Derive the local cubic shape functions.

11. Evaluate the integral $\int_{X_i}^{X_j} S_i \, dX$ for a quadratic shape function using (a) global coordinates, (b) natural coordinates, and (c) local coordinates.

12. Assume that the deflection of a cantilever beam was approximated with linear one-dimensional elements. The nodal deflections and their corresponding positions are shown in the accompanying figure. (a) What is the deflection of the beam at $X = 2$ ft? (b) Evaluate the slope at the endpoint.

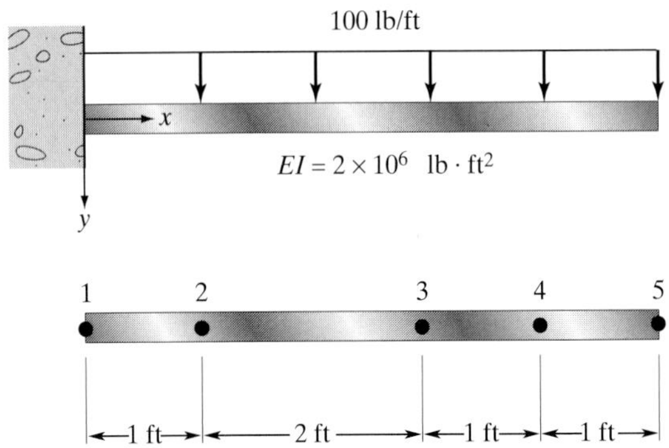

$$\begin{Bmatrix} y_1 \\ y_2 \\ y_3 \\ y_4 \\ y_5 \end{Bmatrix} = \begin{Bmatrix} 0 \\ 0.003275 \\ 0.022275 \\ 0.034400 \\ 0.046875 \end{Bmatrix} \text{ in.}$$

13. We have used linear one-dimensional elements to approximate the temperature distribution inside a metal plate. A heating element is embedded within a plate. The nodal temperatures and their corresponding positions are shown in the accompanying figure. What is the temperature of the plate at $X = 25$ mm? Assume that (a) linear elements were used in obtaining nodal temperatures and (b) quadratic elements were used.

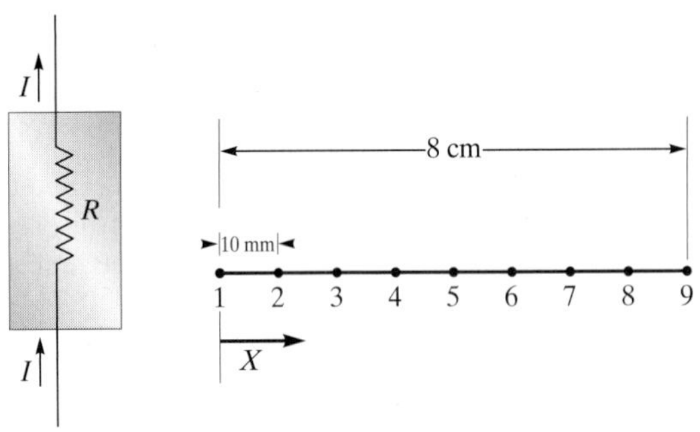

$$\begin{Bmatrix} T_1 \\ T_2 \\ T_3 \\ T_4 \\ T_5 \\ T_6 \\ T_7 \\ T_8 \\ T_9 \end{Bmatrix} = \begin{Bmatrix} 120 \\ 119 \\ 116 \\ 111 \\ 104 \\ 95 \\ 84 \\ 71 \\ 56 \end{Bmatrix} {}^\circ\text{C}$$

14. Quadratic elements are used to approximate the temperature distribution in a straight fin. The nodal temperatures and their corresponding positions are shown in the accompanying figure. What is the temperature of the fin at $X = 7$ cm?

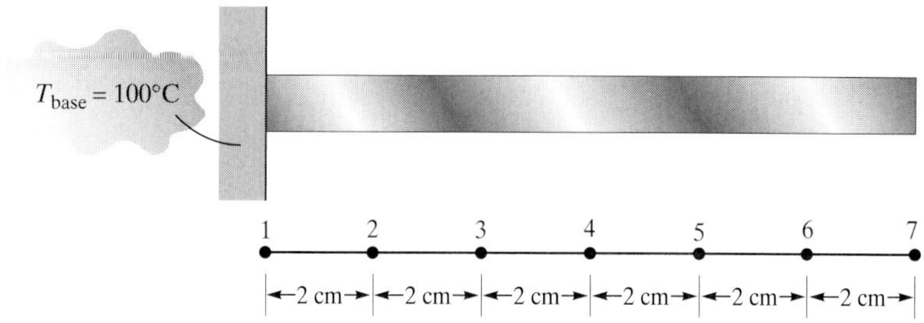

$$\begin{Bmatrix} T_1 \\ T_2 \\ T_3 \\ T_4 \\ T_5 \\ T_6 \\ T_7 \end{Bmatrix} = \begin{Bmatrix} 100 \\ 74 \\ 56 \\ 44 \\ 36 \\ 31 \\ 28 \end{Bmatrix} \text{°C}$$

15. Develop the shape functions for a linear element, shown in the accompanying figure, using the local coordinate x whose origin lies at the one-fourth point of the element.

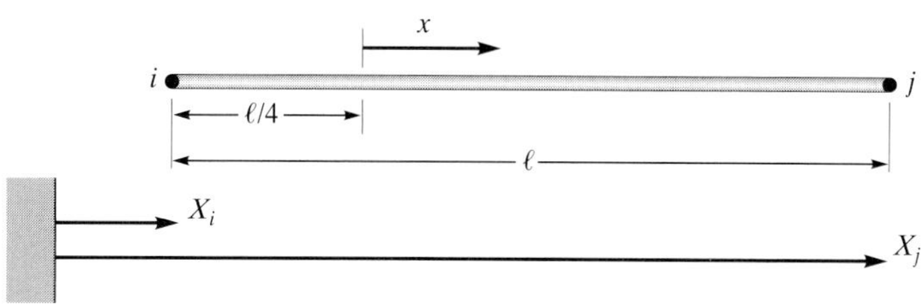

16. Using the natural coordinate system shown in the accompanying figure, develop the natural shape functions for a linear element.

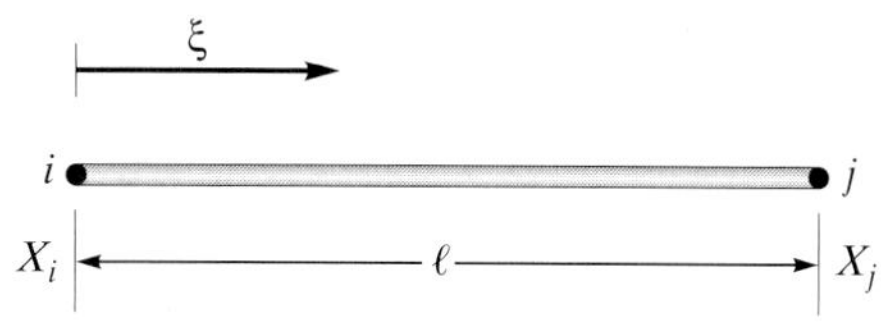

17. In the accompanying figure, the deflection of nodes 2 and 3 are 0.02 mm and 0.025 mm, respectively. What are the deflections at point A and point B, provided that linear elements were used in the analysis?

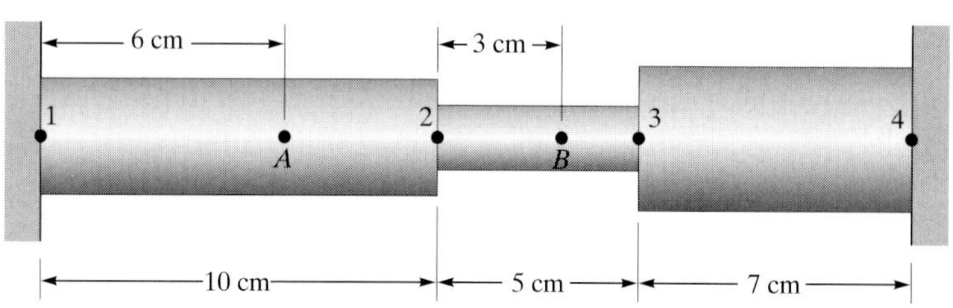

18. Consider the steel column depicted in the accompanying figure. Under the assumption of axial loading, and using linear elements, we determined that the downward vertical displacements of the column at various floor–column connection points are

$$\begin{Bmatrix} u_1 \\ u_2 \\ u_3 \\ u_4 \\ u_5 \end{Bmatrix} = \begin{Bmatrix} 0 \\ 0.03283 \\ 0.05784 \\ 0.07504 \\ 0.08442 \end{Bmatrix} \text{in.}$$

Using local shape functions, determine the deflections of points A and B, located in the middle of elements (3) and (4) respectively.

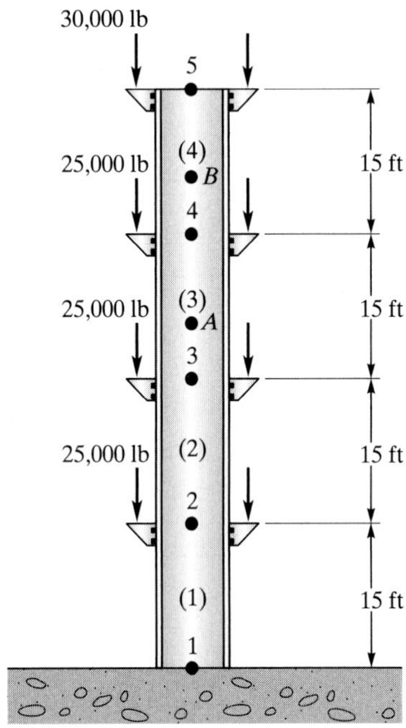

19. Determine the deflection of points A and B on the column in problem 18 using natural coordinates.

20. A 20-ft-tall post is used to support advertisement signs at various locations along its height, as shown in the accompanying figure. The post is made of structural steel with a modulus of elasticity of $E = 29 \times 10^6$ lb/in^2. Not considering wind loading on the signs and using linear elements, we determined that the downward deflections of the post at the points of load application are

$$\begin{Bmatrix} u_1 \\ u_2 \\ u_3 \\ u_4 \end{Bmatrix} = \begin{Bmatrix} 0 \\ 6.312 \times 10^{-4} \\ 8.718 \times 10^{-4} \\ 11.470 \times 10^{-4} \end{Bmatrix} \text{in.}$$

Determine the deflection of point A, located at the midpoint of the middle member, using (a) global shape functions, (b) local shape functions, and (c) natural shape functions.

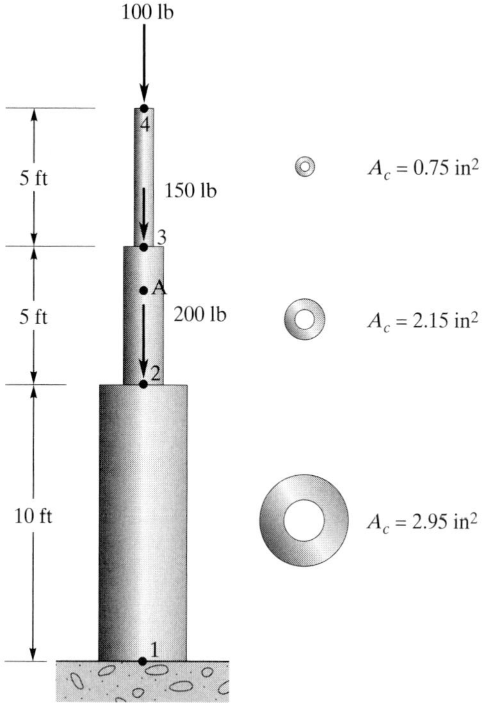

21. Evaluate the integral in problem 11 using Gauss–Legendre two-point formula.

22. Evaluate the given integral analytically and using Gauss–Legendre formula.

$$\int_{1}^{5} (x^3 + 5x^2 + 10)dx$$

23. Evaluate the given integral analytically and using Gauss–Legendre two- and three-point formulae.

$$\int_{-2}^{8} (3x^4 + x^2 - 7x + 10)dx$$

24. In Section 5.6 we mentioned that when the function being integrated is known, the Gauss–Legndre formulae offer a more efficient way of evaluating an integral as compared to the trapezoidal method. The trapezoidal approximation of an integral deals with discrete data at uniformly spaced intervals and is computed from

$$\int_{a}^{b} f(x)dx \approx h\left(\frac{1}{2}y_0 + y_1 + y_2 + \cdots + y_{n-2} + y_{n-1} + \frac{1}{2}y_n\right)$$

The above equation is known as the trapezoidal rule, and h represents the spacing among data points, $y_0, y_1, y_2, \ldots,$ and y_n. For problem 23, use $h = 1$ and generate 11 data points: y_0 at $x = -2$, y_1 at $x = -1, \ldots,$ and y_{10} at $x = 8$. Use the generated data and the trapezoidal rule to approximate the integral. Compare this result to the results of problem 23.

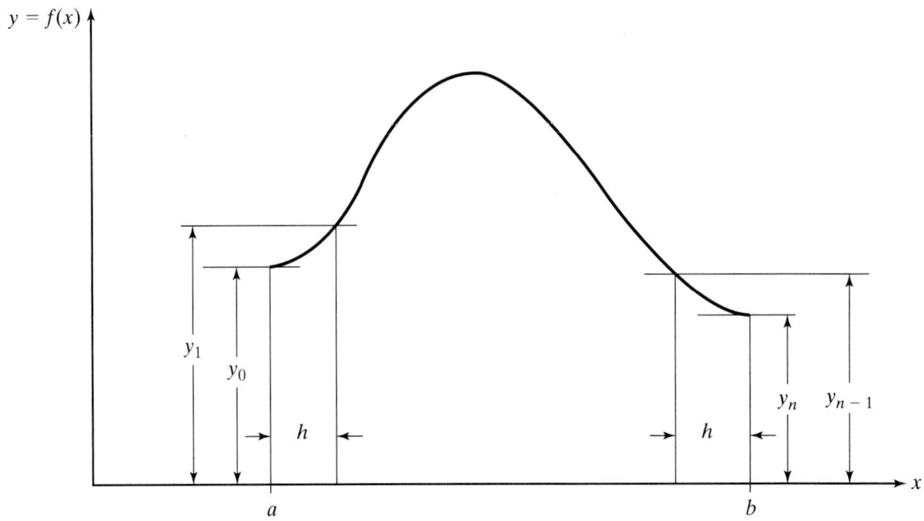

25. Derive and plot the spatial derivatives for linear, quadratic, and cubic elements. Discuss the differences.

26. Temperature distribution along a section of a material is given by $T_0 = 80°\text{C}$, $T_1 = 70°\text{C}$, $T_2 = 62°\text{C}$, $T_3 = 55°\text{C}$ with $X_0 = 0$, $X_1 = 2$ cm, $X_2 = 4$ cm, $X_3 = 6$ cm. Use single linear, quadratic, and cubic elements to approximate the given temperature distribution. Plot the actual data points and compare them to the linear, quadratic, and cubic approximations. Also, estimate the spatial derivative of the given data and compare it to the derivatives of the linear, quadratic, and cubic representations.

27. Use the isoparametric formulation to express the following information: $T_1 = 100°\text{C}$, $T_2 = 58°\text{C}$ with the corresponding global coordinates $X_1 = 2$ cm, $X_2 = 5$ cm. Show the transformation equations between the global coordinate, the local coordinate, and the natural coordinate.

28. Use the isoparametric formulation to express the following deflection information: $U_1 = 0.01$ cm, $U_2 = 0.025$ cm with the corresponding global coordinates $X_1 = 5$cm, $X_2 = 12$ cm. Show the transformation equations between the global coordinate, the local coordinate, and the natural coordinate.

CHAPTER 6

Analysis of One-Dimensional Problems

The main objective of this chapter is to introduce the analysis of one-dimensional problems. Most often, a physical problem is not truly one-dimensional in nature; however, as a starting point, we may model the behavior of a system using one-dimensional approximation. This approach can usually provide some basic insight into a more complex problem. If necessary, as a next step we can always analyze the problem using a two- or three-dimensional approach. This chapter first presents the one-dimensional Galerkin formulation used for heat transfer problems. This presentation is followed by an example demonstrating analysis of a one-dimensional fluid mechanics problem. The main topics discussed in Chapter 6 are

6.1 Heat Transfer Problems
6.2 A Fluid Mechanics Problem
6.3 An Example Using ANSYS
6.4 Verification of Results

6.1 HEAT TRANSFER PROBLEMS

Recall that in Chapter 1 we discussed the basic steps involved in any finite element analysis; to refresh your memory, these steps are repeated here.

PREPROCESSING PHASE

1. Create and discretize the solution domain into finite elements; that is, subdivide the problem into nodes and elements.
2. Assume shape functions to represent the behavior of an element; that is, assume an approximate continuous function to represent the solution for a element. The one-dimensional linear and quadratic shape functions were discussed in Chapter 5.

3. Develop equations for an element. This step is the main focus of the current chapter. We will use the Galerkin approach to formulate elemental descriptions. In Chapter 4, we used the minimum potential energy theorem to generate finite element models for members under axial loading and for beam elements.
4. Assemble the elements to represent the entire problem. Construct the global stiffness or conductance matrix.
5. Apply boundary conditions and loading.

SOLUTION PHASE

6. Solve a set of linear algebraic equations simultaneously to obtain nodal results, such as the temperature at different nodes or displacements.

POSTPROCESSING PHASE

7. Obtain other important information. We may be interested in determining the heat loss or stress in each element.

We now focus our attention on step 3 of the preprocessing phase. We formulate the conductance and the thermal load matrices for a typical one-dimensional fin element. We considered a straight fin of a uniform cross section in Chapter 5. For the sake of convenience, the fin is shown again in Figure 6.1. The fin is modeled using three elements and four nodes. The temperature distribution along the element is interpolated using linear functions. The actual and the approximate piecewise linear temperature distribution along the fin are shown in Figure 6.1. We will concentrate on a typical element belonging to the fin and formulate the conductance matrix and the thermal load matrix for such an element.

One-dimensional heat transfer in a straight fin is governed by the following heat equation, as given in any introductory text on heat transfer:

$$kA\frac{d^2T}{dX^2} - hpT + hpT_f = 0 \tag{6.1}$$

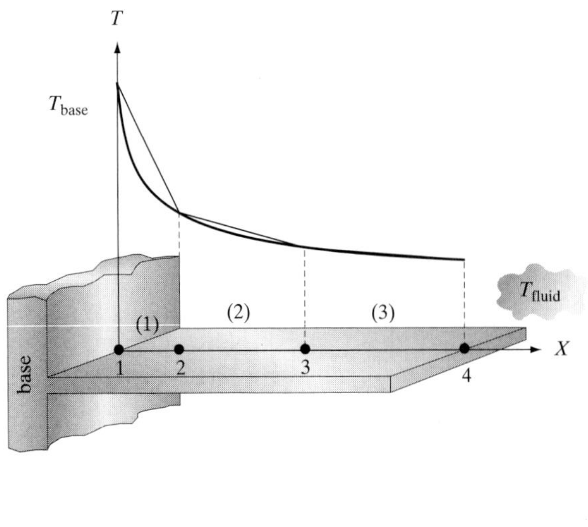

FIGURE 6.1 The actual and approximate temperature distribution for a fin of uniform cross section.

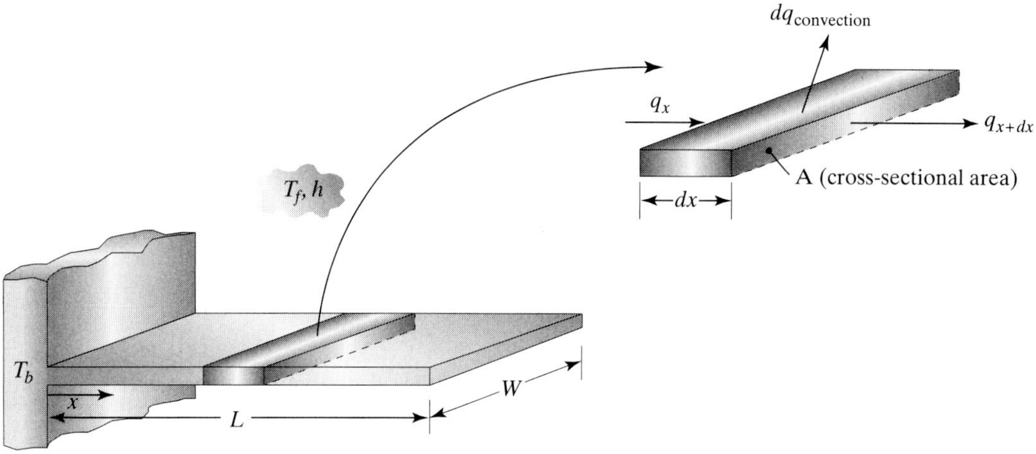

We start by applying the energy balance to a differential element

$$q_x = q_{x+dx} + dq_{\text{convection}}$$

$$q_x = q_x + \frac{dq_x}{dx}dx + dq_{\text{convection}}$$

Next we use Fourier's law

$$q_x = -kA\frac{dT}{dx}$$

and use Newton's law of cooling,

$$dq_{\text{convection}} = h(dA_s)(T - T_f)$$

$$0 = \frac{dq_x}{dx}dx + dq_{\text{convection}} = \frac{d}{dx}\left(-kA\frac{dT}{dx}\right)dx + h(dA_s)(T - T_f)$$

Writing dA_s (differential surface area) in terms of the perimeter of the fin and dx and simplifying, we are left with

$$-kA\frac{d^2T}{dx^2} + hp(T - T_f) = 0$$

FIGURE 6.2 Derivation of the heat equation for a fin.

Equation (6.1) is derived by applying the conservation of energy to a differential section of a fin, as shown in Figure 6.2. The heat transfer in the fin is accomplished by conduction in the longitudinal direction (x-direction) and convection to the surrounding fluid. In Eq. (6.1), k is the thermal conductivity, and A denotes the cross-sectional area of the fin. The convective heat transfer coefficient is represented by h, the perimeter of the fin is denoted by p, and T_f is the temperature of the surrounding fluid. Equation (6.1) is subjected to a set of boundary conditions. First, the temperature of the base is generally known; that is,

$$T(0) = T_b \tag{6.2}$$

The other boundary condition deals with the heat loss at the tip of the fin. In general, there are three possibilities. One possibility is that the tip is long enough so that the temperature of the tip is equal to the temperature of the surrounding fluid temperature. This situation is represented by the condition

$$T(L) = T_f \tag{6.3}$$

The situation in which the heat loss from the tip of the fin may be neglected is represented by the condition

$$-kA\frac{dT}{dX}\bigg|_{X=L} = 0 \tag{6.4}$$

If the heat loss from the tip of the fin should be included in the analysis, then we have the condition

$$-kA\frac{dT}{dX}\bigg|_{X=L} = hA(T_L - T_f) \tag{6.5}$$

Equation (6.5) is obtained by applying the energy balance to the cross-sectional area of the tip. Equation (6.5) simply states that the heat conducted to the tip's surface is convected away by the surrounding fluid. Therefore, we can use one of the boundary conditions given by Eqs. (6.3)–(6.5) and the base temperature to model an actual problem. Before we proceed with the formulation of the conductance matrix and the thermal load matrix for a typical element, let us emphasize the following points: (1) The governing differential equation of the fin represents the balance of energy at any point along the fin and thus governs the balance of energy at all nodes of a finite element model as well, and (2) the exact solution of the governing differential equation (if possible) subject to two appropriate boundary conditions renders the detailed temperature distribution along the fin, and the finite element solution represents an approximation of this solution. We now focus on a typical element and proceed with the formulation of the conductance matrix, recalling that the temperature distribution for a typical element may be approximated using linear shape functions, as discussed in Chapter 5. That is,

$$T^{(e)} = [S_i \quad S_j]\begin{Bmatrix} T_i \\ T_j \end{Bmatrix} \tag{6.6}$$

where the shape functions are given by

$$S_i = \frac{X_j - X}{\ell} \quad \text{and} \quad S_j = \frac{X - X_i}{\ell} \tag{6.7}$$

In order to make this derivation as general as possible and applicable to other type of problems with the same form of differential equations, let $c_1 = kA, c_2 = -hp$, $c_3 = hpT_f$, and $\Psi = T$. Thus, Eq. (6.1) can be written as

$$c_1\frac{d^2\Psi}{dX^2} + c_2\Psi + c_3 = 0 \tag{6.8}$$

Recall from our introductory discussion of weighted residual methods in Chapter 1 that when we substitute an approximate solution into the governing differential equation, the approximate solution does not satisfy the differential equation exactly, and thus, an error, or a residual, is produced. Also recall that the Galerkin formulation requires the error to be orthogonal to some weighting functions. Furthermore, the weighting functions are chosen to be members of the approximate solution. Here we will use the shape functions as the weighting functions because they are members of the approximate solution.

We can obtain the residual equations in one of two ways: using a nodal approach or an elemental approach. Consider three consecutive nodes, *i*, *j*, and *k*, belonging to two adjacent elements (e), and $(e + 1)$, as shown in Figure 6.3. Elements (e) and $(e + 1)$ both contribute to error at node *j*. Realizing this fact, we can write the residual equation for node *j* as

$$R_j = R_j^{(e)} + R_j^{(e+1)} = \int_{X_i}^{X_j} S_j^{(e)} \left[c_1 \frac{d^2\Psi}{dX^2} + c_2\Psi + c_3 \right]^{(e)} dX$$
$$+ \int_{X_j}^{X_k} S_j^{(e+1)} \left[c_1 \frac{d^2\Psi}{dX^2} + c_2\Psi + c_3 \right]^{(e+1)} dX = 0 \tag{6.9}$$

Pay close attention to the subscripts denoting node numbers and superscripts referring to element numbers while following the forthcoming derivation. Writing the residual or the error equations for each node of a finite element model leads to a set of equations of the form

$$\begin{Bmatrix} R_1 \\ R_2 \\ R_3 \\ \cdot \\ R_n \end{Bmatrix} = \begin{Bmatrix} 0 \\ 0 \\ 0 \\ 0 \\ 0 \end{Bmatrix} \tag{6.10}$$

where $R_2 = R_2^{(1)} + R_2^{(2)}$, $R_3 = R_3^{(2)} + R_3^{(3)}$, and so on.

Let us now look at the contribution of each element to the nodal residual equations more closely. Expanding Eq. (6.9) for a number of nodes (1, 2, 3, 4, and so on), we get

$$R_1 = \int_{X_1}^{X_2} S_1^{(1)} \left[c_1 \frac{d^2\Psi}{dX^2} + c_2\Psi + c_3 \right]^{(1)} dX = 0 \tag{6.11}$$

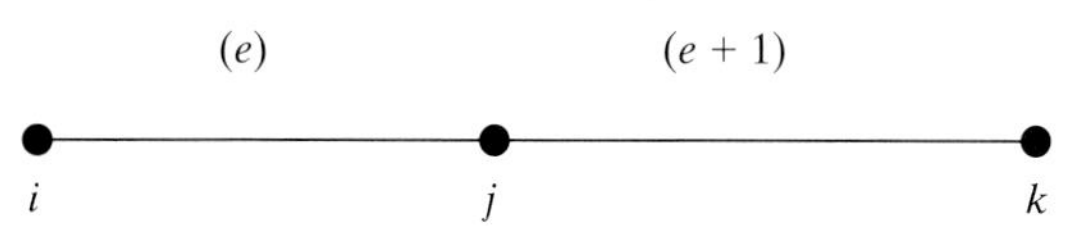

FIGURE 6.3 Elements (e) and $(e + 1)$ and their respective nodes.

$$R_2 = R_2^{(1)} + R_2^{(2)} = \int_{X_1}^{X_2} S_2^{(1)}\left[c_1\frac{d^2\Psi}{dX^2} + c_2\Psi + c_3\right]^{(1)} dX$$

$$+ \overbrace{\int_{X_2}^{X_3} S_2^{(2)}\left[c_1\frac{d^2\Psi}{dX^2} + c_2\Psi + c_3\right]^{(2)} dX}^{\text{element 2 contribution}} = 0 \tag{6.12}$$

$$R_3 = R_3^{(2)} + R_3^{(3)} = \overbrace{\int_{X_2}^{X_3} S_3^{(2)}\left[c_1\frac{d^2\Psi}{dX^2} + c_2\Psi + c_3\right]^{(2)} dX}^{\text{element 2 contribution}}$$

$$+ \overbrace{\int_{X_3}^{X_4} S_3^{(3)}\left[c_1\frac{d^2\Psi}{dX^2} + c_2\Psi + c_3\right]^{(3)} dX}^{\text{element 3 contribution}} = 0 \tag{6.13}$$

$$R_4 = R_4^{(3)} + R_4^{(4)} = \overbrace{\int_{X_3}^{X_4} S_4^{(3)}\left[c_1\frac{d^2\Psi}{dX^2} + c_2\Psi + c_3\right]^{(3)} dX}^{\text{element 3 contribution}}$$

$$+ \overbrace{\int_{X_4}^{X_5} S_4^{(4)}\left[c_1\frac{d^2\Psi}{dX^2} + c_2\Psi + c_3\right]^{(4)} dX}^{\text{element 4 contribution}} = 0 \tag{6.14}$$

and so on. Note that element (2) contributes to Eqs. (6.12) and (6.13), and element (3) contributes to Eqs. (6.13) and (6.14), and so on. In general, an arbitrary element (e) having nodes i and j contributes to the residual equations in the following manner:

$$R_i^{(e)} = \int_{X_i}^{X_j} S_i^{(e)}\left[c_1\frac{d^2\Psi}{dX^2} + c_2\Psi + c_3\right]^{(e)} dX \tag{6.15}$$

$$R_j^{(e)} = \int_{X_i}^{X_j} S_j^{(e)}\left[c_1\frac{d^2\Psi}{dX^2} + c_2\Psi + c_3\right]^{(e)} dX \tag{6.16}$$

This approach leads to the elemental formulation. It is important to note that we have set up the residual equations for an arbitrary element (e). Moreover, the elemental matrices obtained in this manner are then assembled to present the entire problem, and the residuals equations are set equal to zero.

Evaluation of the integrals given by Eqs. (6.15) and (6.16) will result in the elemental formulation. But first, we will manipulate the second-order terms into first-order terms. This manipulation is accomplished by using the chain rule in the following manner:

$$\frac{d}{dX}\left(S_i\frac{d\Psi}{dX}\right) = S_i\frac{d^2\Psi}{dX^2} + \frac{dS_i}{dX}\frac{d\Psi}{dX} \tag{6.17}$$

$$S_i\frac{d^2\Psi}{dX^2} = \frac{d}{dX}\left(S_i\frac{d\Psi}{dX}\right) - \frac{dS_i}{dX}\frac{d\Psi}{dX} \tag{6.18}$$

Substituting Eq. (6.18) into Eq. (6.15), we obtain

$$R_i^{(e)} = \int_{X_i}^{X_j}\left(c_1\left(\frac{d}{dX}\left(S_i\frac{d\Psi}{dX}\right) - \frac{dS_i}{dX}\frac{d\Psi}{dX}\right) + S_i(c_2\Psi + c_3)\right)dX \tag{6.19}$$

We eventually need to follow the same procedure for Eq. (6.16) as well, but for now let us focus only on one of the residual equations. There are four terms in Eq. (6.19) that need to be evaluated:

$$R_i^{(e)} = \int_{X_i}^{X_j} c_1\left(\frac{d}{dX}\left(S_i\frac{d\Psi}{dX}\right)\right)dX + \int_{X_i}^{X_j} c_1\left(-\frac{dS_i}{dX}\frac{d\Psi}{dX}\right)dX$$
$$+ \int_{X_i}^{X_j} S_i(c_2\Psi)\,dX + \int_{X_i}^{X_j} S_i\,c_3\,dX \tag{6.20}$$

Considering and evaluating the first term, we have

$$\int_{X_i}^{X_j} c_1\left(\frac{d}{dX}\left(S_i\frac{d\Psi}{dX}\right)\right)dX = c_1S_i\frac{d\Psi}{dX}\bigg|_{X=X_j} - c_1S_i\frac{d\Psi}{dX}\bigg|_{X=X_i} = -c_1\frac{d\Psi}{dX}\bigg|_{X=X_i} \tag{6.21}$$

It is important to realize that in order for us to obtain the result given by Eq. (6.21), S_i is zero at $X = X_j$ and $S_i = 1$ at $X = X_i$. The second integral in Eq. (6.20) is evaluated as

$$\int_{X_i}^{X_j} c_1\left(-\frac{dS_i}{dX}\frac{d\Psi}{dX}\right)dX = -\frac{c_1}{\ell}(\Psi_i - \Psi_j) \tag{6.22}$$

To obtain the results given by Eq. (6.22), first substitute for $S_i = \dfrac{X_j - X}{\ell}$ and $\Psi = S_i\Psi_i + S_j\Psi_j = \dfrac{X_j - X}{\ell}\Psi_i + \dfrac{X - X_i}{\ell}\Psi_j$ and then proceed with evaluation of integral. Evaluation of the third and the fourth integrals in Eq. (6.20) yields

$$\int_{X_i}^{X_j} S_i(c_2\Psi)\,dX = \frac{c_2\ell}{3}\Psi_i + \frac{c_2\ell}{6}\Psi_j \tag{6.23}$$

$$\int_{X_i}^{X_j} S_ic_3\,dX = c_3\frac{\ell}{2} \tag{6.24}$$

In exactly the same manner, we can evaluate the second residual equation for node j, as given by Eq. (6.16). This evaluation results in the following equations:

$$\int_{X_i}^{X_j} c_1\left(\frac{d}{dX}\left(S_j\frac{d\Psi}{dX}\right)\right)dX = c_1 S_j\frac{d\Psi}{dX}\bigg|_{X=X_j} - c_1 S_j\frac{d\Psi}{dX}\bigg|_{X=X_i} = c_1\frac{d\Psi}{dX}\bigg|_{X=X_j} \tag{6.25}$$

$$\int_{X_i}^{X_j} c_1\left(-\frac{dS_j}{dX}\frac{d\Psi}{dX}\right)dX = -\frac{c_1}{\ell}(-\Psi_i + \Psi_j) \tag{6.26}$$

$$\int_{X_i}^{X_j} S_j(c_2\Psi)\,dX = \frac{c_2\ell}{6}\Psi_i + \frac{c_2\ell}{3}\Psi_j \tag{6.27}$$

$$\int_{X_i}^{X_j} S_j\,c_3\,dX = c_3\frac{\ell}{2} \tag{6.28}$$

It should be clear by now that evaluation of Eqs. (6.15) and (6.16) results in two sets of linear equations, as given by:

$$\begin{Bmatrix} R_i \\ R_j \end{Bmatrix} = \begin{Bmatrix} -c_1\dfrac{d\Psi}{dX}\bigg|_{X=X_i} \\ c_1\dfrac{d\Psi}{dX}\bigg|_{X=X_j} \end{Bmatrix} - \frac{c_1}{\ell}\begin{bmatrix} 1 & -1 \\ -1 & 1 \end{bmatrix}\begin{Bmatrix} \Psi_i \\ \Psi_j \end{Bmatrix}$$

$$+\frac{c_2\ell}{6}\begin{bmatrix} 2 & 1 \\ 1 & 2 \end{bmatrix}\begin{Bmatrix} \Psi_i \\ \Psi_j \end{Bmatrix} + \frac{c_3\ell}{2}\begin{Bmatrix} 1 \\ 1 \end{Bmatrix} \tag{6.29}$$

Note that from here on, for the sake of simplicity of presentation of conductance and load matrices, we will set the residual of an element equal to zero. However, as we mentioned earlier, it is important to realize that the residuals are set equal to zero after all the elements have been assembled. Therefore, we rewrite the Eq. (6.29) as

$$\begin{Bmatrix} c_1\dfrac{d\Psi}{dX}\bigg|_{X=X_i} \\ -c_1\dfrac{d\Psi}{dX}\bigg|_{X=X_j} \end{Bmatrix} + \frac{c_1}{\ell}\begin{bmatrix} 1 & -1 \\ -1 & 1 \end{bmatrix}\begin{Bmatrix} \Psi_i \\ \Psi_j \end{Bmatrix} + \frac{-c_2\ell}{6}\begin{bmatrix} 2 & 1 \\ 1 & 2 \end{bmatrix}\begin{Bmatrix} \Psi_i \\ \Psi_j \end{Bmatrix} = \frac{c_3\ell}{2}\begin{Bmatrix} 1 \\ 1 \end{Bmatrix} \tag{6.30}$$

Combining the unknown nodal parameters, we obtain

$$\begin{Bmatrix} c_1\dfrac{d\Psi}{dX}\bigg|_{X=X_i} \\ -c_1\dfrac{d\Psi}{dX}\bigg|_{X=X_j} \end{Bmatrix} + \{[\mathbf{K}]_{c_1}^{(e)} + [\mathbf{K}]_{c_2}^{(e)}\}\begin{Bmatrix} \Psi_i \\ \Psi_j \end{Bmatrix} = \{\mathbf{F}\}^{(e)} \tag{6.31}$$

where

$$[\mathbf{K}]_{c_1}^{(e)} = \frac{c_1}{\ell}\begin{bmatrix} 1 & -1 \\ -1 & 1 \end{bmatrix}$$

is the elemental conductance for a heat transfer problem (or, it could represent the stiffness for solid mechanics problems), due to the c_1 coefficient,

$$[\mathbf{K}]_{c_2}^{(e)} = \frac{-c_2 \ell}{6}\begin{bmatrix} 2 & 1 \\ 1 & 2 \end{bmatrix}$$

is the elemental conductance (or, for a solid mechanics problem, the stiffness) due to the c_2 coefficient, and

$$\{\mathbf{F}\}^{(e)} = \frac{c_3 \ell}{2}\begin{Bmatrix} 1 \\ 1 \end{Bmatrix}$$

is the load matrix for a given element. The terms

$$\begin{Bmatrix} c_1 \left.\dfrac{d\Psi}{dX}\right|_{X=X_i} \\ -c_1 \left.\dfrac{d\Psi}{dX}\right|_{X=X_j} \end{Bmatrix}$$

contribute to both the conductance (or, for a solid-mechanics problem, the stiffness) matrix and the load matrix. They need to be evaluated for specific boundary conditions. We shall undertake this task shortly. However, let us first write down the conductance matrix for a typical one-dimensional fin in terms of its parameters. The conductance matrices are given by

$$[\mathbf{K}]_{c_1}^{(e)} = \frac{c_1}{\ell}\begin{bmatrix} 1 & -1 \\ -1 & 1 \end{bmatrix} = \frac{kA}{\ell}\begin{bmatrix} 1 & -1 \\ -1 & 1 \end{bmatrix} \tag{6.32}$$

and

$$[\mathbf{K}]_{c_2}^{(e)} = \frac{-c_2 \ell}{6}\begin{bmatrix} 2 & 1 \\ 1 & 2 \end{bmatrix} = \frac{hp\ell}{6}\begin{bmatrix} 2 & 1 \\ 1 & 2 \end{bmatrix} \tag{6.33}$$

In general, the elemental conductance matrix may consist of three terms: The $[\mathbf{K}]_{c_1}^{(e)}$ term is due to conduction heat transfer along the fin (through the cross-sectional area); the $[\mathbf{K}]_{c_2}^{(e)}$ term represents the heat loss through the top, bottom, and side surfaces (periphery) of an element of a fin; and, depending on the boundary condition of the tip, an additional elemental conductance matrix $[\mathbf{K}]_{\text{B.C.}}^{(e)}$ can exist. For the very last element containing the tip surface, and referring to the boundary condition given by Eq. (6.5), the heat loss through the tip surface can be evaluated as

$$\begin{Bmatrix} c_1 \left.\dfrac{d\Psi}{dX}\right|_{X=X_i} \\ -c_1 \left.\dfrac{d\Psi}{dX}\right|_{X=X_j} \end{Bmatrix} = \begin{Bmatrix} kA \left.\dfrac{dT}{dX}\right|_{X=X_i} \\ -kA \left.\dfrac{dT}{dX}\right|_{X=X_j} \end{Bmatrix} = \begin{Bmatrix} 0 \\ hA(T_j - T_f) \end{Bmatrix} \tag{6.34}$$

By rearranging and simplifying, we have

$$\begin{Bmatrix} kA\dfrac{dT}{dX}\Big|_{X=X_i} \\ -kA\dfrac{dT}{dX}\Big|_{X=X_j} \end{Bmatrix} = \begin{Bmatrix} 0 \\ hA(T_j - T_f) \end{Bmatrix} = \begin{bmatrix} 0 & 0 \\ 0 & hA \end{bmatrix}\begin{Bmatrix} T_i \\ T_j \end{Bmatrix} - \begin{Bmatrix} 0 \\ hAT_f \end{Bmatrix} \tag{6.35}$$

$$[\mathbf{K}]^{(e)}_{\text{B.C.}} = \begin{bmatrix} 0 & 0 \\ 0 & hA \end{bmatrix} \tag{6.36}$$

The term $-\begin{Bmatrix} 0 \\ hAT_f \end{Bmatrix}$ belongs to the right side of Eq. (6.31) with the thermal load matrix. It shows the contribution of the boundary condition of the tip to the load matrix:

$$\{\mathbf{F}\}^{(e)}_{\text{B.C.}} = \begin{Bmatrix} 0 \\ hAT_f \end{Bmatrix} \tag{6.37}$$

To summarize, the conductance matrix for all elements, excluding the last element, is given by

$$[\mathbf{K}]^{(e)} = \left\{\frac{kA}{\ell}\begin{bmatrix} 1 & -1 \\ -1 & 1 \end{bmatrix} + \frac{hp\ell}{6}\begin{bmatrix} 2 & 1 \\ 1 & 2 \end{bmatrix}\right\} \tag{6.38}$$

If the heat loss through the tip of the fin must be accounted for, the conductance matrix for the very last element must be computed from the equation

$$[\mathbf{K}]^{(e)} = \left\{\frac{kA}{\ell}\begin{bmatrix} 1 & -1 \\ -1 & 1 \end{bmatrix} + \frac{hp\ell}{6}\begin{bmatrix} 2 & 1 \\ 1 & 2 \end{bmatrix} + \begin{bmatrix} 0 & 0 \\ 0 & hA \end{bmatrix}\right\} \tag{6.39}$$

The thermal load matrix for all elements, excluding the last element, is given by

$$\{\mathbf{F}\}^{(e)} = \frac{hp\ell T_f}{2}\begin{Bmatrix} 1 \\ 1 \end{Bmatrix} \tag{6.40}$$

If the heat loss through the tip of the fin must be included in the analysis, the thermal load matrix for the very last element must be computed from the relation

$$\{\mathbf{F}\}^{(e)} = \frac{hp\ell T_f}{2}\begin{Bmatrix} 1 \\ 1 \end{Bmatrix} + \begin{Bmatrix} 0 \\ hAT_f \end{Bmatrix} \tag{6.41}$$

The next set of examples demonstrates the assembly of elements to present the entire problem and the treatment of other boundary conditions.

EXAMPLE 6.1 A Fin Problem

Aluminum fins of a rectangular profile, shown in Figure 6.4, are used to remove heat from a surface whose temperature is 100°C. The temperature of the ambient air is 20°C. The

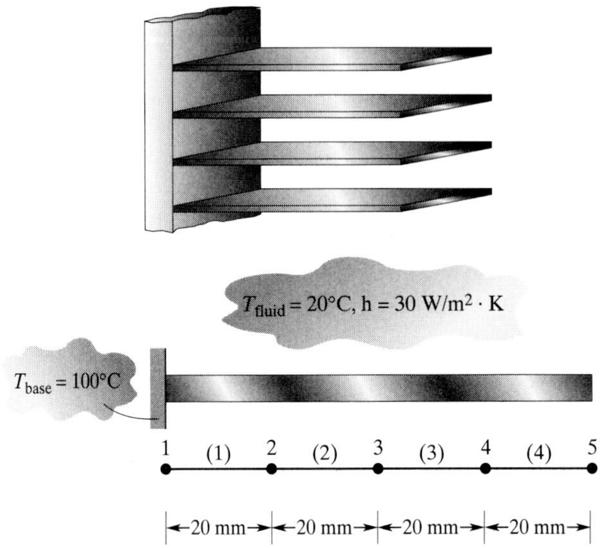

FIGURE 6.4 Finite element model of a straight fin.

thermal conductivity of aluminum is 168 W/m · K (W/m · °C). The natural convective heat transfer coefficient associated with the surrounding air is 30 $W/m^2 \cdot K$ ($W/m^2 \cdot °C$). The fins are 80 mm long, 5 mm wide, and 1 mm thick. (a) Determine the temperature distribution along the fin using the finite element model shown in Figure 6.4. (b) Compute the heat loss per fin.

We will solve this problem using two boundary conditions for the tip. First, let us include the heat transfer from the tip's surface in the analysis. For elements (1), (2), and (3) in the situation, the conductance and thermal load matrices are given by

$$[\mathbf{K}]^{(e)} = \left\{ \frac{kA}{\ell} \begin{bmatrix} 1 & -1 \\ -1 & 1 \end{bmatrix} + \frac{hp\ell}{6} \begin{bmatrix} 2 & 1 \\ 1 & 2 \end{bmatrix} \right\}$$

$$\{\mathbf{F}\}^{(e)} = \frac{hp\ell T_f}{2} \begin{Bmatrix} 1 \\ 1 \end{Bmatrix}$$

Substituting for the properties, we obtain

$$[\mathbf{K}]^{(e)} = \left\{ \frac{(168)(5 \times 1 \times 10^{-6})}{20 \times 10^{-3}} \begin{bmatrix} 1 & -1 \\ -1 & 1 \end{bmatrix} + \frac{30 \times 12 \times 20 \times 10^{-6}}{6} \begin{bmatrix} 2 & 1 \\ 1 & 2 \end{bmatrix} \right\}$$

$$\{\mathbf{F}\}^{(e)} = \frac{30 \times 12 \times 20 \times 10^{-6} \times 20}{2} \begin{Bmatrix} 1 \\ 1 \end{Bmatrix} = \begin{Bmatrix} 0.072 \\ 0.072 \end{Bmatrix}$$

The conductance matrix for elements (1), (2), and (3) is

$$[\mathbf{K}]^{(1)} = [\mathbf{K}]^{(2)} = [\mathbf{K}]^{(3)} = \begin{bmatrix} 0.0444 & -0.0408 \\ -0.0408 & 0.0444 \end{bmatrix} \frac{\text{W}}{°\text{C}}$$

and the thermal-load matrix for elements (1), (2), and (3) is

$$\{\mathbf{F}\}^{(1)} = \{\mathbf{F}\}^{(2)} = \{\mathbf{F}\}^{(3)} = \begin{Bmatrix} 0.072 \\ 0.072 \end{Bmatrix} \text{W}$$

Including the boundary condition of the tip, the conductance and load matrices for element (4) are obtained in the following manner:

$$[\mathbf{K}]^{(e)} = \left\{ \frac{kA}{\ell}\begin{bmatrix} 1 & -1 \\ -1 & 1 \end{bmatrix} + \frac{hp\ell}{6}\begin{bmatrix} 2 & 1 \\ 1 & 2 \end{bmatrix} + \begin{bmatrix} 0 & 0 \\ 0 & hA \end{bmatrix} \right\}$$

$$[\mathbf{K}]^{(4)} = \begin{bmatrix} 0.0444 & -0.0408 \\ -0.0408 & 0.0444 \end{bmatrix} + \begin{bmatrix} 0 & 0 \\ 0 & (30 \times 5 \times 1 \times 10^{-6}) \end{bmatrix} = \begin{bmatrix} 0.0444 & -0.0408 \\ -0.0408 & 0.04455 \end{bmatrix} \frac{\text{W}}{^\circ\text{C}}$$

$$\{\mathbf{F}\}^{(e)} = \frac{hp\ell T_f}{2}\begin{Bmatrix} 1 \\ 1 \end{Bmatrix} + \begin{Bmatrix} 0 \\ hAT_f \end{Bmatrix}$$

$$\{\mathbf{F}\}^{(4)} = \begin{Bmatrix} 0.072 \\ 0.072 \end{Bmatrix} + \begin{Bmatrix} 0 \\ (30 \times 5 \times 1 \times 10^{-6} \times 20) \end{Bmatrix} = \begin{Bmatrix} 0.072 \\ 0.075 \end{Bmatrix} \text{W}$$

Assembly of the elements leads to the global conductance matrix $[\mathbf{K}]^{(G)}$ and the global load matrix $\{\mathbf{F}\}^{(G)}$:

$$[\mathbf{K}]^{(G)} = \begin{bmatrix} 0.0444 & -0.0408 & 0 & 0 & 0 \\ -0.0408 & 0.0444 + 0.0444 & -0.0408 & 0 & 0 \\ 0 & -0.0408 & 0.0444 + 0.0444 & -0.0408 & 0 \\ 0 & 0 & -0.0408 & 0.0444 + 0.0444 & -0.0408 \\ 0 & 0 & 0 & -0.0408 & 0.04455 \end{bmatrix}$$

$$\{\mathbf{F}\}^{(G)} = \begin{Bmatrix} 0.072 \\ 0.072 + 0.072 \\ 0.072 + 0.072 \\ 0.072 + 0.072 \\ 0.075 \end{Bmatrix}$$

Applying the base boundary condition $T_1 = 100^\circ\text{C}$, we find that the final set of linear equations becomes

$$\begin{bmatrix} 1 & 0 & 0 & 0 & 0 \\ -0.0408 & 0.0888 & -0.0408 & 0 & 0 \\ 0 & -0.0408 & 0.0888 & -0.0408 & 0 \\ 0 & 0 & -0.0408 & 0.0888 & -0.0408 \\ 0 & 0 & 0 & -0.0408 & 0.04455 \end{bmatrix} \begin{Bmatrix} T_1 \\ T_2 \\ T_3 \\ T_4 \\ T_5 \end{Bmatrix} = \begin{Bmatrix} 100 \\ 0.144 \\ 0.144 \\ 0.144 \\ 0.075 \end{Bmatrix}$$

We can obtain the nodal temperatures from the solution of the above equation. The nodal solutions are

$$\begin{Bmatrix} T_1 \\ T_2 \\ T_3 \\ T_4 \\ T_5 \end{Bmatrix} = \begin{Bmatrix} 100 \\ 75.03 \\ 59.79 \\ 51.56 \\ 48.90 \end{Bmatrix} {}^\circ\mathrm{C}$$

Note that the nodal temperatures are given in °C and not in °K.

Because the cross-sectional area of the given fin is relatively small, we could have neglected the heat loss from the tip. Under this assumption, the elemental conductance and forcing matrices for all elements are given by:

$$[\mathbf{K}]^{(1)} = [\mathbf{K}]^{(2)} = [\mathbf{K}]^{(3)} = [\mathbf{K}]^{(4)} = \begin{bmatrix} 0.0444 & -0.0408 \\ -0.0408 & 0.0444 \end{bmatrix} \frac{\mathrm{W}}{{}^\circ\mathrm{C}}$$

$$\{\mathbf{F}\}^{(1)} = \{\mathbf{F}\}^{(2)} = \{\mathbf{F}\}^{(3)} = \{\mathbf{F}\}^{(4)} = \begin{Bmatrix} 0.072 \\ 0.072 \end{Bmatrix} \mathrm{W}$$

Assembly of the elements leads to the global conductance matrix $[\mathbf{K}]^G$ and the global load matrix $\{\mathbf{F}\}^G$:

$$[\mathbf{K}]^{(G)} = \begin{bmatrix} 0.0444 & -0.0408 & 0 & 0 & 0 \\ -0.0408 & 0.0444 + 0.0444 & -0.0408 & 0 & 0 \\ 0 & -0.0408 & 0.0444 + 0.0444 & -0.0408 & 0 \\ 0 & 0 & -0.0408 & 0.0444 + 0.0444 & -0.0408 \\ 0 & 0 & 0 & -0.0408 & 0.0444 \end{bmatrix}$$

$$\{\mathbf{F}\}^{(G)} = \begin{Bmatrix} 0.072 \\ 0.072 + 0.072 \\ 0.072 + 0.072 \\ 0.072 + 0.072 \\ 0.072 \end{Bmatrix}$$

Applying the base boundary condition $T_1 = 100^\circ\mathrm{C}$, we find that the final set of linear equations becomes

$$\begin{bmatrix} 1 & 0 & 0 & 0 & 0 \\ -0.0408 & 0.0888 & -0.0408 & 0 & 0 \\ 0 & -0.0408 & 0.0888 & -0.0408 & 0 \\ 0 & 0 & -0.0408 & 0.0888 & -0.0408 \\ 0 & 0 & 0 & -0.0408 & 0.0444 \end{bmatrix} \begin{Bmatrix} T_1 \\ T_2 \\ T_3 \\ T_4 \\ T_5 \end{Bmatrix} = \begin{Bmatrix} 100 \\ 0.144 \\ 0.144 \\ 0.144 \\ 0.072 \end{Bmatrix}$$

which has approximately the same solution as that calculated previously:

$$\begin{Bmatrix} T_1 \\ T_2 \\ T_3 \\ T_4 \\ T_5 \end{Bmatrix} = \begin{Bmatrix} 100 \\ 75.08 \\ 59.89 \\ 51.74 \\ 49.19 \end{Bmatrix} °\mathrm{C}$$

Compared to the previous results, the nodal temperatures are slightly higher because we neglected the heat loss through the end surface of the tip.

The total heat loss Q from the fin can be determined by summing the heat loss through individual elements:

$$Q_{\text{total}} = \Sigma Q^{(e)} \tag{6.42}$$

$$Q^{(e)} = \int_{X_i}^{X_j} hp(T - T_f)\,dX$$

$$= \int_{X_i}^{X_j} hp((S_iT_i + S_jT_j) - T_f)\,dX = hp\ell\left(\left(\frac{T_i + T_j}{2}\right) - T_f\right) \tag{6.43}$$

Applying the temperature results to Eqs. (6.42) and (6.43), we have

$$Q_{\text{total}} = Q^{(1)} + Q^{(2)} + Q^{(3)} + Q^{(4)}$$

$$Q^{(1)} = hp\ell\left(\left(\frac{T_i + T_j}{2}\right) - T_f\right) = 30 \times 12 \times 20 \times 10^{-6}\left(\left(\frac{100 + 75.08}{2}\right) - 20\right) = 0.4862 \text{ W}$$

$$Q^{(2)} = 30 \times 12 \times 20 \times 10^{-6}\left(\left(\frac{75.08 + 59.89}{2}\right) - 20\right) = 0.3418 \text{ W}$$

$$Q^{(3)} = 30 \times 12 \times 20 \times 10^{-6}\left(\left(\frac{59.89 + 51.74}{2}\right) - 20\right) = 0.2578 \text{ W}$$

$$Q^{(4)} = 30 \times 12 \times 20 \times 10^{-6}\left(\left(\frac{51.74 + 49.19}{2}\right) - 20\right) = 0.2193 \text{ W}$$

$$Q_{\text{total}} = 1.3051 \text{ W}$$

EXAMPLE 6.2 A Composite Wall Problem

A wall of an industrial oven consists of three different materials, as depicted in Figure 6.5. The first layer is composed of 5 cm of insulating cement with a clay binder that has a thermal conductivity of 0.08 W/m · K. The second layer is made from 15 cm of 6-ply asbestos board with a thermal conductivity of 0.074 W/m · K (W/m · °C). The exterior consists of 10-cm common brick with a thermal conductivity of 0.72 W/m^2 · K (W/m · °C). The inside wall temperature of the oven is 200°C, and the outside air is 30°C with a convection coefficient of 40 W/m^2 · K (W/m^2 · °C). Determine the temperature distribution along the composite wall.

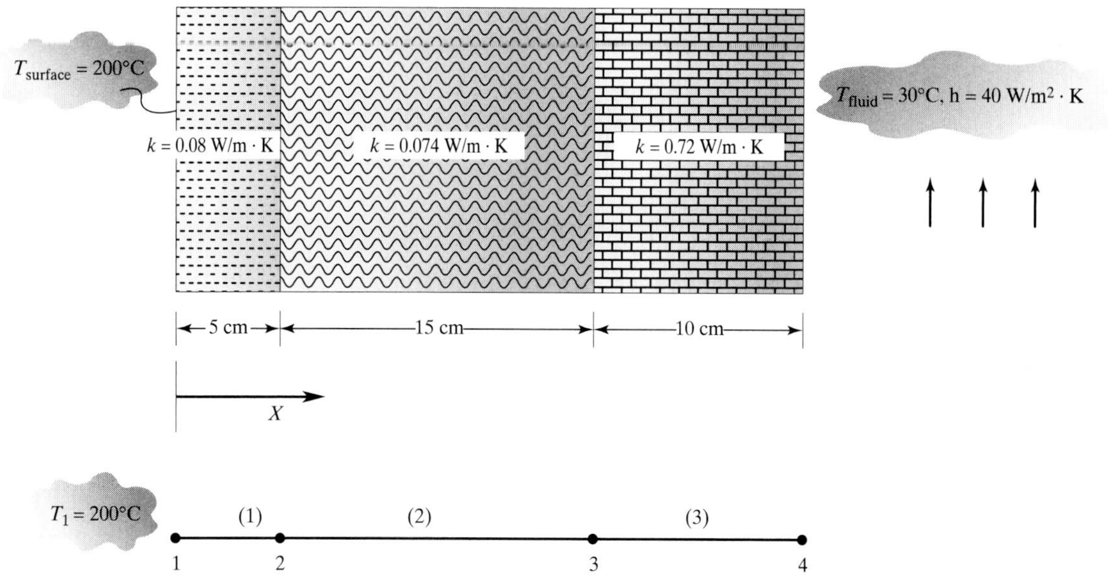

FIGURE 6.5 A composite wall of an industrial oven.

This heat conduction problem is governed by the equation

$$kA\frac{d^2T}{dX^2} = 0 \tag{6.44}$$

and is subjected to the boundary conditions $T_1 = 200°\text{C}$ and $-kA\frac{dT}{dX}\big|_{X=30\,\text{cm}} = hA(T_4 - T_f)$. For this example, we compare Eq. (6.44) to Eq. (6.8), finding that $c_1 = kA$, $c_2 = 0$, $c_3 = 0$, and $\Psi = \text{T}$. Thus, for element (1), we have

$$[\mathbf{K}]^{(1)} = \frac{kA}{\ell}\begin{bmatrix} 1 & -1 \\ -1 & 1 \end{bmatrix} = \frac{0.08 \times 1}{0.05}\begin{bmatrix} 1 & -1 \\ -1 & 1 \end{bmatrix} = \begin{bmatrix} 1.6 & -1.6 \\ -1.6 & 1.6 \end{bmatrix}\frac{\text{W}}{°\text{C}}$$

$$\{\mathbf{F}\}^{(1)} = \begin{Bmatrix} 0 \\ 0 \end{Bmatrix}\text{W}$$

For element (2), we have

$$[\mathbf{K}]^{(2)} = \frac{kA}{\ell}\begin{bmatrix} 1 & -1 \\ -1 & 1 \end{bmatrix} = \frac{0.074 \times 1}{0.15}\begin{bmatrix} 1 & -1 \\ -1 & 1 \end{bmatrix} = \begin{bmatrix} 0.493 & -0.493 \\ -0.493 & 0.493 \end{bmatrix}\frac{\text{W}}{°\text{C}}$$

$$\{\mathbf{F}\}^{(2)} = \begin{Bmatrix} 0 \\ 0 \end{Bmatrix}\text{W}$$

For element (3), including the boundary condition at node 4, we have

$$[\mathbf{K}]^{(3)} = \frac{kA}{\ell}\begin{bmatrix} 1 & -1 \\ -1 & 1 \end{bmatrix} + \begin{bmatrix} 0 & 0 \\ 0 & hA \end{bmatrix} = \frac{0.72 \times 1}{0.1}\begin{bmatrix} 1 & -1 \\ -1 & 1 \end{bmatrix} + \begin{bmatrix} 0 & 0 \\ 0 & (40 \times 1) \end{bmatrix}$$

$$= \begin{bmatrix} 7.2 & -7.2 \\ -7.2 & 47.2 \end{bmatrix} \frac{\text{W}}{°\text{C}}$$

$$\{\mathbf{F}\}^{(3)} = \begin{Bmatrix} 0 \\ hAT_f \end{Bmatrix} = \begin{Bmatrix} 0 \\ (40 \times 1 \times 30) \end{Bmatrix} = \begin{Bmatrix} 0 \\ 1200 \end{Bmatrix} \text{W}$$

Assembling elements, we obtain

$$[\mathbf{K}]^{(G)} = \begin{bmatrix} 1.6 & -1.6 & 0 & 0 \\ -1.6 & 1.6 + 0.493 & -0.493 & 0 \\ 0 & -0.493 & 0.493 + 7.2 & -7.2 \\ 0 & 0 & -7.2 & 47.2 \end{bmatrix}$$

$$\{\mathbf{F}\}^{(G)} = \begin{Bmatrix} 0 \\ 0 \\ 0 \\ 1200 \end{Bmatrix}$$

Applying the boundary condition at the inside furnace wall, we get

$$\begin{bmatrix} 1 & 0 & 0 & 0 \\ -1.6 & 2.093 & -0.493 & 0 \\ 0 & -0.493 & 7.693 & -7.2 \\ 0 & 0 & -7.2 & 47.2 \end{bmatrix} \begin{Bmatrix} T_1 \\ T_2 \\ T_3 \\ T_4 \end{Bmatrix} = \begin{Bmatrix} 200 \\ 0 \\ 0 \\ 1200 \end{Bmatrix}$$

and solving the set of linear equations, we have the following results:

$$\begin{Bmatrix} T_1 \\ T_2 \\ T_3 \\ T_4 \end{Bmatrix} = \begin{Bmatrix} 200 \\ 162.3 \\ 39.9 \\ 31.5 \end{Bmatrix} °\text{C}$$

Note that this type of heat conduction problem can be solved just as easily using fundamental concepts of heat transfer without resorting to finite element formulation. The point of this exercise was to demonstrate the steps involved in finite element analysis using a simple problem.

6.2 A FLUID MECHANICS PROBLEM

EXAMPLE 6.3 A Fluid Mechanics Problem

In a chemical processing plant, aqueous glycerin solution flows in a narrow channel, as shown in Figure 6.6. The pressure drop along the channel is continuously monitored. The upper wall of the channel is maintained at 50°C, while the lower wall is kept at 20°C.

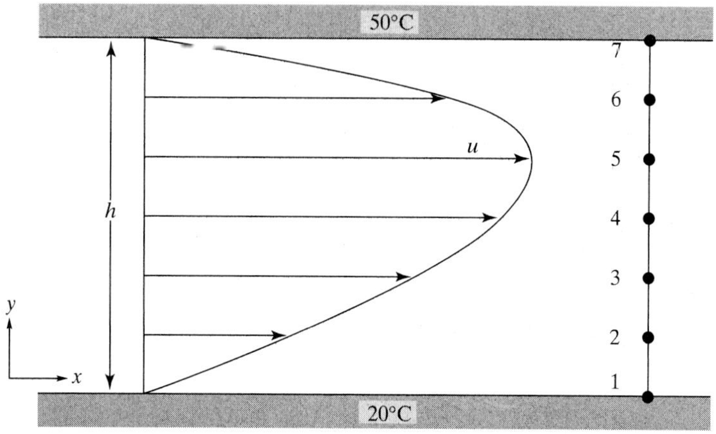

FIGURE 6.6 Laminar flow of aqueous glycerin solution through a channel.

The variation of viscosity and density of the glycerin with the temperature is given in Table 6.1. For a relatively low flow, the pressure drop along the channel is measured to be 120 Pa/m. The channel is 3 m long, 9 cm high, and 40 cm wide. Determine the velocity profile and the mass flow rate of the fluid through the channel.

The laminar flow of a fluid with a constant viscosity inside a channel is governed by the balance between the net shear forces and the net pressure forces acting on a parcel of fluid. The equation of motion is

$$\mu\frac{d^2u}{dy^2} - \frac{dp}{dx} = 0 \tag{6.45}$$

subject to the boundary conditions $u(0) = 0$ and $u(h) = 0$. Here, u represents fluid velocity, μ is the dynamic viscosity of the fluid, and $\frac{dp}{dx}$ is the pressure drop in the direction of the flow. For this problem, when comparing Eq. (6.45) to Eq. (6.8), we find that $c_1 = \mu$, $c_2 = 0$, $c_3 = -\frac{dp}{dx}$, and $\Psi = u$.

TABLE 6.1 Properties of aqueous glycerin solution as a function of temperature

Temperature (°C)	Viscosity (kg/m · s)	Density (kg/m³)
20	0.90	1255
25	0.65	1253
30	0.40	1250
35	0.28	1247
40	0.20	1243
45	0.12	1238
50	0.10	1233

Here, the viscosity of the aqueous glycerin solution varies with the height of the channel. We will use an average value of viscosity over each element when computing the elemental resistance matrices. The average values of viscosity and density associated with each element are given in Table 6.2.

Using the properties from Table 6.2, we can compute the elemental flow-resistance matrices as

$$[\mathbf{K}]^{(1)} = \frac{\mu}{\ell}\begin{bmatrix} 1 & -1 \\ -1 & 1 \end{bmatrix} = \frac{0.775}{1.5 \times 10^{-2}}\begin{bmatrix} 1 & -1 \\ -1 & 1 \end{bmatrix} = \begin{bmatrix} 51.67 & -51.67 \\ -51.67 & 51.67 \end{bmatrix}\frac{kg}{m^2 \cdot s}$$

$$[\mathbf{K}]^{(2)} = \frac{\mu}{\ell}\begin{bmatrix} 1 & -1 \\ -1 & 1 \end{bmatrix} = \frac{0.525}{1.5 \times 10^{-2}}\begin{bmatrix} 1 & -1 \\ -1 & 1 \end{bmatrix} = \begin{bmatrix} 35 & -35 \\ -35 & 35 \end{bmatrix}\frac{kg}{m^2 \cdot s}$$

$$[\mathbf{K}]^{(3)} = \frac{\mu}{\ell}\begin{bmatrix} 1 & -1 \\ -1 & 1 \end{bmatrix} = \frac{0.340}{1.5 \times 10^{-2}}\begin{bmatrix} 1 & -1 \\ -1 & 1 \end{bmatrix} = \begin{bmatrix} 22.67 & -22.67 \\ -22.67 & 22.67 \end{bmatrix}\frac{kg}{m^2 \cdot s}$$

$$[\mathbf{K}]^{(4)} = \frac{\mu}{\ell}\begin{bmatrix} 1 & -1 \\ -1 & 1 \end{bmatrix} = \frac{0.240}{1.5 \times 10^{-2}}\begin{bmatrix} 1 & -1 \\ -1 & 1 \end{bmatrix} = \begin{bmatrix} 16 & -16 \\ -16 & 16 \end{bmatrix}\frac{kg}{m^2 \cdot s}$$

$$[\mathbf{K}]^{(5)} = \frac{\mu}{\ell}\begin{bmatrix} 1 & -1 \\ -1 & 1 \end{bmatrix} = \frac{0.160}{1.5 \times 10^{-2}}\begin{bmatrix} 1 & -1 \\ -1 & 1 \end{bmatrix} = \begin{bmatrix} 10.67 & -10.67 \\ -10.67 & 10.67 \end{bmatrix}\frac{kg}{m^2 \cdot s}$$

$$[\mathbf{K}]^{(6)} = \frac{\mu}{\ell}\begin{bmatrix} 1 & -1 \\ -1 & 1 \end{bmatrix} = \frac{0.110}{1.5 \times 10^{-2}}\begin{bmatrix} 1 & -1 \\ -1 & 1 \end{bmatrix} = \begin{bmatrix} 7.33 & -7.33 \\ -7.33 & 7.33 \end{bmatrix}\frac{kg}{m^2 \cdot s}$$

Since $[\mathbf{K}]$ represents resistance to flow, we have opted to use the term elemental flow-resistance matrix instead of elemental stiffness matrix. Because the flow is fully developed, $\frac{dp}{dx}$ is constant; thus, the forcing matrix has the same value for all elements:

$$\{\mathbf{F}\}^{(1)} = \{\mathbf{F}\}^{(2)} = \cdots = \{\mathbf{F}\}^{(5)} = \{\mathbf{F}\}^{(6)} = \frac{-\frac{dp}{dx}\ell}{2}\begin{Bmatrix} 1 \\ 1 \end{Bmatrix} = \frac{-(-120)(1.5 \times 10^{-2})}{2}\begin{Bmatrix} 1 \\ 1 \end{Bmatrix} = \begin{Bmatrix} 0.9 \\ 0.9 \end{Bmatrix}\frac{N}{m^2}$$

TABLE 6.2 Properties of each element

Element	Average Viscosity (kg/m · s)	Average Density (kg/m^3)
1	0.775	1254
2	0.525	1252
3	0.34	1249
4	0.24	1245
5	0.16	1241
6	0.11	1236

The negative value associated with the pressure drop represents the decreasing nature of the pressure along the direction of flow in the channel. The global resistance matrix is obtained by assembling the elemental resistance matrices:

$$[\mathbf{K}]^{(G)} = \begin{bmatrix} 51.67 & -51.67 & 0 & 0 & 0 & 0 & 0 \\ -51.67 & 51.67 + 35 & -35 & 0 & 0 & 0 & 0 \\ 0 & -35 & 35 + 22.67 & -22.67 & 0 & 0 & 0 \\ 0 & 0 & -22.67 & 22.67 + 16 & -16 & 0 & 0 \\ 0 & 0 & 0 & -16 & 16 + 10.67 & -10.67 & 0 \\ 0 & 0 & 0 & 0 & -10.67 & 10.67 + 7.33 & -7.33 \\ 0 & 0 & 0 & 0 & 0 & -7.33 & 7.33 \end{bmatrix}$$

and the global forcing matrix is

$$\{\mathbf{F}\}^{(G)} = \begin{Bmatrix} 0.9 \\ 0.9 + 0.9 \\ 0.9 + 0.9 \\ 0.9 + 0.9 \\ 0.9 + 0.9 \\ 0.9 + 0.9 \\ 0.9 \end{Bmatrix}$$

Applying the no-slip boundary conditions at the walls leads to the matrix

$$\begin{bmatrix} 1 & 0 & 0 & 0 & 0 & 0 & 0 \\ -51.67 & 86.67 & -35 & 0 & 0 & 0 & 0 \\ 0 & -35 & 57.67 & -22.67 & 0 & 0 & 0 \\ 0 & 0 & -22.67 & 38.67 & -16 & 0 & 0 \\ 0 & 0 & 0 & -16 & 26.67 & -10.67 & 0 \\ 0 & 0 & 0 & 0 & -10.67 & 18 & -7.33 \\ 0 & 0 & 0 & 0 & 0 & 0 & 1 \end{bmatrix} \begin{Bmatrix} u_1 \\ u_2 \\ u_3 \\ u_4 \\ u_5 \\ u_6 \\ u_7 \end{Bmatrix} = \begin{Bmatrix} 0 \\ 1.8 \\ 1.8 \\ 1.8 \\ 1.8 \\ 1.8 \\ 0 \end{Bmatrix}$$

The solution provides the fluid velocities at each node:

$$\begin{Bmatrix} u_1 \\ u_2 \\ u_3 \\ u_4 \\ u_5 \\ u_6 \\ u_7 \end{Bmatrix} = \begin{Bmatrix} 0 \\ 0.1233 \\ 0.2538 \\ 0.3760 \\ 0.4366 \\ 0.3588 \\ 0 \end{Bmatrix} \text{m/s}$$

The mass flow rate through the channel can be determined from

$$\dot{m}_{\text{total}} = \Sigma \dot{m}^{(e)} \tag{6.46}$$

$$\dot{m}^{(e)} = \int_{y_i}^{y_j} \rho u W\,dy = \int_{y_i}^{y_j} \rho W(S_i u_i + S_j u_j)\,dy = \rho W \ell\left(\frac{u_i + u_j}{2}\right) \tag{6.47}$$

In Eq. (6.47), W represents the width of the channel. The elemental and total mass flow rates are given by

$$\dot{m}^{(1)} = \rho W \ell\left(\frac{u_i + u_j}{2}\right) = 1254 \times 0.4 \times 1.5 \times 10^{-2} \times \frac{0 + 0.1233}{2} = 0.4638 \text{ kg/s}$$

$$\dot{m}^{(2)} = 1252 \times 0.4 \times 1.5 \times 10^{-2} \times \frac{0.1233 + 0.2538}{2} = 1.4164 \text{ kg/s}$$

$$\dot{m}^{(3)} = 1249 \times 0.4 \times 1.5 \times 10^{-2} \times \frac{0.2538 + 0.3760}{2} = 2.3598 \text{ kg/s}$$

$$\dot{m}^{(4)} = 1245 \times 0.4 \times 1.5 \times 10^{-2} \times \frac{0.3760 + 0.4366}{2} = 3.0350 \text{ kg/s}$$

$$\dot{m}^{(5)} = 1241 \times 0.4 \times 1.5 \times 10^{-2} \times \frac{0.4366 + 0.3588}{2} = 2.9612 \text{ kg/s}$$

$$\dot{m}^{(6)} = 1236 \times 0.4 \times 1.5 \times 10^{-2} \times \frac{0.3588 + 0}{2} = 1.3304 \text{ kg/s}$$

$$\dot{m}_{\text{total}} = 11.566 \text{ kg/s}$$

6.3 AN EXAMPLE USING ANSYS

EXAMPLE 6.4 Revisited

A wall of an industrial oven consists of three different materials, as shown in Figure 6.4, repeated here as Figure 6.7. The first layer is composed of 5 cm of insulating cement with a clay binder that has a thermal conductivity of 0.08 W/m · K. The second layer is made from 15 cm of 6-ply asbestos board with a thermal conductivity of 0.074 W/m · K. The exterior consists of 10-cm common brick with a thermal conductivity of 0.72 W/m^2 · K. The inside wall temperature of the oven is 200°C, and the outside air is 30°C with a convection coefficient of 40 W/m^2 · K. Determine the temperature distribution along the composite wall.

The following steps demonstrate how to create one-dimensional conduction problems with convective boundary conditions in ANSYS. This task includes choosing appropriate element types, assigning attributes, applying boundary conditions, and obtaining results.

To solve this problem using ANSYS, we employ the following steps:

Enter the **ANSYS** program by using the Launcher.

Type **HeatTran** (or a file name of your choice) in the **Jobname** entry field of the dialog box, and pick **Run** to start the GUI.

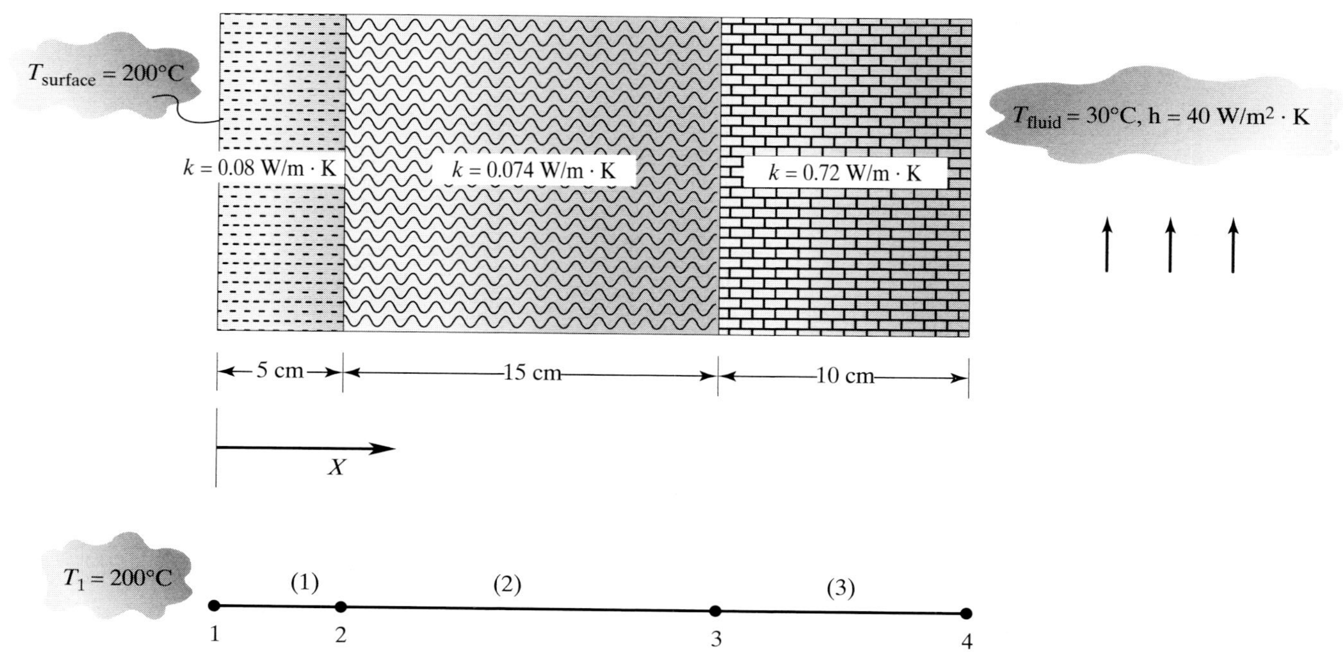

FIGURE 6.7 A composite wall of an industrial oven.

Create a title for the problem. This title will appear on ANSYS display windows to provide a simple way of identifying the displays:

utility menu: **File → Change Title** ...

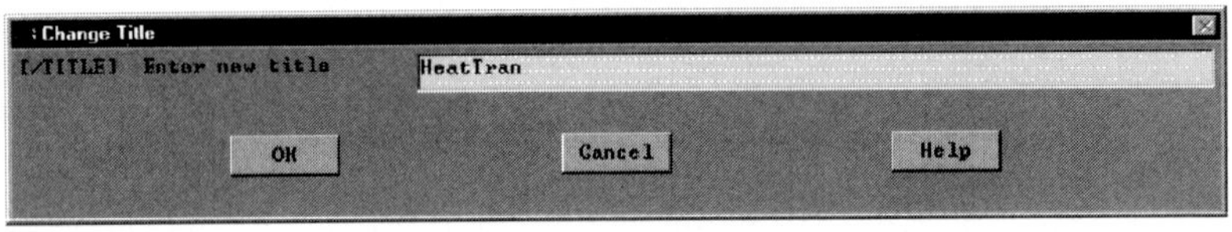

Define the element type and material properties:

main menu: **Preprocessor → Element Type → Add/Edit/Delete**

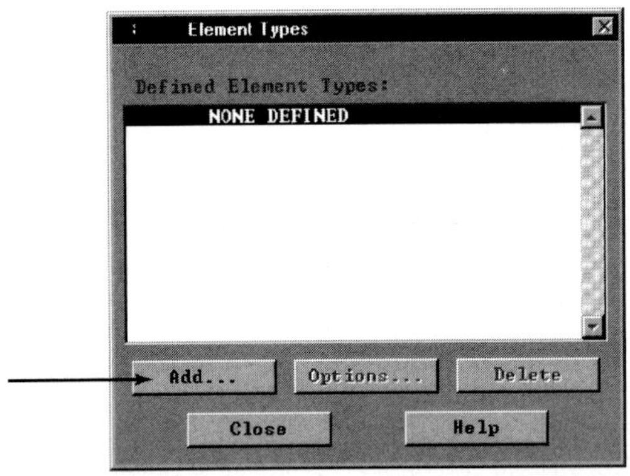

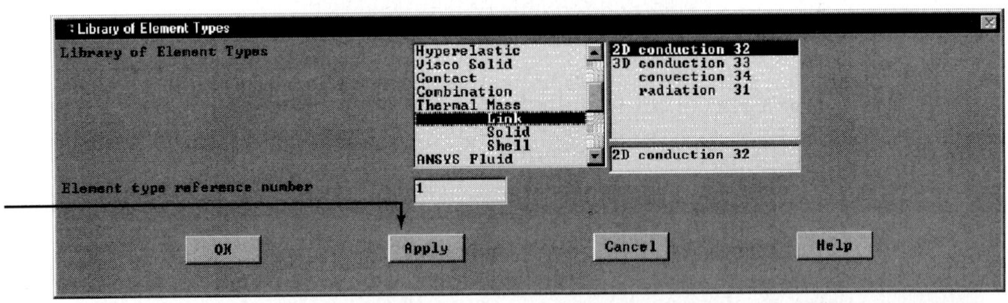

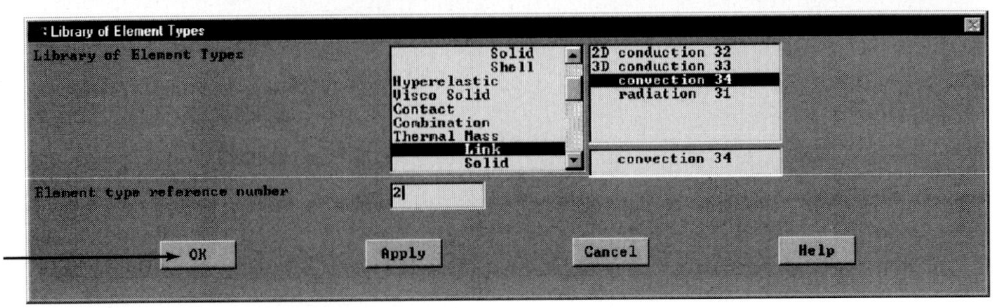

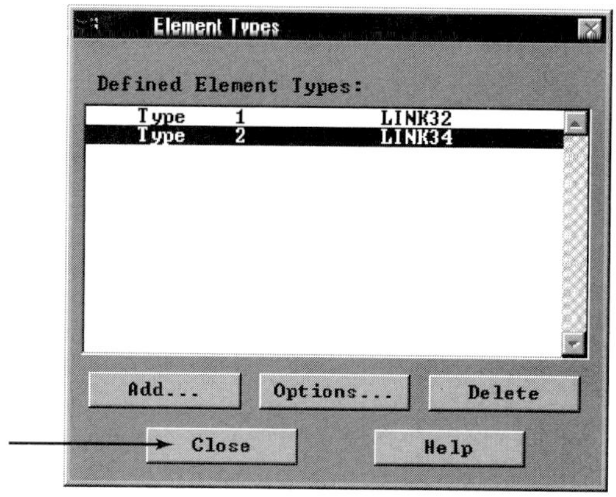

Assign the cross-sectional area of the wall.

main menu: **Preprocessor → Real Constants → Add/Edit/Delete**

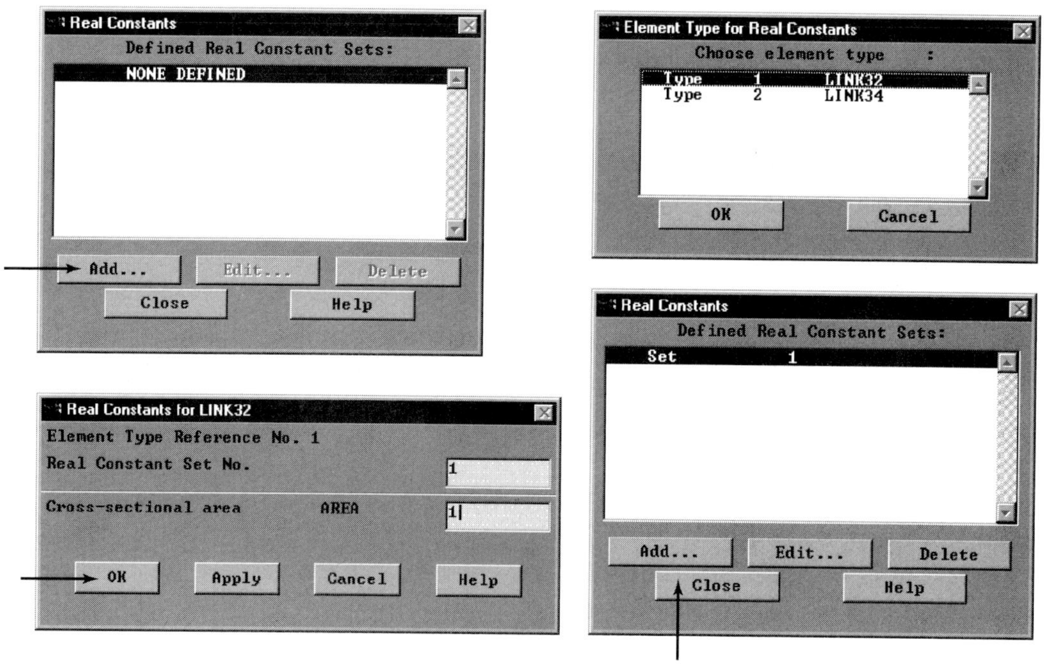

Assign the thermal conductivity values.

main menu: **Preprocessor → Material Props → Material Models → Thermal → Conductivity → Isotropic**

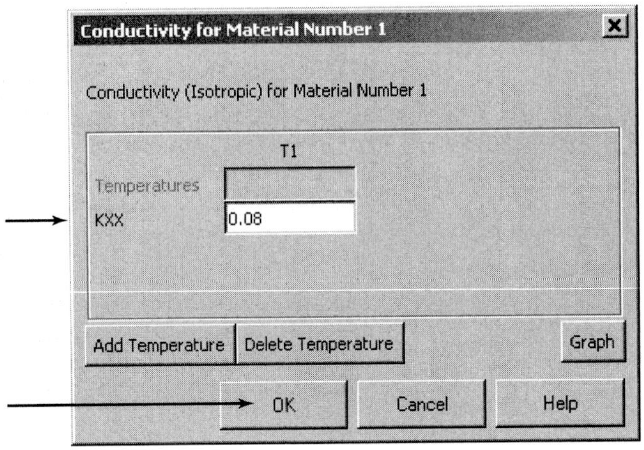

From the Define Material Model Behavior window:

Material → New Model . . .

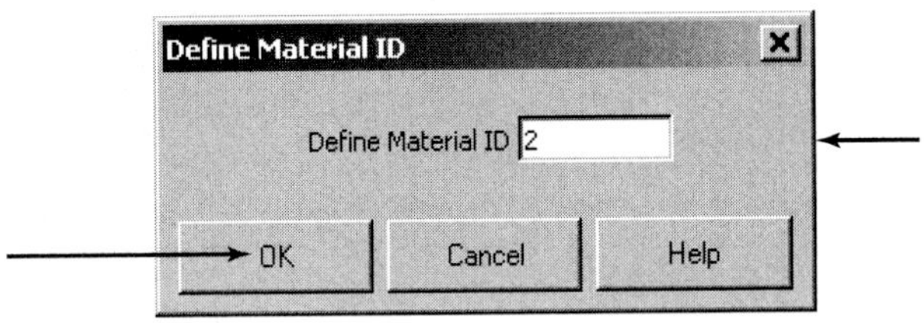

Now double-click on **Isotropic**.

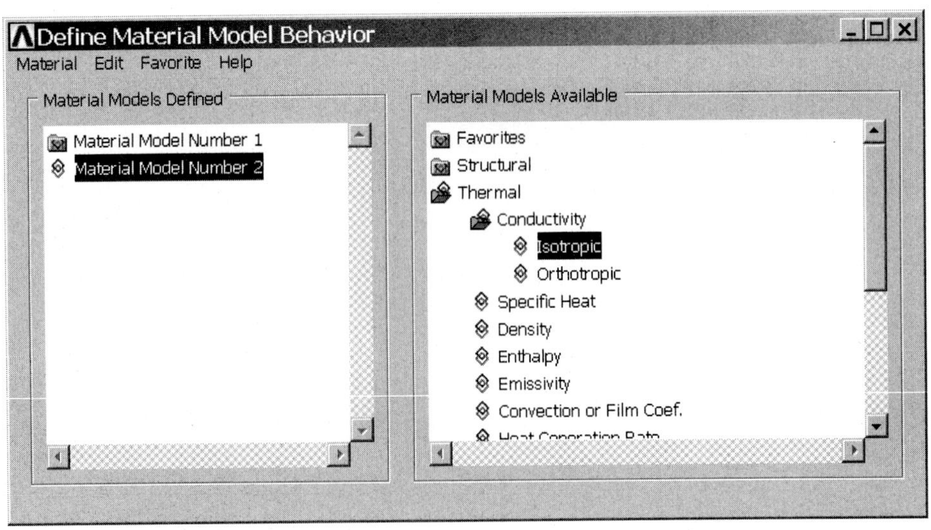

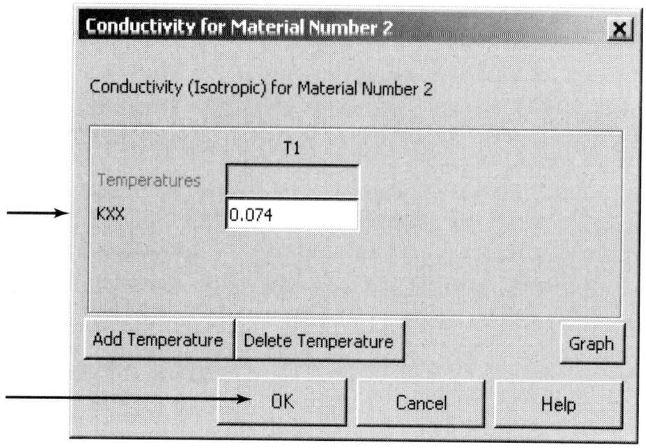

From the Define Material Model Behavior window:

Material → New Model . . .

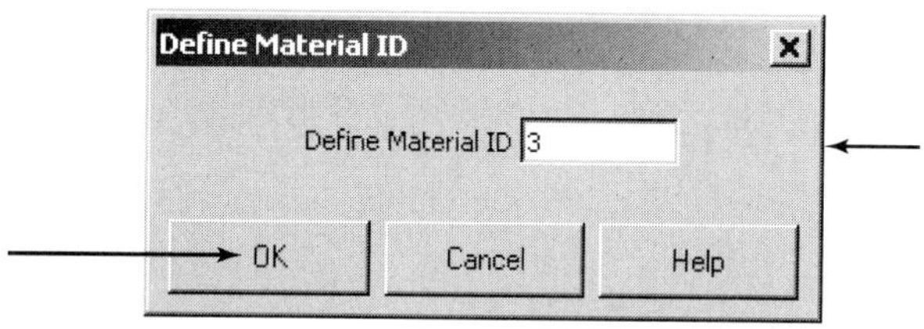

Assign the thermal conductivity of the third layer by double-clicking on **Isotropic** again.

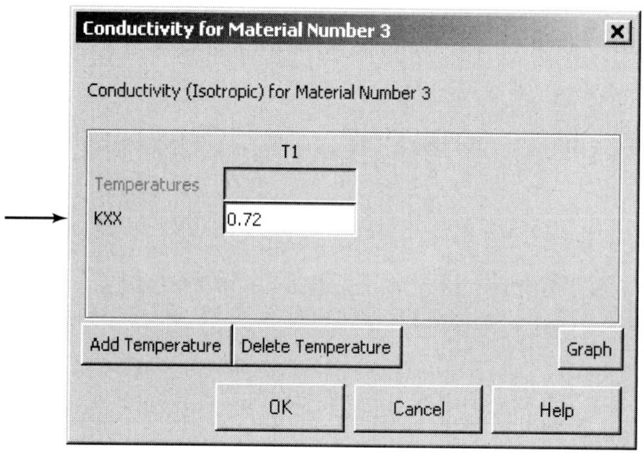

From the Define Material Model Behavior window:

Material → New Model . . .

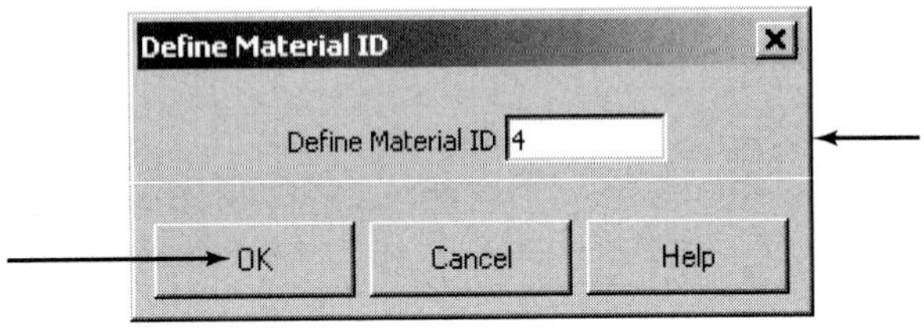

Assign the heat transfer coefficient by double-clicking on **Convection or Film Coef.**

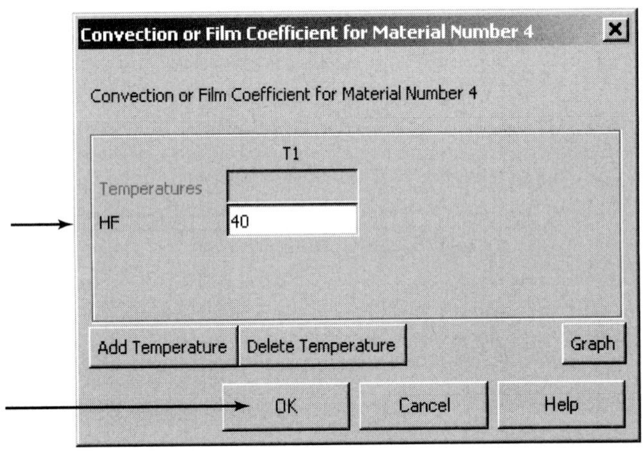

ANSYS Toolbar: **SAVE_DB**

Set up the graphics area (i.e., workplane, zoom, etc.):

utility menu: **Workplane → WP Settings** . . .

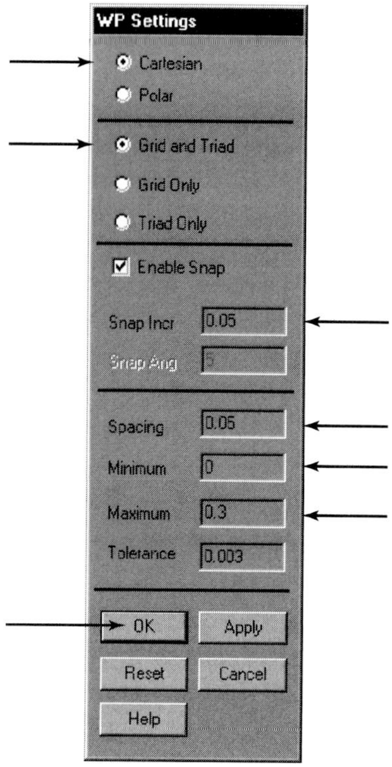

Toggle on the workplane by the following sequence:

utility menu: **Workplane → Display Working Plane**

Bring the workplane to view using the following sequence:

utility menu: **PlotCtrls → Pan, Zoom, Rotate** ...

Create nodes by picking points on the workplane:

main menu: **Preprocessor → Modeling → Create → Nodes → On Working Plane**

On the workplane, pick the location of nodes and apply them:

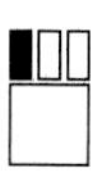

[WP = 0,0]

[WP = 0.05,0]

[WP = 0.2,0]

[WP = 0.3,0]

Create the node for the convection element:

[WP = 0.3,0]

OK

You may want to turn off the workplane now and turn on node numbering:

utility menu: **Workplane → Display Working plane**

utility menu: **PlotCtrls → Numbering** ...

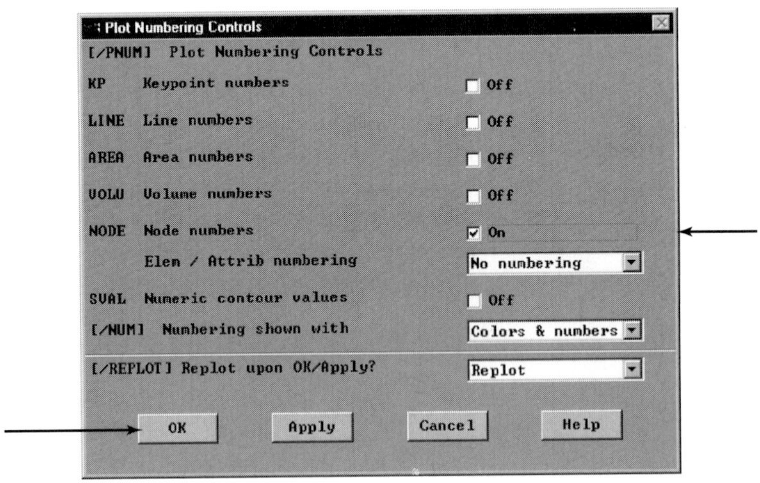

You may want to list nodes at this point in order to check your work:

utility menu: **List → Nodes** ...

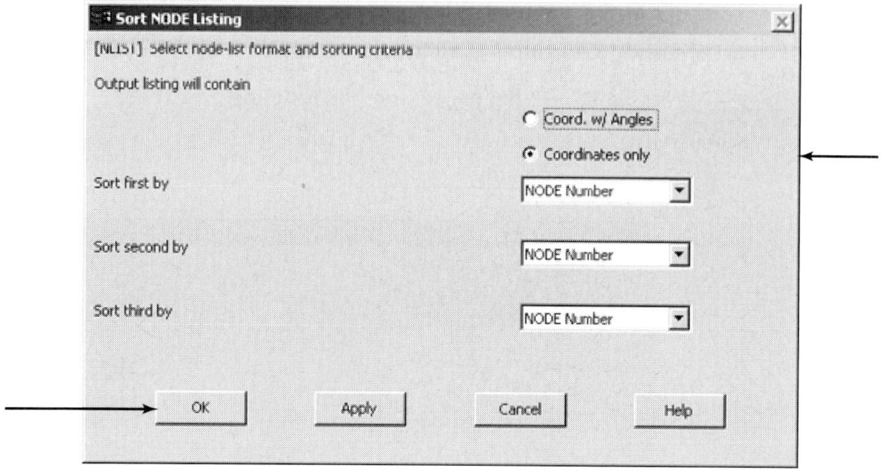

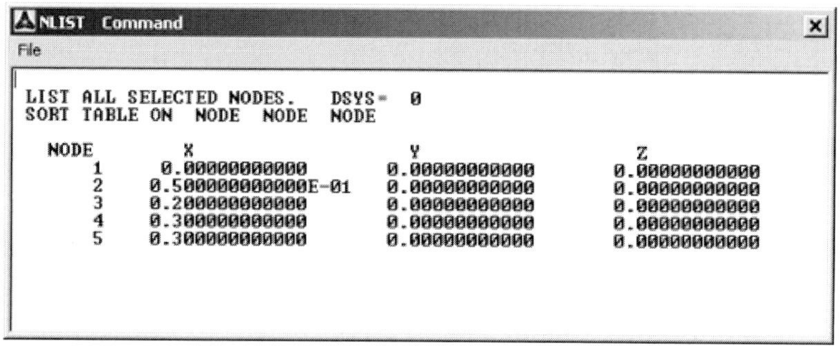

Close

ANSYS Toolbar: **SAVE_DB**

Define elements by picking nodes:

main menu: **Preprocessor** → **Modeling** → **Create** → **Elements** → **Auto Numbered** → **Thru Nodes**

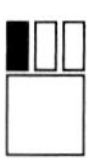 **[node 1 and then node 2]**

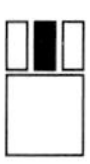 **[use the middle button anywhere in the ANSYS graphics window to apply.]**

OK

Assign the thermal conductivity of the second layer (element), and then connect the nodes to define the element:

main menu: **Preprocessor → Modeling → Create → Elements → Element Attributes**

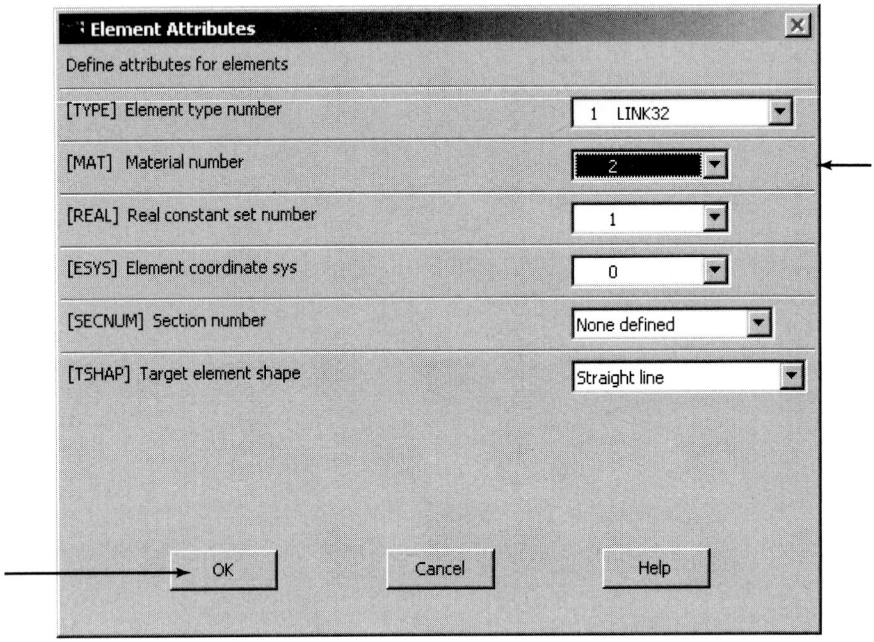

main menu: **Preprocessor → Modeling → Create → Elements → Auto Numbered → Thru Nodes**

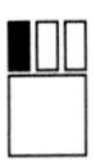 **[node 2 and then node 3]**

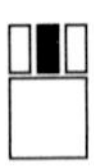 **[anywhere in the ANSYS graphics window]**

OK

Assign the thermal conductivity of third layer (element), and then connect the nodes to define the element:

main menu: **Preprocessor → Modeling → Create → Elements → Element Attributes**

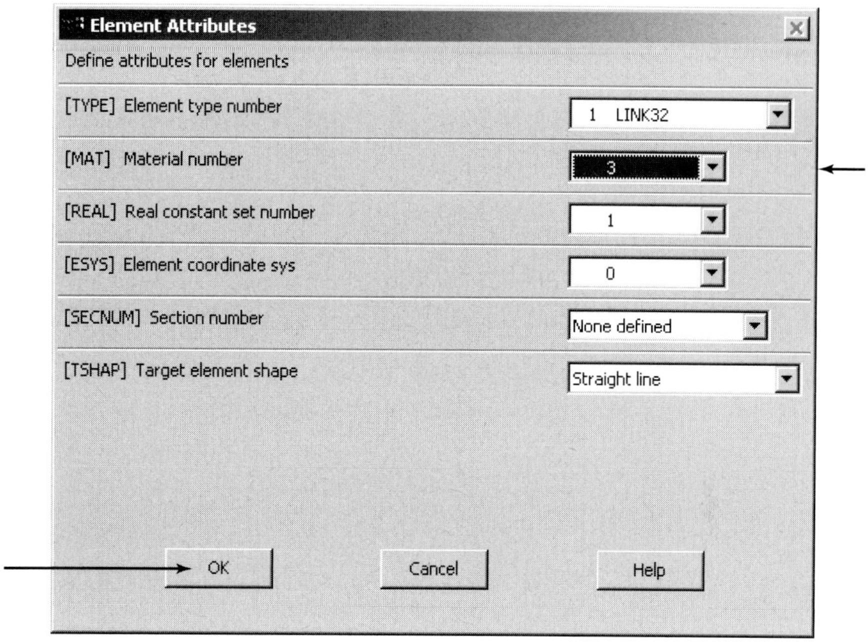

main menu: **Preprocessor → Modeling → Create → Elements → Auto Numbered → Thru Nodes**

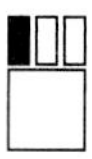

[node 3 and then node 4*]

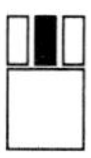

[anywhere in the ANSYS graphics window]

OK

Create the convection link:

main menu: **Preprocessor → Modeling → Create → Elements → Element Attributes**

*Press the **OK** key of the Multiple-Entities window and proceed.

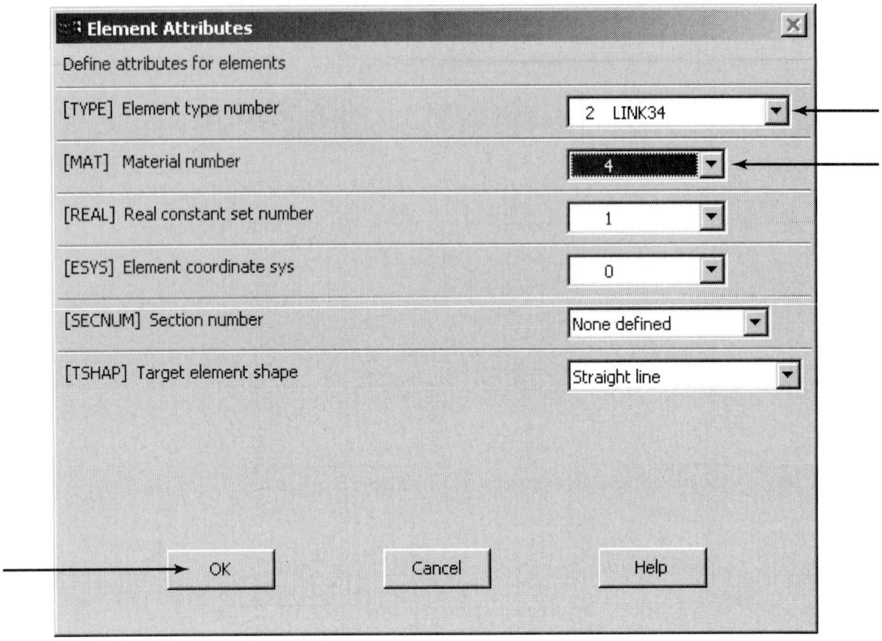

main menu: **Preprocessor → Modeling → Create → Elements → Auto Numbered → Thru Nodes**

[node 4† and then 5‡]

OK

ANSYS Toolbar: **SAVE_DB**

Apply boundary conditions:

main menu: **Solution → Define Loads → Apply → Thermal → Temperature → On Nodes**

[node 1]

[anywhere in the ANSYS graphics window]

---

†Press the OK key of the Multiple-Entities window.

‡Press the **Next** key of the Multiple-Entities window, and then **OK.**

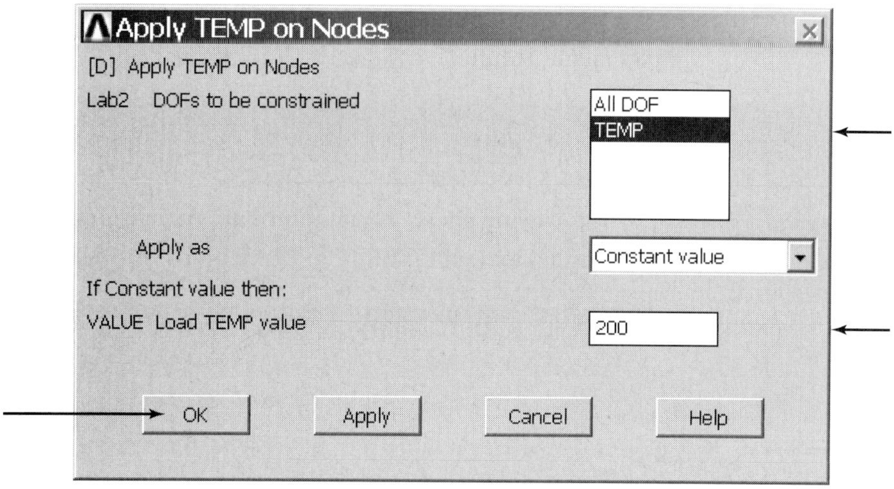

main menu: **Solution** → **Define Loads** → **Apply** → **Thermal** → **Temperature** → **On Nodes**

[node 5*]

[anywhere in the ANSYS graphics window]

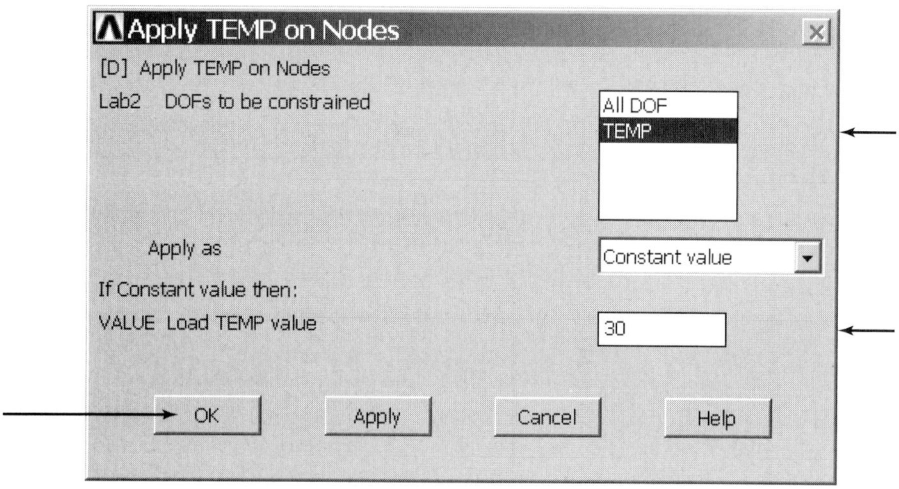

ANSYS Toolbar: **SAVE_DB**

*Press the **Next** key and then the **OK** key of the Multiple-Entities window and proceed.

Solve the problem:

main menu: **Solution → Solve → Current LS**

OK

Close (the solution is done!) window.

Close (the/STAT Command) window.

For the postprocessing phase, obtain information such as nodal temperatures:

main menu: **General Postproc → List Results → Nodal Solution**

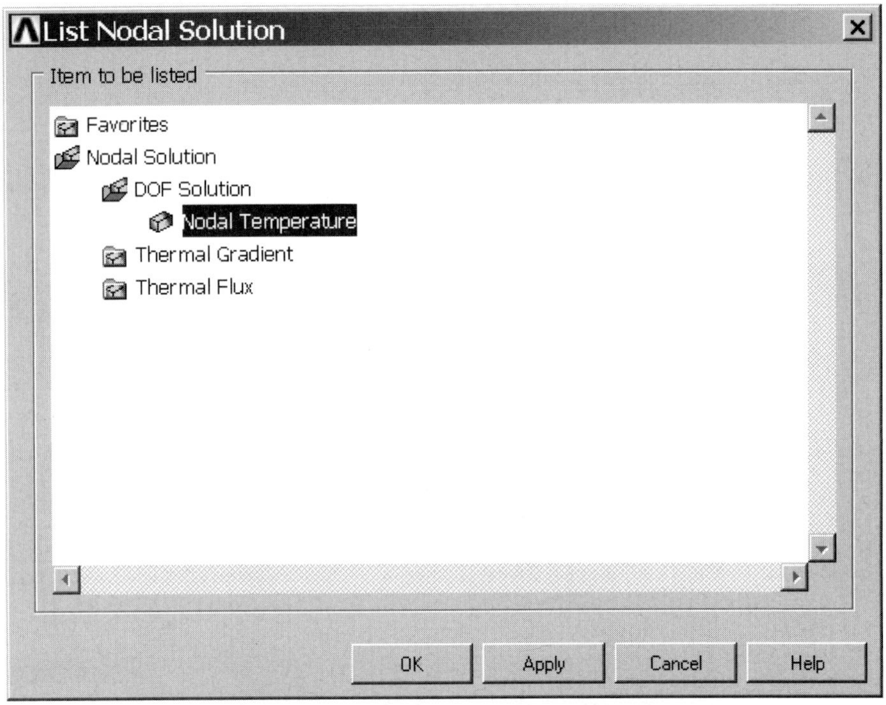

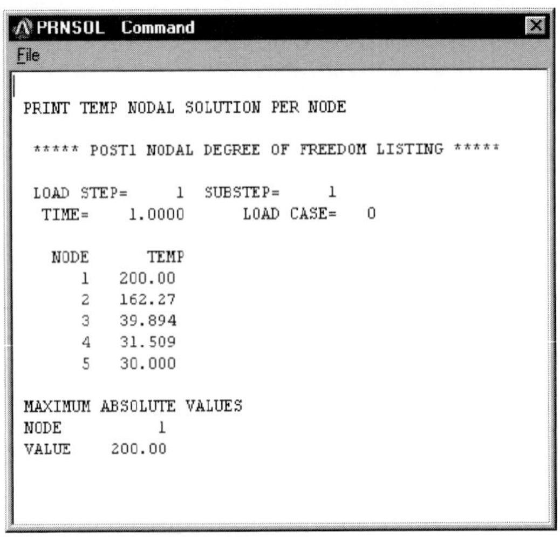


```
PRINT TEMP NODAL SOLUTION PER NODE

 ***** POST1 NODAL DEGREE OF FREEDOM LISTING *****

 LOAD STEP=     1  SUBSTEP=     1
  TIME=    1.0000      LOAD CASE=   0

   NODE       TEMP
      1   200.00
      2   162.27
      3   39.894
      4   31.509
      5   30.000

MAXIMUM ABSOLUTE VALUES
NODE          1
VALUE    200.00
```

Close

Exit ANSYS and save everything.

Toolbar: **QUIT**

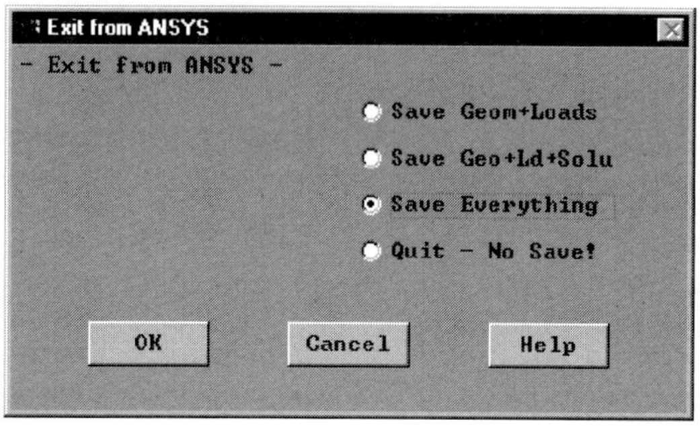

6.4 VERIFICATION OF RESULTS

There are various ways to verify your findings. Consider the nodal temperatures of Example 6.2, as computed by ANSYS and displayed in Table 6.3.

In general, for a heat transfer problem under steady-state conditions, conservation of energy applied to a control volume surrounding an arbitrary node must be satisfied. Are the energies flowing into and out of a node balanced out? Let us use Example 6.2 to demonstrate this important concept. The heat loss through each layer of the composite wall must be equal. Furthermore, heat loss from the last layer must equal the heat removed by the surrounding air. So,

$$Q^{(1)} = Q^{(2)} = Q^{(3)} = Q^{(4)}$$

$$Q^{(1)} = kA\frac{\Delta T}{\ell} = (0.08)(1)\left(\frac{200 - 162.3}{0.05}\right) = 60 \text{ W}$$

$$Q^{(2)} = (0.074)(1)\left(\frac{162.3 - 39.9}{0.15}\right) = 60 \text{ W}$$

$$Q^{(3)} = (0.72)(1)\left(\frac{39.9 - 31.5}{0.1}\right) = 60 \text{ W}$$

TABLE 6.3 Nodal temperatures

Node Number	Temperature (°C)
1	200
2	162.3
3	39.9
4	31.5
5	30

and the heat removal by the fluid is given by

$$Q^{(4)} = hA\Delta T = (40)(1)(31.5 - 30) = 60 \text{ W}$$

For thermal elements, ANSYS provides information such as heat flow through each element. Therefore, we could have compared those values with the one we calculated above.

Another check on the validity of your results could have come from examining the slopes of temperatures in each layer. The first layer has a temperature slope of 754°C/m. For the second layer, the slope of the temperature is 816°C/m. This layer consists of a material with relatively low thermal conductivity and, therefore, a relatively large temperature drop. The slope of the temperature in the exterior wall is 84°C/m. Because the exterior wall is made of a material with relatively high thermal conductivity, we expect the temperature drop through this layer not to be as significant as the other layers.

Now consider the fin problem in Example 6.1. For this problem, recall that all elements have the same length. We determined the temperature distribution and the heat loss from each element. Comparing heat loss results, it is important to realize that element (1) has the highest value because the greatest thermal potential exists at the base of the fin, and as the temperature of the fin drops, so does the rate of heat loss for each element. This outcome is certainly consistent with the results we obtained previously.

These simple problems illustrate the importance of checking for equilibrium conditions when verifying results.

SUMMARY

At this point you should:

1. know how to formulate conductance or resistance matrices, and be able to formulate load matrices for various one-dimensional problems.
2. know how to apply appropriate boundary conditions.
3. have a good understanding of the Galerkin and energy formulations of one-dimensional problems.
4. know how to verify your results.

REFERENCES

ANSYS User's Manual: Procedures, Vol. I, Swanson Analysis Systems, Inc.

ANSYS User's Manual: Commands, Vol. II, Swanson Analysis Systems, Inc.

ANSYS User's Manual: Elements, Vol. III, Swanson Analysis Systems, Inc.

Incropera, F., and Dewitt, D., *Fundamentals of Heat and Mass Transfer*, 2nd ed., New York, John Wiley and Sons, 1985.

Glycerin Producers' Association, "Physical Properties of Glycerin And Its Solutions," New York.

Segrlind, L., *Applied Finite Element Analysis*, 2nd ed., New York, John Wiley and Sons, 1984.

PROBLEMS

1. Aluminum fins, similar to the ones in Example 6.1, with rectangular profiles are used to remove heat from a surface whose temperature is 150°C. The temperature of the ambient air is 20°C. The thermal conductivity of aluminum is 168 W/m · K. The natural convective coefficient associated with the surrounding air is 35 W/m^2 · K. The fins are 150 mm long, 5 mm wide, and 1 mm thick. (a) Determine the temperature distribution along a fin using five equally spaced elements. (b) Approximate the total heat loss for an array of 50 fins.

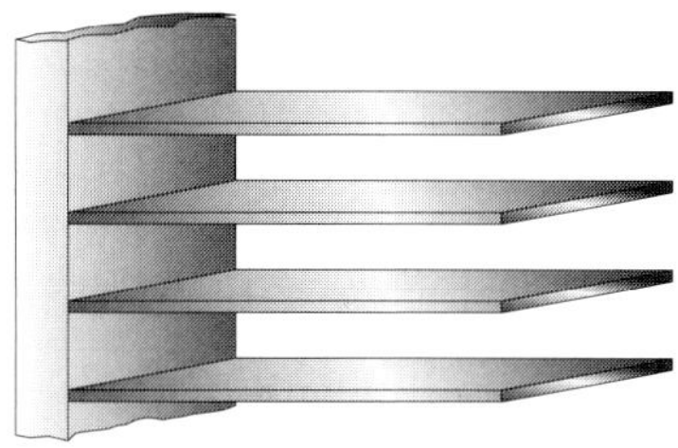

2. For the aluminum fins in problem 1, determine the temperature of a point on the fin 45 mm from the base. Also compute the fraction of the total heat that is lost through this section of the fin.
3. A pin fin, or spine, is a fin with a circular cross section. An array of aluminum pin fins are used to remove heat from a surface whose temperature is 120°C. The temperature of the ambient air is 25°C. The thermal conductivity of aluminum is 168 W/m · K. The natural convective coefficient associated with the surrounding air is 30 W/m^2 · K. The fins are 100 mm long and have respective diameters of 4 mm. (a) Determine the temperature distribution along a fin using five equally spaced elements. (b) Approximate the total heat loss for an array of 100 fins.
4. A rectangular aluminum fin is used to remove heat from a surface whose temperature is 80°C. The temperature of the ambient air varies between 18°C and 25°C. The thermal conductivity of aluminum is 168 W/m · K. The natural convective coefficient associated with the surrounding air is 25 W/m^2 · K. The fin is 100 mm long, 5 mm wide, and 1 mm thick. (a) Determine the temperature distribution along a fin using five equally spaced elements for both ambient conditions. (b) Approximate the total heat loss for an array of 50 fins for each ambient condition. (c) The exact temperature distribution and heat loss for a fin with a negligible heat loss at its tip is given by

$$\frac{T(x) - T_f}{T_b - T_f} = \frac{\cosh\left[\sqrt{\frac{hp}{kA_c}}(L - x)\right]}{\cosh\left[\sqrt{\frac{hp}{kA_c}}(L)\right]}$$

$$Q = \sqrt{hpkA_c}\left(\tanh\left[\sqrt{\frac{hp}{kA_c}}(L)\right]\right)(T_b - T_f)$$

Compare your finite element results to the exact results.

5. Evaluate the integral $\int_{X_i}^{X_j} S_i h p T_f \, dX$ for a situation in which the heat transfer coefficient h varies linearly over a given element.

6. The front window of a car is defogged by supplying warm air at 90°F to its inner surface. The glass has a thermal conductivity of $k = 0.8$ W/m · °C with a thickness of approximately 1/4 in. With the supply fan running at moderate speed, the heat transfer coefficient associated with the air is 50 W/m^2 · K. The outside air is at a temperature of 20°F with an associated heat transfer coefficient of 110 W/m^2 · K. Determine (a) the temperatures of the inner and outer surfaces of the glass and (b) the heat loss through the glass if the area of the glass is approximately 10 ft^2.

7. A wall of an industrial oven consists of three different materials, as shown in the accompanying figure. The first layer is composed of 10 cm of insulating cement with a thermal conductivity of 0.12 W/m · K. The second layer is made from 20 cm of 8-ply asbestos board with a thermal conductivity of 0.071 W/m · K. The exterior consists of 12-cm cement mortar with a thermal conductivity of 0.72 W/m^2 · K. The inside wall temperature of the oven is 250°C, and the outside air is at a temperature of 35°C with a convection coefficient of 45 W/m^2 · K. Determine the temperature distribution along the composite wall.

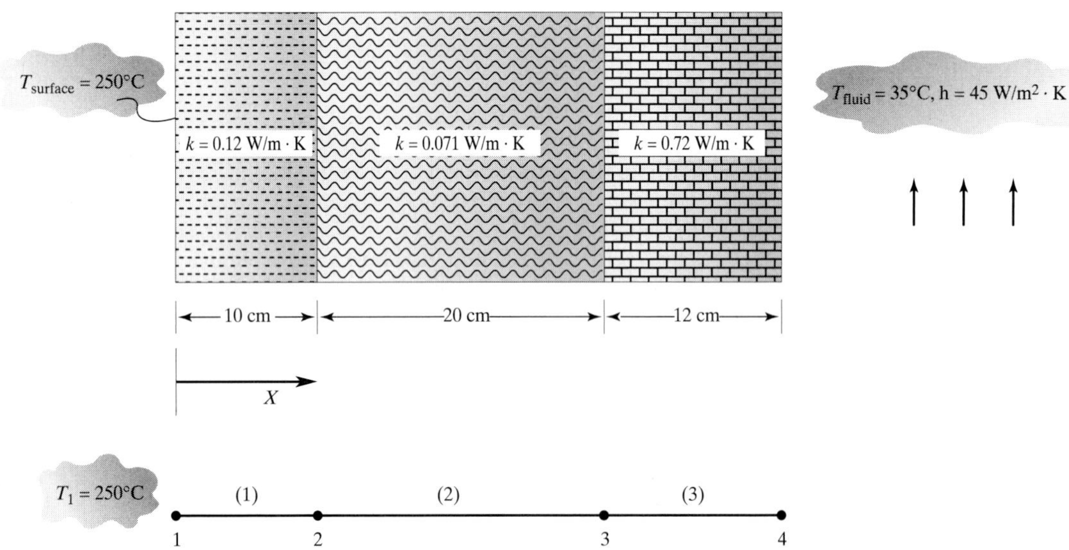

8. Replace the temperature boundary condition of the inside wall of the oven in problem 7 with air temperature of 400°C and an associated convection coefficient of 100 W/m^2 · K. Show the contribution of this boundary condition to the conductance matrix and the forcing matrix of element (1). Also determine the temperature distribution along the composite wall.

9. The equation for the heat diffusion of a one-dimensional system with heat generation in a Cartesian coordinate system is

$$k \frac{\partial^2 T}{\partial X^2} + q^{\cdot} = 0$$

The rate of thermal energy generation $\dot{q}$ represents the conversion of energy from electrical, chemical, nuclear, or electromagnetic forms to thermal energy within the volume of a given system. Derive the contribution of $\dot{q}$ to the load matrix. Consider a strip of heating elements embedded within the rear glass of a car producing a uniform heat generation at a rate of approximately 7000 W/m^3. The glass has a thermal conductivity of $k = 0.8$ W/m · °C with a thickness of approximately 6 mm. The heat transfer coefficient associated with the 20°C air inside the back of the car is approximately 20 W/m^2 · K. The outside air is at a temperature of −5°C with an associated heat transfer coefficient of 50 W/m^2 · K. Determine the temperatures of the inner and outer surfaces of the glass.

10. Verify the evaluation of the integral given by Eq. (6.23):

$$\int_{X_i}^{X_j} S_i(c_2\Psi)\,dX = \frac{c_2\ell}{3}\Psi_i + \frac{c_2\ell}{6}\Psi_j$$

11. Verify the evaluation of the integral given by Eq. (6.26):

$$\int_{X_i}^{X_j} c_1\left(-\frac{dS_j\,d\Psi}{dX\,dX}\right)dX = -\frac{c_1}{\ell}(-\Psi_i + \Psi_j)$$

12. The deformation of an axial element of length ℓ due to the change in its temperature is given by

$$\delta_T = \alpha\ell\Delta T$$

where δ_T is the change in the length of the element, α is the thermal expansion coefficient of the material, and ΔT represents the temperature change. Formulate the contribution of thermal strains to the strain energy of an element. Also formulate the stiffness and the load matrices for such an element.

13. You are to size fins of a rectangular cross section to remove a total of 200 W from a 400-cm^2 surface whose temperature is to be kept at 80°C. The temperature of the surrounding air is 25°C, and you may assume that the natural convection coefficient value is 25 W/m^2 · K. Because of restrictions on the amount of space, the fins cannot be extended more than 100 mm from the hot surface. You may select from the following materials: aluminum, copper, or steel. In a brief report, explain how you came up with your final design.

CHAPTER 7

Two-Dimensional Elements

The objective of this chapter is to introduce the concept of two-dimensional shape functions, along with two-dimensional elements and their properties. Natural coordinates associated with quadrilateral and triangular elements are also presented. We derive the shape functions for rectangular elements, quadratic quadrilateral elements, and triangular elements. Examples of two-dimensional thermal and structural elements in ANSYS are also presented. The main topics discussed in Chapter 7 include the following:

7.1 Rectangular Elements
7.2 Quadratic Quadrilateral Elements
7.3 Linear Triangular Elements
7.4 Quadratic Triangular Elements
7.5 Axisymmetric Elements
7.6 Isoparametric Elements
7.7 Two-Dimensional Integrals: Gauss-Legendre Quadrature
7.8 Examples of Two-Dimensional Elements in ANSYS

7.1 RECTANGULAR ELEMENTS

In Chapter 6, we studied the analysis of one-dimensional problems. We investigated heat transfer in a straight fin. We used one-dimensional linear shape functions to approximate temperature distributions along elements and formulated the conductance matrix and the thermal load matrix. The resulting systems of equations, once solved, yielded the nodal temperatures. In this chapter, we lay the groundwork for the analysis of two-dimensional problems by first studying two-dimensional shape functions and elements. To aid us in presenting this material, let us consider the straight fin shown in Figure 7.1. The dimensions of the fin and thermal boundary conditions are such that we cannot accurately approximate temperature distribution along the fin by a one-dimensional function. The temperature varies in both the X-direction and the Y-direction.

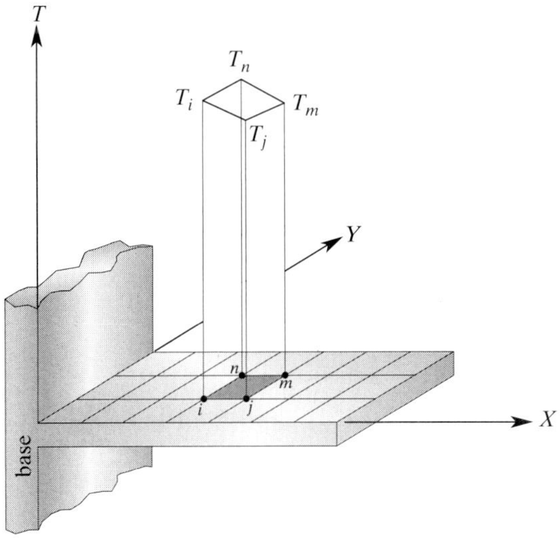

FIGURE 7.1 Using rectangular elements to describe a two-dimensional temperature distribution.

At this point, it is important to understand that the one-dimensional solutions are approximated by line segments, whereas the two-dimensional solutions are represented by plane segments. This point is illustrated in Figure 7.1. A close-up look at a typical rectangular element and its nodal values is shown in Figure 7.2a.

It is clear from examining Figure 7.2a that the temperature distribution over the element is a function of both X- and Y-coordinates. We can approximate the temperature distribution for an arbitrary rectangular element by

$$T^{(e)} = b_1 + b_2x + b_3y + b_4xy \tag{7.1}$$

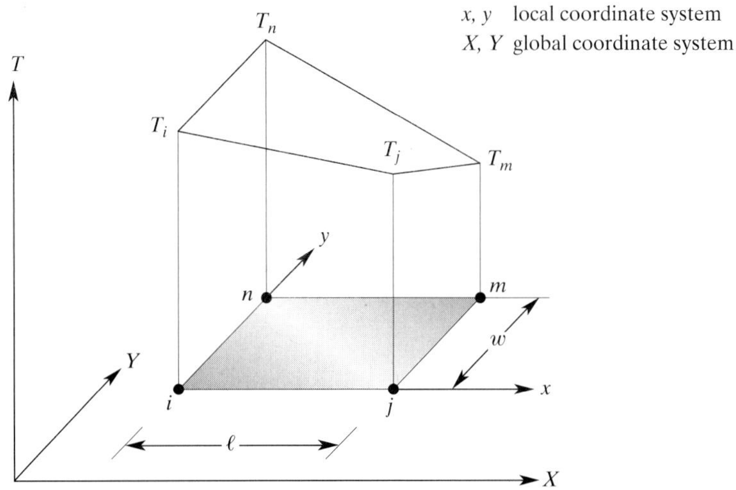

FIGURE 7.2a A typical rectangular element.

Note that there are four unknowns (b_1, b_2, b_3, b_4) in Eq. (7.1), because a rectangular element is defined by four nodes: i, j, m, n. Also note that the function varies linearly along the edges of the element, and it becomes nonlinear inside the element (see Problem 31). Elements with these types of characteristics are commonly referred to as bilinear elements. The procedure for deriving two-dimensional shape functions is essentially the same as that for one-dimensional elements. To obtain b_1, b_2, b_3, and b_4, we use the local coordinates x and y. Considering nodal temperatures, we must satisfy the following conditions:

$$\begin{aligned}
T &= T_i \quad \text{at} \quad x = 0 \quad \text{and} \quad y = 0 \\
T &= T_j \quad \text{at} \quad x = \ell \quad \text{and} \quad y = 0 \\
T &= T_m \quad \text{at} \quad x = \ell \quad \text{and} \quad y = w \\
T &= T_n \quad \text{at} \quad x = 0 \quad \text{and} \quad y = w
\end{aligned} \tag{7.2}$$

Applying the nodal conditions given by Eq. (7.2) to Eq. (7.1) and solving for b_1, b_2, b_3, and b_4, we have

$$\begin{aligned}
b_1 &= T_i & b_2 &= \frac{1}{\ell}(T_j - T_i) \\
b_3 &= \frac{1}{w}(T_n - T_i) & b_4 &= \frac{1}{\ell w}(T_i - T_j + T_m - T_n)
\end{aligned} \tag{7.3}$$

Substituting expressions given for b_1, b_2, b_3, and b_4 into Eq. (7.1) and regrouping parameters will result in the temperature distribution for a typical element in terms of shape functions:

$$T^{(e)} = [S_i \quad S_j \quad S_m \quad S_n] \begin{Bmatrix} T_i \\ T_j \\ T_m \\ T_n \end{Bmatrix} \tag{7.4}$$

The shape functions in the above expression are

$$\begin{aligned}
S_i &= \left(1 - \frac{x}{\ell}\right)\left(1 - \frac{y}{w}\right) \\
S_j &= \frac{x}{\ell}\left(1 - \frac{y}{w}\right) \\
S_m &= \frac{xy}{\ell w} \\
S_n &= \frac{y}{w}\left(1 - \frac{x}{\ell}\right)
\end{aligned} \tag{7.5}$$

It should be clear that we can use these shape functions to represent the variation of any unknown variable Ψ over a rectangular region in terms of its nodal values Ψ_i, Ψ_j, Ψ_m,

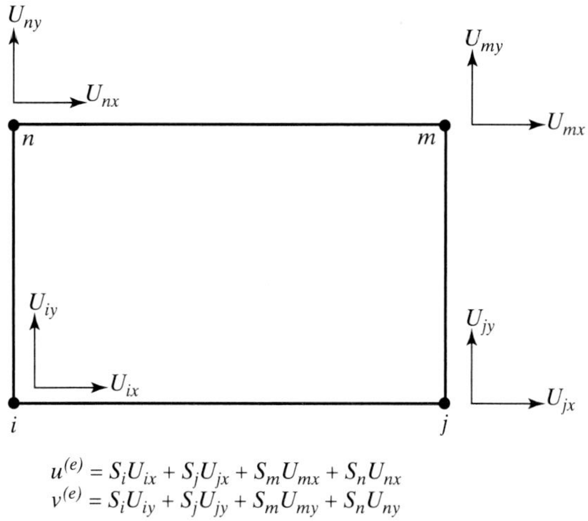

FIGURE 7.2b A rectangular element used in formulating plane-stress problems.

and Ψ_n. Thus, in general, we can write

$$\Psi^{(e)} = [S_i \quad S_j \quad S_m \quad S_n] \begin{Bmatrix} \Psi_i \\ \Psi_j \\ \Psi_m \\ \Psi_n \end{Bmatrix} \tag{7.6}$$

For example, Ψ could represent a solid element displacement in a certain direction as shown in Figure 7.2b.

Natural Coordinates

As was discussed in Chapter 5, natural coordinates are basically local coordinates in a dimensionless form. Moreover, as you may recall, most finite element programs perform element numerical integration by Gaussian quadratures, and as the limits of integration, they use an interval from -1 to 1. The origin of the local coordinate system x, y used earlier coincides with the natural coordinates $\xi = -1$ and $\eta = -1$, as shown in Figure 7.3.

If we let $\xi = \dfrac{2x}{\ell} - 1$ and $\eta = \dfrac{2y}{w} - 1$, then the shape functions in terms of the natural coordinates ξ and η are

$$\begin{aligned} S_i &= \frac{1}{4}(1 - \xi)(1 - \eta) \\ S_j &= \frac{1}{4}(1 + \xi)(1 - \eta) \\ S_m &= \frac{1}{4}(1 + \xi)(1 + \eta) \\ S_n &= \frac{1}{4}(1 - \xi)(1 + \eta) \end{aligned} \tag{7.7}$$

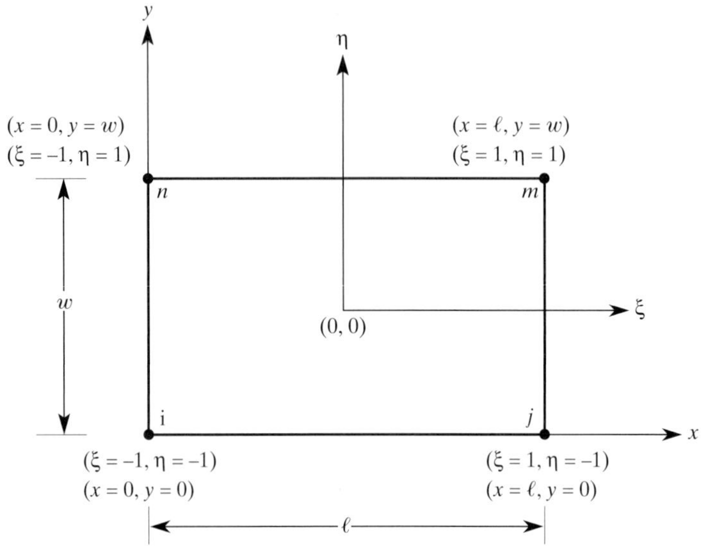

FIGURE 7.3 Natural coordinates used to describe a quadrilateral element.

These shape functions have the same general basic properties as their one-dimensional counterparts. For example, S_i has a value of 1 when evaluated at the coordinates of node *i*, but has a value of zero at all other nodes.

Alternatively, we could have obtained the expressions given in Eq. (7.7) by using a product of linear functions similar to the Lagrange functions as explained in Chapter 5, Section 5.3. For example, for node *i*, we select the functions such that their product will produce a value of zero at other nodes—namely, *j*, *m*, and *n*—and a value of unity at the given node *i*. Along the *j*–*m* side ($\xi = 1$) and *n*–*m* side ($\eta = 1$), so if we choose the product of functions $(1 - \xi)(1 - \eta)$ then the product will produce a value of zero along *j*–*m* side and *n*–*m* side. We then evaluate a_1, an unknown coefficient, such that when the shape function S_i is evaluated at node $i(\xi = -1)$ and $(\eta = -1)$, it will produce a value of unity. That is

$$1 = a_1(1 - \xi)(1 - \eta) = a_1(1-(-1))(1-(-1)) \Rightarrow a_1 = \frac{1}{4}$$

7.2 QUADRATIC QUADRILATERAL ELEMENTS

The eight-node quadratic quadrilateral element is basically a higher order version of the two-dimensional four-node quadrilateral element. This type of element is better suited for modeling problems with curved boundaries. A typical eight-node quadratic element is shown in Figure 7.4. When compared to the linear elements, for the same number of elements, quadratic elements offer better nodal results. In terms of the natural coordinates ξ and η, the eight-node quadratic element has the general

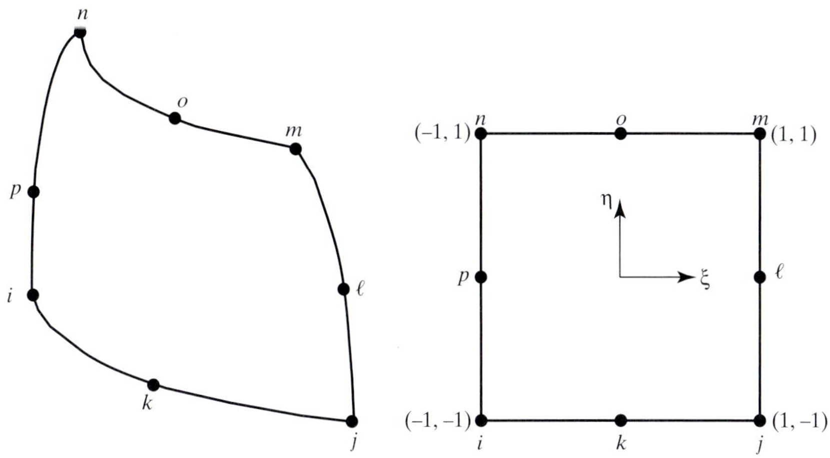

FIGURE 7.4 Eight-node quadratic quadrilateral element.

form of

$$\Psi^{(e)} = b_1 + b_2\xi + b_3\eta + b_4\xi\eta + b_5\xi^2 + b_6\eta^2 + b_7\xi^2\eta + b_8\xi\eta^2 \tag{7.8}$$

To solve for $b_1, b_2, b_3, \ldots, b_8$, we must first apply the nodal conditions and create eight equations from which we can solve for these coefficients. Instead of using this laborious and difficult method, we will follow an alternative approach, which is demonstrated next.

In general, the shape function associated with each node can be represented in terms of the product of two functions F_1 and F_2:

$$S = F_1(\xi, \eta)F_2(\xi, \eta) \tag{7.9}$$

For a given node, we select the first function F_1 such that it will produce a value of zero when evaluated along the sides of the element that the given node does not contact. Moreover, the second function F_2 is selected such that when multiplied by F_1, it will produce a value of unity at the given node and a value of zero at other neighboring nodes. The product of functions F_1 and F_2 must also produce linear and nonlinear terms similar to the ones given by Eq. (7.8). To demonstrate this method, let us consider corner node m, with natural coordinates $\xi = 1$ and $\eta = 1$. First, F_1 must be selected such that along the ij-side ($\eta = -1$) and in-side ($\xi = -1$), the function will produce a value of zero. We select

$$F_1(\xi, \eta) = (1 + \xi)(1 + \eta)$$

which satisfies the condition. We then select

$$F_2(\xi, \eta) = c_1 + c_2\xi + c_3\eta$$

The coefficients in F_2 should be selected such that when F_2 multiplied by F_1, they will produce a value of unity at the given node m, and a value of zero at the adjacent neighboring nodes ℓ and o. Evaluating S_m at node m should give $S_m = 1$ for $\xi = 1$ and $\eta = 1$;

evaluating S_m at node ℓ should give $S_m = 0$ for $\xi = 1$ and $\eta = 0$; and evaluating S_m at node o should give $S_m = 0$ for $\xi = 0$ and $\eta = 1$. Applying these conditions to Eq. (7.9), we obtain

$$\begin{aligned}1 &= \overbrace{(1+1)(1+1)}^{F_1(\xi,\eta)}\overbrace{(c_1 + c_2(1) + c_3(1))}^{F_2(\xi,\eta)} = 4c_1 + 4c_2 + 4c_3\\ 0 &= (1+1)(1+0)(c_1 + c_2(1) + c_3(0)) = 2c_1 + 2c_2\\ 0 &= (1+0)(1+1)(c_1 + c_2(0) + c_3(1)) = 2c_1 + 2c_3\end{aligned}$$

which results in $c_1 = -\frac{1}{4}$, $c_2 = \frac{1}{4}$, and $c_3 = \frac{1}{4}$ with $S_m = (1+\xi)(1+\eta)(-\frac{1}{4} + \frac{1}{4}\xi + \frac{1}{4}\eta)$. The shape functions associated with the other corner nodes are determined in a similar fashion. The corner node shape functions are

$$\begin{aligned}S_i &= -\frac{1}{4}(1-\xi)(1-\eta)(1+\xi+\eta)\\ S_j &= \frac{1}{4}(1+\xi)(1-\eta)(-1+\xi-\eta)\\ S_m &= \frac{1}{4}(1+\xi)(1+\eta)(-1+\xi+\eta)\\ S_n &= -\frac{1}{4}(1-\xi)(1+\eta)(1+\xi-\eta)\end{aligned} \tag{7.10}$$

Let us turn our attention to the shape functions for the middle nodes. As an example, we will develop the shape function associated with node o. First, F_1 must be selected such that along the ij-side ($\eta = -1$), in-side ($\xi = -1$), and jm-side ($\xi = 1$), the function will produce a value of zero. We select

$$F_1(\xi, \eta) = (1-\xi)(1+\eta)(1+\xi)$$

Note that the product of the terms given by F_1 will produce linear and nonlinear terms, as required by Eq. (7.8). Therefore, the second function F_2 must be a constant; otherwise, the product of F_1 and F_2 will produce third-order polynomial terms, which we certainly do not want! So,

$$F_2(\xi, \eta) = c_1$$

Applying the nodal condition

$$S_o = 1 \text{ for } \xi = 0 \text{ and } \eta = 1$$

leads to

$$1 = \overbrace{(1-0)(1+1)(1+0)}^{F_1(\xi,\eta)}\overbrace{c_1}^{F_2(\xi,\eta)} = 2c_1$$

resulting in $c_1 = \frac{1}{2}$ with $S_o = \frac{1}{2}(1 - \xi)(1 + \eta)(1 + \xi) = S_o = \frac{1}{2}(1 + \eta)(1 - \xi^2)$. Using a similar procedure, we can obtain the shape functions for the midpoint nodes k, ℓ, and p. Thus, the midpoint shape functions are

$$S_k = \frac{1}{2}(1 - \eta)(1 - \xi^2) \tag{7.11}$$

$$S_\ell = \frac{1}{2}(1 + \xi)(1 - \eta^2)$$

$$S_o = \frac{1}{2}(1 + \eta)(1 - \xi^2)$$

$$S_p = \frac{1}{2}(1 - \xi)(1 - \eta^2)$$

EXAMPLE 7.1

We have used two-dimensional rectangular elements to model the stress distribution in a thin plate. The nodal stresses for an element belonging to the plate are shown in Figure 7.5. What is the value of stress at the center of this element?

The stress distribution for the element is

$$\sigma^{(e)} = [S_i \quad S_j \quad S_m \quad S_n]\begin{Bmatrix} \sigma_i \\ \sigma_j \\ \sigma_m \\ \sigma_n \end{Bmatrix}$$

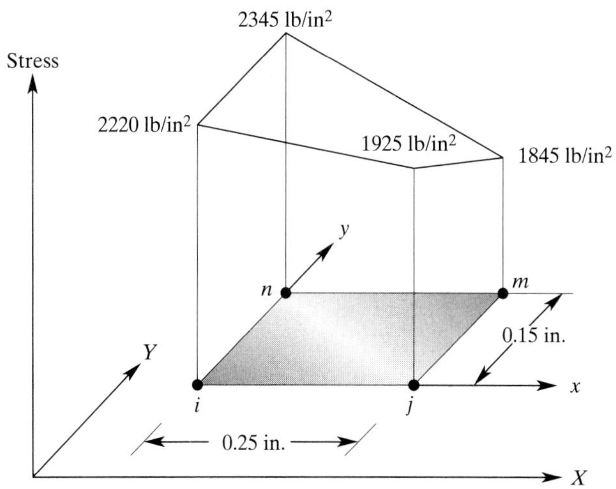

FIGURE 7.5 Nodal stresses for Example 7.1.

where σ_i, σ_j, σ_m, and σ_n are stresses at nodes i, j, m, and n respectively, and the shape functions are given by Eq. (7.5):

$$S_i = \left(1 - \frac{x}{\ell}\right)\left(1 - \frac{y}{w}\right) = \left(1 - \frac{x}{0.25}\right)\left(1 - \frac{y}{0.15}\right)$$

$$S_j = \frac{x}{\ell}\left(1 - \frac{y}{w}\right) = \frac{x}{0.25}\left(1 - \frac{y}{0.15}\right)$$

$$S_m = \frac{xy}{\ell w} = \frac{xy}{(0.25)(0.15)}$$

$$S_n = \frac{y}{w}\left(1 - \frac{x}{\ell}\right) = \frac{y}{0.15}\left(1 - \frac{x}{0.25}\right)$$

For the given element, the stress distribution in terms of the local coordinates x and y is given by

$$\sigma^{(e)} = \overbrace{\left(1 - \frac{x}{0.25}\right)\left(1 - \frac{y}{0.15}\right)}^{S_i}\overbrace{(2220)}^{\sigma_i} + \overbrace{\frac{x}{0.25}\left(1 - \frac{y}{0.15}\right)}^{S_j}\overbrace{(1925)}^{\sigma_j} + \overbrace{\frac{xy}{(0.25)(0.15)}}^{S_m}\overbrace{(1845)}^{\sigma_m}$$
$$+ \overbrace{\frac{y}{0.15}\left(1 - \frac{x}{0.25}\right)}^{S_n}\overbrace{(2345)}^{\sigma_n}$$

We can compute the stress at any point within this element from the aforementioned equation. Here, we are interested in the value of the stress at the midpoint. Substituting $x = 0.125$ and $y = 0.075$ into the equation, we have

$$\sigma(0.125, 0.075) = 555 + 481 + 461 + 586 = 2083 \quad \text{lb/in}^2$$

Note that we could have solved this problem using natural coordinates. This approach may be easier because the point of interest is located at the center of the element $\xi = 0$ and $\eta = 0$. The quadrilateral natural shape functions are given by Eq. (7.7):

$$S_i = \frac{1}{4}(1 - \xi)(1 - \eta) = \frac{1}{4}(1 - 0)(1 - 0) = \frac{1}{4}$$

$$S_j = \frac{1}{4}(1 + \xi)(1 - \eta) = \frac{1}{4}(1 + 0)(1 - 0) = \frac{1}{4}$$

$$S_m = \frac{1}{4}(1 + \xi)(1 + \eta) = \frac{1}{4}(1 + 0)(1 + 0) = \frac{1}{4}$$

$$S_n = \frac{1}{4}(1 - \xi)(1 + \eta) = \frac{1}{4}(1 - 0)(1 + 0) = \frac{1}{4}$$

$$\sigma(0.125, 0.075) = \frac{1}{4}(2220) + \frac{1}{4}(1925) + \frac{1}{4}(1845) + \frac{1}{4}(2345) = 2083 \quad \text{lb/in}^2$$

Thus, the stress at the midpoint of the rectangular element is the average of the nodal stresses.

EXAMPLE 7.2

Confirm the expression given for the quadratic quadrilateral shape function S_n.

Referring to the procedure discussed in Section 7.2, we can represent S_n by

$$S_n = F_1(\xi, \eta)F_2(\xi, \eta)$$

For the shape function S_n, F_1 should be selected such that it will have a value of zero along the *ij*-side ($\eta = -1$) and the *jm*-side ($\xi = 1$) So, we choose

$$F_1(\xi, \eta) = (1 - \xi)(1 + \eta)$$

Furthermore, F_2 is given by

$$F_2(\xi, \eta) = c_1 + c_2\xi + c_3\eta$$

and the coefficients c_1, c_2, and c_3 are determined by applying the following conditions:

$$S_n = 1 \quad \text{for} \quad \xi = -1 \quad \text{and} \quad \eta = 1$$

$$S_n = 0 \quad \text{for} \quad \xi = 0 \quad \text{and} \quad \eta = 1$$

$$S_n = 0 \quad \text{for} \quad \xi = -1 \quad \text{and} \quad \eta = 0$$

Recall from our discussion in the previous section, the coefficients in F_2 are selected such that when F_2 is multiplied by F_1, they will produce a value of unity at node n, and a value of zero at adjacent neighboring nodes o and p. After applying these conditions we get

$$1 = 4c_1 - 4c_2 + 4c_3$$

$$0 = 2c_1 + 2c_3$$

$$0 = 2c_1 - 2c_2$$

resulting in $c_1 = -\frac{1}{4}$, $c_2 = -\frac{1}{4}$, and $c_3 = \frac{1}{4}$, which is identical to the expression previously given for S_n. That is,

$$S_n = -\frac{1}{4}(1 - \xi)(1 + \eta)(1 + \xi - \eta)$$

7.3 LINEAR TRIANGULAR ELEMENTS

A major disadvantage associated with using bilinear rectangular elements is that they do not conform to a curved boundary very well. In contrast, triangular elements, shown describing a two-dimensional temperature distribution in Figure 7.6, are better suited to approximate curved boundaries. A triangular element is defined by three nodes, as shown in Figure 7.7. Therefore, we can represent the variation of a dependent variable, such as temperature, over the triangular region by

$$T^{(e)} = a_1 + a_2X + a_3Y \tag{7.12}$$

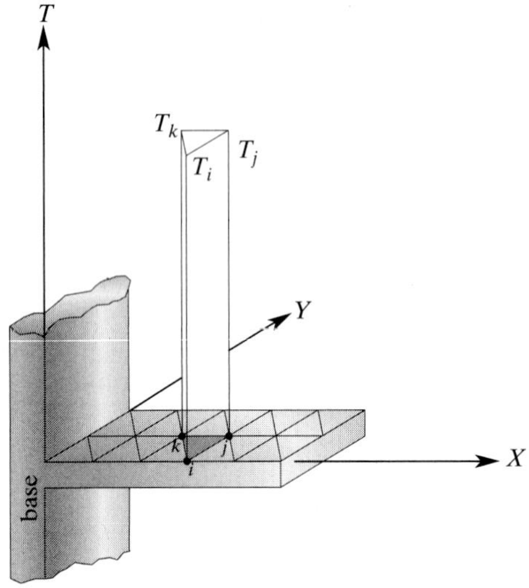

FIGURE 7.6 Using triangular elements to describe a two-dimensional temperature distribution.

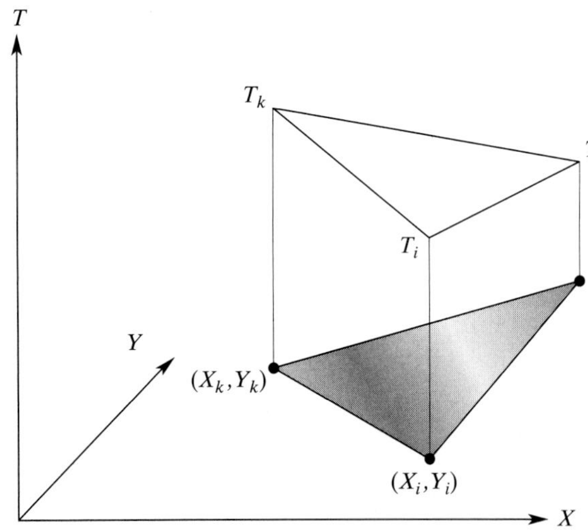

FIGURE 7.7 A triangular element.

Considering the nodal temperatures as shown in Figure 7.7, we must satisfy the following conditions:

$$
\begin{aligned}
T &= T_i \quad \text{at} \quad X = X_i \quad \text{and} \quad Y = Y_i \\
T &= T_j \quad \text{at} \quad X = X_j \quad \text{and} \quad Y = Y_j \\
T &= T_k \quad \text{at} \quad X = X_k \quad \text{and} \quad Y = Y_k
\end{aligned}
\tag{7.13}
$$

Substituting nodal values into Eq. (7.12), we have

$$T_i = a_1 + a_2X_i + a_3Y_i$$
$$T_j = a_1 + a_2X_j + a_3Y_j \tag{7.14}$$
$$T_k = a_1 + a_2X_k + a_3Y_k$$

Solving for a_1, a_2, and a_3, we obtain

$$a_1 = \frac{1}{2A}[(X_jY_k - X_kY_j)T_i + (X_kY_i - X_iY_k)T_j + (X_iY_j - X_jY_i)T_k] \tag{7.15}$$
$$a_2 = \frac{1}{2A}[(Y_j - Y_k)T_i + (Y_k - Y_i)T_j + (Y_i - Y_j)T_k]$$
$$a_3 = \frac{1}{2A}[(X_k - X_j)T_i + (X_i - X_k)T_j + (X_j - X_i)T_k]$$

where A is the area of the triangular element and is computed from the equation

$$2A = X_i(Y_j - Y_k) + X_j(Y_k - Y_i) + X_k(Y_i - Y_j) \tag{7.16}$$

See Example 7.3 for how Eq. (7.16) is derived. Substituting for a_1, a_2, and a_3 into Eq. (7.12) and grouping T_i, T_j, and T_k terms yields

$$T^{(e)} = [S_i \quad S_j \quad S_k]\begin{Bmatrix} T_i \\ T_j \\ T_k \end{Bmatrix} \tag{7.17}$$

where the shape functions S_i, S_j, and S_k are

$$S_i = \frac{1}{2A}(\alpha_i + \beta_iX + \delta_iY) \tag{7.18}$$
$$S_j = \frac{1}{2A}(\alpha_j + \beta_jX + \delta_jY)$$
$$S_k = \frac{1}{2A}(\alpha_k + \beta_kX + \delta_kY)$$

and

$$\alpha_i = X_jY_k - X_kY_j \quad \beta_i = Y_j - Y_k \quad \delta_i = X_k - X_j$$
$$\alpha_j = X_kY_i - X_iY_k \quad \beta_j = Y_k - Y_i \quad \delta_j = X_i - X_k$$
$$\alpha_k = X_iY_j - X_jY_i \quad \beta_k = Y_i - Y_j \quad \delta_k = X_j - X_i$$

Again, keep in mind that triangular shape functions have some basic properties, like other shape functions defined previously. For example, S_i has a value of unity when evaluated at the coordinates of node i and has a value of zero at all other nodes. Or, as another example, the sum of the shape functions has a value of unity. That property is demonstrated by the equation

$$S_i + S_j + S_k = 1 \tag{7.19}$$

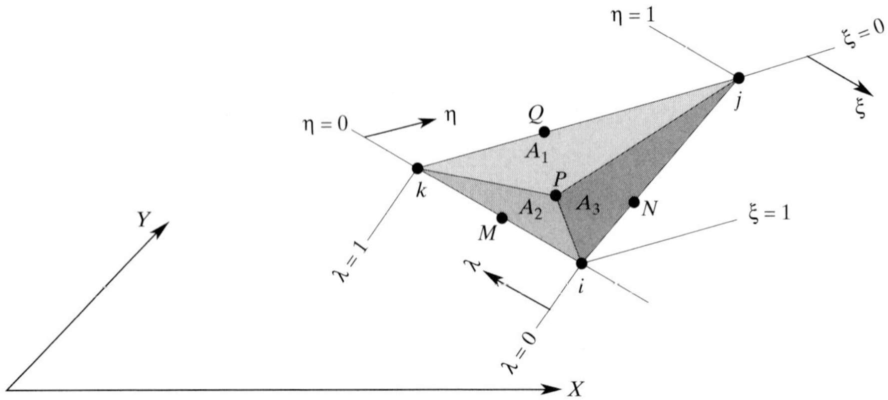

FIGURE 7.8 Natural (area) coordinates for a triangular element.

Natural (Area) Coordinates for Triangular Elements

Consider point P with coordinates (X, Y) inside the triangular region. Connecting this point to nodes, i, j, and k results in dividing the area of the triangle into three smaller areas A_1, A_2, and A_3, as shown in Figure 7.8.

Let us now perform an experiment. We move point P from its inside position to coincide with point Q along the kj-edge of the element. In the process, the value of area A_1 becomes zero. Moving point P to coincide with node i stretches A_1 to fill in the entire area A of the element. Based on the results of our experiment, we can define a natural, or area, coordinate ξ as the ratio of A_1 to the area A of the element so that its values vary from 0 to 1. Similarly, moving point P from its inside position to coincide with point M, along the ki-edge, results in $A_2 = 0$. Moving point P to coincide with node j stretches A_2 such that it fills the entire area of the element; that is, $A_2 = A$. We can define another area coordinate η as the ratio of A_2 to A, and its magnitude varies from 0 to 1. Formally, for a triangular element, the natural (area) coordinates ξ, η, and λ are defined by

$$\xi = \frac{A_1}{A} \qquad (7.20)$$

$$\eta = \frac{A_2}{A}$$

$$\lambda = \frac{A_3}{A}$$

It is important to realize that only two of the natural coordinates are linearly independent, because

$$\frac{A_1}{A} + \frac{A_2}{A} + \frac{A_3}{A} = \frac{A}{A} = 1 = \xi + \eta + \lambda$$

For example, the λ-coordinate can be defined in terms of ξ and η by

$$\lambda = 1 - \xi - \eta \tag{7.21}$$

We can show that the triangular natural (area) coordinates are exactly identical to the shape functions S_i, S_j, and S_k. That is,

$$\begin{aligned} \xi &= S_i \\ \eta &= S_j \\ \lambda &= S_k \end{aligned} \tag{7.22}$$

As an example, consider ξ, which is the ratio of A_1 to A:

$$\xi = \frac{A_1}{A} = \frac{\frac{1}{2}[(X_jY_k - X_kY_j) + X(Y_j - Y_k) + Y(X_k - X_j)]}{\frac{1}{2}[X_i(Y_j - Y_k) + X_j(Y_k - Y_i) + X_k(Y_i - Y_j)]} \tag{7.23}$$

Comparison of Eq. (7.23) to Eq. (7.18)* shows that ξ and S_i are identical. Equation (7.23) was derived by describing triangular areas A_1 and A in terms of the coordinates of their vertexes and using Eq. (7.16). Note that the coordinates of point P is designated by X and Y, because point P could lie anywhere within the area A.

EXAMPLE 7.3

Verify that the area of a triangular element can be computed from Equation (7.16).

As shown in the accompanying diagram, the area of triangle ABD is equal to one half of the area of the parallelogram ABCD—with AB and AD as its sides.

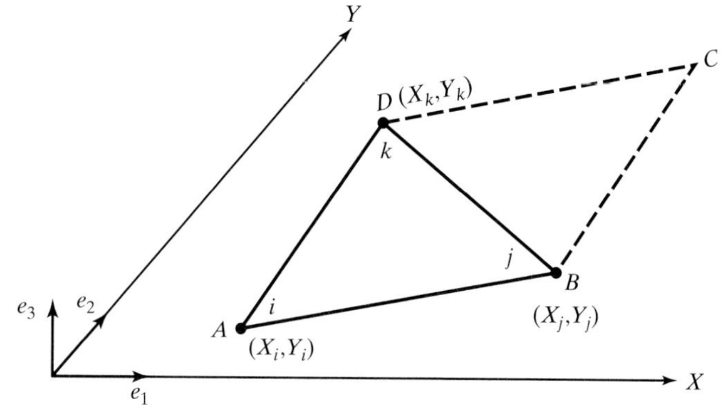

*Substitute for A, α_i, β_i, δ_i in terms of nodal coordinates.

The area of the parallelogram ABCD is equal to the magnitude of $\vec{AB} \times \vec{AD}$; that is,

$$2A = |\vec{AB} \times \vec{AD}|$$

where

$$\vec{AB} = (X_j - X_i)\vec{e}_1 + (Y_j - Y_i)\vec{e}_2$$

$$\vec{AD} = (X_k - X_i)\vec{e}_1 + (Y_k - Y_i)\vec{e}_2$$

and $\vec{e}_1$, $\vec{e}_2$, and $\vec{e}_3$ arc unit vcctors in thc shown dirctions. Carrying out the crossproduct operation in terms of the components of vectors $\vec{AB}$ and $\vec{AD}$,

$$2A = |\vec{AB} \times \vec{AD}| = |[(X_j - X_i)\vec{e}_1 + (Y_j - Y_i)\vec{e}_2] \times [(X_k - X_i)\vec{e}_1 + (Y_k - Y_i)\vec{e}_2]|$$

Noting that $\vec{e}_1 \times \vec{e}_1 = 0$, $\vec{e}_1 \times \vec{e}_2 = \vec{e}_3$, and $\vec{e}_2 \times \vec{e}_1 = -\vec{e}_3$,

$$2A = |(X_j - X_i)(Y_k - Y_i)\vec{e}_3 - (Y_j - Y_i)(X_k - X_i)\vec{e}_3|$$

and simplifying and regrouping terms, we show that the relationship given by Eq. (7.16) is true. That is,

$$2A = X_i(Y_j - Y_k) + X_j(Y_k - Y_i) + X_k(Y_i - Y_j)$$

7.4 QUADRATIC TRIANGULAR ELEMENTS

The spatial variation of a dependent variable, such as temperature, over a region may be approximated more accurately by a quadratic function, such as

$$T^{(e)} = a_1 + a_2X + a_3Y + a_4X^2 + a_5XY + a_6Y^2 \quad (7.24)$$

By now, you should understand how to develop shape functions. Therefore, the shape functions for a quadratic triangular element, shown in Figure 7.9, are given below without proof. The shape functions in terms of natural coordinates are

$$S_i = \xi(2\xi - 1) \quad (7.25)$$

$$S_j = \eta(2\eta - 1)$$

$$S_k = \lambda(2\lambda - 1) = 1 - 3(\xi + \eta) + 2(\xi + \eta)^2$$

$$S_\ell = 4\xi\eta$$

$$S_m = 4\eta\lambda = 4\eta(1 - \xi - \eta)$$

$$S_n = 4\xi\lambda = 4\xi(1 - \xi - \eta)$$

EXAMPLE 7.4

We have used two-dimensional triangular elements to model the temperature distribution in a fin. The nodal temperatures and their corresponding positions for an element are shown in Figure 7.10. (a) What is the value of temperature at $X = 2.15$ cm and

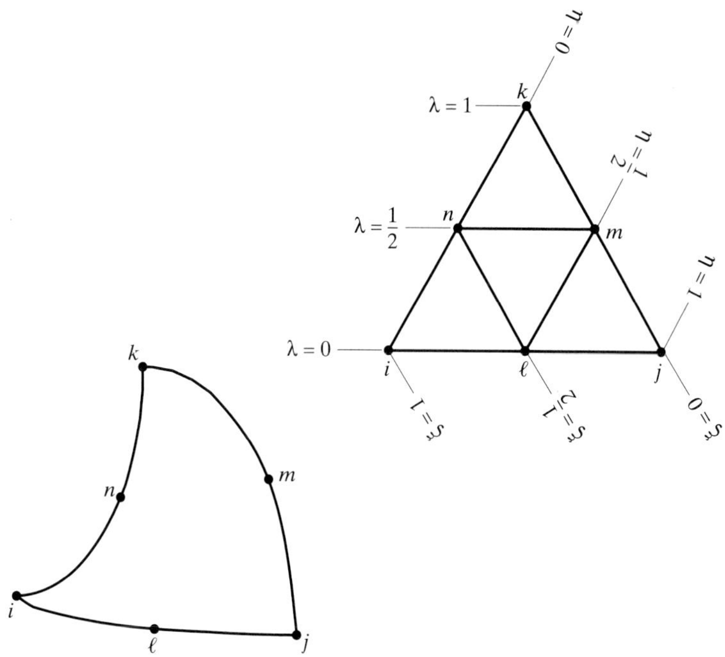

FIGURE 7.9 A quadratic triangular element.

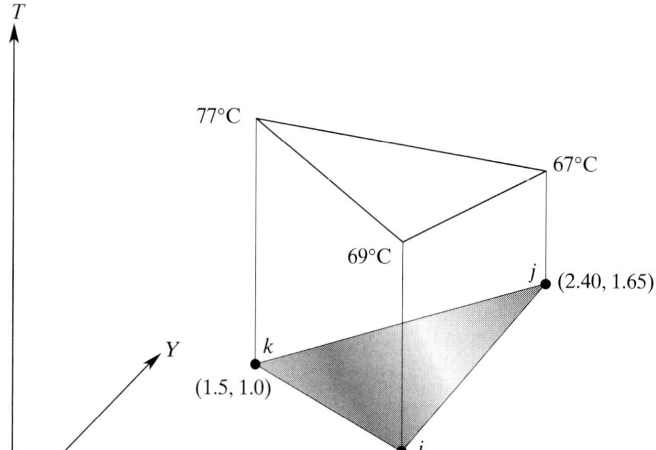

FIGURE 7.10 Nodal temperatures and coordinates for the element in Example 7.4.

$Y = 1.1$ cm? (b) Determine the components of temperature gradients for this element. (c) Determine the location of 70°C and 75°C isotherms.

a. The temperature distribution inside the element is

$$T^{(e)} = [S_i \quad S_j \quad S_k]\begin{Bmatrix} T_i \\ T_j \\ T_k \end{Bmatrix}$$

where the shape functions S_i, S_j, and S_k are

$$S_i = \frac{1}{2A}(\alpha_i + \beta_i X + \delta_i Y)$$

$$S_j = \frac{1}{2A}(\alpha_j + \beta_j X + \delta_j Y)$$

$$S_k = \frac{1}{2A}(\alpha_k + \beta_k X + \delta_k Y)$$

and

$$\alpha_i = X_j Y_k - X_k Y_j = (2.4)(1.0) - (1.5)(1.65) = -0.075$$

$$\alpha_j = X_k Y_i - X_i Y_k = (1.5)(0.75) - (2.25)(1.0) = -1.125$$

$$\alpha_k = X_i Y_j - X_j Y_i = (2.25)(1.65) - (2.40)(0.75) = 1.9125$$

$$\beta_i = Y_j - Y_k = 1.65 - 1.0 = 0.65$$

$$\beta_j = Y_k - Y_i = 1.0 - 0.75 = 0.25$$

$$\beta_k = Y_i - Y_j = 0.75 - 1.65 = -0.9$$

$$\delta_i = X_k - X_j = 1.50 - 2.40 = -0.9$$

$$\delta_j = X_i - X_k = 2.25 - 1.5 = 0.75$$

$$\delta_k = X_j - X_i = 2.40 - 2.25 = 0.15$$

and

$$2A = X_i(Y_j - Y_k) + X_j(Y_k - Y_i) + X_k(Y_i - Y_j)$$

$$2A = 2.25(1.65 - 1.0) + 2.40(1.0 - 0.75) + 1.5(0.75 - 1.65) = 0.7125$$

$$S_i = \frac{1}{2A}(\alpha_i + \beta_i X + \delta_i Y) = \frac{1}{0.7125}(-0.075 + 0.65X - 0.9Y)$$

$$S_j = \frac{1}{2A}(\alpha_j + \beta_j X + \delta_j Y) = \frac{1}{0.7125}(-1.125 + 0.25X + 0.75Y)$$

$$S_k = \frac{1}{2A}(\alpha_k + \beta_k X + \delta_k Y) = \frac{1}{0.7125}(1.9125 - 0.9X + 0.15Y)$$

The temperature distribution for the element is

$$T = \frac{69}{0.7125}(-0.075 + 0.65X - 0.9Y) + \frac{67}{0.7125}(-1.125 + 0.25X + 0.75Y) + \frac{77}{0.7125}(1.9125 - 0.9X + 0.15Y)$$

After simplifying, we have

$$T = 93.632 - 10.808X - 0.421Y$$

Substituting for coordinates of the point $X = 2.15$ and $Y = 1.1$ leads to $T = 69.93°C$.

b. In general, the gradient components of a dependent variable $\Psi^{(e)}$ are computed from

$$\frac{\partial \Psi^{(e)}}{\partial X} = \frac{\partial}{\partial X}[S_i\Psi_i + S_j\Psi_j + S_k\Psi_k]$$

$$\frac{\partial \Psi^{(e)}}{\partial Y} = \frac{\partial}{\partial Y}[S_i\Psi_i + S_j\Psi_j + S_k\Psi_k]$$

$$\begin{Bmatrix} \dfrac{\partial \Psi^{(e)}}{\partial X} \\ \dfrac{\partial \Psi^{(e)}}{\partial Y} \end{Bmatrix} = \frac{1}{2A}\begin{bmatrix} \beta_i & \beta_j & \beta_k \\ \delta_i & \delta_j & \delta_k \end{bmatrix}\begin{Bmatrix} \Psi_i \\ \Psi_j \\ \Psi_k \end{Bmatrix} \tag{7.26}$$

It should be clear from examining Eq. (7.26) that the gradients have constant values. This property is always true for linear triangular elements. The temperature gradients are computed from

$$\begin{Bmatrix} \dfrac{\partial T^{(e)}}{\partial X} \\ \dfrac{\partial T^{(e)}}{\partial Y} \end{Bmatrix} = \frac{1}{2A}\begin{bmatrix} \beta_i & \beta_j & \beta_k \\ \delta_i & \delta_j & \delta_k \end{bmatrix}\begin{Bmatrix} T_i \\ T_j \\ T_k \end{Bmatrix} = \frac{1}{0.7125}\begin{bmatrix} 0.65 & 0.25 & -0.9 \\ -0.9 & 0.75 & 0.15 \end{bmatrix}\begin{Bmatrix} 69 \\ 67 \\ 77 \end{Bmatrix} = \begin{Bmatrix} -10.808 \\ -0.421 \end{Bmatrix}$$

Note that differentiation of the simplified temperature equation $(T = 93.632 - 10.808X - 0.421Y)$ directly would have resulted in exactly the same values.

c. The location of 70°C and 75°C isotherms can be determined from the fact that over a triangular element, temperature varies linearly in both X- and Y-directions. Thus, we can use linear interpolation to calculate coordinates of isotherms. First, let us focus on the 70°C constant temperature line. This isotherm will intersect the 77°C–69°C-edge according to the relations

$$\frac{77 - 70}{77 - 69} = \frac{1.5 - X}{1.5 - 2.25} = \frac{1.0 - Y}{1.0 - 0.75}$$

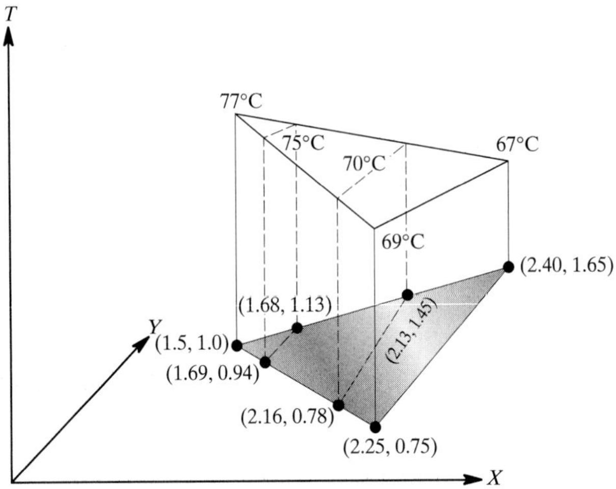

FIGURE 7.11 The isotherms of the element in Example 7.4.

which results in the coordinates $X = 2.16$ cm and $Y = 0.78$ cm. The 70°C isotherm also intersects the 77°C–67°C-edge:

$$\frac{77 - 70}{77 - 67} = \frac{1.5 - X}{1.5 - 2.4} = \frac{1.0 - Y}{1.0 - 1.65}$$

These relations result in the coordinates $X = 2.13$ cm and $Y = 1.45$ cm. Similarly, the location of the 75°C isotherm is determined using the 77°C–69°C-edge:

$$\frac{77 - 75}{77 - 69} = \frac{1.5 - X}{1.5 - 2.25} = \frac{1.0 - Y}{1.0 - 0.75}$$

which results in the coordinates $X = 1.69$ and $Y = 0.94$. Finally, along the 77°C–67°C-edge, we have:

$$\frac{77 - 75}{77 - 67} = \frac{1.5 - X}{1.5 - 2.4} = \frac{1.0 - Y}{1.0 - 1.65}$$

These equations result in the coordinates $X = 1.68$ and $Y = 1.13$. The isotherms and their corresponding locations are shown in Figure 7.11.

7.5 AXISYMMETRIC ELEMENTS

There are special classes of three-dimensional problems whose geometry and loading are symmetrical about an axis, such as the z-axis, as shown in Figure 7.12. These three-dimensional problems may be analyzed using two-dimensional axisymmetric elements. In this section, we discuss the triangular and rectangular axisymmetric elements, and in Chapters 9 and 10 will show finite element formulations employing these types of elements.

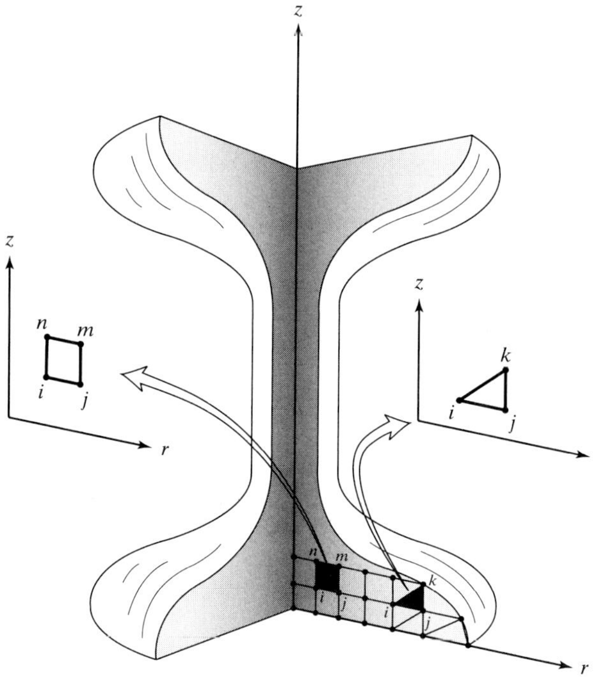

FIGURE 7.12 Examples of axisymmetric elements.

Axisymmetric Triangular Elements

In Section 7.3, we developed the shape functions for a linear triangular element. Recall that the variation of any unknown variable Ψ over a triangular region in terms of its nodal values Ψ_i, Ψ_j, Ψ_k and the shape functions may be represented by

$$\Psi^{(e)} = [S_i \quad S_j \quad S_k]\begin{Bmatrix} \Psi_i \\ \Psi_j \\ \Psi_k \end{Bmatrix}$$

where

$$S_i = \frac{1}{2A}(\alpha_i + \beta_i X + \delta_i Y)$$

$$S_j = \frac{1}{2A}(\alpha_j + \beta_j X + \delta_j Y)$$

$$S_k = \frac{1}{2A}(\alpha_k + \beta_k X + \delta_k Y)$$

We now express the above shape functions in terms of r- and z-coordinates—coordinates that are typically used for axisymmetric triangular elements. A typical axisymmetric triangular element and its coordinates is shown in Figure 7.13.

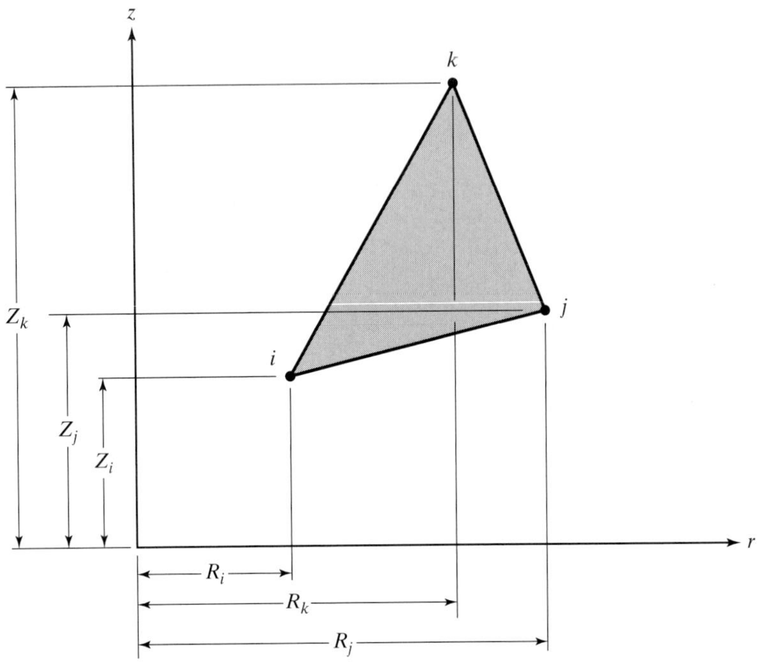

FIGURE 7.13 An axisymmetric triangular element.

Substituting for the spatial coordinates X and Y in terms of r- and z-coordinates and the nodal coordinates X_i, Y_i, X_j, Y_j and X_k, Y_k in terms of R_i, Z_i, R_j, Z_j and R_k, Z_k results in the following set of shape functions

$$
\begin{aligned}
S_i &= \frac{1}{2A}(\alpha_i + \beta_i r + \delta_i z) \\
S_j &= \frac{1}{2A}(\alpha_j + \beta_j r + \delta_j z) \\
S_k &= \frac{1}{2A}(\alpha_k + \beta_k r + \delta_k z)
\end{aligned}
\tag{7.27}
$$

where

$$
\begin{aligned}
\alpha_i &= R_jZ_k - R_kZ_j & \beta_i &= Z_j - Z_k & \delta_i &= R_k - R_j \\
\alpha_j &= R_kZ_i - R_iZ_k & \beta_j &= Z_k - Z_i & \delta_j &= R_i - R_k \\
\alpha_k &= R_iZ_j - R_jZ_i & \beta_k &= Z_i - Z_j & \delta_k &= R_j - R_i
\end{aligned}
$$

Axisymmetric Rectangular Elements

In Section 7.1 we discussed the rectangular element and derived the following shape functions.

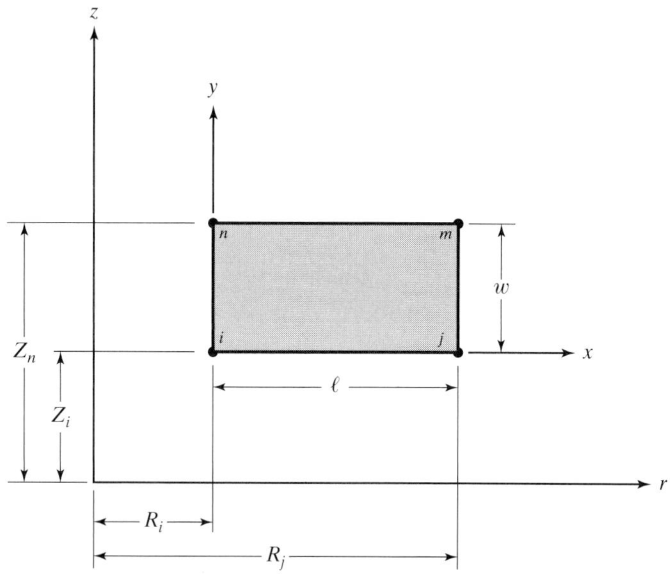

FIGURE 7.14 An axisymmetric rectangular element.

$$S_i = \left(1 - \frac{x}{\ell}\right)\left(1 - \frac{y}{w}\right) \qquad S_j = \frac{x}{\ell}\left(1 - \frac{y}{w}\right)$$

$$S_m = \frac{xy}{\ell w} \qquad S_n = \frac{y}{w}\left(1 - \frac{x}{\ell}\right)$$

Let's now consider the axisymmetric rectangular element shown in Figure 7.14. The relationship between the local coordinates x and y and the axisymmetric coordinates r and z is shown in Figure 7.14.

Substituting for x in terms of r and for y in terms of z, and making use of the following relationships among the nodal coordinates, we get

$$r = R_i + x \qquad \text{or} \qquad x = r - R_i$$

and

$$z = Z_i + y \qquad \text{or} \qquad y = z - Z_i$$

$$S_i = \left(1 - \frac{x}{\ell}\right)\left(1 - \frac{y}{w}\right) = \left(1 - \frac{\overbrace{r - R_i}^{x}}{\ell}\right)\left(1 - \frac{\overbrace{z - Z_i}^{y}}{w}\right) = \left(\frac{\ell - (r - R_i)}{\ell}\right)\left(\frac{w - (z - Z_i)}{w}\right) \tag{7.28}$$

Realizing that $\ell = R_j - R_i$ and $w = Z_n - Z_i$, we can simplify Eq. (7.28) to obtain the shape function S_i in terms of r and z coordinates in the following manner:

$$S_i = \left(\frac{\overbrace{R_j - R_i}^{\ell} - (r - R_i)}{\ell}\right)\left(\frac{\overbrace{Z_n - Z_i}^{w} - (z - Z_i)}{w}\right) = \left(\frac{R_j - r}{\ell}\right)\left(\frac{Z_n - z}{w}\right)$$

Similarly, we can obtain the other shape functions. Thus, the shape functions for an axisymmetric rectangular element are given by

$$S_i = \left(\frac{R_j - r}{\ell}\right)\left(\frac{Z_n - z}{w}\right) \tag{7.29}$$

$$S_j = \left(\frac{r - R_i}{\ell}\right)\left(\frac{Z_n - z}{w}\right)$$

$$S_m = \left(\frac{r - R_i}{\ell}\right)\left(\frac{z - Z_i}{w}\right)$$

$$S_n = \left(\frac{R_j - r}{\ell}\right)\left(\frac{z - Z_i}{w}\right)$$

We discuss the application of these elements in solving heat transfer and solid mechanics problems in Chapters 9 and 10.

EXAMPLE 7.5

We have used axisymmetric rectangular elements to model temperature distribution in a hollow cylinder. The values of nodal temperature for an element belonging to the cylinder are shown in Figure 7.15. What is the value of the temperature at $r = 1.2$ cm and $z = 1.4$ cm?

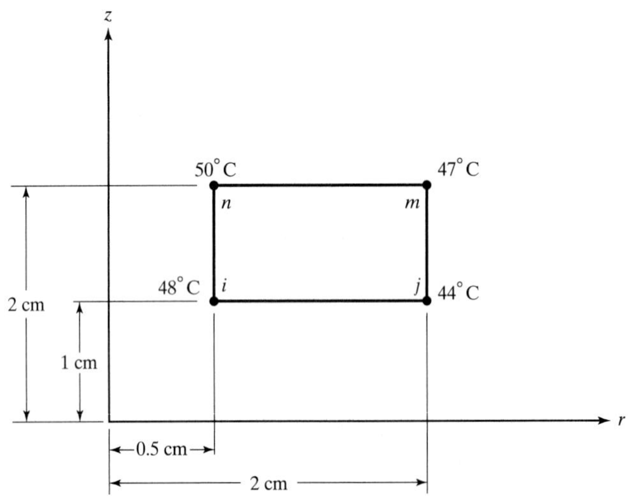

FIGURE 7.15 Nodal temperatures and coordinates for the element in Example 7.5.

The temperature distribution for the element is given by

$$T^{(e)} = [S_i \quad S_j \quad S_m \quad S_n]\begin{Bmatrix} T_i \\ T_j \\ T_m \\ T_n \end{Bmatrix}$$

where T_i, T_j, T_m, and T_n are the values of temperature at nodes i, j, m, and n respectively. Substituting for the coordinates of the point in the shape functions, we get

$$S_i = \left(\frac{R_j - r}{\ell}\right)\left(\frac{Z_n - z}{w}\right) = \left(\frac{2 - 1.2}{1.5}\right)\left(\frac{2 - 1.4}{1}\right) = 0.32$$

$$S_j = \left(\frac{r - R_i}{\ell}\right)\left(\frac{Z_n - z}{w}\right) = \left(\frac{1.2 - 0.5}{1.5}\right)\left(\frac{2 - 1.4}{1}\right) = 0.28$$

$$S_m = \left(\frac{r - R_i}{\ell}\right)\left(\frac{z - Z_i}{w}\right) = \left(\frac{1.2 - 0.5}{1.5}\right)\left(\frac{1.4 - 1}{1}\right) = 0.19$$

$$S_n = \left(\frac{R_j - r}{\ell}\right)\left(\frac{z - Z_i}{w}\right) = \left(\frac{2 - 1.2}{1.5}\right)\left(\frac{1.4 - 1}{1}\right) = 0.21$$

The temperature at the given point is

$$T = (0.32)(48) + (0.28)(44) + (0.19)(47) + (0.21)(50) = 47.11°\text{C}$$

7.6 ISOPARAMETRIC ELEMENTS

As we discussed in Chapter 5, Section 5.5, when we use a single set of parameters (a set of shape functions) to define the unknown variables u, v, T, and so on, and use the same parameters (the same shape functions) to express the geometry, we are using an *isoparametric* formulation. An element expressed in such a manner is called an isoparametric element. Let us turn our attention to the quadrilateral element shown in Figure 7.16. Let us also consider a solid mechanics problem, in which a body undergoes a deformation. Using a quadrilateral element, the displacement field within an element belonging to this solid body can be expressed in terms of its nodal values as

$$\begin{aligned} u^{(e)} &= S_iU_{ix} + S_jU_{jx} + S_mU_{mx} + S_nU_{nx} \\ v^{(e)} &= S_iU_{iy} + S_jU_{jy} + S_mU_{my} + S_nU_{ny} \end{aligned} \tag{7.30}$$

We can write the relations given by Eq. (7.30) in matrix form:

$$\begin{Bmatrix} u \\ v \end{Bmatrix} = \begin{bmatrix} S_i & 0 & S_j & 0 & S_m & 0 & S_n & 0 \\ 0 & S_i & 0 & S_j & 0 & S_m & 0 & S_n \end{bmatrix} \begin{Bmatrix} U_{ix} \\ U_{iy} \\ U_{jx} \\ U_{jy} \\ U_{mx} \\ U_{my} \\ U_{nx} \\ U_{ny} \end{Bmatrix} \tag{7.31}$$

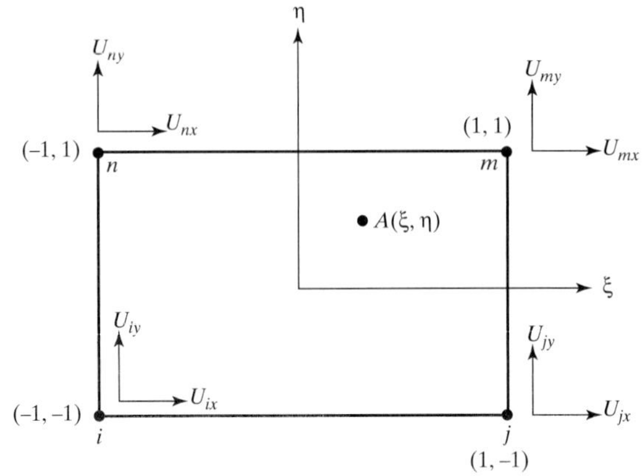

FIGURE 7.16 A quadrilateral element used in formulating plane-stress problems.

Note that using isoparametric formulation, we can use the same shape functions to describe the position of any point, such as A, within the element by the equations

$$x = S_i x_i + S_j x_j + S_m x_m + S_n x_n \qquad (7.32)$$
$$y = S_i y_i + S_j y_j + S_m y_m + S_n y_n$$

As you will see in Chapter 10, the displacement field is related to the components of strains ($\varepsilon_{xx} = \frac{\partial u}{\partial x}, \varepsilon_{yy} = \frac{\partial v}{\partial y}$, and $\gamma_{xy} = \frac{\partial u}{\partial y} + \frac{\partial v}{\partial x}$) and, subsequently, to the nodal displacements using shape functions. In deriving the elemental stiffness matrix from strain energy, we need to take the derivatives of the components of the displacement field with respect to the x- and y-coordinates, which in turn means taking the derivatives of the appropriate shape functions with respect to x and y. At this point, keep in mind that the shape functions are expressed in terms of ξ and η Eq. (7.7). Thus, in general, it is necessary to establish relationships that allow the derivatives of a function $f(x,y)$ to be taken with respect to x and y and to express them in terms of derivatives of the function $f(x,y)$ with respect to ξ and η. This point will become clear soon. Using the chain rule, we can write

$$\frac{\partial f(x,y)}{\partial \xi} = \frac{\partial f(x,y)}{\partial x}\frac{\partial x}{\partial \xi} + \frac{\partial f(x,y)}{\partial y}\frac{\partial y}{\partial \xi} \qquad (7.33)$$
$$\frac{\partial f(x,y)}{\partial \eta} = \frac{\partial f(x,y)}{\partial x}\frac{\partial x}{\partial \eta} + \frac{\partial f(x,y)}{\partial y}\frac{\partial y}{\partial \eta}$$

Expressing Eq. (7.33) in matrix form, we have

$$\begin{Bmatrix} \dfrac{\partial f(x,y)}{\partial \xi} \\ \dfrac{\partial f(x,y)}{\partial \eta} \end{Bmatrix} = \overbrace{\begin{bmatrix} \dfrac{\partial x}{\partial \xi} & \dfrac{\partial y}{\partial \xi} \\ \dfrac{\partial x}{\partial \eta} & \dfrac{\partial y}{\partial \eta} \end{bmatrix}}^{[\mathbf{J}]} \begin{Bmatrix} \dfrac{\partial f(x,y)}{\partial x} \\ \dfrac{\partial f(x,y)}{\partial y} \end{Bmatrix} \qquad (7.34)$$

where the **J** matrix is referred to as the Jacobian of the coordinate transformation. The relationships of Eq. (7.34) can be also presented as

$$\begin{Bmatrix} \dfrac{\partial f(x,y)}{\partial x} \\ \dfrac{\partial f(x,y)}{\partial y} \end{Bmatrix} = [\mathbf{J}]^{-1} \begin{Bmatrix} \dfrac{\partial f(x,y)}{\partial \xi} \\ \dfrac{\partial f(x,y)}{\partial \eta} \end{Bmatrix} \tag{7.35}$$

For a quadrilateral element, the **J** matrix can be evaluated using Eq. (7.32) and (7.7). This evalutation is left as an exercise for you; see problem 7.24. We will discuss the derivation of the element stiffness matrix using the isoparametric formulation in Chapter 10. For the sake of convenience, the results of Sections 7.1 to 7.6 are summarized in Table 7.1.

7.7 TWO-DIMENSIONAL INTEGRALS: GAUSS–LEGENDRE QUADRATURE

As we discussed in Chapter 5, most finite element programs perform numerical integration for elements by Gaussian quadratures, and as the limits of integration, they use an interval from -1 to 1. We now extend the Gauss–Legendre quadrature formulation to two-dimensional problems as follows:

$$I = \int_{-1}^{1}\int_{-1}^{1} f(\xi, \eta)d\xi d\eta \cong \int_{-1}^{1}\left[\sum_{i=1}^{n} w_i f(\xi_i, \eta)\right]d\eta \cong \sum_{i=1}^{n}\sum_{j=1}^{n} w_i w_j f(\xi_i, \eta_j) \tag{7.36}$$

The relationships of Eq. (7.36) should be self-evident. Recall that the weighting factors and the sampling points are given in Chapter 5, Table 5.2.

EXAMPLE 7.6

To demonstrate the steps involved in Gauss–Legendre quadrature computation, let us consider evaluating the integral

$$I = \int_0^2\int_0^2 (3y^2 + 2x)dx\,dy$$

As you know, the given integral can be evaluated analytically as

$$I = \int_0^2\int_0^2 (3y^2 + 2x)dx\,dy = \int_0^2\left[\int_0^2 (3y^2 + 2x)\,dx\right]dy$$

$$= \int_0^2 [(3y^2x + x^2)]_0^2 dy = \int_0^2 (6y^2 + 4)\,dy = 24$$

TABLE 7.1 Two-dimensional shape functions

Linear Rectangular 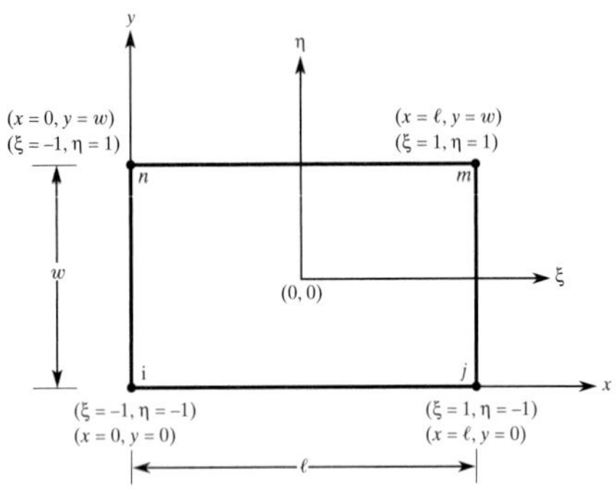	$S_i = \left(1 - \frac{x}{\ell}\right)\left(1 - \frac{y}{w}\right)$ $S_j = \frac{x}{\ell}\left(1 - \frac{y}{w}\right)$ $S_m = \frac{xy}{\ell w}$ $S_n = \frac{y}{w}\left(1 - \frac{x}{\ell}\right)$ $S_i = \frac{1}{4}(1 - \xi)(1 - \eta)$ $S_j = \frac{1}{4}(1 + \xi)(1 - \eta)$ $S_m = \frac{1}{4}(1 + \xi)(1 + \eta)$ $S_n = \frac{1}{4}(1 - \xi)(1 + \eta)$
Quadratic Quadrilateral 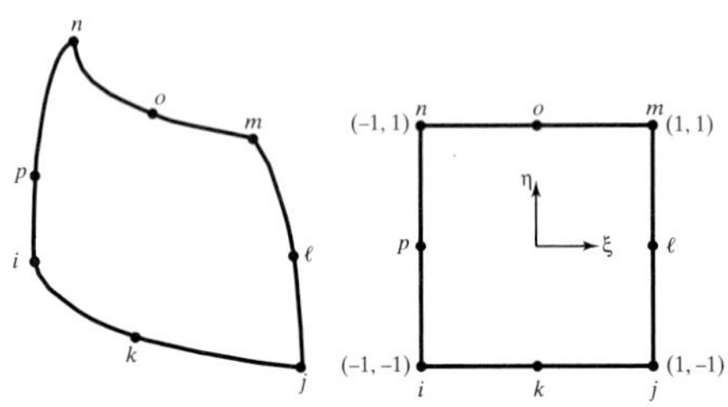	$S_i = -\frac{1}{4}(1 - \xi)(1 - \eta)(1 + \xi + \eta)$ $S_j = \frac{1}{4}(1 + \xi)(1 - \eta)(-1 + \xi - \eta)$ $S_m = \frac{1}{4}(1 + \xi)(1 + \eta)(-1 + \xi + \eta)$ $S_n = -\frac{1}{4}(1 - \xi)(1 + \eta)(1 + \xi - \eta)$ $S_k = \frac{1}{2}(1 - \eta)(1 - \xi^2)$ $S_\ell = \frac{1}{2}(1 + \xi)(1 - \eta^2)$ $S_o = \frac{1}{2}(1 + \eta)(1 - \xi^2)$ $S_p = \frac{1}{2}(1 - \xi)(1 - \eta^2)$

We now evaluate the integral using Gauss–Legendre quadrature. We begin by changing y- and x-variables into ξ and η, using the relationships of Eq. (5.45):

$$x = 1 + \xi \quad \text{and} \quad dx = d\xi$$
$$y = 1 + \eta \quad \text{and} \quad dy = d\eta$$

Thus, the integral I can be expressed by

$$I = \int_0^2 \int_0^2 (3y^2 + 2x)\,dx\,dy = \int_{-1}^{1} \int_{-1}^{1} [3(1 + \eta)^2 + 2(1 + \xi)]\,d\xi\,d\eta$$

TABLE 7.1 *Continued*

Linear Triangular 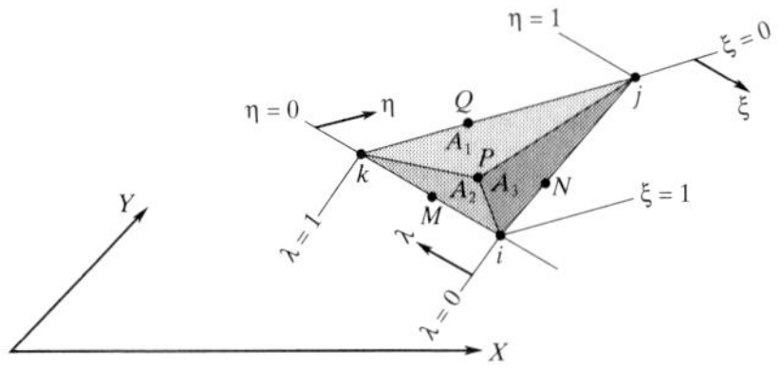	$S_i = \frac{1}{2A}(\alpha_i + \beta_i X + \delta_i Y)$ $S_j = \frac{1}{2A}(\alpha_j + \beta_j X + \delta_j Y)$ $S_k = \frac{1}{2A}(\alpha_k + \beta_k X + \delta_k Y)$ $\alpha_i = X_j Y_k - X_k Y_j \quad \beta_i = Y_j - Y_k \quad \delta_i = X_k - X_j$ $\alpha_j = X_k Y_i - X_i Y_k \quad \beta_j = Y_k - Y_i \quad \delta_j = X_i - X_k$ $\alpha_k = X_i Y_j - X_j Y_i \quad \beta_k = Y_i - Y_j \quad \delta_k = X_j - X_i$
Quadratic Triangular 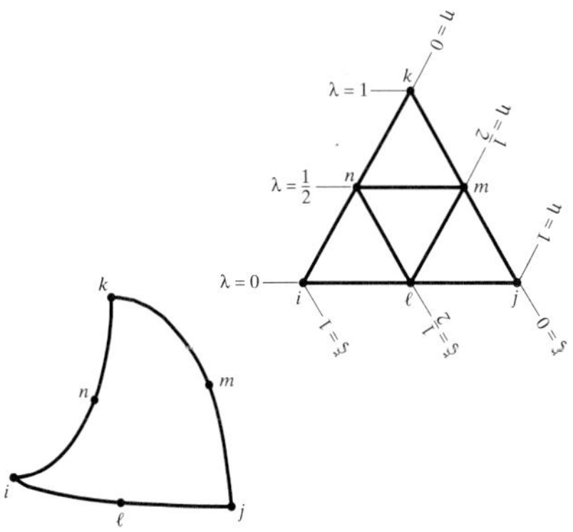	$S_i = \xi(2\xi - 1)$ $S_j = \eta(2\eta - 1)$ $S_k = \lambda(2\lambda - 1) = 1 - 3(\xi + \eta) + 2(\xi + \eta)^2$ $S_l = 4\xi\eta$ $S_m = 4\eta\lambda = 4\eta(1 - \xi - \eta)$ $S_n = 4\xi\lambda = 4\xi(1 - \xi - \eta)$

Using the two-point sampling formula, we have

$$I \cong \sum_{i=1}^{n}\sum_{j=1}^{n} w_i w_j f(\xi_i, \eta_j)$$

$$I \cong \sum_{i=1}^{2}\sum_{j=1}^{2} w_i w_j[3(1 + \eta_j)^2 + 2(1 + \xi_i)]$$

To evaluate the summation, we start with $i = 1$, while changing j from 1 to 2, and we repeat the process with $i = 2$, while changing j from 1 to 2:

$$I \cong [(1)(1)[3(1 + (-0.577350269))^2 + 2(1 + (-0.577350269))]$$

$$+ (1)(1)[3(1 + (0.577350269))^2 + 2(1 + (-0.577350269))]]$$

$$+ [(1)(1)[3(1 + (-0.577350269))^2 + 2(1 + (0.577350269))]$$

$$+ (1)(1)[3(1 + (0.577350269))^2 + 2(1 + (0.577350269))]] = 24.000000000$$

7.8 EXAMPLES OF TWO-DIMENSIONAL ELEMENTS IN ANSYS

ANSYS offers many two-dimensional elements that are based on linear and quadratic quadrilateral and triangular shape functions. We will discuss the formulation of two-dimensional thermal- and solid-structural problems in detail in Chapters 9 and 10. For now, consider some examples of two-dimensional structural-solid and thermal-solid elements.

Plane2 is a six-node triangular structural-solid element. The element has quadratic displacement behavior with two degrees of freedom at each node, translation in the nodal x- and y-directions. The element input data can include thickness if KEYOPTION 3 (plane stress with thickness input) is selected. Surface pressure loads may be applied to element faces. Output data include nodal displacements and element data, such as directional stresses and principal stresses.

Plane35 is a six-node triangular thermal solid element. The element has one degree of freedom at each node, the temperature. Convection and heat fluxes may be input as surface loads at the element faces. The output data for this element include nodal temperatures and element data, such as thermal gradients and thermal fluxes.

Plane42 is a four-node quadrilateral element used in modeling solid problems. The element is defined by four nodes, with two degrees of freedom at each node, the translation in the x- and y-directions. The element input data can include thickness if KEYOPTION 3 (plane stress with thickness input) is selected. Surface pressure loads may be applied to element faces. Output data include nodal displacements and element data, such as directional stresses and principal stresses.

Plane55 is a four-node quadrilateral element used in modeling two-dimensional conduction heat transfer problems. The element has a single degree of freedom, the temperature. Convection or heat fluxes may be input at the element faces. Output data include nodal temperatures and element data, such as thermal gradient and thermal flux components.

Plane77 is an eight-node quadrilateral element used in modeling two-dimensional heat conduction problems. It is basically a higher order version of the two-dimensional, four-node quadrilateral element PLANE55. This element is more capable of modeling problems with curved boundaries. At each node, the element has a single degree of freedom, the temperature. Output data include nodal temperatures and element data, such as thermal gradient and thermal flux components.

Plane82 is an eight-node quadrilateral element used in modeling two-dimensional structural solid problems. It is a higher order version of two-dimensional, four-node quadrilateral element PLANE42. This element offers more accuracy when modeling problems with curved boundaries. At each node, there are two degrees of freedom, the translation in the x- and y-directions. The element input data can include thickness if KEYOPTION 3 (plane stress with thickness input) is selected. Surface pressure loads may be applied to element faces. Output data include nodal displacements and element data, such as directional stresses and principal stresses.

Finally, it may be worth noting that although you generally achieve better results and greater accuracy with higher order elements, these elements require more computational time. This time requirement is because numerical integration of elemental matrices is more involved.

SUMMARY

At this point you should

1. have a good understanding of the linear two-dimensional rectangular and triangular shape functions and of elements, along with their properties and limitations.
2. have a good understanding of the quadratic two-dimensional triangular and quadrilateral elements, as well as shape functions, along with their properties and their advantages over linear elements.
3. know why it is important to use natural coordinate systems.
4. know what is meant by axisymmetric element.
5. know what is meant by isoparametric element and formulation.
6. know how to use Gauss–Legendre quadrature to evaluate two-dimensional integrals.
7. know examples of two-dimensional elements in ANSYS.

REFERENCES

ANSYS User's Manual: Elements, Vol. III, Swanson Analysis Systems, Inc.

Chandrupatla, T., and Belegundu, A., *Introduction to Finite Elements in Engineering*, Englewood Cliffs, NJ, Prentice Hall, 1991.

CRC Standard Mathematical Tables, 25th ed., Boca Raton, FL, CRC Press, 1979.

Segrlind, L., *Applied Finite Element Analysis*, 2nd ed., New York, John Wiley and Sons, 1984.

PROBLEMS

1. We have used two-dimensional rectangular elements to model temperature distribution in a thin plate. The values of nodal temperatures for an element belonging to such a plate are given in the accompanying figure. Using local shape functions, what is the temperature at the center of this element?

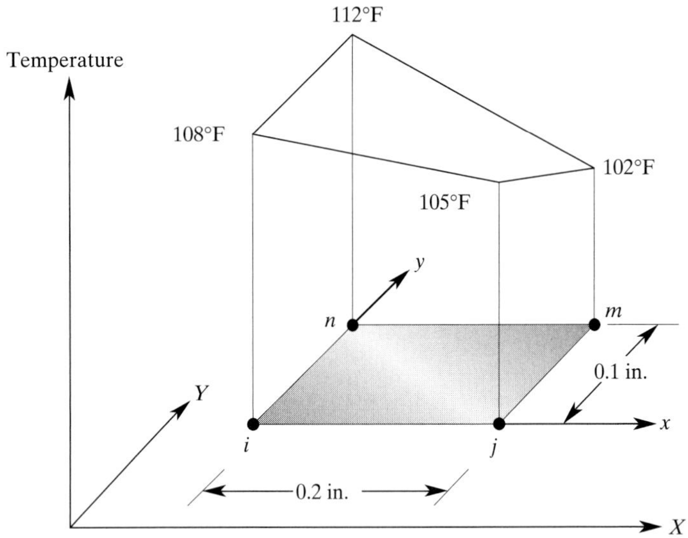

2. Determine the temperature at the center of the element in Problem 1 using natural shape functions.

3. For a rectangular element, derive the x- and y-components of the gradients of a dependent variable Ψ.

4. Determine the components of temperature gradients at the midpoint of the element in Problem 1. Knowing that the element has a thermal conductivity of $k = 92$ Btu/hr · ft · °F, compute the x- and y-components of the heat flux.

5. Compute the location of the 103°F and 107°F isotherms for the element in Problem 1. Also, plot these isotherms.

6. Two-dimensional triangular elements have been used to determine the stress distribution in a machine part. The nodal stresses and their corresponding positions for a triangular element are shown in the accompanying figure. What is the value of stress at $x = 2.15$ cm and $y = 1.1$ cm?

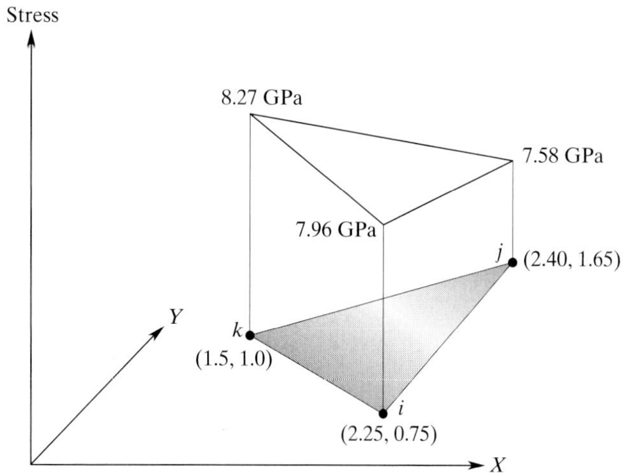

7. Plot the 8.0 GPa and 7.86 GPa stress contour lines for an element of the machine part in Problem 6.

8. For a quadratic quadrilateral element, confirm the expressions given for the shape functions S_i and S_j.

9. For a quadratic quadrilateral element, confirm the expressions given for the shape functions S_k and S_ℓ.

10. For triangular elements, the integral that includes products of area coordinates may be evaluated using the factorial relationship shown below:

$$\int_A \xi^a \eta^b \lambda^c \, dA = \frac{a!b!c!}{(a + b + c + 2)!} 2A$$

Using the above relationship, evaluate the integral $\int_A (S_i^2 + S_j S_k)\, dA$

11. Show that the area A of a triangular element can be computed from the determinant of

$$\begin{vmatrix} 1 & X_i & Y_i \\ 1 & X_j & Y_j \\ 1 & X_k & Y_k \end{vmatrix} = 2A$$

12. In the formulation of two-dimensional heat transfer problems, the need to evaluate the integral $\int_A [\mathbf{S}]^T hT \, dA$ arises; h is the heat transfer coefficient, and T represents the temperature. Using a linear triangular element, evaluate the aforementioned integral, provided that temperature variation is given by

$$T^{(e)} = [S_i \quad S_j \quad S_k] \begin{Bmatrix} T_i \\ T_j \\ T_k \end{Bmatrix}$$

and h is a constant. Also, note that for triangular elements, the integral that includes products of area coordinates may be evaluated using the factorial relationship shown below:

$$\int_A \xi^a \eta^b \lambda^c \, dA = \frac{a!b!c!}{(a+b+c+2)!} 2A$$

13. In the formulation of two-dimensional heat transfer problems, the need to evaluate the integral

$$\int_A k\left(\frac{\partial[\mathbf{S}]^T}{\partial X}\frac{\partial T}{\partial X}\right) dA$$

arises. Using a bilinear rectangular element, evaluate the aforementioned integral, provided temperature is given by

$$T^{(e)} = [S_i \quad S_j \quad S_m \quad S_n]\begin{Bmatrix} T_i \\ T_j \\ T_m \\ T_n \end{Bmatrix}$$

and k is the thermal conductivity of the element and is a constant.

14. Look up the expressions for the nine-node quadratic quadrilateral element (Lagrangian element). Discuss its properties and compare it to the eight-node quadratic quadrilateral element. What is the basic difference between the Lagrangian element and its eight-node quadratic quadrilateral counterpart?

15. For triangular elements, show that the area coordinate $\eta = S_j$ and the area coordinate $\lambda = S_k$.

16. Verify the results given for natural quadrilateral shape functions in Eq. (7.7) by showing that (1) a shape function has a value of unity at its corresponding node and a value of zero at the other nodes and (2) if we sum up the shape functions, we will come up with a value of unity.

17. Verify the results given for natural quadratic triangular shape functions in Eq. (7.25) by showing that a shape function has a value of unity at its corresponding node and a value of zero at the other nodes.

18. For plane stress problems, using triangular elements, we can represent the displacements u and v using a linear triangular element similar to the one shown in the accompanying figure.

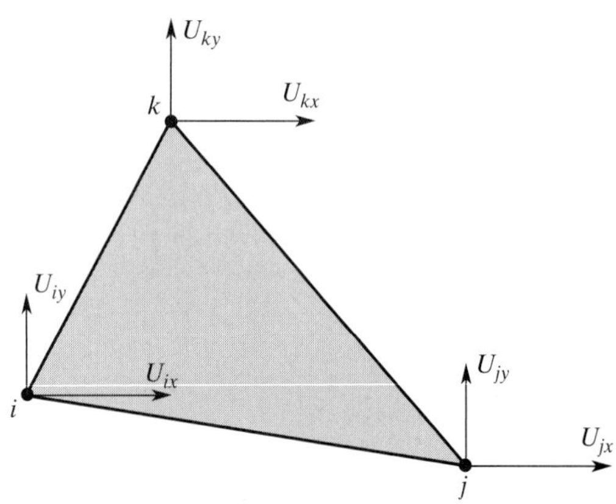

The displacement variables, in terms of linear triangular shape functions and the nodal displacements, are

$$u = S_i U_{ix} + S_j U_{jx} + S_k U_{kx}$$

$$v = S_i U_{iy} + S_j U_{jy} + S_k U_{ky}$$

Moreover, for plane stress situations, the strain displacement relationships are

$$\varepsilon_{xx} = \frac{\partial u}{\partial x} \qquad \varepsilon_{yy} = \frac{\partial v}{\partial y} \qquad \gamma_{xy} = \frac{\partial u}{\partial y} + \frac{\partial v}{\partial x}$$

Show that for a triangular element, strain components are related to the nodal displacements according to the relation

$$\begin{Bmatrix} \varepsilon_{xx} \\ \varepsilon_{yy} \\ \gamma_{xy} \end{Bmatrix} = \frac{1}{2A}\begin{bmatrix} \beta_i & 0 & \beta_j & 0 & \beta_k & 0 \\ 0 & \delta_i & 0 & \delta_j & 0 & \delta_k \\ \delta_i & \beta_i & \delta_j & \beta_j & \delta_k & \beta_k \end{bmatrix}\begin{Bmatrix} U_{ix} \\ U_{iy} \\ U_{jx} \\ U_{iy} \\ U_{kx} \\ U_{ky} \end{Bmatrix}$$

19. Consider point Q along the kj-side of the triangular element shown in the accompanying figure. Connecting this point to node i results in dividing the area of the triangle into two smaller areas A_2 and A_3, as shown.

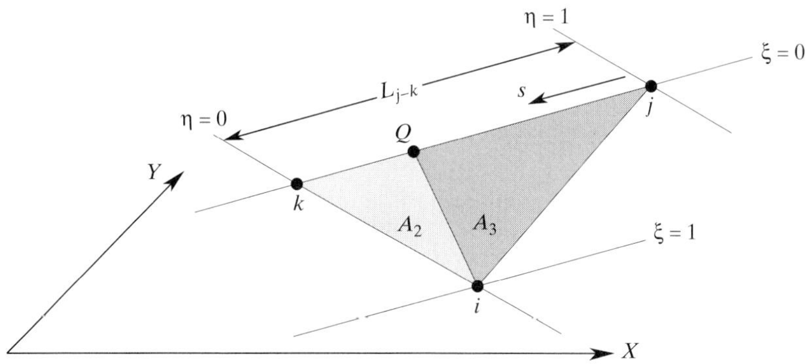

Along the kj-edge, the natural, or area, coordinate ξ has a value of zero. Show that along the kj-edge, the other natural (area) coordinates η and λ reduce to one-dimensional natural coordinates that can be expressed in terms of the local coordinate s according to the equations

$$\eta = \frac{A_2}{A} = 1 - \frac{s}{L_{j-k}}$$

$$\lambda = \frac{A_3}{A} = \frac{s}{L_{j-k}}$$

20. For the element in Problem 19, derive the simplified area coordinates along the *ij* and *ki*-edges using the one-dimensional coordinate *s*.

21. As you will see in Chapters 9 and 10, we need to evaluate integrals along the edges of a triangular element to develop the load matrix in terms of surface loads or derivative boundary conditions. Referring to Problem 19 and making use of the relations

$$\int_0^1 (x)^{m-1}(1-x)^{n-1}dx = \frac{\Gamma(m)\Gamma(n)}{\Gamma(m+n)}$$

$$\Gamma(n) = (n-1)! \quad \text{and} \quad \Gamma(m) = (m-1)!$$

show that

$$L\int_0^1 \left(1-\frac{s}{L}\right)^a \left(\frac{s}{L}\right)^b d\left(\frac{s}{L}\right) = \frac{a!b!}{(a+b+1)!}L$$

and

$$\int_0^{L_{k-j}} (\eta)^a(\lambda)^b ds = L_{j-k}\int_0^1 \left(1-\frac{s}{L_{j-k}}\right)^a \left(\frac{s}{L_{j-k}}\right)^b d\left(\frac{s}{L_{j-k}}\right) = \frac{a!b!}{(a+b+1)!}L_{j-k}$$

22. Consider a triangular element subjected to a distributed load along its *ki*-edge, as shown in the accompanying figure.

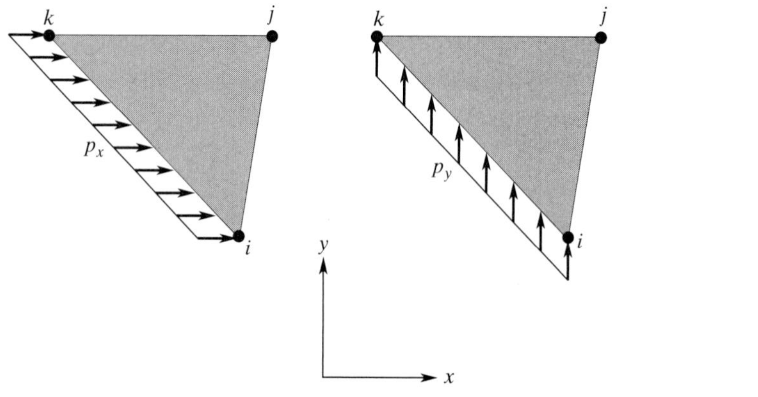

Using the minimum total potential-energy method, the differentiation of the work done by these distributed loads with respect to the nodal displacements gives the load matrix, which is computed from

$$\{\mathbf{F}\}^{(e)} = \int_A [\mathbf{S}]^T\{\mathbf{p}\}dA,$$

where

$$[\mathbf{S}]^T = \begin{bmatrix} S_i & 0 \\ 0 & S_i \\ S_j & 0 \\ 0 & S_j \\ S_k & 0 \\ 0 & S_k \end{bmatrix} \quad \text{and} \quad \{\mathbf{p}\} = \begin{Bmatrix} p_x \\ p_y \end{Bmatrix}$$

Realizing that along the *ki*-edge, $S_j = 0$, evaluate the load matrix for a situation in which the load is applied along the *ki*-edge. Use the results of Problem 21 to help you. Note, in this problem, A is equal to the product of the element thickness and the edge length.

23. For the element in Problem 22, evaluate the load matrices for a situation in which the distributed load is acting along the *ij*-edge and the *jk*-edge.

24. For a quadrilateral element, evaluate the Jacobian matrix J and its inverse $\mathbf{J}^{-1}$ using Eqs. (7.32) and (7.7).

25. Verify the shape function S_j, S_m, and S_n given in Eq. (7. 29) for the axisymmetric rectangular element.

26. We have used axisymmetric triangular elements to model temperature distribution in a system. The values of nodal temperature for an element belonging to the system are shown in the accompanying figure. What is the value of the temperature at $r = 1.8$ cm and $z = 1.9$ cm?

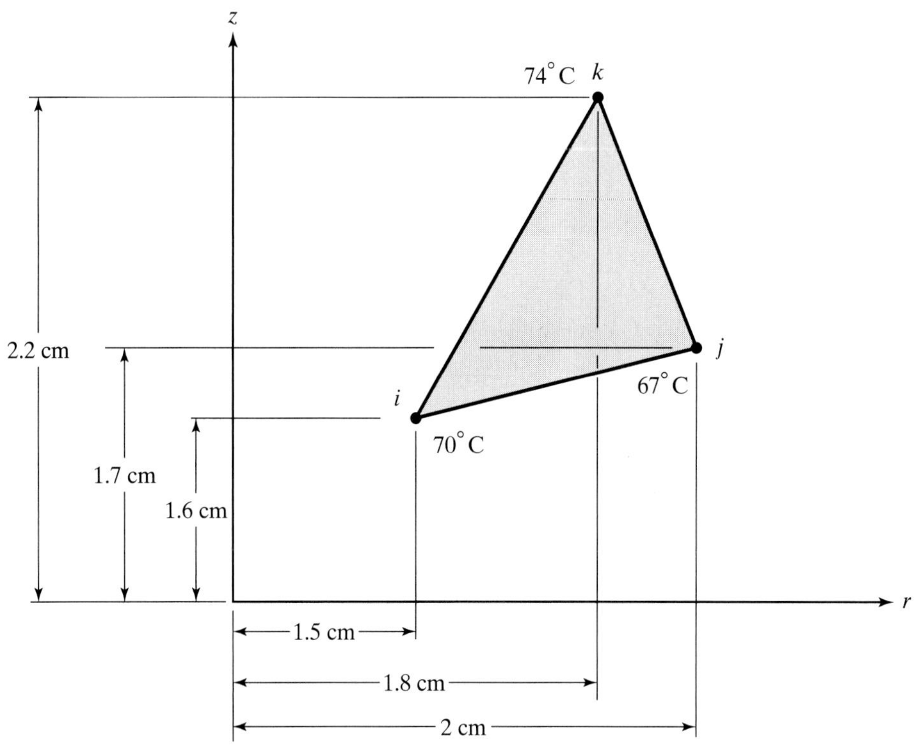

27. We have used axisymmetric rectangular elements to model temperature distribution in a system. The values of nodal temperature for an element belonging to the system are shown in the accompanying figure. What is the value of the temperature at $r = 2.1$ cm and $z = 1.3$ cm?

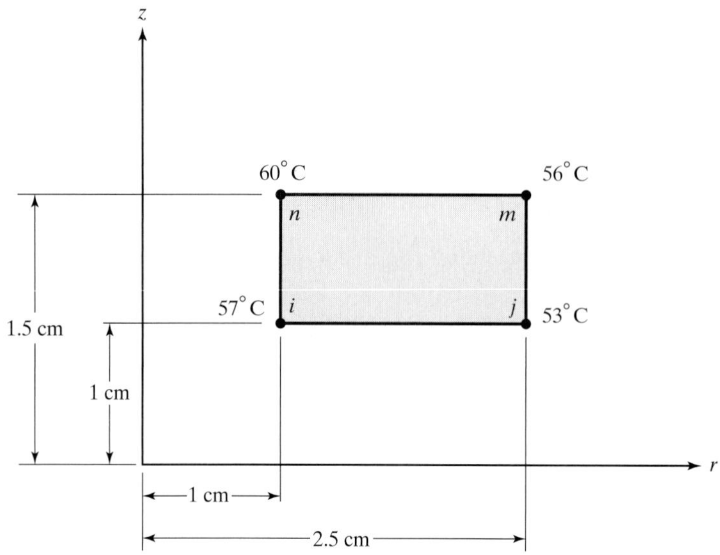

28. For an axisymmetric rectangular element, derive the r- and z-components of the gradients of dependent variable Ψ.

29. Evaluate the given integral analytically and using Gauss–Legendre formula.

$$\int_1^5 \int_0^6 (5y^3 + 2x^2 + 5)\,dx\,dy$$

30. Evaluate the given integral analytically and using Gauss-Legendre formula.

$$\int_{-2}^8 \int_2^{10} (y^3 + 3y^2 + 5y + 2x^2 + x + 10)\,dx\,dy$$

31. Show that for the rectangular element shown in Figure 7.2a and represented by Eq. (7.1), the temperature distribution varies linearly along the edges of the element, and it becomes nonlinear inside the element. (Hint: realize that along the i-j edge, $y = 0$, along i-n, $x = 0$, along j-m, $x = \ell$, and along the n-m edge $y = w$.)

CHAPTER 8

More ANSYS*

The main objective of this chapter is to introduce the essential capabilities and the organization of the ANSYS program. The basic steps in creating and analyzing a model with ANSYS are discussed here, along with an example used to demonstrate these steps at the end of the chapter. The main topics discussed in Chapter 8 include the following:

8.1 ANSYS Program
8.2 ANSYS Database and Files
8.3 Creating a Finite Element Model with ANSYS: Preprocessing
8.4 *h*-Method Versus *p*-Method
8.5 Applying Boundary Conditions, Loads, and the Solution
8.6 Results of Your Finite Element Model: Postprocessing
8.7 Selection Options
8.8 Graphics Capabilities
8.9 Error-Estimation Procedures
8.10 An Example Problem

8.1 ANSYS PROGRAM

The ANSYS program has two basic levels: the *Begin level* and the *Processor level*. When you first enter ANSYS, you are at the Begin level. From the Begin level, you can enter one of the ANSYS processors, as shown in Figure 8.1. A processor is a collection of functions and routines to serve specific purposes. You can clear the database or change a file assignment from the Begin level.

There are three processors that are used most frequently: (1) the *preprocessor* **(PREP7)**, (2) the *processor* **(SOLUTION)**, and (3) the general *postprocessor* **(POST1)**. The preprocessor **(PREP7)** contains the commands needed to build a model:

- define element types and options
- define element real constants

*Materials were adapted with permission from ANSYS documents.

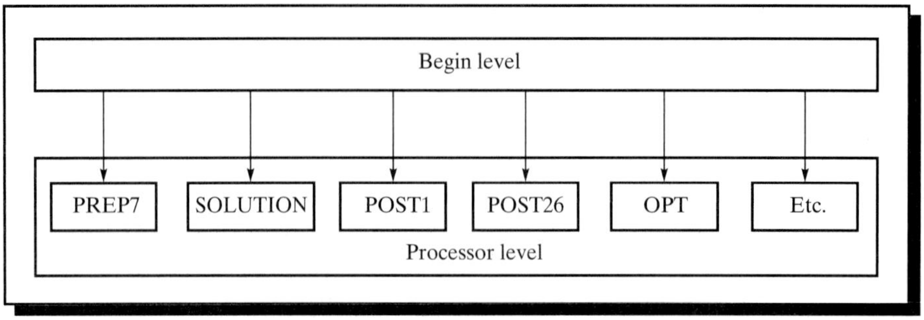

FIGURE 8.1 Organization of ANSYS program.

- define material properties
- create model geometry
- define meshing controls
- mesh the object created

The solution *processor* **(SOLUTION)** has the commands that allow you to apply boundary conditions and loads. For example, for structural problems, you can define displacement boundary conditions and forces, or for heat transfer problems, you can define boundary temperatures or convective surfaces. Once all the information is made available to the solution *processor* **(SOLUTION)**, it solves for the nodal solutions. The general *postprocessor* **(POST1)** contains the commands that allow you to list and display results of an analysis:

- read results data from results file
- read element results data
- plot results
- list results

There are other processors that allow you to perform additional tasks. For example, the *time-history postprocessor* **(POST26)** contains the commands that allow you to review results over time in a transient analysis at a certain point in the model. The *design optimization processor* **(OPT)** allows the user to perform a design optimization analysis.

8.2 ANSYS DATABASE AND FILES

The previous section explained how the ANSYS program is organized. This section discusses the ANSYS database. ANSYS writes and reads many files during a typical analysis. The information you input when modeling a problem (e.g., element type, material property, dimensions, geometry, etc.) is stored as *input data*. During the solution phase, ANSYS computes various results, such as displacements, temperatures, stresses, etc.

This information is stored as *results data*. The input data and the results data are stored in the ANSYS database. The database can be accessed from anywhere in the ANSYS program. The database resides in the memory until the user saves the database to a database file **Jobname.DB**. Jobname is a name that the user specifies upon entering the ANSYS program; this feature will be explained in more detail later. The database can be saved and resumed at any time. When you issue the RESUME_DB command, the database is read into the memory from the database file that was saved most recently. In other words, the database becomes what you saved most recently. When you are uncertain about the next step you should take in your analysis, or if you want to test something, you should issue the SAVE_DB command before proceeding with your test. That way, if you are unhappy with the results of your test, you can issue the RESUME_DB command, which will allow you to go back to the place in your analysis where you started testing. The SAVE_DB, RESUME_DB, and QUIT commands are located in the ANSYS toolbar. In addition, the "Clear & Start New" option, located on the utility menu, allows the user to clear the database. This option is useful when you want to start anew, but do not wish to leave and reenter ANSYS.

When you are ready to exit the ANSYS program, you will be given four options: (1) Save all model data; (2) Save all model data and solution data; (3) Save all model data, solution data, and postprocessing data; or (4) Save nothing.

As previously explained, ANSYS writes and reads many files during a typical analysis. The files are of the form of *Jobname.Ext*. Recall that Jobname is a name you specify when you enter the ANSYS program at the beginning of an analysis. The default jobname is *file*. Files are also given unique extensions to identify their content. Typical files include the following:

- The *log file* (**Jobname.LOG**): This file is opened when ANSYS is first entered. Every command you issue in ANSYS is copied to the log file. Jobname.LOG is closed when you exit ANSYS. Jobname.LOG can be used to recover from a system crash or a serious user error by reading in the file with the **/INPUT** command.
- The *error file* (**Jobname.ERR**): This file is opened when you first enter ANSYS. Every warning and error message given by ANSYS is captured by this file. If Jobname.ERR already exists when you begin a new ANSYS session, all new warnings and error messages will be appended to the bottom of this file.
- The *database file* (**Jobname.DB**): This file is one of the most important ANSYS files because it contains all of your input data and possibly some results. The model portion of the database is automatically saved when you exit the ANSYS program.
- The *output file* (**Jobname.OUT**): This file is opened when you first enter ANSYS. Jobname.OUT is available if you are using the GUI; otherwise, your computer monitor is your output file. Jobname.OUT captures responses given by ANSYS to every command executed by the user. It also records warning and error messages and some results. If you change the Jobname while in a given ANSYS session, the output file name is not changed to the new Jobname.

Other ANSYS files include the *structural analysis results file* (**Jobname.RST**); the *thermal results file* (**Jobname.RTH**); the *magnetic results file* (**Jobname.RMG**); the *graphics file* (**Jobname.GRPH**); and the *element matrices file* (**Jobname.EMAT**).

8.3 CREATING A FINITE ELEMENT MODEL WITH ANSYS: PREPROCESSING

The preprocessor **(PREP7)** contains the commands needed to create a finite element model:

1. define element types and options
2. define element real constants if required for the chosen element type
3. define material properties
4. create model geometry
5. define meshing controls
6. mesh the object created

1. Define element types and options.

ANSYS provides more than 150 various elements to be used to analyze different problems. Selecting the correct element type is a very important part of the analysis process. A good understanding of finite element theory will benefit you the most in this respect, helping you choose the correct element for your analysis. In ANSYS, each element type is identified by a *category name* followed by a *number*. For example, two-dimensional solid elements have the category name **PLANE.** Furthermore, **PLANE42** is a four-node quadrilateral element used to model structural solid problems. The element is defined by four nodes having two degrees of freedom at each node, translation in the x- and y-directions. **PLANE82** is an eight-node (four corner points and four midside nodes) quadrilateral element used to model two-dimensional structural solid problems. It is a higher order version of the two-dimensional, four-node quadrilateral element type, PLANE42. Therefore, the PLANE82 element type offers more accuracy when modeling problems with curved boundaries. At each node, there are two degrees of freedom, translation in the x- and y-directions. Many of the elements used by ANSYS have options that allow you to specify additional information for your analysis. These options are known in ANSYS as *keyoptions* (**KEOPTs**). For example, for PLANE82, with **KEOPT (3)** you can choose plane stress, axisymmetric, plane strain, or plane stress with the thickness analysis option. A list of elements used by ANSYS is shown in Table 8.2 at the end of this chapter. You can define element types and options by choosing

main menu: **Preprocessor → Element Type → Add/Edit/Delete**

You will see the Element Types dialog box next, shown in Figure 8.2.

A **List**: A list of currently defined element types will be shown here. If you have not defined any elements yet, then you need to use the **Add** button to add an element.

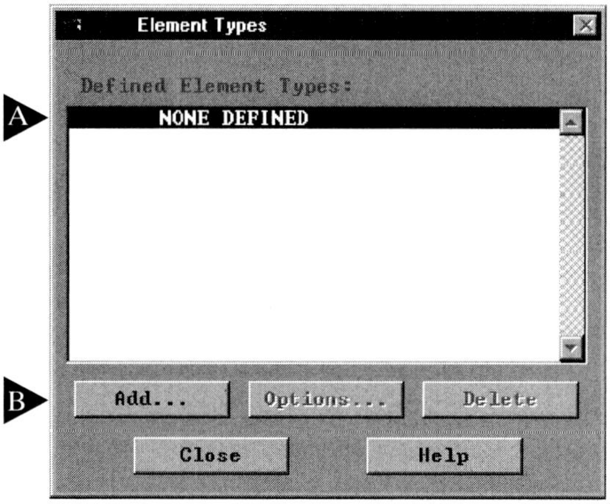

FIGURE 8.2 Element Types dialog box.

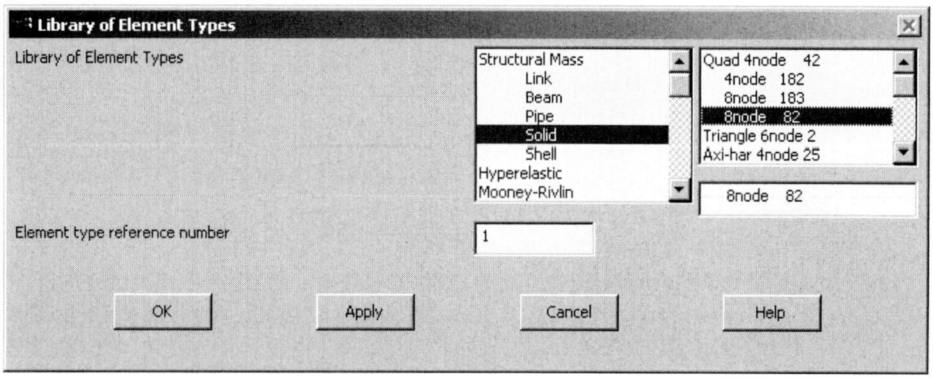

FIGURE 8.3 Library of Element Types dialog box.

The Library of Element Types dialog box will appear next (see Figure 8.3). Then you choose the type of element you desire from the Library.

B **Action Buttons**: The purpose of the **Add** button is to add an element, as we just discussed. The **Delete** button deletes the selected (highlighted) element type. The **Options** button opens the element type options dialog box. You can then choose one of the desired element options for a selected element. For example, if you had selected the element PLANE 82 with KEOPT (3) you could choose plane stress, axisymmetric, plane strain, or plane stress with the thickness analysis option, as shown in Figure 8.4.

2. **Define element real constants.**

   Element *real constants* are quantities that are specific to a particular element. For example, a beam element requires cross-sectional area, second moment of area, and so on. It is important to realize that real constants vary from one element type

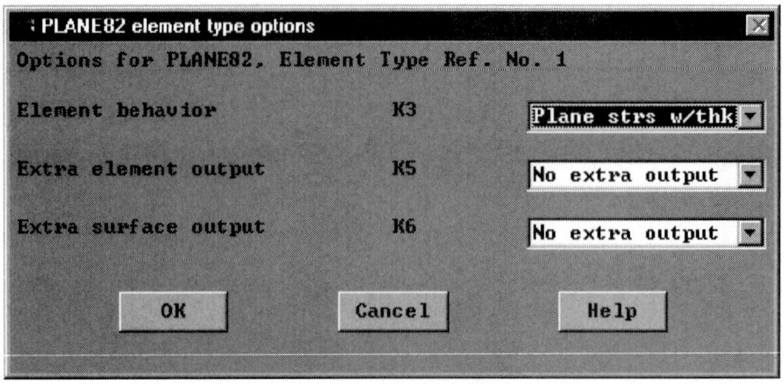

FIGURE 8.4 The element type options dialog box.

to another; furthermore, not all elements require real constants. Real constants can be defined by the command

main menu: **Preprocessor** → **Real Constants** → **Add/Edit/Delete**

You will then see the Real Constants dialog box, as shown in Figure 8.5.

A **List**: A list of currently defined real constants will be shown here. If you have not defined any real constants at this point, you need to use the **Add** button to add real constants. An example of a dialog box for a PLANE 82 element's real constants is shown in Figure 8.6.

B **Action Buttons**: The purpose of the **Add** button has already been explained. The **Delete** Button deletes the selected (highlighted) real constants. The **Edit** button opens a new dialog box that allows you to change the values of existing real constants.

3. Define material properties.

At this point, you define the physical properties of your material. For example, for solid structural problems, you may need to define the modulus of elasticity, Poisson's ratio, or the density of the material, whereas for thermal problems, you

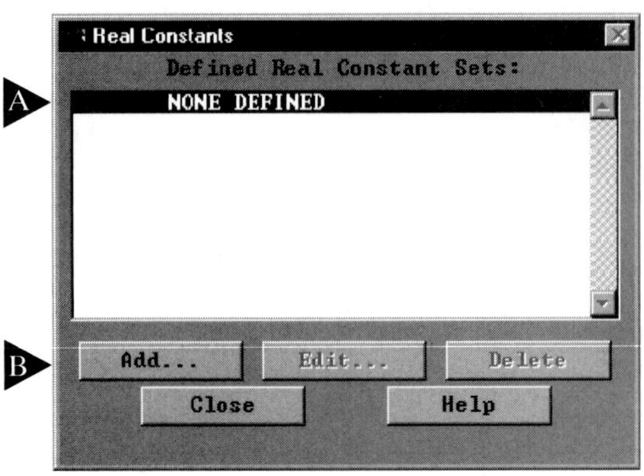

FIGURE 8.5 Real Constants dialog box.

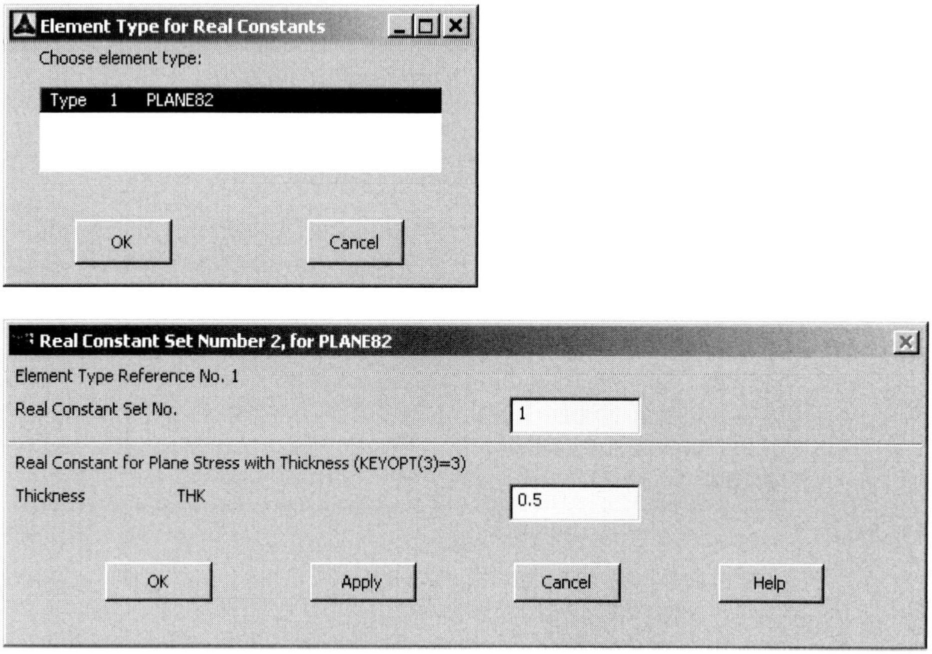

FIGURE 8.6 An example of the dialog box for a PLANE 82 (with options) element's real constants.

may need to define thermal conductivity, specific heat, or the density of the material. You can define material properties by the command

main menu: **Preprocessor** → **Material Props** → **Material Models**

You will then see the Define Material Model Behavior dialog box, as shown in Figure 8.7.

The next dialog box allows you to define the appropriate properties for your analysis, as shown in Figure 8.8. You can use multiple materials in your model if the object you are analyzing is made of different materials. From the Define Material Model Behavior dialog box, click on Material button and then choose New Model.

4. **Create model geometry.**

There are two approaches to constructing a finite element model's geometry: (1) *direct (manual) generation* and (2) the *solid-modeling* approach. Direct generation, or manual generation, is a simple method by which you specify the location of nodes and manually define which nodes make up an element. This approach is generally applied to simple problems that can be modeled with line elements, such as links, beams, and pipes, or if the object is made of simple geometry, such as rectangles. This approach is illustrated in Figure 8.9. Refer back to the truss problem of Example 3.1 in Chapter 3 to refresh your memory about the manual approach, if necessary.

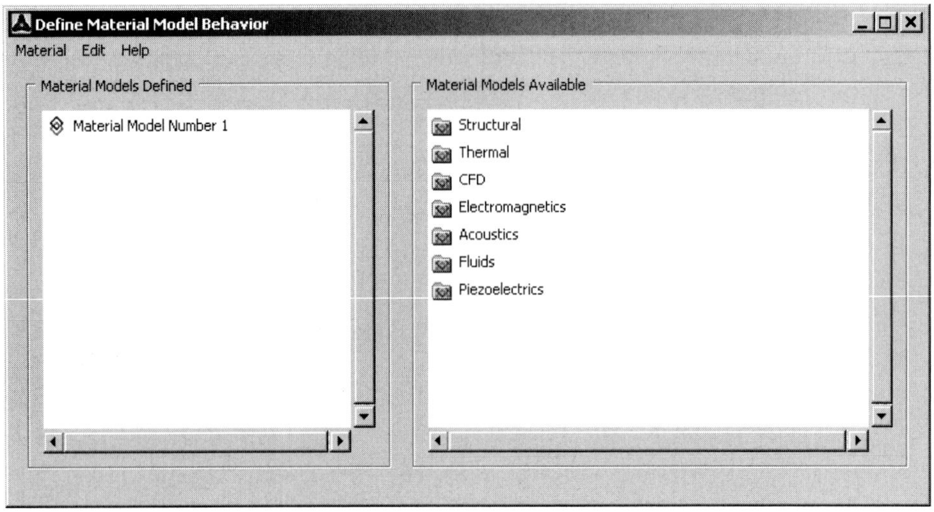

FIGURE 8.7 Material Model dialog box.

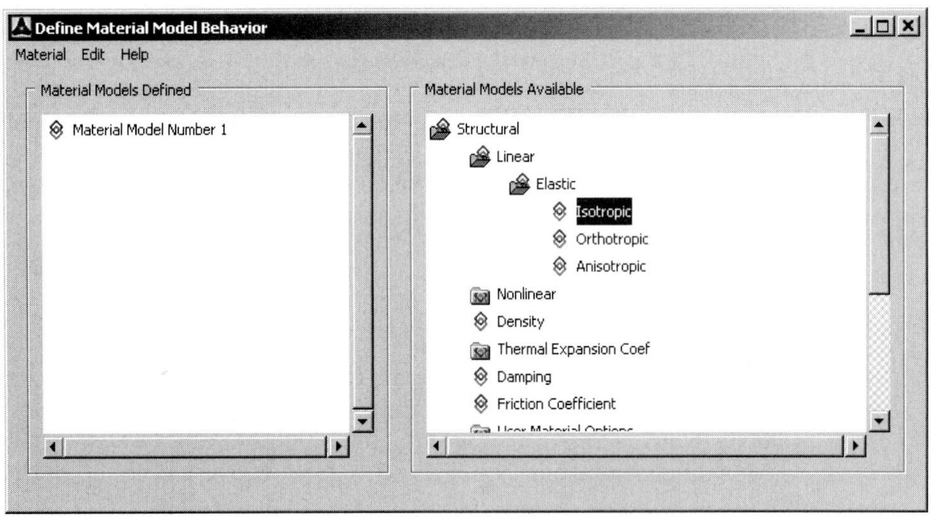

FIGURE 8.8a Isotropic structural material properties dialog box.

With the solid-modeling approach, you use simple *primitives* (simple geometric shapes), such as rectangles, circles, polygons, blocks, cylinders, and spheres, to construct the model. Boolean operations are then used to combine the primitives. Examples of boolean operations include addition, subtraction, and intersection. You then specify the desired element size and shape, and ANSYS will automatically generate all the nodes and the elements. This approach is depicted in Figure 8.10.

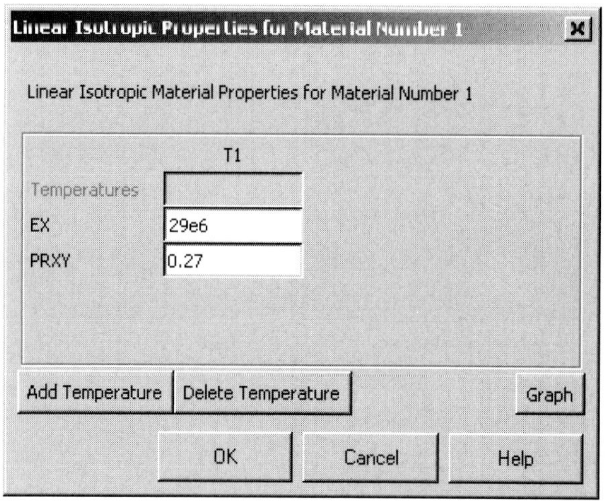

FIGURE 8.8b Linear Isotropic Material Properties dialog box.

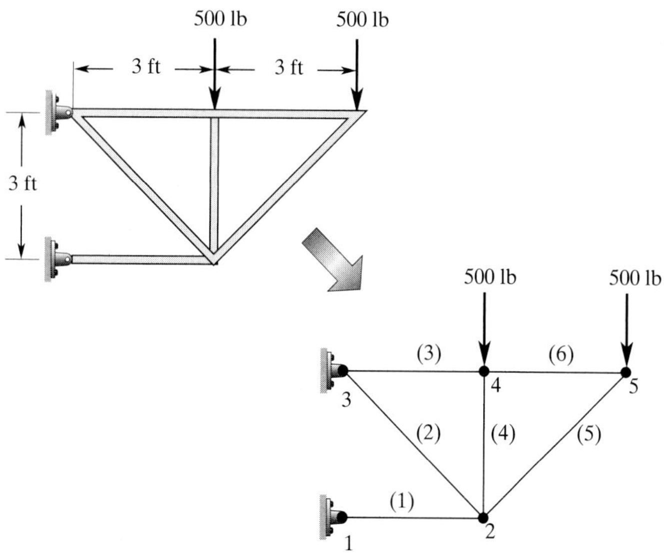

FIGURE 8.9 A truss problem: First nodes 1–5 are created, then nodes are connected to form elements (1)–(6).

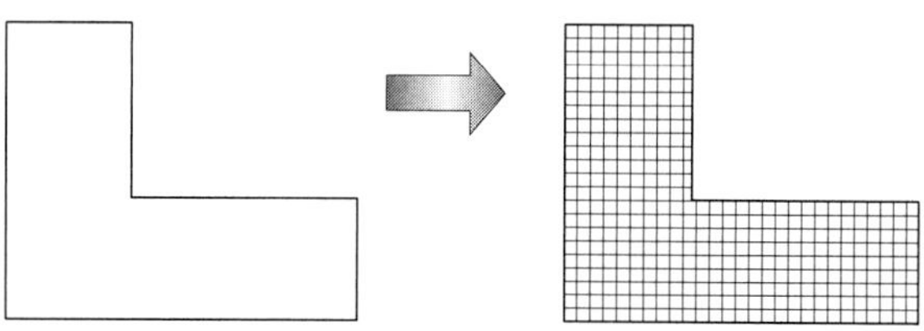

FIGURE 8.10 An example of the solid-modeling approach.

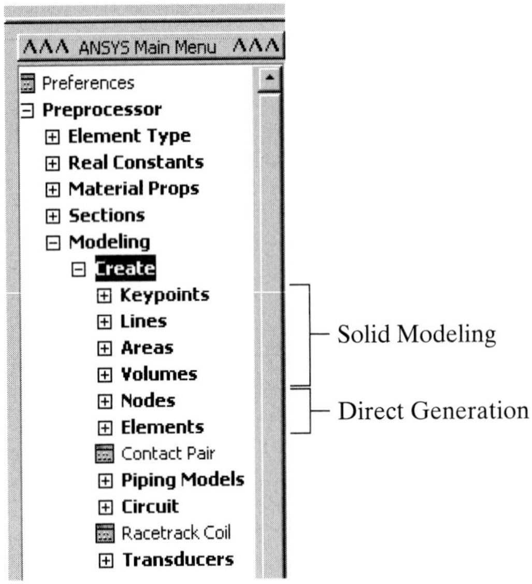

FIGURE 8.11 The Create dialog box.

The construction of the model in ANSYS begins when you choose the **Create** option, as shown in Figure 8.11. You choose this option with the command sequence

main menu: **Preprocessor** → **Modeling** → **Create**

When you create entities such as keypoints, lines, areas, or volumes, they are automatically numbered by the ANSYS program. You use *keypoints* to define the vertices of an object. *Lines* are used to represent the edges of an object. *Areas* are used to represent two-dimensional solid objects. They are also used to define the surfaces of three-dimensional objects. When using primitives to build a model, you need to pay special notice to the hierarchy of the entities. Volumes are bounded by areas, areas are bounded by lines, and lines are bounded by keypoints. Therefore, volumes are considered to be the highest entity, and the keypoints are the lowest entity in solid modeling hierarchy. Remembering this concept is particularly important if you need to delete a primitive. For example, when you define one rectangle, ANSYS automatically creates nine entities: four keypoints, four lines, and one area. The relationship among keypoints, lines, and areas is depicted in Figure 8.12.

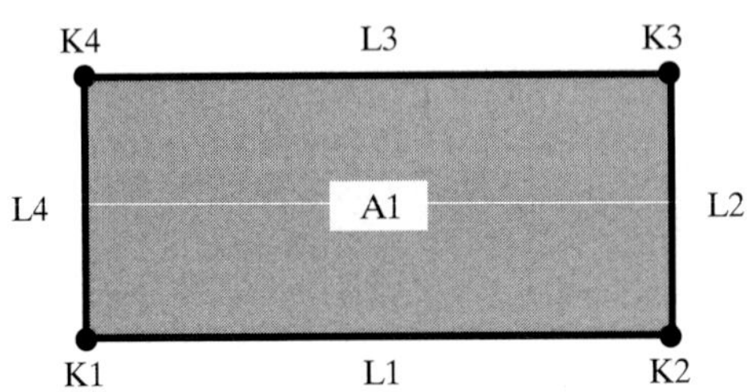

FIGURE 8.12 The relationship among the keypoints, lines, and areas.

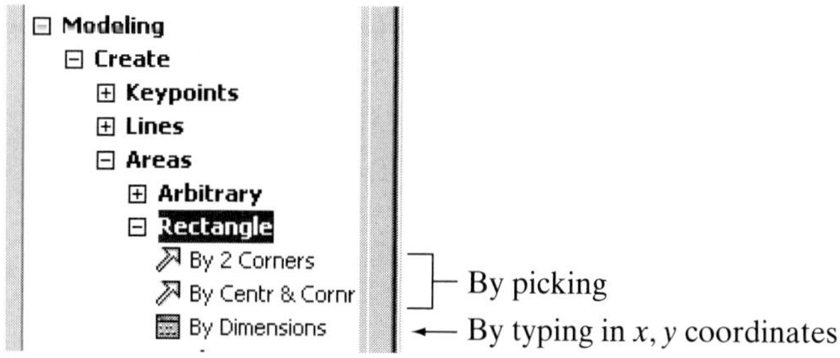

FIGURE 8.13 The Rectangle menu.

Area primitives and volume primitives are grouped under the Areas and Volumes categories in the Create menu. Now let us consider the Rectangle and the Circle menus, because they are commonly used to build two-dimensional models. The Rectangle menu offers three methods for defining a rectangle, as shown in Figure 8.13. The command for accessing the Rectangle menu is

main menu: **Preprocessor → Modeling → Create → Rectangle**

The Circle menu offers several methods for defining a solid circle or annulus, as shown in Figure 8.14.

The Partial Annulus option is limited to circular areas spanning 180° or less. In order to create a partial circle that spans more than 180°, you need to use the By Dimensions option. An example of creating a partial annulus spanning from $\theta = 45°$ to $\theta = 315°$ is shown in Figure 8.15. Note that you can create a solid circle by setting **Rad-1** = 0.

The Working Plane (WP)

In ANSYS, you will use a *working plane (WP)* to create and orient the geometry of the object you are planning to model. All primitives and other modeling entities are

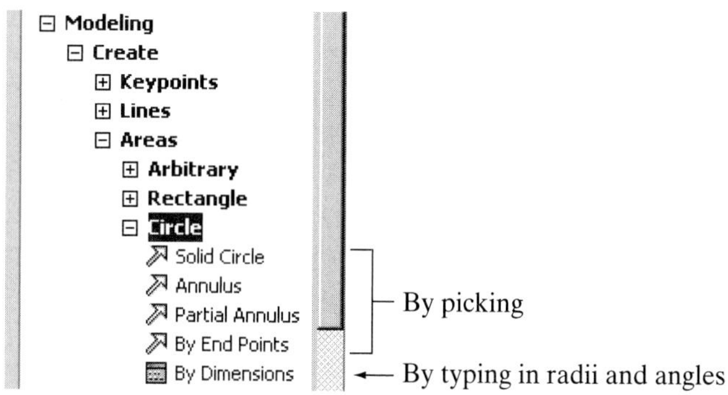

FIGURE 8.14 The Circle menu.

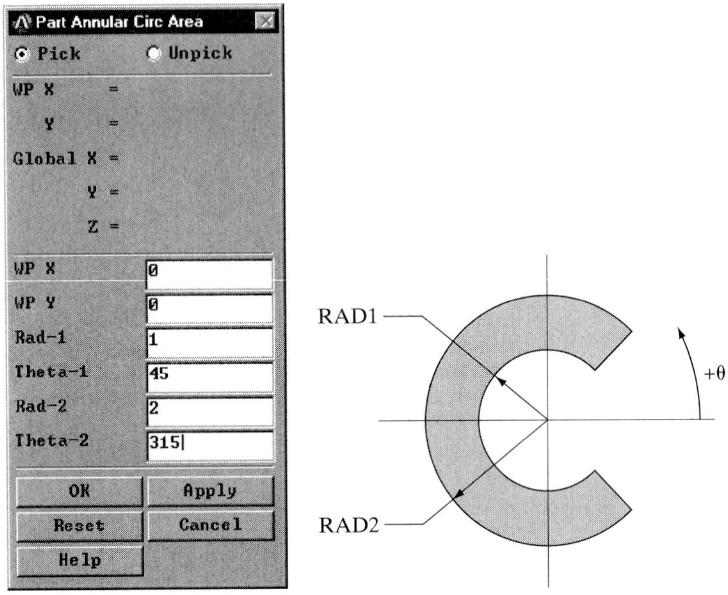

FIGURE 8.15 An example of creating a partial annulus spanning from $\theta = 45°$ to $\theta = 315°$.

defined with respect to this plane. The working plane is basically an infinite plane with a two-dimensional coordinate system. The dimensions of the geometric shapes are defined with respect to the WP. By default, the working plane is a Cartesian plane. You can change the coordinate system to a polar system, if so desired. Other attributes of the working plane may be set by opening the WP settings dialog submenu, as shown in Figure 8.16. To access this dialog box, issue the following sequence of commands:

utility menu: **Work Plane → WP Settings**...

A **Coordinate System**: Choose the working-plane coordinate system you want to use. You locate or define points in terms of *X*- and *Y*-coordinates when using the Cartesian coordinate system. You can also locate or define points with respect to a polar coordinate system using *R*- and Θ-coordinates.

B **Display Options**: This section is where you turn on the grid or grid and triad. The triad appears in the center (0,0 coordinates) of your working plane.

C **Snap Options**: These options control the locations of points that are picked. When activated, these options allow you to pick locations nearest to the snap point. For example, in a Cartesian working plane, **Snap Incr** controls the *X*- and *Y*-increments within the spacing grid. If you have set a spacing of 1.0 and a snap increment of 0.5, then within the *X, Y* grid you can pick coordinates with 0.5 increments. For example, you cannot pick the coordinates 1.25 or 1.75.

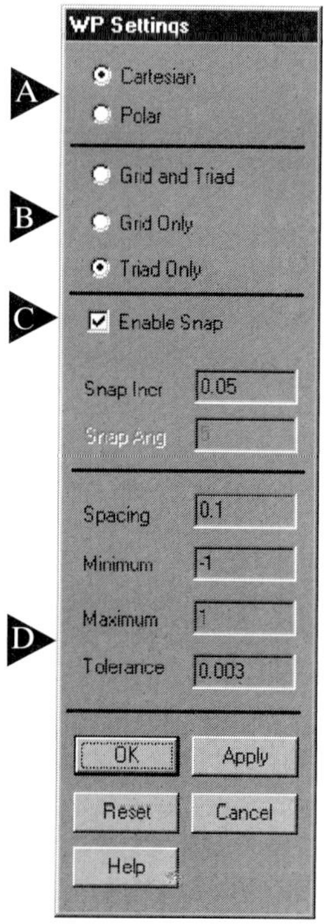

FIGURE 8.16 The WP settings dialog box.

D **Grid Control**

Spacing: This number defines the spacing between the grid lines.

Minimum: This number is the minimum X-location at which you want the grid to be displayed with respect to the Cartesian coordinate system.

Maximum: This number is the maximum X-location at which you want the grid to be displayed with respect to the Cartesian coordinate system.

Radius: This number is the outside radius that you want the grid to be displayed with respect to the polar coordinate system.

Tolerance: This number is the amount that an entity can be off of the current working plane and still be considered as on the plane.

The working plane is always active and, by default, not displayed. To display the working plane, you need to issue the following command:

utility menu: **Work Plane → Display Working Plane**

You can move the WP origin to a different location on the working plane. This feature is useful when you are defining primitives at a location other than the global location. You can move the WP origin by choosing the commands

utility menu: **Work Plane → Offset WP to → XYZ Locations** +

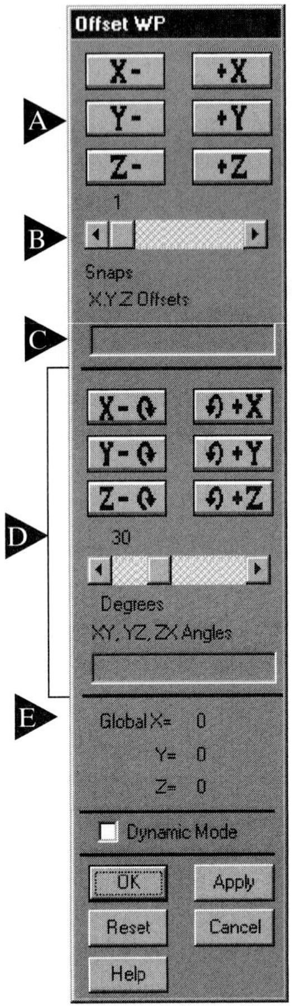

FIGURE 8.17 The dialog box for offsetting the WP.

You can relocate the working plane by offsetting it from its current location, as shown in Figure 8.17. To do so, issue the command

utility menu: **Work Plane → Offset WP by Increments** ...

A **Offset buttons**: Picking these buttons will cause an immediate offset of your working plane in the direction shown on the buttons. The amount of offset is controlled by the Offset slider and the Snap-Incr value on the WP setting dialog box.

B **Offset Slider**: This number controls the amount of offset that occurs with each pick of the offset buttons. If the slider is set to 1, the offset will be one times the Snap-Incr value on the WP setting dialog box.

C **Offset Dialog Input**: This feature allows you to input the exact *X*, *Y*, and *Z* offset values for the working plane. For instance, typing 1,2,2 into this field and pressing the Apply or OK button will move the working plane one unit in the positive X-direction and two units each in the positive Y- and Z-directions.

Location Status: This section displays the current location of the working plane in global Cartesian coordinates. This status is updated each time the working plane is translated.

You can also relocate the working plane by aligning it with specified keypoints, nodes, coordinate locations, and so on, as shown in Figure 8.18. To align the working plane, issue the command

utility menu: **Work Plane → Align WP with**

Plotting Model Entities

You can plot various entities, such as keypoints, lines, areas, volumes, nodes, and elements, using the Plot menu. From the utility menu, you can issue one of the following commands to plot:

utility menu: **Plot → Keypoints**
utility menu: **Plot → Lines**
utility menu: **Plot → Areas**
utility menu: **Plot → Volumes**
utility menu: **Plot → Nodes**
utility menu: **Plot → Elements**

The Plot Numbering Controls menu, shown in Figure 8.19, contains a useful graphics option that allows you to turn on keypoint numbers, line numbers, area numbers, and so on to check your model. To access this option, use the command

utility menu: **PlotCtrls → Numbering** . . .

You may need to replot to see the effects of the numbering command you issue.

5. **Define meshing controls.**

The next step in creating a finite element model is dividing the geometry into nodes and elements. This process is called meshing. The ANSYS program can automatically generate the nodes and elements, provided that you specify the *element attributes* and the *element size*:

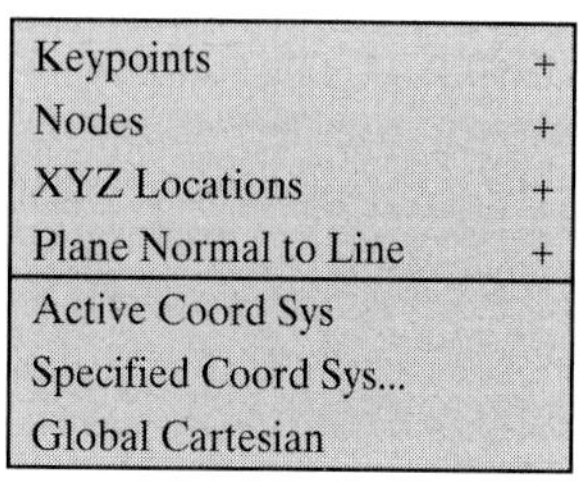

FIGURE 8.18 Working plane-relocation using the Align command.

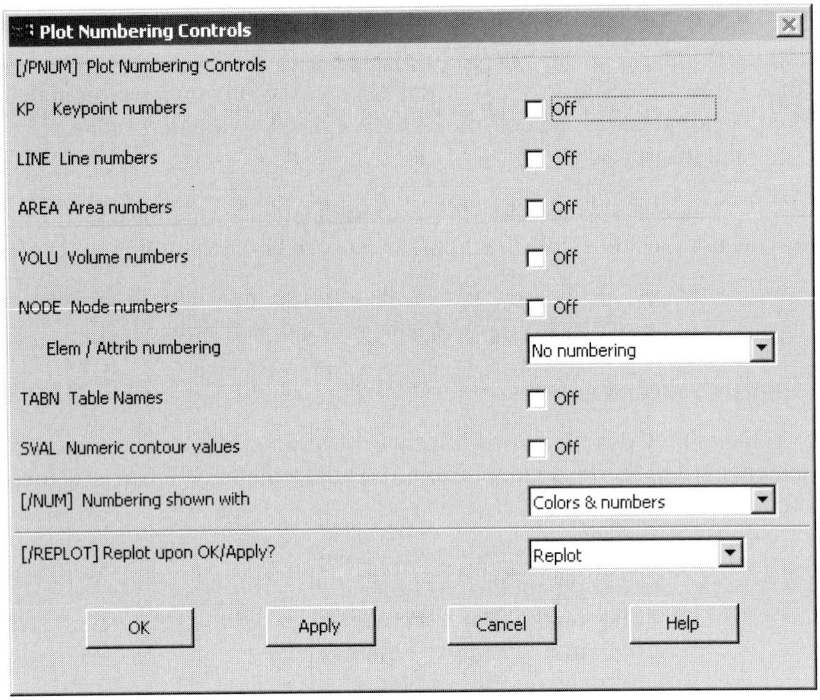

FIGURE 8.19 The Plot Numbering Controls dialog box.

1. The *element attributes* include element type(s), real constants, and material properties.
2. The *element size* controls the fineness of the mesh. The smaller the element size, the finer the mesh. The simplest way to define the element size is by defining a global element size. For example, if you specify an element edge length of 0.1 units, then ANSYS will generate a mesh in which no element edge is larger than 0.1 units. Another way to control the mesh size is by specifying the number of element divisions along a boundary line. The Global Element Sizes dialog box is shown in Figure 8.20. To access this dialog box, issue the following commands:

main menu: **Preprocessor → Meshing → Size Cntrls → Manual Size → Global → Size**

6. **Mesh the object.**

You should get into the habit of saving the database before you initiate meshing. This way, if you are not happy with the mesh generated, you can resume the database and change the element size and remesh the model. To initiate meshing, invoke the commands

main menu: **Preprocessor → Meshing → Mesh → Areas → Free**

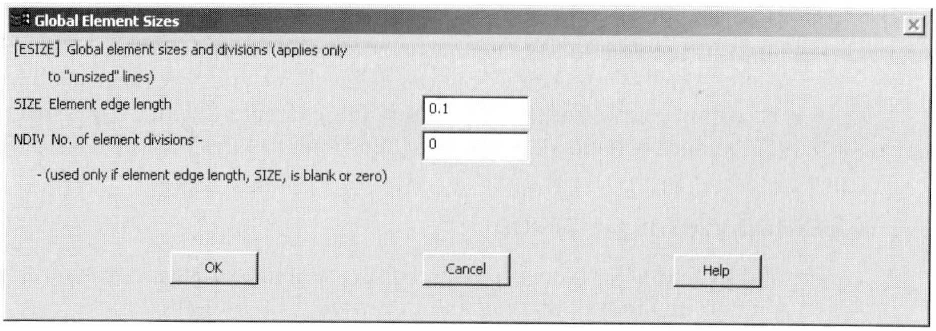

FIGURE 8.20 The Global Element Size dialog box.

Once a picking menu appears, you can pick individual areas or use the Pick All button to select all areas for meshing. Upon selection of the desired areas, Pick the Apply or OK buttons to mesh. The meshing process can take some time, depending on the model complexity and the speed of your computer. During the meshing process, ANSYS periodically writes a meshing status to the output window. Therefore, it is useful to bring the output window to the front to see the meshing status messages.

Free meshing uses either mixed-area element shapes or all-triangular area elements, whereas the *mapped meshing* option uses all quadrilateral area elements and all hexahedral (brick) volume elements. Mapped area mesh requirements include three or four sides, equal numbers of elements on opposite sides, and even numbers of elements for three-sided areas. If you want to mesh an area that is bounded by more than four lines, you can use the concatenate command to combine some of the lines to reduce the total number of lines. Concatenation is usually the last step you take before you start meshing the model. To concatenate, issue the following series of commands:

main menu: **Preprocessor → Meshing → Concatenate → Lines** or **Areas**

Modifying Your Meshed Model

If you want to modify your model, you must keep in mind certain rules enforced by ANSYS:

1. Meshed lines, areas, or volumes may not be deleted or moved.
2. You can delete the nodes and the elements with the meshing Clear command.

Also, areas contained in volumes may not be deleted or changed. Lines contained in areas may not be deleted. Lines can be combined or divided into smaller segments with line operation commands. Keypoints contained in lines may not be deleted. You start the clearing process by issuing the commands

main menu: **Preprocessor → Meshing → Clear**

The clearing process will delete nodes and elements associated with a selected model entity. Then you can use deleting operations to remove all entities associated with an entity. The "... and below" options delete all lower entities associated with the specified entity, as well as the entity itself. For example, deleting "Area and below" will automatically remove the area, the lines, and the keypoints associated with the area.

8.4 *h*-METHOD VERSUS *p*-METHOD

For solid structural problems, to obtain displacements, stresses, or strain, ANSYS offers two solution methods: *h-method* and *p-method*. The *h*-method makes use of elements that are based on shape functions that are typically quadratic. We have already pointed out that the elements based on quadratic shape functions are better than linear elements but not as good as higher order elements, such as cubic. We have also explained that the element size affects the accuracy of your results as well. Whereas the *h*-method makes use of quadratic elements, the *p*-method uses higher order polynomial shape functions to define elements. Because the *h*-method uses quadratic elements, mesh refinement may be necessary if you desire very accurate results. A simple way to determine if the element size is fine enough to produce good results is by solving the problem with a certain number of elements and then comparing its results to the results of a model with twice as many elements. If you detect substantial difference between the results of the two models, then the mesh refinement is necessary. You may have to repeat the process by refining the mesh until you can't detect a substantial difference between the models.

As mentioned above, the *p*-method—which can be used only for linear structural static problems—makes use of polynomial shape functions that are of higher orders than quadratic ones. The user can specify a degree of accuracy, and the *p*-method manipulates the order of the polynomial or the *p*-level to better fit the degree of difficulty associated with the boundaries of the given problem and its behavior to the applied loads. A problem is solved at a given *p*-level, and then the order of the polynomial is increased and the problem is solved again. The results of the iterations are then compared to a set of convergence criteria that the user has specified. In general, a higher *p*-level approximation leads to better results. One of the main benefits of using the *p*-method is that in order to obtain good results, you don't need to manually manipulate the size of elements by creating finer meshes. Depending on the problem, a coarse mesh can provide reasonable results.

Finally, as you will learn in Section 8.9, ANSYS offers error-estimation procedures that calculate the level of solution errors due to meshing employed. It is worth noting that the *p*-method adaptive refinement procedure offers error estimates that are more precise than those of the *h*-method and can be calculated locally at a point or globally.

8.5 APPLYING BOUNDARY CONDITIONS, LOADS, AND THE SOLUTION

The next step of finite element analysis involves applying appropriate boundary conditions and the proper loading. There are two ways to apply the boundary conditions and loading to your model in ANSYS. You can either apply the conditions to the solid model (keypoints, lines, and areas), or the conditions can be directly imposed on the nodes and elements. The first approach may be preferable because should you decide to change the meshing, you will not need to reapply the boundary conditions and the loads to the new finite element model. It is important to note that if you decide to apply the conditions to keypoints, lines, or areas during the solution phase, ANSYS automatically transfers the information to nodes. The solution *processor* (**SOLUTION**) has the commands that allow you to apply boundary conditions and loads. It includes the following options:

for structural problems: displacements, forces, distributed loads (pressures), temperatures for thermal expansion, gravity

for thermal problems: temperatures, heat transfer rates, convection surfaces, internal heat generation

for fluid flow problems: velocities, pressures, temperatures

for electrical problems: voltages, currents

for magnetic problems: potentials, magnetic flux, current density

Degrees of Freedom (DOF) Constraints

In order to constrain a model with fixed (zero displacements) boundary conditions, you need to choose the command sequence

main menu: **Solution** → **Define Loads** → **Apply** → **Structural** → **Displacement**

You can specify the given condition on the keypoints, lines, areas, or nodes. For example, if you choose to constrain certain keypoints, then you need to invoke the commands

main menu: **Solution** → **Define Loads** → **Apply** → **Structural** → **Displacement** → **On Keypoints**

A picking menu will appear. You then pick the keypoints to be constrained and press the OK button. An example of a dialog box for applying displacement constraints on keypoints is shown in Figure 8.21.

The KEXPND field in the dialog box of Figure 8.21 is used to expand the constraint specification to all nodes between the keypoints, as shown in Figure 8.22.

Once you have applied the constraints, you may want to display the constraint symbols graphically. To turn on the boundary condition symbols, open the Symbols dialog box, as shown in Figure 8.23, by choosing the commands

utility menu: **PlotCtrls** → **Symbols** ...

Line or Surface Loads

In order to specify distributed loads on a line or surface of a model, you need to issue the following commands:

main menu: **Solution** → **Define Loads** → **Apply** → **Structural** → **Pressures** → **On Lines** or **On Areas**

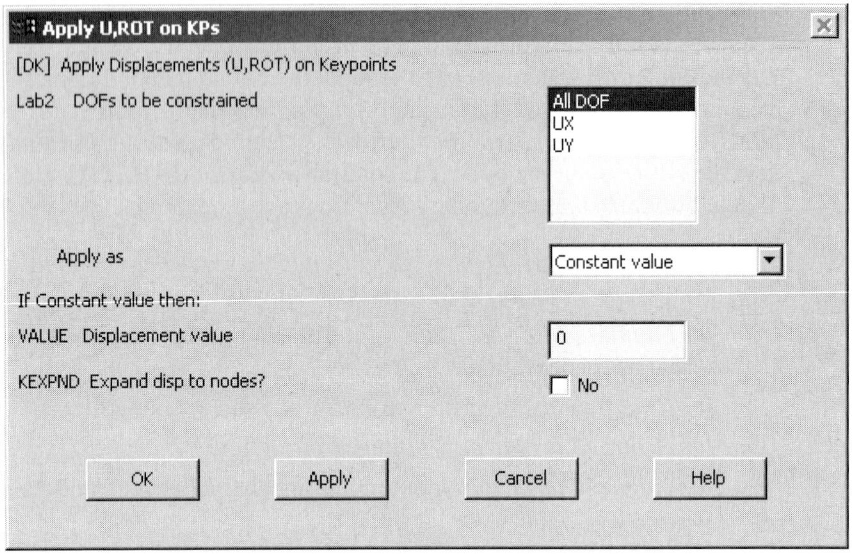

(a)

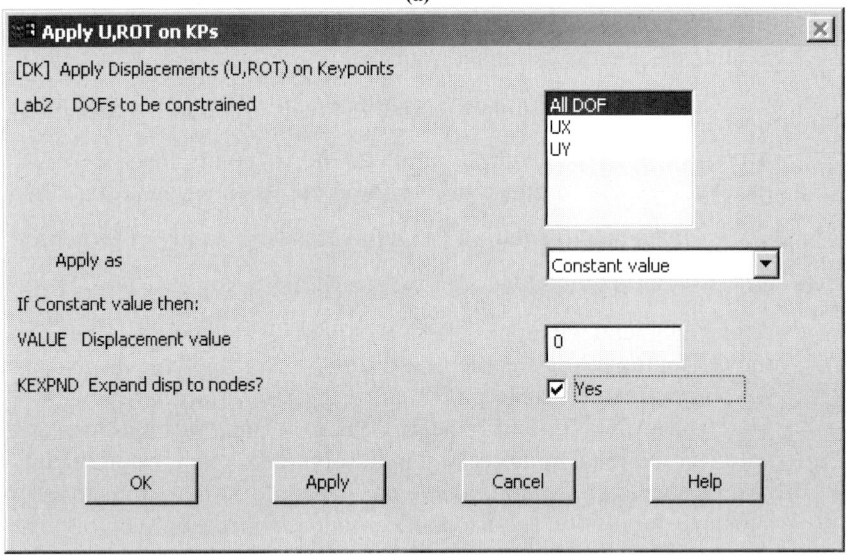

(b)

FIGURE 8.21 The dialog box for applying displacements on keypoints.

A picking menu will appear. You then pick the line(s) or surfaces that require a pressure load and press the OK button. An example of a dialog box for applying pressure loads on line(s) is shown in Figure 8.24.

For uniformly distributed loads, you need to specify only VALI. For a linear distribution, you need to specify both VALI and VALJ, as shown in Figure 8.25. It is important to note that in ANSYS, a positive VALI represents pressure into the surface.

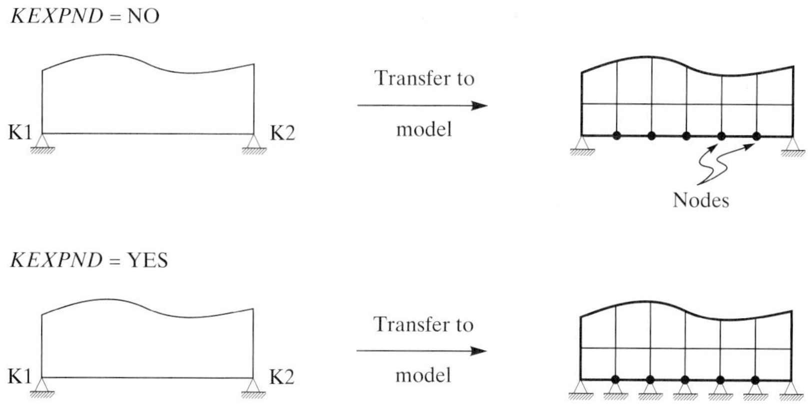

FIGURE 8.22 KEXPND options.

Obtain a Solution

Once you have created the model and have applied the boundary conditions and appropriate loads, then you need to instruct ANSYS to solve the set of equations generated by your model. But first save the database. To initiate the solution, pick the commands

main menu: **Solution → Solve → Current LS**

The next section is about reviewing the results of your analysis.

8.6 RESULTS OF YOUR FINITE ELEMENT MODEL: POSTPROCESSING

There are two postprocessors available for review of your results: (1) **POST1** and (2) **POST26**. The general postprocessor **(POST1)** contains the commands that allow you to list and display results of an analysis:

- Deformed shape displays and contour displays
- Tabular listings of the results data of the analysis
- Calculations for the results data and path operations
- Error estimations

You can read results data from the results file by using one of the choices from the dialog box shown in Figure 8.26. This dialog box may be accessed via the following command:

main menu: **General Postproc**

For example, if you are interested in viewing the deformed shape of a structure under a given loading, you choose the Plot Deformed Shape dialog box, as shown in Figure 8.27. To access this dialog box, issue the following sequence of commands:

main menu: **General Postproc → Plot Results → Deformed Shape**

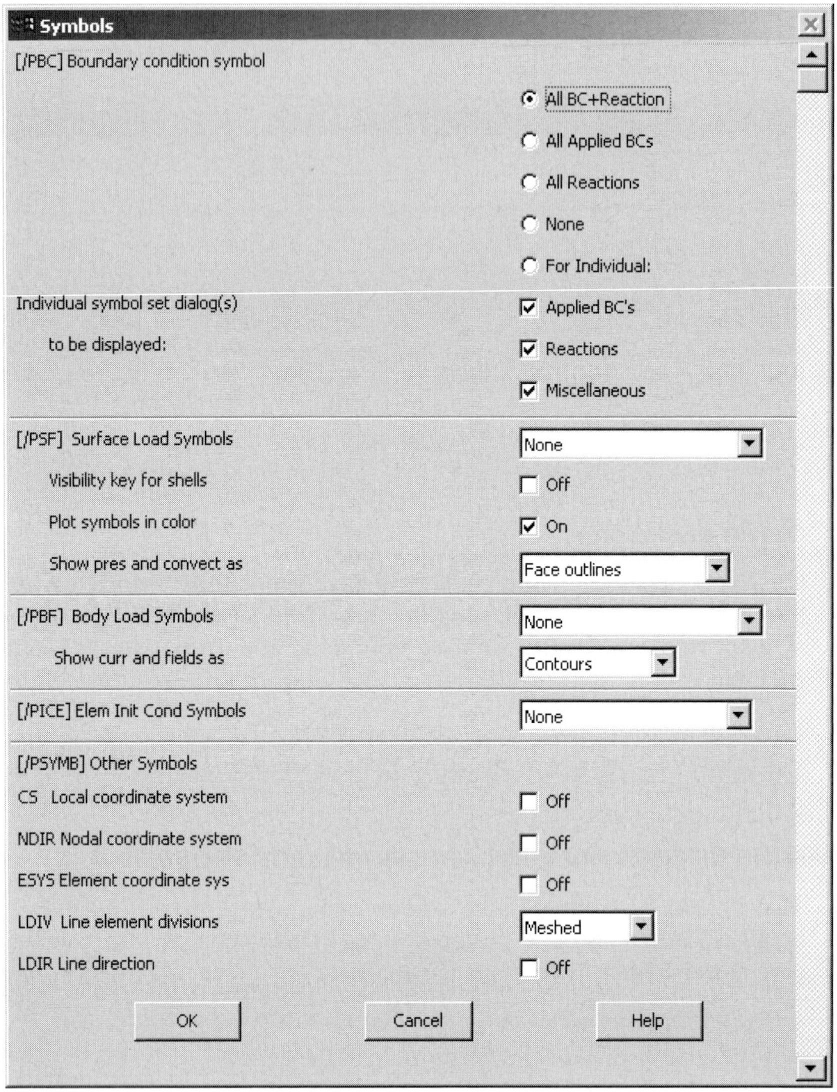

FIGURE 8.23 The Symbols dialog box.

You can also use contour displays to see the distribution of certain variables, such as a component of stress or temperature over the entire model. For example, issue the following command to access the dialog box shown in Figure 8.28.

main menu: **General Postproc → Plot Results →**

Contour Plot → Nodal Solution

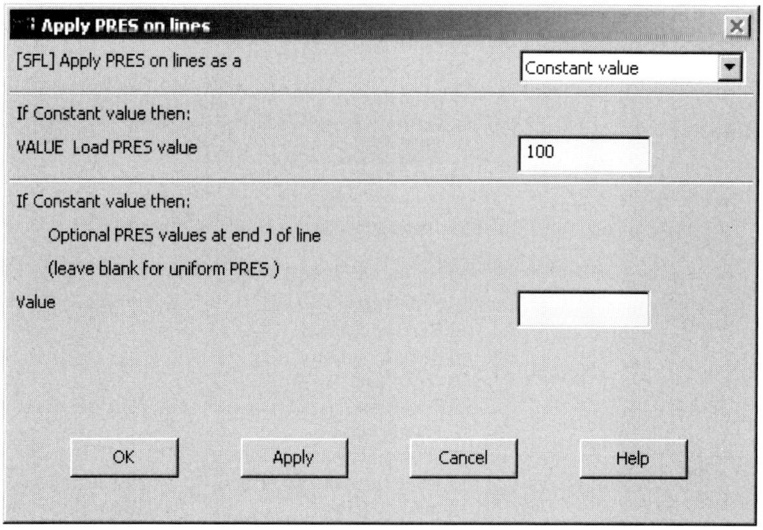

FIGURE 8.24 The dialog box for applying pressure loads on lines.

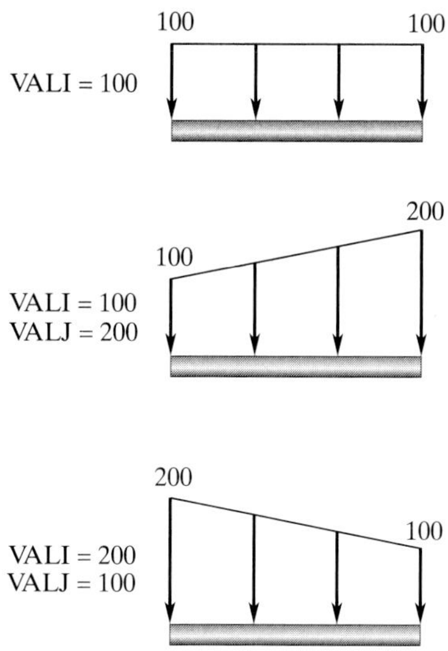

FIGURE 8.25 An example illustrating how to apply uniform and nonuniform loads.

As already mentioned, you can list the results in a tabular form as well. For example, to list the reaction forces, you issue the following command, which gives you a dialog box similar to the one shown in Figure 8.29:

main menu: **General Postproc → List Results → Reaction Solu**

Select the component(s) of your choice and press the OK button.

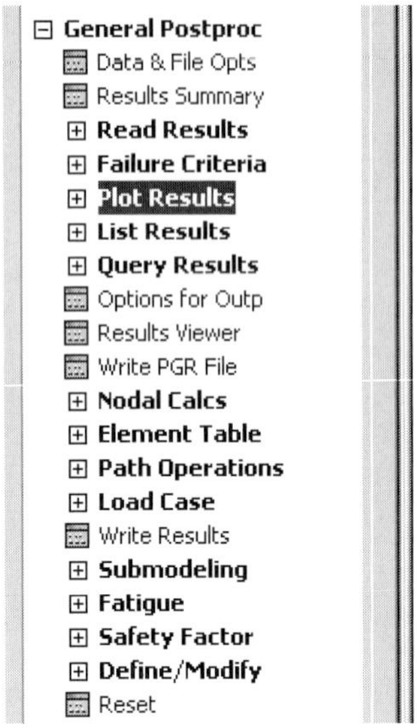

FIGURE 8.26 The General Postprocessing dialog box.

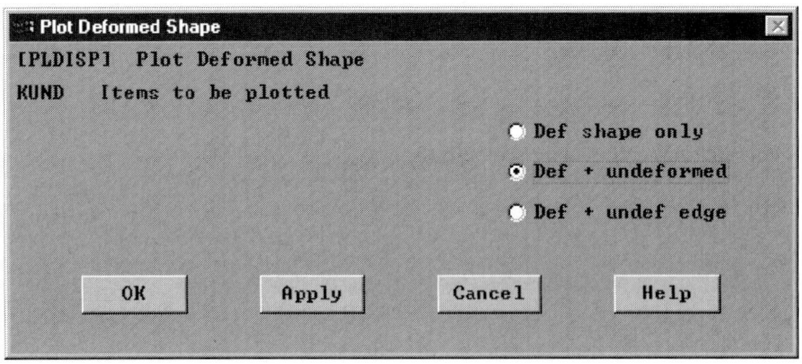

FIGURE 8.27 The Plot Deformed Shape dialog Box.

The *time-history postprocessor* (**POST26**) contains the commands that allow you to review results over time in a transient analysis. These commands will not be discussed here, but you may consult the ANSYS online help for further information about how to use the time-history postprocessor.

Once you have finished reviewing the results and wish to exit the ANSYS program, choose the Quit button from the ANSYS toolbar and pick the option you want. Press the OK button.

If, for any reason, you need to come back to modify a model, first launch ANSYS, and then type the file-name in the **Initial Jobname** entry field of the interactive dialog box. Then press the Run button. From the file menu, choose **Resume Jobname.DB.** Now you have complete access to your model. You can plot keypoints, nodes, elements, and the like to make certain that you have chosen the right problem.

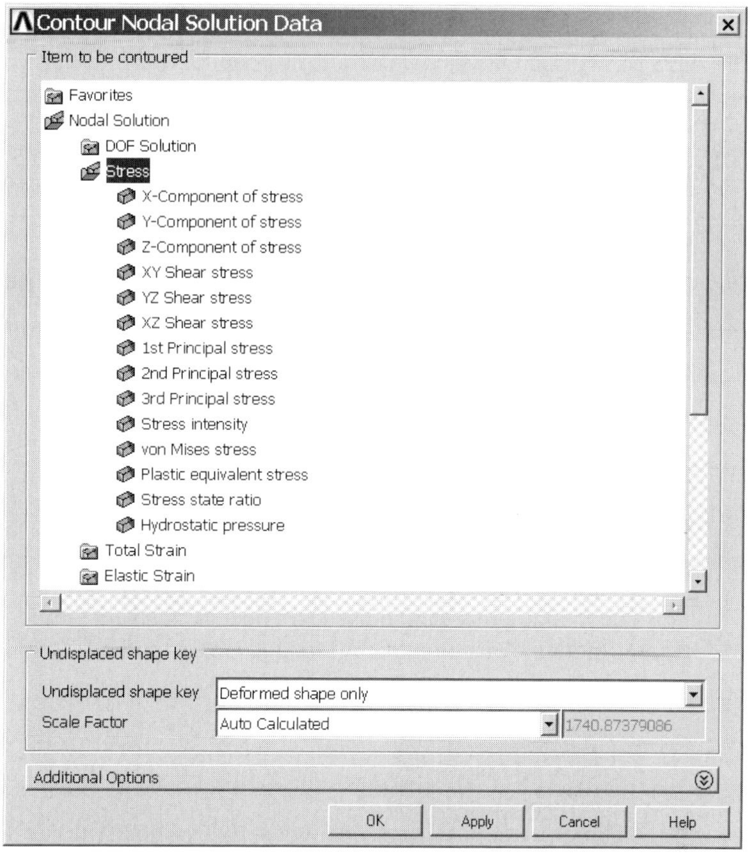

FIGURE 8.28 The Contour Nodal Solution Data dialog box.

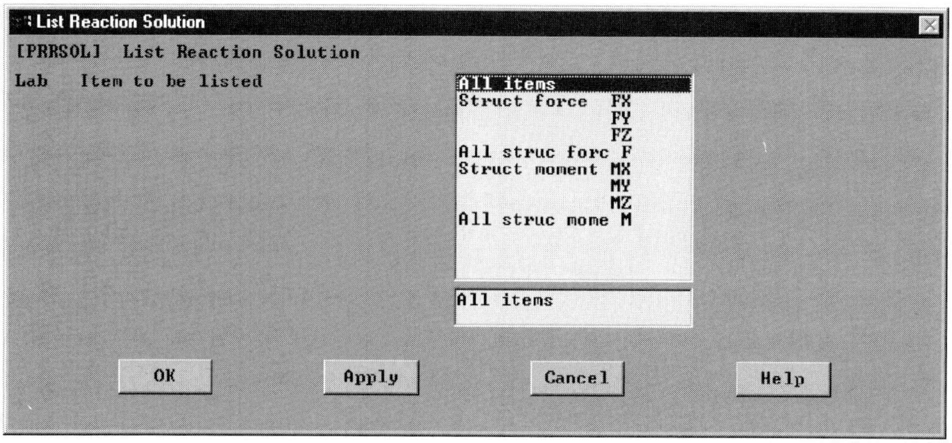

FIGURE 8.29 The List Reaction Solution dialog box.

8.7 SELECTION OPTIONS

The ANSYS program uses a database to store all of the data that you define during an analysis. ANSYS also offers the user the capability to select information about only a portion of the model, such as certain nodes, elements, lines, areas, and volumes for further processing. You can select functions anywhere within ANSYS. To start selecting, issue the following command to bring up the dialogue box shown in Figure 8.30.

utility menu: **Select → Entities** . . .

The various selection commands and their respective uses are as follows:

Select: To select a subset of active items from the full set.

Reselect: To select again from the currently selected subset.

Also Select: To add a different subset to the current subset.

Unselect: To deactivate a portion of the current subset.

Select All: To restore the full set.

Select None: To deactivate the full set (opposite of the Select All command).

Invert: To switch between the active and inactive portions of the set.

The select dialog box can be used to select or unselect entities of your solid or finite element model. You can make selections based on the location of your entities in

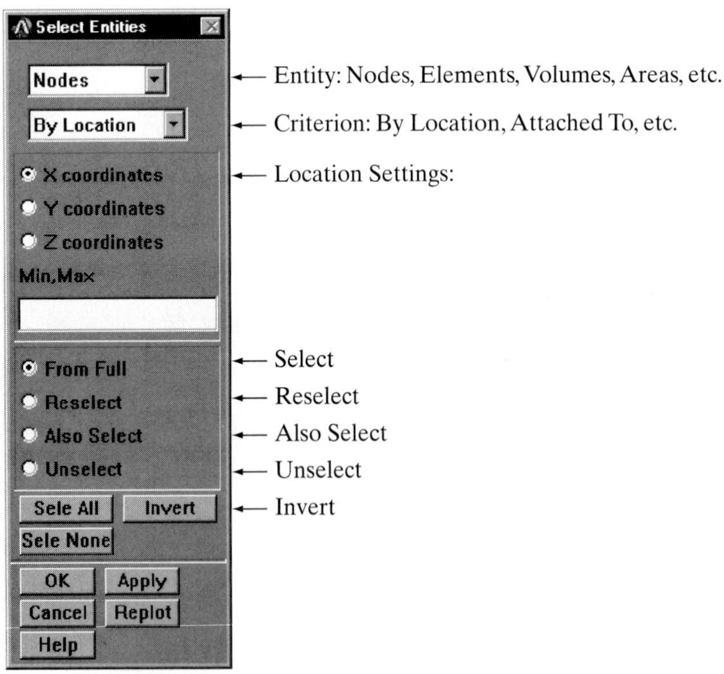

FIGURE 8.30 The Select Entities dialog box.

space, or you can select entities that are attached to other selected entities, such as nodes that are attached to selected elements. Be aware, however, that you must reactivate all entities before solving your model. Unselected entities will not be included in a solution. For example, if you select a subset of nodes on which to apply constraints, you should reactivate all nodes before solving. ANSYS allows the user to activate all entities with one simple operation by the command

utility menu: **Select → Everything** . . .

You can also select a set of related entities in a hierarchical fashion. For example, given a subset of areas, you can select (a) all lines defining the areas, (b) all keypoints defining those lines, (c) all elements belonging to the areas, and so on. To select in this fashion, use the command

utility menu: **Select → Everything Below**

ANSYS also provides the capability to group some selected entities into a component. You can group one type of entity—such as nodes, elements, keypoints, or lines—into a component to be identified by a user-defined name (up to eight characters long).

8.8 GRAPHICS CAPABILITIES

Good graphics are especially important for visualizing and understanding a problem being analyzed. The ANSYS program provides numerous features that allow you to enhance the visual information presented to you. Some examples of the graphics capabilities of ANSYS include deformed shapes, result contours, sectional views, and animation. Consult the ANSYS procedure manual for additional information about more than 100 different graphics functions available to the user.

Up to five ANSYS windows can be opened simultaneously within one graphics window. You can display different information in different windows. ANSYS windows are defined in screen coordinates (-1 to $+1$ in the x-direction and -1 to $+1$ in the y-direction). By default, ANSYS directs all graphics information to one window (window 1). In order to define additional windows, you need to access the window-layout dialog box, as shown in Figure 8.31. To do so, issue the following commands:

utility menu: **PlotCtrls → Window Controls → Window Layout** . . .

There are three important concepts that you need to know with respect to window layout: (1) focus point, (2) distance, and (3) viewpoint. The focus point, with coordinates XF, YF, ZF, is the point on the model that appears at the center of the window. By changing the coordinates of the focus point, you can make a different point on the model appear at the center of the window. Distance determines the magnification of an image. As the distance approaches infinity, the image becomes a point on the screen. As the distance is decreased, the image size increases until the image fills the window. Viewpoint determines the direction from which the object is viewed. A vector is established from the viewpoint to the origin of the display coordinate. The line of sight is parallel to this vector and is directed at the focus point.

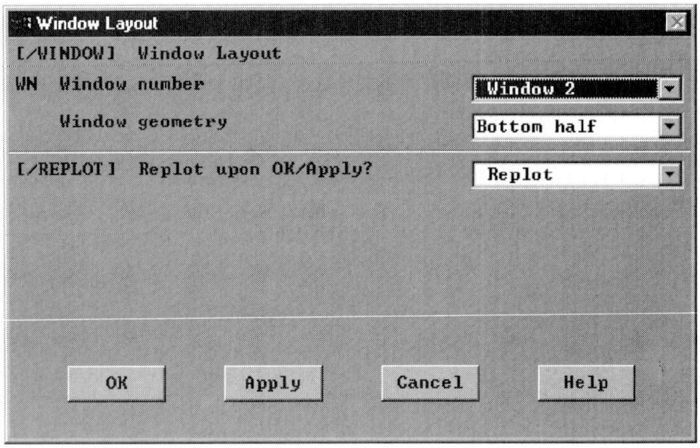

FIGURE 8.31 The Window Layout dialog box.

Next, the Pan–Zoom–Rotate dialog box allows you to change viewing directions, zoom in and out, or rotate your model. You can access this dialog box, shown in Figure 8.32 by the following commands:

utility menu: **PlotCtrls → Pan, Zoom, Rotate** . . .

The various commands within the Pan–Zoom–Rotate dialog box and their respective functions are:

Zoom: Pick the center and the corner of the zoom rectangle.

Box Zoom: Pick the two corners of the zoom rectangle.

Win Zoom: Same as Box Zoom, except the zoom rectangle has the same proportions as the window.

• Zoom out.

● Zoom in.

Dynamic Mode: Allows you to pan, zoom, and rotate the image dynamically.

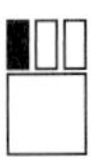
Pan model in *X*- and *Y*-directions.

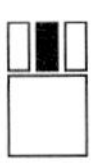
Move the mouse right and left to rotate the model about the *Z*-axis of the screen. Move the mouse up and down to zoom in and out.

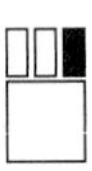
Move the mouse right and left to rotate the model about the *Y*-axis of the screen. Move the mouse up and down to rotate the model about the *X*-axis of the screen.

Fit: Changes the graphics specifications such that the image fits the window exactly.

Reset: Resets the graphics specifications to their default values.

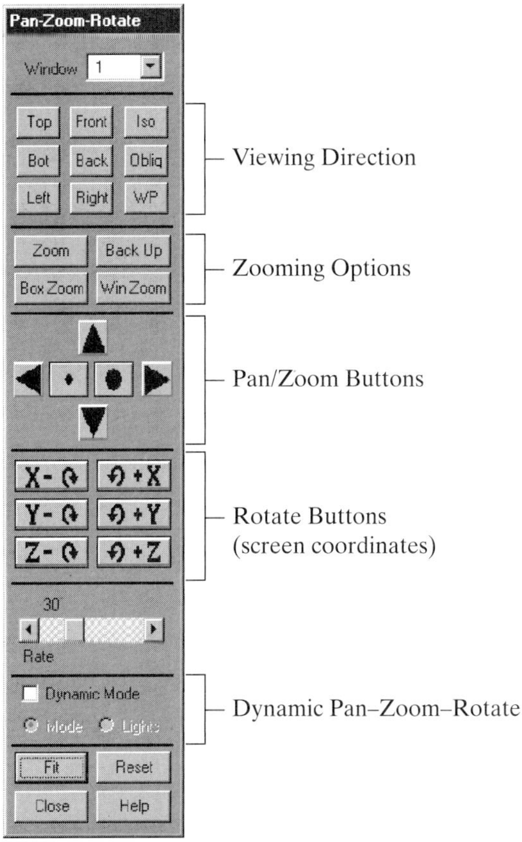

FIGURE 8.32 The Pan–Zoom–Rotate dialog box.

In Section 8.10, an example problem will demonstrate the basic steps in creating and analyzing a model with ANSYS.

8.9 ERROR-ESTIMATION PROCEDURES

In the previous chapters we discussed how to use fundamental principles, such as statics equilibrium conditions or the conservation of energy, to check for the validity of results. We have also noted that when economically feasible or practical, the experimental verification of a finite element model is the best way to check for the validity of results. Moreover, it has been pointed out that the element size affects the accuracy of your results. Now, consider how you know whether the element sizes associated with a meshed model are fine enough to produce good results. A simple way to find out is to first model a problem with a certain number of elements and then compare its results to the results of a model that you create with twice as many elements. In other words, double the number of original elements and compare the results of the analysis. If no significant difference between the results of the two meshes is detected, then the meshing is adequate. If substantially different results are obtained, then further mesh refinement might be necessary.

The ANSYS program offers error-estimation procedures that calculate the level of solution error due to mesh discretization. Error calculations used by ANSYS are based on discontinuity of stresses (or heat fluxes) along interelemental boundaries. Because neighboring elements share common nodes, the difference between the nodal stresses calculated for each element results in a discontinuous stress solution from element to element. The degree of discontinuity is based on both the mesh discretization and the severity of the stress gradient. Therefore, it is the difference in stresses from element to element that forms the basis for error calculation in ANSYS.

Error calculations in ANSYS are presented in three different forms: (1) the elemental-energy error (SERR for structural problems and TERR for thermal problems), which measures the error in each element based on the differences between averaged and unaveraged nodal stress or thermal flux values; (2) the percent error in energy norm (SEPC for structural problems and TEPC for thermal problems), which is a global measure of error energy in the model that is based on the sum of the elemental-error energies; and (3) the nodal-component value deviation (SDSG for structural problems and TDSG for thermal problems), which measures the local error quantity for each element and is determined by computing the difference between the averaged and unaveraged values of stress or heat flux components for an element. To display error distributors, use the following commands:

main menu: **General Postproc → Plot Results → Contour Plot → Element Solu**

You can select and plot the elemental-energy error to observe the high-error regions where mesh refinement may be necessary. You can also plot SDSG (or TDSG) to identify and quantify the region of maximum discretization errors by using the following command:

main menu: **General Postproc → Element Table → Define Table**

The elemental-energy error or the nodal-component deviations can be listed as well by using the following command:

main menu: **General Postproc → Element Table → List Element Table**

Example of ANSYS elements that include error estimations are given in Table 8.1. Note that ANSYS stress-contour plots and listings give the upper and the lower error bounds based on SDSG or TDSG calculations. The estimated-error bound of plotted stresses is denoted by SMXB or SMNB labels in the graphics-status area.

To make the task of mesh evaluation and refinement simpler, ANSYS offers adaptive meshing, which is a process that automatically evaluates mesh-discretization error and performs mesh refinement to reduce the error. The adaptive meshing performed by the ADAPT program of ANSYS will perform the following tasks: (1) it will generate an initial mesh and solve the model; (2) based on error calculations, it will determine if mesh refinement is needed; (3) if mesh refinement is necessary, it will automatically refine the mesh and solve the new model; and (4) it will refine the mesh until a loop limit or an acceptable error limit has been reached. Note that to begin the first run of the adaptive-meshing program, you need to create the initial model by defining the element type, material property, and so on.

TABLE 8.1 Examples of ANSYS elements that include error estimations

Structural Solids	Thermal Solids
PLANE2	PLANE35
PLANE42	PLANE55
PLANE82	PLANE77
SOLID45	SOLID70
SOLID92	SOLID87
SOLID95	SOLID90

8.10 AN EXAMPLE PROBLEM

Consider one of the many steel brackets ($E = 29 \times 10^6$ lb/in^2, $\nu = 0.3$) used to support bookshelves. The dimensions of the bracket are shown in Figure 8.33. The bracket is loaded uniformly along its top surface, and it is fixed along its left edge. Under the given loading and the constraints, plot the deformed shape; also determine the principal stresses and the von Mises stresses in the bracket.

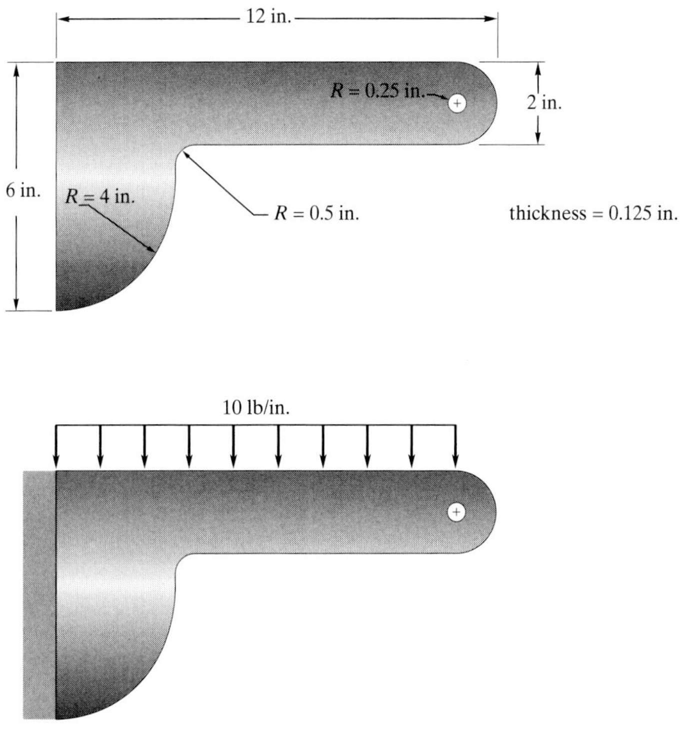

FIGURE 8.33 A schematic of the steel bracket in the example problem.

The following steps demonstrate how to solve this problem using ANSYS:

Enter the **ANSYS** program by using the Launcher.

Type **Bracket** (or a file name of your choice) in the **Jobname** entry field of the dialog box.

Pick **Run** to start the GUI.

Create a title for the problem. This title will appear on ANSYS display windows to provide a simple way of identifying the displays. To create a title, issue the command

utility menu: **File → Change Title** ...

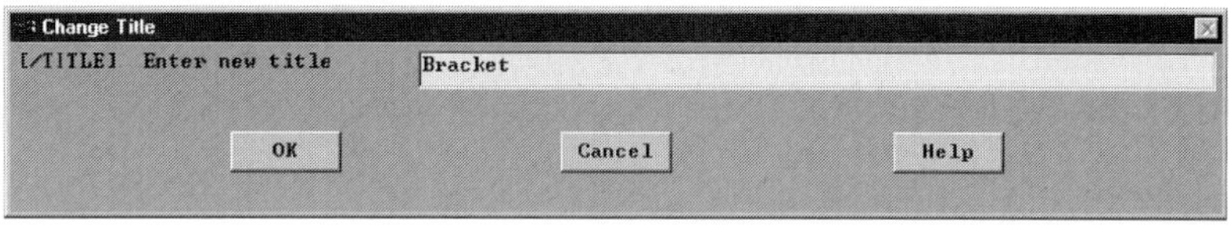

Define the element type and material properties:

main menu: **Preprocessor → Element Type → Add/Edit/Delete**

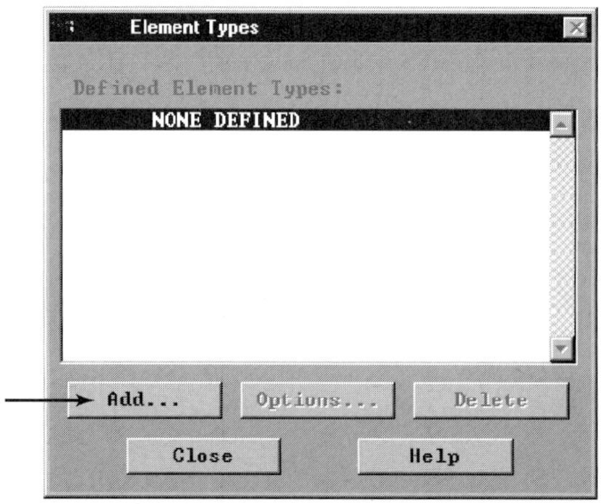

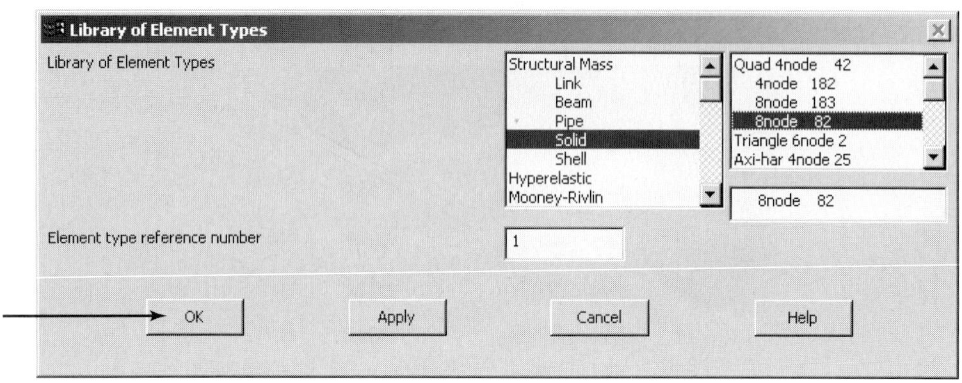

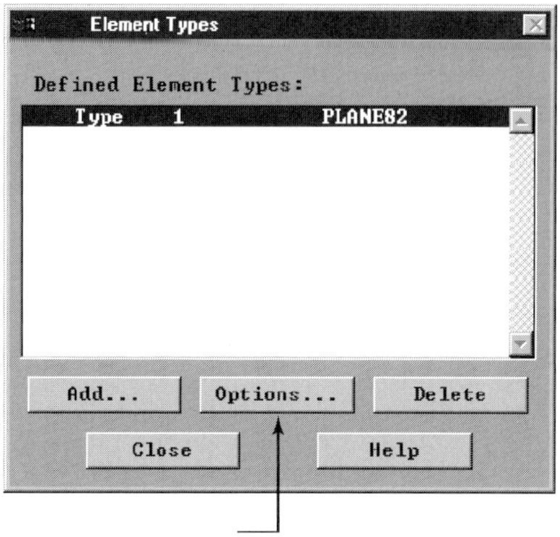
Element Types
Defined Element Types:
Type 1 PLANE82
Add...
Options...
Delete
Close
Help

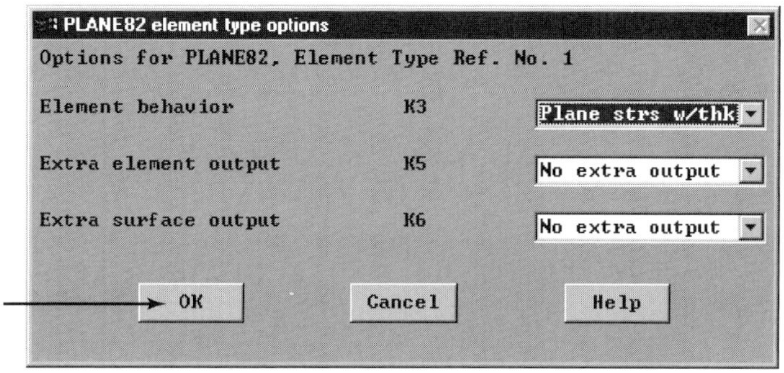
PLANE82 element type options
Options for PLANE82, Element Type Ref. No. 1
Element behavior K3 Plane strs w/thk
Extra element output K5 No extra output
Extra surface output K6 No extra output
OK
Cancel
Help

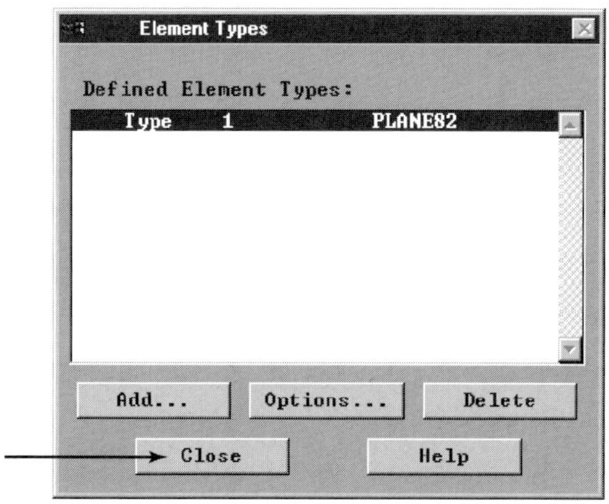
Element Types
Defined Element Types:
Type 1 PLANE82
Add...
Options...
Delete
Close
Help

Assign the thickness of the bracket:

main menu: **Preprocessor → Real Constants → Add/Edit/Delete**

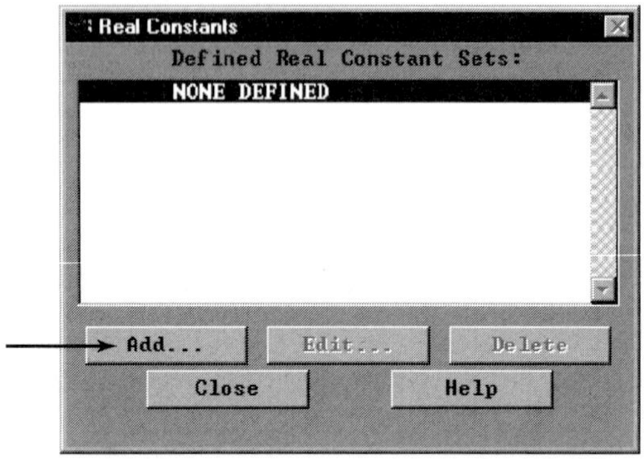

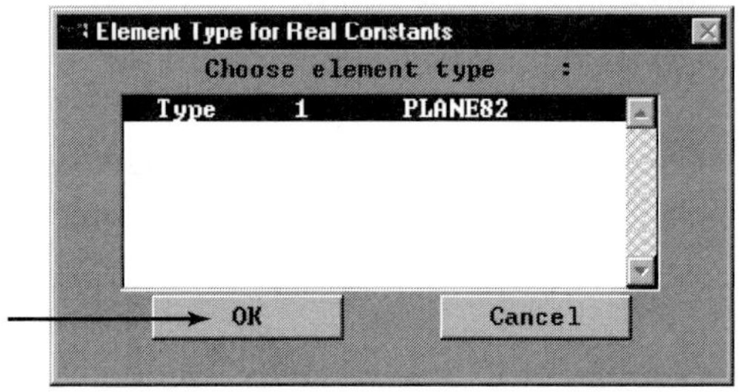

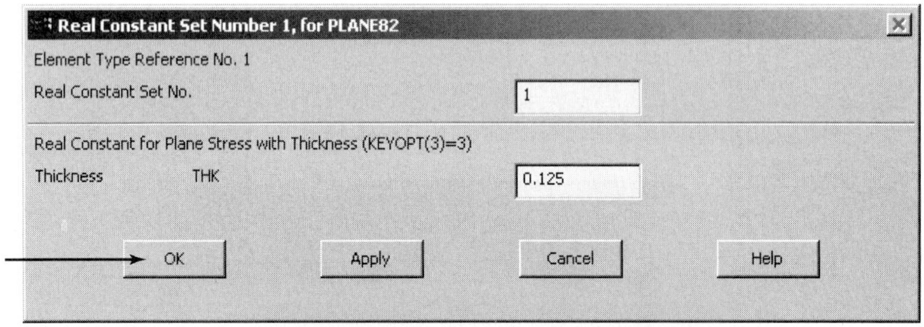

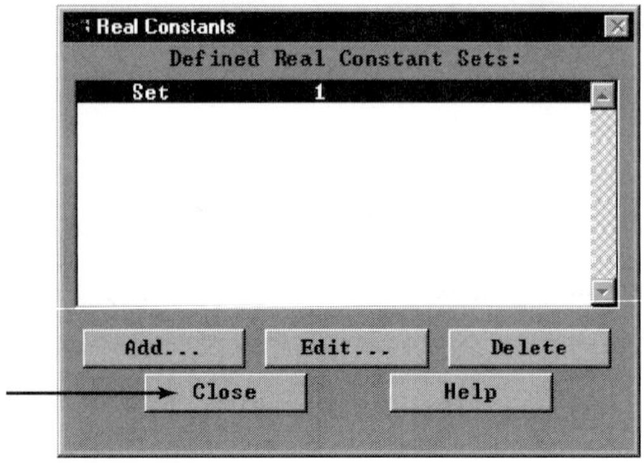

Assign the modulus of elasticity and the Poisson's-ratio values:

main menu: **Preprocessor → Material Props → Material Models → Structural → Linear → Elastic → Isotropic**

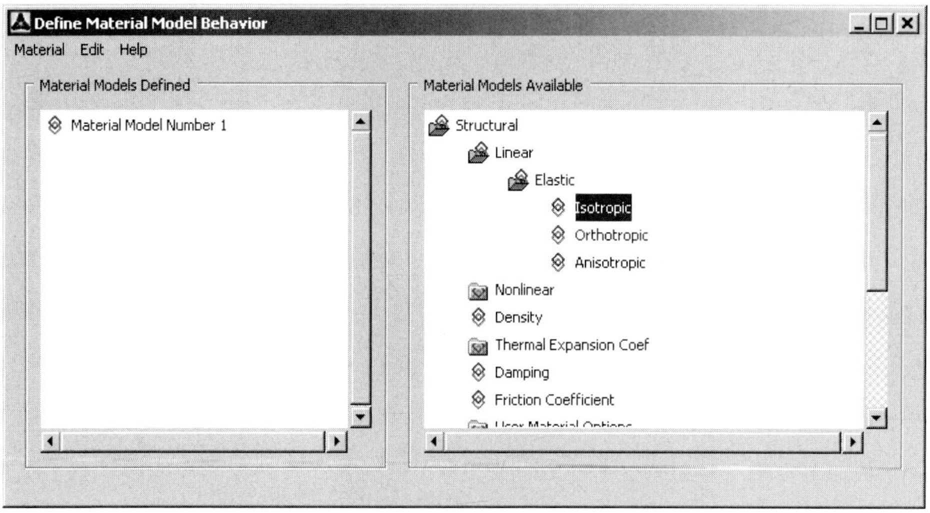

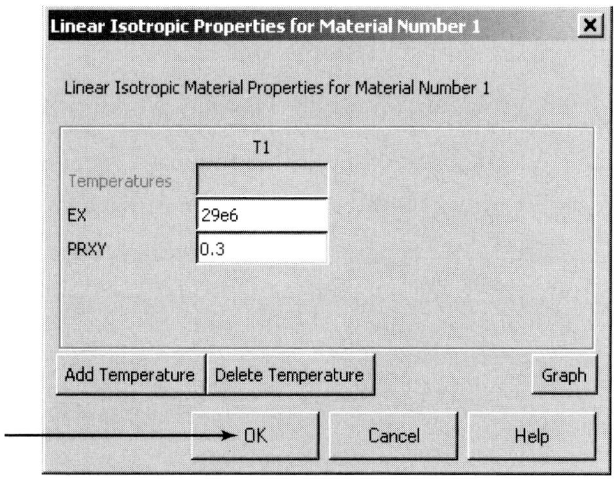

ANSYS Toolbar: **SAVE_DB**

Set up the graphics area—that is, the work plane, zoom, and so on:

utility menu: **Workplane → Wp Settings** . . .

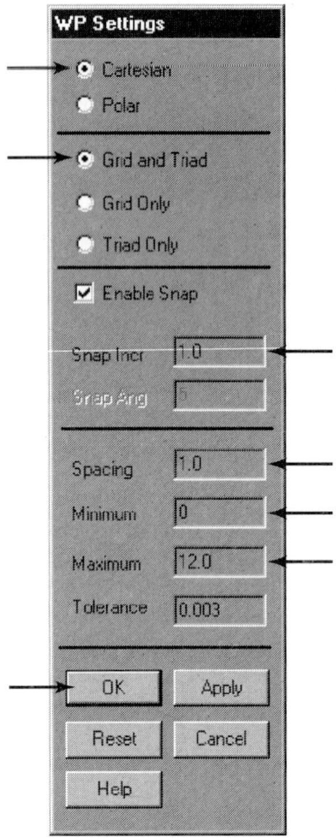

Toggle on the workplane by the command sequence

utility menu: **Workplane → Display Working Plane**

Bring the workplane to view by the command sequence

utility menu: **PlotCtrls → Pan, Zoom, Rotate** . . .

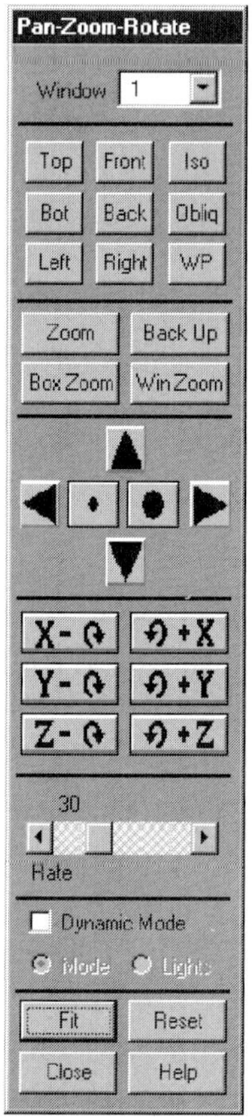

Click on the **small circle** and the **arrows** until you bring the workplane to view, and then create the geometry:

main menu: **Preprocessor** → **Modeling** → **Create** → **Areas** → **Rectangle** → **By 2 Corners**

a) On the workplane, pick the location of the corners of Areas 1 and 2, as shown in Figure 8.34, and apply:

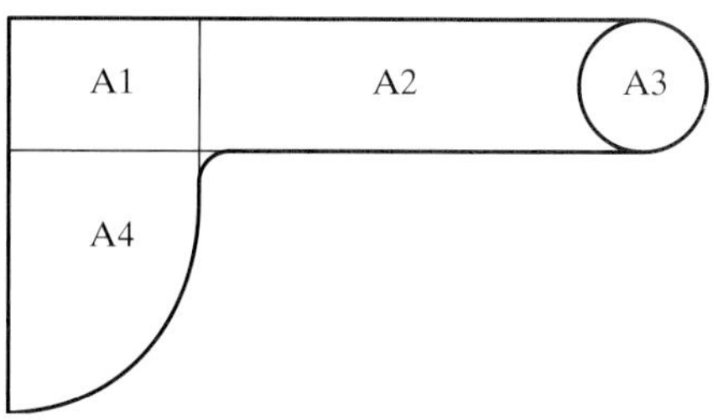

FIGURE 8.34 The areas making up the bracket.

[At WP = 0,12 in the upper left corner of the workplane, press the left button]

[First, expand the rubber band down 2.0 and right 4.0 and, then, press the left button]

[WP = 4,12]

[Expand the rubber band down 2.0 and right 7.0]

OK

b) Create circle A3 by the commands

main menu: **Preprocessor → Modeling → Create → Areas → Circle → Solid Circle**

[WP = 11,11]

[Expand the rubber band to a radius of 1.0]

OK

c) Create quarter-circle A4 by the command

main menu: **Preprocessor → Modeling → Create → Areas → Circle → Partial Annulus**

Type in the following values in the given fields:

[WPX = 0]

[WPY = 10]

[Rad-1 = 0]

[Theta-1 = 0]

[Rad-2 = 4]

[Theta-2 = −90]

OK

d) Before creating the fillet, join the keypoints of Areas 1, 2, and 4 by the commands

main menu: **Preprocessor → Modeling → Operate → Booleans → Glue → Areas**

Pick Areas 1, 2, and 4.

OK

e) Create the fillet by the commands

main menu: **Preprocessor → Modeling → Create → Lines → Line Fillet**

[Pick the bottom edge of rectangular Area 2]

[Pick the curved edge of quarter-circle Area 4]

APPLY

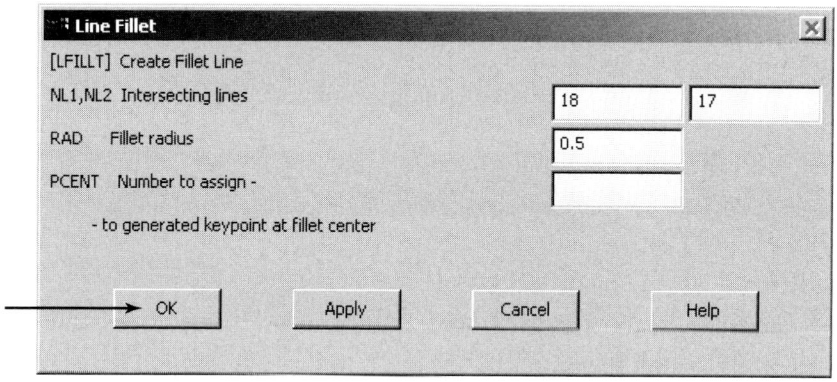

Then, issue the command

utility menu: **PlotCtrls → Pan, Zoom, Rotate** . . .

Use the Box Zoom button to zoom about the fillet region, and issue the command

utility menu: **Plot → Lines**

f) Create an area for the fillet with the commands

main menu: **Preprocessor → Modeling → Create → Areas → Arbitrary → By Lines**

Pick the fillet line and the two intersecting smaller lines.

OK

g) Add the areas together with the commands

main menu: **Preprocessor → Modeling → Operate → Booleans → Add → Areas**

Click on the **Pick All** button and issue the command

utility menu: **PlotCtrls → Pan, Zoom, Rotate** . . .

Click on the **Fit** button and then **Close.**

h) Create the area of the small hole, but first change the **Snap Incr** value in the WP Settings dialog box to 0.25.

OK

Then issue the commands

main menu: **Preprocessor → Modeling → Create → Areas → Circle → Solid Circle**

[WP = 11,11]

Expand the rubber band to a radius of 0.25]

OK

i) Subtract the area of the small hole with the commands

main menu: **Preprocessor → Modeling → Operate → Booleans → Subtract → Areas**

[Pick the bracket area]

[anywhere in the ANSYS graphics area, Apply]

[Pick the small circular area (*r* = 0.25)]

[anywhere in the ANSYS graphics area, Apply]

OK

Now you can toggle off the workplane grids with the command

utility menu: **Workplane → Display Working Plane**

ANSYS Toolbar: **SAVE_DB**

You are now ready to mesh the area of the bracket to create elements and nodes. Issue the commands

main menu: **Preprocessor → Meshing → Size Cntrls → Manual Size → Global → Size**

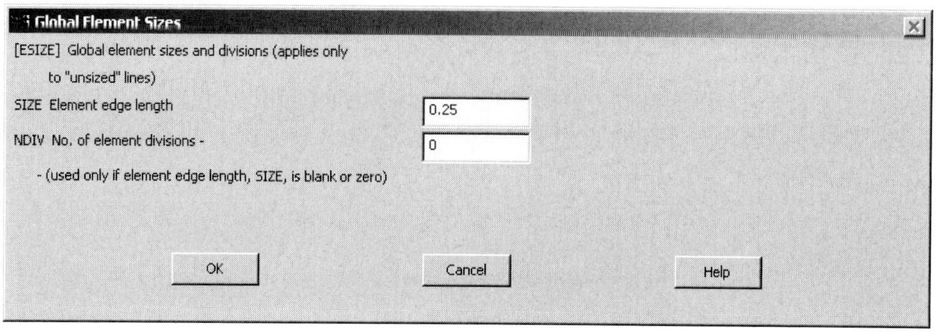

ANSYS Toolbar: **SAVE_DB**

main menu: **Preprocessor → Meshing → Mesh → Areas → Free**

Click on the **Pick All** button.

Apply boundary conditions:

main menu: **Solution → Define Loads → Apply → Structural → Displacement → On Keypoints**

Pick the three keypoints: (1) upper left corner of Area 1, (2) two inches below the keypoint you just picked (i.e., the upper left corner of Area 4), and (3) the lower left corner of Area 4.

OK

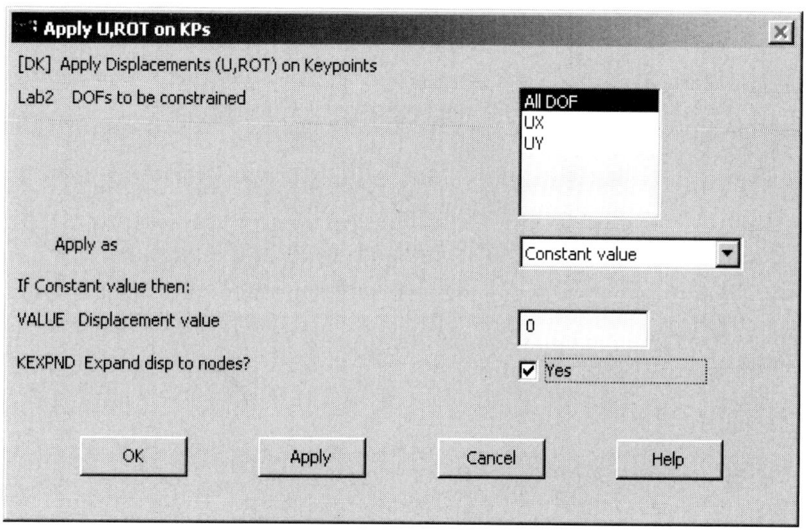

main menu: **Solution → Define Loads → Apply → Structural → Pressure → On Lines**

Pick the upper two horizontal lines associated with Area 1 and Area 2 (on the upper edge of the bracket).

OK

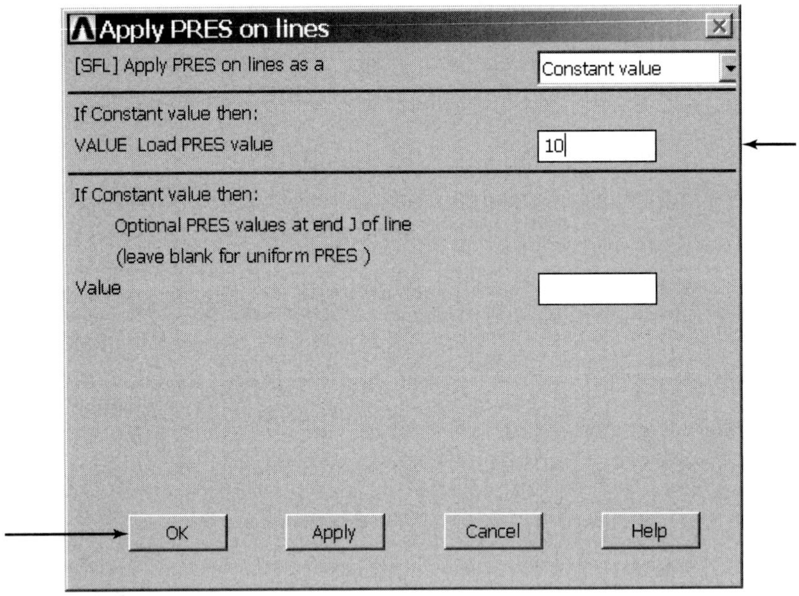

Solve the problem:

main menu: **Solution → Solve → Current LS**

OK

Close (the solution is done!) window.

Close (the/STAT Command) window if it appears.

For the postprocessing phase, first plot the deformed shape by using the commands

main menu: **General Postproc → Plot Results → Deformed Shape**

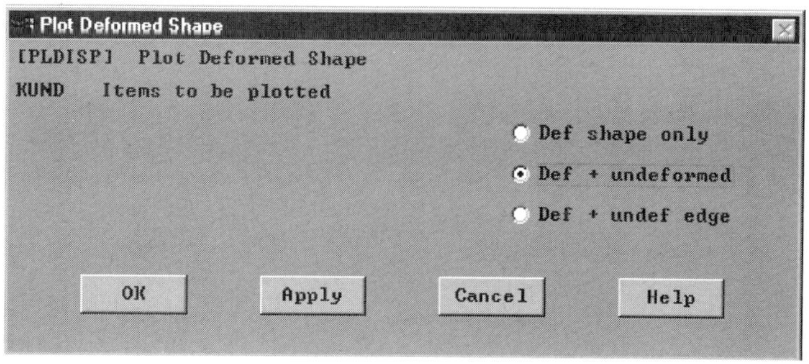

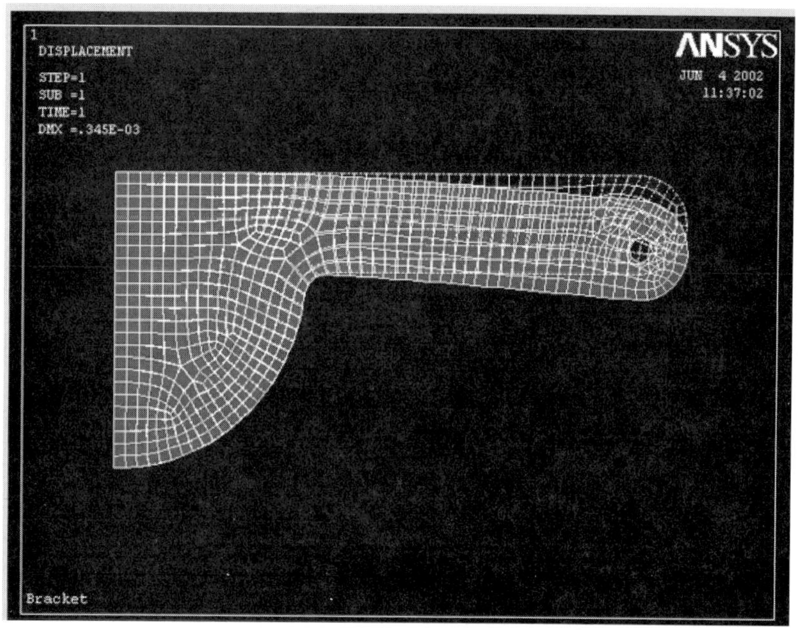

Plot the von Mises stresses with the commands

main menu: **General Postproc → Plot Results**

→ Contour Plot → Nodal Solu

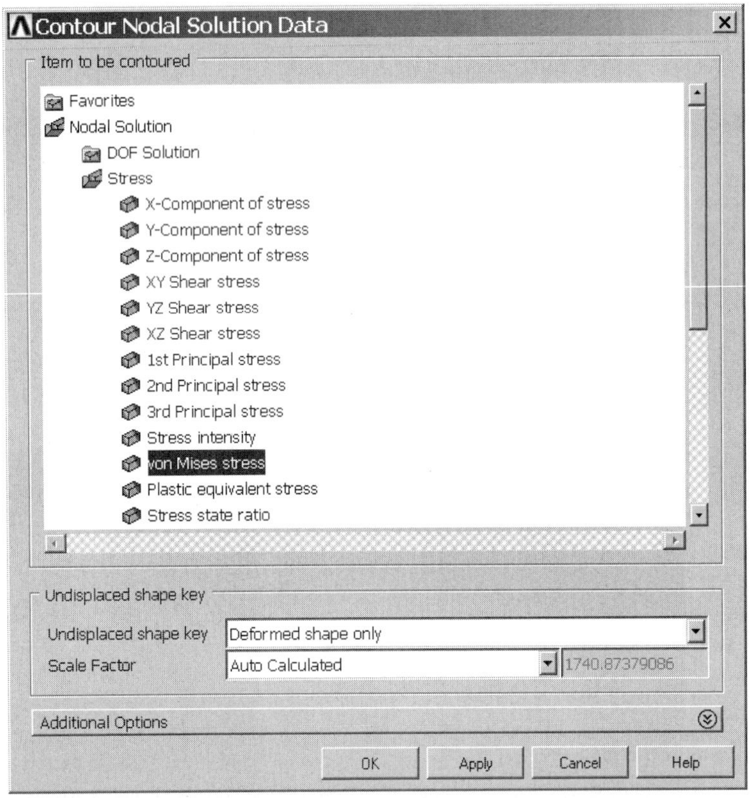
Contour Nodal Solution Data
Item to be contoured
Favorites
Nodal Solution
DOF Solution
Stress
X-Component of stress
Y-Component of stress
Z-Component of stress
XY Shear stress
YZ Shear stress
XZ Shear stress
1st Principal stress
2nd Principal stress
3rd Principal stress
Stress intensity
von Mises stress
Plastic equivalent stress
Stress state ratio
Undisplaced shape key
Undisplaced shape key
Deformed shape only
Scale Factor
Auto Calculated
1740.87379086
Additional Options
OK
Apply
Cancel
Help

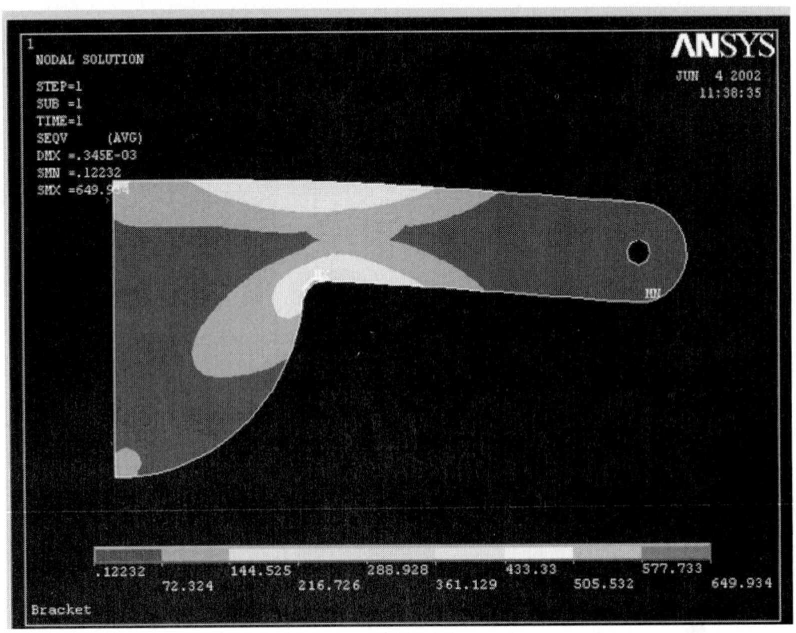
1
NODAL SOLUTION
STEP=1
SUB =1
TIME=1
SEQV (AVG)
DMX =.345E-03
SMN =.12232
ANSYS
JUN 4 2002
11:38:35
.12232
72.324
144.525
216.726
288.928
361.129
433.33
505.532
577.733
649.934
Bracket

Repeat the previous step and pick the principal stresses to be plotted. Then, exit ANSYS and save everything:

Toolbar: **QUIT**

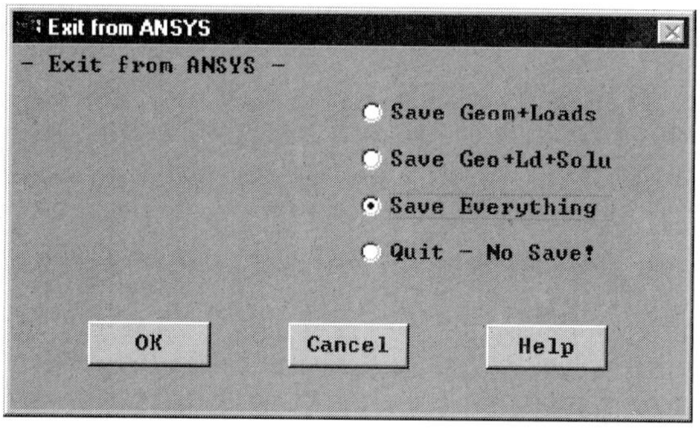

SUMMARY

At this point you should know

1. the basic organization of the ANSYS program. There are three processors that you will use frequently: (1) the *preprocessor* (**PREP7**), (2) the *processor* (**SOLUTION**), and (3) the general *postprocessor* (**POST1**).

2. the commands the preprocessor (**PREP7**) contains that you need to use to build a model:

   - define element types and options
   - define element real constants
   - define material properties
   - create model geometry
   - define meshing controls
   - mesh the object created

3. the commands the Solution *processor* (**SOLUTION**) has that allow you to apply boundary conditions and loads. The solution processor also solves for the nodal solutions and calculates other elemental information.

4. the commands the general *postprocessor* (**POST1**) contains that allow you to list and display results of an analysis:

   - read results data from results file
   - read element results data
   - plot results
   - list results

5. that ANSYS writes and reads many files during a typical analysis.

6. that ANSYS also offers the user the capability to select information about a portion of the model, such as certain nodes, elements, lines, areas, and volumes, for further processing.

7. that the ANSYS program provides numerous features that allow you to enhance the visual information presented to you. Some examples of the graphics capabilities of ANSYS are deformed shapes, result contours, sectional views, and animation.

REFERENCES

ANSYS Manual: Introduction to ANSYS, Vol. I, Swanson Analysis Systems, Inc.

ANSYS User's Manual: Procedures, Vol. I, Swanson Analysis Systems, Inc.

ANSYS User's Manual: Commands, Vol. II, Swanson Analysis Systems, Inc.

ANSYS User's Manual: Elements, Vol. III, Swanson Analysis Systems, Inc.

TABLE 8.2 Element types offered by ANSYS

The ANSYS program offers nearly 150 different element types. For detailed information on a specific element type, see the *Elements* volume (Vol. III) of the ANSYS User's Manual.

Structural Point Structural Mass  MASS21 1 node 3–D space DOF: UX, UY, UZ, ROTX, ROTY, ROTZ	Structural 2–D Line Spar  LINK1 2 nodes 2–D space DOF: UX, UY	Structural 2–D Beam Elastic Beam  BEAM3 2 nodes 2–D space DOF: UX, UY, ROTZ	Plastic Beam  BEAM23 2 nodes 2–D space DOF: UX, UY, ROTZ	Offset Tapered Unsymmetric Beam  BEAM54 2 nodes 2–D space DOF: UX, UY, ROTZ
Structural 3–D Line Spar  LINK8 2 nodes 3–D space DOF: UX, UY, UZ	Tension–Only Spar  LINK10 2 nodes 3–D space DOF: UX, UY, UZ	Linear Actuator 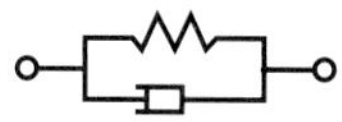 LINK11 2 nodes 3–D space DOF: UX, UY, UZ	Structural 3–D Beam Elastic Beam  BEAM4 2 nodes 3–D space DOF: UX, UY, UZ, ROTX, ROTY, ROTZ	Thin Walled Plastic Beam  BEAM24 2 nodes 3–D space DOF: UX, UY, UZ, ROTX, ROTY, ROTZ
Offset Tapered Unsymmetric Beam  BEAM44 2 nodes 3–D space DOF: UX, UY, UZ, ROTX, ROTY, ROTZ	Structural Pipe Elastic Straight Pipe  PIPE16 2 nodes 3–D space DOF: UX, UY, UZ, ROTX, ROTY, ROTZ	Elastic Pipe Tee  PIPE17 4 nodes 3–D space DOF: UX, UY, UZ, ROTX, ROTY, ROTZ	Curved Pipe (Elbow)  PIPE18 2 nodes 3–D space DOF: UX, UY, UZ, ROTX, ROTY, ROTZ	Plastic Straight Pipe  PIPE20 2 nodes 3–D space DOF: UX, UY, UZ, ROTX, ROTY, ROTZ
Immersed Pipe  PIPE59 2 nodes 3–D space DOF: UX, UY, UZ, ROTX, ROTY, ROTZ	Plastic Curved Pipe  PIPE60 2 nodes 3–D space DOF: UX, UY, UZ, ROTX, ROTY, ROTZ	Structural 2–D Solid Triangular Solid 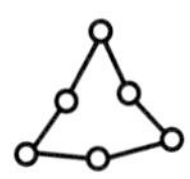 PLANE2 6 nodes 2–D space DOF: UX, UY	Axisymmetric Harmonic Struct. Solid 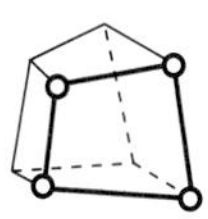 PLANE25 4 nodes 2–D space DOF: UX, UY, UZ	Structural Solid 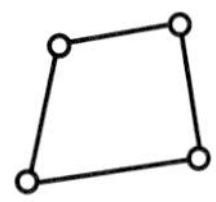 PLANE42 4 nodes 2–D space DOF: UX, UY

TABLE 8.2 (*continued*) Element types offered by ANSYS

The ANSYS program offers nearly 150 different element types. For detailed information on a specific element type, see the *Elements* volume (Vol. III) of the ANSYS User's Manual.

		Structural 3–D Solid		
Structural Solid 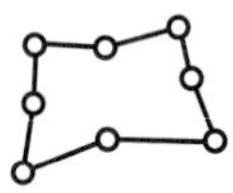 PLANE82 8 nodes 2–D space DOF: UX, UY	Axisymmetric Harmonic Struct. Solid 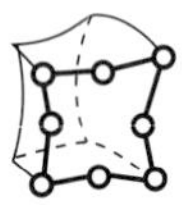 PLANE83 8 nodes 2–D space DOF: UX, UY, UZ	Structural Solid 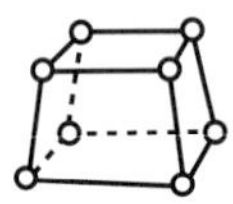 SOLID45 8 nodes 3–D space DOF: UX, UY, UZ	Layered Solid  SOLID46 8 nodes 3–D space DOF: UX, UY, UZ	Anisotropic Solid 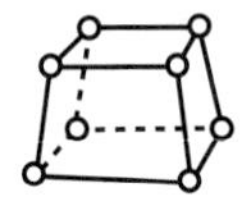 SOLID64 8 nodes 3–D space DOF: UX, UY, UZ
Reinforced Solid 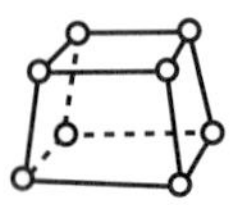 SOLID65 8 nodes 3–D space DOF: UX, UY, UZ	Solid with Rotations 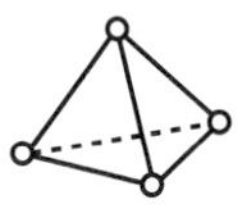 SOLID72 4 nodes 3–D space DOF: UX, UY, UZ, ROTX, ROTY, ROTZ	Solid with Rotations 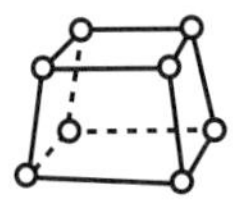 SOLID73 8 nodes 3–D space DOF: UX, UY, UZ, ROTX, ROTY, ROTZ	Tetrahedral Solid 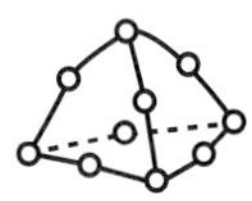 SOLID92 10 nodes 3–D space DOF: UX, UY, UZ	Structural Solid  SOLID95 20 nodes 3–D space DOF: UX, UY, UZ
Structural 2–D Shell		Structural 3–D Shell		
Plastic Axisymmetric Shell with Torsion  SHELL51 2 nodes 2–D space DOF: UX, UY, UZ, ROTZ	Axisymmetric Harmonic Struct. Shell  SHELL61 2 nodes 2–D space DOF: UX, UY, UZ, ROTZ	Shear/Twist Panel 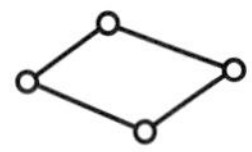 SHELL28 4 nodes 3–D space DOF: UX, UY, UZ or ROTX, ROTY, ROTZ	Membrane Shell 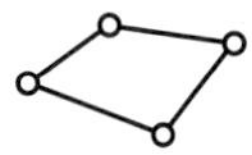 SHELL41 4 nodes 3–D space DOF: UX, UY, UZ	Plastic Shell 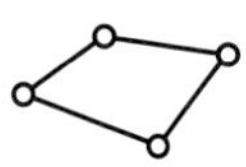 SHELL43 4 nodes 3–D space DOF: UX, UY, UZ, ROTX, ROTY, ROTZ
Elastic Shell 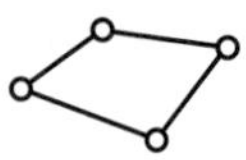 SHELL63 4 nodes 3–D space DOF: UX, UY, UZ, ROTX, ROTY, ROTZ	16–Layer Structural Shell 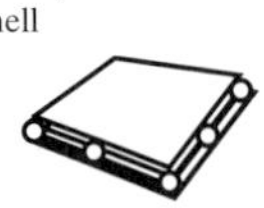 SHELL91 8 nodes 3–D space DOF: UX, UY, UZ, ROTX, ROTY, ROTZ	Structural Shell 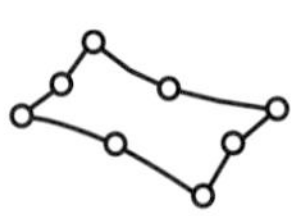 SHELL93 8 nodes 3–D space DOF: UX, UY, UZ, ROTX, ROTY, ROTZ	100–Layer Structural Shell 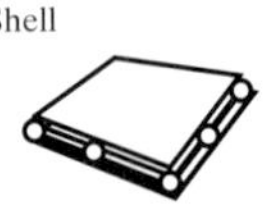 SHELL99 8 nodes 3–D space DOF: UX, UY, UZ, ROTX, ROTY, ROTZ	

TABLE 8.2 (*continued*) Element types offered by ANSYS

The ANSYS program offers nearly 150 different element types. For detailed information on a specific element type, see the *Elements* volume (Vol. III) of the ANSYS User's Manual.

Hyperelastic Solid Hyperelastic Mixed U–P Solid 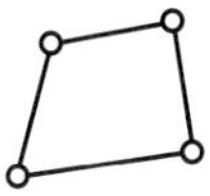 HYPER56 4 nodes 2–D space DOF: UX, UY, UZ	Hyperelastic Mixed U–P Solid 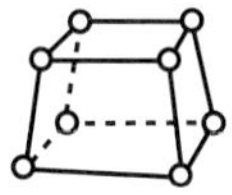 HYPER58 8 nodes 3–D space DOF: UX, UY, UZ	Hyperelastic Mixed U–P Solid 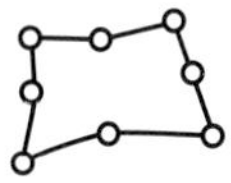 HYPER74 8 nodes 2–D space DOF: UX, UY, UZ	Hyperelastic Solid 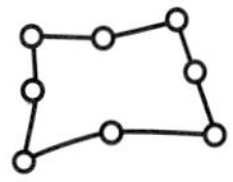 HYPER84 8 nodes 2–D space DOF: UX, UY, UZ	Hyperelastic Solid 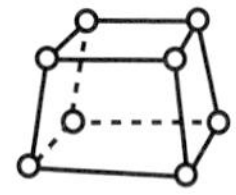 HYPER86 8 nodes 3–D space DOF: UX, UY, UZ
Visco Solid Viscoelastic Solid 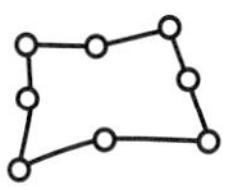 VISCO88 8 nodes 2–D space DOF: UX, UY	Viscoelastic Solid 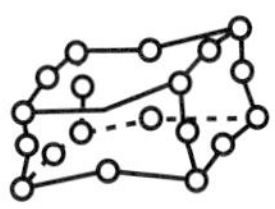 VISCO89 20 nodes 3–D space DOF: UX, UY, UZ	Large Strain Solid 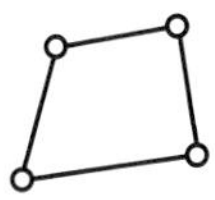 VISCO106 4 nodes 2–D space DOF: UX, UY, UZ	Large Strain Solid 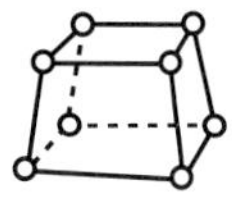 VISCO107 8 nodes 3–D space DOF: UX, UY, UZ	Large Strain Solid 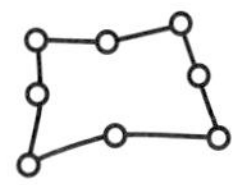 VISCO108 8 nodes 2–D space DOF: UX, UY, UZ
Thermal Point Thermal Mass  MASS71 1 node 3–D space DOF: TEMP	**Thermal Line** Radiation Link  LINK31 2 nodes 3–D space DOF: TEMP	Conduction Bar  LINK32 2 nodes 2–D space DOF: TEMP	Conduction Bar  LINK33 2 nodes 3–D space DOF: TEMP	Convection Link  LINK34 2 nodes 3–D space DOF: TEMP
Thermal 2–D Solid Triangular Thermal Solid 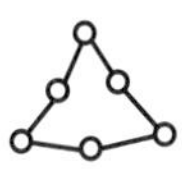 PLANE35 6 nodes 2–D space DOF: TEMP	Thermal Solid 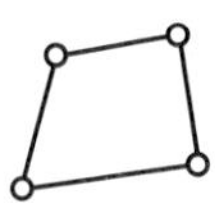 PLANE55 4 nodes 2–D space DOF: TEMP	Axisymmetric Harmonic Thermal Solid 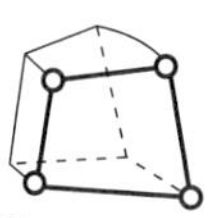 PLANE75 4 nodes 2–D space DOF: TEMP	Thermal Solid 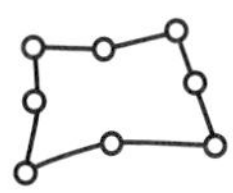 PLANE77 8 nodes 2–D space DOF: TEMP	Axisymmetric Harmonic Thermal Solid 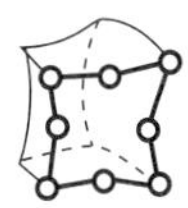 PLANE78 8 nodes 2–D space DOF: TEMP

TABLE 8.2 (*continued*) Element types offered by ANSYS

The ANSYS program offers nearly 150 different element types. For detailed information on a specific element type, see the *Elements* volume (Vol. III) of the ANSYS User's Manual.

Thermal 3–D Solid Thermal Solid 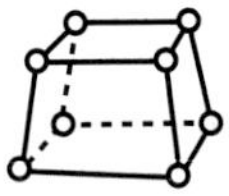 SOLID70 8 nodes 3–D space DOF: TEMP	Tetrahedral Thermal Solid  SOLID87 10 nodes 3–D space DOF: TEMP	Thermal Solid 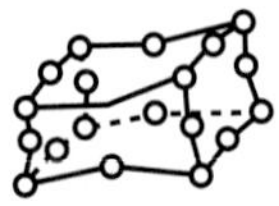 SOLID90 20 nodes 3–D space DOF: TEMP	Thermal Shell Thermal Shell 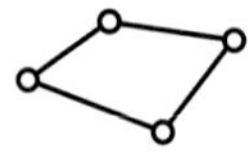 SHELL57 4 nodes 3–D space DOF: TEMP	Fluid Acoustic Fluid 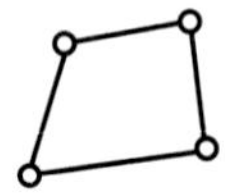 FLUID29 4 nodes 2–D space DOF: UX, UY, PRES
Acoustic Fluid 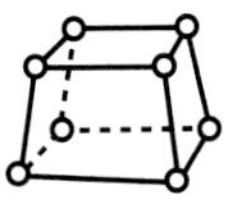 FLUID30 8 nodes 3–D space DOF: UX, UY, UZ, PRES	Dynamic Fluid Coupling  FLUID38 2 nodes 3–D space DOF: UX, UY, UZ	Thermal–Fluid Pipe  FLUID66 2 nodes 3–D space DOF: PRES, TEMP	Contained Fluid 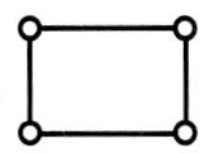 FLUID79 4 nodes 2–D space DOF: UX, UY	Contained Fluid 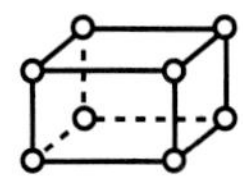 FLUID80 8 nodes 3–D space DOF: UX, UY, UZ
Axisymmetric Harmonic Contained Fluid 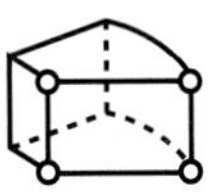 FLUID81 4 nodes 2–D space DOF: UX, UY, UZ	FLOTRAN CFD Fluid–Thermal 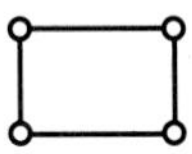 FLUID141 4 nodes 2–D space DOF: VX, VY, VZ, PRES, TEMP, ENKE, ENDS	FLOTRAN CFD Fluid–Thermal 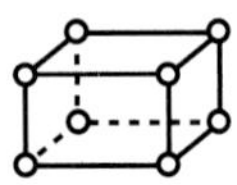 FLUID142 8 nodes 3–D space DOF: VX, VY, VZ, PRES, TEMP, ENKE, ENDS	Thermal Electric Thermal–Electric Solid 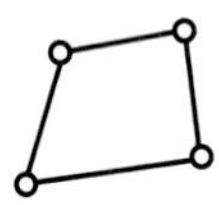 PLANE67 4 nodes 2–D space DOF: TEMP, VOLT	Thermal–Electric Line 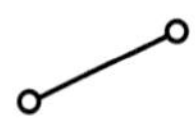 LINK68 2 nodes 3–D space DOF: TEMP, VOLT
Thermal–Electric Solid 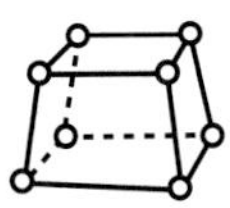 SOLID69 8 nodes 3–D space DOF: TEMP, VOLT	Magnetic Electric Current Source 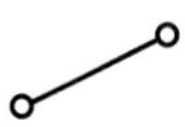 SOURC36 3 nodes 3–D space DOF: MAG	Magnetic Solid 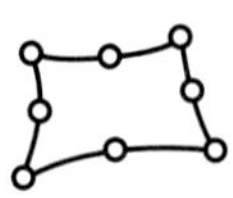 PLANE53 8 nodes 2–D space DOF: VOLT, AZ	Magnetic–Scalar Solid 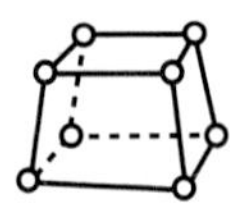 SOLID96 8 nodes 3–D space DOF: MAG	Magnetic Solid 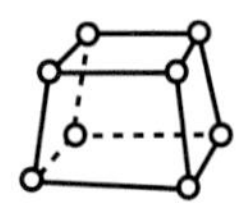 SOLID97 8 nodes 3–D space DOF: VOLT, AX, AY, AZ

TABLE 8.2 (*continued*) Element types offered by ANSYS

The ANSYS program offers nearly 150 different element types. For detailed information on a specific element type, see the *Elements* volume (Vol. III) of the ANSYS User's Manual.

Magnetic Interface 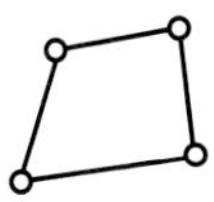 INTER115 4 nodes 3–D space DOF: AX, AY, AZ, MAG	Electrostatic Solid 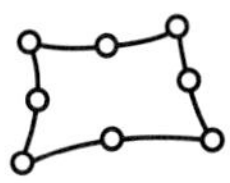 PLANE121 8 nodes 2–D space DOF: VOLT	Electrostatic Solid 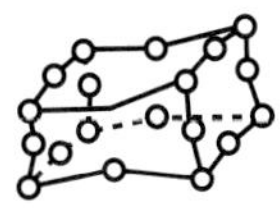 SOLID122 20 nodes 3–D space DOF: VOLT	Tetrahedral Electrostatic Solid 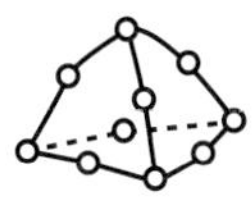 SOLID123 10 nodes 3–D space DOF: VOLT	Coupled–field Coupled–field Solid 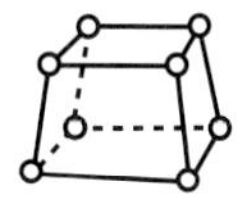 SOLID5 8 nodes 3–D space DOF: UX, UY, UZ, TEMP, VOLT, MAG
Coupled–field Solid 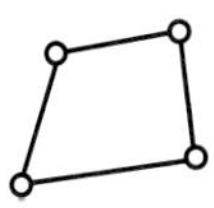 PLANE13 4 nodes 2–D space DOF: UX, UY, TEMP, VOLT, AZ	Coupled–field Solid 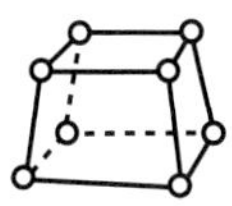 SOLID62 8 nodes 3–D space DOF: UX, UY, UZ, AX, AY, AZ, VOLT	Tetrahedral Couple–field Solid 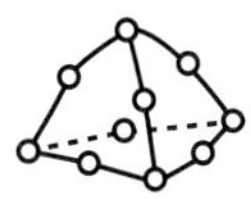 SOLID98 10 nodes 3–D space DOF: UX, UY, UZ, TEMP, VOLT, MAG	Contact Point–to–Point  CONTAC12 2 nodes 2–D space DOF: UX, UY	Point–to–Ground 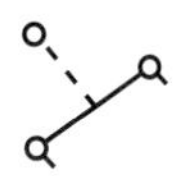 CONTAC26 3 nodes 2–D space DOF: UX, UY
Point–to–Surface 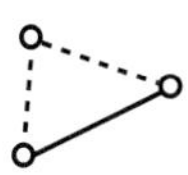 CONTAC48 3 nodes 2–D space DOF: UX, UY, TEMP	Point–to–Surface 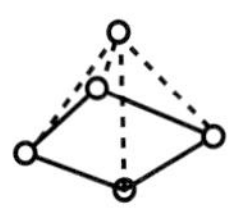 CONTAC49 5 nodes 3–D space DOF: UX, UY, UZ, TEMP	Point–to–Point  CONTAC52 2 nodes 3–D space DOF: UX, UY, UZ	Combination Revolute Joint 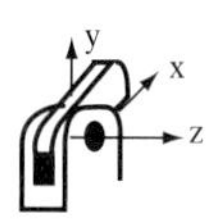 COMBIN7 5 nodes 3–D space DOF: UX, UY, UZ, ROTX, ROTY, ROTZ	Spring–Damper  COMBIN14 2 nodes 3–D space DOF: UX, UY, UZ, ROTX, ROTY, ROTZ PRES, TEMP
Control 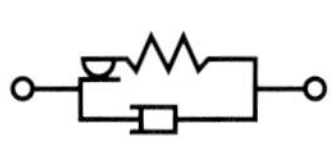 COMBIN37 4 nodes 3–D space DOF: UX, UY, UZ, ROTX, ROTY, ROTZ, PRES, TEMP	Nonlinear Spring  COMBIN39 2 nodes 3–D space DOF: UX, UY, UZ, ROTX, ROTY, ROTZ, PRES, TEMP	Combination 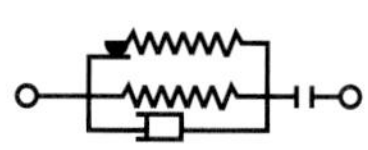 COMBIN40 2 nodes 3–D space DOF: UX, UY, UZ, ROTX, ROTY, ROTZ, PRES, TEMP	Matrix Stiffness, Mass or Damping Matrix 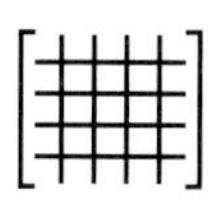 MATRIX27 2 nodes 3–D space DOF: UX, UY, UZ, ROTX, ROTY, ROTZ	Superelement 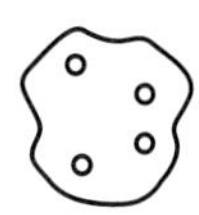 MATRIX50 2–D or 3–D space DOF: Any

TABLE 8.2 *(continued)* Element types offered by ANSYS

The ANSYS program offers nearly 150 different element types. For detailed information on a specific element type, see the *Elements* volume (Vol. III) of the ANSYS User's Manual.

Infinite				Surface
Infinite Boundary  INFIN9 2 nodes 2–D space DOF: AZ, TEMP	Infinite Boundary 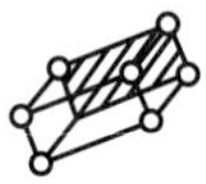 INFIN47 4 nodes 3–D space DOF: MAG, TEMP	Infinite Boundary 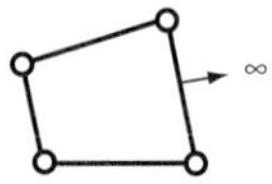 INFIN110 4 nodes 2–D space DOF: AZ, VOLT, TEMP	Infinite Boundary 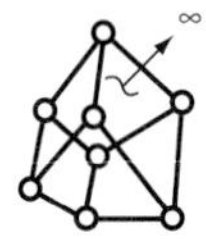 INFIN111 8 nodes 3–D space DOF: MAG, AX, AY, AZ, VOLT, TEMP	Surface Effect  SURF19 3 nodes 2–D space DOF: UX, UY, TEMP
Surface Effect 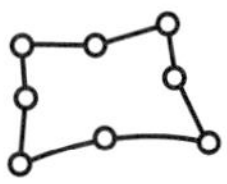 SURF22 8 nodes 3–D space DOF: UX, UY, UZ, TEMP				

CHAPTER 9

Analysis of Two-Dimensional Heat Transfer Problems

The main objective of this chapter is to introduce you to the analysis of two-dimensional heat transfer problems. General conduction problems and the treatment of various boundary conditions are discussed here. The main topics of Chapter 9 include the following:

9.1 GENERAL CONDUCTION PROBLEMS

In this chapter, we are concerned with determining how temperatures may vary with position in a medium as a result of either thermal conditions applied at the boundaries of the medium or heat generation within the medium. We are also interested in determining the heat flux at various points in a system, including its boundaries. Knowledge of temperature and heat flux fields is important in many engineering applications, including, for example, the cooling of electronic equipment, the design of thermal-fluid systems, and material and manufacturing processes. Knowledge of temperature distributions is also useful in determining thermal stresses and corresponding deflections in machine and structural elements. There are three modes of heat transfer: *conduction*, *convection*, and *radiation*. Conduction refers to that mode of heat transfer that occurs when there exists a temperature gradient in a medium. The energy is transported from the high-temperature region to the low-temperature region by molecular activities. Using a twodimensional Cartesian frame of reference, we know that the rate of heat transfer by conduction is given by Fourier's Law:

$$q_X = -kA\frac{\partial T}{\partial X} \tag{9.1}$$

$$q_Y = -kA\frac{\partial T}{\partial Y} \tag{9.2}$$

q_X and q_Y are the X- and the Y-components of the heat transfer rate, k is the thermal conductivity of the medium, A is the cross-sectional area of the medium, and $\frac{\partial T}{\partial X}$ and $\frac{\partial T}{\partial Y}$ are the temperature gradients. Fourier's Law may also be expressed in terms of heat transfer rates per unit area as

$$q''_X = -k\frac{\partial T}{\partial X} \tag{9.3}$$

$$q''_Y = -k\frac{\partial T}{\partial Y} \tag{9.4}$$

where $q''_X = \frac{q_X}{A}$ and $q''_Y = \frac{q_Y}{A}$ are called heat fluxes in the X-direction and the Y-direction, respectively. It is important to realize that the direction of the total heat flow is always perpendicular to the *isotherms* (constant temperature lines or surfaces). This relationship is depicted in Figure 9.1.

Convective heat transfer occurs when a fluid in motion comes into contact with a solid surface whose temperature differs from the moving fluid. The overall heat transfer rate between the fluid and the surface is governed by Newton's Law of Cooling, which is

$$q = hA(T_s - T_f) \tag{9.5}$$

where h is the heat transfer coefficient, T_s is the solid surface temperature, and T_f represents the temperature of the moving fluid. The value of the heat transfer coefficient

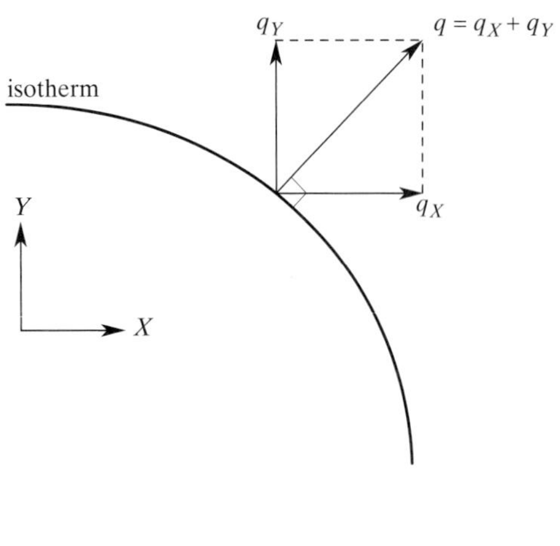

FIGURE 9.1 The heat flux vector is always normal to the isotherms.

for a particular situation is determined from experimental correlations that are available in many books about heat transfer.

All matters emit thermal radiation. This rule is true as long as the body in question is at a finite temperature (expressed in Kelvin or Rankine Scale). Simply stated, the amount of energy emitted by a surface is given by the equation

$$q'' = \varepsilon\sigma T_s^4 \tag{9.6}$$

where q'' represents the rate of thermal energy per unit area emitted by the surface, ε is the emissivity of the surface $0 < \varepsilon < 1$, and σ is the Stefan–Boltzman constant ($\sigma = 5.67 \times 10^{-8}\ \mathrm{W/m^2 \cdot K^4}$). It is important to note here that unlike conduction and convection modes, heat transfer by radiation can occur in a vacuum, and because all objects emit thermal radiation, it is the net energy exchange among the bodies that is of interest to us. The three modes of heat transfer are depicted in Figure 9.2.

In Chapter 1, it was explained that engineering problems are mathematical models of physical situations. Moreover, many of these mathematical models are differential equations that are derived by applying the fundamental laws and principles of nature to a system or a control volume. In heat transfer problems, these governing equations represent the balance of mass, momentum, and energy for a medium. Chapter 1 stated that when possible, the exact solutions of the governing differential equations should be sought because the exact solutions render the detailed behavior of a system. However, for many practical engineering problems, it is impossible to obtain exact solutions to the governing equations because either the geometry is too complex or the boundary conditions are too complicated.

The principle of the conservation of energy plays a significant role in the analysis of heat transfer problems. Consequently, you need to understand this principle fully in

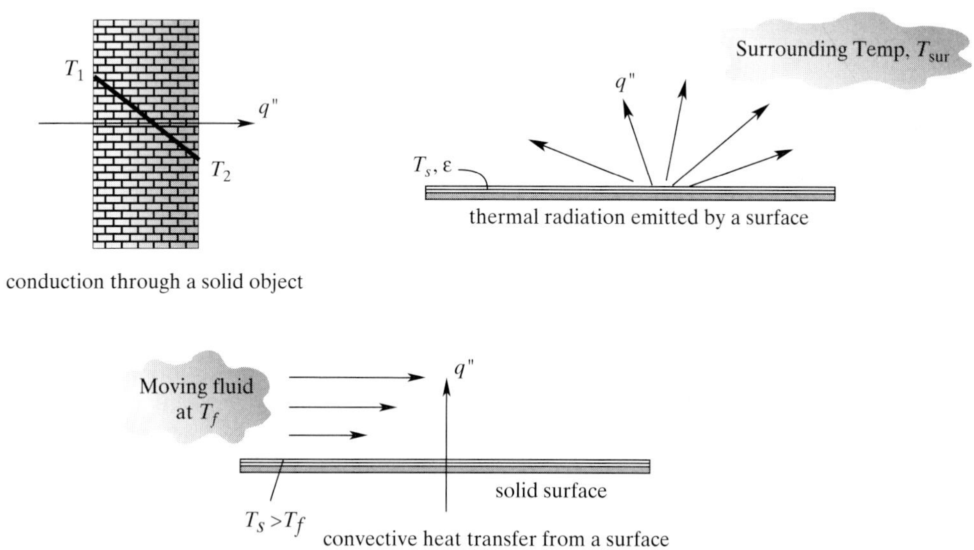

FIGURE 9.2 Various modes of heat transfer.

order to model a physical problem correctly. The principle of the conservation of energy states the following: The rate at which thermal and/or mechanical energy enters a system through its boundaries, minus the rate at which the energy leaves the system through its boundaries, plus the rate of energy generation within the volume of the system, must equal the rate at which energy is stored within the volume of the system. This statement is represented by Figure 9.3 and the equation

$$\dot{E}_{\text{in}} - \dot{E}_{\text{out}} + \dot{E}_{\text{generation}} = \dot{E}_{\text{stored}} \tag{9.7}$$

$\dot{E}_{\text{in}}$ and $\dot{E}_{\text{out}}$ represent the amount of energy crossing into and out of the surfaces of a system. The thermal energy generation rate $\dot{E}_{\text{generation}}$ represents the rate of the conversion of energy from electrical, chemical, nuclear, or electromagnetic forms to thermal energy within the volume of the system. An example of such conversion is the electric current running through a solid conductor. On the other hand, the energy storage term represents the increase or decrease in the amount of thermal internal energy within the volume of the system due to transient processes. It is important to understand the contribution of each term to the overall energy balance of a system in order to model an actual situation properly. A good understanding of the principle of the conservation of energy will also help verify the results of a model.

This chapter focuses on the conduction mode of heat transfer with possible convective or radiative boundary conditions. For now, we will focus on steady-state two-dimensional conduction problems. Applying the principle of the conservation of energy to a system represented in a Cartesian coordinate system results in the following heat diffusion equation:

$$k_X \frac{\partial^2 T}{\partial X^2} + k_Y \frac{\partial^2 T}{\partial Y^2} + \dot{q} = 0 \tag{9.8}$$

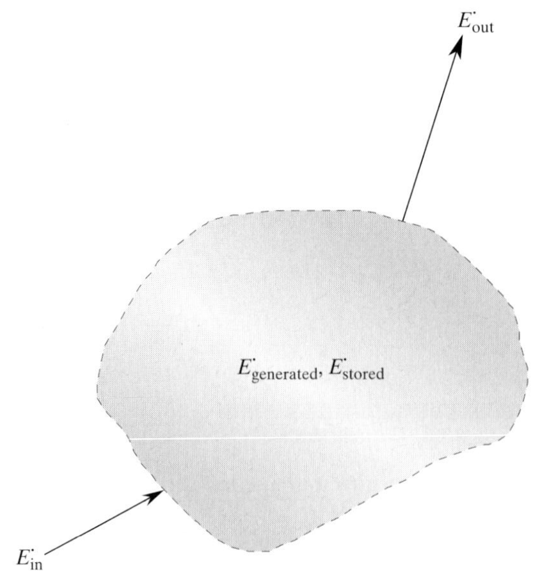

FIGURE 9.3 The principle of the conservation of energy.

The derivation of Eq. (9.8) is shown in Figure 9.4. In Eq. (9.8), $q^{\cdot}$ represents the heat generation per unit volume, within a volume having a unit depth. There are several boundary conditions that occur in conduction problems:

1. A situation wherein heat loss or gain through a surface may be neglected. This situation, shown in Figure 9.5, is commonly referred to as an adiabatic surface or

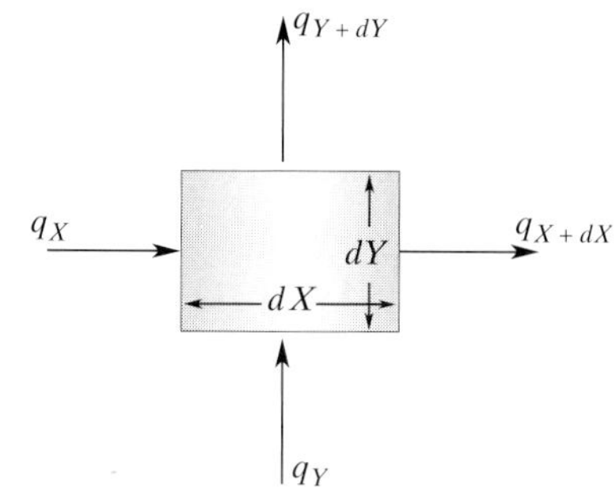

First, we begin by applying the principle of the conservation of energy to a small region (differential volume) in a medium:

$$\dot{E}_{\text{in}} - \dot{E}_{\text{out}} + \dot{E}_{\text{generation}} = \dot{E}_{\text{stored}}$$

$$q_X + q_Y - (q_{X+dX} + q_{Y+dY}) + q^{\cdot} dXdY(1) = \rho c\, dXdY \frac{\partial T}{\partial t}$$

$$q_X + q_Y - \left(q_X + \frac{\partial q_X}{\partial X} dX + q_Y + \frac{\partial q_Y}{\partial Y} dY\right) + q^{\cdot} dXdY = \rho c\, dXdY \frac{\partial T}{\partial t}$$

Simplifying, we get

$$-\frac{\partial q_X}{\partial X} dX - \frac{\partial q_Y}{\partial Y} dY + q^{\cdot} dXdY = \rho c\, dXdY \frac{\partial T}{\partial t}$$

Making use of Fourier's Law, we have

$$q_X = -k_X A \frac{\partial T}{\partial X} = -k_X\, dY(1) \frac{\partial T}{\partial X}$$

$$q_Y = -k_Y A \frac{\partial T}{\partial Y} = -k_Y\, dX(1) \frac{\partial T}{\partial Y}$$

$$-\frac{\partial}{\partial X}\left(-k_X dY \frac{\partial T}{\partial X}\right) dX - \frac{\partial}{\partial Y}\left(-k_Y dX \frac{\partial T}{\partial Y}\right) dY + q^{\cdot} dXdY = \rho c\, dXdY \frac{\partial T}{\partial t}$$

ρ and c are the density and specific heat of the medium, and t represents time. For a steady-state situation, temperature does not change with time, and consequently, the right-hand side is zero. After simplifying, we obtain

$$k_X \frac{\partial^2 T}{\partial X^2} + k_Y \frac{\partial^2 T}{\partial Y^2} + q^{\cdot} = 0$$

FIGURE 9.4 The derivation of the equation of heat conduction under steady-state conditions.

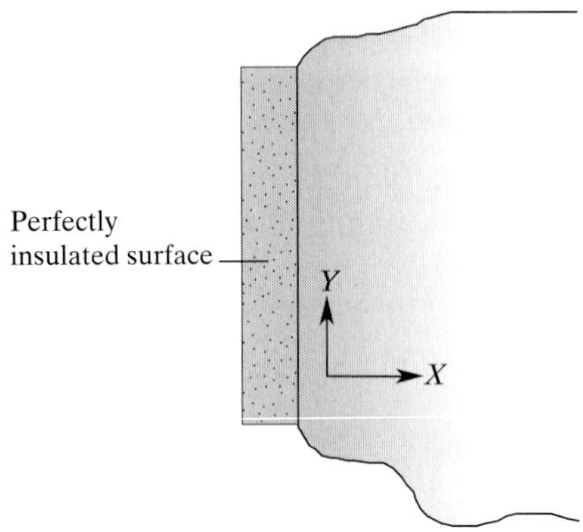

FIGURE 9.5 An adiabatic, or perfectly insulated, surface.

a perfectly insulated surface. In conduction problems, symmetrical lines also represent adiabatic lines. This type of boundary condition is represented by

$$\left.\frac{\partial T}{\partial X}\right|_{(X=0,\,Y)} = 0 \tag{9.9}$$

2. A situation for which a constant heat flux is applied at a surface. This boundary condition, shown in Figure 9.6, is represented by the equation

$$-k\left.\frac{\partial T}{\partial X}\right|_{X=0} = q_0'' \tag{9.10}$$

3. A situation for which cooling or heating is taking place at a surface due to convection processes. This situation, shown in Figure 9.7, is represented by the equation

$$-k\left.\frac{\partial T}{\partial X}\right|_{(X=0,\,Y)} = h[T(0,y) - T_f] \tag{9.11}$$

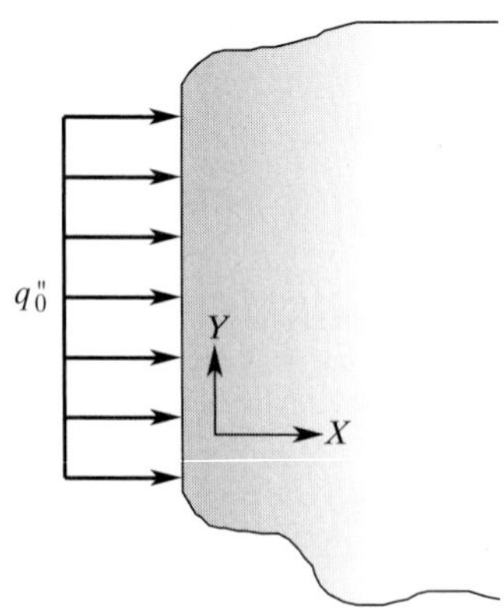

FIGURE 9.6 A constant heat flux applied at a surface.

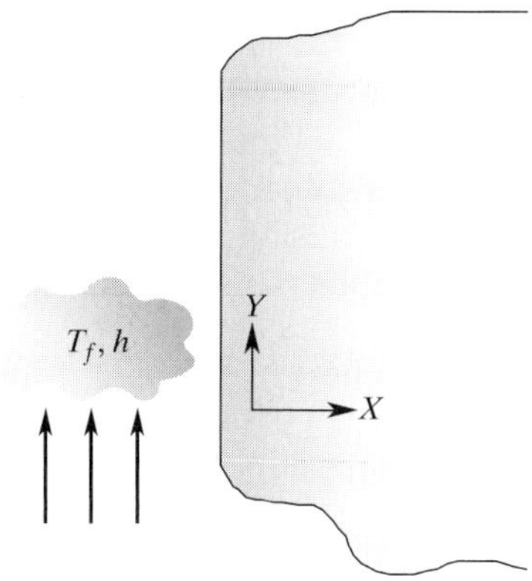

FIGURE 9.7 Convention processes causing cooling or heating to take place at a surface.

4. A situation wherein heating or cooling is taking place at a surface due to net radiation exchange with the surroundings. The expression for this condition will depend on the view factors and the emissivity of the surfaces involved.
5. A situation in which conditions 3 and 4 both exist simultaneously.
6. Constant surface-temperature conditions occur when a fluid in contact with a solid surface experiences phase change, as shown in Figure 9.8. Examples include condensation or evaporation of a fluid at constant pressure. This condition is represented by

$$T(0, Y) = T_0 \tag{9.12}$$

The modeling of actual situations with these boundary conditions will be discussed and illustrated with examples after we consider finite element formulations of two-dimensional heat conduction problems.

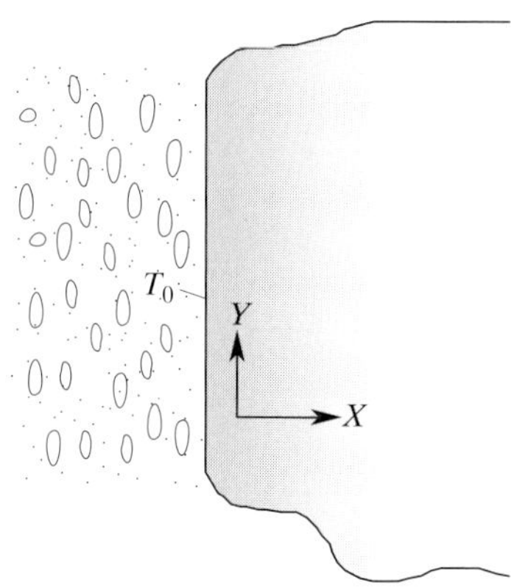

FIGURE 9.8 Constant surface-temperature conditions occur due to phase change of a fluid in contact with a solid surface.

9.2 FORMULATION WITH RECTANGULAR ELEMENTS

Two-dimensional bilinear rectangular elements were covered in detail in Chapter 7. Recall that for problems with straight boundaries, linear rectangular shape functions offer simple means to approximate the spatial variation of a dependent variable, such as temperature. For convenience, the expression for a rectangular element in terms of its nodal temperatures and shape functions is repeated here (also see Figure 9.9). The expression is

$$T^{(e)} = [S_i \quad S_j \quad S_m \quad S_n]\begin{Bmatrix} T_i \\ T_j \\ T_m \\ T_n \end{Bmatrix} \tag{9.13}$$

where the shape functions S_i, S_j, S_m, and S_n are given by

$$\begin{aligned} S_i &= \left(1 - \frac{x}{\ell}\right)\left(1 - \frac{y}{w}\right) \\ S_j &= \frac{x}{\ell}\left(1 - \frac{y}{w}\right) \\ S_m &= \frac{xy}{\ell w} \\ S_n &= \frac{y}{w}\left(1 - \frac{x}{\ell}\right) \end{aligned} \tag{9.14}$$

We now apply the Galerkin approach to the heat diffusion equation, Eq. (9.8) expressed in local coordinates x, y, yielding four residual equations:

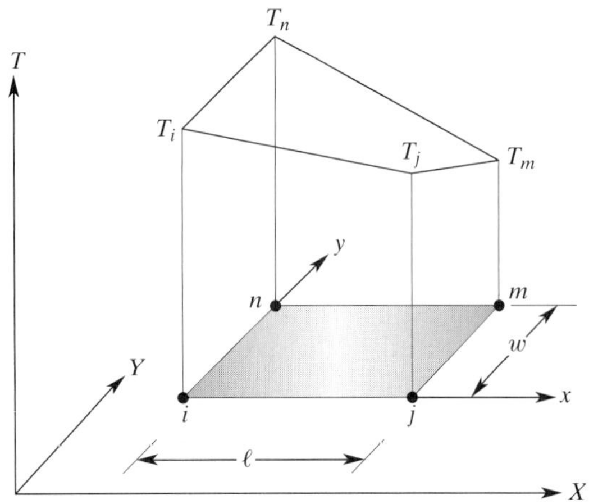

FIGURE 9.9 A typical rectangular element.

$$R_i^{(e)} = \int_A S_i \left(k_x \frac{\partial^2 T}{\partial x^2} + k_y \frac{\partial^2 T}{\partial y^2} + q^{\cdot} \right) dA \tag{9.15}$$
$$R_j^{(e)} = \int_A S_j \left(k_x \frac{\partial^2 T}{\partial x^2} + k_y \frac{\partial^2 T}{\partial y^2} + q^{\cdot} \right) dA$$
$$R_m^{(e)} = \int_A S_m \left(k_x \frac{\partial^2 T}{\partial x^2} + k_y \frac{\partial^2 T}{\partial y^2} + q^{\cdot} \right) dA$$
$$R_n^{(e)} = \int_A S_n \left(k_x \frac{\partial^2 T}{\partial x^2} + k_y \frac{\partial^2 T}{\partial y^2} + q^{\cdot} \right) dA$$

Note that from here on, for the sake of simplicity of presentation of conductance and load matrices, we will set the residual of an element equal to zero. However, as we mentioned earlier, it is important to realize that the residuals are set equal to zero after all the elements have been assembled.

We can rewrite the four equations given by (9.15) in a compact matrix form as

$$\int_A [\mathbf{S}]^T \left(k_x \frac{\partial^2 T}{\partial x^2} + k_y \frac{\partial^2 T}{\partial y^2} + q^{\cdot} \right) dA = 0 \tag{9.16}$$

where the transpose of the shape functions is given by the following matrix:

$$[\mathbf{S}]^T = \begin{Bmatrix} S_i \\ S_j \\ S_m \\ S_n \end{Bmatrix} \tag{9.17}$$

Equation (9.16) consists of three main integrals:

$$\int_A [\mathbf{S}]^T \left(k_x \frac{\partial^2 T}{\partial x^2} \right) dA + \int_A [\mathbf{S}]^T \left(k_y \frac{\partial^2 T}{\partial y^2} \right) dA + \int_A [\mathbf{S}]^T q^{\cdot} dA = 0 \tag{9.18}$$

Let $C_1 = k_x$, $C_2 = k_y$, and $C_3 = q^{\cdot}$ so that we can later apply the results of the forthcoming derivation to other types of problems with similar forms of governing differential equations. As will be demonstrated later in Chapters 10 and 12, we will use the general findings of this chapter to analyze the torsion of solid members and ideal fluid flow problems. So making respective substitutions, we have

$$\int_A [\mathbf{S}]^T \left(C_1 \frac{\partial^2 T}{\partial x^2} \right) dA + \int_A [\mathbf{S}]^T \left(C_2 \frac{\partial^2 T}{\partial y^2} \right) dA + \int_A [\mathbf{S}]^T C_3 \, dA = 0 \tag{9.19}$$

Evaluation of the integrals given by Eq. (9.19) will result in the elemental formulation. We first manipulate the second-order terms into first-order terms by using the chain rule in the following manner:

$$\frac{\partial}{\partial x}\left([\mathbf{S}]^T \frac{\partial T}{\partial x} \right) = [\mathbf{S}]^T \frac{\partial^2 T}{\partial x^2} + \frac{\partial [\mathbf{S}]^T}{\partial x} \frac{\partial T}{\partial x} \tag{9.20}$$

Rearranging Eq. (9.20), we have

$$[\mathbf{S}]^T \frac{\partial^2 T}{\partial x^2} = \frac{\partial}{\partial x}\left([\mathbf{S}]^T \frac{\partial T}{\partial x}\right) - \frac{\partial [\mathbf{S}]^T}{\partial x}\frac{\partial T}{\partial x} \tag{9.21}$$

Applying the results given by Eq. (9.21) to the first and the second terms in Eq. (9.19), we obtain

$$\int_A [\mathbf{S}]^T\left(C_1 \frac{\partial^2 T}{\partial x^2}\right) dA = \int_A C_1 \frac{\partial}{\partial x}\left([\mathbf{S}]^T \frac{\partial T}{\partial x}\right) dA - \int_A C_1\left(\frac{\partial [\mathbf{S}]^T}{\partial x}\frac{\partial T}{\partial x}\right) dA \tag{9.22}$$

$$\int_A [\mathbf{S}]^T\left(C_2 \frac{\partial^2 T}{\partial y^2}\right) dA = \int_A C_2 \frac{\partial}{\partial y}\left([\mathbf{S}]^T \frac{\partial T}{\partial y}\right) dA - \int_A C_2\left(\frac{\partial [\mathbf{S}]^T}{\partial y}\frac{\partial T}{\partial y}\right) dA \tag{9.23}$$

Using Green's theorem, we can write the terms

$$\int_A C_1 \frac{\partial}{\partial x}\left([\mathbf{S}]^T \frac{\partial T}{\partial x}\right) dA$$

and

$$\int_A C_2 \frac{\partial}{\partial y}\left([\mathbf{S}]^T \frac{\partial T}{\partial y}\right) dA$$

in terms of integrals around the element boundary. We will come back to these terms later. For now, let us consider the

$$-\int_A C_1\left(\frac{\partial [\mathbf{S}]^T}{\partial x}\frac{\partial T}{\partial x}\right) dA$$

term in Eq. (9.22). This term can easily be evaluated. Evaluating the derivatives for a rectangular element, we obtain

$$\frac{\partial T}{\partial x} = \frac{\partial}{\partial x}[S_i \quad S_j \quad S_m \quad S_n]\begin{Bmatrix} T_i \\ T_j \\ T_m \\ T_n \end{Bmatrix} = \frac{1}{\ell w}[(-w + y) \ (w - y) \ y \ -y]\begin{Bmatrix} T_i \\ T_j \\ T_m \\ T_n \end{Bmatrix} \tag{9.24}$$

Also evaluating $\dfrac{\partial [\mathbf{S}]^T}{\partial x}$ we have

$$\frac{\partial [\mathbf{S}]^T}{\partial x} = \frac{\partial}{\partial x}\begin{Bmatrix} S_i \\ S_j \\ S_m \\ S_n \end{Bmatrix} = \frac{1}{\ell w}\begin{Bmatrix} -w + y \\ w - y \\ y \\ -y \end{Bmatrix} \tag{9.25}$$

Substituting the results of Eqs. (9.24) and (9.25) into the term

$$-\int_A C_1\left(\frac{\partial [\mathbf{S}]^T}{\partial x}\frac{\partial T}{\partial x}\right) dA$$

we have

$$-\int_A C_1\left(\frac{\partial[\mathbf{S}]^T}{\partial x}\frac{\partial T}{\partial x}\right)dA = -C_1\int_A \frac{1}{(\ell w)^2}\begin{Bmatrix} -w+y \\ w-y \\ y \\ -y \end{Bmatrix}[(-w+y)\ \ (w-y)\ \ y\ \ -y]\begin{Bmatrix} T_i \\ T_j \\ T_m \\ T_n \end{Bmatrix}dA \tag{9.26}$$

Integrating yields

$$-C_1\int_A \frac{1}{(\ell w)^2}\begin{Bmatrix} -w+y \\ w-y \\ y \\ -y \end{Bmatrix}[(-w+y)\ \ (w-y)\ \ y\ \ -y]\begin{Bmatrix} T_i \\ T_j \\ T_m \\ T_n \end{Bmatrix}dA$$

$$= -\frac{C_1 w}{6\ell}\begin{bmatrix} 2 & -2 & -1 & 1 \\ -2 & 2 & 1 & -1 \\ -1 & 1 & 2 & -2 \\ 1 & -1 & -2 & 2 \end{bmatrix}\begin{Bmatrix} T_i \\ T_j \\ T_m \\ T_n \end{Bmatrix} \tag{9.27}$$

To get the results of Eq. (9.27), we carry out the integration for each term in the 4×4 matrix. For example, by integrating the expression in the first row and first column, we get

$$\int_0^\ell\int_0^w \frac{1}{(\ell w)^2}(-w+y)^2\,dy\,dx = \frac{1}{(\ell w)^2}\int_0^\ell\left(w^2 w + \frac{w^3}{3} - 2w\frac{w^2}{2}\right)dx$$

$$= \frac{1}{(\ell w)^2}\int_0^\ell \frac{w^3}{3}dx = \frac{1}{(\ell w)^2}\frac{w^3\ell}{3} = \frac{1}{3}\frac{w}{\ell}$$

or the integration of the expression in the first row and third column yields

$$\int_0^\ell\int_0^w \frac{1}{(\ell w)^2}(-w+y)y\,dy\,dx = \frac{1}{(\ell w)^2}\int_0^\ell\left(-w\frac{w^2}{2} + \frac{w^3}{3}\right)dx$$

$$= \frac{1}{(\ell w)^2}\int_0^\ell -\frac{w^3}{6}dy = -\frac{1}{(\ell w)^2}\frac{w^3\ell}{6} = -\frac{1}{6}\frac{w}{\ell}$$

In the same manner, we can evaluate the term

$$-\int_A C_2\left(\frac{\partial[\mathbf{S}]^T}{\partial y}\frac{\partial T}{\partial y}\right)dA$$

in Eq. (9.23) in the y-direction:

$$\frac{\partial T}{\partial y} = \frac{\partial}{\partial y}[S_i \quad S_j \quad S_m \quad S_n]\begin{Bmatrix} T_i \\ T_j \\ T_m \\ T_n \end{Bmatrix} = \frac{1}{\ell w}[(-\ell+x)\ \ -x\ \ x\ \ (\ell-x)]\begin{Bmatrix} T_i \\ T_j \\ T_m \\ T_n \end{Bmatrix} \tag{9.28}$$

And evaluating $\frac{\partial[\mathbf{S}]^T}{\partial y}$ we have

$$\frac{\partial[\mathbf{S}]^T}{\partial x} = \frac{\partial}{\partial y}\begin{Bmatrix} S_i \\ S_j \\ S_m \\ S_n \end{Bmatrix} = \frac{1}{\ell w}\begin{Bmatrix} -\ell + x \\ -x \\ x \\ \ell - x \end{Bmatrix} \tag{9.29}$$

Substituting the results of Eqs. (9.28) and (9.29) into the term

$$-\int_A C_2\left(\frac{\partial[\mathbf{S}]^T}{\partial y}\frac{\partial T}{\partial y}\right)dA$$

we have

$$-\int_A C_2\left(\frac{\partial[\mathbf{S}]^T}{\partial y}\frac{\partial T}{\partial y}\right)dA = -C_2\int_A \frac{1}{(\ell w)^2}\begin{Bmatrix} -\ell + x \\ -x \\ x \\ \ell - x \end{Bmatrix}[(-\ell + x) \quad -x \quad x \quad (\ell - x)]\begin{Bmatrix} T_i \\ T_j \\ T_m \\ T_n \end{Bmatrix}dA \tag{9.30}$$

Evaluation of the integral yields

$$-C_2\int_A \frac{1}{(\ell w)^2}\begin{Bmatrix} -\ell + x \\ -x \\ x \\ \ell - x \end{Bmatrix}[(-\ell + x) \quad -x \quad x \quad (\ell - x)]\begin{Bmatrix} T_i \\ T_j \\ T_m \\ T_n \end{Bmatrix}dA =$$

$$-\frac{C_2\ell}{6w}\begin{bmatrix} 2 & 1 & -1 & -2 \\ 1 & 2 & -2 & -1 \\ -1 & -2 & 2 & 1 \\ -2 & -1 & 1 & 2 \end{bmatrix}\begin{Bmatrix} T_i \\ T_j \\ T_m \\ T_n \end{Bmatrix} \tag{9.31}$$

Next, we will evaluate the thermal load term $\int_A [\mathbf{S}]^T C_3\, dA$:

$$\int_A [\mathbf{S}]^T C_3\, dA = C_3\int_A \begin{Bmatrix} S_i \\ S_j \\ S_m \\ S_n \end{Bmatrix} dA = \frac{C_3 A}{4}\begin{Bmatrix} 1 \\ 1 \\ 1 \\ 1 \end{Bmatrix} \tag{9.32}$$

We now return to the terms

$$\int_A C_1\frac{\partial}{\partial x}\left([\mathbf{S}]^T\frac{\partial T}{\partial x}\right)dA$$

and

$$\int_A C_2 \frac{\partial}{\partial y}\left([\mathbf{S}]^T \frac{\partial T}{\partial y}\right) dA$$

As mentioned earlier, we can use Green's theorem to rewrite these area integrals in terms of line integrals around the element boundary:

$$\int_A C_1 \frac{\partial}{\partial x}\left([\mathbf{S}]^T \frac{\partial T}{\partial x}\right) dA = \int_\tau C_1 [\mathbf{S}]^T \frac{\partial T}{\partial x} \cos\theta \, d\tau \tag{9.33}$$

$$\int_A C_2 \frac{\partial}{\partial y}\left([\mathbf{S}]^T \frac{\partial T}{\partial y}\right) dA = \int_\tau C_2 [\mathbf{S}]^T \frac{\partial T}{\partial y} \sin\theta \, d\tau \tag{9.34}$$

τ represents the element boundary, and θ measures the angle to the unit normal.

A Review of Green's Theorem

Before we proceed with the evaluation of Eqs. (9.33) and (9.34), let's briefly review Green's theorem, which is given by the following equation:

$$\iint\limits_{Region} \left(\frac{\partial g}{\partial x} - \frac{\partial f}{\partial y}\right) dxdy = \int\limits_{Contour} f dx + g dy \tag{9.35}$$

In the relationship given by Eq. (9.35), $f(x, y)$ and $g(x, y)$ are continuous functions and have continuous partial derivatives.

Next, using simple area examples, we demonstrate how Green's theorem is applied.

We can establish a relationship between the area of a region bounded by its contour τ and a line integral around it by substituting for $f = 0$ and $g = x$ or by letting $f = -y$ and $g = 0$ in Eq. (9.35). This approach leads to the following relationships:

$$A = \iint\limits_{Region} dxdy = \int\limits_{Contour,\tau} xdy \tag{9.36}$$

or

$$A = \iint\limits_{Region} dxdy = -\int\limits_{Contour,\tau} ydx \tag{9.37}$$

We can also combine Eqs. (9.36) and (9.37), which results in yet another relationship between area and line integrals:

$$2A = 2\iint\limits_{Region} dxdy = \int\limits_{Contour,\tau} (xdy - ydx) \tag{9.38}$$

Let's now use the relationship given by Eq. (9.38) to calculate the area of a circle having a radius R. As shown in the accompanying figure, for a circle

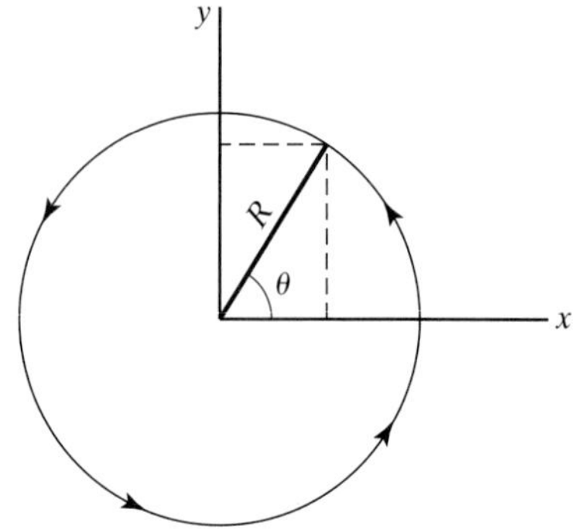

$$x^2 + y^2 = R^2$$

$$x = R\cos\theta \quad \text{and} \quad dx = -R\sin\theta\, d\theta$$

$$y = R\sin\theta \quad \text{and} \quad dy = R\cos\theta\, d\theta$$

Substituting these relationships into Eq. (9.38), we have

$$2A = \int_{Contour,\,\tau} (xdy - ydx) = \int_0^{2\pi} (\overbrace{(R\cos\theta)}^{x}\overbrace{(R\cos\theta d\theta)}^{dy} - \overbrace{(R\sin\theta)}^{y}\overbrace{(-R\sin\theta d\theta)}^{dx})$$

and simplifying, we get

$$2A = \int_0^{2\pi} R^2(\cos^2\theta + \sin^2\theta)d\theta = \int_0^{2\pi} R^2\, d\theta = 2\pi R^2$$

or

$$A = \pi R^2$$

As another example, let's calculate the area of the rectangle shown in the accompanying diagram.

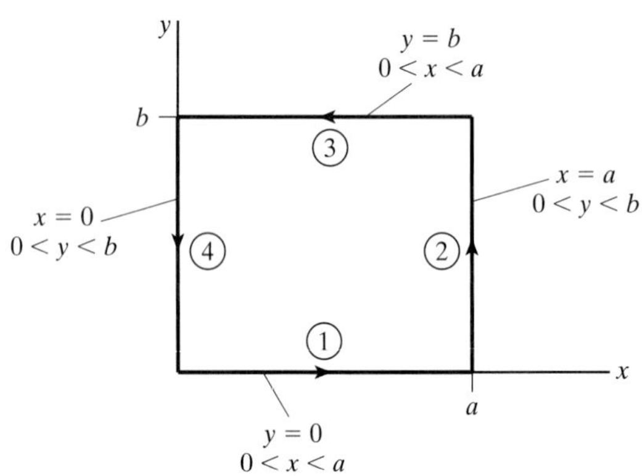

Using Eq. (9.36), we have

$$A = \int_{\tau_1} xdy + \int_{\tau_2} xdy + \int_{\tau_3} xdy + \int_{\tau_4} xdy \tag{9.39}$$

Note that along τ_1, x varies from 0 to a and y is zero $(dy = 0)$, and along τ_2, $x = a$ and y varies from 0 to b. Along τ_3, x varies from a to 0 and y is constant $(dy = 0)$, and along τ_4, $x = 0$ and y varies from b to 0. Then, substituting for these relationships in Eq. (9.39), we get

$$A = 0 + \int_0^b ady + 0 + 0 = ab$$

Let us now return to Eqs. (9.33) and (9.34), which contribute to the derivative boundary conditions. To understand what is meant by derivative boundary conditions, consider an element with a convection boundary condition, as shown in Figure 9.10.

Neglecting radiation, the application of the conservation of energy in the x-direction to the jm edge requires that the energy that reaches the jm edge through conduction must be equal to the energy being convected away (by the fluid adjacent to the jm edge). So,

$$-k\frac{\partial T}{\partial x} = h(T - T_f) \tag{9.40}$$

Substituting the right-hand side of Eq. (9.40) into Eq. (9.33), we get

$$\int_{\tau} C_1[\mathbf{S}]^T\frac{\partial T}{\partial x}\cos\theta\, d\tau = \int_{\tau} k[\mathbf{S}]^T\frac{\partial T}{\partial x}\cos\theta\, d\tau = -\int_{\tau} h[\mathbf{S}]^T(T - T_f)\cos\theta\, d\tau \tag{9.41}$$

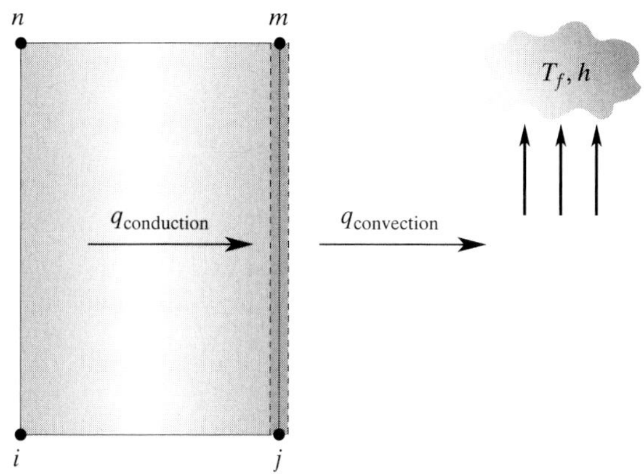

FIGURE 9.10 A rectangular element with a convective boundary condition.

The integral given by Eq. (9.41) has two terms:

$$-\int_{\tau} h[\mathbf{S}]^T(T - T_f)\cos\theta\, d\tau = -\int_{\tau} h[\mathbf{S}]^T T \cos\theta\, d\tau + \int_{\tau} h[\mathbf{S}]^T T_f \cos\theta\, d\tau \qquad (9.42)$$

The terms $\int_{\tau} h[\mathbf{S}]^T T \cos\theta\, d\tau$ and $\int_{\tau} h[\mathbf{S}]^T T \sin\theta\, d\tau$, for convective boundary conditions along different edges of the rectangular element, contribute to conductance matrices:

$$[\mathbf{K}]^{(e)} = \frac{h\ell_{ij}}{6}\begin{bmatrix} 2 & 1 & 0 & 0 \\ 1 & 2 & 0 & 0 \\ 0 & 0 & 0 & 0 \\ 0 & 0 & 0 & 0 \end{bmatrix} \qquad (9.43) \qquad [\mathbf{K}]^{(e)} = \frac{h\ell_{jm}}{6}\begin{bmatrix} 0 & 0 & 0 & 0 \\ 0 & 2 & 1 & 0 \\ 0 & 1 & 2 & 0 \\ 0 & 0 & 0 & 0 \end{bmatrix} \qquad (9.44)$$

$$[\mathbf{K}]^{(e)} = \frac{h\ell_{mn}}{6}\begin{bmatrix} 0 & 0 & 0 & 0 \\ 0 & 0 & 0 & 0 \\ 0 & 0 & 2 & 1 \\ 0 & 0 & 1 & 2 \end{bmatrix} \qquad (9.45) \qquad [\mathbf{K}]^{(e)} = \frac{h\ell_{ni}}{6}\begin{bmatrix} 2 & 0 & 0 & 1 \\ 0 & 0 & 0 & 0 \\ 0 & 0 & 0 & 0 \\ 1 & 0 & 0 & 2 \end{bmatrix} \qquad (9.46)$$

Referring to Figure 9.9, note that in the above matrices, $\ell_{ij} = \ell_{mn} = \ell$ and $\ell_{jm} = \ell_{in} = w$. To get the results of Eqs. (9.43) through (9.46), we need to carry out the integration for each term in the 4×4 matrix.

$$\int_{\tau} h[S]^T T\, d\tau = h\int_{\tau} \begin{Bmatrix} S_i \\ S_j \\ S_m \\ S_n \end{Bmatrix} \overbrace{[S_i \quad S_j \quad S_m \quad S_n] \begin{Bmatrix} T_i \\ T_j \\ T_m \\ T_n \end{Bmatrix}}^{T} d\tau$$

$$= h\int_{\tau} \begin{bmatrix} S_i^2 & S_jS_i & S_mS_i & S_nS_i \\ S_iS_j & S_j^2 & S_mS_j & S_nS_j \\ S_iS_m & S_jS_m & S_m^2 & S_nS_m \\ S_iS_n & S_jS_n & S_mS_n & S_n^2 \end{bmatrix} \begin{Bmatrix} T_i \\ T_j \\ T_m \\ T_n \end{Bmatrix} d\tau \qquad (9.47)$$

For example, along the i–j edge, where $S_m = 0$ and $S_n = 0$, the contribution of Eq. (9.47) to the conductance matrix becomes

$$\int_{\tau} h[S]^T T d\tau = h\int_{\tau} \begin{bmatrix} S_i^2 & S_jS_i & 0 & 0 \\ S_iS_j & S_j^2 & 0 & 0 \\ 0 & 0 & 0 & 0 \\ 0 & 0 & 0 & 0 \end{bmatrix} \begin{Bmatrix} T_i \\ T_j \\ T_m \\ T_n \end{Bmatrix} d\tau \tag{9.48}$$

Expressing the shape functions S_i and S_j in terms of natural coordinates:

$$S_i = \frac{1}{4}(1 - \xi)(1 - \eta)$$

$$S_j = \frac{1}{4}(1 + \xi)(1 - \eta)$$

and noting

$$\xi = \frac{2x}{\ell} - 1$$

$$d\xi = \frac{2}{\ell}dx \quad \text{or} \quad dx = \frac{\ell}{2}d\xi$$

and ξ varies from -1 to 1, Eq. (9.48) then becomes

$$\int_{\tau} h[S]^T T d\tau =$$

$$\frac{h\ell_{ij}}{2}\int_{-1}^{1} \begin{bmatrix} \left(\frac{1}{4}(1-\xi)(1-\eta)\right)^2 & \left(\frac{1}{16}(1-\xi)(1-\eta)(1+\xi)(1-\eta)\right) & 0 & 0 \\ \left(\frac{1}{16}(1-\xi)(1-\eta)(1+\xi)(1-\eta)\right) & \left(\frac{1}{4}(1+\xi)(1-\eta)\right)^2 & 0 & 0 \\ 0 & 0 & 0 & 0 \\ 0 & 0 & 0 & 0 \end{bmatrix}$$

$$\begin{Bmatrix} T_i \\ T_j \\ T_m \\ T_n \end{Bmatrix} d\xi \tag{9.49}$$

Making use of the fact that along i–j edge $\eta = -1$, and integrating Eq. (9.49), we get

$$\int_\tau h[S]^T T\,d\tau = \frac{h\ell_{ij}}{2}\int_{-1}^{1}\begin{bmatrix} \left(\frac{1}{2}(1-\xi)\right)^2 & \left(\frac{1}{4}(1-\xi)(1+\xi)\right) & 0 & 0 \\ \left(\frac{1}{4}(1-\xi)(1+\xi)\right) & \left(\frac{1}{2}(1+\xi)\right)^2 & 0 & 0 \\ 0 & 0 & 0 & 0 \\ 0 & 0 & 0 & 0 \end{bmatrix} d\xi \begin{Bmatrix} T_i \\ T_j \\ T_m \\ T_n \end{Bmatrix}$$

$$= \frac{h\ell_{ij}}{6}\begin{bmatrix} 2 & 1 & 0 & 0 \\ 1 & 2 & 0 & 0 \\ 0 & 0 & 0 & 0 \\ 0 & 0 & 0 & 0 \end{bmatrix}\begin{Bmatrix} T_i \\ T_j \\ T_m \\ T_n \end{Bmatrix}$$

where

$$\int_{-1}^{1}\left(\frac{1}{2}(1-\xi)\right)^2 d\xi = \frac{1}{4}\left[\xi + \frac{\xi^3}{3} - \xi^2\right]_{-1}^{1} = \frac{2}{3}$$

and

$$\int_{-1}^{1}\left(\frac{1}{4}(1-\xi)(1+\xi)\right) d\xi = \frac{1}{4}\left[\xi - \frac{\xi^3}{3}\right]_{-1}^{1} = \frac{1}{3}$$

$$\int_{-1}^{1}\left(\frac{1}{2}(1+\xi)\right)^2 d\xi = \frac{1}{4}\left[\xi + \frac{\xi^3}{3} + \xi^2\right]_{-1}^{1} = \frac{2}{3}$$

Similarly, we can obtain the results given by Eqs. (9.44) through (9.46).

The terms $\int_\tau h[\mathbf{S}]^T T_f \cos\theta\, d\tau$ and $\int_\tau h[\mathbf{S}]^T T_f \sin\theta\, d\tau$ contribute to the elemental thermal load matrix. Evaluating these integrals along the edges of the rectangular element, we obtain

$$\{\mathbf{F}\}^{(e)} = \frac{hT_f\ell_{ij}}{2}\begin{Bmatrix} 1 \\ 1 \\ 0 \\ 0 \end{Bmatrix} \quad (9.50) \qquad \{\mathbf{F}\}^{(e)} = \frac{hT_f\ell_{jm}}{2}\begin{Bmatrix} 0 \\ 1 \\ 1 \\ 0 \end{Bmatrix} \quad (9.51)$$

$$\{\mathbf{F}\}^{(e)} = \frac{hT_f\ell_{mn}}{2}\begin{Bmatrix} 0 \\ 0 \\ 1 \\ 1 \end{Bmatrix} \quad (9.52) \qquad \{\mathbf{F}\}^{(e)} = \frac{hT_f\ell_{ni}}{2}\begin{Bmatrix} 1 \\ 0 \\ 0 \\ 1 \end{Bmatrix} \quad (9.53)$$

Let us summarize what we have done so far. The conductance matrix for a bilinear rectangular element is given by:

$$[\mathbf{K}]^{(e)} = \frac{k_x w}{6\ell}\begin{bmatrix} 2 & -2 & -1 & 1 \\ -2 & 2 & 1 & -1 \\ -1 & 1 & 2 & -2 \\ 1 & -1 & -2 & 2 \end{bmatrix} + \frac{k_y \ell}{6w}\begin{bmatrix} 2 & 1 & -1 & -2 \\ 1 & 2 & -2 & -1 \\ -1 & -2 & 2 & 1 \\ -2 & -1 & 1 & 2 \end{bmatrix}$$

Note that the elemental conductance matrix is composed of (1) a conduction component in the x-direction; (2) a conduction component in the y-direction; and (3) a possible heat transfer term by convection around the edge of a given element, as given by Eqs. (9.43) to (9.46). The thermal load matrix for an element could have two components: (1) a component due to possible heat generation within a given element, and (2) a component due to possible convection heat transfer along an element's edge(s), as given by Eqs. (9.50) to (9.53). The contribution of the heat generation to the element's thermal-load matrix is given by

$$\{\mathbf{F}\}^{(e)} = \frac{q^{\cdot}A}{4}\begin{Bmatrix} 1 \\ 1 \\ 1 \\ 1 \end{Bmatrix}$$

It is worth noting that in situations in which constant heat-flux boundary conditions occur along the edges of a rectangular element, the elemental load matrix is given by (see problem 5)

$$\{\mathbf{F}\}^{(e)} = \frac{q''_o \ell_{ij}}{2}\begin{Bmatrix} 1 \\ 1 \\ 0 \\ 0 \end{Bmatrix} \qquad \{\mathbf{F}\}^{(e)} = \frac{q''_o \ell_{jm}}{2}\begin{Bmatrix} 0 \\ 1 \\ 1 \\ 0 \end{Bmatrix}$$

$$\{\mathbf{F}\}^{(e)} = \frac{q''_o \ell_{mn}}{2}\begin{Bmatrix} 0 \\ 0 \\ 1 \\ 1 \end{Bmatrix} \qquad \{\mathbf{F}\}^{(e)} = \frac{q''_o \ell_{ni}}{2}\begin{Bmatrix} 1 \\ 0 \\ 0 \\ 1 \end{Bmatrix}$$

The next step involves assembling elemental matrices to form the global matrices and solving the set of equations $[\mathbf{K}]\{\mathbf{T}\} = \{\mathbf{F}\}$ to obtain the nodal temperatures. We will demonstrate this step in Example 9.1. For now, let us turn our attention to the derivation of the elemental conductance and load matrices for a triangular element.

9.3 FORMULATION WITH TRIANGULAR ELEMENTS

As we discussed in Chapter 7, a major disadvantage associated with using rectangular elements is that they do not conform to curved boundaries. In contrast, triangular elements are better suited to approximate curved boundaries. For the sake of convenience, a triangular element is shown in Figure 9.11.

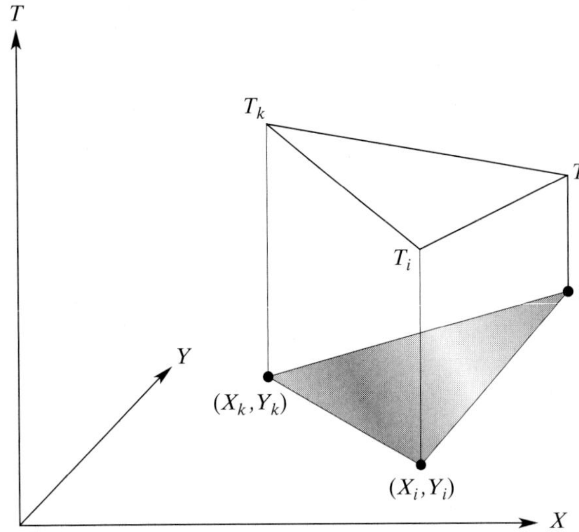

FIGURE 9.11 A triangular element.

Recall that a triangular element is defined by three nodes and that we represent the variation of a dependent variable, such as temperature, over a triangular region using shape functions and the corresponding nodal temperatures by the equation

$$T^{(e)} = [S_i \quad S_j \quad S_k]\begin{Bmatrix} T_i \\ T_j \\ T_k \end{Bmatrix} \tag{9.54}$$

where the shape functions S_i, S_j, and S_k are

$$S_i = \frac{1}{2A}(\alpha_i + \beta_i X + \delta_i Y)$$

$$S_j = \frac{1}{2A}(\alpha_j + \beta_j X + \delta_j Y)$$

$$S_k = \frac{1}{2A}(\alpha_k + \beta_k X + \delta_k Y)$$

A is the area of the element and is computed from the equation

$$2A = X_i(Y_j - Y_k) + X_j(Y_k - Y_i) + X_k(Y_i - Y_j)$$

Also,

$$\begin{aligned} \alpha_i &= X_j Y_k - X_k Y_j \quad \beta_i = Y_j - Y_k \quad \delta_i = X_k - X_j \\ \alpha_j &= X_k Y_i - X_i Y_k \quad \beta_j = Y_k - Y_i \quad \delta_j = X_i - X_k \\ \alpha_k &= X_i Y_j - X_j Y_i \quad \beta_k = Y_i - Y_j \quad \delta_k = X_j - X_i \end{aligned} \tag{9.55}$$

Employing the Galerkin approach, the three residual equations for a triangular element, in matrix form, are given by

$$\int_A [\mathbf{S}]^T\left(k_X\frac{\partial^2 T}{\partial X^2} + k_Y\frac{\partial^2 T}{\partial Y^2} + q^{\cdot}\right)dA = 0 \tag{9.56}$$

where

$$[\mathbf{S}]^T = \begin{Bmatrix} S_i \\ S_j \\ S_m \end{Bmatrix}$$

We now proceed with steps similar to the ones we followed to formulate the conductance and thermal load matrices for rectangular elements. First, we rewrite the second-derivative expressions in terms of the first-derivative expressions using the chain rule. Evaluating the integral

$$-\int_A C_1\left(\frac{\partial[\mathbf{S}]^T}{\partial X}\frac{\partial T}{\partial X}\right)dA$$

for a triangular element, we obtain

$$\frac{\partial[\mathbf{S}]^T}{\partial X} = \frac{\partial}{\partial X}\begin{Bmatrix} S_i \\ S_j \\ S_k \end{Bmatrix} = \frac{1}{2A}\begin{Bmatrix} \beta_i \\ \beta_j \\ \beta_k \end{Bmatrix} \tag{9.57}$$

$$\frac{\partial T}{\partial X} = \frac{\partial}{\partial X}[S_i \quad S_j \quad S_k]\begin{Bmatrix} T_i \\ T_j \\ T_k \end{Bmatrix} = \frac{1}{2A}[\beta_i \quad \beta_j \quad \beta_k]\begin{Bmatrix} T_i \\ T_j \\ T_k \end{Bmatrix} \tag{9.58}$$

Substituting for the derivatives, we get

$$-\int_A C_1\left(\frac{\partial[\mathbf{S}]^T}{\partial X}\frac{\partial T}{\partial X}\right)dA = -C_1\int_A \frac{1}{4A^2}\begin{Bmatrix} \beta_i \\ \beta_j \\ \beta_k \end{Bmatrix}[\beta_i \quad \beta_j \quad \beta_k]\begin{Bmatrix} T_i \\ T_j \\ T_k \end{Bmatrix}dA \tag{9.59}$$

and integrating, we are left with

$$-C_1\int_A \frac{1}{4A^2}\begin{Bmatrix} \beta_i \\ \beta_j \\ \beta_k \end{Bmatrix}[\beta_i \quad \beta_j \quad \beta_k]\begin{Bmatrix} T_i \\ T_j \\ T_k \end{Bmatrix}dA = \frac{C_1}{4A}\begin{bmatrix} \beta_i^2 & \beta_i\beta_j & \beta_i\beta_k \\ \beta_i\beta_j & \beta_j^2 & \beta_j\beta_k \\ \beta_i\beta_k & \beta_j\beta_k & \beta_k^2 \end{bmatrix}\begin{Bmatrix} T_i \\ T_j \\ T_k \end{Bmatrix} \tag{9.60}$$

In the same manner, we can evaluate the term

$$-\int_A C_2\left(\frac{\partial[\mathbf{S}]^T}{\partial Y}\frac{\partial T}{\partial Y}\right)dA$$

as

$$\frac{\partial[\mathbf{S}]^T}{\partial Y} = \frac{\partial}{\partial Y}\begin{Bmatrix} S_i \\ S_j \\ S_k \end{Bmatrix} = \frac{1}{2A}\begin{Bmatrix} \delta_i \\ \delta_j \\ \delta_k \end{Bmatrix} \tag{9.61}$$

$$\frac{\partial T}{\partial Y} = \frac{\partial}{\partial Y}[S_i \quad S_j \quad S_k]\begin{Bmatrix} T_i \\ T_j \\ T_k \end{Bmatrix} = \frac{1}{2A}[\delta_i \quad \delta_j \quad \delta_k]\begin{Bmatrix} T_i \\ T_j \\ T_k \end{Bmatrix} \tag{9.62}$$

Substituting for the derivatives and integrating, we have

$$-C_2\int_A \frac{1}{4A^2}\begin{Bmatrix} \delta_i \\ \delta_j \\ \delta_k \end{Bmatrix}[\delta_i \quad \delta_j \quad \delta_k]\begin{Bmatrix} T_i \\ T_j \\ T_k \end{Bmatrix} dA = -\frac{C_2}{4A}\begin{bmatrix} \delta_i^2 & \delta_i\delta_j & \delta_i\delta_k \\ \delta_i\delta_j & \delta_j^2 & \delta_j\delta_k \\ \delta_i\delta_k & \delta_j\delta_k & \delta_k^2 \end{bmatrix}\begin{Bmatrix} T_i \\ T_j \\ T_k \end{Bmatrix} \tag{9.63}$$

For a triangular element, the thermal load matrix due to the heat generation term C_3 is

$$\int_A [\mathbf{S}]^T C_3\, dA = C_3\int_A \begin{Bmatrix} S_i \\ S_j \\ S_k \end{Bmatrix} dA = \frac{C_3 A}{3}\begin{Bmatrix} 1 \\ 1 \\ 1 \end{Bmatrix} \tag{9.64}$$

Evaluating the terms $\int_\tau h[\mathbf{S}]^T T \cos\theta\, d\tau$ and $\int_\tau h[\mathbf{S}]^T T \sin\theta\, d\tau$ for a convective boundary condition along the edges of the triangular element results in the equations

$$[\mathbf{K}]^{(e)} = \frac{h\ell_{ij}}{6}\begin{bmatrix} 2 & 1 & 0 \\ 1 & 2 & 0 \\ 0 & 0 & 0 \end{bmatrix} \tag{9.65}$$

$$[\mathbf{K}]^{(e)} = \frac{h\ell_{jk}}{6}\begin{bmatrix} 0 & 0 & 0 \\ 0 & 2 & 1 \\ 0 & 1 & 2 \end{bmatrix} \tag{9.66}$$

$$[\mathbf{K}]^{(e)} = \frac{h\ell_{ki}}{6}\begin{bmatrix} 2 & 0 & 1 \\ 0 & 0 & 0 \\ 1 & 0 & 2 \end{bmatrix} \tag{9.67}$$

Note that in the above matrices, ℓ_{ij}, ℓ_{jk}, and ℓ_{ki} represent the respective lengths of the three sides of the triangular element. The terms $\int_\tau h[\mathbf{S}]^T T_f \cos\theta\, d\tau$ and $\int_\tau h[\mathbf{S}]^T T_f \sin\theta\, d\tau$ contribute to the elemental thermal loads. Evaluating these integrals along the edges of the triangular element yields

$$\{\mathbf{F}\}^{(e)} = \frac{hT_f\ell_{ij}}{2}\begin{Bmatrix} 1 \\ 1 \\ 0 \end{Bmatrix} \tag{9.68}$$

$$\{\mathbf{F}\}^{(e)} = \frac{hT_f\ell_{jk}}{2}\begin{Bmatrix} 0 \\ 1 \\ 1 \end{Bmatrix} \tag{9.69}$$

$$\{\mathbf{F}\}^{(e)} = \frac{hT_f\ell_{ki}}{2}\begin{Bmatrix} 1 \\ 0 \\ 1 \end{Bmatrix} \tag{9.70}$$

Let us summarize the triangular formulation. The conductance matrix for a triangular element is

$$[\mathbf{K}]^{(e)} = \frac{k_X}{4A}\begin{bmatrix} \beta_i^2 & \beta_i\beta_j & \beta_i\beta_k \\ \beta_i\beta_j & \beta_j^2 & \beta_j\beta_k \\ \beta_i\beta_k & \beta_j\beta_k & \beta_k^2 \end{bmatrix} + \frac{k_Y}{4A}\begin{bmatrix} \delta_i^2 & \delta_i\delta_j & \delta_i\delta_k \\ \delta_i\delta_j & \delta_j^2 & \delta_j\delta_k \\ \delta_i\delta_k & \delta_j\delta_k & \delta_k^2 \end{bmatrix}$$

Note once again that the elemental conductance matrix for a triangular element is composed of (1) a conduction component in the X-direction; (2) a conduction component in the Y-direction; and (3) a possible heat loss term by convection from the edge(s) of a given element, as given by Eq. (9.65) to (9.67). The thermal load matrix for a triangular element could have two components: (1) a component resulting from a possible heat-generation term within a given element, and (2) a component due to possible convection heat loss from the element's edge(s), as given by Eq. (9.68) to (9.70). The contribution of the heat generation to the element's load matrix is

$$\{\mathbf{F}\}^{(e)} = \frac{\dot{q}A}{3}\begin{Bmatrix} 1 \\ 1 \\ 1 \end{Bmatrix}$$

The development of constant heat flux boundary conditions for triangular elements is left as an exercise. (See problem 6.)

Next, we use an example to demonstrate how to assemble the elemental information to obtain the global conductance matrix and the global load matrix.

EXAMPLE 9.1

Consider a small industrial chimney constructed from concrete with a thermal conductivity value of $k = 1.4\ \text{W/m}\cdot\text{K}$, as shown in Figure 9.12. The inside surface temperature of the chimney is assumed to be uniform at 100°C. The exterior surface is exposed to the surrounding air, which is at 30°C, with a corresponding natural convection heat transfer coefficient of $h = 20\ \text{W/m}^2\cdot\text{K}$. Determine the temperature distribution within the concrete under steady-state conditions.

We can make use of the symmetry of the problem, as shown in Figure 9.12, and only analyze a section of chimney containing 1/8 of the area. The selected section of the chimney is divided into nine nodes with five elements. Elements (1), (2), and (3) are squares, while elements (4) and (5) are triangular elements. Consult Table 9.1 while following the solution.

TABLE 9.1 The relationship between the elements and their corresponding nodes

Element	i	j	m or k	n
(1)	1	2	4	3
(2)	3	4	7	6
(3)	4	5	8	7
(4)	2	5	4	
(5)	5	9	8	

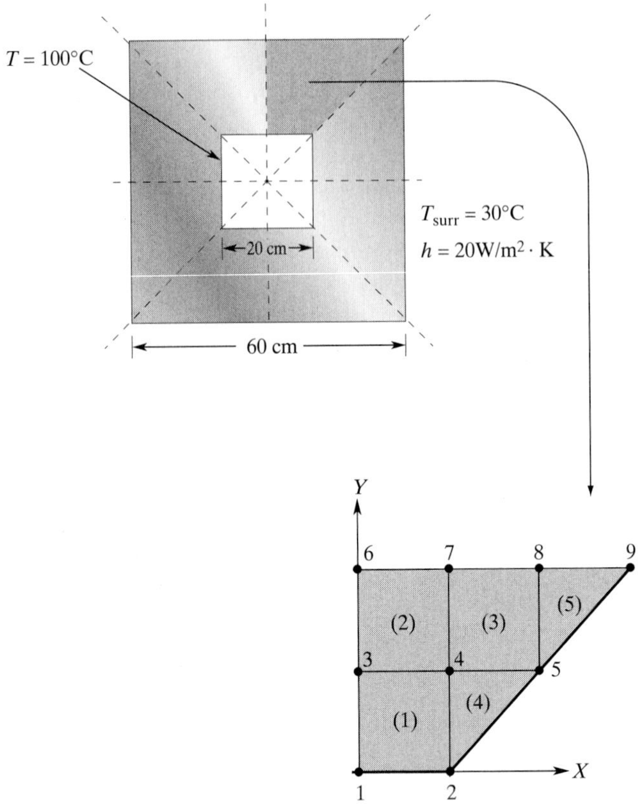

FIGURE 9.12 A schematic of the chimney in Example 9.1.

The conductance matrix due to conduction in a rectangular element is given by

$$[\mathbf{K}]^{(e)} = \frac{kw}{6\ell}\begin{bmatrix} 2 & -2 & -1 & 1 \\ -2 & 2 & 1 & -1 \\ -1 & 1 & 2 & -2 \\ 1 & -1 & -2 & 2 \end{bmatrix} + \frac{k\ell}{6w}\begin{bmatrix} 2 & 1 & -1 & -2 \\ 1 & 2 & -2 & -1 \\ -1 & -2 & 2 & 1 \\ -2 & -1 & 1 & 2 \end{bmatrix}$$

Elements (1), (2), and (3) all have the same dimensions; therefore,

$$[\mathbf{K}]^{(1)} = [\mathbf{K}]^{(2)} = [\mathbf{K}]^{(3)} = \frac{(1.4)(0.1)}{6(0.1)}\begin{bmatrix} 2 & -2 & -1 & 1 \\ -2 & 2 & 1 & -1 \\ -1 & 1 & 2 & -2 \\ 1 & -1 & -2 & 2 \end{bmatrix}$$

$$+ \frac{(1.4)(0.1)}{6(0.1)}\begin{bmatrix} 2 & 1 & -1 & -2 \\ 1 & 2 & -2 & -1 \\ -1 & -2 & 2 & 1 \\ -2 & -1 & 1 & 2 \end{bmatrix}$$

To help with assembly of the elements later, the corresponding node numbers for each element are shown on the top and the side of each matrix:

$$[\mathbf{K}]^{(1)} = \begin{array}{c} \begin{matrix} 1(i) & 2(j) & 4(m) & 3(n) \end{matrix} \\ \begin{bmatrix} 0.933 & -0.233 & -0.466 & -0.233 \\ -0.233 & 0.933 & -0.233 & -0.466 \\ -0.466 & -0.233 & 0.933 & -0.233 \\ -0.233 & -0.466 & -0.233 & 0.933 \end{bmatrix} \begin{matrix} 1 \\ 2 \\ 4 \\ 3 \end{matrix} \end{array}$$

$$[\mathbf{K}]^{(2)} = \begin{array}{c} \begin{matrix} 3(i) & 4(j) & 7(m) & 6(n) \end{matrix} \\ \begin{bmatrix} 0.933 & -0.233 & -0.466 & -0.233 \\ -0.233 & 0.933 & -0.233 & -0.466 \\ -0.466 & -0.233 & 0.933 & -0.233 \\ -0.233 & -0.466 & -0.233 & 0.933 \end{bmatrix} \begin{matrix} 3 \\ 4 \\ 7 \\ 6 \end{matrix} \end{array}$$

$$[\mathbf{K}]^{(3)} = \begin{array}{c} \begin{matrix} 4(i) & 5(j) & 8(m) & 7(n) \end{matrix} \\ \begin{bmatrix} 0.933 & -0.233 & -0.466 & -0.233 \\ -0.233 & 0.933 & -0.233 & -0.466 \\ -0.466 & -0.233 & 0.933 & -0.233 \\ -0.233 & -0.466 & -0.233 & 0.933 \end{bmatrix} \begin{matrix} 4 \\ 5 \\ 8 \\ 7 \end{matrix} \end{array}$$

For triangular elements (4) and (5), the conductance matrix is

$$[\mathbf{K}]^{(e)} = \frac{k}{4A}\begin{bmatrix} \beta_i^2 & \beta_i\beta_j & \beta_i\beta_k \\ \beta_i\beta_j & \beta_j^2 & \beta_j\beta_k \\ \beta_i\beta_k & \beta_j\beta_k & \beta_k^2 \end{bmatrix} + \frac{k}{4A}\begin{bmatrix} \delta_i^2 & \delta_i\delta_j & \delta_i\delta_k \\ \delta_i\delta_j & \delta_j^2 & \delta_j\delta_k \\ \delta_i\delta_k & \delta_j\delta_k & \delta_k^2 \end{bmatrix}$$

where the β- and δ-terms are given by the relations of Eq. (9.55). Because the β- and δ-terms are calculated from the difference of the coordinates of the involved nodes, it does not matter where we place the origin of the coordinate system X, Y. Evaluating the coefficients for element (4), we have

$$\begin{aligned} \beta_i &= Y_j - Y_k = 0.1 - 0.1 = 0 & \delta_i &= X_k - X_j = 0 - 0.1 = -0.1 \\ \beta_j &= Y_k - Y_i = 0.1 - 0 = 0.1 & \delta_j &= X_i - X_k = 0 - 0 = 0 \\ \beta_k &= Y_i - Y_j = 0 - 0.1 = -0.1 & \delta_k &= X_j - X_i = 0.1 - 0 = 0.1 \end{aligned}$$

Evaluating the coefficients for element (5) renders the same results because the difference between the coordinates of its nodes is identical to that of element (4). Therefore, elements (4) and (5) will both have the following conductance matrix:

$$[\mathbf{K}]^{(4)} = [\mathbf{K}]^{(5)} = \frac{1.4}{4(0.005)}\begin{bmatrix} 0 & 0 & 0 \\ 0 & (0.1)^2 & (0.1)(-0.1) \\ 0 & (0.1)(-0.1) & (-0.1)^2 \end{bmatrix}$$

$$+ \frac{1.4}{4(0.005)}\begin{bmatrix} (-0.1)^2 & 0 & (-0.1)(0.1) \\ 0 & 0 & 0 \\ (-0.1)(0.1) & 0 & (0.1)^2 \end{bmatrix}$$

Showing the corresponding node numbers on the top and the side of each respective conductance matrix for elements (4) and (5), we obtain

$$[\mathbf{K}]^{(4)} = \begin{array}{c} \begin{matrix} 2(i) & 5(j) & 4(k) \end{matrix} \\ \begin{bmatrix} 0.7 & 0 & -0.7 \\ 0 & 0.7 & -0.7 \\ -0.7 & -0.7 & 1.4 \end{bmatrix} \begin{matrix} 2 \\ 5 \\ 4 \end{matrix} \end{array}$$

$$[\mathbf{K}]^{(5)} = \begin{array}{c} \begin{matrix} 5(i) & 9(j) & 8(k) \end{matrix} \\ \begin{bmatrix} 0.7 & 0 & -0.7 \\ 0 & 0.7 & -0.7 \\ -0.7 & -0.7 & 1.4 \end{bmatrix} \begin{matrix} 5 \\ 9 \\ 8 \end{matrix} \end{array}$$

As explained earlier, the convective boundary condition contributes to both the conductance matrix and the load matrix. The convective boundary condition contributes to the conductance matrices of elements (2) and (3) according to the relationship

$$[\mathbf{K}]^{(e)} = \frac{h\ell_{mn}}{6} \begin{bmatrix} 0 & 0 & 0 & 0 \\ 0 & 0 & 0 & 0 \\ 0 & 0 & 2 & 1 \\ 0 & 0 & 1 & 2 \end{bmatrix} \begin{matrix} i \\ j \\ m \\ n \end{matrix}$$

$$[\mathbf{K}]^{(2)} = [\mathbf{K}]^{(3)} = \frac{(20)(0.1)}{6} \begin{bmatrix} 0 & 0 & 0 & 0 \\ 0 & 0 & 0 & 0 \\ 0 & 0 & 2 & 1 \\ 0 & 0 & 1 & 2 \end{bmatrix} = \begin{bmatrix} 0 & 0 & 0 & 0 \\ 0 & 0 & 0 & 0 \\ 0 & 0 & 0.666 & 0.333 \\ 0 & 0 & 0.333 & 0.666 \end{bmatrix}$$

Including the nodal information, the conductance matrices for elements (2) and (3) are

$$[\mathbf{K}]^{(2)} = \begin{array}{c} \begin{matrix} 3 & 4 & 7 & 6 \end{matrix} \\ \begin{bmatrix} 0 & 0 & 0 & 0 \\ 0 & 0 & 0 & 0 \\ 0 & 0 & 0.666 & 0.333 \\ 0 & 0 & 0.333 & 0.666 \end{bmatrix} \begin{matrix} 3 \\ 4 \\ 7 \\ 6 \end{matrix} \end{array}$$

$$[\mathbf{K}]^{(3)} = \begin{array}{c} \begin{matrix} 4 & 5 & 8 & 7 \end{matrix} \\ \begin{bmatrix} 0 & 0 & 0 & 0 \\ 0 & 0 & 0 & 0 \\ 0 & 0 & 0.666 & 0.333 \\ 0 & 0 & 0.333 & 0.666 \end{bmatrix} \begin{matrix} 4 \\ 5 \\ 8 \\ 7 \end{matrix} \end{array}$$

Heat loss by convection also occurs along jk edge of element (5); thus,

$$[\mathbf{K}]^{(e)} = \frac{h\ell_{jk}}{6}\begin{bmatrix} 0 & 0 & 0 \\ 0 & 2 & 1 \\ 0 & 1 & 2 \end{bmatrix}\begin{matrix} i \\ j \\ k \end{matrix}$$

$$[\mathbf{K}]^{(e)} = \frac{(20)(0.1)}{6}\begin{bmatrix} 0 & 0 & 0 \\ 0 & 2 & 1 \\ 0 & 1 & 2 \end{bmatrix} = \begin{bmatrix} 0 & 0 & 0 \\ 0 & 0.666 & 0.333 \\ 0 & 0.333 & 0.666 \end{bmatrix}$$

$$[\mathbf{K}]^{(5)} = \begin{matrix} 5 & 9 & 8 \\ \end{matrix}\begin{bmatrix} 0 & 0 & 0 \\ 0 & 0.666 & 0.333 \\ 0 & 0.333 & 0.666 \end{bmatrix}\begin{matrix} 5 \\ 9 \\ 8 \end{matrix}$$

The convective boundary condition contributes to the thermal load matrices for elements (2) and (3) along their mn edge according to the relationship

$$\{\mathbf{F}\}^{(e)} = \frac{hT_f\,\ell_{mn}}{2}\begin{Bmatrix} 0 \\ 0 \\ 1 \\ 1 \end{Bmatrix} = \frac{(20)(30)(0.1)}{2}\begin{Bmatrix} 0 \\ 0 \\ 1 \\ 1 \end{Bmatrix} = \begin{Bmatrix} 0 \\ 0 \\ 30 \\ 30 \end{Bmatrix}$$

Including the nodal information, we have

$$\{\mathbf{F}\}^{(2)} = \begin{Bmatrix} 0 \\ 0 \\ 30 \\ 30 \end{Bmatrix}\begin{matrix} 3 \\ 4 \\ 7 \\ 6 \end{matrix}$$

$$\{\mathbf{F}\}^{(3)} = \begin{Bmatrix} 0 \\ 0 \\ 30 \\ 30 \end{Bmatrix}\begin{matrix} 4 \\ 5 \\ 8 \\ 7 \end{matrix}$$

The convective boundary condition contributes to the load matrix for element (5) along its jk edge according to the matrix

$$\{\mathbf{F}\}^{(e)} = \frac{hT_f\,\ell_{jk}}{2}\begin{Bmatrix} 0 \\ 1 \\ 1 \end{Bmatrix} = \frac{(20)(30)(0.1)}{2}\begin{Bmatrix} 0 \\ 1 \\ 1 \end{Bmatrix} = \begin{Bmatrix} 0 \\ 30 \\ 30 \end{Bmatrix}$$

Again, including the nodal information, we have

$$\{\mathbf{F}\}^{(5)} = \begin{Bmatrix} 0 \\ 30 \\ 30 \end{Bmatrix} \begin{matrix} 5 \\ 9 \\ 8 \end{matrix}$$

Next, we need to assemble all of the elemental matrices. Using the nodal information presented next to each element, the global conductance matrix becomes

$$[\mathbf{K}]^{(G)} = \begin{matrix} \begin{matrix} 1 & 2 & 3 & 4 & 5 & 6 & 7 & 8 & 9 \end{matrix} & \\ \begin{bmatrix} 0.933 & -0.233 & -0.233 & -0.466 & 0 & 0 & 0 & 0 & 0 \\ -0.233 & 1.633 & -0.466 & -0.933 & 0 & 0 & 0 & 0 & 0 \\ -0.233 & -0.466 & 1.866 & -0.466 & 0 & -0.233 & -0.466 & 0 & 0 \\ -0.466 & -0.933 & -0.466 & 4.199 & -0.933 & -0.466 & -0.466 & -0.466 & 0 \\ 0 & 0 & 0 & -0.933 & 2.333 & 0 & -0.466 & -0.933 & 0 \\ 0 & 0 & -0.233 & -0.466 & 0 & 1.599 & 0.1 & 0 & 0 \\ 0 & 0 & -0.466 & -0.466 & -0.466 & 0.1 & 3.198 & 0.1 & 0 \\ 0 & 0 & 0 & -0.466 & -0.933 & 0 & 0.1 & 3.665 & -0.367 \\ 0 & 0 & 0 & 0 & 0 & 0 & 0 & -0.367 & 1.366 \end{bmatrix} & \begin{matrix} 1 \\ 2 \\ 3 \\ 4 \\ 5 \\ 6 \\ 7 \\ 8 \\ 9 \end{matrix} \end{matrix}$$

Applying the constant temperature boundary condition at nodes 1 and 2 results in the global matrix

$$[\mathbf{K}]^{(G)} = \begin{bmatrix} 1 & 0 & 0 & 0 & 0 & 0 & 0 & 0 & 0 \\ 0 & 1 & 0 & 0 & 0 & 0 & 0 & 0 & 0 \\ -0.233 & -0.466 & 1.866 & -0.466 & 0 & -0.233 & -0.466 & 0 & 0 \\ -0.466 & -0.933 & -0.466 & 4.199 & -0.933 & -0.466 & -0.466 & -0.466 & 0 \\ 0 & 0 & 0 & -0.933 & 2.333 & 0 & -0.466 & -0.933 & 0 \\ 0 & 0 & -0.233 & -0.466 & 0 & 1.599 & 0.1 & 0 & 0 \\ 0 & 0 & -0.466 & -0.466 & -0.466 & 0.1 & 3.198 & 0.1 & 0 \\ 0 & 0 & 0 & -0.466 & -0.933 & 0 & 0.1 & 3.665 & -0.367 \\ 0 & 0 & 0 & 0 & 0 & 0 & 0 & -0.367 & 1.366 \end{bmatrix}$$

Assembling the thermal load matrix, we have

$$\{\mathbf{F}\}^{(G)} = \begin{Bmatrix} 0 \\ 0 \\ 0 \\ 0 \\ 0 \\ 30 \\ 30 + 30 \\ 30 + 30 \\ 30 \end{Bmatrix}$$

and applying the constant temperature boundary condition at nodes 1 and 2 leads to the following final form of the thermal load matrix:

$$\{\mathbf{F}\}^{(G)} = \begin{Bmatrix} 100 \\ 100 \\ 0 \\ 0 \\ 0 \\ 30 \\ 60 \\ 60 \\ 30 \end{Bmatrix}$$

The final set of nodal equations is given by

$$\begin{bmatrix} 1 & 0 & 0 & 0 & 0 & 0 & 0 & 0 & 0 \\ 0 & 1 & 0 & 0 & 0 & 0 & 0 & 0 & 0 \\ -0.233 & -0.466 & 1.866 & -0.466 & 0 & -0.233 & -0.466 & 0 & 0 \\ -0.466 & -0.933 & -0.466 & 4.199 & -0.933 & -0.466 & -0.466 & -0.466 & 0 \\ 0 & 0 & 0 & -0.933 & 2.333 & 0 & -0.466 & -0.933 & 0 \\ 0 & 0 & -0.233 & -0.466 & 0 & 1.599 & 0.1 & 0 & 0 \\ 0 & 0 & -0.466 & -0.466 & -0.466 & 0.1 & 3.198 & 0.1 & 0 \\ 0 & 0 & 0 & -0.466 & -0.933 & 0 & 0.1 & 3.665 & -0.367 \\ 0 & 0 & 0 & 0 & 0 & 0 & 0 & -0.367 & 1.366 \end{bmatrix}$$

$$\times \begin{Bmatrix} T_1 \\ T_2 \\ T_3 \\ T_4 \\ T_5 \\ T_6 \\ T_7 \\ T_8 \\ T_9 \end{Bmatrix} = \begin{Bmatrix} 100 \\ 100 \\ 0 \\ 0 \\ 0 \\ 30 \\ 60 \\ 60 \\ 30 \end{Bmatrix}$$

Solving the set of linear equations simultaneously leads to the following nodal solution:

$$[T]^T = [100 \quad 100 \quad 70.83 \quad 67.02 \quad 51.56 \quad 45.88 \quad 43.67 \quad 40.10 \quad 32.73]°\text{C}$$

To check for the accuracy of the results, first note that nodal temperatures are within the imposed boundary temperatures. Moreover, all temperatures at the outer edge are slightly above 30°C, with node 9 having the smallest value. This condition makes physical sense because node 9 is the outermost cornerpoint. As another check on the validity of the results, we can make sure that the conservation of energy, as applied to a control volume surrounding an arbitrary node, is satisfied. Are the energies flowing into

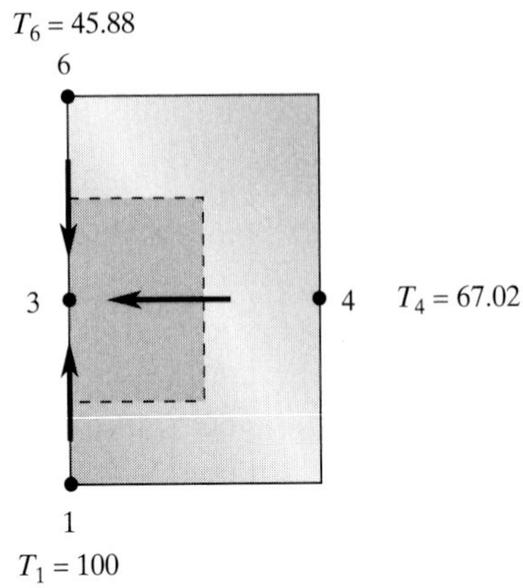

FIGURE 9.13 Applying the principle of energy balance to node 3 to check the validity of our results.

and out of a node balanced out? As an example, let us consider node 3. Figure 9.13 shows the control volume surrounding node 3 used to apply the conservation of energy principle.

We start with the equation

$$\sum q = 0$$

and using Fourier's law we have

$$k(0.1)\left(\frac{67.02 - T_3}{0.1}\right) + k(0.05)\left(\frac{45.88 - T_3}{0.1}\right) + k(0.05)\left(\frac{100 - T_3}{0.1}\right) = 0$$

Solving for T_3, we find that $T_3 = 69.98°C$. This value is reasonably close to the value of 70.83°C, particularly considering the coarseness of the element sizes. We will discuss the verification of results further with another example problem solved using ANSYS.

9.4 AXISYMMETRIC FORMULATION OF THREE-DIMENSIONAL PROBLEMS

As we explained in Chapter 7, Section 7.5, there is a special class of three-dimensional problems whose geometry and loading are symmetrical about an axis, such as a z-axis, as shown in Figure 7.12. These three-dimensional problems may be analyzed using two-dimensional axisymmetric elements. We discussed the formulation of axisymmetric elements in Section 7.5; in this section, we discuss the finite element formulation of axisymmetric conduction problems using triangular axisymmetric elements.

We begin our finite element formulation by applying the principle of the conservation of energy to a differential volume represented in a cylindrical coordinate system. The resulting heat conduction equation is given by

$$\frac{1}{r}\frac{\partial}{\partial r}\left(k_r r\frac{\partial T}{\partial r}\right) + \frac{1}{r^2}\frac{\partial}{\partial \theta}\left(k_\theta\frac{\partial T}{\partial \theta}\right) + \frac{\partial}{\partial z}\left(k_z\frac{\partial T}{\partial z}\right) + q^{\cdot} = \rho c\frac{\partial T}{\partial t} \tag{9.71}$$

In Eq. (9.71), k_r, k_θ, k_z represent thermal conductivities in r-, θ-, and z-directions, $\boldsymbol{q}^{\cdot}$ is the heat generation per unit volume, and c represents the specific heat of the material. The derivation of Eq. (9.71) is shown in Figure 9.14. For steady-state problems, the right-hand side of Eq. (9.71) vanishes. Moreover, for axisymmetric situations, there is no variation of temperature in the θ-direction, and assuming thermal conductivity is constant $(k_r = k_\theta = k_z = k)$,Eq. (9.71) reduces to

$$\frac{k}{r}\frac{\partial}{\partial r}\left(r\frac{\partial T}{\partial r}\right) + k\frac{\partial^2 T}{\partial z^2} + q^{\cdot} = 0 \tag{9.72}$$

The Galerkin residuals for an arbitrary triangular element become

$$\{\boldsymbol{R}^{(e)}\} = \int_V [\boldsymbol{S}]^T\left(\frac{k}{r}\frac{\partial}{\partial r}\left(r\frac{\partial T}{\partial r}\right) + k\frac{\partial^2 T}{\partial z^2} + q^{\cdot}\right)dV \tag{9.73}$$

where

$$S_i = \frac{1}{2A}(\alpha_i + \beta_i r + \delta_i z)$$

$$[\boldsymbol{S}]^T = \begin{Bmatrix} S_i \\ S_j \\ S_k \end{Bmatrix} \quad \text{and} \quad S_j = \frac{1}{2A}(\alpha_j + \beta_j r + \delta_j z) \quad \text{and}$$

$$S_k = \frac{1}{2A}(\alpha_k + \beta_k r + \delta_k z)$$

$$\begin{array}{lll} \alpha_i = R_j Z_k - R_k Z_j & \beta_i = Z_j - Z_k & \delta_i = R_k - R_j \\ \alpha_j = R_k Z_i - R_i Z_k & \beta_j = Z_k - Z_i & \delta_j = R_i - R_k \\ \alpha_k = R_i Z_j - R_j Z_i & \beta_k = Z_i - Z_j & \delta_k = R_j - R_i \end{array}$$

As marked below, Eq. (9.73) has three main parts

$$\{\boldsymbol{R}^{(e)}\} = \overbrace{\int_V [\boldsymbol{S}]^T\left(\frac{k}{r}\frac{\partial}{\partial r}\left(r\frac{\partial T}{\partial r}\right)\right)dV}^{\text{part one}} + \overbrace{\int_V [\boldsymbol{S}]^T\left(k\frac{\partial^2 T}{\partial z^2}\right)dV}^{\text{part two}} + \overbrace{\int_V [\boldsymbol{S}]^T q^{\cdot} dV}^{\text{part three}} \tag{9.74}$$

Using the chain rule, we can rearrange the terms in part one in the following manner:

$$\frac{k}{r}\frac{\partial}{\partial r}\left([\boldsymbol{S}]^T r\frac{\partial T}{\partial r}\right) = \frac{k}{r}\frac{\partial [\boldsymbol{S}]^T}{\partial r} r\frac{\partial T}{\partial r} + \frac{k}{r}[\boldsymbol{S}]^T\frac{\partial}{\partial r}\left(r\frac{\partial T}{\partial r}\right) \tag{9.75}$$

Now, using the relationship given by Eq. (9.75), the part one in Eq. (9.74) may be expressed as

$$[\boldsymbol{S}]^T\left(\frac{k}{r}\frac{\partial}{\partial r}\left(r\frac{\partial T}{\partial r}\right)\right) = \frac{k}{r}\frac{\partial}{\partial r}\left([\boldsymbol{S}]^T r\frac{\partial T}{\partial r}\right) - k\frac{\partial [\boldsymbol{S}]^T}{\partial r}\frac{\partial T}{\partial r} \tag{9.76}$$

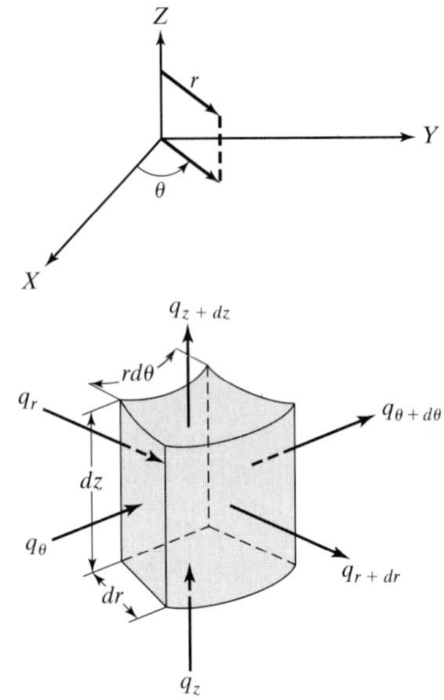

Applying the conservation of energy to a differential cylindrical segment,

$$\dot{E}_{\text{in}} - \dot{E}_{\text{out}} + \dot{E}_{\text{generated}} = \dot{E}_{\text{stored}}$$

The r-, θ-, and z-components of Fourier's law in cylindrical coordinates are

$$q_r = -k_r A_r \frac{\partial T}{\partial r} = -k_r \overbrace{r\,d\theta\,dz}^{A_r} \frac{\partial T}{\partial r}$$

$$q_\theta = -\frac{k_\theta}{r} A_\theta \frac{\partial T}{\partial \theta} = -\frac{\overbrace{k_\theta drdz}^{A_\theta}}{r} \frac{\partial T}{\partial \theta}$$

$$q_z = -k_z A_z \frac{\partial T}{\partial z} = -k_z \overbrace{rd\theta dr}^{A_r} \frac{\partial T}{\partial z}$$

$$(q_r + q_\theta + q_z) - (q_{r+dr} + q_{\theta+d\theta} + q_{z+dz}) + \dot{q}\overbrace{drdzrd\theta}^{dV} = \rho c\,\overbrace{drdzrd\theta}^{dV}\frac{\partial T}{\partial t}$$

$$(q_r + q_\theta + q_z) - \left(\overbrace{q_r + \frac{\partial q_r}{\partial r}dr}^{q_{r+dr}} + \overbrace{q_\theta + \frac{\partial q_\theta}{\partial \theta}d\theta}^{q_{\theta+d\theta}} + \overbrace{q_z + \frac{\partial q_z}{\partial z}dz}^{q_{z+dz}}\right) + \dot{q}\,drdzrd\theta = \rho c\; drdzrd\theta\frac{\partial T}{\partial t}$$

$$-\frac{\partial q_r}{\partial r}dr - \frac{\partial q_\theta}{\partial \theta}d\theta - \frac{\partial q_z}{\partial z}dz + \dot{q}\,drdzrd\theta = \rho c\,drdzrd\theta\frac{\partial T}{\partial t}$$

Substituting for q_r, q_θ, and q_z in the above equation

$$-\frac{\partial}{\partial r}\left(-k_r rd\theta dz\frac{\partial T}{\partial r}\right)dr - \frac{\partial}{\partial \theta}\left(-\frac{k_\theta drdz}{r}\frac{\partial T}{\partial \theta}\right)d\theta - \frac{\partial}{\partial z}\left(-k_z rd\theta dr\frac{\partial T}{\partial z}\right)dz + \dot{q}\,drdzrd\theta = \rho c\,drdzrd\theta\frac{\partial T}{\partial t}$$

and simplifying the differential volume terms, we have

$$\frac{1}{r}\frac{\partial}{\partial r}\left(k_r r\frac{\partial T}{\partial r}\right) + \frac{1}{r^2}\frac{\partial}{\partial \theta}\left(k_\theta\frac{\partial T}{\partial \theta}\right) + \frac{\partial}{\partial z}\left(k_z\frac{\partial T}{\partial z}\right) + \dot{q} - \rho c\frac{\partial T}{\partial t}$$

FIGURE 9.14 The derivation of equation of heat conduction in cylindrical coordinates under steady-state conditions.

Similarly, using the chain rule and rearranging terms in part two of Eq. (9.74), we have

$$k\frac{\partial}{\partial z}\left([\mathbf{S}]^T\frac{\partial T}{\partial z}\right) = k[\mathbf{S}]^T\frac{\partial^2 T}{\partial z^2} + k\frac{\partial[\mathbf{S}]^T}{\partial z}\frac{\partial T}{\partial z} \tag{9.77}$$

and

$$k[\mathbf{S}]^T\frac{\partial^2 T}{\partial z^2} = k\frac{\partial}{\partial z}\left([\mathbf{S}]^T\frac{\partial T}{\partial z}\right) - k\frac{\partial[\mathbf{S}]^T}{\partial z}\frac{\partial T}{\partial z} \tag{9.78}$$

Applying the results given by Eqs. (9.76) and (9.78), the contributions of part one and part two to the residual matrix is

$$\overbrace{\int_V [\mathbf{S}]^T\left(\frac{k}{r}\frac{\partial}{\partial r}\left(r\frac{\partial T}{\partial r}\right)\right)dV}^{\text{part one}} = \int_V \frac{k}{r}\frac{\partial}{\partial r}\left([\mathbf{S}]^T r\frac{\partial T}{\partial r}\right)dV - \int_V k\frac{\partial[\mathbf{S}]^T}{\partial r}\frac{\partial T}{\partial r}dV \tag{9.79}$$

$$\overbrace{\int_V [\mathbf{S}]^T\left(k\frac{\partial^2 T}{\partial z^2}\right)dV}^{\text{part two}} = \int_V \frac{\partial}{\partial z}\left([\mathbf{S}]^T\frac{\partial T}{\partial z}\right)dV - \int_V \frac{\partial[\mathbf{S}]^T}{\partial z}\frac{\partial T}{\partial z}dV \tag{9.80}$$

Similar to what we did in Section 9.2, using Green's theorem, we can rewrite the volume integrals, $\int_V \frac{k}{r}\frac{\partial}{\partial r}\left([\mathbf{S}]^T r\frac{\partial T}{\partial r}\right)dV$ and $\int_V \frac{\partial}{\partial z}\left([\mathbf{S}]^T\frac{\partial T}{\partial z}\right)dV$ in terms of area integrals around the element surface boundaries. We will come back to these terms later; for now let us consider the $-\int_V k\frac{\partial[\mathbf{S}]^T}{\partial r}\frac{\partial T}{\partial r}dV$ term in Eq. (9.79). This term can easily be evaluated.

Evaluating the derivative for a triangular element, we obtain

$$\frac{\partial T}{\partial r} = \frac{\partial}{\partial r}[S_i \quad S_j \quad S_k]\begin{Bmatrix} T_i \\ T_j \\ T_k \end{Bmatrix} = \frac{1}{2A}[\beta_i \quad \beta_j \quad \beta_k]\begin{Bmatrix} T_i \\ T_j \\ T_k \end{Bmatrix} \tag{9.81}$$

Also, evaluating $\frac{\partial[\mathbf{S}]^T}{\partial r}$, we have

$$\frac{\partial[\mathbf{S}]^T}{\partial r} = \frac{\partial}{\partial r}\begin{Bmatrix} S_i \\ S_j \\ S_k \end{Bmatrix} = \frac{1}{2A}\begin{Bmatrix} \beta_i \\ \beta_j \\ \beta_k \end{Bmatrix} \tag{9.82}$$

Substituting the results of Eqs. (9.81) and (9.82) into the term $-\int_V k\frac{\partial[\mathbf{S}]^T}{\partial r}\frac{\partial T}{\partial r}dV$ and integrating, we have

$$-\int_V k\frac{\partial[\mathbf{S}]^T}{\partial r}\frac{\partial T}{\partial r}dV = -k\int_V \frac{1}{4A^2}\begin{Bmatrix}\beta_i\\ \beta_j\\ \beta_k\end{Bmatrix}[\beta_i \quad \beta_j \quad \beta_k]dV = \frac{k}{4A^2}\begin{bmatrix}\beta_i^2 & \beta_i\beta_j & \beta_i\beta_k\\ \beta_i\beta_j & \beta_j^2 & \beta_j\beta_k\\ \beta_i\beta_k & \beta_j\beta_k & \beta_k^2\end{bmatrix}V \quad (9.83)$$

Next, we use the *Pappus–Guldinus* theorem to compute the volume of the element that is generated by revolving the area element about the z-axis. You may recall from your Statics class that the Pappus–Guldinus theorem is used to calculate the volume of a body that is generated by revolving a sectional area about an axis similar to the one shown in Figure 9.15.

$$V = 2\pi\bar{r}A \quad (9.84)$$

In Equation (9.84), $2\pi\bar{r}$ represents the distance traveled by the centroid of the sectional area. See Example 9.2 to refresh your memory about how the Pappus–Guldinus theorem

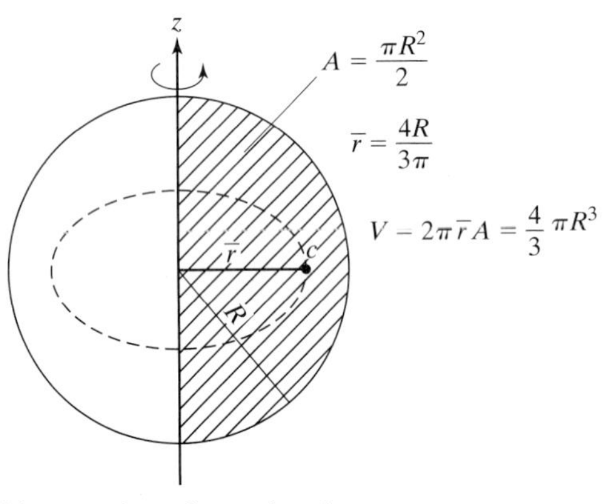

(a) generating volume of a sphere

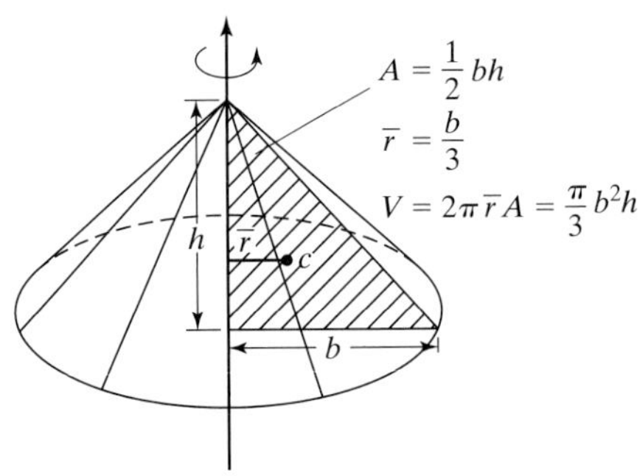

(b) generating volume of a cone

FIGURE 9.15 The Pappus–Guldinus theorem—A volume generated by revolving an area about a fixed axis.

is used to calculate a volume of an object. Substituting in Eq. (9.83) and simplifying, we get

$$-\int_V \boldsymbol{k}\frac{\partial[\mathbf{S}]^T}{\partial r}\frac{\partial T}{\partial r}dV = \frac{\pi\bar{r}k}{2A}\begin{bmatrix} \beta_i^2 & \beta_i\beta_j & \beta_i\beta_k \\ \beta_i\beta_j & \beta_j^2 & \beta_j\beta_k \\ \beta_i\beta_k & \beta_j\beta_k & \beta_k^2 \end{bmatrix} \tag{9.85}$$

As we discussed in Chapter 7, we can use shape functions to describe the position of any point within an element. Therefore, the radial position of the centroid of the element for which $S_i = S_j = S_k = 1/3$ can be expressed by

$$\bar{r} = S_iR_i + S_jR_j + S_kR_k = \frac{R_i + R_j + R_k}{3} \tag{9.86}$$

Equation (9.86) can be used to calculate $\bar{r}$. In a manner shown previously, we can also evaluate the term $-\int_V \frac{\partial[\mathbf{S}]^T}{\partial z}\frac{\partial T}{\partial z}dV$

$$\frac{\partial T}{\partial z} = \frac{\partial}{\partial z}[S_i \quad S_j \quad S_k]\begin{Bmatrix} T_i \\ T_j \\ T_k \end{Bmatrix} = \frac{1}{2A}[\delta_i \quad \delta_j \quad \delta_k]\begin{Bmatrix} T_i \\ T_j \\ T_k \end{Bmatrix} \tag{9.87}$$

$$\frac{\partial[\mathbf{S}]^T}{\partial z} = \frac{\partial}{\partial z}\begin{Bmatrix} S_i \\ S_j \\ S_k \end{Bmatrix} = \frac{1}{2A}\begin{Bmatrix} \delta_i \\ \delta_j \\ \delta_k \end{Bmatrix} \tag{9.88}$$

Substituting for the derivatives and integrating, we have

$$-\int_V \boldsymbol{k}\frac{\partial[\mathbf{S}]^T}{\partial z}\frac{\partial T}{\partial z}dV = -k\int_V \frac{1}{4A^2}\begin{Bmatrix} \delta_i \\ \delta_j \\ \delta_k \end{Bmatrix}[\delta_i \quad \delta_j \quad \delta_k]dV = \frac{k}{4A^2}\begin{bmatrix} \delta_i^2 & \delta_i\delta_j & \delta_i\delta_k \\ \delta_i\delta_j & \delta_j^2 & \delta_j\delta_k \\ \delta_i\delta_k & \delta_j\delta_k & \delta_k^2 \end{bmatrix}V \tag{9.89}$$

And using Pappus–Guldinus theorem, substituting for $V = 2\pi\bar{r}A$ in Eq. (9.89), we have

$$-\int_V \boldsymbol{k}\frac{\partial[\mathbf{S}]^T}{\partial z}\frac{\partial T}{\partial z}dV = \frac{k}{4A^2}\begin{bmatrix} \delta_i^2 & \delta_i\delta_j & \delta_i\delta_k \\ \delta_i\delta_j & \delta_j^2 & \delta_j\delta_k \\ \delta_i\delta_k & \delta_j\delta_k & \delta_k^2 \end{bmatrix}2\pi\bar{r}A = \frac{\pi\bar{r}k}{2A}\begin{bmatrix} \delta_i^2 & \delta_i\delta_j & \delta_i\delta_k \\ \delta_i\delta_j & \delta_j^2 & \delta_j\delta_k \\ \delta_i\delta_k & \delta_j\delta_k & \delta_k^2 \end{bmatrix} \tag{9.90}$$

The thermal load due to the heat generation term is evaluated by first substituting for $dV = 2\pi\bar{r}dA$ and $\bar{r} = \frac{R_i + R_j + R_k}{3}$, then integrating the resulting expression

$$\int_V [\mathbf{S}]^T q^{\cdot} dV = q^{\cdot}\int_V [\mathbf{S}]^T dV = q^{\cdot}\int_V \begin{Bmatrix} S_i \\ S_j \\ S_k \end{Bmatrix}dV = q^{\cdot}2\pi\bar{r}\int_V \begin{Bmatrix} S_i \\ S_j \\ S_k \end{Bmatrix}dA \tag{9.91}$$

and substituting for S_i, S_j, and S_k and integrating, we get

$$\int_V [\mathbf{S}]^T \dot{q}\, dV = \dot{q} 2\pi \bar{r} \int_V \begin{Bmatrix} S_i \\ S_j \\ S_k \end{Bmatrix} dA = \frac{2\pi \dot{q} A}{12} \begin{bmatrix} 2 & 1 & 1 \\ 1 & 2 & 1 \\ 1 & 1 & 2 \end{bmatrix} \begin{Bmatrix} R_i \\ R_j \\ R_k \end{Bmatrix} \tag{9.92}$$

We now return to the terms $\int_V \frac{k}{r} \frac{\partial}{\partial r}\left([\mathbf{S}]^T r \frac{\partial T}{\partial r} \right) dV$ and $\int_V \frac{\partial}{\partial z}\left([\mathbf{S}]^T \frac{\partial T}{\partial z} \right) dV$.

As mentioned earlier, we can use Green's theorem to rewrite these volume integrals in terms of area integrals around the boundaries similar to the approach we discussed in Section 9.2. For convective boundary conditions along the areas generated by revolving the edges of the triangular element, the above terms contribute to both the conductance matrix and the thermal load matrix in the following manner:

$$\text{Along } i\text{–}j, \quad [\mathbf{K}]^{(e)} = \frac{2\pi \ell_{ij}}{12} \begin{bmatrix} 3R_i + R_j & R_i + R_j & 0 \\ R_i + R_j & R_i + 3R_j & 0 \\ 0 & 0 & 0 \end{bmatrix} \tag{9.93}$$

and

$$\{\boldsymbol{f}\}^{(e)} = \frac{2\pi h T_f \ell_{ij}}{6} \begin{Bmatrix} 2R_i + R_j \\ R_i + 2R_j \\ 0 \end{Bmatrix} \tag{9.94}$$

$$\text{Along } j\text{–}k, \quad [\mathbf{K}]^{(e)} = \frac{2\pi \ell_{jk}}{12} \begin{bmatrix} 0 & 0 & 0 \\ 0 & 3R_j + R_k & R_j + R_k \\ 0 & R_j + R_k & R_j + 3R_k \end{bmatrix} \tag{9.95}$$

and

$$\{f\}^{(e)} = \frac{2\pi h T_f \ell_{jk}}{6} \begin{Bmatrix} 0 \\ 2R_j + R_k \\ R_j + 2R_k \end{Bmatrix} \tag{9.96}$$

$$\text{Along } k\text{–}i, \quad [\mathbf{K}]^{(e)} = \frac{2\pi \ell_{ki}}{12} \begin{bmatrix} 3R_i + R_k & 0 & R_i + R_k \\ 0 & 0 & 0 \\ R_i + R_k & 0 & R_i + 3R_k \end{bmatrix} \tag{9.97}$$

and

$$\{f\}^{(e)} = \frac{2\pi h T_f \ell_{ki}}{6} \begin{Bmatrix} 2R_i + R_k \\ 0 \\ R_i + 2R_k \end{Bmatrix} \tag{9.98}$$

Let us now summarize the axisymmetric triangular formulation. The conductance matrix for an axisymmetric triangular element is

$$[\boldsymbol{K}]^{(e)} = \frac{\pi \bar{r} k}{2A}\begin{bmatrix} \beta_i^2 & \beta_i\beta_j & \beta_i\beta_k \\ \beta_i\beta_j & \beta_j^2 & \beta_j\beta_k \\ \beta_i\beta_k & \beta_j\beta_k & \beta_k^2 \end{bmatrix} + \frac{\pi \bar{r} k}{2A}\begin{bmatrix} \delta_i^2 & \delta_i\delta_j & \delta_i\delta_k \\ \delta_i\delta_j & \delta_j^2 & \delta_j\delta_k \\ \delta_i\delta_k & \delta_j\delta_k & \delta_k^2 \end{bmatrix}$$

Note that the elemental conductance matrix for an axisymmetric triangular element is composed of (1) a conduction component in the radial direction; (2) a conduction component in the z-direction; and (3) a possible heat loss or gain term by convection from the edge(s) of a given element as given by Eqs. (9.93), (9.95), and (9.97). The thermal load matrix for an axisymmetric triangular element could have two components: (1) a component resulting from a possible heat generation term within a given element, Eq. (9.92), and (2) a component due to possible convection heat loss or gain from the element's edge(s) as given by Eqs. (9.94), (9.96), and (9.98). The development of constant heat flux boundary conditions for an axisymmetric element is left as an exercise. (See problem 25.)

EXAMPLE 9.2

Use the Pappus–Guldinus theorem to calculate the volume of the solid portion of the annulus generated by revolving the rectangle about the z-axis, as shown in the accompanying figure.

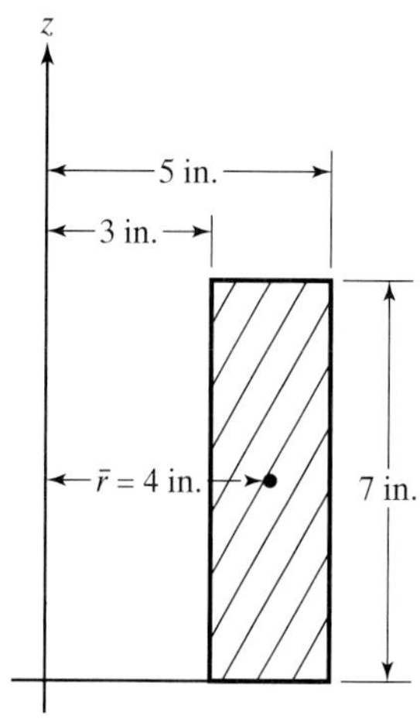

The centroid of the revolving rectangle is located 4 in from the z-axis.

$$V = 2\pi \bar{r} A = 2\pi(4 \text{ in})(2 \text{ in})(7 \text{ in}) = 112\pi \text{ in}^3$$

Alternatively, the volume of the solid portion of the annulus can be calculated from

$$V = \pi R_2^2 h - \pi R_1^2 h = \pi((5\text{ in})^2 - (3\text{ in})^2)(7\text{ in}) = 112\pi\text{ in}^3$$

As you can see, the results are in agreement.

9.5 UNSTEADY HEAT TRANSFER

In this section, we are concerned with determining how temperatures may vary with position and time as a result of applied or existing thermal conditions. First, we review some important concepts dealing with transient heat transfer problems. A good understanding of these concepts will better assist you to model a physical problem using ANSYS. We begin our review with problems for which the spatial variation of temperature is negligible during the transient process and consequently, the variation of temperature with time is only important. Consider the cooling of a small spherical steel pellet as shown in Figure 9.16.

The thermophysical variables affecting the solution are shown in Figure 9.16. Because of the relatively small size of the pellet, the temperature variation with position inside the sphere is negligible. We can obtain an expression for variation of temperature with time starting with Eq. (9.7), $\dot{E}_{in} - \dot{E}_{out} + \dot{E}_{gen} = \dot{E}_{stored}$, and realizing that for this problem $\dot{E}_{in} = 0$, $\dot{E}_g = 0$. Moreover, convection heat transfer occurs between the sphere and the surrounding fluid according to Eq. (9.5), $\dot{E}_{out} = -hA(T - T_f)$. The $\dot{E}_{stored}$ term represents the decrease in the thermal energy of the sphere due to the cooling process and is given by $\dot{E}_{stored} = \rho c V \frac{dT}{dt}$. Substituting for $\dot{E}_{in} = 0$, $\dot{E}_g = 0$, $\dot{E}_{out} = -hA(T - T_f)$, and $\dot{E}_{stored} = \rho c V \frac{dT}{dt}$ in Eq. (9.7), we get

$$-hA(T - T_f) = \rho c V \frac{dT}{dt} \tag{9.99}$$

Assuming a relatively large cooling reservoir, it is reasonable to expect the temperature of fluid not to change with time. Thus, in Equation (9.99) we have assumed that T_f is constant. Equation (9.99) is a first order differential equation and to solve it,

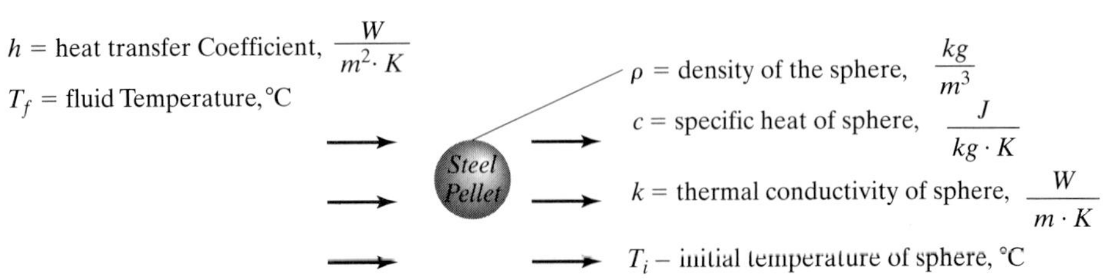

FIGURE 9.16 Cooling of a small spherical pellet.

we let $\Theta = T - T_f$, then we separate variables and integrate, which results in

$$-hA\Theta = \rho cV\frac{d\Theta}{dt} \Rightarrow \frac{\rho cV}{hA}\int_{\Theta_i}^{\Theta}\frac{d\Theta}{\Theta} = -\int_0^t dt \Rightarrow t = \frac{\rho cV}{hA}\ln\frac{\Theta_i}{\Theta} \qquad (9.100)$$

Thus, the time that it takes for a sphere to reach a certain temperature T is computed from

$$t = \frac{\rho cV}{hA}\ln\frac{T_i - T_f}{T - T_f} \qquad (9.101)$$

And if we were interested in knowing what the temperature of a sphere was at a given time, we could rearrange Eq. (9.101) in the following manner

$$\frac{T - T_f}{T_i - T_f} = \exp\left(-\frac{hA}{\rho cV}t\right) \qquad (9.102)$$

For a transient problem, we may also be interested in knowing, the amount of heat transferred to the fluid (or removed from a solid object) up to a certain time t. This information is obtained from

$$Q = \int_0^t hA(T - T_f)dt \qquad (9.103)$$

Substituting for $T - T_f$ from Eq. (9.102) in Eq. (9.103) and integrating it, we get

$$Q = \rho cV(T_i - T_f)\left[1 - \exp\left(-\frac{hA}{\rho cV}t\right)\right] \qquad (9.104)$$

When dealing with the analysis of transient heat transfer problems, there are two dimensionless quantities, the Biot and Fourier numbers, that are very useful. Biot number provides a measure of the thermal resistance within the solid, being cooled (or heated in some problems), relative to the thermal resistance offered by the cooling (heating) fluid. Biot number is defined according to

$$Bi = \frac{hL_c}{k_{\text{solid}}} \qquad (9.105)$$

where L_c is a characteristic length, and is typically defined as the ratio of the volume of the object to its exposed surface areas. Small Biot numbers (typically, $Bi < 0.1$) implies that thermal resistance within the solid object is negligible and consequently, the temperature distribution is approximately uniform at any instant inside the solid object.

Another important variable is Fourier number, which is a dimensionless time parameter. It provides a measure of the rate of conduction within the solid relative to the rate of the thermal storage. Fourier number is defined as

$$Fo = \frac{\alpha t}{L_c^2} \qquad (9.106)$$

In Eq. (9.106), α is called thermal diffusivity and is equal to $\alpha = \dfrac{k}{\rho c}$. Thermal diffusivity value represents a measure of how well the material conducts heat in comparison to storing the heat. Thus, materials with low thermal diffusivity are better in storing thermal energy than they are in conducting it. Thc mcthod which we described in this section to obtain solutions to problems wherein spatial variation of temperature at a given time step is neglected is called *Lumped Capacitance method*. For problems in which $Bi < 0.1$, lumped capacitance method renders accurate results.

Exact Solutions

We now consider problems for which spatial and temporal variations of temperature must be considered. One-dimensional transient heat transfer, described in a Cartesian coordinate system, is governed by

$$\frac{\partial^2 T}{\partial x^2} = \frac{1}{\alpha}\frac{\partial T}{\partial t} \tag{9.107}$$

The solution of the governing differential heat equation requires application of two boundary conditions and one initial condition. Infinite series solutions exist for simple geometries with convective boundary conditions and uniform initial temperatures. These solutions provide temperature distribution as a function of position x inside the medium and time t. The infinite series solutions are also available for long cylinders and spheres. For heat transfer problems wherein $Fo > 0.2$, the one term approximate solution provides accurate results and are presented in the form of charts called *Heisler charts*. For simple three-dimensional problems, the solution may be represented in terms of the product of one the dimensional solutions. For an in-depth review of exact solutions, you are encouraged to study a good text on the heat transfer. For example of such a text, see Incropera and DeWitt (1996).

EXAMPLE 9.3

In an annealing process, thin steel plates ($k = 40$ W/m.K, $\rho = 7800$ kg/m^3, $c = 400$ J/kg.K) are heated to temperatures of 900°C and then cooled in an environment with temperature of 35°C and $h = 25$ W/m^2.K. Each plate has a thickness of 5 cm. We are interested in determining how long it would take for a plate to reach a temperature of 50°C and what the temperature of the plate is after one hour.

First, we need to calculate the Biot number to see if the lumped capacitance method is applicable. The characteristic length for this problem is equal to half of the plate thickness. Recall that characteristic length is defined as the ratio of volume of the object to its exposed surface areas. Thus

$$L_c = \frac{V}{A_{\text{exposed}}} = \frac{(\text{area})(\text{thickness})}{2(\text{area})} = \frac{0.05\text{ m}}{2} = 0.025\text{ m}$$

$$Bi = \frac{hL_c}{k_{\text{solid}}} = \frac{(25\text{ W/m}^2\text{.K})(0.025\text{ m})}{40\text{ W/m.k}} = 0.015$$

Because the Biot number is less than 0.1, the lumped capacitance method is applicable. We use Eq. (9.101) to determine the time that it takes for a plate to reach a temperature of 50°C.

$$t = \frac{\rho c V}{hA}\ln\frac{T_i - T_f}{T - T_f} = \frac{(7800\ \text{kg/m}^3)(400\ \text{J/kg.K})(0.025\ \text{m})}{(25\ \text{W/m.K})}\ln\frac{900 - 35}{50 - 35} = 12650\ \text{sec} = 3.5\ \text{hr}$$

We use Eq. (9.102) to determine the temperature of a plate after one hour.

$$\frac{T - T_f}{T_i - T_f} = \exp\left(-\frac{hA}{\rho c V}t\right) = \frac{T - 35}{900 - 35} = \exp\left(-\frac{(25\ \text{W/m.K})(3600\ \text{sec})}{(7800\ \text{kg/m}^3)(400\ \text{J/kg.K})(0.025\ \text{m})}\right)$$

$$\Rightarrow T = 304.4°\text{C}$$

Finite Difference Approach

There are many practical heat transfer problems for which we cannot obtain exact solutions, consequently we resort to numerical approximations. As we explained in Chapter 1, there are two common classes of numerical methods: (1) finite difference methods and (2) finite element methods. To better understand the finite element formulation of transient problems, let us first review two common finite difference procedures, the *explicit method* and *Implicit method*. The first step of any finite difference scheme is discretization. We divide the medium of interest into a number of subregions and nodes as shown in Figure 9.17. The governing differential heat equation (e.g., Eq. (9.107)) is then written for each node, and the derivatives in the governing equations are replaced by difference equations.

Using the explicit method, the temperature of an arbitrary node n at time step $p + 1$, T_n^{p+1} is determined from the knowledge of temperatures of node n and its neighboring nodes $n - 1$ and $n + 1$ at the previous time step p. Starting with expressing Eq. (9.107) in a finite difference form,

$$\frac{T_{n-1}^p - 2T_n^p + T_{n+1}^p}{(\Delta x)^2} = \frac{1}{\alpha}\frac{T_n^{p+1} - T_n^p}{\Delta t}$$

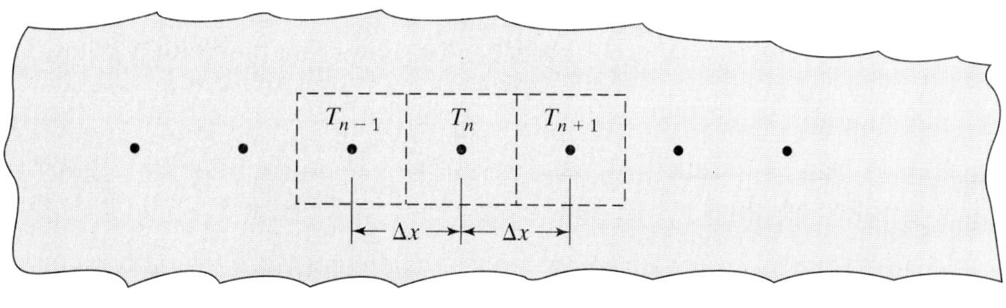

FIGURE 9.17 Discretization of the medium into smaller subregions and nodes.

and simplifying,

$$T_n^{p+1} = \frac{\alpha \Delta t}{(\Delta x)^2}(T_{n-1}^p + T_{n+1}^p) + \left[1 - 2\left(\frac{\alpha \Delta t}{(\Delta x)^2}\right)\right]T_n^p$$

we get

$$T_n^{p+1} = Fo(T_{n-1}^p + T_{n+1}^p) + [1 - 2Fo]T_n^p \tag{9.108}$$

where

$$Fo = \frac{\alpha \Delta t}{(\Delta x)^2} \tag{9.109}$$

We start at time $t = 0$ corresponding to time step $p = 0$ and use the initial temperature values in Eq. (9.108) to determine the temperature of node n at the next time step corresponding to time $t = \Delta t$. We then proceed by using the newly calculated temperatures as values for time step p and compute the temperatures at next time step $p + 1$. The march in time continues until the nodal temperature distribution corresponding to the desired time is obtained. As the name implies, explicit method makes explicit use of the knowledge of the temperature at previous time steps to compute the nodal values at the next time step. Although the explicit procedure is simple, there is a restriction on the size of time step that must be followed. Failure to follow the restriction will lead to unstable solutions that will not make physical sense. In Eq. (9.108) the term $(1 - 2Fo)$ must be positive, that is $1 - 2Fo \geq 0$ or $Fo \leq \frac{1}{2}$ to ensure stable solutions. To better understand this restriction, consider a situation for which temperature of node n at time step p is, say, $T_n^p = 60°\text{C}$, and the temperatures of neighboring nodes $n - 1$ and $n + 1$ at time step p are $T_{n-1}^p = T_{n+1}^p = 80°\text{C}$. If we were to use a Fourier value greater than 0.5, say 0.6, then using Equation (9.108), we would get

$$T_n^{p+1} = Fo(T_{n-1}^p + T_{n+1}^p) + [1 - 2Fo]T_n^p = 0.6(80 + 80) + [1 - 2(0.6)](60) = 84°\text{C}$$

As you can see, the resulting temperature is not physically meaningful, because the temperature of node n at time step $p + 1$ can not exceed the temperature of its neighboring nodes. This violates the second law of thermodynamics. Therefore, for one-dimensional problems, the stability criterion is given by

$$Fo \leq \frac{1}{2} \tag{9.110}$$

For two-dimensional problems, it can be shown that the stability criterion becomes

$$Fo \leq \frac{1}{4} \tag{9.111}$$

Implicit Method

In order to overcome the time step restriction imposed by stability requirement, we can resort to implicit method. Using the *implicit method*, the temperature of an arbitrary node n at time step $p + 1$, T_n^{p+1} is determined from its value at the previous time p and the value of temperatures at its neighboring nodes $n - 1$ and $n + 1$ at time step $p + 1$. The heat equation, Eq. (9.107) for an arbitrary node n is expressed by

$$\frac{T_{n-1}^{p+1} - 2T_n^{p+1} + T_{n+1}^{p+1}}{(\Delta x)^2} = \frac{1}{\alpha}\frac{T_n^{p+1} - T_n^p}{\Delta t}$$

and when simplified it results in

$$T_n^{p+1} - T_n^p = Fo(T_{n-1}^{p+1} - 2T_n^{p+1} + T_{n+1}^{p+1}) \tag{9.112}$$

Separating the unknown temperatures at time step $p + 1$ from the known temperature at time step p in Eq. (9.112), we get

$$T_n^{p+1} - Fo(T_{n-1}^{p+1} - 2T_n^{p+1} + T_{n+1}^{p+1}) = T_n^p \tag{9.113}$$

The implicit method leads to a set of linear equations that are solved simultaneously to obtain the nodal temperatures at time step $p + 1$. Example 9.4 demonstrate the steps involved in formulating and obtaining solutions using explicit and implicit schemes.

EXAMPLE 9.4

Consider the long thin plate shown in Figure 9.18, which is initially at a uniform temperature of $T = 250°\text{C}$. The plate is made of Silicon carbide with the following properties: $k = 510$ W/m. K, $\rho = 3100$ kg/m^3, and $c = 655$ J/kg.K. The plate is suddenly subjected to moving cold water with a very high convective heat transfer coefficient approximating a constant surface temperature of 0°C. We are interested in determining the variation of temperature within the plate.

Explicit Solution

We can make use of the symmetry of the problem, as shown in Figure 9.18, and only analyze half of the plate. The selected section of the plate is divided into six nodes as shown in Figure 9.18. The thermal diffusivity of the plate is

$$\alpha = \frac{k}{\rho c} = \frac{510 \text{ W/m.K}}{(3100 \text{ kg/m}^3)(655 \text{ J/kg.K})} = 2.5 \times 10^{-4}\text{m}^2\text{/s}$$

To obtain meaningful results, we set $Fo = \frac{1}{2}$ and solve for the appropriate time step

$$Fo = \frac{\alpha \Delta t}{(\Delta x)^2} = \frac{1}{2} = \frac{(2.5 \times 10^{-4}\text{m}^2\text{/s})\Delta t}{(0.01 \text{ m})^2} \Rightarrow \Delta t = 0.2 \text{ sec}$$

The nodal equations are:

$$\text{For node 1:} \quad \left.\frac{\partial T}{\partial x}\right|_{x=0} = 0 \quad \Rightarrow T_1^{p+1} = T_2^{p+1}$$

Substituting for $Fo = 0.5$ in Eq. (9.108), for nodes 2, 3, 4, and 5, we have

$$T_n^{p+1} = 0.5(T_{n-1}^p + T_{n+1}^p)$$

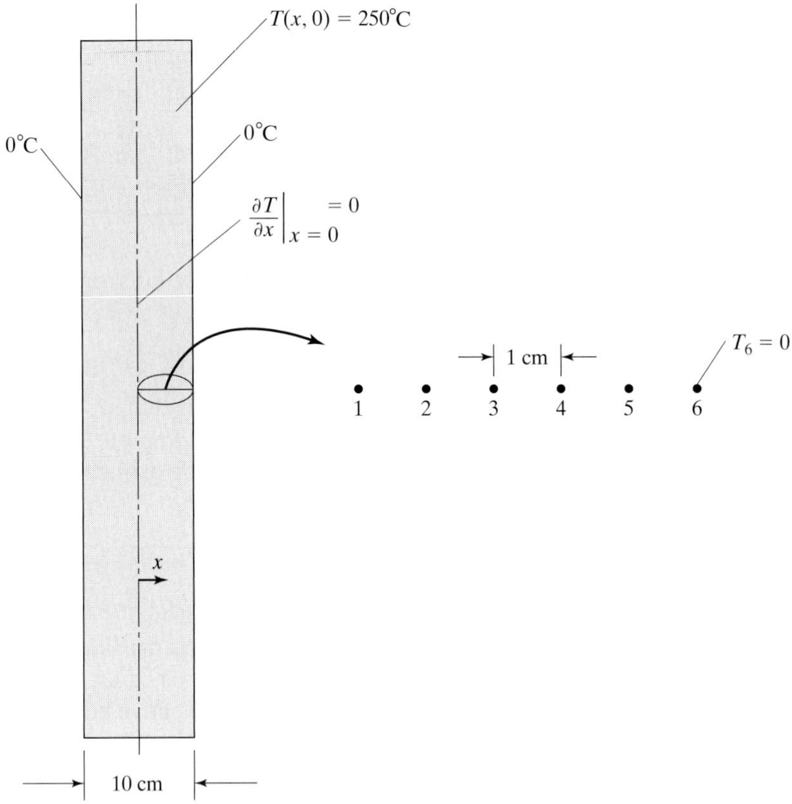

FIGURE 9.18 A schematic of the plate in Example 9.4.

and for node 6: $T_6^{p+1} = 0°\text{C}$

Starting with the initial condition and marching out in time, we obtain the results shown in Table 9.2.

Implicit Solution

The boundary condition of node 1 requires that

For node 1: $T_1^{p+1} = T_2^{p+1}$

We can use Eq. (9.113) to formulate the expressions for nodes 2, 3, 4, and 5:

$$-T_{n+1}^{p+1} + 4T_n^{p+1} - T_{n-1}^{p+1} = 2T_m^p$$

and for node 6: $T_6^{p+1} = 0°\text{C}$

TABLE 9.2 Temperatures for Example 9.4, using the explicit method

		Temperature (°C)					
		$n = 1$	2	3	4	5	6
p	t (sec)	$x = 0$	0.01	0.02	0.03	0.04	0.05
0	0	250	250	250	250	250	250
1	0.2	250	250	250	250	250	0
2	0.4	250	250	250	250	125	0
3	0.6	250	250	250	187.5	125	0
4	0.8	250	250	218.7	187.5	93.7	0
5	1	234.3	234.3	218.7	156.2	93.7	0
6	1.2	226.5	226.5	195.2	156.2	78.1	0
7	1.4	210.8	210.8	191.3	136.6	78.1	0
8	1.6	201.1	201.1	173.7	134.7	68.3	0
9	1.8	187.4	187.4	167.9	121	67.3	0
10	2	177.6	177.6	154.2	117.6	60.5	0
11	2.2	165.9	165.9	147.6	107.3	58.8	0
12	2.4	156.7	156.7	136.6	103.2	53.6	0
13	2.6	146.6	146.6	129.9	95.1	51.6	0
14	2.8	138.2	138.2	120.8	90.7	47.5	0
15	3	129.5	129.5	114.4	84.1	45.3	0

Expressing the nodal equations in a matrix form, we have

$$\begin{bmatrix} 1 & -1 & 0 & 0 & 0 & 0 \\ -1 & 4 & -1 & 0 & 0 & 0 \\ 0 & -1 & 4 & -1 & 0 & 0 \\ 0 & 0 & -1 & 4 & -1 & 0 \\ 0 & 0 & 0 & -1 & 4 & -1 \\ 0 & 0 & 0 & 0 & 0 & 1 \end{bmatrix} \begin{Bmatrix} T_1^{P+1} \\ T_2^{P+1} \\ T_3^{P+1} \\ T_4^{P+1} \\ T_5^{P+1} \\ T_6^{P+1} \end{Bmatrix} = \begin{Bmatrix} 0 \\ 2T_2^P \\ 2T_3^P \\ 2T_4^P \\ 2T_5^P \\ 0 \end{Bmatrix}$$

Initially the plate is at a uniform temperature of 250°C. Using the initial conditions in the right hand side of the above matrix, we get

$$\begin{bmatrix} 1 & -1 & 0 & 0 & 0 & 0 \\ -1 & 4 & -1 & 0 & 0 & 0 \\ 0 & -1 & 4 & -1 & 0 & 0 \\ 0 & 0 & -1 & 4 & -1 & 0 \\ 0 & 0 & 0 & -1 & 4 & -1 \\ 0 & 0 & 0 & 0 & 0 & 1 \end{bmatrix} \begin{Bmatrix} T_1^1 \\ T_2^1 \\ T_3^1 \\ T_4^1 \\ T_5^1 \\ T_6^1 \end{Bmatrix} = \begin{Bmatrix} 0 \\ 500 \\ 500 \\ 500 \\ 500 \\ 0 \end{Bmatrix}$$

Solving the set of linear equations we obtain the nodal temperature values corresponding to $t = 0.2$ sec.

$$[\mathbf{T}^1] = [248.36 \quad 248.36 \quad 245.09 \quad 232.02 \quad 183.00 \quad 0]$$

Using the above results we can now solve for nodal values at $t = 0.4$ sec.

$$\begin{bmatrix} 1 & -1 & 0 & 0 & 0 & 0 \\ -1 & 4 & -1 & 0 & 0 & 0 \\ 0 & -1 & 4 & -1 & 0 & 0 \\ 0 & 0 & -1 & 4 & -1 & 0 \\ 0 & 0 & 0 & -1 & 4 & -1 \\ 0 & 0 & 0 & 0 & 0 & 1 \end{bmatrix} \begin{Bmatrix} T_1^1 \\ T_2^1 \\ T_3^1 \\ T_4^1 \\ T_5^1 \\ T_6^1 \end{Bmatrix} = \begin{Bmatrix} 0 \\ 2 \times 248.36 \\ 2 \times 245.09 \\ 2 \times 232.02 \\ 2 \times 183.00 \\ 0 \end{Bmatrix}$$

$$[\mathbf{T}^2] = [244.38 \quad 244.38 \quad 236.44 \quad 211.19 \quad 144.29 \quad 0]$$

Using similar steps, we can march in time to obtain solutions at any desired time. Example 9.4 demonstrated the difference between the explicit and implicit schemes. Next, we will look at the finite element formulation of transient heat transfer problems.

Finite Element Approach

In sections 9.2 and 9.3 we derived the conductance and thermal load matrices for steady state heat problems. The finite element formulation of these problems leads to the following general form:

$$[\mathbf{K}]\{\mathbf{T}\} = \{\mathbf{F}\} \tag{9.114}$$

$$[\textbf{conductance matrix}]\{\textbf{temperature matrix}\} = \{\textbf{thermal load matrix}\}$$

For transient problems, we must also account for the thermal energy storage term, which leads to

$$[\mathbf{C}]\{\dot{\mathbf{T}}\} + [\mathbf{K}]\{\mathbf{T}\} = \{\mathbf{F}\} \tag{9.115}$$

$$[\textbf{heat storage matrix}] + [\textbf{conductance matrix}]\{\textbf{temperature matrix}\} = \{\textbf{thermal load matrix}\}$$

To obtain the solution, at discrete points in time, to the system of equations given by Eq. (9.115), we use a time integration procedure. The time step used in this procedure plays an important role in the accuracy of results. If we select a time step that is too small, then spurious oscillation may occur in the temperature solutions leading to meaningless results. On the other hand, if the time step is too large, then the temperature gradients cannot be accurately computed. We can use Biot number $\left(Bi = \dfrac{h\Delta x}{k_{\text{solid}}}\right)$ and Fourier number $\left(Fo = \dfrac{\alpha \Delta t}{(\Delta x)^2}\right)$ to arrive at a reasonable time step. Δx and Δt represent the mean element width and the time step, respectively. For problems for which the $Bi < 1$, the time step size my be estimated by setting the Fourier number equal to a scaling value b whose magnitude varies between 0.1 and 0.5 as shown below.

$$Fo = \frac{\alpha \Delta t}{(\Delta x)^2} = b$$

and solving for Δt, we get

$$\Delta t = b\frac{(\Delta x)^2}{\alpha} \quad \text{where } 0.1 \leq b \leq 0.5 \tag{9.116}$$

If for a problem $Bi > 1$, then the time step may be estimated from the product of the Fourier and Biot numbers in the following manner:

$$(Fo)(Bi) = \left[\frac{\alpha\Delta t}{(\Delta x)^2}\right]\left[\frac{h\Delta x}{k_{\text{solid}}}\right] = b$$

and solving for Δt, we get

$$\Delta t = b\frac{(\Delta x)k_{\text{solid}}}{h\alpha} = b\frac{(\Delta x)\rho c}{h} \quad \text{where } 0.1 \leq b \leq 0.5 \tag{9.117}$$

ANSYS uses the following generalized Euler scheme for time integration.

$$\{\mathbf{T}^{p+I}\} = \{\mathbf{T}^p\} + (1 - \theta)\Delta t\{\dot{\mathbf{T}}^p\} + \theta\Delta t\{\dot{\mathbf{T}}^{p+1}\} \tag{9.118}$$

In Equation (9.118), θ is called the Euler parameter. Temperature solutions are obtained for time step $p + 1$ from the knowledge of temperature values at previous time step p, starting at time $t = 0$, corresponding to the time step $p = 0$, when the initial temperatures are known. For implicit scheme, which is unconditionally stable, the value of θ is limited to $\frac{1}{2} \leq \theta \leq 1$. When $\theta = \frac{1}{2}$, the integration scheme is commonly referred to as Crank-Nicolson. It provides accurate results for most transient heat transfer problems. When $\theta = 1$, the integration technique is called Backward Euler and is the default setting in ANSYS. Substituting Eq. (9.118) into Eq. (9.115), gives

$$\overbrace{\left(\frac{1}{\theta\Delta t}[\mathbf{C}] + [\mathbf{K}]\right)}^{\text{Equivalent } K \text{ matrix}}\{\mathbf{T}^{p+1}\} = \overbrace{\{\mathbf{F}\} + [\mathbf{C}]\left(\frac{1}{\theta\Delta t}\{\mathbf{T}^p\} + \frac{1-\theta}{1}\{\dot{\mathbf{T}}^p\}\right)}^{\text{Equivalent } F \text{ matrix}} \tag{9.119}$$

System of equations represented by Eq. (9.119) are solved to obtain nodal temperatures at a discrete points in time. Now, a few words about how the thermal loads should be applied. Thermal loads can be applied either suddenly as a step function or increased (or decreased) over a period of time as a ramp function as shown in Figure 9.19. Note that when using ANSYS, the value of the step load is typically applied at the first time substep. In Section 9.7 we will use ANSYS to solve a transient heat transfer problem.

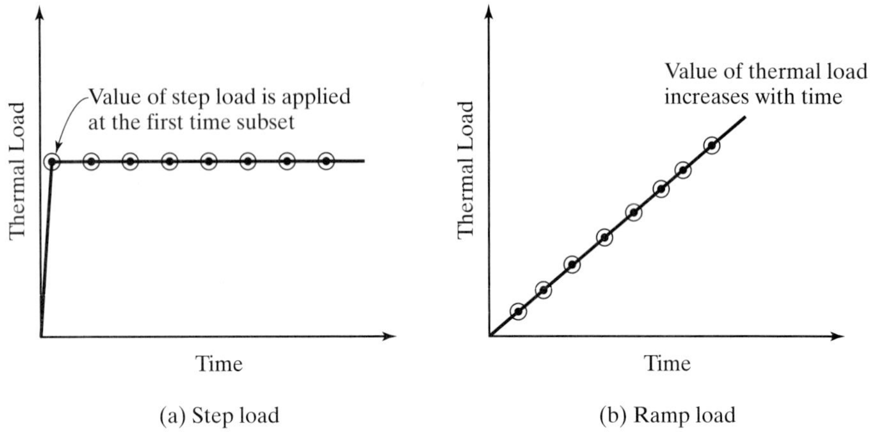

FIGURE 9.19 Thermal load options for a transient problem.

9.6 CONDUCTION ELEMENTS USED BY ANSYS

ANSYS offers many two-dimensional thermal-solid elements that are based on linear and quadratic quadrilateral and triangular shape functions:

PLANE35 is a six-node triangular thermal-solid element. The element has one degree of freedom at each node—namely, temperature. Convection and heat fluxes may be input as surface loads at the element's faces. The output data for this element include nodal temperatures and other data, such as thermal gradients and thermal fluxes. This element is compatible with the eight-node PLANE77 element.

PLANE55 is a four-node quadrilateral element used in modeling two-dimensional conduction heat transfer problems. The element has a single degree of freedom, which is temperature. Convection or heat fluxes may be input at the element's faces. Output data include nodal temperatures and element data, such as thermal gradient and thermal flux components.

PLANE77 is an eight-node quadrilateral element used in modeling two-dimensional heat conduction problems. It is basically a higher order version of the two-dimensional four-node quadrilateral PLANE55 element. This element is better capable of modeling problems with curved boundaries. At each node, the element has a single degree of freedom, which is temperature. Output data include nodal temperatures and element data, such as thermal gradient and thermal flux components.

Keep in mind that although you generally achieve more accuracy of results with higher order elements, they require more computational time because numerical integration of elemental matrices is more involved.

9.7 EXAMPLES USING ANSYS

Steady State Example

Consider a small chimney constructed from two different materials. The inner layer is constructed from concrete with a thermal conductivity $k = 0.07$ Btu/hr · in · °F. The outer layer of the chimney is constructed from bricks with a thermal conductivity value $k = 0.04$ Btu/hr · in · °F. The temperature of the hot gases on the inside surface of the chimney is assumed to be 140°F, with a convection heat transfer coefficient of 0.037 Btu/hr · in^2 · °F. The outside surface is exposed to the surrounding air, which is at 10°F, with a corresponding convection heat transfer coefficient $h = 0.012$ Btu/hr · in^2 · °F. The dimensions of the chimney are shown in Figure 9.20. Determine the temperature distribution within the concrete and within the brick layers under steady-state conditions. Also, plot the heat fluxes through each layer.

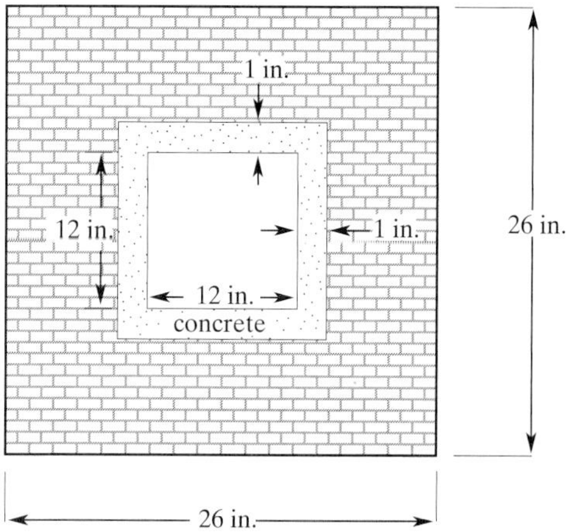

FIGURE 9.20 A schematic of the chimney in the example problem of Section 9.7.

The following steps demonstrate how to choose the appropriate element type, create the geometry of the problem, apply boundary conditions, and obtain nodal results for this problem using ANSYS.

Enter the **ANSYS** program by using the Launcher. Type **Chimney** (or a file name of your choice) in the **Jobname** entry field of the dialog box. Pick **Run** to start the GUI.

Create a title for the problem. This title will appear on ANSYS display windows to provide a simple way of identifying the displays. To create a title, issue the following command:

utility menu: **File → Change Title** . . .

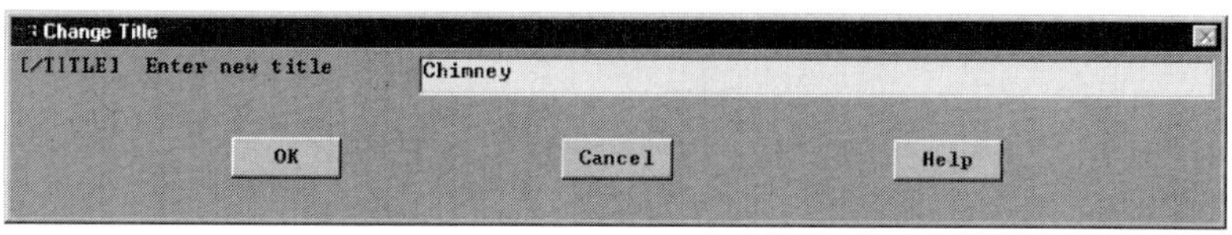

Define the element type and material properties with the commands

main menu: **Preprocessor → Element Type → Add/Edit/Delete**

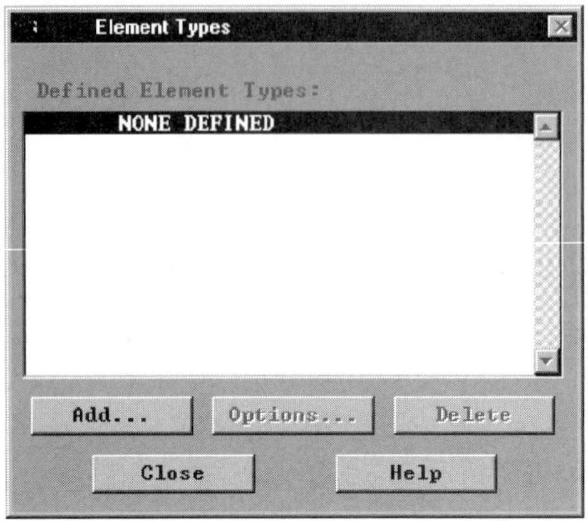

From the Library of Element Types, under Thermal Mass, choose **Solid**, then choose **Quad 4node 55**:

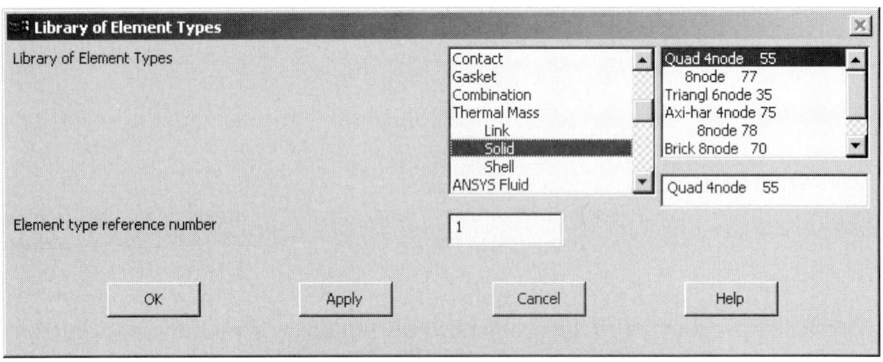

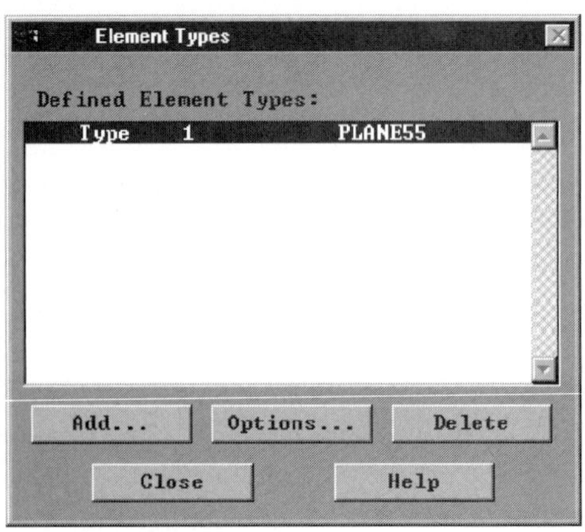

Assign the thermal conductivity values for concrete and brick. First, assign the value for concrete with the commands

main menu: **Preprocessor → Material Props → Material Models → Thermal → Conductivity → Isotropic**

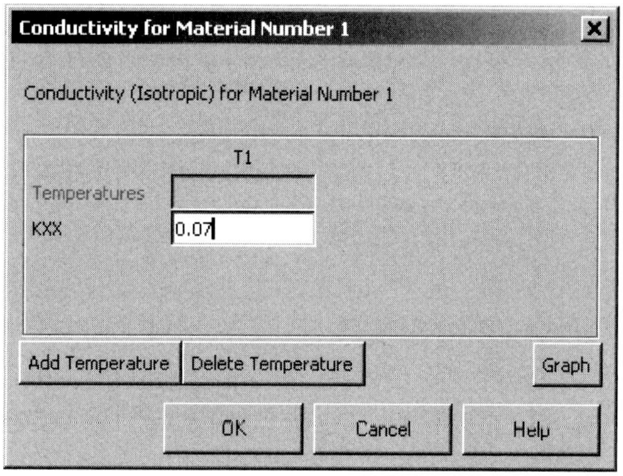

From the Define Material Model behavior window:

Material → New Model . . .

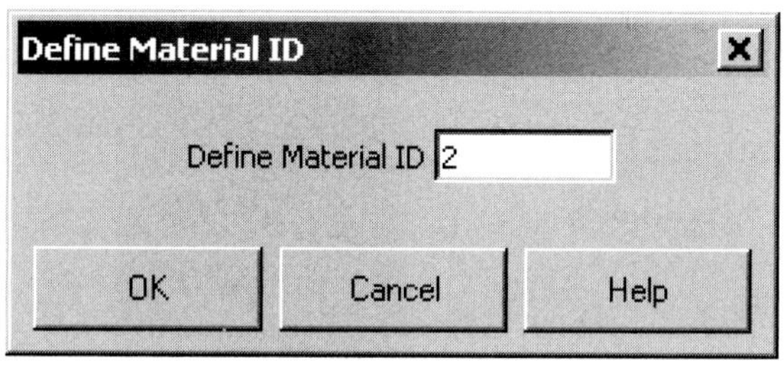

Double click on **Isotropic** and assign thermal Conductivity for brick.

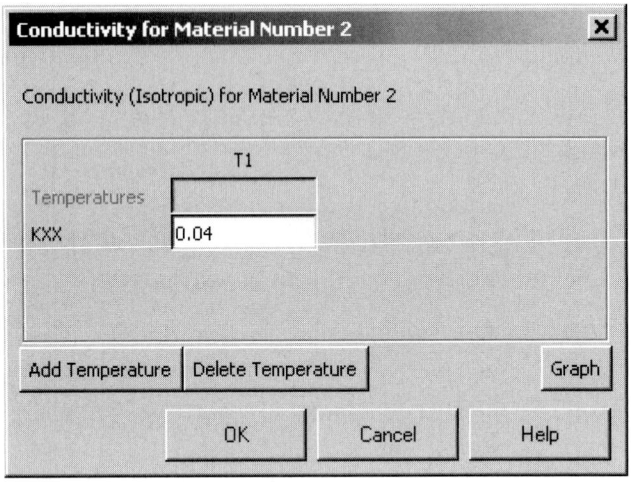

ANSYS Toolbar: **SAVE_DB**

Set up the graphics area (i.e., the workplane, zoom, etc.) with the commands

utility menu: **Workplane → WP Settings** . . .

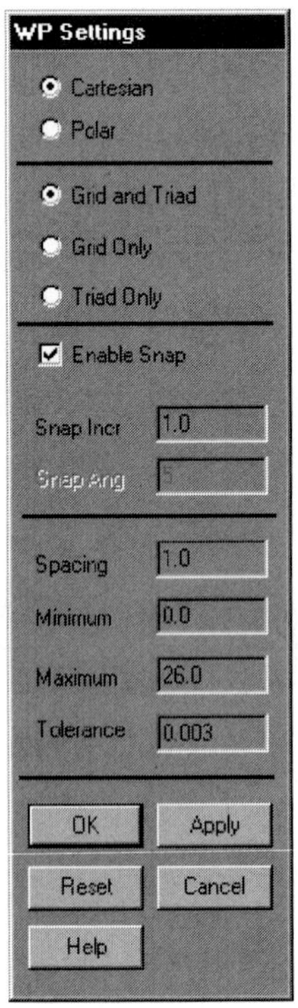

Toggle on the workplane by the issuing the command

utility menu: **Workplane → Display Working Plane**

Bring the workplane to view with the command

utility menu: **PlotCtrls → Pan, Zoom, Rotate** . . .

Click on the **small circle** until you bring the workplane to view. Then create the brick section of the chimney by issuing the commands

main menu: **Preprocessor → Modeling → Create → Areas → Rectangle → By 2 Corners**

On the workplane, pick the respective locations of the corners of areas and apply:

[WP = 0,0 lower left corner of the workplane]

[Expand the rubber band up 26.0 and right 26.0]

[WP = 6,6]

[Expand the rubber band up 14.0 and right 14.0]

OK

To create the brick area of the chimney, subtract the two areas you have created with the commands

main menu: **Preprocessor → Modeling → Operate → Booleans → Subtract → Areas**

[Pick area 1]

[Use the middle button anywhere in the ANSYS graphics window to apply]

[Pick area 2]

[anywhere in the ANSYS graphics window]

OK

Next, create the area of concrete by issuing the following commands:

main menu: **Preprocessor → Modeling → Create → Areas → Rectangles → By 2 Corners**

On the workplane, pick the respective locations of the corners of areas and apply:

[WP = 6,6]

[Expand the rubber band up 14.0 and right 14.0]

This is our area number 4.

[WP = 7,7]

[Expand the rubber band up 12.0 and right 12.0]

Apply. This is our area number 5

OK

Next, subtract the two inside areas with the commands

main menu: **Preprocessor → Modeling → Operate → Booleans → Subtract → Areas**

[Pick the area number 4]

[Use the middle button anywhere in the ANSYS graphics window to apply]

[Pick the area number 5]

[anywhere in the ANSYS graphics window]

OK

To check your work thus far, plot the areas. First, toggle off the workplane and turn on area numbering by the commands

utility menu: **Workplane → Display Working Plane**

utility menu: **PlotCtrls → Numbering** . . .

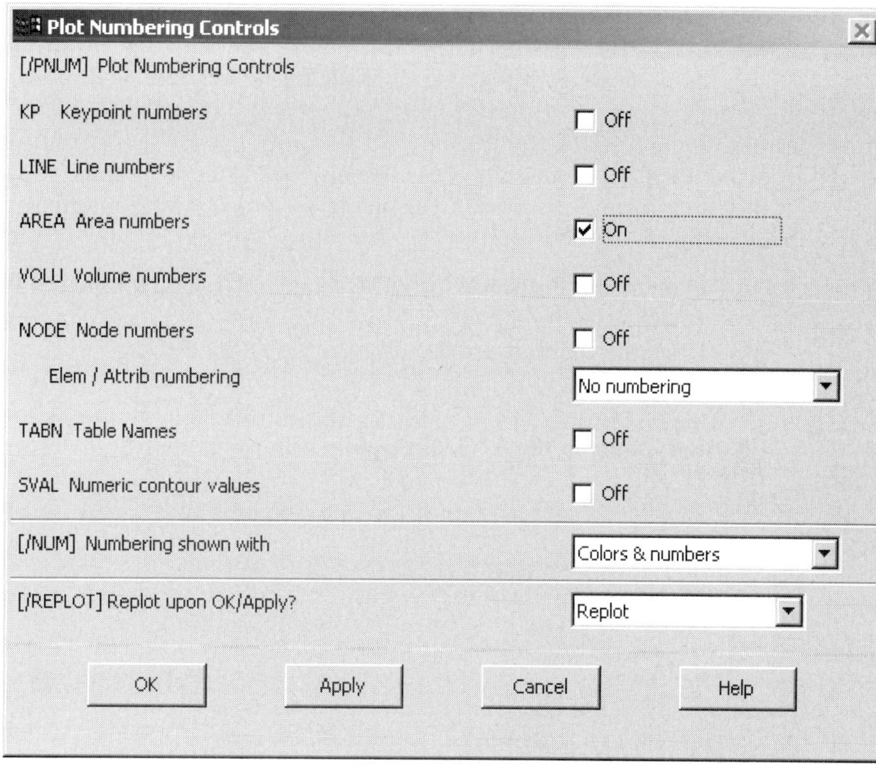

utility menu: **Plot → Areas**

ANSYS Toolbar: **SAVE_DB**

We now want to mesh the areas to create elements and nodes. But first, we need to specify the element sizes. So issue the commands

main menu: **Preprocessor → Meshing → Size Cntrls**

→ Manualsize → Global → Size

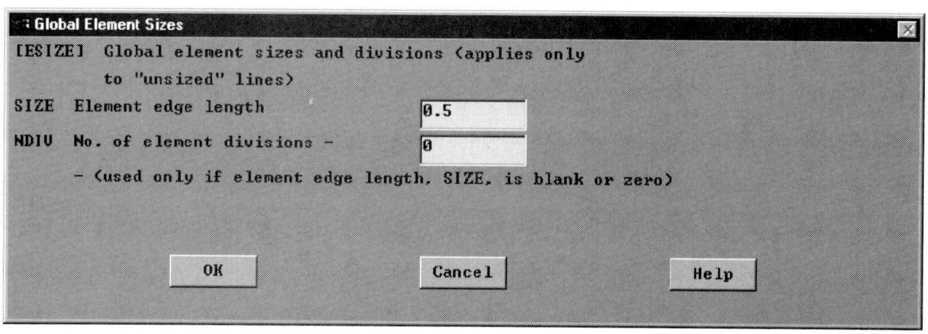

Next, glue areas to merge keypoints with the commands

main menu: **Preprocessor → Modeling → Operate → Booleans → Glue → Areas**

Select **Pick All** to glue the areas. We also need to specify material attributes for the concrete and the brick areas before we proceed with meshing. So, issue the commands

main menu: **Preprocessor → Meshing → Mesh Attributes → Picked Areas**

[Pick the concrete area]

[anywhere in the ANSYS graphics window to apply]

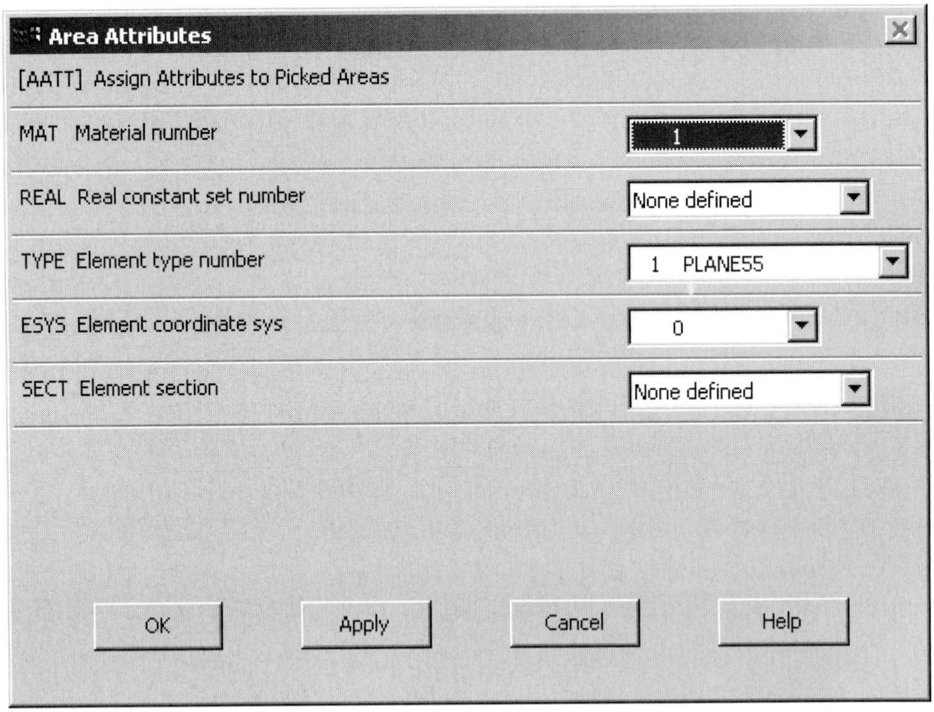

main menu: **Preprocessor → Meshing → Mesh Attributes → Picked Areas**

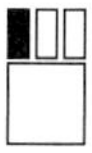 **[Pick the brick area]**

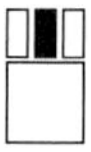 **[anywhere in the ANSYS graphics window to apply]**

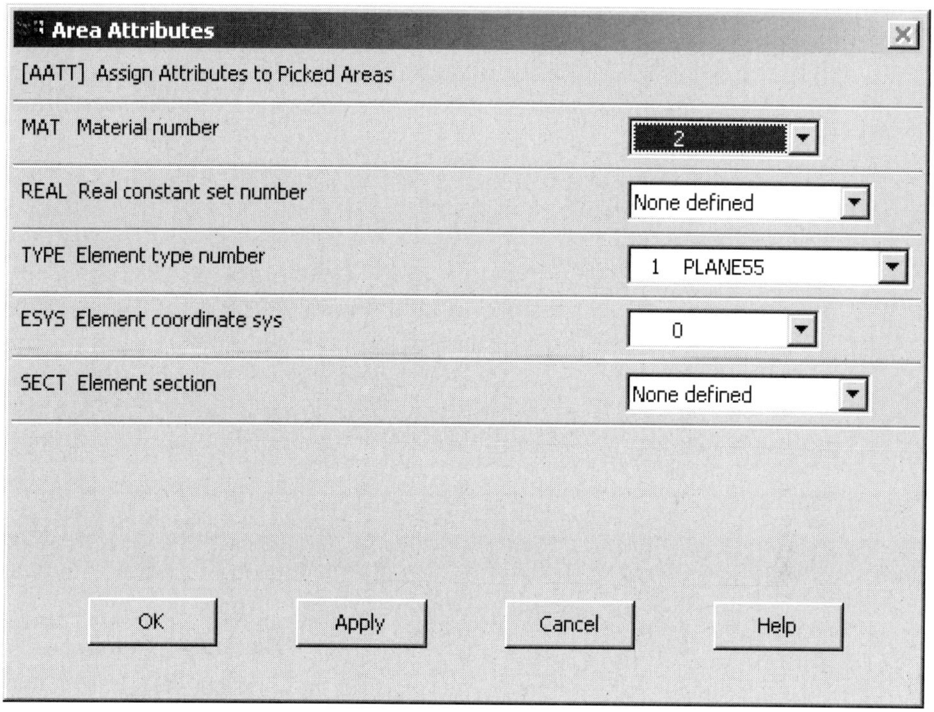

ANSYS Toolbar: **SAVE_DB**

We can proceed with meshing now. So, issue the following commands:

main menu: **Preprocessor → Meshing → Mesh → Areas → Free**

Select **Pick All** and proceed. Then issue the command

utility menu: **PlotCtrls → Numbering** . . .

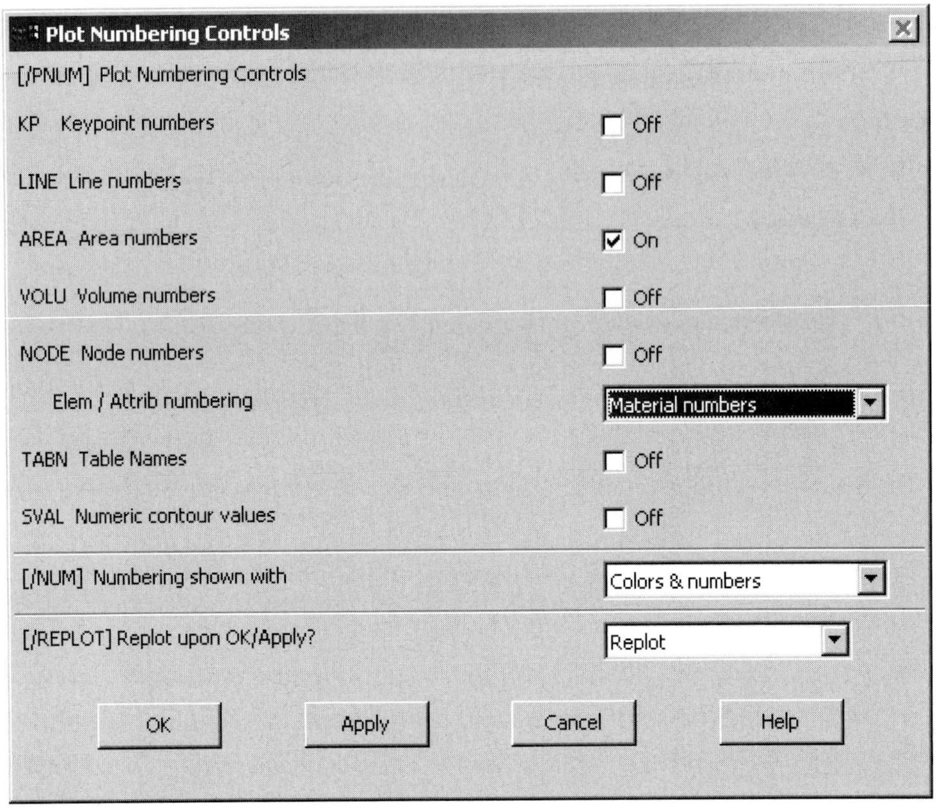

Apply boundary conditions using the command

main menu: **Solution** → **Define Loads** → **Apply** → **Thermal** → **Convection** → **On lines**

Pick the convective lines of the concrete, and press the **OK** button to specify the convection coefficient and the temperature:

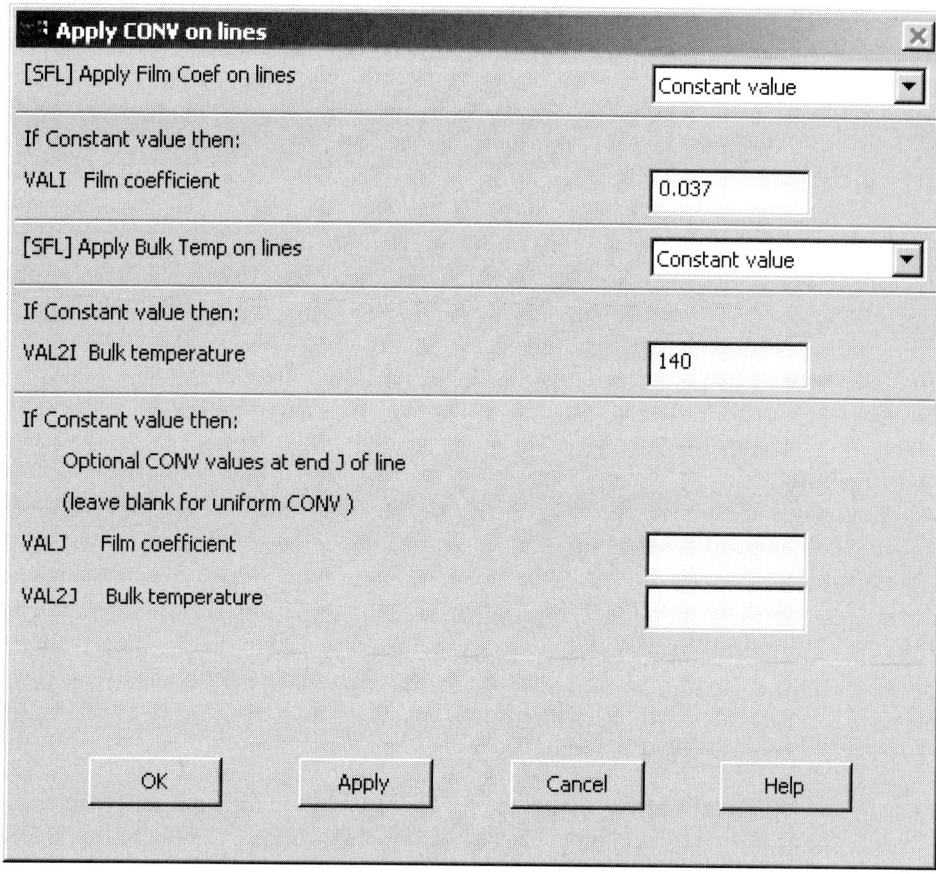

main menu: **Solution** → **Define Loads** → **Apply**
→ **Thermal** → **Convection** → **On lines**

Pick the exterior lines of the brick layer, and press the **OK** button to specify the convection coefficient and the temperature.

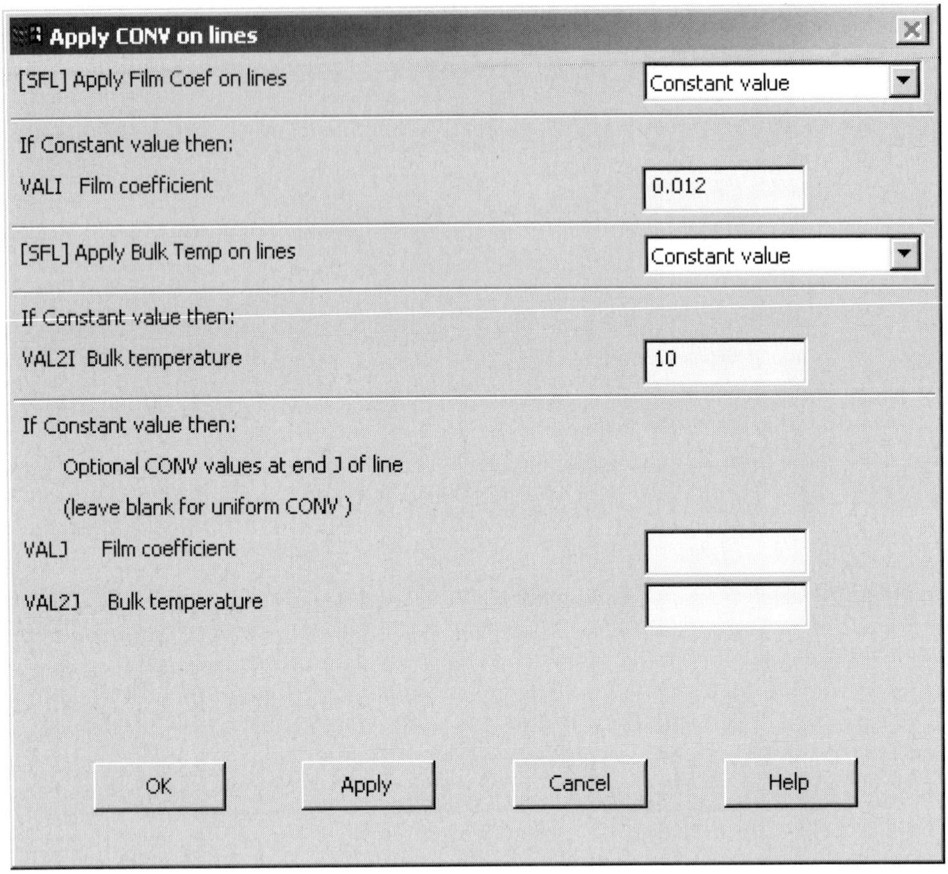

To see the applied convective boundary conditions, issue the command

utility menu: **PlotCtrls → Symbols** . . .

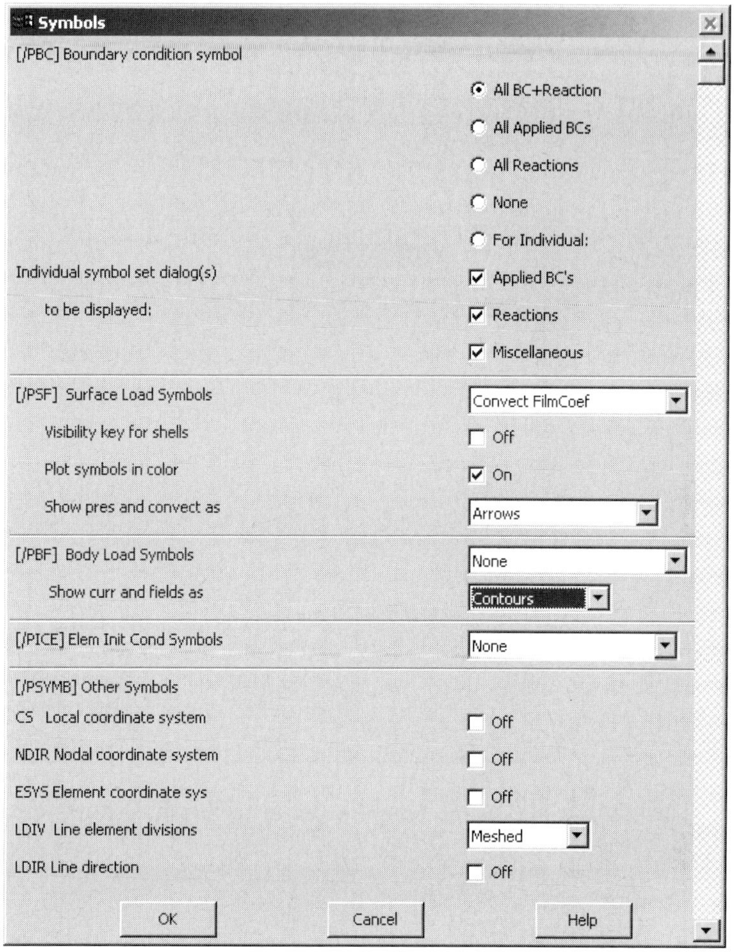

utility menu: **Plot → Lines**

ANSYS Toolbar: **SAVE_DB**

Now, solve the problem with the following commands:

main menu: **Solution → Solve → Current LS**

OK

Close (the solution is done!) window.

Close (the/STAT Command) window.

Begin the postprocessing phase. First obtain information, such as nodal temperatures and heat fluxes with the command main menu: **General Postproc → Plot Results → Contour Plot → Nodal Solu**

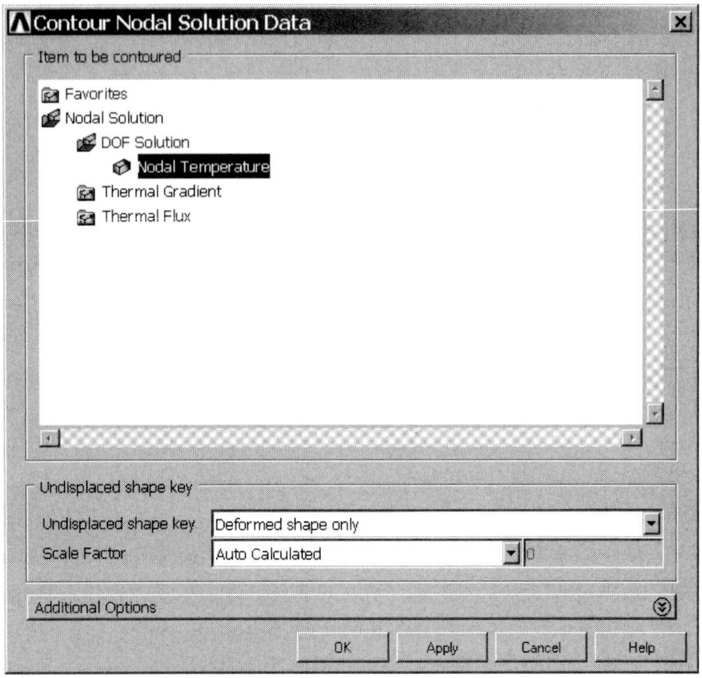

The contour plot of the temperature distribution is shown in Figure 9.21.

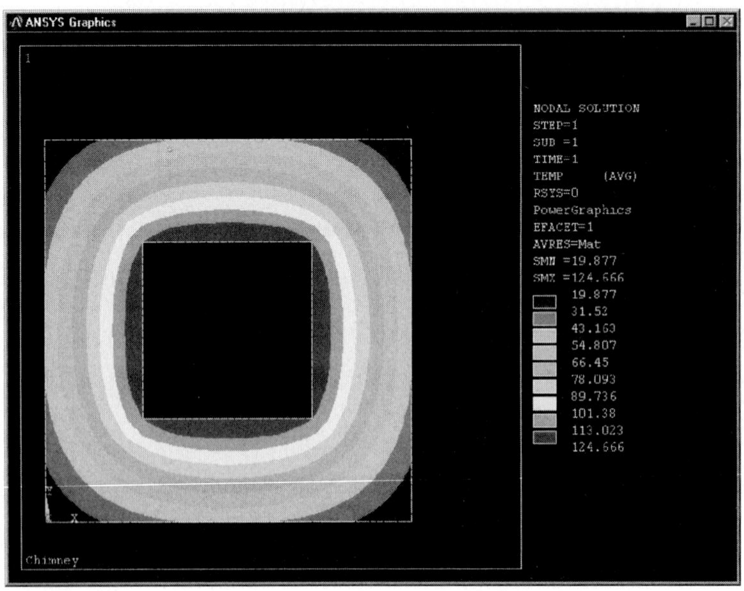

FIGURE 9.21 Temperature contour plot.

Now use the following command to plot the heat flow vectors (the plot is shown in Figure 9.22):

main menu: **General Postproc → Plot Results**

→ Vector Plot → Predefined

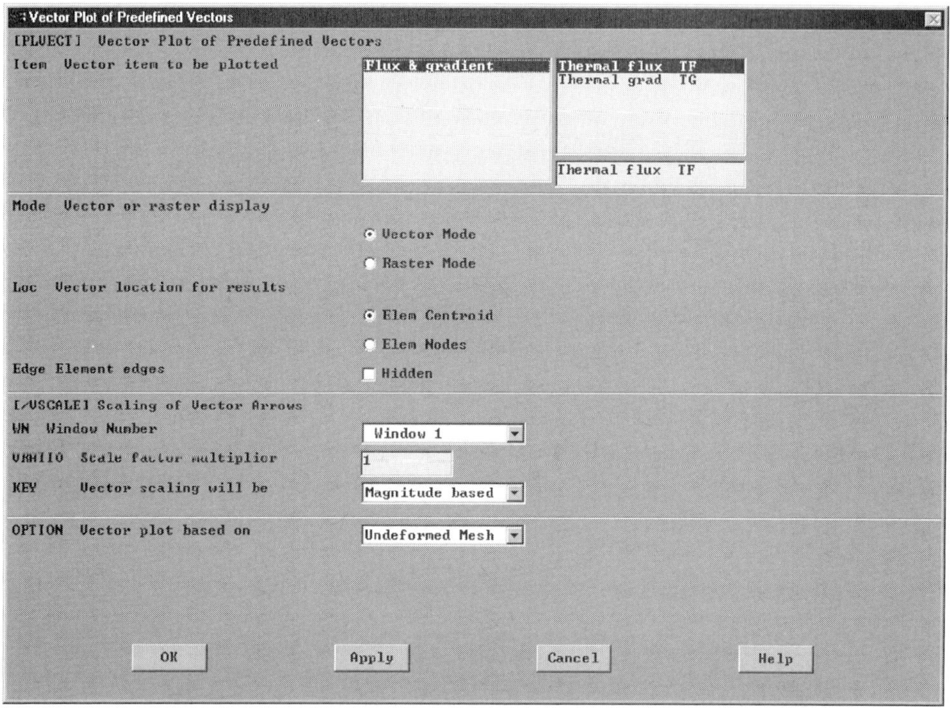

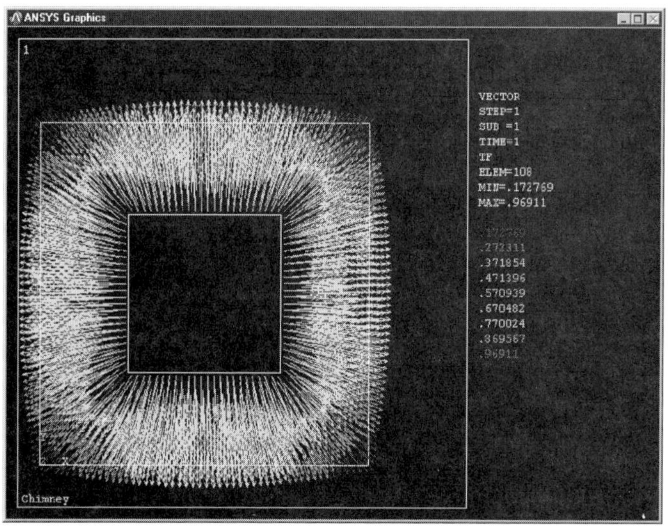

FIGURE 9.22 Plot of the heat flow vectors.

Next, issue the following commands:

utility menu: **Plot → Areas**

main menu: **General Postproc → Path Operations → Define Path → On Working Plane**

Pick the two points along the line marked as *A–A*, as shown in Figure 9.23, and press the **OK** button.

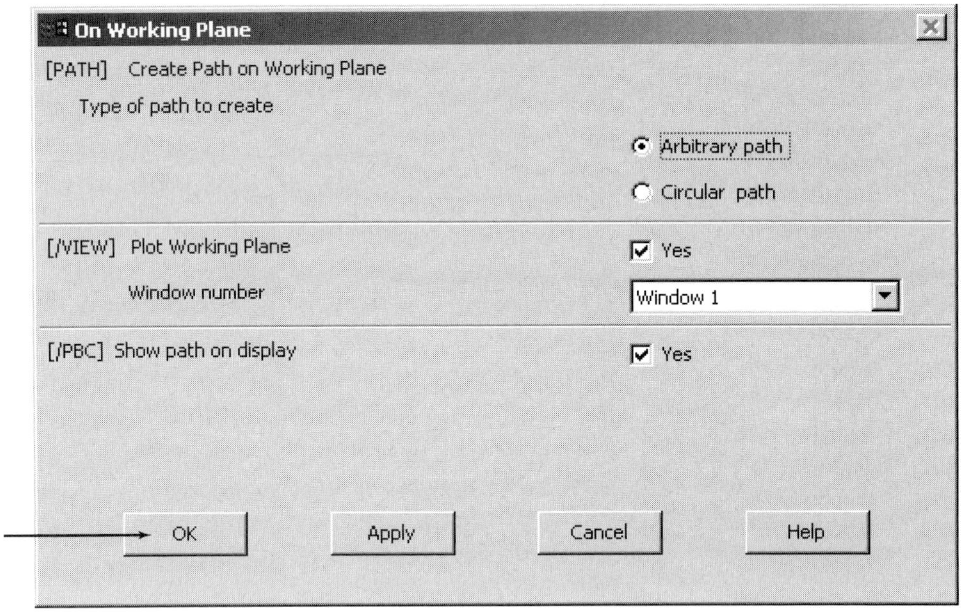

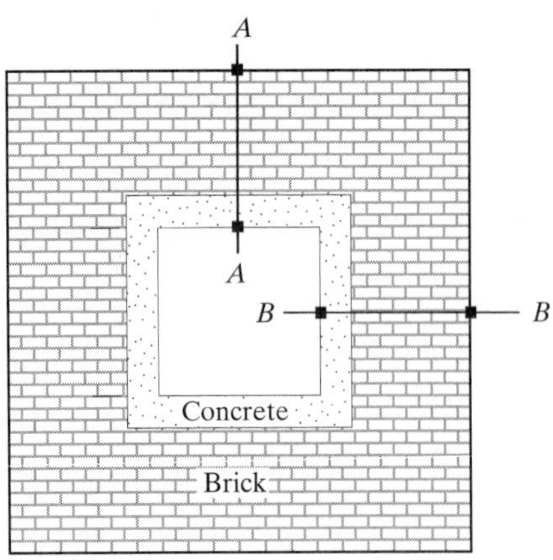

FIGURE 9.23 Defining the path for path operation.

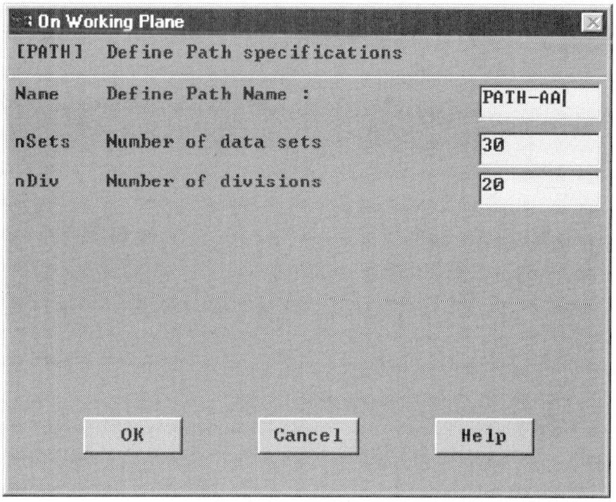

Then, issue the commands

main menu: **General Postproc → Path Operations → Map onto Path**

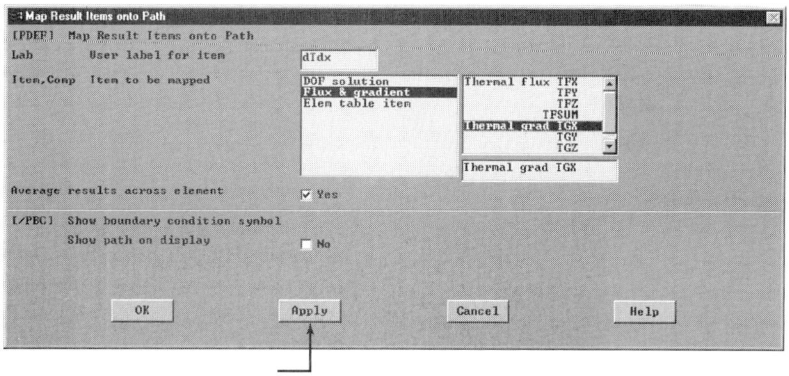

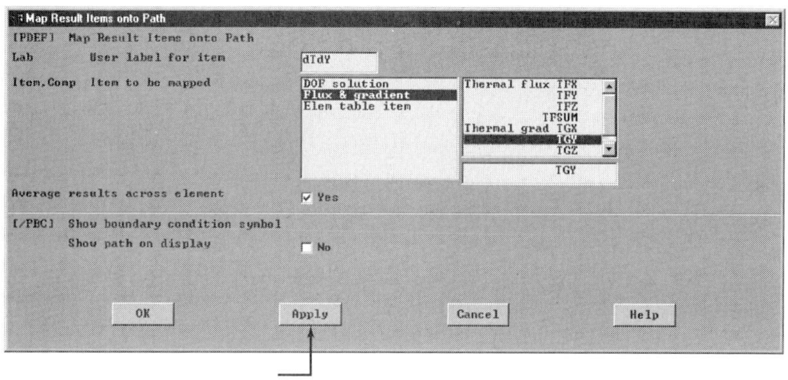

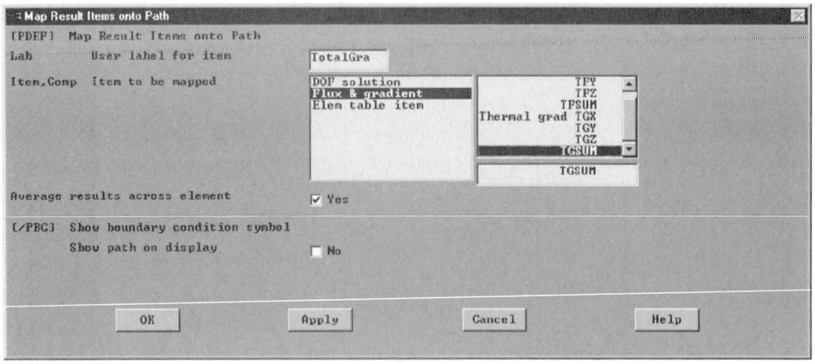

main menu: **General Postproc → Path Operations → Plot Path Item → On Graph**

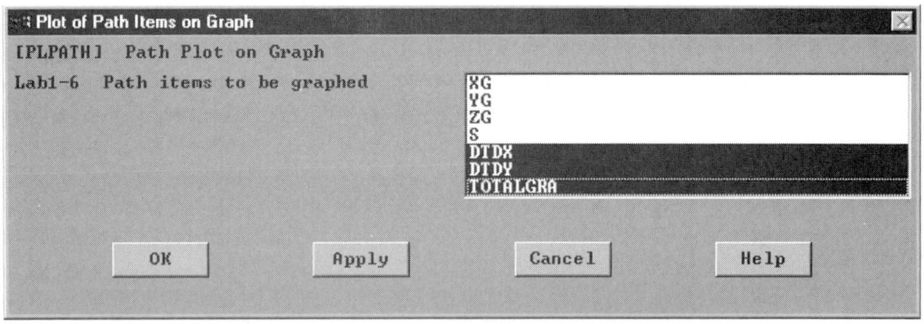

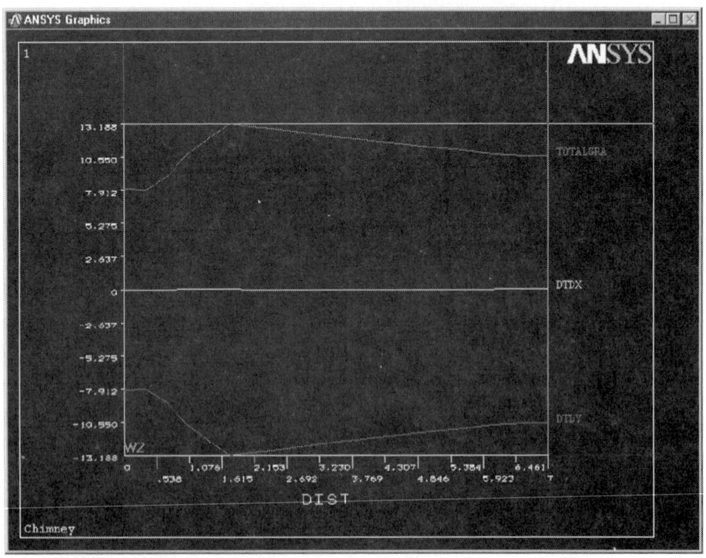

FIGURE 9.24 The variation of temperature gradients along path A–A.

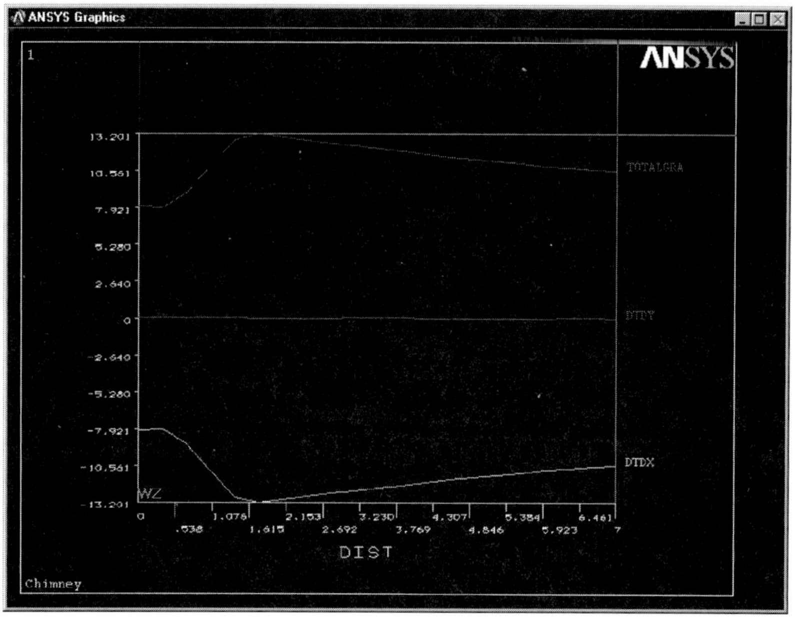

FIGURE 9.25 The variation of temperature gradients along path *B–B*.

In a similar fashion, plot the variation of temperature gradients along path B-B.

Finally, exit ANSYS and save everything:

ANSYS Toolbar: **Quit**

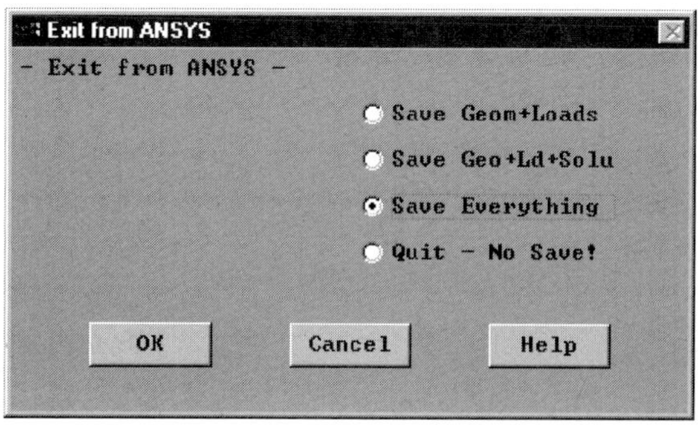

Transient Example

In this problem we will use ANSYS to study the transient response of a fin. Aluminum fins ($k = 170$ W/m.K, $\rho = 2800$ kg/m^3, $c = 870$ J/kg.K) are commonly used to dissipate heat from various devices. An example of a section of a fin is shown in Figure 9.26. The fin is initially at a uniform temperature of 28°C. Assume that shortly after the device is turned on, the temperature of the base of the fin is suddenly increased to 90°C. The temperature of the surrounding air is 28°C with a corresponding heat transfer coefficient of $h = 30$ W/m^2.K.

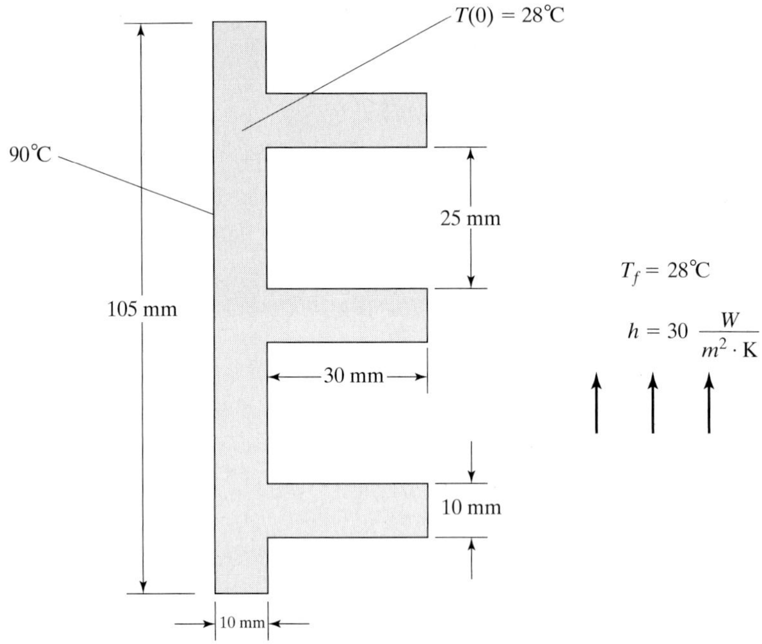

FIGURE 9.26 A schematic of the fin in the example problem.

Enter **ANSYS** program by using the Launcher. Type **TranFin** (or a file name of your choice) in the Jobname entry field of the dialog box. Pick **Run** to start the Graphic User Interface (GUI).

Create a title for the problem.

utility menu: **File → Change Title** . . .

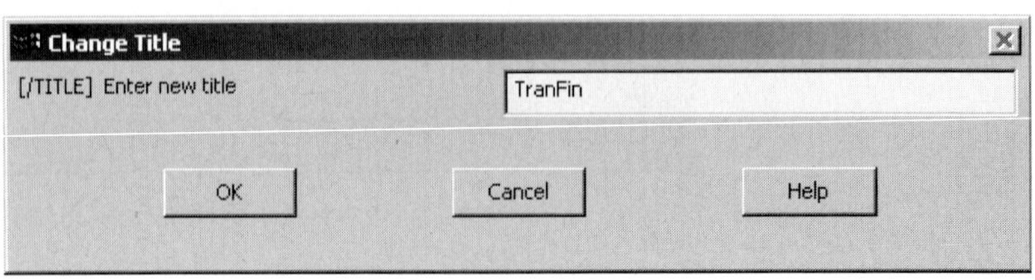

main menu: **Preprocessor → Element Type → Add/Edit/Delete**

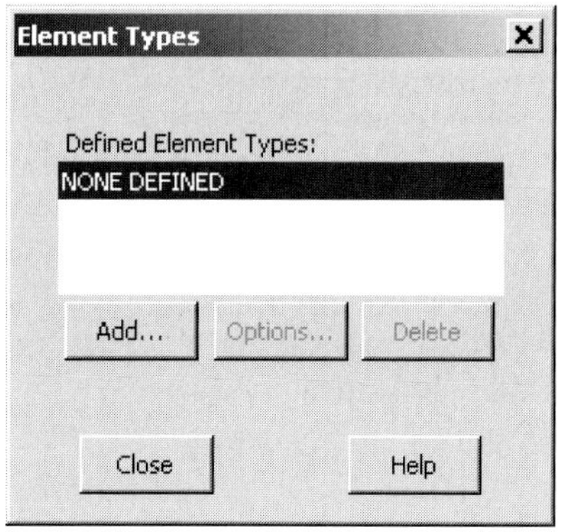

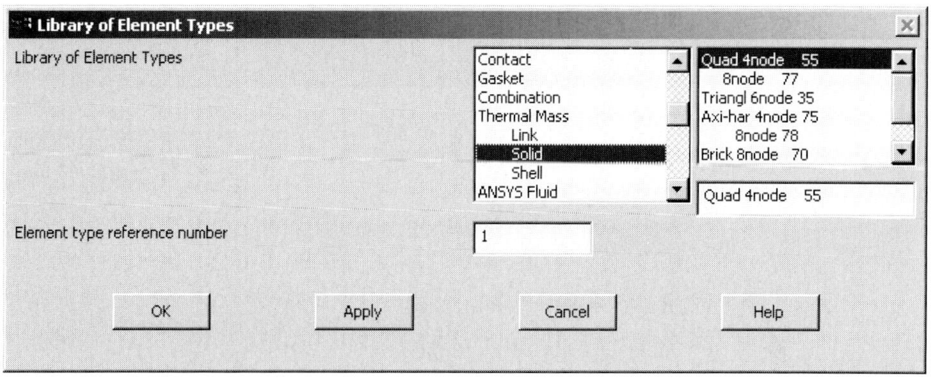

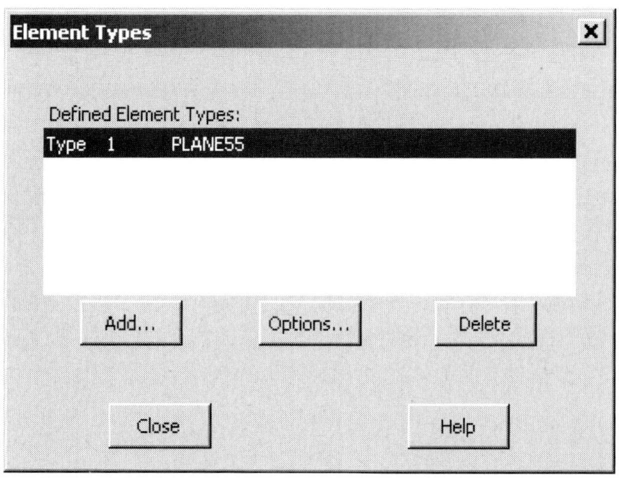

main menu: **Preprocessor → Material Props → Material Models → Thermal → Conductivity → Isotropic**

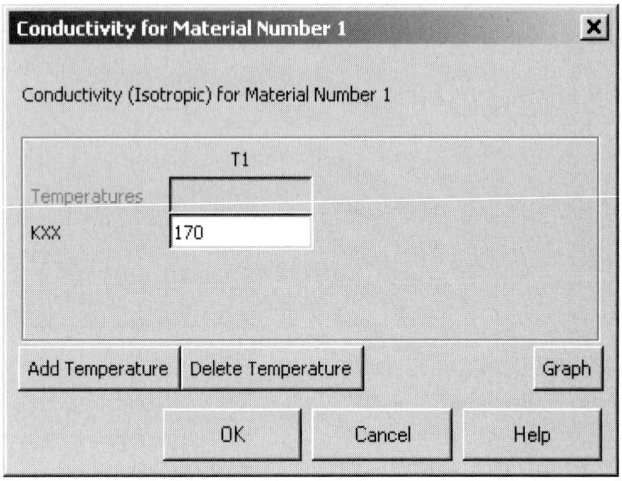

main menu: **Preprocessor → Material Props → Material Models → Thermal → Specific Heat**

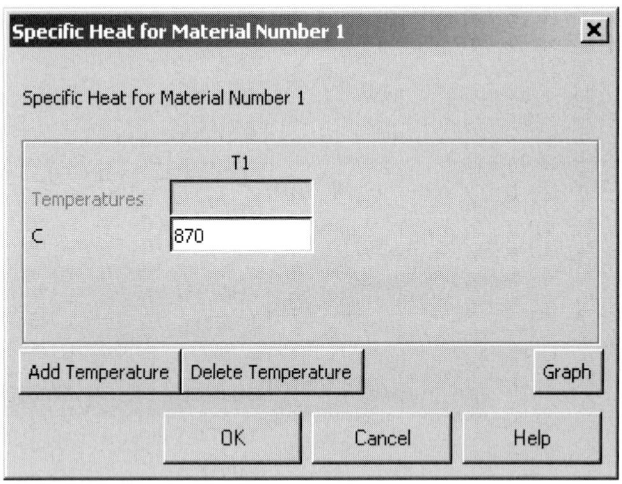

main menu: **Preprocessor → Material Props → Material Models → Thermal → Density**

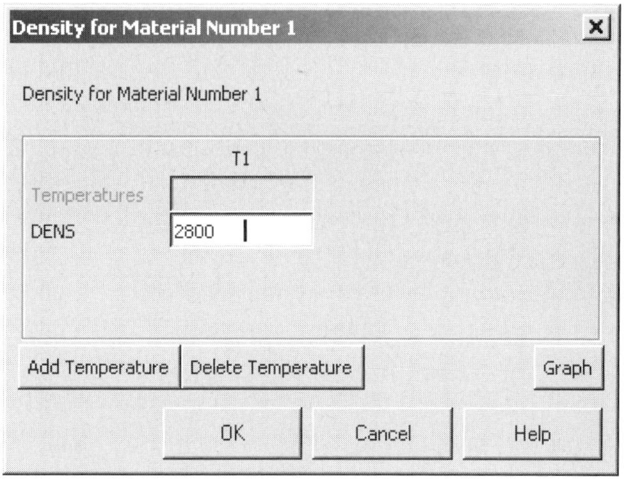

ANSYS Toolbar: **SAVE_DB**

main menu: **Preprocessor → Modeling → Create → Areas → Rectangle → By 2 Corners**

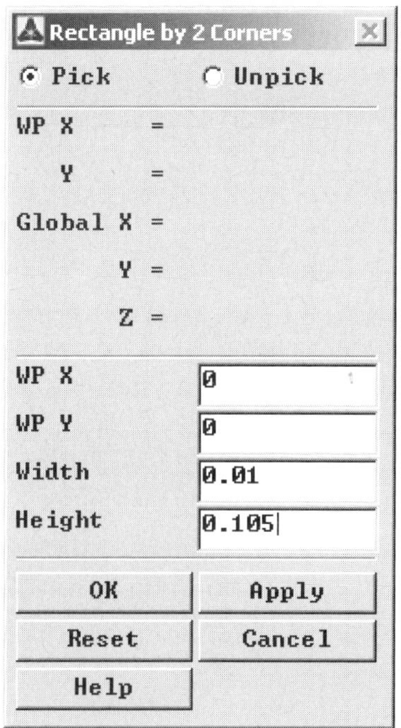

Apply

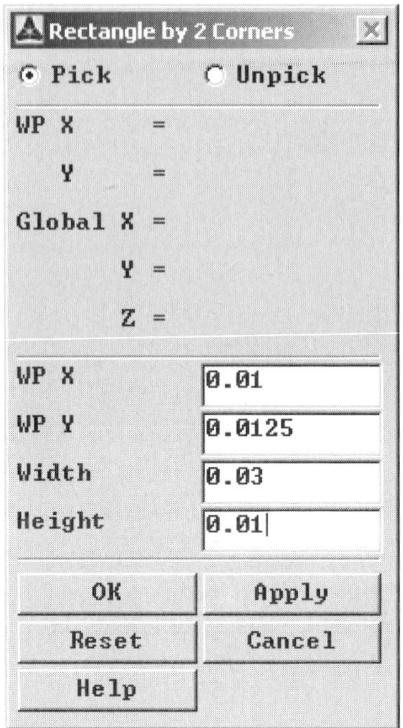
Rectangle by 2 Corners
Pick
Unpick
WP X =
Y =
Global X =
Y =
Z =
WP X
0.01
WP Y
0.0125
Width
0.03
Height
0.01
OK
Apply
Reset
Cancel
Help

Apply

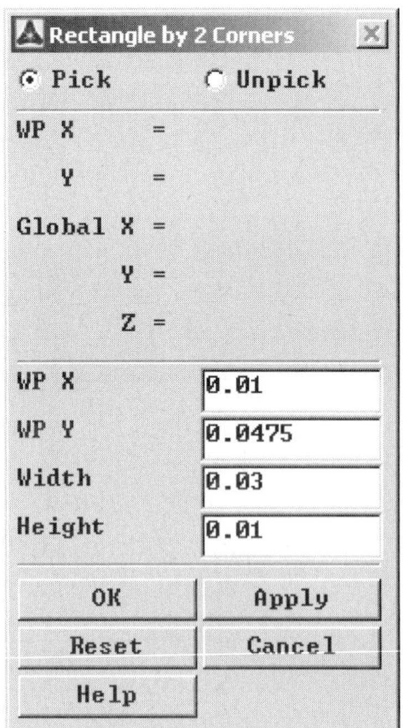
Rectangle by 2 Corners
Pick
Unpick
WP X =
Y =
Global X =
Y =
Z =
WP X
0.01
WP Y
0.0475
Width
0.03
Height
0.01
OK
Apply
Reset
Cancel
Help

Apply

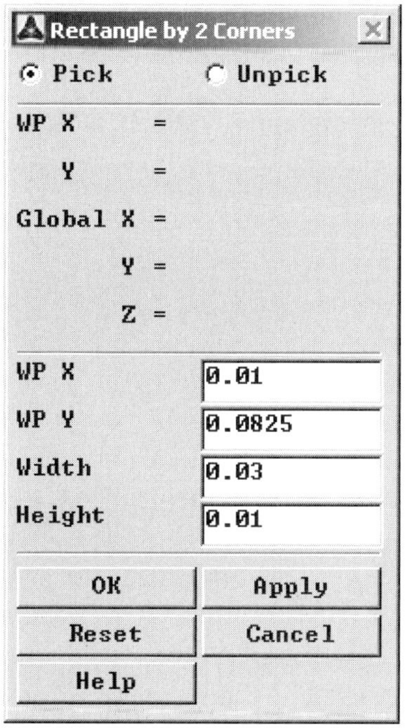

OK

main menu: **Preprocessor → Modeling → Operate → Booleans → Add → Areas**

Pick All

main menu: **Preprocessor → Meshing → Size Cntrls → Smart Size → Basic**

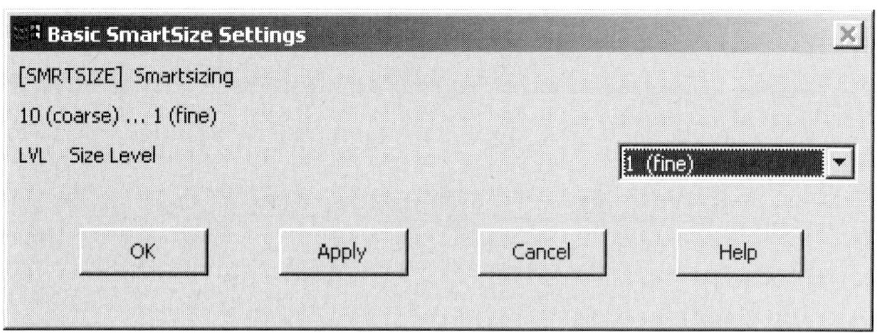

main menu: **Preprocessor → Meshing → Mesh → Areas → Free**

Pick All

main menu: **Solution → Analysis Type → New Analysis**

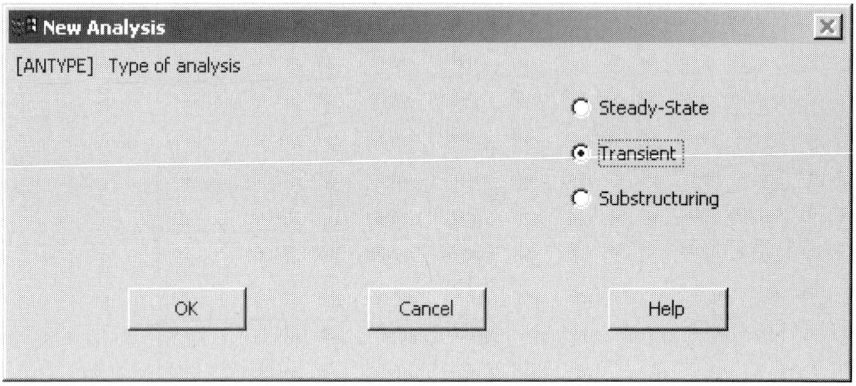

OK

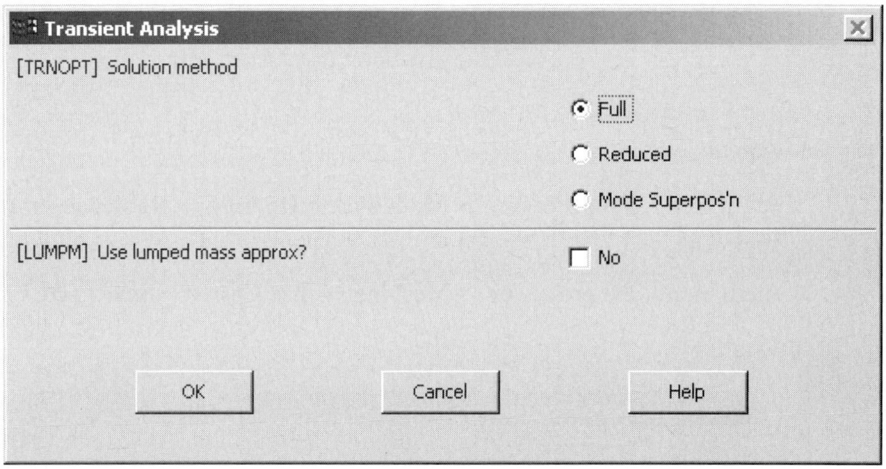

main menu: **Solution → Define Loads → Settings → Uniform Temp**

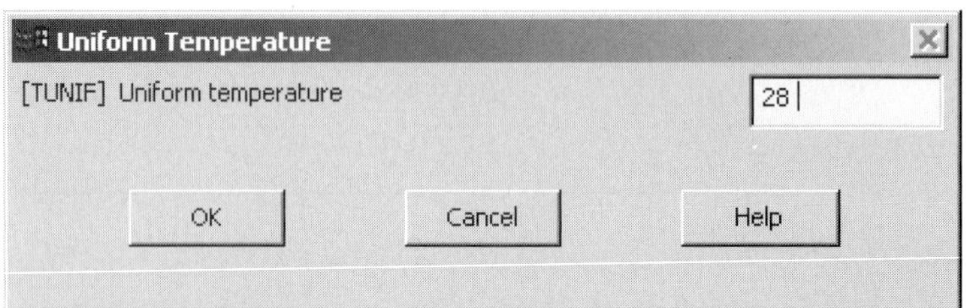

main menu: **Solution → Define Loads → Apply → Thermal → Convection → On Lines**

Pick the edges of the fin which are exposed to the convective environment, and **Apply**.

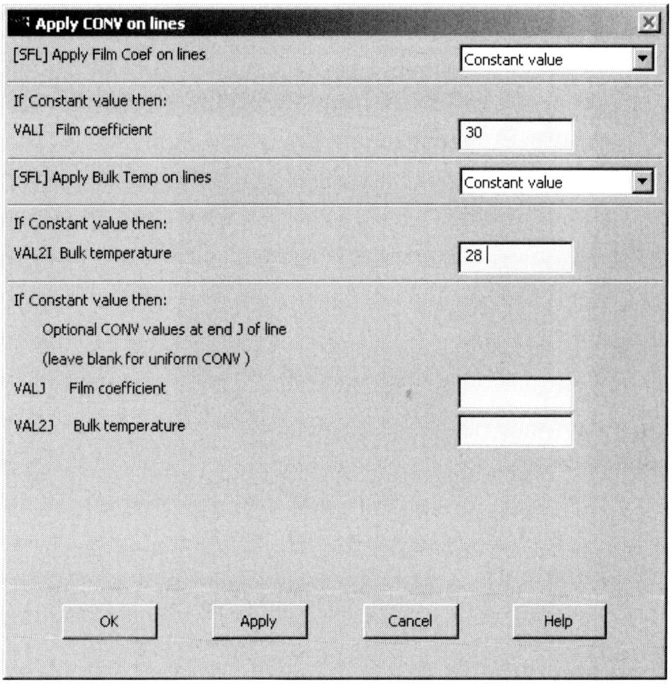

main menu: **Solution → Define Loads → Apply → Thermal → Temperature → On Lines**

Pick the line representing the base of the fin.

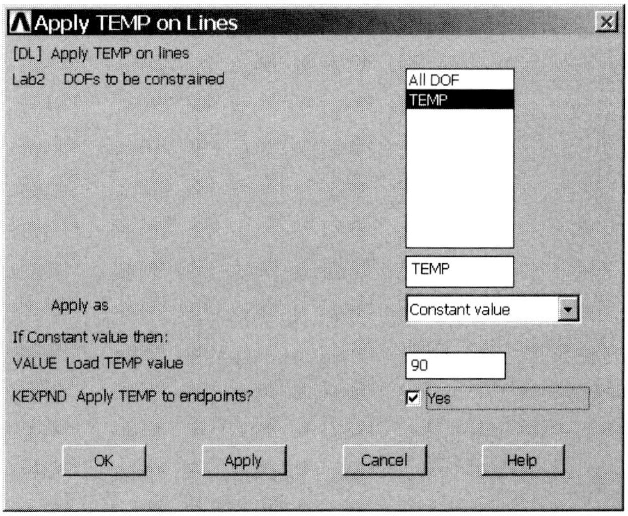

OK

Next, we will set a time duration of 300 seconds and a time step size of 1 second. We will also use ANSYS's automatic time stepping capabilities to modify the time step size as needed.

main menu: **Solution → Load Step Opts → Time/Frequenc → Time-Time Step**

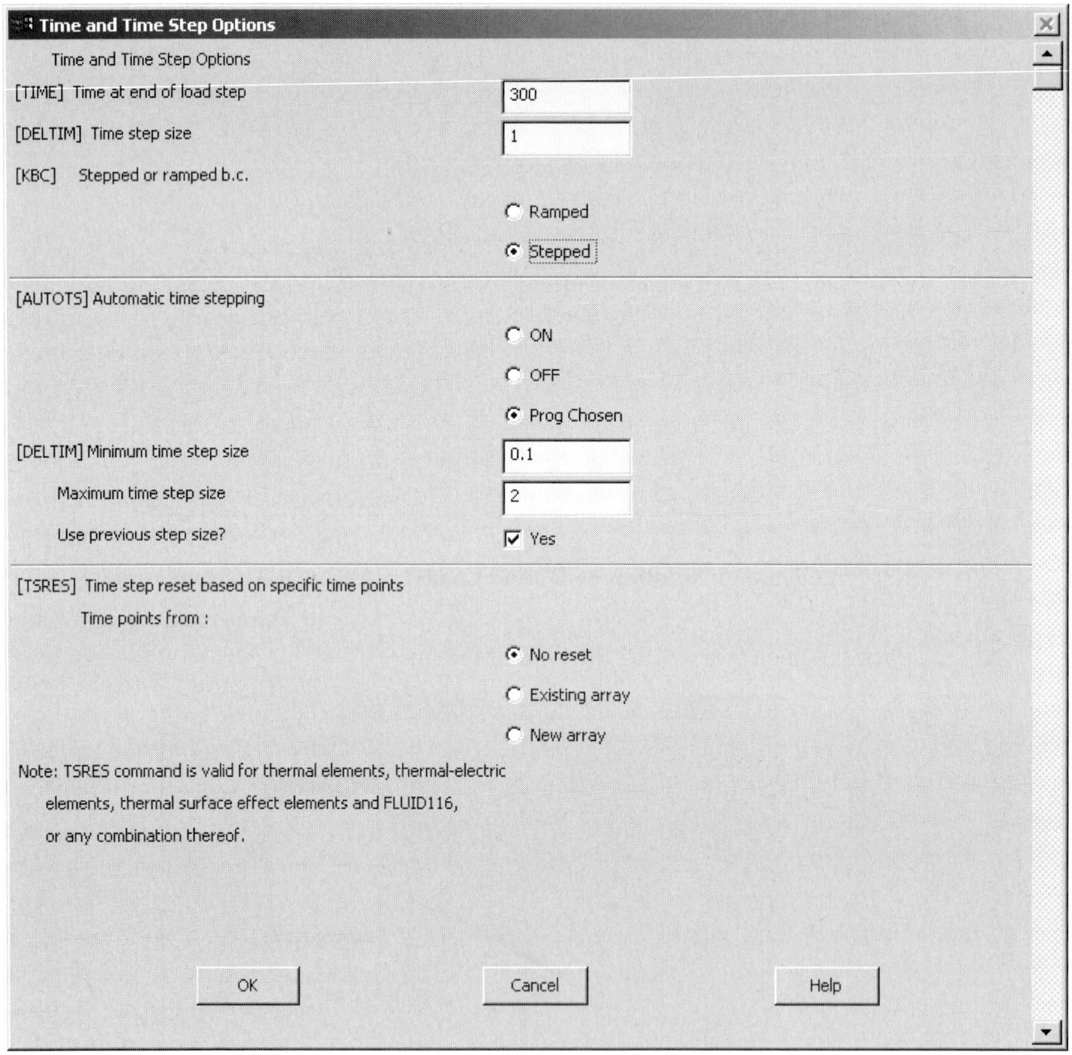

Now, we set the file write frequency to every subset.

main menu: **Solution → Load Step Opts → Output Ctrls → DB/Results File**

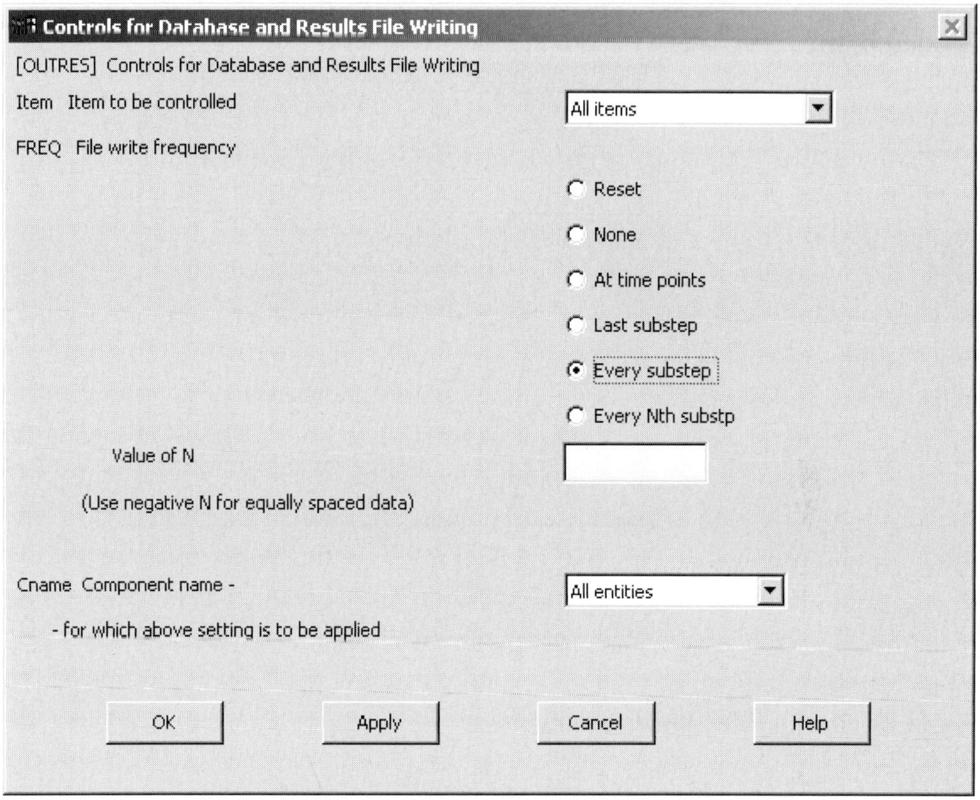

main menu: **Solution → Solve → Current LS**

Close (the solution is done!) window.

Close (the /STAT Command) window.

Next, we define a variable called *corner_pt* to store the temperature of the corner node with the following location $X = 0.01$ m, $Y = 0.0475$ m, $Z = 0$. You can choose some other node if you wish or you can directly pick the node of interest when asked in the **Time History Postprocessing.**

utility menu: **Parameters → Scalar Parameters** . . .

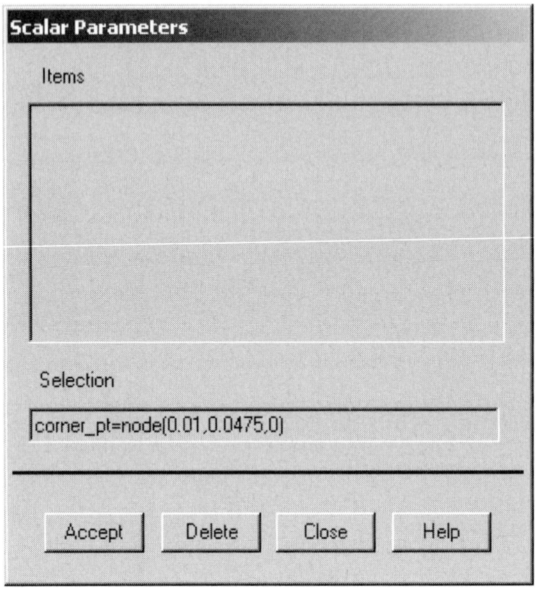

Press the Accept button.

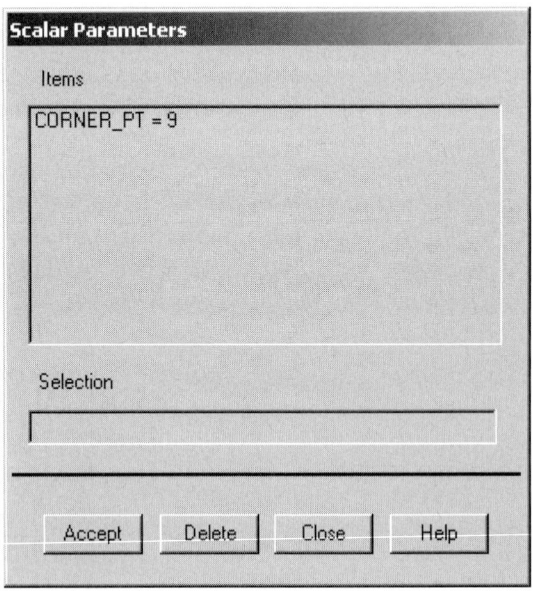

Close

main menu: **TimeHist Postproc**

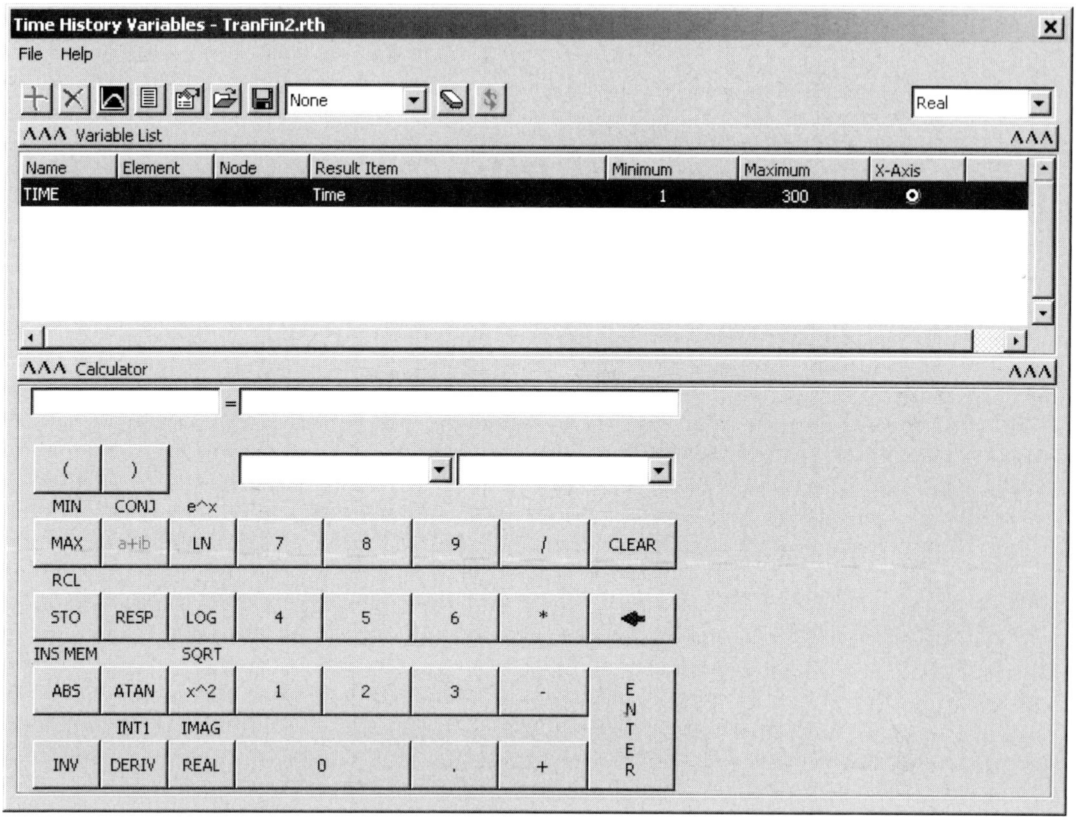

Press the Add Data button, the green plus button. Next, double-click on Nodal Solution, then DOF Solution, and Temperature.

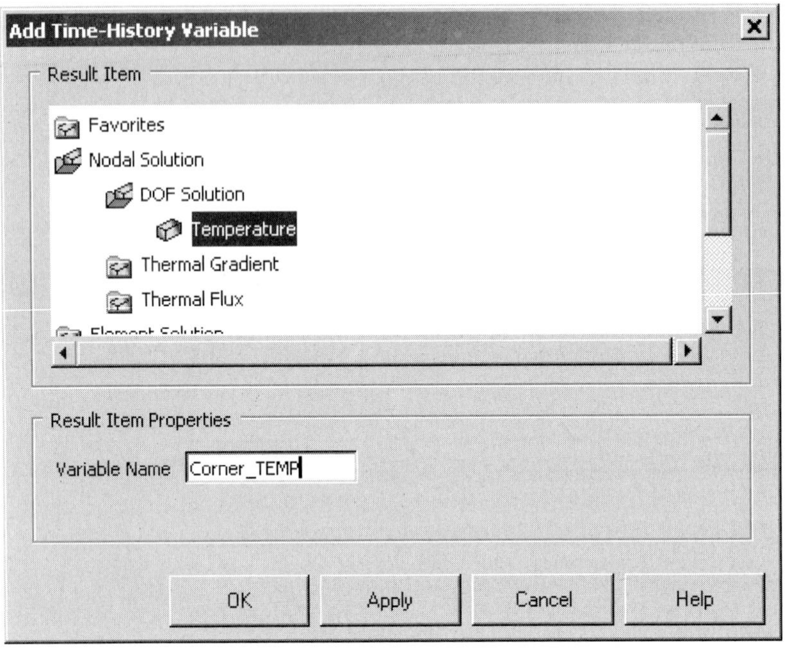

Type *Corner_Temp* or a name of your choice in the Variable Name field as shown.

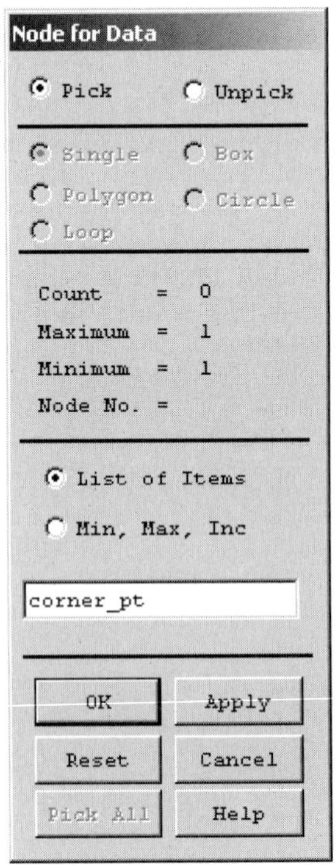

Type *corner_pt* in the field shown. Recall corner_point corresponds to the node with the location X = 0.01 m, Y = 0.0475 m, Z = 0. Press the Enter key and then the OK button. This is the step where you could pick another node by manual picking instead of using the defined parameter *corner_pt*.

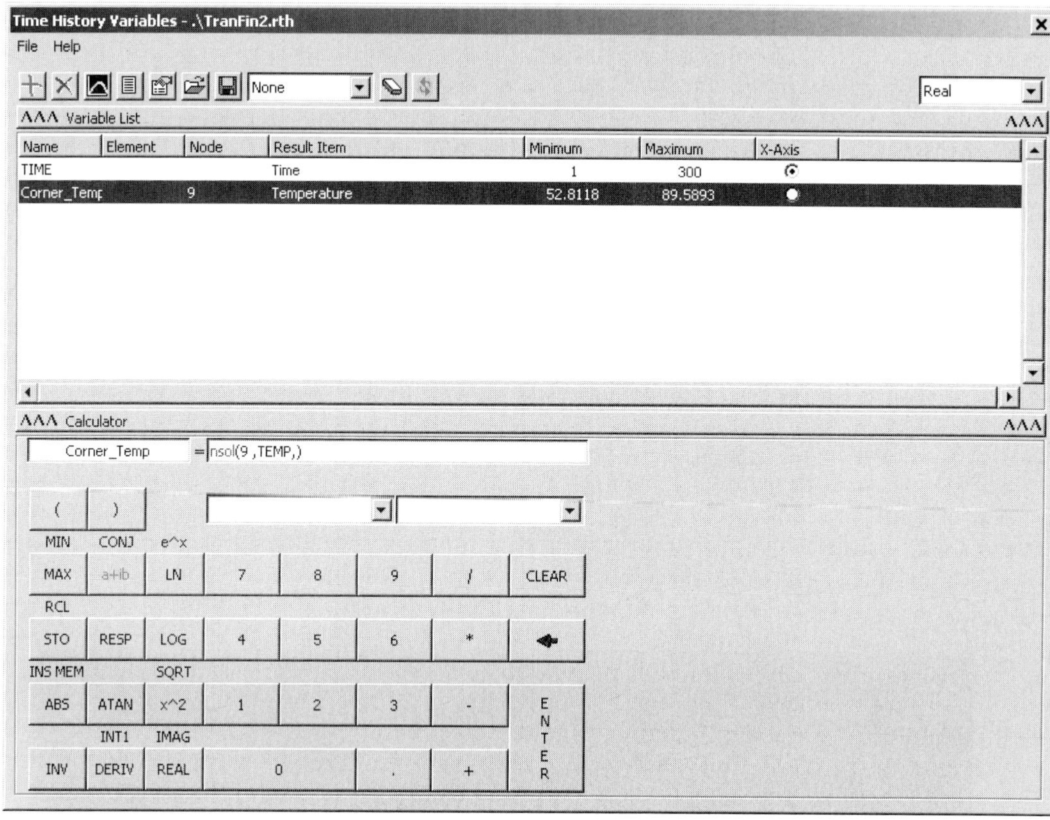

Close

main menu: **TimeHist Postproc → Graph Variables**

Type 2 in the NVAR1 1st variable to graph field.

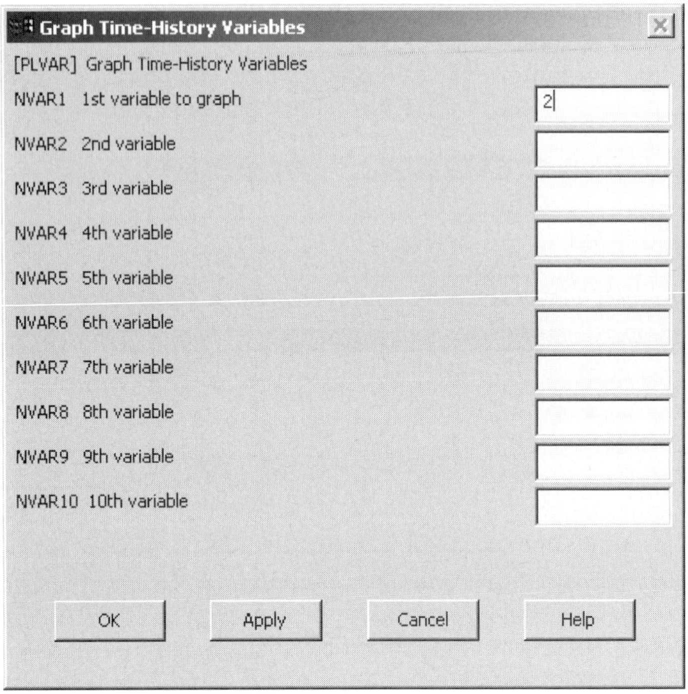
Graph Time-History Variables
[PLVAR] Graph Time-History Variables
NVAR1 1st variable to graph
2
NVAR2 2nd variable
NVAR3 3rd variable
NVAR4 4th variable
NVAR5 5th variable
NVAR6 6th variable
NVAR7 7th variable
NVAR8 8th variable
NVAR9 9th variable
NVAR10 10th variable
OK
Apply
Cancel
Help

OK

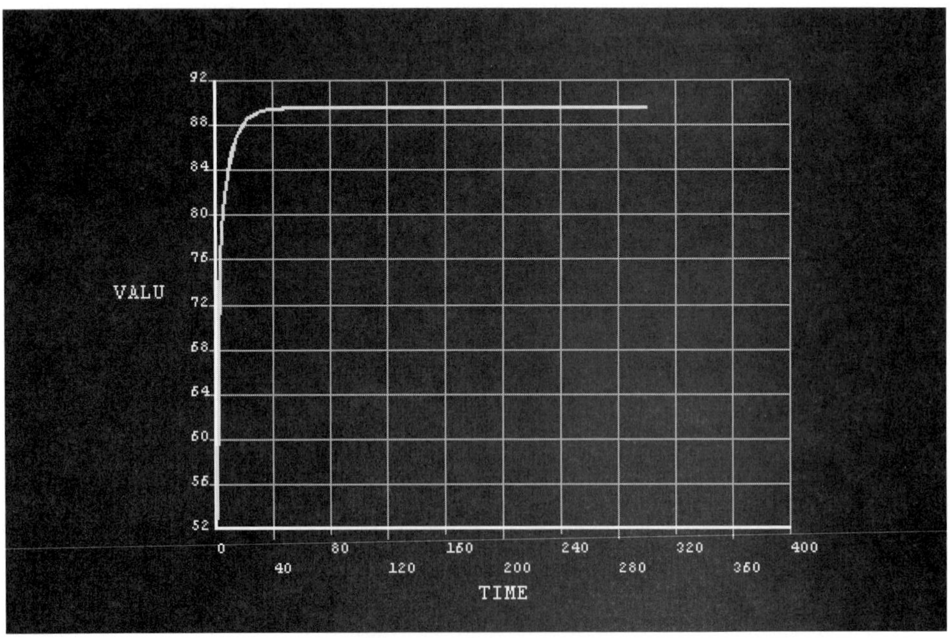
VALU
92
88
84
80
76
72
68
64
60
56
52
0
40
80
120
160
200
240
280
320
360
400
TIME

You can now see how the temperature of the *corner_pt* changes with time.

The next few steps will show how to animate the temperature as it changes with time.

main menu: **General Postproc → Read Results → First Set**

Utility menu: **Plot Ctrls → Numbering** . . .

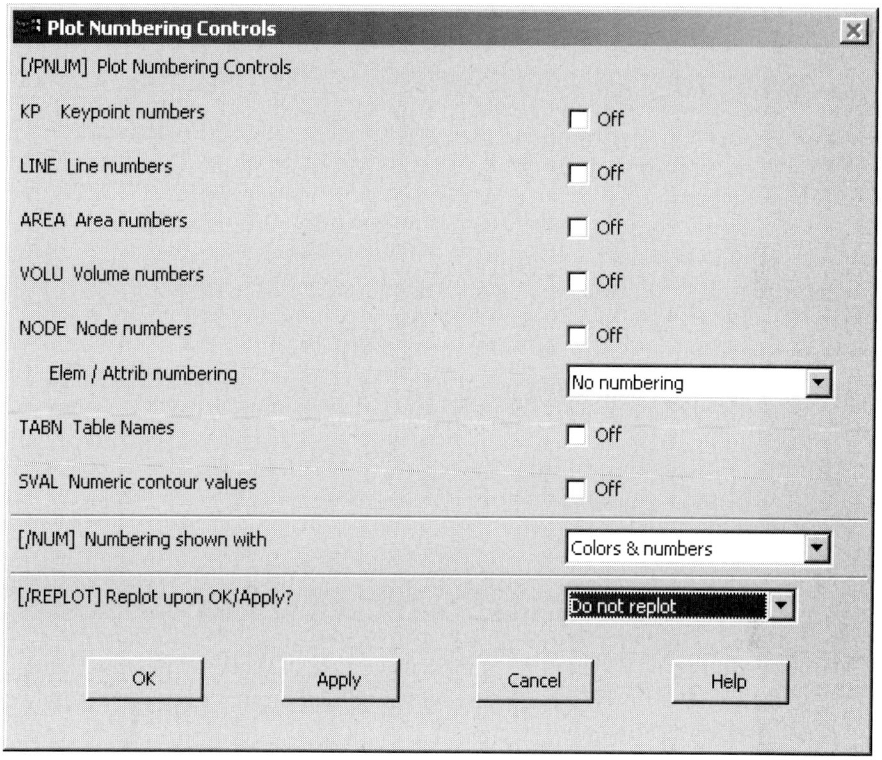

Utility menu: **Plot Ctrls → Style → Contours → Non_uniform Contours** . . .

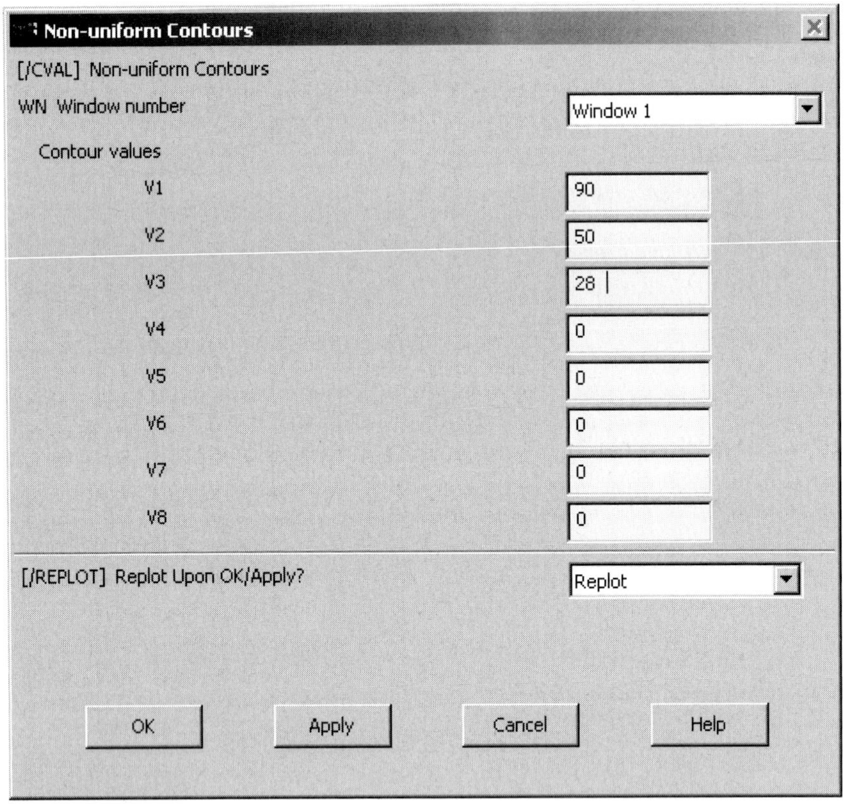

Enter 90 in V1, 50 in V2, and 28 in V3 fields to specify the upper bounds of the first, second, and third contours. Now you can animate the results by issuing the following command:

utility menu: **Plot Ctrls → Animate → Over Time** . . .

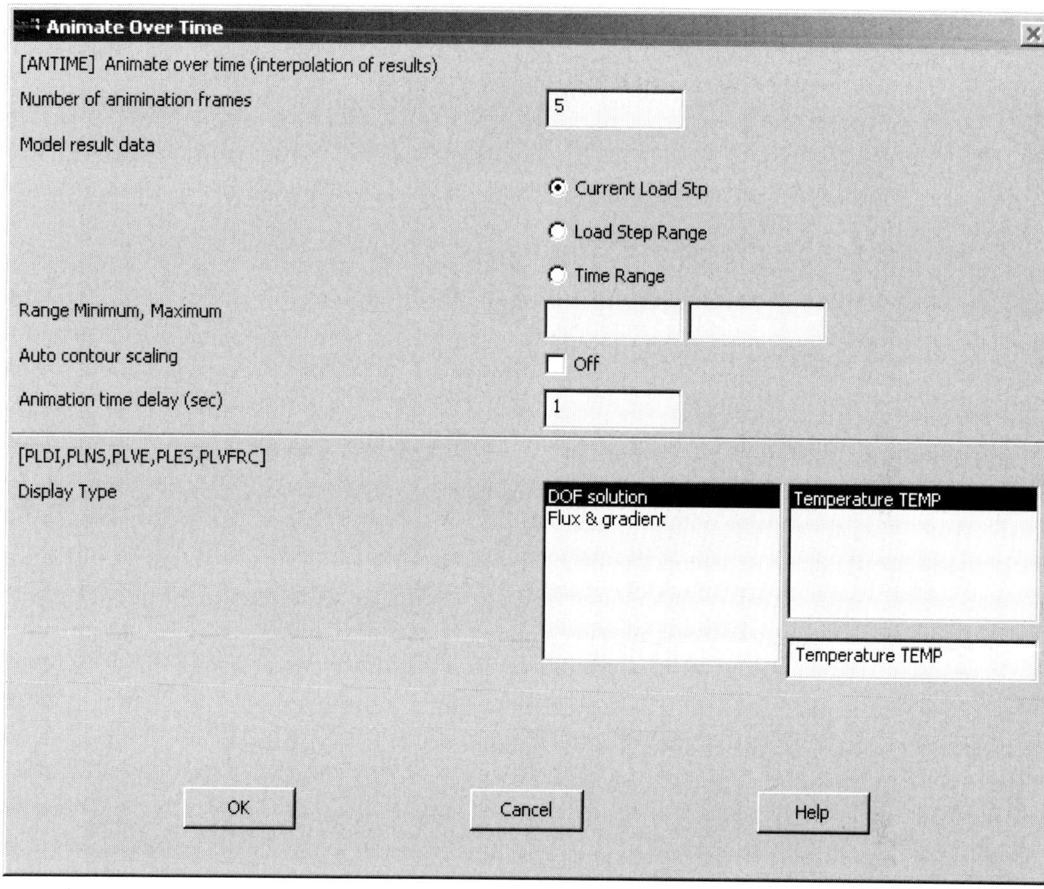
Animate Over Time
[ANTIME] Animate over time (interpolation of results)
Number of animination frames
5
Model result data
Current Load Stp
Load Step Range
Time Range
Range Minimum, Maximum
Auto contour scaling
Off
Animation time delay (sec)
1
[PLDI,PLNS,PLVE,PLES,PLVFRC]
Display Type
DOF solution
Flux & gradient
Temperature TEMP
Temperature TEMP
OK
Cancel
Help

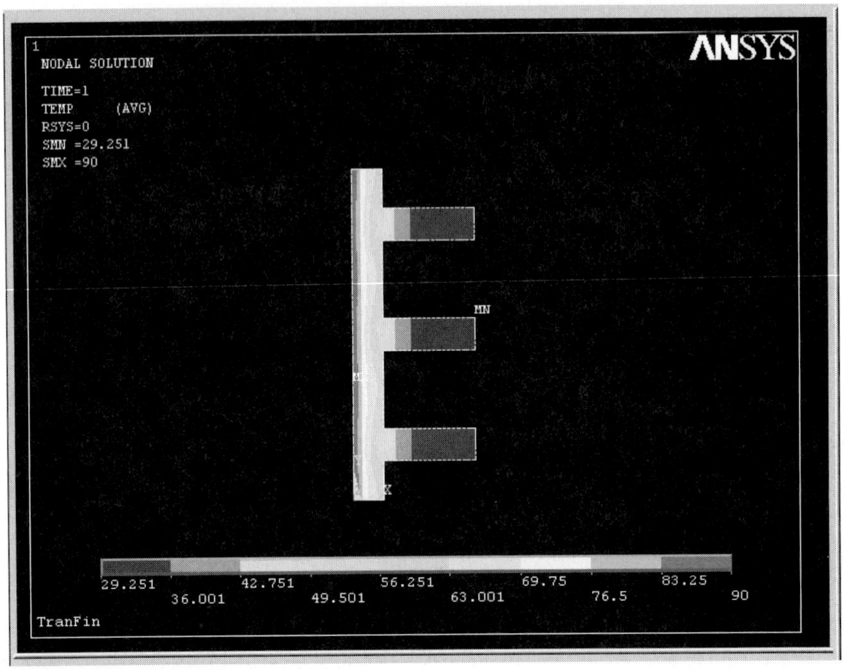

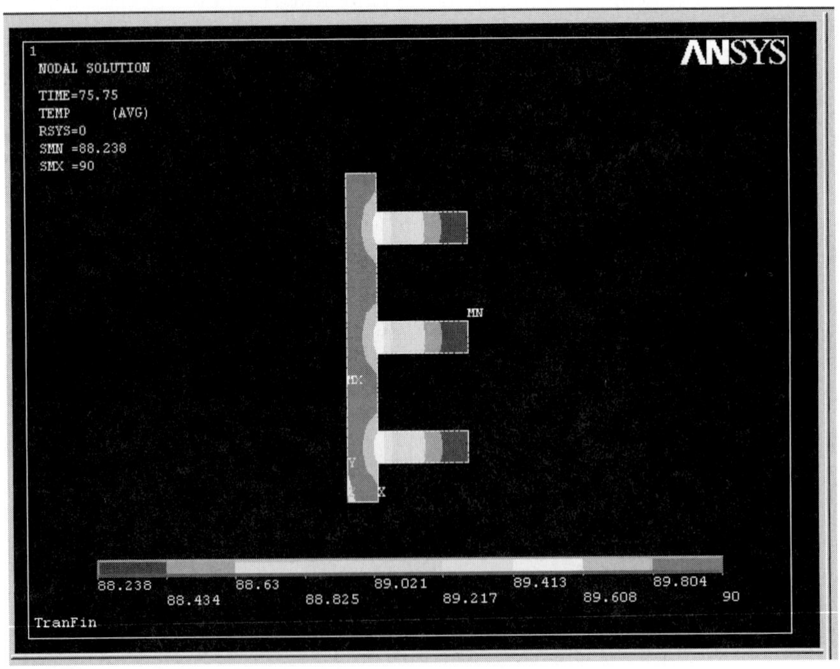

Then you can exit ANSYS.

9.8 VERIFICATION OF RESULTS

First, let us discuss some simple yet powerful ways to verify your results visually. For symmetrical problems, you should always identify lines of symmetry created by geometrical and thermal conditions. Lines of symmetry are always adiabatic lines, meaning that no heat flows in the directions perpendicular to these lines. Because no heat flows in the directions perpendicular to lines of symmetry, they constitute heat flow lines. In other words, heat flows parallel to these lines. Consider the variation of the temperature gradients $\frac{\partial T}{\partial X}$ and $\frac{\partial T}{\partial Y}$ and their vector sum along path A–A, as shown in Figure 9.25. Note that path A–A (in Figure 9.23) is a line of symmetry and, therefore, constitutes an adiabatic line. Because of this fact, the magnitude of $\frac{\partial T}{\partial X}$ is zero along path A–A and $\frac{\partial T}{\partial Y}$ equals vector sum as shown in Figure 9.25. Comparing the variation of the temperature gradients $\frac{\partial T}{\partial X}$ and $\frac{\partial T}{\partial Y}$ and their vector sum along path B–B, as shown in Figure 9.26, renders the conclusion that the magnitude of $\frac{\partial T}{\partial Y}$ is now zero and $\frac{\partial T}{\partial X}$ equals vector sum.

Another important visual inspection of the results requires that the isotherms (lines of constant temperatures) always be perpendicular to the adiabatic lines, or lines of symmetry. You can see this orthogonal relationship in the temperature contour plot of the chimney, as shown in Figure 9.21.

We can also perform a quantitative check on the validity of the results. For example, the conservation of energy applied to a control volume surrounding an arbitrary node must be satisfied. Are the energies flowing into and out of a node balanced out? This approach was demonstrated earlier with Example 9.1.

SUMMARY

At this point you should

1. understand the fundamental concepts of the three modes of heat transfer. You should also know the various types of boundary conditions that could occur in a conduction problem.
2. know how the conductance matrices and the load matrices for two-dimensional conduction problems were obtained. The conductance matrix for a bilinear rectangular element is

$$[\mathbf{K}]^{(e)} = \frac{k_x w}{6\ell}\begin{bmatrix} 2 & -2 & -1 & 1 \\ -2 & 2 & 1 & -1 \\ -1 & 1 & 2 & -2 \\ 1 & -1 & -2 & 2 \end{bmatrix} + \frac{k_y \ell}{6w}\begin{bmatrix} 2 & 1 & -1 & -2 \\ 1 & 2 & -2 & -1 \\ -1 & -2 & 2 & 1 \\ -2 & -1 & 1 & 2 \end{bmatrix}$$

Heat loss by convection around the edge of a rectangular element can also contribute to the conductance matrix

$$[\mathbf{K}]^{(e)} = \frac{h\ell_{ij}}{6}\begin{bmatrix} 2 & 1 & 0 & 0 \\ 1 & 2 & 0 & 0 \\ 0 & 0 & 0 & 0 \\ 0 & 0 & 0 & 0 \end{bmatrix} \qquad [\mathbf{K}]^{(e)} = \frac{h\ell_{jm}}{6}\begin{bmatrix} 0 & 0 & 0 & 0 \\ 0 & 2 & 1 & 0 \\ 0 & 1 & 2 & 0 \\ 0 & 0 & 0 & 0 \end{bmatrix}$$

$$[\mathbf{K}]^{(e)} = \frac{h\ell_{mn}}{6}\begin{bmatrix} 0 & 0 & 0 & 0 \\ 0 & 0 & 0 & 0 \\ 0 & 0 & 2 & 1 \\ 0 & 0 & 1 & 2 \end{bmatrix} \qquad [\mathbf{K}]^{(e)} = \frac{h\ell_{ni}}{6}\begin{bmatrix} 2 & 0 & 0 & 1 \\ 0 & 0 & 0 & 0 \\ 0 & 0 & 0 & 0 \\ 1 & 0 & 0 & 2 \end{bmatrix}$$

The load vector for a rectangular element could have many components. It could have a component due to a possible heat generation term within a given element:

$$\{\mathbf{F}\}^{(e)} = \frac{q^{\cdot} A}{4}\begin{Bmatrix} 1 \\ 1 \\ 1 \\ 1 \end{Bmatrix}$$

It could also have a possible convection heat loss term(s) along the edge(s):

$$\{\mathbf{F}\}^{(e)} = \frac{hT_f\ell_{ij}}{2}\begin{Bmatrix} 1 \\ 1 \\ 0 \\ 0 \end{Bmatrix} \qquad \{\mathbf{F}\}^{(e)} = \frac{hT_f\ell_{jm}}{2}\begin{Bmatrix} 0 \\ 1 \\ 1 \\ 0 \end{Bmatrix}$$

$$\{\mathbf{F}\}^{(e)} = \frac{hT_f\ell_{mn}}{2}\begin{Bmatrix} 0 \\ 0 \\ 1 \\ 1 \end{Bmatrix} \qquad \{\mathbf{F}\}^{(e)} = \frac{hT_f\ell_{ni}}{2}\begin{Bmatrix} 1 \\ 0 \\ 0 \\ 1 \end{Bmatrix}$$

The conductance matrix for a triangular element is

$$[\mathbf{K}]^{(e)} = \frac{k_x}{4A}\begin{bmatrix} \beta_i^2 & \beta_i\beta_j & \beta_i\beta_k \\ \beta_i\beta_j & \beta_j^2 & \beta_j\beta_k \\ \beta_i\beta_k & \beta_j\beta_k & \beta_k^2 \end{bmatrix} + \frac{k_y}{4A}\begin{bmatrix} \delta_i^2 & \delta_i\delta_j & \delta_i\delta_k \\ \delta_i\delta_j & \delta_j^2 & \delta_j\delta_k \\ \delta_i\delta_k & \delta_j\delta_k & \delta_k^2 \end{bmatrix}$$

Heat loss by convection around the edge of a triangular element can also contribute to the conductance matrix according to the equations

$$[\mathbf{K}]^{(e)} = \frac{h\ell_{ij}}{6}\begin{bmatrix} 2 & 1 & 0 \\ 1 & 2 & 0 \\ 0 & 0 & 0 \end{bmatrix} \qquad [\mathbf{K}]^{(e)} = \frac{h\ell_{jk}}{6}\begin{bmatrix} 0 & 0 & 0 \\ 0 & 2 & 1 \\ 0 & 1 & 2 \end{bmatrix}$$

$$[\mathbf{K}]^{(e)} = \frac{h\ell_{ki}}{6}\begin{bmatrix} 2 & 0 & 1 \\ 0 & 0 & 0 \\ 1 & 0 & 2 \end{bmatrix}$$

The load matrix for a triangular element could have many components. It could have a component due to a possible heat generation term within a given element:

$$\{\mathbf{F}\}^{(e)} = \frac{q^{\cdot} A}{3}\begin{Bmatrix}1\\1\\1\end{Bmatrix}$$

Also, it could have a possible convection heat loss term(s) along the edge(s):

$$\{\mathbf{F}\}^{(e)} = \frac{hT_f\ell_{ij}}{2}\begin{Bmatrix}1\\1\\0\end{Bmatrix} \qquad [\mathbf{F}]^{(e)} = \frac{hT_f\ell_{jk}}{2}\begin{Bmatrix}0\\1\\1\end{Bmatrix}$$

$$[\mathbf{F}]^{(e)} = \frac{hT_f\ell_{ki}}{2}\begin{Bmatrix}1\\0\\1\end{Bmatrix}$$

3. understand the contribution of convective boundary conditions to the conductance matrix and the forcing matrix.
4. how the conductance and load matrices were obtained for axisymmetric problems using triangular elements.
5. be familiar with formulation of transient heat transfer problems.
6. always find ways to verify your results.

REFERENCES

ANSYS User's Manual: Procedures, Vol. I, Swanson Analysis Systems, Inc.

ANSYS User's Manual: Commands, Vol. II, Swanson Analysis Systems, Inc.

ANSYS User's Manual: Elements, Vol. III, Swanson Analysis Systems, Inc.

Incropera, F., and Dewitt, D., *Fundamentals of Heat and Mass Transfer*, 2nd ed., New York, John Wiley and Sons, 1985.

Segrlind, L., *Applied Finite Element Analysis*, 2nd ed., New York, John Wiley and Sons, 1984.

PROBLEMS

1. Construct the conductance matrices for the elements shown in the accompanying figure. Also, assemble the elements to obtain the global conductance matrix. The properties and the boundary conditions for each element are shown in the figure.

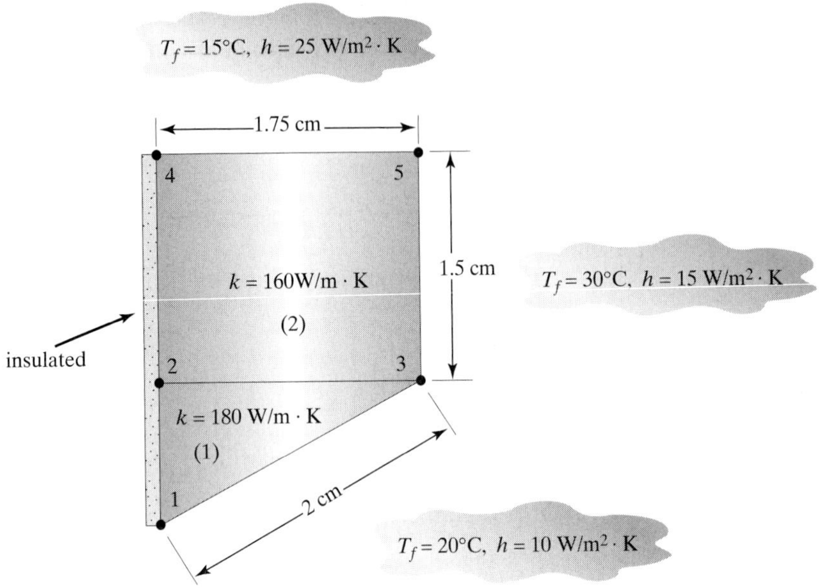

2. Construct the load matrix for each element in problem 1. Also, assemble the elemental load matrices to construct the global load matrix.
3. Construct the conductance matrices shown in the accompanying figure. Also, assemble the elements to obtain the global conductance matrix. The properties and the boundary conditions for each element are shown in the figure.

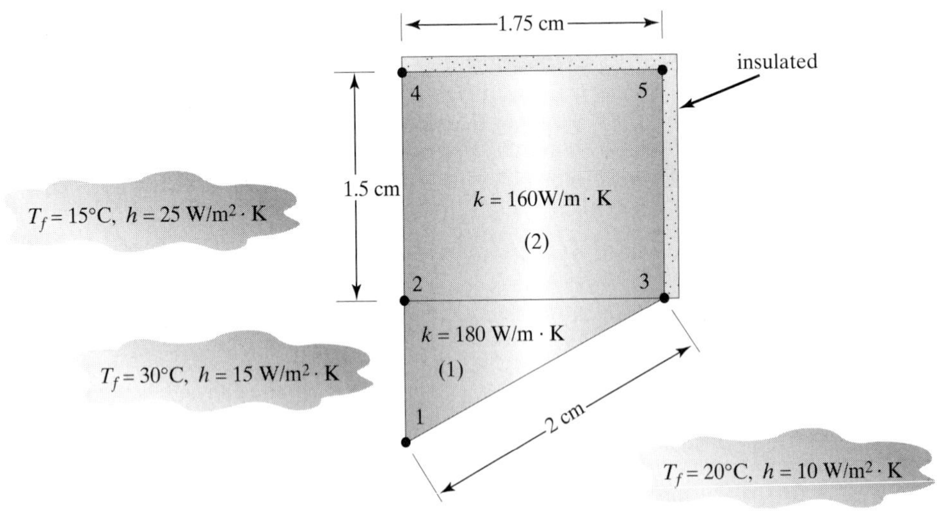

4. Construct the load matrix for each element in problem 3. Also, assemble the elemental load matrices to construct the global load matrix.

5. Show that for a constant heat flux boundary condition q_o'', evaluation of the terms $\int_\tau [\mathbf{S}]^T q_o'' \cos\theta \, d\tau$ and $\int_\tau [\mathbf{S}]^T q_o'' \sin\theta \, d\tau$ along the edges of the rectangular element results in the elemental load matrices

$$\{\mathbf{F}\}^{(e)} = \frac{q_o''\ell_{ij}}{2}\begin{Bmatrix}1\\1\\0\\0\end{Bmatrix} \qquad \{\mathbf{F}\}^{(e)} = \frac{q_o''\ell_{jm}}{2}\begin{Bmatrix}0\\1\\1\\0\end{Bmatrix}$$

$$\{\mathbf{F}\}^{(e)} = \frac{q_o''\ell_{mn}}{2}\begin{Bmatrix}0\\0\\1\\1\end{Bmatrix} \qquad \{\mathbf{F}\}^{(e)} = \frac{q_o''\ell_{ni}}{2}\begin{Bmatrix}1\\0\\0\\1\end{Bmatrix}$$

6. Evaluate the constant heat flux boundary condition in problem 5 for a triangular element.

7. Using the results of problem 5, construct the load matrix for each element shown in the accompanying figure. Also, assemble the elemental matrices to construct the global load matrix. The boundary conditions are shown in the figure. Note that if all elements are made from the same material, then heat flow becomes one-dimensional.

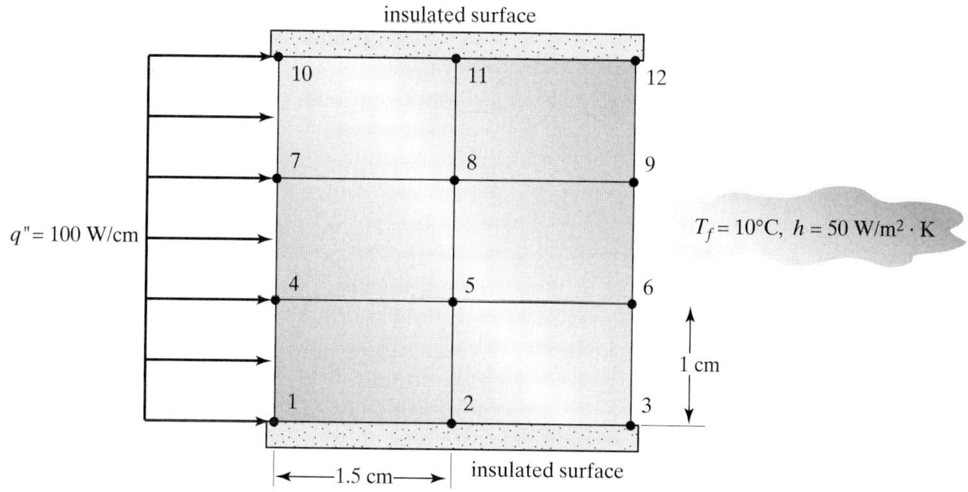

8. In the Galerkin formulation of two-dimensional fins, the convection heat loss from the periphery of the extended surface gives rise to the term $\int_A [S]^T hT\,dA$. The term contributes to the elemental conductance matrix. Show that for a bilinear rectangular element, the integral yields

$$\int_A [\mathbf{S}]^T hT\,dA = \int_A [\mathbf{S}]^T h[S_i \quad S_j \quad S_m \quad S_n]\begin{Bmatrix}T_i\\T_j\\T_m\\T_n\end{Bmatrix} dA = \frac{hA}{36}\begin{bmatrix}4 & 2 & 1 & 2\\2 & 4 & 2 & 1\\1 & 2 & 4 & 2\\2 & 1 & 2 & 4\end{bmatrix}\begin{Bmatrix}T_i\\T_j\\T_m\\T_n\end{Bmatrix}$$

9. Evaluate the integral in problem 8 for a triangular element. Show that

$$\int_A [\mathbf{S}]^T hT\,dA = \int_A [\mathbf{S}]^T h[S_i \quad S_j \quad S_k]\begin{Bmatrix} T_i \\ T_j \\ T_k \end{Bmatrix} dA = \frac{hA}{12}\begin{bmatrix} 2 & 1 & 1 \\ 1 & 2 & 1 \\ 1 & 1 & 2 \end{bmatrix}\begin{Bmatrix} T_i \\ T_j \\ T_k \end{Bmatrix}$$

10. Consider a small rectangular aluminum plate with dimensions of 20 cm × 10 cm and a thermal conductivity value of $k = 168\ \mathrm{W/m \cdot K}$, as shown in the accompanying figure. The plate is exposed to the boundary conditions shown in the figure. Using manual calculations, determine the temperature distribution within the plate, under steady-state conditions.(*Hint*: Because there are two axes of symmetry, you should model only a quarter of the plate.)

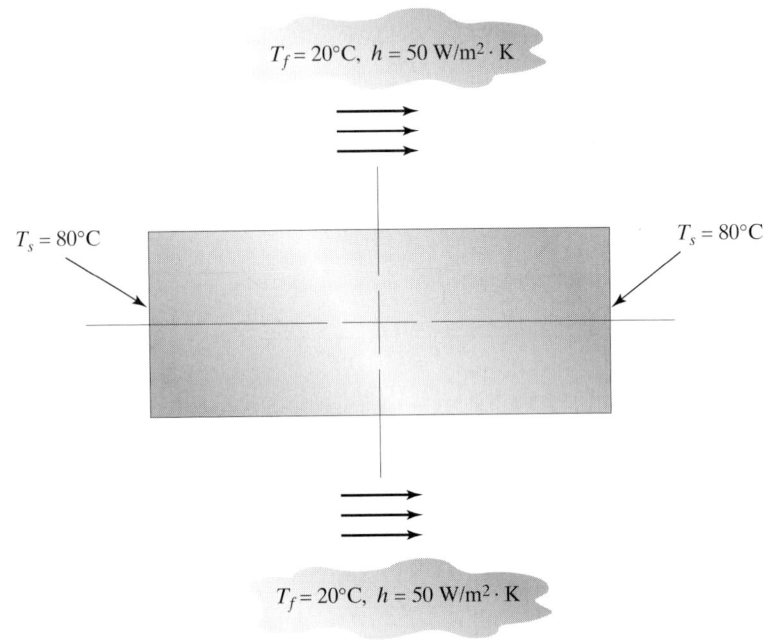

11. Aluminum fins with triangular profiles, shown in the accompanying figure, are used to remove heat from a surface whose temperature is 150°C. The temperature of the surrounding air is 20°C. The natural heat transfer coefficient associated with the surrounding air is $30\ \mathrm{W/m^2 \cdot K}$. The thermal conductivity of aluminum is $k = 168\ \mathrm{W/m \cdot K}$. Using manual calculations, determine the temperature distribution along a fin. Approximate the heat loss for one such fin.

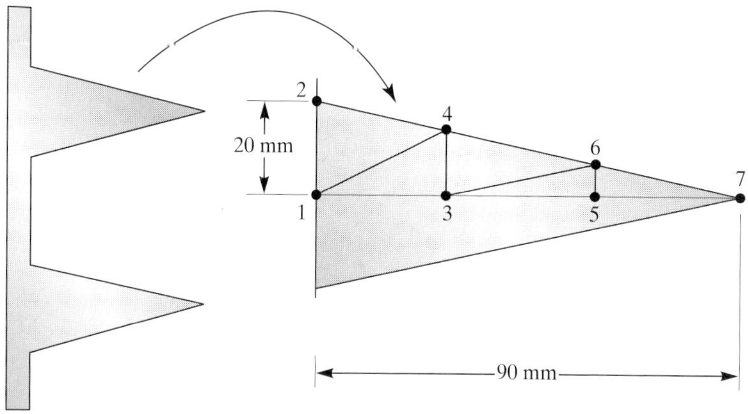

12. For the fin in problem 11, use ANSYS to determine the temperature distribution within the fin. What is the overall heat loss through the fin? Compare these results to the results of your manual calculations.

13. Aluminum fins with parabolic profiles, shown in the accompanying figure, are used to remove heat from a surface whose temperature is 120°C. The temperature of the surrounding air is 20°C. The natural heat transfer coefficient associated with the surrounding air is 25 W/m^2 · K. The thermal conductivity of aluminum is k = 168 W/m · K. Using ANSYS, determine the temperature distribution along a fin. Approximate the heat loss for one such fin.

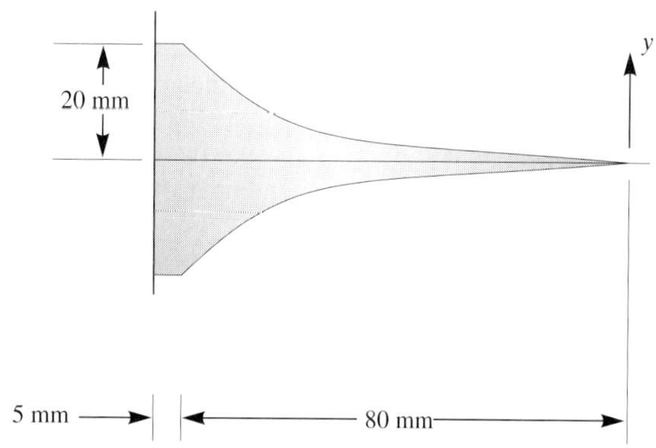

14. Using ANSYS, determine the temperature distribution in the window assembly shown in the accompanying figure. During the winter months, the inside air temperature is kept at 68°F, with a corresponding heat transfer coefficient of h = 1.46 Btu/hr · ft^2 · °F. Assume an outside air temperature of 10°F and a corresponding heat transfer coefficient of h = 6 Btu/hr · ft^2 · °F. What is the overall heat loss through the window assembly?

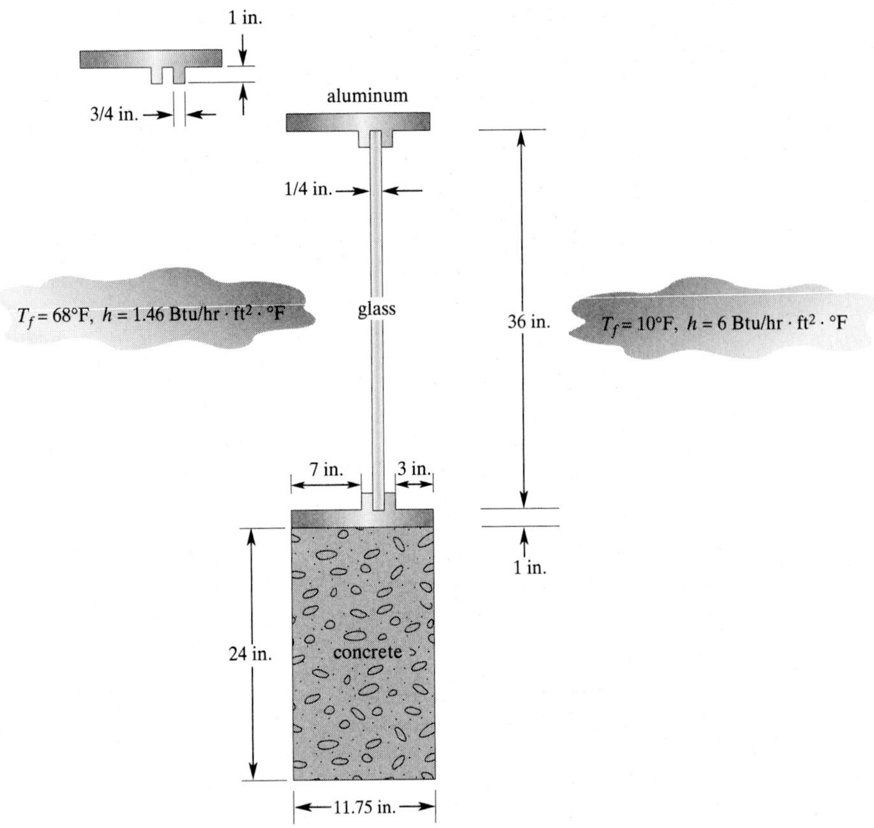

15. Aluminum fins ($k = 170$ W/m · K) are commonly used to dissipate heat from electronic devices. An example of such a fin is shown in the accompanying figure. Using ANSYS, determine the temperature distribution within the fin. The base of the fin experiences a constant flux of $q' = 1000$ W/m. A fan forces air over the surfaces of the fin. The temperature of the surrounding air is 20°C with a corresponding heat transfer coefficient of $h = 40$ W/m² · K.

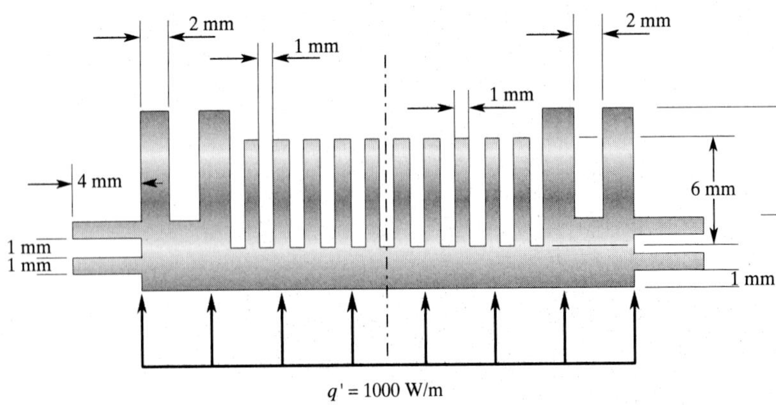

16. Hot water flows through pipes that are embedded in a concrete slab. A section of the slab is shown in the accompanying figure. The temperature of the water inside the pipe is 50°C, with a corresponding heat transfer coefficient of 200 W/m^2 · K. With the conditions shown at the surface, use ANSYS to determine the temperature of the surface. Assuming that the heat transfer coefficient associated with the hot-water flow remains constant, find the water temperature at which the surface freezes. Neglect the thermal resistance through the pipe walls.

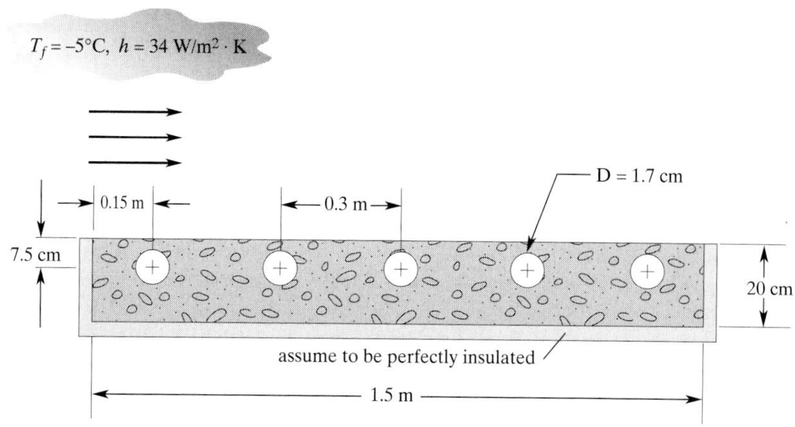

17. Consider the heat transfer through a basement wall with the dimensions given in the accompanying figure. The wall is constructed from concrete and has a thermal conductivity of $k = 1.0$ Btu/hr · ft · °F. The nearby ground has an average thermal conductivity of $k = 0.85$ Btu/hr · ft · °F. Using ANSYS, determine the temperature distribution within the wall and the heat loss from the wall. The inside air is kept at 68°F with a corresponding heat transfer coefficient of $h = 1.46$ Btu/hr · ft^2 · °F. Assume an outside air temperature of 15°F, and a corresponding heat transfer coefficient of $h = 6$ Btu/hr · ft^2 · °F. Assume that at about four feet away from the wall, the horizontal component of the heat transfer in the soil becomes negligible.

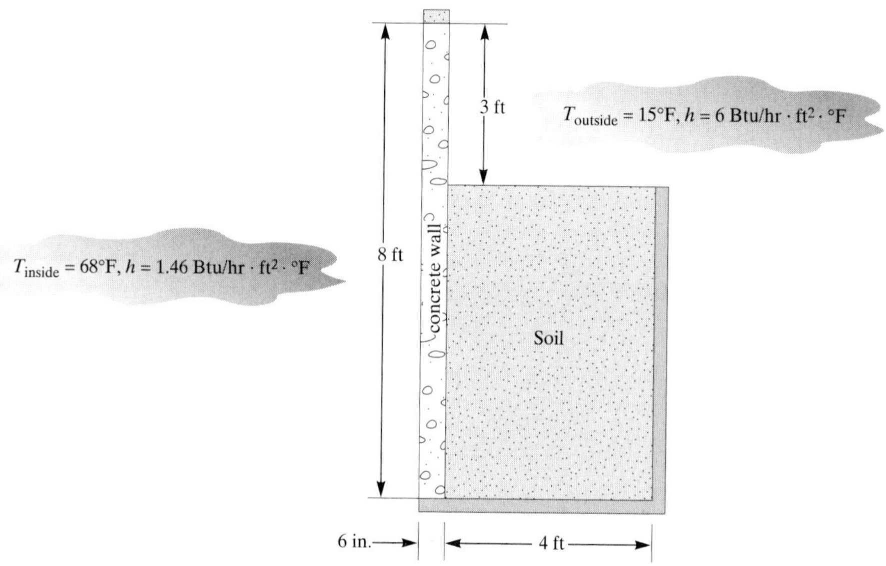

18. We would like to include the heat transfer rates through an uninsulated basement floor in our model in problem 17. Considering the heat transfer model shown in the accompanying figure, determine the temperature distributions in the wall, the floor, and the soil, and the heat loss from the floor and the wall. As shown in the figure, assume that at about four feet away from the wall and the floor, the horizontal and the vertical components of the heat transfer in the soil become negligible.

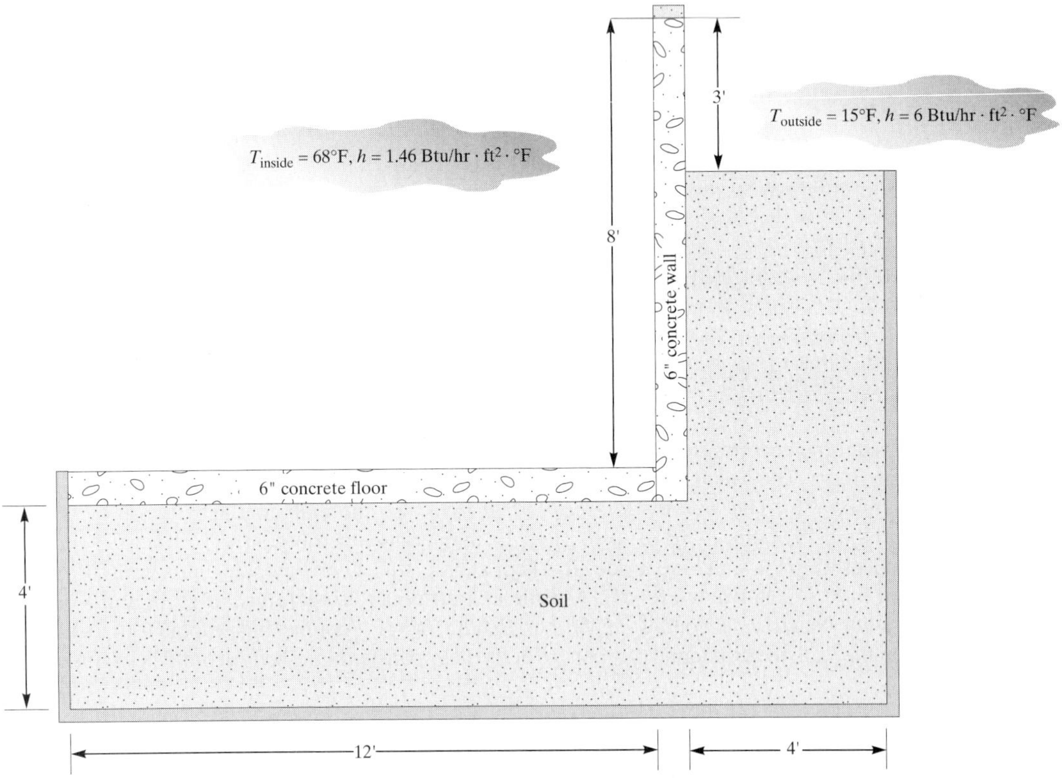

19. In order to enhance heat transfer rates, the inside surface of a tube is extended to form longitudinal fins, as shown in the accompanying figure. Determine the temperature distribution inside the tube wall, given the following data:

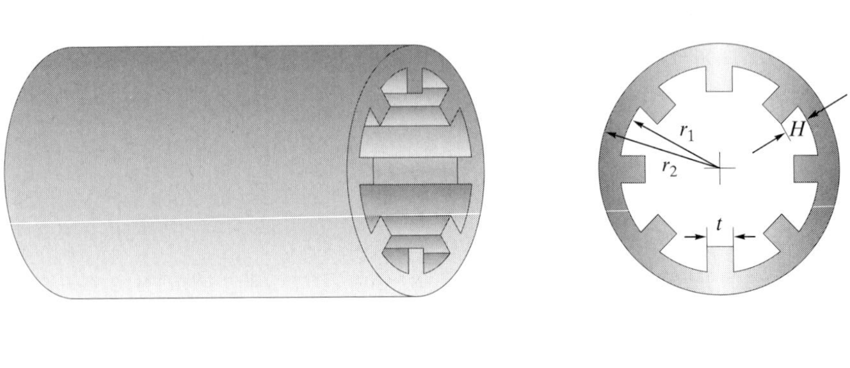

$r_1 = 2$ in	$k = 400$ W/m·K
$r_2 = 2\frac{1}{4}$ in	$T_{\text{inside}} = 80°\text{C}$
$t = \frac{3}{4}$ in	$h_{\text{inside}} = 150$ W/m^2·K
$H = \frac{3}{4}$ in	$T_{\text{outside}} = 15°\text{C}$
	$h_{\text{outside}} = 30$ W/m^2·K

20. Consider the concentric-tube heat exchanger shown in the accompanying figure. A mixture of aqueous ethylene glycol solution arriving from a solar collector is passing through the inner tube. Water flows through the annulus as shown in the figure. The average temperature of the water at the section shown is 15°C, with a corresponding heat transfer coefficient of $h = 200$ W/m^2·K. The average temperature of the ethylene glycol mixture is 48°C, with an associated heat transfer coefficient of $h = 150$ W/m^2·K. In order to enhance the heat transfer rates between the fluids, the outside surface of the inner tube is extended to form longitudinal fins, as shown in the figure. Determine the temperature distribution inside the heat exchanger's walls, assuming that the outside of the heat exchanger is perfectly insulated. Also, determine the heat transfer rate between the fluids.

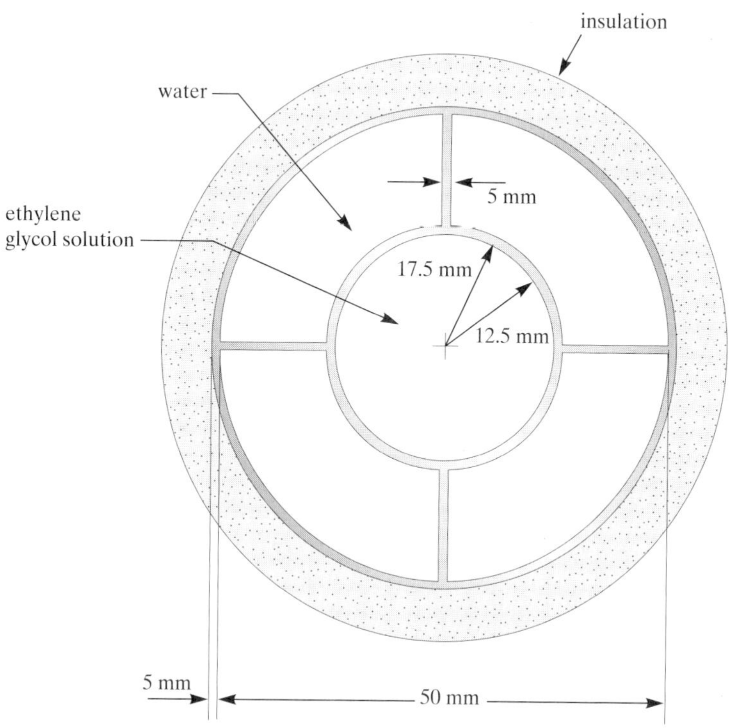

21. Use Green's theorem Eq. (9.36) to calculate the area of the triangle shown in the accompanying figure.

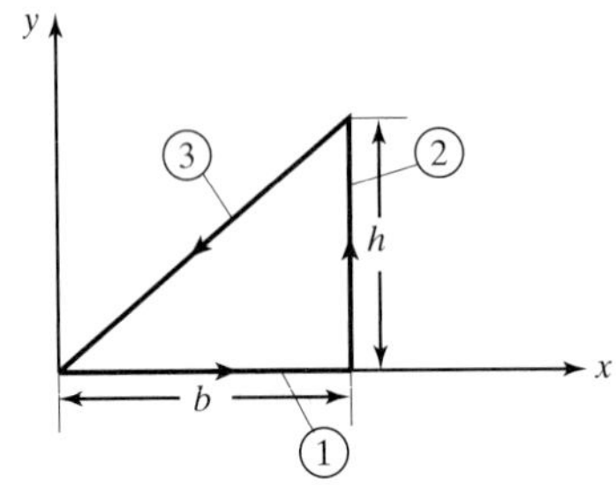

22. Calculate the conductance matrix for the axisymmetric triangular element shown. The element belongs to a solid body made of aluminum alloy with a thermal conductivity $k = 170\ \mathrm{W/m \cdot K}$.

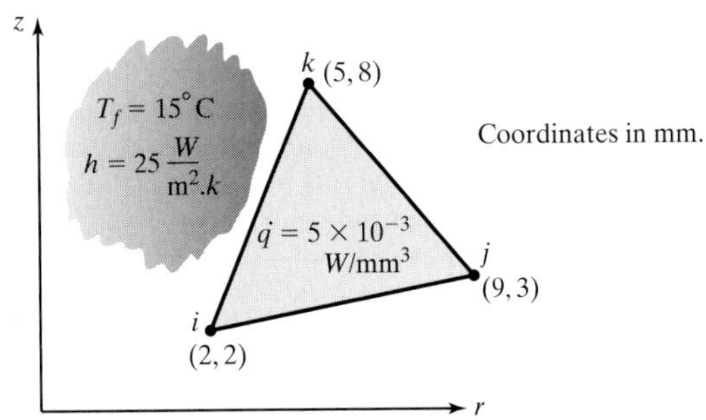

23. Construct the load matrix for problem 22.

24. Construct the load matrix for each axisymmetric triangular element shown. Also, assemble the elemental load matrices to construct the global load matrix.

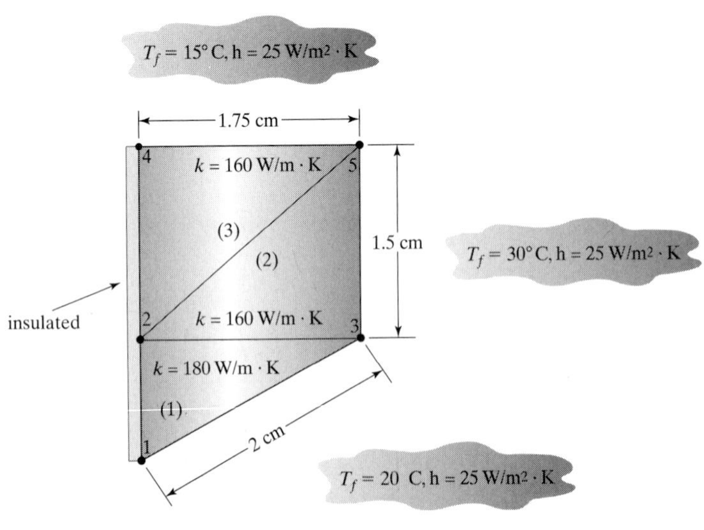

25. Evaluate the constant heat-flux boundary condition for an axisymmetric triangular element.

26. Evaluate the integral $-\int_V \frac{k}{r}\frac{\partial[\mathbf{S}]^T}{\partial r}\frac{\partial T}{\partial r}dV$ for an axisymmetric rectangular element.

27. Evaluate the integral $-\int_V \frac{\partial[\mathbf{S}]^T}{\partial z}\frac{\partial T}{\partial z}dV$ for an axisymmetric rectangular element.

28. Under certain conditions, heat transfer rate q in a two-dimensional system may be computed using the equation $q = kS(T_1 - T_2)$, where k is the thermal conductivity of the medium, S is the shape factor, and T_1 and T_2 are temperatures of the surfaces, as shown in accompanying diagrams. Using ANSYS plot the temperature distribution in the soil for each of the situations shown. Also, using ANSYS calculate the heat-transfer rate for each case and compare it to the shape factor solution.

First situation deals with an isothermal sphere that is buried in soil with $k = 0.5$ W/m · K, $z = 10$ m, $D = 1$ m, $T_1 = 300°C$, $T_2 = 27°C$, and the corresponding shape factor is given by

$$S = \frac{2\pi D}{1 - \frac{D}{4z}}.$$

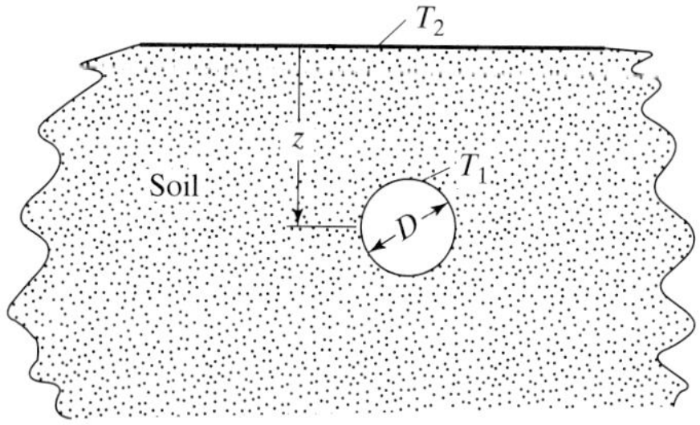

Second situation deals with an isothermal horizontal pipe is buried in soil with $k = 0.5$ W/m · K, $z = 2$ m, $D = 0.5$ m, $L = 50$ m, $T_1 = 100°C$, $T_2 = 5°C$ and the corresponding shape factor is given by $S = \frac{2\pi L}{\ln\left(\frac{4z}{D}\right)}$.

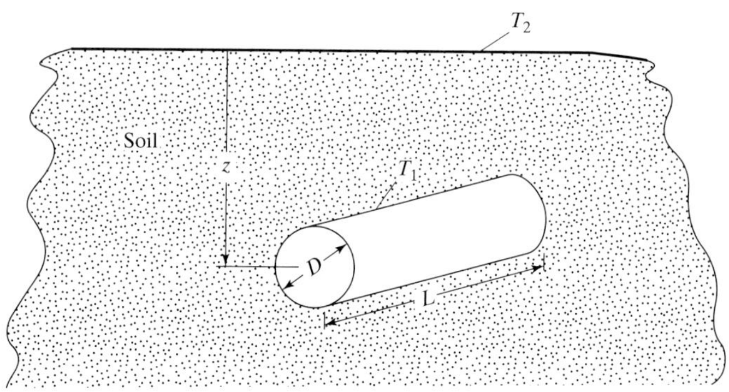

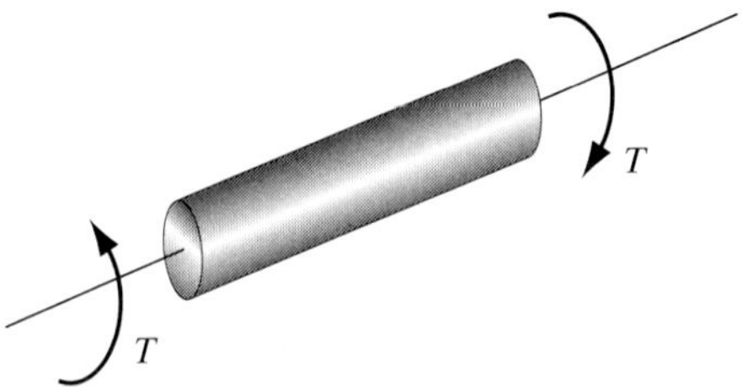

FIGURE 10.1 Torsion of a shaft.

problem when the applied moment or torque twists the member about its longitudinal axis, as shown in Figure 10.1.

Over the elastic limit, the shear stress (τ) distribution within a member with a circular cross section, such as a shaft or a tube, is given by the equation

$$\tau = \frac{Tr}{J} \tag{10.1}$$

where T is the applied torque, r is the radial distance measured from the center of the shaft to a desired point in the cross section, and J represents the polar moment of inertia of the cross-sectional area. It should be clear from examination of Eq. (10.1) that the maximum shear stress occurs at the outer surface of the shaft, where r is equal to the radius of the shaft. Also, recall that the angle of twist (θ) caused by the applied torque can be determined from the equation

$$\theta = \frac{TL}{JG} \tag{10.2}$$

in which L is the length of the member and G is the shear modulus (modulus of rigidity) of the material. Furthermore, there are analytical solutions that can be applied to torsion of members with rectangular cross-sectional areas.* When a torque is applied to a straight bar with a rectangular cross-sectional area, within the elastic region of the material, the maximum shearing stress and an angle of twist caused by the torque are given by

$$\tau_{max} = \frac{T}{c_1 wh^2} \tag{10.3}$$

$$\theta = \frac{TL}{c_2 Gwh^3} \tag{10.4}$$

L is the length of the bar and w and h are the larger and smaller sides of the cross-section, respectively. (See Figure 10.2.) The values of coefficients c_1 and c_2 (given in Table 10.1) are dependent on the aspect ratio of the cross section. As the aspect ratio approaches large numbers ($w/h \rightarrow \infty$), $c_1 = c_2 = 0.3333$. This relationship is demonstrated in Table 10.1.

*See Timoshenko and Goadier (1970) for more detail.

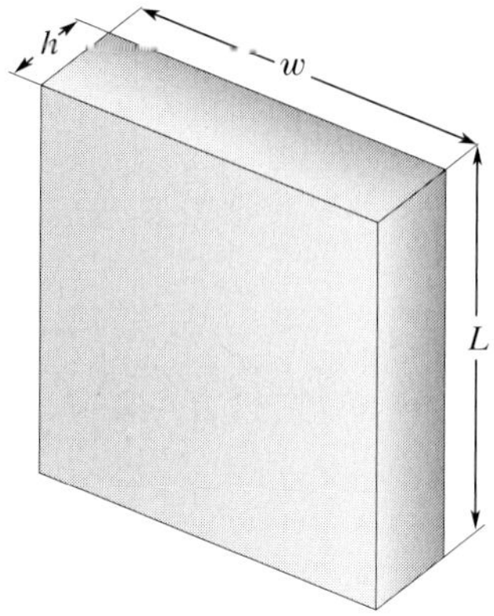

FIGURE 10.2 A straight rectangular bar in torsion.

The maximum shear stress and the angle of twist for cross-sectional geometries with high aspect ratios ($w/h > 10$) are given by:

$$\tau_{max} = \frac{T}{0.333\,wh^2} \tag{10.5}$$

$$\theta = \frac{TL}{0.333\,Gwh^3} \tag{10.6}$$

These types of members are commonly referred to as thin-wall members. Examples of some thin-wall members are shown in Figure 10.3.

Therefore, if you come across a problem that fits these categories, solve it using the torsional formulae. Do not spend a great deal of time generating a finite element model.

Finite Element Formulation of Torsional Problems

Fung (1965) discusses the elastic torsional behavior of noncircular shafts in detail. There are two basic theories: (1) St. Venant's formulation and (2) the Prandtl formulation.

TABLE 10.1 c_1 and c_2 values for a bar with a rectangular cross section

w/h	c_1	c_2
1.0	0.208	0.141
1.2	0.219	0.166
1.5	0.231	0.196
2.0	0.246	0.229
2.5	0.258	0.249
3.0	0.267	0.263
4.0	0.282	0.281
5.0	0.291	0.291
10.0	0.312	0.312
∞	0.333	0.333

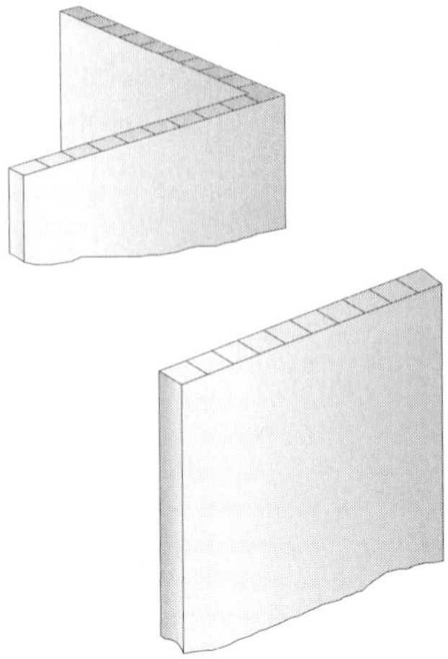

FIGURE 10.3 Examples of thin-wall members.

Here, we will use the Prandtl formulation. The governing differential equation for the elastic torsion of a shaft in terms of the stress function ϕ is

$$\frac{\partial^2\phi}{\partial x^2} + \frac{\partial^2\phi}{\partial y^2} + 2G\theta = 0 \tag{10.7}$$

where G is the shear modulus of elasticity of the bar and θ represents the angle of twist per unit length. The shear stress components are related to the stress function ϕ according to the equations

$$\tau_{zx} = \frac{\partial\phi}{\partial y} \tag{10.8}$$

$$\tau_{zy} = -\frac{\partial\phi}{\partial x} \tag{10.9}$$

Note that with Prandtl's formulation, the applied torque does not directly appear in the governing equation. Instead, the applied torque is related to the stress function and is

$$T = 2\int_A \phi\, dA \tag{10.10}$$

In Eq. (10.10), A represents the cross-sectional area of the shaft. Comparing the differential equation governing the torsional behavior of a member, Eq. (10.7), to the heat diffusion equation, Eq. (9.8), we note that both of these equations have the same form. Therefore, we can apply the results of Section 9.2 and Section 9.3 to torsional problems. However, when comparing the differential equations for torsional problems, we let $c_1 = 1$ and $c_2 = 1$, $c_3 = 2G\theta$. The stiffness matrix for a rectangular element

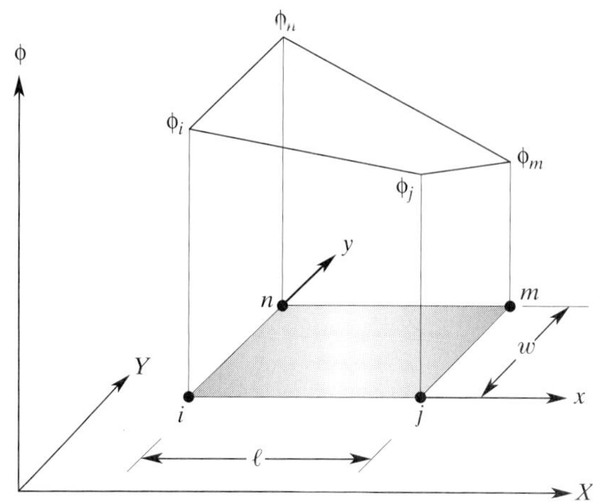

FIGURE 10.4 Nodal values of the stress function for a rectangular element.

then becomes

$$[\mathbf{K}]^{(e)} = \frac{w}{6\ell}\begin{bmatrix} 2 & -2 & -1 & 1 \\ -2 & 2 & 1 & -1 \\ -1 & 1 & 2 & 2 \\ 1 & -1 & -2 & 2 \end{bmatrix} + \frac{\ell}{6w}\begin{bmatrix} 2 & 1 & -1 & -2 \\ 1 & 2 & -2 & -1 \\ -1 & -2 & 2 & 1 \\ -2 & -1 & 1 & 2 \end{bmatrix} \quad (10.11)$$

where w and ℓ are the length and the width, respectively, of the rectangular element, as shown in Figure 10.4. The load matrix for an element is

$$\{\mathbf{F}\}^{(e)} = \frac{2G\theta A}{4}\begin{Bmatrix} 1 \\ 1 \\ 1 \\ 1 \end{Bmatrix} \quad (10.12)$$

and for triangular elements, shown in Figure 10.5, the stiffness and load matrices are

$$[\mathbf{K}]^{(e)} = \frac{1}{4A}\begin{bmatrix} \beta_i^2 & \beta_i\beta_j & \beta_i\beta_k \\ \beta_i\beta_j & \beta_j^2 & \beta_j\beta_k \\ \beta_i\beta_k & \beta_j\beta_k & \beta_k^2 \end{bmatrix} + \frac{1}{4A}\begin{bmatrix} \delta_i^2 & \delta_i\delta_j & \delta_i\delta_k \\ \delta_i\delta_j & \delta_j^2 & \delta_j\delta_k \\ \delta_i\delta_k & \delta_j\delta_k & \delta_k^2 \end{bmatrix} \quad (10.13)$$

$$\{\mathbf{F}\}^{(e)} = \frac{2G\theta A}{3}\begin{Bmatrix} 1 \\ 1 \\ 1 \end{Bmatrix} \quad (10.14)$$

where the area A of the triangular element and the α, β, and δ-terms are given by

$$2A = X_i(Y_j - Y_k) + X_j(Y_k - Y_i) + X_k(Y_i - Y_j)$$
$$\alpha_i = X_jY_k - X_kY_j \quad \beta_i = Y_j - Y_k \quad \delta_i = X_k - X_j$$
$$\alpha_j = X_kY_i - X_iY_k \quad \beta_j = Y_k - Y_i \quad \delta_j = X_i - X_k$$
$$\alpha_k = X_iY_j - X_jY_i \quad \beta_k = Y_i - Y_j \quad \delta_k = X_j - X_i$$

Next, we will consider an example problem dealing with torsion of a steel bar.

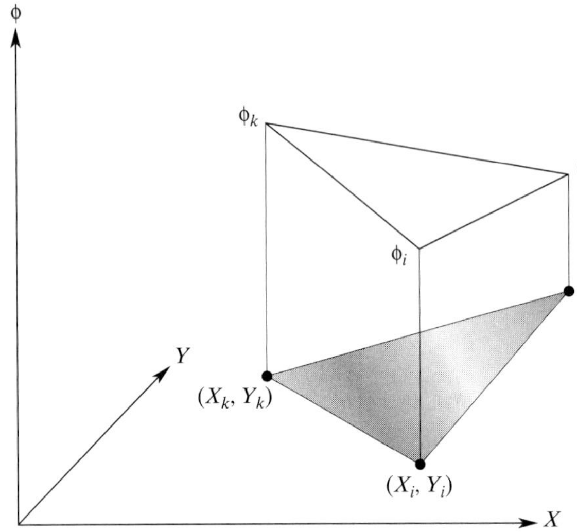

FIGURE 10.5 Nodal values of the stress function for a triangular element.

EXAMPLE 10.1

Consider the torsion of a steel bar $(G = 11 \times 10^3 \text{ksi})$ having a rectangular cross section, as shown in the accompanying figure. Assume that $\theta = 0.0005$ rad/in. Using the finite element procedure discussed above, we are interested in determining the shear stress distribution within the bar.

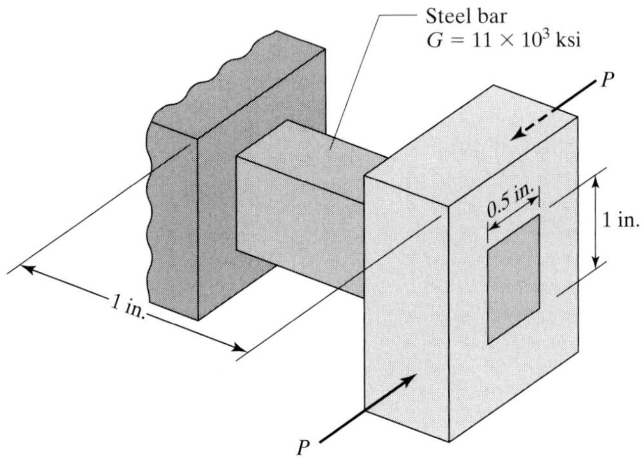

Before we proceed with the solution, note that the point of this exercise is to demonstrate the finite element steps for a torsional problem. As we mentioned earlier, this problem has a simple analytical solution which we will present when we revisit this problem later in this chapter.

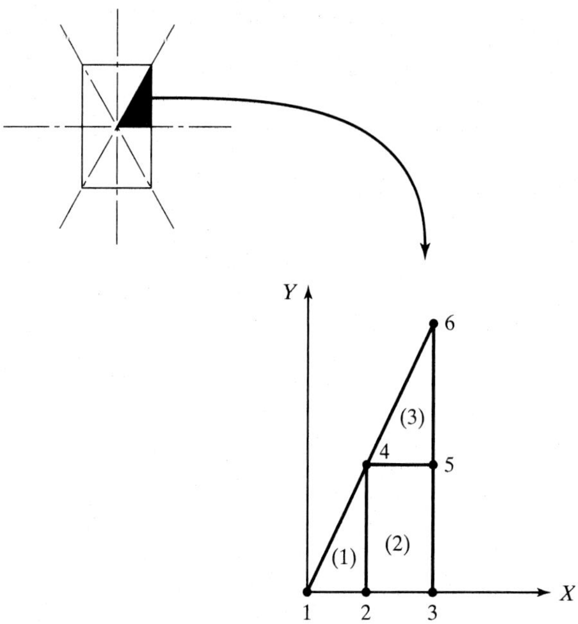

FIGURE 10.6 A schematic of the bar in Example 10.1.

We can make use of the symmetry of the problem, as shown in Figure 10.6, and only analyze a section of the shaft containing 1/8 of the area. The selected section of the shaft is divided into six nodes and three elements (a crude model). Elements (1) and (3) are triangular, while element (2) is rectangular. Consult Table 10.2 while following the solution.

For triangular elements (1) and (3), the stiffness matrix is

$$[\boldsymbol{K}]^{(1)} = [\boldsymbol{K}]^{(3)} = \frac{1}{4A}\begin{bmatrix} \beta_i^2 & \beta_i\beta_j & \beta_i\beta_k \\ \beta_i\beta_j & \beta_j^2 & \beta_j\beta_k \\ \beta_i\beta_k & \beta_j\beta_k & \beta_k^2 \end{bmatrix} + \frac{1}{4A}\begin{bmatrix} \delta_i^2 & \delta_i\delta_j & \delta_i\delta_k \\ \delta_i\delta_j & \delta_j^2 & \delta_j\delta_k \\ \delta_i\delta_k & \delta_j\delta_k & \delta_k^2 \end{bmatrix}$$

Evaluating the β- and δ- coefficients for element (1), we get

$$\beta_i = Y_j - Y_k = 0 - 0.25 = -0.25 \quad \delta_i = X_k - X_j = 0.125 - 0.125 = 0$$
$$\beta_j = Y_k - Y_i = 0.25 - 0 = 0.25 \quad \delta_j = X_i - X_k = 0 - 0.125 = -0.125$$
$$\beta_k = Y_i - Y_j = 0 - 0 = 0 \quad \delta_k = X_j - X_i = 0.125 - 0 = 0.125$$

TABLE 10.2 The relationship between elements and their corresponding nodes and their [coordinates]

Element	i	j	m or k	n
(1)	**1** [0, 0]	**2** [0.125, 0]	**4** [0.125, 0.25]	
(2)	**2** [0.125, 0]	**3** [0.25, 0]	**5** [0.25, 0.25]	**4** [0.125, 0.25]
(3)	**4** [0.125, 0.25]	**5** [0.25, 0.25]	**6** [0.25, 0.5]	

Note that evaluating the β- and δ-coefficients for element (3) renders the same values because the difference between the coordinates of its nodes is identical to that of element (1). Therefore, elements (1) and (3) will both have the following stiffness matrix:

$$[\boldsymbol{K}]^{(1)} = [\boldsymbol{K}]^{(3)} = \frac{1}{4(0.015625)}\begin{bmatrix} 0.0625 & -0.0625 & 0 \\ -0.0625 & 0.0625 & 0 \\ 0 & 0 & 0 \end{bmatrix}$$

$$+ \frac{1}{4(0.015625)}\begin{bmatrix} 0 & 0 & 0 \\ 0 & 0.015625 & -0.015625 \\ 0 & -0.015625 & 0.015625 \end{bmatrix}$$

To help with assembly of the elements later, the corresponding node numbers for the elements are shown on the top and the side of the stiffness matrices. The stiffness matrices for elements (1) and (3), are

$$[\boldsymbol{K}]^{(1)} = \begin{array}{c} \begin{matrix} 1(i) & 2(j) & 4(k) \end{matrix} \\ \begin{bmatrix} 1 & -1 & 0 \\ -1 & 1.25 & -0.25 \\ 0 & -0.25 & 0.25 \end{bmatrix} \end{array} \begin{matrix} 1 \\ 2 \\ 4 \end{matrix}$$

$$[\boldsymbol{K}]^{(3)} = \begin{array}{c} \begin{matrix} 4(i) & 5(j) & 6(k) \end{matrix} \\ \begin{bmatrix} 1 & -1 & 0 \\ -1 & 1.25 & -0.25 \\ 0 & -0.25 & 0.25 \end{bmatrix} \end{array} \begin{matrix} 4 \\ 5 \\ 6 \end{matrix}$$

The stiffness matrix for element (2) is

$$[\boldsymbol{K}]^{(2)} = \frac{w}{6\ell}\begin{bmatrix} 2 & -2 & -1 & 1 \\ -2 & 2 & 1 & -1 \\ -1 & 1 & 2 & -2 \\ 1 & -1 & -2 & 2 \end{bmatrix} + \frac{\ell}{6w}\begin{bmatrix} 2 & 1 & -1 & -2 \\ 1 & 2 & -2 & -1 \\ -1 & -2 & 2 & 1 \\ -2 & -1 & 1 & 2 \end{bmatrix}$$

$$[\boldsymbol{K}]^{(2)} = \frac{0.25}{6(0.125)}\begin{bmatrix} 2 & -2 & -1 & 1 \\ -2 & 2 & 1 & -1 \\ -1 & 1 & 2 & -2 \\ 1 & -1 & -2 & 2 \end{bmatrix} + \frac{0.125}{6(0.25)}\begin{bmatrix} 2 & 1 & -1 & -2 \\ 1 & 2 & -2 & -1 \\ -1 & -2 & 2 & 1 \\ -2 & -1 & 1 & 2 \end{bmatrix}$$

The stiffness matrix for element (2) with the corresponding node numbers is

$$[\boldsymbol{K}]^{(2)} = \begin{array}{c} \begin{matrix} 2(i) & 3(j) & 5(m) & 4(n) \end{matrix} \\ \begin{bmatrix} 0.83333333 & -0.58333333 & -0.4166666 & 0.16666667 \\ -0.58333333 & 0.83333333 & 0.16666667 & -0.41666667 \\ -0.41666667 & 0.16666667 & 0.83333333 & -0.58333333 \\ 0.16666667 & -0.41666667 & -0.58333333 & 0.83333333 \end{bmatrix} \end{array} \begin{matrix} 2 \\ 3 \\ 5 \\ 4 \end{matrix}$$

The load matrices for the triangular elements (1) and (3) are computed from

$$\{\boldsymbol{F}\}^{(e)} = \frac{2G\theta A}{3}\begin{Bmatrix}1\\1\\1\end{Bmatrix} \text{ leading to: } \{\boldsymbol{F}\}^{(1)} = \begin{Bmatrix}57.29166667\\57.29166667\\57.29166667\end{Bmatrix}\begin{matrix}1\\2\\4\end{matrix} \text{ and } \{\boldsymbol{F}\}^{(3)} = \begin{Bmatrix}57.29166667\\57.29166667\\57.29166667\end{Bmatrix}\begin{matrix}4\\5\\6\end{matrix}$$

The load matrix for the rectangular element (2) including the nodal information is

$$\{\boldsymbol{F}\}^{(2)} = \frac{2G\theta A}{4}\begin{Bmatrix}1\\1\\1\\1\end{Bmatrix} = \begin{Bmatrix}85.9375\\85.9375\\85.9375\\85.9375\end{Bmatrix}\begin{matrix}2\\3\\5\\4\end{matrix}$$

Using the nodal information presented next to each element, the global stiffness matrix becomes

$$[\boldsymbol{K}]^{(G)} = \begin{matrix} 1 & 2 & 3 & 4 & 5 & 6 & \\ \end{matrix}$$
$$\left[\begin{matrix} 1 & -1 & 0 & 0 & 0 & 0 \\ -1 & 2.08333333 & -0.58333333 & -0.08333333 & -0.41666666 & 0 \\ 0 & -0.58333333 & 0.83333333 & -0.41666667 & 0.16666667 & 0 \\ 0 & -0.08333333 & -0.41666667 & 2.08333333 & -1.58333333 & 0 \\ 0 & -0.41666666 & 0.16666667 & -1.58333333 & 2.08333333 & -0.25 \\ 0 & 0 & 0 & 0 & -0.25 & 0.25 \end{matrix}\right]\begin{matrix}1\\2\\3\\4\\5\\6\end{matrix}$$

Assembling the load matrices, we get

$$\{\boldsymbol{F}\}^{(G)} = \begin{Bmatrix}57.3\\57.3 + 85.9\\85.9\\57.3 + 57.3 + 85.9\\57.3 + 85.9\\57.3\end{Bmatrix} = \begin{Bmatrix}57.3\\143.2\\85.93\\200.5\\143.2\\57.3\end{Bmatrix}$$

Applying the boundary condition of $\phi = 0$ to nodes 3, 5, and 6, we get the following 3×3 matrix:

$$\begin{bmatrix}1 & -1 & 0\\-1 & 2.083 & -0.083\\0 & -0.083 & 2.083\end{bmatrix}\begin{Bmatrix}\phi_1\\\phi_2\\\phi_3\end{Bmatrix} = \begin{Bmatrix}57.3\\143.2\\200.5\end{Bmatrix}$$

Solving the set of linear equations simultaneously leads to the following nodal solution:

$$[\phi]^T = [250 \quad 193 \quad 0 \quad 104 \quad 0 \quad 0]$$

We can compute the shear stress components (see Figure 10.7) from Eqs. (10.8) and (10.9) in the following manner. For elements (1) and (3)

$$\tau_{ZX} = \frac{\partial\phi}{\partial Y} = \frac{\partial}{\partial Y}[S_i\phi_i + S_j\phi_j + S_k\phi_k] = \frac{\partial}{\partial Y}[S_i \quad S_j \quad S_k]\begin{Bmatrix}\phi_i\\\phi_j\\\phi_k\end{Bmatrix} = [\delta_i \quad \delta_j \quad \delta_k]\begin{Bmatrix}\phi_i\\\phi_j\\\phi_k\end{Bmatrix}$$

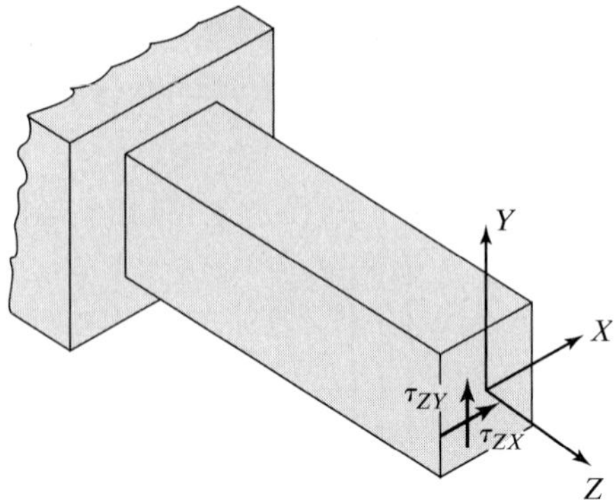

FIGURE 10.7 The directions of shear stress components.

$$\tau_{ZY} = -\frac{\partial \phi}{\partial X} = -\frac{\partial}{\partial X}[S_i\phi_i + S_j\phi_j + S_k\phi_k] = -\frac{\partial}{\partial X}[S_i \quad S_j \quad S_k]\begin{Bmatrix} \phi_i \\ \phi_j \\ \phi_k \end{Bmatrix} =$$

$$-[\beta_i \quad \beta_j \quad \beta_k]\begin{Bmatrix} \phi_i \\ \phi_j \\ \phi_k \end{Bmatrix}$$

The shear stress components for element (1) are

$$\tau_{ZX}^{(1)} = [\delta_i \quad \delta_j \quad \delta_k]\begin{Bmatrix} \phi_i \\ \phi_j \\ \phi_k \end{Bmatrix} = [0 \quad -0.125 \quad 0.125]\begin{Bmatrix} 250 \\ 193 \\ 104 \end{Bmatrix} = -11 \text{ lb/in}^2$$

$$\tau_{ZY}^{(1)} = -[\beta_i \quad \beta_j \quad \beta_k]\begin{Bmatrix} \phi_i \\ \phi_j \\ \phi_k \end{Bmatrix} = -[-0.25 \quad 0.25 \quad 0]\begin{Bmatrix} 250 \\ 193 \\ 104 \end{Bmatrix} = 14 \text{ lb/in}^2$$

Similarly, we can compute the shear stress components for element (3).

$$\tau_{ZX}^{(3)} = [0 \quad -0.125 \quad 0.125]\begin{Bmatrix} 104 \\ 0 \\ 0 \end{Bmatrix} = 0 \text{ lb/in}^2$$

$$\tau_{ZY}^{(3)} = -[-0.25 \quad 0.25 \quad 0]\begin{Bmatrix} 104 \\ 0 \\ 0 \end{Bmatrix} = 26 \text{ lb/in}^2$$

Because the spatial derivatives of linear triangular elements are constants, note that the computed shear stress distribution over the region represented by the triangular element is constant. This is one of the drawbacks of using linear triangular elements. We could have divided the selected section of the shaft into many more elements or used higher order elements, as discussed in Chapter 7, to come up with better results. The shear stress components over the rectangular element is obtained in the same manner.

$$\tau_{ZX} = \frac{\partial \phi}{\partial Y} = \frac{\partial}{\partial Y}[S_i\phi_i + S_j\phi_j + S_m\phi_m + S_n\phi_n] = \frac{\partial}{\partial Y}[S_i \quad S_j \quad S_m \quad S_n]\begin{Bmatrix} \phi_i \\ \phi_j \\ \phi_m \\ \phi_n \end{Bmatrix}$$

$$\tau_{ZX} = \frac{1}{\ell w}[(-\ell + x) \quad -x \quad x \quad (\ell - x)]\begin{Bmatrix} \phi_i \\ \phi_j \\ \phi_m \\ \phi_n \end{Bmatrix}$$

Similarly,

$$\tau_{ZY} = -\frac{\partial \phi}{\partial X} = -\frac{1}{\ell w}[(-w + y) \quad (w - y) \quad y \quad -y]\begin{Bmatrix} \phi_i \\ \phi_j \\ \phi_m \\ \phi_n \end{Bmatrix}$$

Note that for the bilinear rectangular element, the shear stress components vary with position and can be computed for a specific location within the element. It is left as an exercise for you to substitute for the local coordinates of a point within the boundaries of element (2) to obtain the shear stress components. We will revisit this problem later and solve it using ANSYS.

10.2 PLANE-STRESS FORMULATION

We begin by reviewing some of the fundamental concepts dealing with the elastic behavior of materials. Consider an infinitesimally small cube volume surrounding a point within a material. An enlarged version of this volume is shown in Figure 10.8. The faces

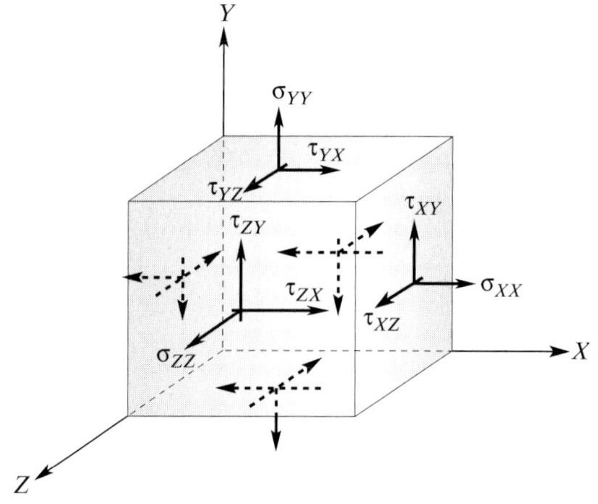

FIGURE 10.8 The components of stress at a point.

of the cube are oriented in the directions of (X, Y, Z) coordinate system.* The application of external forces creates internal forces and, subsequently, stresses within the material. The state of stress at a point can be defined in terms of the nine components on the positive faces and their counterparts on the negative surfaces, as shown in the figure. However, recall that because of equilibrium requirements, only six independent stress components are needed to characterize the general state of stress at a point. Thus the general state of stress at a point is defined by:

$$[\boldsymbol{\sigma}]^T = [\sigma_{XX} \quad \sigma_{YY} \quad \sigma_{ZZ} \quad \tau_{XY} \quad \tau_{YZ} \quad \tau_{XZ}] \tag{10.15}$$

where σ_{XX}, σ_{YY}, and σ_{ZZ} are the normal stresses and τ_{XY}, τ_{YZ}, and τ_{XZ} are the shear stress components, and they provide a measure of the intensity of the internal forces acting over areas of the cube faces. In many practical problems, we come across situations where there are no forces acting in the Z-direction and, consequently, no internal forces acting on the Z- faces. This situation is commonly referred to as a *plane-stress* situation, as shown in Figure 10.9.

For a plane-stress situation, the state of stress reduces to three components:

$$[\boldsymbol{\sigma}]^T = [\sigma_{XX} \quad \sigma_{YY} \quad \tau_{XY}] \tag{10.16}$$

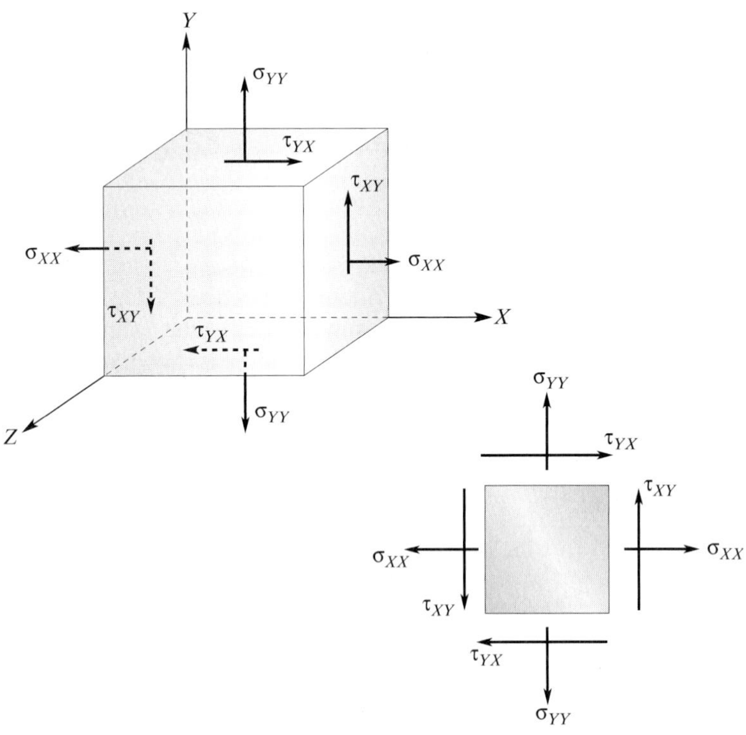

FIGURE 10.9 Plane state of stress.

*Note that throughout this section X, Y, Z and x, y, z coordinate systems are aligned.

We have just considered how an applied force can create stresses within a body. As you know, the applied force will also cause a body to undergo deformation, or change in its shape. We can use a displacement vector to measure the changes that occur in the position of a point within a body. The displacement vector $\vec{\delta}$ can be written in terms of its Cartesian components as

$$\vec{\delta} = u(x, y, z)\vec{i} + v(x, y, z)\vec{j} + w(x, y, z)\vec{k}$$

where the i, j, and k components of the displacement vector represent the difference in the coordinates of the displacement of the point from its original position (x, y, z) to a new position (x', y', z') caused by loading, as given by the equations

$$\begin{aligned} u(x, y, z) &= x' - x \\ v(x, y, z) &= y' - y \\ w(x, y, z) &= z' - z \end{aligned}$$

To better measure the size and shape changes that occur locally within the material, we define normal and shear strains. The state of strain at a point is, therefore, characterized by six independent components:

$$[\boldsymbol{\varepsilon}]^T = [\varepsilon_{xx} \quad \varepsilon_{yy} \quad \varepsilon_{zz} \quad \gamma_{xy} \quad \gamma_{yz} \quad \gamma_{xz}] \tag{10.17}$$

ε_{xx}, ε_{yy}, and ε_{zz} are the normal strains, and γ_{xy}, γ_{yz}, and γ_{xz} are the shear-strain components. These components provide information about the size and shape changes that occur locally in a given material due to loading. The situation in which no displacements occur in the z-direction is known as a *plane-strain* situation. As you may recall from your study of the mechanics of materials, there exists a relationship between the strain and the displacement. These relationships are

$$\varepsilon_{xx} = \frac{\partial u}{\partial x} \qquad \varepsilon_{yy} = \frac{\partial v}{\partial y} \qquad \varepsilon_{zz} = \frac{\partial w}{\partial z} \tag{10.18}$$

$$\gamma_{xy} = \frac{\partial u}{\partial y} + \frac{\partial v}{\partial x} \qquad \gamma_{yz} = \frac{\partial v}{\partial z} + \frac{\partial w}{\partial y} \qquad \gamma_{xz} = \frac{\partial u}{\partial z} + \frac{\partial w}{\partial x}$$

Over the elastic region of a material, there also exists a relationship between the state of stresses and strains, according to the generalized Hooke's law. These relationships are

$$\varepsilon_{xx} = \frac{1}{E}[\sigma_{xx} - \nu(\sigma_{yy} + \sigma_{zz})] \tag{10.19}$$

$$\varepsilon_{yy} = \frac{1}{E}[\sigma_{yy} - \nu(\sigma_{xx} + \sigma_{zz})]$$

$$\varepsilon_{zz} = \frac{1}{E}[\sigma_{zz} - \nu(\sigma_{xx} + \sigma_{yy})]$$

$$\gamma_{xy} = \frac{1}{G}\tau_{xy} \qquad \gamma_{yz} = \frac{1}{G}\tau_{yz} \qquad \gamma_{zx} = \frac{1}{G}\tau_{zx}$$

where E is the modulus of elasticity (Young's modulus), ν is Poisson's ratio, and G is the shear modulus of elasticity (modulus of rigidity). For a plane-stress situation, the generalized Hooke's law reduces to

$$\begin{Bmatrix} \sigma_{xx} \\ \sigma_{yy} \\ \tau_{xy} \end{Bmatrix} = \frac{E}{1-\nu^2}\begin{bmatrix} 1 & \nu & 0 \\ \nu & 1 & 0 \\ 0 & 0 & \dfrac{1-\nu}{2} \end{bmatrix}\begin{Bmatrix} \varepsilon_{xx} \\ \varepsilon_{yy} \\ \gamma_{xy} \end{Bmatrix} \tag{10.20}$$

or, in a compact matrix form,

$$\{\boldsymbol{\sigma}\} = [\boldsymbol{\nu}]\{\boldsymbol{\varepsilon}\} \tag{10.21}$$

where

$$[\boldsymbol{\sigma}]^T = [\sigma_{xx} \quad \sigma_{yy} \quad \tau_{xy}]$$

$$[\boldsymbol{\nu}] = \frac{E}{1-\nu^2}\begin{bmatrix} 1 & \nu & 0 \\ \nu & 1 & 0 \\ 0 & 0 & \dfrac{1-\nu}{2} \end{bmatrix}$$

$$\{\boldsymbol{\varepsilon}\} = \begin{Bmatrix} \varepsilon_{xx} \\ \varepsilon_{yy} \\ \gamma_{xy} \end{Bmatrix}$$

For a plane-strain situation, the generalized Hooke's law becomes

$$\begin{Bmatrix} \sigma_{xx} \\ \sigma_{yy} \\ \tau_{xy} \end{Bmatrix} = \frac{E}{(1+\nu)(1-2\nu)}\begin{bmatrix} 1-\nu & \nu & 0 \\ \nu & 1-\nu & 0 \\ 0 & 0 & \dfrac{1}{2}-\nu \end{bmatrix}\begin{Bmatrix} \varepsilon_{xx} \\ \varepsilon_{yy} \\ \gamma_{xy} \end{Bmatrix} \tag{10.22}$$

Furthermore, for plane stress situations, the strain–displacement relationship becomes

$$\varepsilon_{xx} = \frac{\partial u}{\partial x} \qquad \varepsilon_{yy} = \frac{\partial v}{\partial y} \qquad \gamma_{xy} = \frac{\partial u}{\partial y} + \frac{\partial v}{\partial x} \tag{10.23}$$

We have discussed throughout this text that the minimum total potential energy approach is very commonly used to generate finite element models in solid mechanics. External loads applied to a body will cause the body to deform. During the deformation, the work done by the external forces is stored in the material in the form of elastic energy, which is called strain energy. For a solid material under biaxial loading, the strain energy Λ is

$$\Lambda^{(e)} = \frac{1}{2}\int_V (\sigma_{xx}\varepsilon_{xx} + \sigma_{yy}\varepsilon_{yy} + \tau_{xy}\gamma_{xy})\, dV \tag{10.24}$$

Or, in a compact matrix form,

$$\Lambda^{(e)} = \frac{1}{2}\int_V [\boldsymbol{\sigma}]^T\{\boldsymbol{\varepsilon}\}\,dV \tag{10.25}$$

Substituting for stresses in terms of strains using Hooke's law, Eq. (10.25) can be written as

$$\Lambda^{(e)} = \frac{1}{2}\int_V ([\boldsymbol{\nu}]\{\boldsymbol{\varepsilon}\})^T\{\boldsymbol{\varepsilon}\} = \frac{1}{2}\int_V \{\boldsymbol{\varepsilon}\}^T[\boldsymbol{\nu}]^T\{\boldsymbol{\varepsilon}\} = \frac{1}{2}\int_V \{\boldsymbol{\varepsilon}\}^T[\boldsymbol{\nu}]\{\boldsymbol{\varepsilon}\}\,dV \tag{10.26}$$

Note in arriving at the final form of integral, we have used Eq. (2.12) and $[v]^T = [v]$. We are now ready to look at finite element formulation of plane stress problems using triangular elements. We can represent the displacements u and v using a linear triangular element similar to the one shown in Figure 10.10.

The displacement variable, in terms of linear triangular shape functions and the nodal displacements, is

$$\begin{aligned} u &= S_iU_{ix} + S_jU_{jx} + S_kU_{kx} \\ v &= S_iU_{iy} + S_jU_{jy} + S_kU_{ky} \end{aligned} \tag{10.27}$$

We can write the relations given by Eq. (10.27) in a matrix form:

$$\begin{Bmatrix} u \\ v \end{Bmatrix} = \begin{bmatrix} S_i & 0 & S_j & 0 & S_k & 0 \\ 0 & S_i & 0 & S_j & 0 & S_k \end{bmatrix} \begin{Bmatrix} U_{ix} \\ U_{iy} \\ U_{jx} \\ U_{jy} \\ U_{kx} \\ U_{ky} \end{Bmatrix} \tag{10.28}$$

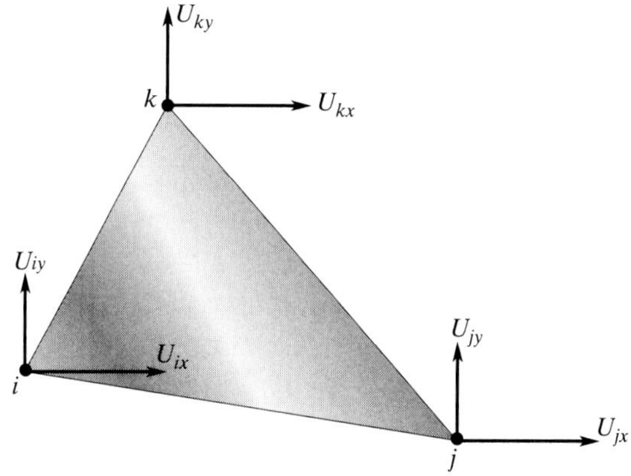

FIGURE 10.10 A triangular element used in formulating plane stress problems.

The next step involves relating the strains to the displacement field and, subsequently, relating the strains to the nodal displacements using shape functions. Referring to the strain–displacement relations as given by Eq. (10.23), we need to take the derivatives of the components of the displacement field with respect to the x and y coordinates, which in turn means taking the derivatives of the appropriate shape functions with respect to x and y. Performing these operations results in the following relations:

$$\begin{aligned}
\varepsilon_{xx} &= \frac{\partial u}{\partial x} = \frac{\partial}{\partial x}(S_i U_{ix} + S_j U_{jx} + S_k U_{kx}) = \frac{1}{2A}[\beta_i U_{ix} + \beta_j U_{jx} + \beta_k U_{kx}] \\
\varepsilon_{yy} &= \frac{\partial v}{\partial y} = \frac{\partial}{\partial y}(S_i U_{iy} + S_j U_{jy} + S_k U_{ky}) = \frac{1}{2A}[\delta_i U_{iy} + \delta_j U_{jy} + \delta_k U_{ky}] \\
\gamma_{xy} &= \frac{\partial u}{\partial y} + \frac{\partial v}{\partial x} = \frac{1}{2A}[\delta_i U_{ix} + \beta_i U_{iy} + \delta_j U_{jx} + \beta_j U_{jy} + \delta_k U_{kx} + \beta_k U_{ky}]
\end{aligned} \tag{10.29}$$

Representing the relations of Eq. (10.29) in a matrix form, we have

$$\begin{Bmatrix} \varepsilon_{xx} \\ \varepsilon_{yy} \\ \gamma_{xy} \end{Bmatrix} = \frac{1}{2A}\begin{bmatrix} \beta_i & 0 & \beta_j & 0 & \beta_k & 0 \\ 0 & \delta_i & 0 & \delta_j & 0 & \delta_k \\ \delta_i & \beta_i & \delta_j & \beta_j & \delta_k & \beta_k \end{bmatrix}\begin{Bmatrix} U_{ix} \\ U_{iy} \\ U_{jx} \\ U_{jy} \\ U_{kx} \\ U_{ky} \end{Bmatrix} \tag{10.30}$$

and in a compact matrix form, Eq. (10.30) becomes

$$\{\boldsymbol{\varepsilon}\} = [\mathbf{B}]\{\mathbf{U}\} \tag{10.31}$$

where

$$\{\boldsymbol{\varepsilon}\} = \begin{Bmatrix} \varepsilon_{xx} \\ \varepsilon_{yy} \\ \gamma_{xy} \end{Bmatrix} \quad [\mathbf{B}] = \frac{1}{2A}\begin{bmatrix} \beta_i & 0 & \beta_j & 0 & \beta_k & 0 \\ 0 & \delta_i & 0 & \delta_j & 0 & \delta_k \\ \delta_i & \beta_i & \delta_j & \beta_j & \delta_k & \beta_k \end{bmatrix} \quad \{\mathbf{U}\} = \begin{Bmatrix} U_{ix} \\ U_{iy} \\ U_{jx} \\ U_{jy} \\ U_{kx} \\ U_{ky} \end{Bmatrix}$$

Substituting into the strain energy equation for the strain components in terms of the displacements, we obtain

$$\Lambda^{(e)} = \frac{1}{2}\int_V \{\boldsymbol{\varepsilon}\}^T[\boldsymbol{\nu}]\{\boldsymbol{\varepsilon}\}\,dV = \frac{1}{2}\int_V [\mathbf{U}]^T[\mathbf{B}]^T[\boldsymbol{\nu}][\mathbf{B}][\mathbf{U}]\,dV \tag{10.32}$$

Differentiating with respect to the nodal displacements, we obtain

$$\frac{\partial \Lambda^{(e)}}{\partial U_k} = \frac{\partial}{\partial U_k}\left(\frac{1}{2}\int_V [\mathbf{U}]^T[\mathbf{B}]^T[\boldsymbol{\nu}][\mathbf{B}][\mathbf{U}]\,dV\right) \quad \text{for } k = 1, 2, \ldots, 6 \tag{10.33}$$

Evaluation of Eq. (10.33) results in the expression $[\mathbf{K}]^{(e)}\{\mathbf{U}\}$. The expression for the stiffness matrix is thus

$$[\mathbf{K}]^{(e)} = \int_V [\mathbf{B}]^T[\boldsymbol{\nu}][\mathbf{B}]\, dV = V[\mathbf{B}]^T[\boldsymbol{\nu}][\mathbf{B}] \tag{10.34}$$

Here, V is the volume of the element and is the product of the area of the element and its thickness. Example 10.2 will show how Eq. (10.34) is used to evaluate the stiffness matrix for a two-dimensional triangular plane stress element.

Load Matrix

To obtain the load matrix for a two-dimensional plane stress element, we must first compute the work done by the external forces, such as distributed loads or point loads. The work done by a concentrated load Q is the product of the load component and the corresponding displacement component. We can represent the work done by concentrated loads in a compact matrix form as

$$W^{(e)} = \{\mathbf{U}\}^T\{\mathbf{Q}\} \tag{10.35}$$

A distributed load with p_x and p_y components does work according to the relationship

$$W^{(e)} = \int_A (up_x + vp_y)\, dA \tag{10.36}$$

where u and v are the displacements in the x and y directions respectively, and A represents the surface over which the distributed load components are acting. The magnitude of the surface A is the product of the element thickness t and the length of the edge over which the distributed load is applied. Using triangular elements to represent the displacements, we find that the work done by distributed loads becomes

$$W^{(e)} = \int_A \{\mathbf{U}\}^T[\mathbf{S}]^T\{\mathbf{p}\}\, dA \tag{10.37}$$

where

$$\{\mathbf{p}\} = \begin{Bmatrix} p_x \\ p_y \end{Bmatrix}$$

The next step in evaluating the load matrix involves the minimization process. In the case of the concentrated load, differentiation of Eq. (10.35) with respect to nodal displacements yields the components of the loads:

$$\{\mathbf{F}\}^{(e)} = \begin{Bmatrix} Q_{ix} \\ Q_{iy} \\ Q_{jx} \\ Q_{jy} \\ Q_{kx} \\ Q_{ky} \end{Bmatrix} \tag{10.38}$$

The differentiation of the work done by the distributed load with respect to the nodal displacements gives the load matrix

$$\{\mathbf{F}\}^{(e)} = \int_A [\mathbf{S}]^T \{\mathbf{p}\}\, dA \tag{10.39}$$

where

$$[\mathbf{S}]^T = \begin{bmatrix} S_i & 0 \\ 0 & S_i \\ S_j & 0 \\ 0 & S_j \\ S_k & 0 \\ 0 & S_k \end{bmatrix}$$

Consider an element subjected to a distributed load along its *ki*-edge, as shown in Figure 10.11.

Evaluating Eq. (10.39) along the *ki*-edge and realizing that along the *ki*-edge, $S_j = 0$, we have

$$\{\mathbf{F}\}^{(e)} = \int_A \begin{bmatrix} S_i & 0 \\ 0 & S_i \\ S_j & 0 \\ 0 & S_j \\ S_k & 0 \\ 0 & S_k \end{bmatrix} \begin{Bmatrix} p_x \\ p_y \end{Bmatrix} dA = t \int_{\ell_{ki}} \begin{bmatrix} S_i & 0 \\ 0 & S_i \\ 0 & 0 \\ 0 & 0 \\ S_k & 0 \\ 0 & S_k \end{bmatrix} \begin{Bmatrix} p_x \\ p_y \end{Bmatrix} d\ell = \frac{tL_{ik}}{2} \begin{Bmatrix} p_x \\ p_y \\ 0 \\ 0 \\ p_x \\ p_y \end{Bmatrix} \tag{10.40}$$

Note that the effect of the distributed load in Figure 10.11 along the *ki*-edge is represented by two equal nodal forces at *i* and *k*, with each force having *x* and *y* components.

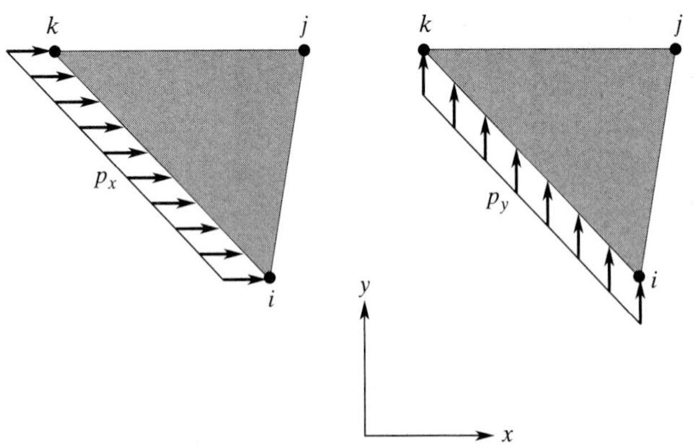

FIGURE 10.11 A distributed load acting over the *ki*-edge of a triangular element.

In a similar fashion, we can formulate the load matrix for a distributed load acting along other sides of the triangular element. Evaluation of the integral in Eq. (10.39) along the ij-edge and the jk-edge results in

$$\{\boldsymbol{F}\}^{(e)} = \frac{tL_{ij}}{2}\begin{Bmatrix} p_x \\ p_y \\ p_x \\ p_y \\ 0 \\ 0 \end{Bmatrix} \qquad \{\boldsymbol{F}\}^{(e)} = \frac{tL_{jk}}{2}\begin{Bmatrix} 0 \\ 0 \\ p_x \\ p_y \\ p_x \\ p_y \end{Bmatrix} \tag{10.41}$$

It is worth noting that, generally speaking, linear triangular elements do not offer as accurate results as do the higher order elements. The purpose of the above derivation was to demonstrate the general steps involved in obtaining the elemental stiffness and load matrices. Next, we will derive the stiffness matrix for a quadrilateral element using isoparametric formulation.

10.3 ISOPARAMETRIC FORMULATION: USING A QUADRILATERAL ELEMENT

As we discussed in Chapters 5 and 7, when we use a *single* set of parameters (a set of shape functions) to define the unknown variables u, v, T, and so on, and to express the position of any point within the element, we are using isoparametric formulation. An element expressed in such manner is called an isoparametric element. We will now turn our attention to the quadrilateral element previously shown as Figure 7.16 (repeated here

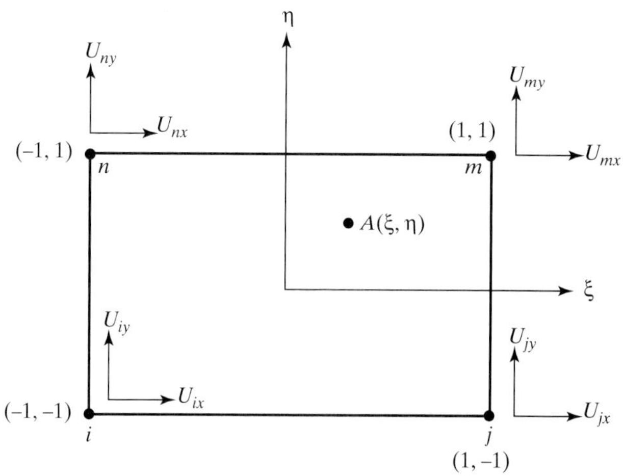

FIGURE 7.16 A quadrilateral element used in formulating plane stress problems.

for convenience). Using a quadrilateral element, we can express the displacement field within an element by Eq. (7.30):

$$\begin{aligned} u &= S_iU_{ix} + S_jU_{jx} + S_mU_{mx} + S_nU_{nx} \\ v &= S_iU_{iy} + S_jU_{jy} + S_mU_{my} + S_nU_{ny} \end{aligned} \tag{7.30}$$

We can write the relations given by Eq. (7.30) in matrix form, given previously in Eq. (7.31):

$$\begin{Bmatrix} u \\ v \end{Bmatrix} = \begin{bmatrix} S_i & 0 & S_j & 0 & S_m & 0 & S_n & 0 \\ 0 & S_i & 0 & S_j & 0 & S_m & 0 & S_n \end{bmatrix} \begin{Bmatrix} U_{ix} \\ U_{iy} \\ U_{jx} \\ U_{jy} \\ U_{mx} \\ U_{my} \\ U_{nx} \\ U_{ny} \end{Bmatrix} \tag{7.31}$$

Note that using isoparametric formulation, we can use the same shape functions to describe the position of any point within the element by the relationships in Eq. (7.32):

$$\begin{aligned} x &= S_ix_i + S_jx_j + S_mx_m + S_nx_n \\ y &= S_iy_i + S_jy_j + S_my_m + S_ny_n \end{aligned} \tag{7.32}$$

The displacement field is related to the components of strains ($\varepsilon_{xx} = \frac{\partial u}{\partial x}$, $\varepsilon_{yy} = \frac{\partial v}{\partial y}$, and $\gamma_{xy} = \frac{\partial u}{\partial y} + \frac{\partial v}{\partial x}$) and subsequently to the nodal displacements through shape functions.

In Chapter 7, we also showed that using the Jacobian of the coordinate transformation, we can write the following, previously presented as Eq. (7.34):

$$\begin{Bmatrix} \dfrac{\partial f(x,y)}{\partial \xi} \\ \dfrac{\partial f(x,y)}{\partial \eta} \end{Bmatrix} = \overbrace{\begin{bmatrix} \dfrac{\partial x}{\partial \xi} & \dfrac{\partial y}{\partial \xi} \\ \dfrac{\partial x}{\partial \eta} & \dfrac{\partial y}{\partial \eta} \end{bmatrix}}^{[\mathbf{J}]} \begin{Bmatrix} \dfrac{\partial f(x,y)}{\partial x} \\ \dfrac{\partial f(x,y)}{\partial y} \end{Bmatrix} \tag{7.34}$$

The relationship of Eq. (7.34) was also presented as the following, previously shown as Eq. (7.35):

$$\begin{Bmatrix} \dfrac{\partial f(x,y)}{\partial x} \\ \dfrac{\partial f(x,y)}{\partial y} \end{Bmatrix} = [\mathbf{J}]^{-1} \begin{Bmatrix} \dfrac{\partial f(x,y)}{\partial \xi} \\ \dfrac{\partial f(x,y)}{\partial \eta} \end{Bmatrix} \tag{7.35}$$

For a quadrilateral element, the **J** matrix can be evaluated using Eqs. (7.32) and (7.7):

$$[\mathbf{J}] = \begin{bmatrix} \dfrac{\partial x}{\partial \xi} & \dfrac{\partial y}{\partial \xi} \\ \dfrac{\partial x}{\partial \eta} & \dfrac{\partial y}{\partial \eta} \end{bmatrix} = \begin{bmatrix} \dfrac{\partial}{\partial \xi}[S_i x_i + S_j x_j + S_m x_m + S_n x_n] & \dfrac{\partial}{\partial \xi}[S_i y_i + S_j y_j + S_m y_m + S_n y_n] \\ \dfrac{\partial}{\partial \eta}[S_i x_i + S_j x_j + S_m x_m + S_n x_n] & \dfrac{\partial}{\partial \eta}[S_i y_i + S_j y_j + S_m y_m + S_n y_n] \end{bmatrix} \tag{10.42}$$

$$[\mathbf{J}] = \frac{1}{4}\begin{bmatrix} [-(1-\eta)x_i + (1-\eta)x_j + (1+\eta)x_m - (1+\eta)x_n] & [-(1-\eta)y_i + (1-\eta)y_j + (1+\eta)y_m - (1+\eta)y_n] \\ [-(1-\xi)x_i - (1+\xi)x_j + (1+\xi)x_m + (1-\xi)x_n] & [-(1-\xi)y_i - (1+\xi)y_j + (1+\xi)y_m + (1-\xi)y_n] \end{bmatrix} = \begin{bmatrix} J_{11} & J_{12} \\ J_{21} & J_{22} \end{bmatrix} \tag{10.43}$$

Also recall that the inverse of a two-dimensional square matrix is given by

$$[\mathbf{J}]^{-1} = \frac{1}{J_{11}J_{22} - J_{12}J_{21}}\begin{bmatrix} J_{22} & -J_{12} \\ -J_{21} & J_{11} \end{bmatrix} = \frac{1}{\det \mathbf{J}}\begin{bmatrix} J_{22} & -J_{12} \\ -J_{21} & J_{11} \end{bmatrix} \tag{10.44}$$

We can now proceed with the formulation of the stiffness matrix. The strain energy of an element is

$$\Lambda^{(e)} = \frac{1}{2}\int_V \{\boldsymbol{\varepsilon}\}^T[\boldsymbol{\nu}]\{\boldsymbol{\varepsilon}\}\,dV = \frac{1}{2}(t_e)\int_A \{\boldsymbol{\varepsilon}\}^T[\boldsymbol{\nu}]\{\boldsymbol{\varepsilon}\}\,dA \tag{10.45}$$

where t_e is the thickness of the element. Recall the strain–displacement relationships in matrix form

$$\{\boldsymbol{\varepsilon}\} = \begin{Bmatrix} \varepsilon_{xx} \\ \varepsilon_{yy} \\ \gamma_{xy} \end{Bmatrix} = \begin{Bmatrix} \dfrac{\partial u}{\partial x} \\ \dfrac{\partial v}{\partial y} \\ \dfrac{\partial u}{\partial y} + \dfrac{\partial v}{\partial x} \end{Bmatrix} \tag{10.46}$$

Evaluating the derivatives, we obtain

$$\begin{Bmatrix} \dfrac{\partial u}{\partial x} \\ \dfrac{\partial u}{\partial y} \end{Bmatrix} = \frac{1}{\det \mathbf{J}}\begin{bmatrix} J_{22} & -J_{12} \\ -J_{21} & J_{11} \end{bmatrix}\begin{Bmatrix} \dfrac{\partial u}{\partial \xi} \\ \dfrac{\partial u}{\partial \eta} \end{Bmatrix} \tag{10.47}$$

and

$$\begin{Bmatrix} \dfrac{\partial v}{\partial x} \\ \dfrac{\partial v}{\partial y} \end{Bmatrix} = \frac{1}{\det \mathbf{J}} \begin{bmatrix} J_{22} & -J_{12} \\ -J_{21} & J_{11} \end{bmatrix} \begin{Bmatrix} \dfrac{\partial v}{\partial \xi} \\ \dfrac{\partial v}{\partial \eta} \end{Bmatrix} \tag{10.48}$$

Combining Eqs. (10.46), (10.47), and (10.48) into a single relationship, we have

$$\{\boldsymbol{\varepsilon}\} = \begin{Bmatrix} \dfrac{\partial u}{\partial x} \\ \dfrac{\partial v}{\partial y} \\ \dfrac{\partial u}{\partial y} + \dfrac{\partial v}{\partial x} \end{Bmatrix} = \overbrace{\frac{1}{\det \mathbf{J}} \begin{bmatrix} J_{22} & -J_{12} & 0 & 0 \\ 0 & 0 & -J_{21} & J_{11} \\ -J_{21} & J_{11} & J_{22} & -J_{12} \end{bmatrix}}^{[\mathbf{A}]} \begin{Bmatrix} \dfrac{\partial u}{\partial \xi} \\ \dfrac{\partial u}{\partial \eta} \\ \dfrac{\partial v}{\partial \xi} \\ \dfrac{\partial v}{\partial \eta} \end{Bmatrix} \tag{10.49}$$

Note how we defined the [**A**] matrix, to be used later. Using Eq. (7.30), we can perform the following evaluation:

$$\begin{Bmatrix} \dfrac{\partial u}{\partial \xi} \\ \dfrac{\partial u}{\partial \eta} \\ \dfrac{\partial v}{\partial \xi} \\ \dfrac{\partial v}{\partial \eta} \end{Bmatrix} = \overbrace{\frac{1}{4} \begin{bmatrix} -(1-\eta) & 0 & (1-\eta) & 0 & (1+\eta) & 0 & -(1+\eta) & 0 \\ -(1-\xi) & 0 & -(1+\xi) & 0 & (1+\xi) & 0 & (1-\xi) & 0 \\ 0 & -(1-\eta) & 0 & (1-\eta) & 0 & (1+\eta) & 0 & -(1+\eta) \\ 0 & -(1-\xi) & 0 & -(1+\xi) & 0 & (1+\xi) & 0 & (1-\xi) \end{bmatrix}}^{[\mathbf{D}]} \overbrace{\begin{Bmatrix} U_{ix} \\ U_{iy} \\ U_{jx} \\ U_{jy} \\ U_{mx} \\ U_{my} \\ U_{nx} \\ U_{ny} \end{Bmatrix}}^{\{\mathbf{U}\}} \tag{10.50}$$

We can express the relationship in Eq. (10.50) in a compact matrix form as

$$\{\boldsymbol{\varepsilon}\} = [\mathbf{A}][\mathbf{D}]\{\mathbf{U}\} \tag{10.51}$$

Next, we need to transform the dA term ($dA = dx\,dy$) in the strain energy integral into a product of natural coordinates. This transformation is achieved in the following manner:

$$\Lambda^{(e)} = \frac{1}{2}(t_e)\int_A \{\boldsymbol{\varepsilon}\}^T[\boldsymbol{\nu}]\{\boldsymbol{\varepsilon}\}\,dA = \frac{1}{2}(t_e)\int_{-1}^{1}\int_{-1}^{1}\{\boldsymbol{\varepsilon}\}^T[\boldsymbol{\nu}]\{\boldsymbol{\varepsilon}\}\overbrace{\det \mathbf{J}d\xi d\eta}^{dA} \tag{10.52}$$

Substituting for the strain matrix $\{\boldsymbol{\varepsilon}\}$ and the properties of the material matrix $[\boldsymbol{\nu}]$ into Eq. (10.52) and differentiating the strain energy of the element with respect to its nodal displacements, we find that the expression for the element stiffness matrix becomes:

$$[\mathbf{K}]^{(e)} = t_e\int_{-1}^{1}\int_{-1}^{1}[[\mathbf{A}][\mathbf{D}]]^T[\boldsymbol{\nu}][\mathbf{A}][\mathbf{D}]\det \mathbf{J}d\xi d\eta \tag{10.53}$$

Note that the resulting stiffness matrix is an 8×8 matrix. Furthermore, as discussed in Chapter 7, the integral of Eq. (10.53) is to be evaluated numerically, using the Gauss–Legendre formula.

EXAMPLE 10.2

A two-dimensional triangular plane stress element made of steel, with modulus of elasticity $E = 200$ GPa and Poisson's ratio $v = 0.32$, is shown in Figure 10.12. The element is 3 mm thick, and the coordinates of nodes i, j, and k are given in centimeters in Figure 10.12. Determine the stiffness and load matrices under the given conditions.

The element stiffness matrix is

$$[\mathbf{K}]^{(e)} = V[\mathbf{B}]^T[\boldsymbol{\nu}][\mathbf{B}]$$

where

$$V = tA$$

$$[\mathbf{B}] = \frac{1}{2A}\begin{bmatrix} \beta_i & 0 & \beta_j & 0 & \beta_k & 0 \\ 0 & \delta_i & 0 & \delta_j & 0 & \delta_k \\ \delta_i & \beta_i & \delta_j & \beta_j & \delta_k & \beta_k \end{bmatrix}$$

$$[\boldsymbol{\nu}] = \frac{E}{1-\nu^2}\begin{bmatrix} 1 & \nu & 0 \\ \nu & 1 & 0 \\ 0 & 0 & \dfrac{1-\nu}{2} \end{bmatrix}$$

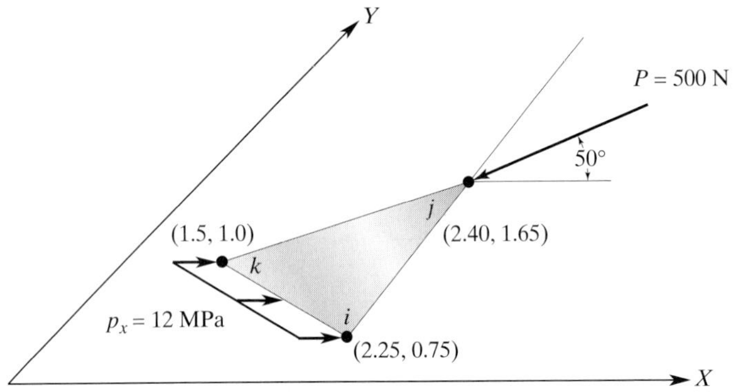

FIGURE 10.12 The loading and nodal coordinates for the element in Example 10.2.

Thus,

$$\beta_i = Y_j - Y_k = 1.65 - 1.0 = 0.65 \qquad \delta_i = X_k - X_j = 1.50 - 2.40 = -0.9$$
$$\beta_j = Y_k - Y_i = 1.0 - 0.75 = 0.25 \qquad \delta_j = X_i - X_k = 2.25 - 1.5 = 0.75$$
$$\beta_k = Y_i - Y_j = 0.75 - 1.65 = -0.9 \quad \delta_k = X_j - X_i = 2.40 - 2.25 = 0.15$$

and

$$2A = X_i(Y_j - Y_k) + X_j(Y_k - Y_i) + X_k(Y_i - Y_j)$$
$$2A = 2.25(1.65 - 1.0) + 2.40(1.0 - 0.75) + 1.5(0.75 - 1.65) = 0.7125$$

Substituting appropriate values into the above matrices, we have

$$[\mathbf{B}] = \frac{1}{0.7125}\begin{bmatrix} 0.65 & 0 & 0.25 & 0 & -0.9 & 0 \\ 0 & -0.9 & 0 & 0.75 & 0 & 0.15 \\ -0.9 & 0.65 & 0.75 & 0.25 & 0.15 & -0.9 \end{bmatrix}$$

$$[\mathbf{B}]^T = \frac{1}{0.7125}\begin{bmatrix} 0.65 & 0 & -0.9 \\ 0 & -0.9 & 0.65 \\ 0.25 & 0 & 0.75 \\ 0 & 0.75 & 0.25 \\ -0.9 & 0 & 0.15 \\ 0 & 0.15 & -0.9 \end{bmatrix}$$

$$[\boldsymbol{\nu}] = \frac{200 \times 10^5 \dfrac{\text{N}}{\text{cm}^2}}{1 - (0.32)^2}\begin{bmatrix} 1 & 0.32 & 0 \\ 0.32 & 1 & 0 \\ 0 & 0 & \dfrac{1 - 0.32}{2} \end{bmatrix} = \begin{bmatrix} 22281640 & 7130125 & 0 \\ 7130125 & 22281640 & 0 \\ 0 & 0 & 7575758 \end{bmatrix}$$

Carrying out the matrix operations results in the element stiffness matrix:

$$[\mathbf{K}]^{(e)} = \frac{(0.3)\left(\frac{0.7125}{2}\right)}{(0.7125)^2}\begin{bmatrix} 0.65 & 0 & -0.9 \\ 0 & -0.9 & 0.65 \\ 0.25 & 0 & 0.75 \\ 0 & 0.75 & 0.25 \\ -0.9 & 0 & 0.15 \\ 0 & 0.15 & -0.9 \end{bmatrix}\begin{bmatrix} 22281640 & 7130125 & 0 \\ 7130125 & 22281640 & 0 \\ 0 & 0 & 7575758 \end{bmatrix}$$

$$\begin{bmatrix} 0.65 & 0 & 0.25 & 0 & -0.9 & 0 \\ 0 & -0.9 & 0 & 0.75 & 0 & 0.15 \\ -0.9 & 0.65 & 0.75 & 0.25 & 0.15 & -0.9 \end{bmatrix}$$

Simplifying, we obtain

$$[\mathbf{K}]^{(e)} = \begin{bmatrix} 3273759 & -1811146 & -314288 & 372924 & -2959471 & 1438221 \\ -1811146 & 4473449 & 439769 & -2907167 & 1371376 & -1566282 \\ -314288 & 439769 & 1190309 & 580495 & -876020 & -1020265 \\ 372924 & -2907167 & 580495 & 2738296 & -953420 & 168871 \\ -2959471 & 1371376 & -876020 & -953420 & 3835491 & -417957 \\ 1438221 & -1566282 & -1020265 & 168871 & -417957 & 1397411 \end{bmatrix} (\text{N/cm})$$

The load matrix due to the distributed load is

$$\{\boldsymbol{F}\}^{(e)} = \frac{tL_{ik}}{2}\begin{Bmatrix} p_x \\ p_y \\ 0 \\ 0 \\ p_x \\ p_y \end{Bmatrix} = \frac{(0.3)\sqrt{(2.25-1.5)^2+(0.75-1.0)^2}}{2}\begin{Bmatrix} 1200 \\ 0 \\ 0 \\ 0 \\ 1200 \\ 0 \end{Bmatrix} = \begin{Bmatrix} 142 \\ 0 \\ 0 \\ 0 \\ 142 \\ 0 \end{Bmatrix}$$

The load matrix due to the concentrated load is

$$\{\boldsymbol{F}\}^{(e)} = \begin{Bmatrix} 0 \\ 0 \\ Q_{jx} \\ Q_{jy} \\ 0 \\ 0 \end{Bmatrix} = \begin{Bmatrix} 0 \\ 0 \\ -500\cos(50) \\ -500\sin(50) \\ 0 \\ 0 \end{Bmatrix} = \begin{Bmatrix} 0 \\ 0 \\ -321 \\ -383 \\ 0 \\ 0 \end{Bmatrix}$$

The complete load matrix for the element is

$$\{F\}^{(e)} = \begin{Bmatrix} 142 \\ 0 \\ -321 \\ -383 \\ 142 \\ 0 \end{Bmatrix} (\text{N})$$

10.4 AXISYMMETRIC FORMULATION

In this section we briefly discuss the formulation of a stiffness matrix using axisymmetric triangular elements. The steps that we are about to take are similar to the steps shown in Section 10.2, where we expressed the state of the stress at a point using a Cartesian coordinate system. However, as you saw in Chapter 9, Section 9.4, axisymmetric formulations require the use of a cylindrical coordinate system. Moreover, because geometry and loading are symmetrical about the z-axis, the state of stress and strain at a point are defined by only the following components:

$$[\sigma]^T = [\sigma_{rr} \quad \sigma_{zz} \quad \tau_{rz} \quad \sigma_{\theta\theta}] \tag{10.54}$$

$$[\varepsilon]^T = [\varepsilon_{rr} \quad \varepsilon_{zz} \quad \gamma_{rz} \quad \varepsilon_{\theta\theta}] \tag{10.55}$$

The relationship between the strains and the displacements are given by

$$\varepsilon_{rr} = \frac{\partial u}{\partial r} \quad \varepsilon_{zz} = \frac{\partial w}{\partial z} \quad \gamma_{rz} = \frac{\partial u}{\partial z} + \frac{\partial w}{\partial r} \quad \varepsilon_{\theta\theta} = \frac{u}{r} \tag{10.56}$$

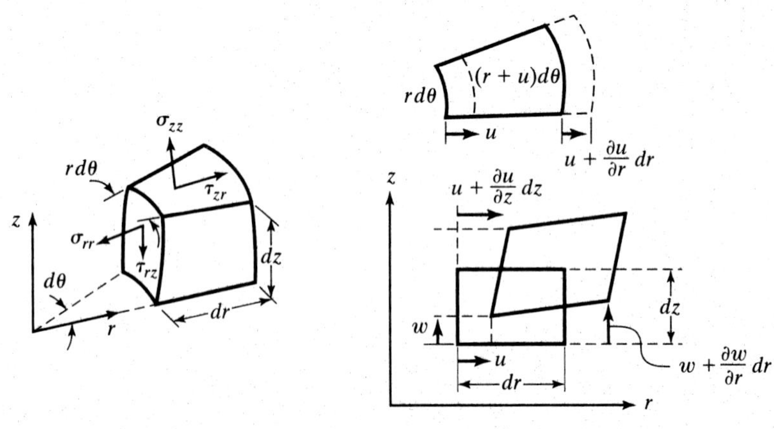

FIGURE 10.13 The component of the stress and displacements in a cylindrical coordinate system.

See Figure 10.13 to help you visualize the deformations. The relationship between the stress and the strain is given by generalized Hooke's law according to

$$\begin{Bmatrix} \sigma_{rr} \\ \sigma_{zz} \\ \tau_{rz} \\ \sigma_{\theta\theta} \end{Bmatrix} = \frac{E(1-\nu)}{(1+\nu)(1-2\nu)} \begin{bmatrix} 1 & \dfrac{\nu}{1-\nu} & 0 & \dfrac{\nu}{1-\nu} \\ \dfrac{\nu}{1-\nu} & 1 & 0 & \dfrac{\nu}{1-\nu} \\ 0 & 0 & \dfrac{1-2\nu}{2(1-\nu)} & 0 \\ \dfrac{\nu}{1-\nu} & \dfrac{\nu}{1-\nu} & 0 & 1 \end{bmatrix} \begin{Bmatrix} \varepsilon_{rr} \\ \varepsilon_{zz} \\ \gamma_{rz} \\ \varepsilon_{\theta\theta} \end{Bmatrix} \quad (10.57)$$

We can express Eq. (10.57) in a compact matrix form by

$$\{\boldsymbol{\sigma}\} = [\boldsymbol{\nu}]\{\boldsymbol{\varepsilon}\} \quad (10.58)$$

The next step involves writing the displacement variables u and w in terms of axisymmetric triangular shape functions. The nodal displacements of the axisymmetric triangular element are shown in Figure 10.14.

$$\begin{Bmatrix} u \\ w \end{Bmatrix} = \begin{bmatrix} S_i & 0 & S_j & 0 & S_k & 0 \\ 0 & S_i & 0 & S_j & 0 & S_k \end{bmatrix} \begin{Bmatrix} U_{ir} \\ U_{iz} \\ U_{jr} \\ U_{jz} \\ U_{kr} \\ U_{kz} \end{Bmatrix} \quad (10.59)$$

The expressions for the shape functions are given in Eq. (7.27). As a next step we need to evaluate the strains from the strain displacement relationships.

$$\varepsilon_{rr} = \frac{\partial u}{\partial r} = \frac{\partial}{\partial r}[S_i U_{ir} + S_j U_{jr} + S_k U_{kr}] \quad (10.60)$$

$$\varepsilon_{zz} = \frac{\partial w}{\partial z} = \frac{\partial}{\partial z}[S_i U_{iz} + S_j U_{jz} + S_k U_{kz}] \quad (10.61)$$

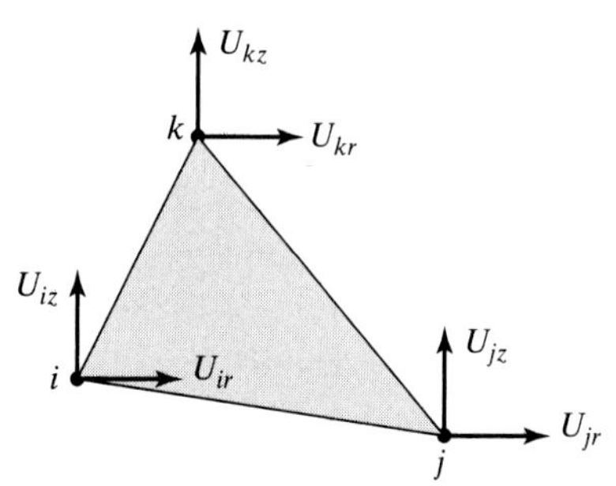

FIGURE 10.14 The nodal displacements of an axisymmetric triangular element.

$$\gamma_{rz} = \frac{\partial u}{\partial z} + \frac{\partial w}{\partial r} = \frac{\partial}{\partial z}[S_iU_{ir} + S_jU_{jr} + S_kU_{kr}] + \frac{\partial}{\partial r}[S_iU_{iz} + S_jU_{jz} + S_kU_{kz}] \quad (10.62)$$

$$\varepsilon_{\theta\theta} = \frac{u}{r} = \frac{S_iU_{ir} + S_jU_{jr} + S_kU_{kr}}{r} \quad (10.63)$$

As we did in section 10.2, after taking the derivatives of the components of the displacement fields we can express the relationship between the strain components and the nodal displacements in a compact matrix form by

$$\{\boldsymbol{\varepsilon}\} = [\boldsymbol{B}]\{\boldsymbol{U}\} \quad (10.64)$$

We then substitute for the strain matrix (in terms of displacements) in the strain energy of the axisymmetric element that eventually leads to the stiffness matrix, which is computed from

$$[\boldsymbol{K}]^{(e)} = 2\pi \int [\boldsymbol{B}]^T[\nu][\boldsymbol{B}]r\,dA \quad (10.65)$$

Note that the resulting matrix is a 6×6 matrix.

10.5 BASIC FAILURE THEORIES

One of the goals of most structural solid analyses is to check for failure. The prediction of failure is quite complex in nature; consequently, many investigators have been studying this topic. This section presents a brief overview of some failure theories. For an in-depth review of failure theories, you are encouraged to study a good text on the mechanics of materials or on machine design. (For a good example of such a text, see *Shigley and Mischke, 1989*).

Using ANSYS, you can calculate the distribution of the stress components σ_x, σ_y, and τ_{xy}, and the principal stresses σ_1, and σ_2 within the material. But how would you decide whether or not the solid part you are analyzing will permanently deform or fail under the applied loading? You may recall from your previous study of the mechanics of materials that to compensate for what we do not know about the exact behavior of a material and/or to account for future loading for which we may have not accounted, but to which someone may subject the part, we introduce a *Factor of Safety* (F.S.), which is defined as

$$\text{F.S.} = \frac{P_{\text{max}}}{P_{\text{allowable}}} \quad (10.66)$$

where P_{max} is the load that can cause failure. For certain situations, it is also customary to define the factor of safety in terms of the ratio of maximum stress that causes failure to the allowable stresses if the applied loads are linearly related to the stresses. But how do we apply the knowledge of stress distributions in a material to predict failure? Let us begin by reviewing how the *principal stresses* and *maximum shear stresses* are computed. The in-plane principal stresses at a point are determined from the values of

σ_{xx}, σ_{yy}, and τ_{xy} at that point using the equation

$$\sigma_{1,2} = \frac{\sigma_x + \sigma_y}{2} \pm \sqrt{\left(\frac{\sigma_x - \sigma_y}{2}\right)^2 + \tau_{xy}^2} \tag{10.67}$$

The maximum in-plane shear stress at the point is determined from the relationship

$$\tau_{max} = \sqrt{\left(\frac{\sigma_x - \sigma_y}{2}\right)^2 + \tau_{xy}^2} \tag{10.68}$$

There are a number of failure criteria, including the maximum-normal-stress theory, the maximum-shear-stress theory, and the distortion-energy theory. The distortion-energy theory, often called the von Mises–Hencky theory, is one of the most commonly used criteria to predict failure of ductile materials. This theory is used to define the start of yielding. For design purposes, the von Mises stress σ_v is calculated according to the equation

$$\sigma_v = \sqrt{\sigma_1^2 - \sigma_1\sigma_2 + \sigma_2^2} \tag{10.69}$$

A safe design is one that keeps the von Mises stresses in the material below the yield strength of the material. The relationship among the von Mises stress, the yield strength, and the factor of safety is

$$\sigma_v = \frac{S_Y}{\text{F.S.}} \tag{10.70}$$

where S_Y is the yield strength of the material, obtained from a tension test. Most brittle materials have a tendency to fail abruptly without any yielding. For a brittle material under plane-stress conditions, the maximum-normal-stress theory states that the material will fail if any point within the material experiences principal stresses exceeding the ultimate normal strength of the material. This idea is represented by the equations

$$|\sigma_1| = S_{\text{ultimate}} \qquad |\sigma_2| = S_{\text{ultimate}} \tag{10.71}$$

where S_{ultimate} is the ultimate strength of the material, obtained from a tension test. The maximum-normal-stress theory may not produce reasonable predictions for materials with different tension and compression properties; in such structures, consider using the *Mohr failure criteria* instead.

10.6 EXAMPLES USING ANSYS

ANSYS offers a number of elements that can be used to model two-dimensional solid-structural problems. Some of these elements were introduced in Chapter 7. The two-dimensional solid-structural elements in ANSYS include PLANE2, PLANE42, and PLANE82.

PLANE2 is a six-node triangular structural-solid element. The element has quadratic displacement behavior, with two degrees of freedom at each node: translation in the nodal x- and y-directions. The element input data can include thickness if the KEYOPTION 3 (plane stress with thickness input) is selected. Surface-pressure loads may be applied to element faces. Output data include nodal displacements and elemental data, such as directional stresses and principal stresses.

PLANE42 is a four-node quadrilateral element used to model solid problems. The element is defined by four nodes, with two degrees of freedom at each node: translation in the x- and y-directions. The element input data can include thickness if the KEYOPTION 3 (plane stress with thickness input) is selected. Surface-pressure loads may be applied to element faces. Output data include nodal displacements and elemental data, such as directional stresses and principal stresses.

PLANE82 is an eight-node quadrilateral element used to model two-dimensional structural-solid problems. It is a higher order version of the two-dimensional, four-node quadrilateral PLANE42 element. This element offers more accuracy when modeling problems with curved boundaries. At each node, there are two degrees of freedom: translation in the x- and y-directions. The element input data can include thickness if the KEYOPTION 3 (plane stress with thickness input) is selected. Surface-pressure loads may be applied to element faces. Output data include nodal displacements and elemental data, such as directional stresses and principal stresses.

As the theory in Section 10.1 suggested, because of similarities between the governing equation of torsional problems and heat transfer, in addition to the elements listed above, you can use thermal solid elements (e.g., PLANE35, a six-node triangular element; PLANE55, a four-node quadrilateral element; or PLANE77, an eight-node quadrilateral element) to model torsional problems. However, when using the solid thermal elements, make sure that the appropriate values are supplied to the property fields and the boundary conditions. Example 10.1 (revisited) demonstrates this point.

EXAMPLE 10.1 Revisited

Consider Example 10.1, which deals with the torsion of a steel bar ($G = 11 \times 10^3$ ksi) having a rectangular cross section, as shown in the accompanying figure. Assuming that $\theta = 0.0005$ rad/in, and using ANSYS, we are interested in determining the location and magnitude of the maximum shear stress. We will then compare the solution generated with ANSYS to the exact solution for a straight rectangular bar, as discussed in Section 10.1.

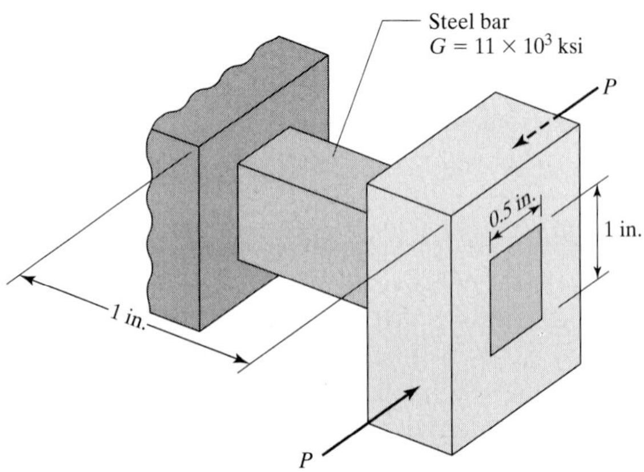

Enter the **ANSYS** program by using the Launcher. Type **Torsion** (or a file name of your choice) in the **Jobname** entry field of the dialog box. Pick **Run** to start the GUI.

Create a title for the problem.

utility menu: **File → Change Title** . . .

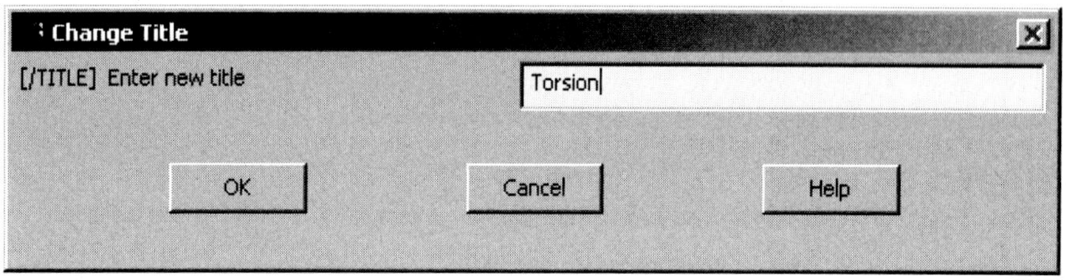

main menu: **Preprocessor → Element Type → Add/Edit/Delete**

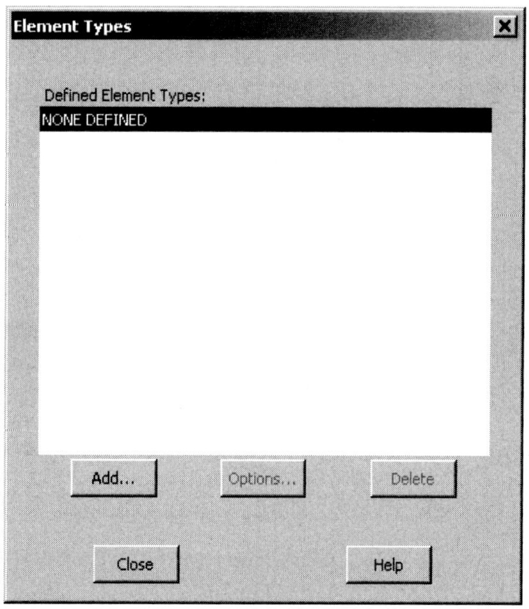

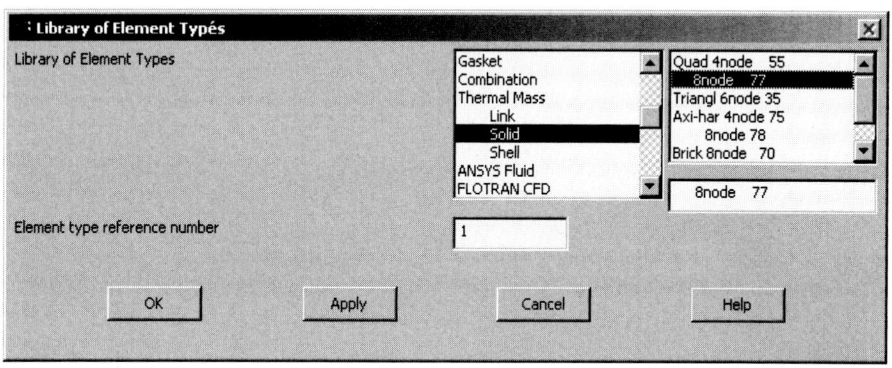

main menu: **Preprocessor → Material Props → Material Models → Thermal → Conductivity → Isotropic**

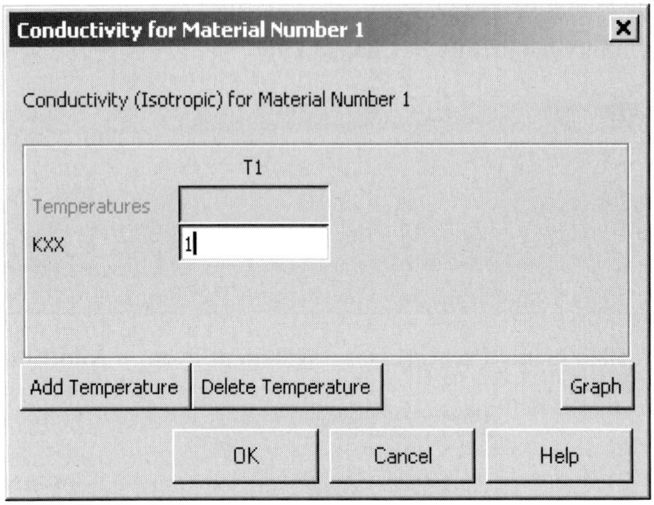

ANSYS Toolbar: **SAVE_DB**

main menu: **Preprocessor → Modeling → Create → Areas → Rectangle → by 2 Corners**

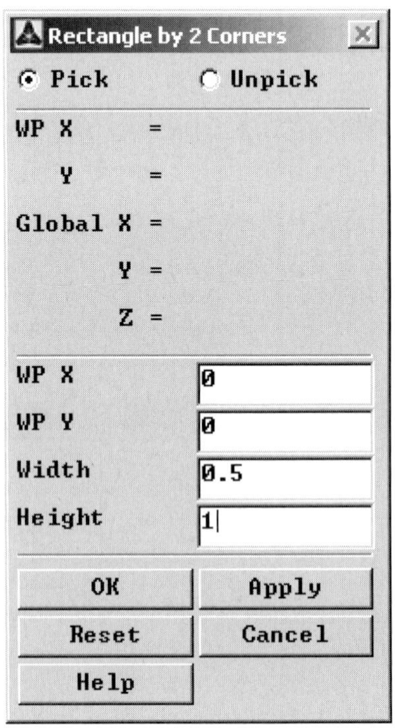

main menu: **Preprocessor → Meshing → Size Cntrls → Smart Size → Basic**

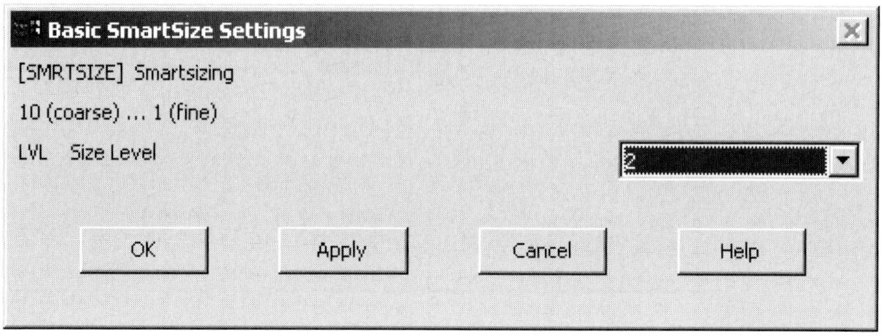

main menu: **Preprocessor → Meshing → Mesh → Areas → Free**

Pick All

main menu: **Solution → Define Loads → Apply → Thermal → Temperature → On Line**

Pick the edges of the rectangle and apply a zero constant temperature boundary condition on all four edges.

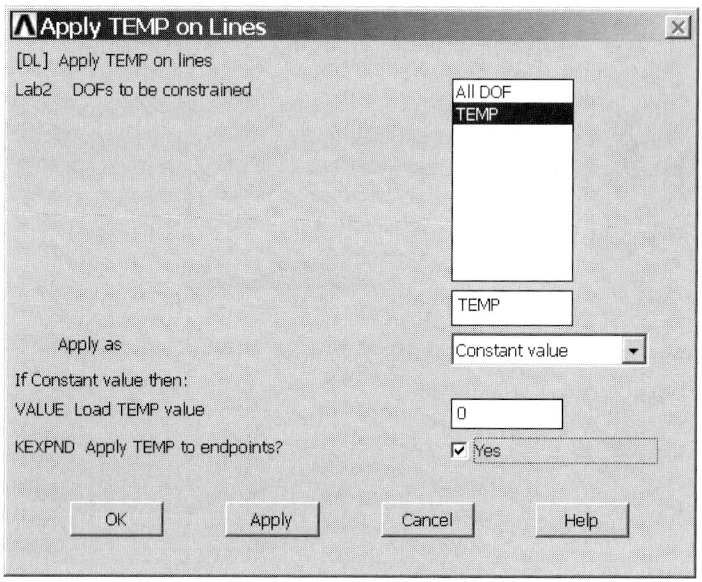

main menu: **Solution → Define Loads → Apply → Thermal → Heat Generat → On Area**

Pick the area of the rectangle and apply a constant value of $2G\theta = 11000$.

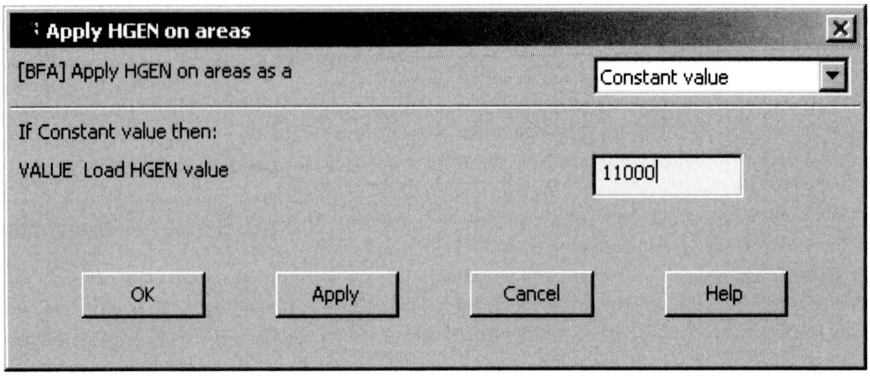

main menu: **Solution → Solve → Current LS**

OK
Close (the solution is done!) window.
Close (the /STAT Command) window.

main menu: **General Postproc → Plot Results → Contour Plot → Nodal Solu**

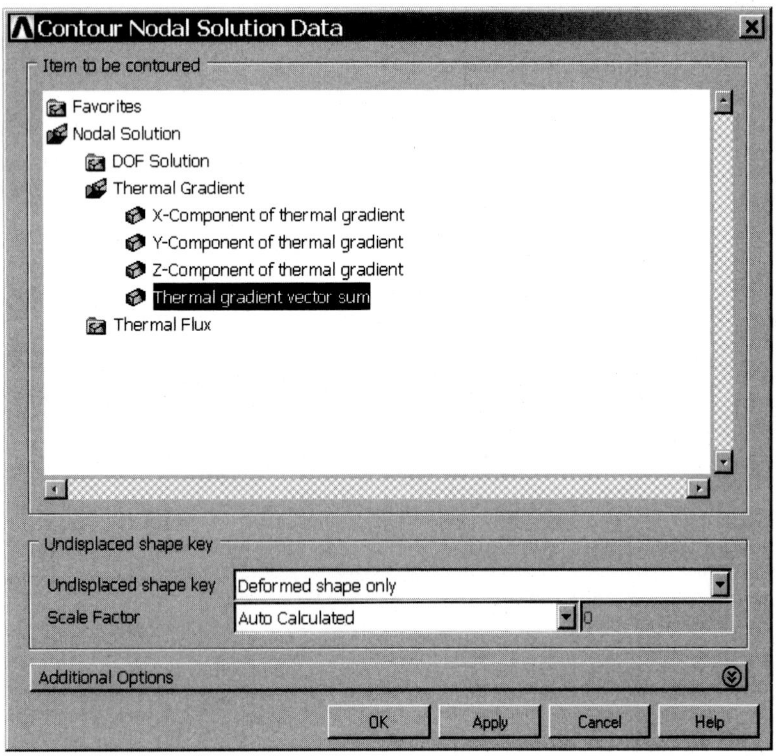

The solution is then given in the following figure.

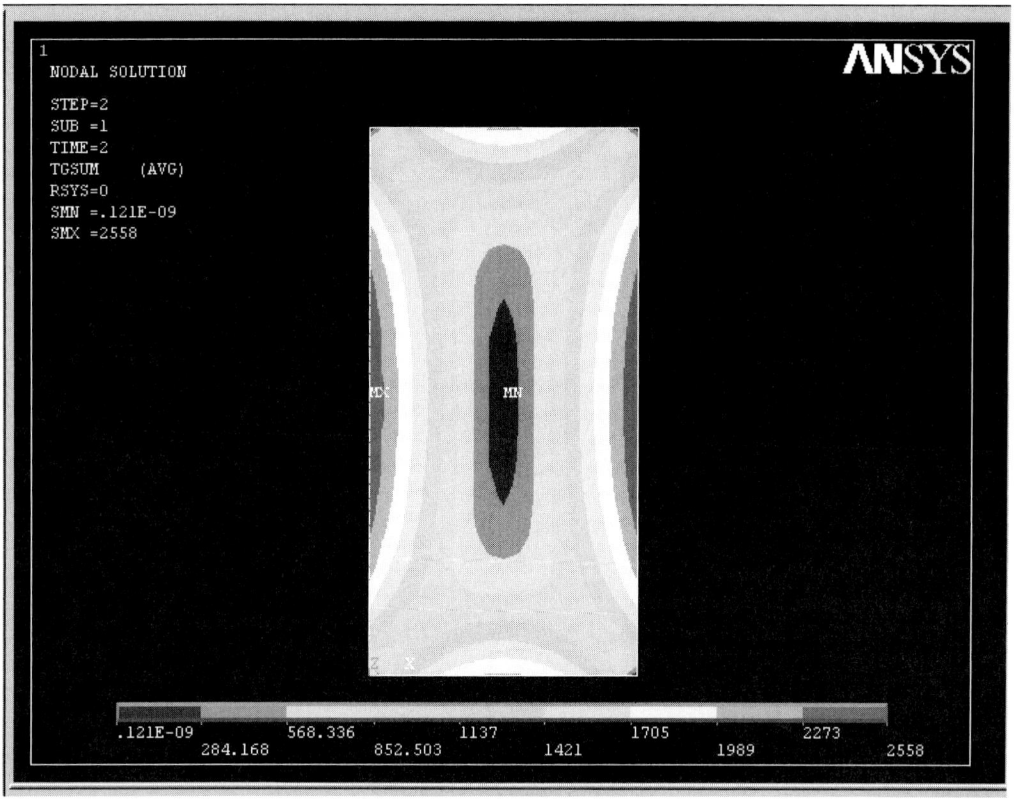

By calculating the aspect ratio of the cross section $\frac{w}{h} = \frac{1.0 \text{ in.}}{0.5 \text{ in.}} = 2.0$ and consulting Table 10.1, we find that $c_1 = 0.246$ and $c_2 = 0.229$. Substituting for G, w, h, θ, c_1, and c_2 in Eqs. (10.3) and (10.4), we get

$$\theta = \frac{TL}{c_2 G w h^3} = 0.0005 \text{ rad/in.} = \frac{T(1 \text{ in.})}{0.229(11 \times 10^6 \text{lb/in}^2)(1 \text{ in.})(0.5 \text{ in.})^3} \Rightarrow T = 157.5 \text{ lb.in}$$

$$\tau_{max} = \frac{T}{c_1 w h^2} = \frac{157.5 \text{ lb.in}}{0.246(1 \text{ in.})(0.5 \text{ in.})^2} = 2560 \text{ lb/in}^2$$

When comparing 2560 lb/in^2 to the FEA results of 2558 lb/in^2, you see that we could have saved lots of time by calculating the maximum shear stress using the analytical solution and avoided generating a finite element model.

EXAMPLE 10.3

The bicycle wrench shown in Figure 10.15 is made of steel with a modulus of elasticity $E = 200\text{ GPa}\left(200 \times 10^5 \dfrac{N}{\text{cm}^2}\right)$ and a Poisson's ratio $\nu = 0.32$. The wrench is 3 mm (0.3 cm) thick. Determine the von Mises stresses under the given distributed load and boundary conditions.

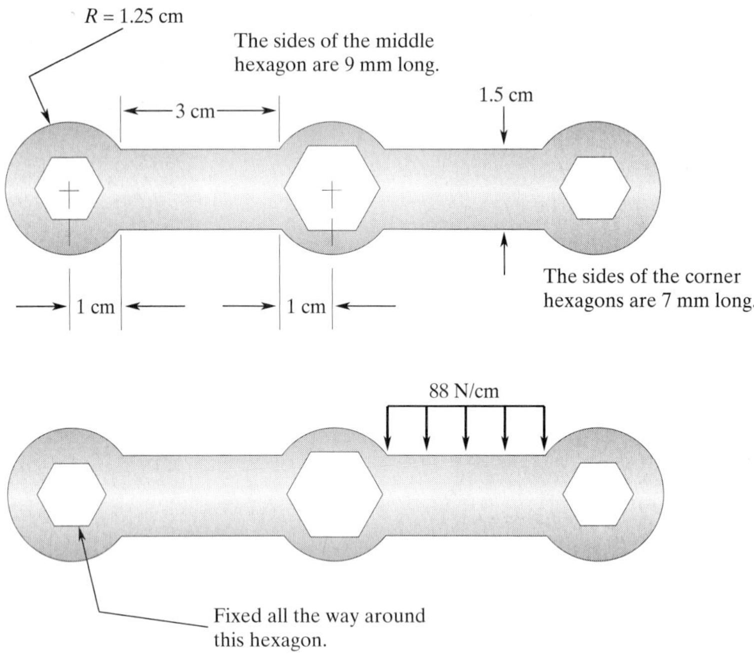

FIGURE 10.15 A schematic for the bicycle wrench in Example 10.3.

The following steps demonstrate how to (1) create the geometry of the problem, (2) choose the appropriate element type, (3) apply boundary conditions, and (4) obtain nodal results:

Enter the **ANSYS** program by using the Launcher. Type **Bikewh** (or a file name of your choice) in the **Jobname** entry field of the dialog box. Pick **Run** to start the GUI.

Create a title for the problem. This title will appear on ANSYS display windows to provide a simple way to identify the displays. So, issue the command utility menu: **File → Change Title** . . .

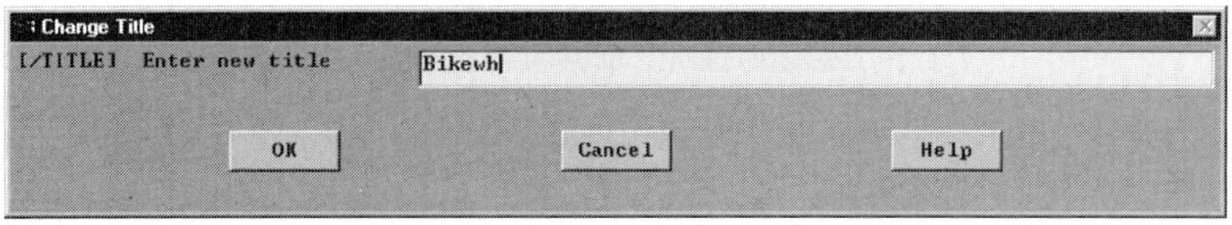

Define the element type and material properties with the following commands:

main menu: **Preprocessor → Element Type → Add/Edit/Delete**

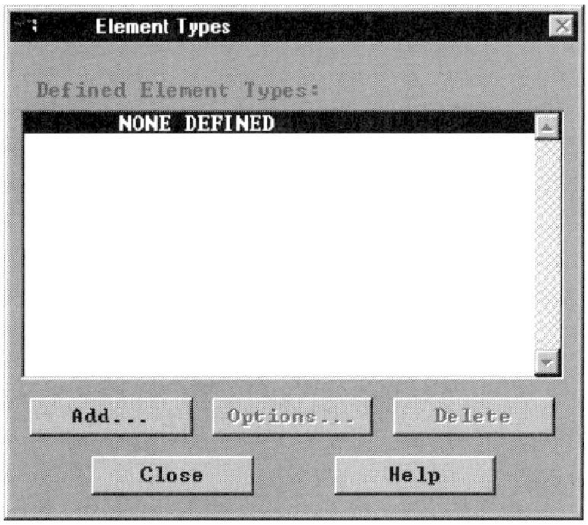

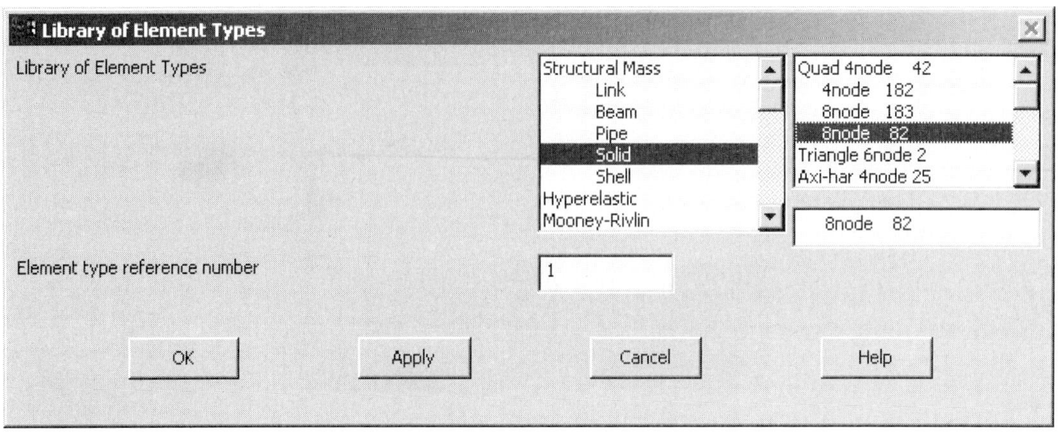

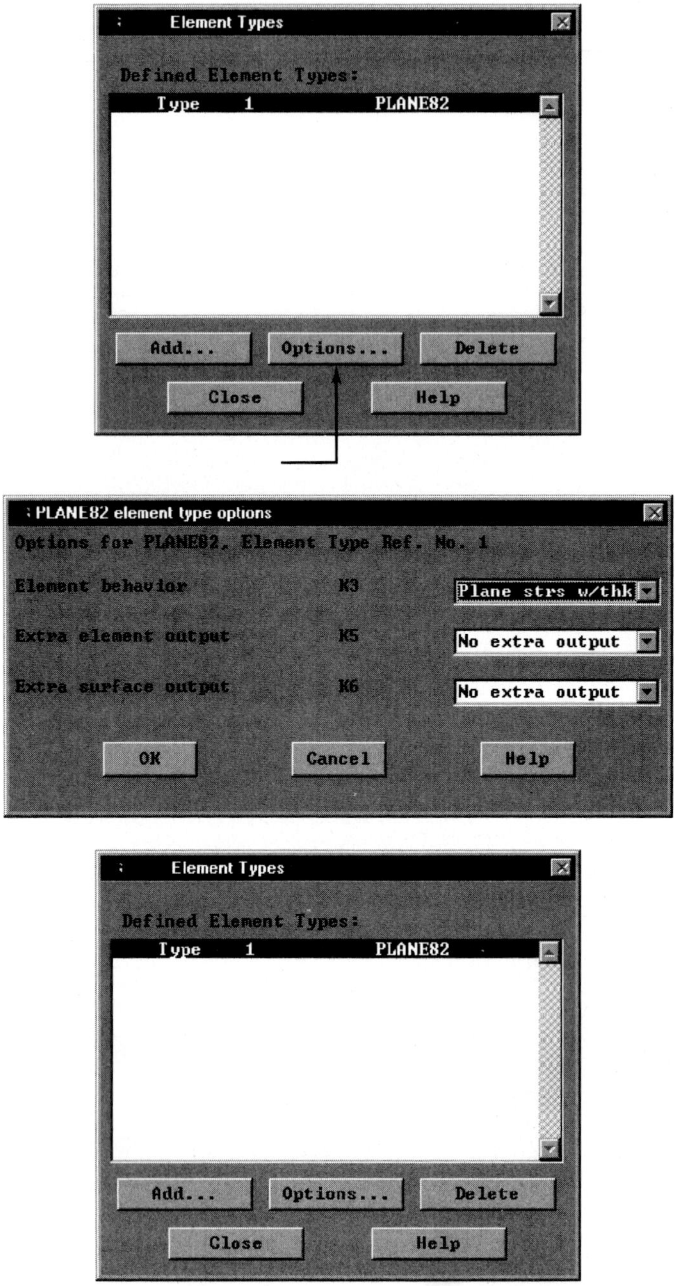

Assign the thickness of the wrench with the following commands:

main menu: **Preprocessor → Real Constants → Add/Edit/Delete**

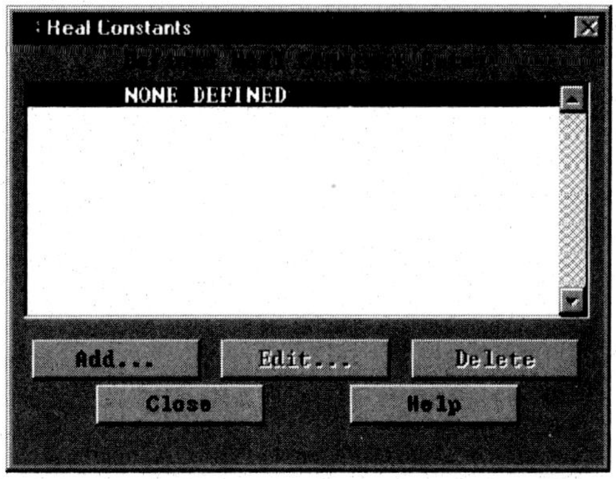
Real Constants
NONE DEFINED
Add...
Edit...
Delete
Close
Help

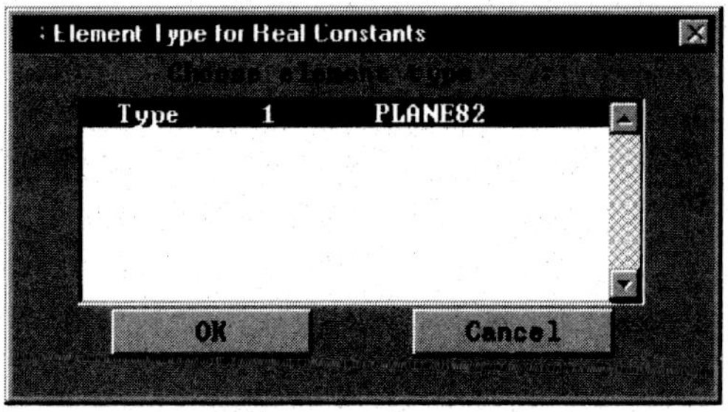
Element Type for Real Constants
Type 1 PLANE82
OK
Cancel

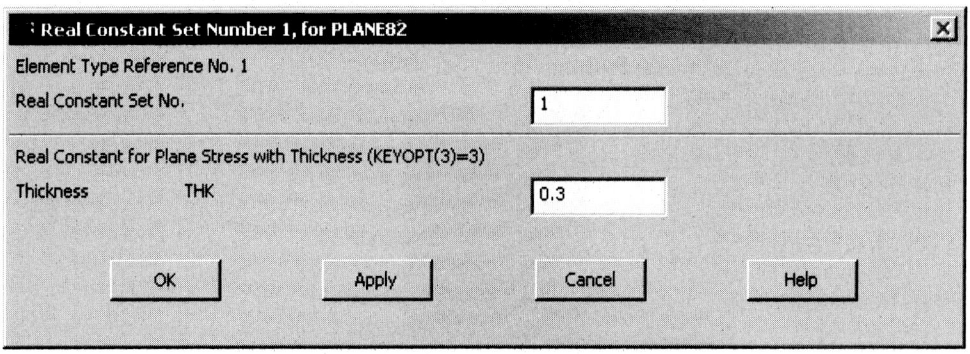
Real Constant Set Number 1, for PLANE82
Element Type Reference No. 1
Real Constant Set No.
1
Real Constant for Plane Stress with Thickness (KEYOPT(3)=3)
Thickness THK
0.3
OK
Apply
Cancel
Help

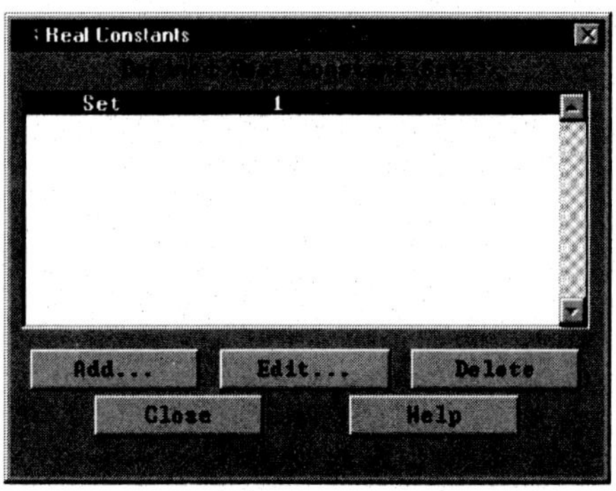
Real Constants
Set 1
Add...
Edit...
Delete
Close
Help

Assign the modulus of elasticity and Poisson's ratio values by using the following commands:

main menu: **Preprocessor → Material Props → Material Models → Structural → Linear → Elastic → Isotropic**

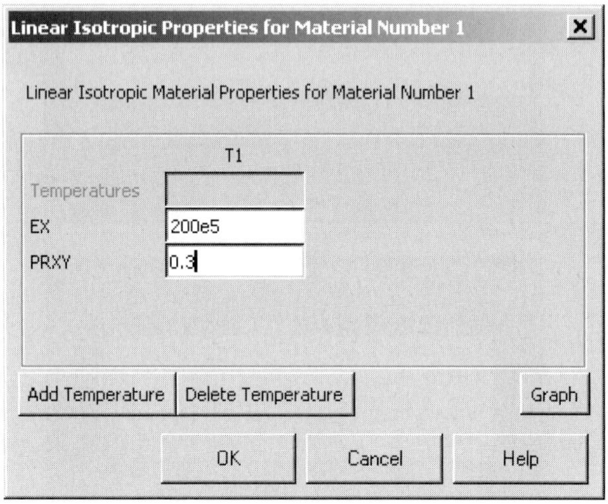

ANSYS Toolbar: **SAVE_DB**

Set up the graphics area (i.e., workplane, zoom, etc.) with the following commands:

utility menu: **Workplane → Wp Settings** . . .

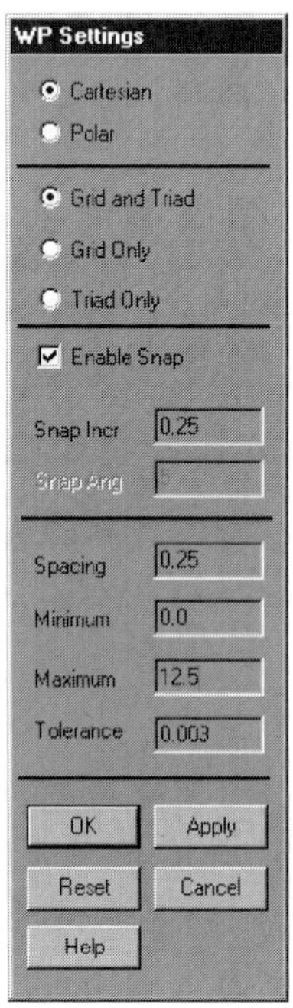

Toggle on the workplane by using the command

utility menu: **Workplane → Display Working Plane**

Bring the workplane to view by using the command

utility menu: **PlotCtrls → Pan, Zoom, Rotate** . . .

Click on the small circle until you bring the work plane to view. Then, create the geometry with the following commands:

main menu: **Preprocessor → Modeling → Create →**

Areas → Rectangles → By 2 Corners

On the workplane, create the two rectangles:

Use the mouse buttons as shown below, or type the values in the appropriate fields.

[WP = 2.25, 0.5]

[Expand the rubber band up 1.5 and right 3.0]

[WP = 7.25, 0.5]

[Expand the rubber band up 1.5 and right 3.0]

OK

Create the circles with the following commands:

main menu: **Preprocessor → Modeling → Create → Areas →**

Circle → Solid Circle

[WP = 1.25,1.25]

[Expand the rubber band to a radius of 1.25]

[WP = 6.25,1.25]

[Expand the rubber band to a radius of 1.25]

[WP = 11.25, 1.25]

[Expand the rubber band to a radius of 1.25]

OK

Add the areas together with the commands:

main menu: **Preprocessor → Modeling → Operate → Booleans → Add → Areas**

Click on the **Pick All** button, and then create the hexagons. First, change the **Snap Incr** in the WP Settings dialog box to 0.1 with the command

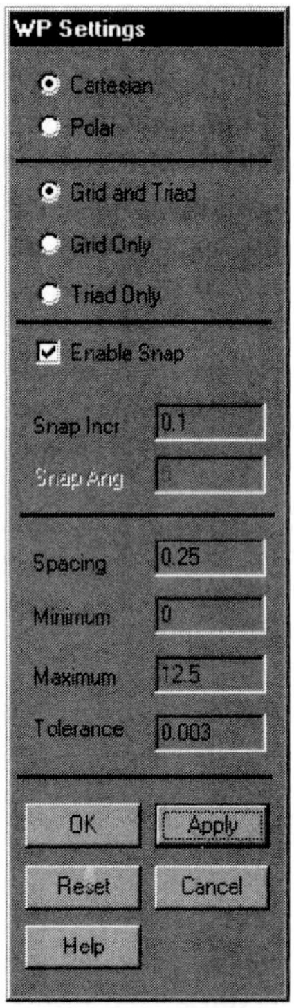

utility menu: **PlotCtrls → Pan, Zoom, Rotate . . .**

Click on the **Box Zoom**, and put a box zoom around the left circle, then using the following commands create the hexagon:

main menu: **Preprocessor → Modeling → Create → Areas → Polygon → Hexagon**

Use the mouse buttons as shown below, or type the values in the appropriate fields:

[1.25, 1.25]

[Expand the hexagon to WP Rad = 0.7, Ang = 120]

or

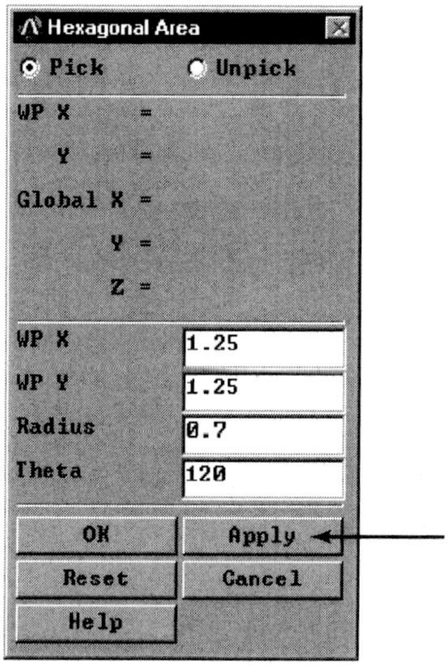

Then, issue the command

utility menu: **PlotCtrls → Pan, Zoom, Rotate** . . .

Click on the **Fit** button. Then, click on the **Box Zoom**, and put a box zoom around the center circle. Use the mouse buttons as shown below, or type the values in the appropriate fields:

[6.25, 1.25]

[Expand the hexagon to WP Rad = 0.9, Ang = 120]

or

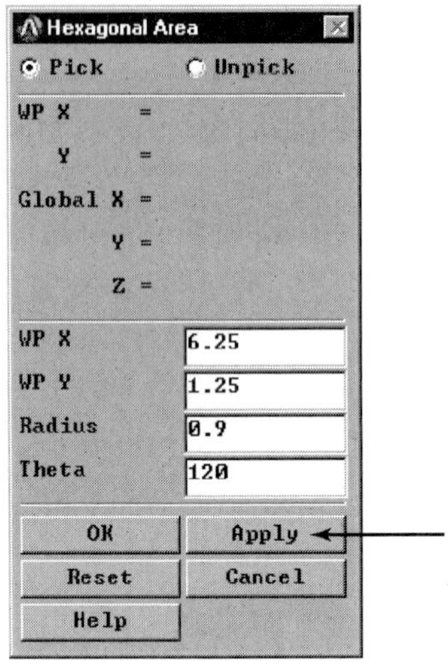

utility menu: **PlotCtrls → Pan, Zoom, Rotate** . . .

Click on the **Fit** button. Then, click on the **Box Zoom**, and put a box zoom around the right-end circle. Use the mouse buttons as shown below, or type the values in appropriate fields:

[11.25, 1.25]

[Expand the hexagon to WP Rad = 0.7, Ang = 120]

or

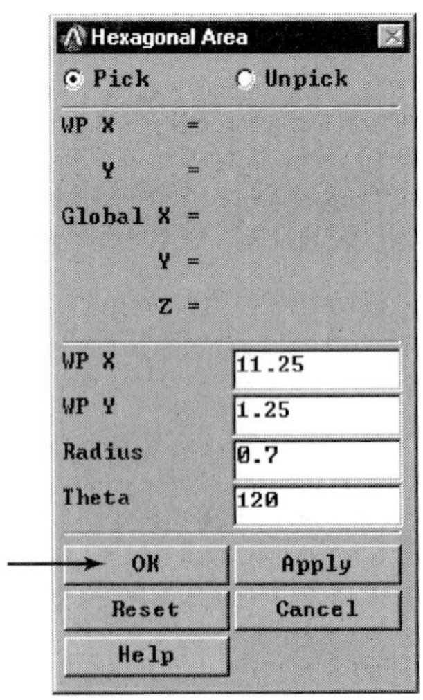

ANSYS Toolbar: **SAVE_DB**

Subtract the areas of the hexagons to create the driver holes:

main menu: **Preprocessor → Modeling → Operate → Booleans → Subtract → Areas**

[Pick the solid area of the wrench]

[Apply anywhere in the ANSYS graphics area]

[pick the left hexagon area]

[pick the center hexagon area]

[pick the right hexagon area]

[Apply anywhere in the ANSYS graphics area]

OK

Now you can toggle off the workplane grids with the following command:

utility menu: **Workplane → Display Working Plane**

ANSYS Toolbar: **SAVE_DB**

You are now ready to mesh the area of the bracket to create elements and nodes. So, issue the following commands:

main menu: **Preprocessor → Meshing → Size Cntrls → Manual Size → Global → Size**

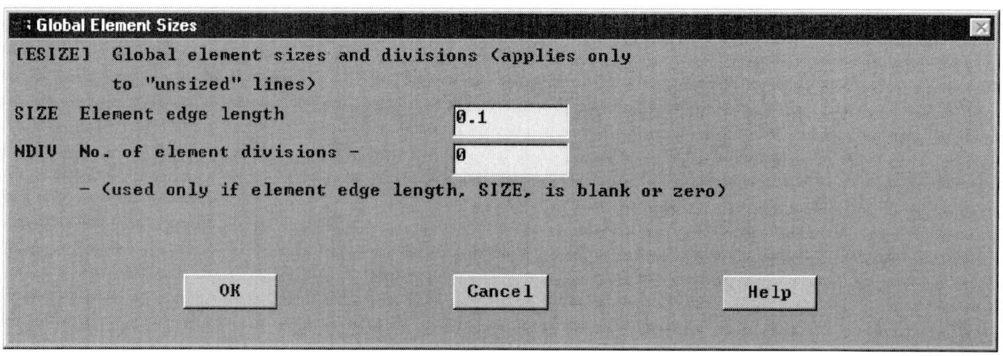

ANSYS Toolbar: **SAVE_DB**

main menu: **Preprocessor → Meshing → Mesh → Areas → Free**

Click on the **Pick All** button.

OK

Apply the boundary conditions and the load:

main menu: **Solution → Define Loads → Apply → Structural → Displacements → On Keypoints**

Pick the six corner keypoints of the left hexagon:

OK

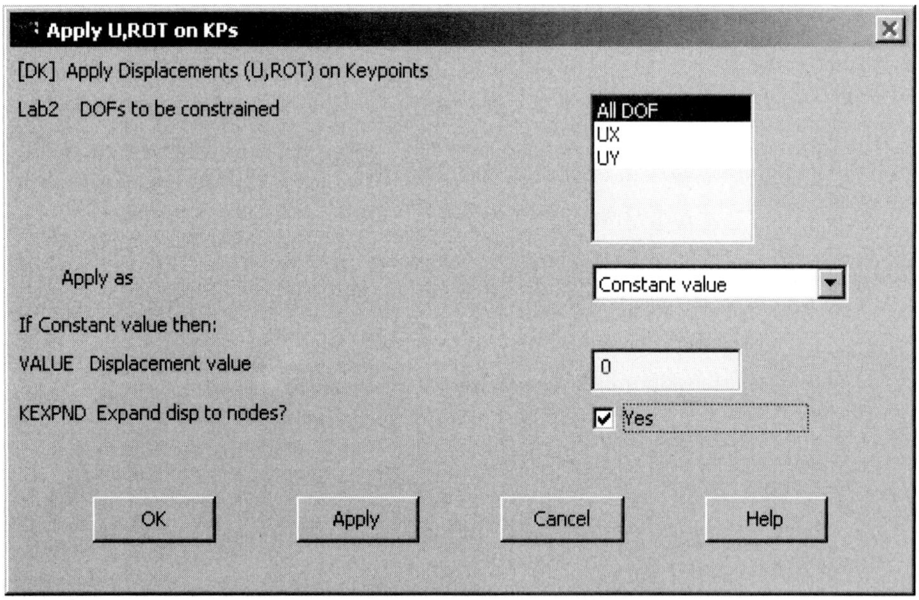

main menu: **Solution → Define Loads → Apply → Structural → Pressure → On Lines**

Pick the appropriate horizontal line, as shown in the problem statement:

OK

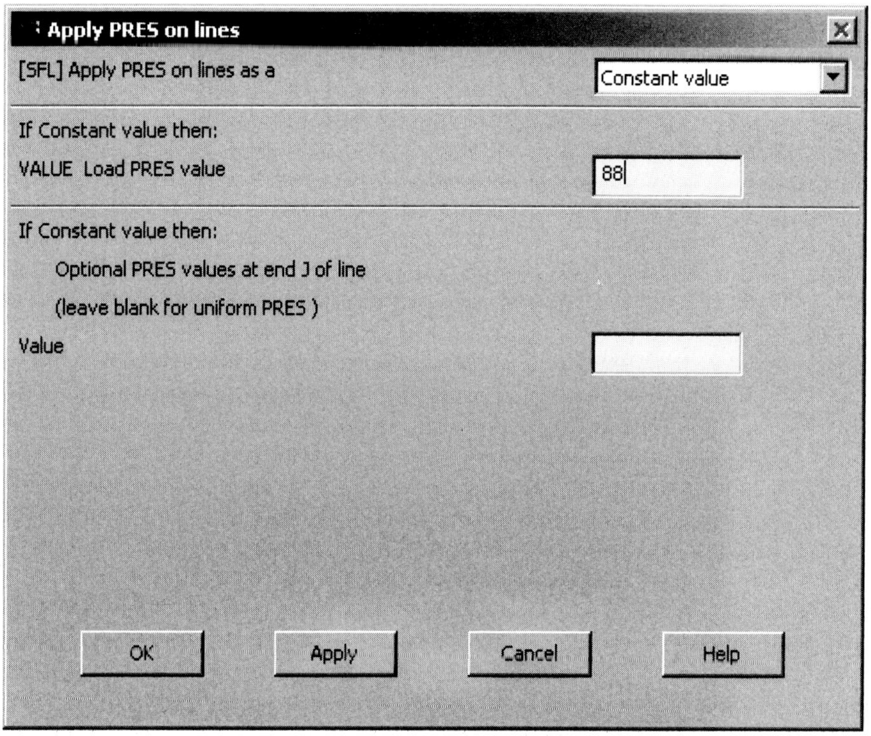

Solve the problem:

main menu: **Solution → Solve → Current LS**

OK

Close (the solution is done!) window.

Close (the STAT Command) window.

Begin the postprocessing phase and plot the deformed shape with the following commands:

main menu: **General Postproc → Plot Results → Deformed Shape**

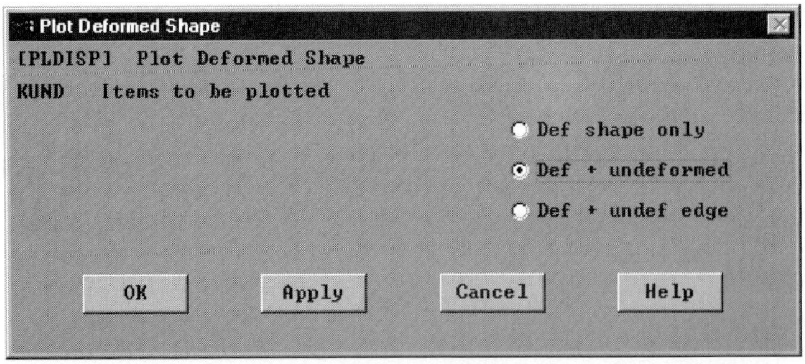

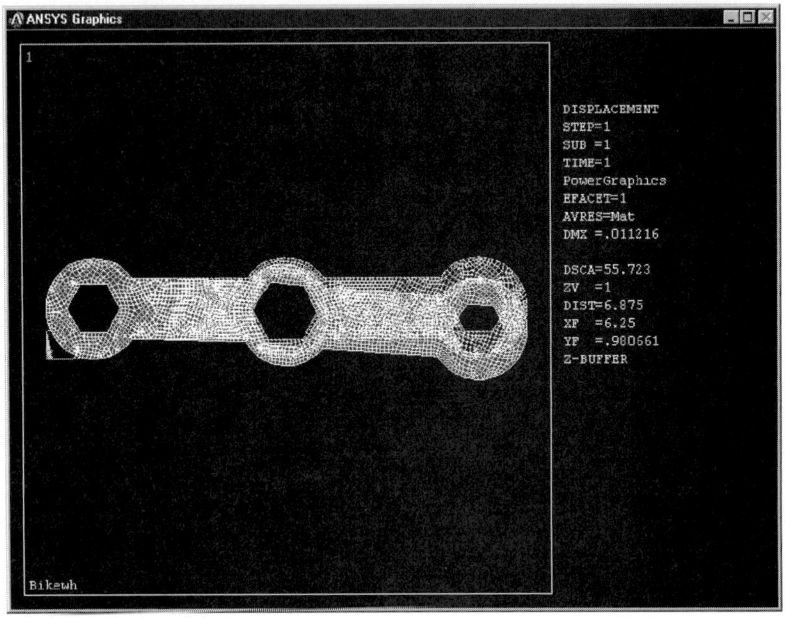

Plot the von Mises stresses with the following commands:

main menu: **General Postproc → Plot Results →**

Contour Plot → Nodal Solu

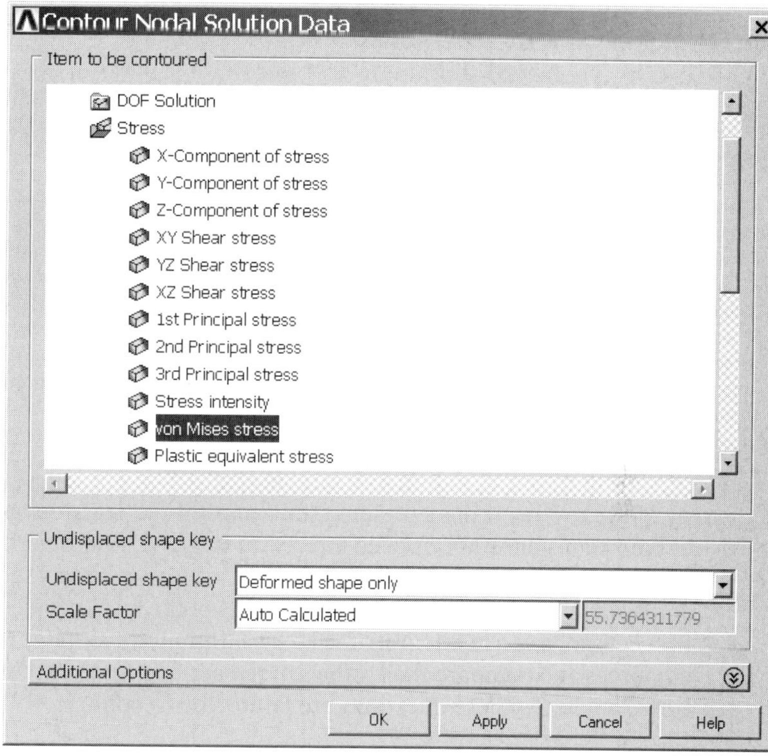

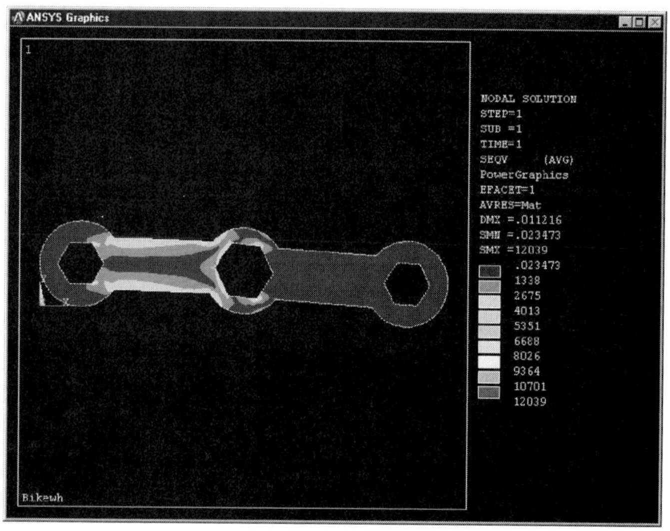

Exit ANSYS and save everything:

ANSYS Toolbar: **QUIT**

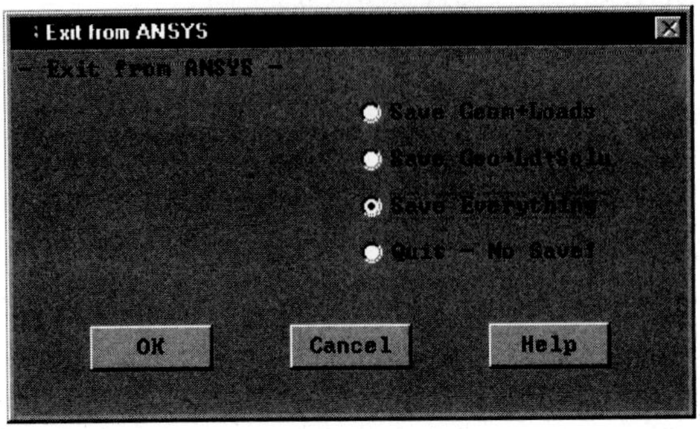

10.7 VERIFICATION OF RESULTS

Now we turn our attention to Example 10.3. There are several ways to check the validity of the results of this problem. You can print the reaction forces and check the value of their sum against the applied force. Are statics equilibrium conditions satisfied? Using the path operations of ANSYS, you can also cut an arbitrary section through the wrench and visually assess the x- and y-components of the local stresses and shear stresses along the section. You can integrate the stress information along the path to obtain the internal forces and compare their values to the applied force. Are statics equilibrium conditions satisfied? These questions are left to you to confirm.

SUMMARY

At this point you should

1. know that it is wise to use simple analytical solutions rather than finite element modeling for a simple problem whenever appropriate. Use finite element modeling only when it is necessary to do so. Simple analytical solutions are particularly appropriate when you are solving basic torsional problems.
2. know that the stiffness matrix for torsional problems is similar to the conductance matrix obtained for two-dimensional conduction problems. The stiffness matrix and the load matrix using a rectangular element are

$$[\mathbf{K}]^{(e)} = \frac{w}{6\ell}\begin{bmatrix} 2 & -2 & -1 & 1 \\ -2 & 2 & 1 & -1 \\ -1 & 1 & 2 & -2 \\ 1 & -1 & -2 & 2 \end{bmatrix} + \frac{\ell}{6w}\begin{bmatrix} 2 & 1 & -1 & -2 \\ 1 & 2 & -2 & -1 \\ -1 & -2 & 2 & 1 \\ -2 & -1 & 1 & 2 \end{bmatrix}$$

$$\{\mathbf{F}\}^{(e)} = \frac{2G\theta A}{4}\begin{Bmatrix} 1 \\ 1 \\ 1 \\ 1 \end{Bmatrix}$$

For triangular elements, the stiffness and load matrices are, respectively

$$[\mathbf{K}]^{(e)} = \frac{1}{4A}\begin{bmatrix} \beta_i^2 & \beta_i\beta_j & \beta_i\beta_k \\ \beta_i\beta_j & \beta_j^2 & \beta_j\beta_k \\ \beta_i\beta_k & \beta_j\beta_k & \beta_k^2 \end{bmatrix} + \frac{1}{4A}\begin{bmatrix} \delta_i^2 & \delta_i\delta_j & \delta_i\delta_k \\ \delta_i\delta_j & \delta_j^2 & \delta_j\delta_k \\ \delta_i\delta_k & \delta_j\delta_k & \delta_k^2 \end{bmatrix}$$

$$\{\mathbf{F}\}^{(e)} = \frac{2G\theta A}{3}\begin{Bmatrix} 1 \\ 1 \\ 1 \end{Bmatrix}$$

3. know that the stiffness matrix for a plane-stress triangular element is

$$[\mathbf{K}]^{(e)} = V[\mathbf{B}]^T[\boldsymbol{\nu}][\mathbf{B}]$$

where

$$V = tA$$

$$[\mathbf{B}] = \frac{1}{2A}\begin{bmatrix} \beta_i & 0 & \beta_j & 0 & \beta_k & 0 \\ 0 & \delta_i & 0 & \delta_j & 0 & \delta_k \\ \delta_i & \beta_i & \delta_j & \beta_j & \delta_k & \beta_k \end{bmatrix} \quad [\boldsymbol{\nu}] = \frac{E}{1-\nu^2}\begin{bmatrix} 1 & \nu & 0 \\ \nu & 1 & 0 \\ 0 & 0 & \dfrac{1-\nu}{2} \end{bmatrix}$$

and

$$\begin{aligned} \beta_i &= Y_j - Y_k \qquad \delta_i = X_k - X_j \\ \beta_j &= Y_k - Y_i \qquad \delta_j = X_i - X_k \\ \beta_k &= Y_i - Y_j \qquad \delta_k = X_j - X_i \\ 2A &= X_i(Y_j - Y_k) + X_j(Y_k - Y_i) + X_k(Y_i - Y_j) \end{aligned}$$

4. know that the load matrix due to a distributed load along the element's edges is

$$\{\mathbf{F}\}^{(e)} = \frac{tL_{ij}}{2}\begin{Bmatrix} p_x \\ p_y \\ p_x \\ p_y \\ 0 \\ 0 \end{Bmatrix} \quad \{\mathbf{F}\}^{(e)} = \frac{tL_{jk}}{2}\begin{Bmatrix} 0 \\ 0 \\ p_x \\ p_y \\ p_x \\ p_y \end{Bmatrix} \quad \{\mathbf{F}\}^{(e)} = \frac{tL_{ik}}{2}\begin{Bmatrix} p_x \\ p_y \\ 0 \\ 0 \\ p_x \\ p_y \end{Bmatrix}$$

5. understand how an element's stiffness matrix is obtained through the isoparametric formulation.

6. understand how an element's stiffness matrix is obtained through axisymmetric formulation.

REFERENCES

ANSYS User's Manual: Procedures, Vol. I, Swanson Analysis Systems, Inc.

ANSYS User's Manual: Commands, Vol. II, Swanson Analysis Systems, Inc.

ANSYS User's Manual: Elements, Vol. III, Swanson Analysis Systems, Inc.

Beer, P., and Johnston, E. R., *Mechanics of Materials*, 2nd ed., New York, McGraw-Hill, 1992.

Fung, Y. C., *Foundations of Solid Mechanics*, Englewood Cliffs, NJ, Prentice-Hall, 1965.

Hibbleer, R. C., *Mechanics of Materials*, 2nd ed., New York, Macmillan, 1994.

Segrlind, L., *Applied Finite Element Analysis*, 2nd ed., New York, John Wiley and Sons, 1984.

Shigley, J. E., and Mischke, C. R., *Mechanical Engineering Design*, 5th ed., New York, McGraw-Hill, 1989.

Timoshenko, S. P., and Goodier J. N., *Theory of Elasticity*, 3rd ed., New York, McGraw-Hill, 1970.

PROBLEMS

1. Using ANSYS, verify the stress-concentration chart for a flat bar with a circular hole under axial loading. Refer to a textbook on the mechanics of materials or textbook on machine design for the appropriate chart. Recall that the stress-concentration factor k is defined as

$$k = \frac{\sigma_{max}}{\sigma_{avg}}$$

and for this case, its value changes from approximately 3.0 to 2.0, depending on the size of the hole. Use ANSYS's selection options to list the value of σ_{max} at point A or B.

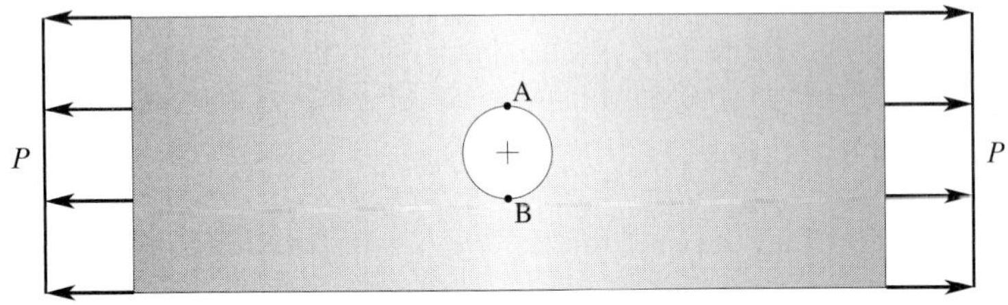

2. Consider one of the many steel brackets ($E = 29 \times 10^6$ lb/in^2, $\nu = 0.3$) used to support bookshelves. The thickness of the bracket is 1/8 in. The dimensions of the bracket are shown in the accompanying figure. The bracket is loaded uniformly along its top surface, and it is fixed along its left edge. Under the given loading and the constraints, plot the deformed shape; also, determine the von Mises stresses in the bracket.

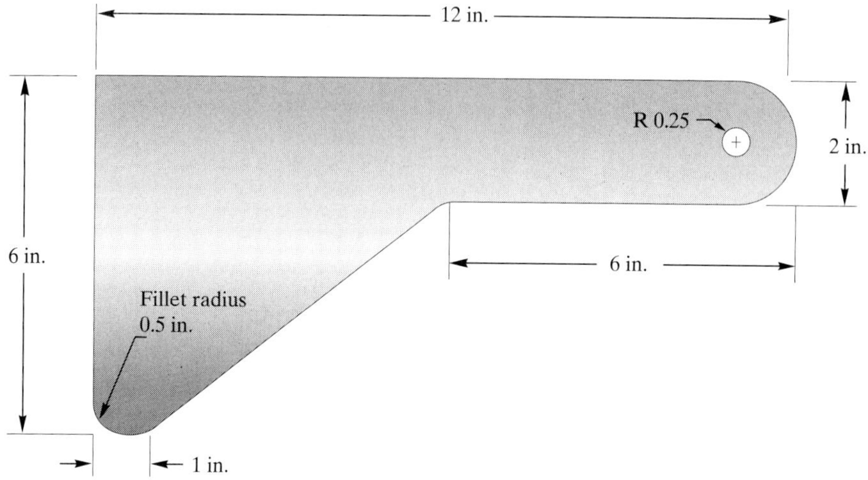

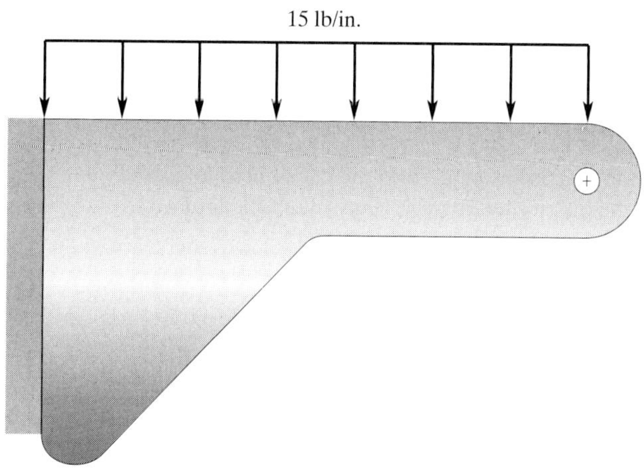

3. A $\frac{1}{8}$-in-thick plate supports a load of 100 lb, as shown in the accompanying figure. The plate is made of steel, with $E = 29 \times 10^6$ lb/in^2 and $\nu = 0.3$. Using ANSYS, determine the principal stresses in the plate. When modeling, distribute the load over part of the bottom portion of the hole.

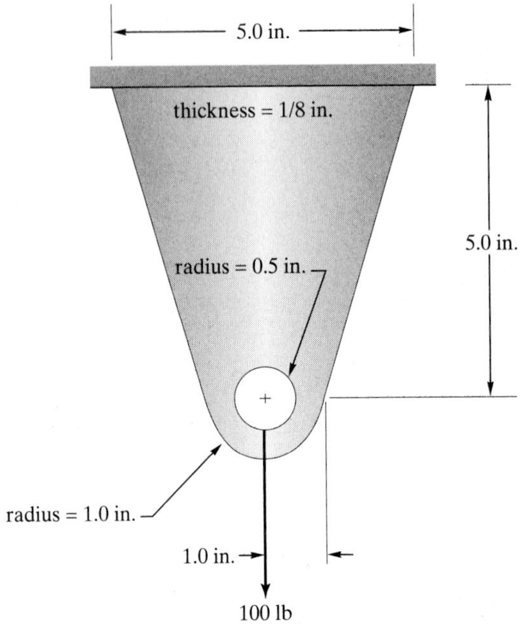

4. Elements (1) and (2) are subjected to the distributed loads shown in the accompanying figure. Replace the distributed loads by equivalent loads at nodes 3, 4, and 5.

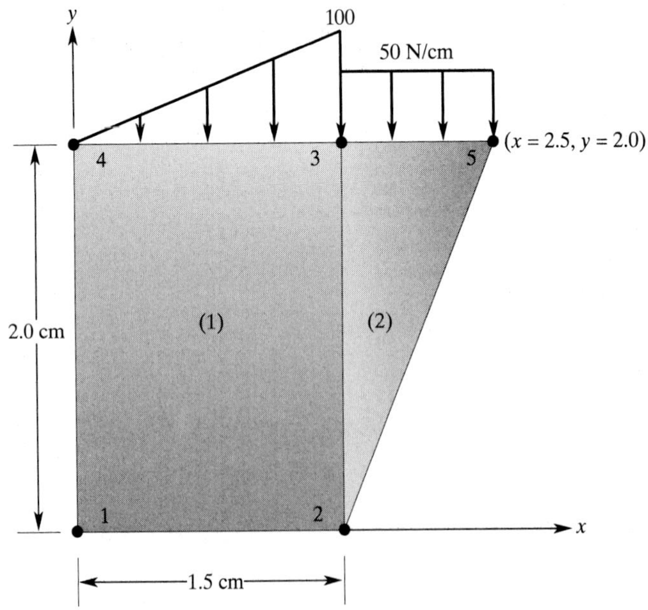

5. Using a steel sample similar to the one shown in the accompanying figure, perform a numerical tension test over the elastic region of the material. Plot the stress-strain diagram over the elastic region. Use ANSYS's selection options to list the stress and strain values at the mid-section of the steel sample.

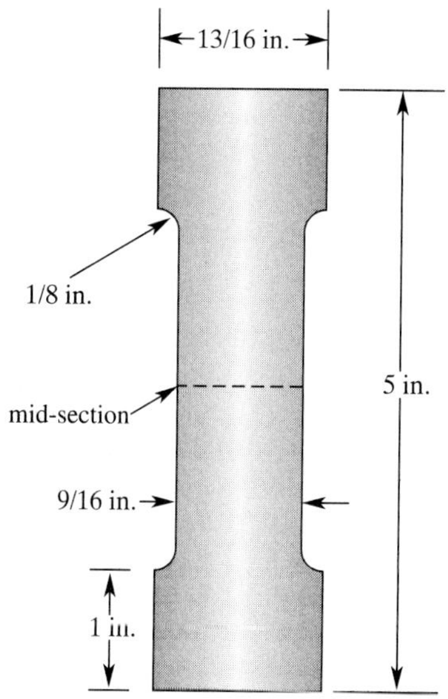

6. **Example 1.4 (revisited)**. A steel plate is subjected to an axial load, as shown in the accompanying figure. The plate is 1/16 in thick, and it has a modulus of elasticity $E = 29 \times 10^6$ lb/in^2. Recall that we approximated the deflections and average stresses along the plate using the concept of one-dimensional direct formulation. Using ANSYS, determine the deflection and the x- and y-components of the stress distributions in the plate. Also, determine the location of the maximum-stress-concentration regions. Plot the variation of the x-component of the stress at sections A–A, B–B, and C–C. Compare the results of the direct-formulation model to the results obtained from ANSYS. Furthermore, recall that for the given problem, it was mentioned that the way in which you apply the external load to your finite element model will influence the stress-distribution results. Experiment with applying the load over an increasingly large load-contact surface area. Discuss your results.

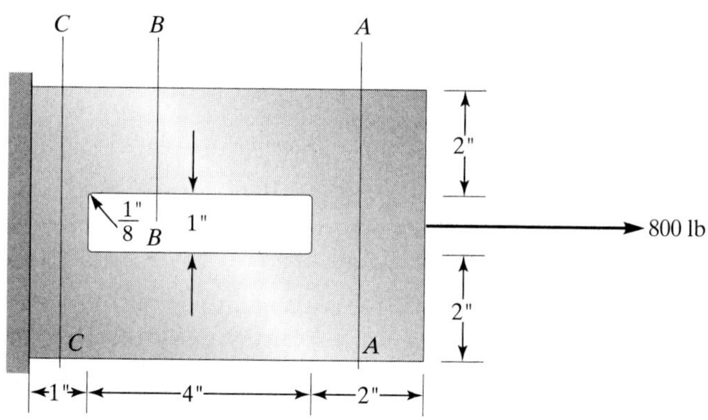

7. Consider a plate with a variable cross section supporting a load of 1500 lb, as shown in the accompanying figure. Using ANSYS, determine the deflection and the x- and y-components of the stress distribution in the plate. The plate is made of a material with a modulus of elasticity $E = 10.6 \times 10^3$ ksi. In problem 24 of Chapter 1, you were asked to analyze this problem using simple direct formulation. Compare the results of your direct-formulation model to the results obtained from ANSYS. Experiment with applying the load over an increasingly large load-contact surface area. Discuss your results.

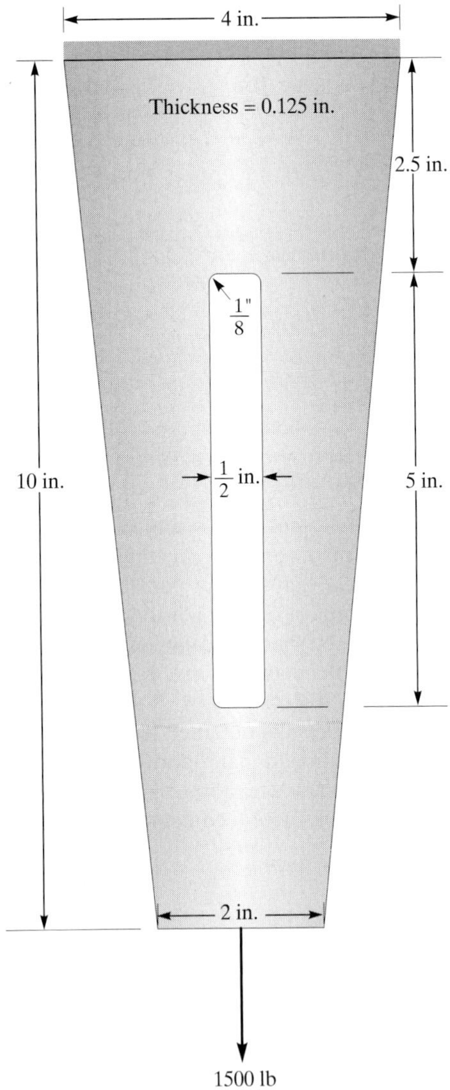

8. A thin steel plate with the profile given in the accompanying figure is subjected to an axial load. Using ANSYS, determine the deflection and the x- and y-components of the stress distributions in the plate. The plate has a thickness of 0.125 in and a modulus of elasticity of

$E = 28 \times 10^3$ ksi. In problem 4 of Chapter 1, you were asked to analyze this problem using simple direct formulation. Compare the results of your direct-formulation model to the results obtained from ANSYS. Experiment with applying the load over an increasingly larger load-contact surface area. Discuss your results.

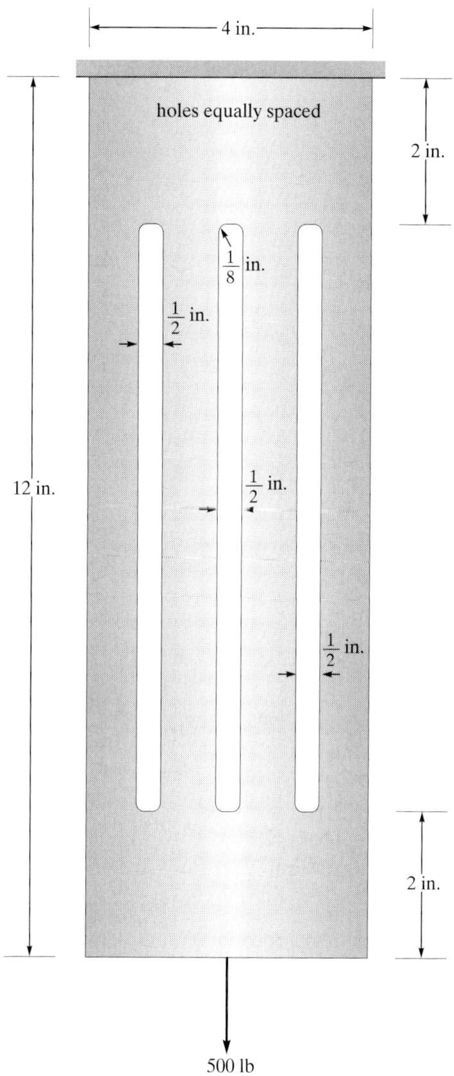

9. Consider the torsion of a steel bar ($G = 11 \times 10^3$ ksi) having an equilateral-triangular cross section, as shown in the accompanying figure. Assuming that $\theta = 0.0005$ rad/in and using ANSYS, determine the location(s) and magnitude of the maximum shear stress. Compare the solution generated with ANSYS to the exact solution obtained from the equation

$$\tau_{\max} = \frac{GL\theta}{2 \cdot 31}.$$

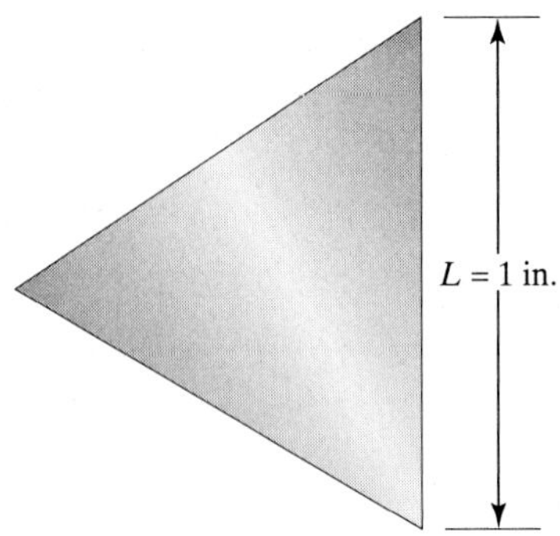

10. Consider the torsion of a steel wide-flange member (W4 × 13 and $G = 11 \times 10^3$ ksi) with dimensions shown in the accompanying figure. Assuming $\theta = 0.00035$ rad/in and using ANSYS, plot the shear stress distributions. Could you have solved this problem using the thin-wall member assumption and, thus, avoid resorting to a finite element model?

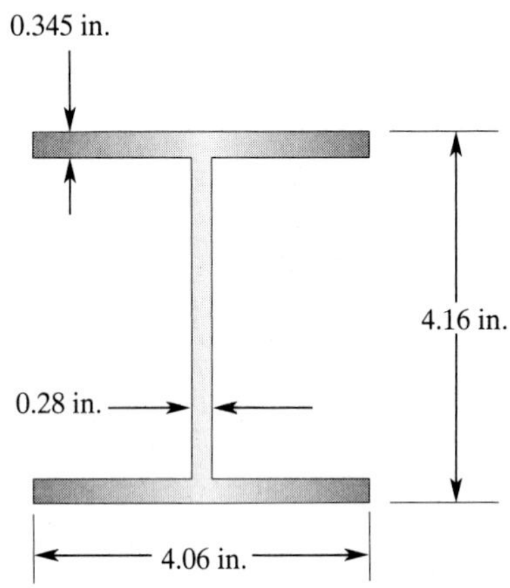

11. Consider the torsion of a steel bar ($G = 11 \times 10^3$ ksi) having a square cross section, as shown in the accompanying figure. Assuming that $\theta = 0.0005$ rad/in and using ANSYS, determine the location(s) and magnitude of the maximum shear stress. Compare the solution generated with ANSYS to the exact solution obtained from the equation $\tau_{max} = \dfrac{Gh\theta}{1.6}$.

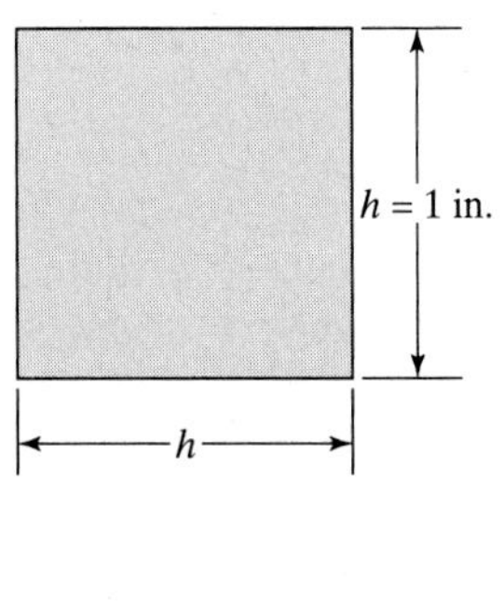

12. Consider the torsion of a steel bar ($G = 11 \times 10^3$ ksi) having a polygon cross section, as shown in the accompanying figure. Assuming that $\theta = 0.0005$ rad/in and using ANSYS, determine the location(s) and magnitude of the maximum shear stress. Compare the solution generated with ANSYS to the exact solution obtained from the equation $\tau_{max} = \dfrac{GL\theta}{0.9}$.

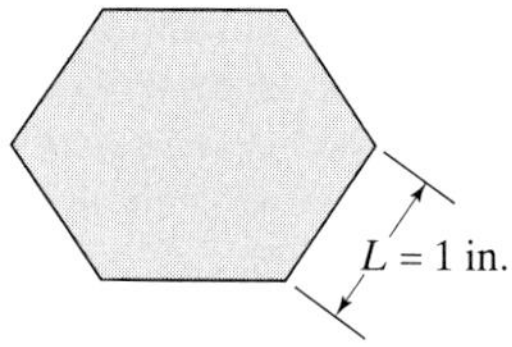

13. Consider the torsion of a steel bar ($G = 11 \times 10^3$ ksi) having an elliptical cross section, as shown in the accompanying figure. Assuming that $\theta = 0.0005$ rad/in and using ANSYS, determine the location(s) and magnitude of the maximum shear stress. Compare the solution generated with ANSYS to the exact solution obtained from the equation $\tau_{max} = \dfrac{Gbh^2\theta}{b^2 + h^2}$.

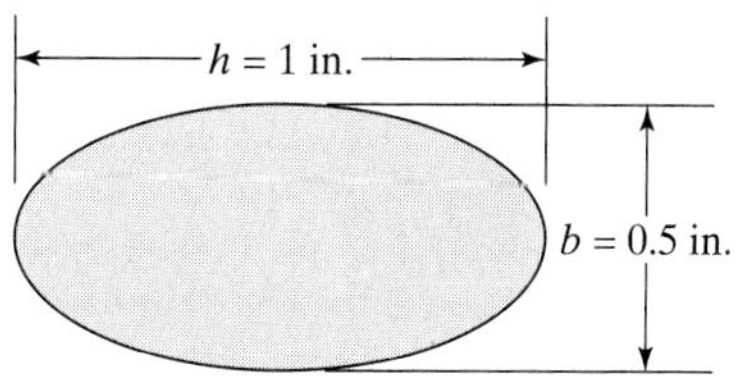

14. Consider the torsion of a steel bar ($G = 11 \times 10^3$ ksi) having a hollow, circular cross section, as shown in the accompanying figure. Assuming that $\theta = 0.0005$ rad/in and using ANSYS, determine the location(s) and magnitude of the maximum shear stress. Compare the solution generated with ANSYS to the exact solution obtained from the equation $\tau_{max} = \dfrac{GD\theta}{2}$.

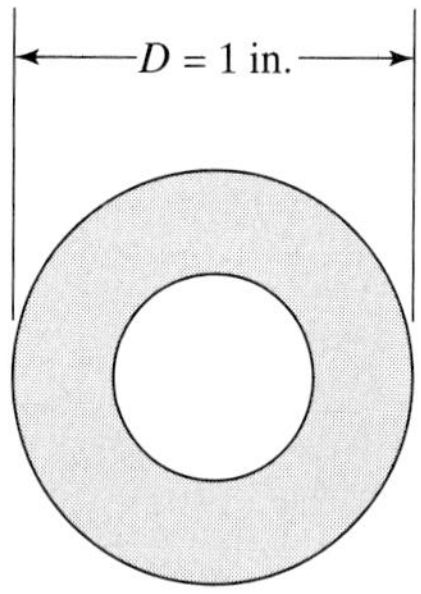

15. **Design Project** The purpose of this project is twofold: (1) to provide a basis for the application of solid-design principles using finite element methods and (2) to foster competitiveness among students. Each student is to design and construct a structural model from a $\frac{3}{8} \times 6 \times 6$ in sheet of plexiglas material that adheres to the specifications and rules given later in this problem and that is capable of competing in three areas: (1) maximum failure load per model weight, (2) predication of failure load using ANSYS, and (3) workmanship. A

sketch of a *possible* model is shown in the accompanying figure. Each end of the model will have a diameter hole (eye) of $d > 1/2''$ drilled through it perpendicular to the axis of loading, for which pins can be inserted and the model loaded in tension. The dimension a must also be set such that $a > 1''$. The distance between the eyes will be $\ell > 2''$. The maximum thickness of the member in the region of the eye will be $t < 3/8''$. This requirement will ensure that the model fits into the loading attachment. A dimension of $b < 1''$ from the center of the eyes to the outer edge in the direction of loading must be maintained so that the loading attachment can be used. The maximum width is limited to $w < 6''$, and the maximum height is limited to $h < 6''$. Any configuration may be used. Two sheets of $\frac{3}{8} \times 6 \times 6$ in Plexiglas™ will be provided. You can use one sheet to experiment and one sheet for your final design. Write a brief report discussing the evolution of your final design.

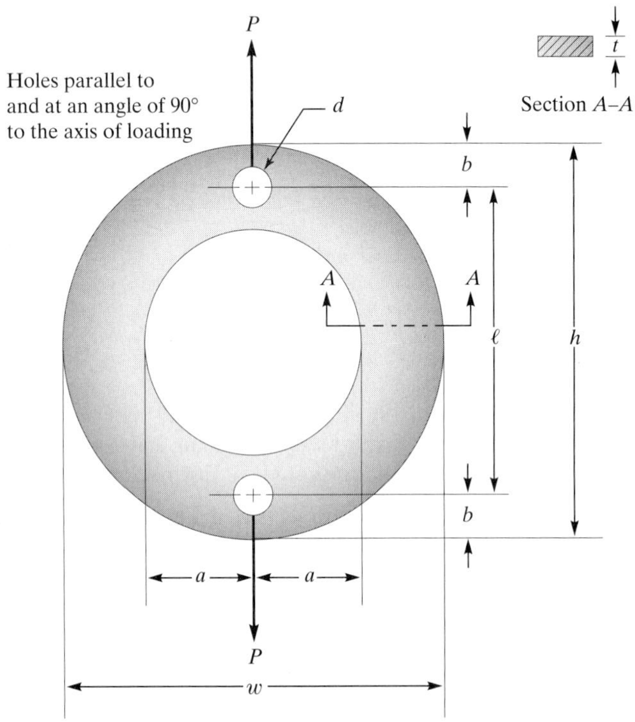

CHAPTER 11

Dynamic Problems

The main objective of this chapter is to introduce analysis of dynamic systems. A dynamic system is defined as a system that has mass and components, or parts, that are capable of relative motion. Examples of dynamic systems include structures such as buildings, bridges, water towers, planes, and machine components. In most engineering applications mechanical vibration is undesirable. However, there are systems, such as shakers, mixers, and vibrators, that are intentionally designed to vibrate. Before discussing finite element formulations of dynamic problems, we review dynamics of particles and rigid bodies, and examine basic concepts dealing with vibration of mechanical and structural systems. A good understanding of the fundamental concepts of dynamics and vibration is necessary for an accurate finite element modeling of actual physical situations. After we establish the basic foundation, we consider finite element formulation of an axial member, a beam, and frame elements. The main topics discussed in Chapter 11 include the following:

11.1 Review of Dynamics
11.2 Review of Vibration of Mechanical and Structural Systems
11.3 Lagrange's Equations
11.4 Finite Element Formulation of Axial Members
11.5 Finite Element Formulation of Beams and Frames
11.6 Examples Using ANSYS

11.1 REVIEW OF DYNAMICS

The subject of dynamics is traditionally divided into two broad areas: *kinematics* and *kinetics*. Kinematics deals with space and time relationships—in other words, the geometric aspect of motion. Kinematics is the study of variables such as distance traveled by an object, its speed, and its acceleration. The fundamental dimensions associated with all of these variables are length and time. When studying kinematics, we are not concerned with causes of motion but focus instead on the motion itself. Study of kinetics, on the

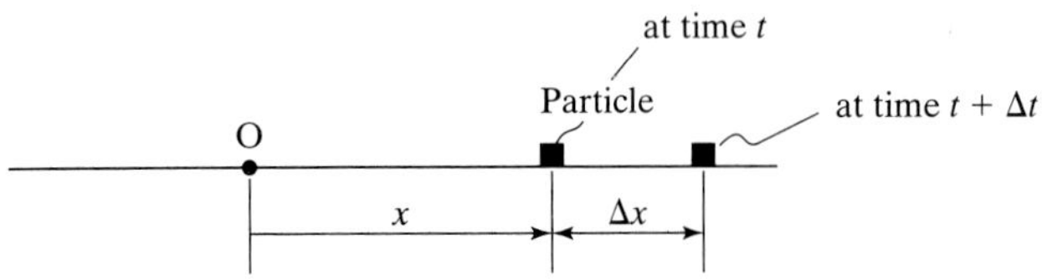

FIGURE 11.1 Rectilinear motion of a particle.

other hand, deals with the relationship between forces and moments, and the resulting motion.

Kinematics of a Particle

Let us first define a particle and explain when it is appropriate to model a problem using a particle. For an object to be considered as *a particle,* all forces acting on the object must act at the same point and create no rotation. Moreover, an object is modeled as a particle when its size does not play a significant role in the way it behaves. In general, the motion of a particle is described by its position, instantaneous velocity, and the instantaneous acceleration.

The motion of an object along a straight path is called *rectilinear motion*, which is the simplest form of motion. The kinematical relationships for an object moving along a straight line is given by Eqs. (11.1) through (11.3). In these equations, x represents the position of the object, t is time, v is velocity, and a represents the acceleration of the particle.

$$v = \frac{dx}{dt} \tag{11.1}$$

$$a = \frac{dv}{dt} \tag{11.2}$$

$$vdv = adx \tag{11.3}$$

The position and displacement of an object moving along a straight path are shown in Figure 11.1.

Plane Curvilinear Motion When a particle moves along a curved path, its motion can be described using a number of coordinate systems. As shown in Figure 11.2,

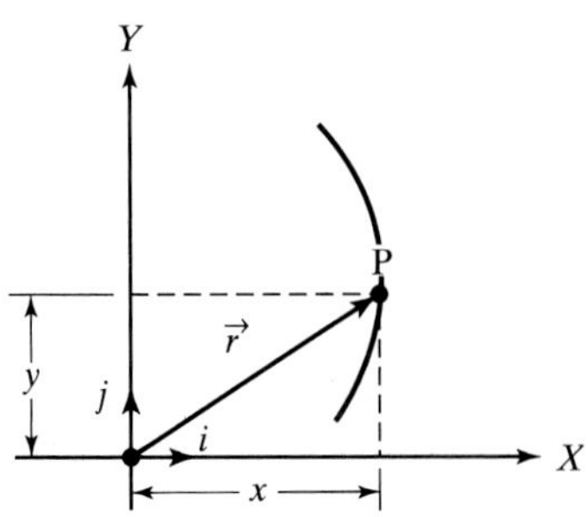

FIGURE 11.2 Rectangular components of motion of a particle.

using a rectangular coordinate system (x, y), we can describe the position $\vec{r}$, velocity $\vec{v}$, and acceleration $\vec{a}$ of the particle with

$$\vec{r} = x\vec{i} + y\vec{j} \tag{11.4}$$

$$\vec{v} = v_x\vec{i} + v_y\vec{j} \quad \text{where} \quad v_x = \frac{dx}{dt} \quad \text{and} \quad v_y = \frac{dy}{dt} \tag{11.5}$$

$$\vec{a} = a_x\vec{i} + a_y\vec{j} \quad \text{where} \quad a_x = \frac{dv_x}{dt} \quad \text{and} \quad a_y = \frac{dv_y}{dt} \tag{11.6}$$

In Eqs. (11.4) through (11.6) x and y are rectangular components of the position vector; v_x, v_y, a_x, and a_y are the Cartesian components of the velocity and acceleration vectors.

Normal and Tangential Coordinates The plane motion of a particle can also be described using normal and tangential unit vectors, as shown in Figure 11.3. It is important to note that unlike the previously employed Cartesian unit vectors $\vec{i}$ and $\vec{j}$, whose directions are fixed in space, the directions of unit vectors $\vec{e}_t$ and $\vec{e}_n$ change as they move along with the particle. The changes in the direction of $\vec{e}_t$ and $\vec{e}_n$, as the particle moves from position 1 to position 2, are shown in Figure 11.3. The velocity and acceleration of a particle in terms of unit vectors $\vec{e}_t$ and $\vec{e}_n$ are

$$\vec{v} = v\vec{e}_t \tag{11.7}$$

$$\vec{a} = a_n\vec{e}_n + a_t\vec{e}_t \quad \text{where} \quad a_n = \frac{v^2}{r} \quad \text{and} \quad a_t = \frac{dv}{dt} \tag{11.8}$$

Polar Coordinates A polar coordinate (or radial and transverse) system offers yet another way of describing motion of an object along a curved path. To locate the object, two pieces of information are used: a radial distance r in a direction specified by the unit vector $\vec{e}_r$ and the angular coordinate θ in the $\vec{e}_\theta$ direction, as shown in Figure 11.4. The position, velocity, and acceleration of the object are given by

$$\vec{r} = r\vec{e}_r \tag{11.9}$$

$$\vec{v} = v_r\vec{e}_r + v_\theta\vec{e}_\theta \quad \text{where} \quad v_r = \frac{dr}{dt} \quad \text{and} \quad v_\theta = r\frac{d\theta}{dt} \tag{11.10}$$

$$\vec{a} = a_r\vec{e}_r + a_t\vec{e}_t \quad \text{where} \quad a_r = \frac{d^2r}{dt^2} - r\left(\frac{d\theta}{dt}\right)^2 \quad \text{and} \quad a_\theta = r\frac{d^2\theta}{dt^2} + 2\left(\frac{dr}{dt}\right)\left(\frac{d\theta}{dt}\right) \tag{11.11}$$

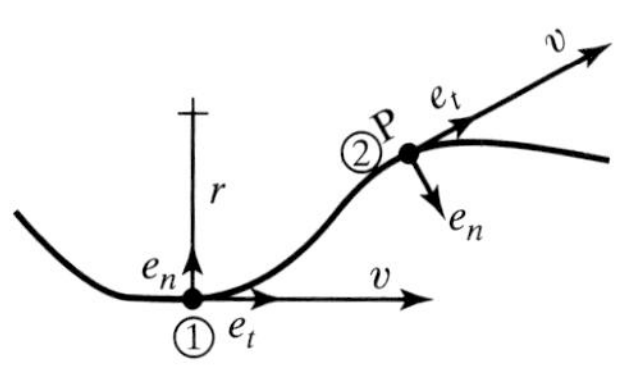

FIGURE 11.3 Normal and tangential components of motion of a particle.

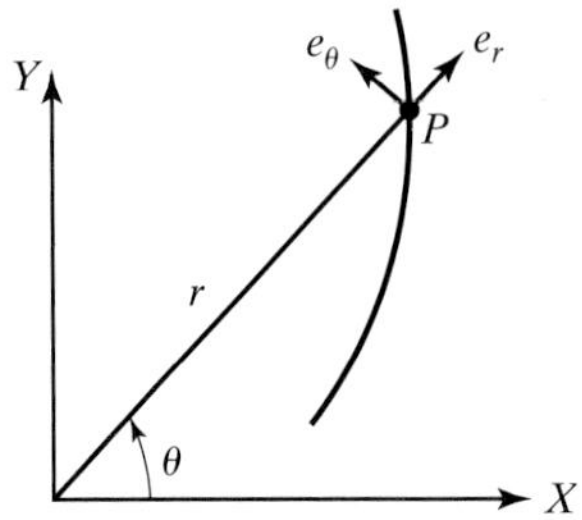

FIGURE 11.4 Polar components of a curvilinear motion.

Note similar to normal and tangential unit vectors, the directions of $\vec{e}_\theta$ and $\vec{e}_r$ change as they move along with the particle.

Relative Motion The relationship between the position of two particles moving along different paths is shown in Figure 11.5. When studying Figure 11.5, note that the X,Y coordinate system is fixed, and an observer located at its origin (point O) measures the absolute motion of the particles A and B. Therefore, vectors $\vec{r}_A$ and $\vec{r}_B$ represent the absolute positions of particle A and B with respect to the observer. On the other hand, vector $\vec{r}_{B/A}$ represents the position of particle B with respect to A. The relationship among these vectors is given by

$$\vec{r}_B = \vec{r}_A + \vec{r}_{B/A} \tag{11.12}$$

Taking the time derivative of Eq. (11.12), we get

$$\vec{v}_B = \vec{v}_A + \vec{v}_{B/A} \tag{11.13}$$

where $\vec{v}_B$ and $\vec{v}_A$ are absolute velocities of particle A and B measured by the observer at *O*, and $\vec{v}_{B/A}$ is the velocity of particle B relative to A (measured by someone moving along with A). By taking the time derivative of Eq. (11.13), we can obtain the relationship between the absolute accelerations of particle A and B and acceleration of particle B relative to A.

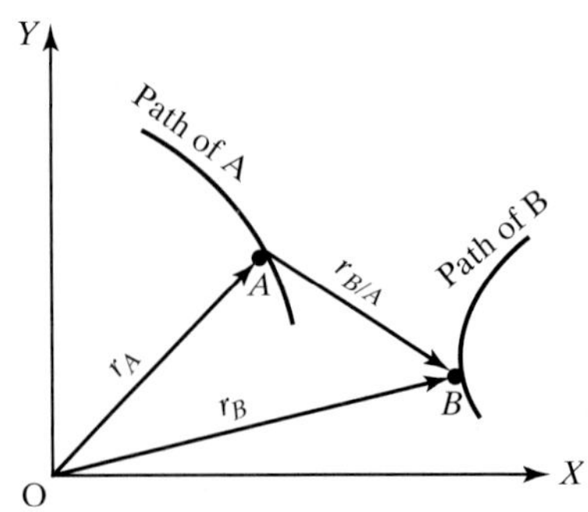

FIGURE 11.5 Relationship between two particles moving along different paths.

Kinetics of a Particle

The equations of motion for a particle of constant mass m under the influence of forces $F_1, F_2, F_3, \ldots, F_n$ is governed by Newton's second law according to

$$\sum_{i=1}^{n} \vec{F}_i = m\vec{a} \tag{11.14}$$

We can describe the equations of motion using either a rectangular, normal and tangential, or polar coordinate system. Using a rectangular coordinate system, Eq. (11.14) is expressed by

$$\sum F_x = ma_x \tag{11.15}$$

$$\sum F_y = ma_y \tag{11.16}$$

In a normal and tangential coordinate system, Eq. (11.14) becomes

$$\sum F_n = ma_n \tag{11.17}$$

$$\sum F_t = ma_t \tag{11.18}$$

And in a polar coordinate system, we have

$$\sum F_r = ma_r \tag{11.19}$$

$$\sum F_\theta = ma_\theta \tag{11.20}$$

where the respective components of accelerations are given in the previous sections.

It is important to point out that the only way to account correctly for every force acting on a particle is to draw a free-body diagram. A free-body diagram represents the interaction of the particle with its surroundings. To draw a free-body diagram, as the name implies, you must free the body from its surroundings and show all the interaction of the body with its surroundings with forces of appropriate magnitudes and directions (see Figure 11.6).

Newton's second law of motion is a vectorial equation that relates forces acting on the object to its mass and acceleration. If the solution to a problem requires position and velocity information, then kinematical relationships are used to obtain such information from the knowledge of the object's acceleration.

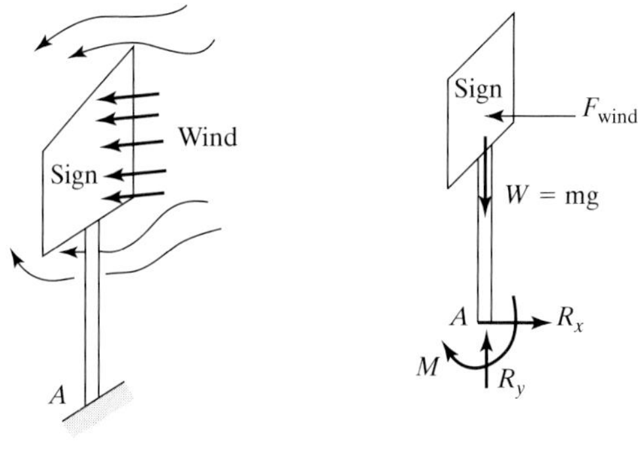

FIGURE 11.6 A free-body diagram of the object shown. Note the object shown can not be considered as a particle. The sign is merely used to demonstrate the concept of free-body diagram.

Work and Energy Principle Whereas Newton's second law is a vectorial equation, the work–energy principle is a scalar relationship. The work–energy principle relates the work done by forces over a distance to the object's mass and speed. The work–energy principle becomes very useful for situations for which we are interested in determining the change in speed of an object due to applied forces. The work done by a force moving an object from position 1 to position 2, as shown in Figure 11.7, is defined by

$$W_{1-2} = \int \vec{F}.d\vec{r} \tag{11.21}$$

The work–energy principle simply states that net work done by all forces acting on an object will bring about a change in its kinetic energy according to

$$\int \vec{F}.d\vec{r} = \frac{1}{2}mv_2^2 - \frac{1}{2}mv_1^2 \tag{11.22}$$

The terms $\frac{1}{2}mv_2^2$ and $\frac{1}{2}mv_1^2$ represent the kinetic energy of the object and correspond to its 2 and 1 positions. When using the work–energy principle, you must pay close attention to two important things: (1) work is done only when a force undergoes a displacement; (2) if the tangential component of the force (the component that moves the object) and displacement have the same direction, then the work done is considered positive, and if they have opposite directions, work done is a negative quantity.

Linear Impulse and Linear Momentum For problems wherein the time history of forces is known (i.e., how forces act on an object over a time period), the impulse and momentum approach may be used to determine the resulting changes in the velocities of the object. Newton's second law can be rearranged and integrated over time in the following manner:

$$\sum \vec{F} = \frac{dm\vec{v}}{dt} \Rightarrow \sum \vec{F}dt = dm\vec{v} \tag{11.23}$$

$$\int_{t_1}^{t_2} \sum \vec{F}dt = m\vec{v}_2 - m\vec{v}_1 \tag{11.24}$$

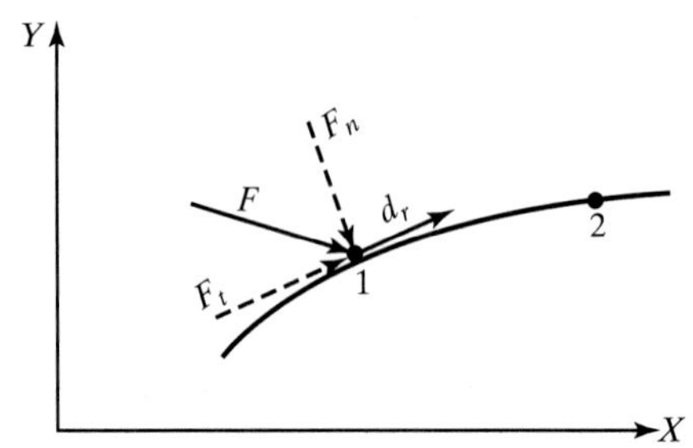

FIGURE 11.7 Work done by a force moving an object.

Equation (11.24) may be expressed in any of the previously mentioned coordinate systems. For example, using a Cartesian coordinate system, Eq. (11.24) becomes

$$\int_{t_1}^{t_2} \sum F_x dt = m(v_x)_2 - m(v_x)_1 \tag{11.25}$$

$$\int_{t_1}^{t_2} \sum F_y dt = m(v_y)_2 - m(v_y)_1 \tag{11.26}$$

Next, we will use Example 11.1 to demonstrate some of the ideas we have discussed so far.

EXAMPLE 11.1

A small sphere having mass m is released from position 1, as shown in Figure 11.8. We are interested in determining the velocity of the sphere as a function of θ. We solve this problem using both Newton's law and the work–energy principle.

Using Newton's second law and referring to the free-body diagram shown in the accompanying figure, applying Eq. (11.18), we have

$$\sum F_t = ma_t$$

$$mg \cos \theta = ma_t \Rightarrow a_t = g \cos \theta$$

Now, using Eq. (11.3), the kinematical relationship between velocity and acceleration, and noting that $ds = Ld\theta$,

$$vdv = a_t ds$$

$$\int_0^v vdv = \int_0^\theta g \cos \theta L d\theta \quad \Rightarrow \quad v = \sqrt{2Lg \sin \theta}$$

We now solve the problem using the work–energy principle. The force due to the tension in the string does not perform any work because it acts normal to the path of the sphere. Relating the work done by the weight of the sphere $W_{1-2} = mgL \sin \theta$ to the change in its kinetic energy, we get

$$mgL \sin \theta = \frac{1}{2}mv^2 - 0 \quad \Rightarrow \quad v = \sqrt{2Lg \sin \theta}$$

You realize that it is easier to solve this problem using the work–energy principle, because we were interested in determining the speed of the sphere and not its acceleration.

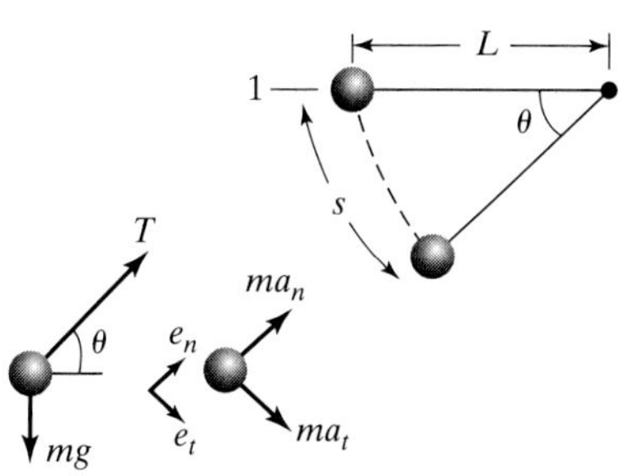

FIGURE 11.8 The sphere of Example 11.1.

Kinematics of a Rigid Body

In this section we discuss kinematics of a rigid body. Unlike a particle model, a rigid body is an object whose size affects its dynamic behavior and forces can act anywhere on the body. Moreover, as the name implies, a rigid body is considered rigid—no deformation due to application of forces. A rigid body is an idealization of an actual situation in which the magnitude of motion caused by forces and moments is much larger than internal displacements. Motion of a rigid body may be classified as pure translation, pure rotation, or a combination of translation and rotation called general plane motion.

Translation of a Rigid Body When a rigid body undergoes a pure translational motion, all constituent particles move with the same velocity and acceleration. As shown in Figure 11.9, points A and B have the same velocity.

Rotation of a Rigid Body about a Fixed Axis When a rigid body rotates about a fixed axis, the particles of the rigid body follow circular paths, as shown in Figure 11.10. The relationships between the velocity of a point A such as v_A and its acceleration and the angular velocity and acceleration of the rigid body, ω and α, are given by

$$v_A = r_A\omega \tag{11.27}$$

$$a_n = r_A\omega^2 = \frac{v_A^2}{r_A} = v_A\omega \tag{11.28}$$

$$a_t = r_A\alpha \tag{11.29}$$

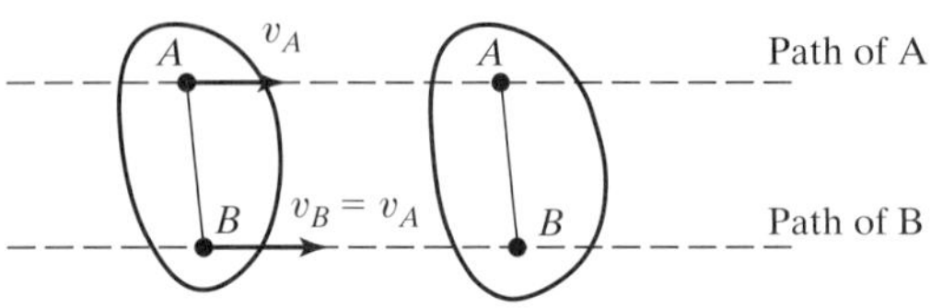

FIGURE 11.9 A pure translation of a rigid body.

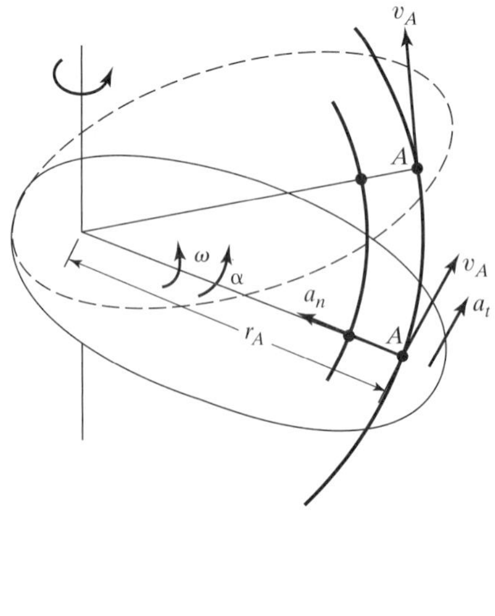

FIGURE 11.10 Pure rotation motion.

In general, the velocity and the acceleration components of a point (such as A) located on a rigid body in terms of the position vector $\vec{r}_A$, the angular velocity $\vec{\omega}$, and the angular acceleration $\vec{\alpha}$ are given by

$$\vec{v}_A = \vec{\omega} \times \vec{r}_A \tag{11.30}$$

$$a_n = \vec{\omega} \times (\vec{\omega} \times \vec{r}_A) \tag{11.31}$$

$$a_t = \vec{\alpha} \times \vec{r}_A \tag{11.32}$$

General Plane Motion The motion of a rigid body undergoing simultaneous rotation and translation is commonly referred to as general plane motion. The relationship between the velocities of two points A and B at an instant is given by

$$\vec{v}_A = \vec{v}_B + \vec{v}_{A/B} \tag{11.33}$$

where $\vec{v}_A$ and $\vec{v}_B$ are the absolute velocities of points A and B and $v_{A/B}$ represents the velocity of point A relative to point B. The magnitude of $v_{A/B}$ is expressed by $v_{A/B} = r_{A/B}\,\omega$, and its direction is normal to the position vector $r_{B/A}$, as shown in Figure 11.11.

The acceleration of points A and B are related according to

$$\vec{a}_A = \vec{a}_B + \vec{a}_{A/B} = \vec{a}_B + (\vec{a}_{A/B})_n + (\vec{a}_{A/B})_t \tag{11.34}$$

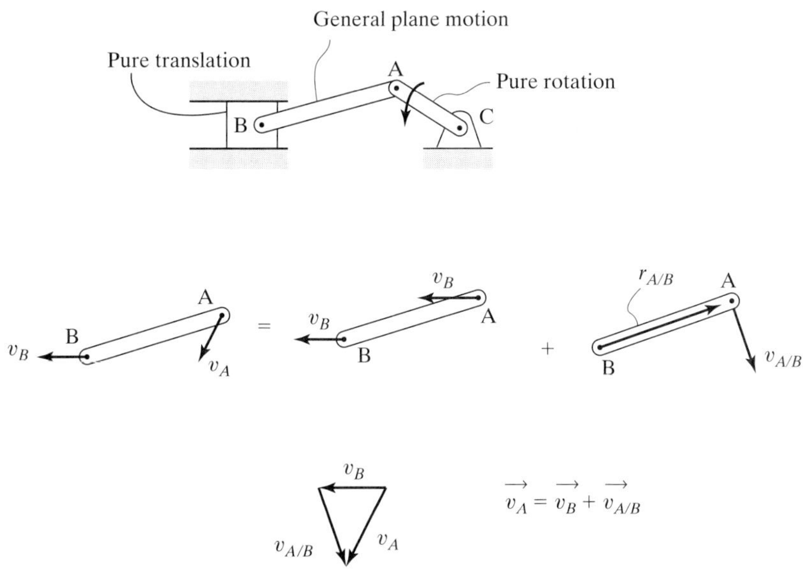

FIGURE 11.11 The relationship between the velocities of two points belonging to a rigid body undergoing a general plane motion.

where

$$(\vec{a}_{A/B})_n = \vec{\omega} \times (\vec{\omega} \times \vec{r}_{A/B}) \tag{11.35}$$

$$(\vec{a}_{A/B})_t = \vec{\alpha} \times \vec{r}_{A/B} \tag{11.36}$$

and the magnitudes of the normal and tangential components of the acceleration of A relative to B are given by

$$(a_{A/B})_n = \frac{v_{A/B}^2}{r_{A/B}} = r_{A/B}\,\omega^2 \tag{11.37}$$

$$(a_{A/B})_t = r_{A/B}\,\alpha \tag{11.38}$$

The directions of these components are shown in Figure 11.12.

Kinetics of a Rigid Body

The study of kinetics of a rigid body includes forces and moments that create the motion of the rigid body.

Rectilinear Translation The rectilinear translation of a rigid body under the influence of forces $F_1, F_2, F_3, \ldots, F_n$ is governed by Newton's second law according to

$$\sum F_x = m(a_G)_x \tag{11.39}$$

$$\sum F_y = m(a_G)_y \tag{11.40}$$

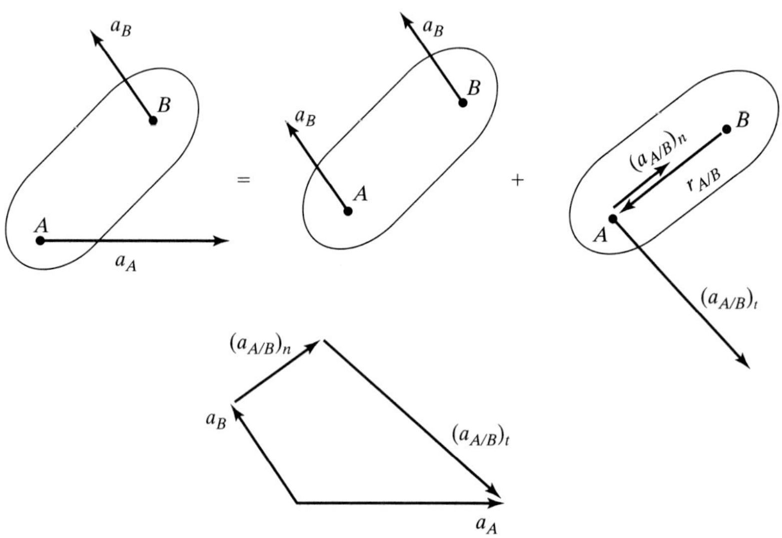

FIGURE 11.12 The directions of the normal and tangential components of $\vec{a}_{A/B}$.

Although we have related the sum of forces to the acceleration of the mass center $\vec{a}_G$, it is important to realize that for a rigid body undergoing translation, all constituent particles have the same velocity and acceleration. Moreover, since the body is not rotating, the sum of the moments of the forces about mass center G must be zero.

$$\circlearrowleft + \sum M_G = 0 \tag{11.41}$$

However, if we were to take the sum of the moment about another point, such as O, then the sum of the moments about that point is not zero, because the inertia forces $m(a_G)_x$ and $m(a_G)_y$ create moments about that point. That is,

$$\circlearrowleft + \sum M_O = m(a_G)_x d_1 - m(a_G)_y d_2 \tag{11.42}$$

The free-body and inertia diagrams for a rigid body undergoing pure translational motion are shown in Figure 11.13.

Rotation about a Fixed Axis The rotation of a rigid body is governed by the following equations:

$$\sum F_n = m r_{G/O}\,\omega^2 \tag{11.43}$$

$$\sum F_t = m r_{G/O}\,\alpha \tag{11.44}$$

$$\circlearrowleft + \sum M_O = I_O\,\alpha \tag{11.45}$$

In Eq. (11.45), I_o is the mass moment of inertia of the body about point O, as shown in Figure 11.14. Whereas mass provides a measure of resistance to translational motion, the mass moment of inertia represents the amount of the body's inherent resistance to rotational motion. For a special case when the rigid body is rotating about its mass center G, the equations of motion become

$$\sum F_x = 0 \tag{11.46}$$

$$\sum F_y = 0 \tag{11.47}$$

$$\circlearrowleft + \sum M_G = I_G\,\alpha \tag{11.48}$$

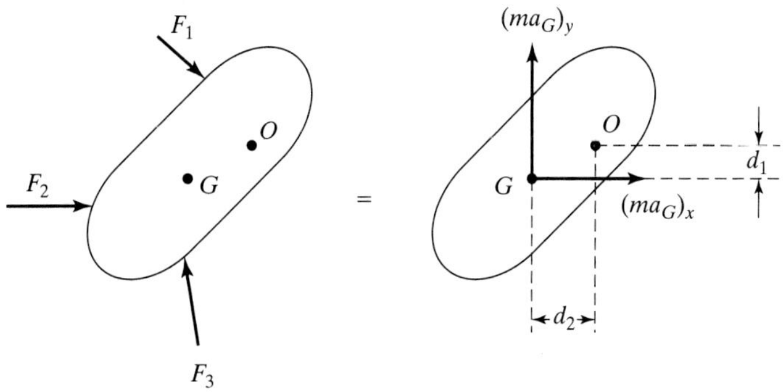

FIGURE 11.13 The kinetics of a rigid body under translational motion.

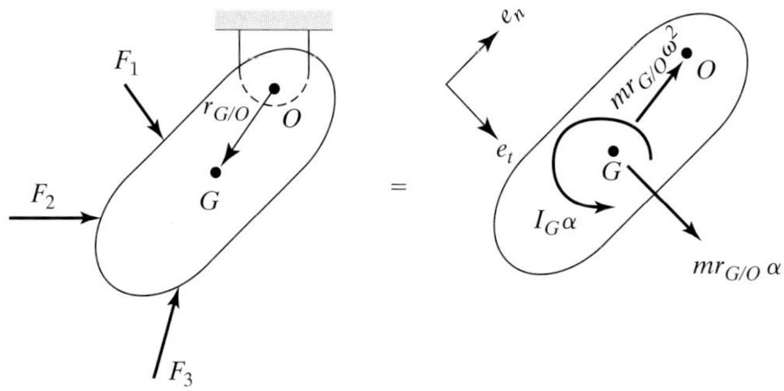

FIGURE 11.14 The rotational motion of a rigid body.

General Plane Motion For a rigid body that is translating and rotating simultaneously, the equations of motion become

$$\sum F_x = m(a_G)_x \tag{11.49}$$

$$\sum F_y = m(a_G)_y \tag{11.50}$$

$$\circlearrowleft+ \sum M_G = I_G\alpha \tag{11.51}$$

or as shown in Figure 11.15, the sum of the moments about another point such as O must include the moments created by inertia forces.

$$\circlearrowleft+ \sum M_O = I_G\alpha + m(a_G)_x d_1 - m(a_G)_y d_2 \tag{11.52}$$

Work–Energy Relations The work–energy principle relates the work done by forces and moments to the change in the kinetic energy of the rigid body as shown in

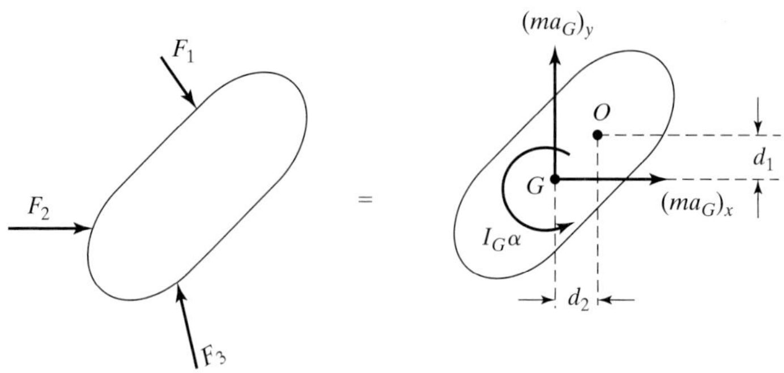

FIGURE 11.15 General plane motion of a rigid body.

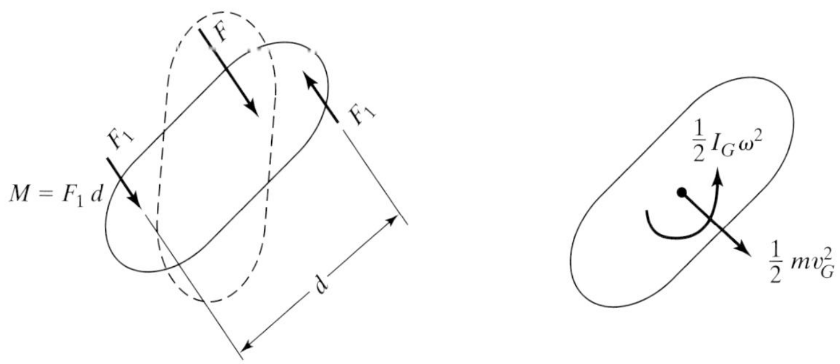

FIGURE 11.16 Work–energy principle for a rigid body.

Figure 11.16. The forces and moments acting on a rigid body do work according to

$$W_{1-2} = \int \vec{F}.d\vec{r} \tag{11.53}$$

$$W_{1-2} = \int M d\theta \tag{11.54}$$

where dr and $d\theta$ are the infinitesimal translation and rotational displacements. Because the rigid body can translate and rotate, its kinetic energy has two parts: a translational and a rotational part. The rigid body's translational part of kinetic energy is given by

$$T = \frac{1}{2} m v_G^2 \tag{11.55}$$

The rotational kinetic energy of a rigid body about it mass center G is given by

$$T = \frac{1}{2} I_G \omega^2 \tag{11.56}$$

The total kinetic energy of a rigid body undergoing a general plane motion is then given by

$$T = \frac{1}{2} m v_G^2 + \frac{1}{2} I_G \omega^2 \tag{11.57}$$

The rotational kinetic energy of a rigid body about an arbitrary point O is

$$T = \frac{1}{2} I_O \omega^2$$

Impulse and Momentum For problems in which time history of forces and moments are known, the impulse and momentum approach is used to determine the changes in the velocities of the rigid body. The linear impulse and momentum equations for a rigid body are

$$\int_{t_1}^{t_2} \sum \vec{F} dt = m(\vec{v}_G)_2 - m(\vec{v}_G)_1 \tag{11.58a}$$

$$\int_{t_1}^{t_2} \sum M_G \, dt = I_G(\omega)_2 - I_G(\omega)_1 \tag{11.58b}$$

In Eq. (11.58b), $\sum M_G$ represents the sum of all moments about mass center G, and I_G is the mass moment of inertia of the rigid body about its mass center G.

Next, we use Example 11.2 to show how the concepts discussed previously are applied to formulate the equations of motion for a rigid body.

EXAMPLE 11.2

In this example we derive the equations of motion for the system shown in the accompanying figure. When studying this example, keep in mind that the rod can rotate and translate.

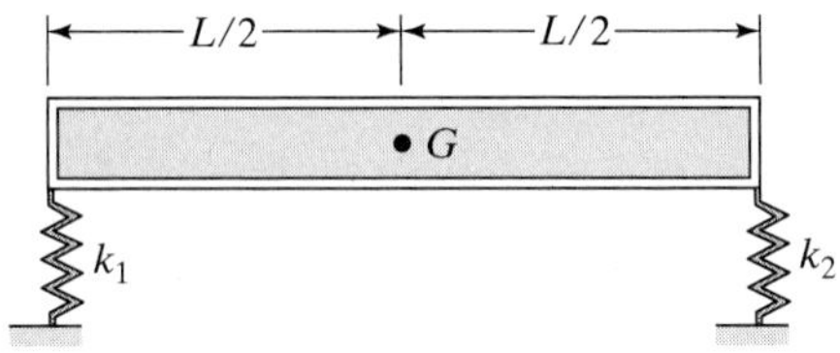

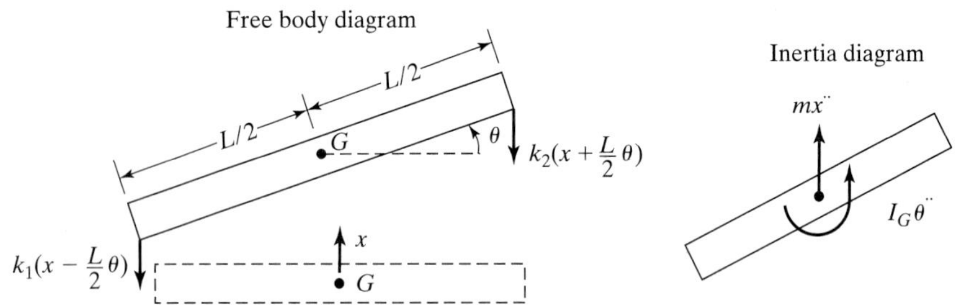

The free-body diagram of the rod is shown above. This is a snapshot of the rod at a point in time as it oscillates. One way of arriving at the free-body diagram shown is by imagining that you pull the rod straight up and turn it counterclockwise and then let it go. To keep track of the motion of the rod, we will use x and $\ddot{x}$ to locate translation of the mass center and use θ and $\ddot{\theta}$ to measure the rotational motion. Note x is measured from the static equilibrium position and as a result the weight of the rod and the spring forces in static position cancel each other out. Applying Eqs. (11.50) and (11.51), we have

$$\sum F_x = m(a_G)_x = m\ddot{x}$$

$$-k_1\left(x - \frac{L}{2}\theta\right) - k_2\left(x + \frac{L}{2}\theta\right) = m\ddot{x}$$

and simplifying, we get the equation of translational motion

$$m\ddot{x} + (k_1 + k_2)x - (k_1 - k_2)\frac{L}{2}\theta = 0$$

The equation for rotational component is obtained from

$$\overset{+}{\circlearrowleft}\Sigma M_G = I_G\alpha = I_G\ddot{\theta}$$

$$k_1\left(x - \frac{L}{2}\theta\right)\frac{L}{2} - k_2\left(x + \frac{L}{2}\theta\right)\frac{L}{2} = I_G\ddot{\theta}$$

and simplifying we get the equation for rotational motion, which is given by

$$I_G\ddot{\theta} - (k_1 - k_2)\frac{L}{2}x + (k_1 + k_2)\left(\frac{L}{2}\right)^2\theta = 0$$

Now that we have reviewed dynamics of particles and rigid bodies, we are ready to review the basic definitions, fundamental principles, and governing equations in vibrations.

11.2 REVIEW OF VIBRATION OF MECHANICAL AND STRUCTURAL SYSTEMS

A dynamic system is defined as a system that has mass and components (parts) that are capable of relative motion. A dynamic system has the following properties:

Because of the mass of the system and the changes in the velocity of the system, the kinetic energy of the system can increase or decrease with time.

Elastic members of the system are capable of storing elastic energy.

The materials making up the system have damping characteristics that could convert a portion of the work or energy input (into the system) into heat.

Work or energy enters the system through excitation of support or direct application of forces.

Examples of dynamic systems are given in Table 11.1.

Degrees of Freedom

The degree of freedom is defined as the number of spatial coordinates required to completely describe the motion of a system—in other words, the number of coordinates needed to keep track of all components or lumped masses making up the system. For example, the four-story building shown in Table 11.1 requires four spatial coordinates to locate the temporal position of each floor mass. As another example we may use a model with three degrees of freedom to study the dynamic behavior of an airplane as shown in Table 11.1.

Simple Harmonic Motion

Consider the simple single degree of freedom system consisting of a linear spring and mass, as shown in Figure 11.17.

TABLE 11.1 Examples of dynamic systems

Actual System	Possible Dynamic Model
A water tower 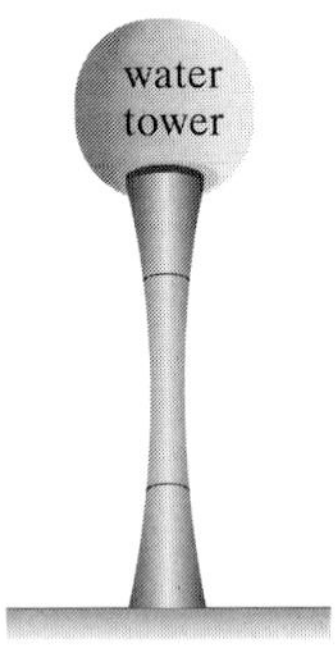	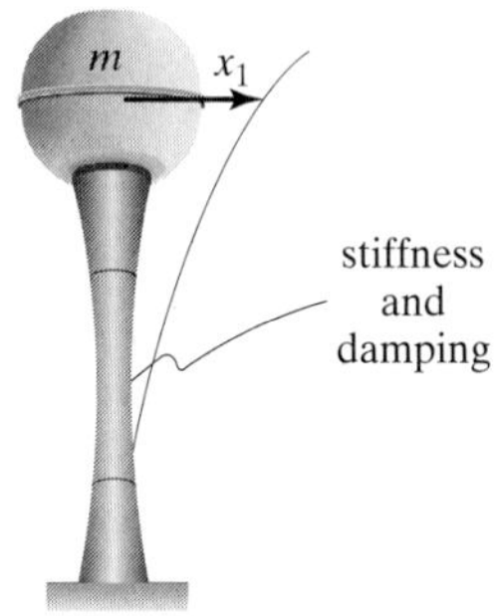
A four-story building 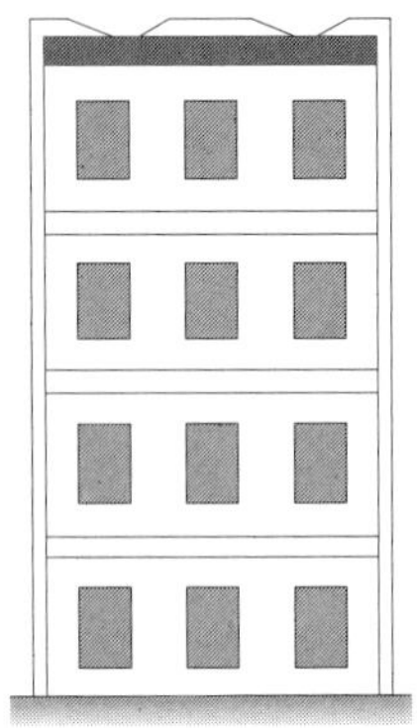	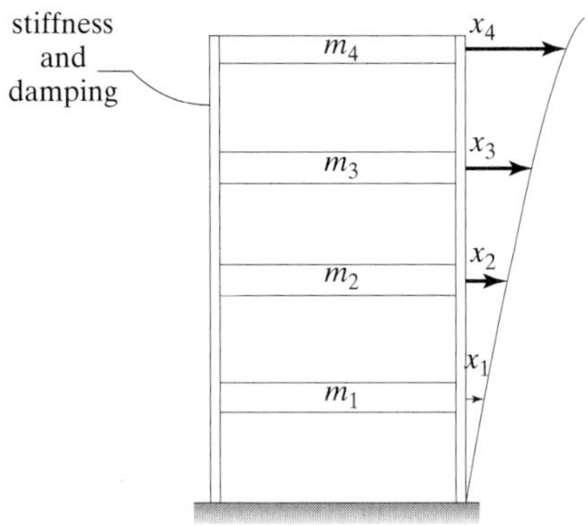
An airplane 	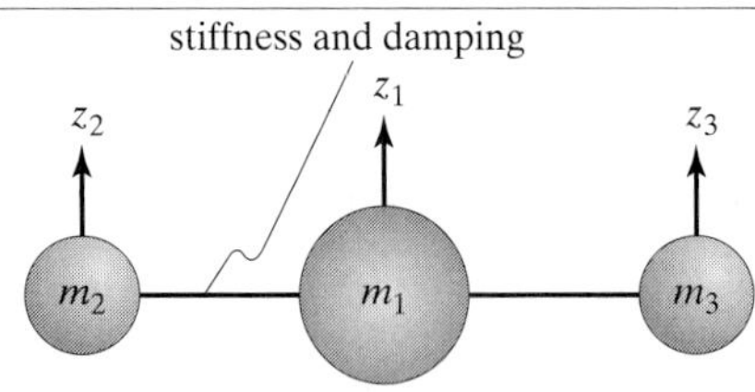

The free-body diagram for the static equilibrium is shown in Figure 11.17(b). As you can see, at static equilibrium, the weight of the mass is supported by the spring force, $k\delta_{static} = W$. In the forthcoming analysis, k represents the stiffness of the spring in N/mm (or lb/in), and δ_{static} is the static deflection in mm (or in). To obtain the governing equation of motion, we displace the mass and then release it, as shown in Figure 11.17(c).

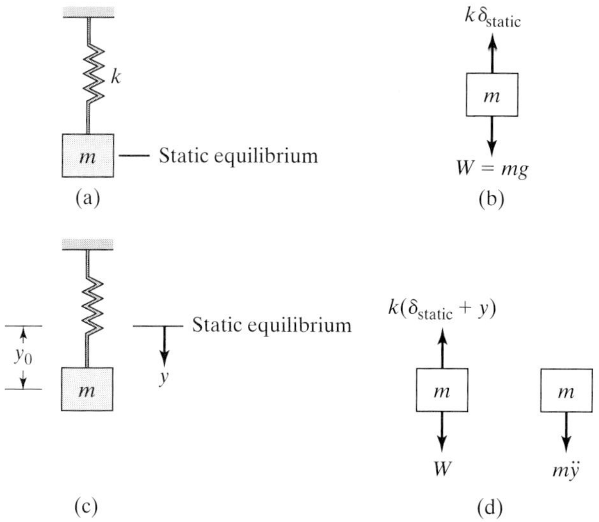

FIGURE 11.17 A simple single degree of freedom system: (a) the system, (b) the free-body diagram of the system in static equilibrium, (c) displacing mass by y_0, and (d) the free-body diagram for the oscillating mass.

We then apply Newton's second law, which leads to

$$\sum F_y - m\ddot{y} \tag{11.59}$$

$$-k\delta_{\text{static}} - ky + W = m\ddot{y} \tag{11.60}$$

We previously showed that $W = k\delta_{\text{static}}$; therefore, Eq. (11.60) reduces to

$$m\ddot{y} + ky = 0 \tag{11.61}$$

Referring to Figure 11.17(c), you are reminded that y is measured from the static equilibrium position of the system. It is customary to write Eq. (11.61) in the following manner:

$$\ddot{y} + \omega_n^2 y = 0 \tag{11.62}$$

where

$$\omega_n^2 = \frac{k}{m} = \text{undamped natural circular frequency of the system (rad/s)} \tag{11.63}$$

In order to solve the governing differential equation of motion (Eq. 11.62), we need to first define the initial conditions. Since Eq. (11.62) is a second-order differential equation, two initial conditions are required. Let's assume that at time $t = 0$, we pull the mass down by a distance of y_0 and then release it without giving the mass any initial velocity. Then, for time $t = 0$, we can write the following initial conditions:

$$y = y_0 \text{ (or written alternatively as } y(0) = y_0) \tag{11.64}$$

and

$$\dot{y} = 0 \text{ (or } \dot{y}(0) = 0) \tag{11.65}$$

For this simple degree of freedom system, the general solution to the governing differential equation is then given by

$$y(t) = c_1 \sin \omega_n t + c_2 \cos \omega_n t \tag{11.66}$$

You should know from your studies in differential equation class how to obtain Eq. (11.66). Moreover, you should recall that the solution to a differential equation must satisfy the differential equation. In other words, if we were to substitute the solution, Eq. (11.66), back into the governing differential equation of motion, Eq. (11.62), the outcome must be zero. To demonstrate this point, we substitute the solution back into the governing differential equation:

$$\overbrace{(-c_1\omega_n^2 \sin \omega_n t - c_2\omega_n^2 \cos \omega_n t)}^{\ddot{y}} + \omega_n^2 \overbrace{(c_1 \sin \omega_n t + c_2 \cos \omega_n t)}^{y} = 0$$
$$0 = 0 \qquad \text{Q.E.D.}$$

Applying the initial condition $y(0) = y_0$, we have

$$y_0 = c_1 \sin (0) + c_2 \cos (0)$$

which leads to $c_2 = y_0$, and applying $\dot{y}(0) = 0$ leads to

$$\dot{y} = c_1\omega_n \cos \omega_n t - c_2 \omega_n \sin \omega_n t$$
$$0 = c_1 \omega_n \cos(0) - c_2\omega_n \sin(0)$$
$$c_1 = 0$$

After substituting for c_1 and c_2 in Eq. (11.66), the solution representing the location of the mass (from the static equilibrium position) as a function of time is given by

$$y(t) = y_0 \cos \omega_n t \tag{11.67}$$

and the velocity and acceleration of the mass are given by

$$\dot{y}(t) = \frac{dy}{dt} = -y_0\, \omega_n \sin \omega_n t \tag{11.68}$$

$$\ddot{y}(t) = \frac{d^2y}{dt^2} = -y_0\, \omega_n^2 \cos \omega_n t \tag{11.69}$$

In order to look at the harmonic behavior of the system, we have plotted the position of the mass in Figure 11.18.

Using Figure 11.18, we can now define period and frequency for a dynamic system. As shown in Figure 11.18, period T is defined as the time that it takes for the mass m to complete one cycle, typically measured in seconds. On the other hand, frequency f is defined as the number of cycles per second and is expressed in Hertz. The relationship between frequency and period is given by

$$f = \frac{1}{T} \tag{11.70}$$

This is a good place to point out the difference between the circular frequency ω, which is expressed in radians per second (rad/s), and frequency f, which is expressed in cycles

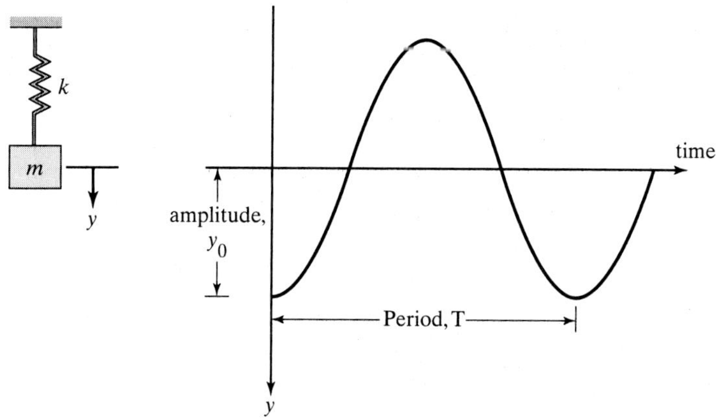

FIGURE 11.18 The behavior of a simple single degree of freedom system.

per second (hz). The relationship between ω and *f* is given by

$$\omega = 2\pi f \tag{11.71}$$

$$\omega\left(\frac{\text{radians}}{\text{second}}\right) = \left(\frac{2\pi\ \text{radians}}{\text{cycle}}\right) f\left(\frac{\text{cycles}}{\text{second}}\right)$$

EXAMPLE 11.3

Consider the simple degree of freedom system shown in the accompanying figure. We are interested in determining the position and the velocity of the 4 kg mass as a function of time. The spring has a stiffness value of 39.5 N/cm. To start the vibration, we will pull down the mass by 2 cm and then release it with no initial velocity.

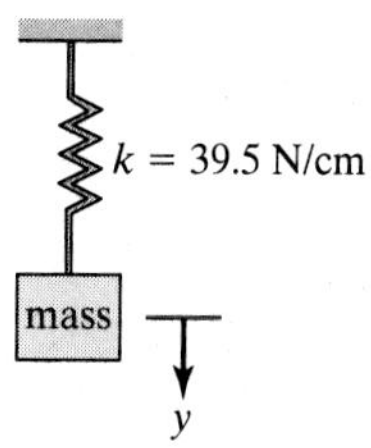

The natural circular frequency is calculated from Eq. (11.63),

$$\omega_n = \sqrt{\frac{k}{m}} = \sqrt{\frac{(39.5\ \text{N/cm})(100\text{cm}/1\text{m})}{4\text{kg}}} = 31.42\ \text{rad/s}$$

and the frequency and period are

$$f = \frac{\omega_n}{2\pi} = \frac{31.42}{2\pi} = 5\ \text{hz}$$

$$T = \frac{1}{f} = \frac{1}{5} = 0.2\ \text{s}$$

It takes 0.2 seconds for the mass to complete one cycle. When designing for dynamic systems, the knowledge of magnitude of forces transmitted to the support, or the foundation, is important to ensure the integrity of the foundation. For Example 11.3, the magnitude of the force transmitted to the support is determined from

$$R(t) = ky + W = ky_0 \cos \omega_n t + W$$

Substituting for $W = mg = (4\text{ kg})(9.81\text{ m/s}^2) = 39.2\text{ N}$, we get

$$R(t) = (39.5)(2)\cos(31.42t) + 39.2$$

The maximum reaction force occurs when the spring is stretched the maximum amount and whenever the value of $\cos(31.42t)$ becomes 1.

$$R_{max} = (39.5)(2) + 39.2 = 118.2\text{ N}$$

Note that for the given system, the value of R_{max} depends on the initial displacement. Also note that the support reaction should be equal to the weight of the system when $y = 0$.

Forced Vibration of a Single Degree of Freedom System

The vibration of a structural system may be induced by a number of sources such as wind, earthquake, or an unbalanced rotating machinery housed on a floor of a building. These sources of excitations may create forces that are sinusoidal, sudden, random, or that vary with time in a certain manner. We can learn a lot about the behavior of a system by considering basic excitation functions, such as a sinusoidal, step, or a ramp function. Moreover, to simplify presentation and since for most structures damping is relatively small, we will not consider it in our forthcoming presentations. Let us now consider a spring mass system subjected to a sinusoidal force, as shown in Figure 11.19. The motion of the system is governed by

$$m\ddot{y} + ky = F_0 \sin \omega t \tag{11.72}$$

Dividing both sides of Eq. (11.72) by mass m,

$$\ddot{y} + \frac{k}{m}y = \frac{F_0}{m}\sin \omega t \tag{11.73}$$

and by letting $\omega_n^2 = \dfrac{k}{m}$, we have

$$\ddot{y} + \omega_n^2 y = \frac{F_0}{m}\sin \omega t \tag{11.74}$$

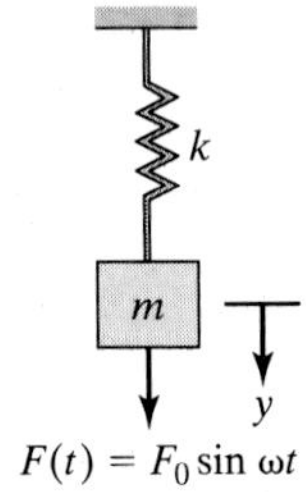

FIGURE 11.19 A spring mass system subjected to a harmonic forcing function.

The general solution to Eq. (11.74) has two parts: a homogenous solution and a particular solution. As shown previously, the homogenous solution y_h is given by

$$y_h(t) = A \sin \omega_n t + B \cos \omega_n t \tag{11.75}$$

For the particular solution y_p, we assume a function of the form

$$y_p(t) = Y_0 \sin \omega t \tag{11.76}$$

and differentiating it to get $\ddot{y}_p$, we get

$$\dot{y}_p(t) = Y_0 \omega \cos \omega t \tag{11.77}$$

$$\ddot{y}_p(t) = -Y_0 \omega^2 \sin \omega t$$

Substituting for $\ddot{y}_p$ and y_p into Eq. (11.74)

$$-Y_0 \omega^2 \sin \omega t + \omega_n^2 Y_0 \sin \omega t = \frac{F_0}{m} \sin \omega t$$

and solving for Y_0, we have

$$Y_0 = \frac{\frac{F_0}{m}}{-\omega^2 + \omega_n^2} = \frac{\frac{\frac{F_0}{m}}{\omega_n^2}}{\frac{-\omega^2 + \omega_n^2}{\omega_n^2}} = \frac{\frac{F_0}{k}}{1 - \left(\frac{\omega}{\omega_n}\right)^2} \tag{11.78}$$

Note that upon arriving at the final expression for Y_0, we divided the numerator and the denominator by ω_n^2, and we substituted for $\omega_n^2 = \frac{k}{m}$.

The forced behavior of the spring mass system to the harmonic forcing function is then given by

$$y(t) = \overbrace{A \cos \omega_n t + B \sin \omega_n t}^{\text{Natural Response}} + \overbrace{\frac{\frac{F_0}{k}}{1 - \left(\frac{\omega}{\omega_n}\right)^2} \sin \omega t}^{\text{Forced Response}} \tag{11.79}$$

It is important to point out that the natural response of the system will eventually die out because every system has some inherent damping. Consequently, in the upcoming discussions we will focus only on the forced response. To shed more light on the response of a single degree of freedom system to a sinusoidal force, using Eq. (11.78), we have plotted the ratio of the amplitude of the forced response Y_0 to the static deflection $\frac{F_0}{k}$ (that is caused if the force is applied statically) as a function of frequency ratio $\frac{\omega}{\omega_n}$. The response is shown in Figure 11.20.

It is clear from examining Figure 11.20 that as the frequency ratio approaches unity, the amplitude of forced oscillation becomes very large. This condition, known as resonance, is highly undesirable.

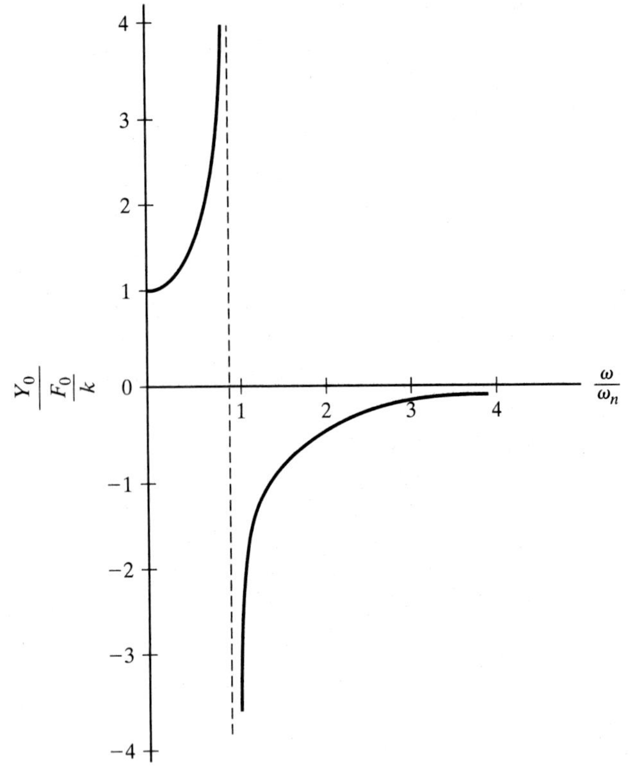

FIGURE 11.20 Amplitude of the forced behavior as a function of the frequency ratio.

Forced Vibration Caused by an Unbalanced Rotating Mass

As we mentioned earlier, vibration of a system may be induced by an unbalanced rotating mass within a machine. In machines the vibration occurs when the center of mass of the rotating component does not coincide with its center of rotation. The vibration of an unbalanced machine may be modeled by a single degree of freedom system that is excited by a sinusoidal forcing function. The sinusoidal forcing function results from the vertical component of the normal acceleration of unbalanced mass, as shown in Figure 11.21.

For these situations, the forcing function is represented by

$$F(t) = m_0 e\,\omega^2 \sin \omega t \tag{11.80}$$

where m_0 is the unbalanced mass, e is eccentricity, and ω is the angular speed of the rotating component. Comparing Eq. (11.80) to Eq. (11.72), we note that $F_0 = m_0 e\,\omega^2$ and substituting for F_0 in Eq. (11.79), and considering only the forced response we get

$$y(t) = \frac{\overbrace{\dfrac{m_0 e \omega^2}{k}}^{Y_0}}{1 - \left(\dfrac{\omega}{\omega_n}\right)^2} \sin \omega t \tag{11.81}$$

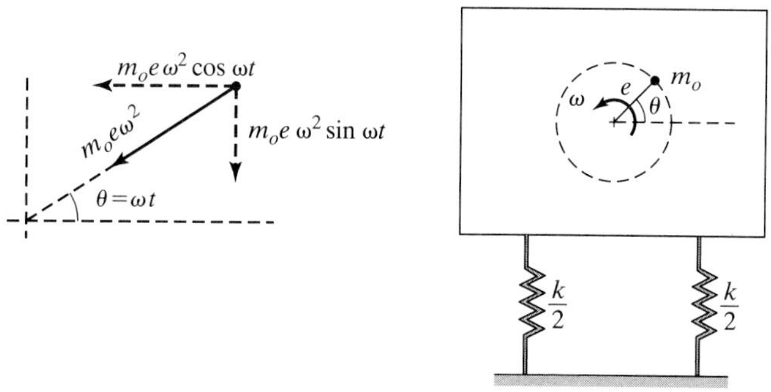

FIGURE 11.21 An unbalanced rotating machine.

and rewriting the amplitude of the vibration Y_0, by substituting $k = m\omega_n^2$, and regrouping,

$$Y_0 = \frac{\dfrac{m_0 e\omega^2}{k}}{1 - \left(\dfrac{\omega}{\omega_n}\right)^2} = \frac{\dfrac{m_0 e\omega^2}{m\omega_n^2}}{1 - \left(\dfrac{\omega}{\omega_n}\right)^2} \Rightarrow \frac{Y_0 m}{m_0 e} = \frac{\dfrac{\omega^2}{\omega_n^2}}{1 - \left(\dfrac{\omega}{\omega_n}\right)^2} \tag{11.82}$$

From examining the numerator $\dfrac{m_0 e\omega^2}{m}$ in Eq. (11.82), we note that by increasing the mass of the system m, we can reduce the amplitude of vibration. You may have seen turbine or large pumps mounted on heavy concrete blocks to reduce the amplitude of unexpected vibration. Moreover, we can show that as

$$\frac{\omega}{\omega_n} \ll 1 \quad \Rightarrow \quad \frac{Y_0 m}{m_0 e} = 0 \tag{11.83}$$

and

$$\frac{\omega}{\omega_n} \gg 1 \quad \Rightarrow \quad \frac{Y_0 m}{m_0 e} = -1 \tag{11.84}$$

Using Eq. (11.82), we have plotted $\dfrac{Y_0 m}{m_0 e}$ as a function of frequency ratio $\dfrac{\omega}{\omega_n}$ and have shown it in Figure 11.22. The behavior of the system as a function of frequency ratio is self-evident when examining Figure 11.22.

The derivation of the response of a single degree of freedom mass spring system to a suddenly applied force, a ramp function, or a suddenly applied force that decays with time is left as an exercise. See problems 4 and 5 at the end of this chapter.

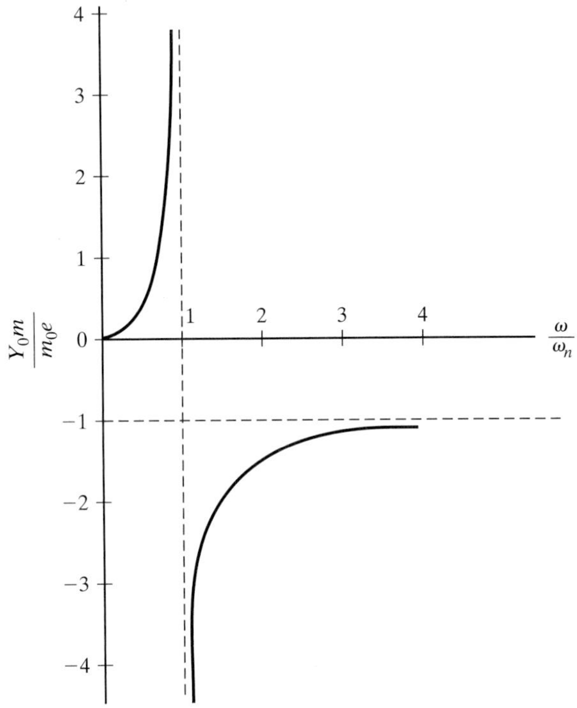

FIGURE 11.22 A plot of $\frac{Y_0 m}{m_0 e}$ as a function of frequency ratio $\frac{\omega}{\omega_n}$.

Forces Transmitted to Foundation

As we mentioned earlier, when designing for dynamic systems, the knowledge of the forces transmitted to the support or the foundation is important to ensure the integrity of the foundation. The relationship between the vibration of the mass and the forces transmitted to the foundation is depicted in Figure 11.23.

As shown in Figure 11.23, forces are transmitted to the foundation through the springs. The magnitude of these forces, which varies with time, is given by

$$F(t) = ky(t) = \frac{k\dfrac{F_0}{k}}{1 - \left(\dfrac{\omega}{\omega_n}\right)^2} \sin \omega t = \frac{F_0}{1 - \left(\dfrac{\omega}{\omega_n}\right)^2} \sin \omega t \tag{11.85}$$

However, for most engineering applications, we are interested in determining maximum magnitude of forces transmitted to the foundation, which occurs when $\sin \omega t$ has a value of 1 in Eq. (11.85), and is given by

$$F_{\max} = \frac{F_0}{1 - \left(\dfrac{\omega}{\omega_n}\right)^2} \tag{11.86}$$

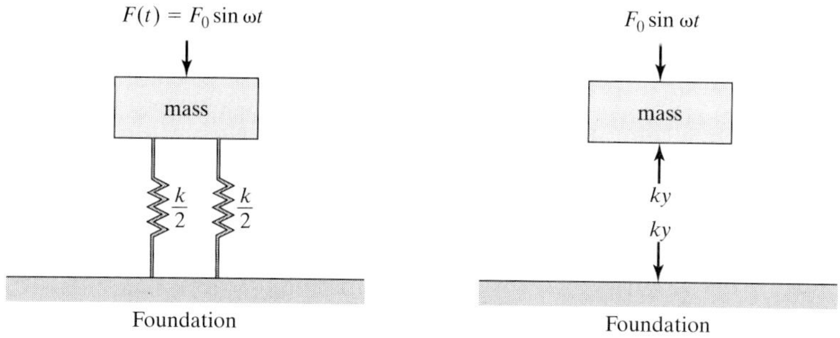

FIGURE 11.23 Forces transmitted to a foundation.

In the field of vibration, it is customary to define a transmission ratio or transmissibility TR as the ratio of F_{max} to the static magnitude F_0:

$$TR = \left| \frac{F_{max}}{F_0} \right| = \left| \frac{\dfrac{F_0}{1 - \left(\dfrac{\omega}{\omega_n}\right)^2}}{F_0} \right| = \left| \frac{1}{1 - \left(\dfrac{\omega}{\omega_n}\right)^2} \right| \tag{11.87}$$

Figure 11.24 shows the relationship between transmissibility and the frequency ratio, as given by Eq. (11.87).

It is clear from examining Figure 11.24, as the frequency ratio approaches unity, the transmissibility approaches very large values, which could have catastrophic consequences. Furthermore, Figure 11.24 shows that in order to keep transmissibility

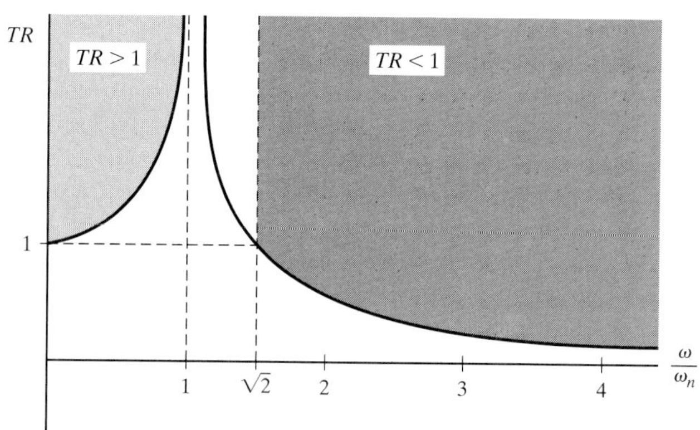

FIGURE 11.24 Transmissibility as a function of frequency ratio.

low, machines should operate at frequencies that are much greater than their natural frequencies.

Support Excitation

In this section we review support excitation. A spring mass system that is being excited by its support is shown in Figure 11.25. Using Newton's second law, we can formulate the governing equation of motion of the oscillating mass as follows:

$$-k(y_1 - y_2) = m\ddot{y}_1 \tag{11.88}$$

Note that the magnitude of the spring force depends on the relative position of the mass with respect to its support. Separating the excitation term from the response, we have

$$m\ddot{y}_1 + ky_1 = ky_2 \tag{11.89}$$

Dividing both sides of Eq. (11.89) by m, we get

$$\ddot{y}_1 + \omega_n^2 y_1 = \omega_n^2 y_2 = \omega_n^2 Y_2 \sin \omega t \tag{11.90}$$

We can use the solution we obtained earlier to the differential equation given by Eq. (11.74). By comparing Eq. (11.90) to Eq. (11.74), and noting that $\omega_n^2 Y_2 = \frac{F_0}{m}$, the solution to Eq. (11.90) then becomes

$$y_1(t) = \frac{\dfrac{\omega_n^2 Y_2}{\omega_n^2}}{1 - \left(\dfrac{\omega}{\omega_n}\right)^2} \sin \omega t = \overbrace{\frac{Y_2}{1 - \left(\dfrac{\omega}{\omega_n}\right)^2}}^{\text{amplitude of mass}} \sin \omega t \tag{11.91}$$

To get a physical feeling for what the solution represents, let's turn our attention to the amplitude of mass, as given by Eq. 11.91. If we replace the spring with a stiff rod and excite the support, as shown in Figure 11.26(a), we note that because of the rod's large k

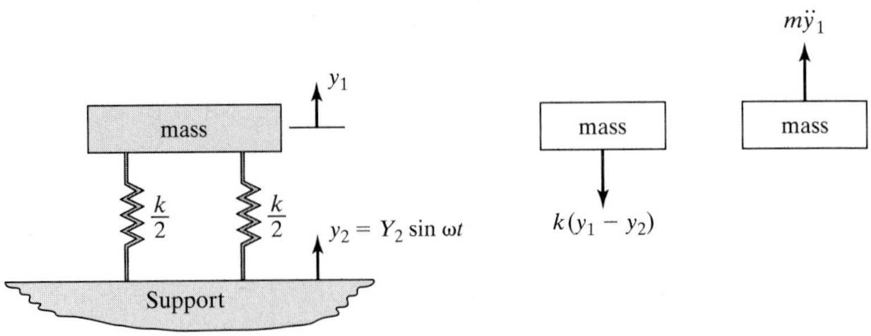

FIGURE 11.25 Support excitation.

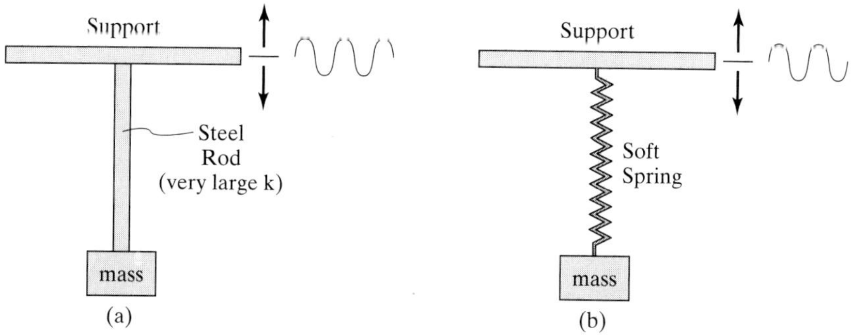

FIGURE 11.26 Experiments demonstrating the relationship between the amplitude of support excitation and the amplitude of the mass.

value and consequently large ω_n, the frequency ratio would be very small, $\frac{\omega}{\omega_n} \ll 1$, and consequently the amplitude of oscillation of mass would equal Y_2. As you would expect, the mass will move with the same amplitude as the support. Now, if we replace the rod with a soft spring having a small k value (thus a small ω_n), the frequency ratio becomes $\frac{\omega}{\omega_n} \gg 1$. Substituting a large value for frequency ratio in Eq. (11.91) will show that the mass will vibrate with very small amplitude and will appear nearly stationary.

Multiple Degrees of Freedom

In the previous sections we considered the natural and forced behavior of a single degree of freedom system. Next, we will demonstrate some of the important characteristics of a multiple degrees of freedom system using a simple two degrees of freedom system. Consider the two degrees of freedom system shown in Figure 11.27. We are interested in determining the natural frequencies of the system shown. We begin by formulating the governing equations of motion for each mass. To write the equations of motion, we will initiate the free vibration by displacing the masses such that $x_2 > x_1$.

Using the free-body diagrams shown, the equations of motion are

$$m_1\ddot{x}_1 + 2kx_1 - kx_2 = 0 \tag{11.92}$$

$$m_2\ddot{x}_2 - kx_1 + 2kx_2 = 0 \tag{11.93}$$

or, in a matrix form,

$$\begin{bmatrix} m_1 & 0 \\ 0 & m_2 \end{bmatrix} \begin{Bmatrix} \ddot{x}_1 \\ \ddot{x}_2 \end{Bmatrix} + \begin{bmatrix} 2k & -k \\ -k & 2k \end{bmatrix} \begin{Bmatrix} x_1 \\ x_2 \end{Bmatrix} = \begin{Bmatrix} 0 \\ 0 \end{Bmatrix}$$

Note that Eqs. (11.92) and (11.93) are second-order homogenous differential equations. Also note that these equations are coupled because both x_1 and x_2 appear in each equation. This type of system is called elastically coupled and may be represented in the general matrix form by

$$[\boldsymbol{M}]\{\ddot{\boldsymbol{x}}\} + [\boldsymbol{K}]\{\boldsymbol{x}\} = 0 \tag{11.94}$$

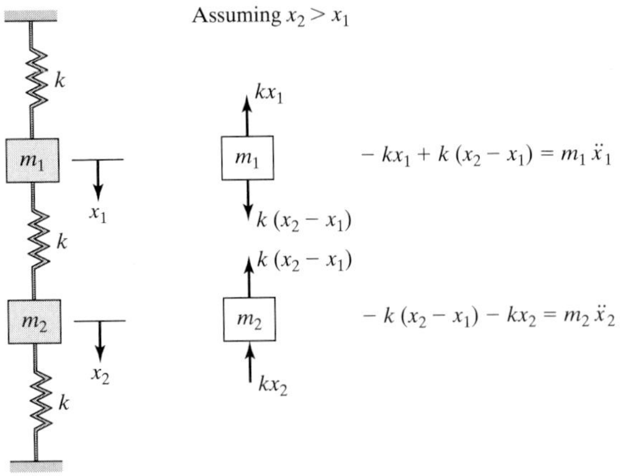

FIGURE 11.27 A schematic diagram of an elastic system with two degrees of freedom.

where $[M]$ and $[K]$ are the mass and the stiffness matrices respectively. We can simplify Eqs. (11.92) and (11.93) by dividing both sides of each equation by the values of the respective masses:

$$\ddot{x}_1 + \frac{2k}{m_1}x_1 - \frac{k}{m_1}x_2 = 0 \tag{11.95}$$

$$\ddot{x}_2 - \frac{k}{m_2}x_1 + \frac{2k}{m_2}x_2 = 0 \tag{11.96}$$

Using matrix notation, we premultiply the matrix form of the equations of motion by the inverse of the mass matrix $[M]^{-1}$, which leads to

$$\{\ddot{x}\} + [M]^{-1}[K]\{x\} = 0 \tag{11.97}$$

As a next step, we assume a harmonic solution of the form $x_1(t) = X_1 \sin(\omega t + \phi)$ and $x_2(t) = X_2 \sin(\omega t + \phi)$ [or in matrix form, $\{x\} = \{X\} \sin(\omega t + \phi)$] and substitute the assumed solutions into the differential equations of motion, Eqs. (11.95) and (11.96), to create a set of linear algebraic equations.

This step leads to

$$-\omega^2 X_1 \sin(\omega t + \phi) + \frac{2k}{m_1}X_1\sin(\omega t + \phi) - \frac{k}{m_1}X_2\sin(\omega t + \phi) = 0$$

$$-\omega^2 X_2 \sin(\omega t + \phi) - \frac{k}{m_2}X_1\sin(\omega t + \phi) + \frac{2k}{m_2}X_2\sin(\omega t + \phi) = 0$$

After simplifying the $\sin(\omega t + \phi)$ terms, we get

$$-\omega^2 \begin{Bmatrix} X_1 \\ X_2 \end{Bmatrix} + \begin{bmatrix} \dfrac{2k}{m_1} & -\dfrac{k}{m_1} \\ -\dfrac{k}{m_2} & \dfrac{2k}{m_2} \end{bmatrix} \begin{Bmatrix} X_1 \\ X_2 \end{Bmatrix} = \begin{Bmatrix} 0 \\ 0 \end{Bmatrix} \tag{11.98}$$

or, in a general matrix form,

$$-\omega^2\{\boldsymbol{X}\} + [\boldsymbol{M}]^{-1}[\boldsymbol{K}]\{\boldsymbol{X}\} = 0 \qquad (11.99)$$

Note that $\{\boldsymbol{x}\} = \begin{Bmatrix} x_1(t) \\ x_2(t) \end{Bmatrix}$ represents the position of each mass as the function of time, the $\{\boldsymbol{X}\} = \begin{Bmatrix} X_1 \\ X_2 \end{Bmatrix}$ matrix denotes the amplitudes of each oscillating mass, and ϕ is the phase angle. Equation (11.98) may be written as

$$-\omega^2\begin{bmatrix} 1 & 0 \\ 0 & 1 \end{bmatrix}\begin{Bmatrix} X_1 \\ X_2 \end{Bmatrix} + \begin{bmatrix} \frac{1}{m_1} & 0 \\ 0 & \frac{1}{m_2} \end{bmatrix}\begin{bmatrix} 2k & -k \\ -k & 2k \end{bmatrix}\begin{Bmatrix} X_1 \\ X_2 \end{Bmatrix} = 0 \qquad (11.100)$$

or by

$$\left[\begin{bmatrix} \frac{2k}{m_1} & -\frac{k}{m_1} \\ -\frac{k}{m_2} & \frac{2k}{m_2} \end{bmatrix} - \omega^2\begin{bmatrix} 1 & 0 \\ 0 & 1 \end{bmatrix}\right]\begin{Bmatrix} X_1 \\ X_2 \end{Bmatrix} = \begin{Bmatrix} 0 \\ 0 \end{Bmatrix} \qquad (11.101)$$

Simplifying Eq. (11.101) further, we have

$$\begin{bmatrix} -\omega^2 + \frac{2k}{m_1} & -\frac{k}{m_1} \\ -\frac{k}{m_2} & -\omega^2 + \frac{2k}{m_2} \end{bmatrix}\begin{Bmatrix} X_1 \\ X_2 \end{Bmatrix} = 0 \qquad (11.102)$$

Problems with governing equations of the type (11.99) or (11.102) have nontrivial solutions only when the determinant of the coefficient matrix is zero. Let's assign some numerical values to the above example problem and proceed with the solution. Let $m_1 = m_2 = 0.1$ kg and $k = 100$ N/m. Forming the determinant of the coefficient matrix and setting it equal to zero, we get

$$\begin{vmatrix} -\omega^2 + 2000 & -1000 \\ -1000 & -\omega^2 + 2000 \end{vmatrix} = 0 \qquad (11.103)$$

$$(-\omega^2 + 2000)(-\omega^2 + 2000) - (-1000)(-1000) = 0 \qquad (11.104)$$

Simplifying Eq. (11.104), we have

$$\omega^4 - 4000\omega^2 + 3{,}000{,}000 = 0 \qquad (11.105)$$

Equation (11.105) is called the characteristic equation, and its roots are the natural frequencies of the system.

$$\omega_1^2 = \lambda_1 = 1000\,(\text{rad/s})^2 \quad \text{and} \quad \omega_1 = 31.62 \text{ rad/s}$$
$$\omega_2^2 = \lambda_2 = 3000\,(\text{rad/s})^2 \quad \text{and} \quad \omega_2 = 54.77 \text{ rad/s}$$

Once the ω^2 values are known, they can be substituted back into Equation (11.102) to solve for the relationship between X_1 and X_2. The relationship between the amplitudes of mass oscillating at natural frequencies is called natural modes. We can use either relationship (rows) in Eq. (11.102).

$$(-\omega^2 + 2000)\,\boldsymbol{X}_1 - 1000\boldsymbol{X}_2 = 0 \quad \text{and substituting for } \omega_1^2 = 1000$$
$$(-1000 + 2000)\boldsymbol{X}_1 - 1000\,\boldsymbol{X}_2 = 0 \quad \rightarrow \frac{\boldsymbol{X}_2}{\boldsymbol{X}_1} = 1$$

Or, using the second row,

$$-1000\,\boldsymbol{X}_1 + (-\omega^2 + 2000)\boldsymbol{X}_2 = 0 \quad \text{and substituting for } \omega_1^2 = 1000$$
$$-1000\boldsymbol{X}_1 + (-1000 + 2000)\boldsymbol{X}_2 = 0 \rightarrow \frac{\boldsymbol{X}_2}{\boldsymbol{X}_1} = 1$$

As expected, the results are identical. The second mode is obtained in a similar manner by substituting for $\omega_2^2 = 3000$ in Eq. (11.102).

$$(-\omega^2 + 2000)\,\boldsymbol{X}_1 - 1000\boldsymbol{X}_2 = 0 \quad \text{and substituting for } \omega_2^2 = 3000$$
$$(-3000 + 2000)\boldsymbol{X}_1 - 1000\boldsymbol{X}_2 = 0 \rightarrow \frac{\boldsymbol{X}_2}{\boldsymbol{X}_1} = -1$$

It is important to note again that the solution of the eigenvalue problems leads to establishing a relationship among the unknowns, not specific values. To shed more light on what the above natural frequencies and modes represent, consider the following experiment. Pull down both mass one and two by, say, 1 inch ($X_1 = X_2 = 1$), and then release the system. Under these initial conditions, the system will oscillate at the first natural frequency ($\omega_1 = 31.62$ rad/s). However, if you were to give the system the following initial conditions: pull mass one up by 1 inch and pull mass two down by 1 inch ($X_2 = -X_1 = 1$), then release them, the system would oscillate at the second natural frequency ($\omega_2 = 54.77$ rad/s). Any other initial conditions will result in the system oscillating such that both natural frequencies affect its behavior.

Equations of Motion for Forced Vibration of Multiple Degrees of Freedom

In the previous section we showed that the general form of equations of motion for free vibration of multiple degrees of freedom is

$$[\boldsymbol{M}]\{\ddot{\boldsymbol{x}}\} + [\boldsymbol{K}]\{\boldsymbol{x}\} = 0 \tag{11.106}$$

With damping $[\boldsymbol{C}]$ included, the equations of motion for free vibration of multiple degrees of freedom becomes

$$[\boldsymbol{M}]\{\ddot{\boldsymbol{x}}\} + [\boldsymbol{C}]\{\dot{\boldsymbol{x}}\} + [\boldsymbol{K}]\{\boldsymbol{x}\} = 0 \tag{11.107}$$

The natural response of a system with a multiple degrees of freedom can also be obtained using modal analysis. Modal analysis involves the uncoupling of the differential equations of motion using what is called *principal coordinates*. The basic idea is to represent the motion of each part (mass) by a single coordinate that makes no reference to any other coordinate. Once the equations of motion are uncoupled, then each independent equation is treated as a single degree of freedom system. To better understand the idea of principal coordinates, consider the two degrees of freedom system shown in Figure 11.27 and assume that $m_1 = m_2 = m$, then the equations of motion become

$$\ddot{x}_1 + \frac{2k}{m}x_1 - \frac{k}{m}x_2 = 0 \tag{11.95b}$$

$$\ddot{x}_2 - \frac{k}{m}x_1 + \frac{2k}{m}x_2 = 0 \tag{11.96b}$$

Substitute in Eqs. (11.95) and (11.96) $m_1 = m_2 = m$ to arrive at Eqs. (11.95b) and (11.96b). Now consider first adding Eqs. (11.95b) and (11.96b) and then subtracting Eq. (11.96b) from Eq. (11.95 b). These operations result in

$$\ddot{x}_1 + \ddot{x}_2 + \frac{k}{m}(x_1 + x_2) = 0 \qquad \Rightarrow \ddot{p}_1 + \frac{k}{m}p_1 = 0$$

$$\ddot{x}_1 - \ddot{x}_2 + \frac{3k}{m}(x_1 - x_2) = 0 \qquad \Rightarrow \ddot{p}_2 + \frac{3k}{m}p_2 = 0$$

where $p_1 = x_1 + x_2$ and $p_2 = x_1 - x_2$. As you can see by using the principal coordinates p_1 and p_2, we were able to decouple the equations of motion. The natural frequencies of the system are $\omega_1 = \sqrt{\frac{k}{m}}$ and $\omega_2 = \sqrt{\frac{3k}{m}}$. It is left as an exercise for you to verify these results by determining the roots of characteristics equation. Although the decoupling of equations of motion are much more involved than what is shown here, the above example demonstrates the basic idea of principal coordinates and decoupling. The modal analysis is also used in determining the natural and the forced response of multiple degrees of freedom systems with damping.

The matrix form of equations of motion for multiple degrees of freedom subjected to forces is given by

$$[\boldsymbol{M}]\{\ddot{\boldsymbol{x}}\} + [\boldsymbol{K}]\{\boldsymbol{x}\} = \{\boldsymbol{F}\} \tag{11.108}$$

where $\{\boldsymbol{F}\}$ is the force matrix. And with damping, we have the following relationship:

$$[\boldsymbol{M}]\{\ddot{\boldsymbol{x}}\} + [\boldsymbol{C}]\{\dot{\boldsymbol{x}}\} + [\boldsymbol{K}]\{\boldsymbol{x}\} = \{\boldsymbol{F}\} \tag{11.109}$$

Up to this point, we have explained how to approximate the behavior of systems with distributed mass and elastic properties by discrete models that consist of lumped masses and equivalent (or bulk) stiffness, and a finite number of degrees of freedom. Moreover, as you have seen, these models are represented by ordinary differential equations, subject to initial conditions, whose solutions render the natural or the forced responses of the system.

Because rods and beams play significant role in many engineering applications, we will discuss their finite element formulations in detail in Sections 11.4 and 11.5. Rods and beams are continuous systems which theoretically possess infinite numbers of degrees of freedom and natural frequencies. However, for most practical problems, only the first few natural frequencies are important. In general, the governing equations of motions of continuous systems are partial differential equations whose exact solutions require both boundary and initial conditions. Except for a few simple problems, the solutions are very complex and hard to find. Therefore, we resort to numerical approximation of discrete models to solve many practical problems. We will discuss the finite element formulation of bars, beams, and frames using Lagrange's equations in the forthcoming sections.

11.3 LAGRANGE'S EQUATIONS

In the previous sections we derived the governing equations of motions of vibrating systems using Newton's second law. In this section we introduce another approach, which uses Lagrange's equations to formulate equations of motions. Lagrange's equations are given by

$$\frac{d}{dt}\left(\frac{\partial T}{\partial \dot{q}_i}\right) - \frac{\partial T}{\partial q_i} + \frac{\partial \Lambda}{\partial q_i} = Q_i \quad (i = 1, 2, 3, \ldots, n) \tag{11.110}$$

where

t = time

T = kinetic energy of the system

q_i = coordinate system

$\dot{q}_i$ = time derivate of the coordinate system representing velocity

Λ = potential energy of the system

Q_i = nonconservative forces or moments

We will use Example 11.4 to demonstrate how to use Lagrange's equations to formulate equations of motion for a dynamic system.

EXAMPLE 11.4

Use Lagrange's equations to formulate the governing equations of motion for the systems shown in Figure 11.28.

The spring mass system shown in Figure 11.28(a) is a single degree of freedom system that requires only one coordinate q to describe its behavior. To apply Eq. (11.110), we must first express the kinetic and potential energies of the system in terms of the coordinate q and its derivative $\dot{q}$. Note that for this problem, we let $q = x$. The kinetic

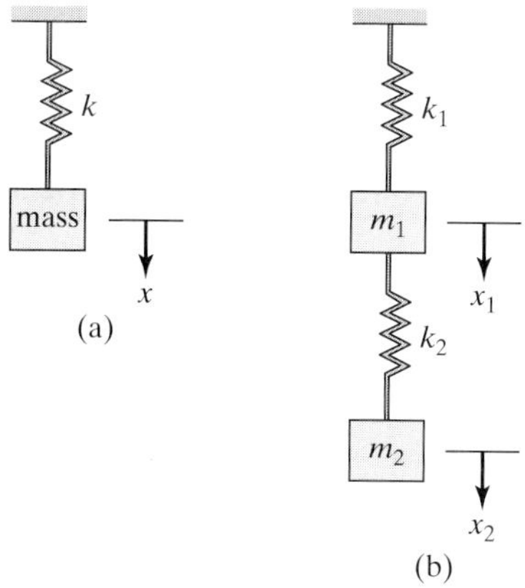

FIGURE 11.28 Spring mass systems of Example 11.4.

and potential energies of the system are

$$T = \frac{1}{2}m\dot{x}^2$$

$$\Lambda = \frac{1}{2}kx^2$$

Next, as required by Lagrange's equation, we differentiate the kinetic energy term with respect to $\dot{q}$, or in our case, $\dot{x}$:

$$\frac{\partial T}{\partial \dot{q}} = \frac{\partial T}{\partial \dot{x}} = \frac{\partial}{\partial \dot{x}}\left(\frac{1}{2}m\dot{x}^2\right) = (2)\left(\frac{1}{2}\right)m\dot{x} = m\dot{x}$$

Then, taking the time derivative of $\frac{\partial T}{\partial \dot{x}}$, we get

$$\frac{d}{dt}\left(\frac{\partial T}{\partial \dot{x}}\right) = \frac{d}{dt}(m\dot{x}) = m\ddot{x}$$

Because T is a function of $\dot{x}$ and not x, the $\frac{\partial T}{\partial x} = 0$. Evaluating the potential term, $\frac{\partial \Lambda}{\partial q}$, in Eq. (11.110), we have

$$\frac{\partial \Lambda}{\partial q} = \frac{\partial \Lambda}{\partial x} = \frac{\partial}{\partial x}\left(\frac{1}{2}kx^2\right) = kx$$

Finally, substituting for each term in Eq. (11.110), we get

$$\overbrace{\frac{d}{dt}\left(\frac{\partial T}{\partial \dot{q}_i}\right)}^{m\ddot{x}} - \overbrace{\frac{\partial T}{\partial q_i}}^{0} + \overbrace{\frac{\partial \Lambda}{\partial q_i}}^{kx} = \overbrace{Q_i}^{0}$$

As expected, the governing differential equation of motion is given by

$$m\ddot{x} + kx = 0$$

The system shown in Figure 11.28(b) has two degrees of freedom; consequently, we need two coordinates, x_1 and x_2, to formulate the kinetic and potential energies.

$$T = \frac{1}{2}m_1\dot{x}_1^2 + \frac{1}{2}m_2\dot{x}_2^2$$

$$\Lambda = \frac{1}{2}k_1x_1^2 + \frac{1}{2}k_2(x_2 - x_1)^2$$

Taking the derivatives of each term, as required by Eq. (11.110), we get

$$\frac{\partial T}{\partial \dot{x}_1} = m_1\dot{x}_1 \quad \text{and} \quad \frac{d}{dt}\left(\frac{\partial T}{\partial \dot{x}_1}\right) = m_1\ddot{x}_1$$

$$\frac{\partial T}{\partial \dot{x}_2} = m_2\dot{x}_2 \quad \text{and} \quad \frac{d}{dt}\left(\frac{\partial T}{\partial \dot{x}_2}\right) = m_2\ddot{x}_2$$

$$\frac{\partial T}{\partial x_1} = \frac{\partial T}{\partial x_2} = 0$$

$$\frac{\partial \Lambda}{\partial x_1} = k_1x_1 + k_2(x_1 - x_2)$$

$$\frac{\partial \Lambda}{\partial x_2} = k_2(x_2 - x_1)$$

Substituting each term in Lagrange's equation, Eq. (11.110), we have

$$m_1\ddot{x}_1 + (k_1 + k_2)x_1 - k_2x_2 = 0$$

$$m_2\ddot{x}_2 - k_2x_1 + k_2x_2 = 0$$

or we can express the equations of motion in a matrix form by

$$\begin{bmatrix} m_1 & 0 \\ 0 & m_2 \end{bmatrix}\begin{Bmatrix} \ddot{x}_1 \\ \ddot{x}_2 \end{Bmatrix} + \begin{bmatrix} k_1 + k_2 & -k_2 \\ -k_2 & k_2 \end{bmatrix}\begin{Bmatrix} x_1 \\ x_2 \end{Bmatrix} = \begin{Bmatrix} 0 \\ 0 \end{Bmatrix}$$

11.4 FINITE ELEMENT FORMULATION OF AXIAL MEMBERS

In this section, using Lagrange's equations, we first formulate the mass matrix for an axial member and then use it to obtain the natural frequencies of an axial member. Recall that the displacement of an axial member may be expressed using one-dimensional shape functions S_i and S_j, as shown below:

$$u = S_iU_i + S_jU_j \tag{11.111}$$

The shape functions in terms of the local coordinate x shown in Figure 11.29 are given by

$$S_i = 1 - \frac{x}{L} \tag{11.112}$$

$$S_j = \frac{x}{L} \tag{11.113}$$

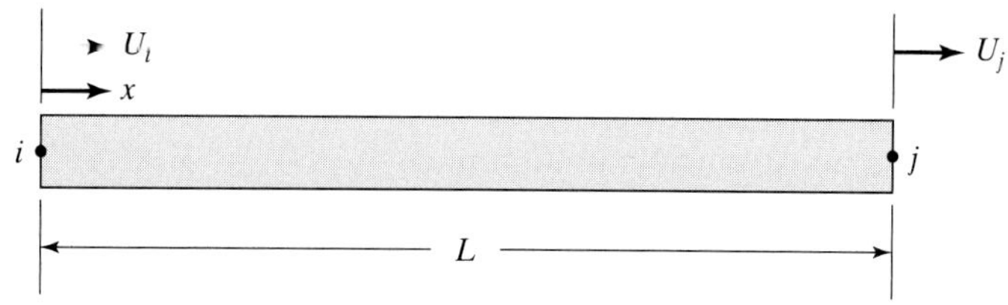

FIGURE 11.29 An axial member.

It is important to note that whereas the displacement function for a static problem depends on coordinate x only, for a dynamic problem the displacement function is a function of x and time t; that is, $u = u(x, t)$. The total kinetic energy of the member is the sum of the kinetic energies of its constituent particles (or smaller chunks).

$$T = \int_0^L \frac{\gamma}{2} \dot{u}^2 dx \qquad (11.114)$$

In Eq. (11.114), $\dot{u}$ represents the velocity of the particles along the member, and γ is mass per unit length. The velocity of the member can be expressed in terms of its nodal velocities $\dot{U}_i$ and $\dot{U}_j$ and is given by

$$\dot{u} = S_i\dot{U}_i + S_j\dot{U}_j \qquad (11.115)$$

Substituting Eq. (11.115) into Eq. (11.114), we have

$$T = \frac{\gamma}{2}\int_0^L (S_i\dot{U}_i + S_j\dot{U}_j)^2 dx \qquad (11.116)$$

And taking the derivatives as required by Lagrange's equation, Eq. (11.110), we get

$$\frac{\partial T}{\partial \dot{U}_i} = \frac{\gamma}{2}\int_0^L 2S_i(S_i\dot{U}_i + S_j\dot{U}_j)dx \qquad (11.117)$$

$$\frac{\partial T}{\partial \dot{U}_j} = \frac{\gamma}{2}\int_0^L 2S_j(S_i\dot{U}_i + S_j\dot{U}_j)dx \qquad (11.118)$$

$$\frac{d}{dt}\left(\frac{\partial T}{\partial \dot{U}_i}\right) = \gamma\left[\int_0^L S_i^2\ddot{U}_i dx + \int_0^L S_iS_j\ddot{U}_j dx\right] \qquad (11.119)$$

$$\frac{d}{dt}\left(\frac{\partial T}{\partial \dot{U}_j}\right) = \gamma\left[\int_0^L S_iS_j\ddot{U}_i dx + \int_0^L S_j^2\ddot{U}_j dx\right] \qquad (11.120)$$

Note that S_i and S_j are functions of x alone, whereas $\ddot{U}_i$ and $\ddot{U}_j$ represent accelerations at nodes i and j respectively, which are functions of time. Evaluating the integrals of Eqs. (11.119) and (11.120), we get

$$\gamma\int_0^L S_i^2 dx = \gamma\int_0^L \left(1 - \frac{x}{L}\right)^2 dx = \frac{\gamma L}{3} \qquad (11.121)$$

$$\gamma \int_0^L S_i S_j dx = \gamma \int_0^L \left(1 - \frac{x}{L}\right)\left(\frac{x}{L}\right) dx = \frac{\gamma L}{6} \tag{11.122}$$

$$\gamma \int_0^L S_j^2 dx = \gamma \int_0^L \left(\frac{x}{L}\right)^2 dx = \frac{\gamma L}{3} \tag{11.123}$$

Substituting the results of the integrals given by Eqs. (11.121) through (11.123) into Eqs. (11.119) and (11.120), leads to $[\boldsymbol{M}]\{\ddot{\boldsymbol{u}}\}$. Then for an axial member, the mass matrix becomes

$$[\boldsymbol{M}]^{(e)} = \frac{\gamma L}{6}\begin{bmatrix} 2 & 1 \\ 1 & 2 \end{bmatrix} \tag{11.124}$$

We derived the stiffness matrix for an axial element in Chapter 4, Section 4.1, which is given by

$$[\boldsymbol{K}]^{(e)} = \frac{AE}{L}\begin{bmatrix} 1 & -1 \\ -1 & 1 \end{bmatrix}$$

Next, we will use Example 11.5 to demonstrate how to use the results of this section.

EXAMPLE 11.5

Consider the 30-cm long aluminum rod shown in Figure 11.30. The rod has a modulus of elasticity $E = 70$ GPa and density $\rho = 2700\ \text{kg/m}^3$ ($\gamma = 5.4$ kg/m). The rod is fixed at one end, as shown in the figure, and we are interested in approximating the natural frequencies of the rod using the three-element model shown.

The mass matrix for each element is computed from Eq. (11.124)

$$[\boldsymbol{M}]^{(1)} = [\boldsymbol{M}]^{(2)} = [\boldsymbol{M}]^{(3)} = \frac{\gamma L}{6}\begin{bmatrix} 2 & 1 \\ 1 & 2 \end{bmatrix} = \frac{(5.4)(0.1)}{6}\begin{bmatrix} 2 & 1 \\ 1 & 2 \end{bmatrix} = \begin{bmatrix} 0.18 & 0.09 \\ 0.09 & 0.18 \end{bmatrix}$$

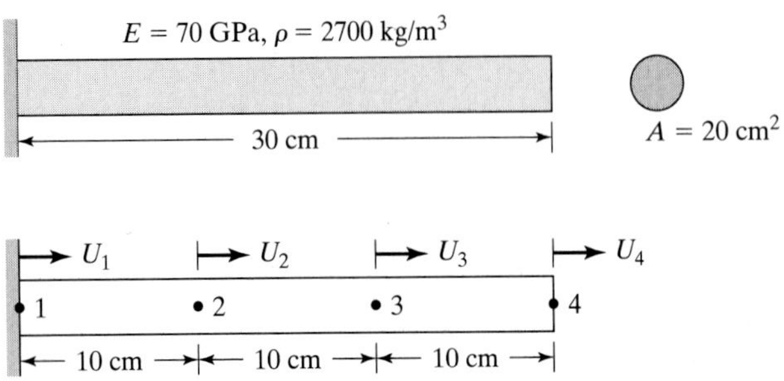

FIGURE 11.30 The aluminum rod of Example 11.5.

and the elemental stiffness matrix is

$$[\boldsymbol{K}]^{(1)} = [\boldsymbol{K}]^{(2)} = [\boldsymbol{K}]^{(3)} = \frac{AE}{L}\begin{bmatrix} 1 & -1 \\ -1 & 1 \end{bmatrix} =$$

$$\frac{(20 \times 10^{-4})(70 \times 10^{9})}{0.1}\begin{bmatrix} 1 & -1 \\ -1 & 1 \end{bmatrix} = 1.4 \times 10^{9}\begin{bmatrix} 1 & -1 \\ -1 & 1 \end{bmatrix}$$

Assembling the mass and the stiffness matrices,

$$[\boldsymbol{M}]^{(G)} = \begin{bmatrix} 0.18 & 0.09 & 0 & 0 \\ 0.09 & 0.36 & 0.09 & 0 \\ 0 & 0.09 & 0.36 & 0.09 \\ 0 & 0 & 0.09 & 0.18 \end{bmatrix}$$

$$[\boldsymbol{K}]^{(G)} = 10^{9}\begin{bmatrix} 1.4 & -1.4 & 0 & 0 \\ -1.4 & 2.8 & -1.4 & 0 \\ 0 & -1.4 & 2.8 & -1.4 \\ 0 & 0 & -1.4 & 1.4 \end{bmatrix}$$

Applying the boundary condition—because node 1 is fixed—the first rows and columns of mass and stiffness matrices are eliminated. Application of the boundary condition at node 1 reduces the mass and the stiffness matrices to

$$[\boldsymbol{M}]^{(G)} = \begin{bmatrix} 0.36 & 0.09 & 0 \\ 0.09 & 0.36 & 0.09 \\ 0 & 0.09 & 0.18 \end{bmatrix}$$

$$[\boldsymbol{K}]^{(G)} = 10^{9}\begin{bmatrix} 2.8 & -1.4 & 0 \\ -1.4 & 2.8 & -1.4 \\ 0 & -1.4 & 1.4 \end{bmatrix}$$

As previously discussed in the section dealing with the natural vibration of multiple degrees of freedom systems, to get the natural frequencies of the system, we need to solve $[\boldsymbol{M}]^{-1}[\boldsymbol{K}]\{\boldsymbol{X}\} = \omega^2\{\boldsymbol{X}\}$ or, in the case of the rod, $[\boldsymbol{M}]^{-1}[\boldsymbol{K}]\{\boldsymbol{U}\} = \omega^2\{\boldsymbol{U}\}$. Computing the inverse of the mass matrix, we get

$$[\boldsymbol{M}]^{-1} = \begin{bmatrix} 2.9915 & -0.8547 & 0.4274 \\ -0.8547 & 3.4188 & -1.7094 \\ 0.4274 & -1.7094 & 6.4103 \end{bmatrix}$$

And calculating

$$[\boldsymbol{M}]^{-1}[\boldsymbol{K}] = 10^{10}\begin{bmatrix} 0.9573 & -0.7179 & 0.1795 \\ -0.7179 & 1.3162 & -0.7179 \\ 0.3590 & -1.4359 & 1.1368 \end{bmatrix}$$

and solving for the eigenvlaues, we obtain the natural frequencies: $\omega_1 = 1.5999 \times 10^5$ rad/s, $\omega_2 = 0.8819 \times 10^5$ rad/s, $\omega_3 = 0.2697 \times 10^5$ rad/s.

11.5 FINITE ELEMENT FORMULATION OF BEAMS AND FRAMES

In this section, using Lagrange's equations, we will formulate the vibration of a beam and a frame element. Recall from Chapter 4, Section 4.2 that the deflection of a beam element may be represented using shape functions $S_{i1}, S_{i2}, S_{j1}, S_{j2}$ and the nodal displacements $U_{i1}, U_{i2}, U_{j1}, U_{j2}$ by

$$v = S_{i1}U_{i1} + S_{i2}U_{i2} + S_{j1}U_{j1} + S_{j2}U_{j2}$$

where the shape functions are given by

$$S_{i1} = 1 - \frac{3x^2}{L^2} + \frac{2x^3}{L^3}$$

$$S_{i2} = x - \frac{2x^2}{L} + \frac{x^3}{L^2}$$

$$S_{j1} = \frac{3x^2}{L^2} - \frac{2x^3}{L^3}$$

$$S_{j2} = -\frac{x^2}{L} + \frac{x^3}{L^2}$$

For the sake of presentation continuity, the beam element shown previously in Figure 4.8 is repeated here.

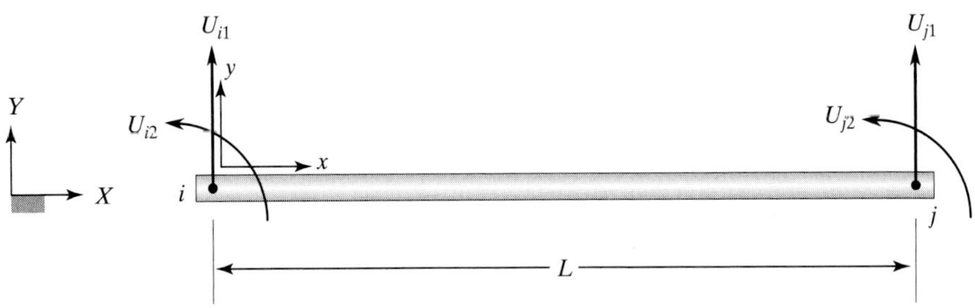

FIGURE 4.8 A beam element.

The kinetic energy of a beam element is determined by adding the kinetic energy of its constituent particles according to

$$T = \int_0^L \frac{\gamma}{2} \dot{v}^2 \, dx \tag{11.125}$$

In Eq. (11.125), $\dot{v}$ represents the velocity distribution within the beam and is a function of time and position. We can represent the velocity of the beam in terms of shape functions and the lateral and rotational velocities of nodes i and j according to

$$v^{\cdot} = S_{i1}\dot{U}_{i1} + S_{i2}\dot{U}_{i2} + S_{j1}\dot{U}_{j1} + S_{j2}\dot{U}_{j2} \tag{11.126}$$

Substituting for the velocity distribution in the kinetic energy equation, we have

$$T = \int_0^L \frac{\gamma}{2}\dot{v}^2 dx = \frac{\gamma}{2}\int_0^L (S_{i1}\dot{U}_{i1} + S_{i2}\dot{U}_{i2} + S_{j1}\dot{U}_{j1} + S_{j2}\dot{U}_{j2})^2 dx \tag{11.127}$$

Although $\dot{v}$ is a function of time and position, note that the nodal velocities are only a function of time. The shape functions account for the spatial variations in Eqs. (11.26) and (11.27). Evaluating the derivative terms as required by Lagrange's equations, we get

$$\frac{\partial T}{\partial \dot{U}_{i1}} = \frac{\gamma}{2}\int_0^L 2S_{i1}(S_{i1}\dot{U}_{i1} + S_{i2}\dot{U}_{i2} + S_{j1}\dot{U}_{j1} + S_{j2}\dot{U}_{j2})dx \tag{11.128}$$

$$\frac{\partial T}{\partial \dot{U}_{i2}} = \frac{\gamma}{2}\int_0^L 2S_{i2}(S_{i1}\dot{U}_{i1} + S_{i2}\dot{U}_{i2} + S_{j1}\dot{U}_{j1} + S_{j2}\dot{U}_{j2})dx \tag{11.129}$$

$$\frac{\partial T}{\partial \dot{U}_{j1}} = \frac{\gamma}{2}\int_0^L 2S_{j1}(S_{i1}\dot{U}_{i1} + S_{i2}\dot{U}_{i2} + S_{j1}\dot{U}_{j1} + S_{j2}\dot{U}_{j2})dx \tag{11.130}$$

$$\frac{\partial T}{\partial \dot{U}_{j2}} = \frac{\gamma}{2}\int_0^L 2S_{j2}(S_{i1}\dot{U}_{i1} + S_{i2}\dot{U}_{i2} + S_{j1}\dot{U}_{j1} + S_{j2}\dot{U}_{j2})dx \tag{11.131}$$

and evaluating the $\frac{d}{dt}\left(\frac{\partial T}{\partial \dot{q}_i}\right)$ terms, we have

$$\frac{d}{dt}\left(\frac{\partial T}{\partial \dot{U}_{i1}}\right) = \gamma\left[\int_0^L S_{i1}(S_{i1}\ddot{U}_{i1} + S_{i2}\ddot{U}_{i2} + S_{j1}\ddot{U}_{j1} + S_{j2}\ddot{U}_{j2})\,dx\right] \tag{11.132}$$

$$\frac{d}{dt}\left(\frac{\partial T}{\partial \dot{U}_{i2}}\right) = \gamma\left[\int_0^L S_{i2}(S_{i1}\ddot{U}_{i1} + S_{i2}\ddot{U}_{i2} + S_{j1}\ddot{U}_{j1} + S_{j2}\ddot{U}_{j2})\,dx\right] \tag{11.133}$$

$$\frac{d}{dt}\left(\frac{\partial T}{\partial \dot{U}_{j1}}\right) = \gamma\left[\int_0^L S_{j1}(S_{i1}\ddot{U}_{i1} + S_{i2}\ddot{U}_{i2} + S_{j1}\ddot{U}_{j1} + S_{j2}\ddot{U}_{j2})\,dx\right] \tag{11.134}$$

$$\frac{d}{dt}\left(\frac{\partial T}{\partial \dot{U}_{j2}}\right) = \gamma\left[\int_0^L S_{j2}(S_{i1}\ddot{U}_{i1} + S_{i2}\ddot{U}_{i2} + S_{j1}\ddot{U}_{j1} + S_{j2}\ddot{U}_{j2})\,dx\right] \tag{11.135}$$

Like nodal velocities, note that the lateral and rotational accelerations of nodes i and j $\ddot{U}_{i1}$, $\ddot{U}_{i2}$, $\ddot{U}_{j1}$, $\ddot{U}_{j1}$ are independent of coordinate x and are only functions of time. This realization allows us to pull the nodal accelerations out of the integrals in Eqs. (11.132) through (11.135), and only integrate the products of the shape functions. It is important to point out that when integrating Eqs. (11.132) to (11.135), we need not evaluate all

16 integrals. Some of these integrals are identical. The integrals that must be evaluated follow:

$$\gamma \int_0^L S_{i1}^2 \, dx = \gamma \int_0^L \left(1 - \frac{3x^2}{L^2} + \frac{2x^3}{L^3}\right)^2 dx = \frac{13\gamma L}{35} = \frac{13}{35} m \tag{11.136}$$

$$\gamma \int_0^L S_{i1}S_{i2} \, dx = \gamma \int_0^L \left(1 - \frac{3x^2}{L^2} + \frac{2x^3}{L^3}\right)\left(x - \frac{2x^2}{L} + \frac{x^3}{L^2}\right) dx = \frac{11\gamma L^2}{210} = \frac{11}{210} mL \tag{11.137}$$

$$\gamma \int_0^L S_{i1}S_{j1} \, dx = \gamma \int_0^L \left(1 - \frac{3x^2}{L^2} + \frac{2x^3}{L^3}\right)\left(\frac{3x^2}{L^2} - \frac{2x^3}{L^3}\right) dx = \frac{9\gamma L}{70} \tag{11.138}$$

$$\gamma \int_0^L S_{i1}S_{j2} \, dx = \gamma \int_0^L \left(1 - \frac{3x^2}{L^2} + \frac{2x^3}{L^3}\right)\left(-\frac{x^2}{L} + \frac{x^3}{L^2}\right) dx = -\frac{13\gamma L^2}{420} \tag{11.139}$$

$$\gamma \int_0^L S_{j2}^2 \, dx = \gamma \int_0^L \left(-\frac{x^2}{L} + \frac{x^3}{L^2}\right)^2 dx = \frac{\gamma L^3}{105} \tag{11.140}$$

$$\gamma \int_0^L S_{j2}S_{j1} \, dx = \gamma \int_0^L \left(-\frac{x^2}{L} + \frac{x^3}{L^2}\right)\left(\frac{3x^2}{L^2} - \frac{2x^3}{L^3}\right) dx = -\frac{11\gamma L^2}{210} \tag{11.141}$$

Incorporating the results of the integrations leads to $[\boldsymbol{M}]\{\ddot{\boldsymbol{v}}\}$, then the mass matrix for a beam element is given by

$$[\boldsymbol{M}]^{(e)} = \frac{\gamma L}{420}\begin{bmatrix} 156 & 22L & 54 & -13L \\ 22L & 4L^2 & 13L & -3L^2 \\ 54 & 13L & 156 & -22L \\ -13L & -3L^2 & -22L & 4L^2 \end{bmatrix} \tag{11.142}$$

We now turn our attention to the stiffness matrix. Recall from Section 4.2 that the stiffness matrix for a beam element is

$$[\boldsymbol{K}]^{(e)} = \frac{EI}{L^3}\begin{bmatrix} 12 & 6L & -12 & 6L \\ 6L & 4L^2 & -6L & 2L^2 \\ -12 & -6L & 12 & -6L \\ 6L & 2L^2 & -6L & 4L^2 \end{bmatrix} \tag{11.143}$$

We discuss the finite element formulation of the mass matrix for a frame element next, and then demonstrate finite element modeling of oscillating frames with Example 11.7.

Frame Element

You may recall from our discussion of frames in Chapter 4 that frames represent structural members that may be rigidly connected with welded joints or bolted joints. For such structures, in addition to rotation and lateral displacement, we also need to be concerned about axial deformations. For the sake of presentation continuity, the frame element, Figure 4.12, is repeated here.

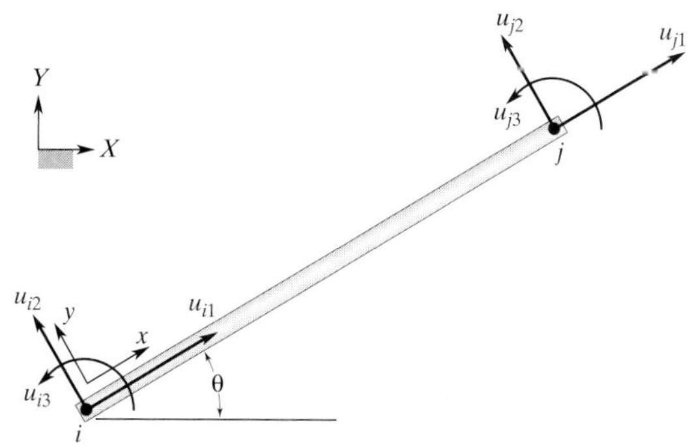

FIGURE 4.12 A frame element.

In the previous section we developed the mass matrix for a beam element. This matrix accounts for lateral displacements and rotations at each node and is

$$[\boldsymbol{M}]^{(e)} = \frac{\gamma L}{420}\begin{bmatrix} 0 & 0 & 0 & 0 & 0 & 0 \\ 0 & 156 & 22L & 0 & 54 & -13L \\ 0 & 22L & 4L^2 & 0 & 13L & -3L^2 \\ 0 & 0 & 0 & 0 & 0 & 0 \\ 0 & 54 & 13L & 0 & 156 & -22L \\ 0 & -13L & -3L^2 & 0 & -22L & 4L^2 \end{bmatrix} \quad (11.144)$$

The mass matrix for a member under axial movement was developed in Section 11.4 and is given by

$$[\boldsymbol{M}]^{(e)} = \frac{\gamma L}{6}\begin{bmatrix} 2 & 0 & 0 & 1 & 0 & 0 \\ 0 & 0 & 0 & 0 & 0 & 0 \\ 0 & 0 & 0 & 0 & 0 & 0 \\ 1 & 0 & 0 & 2 & 0 & 0 \\ 0 & 0 & 0 & 0 & 0 & 0 \\ 0 & 0 & 0 & 0 & 0 & 0 \end{bmatrix} = \frac{\gamma L}{420}\begin{bmatrix} 140 & 0 & 0 & 70 & 0 & 0 \\ 0 & 0 & 0 & 0 & 0 & 0 \\ 0 & 0 & 0 & 0 & 0 & 0 \\ 70 & 0 & 0 & 140 & 0 & 0 \\ 0 & 0 & 0 & 0 & 0 & 0 \\ 0 & 0 & 0 & 0 & 0 & 0 \end{bmatrix} \quad (11.145)$$

Adding Eqs. (11.144) and (11.145) results in the mass matrix for a frame element:

$$[\boldsymbol{M}]^{(e)} = \frac{\gamma L}{420}\begin{bmatrix} 140 & 0 & 0 & 70 & 0 & 0 \\ 0 & 156 & 22L & 0 & 54 & -13L \\ 0 & 22L & 4L^2 & 0 & 13L & -3L^2 \\ 70 & 0 & 0 & 140 & 0 & 0 \\ 0 & 54 & 13L & 0 & 156 & -22L \\ 0 & -13L & -3L^2 & 0 & -22L & 4L^2 \end{bmatrix} \quad (11.146)$$

In Section 4.2, we derived the stiffness matrix for the frame element, which is given by

$$[\boldsymbol{K}]_{xy}^{(e)} = \begin{bmatrix} \frac{AE}{L} & 0 & 0 & -\frac{AE}{L} & 0 & 0 \\ 0 & \frac{12EI}{L^3} & \frac{6EI}{L^2} & 0 & -\frac{12EI}{L^3} & \frac{6EI}{L^2} \\ 0 & \frac{6EI}{L^2} & \frac{4EI}{L} & 0 & -\frac{6EI}{L^2} & \frac{2EI}{L} \\ -\frac{AE}{L} & 0 & 0 & \frac{AE}{L} & 0 & 0 \\ 0 & -\frac{12EI}{L^3} & -\frac{6EI}{L^2} & 0 & \frac{12EI}{L^3} & -\frac{6EI}{L^2} \\ 0 & \frac{6EI}{L^2} & \frac{2EI}{L} & 0 & -\frac{6EI}{L^2} & \frac{4EI}{L} \end{bmatrix} \tag{11.147}$$

In Chapter 4 we also discussed the role and the importance of using local and global coordinate systems to formulate and analyze finite element models. You may recall that the local degrees of freedom are related to the global degrees of freedom through the transformation matrix, according to the relationship

$$\{\boldsymbol{u}\} = [\boldsymbol{T}]\{\boldsymbol{U}\} \tag{11.148}$$

where the transformation matrix is

$$[\boldsymbol{T}] = \begin{bmatrix} \cos\theta & \sin\theta & 0 & 0 & 0 & 0 \\ -\sin\theta & \cos\theta & 0 & 0 & 0 & 0 \\ 0 & 0 & 1 & 0 & 0 & 0 \\ 0 & 0 & 0 & \cos\theta & \sin\theta & 0 \\ 0 & 0 & 0 & -\sin\theta & \cos\theta & 0 \\ 0 & 0 & 0 & 0 & 0 & 1 \end{bmatrix} \tag{11.149}$$

The equations of motion in the element's local coordinate system is given by

$$[\boldsymbol{M}]_{xy}^{(e)}\{\ddot{\boldsymbol{u}}\} + [\boldsymbol{K}]_{xy}^{(e)}\{\boldsymbol{u}\} = \{\boldsymbol{f}\}^{(e)} \tag{11.150}$$

We can now make use of the relationships between local and global displacements and accelerations $\{\boldsymbol{u}\} = [\boldsymbol{T}]\{\boldsymbol{U}\}$ and $\{\ddot{\boldsymbol{u}}\} = [\boldsymbol{T}]\{\ddot{\boldsymbol{U}}\}$ and the local and global description of forces $\{\boldsymbol{f}\} = [\boldsymbol{T}]\{\boldsymbol{F}\}$ and substitute for $\{\boldsymbol{u}\}$, $\{\ddot{\boldsymbol{u}}\}$, $\{\boldsymbol{f}\}$ in Eq. (11.150). These substitutions result in

$$[\boldsymbol{M}]_{xy}^{(e)}\overbrace{[\boldsymbol{T}]\{\ddot{\boldsymbol{U}}\}}^{\{\ddot{\boldsymbol{u}}\}} + [\boldsymbol{K}]_{xy}^{(e)}\overbrace{[\boldsymbol{T}]\{\boldsymbol{U}\}}^{\{\boldsymbol{u}\}} = \overbrace{[\boldsymbol{T}]\{\boldsymbol{F}\}}^{\{\boldsymbol{f}\}}{}^{(e)} \tag{11.151}$$

Premultiplying Eq. (11.151) by $[\boldsymbol{T}]^{-1}$,

$$[\boldsymbol{T}]^{-1}[\boldsymbol{M}]_{xy}^{(e)}[\boldsymbol{T}]\{\ddot{\boldsymbol{U}}\} + [\boldsymbol{T}]^{-1}[\boldsymbol{K}]_{xy}^{(e)}[\boldsymbol{T}]\{\boldsymbol{U}\} = [\boldsymbol{T}]^{-1}[\boldsymbol{T}]\{\boldsymbol{F}\}^{(e)} \tag{11.152}$$

It can be readily shown that for the transformation matrix, $[\boldsymbol{T}]^{-1} = [\boldsymbol{T}]^T$ (see Example 11.6). Using this relationship and simplifying Eq. (11.152), we get

$$\overbrace{[\boldsymbol{T}]^T[\boldsymbol{M}]_{xy}^{(e)}[\boldsymbol{T}]}^{[\boldsymbol{M}]^{(e)}}\{\ddot{\boldsymbol{U}}\} + \overbrace{[\boldsymbol{T}]^T[\boldsymbol{K}]_{xy}^{(e)}[\boldsymbol{T}]}^{[\boldsymbol{K}]^{(e)}}\{\boldsymbol{U}\} = \{\boldsymbol{F}\}^{(e)} \tag{11.153}$$

Finally, the equations of motion in terms of global coordinates become

$$[\boldsymbol{M}]^{(e)}\{\ddot{\boldsymbol{U}}\} + [\boldsymbol{K}]^{(e)}\{\boldsymbol{U}\} = \{\boldsymbol{F}\}^{(e)} \tag{11.154}$$

where

$$[\boldsymbol{M}]^{(e)} = [\boldsymbol{T}]^T[\boldsymbol{M}]_{xy}^{(e)}[\boldsymbol{T}] \tag{11.155}$$

and

$$[\boldsymbol{K}]^{(e)} = [\boldsymbol{T}]^T[\boldsymbol{K}]_{xy}^{(e)}[\boldsymbol{T}] \tag{11.156}$$

We will demonstrate how to use these equations to determine the natural frequencies of beams and frames with Example 11.7.

EXAMPLE 11.6

In this example we show that $[\boldsymbol{T}]^{-1} = [\boldsymbol{T}]^T$. We begin by showing that

$$[\boldsymbol{T}]^T[\boldsymbol{T}] = [\boldsymbol{I}]$$

$$\begin{bmatrix} \cos\theta & -\sin\theta & 0 & 0 & 0 & 0 \\ \sin\theta & \cos\theta & 0 & 0 & 0 & 0 \\ 0 & 0 & 1 & 0 & 0 & 0 \\ 0 & 0 & 0 & \cos\theta & -\sin\theta & 0 \\ 0 & 0 & 0 & \sin\theta & \cos\theta & 0 \\ 0 & 0 & 0 & 0 & 0 & 1 \end{bmatrix}\begin{bmatrix} \cos\theta & \sin\theta & 0 & 0 & 0 & 0 \\ -\sin\theta & \cos\theta & 0 & 0 & 0 & 0 \\ 0 & 0 & 1 & 0 & 0 & 0 \\ 0 & 0 & 0 & \cos\theta & \sin\theta & 0 \\ 0 & 0 & 0 & -\sin\theta & \cos\theta & 0 \\ 0 & 0 & 0 & 0 & 0 & 1 \end{bmatrix} =$$

$$\begin{bmatrix} 1 & 0 & 0 & 0 & 0 & 0 \\ 0 & 1 & 0 & 0 & 0 & 0 \\ 0 & 0 & 1 & 0 & 0 & 0 \\ 0 & 0 & 0 & 1 & 0 & 0 \\ 0 & 0 & 0 & 0 & 1 & 0 \\ 0 & 0 & 0 & 0 & 0 & 1 \end{bmatrix}$$

Note that the relationship $\cos^2\theta + \sin^2\theta = 1$ is used to simplify the results. Because $[\boldsymbol{T}]^{-1}[\boldsymbol{T}] = [\boldsymbol{I}]$ and we just showed that $[\boldsymbol{T}]^T[\boldsymbol{T}] = [\boldsymbol{I}]$, then $[\boldsymbol{T}]^{-1} = [\boldsymbol{T}]^T$ must be true.

EXAMPLE 11.7

Consider the frame shown in Figure 11.31. The frame is made of steel, with $E = 30 \times 10^6$ lb/in^2. The cross-sectional areas and second moment of areas for the members are shown in the figure. The frame is fixed as shown, and we are interested in determining the natural frequencies using the three-element model shown. Members (1) and (3) are W12 × 26 steel beams, while member (2) is a W16 × 26 steel beam.

The mass per unit length is

$$\gamma = \frac{26 \text{ lb}}{(12 \text{ in})(32.2 \text{ ft/s}^2)(12 \text{ in/ft})} = 0.0056 \text{ lb.s}^2/\text{in}^2$$

For each element, the relationship between the local and global coordinate systems is shown in Figure 11.32.

The stiffness values to be used in the stiffness matrix for elements (1) and (3) are

$$\frac{AE}{L} = \frac{(7.65 \text{ in}^2)(30 \times 10^6 \text{ lb/in}^2)}{(15 \text{ ft})(12 \text{ in/ft})} = 1{,}275{,}000 \text{ lb/in}$$

$$\frac{12EI}{L^3} = \frac{(12)(30 \times 10^6 \text{ lb/in}^2)(204 \text{ in}^4)}{((15 \text{ ft})(12 \text{ in/ft}))^3} = 12{,}592 \text{ lb/in}$$

$$\frac{6EI}{L^2} = \frac{(6)(30 \times 10^6 \text{ lb/in}^2)(204 \text{ in}^4)}{((15 \text{ ft})(12 \text{ in/ft}))^2} = 1{,}133{,}333 \text{ lb}$$

$$\frac{2EI}{L} = \frac{(2)(30 \times 10^6 \text{ lb/in}^2)(204 \text{ in}^4)}{(15 \text{ ft})(12 \text{ in/ft})} = 68{,}000{,}000 \text{ lb.in}$$

$$\frac{4EI}{L} = \frac{(4)(30 \times 10^6 \text{ lb/in}^2)(204 \text{ in}^4)}{(15 \text{ ft})(12 \text{ in/ft})} = 136{,}000{,}000 \text{ lb.in}$$

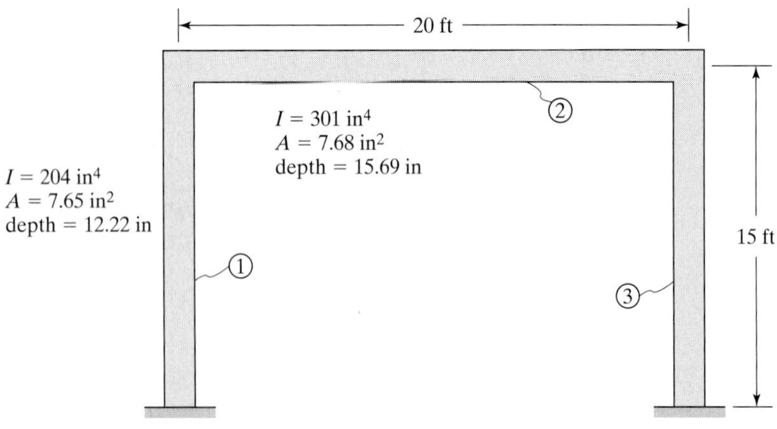

FIGURE 11.31 The frame of Example 11.7.

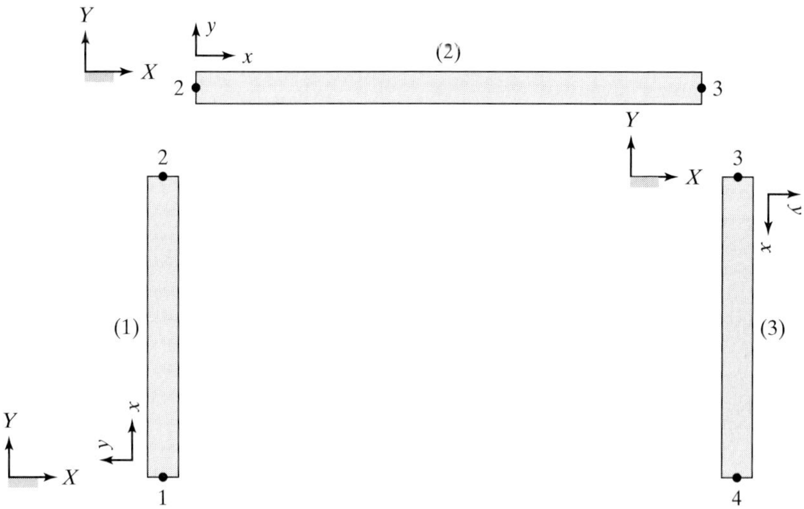

FIGURE 11.32 The configuration of elements (1), (2), and (3).

The local stiffness matrices for elements (1) and (3) are

$$[K]^{(1)}_{xy} = [K]^{(3)}_{xy} = \begin{bmatrix} \frac{AE}{L} & 0 & 0 & -\frac{AE}{L} & 0 & 0 \\ 0 & \frac{12EI}{L^3} & \frac{6EI}{L^2} & 0 & -\frac{12EI}{L^3} & \frac{6EI}{L^2} \\ 0 & \frac{6EI}{L^2} & \frac{4EI}{L} & 0 & -\frac{6EI}{L^2} & \frac{2EI}{L} \\ -\frac{AE}{L} & 0 & 0 & \frac{AE}{L} & 0 & 0 \\ 0 & -\frac{12EI}{L^3} & -\frac{6EI}{L^2} & 0 & \frac{12EI}{L^3} & -\frac{6EI}{L^2} \\ 0 & \frac{6EI}{L^2} & \frac{2EI}{L} & 0 & -\frac{6EI}{L^2} & \frac{4EI}{L} \end{bmatrix}$$

$$= 10^3 \begin{bmatrix} 1275 & 0 & 0 & -1275 & 0 & 0 \\ 0 & 12.592 & 1133.333 & 0 & -12.592 & 1133.333 \\ 0 & 1133.333 & 136000 & 0 & -1133.333 & 68000 \\ -1275 & 0 & 0 & 1275 & 0 & 0 \\ 0 & -12.592 & -1133.333 & 0 & 12.592 & -1133.333 \\ 0 & 1133.333 & 68000 & 0 & -1133.333 & 136000 \end{bmatrix}$$

The local mass matrix for elements (1) and (3) is

$$[\boldsymbol{M}]_{xy}^{(1)} = [\boldsymbol{M}]_{xy}^{(3)} = \frac{\gamma L}{420}\begin{bmatrix} 140 & 0 & 0 & 70 & 0 & 0 \\ 0 & 156 & 22L & 0 & 54 & -13L \\ 0 & 22L & 4L^2 & 0 & 13L & -3L^2 \\ 70 & 0 & 0 & 140 & 0 & 0 \\ 0 & 54 & 13L & 0 & 156 & -22L \\ 0 & -13L & -3L^2 & 0 & -22L & 4L^2 \end{bmatrix}$$

$$= \frac{(0.0056 \text{ lb.s}^2/\text{in}^2)(15 \text{ ft})(12 \text{ in/ft})}{420} \times$$

$$\begin{bmatrix} 140 & 0 & 0 & 70 & 0 & 0 \\ 0 & 156 & (22)(15)(12) & 0 & 54 & -(13)(15)(12) \\ 0 & (22)(15)(12) & (4)((15)(12))^2 & 0 & (13)(15)(12) & -(3)((15)(12))^2 \\ 70 & 0 & 0 & 140 & 0 & 0 \\ 0 & 54 & (13)(15)(12) & 0 & 156 & -(22)(15)(12) \\ 0 & -(13)(15)(12) & -(3)((15)(12))^2 & 0 & -(22)(15)(12) & (4)((15)(12))^2 \end{bmatrix}$$

$$[\boldsymbol{M}]_{xy}^{(1)} = [\boldsymbol{M}]_{xy}^{(3)} = 0.0024\begin{bmatrix} 140 & 0 & 0 & 70 & 0 & 0 \\ 0 & 156 & 3960 & 0 & 54 & -2340 \\ 0 & 3960 & 129600 & 0 & 2340 & -97200 \\ 70 & 0 & 0 & 140 & 0 & 0 \\ 0 & 54 & 2340 & 0 & 156 & -3960 \\ 0 & -2340 & -97200 & 0 & -3960 & 129600 \end{bmatrix}$$

For element (1), the transformation matrix and its transpose are

$$[\boldsymbol{T}] = \begin{bmatrix} \cos(90) & \sin(90) & 0 & 0 & 0 & 0 \\ -\sin(90) & \cos(90) & 0 & 0 & 0 & 0 \\ 0 & 0 & 1 & 0 & 0 & 0 \\ 0 & 0 & 0 & \cos(90) & \sin(90) & 0 \\ 0 & 0 & 0 & -\sin(90) & \cos(90) & 0 \\ 0 & 0 & 0 & 0 & 0 & 1 \end{bmatrix} = \begin{bmatrix} 0 & 1 & 0 & 0 & 0 & 0 \\ -1 & 0 & 0 & 0 & 0 & 0 \\ 0 & 0 & 1 & 0 & 0 & 0 \\ 0 & 0 & 0 & 0 & 1 & 0 \\ 0 & 0 & 0 & -1 & 0 & 0 \\ 0 & 0 & 0 & 0 & 0 & 1 \end{bmatrix}$$

$$[\boldsymbol{T}]^T = \begin{bmatrix} 0 & -1 & 0 & 0 & 0 & 0 \\ 1 & 0 & 0 & 0 & 0 & 0 \\ 0 & 0 & 1 & 0 & 0 & 0 \\ 0 & 0 & 0 & 0 & -1 & 0 \\ 0 & 0 & 0 & 1 & 0 & 0 \\ 0 & 0 & 0 & 0 & 0 & 1 \end{bmatrix}$$

For element (3), the transformation matrix and its transpose are

$$[\boldsymbol{T}]=\begin{bmatrix} \cos(270) & \sin(270) & 0 & 0 & 0 & 0 \\ -\sin(270) & \cos(270) & 0 & 0 & 0 & 0 \\ 0 & 0 & 1 & 0 & 0 & 0 \\ 0 & 0 & 0 & \cos(270) & \sin(270) & 0 \\ 0 & 0 & 0 & -\sin(270) & \cos(270) & 0 \\ 0 & 0 & 0 & 0 & 0 & 1 \end{bmatrix} = \begin{bmatrix} 0 & -1 & 0 & 0 & 0 & 0 \\ 1 & 0 & 0 & 0 & 0 & 0 \\ 0 & 0 & 1 & 0 & 0 & 0 \\ 0 & 0 & 0 & 0 & -1 & 0 \\ 0 & 0 & 0 & 1 & 0 & 0 \\ 0 & 0 & 0 & 0 & 0 & 1 \end{bmatrix}$$

$$[\boldsymbol{T}]^T = \begin{bmatrix} 0 & 1 & 0 & 0 & 0 & 0 \\ -1 & 0 & 0 & 0 & 0 & 0 \\ 0 & 0 & 1 & 0 & 0 & 0 \\ 0 & 0 & 0 & 0 & 1 & 0 \\ 0 & 0 & 0 & -1 & 0 & 0 \\ 0 & 0 & 0 & 0 & 0 & 1 \end{bmatrix}$$

Substituting for $[\boldsymbol{T}]^T$, $[\boldsymbol{K}]^{(1)}_{xy}$, and $[\boldsymbol{T}]$ into Eq. (11.156), we have:

$$[\boldsymbol{K}]^{(1)} = 10^3 \begin{bmatrix} 0 & -1 & 0 & 0 & 0 & 0 \\ 1 & 0 & 0 & 0 & 0 & 0 \\ 0 & 0 & 1 & 0 & 0 & 0 \\ 0 & 0 & 0 & 0 & -1 & 0 \\ 0 & 0 & 0 & 1 & 0 & 0 \\ 0 & 0 & 0 & 0 & 0 & 1 \end{bmatrix} \begin{bmatrix} 1275 & 0 & 0 & -1275 & 0 & 0 \\ 0 & 12.592 & 1133.333 & 0 & -12.592 & 1133.333 \\ 0 & 1133.333 & 136000 & 0 & -1133.333 & 68000 \\ -1275 & 0 & 0 & 1275 & 0 & 0 \\ 0 & -12.592 & -1133.333 & 0 & 12.592 & -1133.333 \\ 0 & 1133.333 & 68000 & 0 & -1133.333 & 136000 \end{bmatrix}$$

$$\begin{bmatrix} 0 & 1 & 0 & 0 & 0 & 0 \\ -1 & 0 & 0 & 0 & 0 & 0 \\ 0 & 0 & 1 & 0 & 0 & 0 \\ 0 & 0 & 0 & 0 & 1 & 0 \\ 0 & 0 & 0 & -1 & 0 & 0 \\ 0 & 0 & 0 & 0 & 0 & 1 \end{bmatrix}$$

$$[\boldsymbol{K}]^{(1)} = 10^3 \begin{bmatrix} 12.592 & 0 & -1133.33 & -12.592 & 0 & -1133.333 \\ 0 & 1275 & 0 & 0 & -1275 & 0 \\ -1133.33 & 0 & 136000 & 1133.333 & 0 & 68000 \\ -12.592 & 0 & 133.333 & 12.59 & 0 & 1133.333 \\ 0 & -1275 & 0 & 0 & 1275 & 0 \\ -1133.333 & 0 & 68000 & 1133.33 & 0 & 136000 \end{bmatrix}$$

Substituting for $[\boldsymbol{T}]^T$, $[\boldsymbol{M}]^{(1)}_{xy}$, and $[\boldsymbol{T}]$ into Eq. (11.155), we have

$$[\boldsymbol{M}]^{(1)} = 0.0024\begin{bmatrix} 0 & -1 & 0 & 0 & 0 & 0 \\ 1 & 0 & 0 & 0 & 0 & 0 \\ 0 & 0 & 1 & 0 & 0 & 0 \\ 0 & 0 & 0 & 0 & -1 & 0 \\ 0 & 0 & 0 & 1 & 0 & 0 \\ 0 & 0 & 0 & 0 & 0 & 1 \end{bmatrix}$$

$$\begin{bmatrix} 140 & 0 & 0 & 70 & 0 & 0 \\ 0 & 156 & 3960 & 0 & 54 & -2340 \\ 0 & 3960 & 129600 & 0 & 2340 & -97200 \\ 70 & 0 & 0 & 140 & 0 & 0 \\ 0 & 54 & 2340 & 0 & 156 & -3960 \\ 0 & -2340 & -97200 & 0 & -3960 & 129600 \end{bmatrix}\begin{bmatrix} 0 & 1 & 0 & 0 & 0 & 0 \\ -1 & 0 & 0 & 0 & 0 & 0 \\ 0 & 0 & 1 & 0 & 0 & 0 \\ 0 & 0 & 0 & 0 & 1 & 0 \\ 0 & 0 & 0 & -1 & 0 & 0 \\ 0 & 0 & 0 & 0 & 0 & 1 \end{bmatrix}$$

$$[\boldsymbol{M}]^{(1)} = 0.0024\begin{bmatrix} 156 & 0 & -3960 & 54 & 0 & 2340 \\ 0 & 140 & 0 & 0 & 70 & 0 \\ -3960 & 0 & 129600 & -2340 & 0 & -97200 \\ 54 & 0 & -2340 & 156 & 0 & 3960 \\ 0 & 70 & 0 & 0 & 140 & 0 \\ 2340 & 0 & -97200 & 3960 & 0 & 129600 \end{bmatrix}$$

Similarly, the stiffness matrix and the mass matrix for element (3) are

$$[\boldsymbol{K}]^{(3)} = 10^3\begin{bmatrix} 0 & 1 & 0 & 0 & 0 & 0 \\ -1 & 0 & 0 & 0 & 0 & 0 \\ 0 & 0 & 1 & 0 & 0 & 0 \\ 0 & 0 & 0 & 0 & 1 & 0 \\ 0 & 0 & 0 & -1 & 0 & 0 \\ 0 & 0 & 0 & 0 & 0 & 1 \end{bmatrix}\begin{bmatrix} 1275 & 0 & 0 & -1275 & 0 & 0 \\ 0 & 12.592 & 1133.333 & 0 & -12.592 & 1133.333 \\ 0 & 1133.333 & 136000 & 0 & -1133.333 & 68000 \\ -1275 & 0 & 0 & 1275 & 0 & 0 \\ 0 & -12.592 & -1133.333 & 0 & 12.592 & -1133.333 \\ 0 & 1133.333 & 68000 & 0 & -1133.333 & 136000 \end{bmatrix}$$

$$\begin{bmatrix} 0 & -1 & 0 & 0 & 0 & 0 \\ 1 & 0 & 0 & 0 & 0 & 0 \\ 0 & 0 & 1 & 0 & 0 & 0 \\ 0 & 0 & 0 & 0 & -1 & 0 \\ 0 & 0 & 0 & 1 & 0 & 0 \\ 0 & 0 & 0 & 0 & 0 & 1 \end{bmatrix}$$

$$[\boldsymbol{K}]^{(3)} = 10^3\begin{bmatrix} 12.592 & 0 & 1133.33 & -12.592 & 0 & 1133.333 \\ 0 & 1275 & 0 & 0 & -1275 & 0 \\ 1133.33 & 0 & 136000 & -1133.333 & 0 & 68000 \\ -12.592 & 0 & -133.333 & 12.59 & 0 & -1133.333 \\ 0 & -1275 & 0 & 0 & 1275 & 0 \\ 1133.333 & 0 & 68000 & -1133.33 & 0 & 136000 \end{bmatrix}$$

$$[\boldsymbol{M}]^{(3)} = 0.0024\begin{bmatrix} 0 & 1 & 0 & 0 & 0 & 0 \\ -1 & 0 & 0 & 0 & 0 & 0 \\ 0 & 0 & 1 & 0 & 0 & 0 \\ 0 & 0 & 0 & 0 & 1 & 0 \\ 0 & 0 & 0 & -1 & 0 & 0 \\ 0 & 0 & 0 & 0 & 0 & 1 \end{bmatrix}$$

$$\begin{bmatrix} 140 & 0 & 0 & 70 & 0 & 0 \\ 0 & 156 & 3960 & 0 & 54 & -2340 \\ 0 & 3960 & 129600 & 0 & 2340 & -97200 \\ 70 & 0 & 0 & 140 & 0 & 0 \\ 0 & 54 & 2340 & 0 & 156 & -3960 \\ 0 & -2340 & -97200 & 0 & -3960 & 129600 \end{bmatrix}\begin{bmatrix} 0 & -1 & 0 & 0 & 0 & 0 \\ 1 & 0 & 0 & 0 & 0 & 0 \\ 0 & 0 & 1 & 0 & 0 & 0 \\ 0 & 0 & 0 & 0 & -1 & 0 \\ 0 & 0 & 0 & 1 & 0 & 0 \\ 0 & 0 & 0 & 0 & 0 & 1 \end{bmatrix}$$

$$[\boldsymbol{M}]^{(3)} = 0.0024\begin{bmatrix} 156 & 0 & 3960 & 54 & 0 & -2340 \\ 0 & 140 & 0 & 0 & 70 & 0 \\ 3960 & 0 & 129600 & 2340 & 0 & -97200 \\ 54 & 0 & 2340 & 156 & 0 & -39600 \\ 0 & 70 & 0 & 0 & 140 & 0 \\ -2340 & 0 & -97200 & -3960 & 0 & 129600 \end{bmatrix}$$

The stiffness values for element (2) are

$$\frac{AE}{L} = \frac{(7.68 \text{ in}^2)(30 \times 10^6 \text{ lb/in}^2)}{(20 \text{ ft})(12 \text{ in/ft})} = 960{,}000 \text{ lb/in}$$

$$\frac{12EI}{L^3} = \frac{(12)(30 \times 10^6 \text{ lb/in}^2)(301 \text{ in}^4)}{((20 \text{ ft})(12 \text{ in/ft}))^3} = 7838 \text{ lb/in}$$

$$\frac{6EI}{L^2} = \frac{(6)(30 \times 10^6 \text{ lb/in}^2)(301 \text{ in}^4)}{((20 \text{ ft})(12 \text{ in/ft}))^2} = 940{,}625 \text{ lb}$$

$$\frac{2EI}{L} = \frac{(2)(30 \times 10^6 \text{ lb/in}^2)(301 \text{ in}^4)}{(20 \text{ ft})(12 \text{ in/ft})} = 75{,}250{,}000 \text{ lb.in}$$

$$\frac{4EI}{L} = \frac{(4)(30 \times 10^6 \text{ lb/in}^2)(301 \text{ in}^4)}{(20 \text{ ft})(12 \text{ in/ft})} = 150{,}500{,}000 \text{ lb.in}$$

For element (2), the local and the global frames of reference are aligned in the same direction; therefore, the stiffness matrix is

$$[\boldsymbol{K}]^{(2)} = \begin{bmatrix} \frac{AE}{L} & 0 & 0 & -\frac{AE}{L} & 0 & 0 \\ 0 & \frac{12EI}{L^3} & \frac{6EI}{L^2} & 0 & -\frac{12EI}{L^3} & \frac{6EI}{L^2} \\ 0 & \frac{6EI}{L^2} & \frac{4EI}{L} & 0 & -\frac{6EI}{L^2} & \frac{2EI}{L} \\ -\frac{AE}{L} & 0 & 0 & \frac{AE}{L} & 0 & 0 \\ 0 & -\frac{12EI}{L^3} & -\frac{6EI}{L^2} & 0 & \frac{12EI}{L^3} & -\frac{6EI}{L^2} \\ 0 & \frac{6EI}{L^2} & \frac{2EI}{L} & 0 & -\frac{6EI}{L^2} & \frac{4EI}{L} \end{bmatrix} =$$

$$10^3 \begin{bmatrix} 960 & 0 & 0 & -960 & 0 & 0 \\ 0 & 7.838 & 940.625 & 0 & -7.838 & 940.625 \\ 0 & 940.625 & 150500 & 0 & -940.625 & 75250 \\ -960 & 0 & 0 & 960 & 0 & 0 \\ 0 & -7.838 & -940.625 & 0 & 7.838 & -940.625 \\ 0 & 940.625 & 75250 & 0 & -940.625 & 150500 \end{bmatrix}$$

The mass matrix for element (2) is

$$[\boldsymbol{M}]^{(2)} = \frac{\gamma L}{420} \begin{bmatrix} 140 & 0 & 0 & 70 & 0 & 0 \\ 0 & 156 & 22L & 0 & 54 & -13L \\ 0 & 22L & 4L^2 & 0 & 13L & -3L^2 \\ 70 & 0 & 0 & 140 & 0 & 0 \\ 0 & 54 & 13L & 0 & 156 & -22L \\ 0 & -13L & -3L^2 & 0 & -22L & 4L^2 \end{bmatrix} = \frac{(0.0056 \text{ lb.s}^2/\text{in}^2)(20 \text{ ft})(12 \text{ in/ft})}{420}$$

$$\begin{bmatrix} 140 & 0 & 0 & 70 & 0 & 0 \\ 0 & 156 & (22)(20)(12) & 0 & 54 & -(13)(20)(12) \\ 0 & (22)(20)(12) & (4)((20)(12))^2 & 0 & (13)(20)(12) & -(3)((20)(12))^2 \\ 70 & 0 & 0 & 140 & 0 & 0 \\ 0 & 54 & (13)(20)(12) & 0 & 156 & -(22)(20)(12) \\ 0 & -(13)(20)(12) & -(3)((20)(12))^2 & 0 & -(22)(20)(12) & (4)((20)(12))^2 \end{bmatrix}$$

$$[\boldsymbol{M}]^{(2)} = 0.0032 \begin{bmatrix} 140 & 0 & 0 & 70 & 0 & 0 \\ 0 & 156 & 5280 & 0 & 54 & -3120 \\ 0 & 5280 & 230400 & 0 & 3120 & -172800 \\ 70 & 0 & 0 & 140 & 0 & 0 \\ 0 & 54 & 3120 & 0 & 156 & -5280 \\ 0 & -3120 & -172800 & 0 & -5280 & 230400 \end{bmatrix}$$

Next, we will construct the global stiffness and mass matrices.

$$[K]^{(G)} = 10^3 \begin{bmatrix} 12.59 & 0 & -1133.333 & -12.59 & 0 & -1133.333 & 0 & 0 & 0 & 0 & 0 & 0 \\ 0 & 1275 & 0 & 0 & -1275 & 0 & 0 & 0 & 0 & 0 & 0 & 0 \\ -1133.333 & 0 & 136000 & 1133.333 & 0 & 68000 & 0 & 0 & 0 & 0 & 0 & 0 \\ -12.59 & 0 & 113.333 & 972.59 & 0 & 1133.33 & -960 & 0 & 0 & 0 & 0 & 0 \\ 0 & -1275 & 0 & 0 & 1282.84 & 940.63 & 0 & -7.84 & 940.63 & 0 & 0 & 0 \\ -1133.333 & 0 & 68000 & 1133.333 & 940.63 & 286500 & 0 & -940.63 & 75250 & 0 & 0 & 0 \\ 0 & 0 & 0 & -960 & 0 & 0 & 972.59 & 0 & 1133.33 & -12.59 & 0 & 1133.333 \\ 0 & 0 & 0 & 0 & -7.84 & -940.63 & 0 & 1282.84 & -940.63 & 0 & -1275 & 0 \\ 0 & 0 & 0 & 0 & 940.63 & 75250 & 1133.33 & -940.63 & 286500 & -1133.33 & 0 & 68000 \\ 0 & 0 & 0 & 0 & 0 & 0 & -12.59 & 0 & -1133.33 & 12.59 & 0 & -1133.333 \\ 0 & 0 & 0 & 0 & 0 & 0 & 0 & -1275 & 0 & 0 & 1275 & 0 \\ 0 & 0 & 0 & 0 & 0 & 0 & 1133.33 & 0 & 68000 & -1133.33 & 0 & 136000 \end{bmatrix}$$

$$[M]^{(G)} = \begin{bmatrix} 0.37 & 0 & -9.50 & 0.13 & 0 & 5.62 & 0 & 0 & 0 & 0 & 0 & 0 \\ 0 & 0.34 & 0 & 0 & 0.17 & 0 & 0 & 0 & 0 & 0 & 0 & 0 \\ -9.50 & 0 & 311.04 & -5.62 & 0 & -233.28 & 0 & 0 & 0 & 0 & 0 & 0 \\ 0.13 & 0 & -5.62 & 0.82 & 0 & 9.50 & 0.22 & 0 & 0 & 0 & 0 & 0 \\ 0 & 0.17 & 0 & 0 & 0.84 & 16.90 & 0 & 0.17 & -9.98 & 0 & 0 & 0 \\ 5.62 & 0 & -233.28 & 9.50 & 16.90 & 1048.32 & 0 & 9.98 & -552.96 & 0 & 0 & 0 \\ 0 & 0 & 0 & 0.22 & 0 & 0 & 0.82 & 0 & 9.50 & 0.13 & 0 & -5.62 \\ 0 & 0 & 0 & 0 & 0.17 & 9.98 & 0 & 0.84 & -16.90 & 0 & 0.17 & 0 \\ 0 & 0 & 0 & 0 & -9.98 & -552.96 & 9.50 & -16.90 & 1048.32 & 5.62 & 0 & -233.28 \\ 0 & 0 & 0 & 0 & 0 & 0 & 0.13 & 0 & 5.62 & 0.37 & 0 & -9.50 \\ 0 & 0 & 0 & 0 & 0 & 0 & 0 & 0.17 & 0 & 0 & 0.34 & 0 \\ 0 & 0 & 0 & 0 & 0 & 0 & -5.62 & 0 & -233.28 & -9.50 & 0 & 311.04 \end{bmatrix}$$

Applying the boundary conditions, the global stiffness and mass matrices reduce to

$$[K]^{(G)} = 10^3 \begin{bmatrix} 972.59 & 0 & 1133.33 & -960 & 0 & 0 \\ 0 & 1282.84 & 940.63 & 0 & -7.84 & 940.63 \\ 1133.33 & 940.63 & 286500 & 0 & -940.63 & 75250 \\ -960 & 0 & 0 & 972.59 & 0 & 1133.33 \\ 0 & -7.84 & -940.63 & 0 & 1282.84 & -940.63 \\ 0 & 940.63 & 75250 & 1133.33 & -940.63 & 286500 \end{bmatrix}$$

$$[\mathbf{M}]^{(G)} = \begin{bmatrix} 0.82 & 0 & 9.50 & 0.22 & 0 & 0 \\ 0 & 0.84 & 16.90 & 0 & 0.17 & -9.98 \\ 9.50 & 16.90 & 1048.32 & 0 & 9.98 & -552.96 \\ 0.22 & 0 & 0 & 0.82 & 0 & 9.50 \\ 0 & 0.17 & 9.98 & 0 & 0.84 & -16.90 \\ 0 & -9.98 & -552.96 & 9.50 & -16.90 & 1048.32 \end{bmatrix}$$

And solving $[M]^{-1}[K]\{U\} = \omega^2\{U\}$ for the eigenvalues, we get

$$\omega_1 = 95 \text{ rad/s} \qquad \omega_2 = 355 \text{ rad/s} \qquad \omega_3 = 893 \text{ rad/s}$$
$$\omega_4 = 1460 \text{ rad/s} \qquad \omega_5 = 1570 \text{ rad/s} \qquad \omega_6 = 2100 \text{ rad/s}$$

11.6 EXAMPLES USING ANSYS

In this section, ANSYS is used to solve two problems. First we revisit Example 11.7 and then consider the natural oscillation of a straight member with a rectangular cross-section. As the formulations in the previous sections showed, the stiffness matrix for oscillating rods, beam, and frame elements are the same as those for static problems. However, you must include density values in the material models for calculation of mass matrices. Therefore, when modeling dynamic problems, the preprocessing phase including element selection is identical to that of static problems. It is during the solution phase that the correct dynamic analysis type must be selected.

EXAMPLE 11.7 Revisited

Consider the frame from Example 11.7 shown in Figure 11.31. The frame is made of steel, with $E = 30 \times 10^6 \text{lb/in}^2$. The cross-sectional areas and second moment of areas for the members are shown in the figure. The frame is fixed as shown, and we are interested in determining the natural frequencies using the three-element model shown. Members (1) and (3) are W12 × 26 steel beams, while member (2) is a W16 × 26 steel beam.

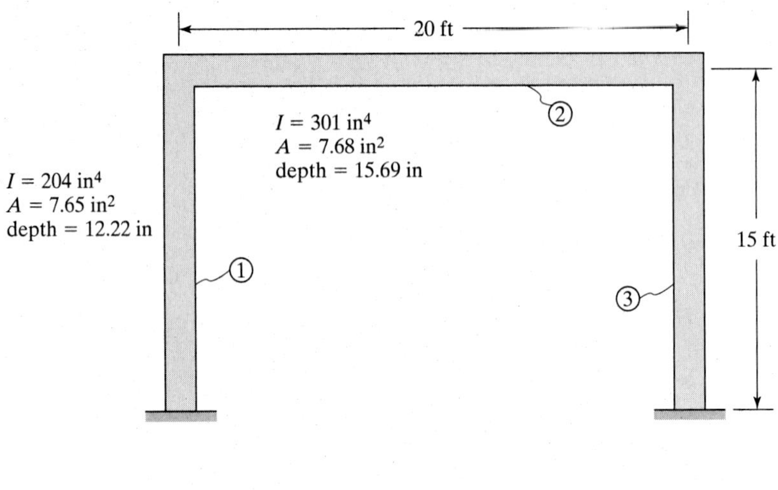

Enter **ANSYS** program by using the Launcher. Type **Osciframe** (or a file name of your choice) in the **Jobname** entry field of the dialog box. Pick **Run** to start the Graphic User Interface (GUI).

Create a title for the problem.

utility menu: **File → ChangeTitle** . . .

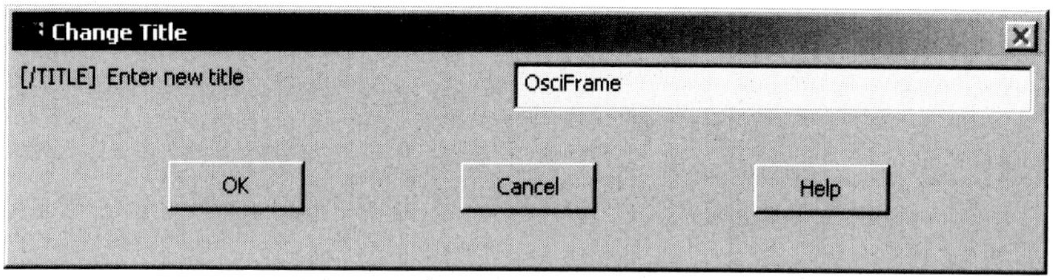

main menu: **Preprocessor → Element Type → Add/Edit/Delete**

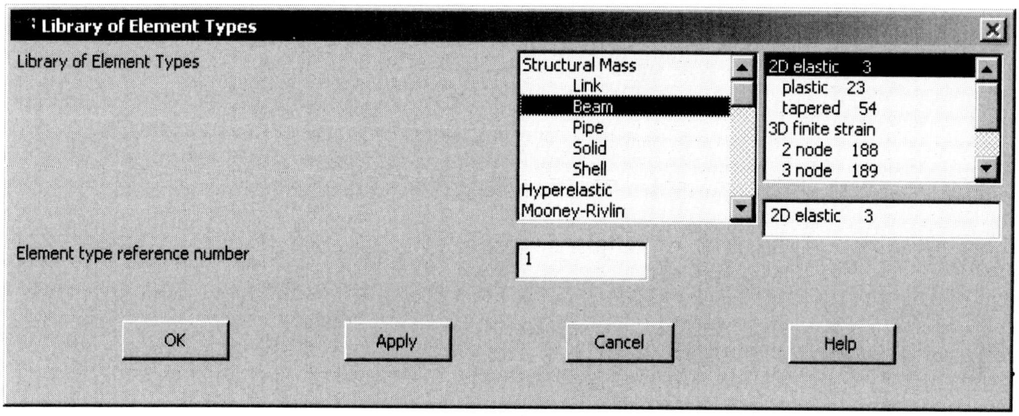

OK

main menu: **Preprocessor → Real Constants → Add/Edit/Delete**

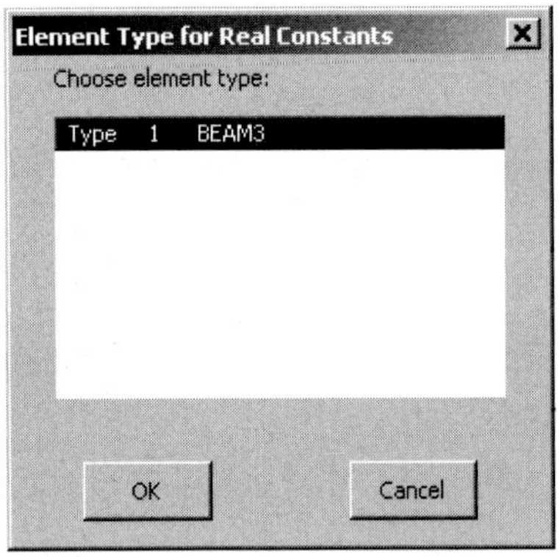
Element Type for Real Constants
Choose element type:
Type 1 BEAM3
OK
Cancel

OK

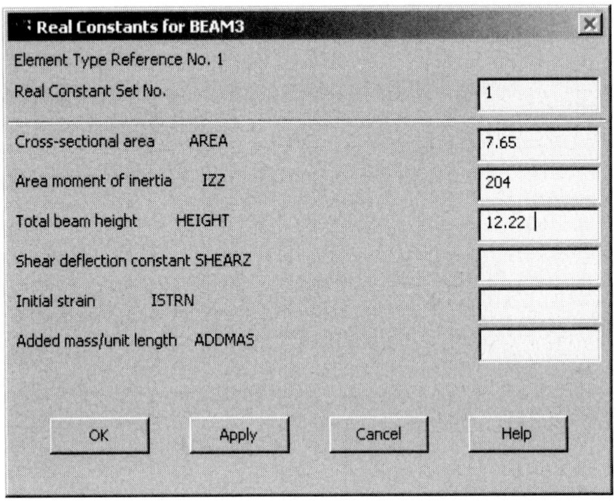
Real Constants for BEAM3
Element Type Reference No. 1
Real Constant Set No.
1
Cross-sectional area AREA
7.65
Area moment of inertia IZZ
204
Total beam height HEIGHT
12.22
Shear deflection constant SHEARZ
Initial strain ISTRN
Added mass/unit length ADDMAS
OK
Apply
Cancel
Help

Apply

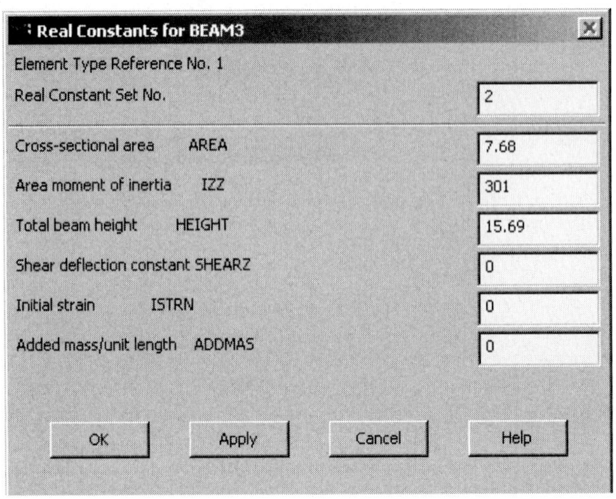
Real Constants for BEAM3
Element Type Reference No. 1
Real Constant Set No.
2
Cross-sectional area AREA
7.68
Area moment of inertia IZZ
301
Total beam height HEIGHT
15.69
Shear deflection constant SHEARZ
0
Initial strain ISTRN
0
Added mass/unit length ADDMAS
0
OK
Apply
Cancel
Help

OK

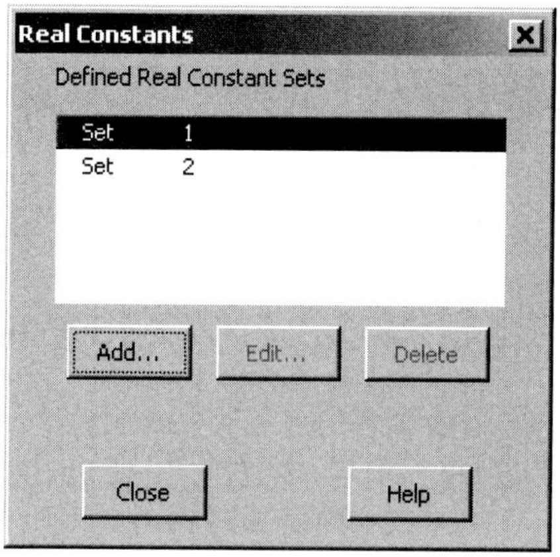

Close

main menu: **Preprocessor** → **Material Props** → **Material Models** → **Structural** → **Linear** → **Elastic** → **Isotropic**

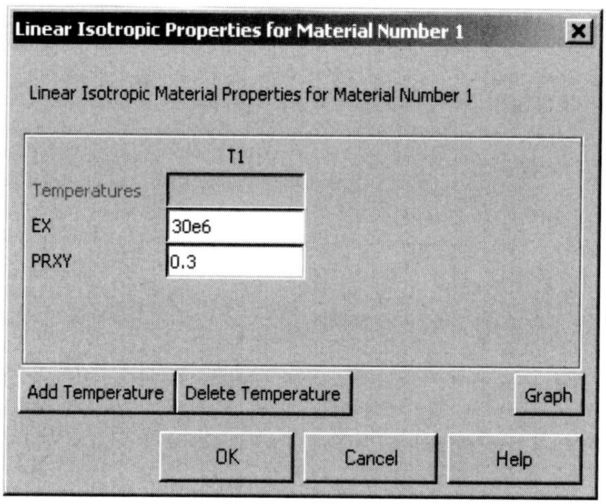

OK

main menu: **Preprocessor** → **Material Props** → **Material Models** → **Structural** → **Density**

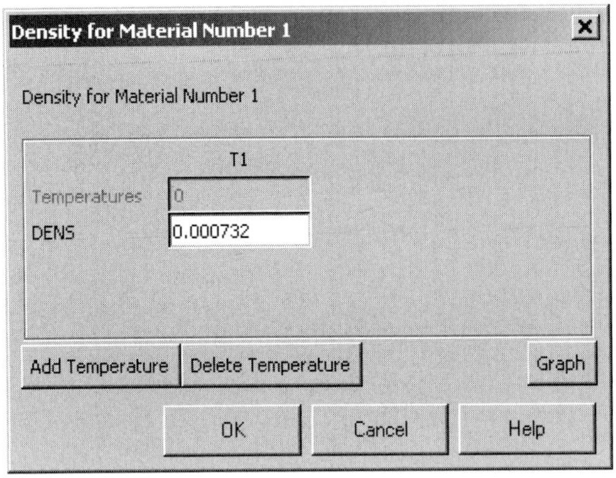

Note, here, density is equal to mass per unit volume.

OK

ANSYS Toolbar: **SAVE_DB**

main menu: **Preprocessor → Modeling → Create → Nodes → In Active CS**

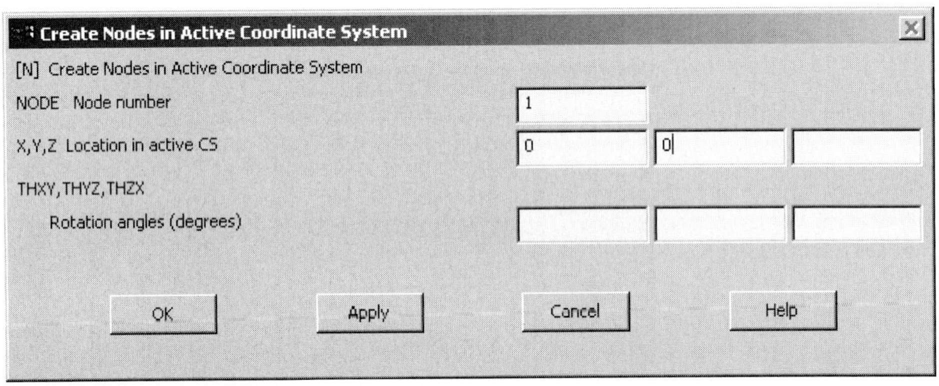

Apply

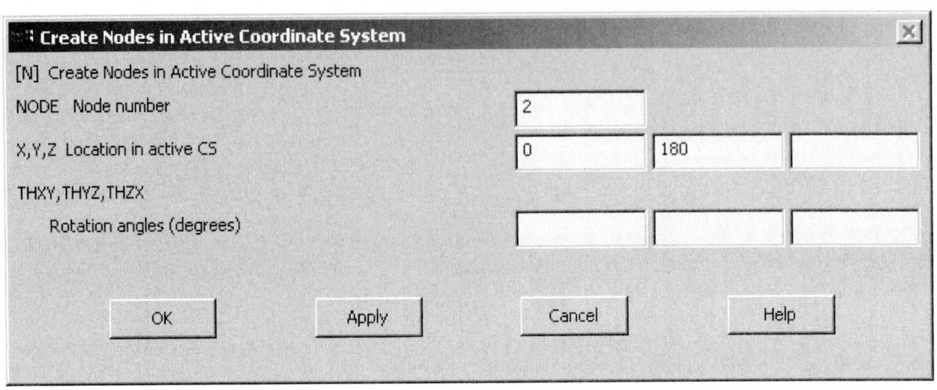

Apply

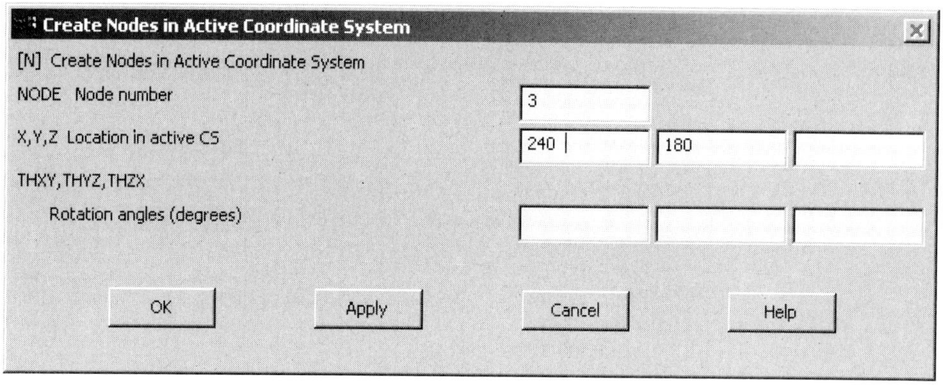

Apply

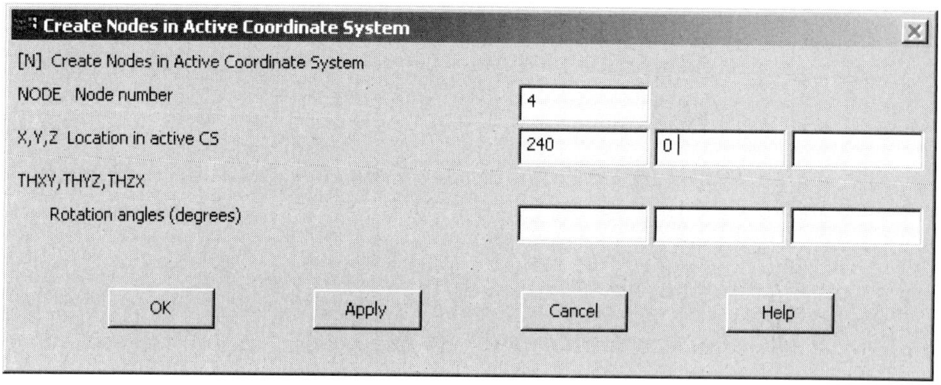

OK

main menu: **Preprocessor → Modeling → Create → Elements → Auto Numbered → Thru Nodes**

Pick Nodes 1 and 2 and **Apply**, then pick Nodes 4 and 3. **OK**

main menu: **Preprocessor → Modeling → Create → Elements → Elem Attributes**

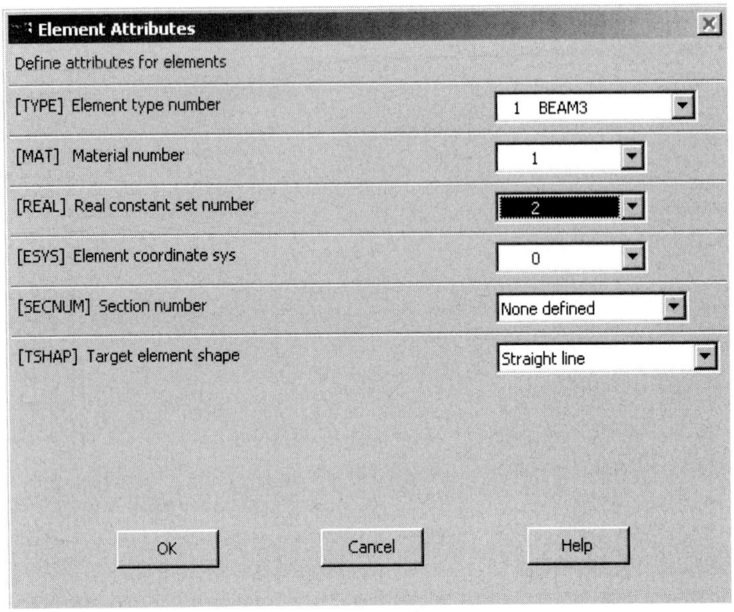

OK

main menu: **Preprocessor → Modeling → Create → Elements → Auto Numbered → Thru Nodes**

Pick nodes 2 and 3

main menu: **Solution → Define Loads → Apply → Structural → Displacement → On Nodes**

Pick nodes 1 and 4

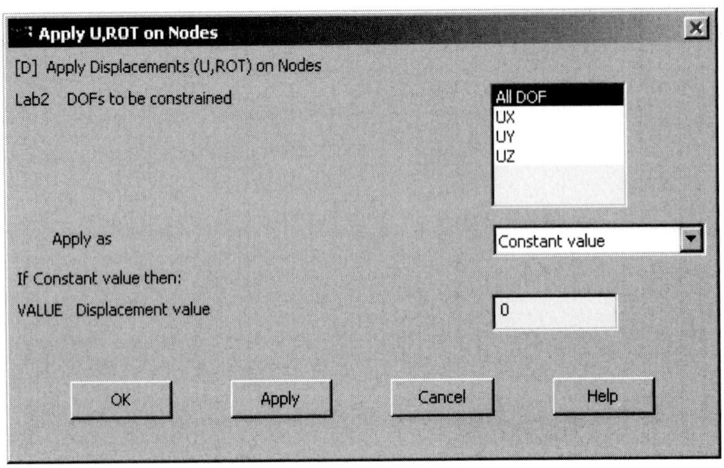

main menu: **Solution → Analysis Type → New Analysis**

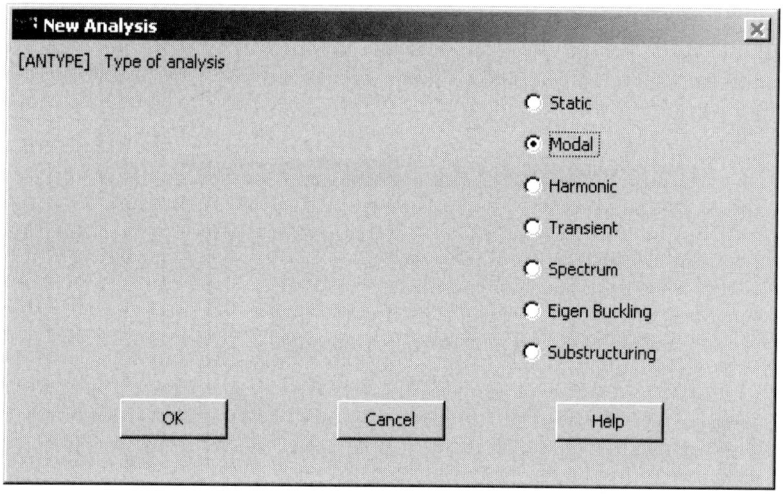

main menu: **Solution → Analysis Type → Analysis Options**

As we mentioned previously, we use modal analysis to obtain the natural frequencies and the mode shapes of a vibrating system. Moreover, you should recall that when a natural frequency of a system matches the excitation frequency, resonance will occur.

You must perform a new analysis each time you change boundary conditions. After you select the Modal in the New Analysis window, you need to choose a mode extraction method. ANSYS offers the following extraction methods: Block Lanczos (default), Subspace, Powerdynamics, Reduced, Unsymmetric, Damped, and QR Damped. You use the Block Lanczos method to solve large symmetrical eigenvalue problems. It uses sparse matrix solver and has a fast convergence rate. You can also use the Subspace method to solve large symmetrical problems. This method offers several solution controls to manage the iteration process. The Powerdynamics method is used for models with over 100,000 degrees of freedom and makes use of lumped mass approximation. As the name implies the Reduced method uses reduced system matrices to extract frequencies. It is not as accurate as the Subspace method but it has a faster convergence rate. Typically, you use the Unsymmetric method for problems dealing with fluid-structure interactions whose matrices are unsymmetrical. The Damped method is used for problems for which damping must be included in the model. The QR Damped method has a faster convergence rate than the Damped method. It uses the reduced modal damped matrix to calculate frequencies. If you choose the Reduced, Unsymmetic, or the Damped methods, then you must specify the number of modes to expand—the reduced solution is expanded to include the full degrees of freedom. In ANSYS, expansion also means writing the mode shapes to result files. When using the Reduced method, you need to define the master degrees of freedom—you choose at least twice as many master degrees of freedom as there are number of modes of interest.

Additional modal analysis options include specifying a frequency range for mode extraction. For most cases you don't need to specify the range and use the ANSYS's default setting. When using Reduced method you need to specify the number of reduced modes to be listed. Finally, if you are planning to perform spectrum analysis, then you need to normalize the mode shapes with respect to mass—the default setting.

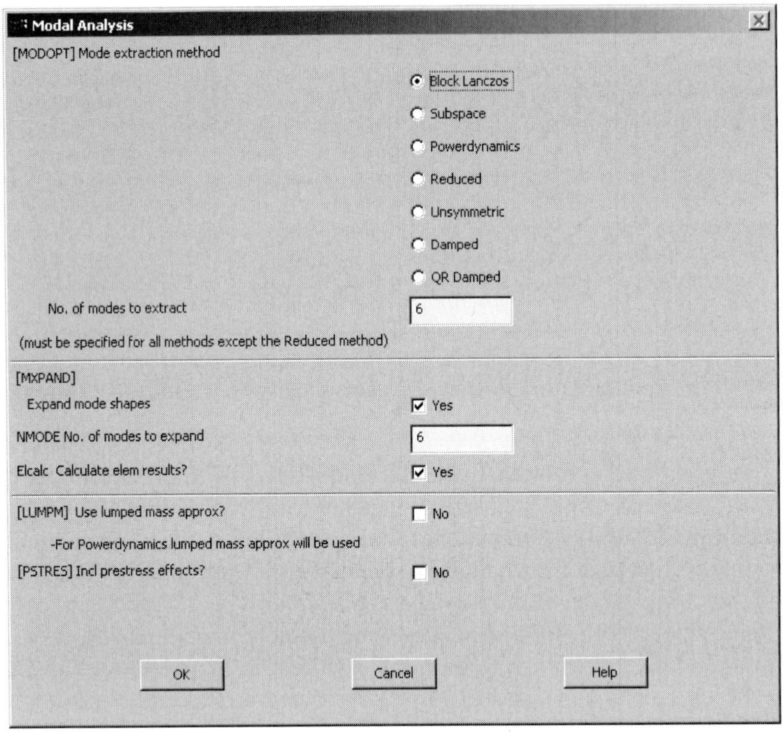

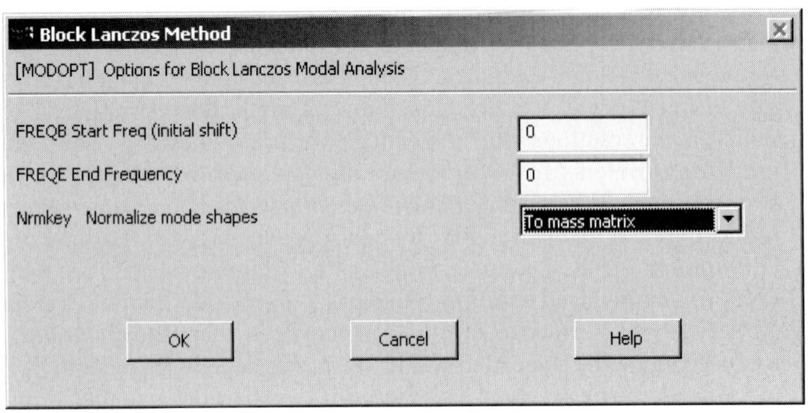

main menu: **Solution → Solve → Current LS**

Close (the solution is done!) window.

Close (the/STAT Command) window.

main menu: **General Postproc → Results Summary**

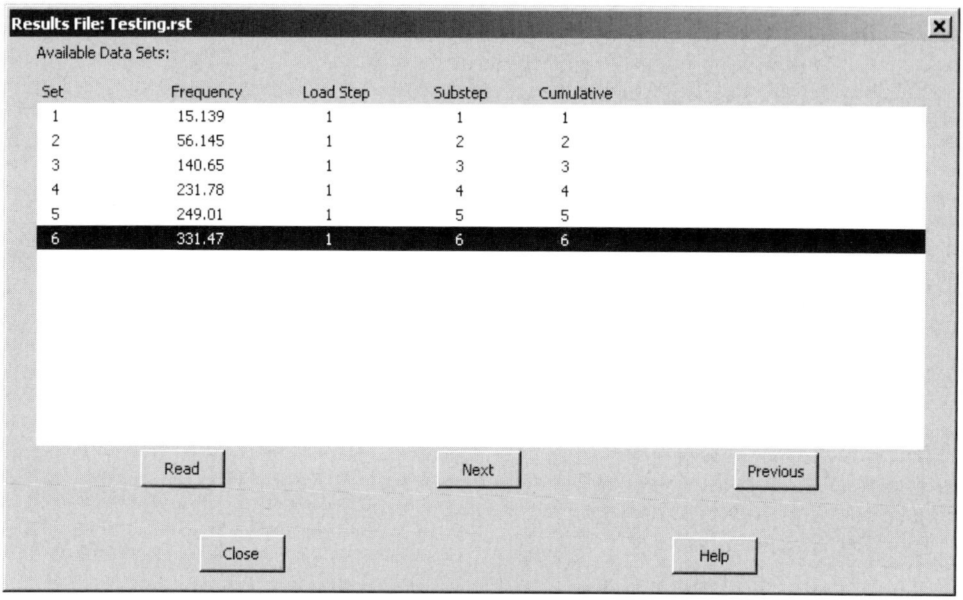

Note frequency values are given in hertz.

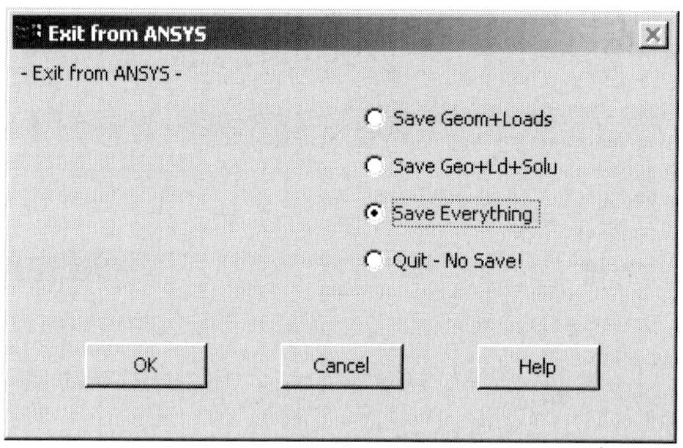

EXAMPLE 11.8

In this example problem we use ANSYS to study the natural oscillation of an aluminum strip with a rectangular cross section. The strip is 3 cm wide, 0.5 mm thick, and 10 cm long. It has a density of 2800 kg/m^3, a modulus of elasticity of $E = 73$ GPa, and a Poisson's ratio of 0.33. We assume the strip to be fixed at one end.

Enter the **ANSYS** program by using the Launcher. Type **Oscistrip** (or a file name of your choice) in the **Jobname** entry field of the dialog box. Pick **Run** to start the GUI.

Create a title for the problem.

utility menu: **File → Change Title** . . .

Type in the title of your choice

main menu: **Preprocessor → Element Type → Add/Edit/Delete**

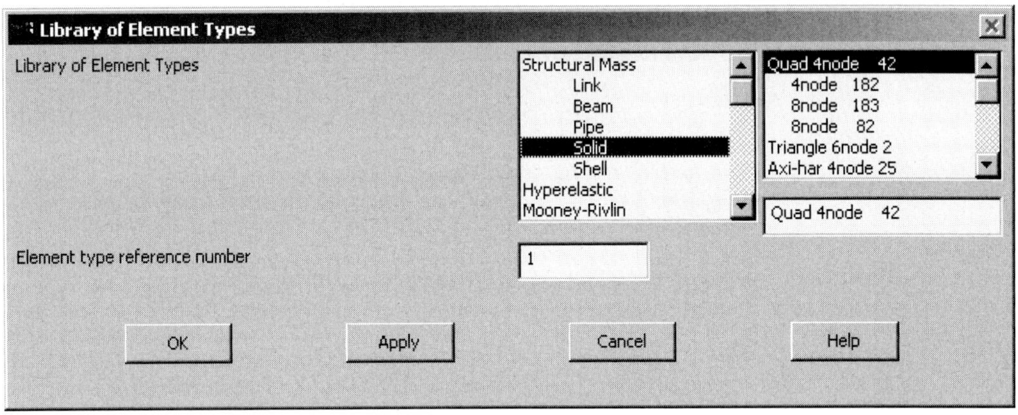

Apply

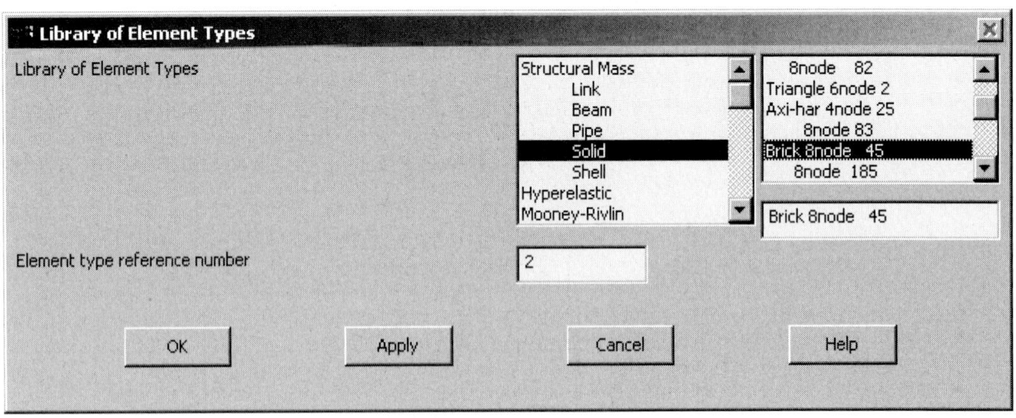

OK

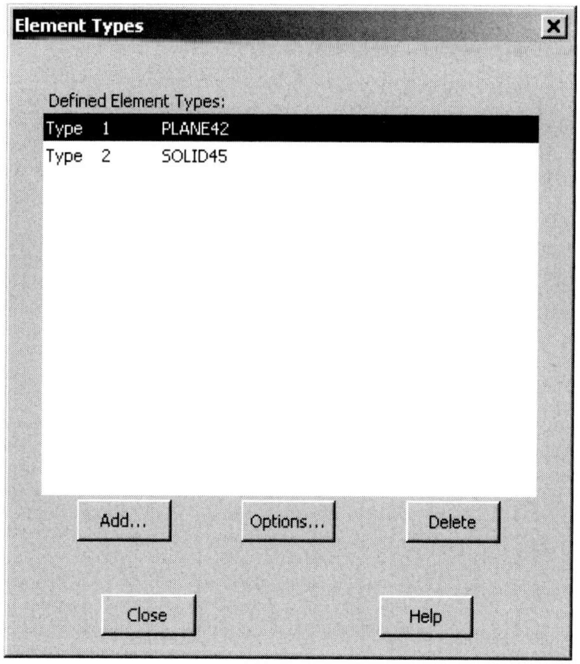

Close

main menu: **Preprocessor → Material Props → Material Models → Structural → Linear → Elastic → Isotropic**

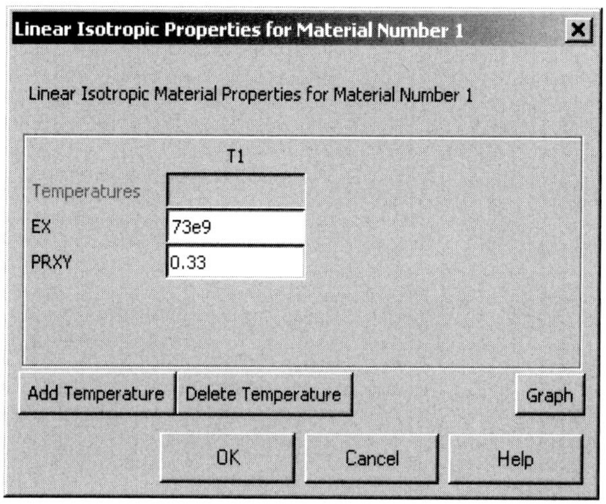

OK

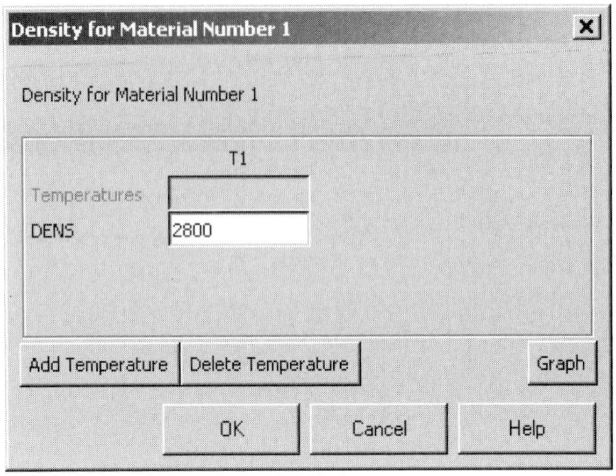

OK

ANSYS Toolbar: **SAVE_DB**

main menu: **Preprocessor → Modeling → Create → Areas → Rectangle → By 2 Corners**

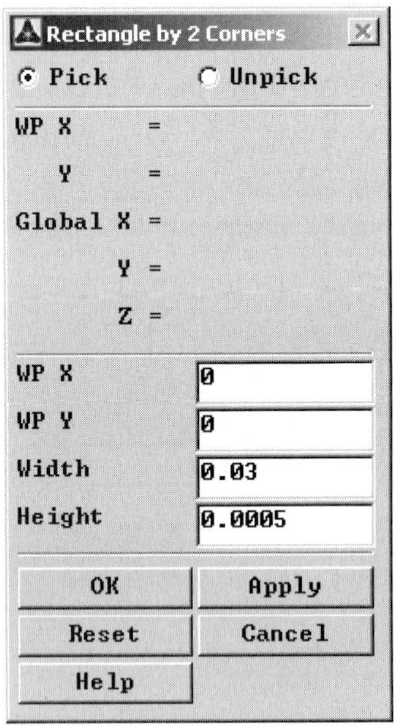

OK

main menu: **Preprocessor → Meshing → Size Cntrls → Smart Size → Basic**

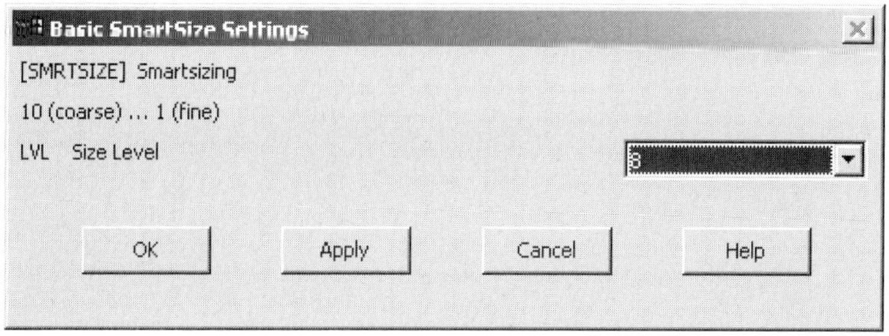

OK

main menu: **Preprocessor → Meshing → Mesh → Areas → Free**

First click on **Pick All,** then press **OK.**

main menu: **Preprocessor → Modeling → Operate → Extrude → Elem Ext Opts**

Choose Element type number 2 **SOLID45**. Also in the Element sizing options for extrusion, **VAL1 No. Elem divs** field type the value **10**.

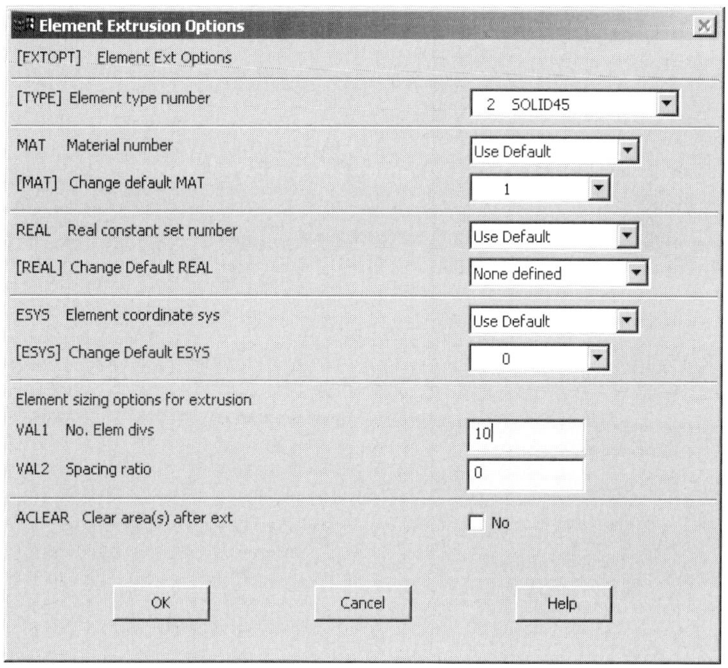

OK

main menu: **Preprocessor → Modeling → Operate → Extrude → Areas → By XYZ Offset**

Pick All

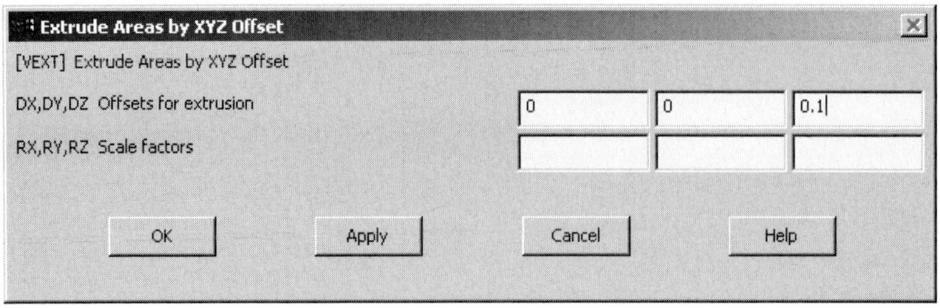

OK
utility menu: **Plot Ctrls → Pan, Zoom, Rotate** . . .

Choose **ISO**

utility menu: **Select → Entities**

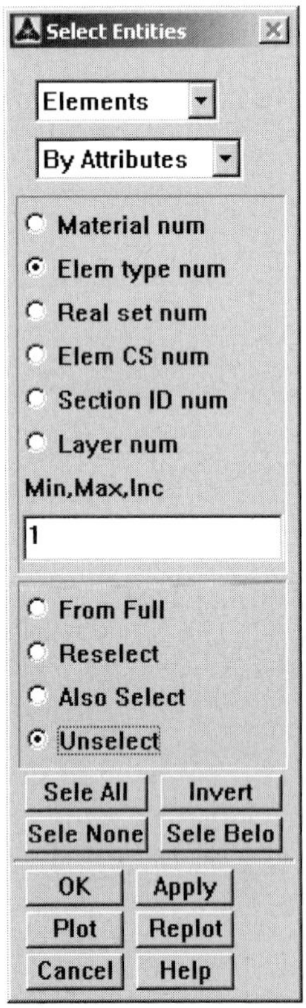

Select **Elements, By Attributes,** pick **Elem type num,** type **1** in the **Min, Max, Inc** field, and pick **Unselect.**

Apply

Then select the nodes to apply the zero displacement boundary conditions. Select **Nodes, By Location,** pick **Z coordinates,** type 0 in the **Min, Max,** and pick **From Full** set.

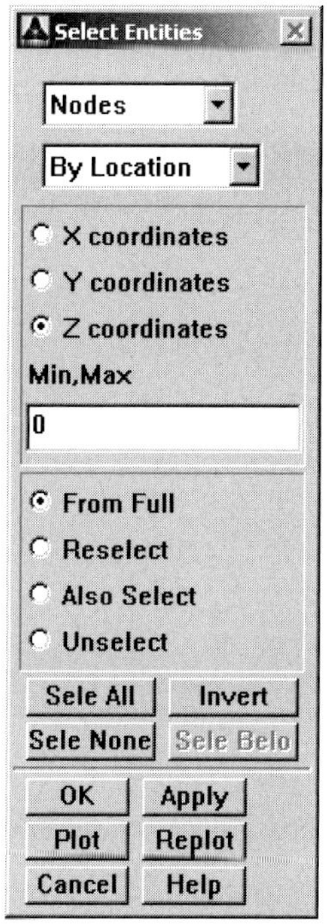

Apply

main menu: **Solution → Define Loads → Apply → Structural → Displacement → On Nodes**

Pick All

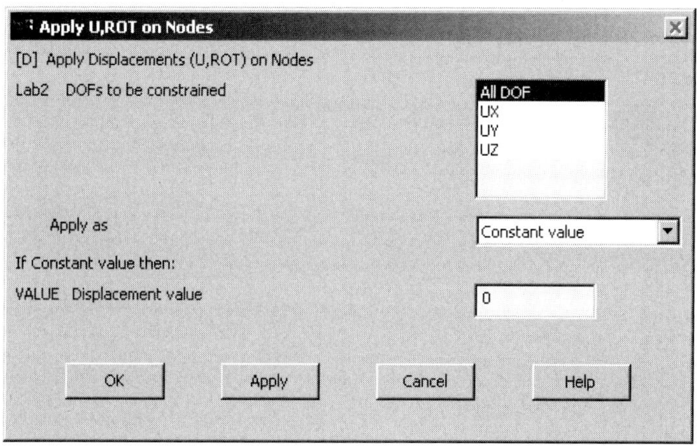

OK

utility menu: **Select → Everything . . .**

main menu: **Solution → Analysis Type → New Analysis**

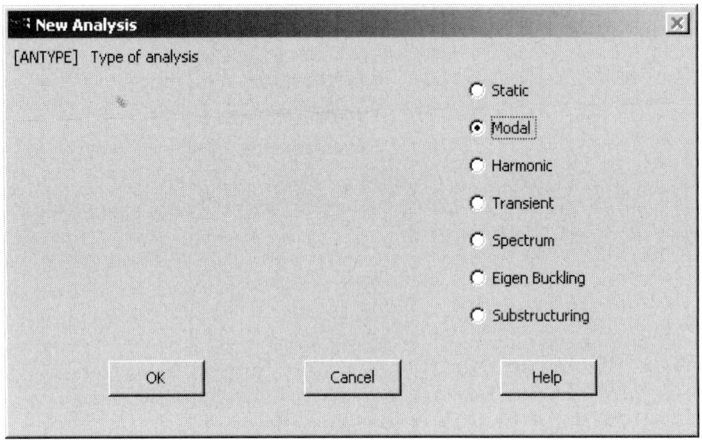

OK

main menu: **Solution → Analysis Type → Analysis Options**

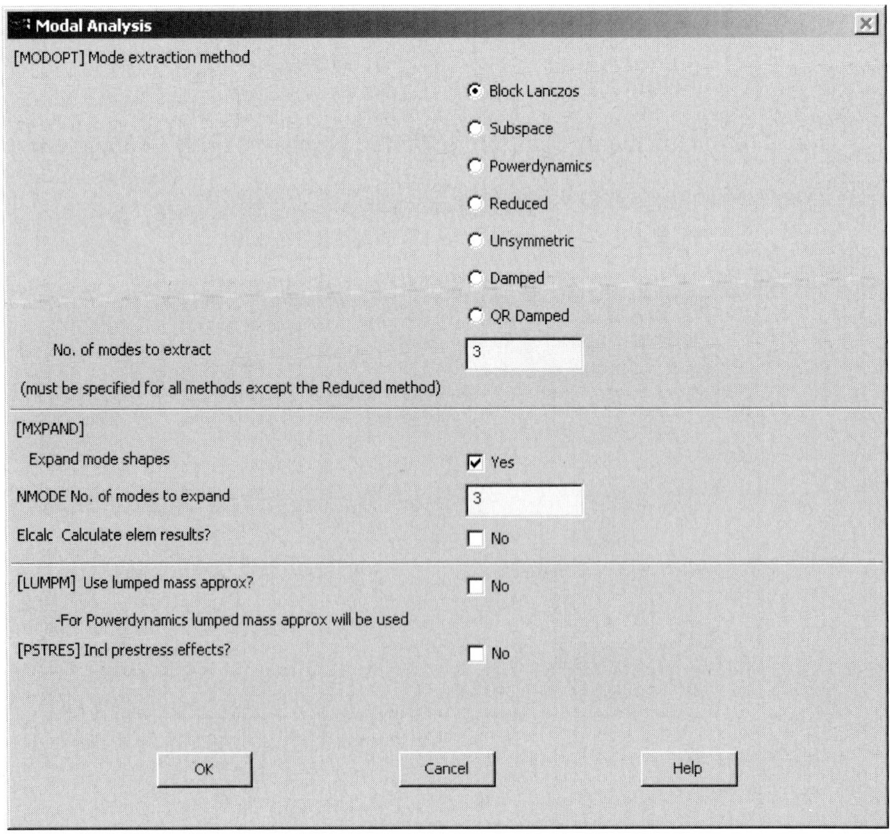

OK

Use the help menu to obtain information on Block Lanczos method.

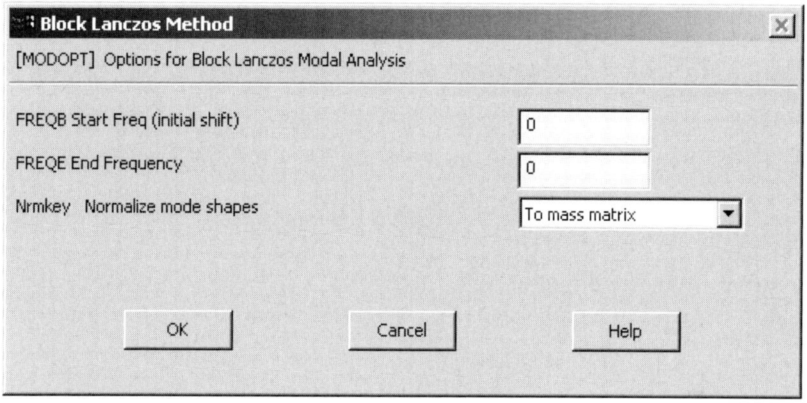

main menu: **Solution → Solve → Current LS**

Close (the solution is done!) window.

Close (the /STAT Command) window.

main menu: **General Postproc → Results Summary**

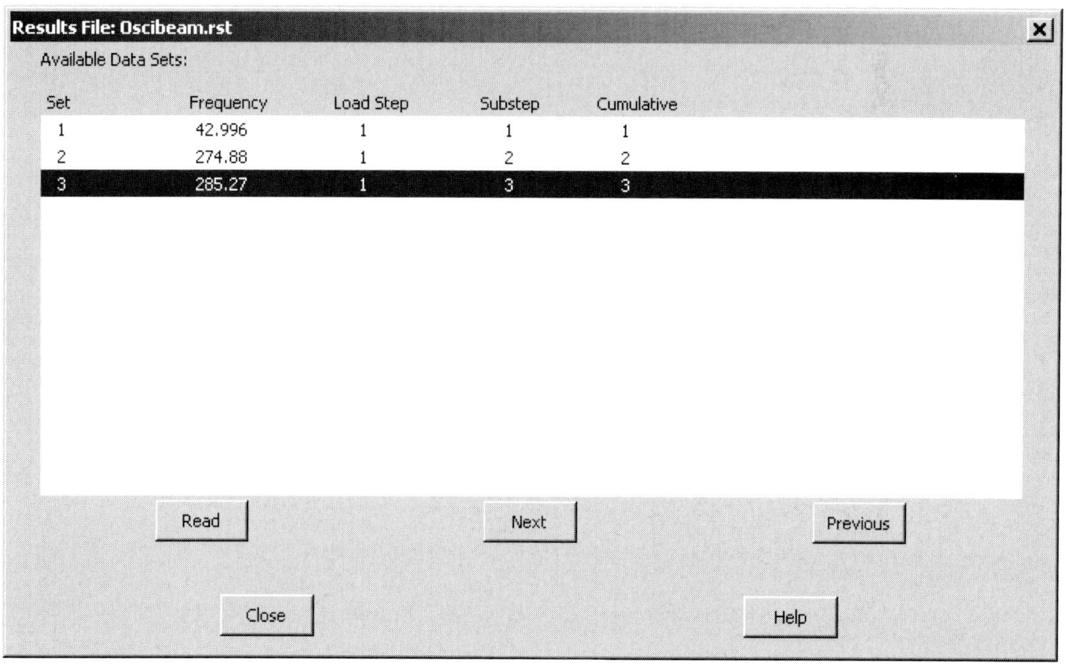

Close

main menu: **General Postproc → Read Results → First Set**

utility menu: **PlotCtrls → Animate → Mode Shape**

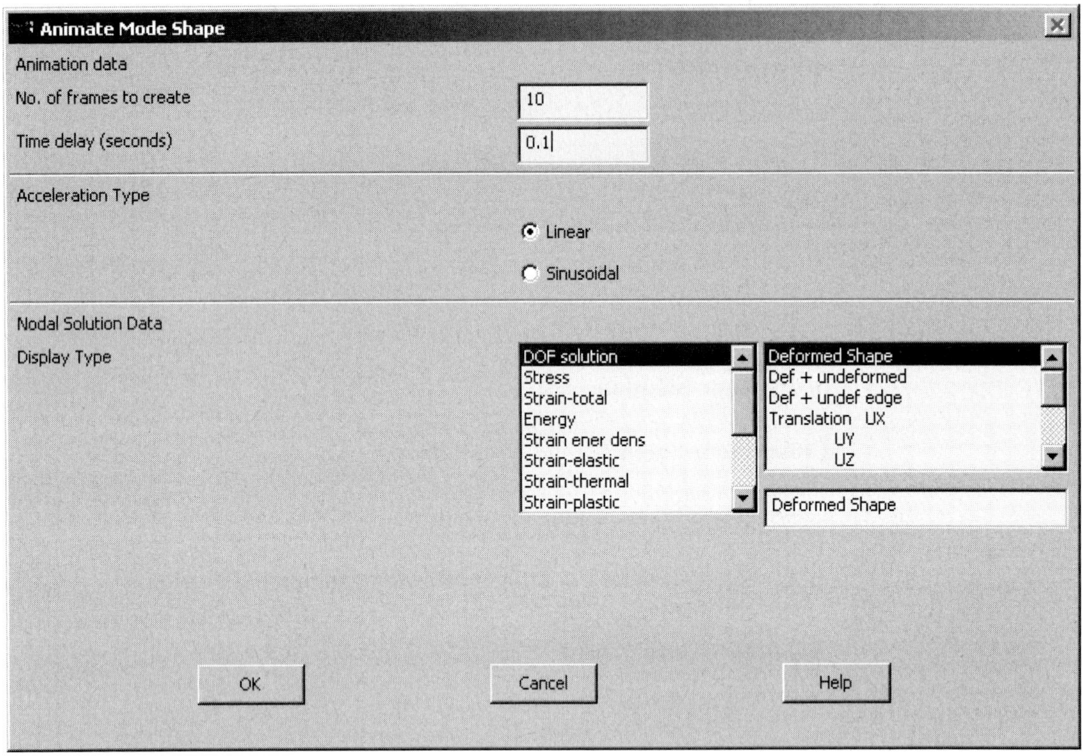

OK

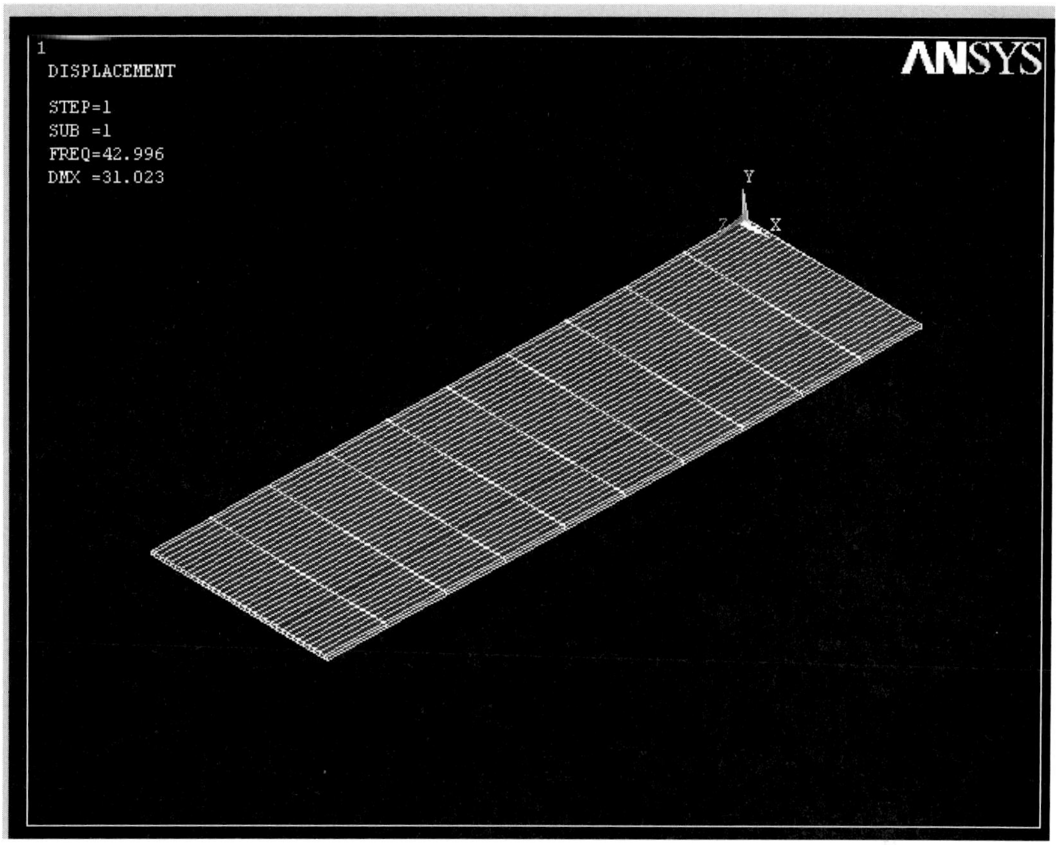

Look at the next mode by issuing the commands

main menu: **General Postproc → Read Results → Next Set**

utility menu: **PlotCtrls → Animate → Mode Shape**

Exit from ANSYS.

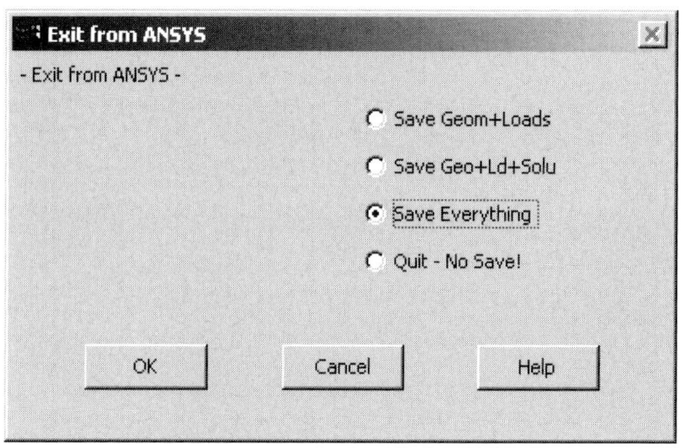

SUMMARY

At this point you should

1. have a good grasp of the fundamental concepts and governing equations of motions for a particle, a rigid body, and a dynamic system.
2. have a good understanding of fundamental definitions and concepts of vibration of mechanical and structural systems.
3. understand the formulation and natural vibration behavior for a system with single and multiple degrees of freedom.
4. understand the formulation and forced vibration behavior for a system with single and multiple degrees of freedom system.
5. understand the finite element formulation of axial members, beams, and frames.
6. know how to use ANSYS to solve some dynamic problems.

REFERENCES

Beer, F. P., and Johnston, E. R., *Vector Mechanics for Engineers*, 5th ed., New York, McGraw-Hill, 1988.

Steidel, R., *An Introduction to Mechanical Vibrations*, 3rd ed., New York, John Wiley and Sons, 1971.

Timoshenko, S., Young, D. H., and Weaver, W., *Vibration Problems in Engineering*, 4th ed., New York, John Wiley and Sons, 1974.

PROBLEMS

1. A simple dynamic system is modeled by a single degree of freedom with $m = 10$ kg and $k_{\text{equivalent}} = 100$ N/cm. Calculate the frequency and period of oscillation. Also, determine the maximum velocity and acceleration of the system. $(y(0) = 5 \text{ mm})$
2. The system described in problem 1 is subjected to a sinusoidal forcing function $F(t) = 30\sin(20t)$ in Newtons. Calculate the amplitude, maximum velocity, and acceleration of the system.
3. Derive the equations of motion for the system shown in the accompanying figure. What is the natural frequency of the system?

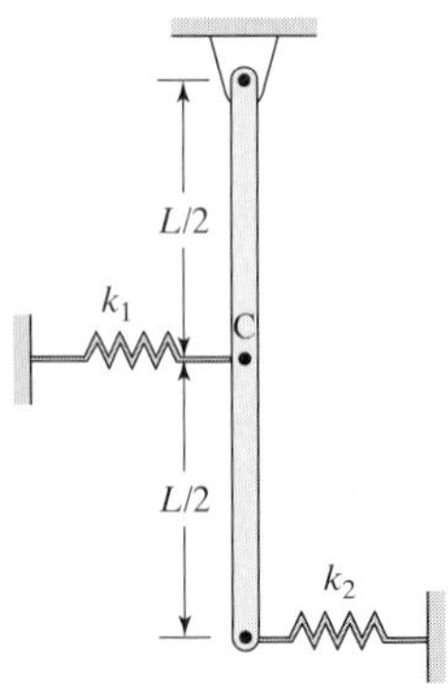

4. Derive the response of a single degree of freedom mass spring system to a *suddenly* applied force F_0, as shown in the accompanying figure. Plot the response. Compare the response of the system to a situation where the force F_0 is applied as a ramp function. Also compare the response of the system to a suddenly applied force to a situtation where the force F_0 is applied statically.

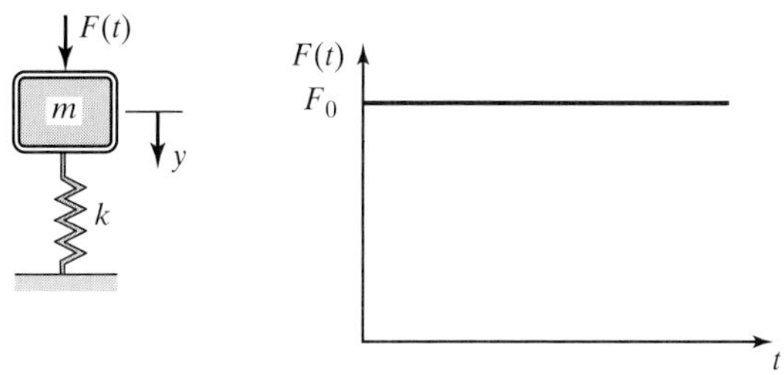

5. Derive the response of a single degree of freedom mass spring system to a *suddenly* applied force F_0 that decays with time according to $F(t) = F_0 e^{-c_1 t}$, as shown in the accompanying figure. The value of c_1 defines the rate of decay. Plot the response of the system for different values of c_1. Compare the response of the system to a situation where the force F_0 is applied statically.

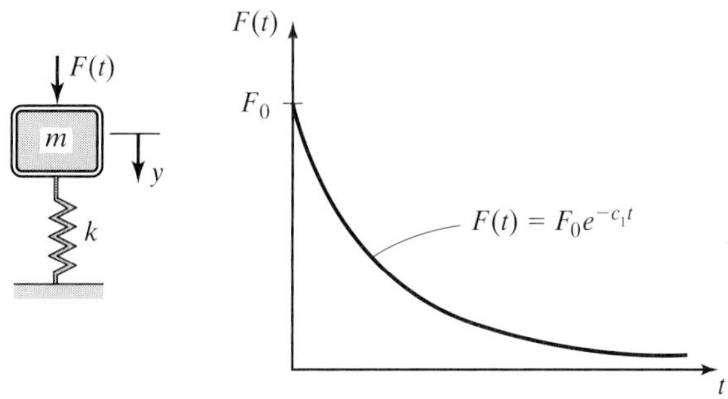

6. Derive the response of the system shown in the accompanying figure to the excitation shown.

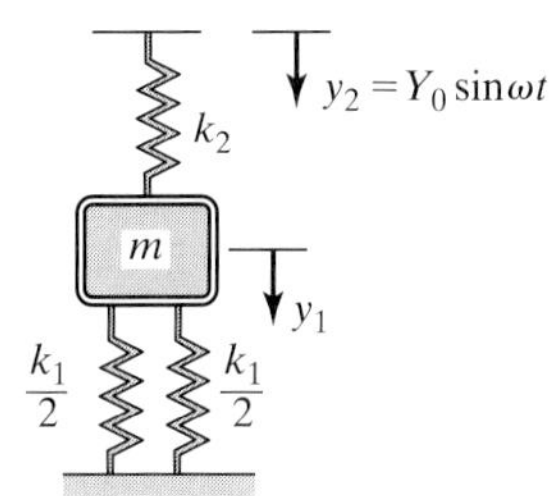

7. Using both Newton's second law and Lagrange's equations, formulate the equations of motion for the system shown in the accompanying figure.

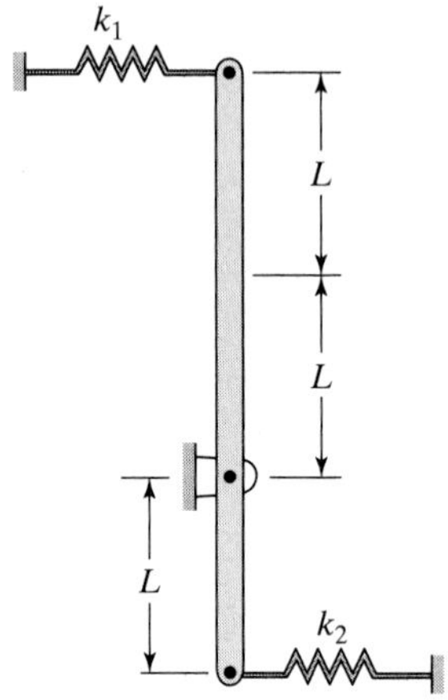

8. Using both Newton's second law and Lagrange's equations, formulate the equations of motion for the system shown in the accompanying figure.

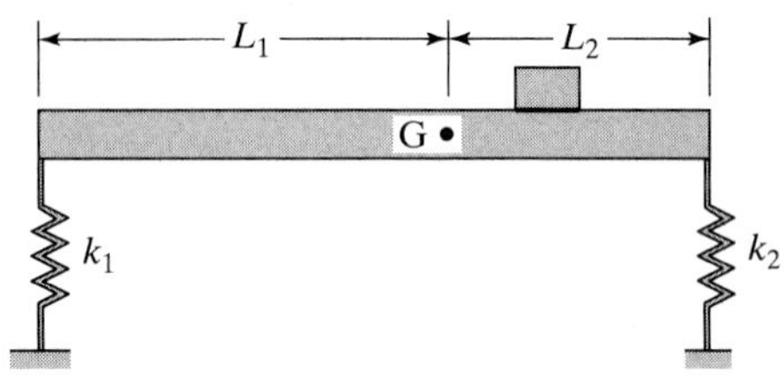

9. Using both Newton's second law and Lagrange's equations, formulate the equations of motion for the system shown in the accompanying figure.

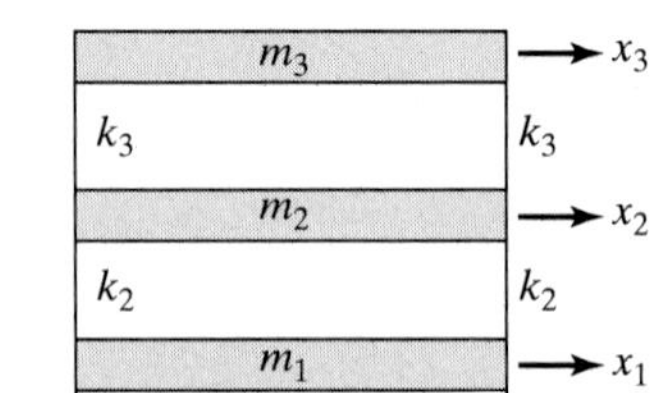

10. Determine the first two natural frequencies of the axial members shown in the accompanying figure. Members are made from structural steel with a modulus of elasticity of $E = 29 \times 10^6$ lb/in^2 and a mass density of 15.2 slugs/ft^3.

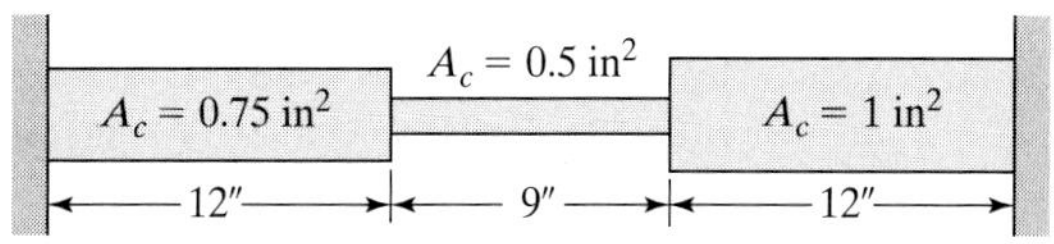

11. Determine the first two natural frequencies of the axial member shown in the accompanying figure. Member is made from aluminum alloy with a modulus of elasticity of $E = 10 \times 10^6$ lb/in^2 and a mass density of 5.4 slugs/ft^3.

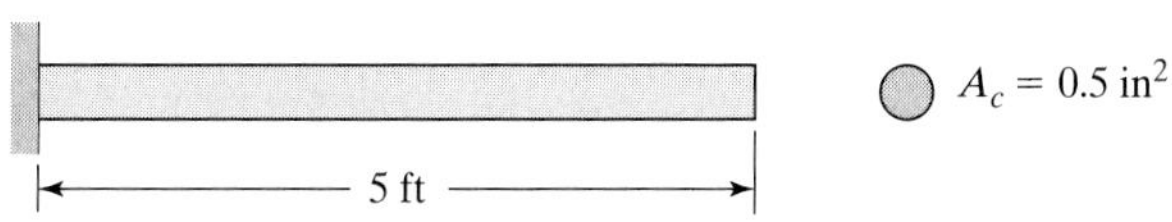

12. Determine the first three natural frequencies of the post shown. The post is made of structural steel with a modulus of elasticity of $E = 29 \times 10^6$ lb/in^2 and a mass density of 15.2 slugs/ft^3. Consider only the axial oscillation.

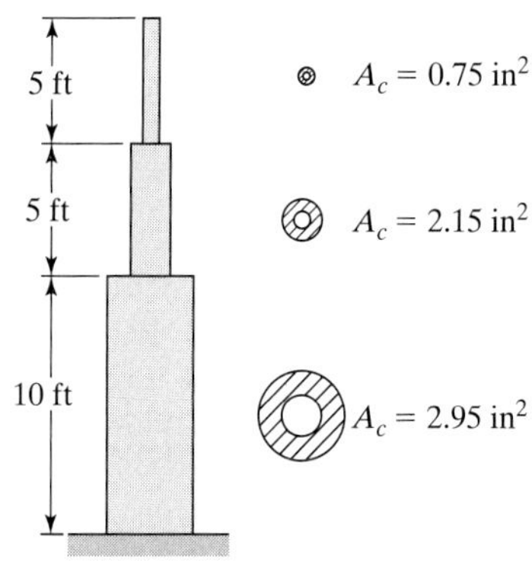

13. The cantilevered beam shown in the accompanying figure is a W18 $\times$ 35. Determine its first two natural frequencies.

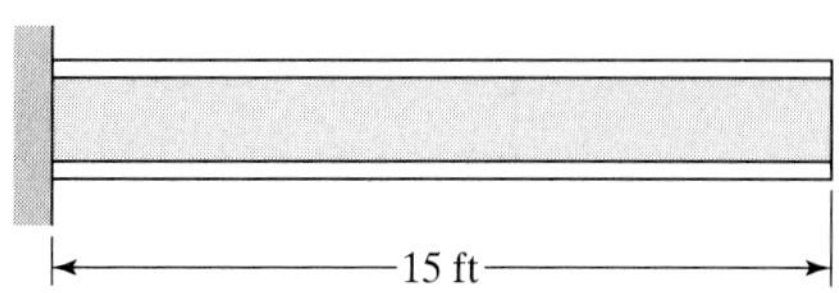

14. The cantilevered beam shown in the accompanying figure is a W16 × 31. Determine its first three natural frequencies.

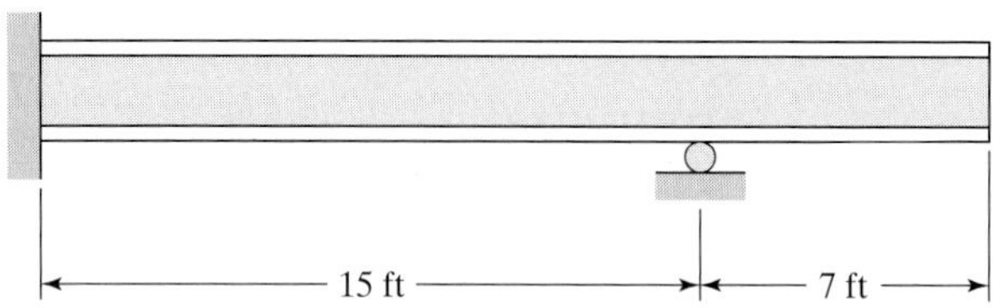

15. The simply supported beam shown in the accompanying figure is a W4 × 13. Determine its first two natural frequencies.

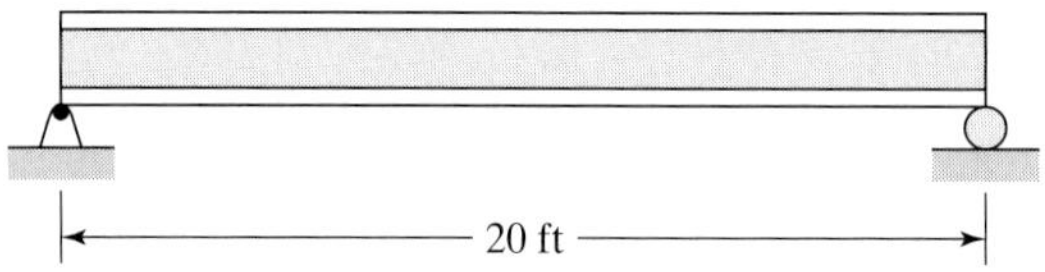

16. Determine the first two natural frequencies of the simply supported beam with the rectangular cross section shown in the accompanying figure.

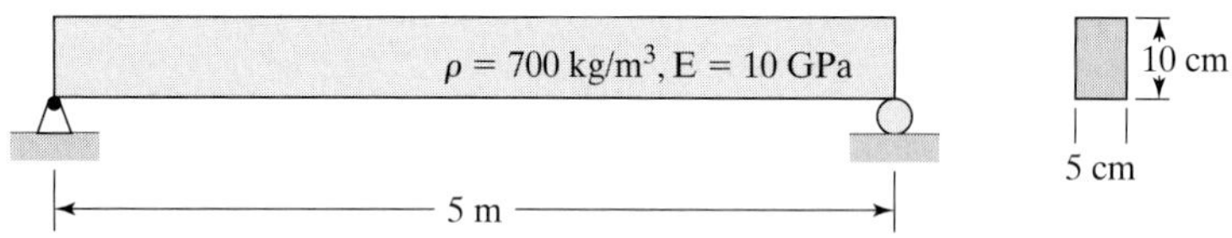

17. Consider the overhang frame shown in the accompanying figure. The cross-sectional areas and second moment of areas for each W12 × 26 member are shown in the figure. Determine the first three natural frequencies of the system.

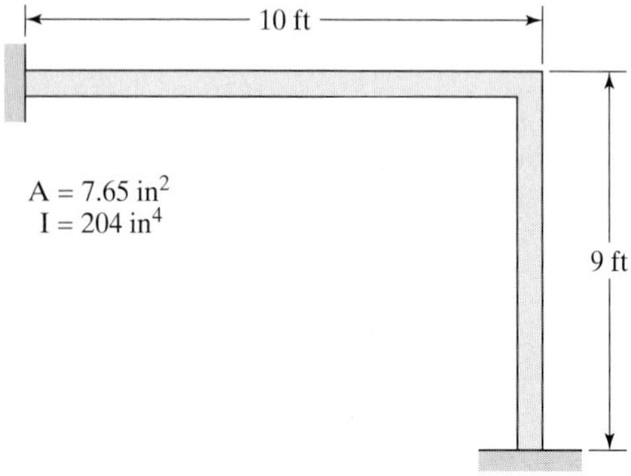

18. Consider the overhang frame shown in the accompanying figure. The members of the frame are W5 × 16. Determine the first three natural frequencies of the system.

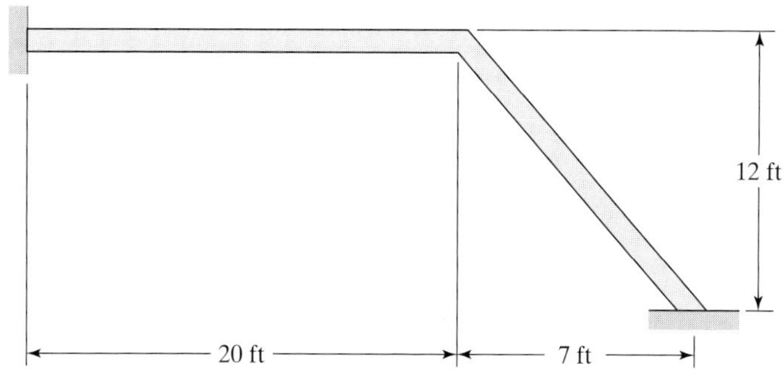

19. Re-solve Example 11.7 for the case where all members of the frame are W12 × 26.

20. Using ANSYS, determine the first three natural frequencies of the frame shown in the accompanying figure.

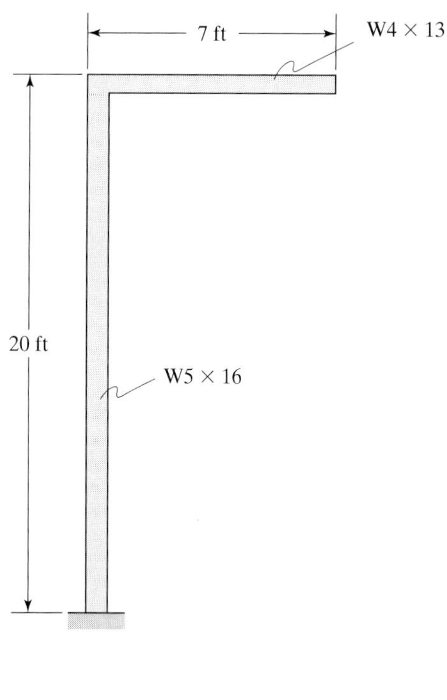

21. Using ANSYS, determine the first three natural frequencies of the frame shown in the accompanying figure.

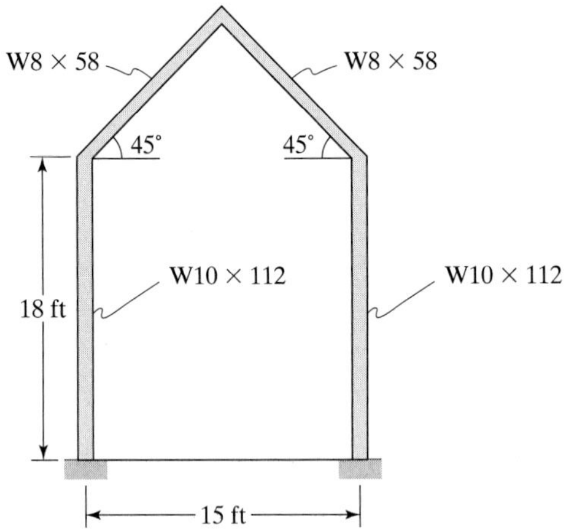

22. Re-solve Example 11.8 for the case where thickness of the strip is 1 mm, and 2 mm. Compare results and discuss your findings.

CHAPTER 12

Analysis of Fluid Mechanics Problems

The main objective of this chapter is to introduce you to the analysis of fluid mechanics problems. First, we will discuss the direct formulation of pipe-network problems. Then, we consider finite element formulation of ideal fluid behavior (inviscid flow). Finally, we briefly look at the flow of fluid through porous media and finite element formulation of underground seepage flows. The main topics discussed in Chapter 12 include the following:

12.1 Direct Formulation of Flow Through Pipes

12.2 Ideal Fluid Flow

12.3 Groundwater Flow

12.4 Examples Using ANSYS

12.5 Verification of Results

12.1 DIRECT FORMULATION OF FLOW THROUGH PIPES

We begin by reviewing fundamental concepts of fluid flow through pipes. The internal flow through a conduit may be classified as laminar or turbulent flow. In laminar flow situations, a thin layer of dye injected into a pipe will show as a straight line. No mixing of fluid layers will be visible. This situation does not hold for turbulent flow, in which the bulk mixing of adjacent fluid layers will occur. Laminar and turbulent flow are depicted in Figure 12.1. Laminar flow typically occurs when the Reynolds number of the flowing fluid is less than 2100. The Reynolds number is defined as

$$\mathrm{Re} = \frac{\rho V D}{\mu} \qquad (12.1)$$

where ρ and μ are the density and the dynamic viscosity of the fluid respectively. V represents the average fluid velocity, and D represents the diameter of the pipe. The flow

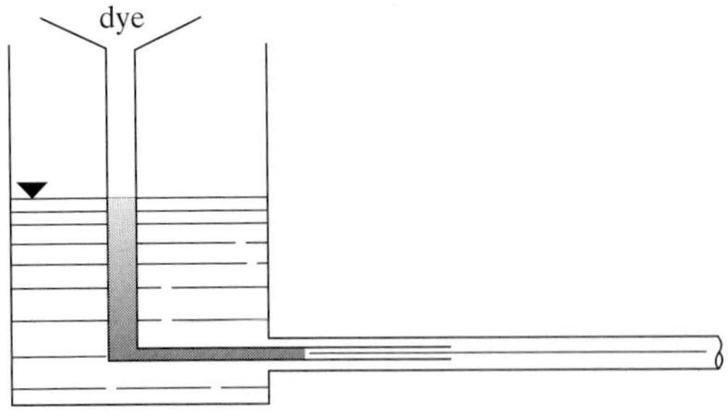

(a) Laminar flow in a pipe.

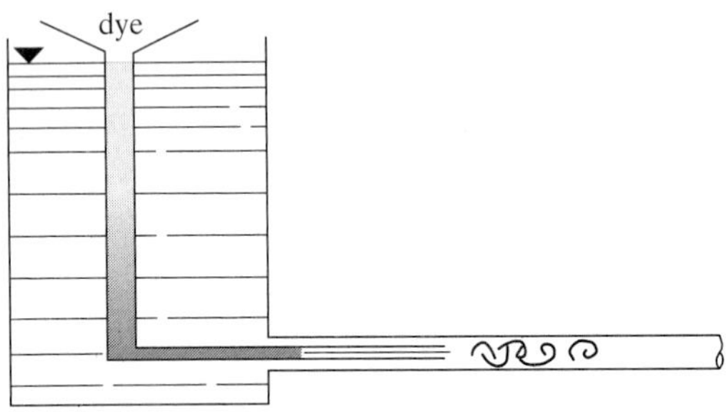

(b) Turbulent flow in a pipe.

FIGURE 12.1 Laminar and turbulent flows.

is said to be in a transition region when the Reynolds number is typically between 2100 and 4000. The behavior of the fluid flow is unpredictable in the transition region. The flow is generally considered to be turbulent when the Reynolds number is greater than 4000. The conservation of mass for a steady flow requires that the mass flow rate at any section of the pipe remains constant according to the equation

$$\dot{m_1} = \dot{m_2} = \rho_1 V_1 A_1 = \rho_2 V_2 A_2 = \text{constant} \tag{12.2}$$

Again, ρ is the density of the fluid, V is the average fluid velocity at a section, and A represents the cross-sectional area of the flow as shown in Figure 12.2.

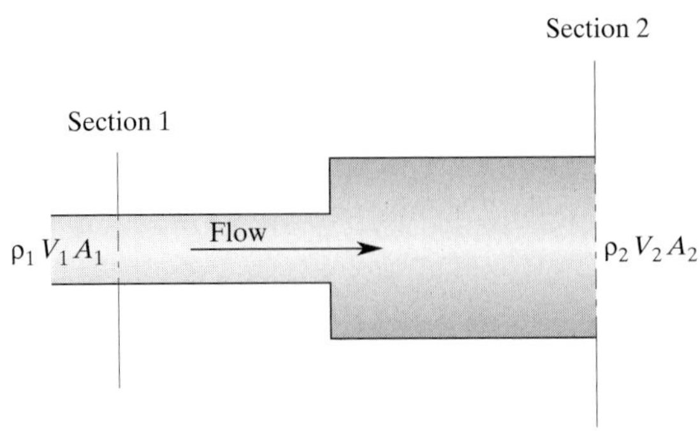

FIGURE 12.2 Flow of fluid through a conduit with variable cross section.

For an incompressible flow—a flow situation where the density of the fluid remains constant—the volumetric flow rate Q through a conduit at any section of the conduit is also constant:

$$Q_1 = Q_2 = V_1A_1 = V_2A_2 \tag{12.3}$$

For a fully developed laminar flow, there exists a relationship between the volumetric flow rate and the pressure drop $P_1 - P_2$ along a pipe of length L. This relationship is given by

$$Q = \frac{\pi D^4}{128\mu}\left(\frac{P_1 - P_2}{L}\right) \tag{12.4}$$

The pressure drop for a turbulent flow is commonly expressed in terms of head loss, which is defined as

$$H_{loss} = \frac{P_1 - P_2}{\rho g} = f\frac{L}{D}\frac{V^2}{2g} \tag{12.5}$$

where f is the friction factor, which depends on the surface roughness of the pipe and the Reynolds number. For turbulent flows, we can also obtain a relationship between the volumetric flow rate and the pressure drop by substituting for V in terms of the flow rate in Eq. (12.5) and rearranging terms:

$$Q^2 = \frac{1}{f}\frac{\pi^2 D^5}{8\rho}\left(\frac{P_1 - P_2}{L}\right) \tag{12.6}$$

When we compare turbulent flow to laminar flow, we note that for turbulent flow, the relationship between the flow rate and pressure drop is nonlinear.

Pipes in Series

For flow of a fluid through a piping network consisting of a series of pipes with respective diameters $D_1, D_2, D_3, \ldots$, as shown in Figure 12.3, the conservation of mass (continuity equation) requires that under steady-state conditions, the mass flow rate through each pipe be the same:

$$\dot{m}_1 = \dot{m}_2 = \dot{m}_3 = \ldots = \text{constant} \tag{12.7}$$

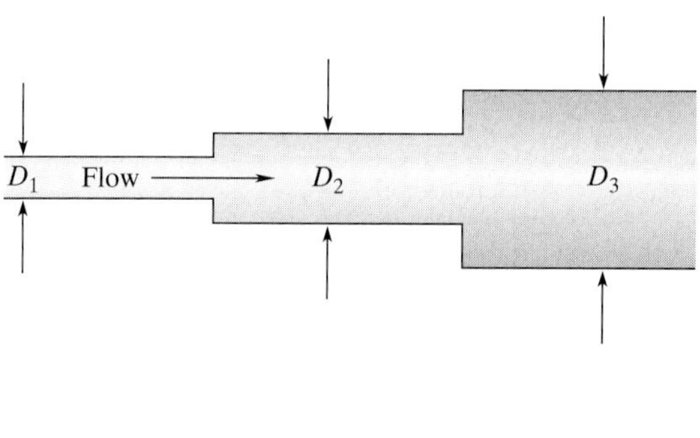

FIGURE 12.3 Pipes in series.

Moreover, for an incompressible flow, the volumetric flow rate through each pipe that is part of a piping network in series is constant. That is,

$$Q_1 = Q_2 = Q_3 = \ldots = \text{constant} \quad (12.8)$$

Expressing the flow rates in terms of the average fluid velocity in each pipe, we obtain

$$V_1 D_1^2 = V_2 D_2^2 = V_3 D_3^2 = \ldots = \text{constant} \quad (12.9)$$

For pipes in series, the total pressure drop through a network is determined from the sum of the pressure drops in each pipe:

$$\Delta P_{\text{total}} = \Delta P_1 + \Delta P_2 + \Delta P_3 + \ldots \quad (12.10)$$

Pipes in Parallel

For flow of a fluid through a piping network consisting of pipes in parallel arrangement, as shown in Figure 12.4, the conservation of mass (continuity equation) requires that

$$\dot{m}_{\text{total}} = \dot{m}_1 + \dot{m}_2 \quad (12.11)$$

Moreover, for an incompressible flow,

$$Q_{\text{total}} = Q_1 + Q_2 \quad (12.12)$$

For pipes in parallel configuration, the pressure drop in each parallel branch is the same, and is related according to

$$\Delta P_{\text{total}} = \Delta P_1 = \Delta P_2 \quad (12.13)$$

Finite Element Formulation

Consider an incompressible laminar flow of a viscous fluid through a network of piping systems, as shown in Figure 12.5. We start by subdividing the problem into nodes and elements. This example may be represented by a model that has four nodes and four elements.

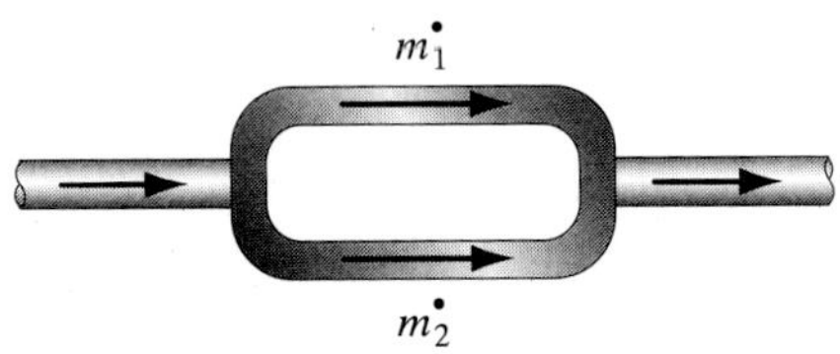

FIGURE 12.4 Pipes in parallel.

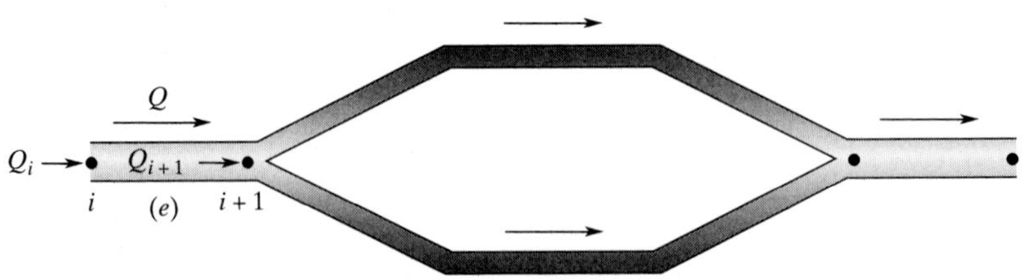

FIGURE 12.5 A network problem: an incompressible laminar flow of a viscous fluid through a network of piping systems.

The behavior of the fluid flow inside a pipe section is modeled by an element with two nodes. The elemental description is given by the relationship between the flow rate and the pressure drop as given by Eq. (12.4), such that

$$Q = \frac{\pi D^4}{128\mu}\left(\frac{P_i - P_{i+1}}{L}\right) = C(P_i - P_{i+1}) \tag{12.14}$$

where the flow-resistance coefficient C is given by

$$C = \frac{\pi D^4}{128L\mu} \tag{12.15}$$

Because there are two nodes associated with each element, we need to create two equations for each element. These equations must involve nodal pressure and the element's flow resistance. Consider the flow rates Q_i and Q_{i+1} and the nodal pressures P_i and P_{i+1} of an element, which are related according to the equations

$$\begin{aligned} Q_i &= C(P_i - P_{i+1}) \\ Q_{i+1} &= C(P_{i+1} - P_i) \end{aligned} \tag{12.16}$$

The equations given by (12.16) were formulated such that the conservation of mass is satisfied as well. The sum of Q_i and Q_{i+1} is zero, which implies that under steady-state element conditions, what flows into a given element must flow out. Equations (12.16) can be expressed in matrix form by

$$\begin{Bmatrix} Q_i \\ Q_{i+1} \end{Bmatrix} = \begin{bmatrix} C & -C \\ -C & C \end{bmatrix}\begin{Bmatrix} P_i \\ P_{i+1} \end{Bmatrix} = \begin{bmatrix} \frac{\pi D^4}{128L\mu} & -\frac{\pi D^4}{128L\mu} \\ -\frac{\pi D^4}{128L\mu} & \frac{\pi D^4}{128L\mu} \end{bmatrix}\begin{Bmatrix} P_i \\ P_{i+1} \end{Bmatrix} \tag{12.17}$$

The element's flow-resistance matrix is then given by

$$[\mathbf{R}]^{(e)} = \begin{bmatrix} \frac{\pi D^4}{128L\mu} & -\frac{\pi D^4}{128L\mu} \\ -\frac{\pi D^4}{128L\mu} & \frac{\pi D^4}{128L\mu} \end{bmatrix} \tag{12.18}$$

Applying the elemental description given by Eq. (12.17) to all elements and assembling them will lead to the formation of the global flow matrix, the flow-resistance matrix, and the pressure matrix.

EXAMPLE 12.1

Oil with dynamic viscosity of $\mu = 0.3\text{N}\cdot\text{s/m}^2$ and density of $\rho = 900\ \text{kg/m}^3$ flows through the piping network shown in Figure 12.6. The 2–4–5 branch was added in parallel to the 2–3–5 branch to allow for the flexibility of performing maintenance on one

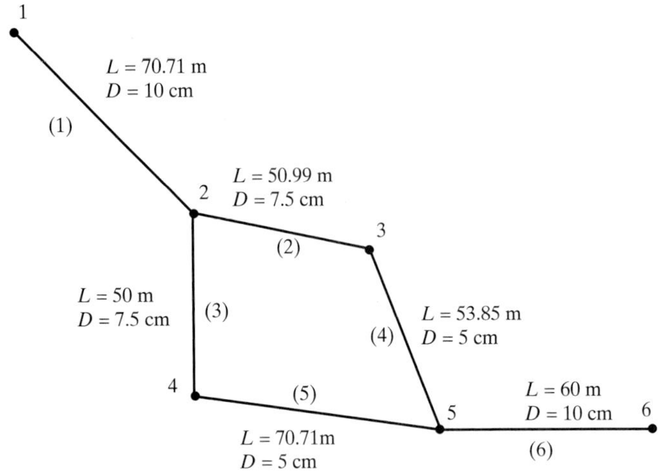

FIGURE 12.6 The piping network of Example 12.1.

branch while the oil flows through the other branch. The dimensions of the piping system are shown in Figure 12.6. Determine the pressure distribution in the system if both branches are on line. The flow rate at node 1 is $5 \times 10^{-4}\ \text{m}^3/\text{s}$. The pressure at node 1 is 39182 Pa (g) and the pressure at node 6 is -3665 Pa (g). For the given conditions, the flow is laminar throughout the system. How does the flow divide in each branch?

The elemental flow resistance is given by Eq. (12.18) as

$$[\mathbf{R}]^{(e)} = \begin{bmatrix} \dfrac{\pi D^4}{128L\mu} & -\dfrac{\pi D^4}{128L\mu} \\ -\dfrac{\pi D^4}{128L\mu} & \dfrac{\pi D^4}{128L\mu} \end{bmatrix}$$

We model the given network using six elements and six nodes. Evaluating the respective resistance matrices for elements (1)–(6), we obtain

$$[\mathbf{R}]^{(1)} = 10^{-9}\begin{bmatrix} 115.70 & -115.70 \\ -115.70 & 115.70 \end{bmatrix}\begin{matrix} 1 \\ 2 \end{matrix} \qquad [\mathbf{R}]^{(2)} = 10^{-9}\begin{bmatrix} 50.76 & -50.76 \\ -50.76 & 50.76 \end{bmatrix}\begin{matrix} 2 \\ 3 \end{matrix}$$

$$[\mathbf{R}]^{(3)} = 10^{-9}\begin{bmatrix} 51.77 & -51.77 \\ -51.77 & 51.77 \end{bmatrix}\begin{matrix} 2 \\ 4 \end{matrix} \qquad [\mathbf{R}]^{(4)} = 10^{-9}\begin{bmatrix} 9.50 & -9.50 \\ -9.50 & 9.50 \end{bmatrix}\begin{matrix} 3 \\ 5 \end{matrix}$$

$$[\mathbf{R}]^{(5)} = 10^{-9}\begin{bmatrix} 7.23 & -7.23 \\ -7.23 & 7.23 \end{bmatrix}\begin{matrix} 4 \\ 5 \end{matrix} \qquad [\mathbf{R}]^{(6)} = 10^{-9}\begin{bmatrix} 136.35 & -136.35 \\ -136.35 & 136.35 \end{bmatrix}\begin{matrix} 5 \\ 6 \end{matrix}$$

Note that in order to aid us in assembling the elemental resistance matrices into the global resistance matrix, the corresponding nodes are shown alongside of each element's resistance matrix. So, we have

$$10^{-9}\begin{bmatrix} 115.7 & -115.7 & 0 & 0 & 0 & 0 \\ -115.7 & 115.7+50.76+51.77 & -50.76 & -51.77 & 0 & 0 \\ 0 & -50.76 & 50.76+9.50 & 0 & -9.50 & 0 \\ 0 & -51.77 & 0 & 51.77+7.23 & -7.23 & 0 \\ 0 & 0 & -9.50 & -7.23 & 9.50+7.23+136.35 & -136.35 \\ 0 & 0 & 0 & 0 & -136.35 & 136.35 \end{bmatrix}\begin{matrix}1\\2\\3\\4\\5\\6\end{matrix}$$

Applying the boundary conditions $P_1 = 39182$ and $P_2 = -3665$, we obtain

$$\begin{bmatrix} 1 & 0 & 0 & 0 & 0 & 0 \\ -115.7 & 218.23 & -50.76 & -51.77 & 0 & 0 \\ 0 & -50.76 & 60.26 & 0 & -9.50 & 0 \\ 0 & -51.77 & 0 & 59.0 & -7.23 & 0 \\ 0 & 0 & -9.50 & -7.23 & 153.08 & -136.35 \\ 0 & 0 & 0 & 0 & 0 & 1 \end{bmatrix}\begin{Bmatrix} P_1 \\ P_2 \\ P_3 \\ P_4 \\ P_5 \\ P_6 \end{Bmatrix} = \begin{Bmatrix} 39182 \\ 0 \\ 0 \\ 0 \\ 0 \\ -3665 \end{Bmatrix}$$

Solving the systems of equations simultaneously results in the nodal pressure values:

$$[\mathbf{P}]^T = [39182 \quad 34860 \quad 29366 \quad 30588 \quad 2 \quad -3665]\text{Pa}$$

The flow rate in each branch is determined from Eq. (12.14):

$$Q = \frac{\pi D^4}{128\mu}\left(\frac{P_i - P_{i+1}}{L}\right) = C(P_i - P_{i+1})$$

$$Q^{(2)} = 50.76 \times 10^{-9}(34860 - 29366) = 2.79 \times 10^{-4}\ \text{m}^3/\text{s}$$

$$Q^{(3)} = 51.77 \times 10^{-9}(34860 - 30588) = 2.21 \times 10^{-4}\ \text{m}^3/\text{s}$$

$$Q^{(4)} = 9.50 \times 10^{-9}(29366 - 2) = 2.79 \times 10^{-4}\ \text{m}^3/\text{s}$$

$$Q^{(5)} = 7.23 \times 10^{-9}(30588 - 2) = 2.21 \times 10^{-4}\ \text{m}^3/\text{s}$$

The verification of these results is discussed in Section 12.5.

12.2 IDEAL FLUID FLOW

All fluids have some viscosity; however, in certain flow situations it may be reasonable to neglect the effects of viscosity and the corresponding shear stresses. The assumption may be made as a first approximation to simplify the behavior of real fluids with relatively small viscosity. Also, in many external viscous flow situations, we can divide the flow into two regions: (1) a thin layer close to a solid boundary—called the boundary layer region—where the effects of viscosity are important and (2) a region outside the boundary layer where the viscous effects are negligible, in which the fluid is considered to be inviscid. This concept is demonstrated for the case of the flow of air over an airfoil

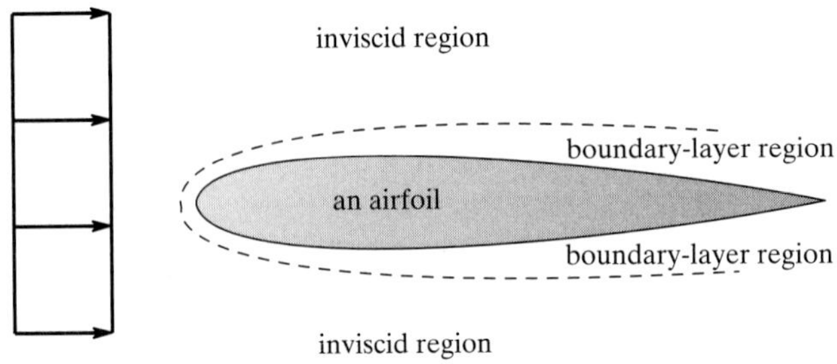

FIGURE 12.7 The flow of air over an airfoil.

in Figure 12.7. For inviscid flow situations, the only forces considered are those resulting from pressure and the inertial forces acting on a fluid element.

Before discussing finite element formulation of ideal fluid problems, let us review some fundamental information. For a two-dimensional flow field, the fluid velocity is

$$\vec{V} = \nu_x \vec{i} + \nu_y \vec{j} \tag{12.19}$$

where ν_x and ν_y are the x- and y-components of the fluid's velocity vector, respectively. The conservation of mass (continuity equation) for a two-dimensional incompressible fluid can be expressed in the differential form in terms of fluid's velocity components as

$$\frac{\partial \nu_x}{\partial x} + \frac{\partial \nu_y}{\partial y} = 0 \tag{12.20}$$

The derivation of Equation (12.20) is shown in Figure 12.8.

The Stream Function and Stream Lines

For a steady flow, a streamline represents the trajectory of a fluid particle. The streamline is a line that is tangent to the velocity of a fluid particle. Streamlines provide a means for visualizing the flow patterns. The stream function $\psi(x, y)$ is defined such that it will satisfy the continuity equation Eq. (12.20) according to the following relationships:

$$\nu_x = \frac{\partial \psi}{\partial y} \quad \text{and} \quad \nu_y = -\frac{\partial \psi}{\partial x} \tag{12.21}$$

Note that upon substitution of Eq. (12.21) into Eq. (12.20), the conservation of mass is satisfied. Along a line of constant $\psi(x, y)$, we have

$$d\psi = 0 = \frac{\partial \psi}{\partial x} dx + \frac{\partial \psi}{\partial y} dy = -\nu_y\, dx + \nu_x\, dy \tag{12.22}$$

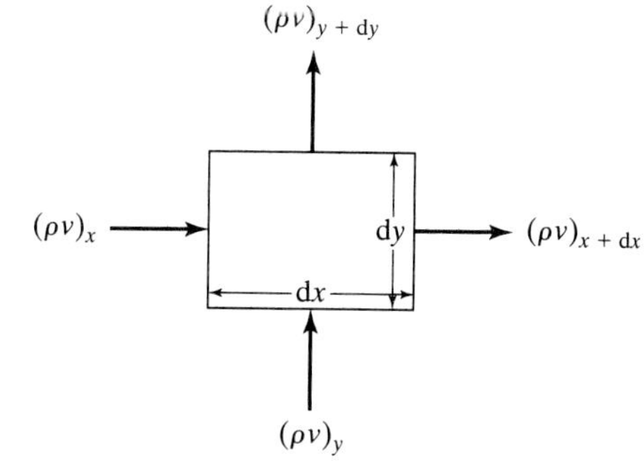

First, we begin by applying the conservation of mass to a small region (differential control volume) in the flow field.

$$\dot{m}_{\text{in}} - \dot{m}_{\text{out}} = \frac{dm_{\text{control volume}}}{dt}$$

$$\rho v_x\,dy - \left(\rho + \frac{\partial \rho}{\partial x}dx\right)\left(v_x + \frac{\partial v_x}{\partial x}dx\right)dy + \rho v_y\,dx -$$

$$\left(\rho + \frac{\partial \rho}{\partial y}dy\right)\left(v_y + \frac{\partial v_y}{\partial y}dy\right)dx = \frac{\partial \rho}{\partial t}dxdy$$

Simplifying, we get

$$-\left(v_x\frac{\partial \rho}{\partial x}dx + \rho\frac{\partial v_x}{\partial x}dx\right)dy - \left(v_y\frac{\partial \rho}{\partial y}dy + \rho\frac{\partial v_y}{\partial y}dy\right)dx = \frac{\partial \rho}{\partial t}dxdy$$

$$-\frac{\partial \rho v_x}{\partial x}dxdy - \frac{\partial \rho v_y}{\partial y}dxdy = \frac{\partial \rho}{\partial t}dxdy$$

Canceling out the *dxdy* terms and assuming incompressible fluid (ρ = constant),

$$\frac{\partial v_x}{\partial x} + \frac{\partial v_y}{\partial y} = 0$$

FIGURE 12.8 The derivation of continuity equation for an incompressible fluid.

or

$$\frac{dy}{dx} = \frac{v_y}{v_x} \tag{12.23}$$

Equation (12.23) can be used to determine the stream function for a specific flow. Moreover, Eq. (12.23) gives the relationship between the slope at any point along a stream line and the fluid velocity components. This relationship is shown in Figure 12.9. To shed more light on the physical meaning of Eq. (12.23), consider Figure 12.10 in which

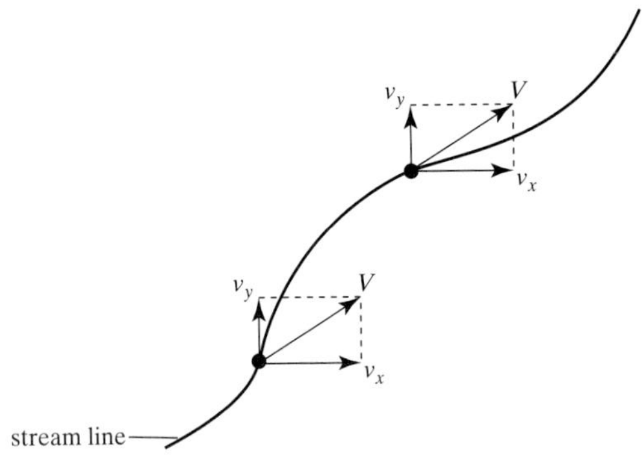

FIGURE 12.9 The relationship between the slope at any point along a stream line and the fluid velocity components.

the flow of a fluid around a sharp corner is shown. For this flow situation, the velocity field is represented by

$$\vec{V} = cx\vec{i} - cy\vec{j}$$

To obtain the expression for the stream function, we make use of Eq. (12.23):

$$\frac{v_y}{v_x} = \frac{dy}{dx} = \frac{-cy}{cx}$$

Integrating, we have

$$\int \frac{dy}{y} = -\int \frac{dx}{x}$$

Evaluating the integral results in the stream function, which is given by

$$xy = \text{constant}$$

or

$$\psi - xy$$

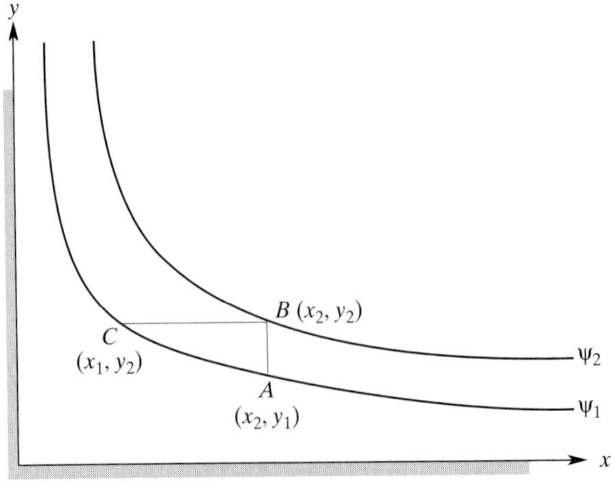

FIGURE 12.10 The flow of fluid around a sharp corner.

To visualize the trajectory of fluid particles we can plot the streamlines by assigning various values to ψ as shown in Figure 12.10. Note that the individual assigned values of streamlines are not important; it is the difference between their values that is important. The difference between the values of two streamlines provides a measure of volumetric flow rate between the streamlines. To demonstrate this idea, let us refer back to Figure 12.10. Along, the A–B section, we can write

$$\frac{Q}{w} = \int_{y_1}^{y_2} v_x\,dy = \int_{y_1}^{y_2} \frac{\partial \psi}{\partial y} dy = \int_{\psi_1}^{\psi_2} d\psi = \psi_2 - \psi_1 \tag{12.24}$$

Similarly, along the B–C section, we have

$$\frac{Q}{w} = \int_{x_1}^{x_2} -v_y\,dx = \int_{x_1}^{x_2} \frac{\partial \psi}{\partial x} dx = \int_{\psi_1}^{\psi_2} d\psi = \psi_2 - \psi_1 \tag{12.25}$$

Therefore, the difference between the values of the streamlines represents the volumetric flow rate per unit width w.

The Irrotational Flow, Potential Function, and Potential Lines

As mentioned earlier, there are many flow situations for which the effects of viscosity may be neglected. Moreover at low speeds, the fluid elements within inviscid flow situations may have an angular velocity of zero (no rotation). These types of flow situations are referred to as irrotational flows. A two-dimensional flow is considered to be irrotational when

$$\frac{\partial v_y}{\partial x} - \frac{\partial v_x}{\partial y} = 0 \tag{12.26}$$

We can also define a potential function ϕ such that the spatial gradients of the potential function are equal to the components of the velocity field:

$$v_x = \frac{\partial \phi}{\partial x} \qquad v_y = \frac{\partial \phi}{\partial y} \tag{12.27}$$

Along a line of constant potential function, we have

$$d\phi = 0 = \frac{\partial \phi}{\partial x} dx + \frac{\partial \phi}{\partial y} dy = v_x\,dx + v_y\,dy = 0 \tag{12.28}$$

$$\frac{dy}{dx} = -\frac{v_x}{v_y} \tag{12.29}$$

By comparing Eqs. (12.29) and (12.23), we can see that the streamlines and the velocity potential lines are orthogonal to each other. It is clear that the potential function complements the stream function. Using the relationships in Eq. (12.27) to substitute for v_x and v_y in the continuity equation Eq. (12.20), we have

$$\frac{\partial^2 \phi}{\partial x^2} + \frac{\partial^2 \phi}{\partial y^2} = 0 \tag{12.30}$$

Using the definitions of stream functions as given by Eq. (12.21) and substituting for v_x and v_y in Eq. (12.26), we have

$$\frac{\partial^2 \psi}{\partial x^2} + \frac{\partial^2 \psi}{\partial y^2} = 0 \tag{12.31}$$

Equations (12.30) and (12.31), which are forms of Laplace's equation, govern the motion of an ideal irrotational flow. Typically, for potential flow situations, the boundary conditions are the known free-stream velocities, and at the solid surface boundary, the fluid cannot have velocity normal to the surface. The latter condition is given by the equation

$$\frac{\partial \phi}{\partial n} = 0 \tag{12.32}$$

Here, n represents a direction normal to the solid surface. Comparing the differential equation governing the irrotational flow behavior of an inviscid fluid, Eq. (12.30), to the heat diffusion equation, Eq. (9.8), we note that both of these equations have the same form; therefore, we can apply the results of Sections 9.2 and 9.3 to the potential flow problems. However, when comparing the differential equations for irrotational flow problems, we let $C_1 = 1, C_2 = 1$, and $C_3 = 0$. Later in this chapter, we will use ANSYS to analyze the flow of an ideal fluid around a cylinder.

Now, let us briefly discuss the analysis of viscous flows. As mentioned earlier, all real fluids have viscosity. The analysis of a complex viscous flow is generally performed by solving the governing equations of motion for a specific boundary condition using the finite differencing approach. However, in recent years, we have made some advances in the finite element formulation of viscous fluid flow problems. *Bathe* (*1996*) discusses a Galerkin procedure for the analysis of the two-dimensional laminar flow of an incompressible fluid. For more details on the formulation of viscous laminar flows, also see Section 7.1 of the theory volume of ANSYS documents.

EXAMPLE 12.2

Consider Figure 12.11, in which a simple uniform flow (constant velocity) in the horizontal direction is shown. As you would expect, for such flow, the fluid particles follow straight horizontal lines and thus the stream lines should be straight parallel lines. We are interested in using the theory discussed in the previous sections to obtain the equations

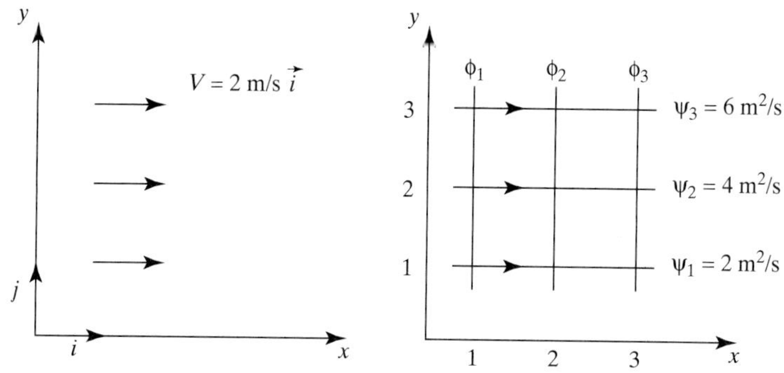

FIGURE 12.11 The velocity, stream lines, and the potential lines for Example 12.2.

for the stream and potential functions and plot a number of stream lines and potential lines.

Recognizing that for this problem, $\nu_x = 2$ m/*s* and $\nu_y = 0$, we can then obtain the equation for the stream function from Eq. (12.21).

$$\nu_x = \frac{\partial \psi}{\partial y} = 2 \quad \Rightarrow \quad \psi = 2y \text{ m}^2/\text{s}$$

We can find the equation for the potential function using Eq. (12.27)

$$\nu_x = \frac{\partial \phi}{\partial x} = 2 \quad \Rightarrow \quad \phi = 2x \text{ m}^2/\text{s}$$

We have plotted three stream lines and three potential lines by letting $x = 1, x = 2,$ and $x = 3$, and $y = 1, y = 2,$ and $y = 3$, as shown in Figure 12.11. Note that as we discussed earlier, the individual values of stream lines are not significant; it is the difference between their values that is significant. Recall that the difference between the values of stream lines represents the volumetric flow rate per unit width, for example, $\psi_3 - \psi_1 = 6 - 2 = 4\ \text{m}^2/\text{s}$ represents the volumetric flow per unit width between stream lines ψ_3 and ψ_1. Also note that the streamlines and the velocity potential lines are orthogonal to each other.

12.3 GROUNDWATER FLOW

The study of fluid flow and heat transfer in porous media is important in many engineering applications, including problems related to oil-recovery methods, groundwater hydrology, solar energy storage, and geothermal energy. The flow of fluid through an unbounded porous medium is governed by Darcy's law. Darcy's law relates the pressure drop to the mean fluid velocity according to the relationship

$$U_D = -\frac{k}{\mu}\frac{dP}{dx} \tag{12.33}$$

where U_D is the mean fluid velocity, k is the permeability of the porous medium, μ is the viscosity of the fluid, and $\frac{dP}{dx}$ is the pressure gradient. For two-dimensional flows, it is customary to use the hydraulic head ϕ to define the components of the fluid velocities. Consider the seepage flow of water under a dam, as shown in Figure 12.12.

The two-dimensional flow of fluid through the soil is governed by Darcy's law, which is given by

$$k_x \frac{\partial^2 \phi}{\partial x^2} + k_y \frac{\partial^2 \phi}{\partial y^2} = 0 \tag{12.34}$$

The components of the seepage velocity are

$$v_x = -k_x \frac{\partial \phi}{\partial x} \quad \text{and} \quad v_y = -k_y \frac{\partial \phi}{\partial y} \tag{12.35}$$

where k_x and k_y are the permeability coefficients and ϕ represents the hydraulic head. Comparing the differential equation governing the groundwater seepage flow, Eq. (12.34), to the heat diffusion equation, Eq. (9.8), we note that both of these equations have the same form; therefore, we can apply the results of Sections 9.2 and 9.3 to the groundwater flow problems. However, when comparing the differential equations for the groundwater seepage flow problems, we let $C_1 = k_x$, $C_2 = k_y$, and $C_3 = 0$.

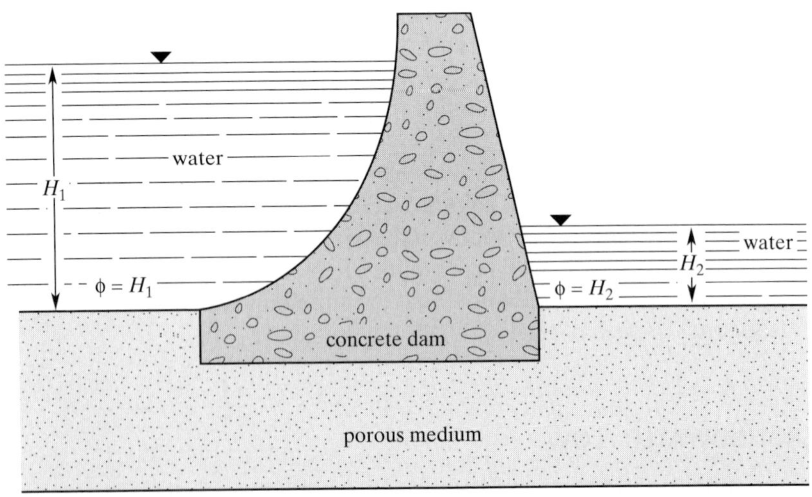

FIGURE 12.12 The seepage flow of water through a porous medium under a dam.

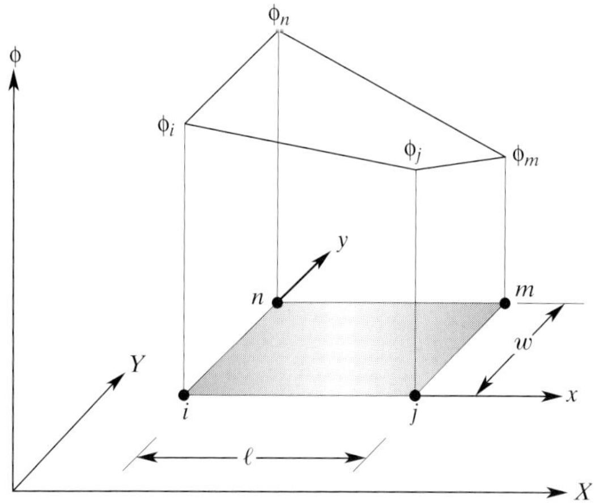

FIGURE 12.13 Nodal values of a hydraulic head for a rectangular element.

The permeability matrix for a rectangular element is

$$[\mathbf{K}]^{(e)} = \frac{k_x w}{6\ell}\begin{bmatrix} 2 & -2 & -1 & 1 \\ -2 & 2 & 1 & -1 \\ -1 & 1 & 2 & -2 \\ 1 & -1 & -2 & 2 \end{bmatrix} + \frac{k_y \ell}{6w}\begin{bmatrix} 2 & 1 & -1 & -2 \\ 1 & 2 & -2 & -1 \\ -1 & -2 & 2 & 1 \\ -2 & -1 & 1 & 2 \end{bmatrix} \quad (12.36)$$

where w and ℓ are the length and the width, respectively, of the rectangular element, as shown in Figure 12.13. In addition, for a typical seepage flow problem, the magnitude of the hydraulic head is generally known at certain surfaces, as shown in Figure 12.12. The known hydraulic head will then serve as a given boundary condition.

The nodal values of a hydraulic head for a triangular element are depicted in Figure 12.14. For triangular elements, the permeability matrix is

$$[\mathbf{K}]^{(e)} = \frac{k_x}{4A}\begin{bmatrix} \beta_i^2 & \beta_i\beta_j & \beta_i\beta_k \\ \beta_i\beta_j & \beta_j^2 & \beta_j\beta_k \\ \beta_i\beta_k & \beta_j\beta_k & \beta_k^2 \end{bmatrix} + \frac{k_y}{4A}\begin{bmatrix} \delta_i^2 & \delta_i\delta_j & \delta_i\delta_k \\ \delta_i\delta_j & \delta_j^2 & \delta_j\delta_k \\ \delta_i\delta_k & \delta_j\delta_k & \delta_k^2 \end{bmatrix} \quad (12.37)$$

where the area A of the triangular element and the α-, β-, and δ-terms are given by

$$\begin{aligned} 2A &= X_i(Y_j - Y_k) + X_j(Y_k - Y_i) + X_k(Y_i - Y_j) \\ \alpha_i &= X_jY_k - X_kY_j \quad \beta_i = Y_j - Y_k \quad \delta_i = X_k - X_j \\ \alpha_j &= X_kY_i - X_iY_k \quad \beta_j = Y_k - Y_i \quad \delta_j = X_i - X_k \\ \alpha_k &= X_iY_j - X_jY_i \quad \beta_k = Y_i - Y_j \quad \delta_k = X_j - X_i \end{aligned}$$

Next, we discuss ANSYS elements.

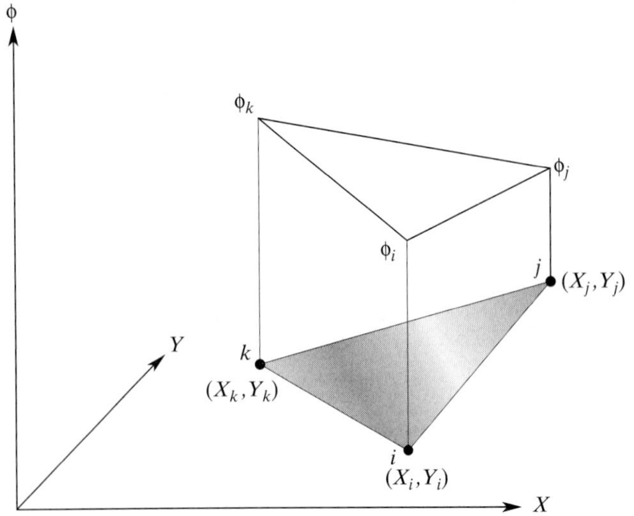

FIGURE 12.14 Nodal values of a hydraulic head for a triangular element.

12.4 EXAMPLES USING ANSYS

ANSYS offers a number of elements for modeling fluid mechanics problems. Examples of those elements include FLUID15, FLUID66, and FLUID79.

FLUID15 is a two-dimensional plane fluid flow element with heat transfer capability. The Navier–Stokes equations, the continuity equation, and the energy equation for incompressible laminar flow are discretized. The element has four corner nodes with three degrees of freedom at each node. The degrees of freedom are the respective velocities in the nodal x- and y-directions and the temeperature. Pressure is also computed at the centroid of each element. The elemental input data include the node locations, the fluid density, thermal conductivity, and viscosity.

FLUID66 is a thermal-flow element with the ability to conduct heat and transport fluid between its two primary nodes. FLUID66 has two degrees of freedom at each node: temperature and pressure. It can also account for convections taking place with two additional optional nodes. The element is defined by its two primary nodes. The elemental input data include the node locations, the fluid density, the convective heat transfer coefficient, thermal conductivity, specific heat, and viscosity.

FLUID79 is a modification of the two-dimensional structural solid element PLANE42. This element is used to model fluids contained within vessels having no net flow rate. This element is defined by four nodes, with two degrees of freedom at each node: translation in the nodal x- and y-directions. The elemental input data include the node locations, the fluid's elastic (bulk) modulus, and viscosity. The viscosity is used to compute a damping matrix for dynamic analysis. Pressure may be input as surface loads on the element faces.

As the theory in the previous sections suggested, because of the similarities among the governing differential equations, in addition to the elements listed above, you can use thermal solid elements (e.g., **PLANE35**, a six-node triangular element; **PLANE55**, a four-node quadrilateral element; or **PLANE77**, an eight-node quadrilateral element) to model irrotational fluid flow or groundwater flow problems. However, when using the solid thermal elements, make sure that the appropriate values are supplied to the property fields. Examples 12.3 and 12.4 demonstrate this point.

EXAMPLE 12.3

Consider an ideal flow of air around a cylinder, as shown in Figure 12.15. The radius of the cylinder is 5 cm, and the velocity of the approach is $U = 10$ cm/s. Using ANSYS, determine the velocity distribution around the cylinder. Assume that the free-stream velocity remains constant at a distance of five diameters downstream and upstream of the cylinder.

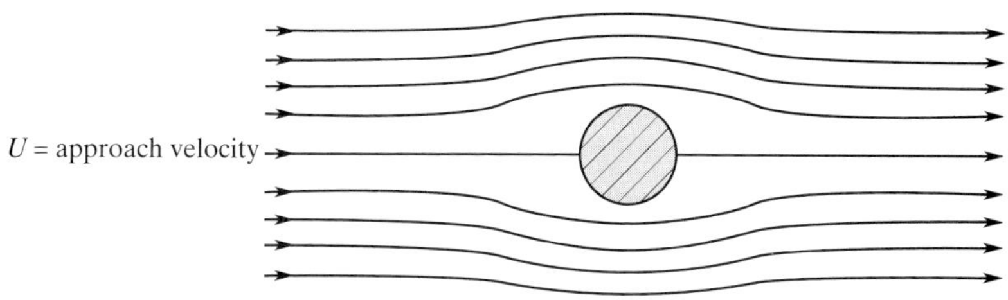

FIGURE 12.15 An ideal flow of air around a cylinder.

Enter the **ANSYS** program by using the Launcher. Type **FlowCYL** (or a file name of your choice) in the **Jobname** entry field of the dialog box. Pick **Run** to start the GUI.

Create a title for the problem. This title will appear on ANSYS display windows to provide a simple way of identifying the displays. So, issue the following command sequence:

utility menu: **File** → **Change Title** ...

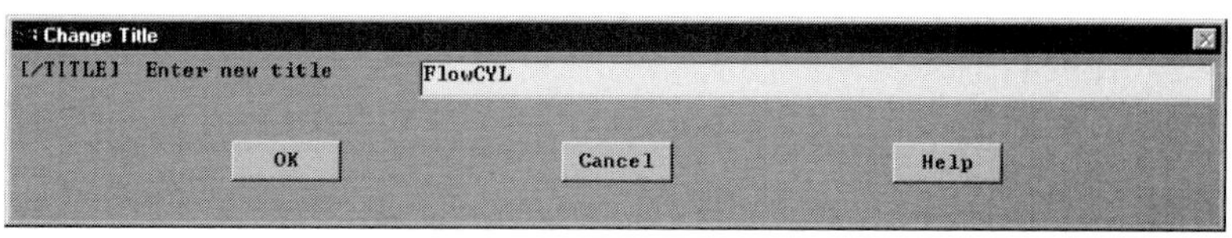

main menu: **Preprocessor → Element Type → Add/Edit/Delete**

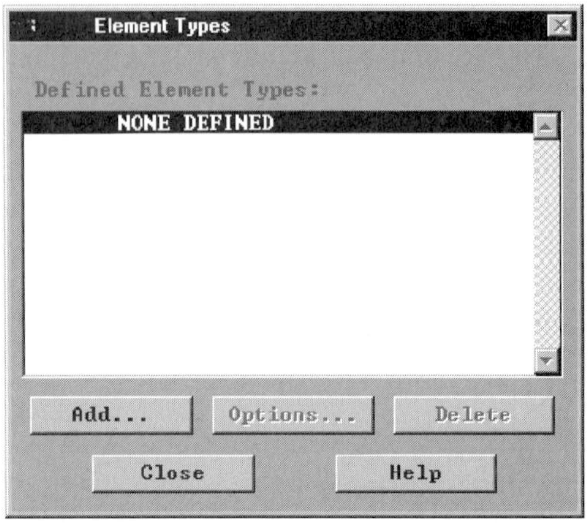

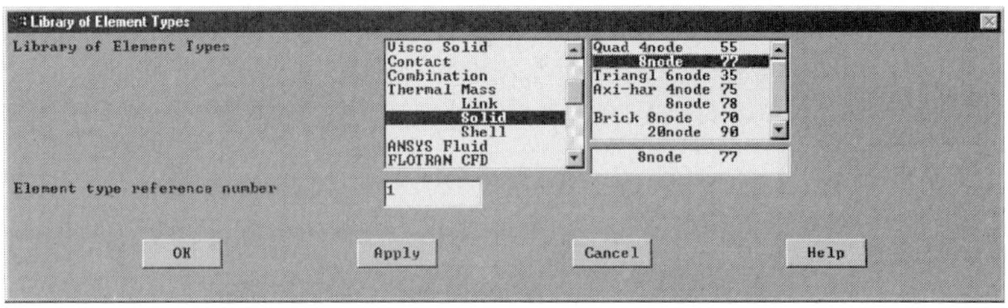

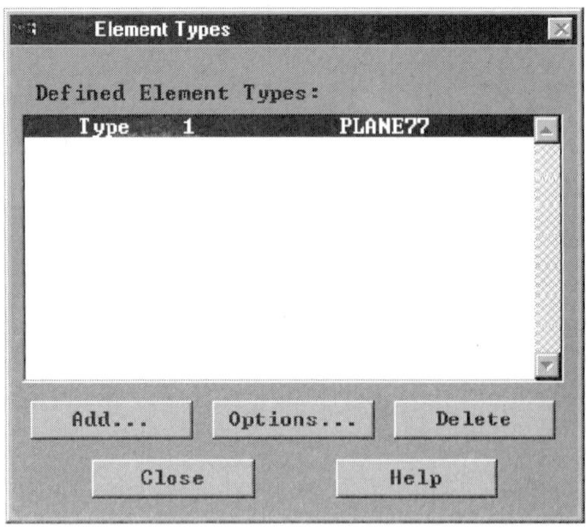

main menu: **Preprocessor → Material Props → Material Models → Thermal → Conductivity → Isotropic**

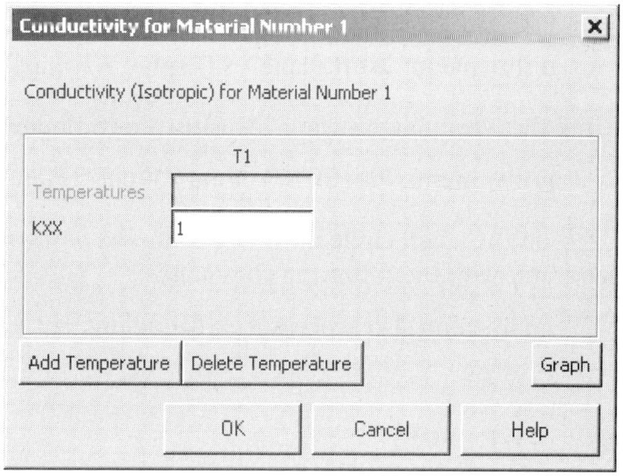

ANSYS Toolbar: **SAVE_DB**

Set up the graphics area (i.e., workplane, zoom, etc.) with the following commands:

utility menu: **Workplane → WP Settings** . . .

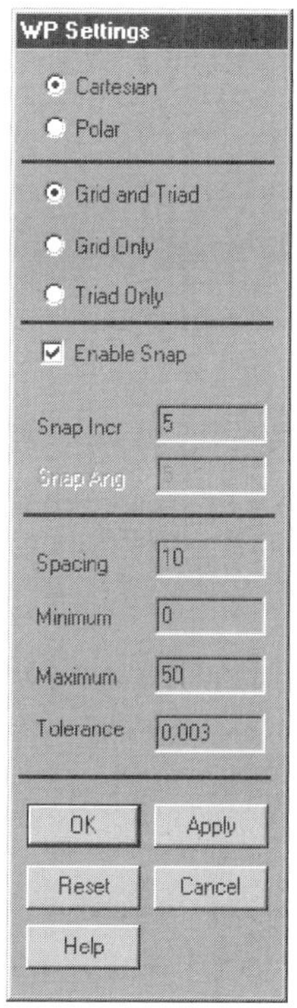

Toggle on the workplane by using the command:

utility menu: **Workplane → Display Working Plane**

Bring the workplane to view by using the command:

utility menu: **PlotCtrls → Pan, Zoom, Rotate** . . .

Click on the **small circle** until you bring the workplane to view. Then, create the geometry with the following commands:

main menu: **Preprocessor → Modeling → Create → Areas → Rectangle → By 2 Corners**

[WP = 0,0]

[Expand the rubber up 50 and right 50]

OK

Create the cross section of the cylinder to be removed later:

main menu: **Preprocessor → Modeling → Create → Areas → Circle → Solid Circle**

[WP = 25, 25]

[Expand the rubber to r = 5.0]

OK

main menu: **Preprocessor → Modeling → Operate → Booleans → Subtract → Areas**

Pick Area1 (the rectangle) and apply; then, pick Area2 (the circle) and apply.
OK

We now want to mesh the area to create elements and nodes, but first, we need to specify the element sizes. So, issue the following commands:

main menu: **Preprocessor → Meshing → Size Cntrls → Manual Size → Global → Size**

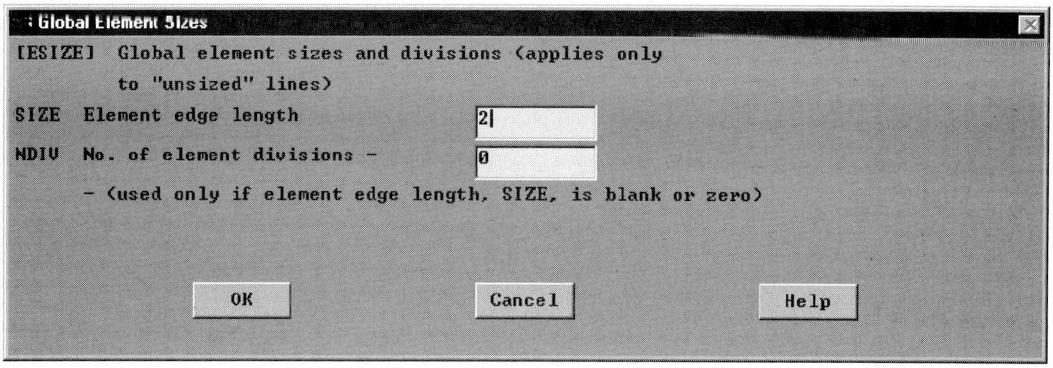

main menu: **Preprocessor → Meshing → Mesh → Areas → Free**

Pick All

Apply boundary conditions with the following commands:

main menu: **Solution → Define Loads → Apply → Thermal → Heat Flux → On Lines**

Pick the left vertical edge of the rectangle.

OK

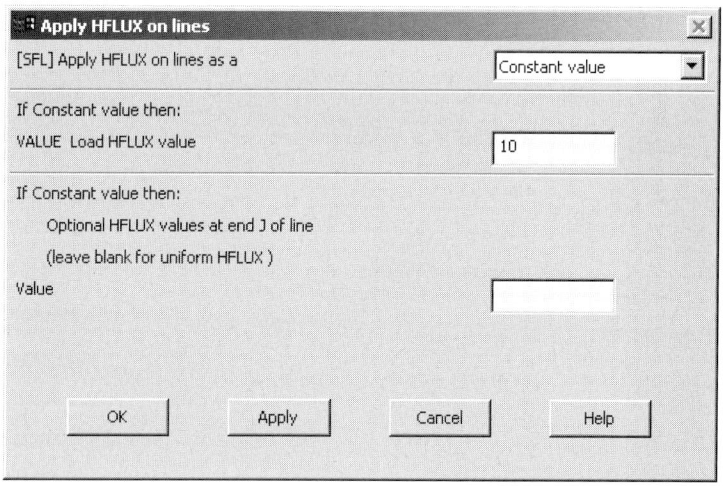

OK

main menu: **Solution → Define Loads → Apply → Thermal → Heat Flux → On Lines**

Pick the right vertical edge of the rectangle.

OK

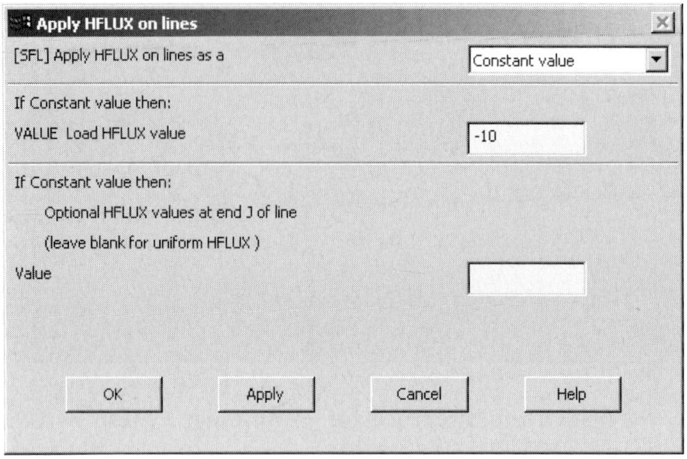

OK

utility menu: **PlotCtrls → Symbols** . . .

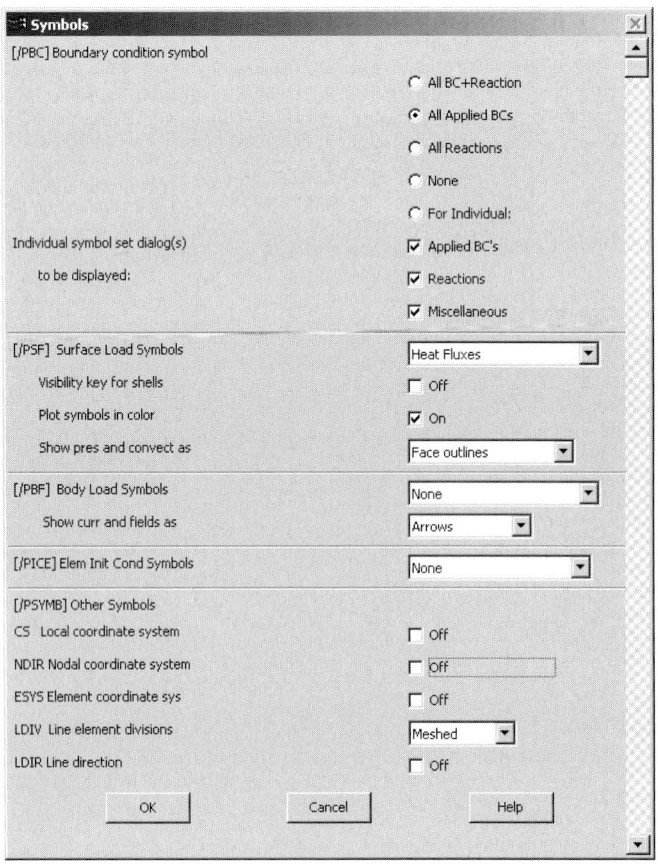

utility menu: **Plot → Lines**

ANSYS Toolbar: **SAVE_DB**

Solve the problem:

main menu: **Solution → Solve → Current LS**

OK

Close (the solution is done!) window.

Close (the/STAT Command) window.

postprocessing phase, obtain information such as velocities (see Figure 12.16):

main menu: **General Postproc → Plot Results → Vector Plot → Predefined**

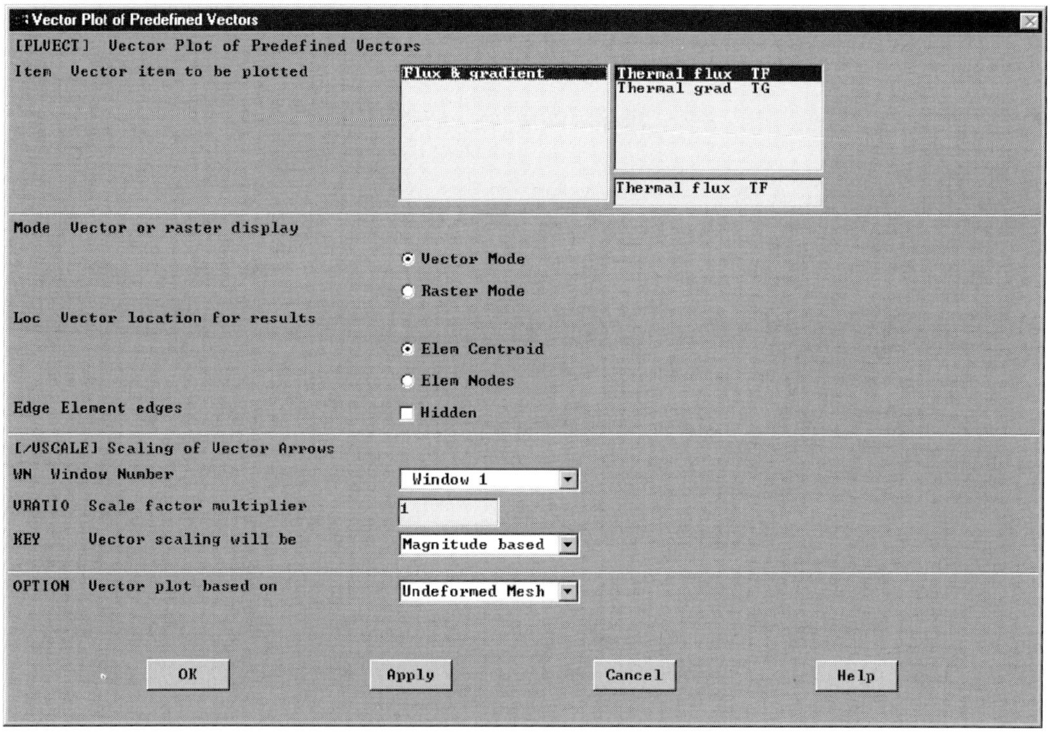

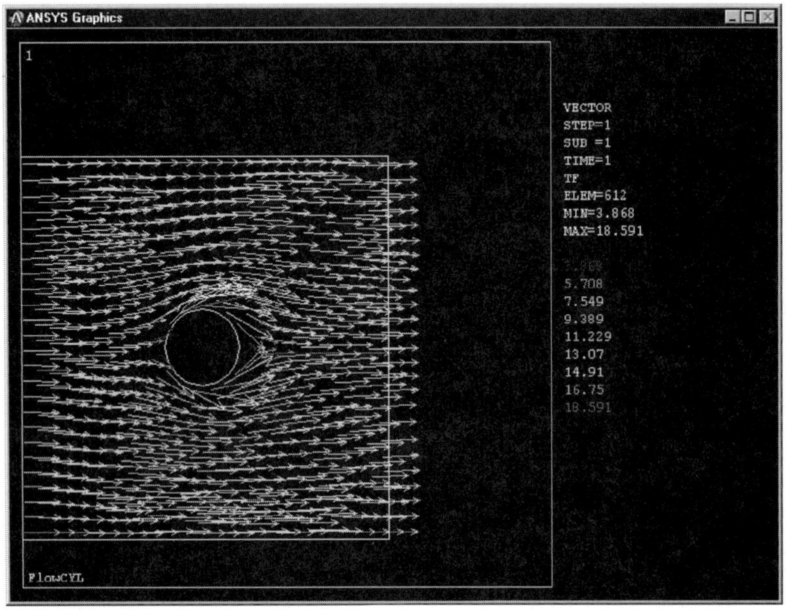

FIGURE 12.16 The velocity vectors.

utility menu: **Plot → Areas**

main menu: **General Postproc → Path Operations → Define Path → On Working Plane**

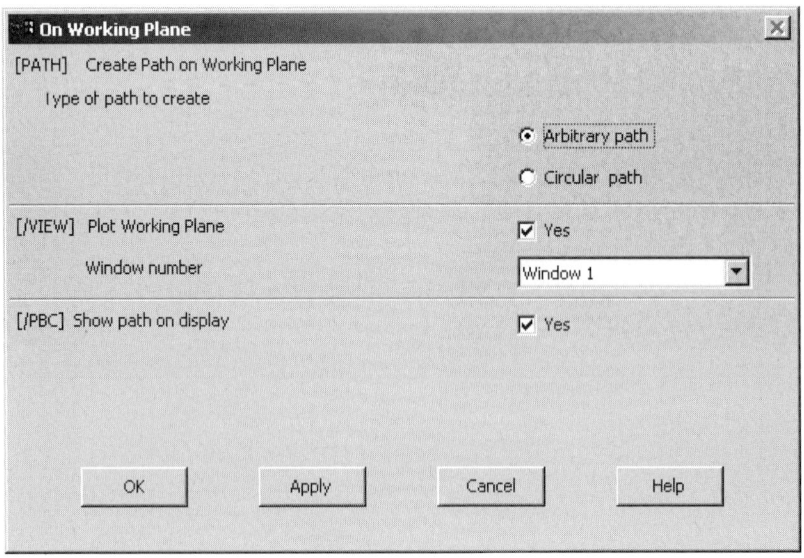

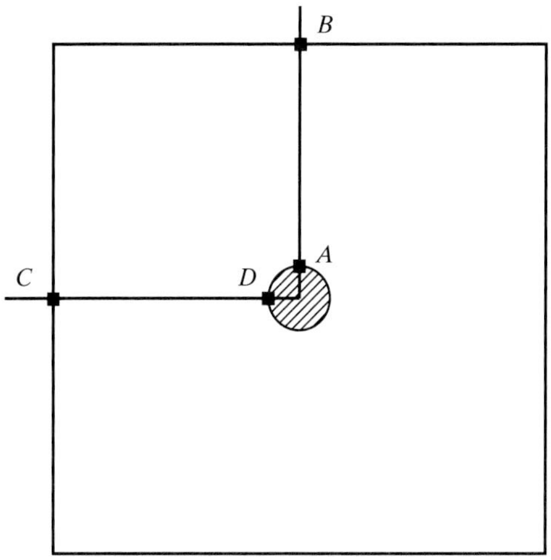

FIGURE 12.17 Defining the path for path operation.

Pick the two points along the line marked A–B, as shown in Figure 12.17.

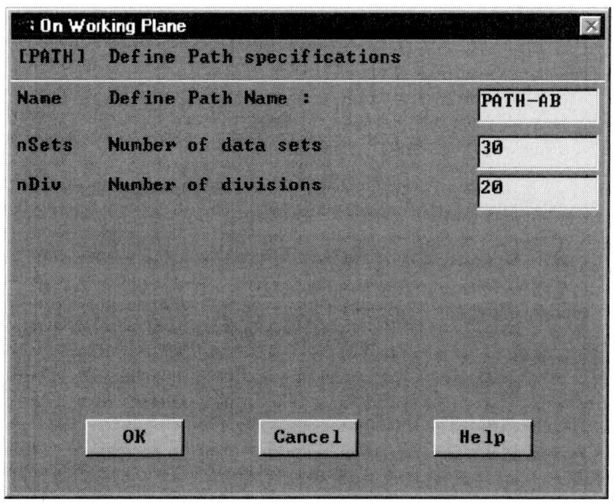

main menu: **General Postproc → Path Operations → Map onto Path**

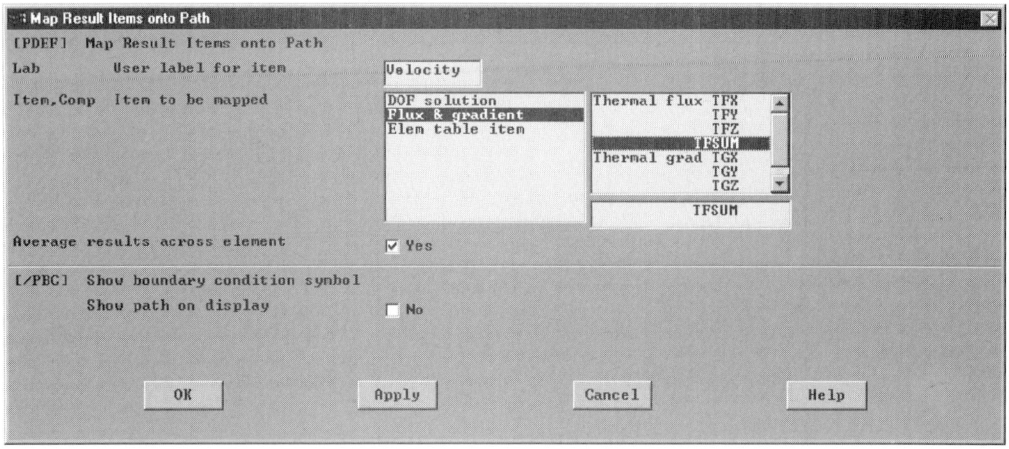

Now, plot the results (see Figure 12.18):

main menu: **General Postproc → Path Operations → Plot Path Item → On Graph**

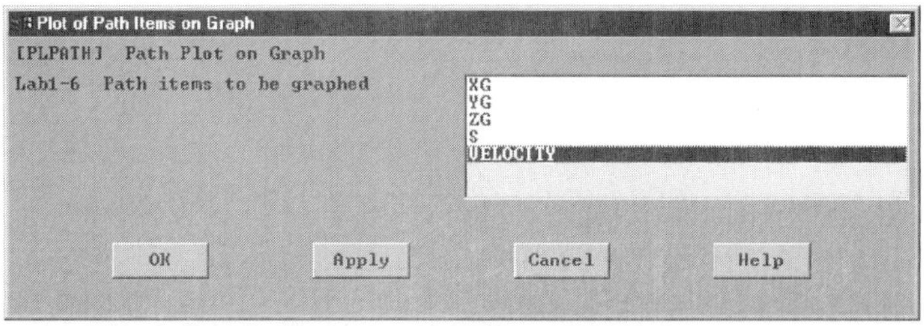

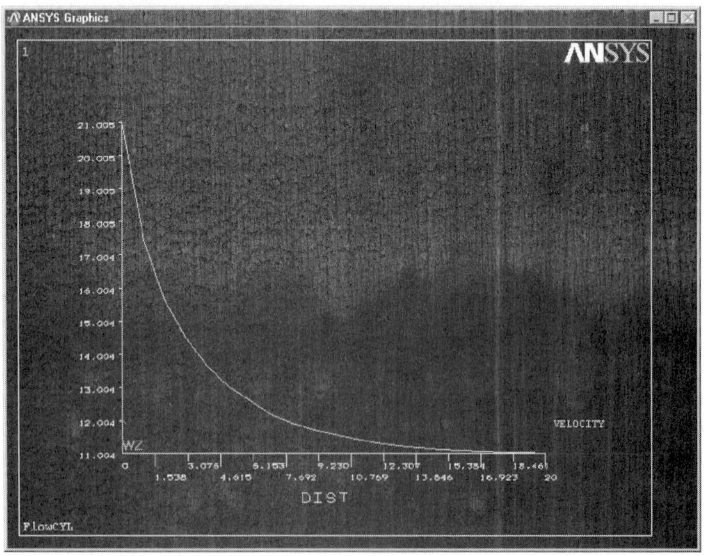

FIGURE 12.18 The variation of fluid velocity along path *A–B*.

utility menu: **Plot → Areas**

main menu: **General Postproc → Path Operations → Define Path → On Working Plane**

Pick the two points along the line marked as *C–D*, as shown in Figure 12.17.

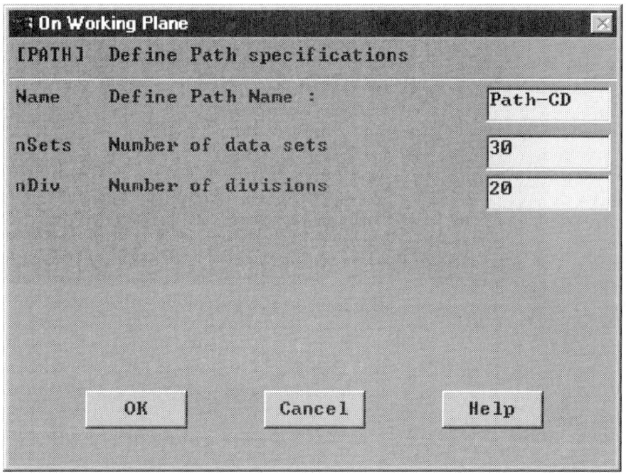

OK

main menu: **General Postproc → Path Operations → Map onto Path**

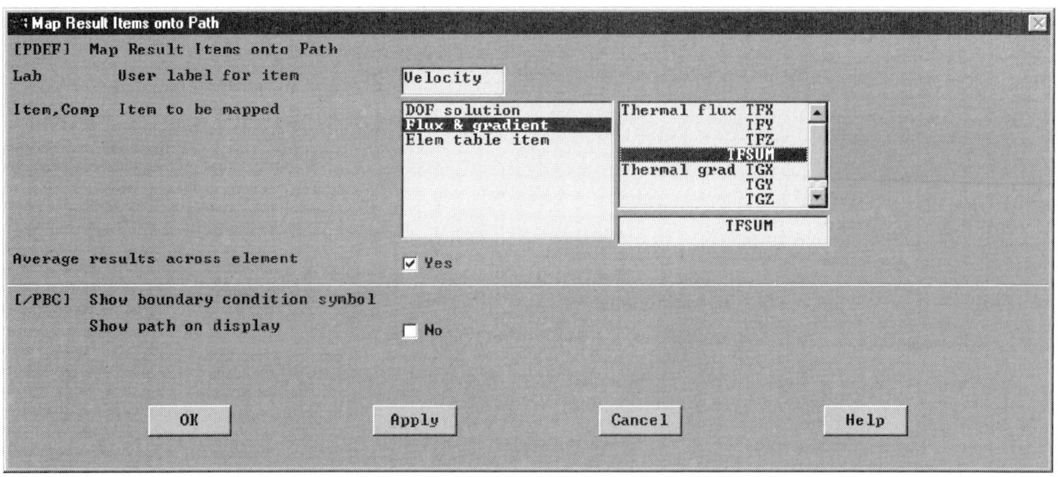

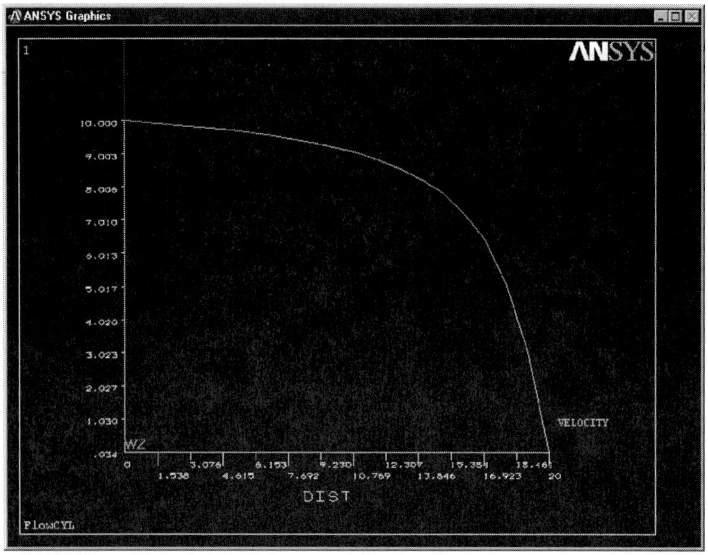

FIGURE 12.19 The variation of fluid velocity along the *C–D* section.

Now, plot the results (see Figure 12.19):

main menu: **General Postproc → Path Operations → Plot Path Item → On Graph**

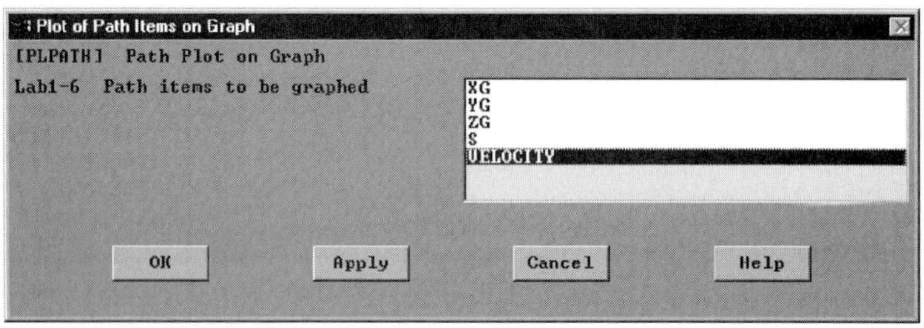

Exit and save the results:

ANSYS Toolbar: **QUIT**

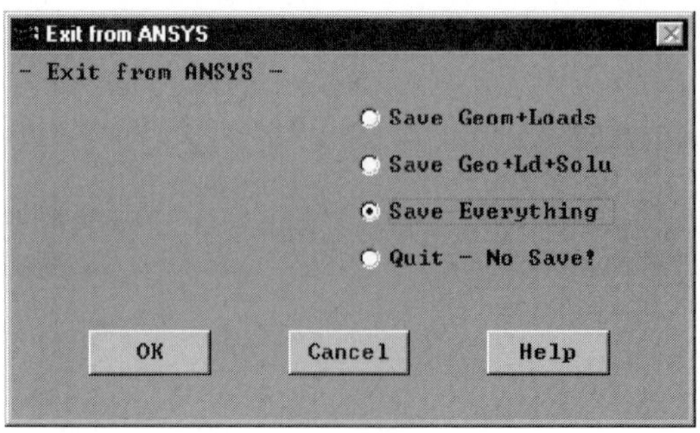

EXAMPLE 12.4

Consider the seepage flow of water under the concrete dam shown in Figure 12.20. The permeability of the porous soil under the dam is approximated as $k = 15$ m/day. Determine the seepage velocity distribution in the porous soil.

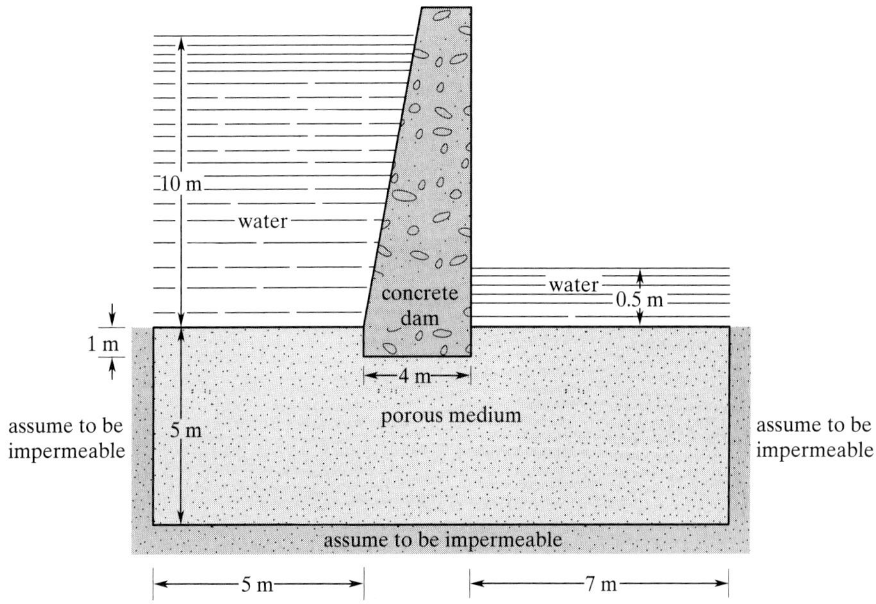

FIGURE 12.20 The seepage flow of water through a porous medium under a concrete dam.

Enter the **ANSYS** program by using the Launcher. Type **DAM** (or a file name of your choice) in the **Jobname** entry field of the dialog box. Pick **Run** to start the GUI.

Create a title for the problem. This title will appear on ANSYS display windows to provide a simple way of identifying the displays. So, issue the following commands:

utility menu: **File → Change Title** . . .

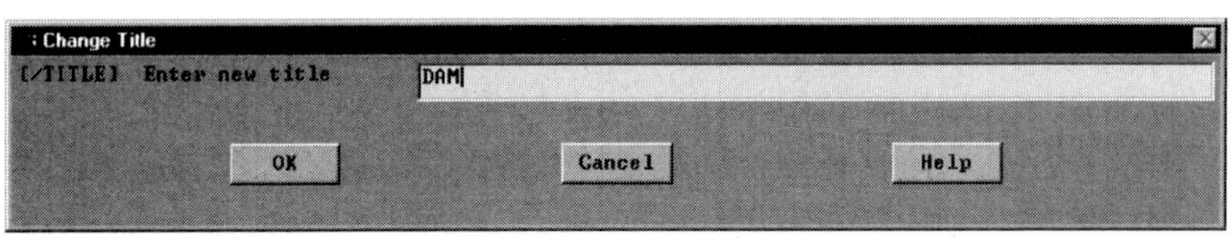

main menu: **Preprocessor → Element Type → Add/Edit/Delete**

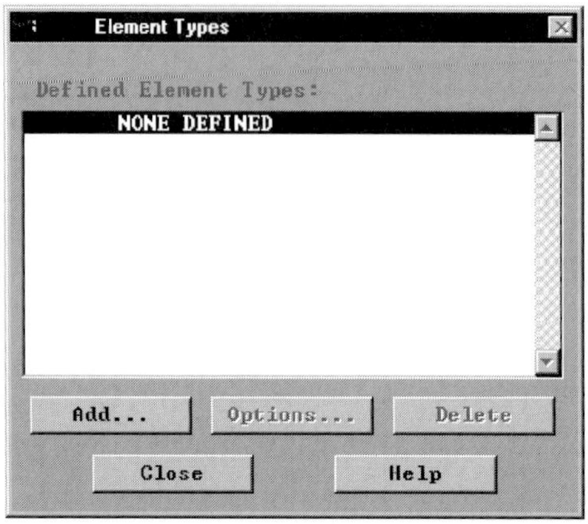

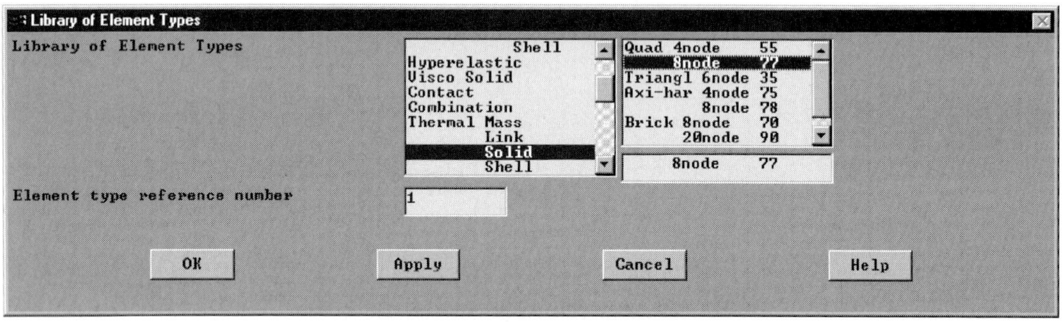

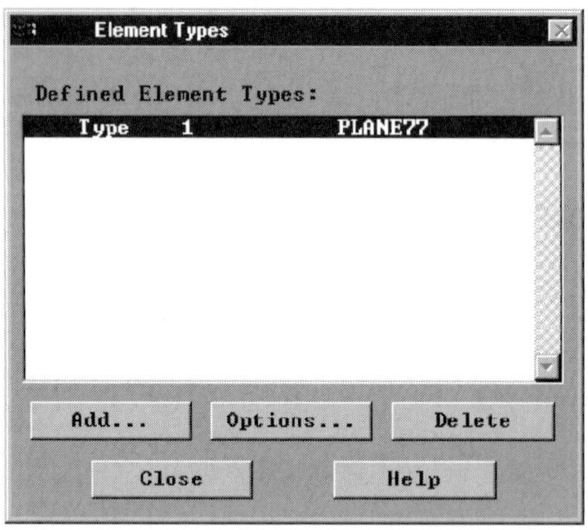

Assign the permeability of the soil with the following commands:

main menu: **Preprocessor → Material Props → Material Models → Thermal → Conductivity → Isotropic**

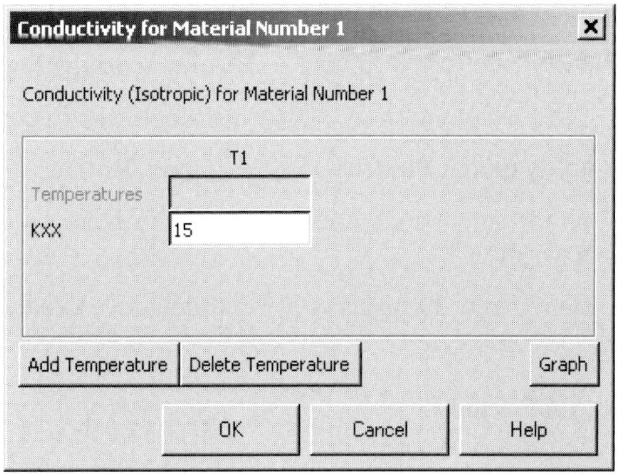

ANSYS Toolbar: **SAVE_DB**

Set up the graphics area (i.e., workplane, zoom, etc.) with the commands

utility menu: **Workplane → WP Settings** . . .

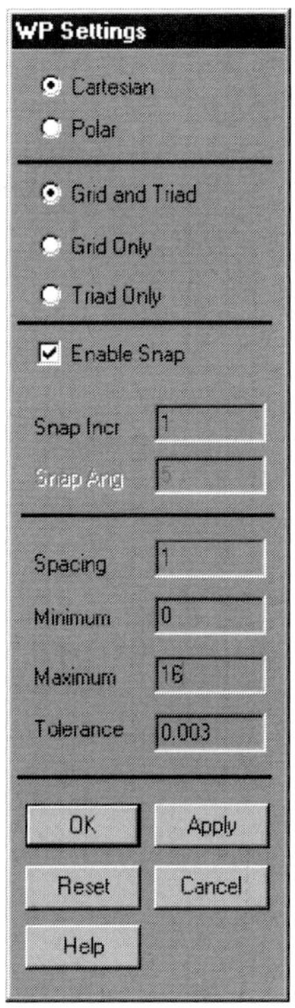

Toggle on the workplane by using the following command:

utility menu: **Workplane → Display Working Plane**

Bring the workplane to view by using the following command:

utility menu: **PlotCtrls → Pan, Zoom, Rotate** . . .

Click on the small circle until you bring the workplane to view. Then, create the geometry:

main menu: **Preprocessor → Modeling → Create → Areas → Rectangle → By 2 Corners**

[WP = 0,0]

[Expand the rubber up 5 and right 16]

[WP = 5,4]

[Expand the rubber up 1 and right 4]

OK

main menu: **Preprocessor → Modeling → Operate → Booleans → Subtract → Areas**

Pick Area1 (the large rectangle) and Apply; then, pick Area2 (the small rectangle) and Apply.

OK

We now want to mesh the areas to create elements and nodes, but first, we need to specify the element sizes:

main menu: **Preprocessor → Meshing → Size Cntrls → Manual Size → Global → Size**

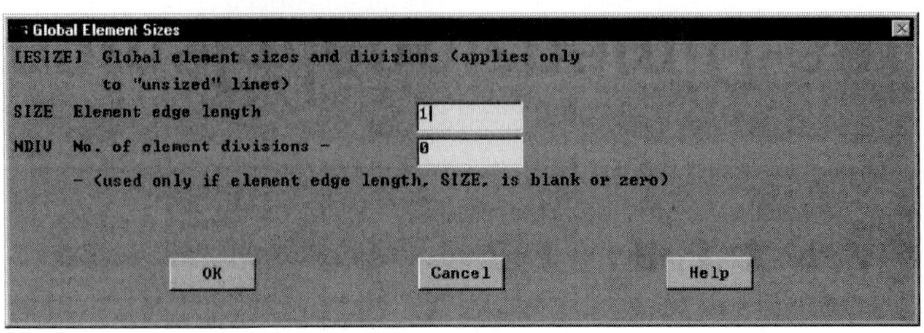

main menu: **Preprocessor → Meshing → Mesh → Areas → Free**

Pick All

Apply boundary conditions with the following commands:

main menu: **Solution → Define Loads → Apply → Thermal → Temperature → On Nodes**

Using the *box* picking mode, pick all of the nodes attached to the left top edge of the rectangle. Hold down the left button while picking.

OK

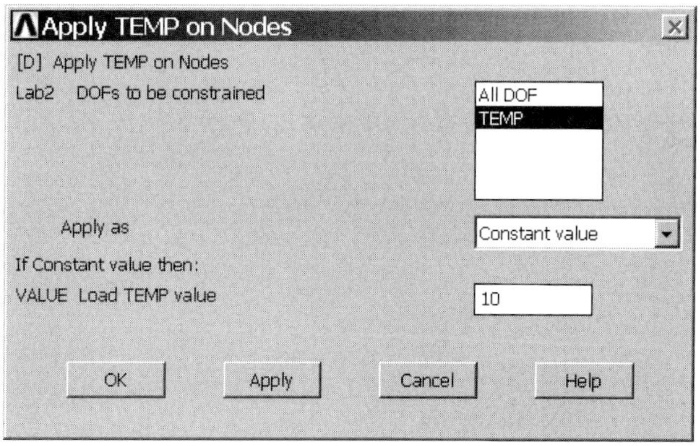

main menu: **Solution → Define Loads → Apply → Thermal → Temperature → On Nodes**

Using the box picking mode, pick all of the nodes attached to the right top edge of the rectangle:

OK

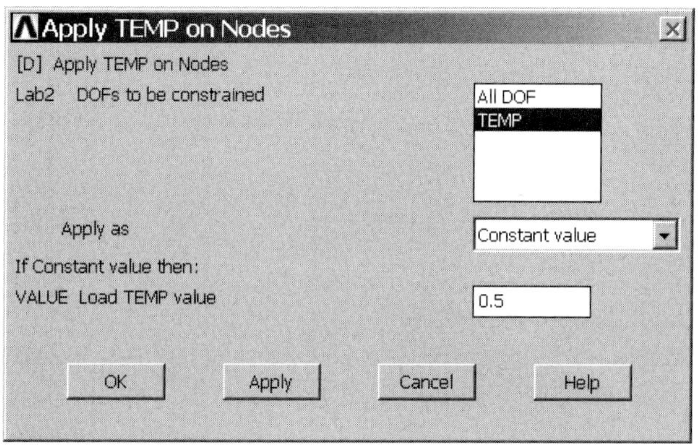

ANSYS Toolbar: **SAVE_DB**

Solve the problem:

main menu: **Solution → Solve → Current LS**

OK

Close (the solution is done!) window.

Close (the/STAT Command) window.

For the postprocessing phase, obtain information such as velocities (see Figure 12.21):

main menu: **General Postproc → Plot Results → Vector Plot → Predefined**

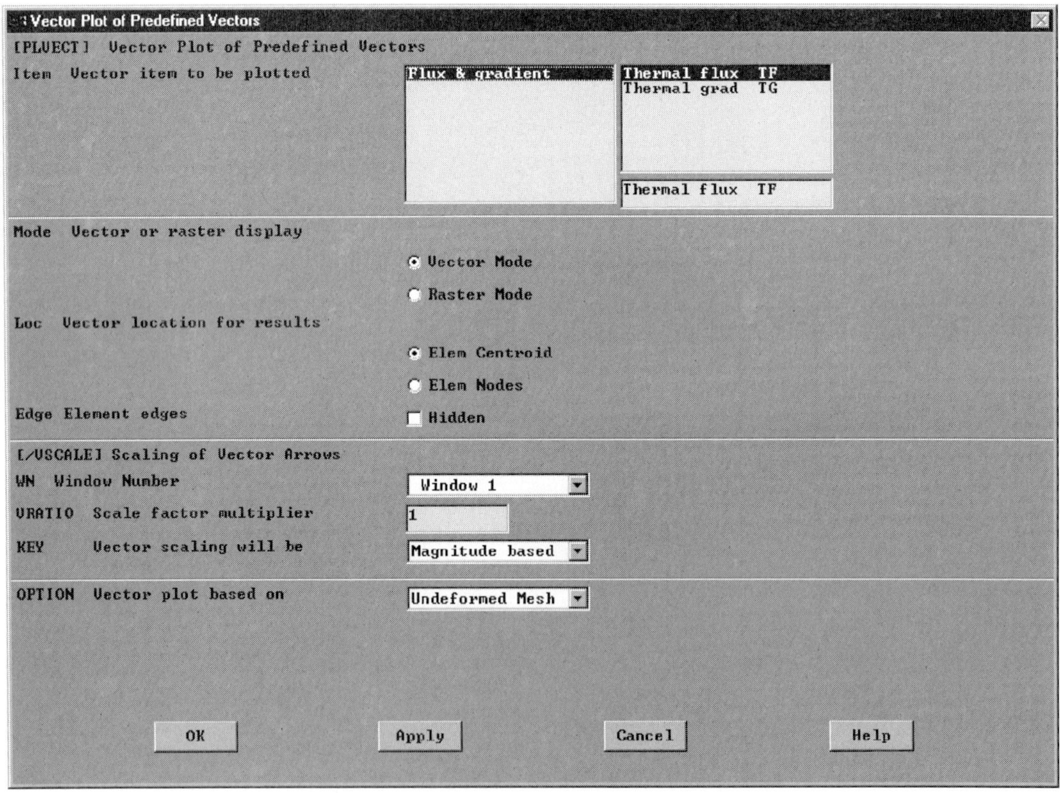

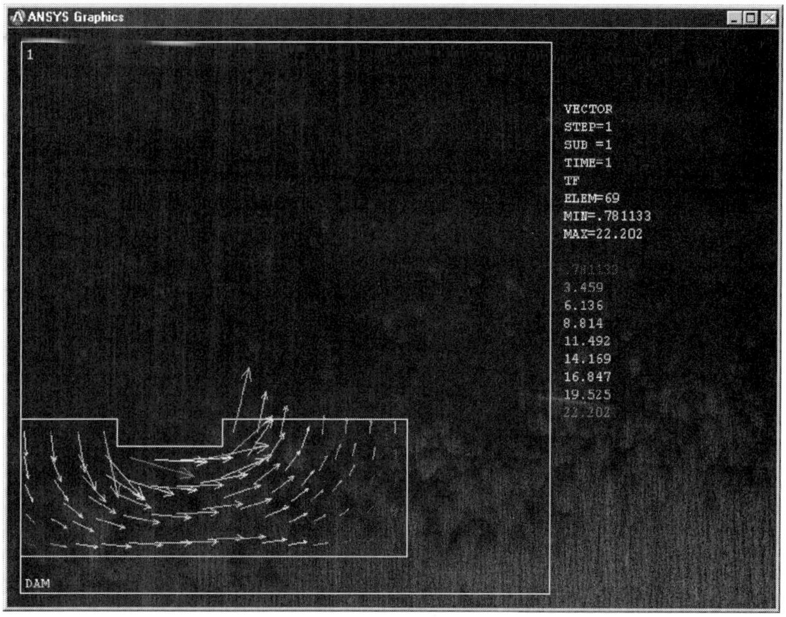

FIGURE 12.21 The seepage-velocity distribution within the soil.

utility menu: **Plot → Areas**

main menu: **General Postproc → Path Operations → Define Path → On Working Plane**

Pick the two points along the line marked as A–B, as shown in Figure 12.22.

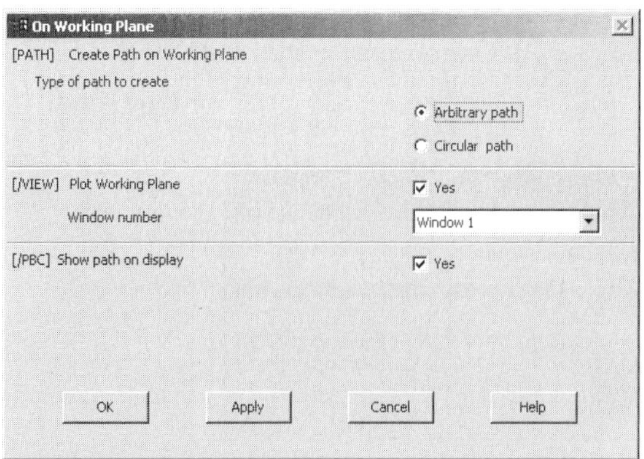

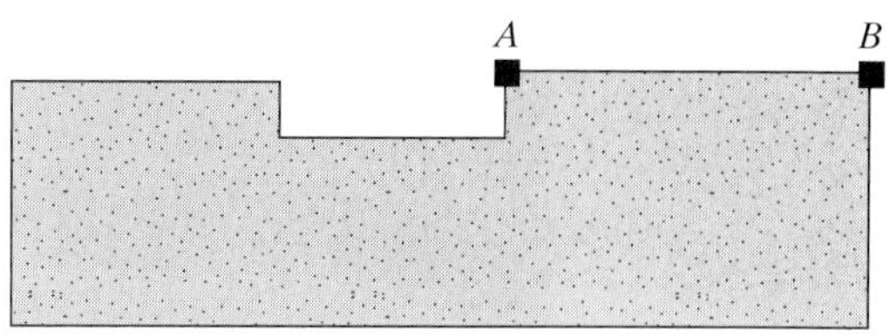

FIGURE 12.22 Defining the path for path operation.

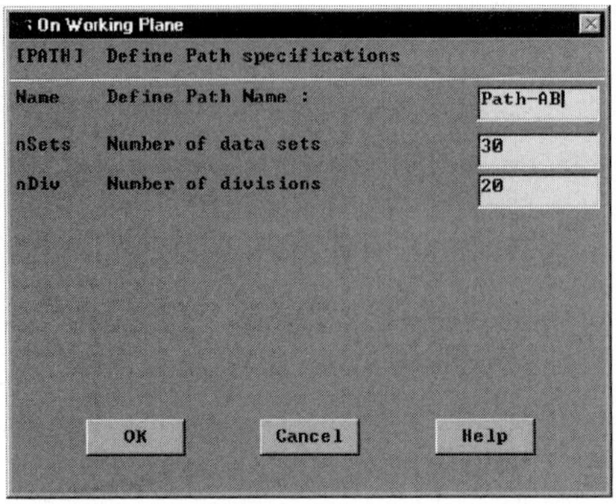

OK

main menu: **General Postproc → Path Operations → Map onto Path**

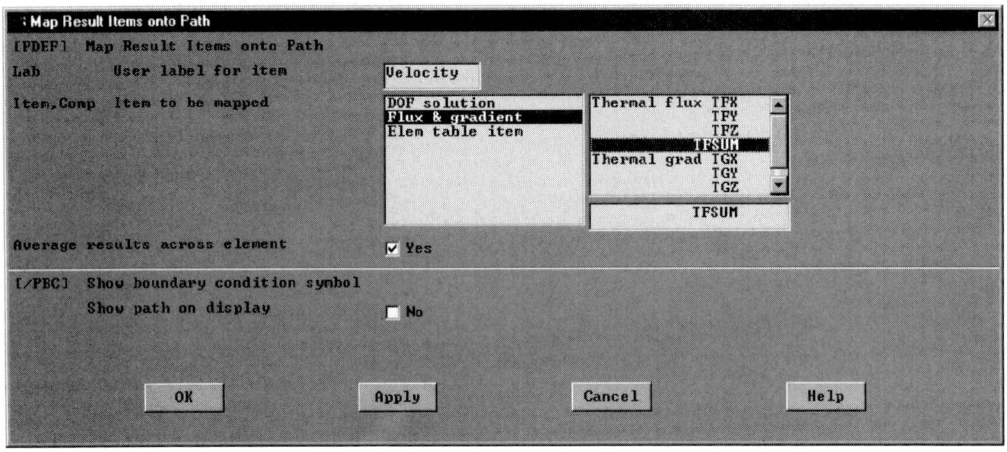

Now, plot the results (see Figure 12.23):

main menu: **General Postproc → Path Operations → Plot Path Item → On Graph**

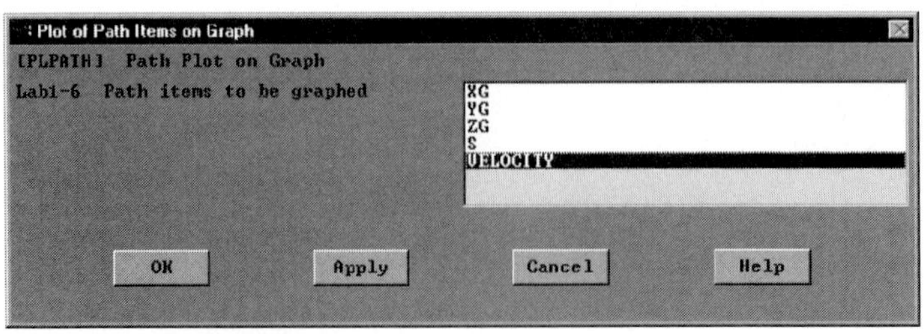

Exit and save the results:

ANSYS Toolbar: **QUIT**

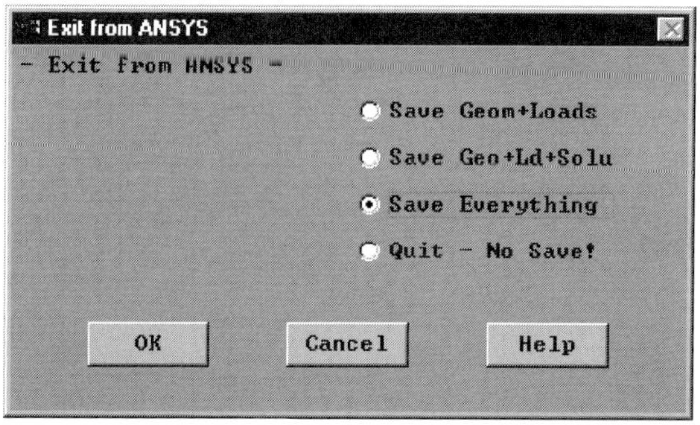

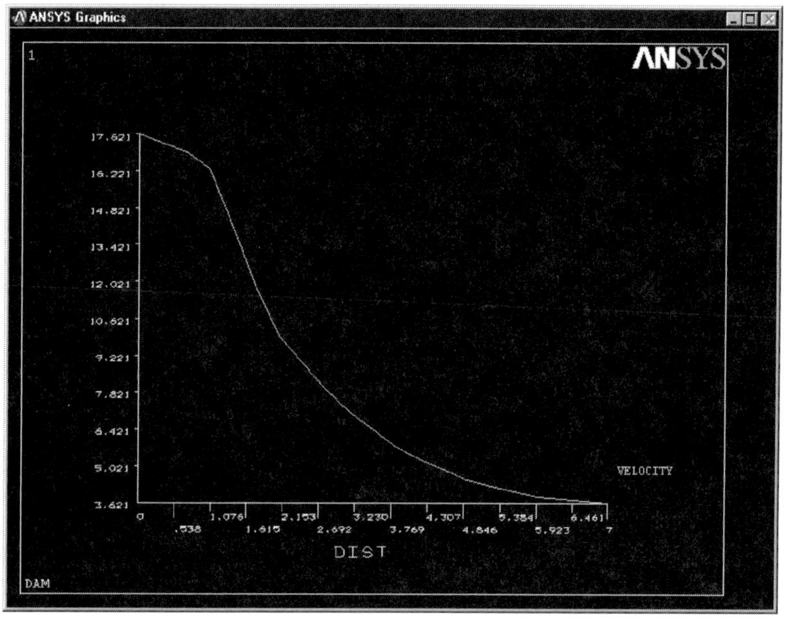

FIGURE 12.23 The variation of the seepage velocity along path A–B.

12.5 VERIFICATION OF RESULTS

There are various ways by which you can verify your findings. Consider the flow rate results of Example 12.1, shown in Table 12.1.

Referring to Figure 12.6, elements (2) and (4) are in series; therefore, the flow rate through each element should be equal. Comparing $Q^{(2)}$ to $Q^{(4)}$, we find that this condition is true. Elements (3) and (5) are also in series, and the computed flow rates for these elements are also equal. Moreover, the sum of the flow rates in elements (2) and (3) should equal the flow rate in element (1). This condition is also true.

Let us now turn our attention to Example 12.3. One way of checking for the validity of your FEA findings is to consider the variation of air velocity along path A–B,

TABLE 12.1 Summary of flow rate results for Example 12.1

Element	Flow Rate (m^3/s)
1	5.0×10^{-4}
2	2.79×10^{-4}
3	2.21×10^{-4}
4	2.79×10^{-4}
5	2.21×10^{-4}
6	5.0×10^{-4}

as shown in Figure 12.18. The fluid velocity is at its maximum value at point A, and it decreases along path A–B, approaching the free-stream value. Another check on the validity of our results could come from examining the fluid velocity variation along path C–D, as shown in Figure 12.19. The air velocity changes from its free-stream value to zero at the forward stagnation point of the cylinder. These results are certainly consistent with the results obtained from applying Euler's equation to an inviscid flow of air around a cylinder.

The results of Example 12.4 can be visually verified in a similar fashion. Consider Figure 12.23, which shows the variation of the seepage velocity along path A–B. It is clear that the seepage velocities are higher near point A than they are near point B. This difference is attributed to the fact that point A lies on the path of the least resistance to the flow, and consequently, more fluid flows near point A than near point B. The other check on the validity of the result could come from comparing the seepage flow rates on the dam's upstream side to the seepage flow on the downstream side; of course, they must be equal.

SUMMARY

At this point you should

1. know how to solve laminar flow network problems. You should also know that the resistance matrix for laminar pipe flow is given by

$$[\mathbf{R}]^{(e)} = \begin{bmatrix} \dfrac{\pi D^4}{128L\mu} & -\dfrac{\pi D^4}{128L\mu} \\ -\dfrac{\pi D^4}{128L\mu} & \dfrac{\pi D^4}{128L\mu} \end{bmatrix}$$

2. know the definitions of streamline and stream function, as well as what they physically represent.
3. know what an irrotational flow is.
4. know that the inviscid flow matrix for a rectangular element is

$$[\mathbf{K}]^{(e)} = \frac{w}{6\ell}\begin{bmatrix} 2 & -2 & -1 & 1 \\ -2 & 2 & 1 & -1 \\ -1 & 1 & 2 & -2 \\ 1 & -1 & -2 & 2 \end{bmatrix} + \frac{\ell}{6w}\begin{bmatrix} 2 & 1 & -1 & -2 \\ 1 & 2 & -2 & -1 \\ -1 & -2 & 2 & 1 \\ -2 & -1 & 1 & 2 \end{bmatrix}$$

and that the inviscid flow matrix for a triangular element is

$$[\mathbf{K}]^{(e)} = \frac{1}{4A}\begin{bmatrix} \beta_i^2 & \beta_i\beta_j & \beta_i\beta_k \\ \beta_i\beta_j & \beta_j^2 & \beta_j\beta_k \\ \beta_i\beta_k & \beta_j\beta_k & \beta_k^2 \end{bmatrix} + \frac{1}{4A}\begin{bmatrix} \delta_i^2 & \delta_i\delta_j & \delta_i\delta_k \\ \delta_i\delta_j & \delta_j^2 & \delta_j\delta_k \\ \delta_i\delta_k & \delta_j\delta_k & \delta_k^2 \end{bmatrix}$$

5. know that the permeability matrix for seepage flow problems is similar to the conductance matrix for two-dimensional conduction problems. The permeability matrix for a rectangular element is

$$[\boldsymbol{K}]^{(e)} = \frac{k_x w}{6\ell}\begin{bmatrix} 2 & -2 & -1 & 1 \\ -2 & 2 & 1 & -1 \\ -1 & 1 & 2 & -2 \\ 1 & -1 & -2 & 2 \end{bmatrix} + \frac{k_y \ell}{6w}\begin{bmatrix} 2 & 1 & -1 & -2 \\ 1 & 2 & -2 & -1 \\ -1 & -2 & 2 & 1 \\ -2 & -1 & 1 & 2 \end{bmatrix}$$

and the permeability matrix for a triangular element is

$$[\boldsymbol{K}]^{(e)} = \frac{k_x}{4A}\begin{bmatrix} \beta_i^2 & \beta_i\beta_j & \beta_i\beta_k \\ \beta_i\beta_j & \beta_j^2 & \beta_j\beta_k \\ \beta_j\beta_k & \beta_j\beta_k & \beta_k^2 \end{bmatrix} + \frac{k_y}{4A}\begin{bmatrix} \delta_i^2 & \delta_i\delta_j & \delta_i\delta_k \\ \delta_i\delta_j & \delta_j^2 & \delta_j\delta_k \\ \delta_i\delta_k & \delta_j\delta_k & \delta_k^2 \end{bmatrix}$$

REFERENCES

Abbot, I. H., and Von Doenhoff, A. E., *Theory of Wing Sections*, New York, Dover Publications, 1959.

ANSYS User's Manual: Elements, Vol. III, Swanson Analysis Systems, Inc.

Bathe, K., *Finite Element Procedures*, Englewood Cliffs, NJ, Prentice Hall, 1996.

Fox, R. W., and McDonald, A. T., *Introduction to Fluid Mechanics*, 4th ed., New York, John Wiley and Sons, 1992.

Segrlind, L., *Applied Finite Element Analysis*, 2nd ed., New York, John Wiley and Sons, 1984.

PROBLEMS

1. Oil with dynamic viscosity of $\mu = 0.3\ \text{N}\cdot\text{s/m}^2$ and density of $\rho = 900\ \text{kg/m}^3$ flows through the piping network shown in the accompanying figure. Determine the pressure distribution in the system if the flow rate at node 1 is $20 \times 10^{-4}\ \text{m}^3/\text{s}$. For the given conditions, the flow is laminar throughout the system. How does the flow divide in each branch?

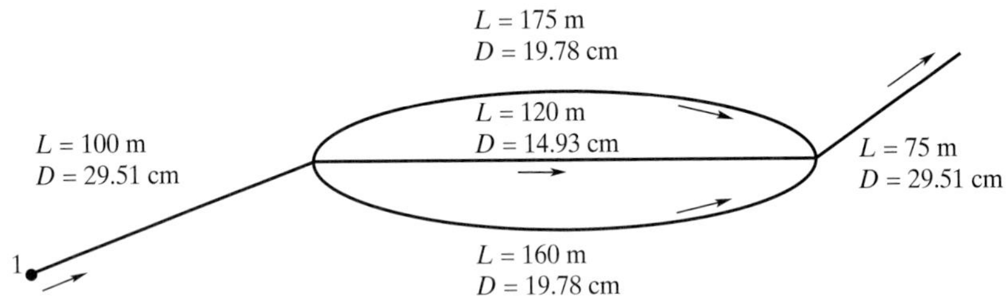

2. Consider the flow of air through the diffuser shown in the accompanying figure. Neglecting the viscosity of air and assuming uniform velocities at the inlet and exit of the diffuser, use ANSYS to compute and plot the velocity distribution within the diffuser.

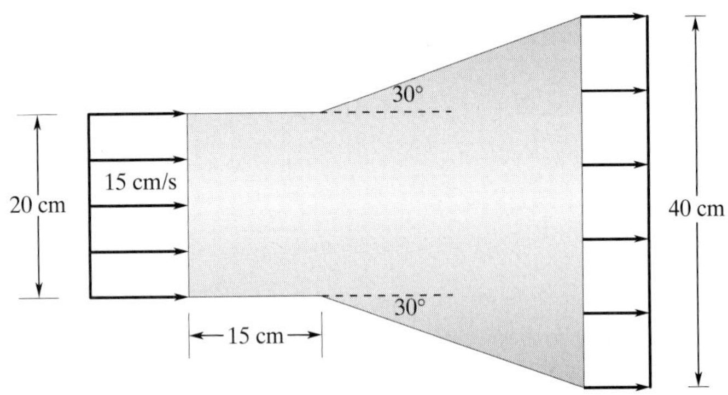

3. Consider the flow of air through a 90° elbow. Assuming ideal flow behavior for air and uniform velocities at the inlet and outlet sections, use ANSYS to compute and plot the velocity distribution within the elbow. The elbow has a uniform depth. Use the continuity equation to obtain the velocity at the outlet.

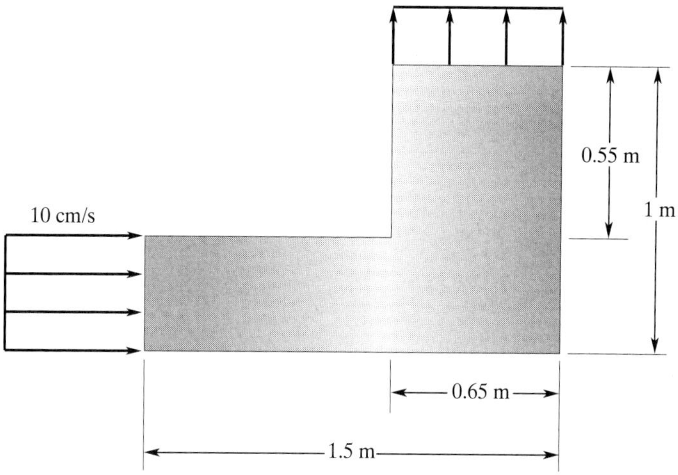

4. Consider the flow of air through the elbow in the accompanying figure. The corners of the elbow are rounded as shown in the figure. Assuming ideal flow behavior for air and uniform velocities at the inlet and outlet sections, use ANSYS to compute and plot the velocity distribution within the elbow. The elbow has a uniform depth. Use the continuity equation to obtain the velocity at the outlet.

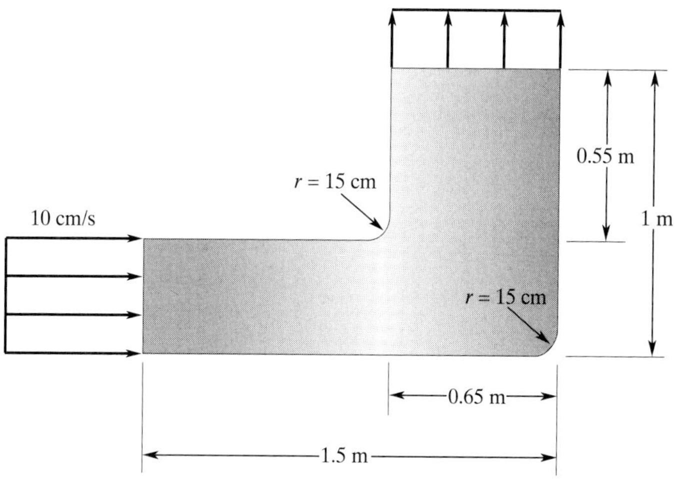

5. Using ANSYS, plot the velocity distributions for the transition-duct fitting shown in the accompanying figure. Plot the results for the combinations of the area ratios and transition angles given in the accompanying table.

A_2/A_1		θ	
0.1	10	20	45
0.25	10	45	60
0.5	20	45	60

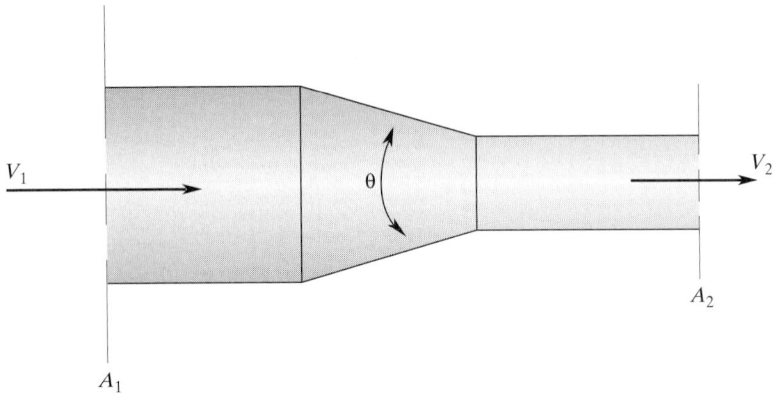

6. Consider the inviscid flow of air past the rounded equilateral triangle shown in the accompanying figure. Perform numerical experiments by changing the velocity of the upstream air and the r/L ratio (for $r/L = 0$, $r/L = 0.1$, and $r/L = 0.25$) and obtaining the corresponding air velocity distributions over the triangle. Discuss your results.

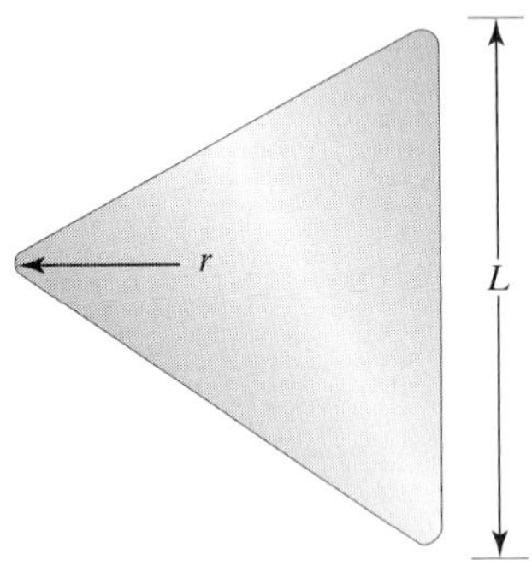

7. Consider the inviscid flow of air past the square rod with rounded corners shown in the accompanying figure. Perform numerical experiments by changing the velocity of the upstream air and the r/L ratio (for $r/L = 0$, $r/L = 0.1$, and $r/L = 0.25$) and obtaining the corresponding air velocity distributions over the square. Discuss your results.

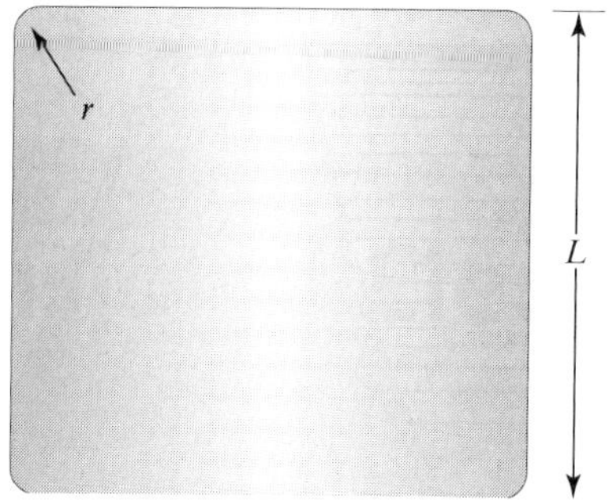

8. Consider the inviscid flow of air past a NACA symmetric airfoil. *Abbot and Von Doenhoff (1959)* provide detailed information about NACA symmetric airfoil shapes, including geometric data for NACA symmetric airfoils. Using their geometric data, obtain the velocity distribution over the NACA 0012-airfoil shown in the accompanying figure. Perform numerical experiments by changing the angle of attack and obtaining the corresponding air velocity distributions over the airfoil. Discuss your results.

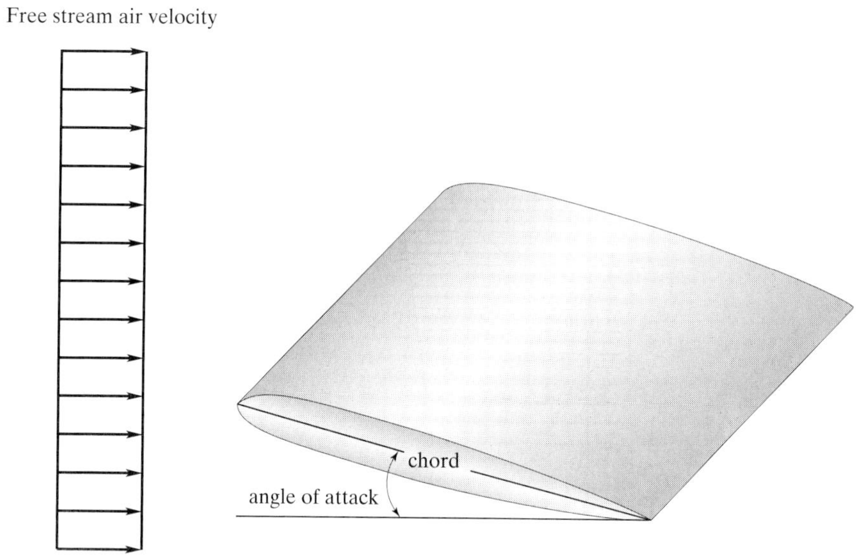

9. Consider the seepage flow of water under the concrete dam shown in the accompanying figure. The permeability of the porous soil under the dam is approximated as $k = 45$ ft/day. Determine the seepage velocity distribution in the porous soil.

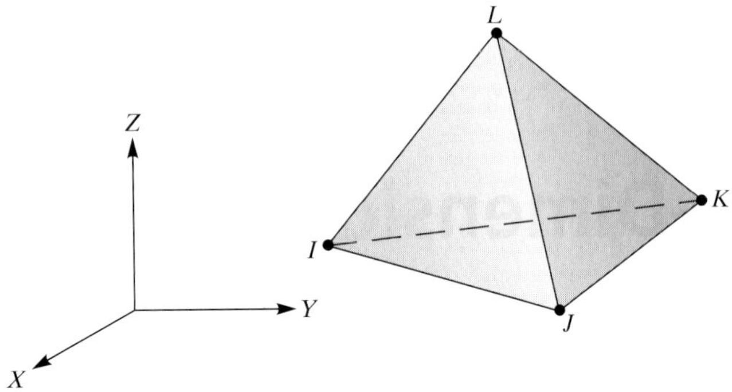

FIGURE 13.1 A four-node tetrahedral element.

gular shape functions for two-dimensional problems. We begin by representing the displacement field by the following equations:

$$\begin{aligned} u &= C_{11} + C_{12}X + C_{13}Y + C_{14}Z \\ v &= C_{21} + C_{22}X + C_{23}Y + C_{24}Z \\ w &= C_{31} + C_{32}X + C_{33}Y + C_{34}Z \end{aligned} \tag{13.1}$$

Considering the nodal displacements, we must satisfy the following conditions:

$$\begin{aligned} &u = u_I \quad \text{at} \quad X = X_I \quad Y = Y_I \quad \text{and} \quad Z = Z_I \\ &u = u_J \quad \text{at} \quad X = X_J \quad Y = Y_J \quad \text{and} \quad Z = Z_J \\ &u = u_K \quad \text{at} \quad X = X_K \quad Y = Y_K \quad \text{and} \quad Z = Z_K \\ &u = u_L \quad \text{at} \quad X = X_L \quad Y = Y_L \quad \text{and} \quad Z = Z_L \end{aligned}$$

Similarly, we must satisfy the following requirements:

$$\begin{aligned} &v = v_I \quad \text{at} \quad X = X_I \quad Y = Y_I \quad \text{and} \quad Z = Z_I \\ &\vdots \qquad\quad \vdots \qquad\quad \vdots \qquad\qquad \vdots \qquad\quad \vdots \qquad\quad \vdots \\ &w = w_L \quad \text{at} \quad X = X_I \quad Y = Y_I \quad \text{and} \quad Z = Z_I \end{aligned}$$

Substitution of respective nodal values into Eqs. (13.1) results in 12 equations and 12 unknowns:

$$\begin{aligned} u_I &= C_{11} + C_{12}X_I + C_{13}Y_I + C_{14}Z_I \\ u_J &= C_{11} + C_{12}X_J + C_{13}Y_J + C_{14}Z_J \\ &\vdots \\ w_L &= C_{31} + C_{32}X_L + C_{33}Y_L + C_{34}Z_L \end{aligned} \tag{13.2}$$

Solving for the unknown C-coefficients, substituting the results back into Eq. (13.1), and regrouping the parameters, we obtain

$$
\begin{aligned}
u &= S_1 u_I + S_2 u_J + S_3 u_K + S_4 u_L \\
v &= S_1 v_I + S_2 v_J + S_3 v_K + S_4 v_L \\
w &= S_1 w_I + S_2 w_J + S_3 w_K + S_4 w_L
\end{aligned}
\tag{13.3}
$$

The shape functions are

$$
\begin{aligned}
S_1 &= \frac{1}{6V}(a_I + b_I X + c_I Y + d_I Z) \\
S_2 &= \frac{1}{6V}(a_J + b_J X + c_J Y + d_J Z) \\
S_3 &= \frac{1}{6V}(a_K + b_K X + c_K Y + d_K Z) \\
S_4 &= \frac{1}{6V}(a_L + b_L X + c_L Y + d_L Z)
\end{aligned}
\tag{13.4}
$$

where V, the volume of the tetrahedral element, is computed from

$$
6V = \det\begin{vmatrix} 1 & X_I & Y_I & Z_I \\ 1 & X_J & Y_J & Z_J \\ 1 & X_K & Y_K & Z_K \\ 1 & X_L & Y_L & Z_L \end{vmatrix}
\tag{13.5}
$$

the $a_I, b_I, c_I, d_I, \ldots,$ and d_L-terms are

$$
a_I = \det\begin{vmatrix} X_J & Y_J & Z_J \\ X_K & Y_K & Z_K \\ X_L & Y_L & Z_L \end{vmatrix} \qquad b_I = -\det\begin{vmatrix} 1 & Y_J & Z_J \\ 1 & Y_K & Z_K \\ 1 & Y_L & Z_L \end{vmatrix}
\tag{13.6}
$$

$$
c_I = \det\begin{vmatrix} X_J & 1 & Z_J \\ X_K & 1 & Z_K \\ X_L & 1 & Z_L \end{vmatrix} \qquad d_I = -\det\begin{vmatrix} X_J & Y_J & 1 \\ X_K & Y_K & 1 \\ X_L & Y_L & 1 \end{vmatrix}
$$

We can represent the $a_J, b_J, c_J, d_J, \ldots,$ and d_L-terms using similar determinants by rotating through the I, J, K, and L subscripts using the right-hand rule. For example,

$$
a_J = \det\begin{vmatrix} X_K & Y_K & Z_K \\ X_L & Y_L & Z_L \\ X_I & Y_I & Z_I \end{vmatrix}
$$

It is important to note here that for thermal problems, we associate only a single degree of freedom with each node of the four-node tetrahedral element—namely, temperature. The variation of temperature over a four-node tetrahedral element is expressed by

$$T = T_I S_1 + T_J S_2 + T_K S_3 + T_L S_4 \tag{13.7}$$

13.2 ANALYSIS OF THREE-DIMENSIONAL SOLID PROBLEMS USING FOUR-NODE TETRAHEDRAL ELEMENTS

You may recall from Chapter 10 that only six independent stress components are needed to characterize the general state of stress at a point. These components are

$$[\boldsymbol{\sigma}]^T = \begin{bmatrix} \sigma_{xx} & \sigma_{yy} & \sigma_{zz} & \tau_{xy} & \tau_{yz} & \tau_{xz} \end{bmatrix} \tag{13.8}$$

where σ_{xx}, σ_{yy}, and σ_{zz} are the normal stresses and τ_{xy}, τ_{yz}, and τ_{xz} are the shear-stress components. Moreover, we discussed the displacement vector that measures the changes occurring in the position of a point within a body when the body is subjected to a load. You may also recall that the displacement vector $\vec{\delta}$ can be written in terms of its Cartesian components as

$$\vec{\delta} = u(x, y, z)\vec{i} + v(x, y, z)\vec{j} + w(x, y, z)\vec{k} \tag{13.9}$$

The corresponding state of strain at a point was also discussed in Chapter 10. The general state of strain is characterized by six independent components as given by

$$[\boldsymbol{\varepsilon}]^T = \begin{bmatrix} \varepsilon_{xx} & \varepsilon_{yy} & \varepsilon_{zz} & \gamma_{xy} & \gamma_{yz} & \gamma_{xz} \end{bmatrix} \tag{13.10}$$

where ε_{xx}, ε_{yy}, and ε_{zz} are the normal strains and γ_{xy}, γ_{yz}, and γ_{xz} are the shear-strain components. As previously discussed, the relationship between the strain and the displacement is represented by

$$\varepsilon_{xx} = \frac{\partial u}{\partial x} \quad \varepsilon_{yy} = \frac{\partial v}{\partial y} \quad \varepsilon_{zz} = \frac{\partial w}{\partial z} \tag{13.11}$$

$$\gamma_{xy} = \frac{\partial u}{\partial y} + \frac{\partial v}{\partial x} \quad \gamma_{yz} = \frac{\partial v}{\partial z} + \frac{\partial w}{\partial y} \quad \gamma_{xz} = \frac{\partial u}{\partial z} + \frac{\partial w}{\partial x}$$

Equations (13.11) can be represented in matrix form as

$$\{\boldsymbol{\varepsilon}\} = LU \tag{13.12}$$

where

$$\{\boldsymbol{\varepsilon}\} = \begin{Bmatrix} \varepsilon_{xx} \\ \varepsilon_{yy} \\ \varepsilon_{zz} \\ \gamma_{xy} \\ \gamma_{yz} \\ \gamma_{xz} \end{Bmatrix}$$

and

$$LU = \begin{Bmatrix} \dfrac{\partial u}{\partial x} \\ \dfrac{\partial v}{\partial y} \\ \dfrac{\partial w}{\partial z} \\ \dfrac{\partial u}{\partial y} + \dfrac{\partial v}{\partial x} \\ \dfrac{\partial v}{\partial z} + \dfrac{\partial w}{\partial y} \\ \dfrac{\partial w}{\partial x} + \dfrac{\partial u}{\partial z} \end{Bmatrix}$$

L is commonly referred to as the linear-differential operator.

Over the elastic region of a material, there also exists a relationship between the state of stresses and strains, according to the generalized Hooke's law. This relationship is given by the following equations:

$$\varepsilon_{xx} = \frac{1}{E}[\sigma_{xx} - \nu(\sigma_{yy} + \sigma_{zz})] \tag{13.13}$$

$$\varepsilon_{yy} = \frac{1}{E}[\sigma_{yy} - \nu(\sigma_{xx} + \sigma_{zz})]$$

$$\varepsilon_{zz} = \frac{1}{E}[\sigma_{zz} - \nu(\sigma_{xx} + \sigma_{yy})]$$

$$\gamma_{xy} = \frac{1}{G}\tau_{xy} \quad \gamma_{yz} = \frac{1}{G}\tau_{yz} \quad \gamma_{zx} = \frac{1}{G}\tau_{zx}$$

The relationship between the stress and strain can be expressed in a compact-matrix form as

$$\{\boldsymbol{\sigma}\} = [\boldsymbol{\nu}]\{\boldsymbol{\varepsilon}\} \tag{13.14}$$

where

$$\{\boldsymbol{\sigma}\} = \begin{Bmatrix} \sigma_{xx} \\ \sigma_{yy} \\ \sigma_{zz} \\ \tau_{xy} \\ \tau_{yz} \\ \tau_{xz} \end{Bmatrix}$$

$$[\boldsymbol{\nu}] = \frac{E}{1+\nu}\begin{bmatrix} \frac{1-\nu}{1-2\nu} & \frac{\nu}{1-2\nu} & \frac{\nu}{1-2\nu} & 0 & 0 & 0 \\ \frac{\nu}{1-2\nu} & \frac{1-\nu}{1-2\nu} & \frac{\nu}{1-2\nu} & 0 & 0 & 0 \\ \frac{\nu}{1-2\nu} & \frac{\nu}{1-2\nu} & \frac{1-\nu}{1-2\nu} & 0 & 0 & 0 \\ 0 & 0 & 0 & \frac{1}{2} & 0 & 0 \\ 0 & 0 & 0 & 0 & \frac{1}{2} & 0 \\ 0 & 0 & 0 & 0 & 0 & \frac{1}{2} \end{bmatrix}$$

$$\{\boldsymbol{\varepsilon}\} = \begin{Bmatrix} \varepsilon_{xx} \\ \varepsilon_{yy} \\ \varepsilon_{zz} \\ \gamma_{xy} \\ \gamma_{yz} \\ \gamma_{xz} \end{Bmatrix}$$

For a solid material under triaxial loading, the strain energy Λ is

$$\Lambda^{(e)} = \frac{1}{2}\int_V (\sigma_{xx}\,\varepsilon_{xx} + \sigma_{yy}\,\varepsilon_{yy} + \sigma_{zz}\,\varepsilon_{zz} + \tau_{xy}\gamma_{xy} + \tau_{xz}\gamma_{xz} + \tau_{yz}\gamma_{yz})\,dV \qquad (13.15)$$

Or, in a compact-matrix form,

$$\Lambda^{(e)} = \frac{1}{2}\int_V [\boldsymbol{\sigma}]^T\{\boldsymbol{\varepsilon}\}\,dV \qquad (13.16)$$

Substituting for stresses in terms of strains using Hooke's law, Eq. (13.15) can be written as

$$\Lambda^{(e)} = \frac{1}{2}\int_V \{\boldsymbol{\varepsilon}\}^T[\boldsymbol{\nu}]\{\boldsymbol{\varepsilon}\}\, dV \qquad (13.17)$$

We will now use the four-node tetrahedral element to formulate the stiffness matrix. Recall that this element has four nodes, with each node having three translational degrees of freedom in the nodal x-, y-, and z-directions. The displacements u, v, and w in terms of the nodal values and the shape functions are represented by

$$\{\mathbf{u}\} = [\mathbf{S}]\{\mathbf{U}\} \qquad (13.18)$$

where

$$\{\mathbf{u}\} = \begin{Bmatrix} u \\ v \\ w \end{Bmatrix}$$

$$[\mathbf{S}] = \begin{bmatrix} S_1 & 0 & 0 & S_2 & 0 & 0 & S_3 & 0 & 0 & S_4 & 0 & 0 \\ 0 & S_1 & 0 & 0 & S_2 & 0 & 0 & S_3 & 0 & 0 & S_4 & 0 \\ 0 & 0 & S_1 & 0 & 0 & S_2 & 0 & 0 & S_3 & 0 & 0 & S_4 \end{bmatrix}$$

$$\{\mathbf{U}\} = \begin{Bmatrix} u_I \\ v_I \\ w_I \\ u_J \\ v_J \\ w_J \\ u_K \\ v_K \\ w_K \\ u_L \\ v_L \\ w_L \end{Bmatrix}$$

The next few steps are similar to the steps we took to derive the stiffness matrix for plane-stress situations in Chapter 10, except that more terms are involved in this case. We begin by relating the strains to the displacement field and, in turn, to the nodal displacements through the shape functions. We need to take the derivatives of the components of the displacement field with respect to the x-, y-, and z-coordinates according to the strain-displacement relations given by Eq. (13.12). The operation results in:

$$\begin{Bmatrix} \varepsilon_{xx} \\ \varepsilon_{yy} \\ \varepsilon_{zz} \\ \gamma_{xy} \\ \gamma_{yz} \\ \gamma_{xz} \end{Bmatrix} = \begin{bmatrix} \frac{\partial S_1}{\partial x} & 0 & 0 & \frac{\partial S_2}{\partial x} & 0 & 0 & \frac{\partial S_3}{\partial x} & 0 & 0 & \frac{\partial S_4}{\partial x} & 0 & 0 \\ 0 & \frac{\partial S_1}{\partial y} & 0 & 0 & \frac{\partial S_2}{\partial y} & 0 & 0 & \frac{\partial S_3}{\partial y} & 0 & 0 & \frac{\partial S_4}{\partial y} & 0 \\ 0 & 0 & \frac{\partial S_1}{\partial z} & 0 & 0 & \frac{\partial S_2}{\partial z} & 0 & 0 & \frac{\partial S_3}{\partial z} & 0 & 0 & \frac{\partial S_3}{\partial z} \\ \frac{\partial S_1}{\partial y} & \frac{\partial S_1}{\partial x} & 0 & \frac{\partial S_2}{\partial y} & \frac{\partial S_2}{\partial x} & 0 & \frac{\partial S_3}{\partial y} & \frac{\partial S_3}{\partial x} & 0 & \frac{\partial S_4}{\partial y} & \frac{\partial S_4}{\partial x} & 0 \\ 0 & \frac{\partial S_1}{\partial z} & \frac{\partial S_1}{\partial y} & 0 & \frac{\partial S_2}{\partial z} & \frac{\partial S_1}{\partial y} & 0 & \frac{\partial S_3}{\partial z} & \frac{\partial S_3}{\partial y} & 0 & \frac{\partial S_4}{\partial z} & \frac{\partial S_4}{\partial y} \\ \frac{\partial S_1}{\partial z} & 0 & \frac{\partial S_1}{\partial x} & \frac{\partial S_2}{\partial z} & 0 & \frac{\partial S_2}{\partial x} & \frac{\partial S_3}{\partial z} & 0 & \frac{\partial S_3}{\partial x} & \frac{\partial S_4}{\partial z} & 0 & \frac{\partial S_4}{\partial x} \end{bmatrix} \begin{Bmatrix} u_I \\ v_I \\ w_I \\ u_J \\ v_J \\ w_J \\ u_K \\ v_K \\ w_K \\ u_L \\ v_L \\ w_L \end{Bmatrix} \tag{13.19}$$

Substituting for the shape functions using the relations of Eq. (13.4) and differentiating, we have

$$\{\boldsymbol{\varepsilon}\} = [\mathbf{B}]\{\mathbf{U}\} \tag{13.20}$$

where

$$[\mathbf{B}] = \frac{1}{6V}\begin{bmatrix} b_I & 0 & 0 & b_J & 0 & 0 & b_K & 0 & 0 & b_L & 0 & 0 \\ 0 & c_I & 0 & 0 & c_J & 0 & 0 & c_K & 0 & 0 & c_L & 0 \\ 0 & 0 & d_I & 0 & 0 & d_J & 0 & 0 & d_K & 0 & 0 & d_L \\ c_I & b_I & 0 & c_J & b_J & 0 & c_K & b_K & 0 & c_L & b_L & 0 \\ 0 & d_I & c_I & 0 & d_J & c_J & 0 & d_K & c_K & 0 & d_L & c_L \\ d_I & 0 & b_I & d_J & 0 & b_J & d_K & 0 & b_K & d_L & 0 & b_L \end{bmatrix}$$

and the volume V and the b-, c-, and d-terms are given by Eqs. (13.5) and (13.6). Substituting into the strain energy equation for the strain components in terms of the displacements, we obtain

$$\Lambda^{(e)} = \frac{1}{2}\int_V \{\boldsymbol{\varepsilon}\}^T[\boldsymbol{\nu}]\{\boldsymbol{\varepsilon}\}\,dV = \frac{1}{2}\int_V [\mathbf{U}]^T[\mathbf{B}]^T[\boldsymbol{\nu}][\mathbf{B}][\mathbf{U}]\,dV \tag{13.21}$$

Differentiating with respect to the nodal displacements yields

$$\frac{\partial \Lambda^{(e)}}{\partial U_k} = \frac{\partial}{\partial U_k}\left(\frac{1}{2}\int_V [\mathbf{U}]^T[\mathbf{B}]^T[\boldsymbol{\nu}][\mathbf{B}][\mathbf{U}]\,dV\right) \quad \text{for } k = 1, 2, \ldots, 12 \tag{13.22}$$

Evaluation of Eq. (13.22) results in the expression $[\mathbf{K}]^{(e)}\{\mathbf{U}\}$ and, subsequently, the expression for the stiffness matrix, which is

$$[\mathbf{K}]^{(e)} = \int_V [\mathbf{B}]^T[\boldsymbol{\nu}][\mathbf{B}]\,dV = V[\mathbf{B}]^T[\boldsymbol{\nu}][\mathbf{B}] \tag{13.23}$$

where V is the volume of the element. Note that the resulting stiffness matrix will have the dimensions of 12 × 12.

Load Matrix

The load matrix for three-dimensional problems is obtained by using a procedure similar to the one described in Section 10.2. The load matrix for a tetrahedral element is a 12 × 1 matrix. For a concentrated-loading situation, the load matrix is formed by placing the components of the load at appropriate nodes in appropriate directions. For a distributed load, the load matrix is computed from the equation

$$\{\mathbf{F}\}^{(e)} = \int_A [\mathbf{S}]^T \{\mathbf{p}\}\, dA \tag{13.24}$$

where

$$\{\mathbf{p}\} = \begin{Bmatrix} p_x \\ p_y \\ p_z \end{Bmatrix}$$

and A represents the surface over which the distributed-load components are acting. The surfaces of the tetrahedral element are triangular in shape. Assuming that the distributed load acts on the I–J–K surface, the load matrix becomes

$$\{\mathbf{F}\}^{(e)} = \frac{A_{I\text{-}J\text{-}K}}{3} \begin{Bmatrix} p_x \\ p_y \\ p_z \\ p_x \\ p_y \\ p_z \\ p_x \\ p_y \\ p_z \\ 0 \\ 0 \\ 0 \end{Bmatrix} \tag{13.25}$$

The load matrix for a distributed load acting on the other surfaces of the tetrahedral element is obtained in a similar fashion.

13.3 THE EIGHT-NODE BRICK ELEMENT

The eight-node brick element is the next simple three-dimensional element used in the analysis of solid mechanics problems. Each of the eight nodes of this element has three translational degrees of freedom in the nodal x-, y-, and z-directions. A typical eight-node brick element is shown in Figure 13.2.

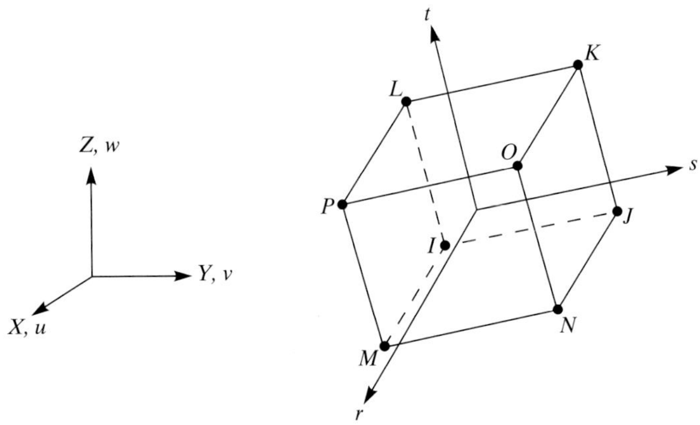

FIGURE 13.2 An eight-node brick element.

The element's displacement field in terms of the nodal displacements and the shape functions can be written as

$$u = \frac{1}{8}(u_I(1-s)(1-t)(1-r) + u_J(1+s)(1-t)(1-r)) \tag{13.26}$$

$$+ \frac{1}{8}(u_K(1+s)(1+t)(1-r) + u_L(1-s)(1+t)(1-r))$$

$$+ \frac{1}{8}(u_M(1-s)(1-t)(1+r) + u_N(1+s)(1-t)(1+r))$$

$$+ \frac{1}{8}(u_O(1+s)(1+t)(1+r) + u_P(1-s)(1+t)(1+r))$$

$$v = \frac{1}{8}(v_I(1-s)(1-t)(1-r) + v_J(1+s)(1-t)(1-r)) \tag{13.27}$$

$$+ \frac{1}{8}(v_K(1+s)(1+t)(1-r) + v_L(1-s)(1+t)(1-r))$$

$$+ \frac{1}{8}(v_M(1-s)(1-t)(1+r) + v_N(1+s)(1-t)(1+r))$$

$$+ \frac{1}{8}(v_O(1+s)(1+t)(1+r) + v_P(1-s)(1+t)(1+r))$$

$$w = \frac{1}{8}(w_I(1-s)(1-t)(1-r) + w_J(1+s)(1-t)(1-r)) \tag{13.28}$$

$$+ \frac{1}{8}(w_K(1+s)(1+t)(1-r) + w_L(1-s)(1+t)(1-r))$$

$$+\frac{1}{8}(w_M(1-s)(1-t)(1+r)+w_N(1+s)(1-t)(1+r))$$

$$+\frac{1}{8}(w_O(1+s)(1+t)(1+r)+w_P(1-s)(1+t)(1+r))$$

In a similar fashion, for thermal problems, the spatial variation of temperature over an element is represented by

$$T=\frac{1}{8}(T_I(1-s)(1-t)(1-r)+T_J(1+s)(1-t)(1-r)) \tag{13.29}$$

$$+\frac{1}{8}(T_K(1+s)(1+t)(1-r)+T_L(1-s)(1+t)(1-r))$$

$$+\frac{1}{8}(T_M(1-s)(1-t)(1+r)+T_N(1+s)(1-t)(1+r))$$

$$+\frac{1}{8}(T_O(1+s)(1+t)(1+r)+T_P(1-s)(1+t)(1+r))$$

13.4 THE TEN-NODE TETRAHEDRAL ELEMENT

The ten-node tetrahedral element, shown in Figure 13.3, is a higher order version of the three-dimensional linear tetrahedral element. When compared to the four-node tetrahedral element, the ten-node tetrahedral element is better suited for and more accurate in modeling problems with curved boundaries.

For solid problems, the displacement field is represented by

$$u=u_I(2S_1-1)S_1+u_J(2S_2-1)S_2+u_K(2S_3-1)S_3+u_L(2S_4-1)S_4 \tag{13.30}$$

$$+4(u_MS_1S_2+u_NS_2S_3+u_OS_1S_3+u_PS_1S_4+u_QS_2S_4+u_RS_3S_4)$$

$$v=v_I(2S_1-1)S_1+v_J(2S_2-1)S_2+v_K(2S_3-1)S_3+v_L(2S_4-1)S_4 \tag{13.31}$$

$$+4(v_MS_1S_2+v_NS_2S_3+v_OS_1S_3+v_PS_1S_4+v_QS_2S_4+v_RS_3S_4)$$

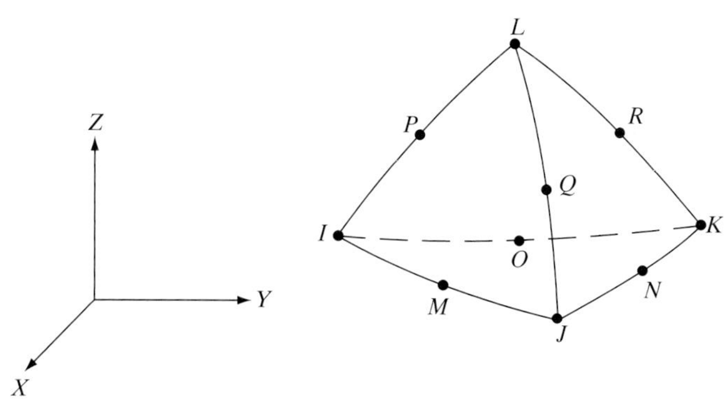

FIGURE 13.3 A ten-node tetrahedral element.

$$w = w_I(2S_1 - 1)S_1 + w_J(2S_2 - 1)S_2 + w_K(2S_3 - 1)S_3 + w_L(2S_4 - 1)S_4 \qquad (13.32)$$
$$+ 4(w_M S_1 S_2 + w_N S_2 S_3 + w_O S_1 S_3 + w_P S_1 S_4 + w_Q S_2 S_4 + w_R S_3 S_4)$$

In similar fashion, the spatial distribution of temperature over an element is given by

$$T = T_I(2S_1 - 1)S_1 + T_J(2S_2 - 1)S_2 + T_K(2S_3 - 1)S_3 + T_L(2S_4 - 1)S_4 \qquad (13.33)$$
$$+ 4(T_M S_1 S_2 + T_N S_2 S_3 + T_O S_1 S_3 + T_P S_1 S_4 + T_Q S_2 S_4 + T_R S_3 S_4)$$

13.5 THE TWENTY-NODE BRICK ELEMENT

The twenty-node brick element, shown in Figure 13.4, is a higher order version of the three-dimensional eight-node brick element. This element is more capable and more accurate for modeling problems with curved boundaries than the eight-node brick element.

For solid mechanics problems, the displacement field is given by

$$u = \frac{1}{8}(u_I(1-s)(1-t)(1-r)(-s-t-r-2) + u_J(1+s)(1-t)(1-r)(s-t-r-2))$$
$$+ \frac{1}{8}(u_K(1+s)(1+t)(1-r)(s+t-r-2) + u_L(1-s)(1+t)(1-r)(-s+t-r-2))$$
$$+ \frac{1}{8}(u_M(1-s)(1-t)(1+r)(-s-t+r-2) + u_N(1+s)(1-t)(1+r)(s-t+r-2))$$
$$+ \frac{1}{8}(u_O(1+s)(1+t)(1+r)(s+t+r-2) + u_P(1-s)(1+t)(1+r)(-s+t+r-2))$$

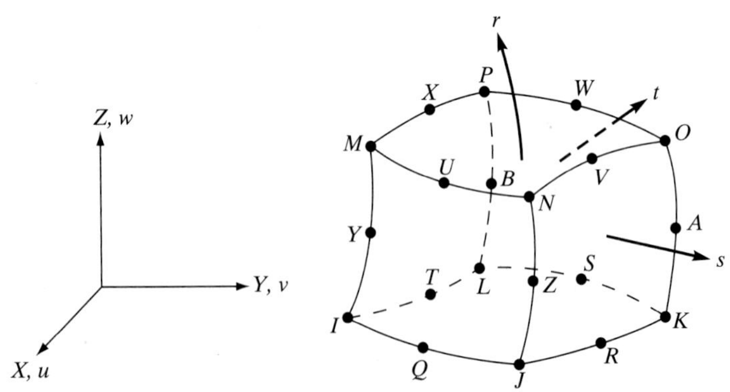

FIGURE 13.4 A twenty-node brick element.

$$+\frac{1}{4}(u_Q(1-s^2)(1-t)(1-r)+u_R(1+s)(1-t^2)(1-r))$$

$$+\frac{1}{4}(u_S(1-s^2)(1+t)(1-r)+u_T(1-s)(1-t^2)(1-r))$$

$$+\frac{1}{4}(u_U(1-s^2)(1-t)(1+r)+u_V(1+s)(1-t^2)(1+r))$$

$$+\frac{1}{4}(u_W(1-s^2)(1+t)(1+r)+u_X(1-s)(1-t^2)(1+r))$$

$$+\frac{1}{4}(u_Y(1-s)(1-t)(1-r^2)+u_Z(1+s)(1-t)(1-r^2))$$

$$+\frac{1}{4}(u_A(1+s)(1+t)(1-r^2)+u_B(1-s)(1+t)(1-r^2)) \quad (13.34)$$

The v- and w-components of the displacement are similar to the u-component:

$$v=\frac{1}{8}(v_I(1-s)(1-t)(1-r)(-s-t-r-2)+v_J(1+s)(1-t)(1-r)(s-t-r-2))$$

$$+\frac{1}{8}(v_K(1+s)(1+t)(1-r)(s+t-r-2)+\dots)$$

$$\dots$$

$$w=\frac{1}{8}(w_I(1-s)(1-t)(1-r)(-s-t-r-2)+w_J(1+s)(1-t)(1-r)(s-t-r-2))$$

$$+\frac{1}{8}(w_K(1+s)(1+t)(1-r)(s+t-r-2)+\dots) \quad (13.35)$$

$$\dots$$

For heat transfer problems, the spatial variation of temperature over an element is given by

$$T=\frac{1}{8}(T_I(1-s)(1-t)(1-r)(-s-t-r-2)+T_J(1+s)(1-t)(1-r)(s-t-r-2))$$

$$+\frac{1}{8}(T_K(1+s)(1+t)(1-r)(s+t-r-2)+\dots) \quad (13.36)$$

$$\dots$$

13.6 EXAMPLES OF THREE-DIMENSIONAL ELEMENTS IN ANSYS*

ANSYS offers a broad variety of elements for the analysis of three-dimensional problems. Some examples of three-dimensional elements in ANSYS are presented next.

Thermal-Solid Elements

SOLID70 is a three-dimensional element used to model conduction heat transfer problems. It has eight nodes, with each node having a single degree of freedom—temperature—as shown in Figure 13.5. The element's faces are shown by the circled numbers. Convection or heat fluxes may be applied to the element's surfaces. In addition, heat-generation rates may be applied at the nodes. This element may be used to analyze steady-state or transient problems.

The solution output consists of nodal temperatures and other information, such as average face temperature, temperature-gradient components, the vector sum at the centroid of the element, and the heat-flux components.

SOLID90 is a twenty-node brick element used to model steady-state or transient conduction heat transfer problems. This element is more accurate than the SOLID70 element, but it requires more solution time. Each node of the element has a single degree of freedom—temperature—as shown in Figure 13.6. This element is well suited to model problems with curved boundaries. The required input data and the solution output are similar to the data format of the SOLID70 elements.

Structural-Solid Elements

SOLID45 is a three-dimensional brick element used to model isotropic solid problems. It has eight nodes, with each node having three translational degrees of freedom in the

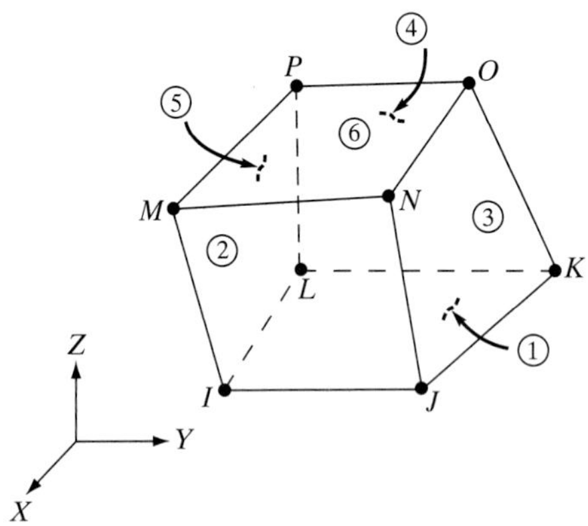

FIGURE 13.5 The SOLID70 element used by ANSYS.

*Materials were adapted with permission from ANSYS documents.

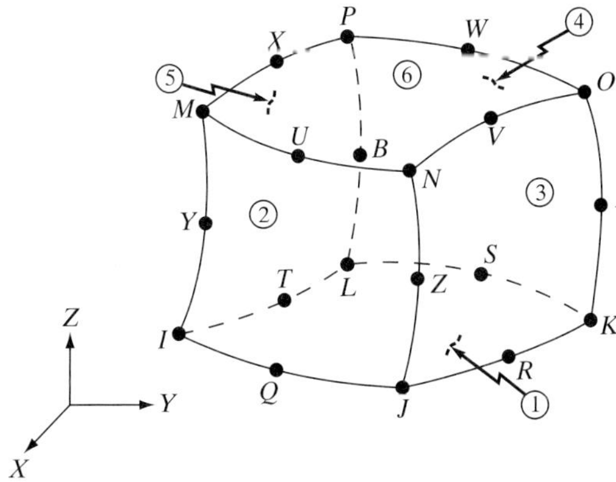

FIGURE 13.6 The SOLID90 element used by ANSYS.

nodal x-, y-, and z-directions, as shown in Figure 13.7 (The element's faces are shown by the circled numbers.) Distributed surface loads (pressures) may be applied to the element's surfaces. This element may be used to analyze large-deflection, large-strain, plasticity, and creep problems.

The solution output consists of nodal displacements. Examples of additional elemental output include normal components of the stresses in x-, y-, and z-directions; shear stresses; and principal stresses. The element's stress directions are parallel to the element's coordinate systems.

SOLID65 is used to model reinforced-concrete problems or reinforced composite materials, such as fiberglass. This element is similar to the SOLID45 elements, and it has eight nodes, with each node having three translational degrees of freedom in the nodal x-, y-, and z-directions, as shown in Figure 13.8. The element may be used to analyze cracking in tension or crushing in compression. The element can also be used to

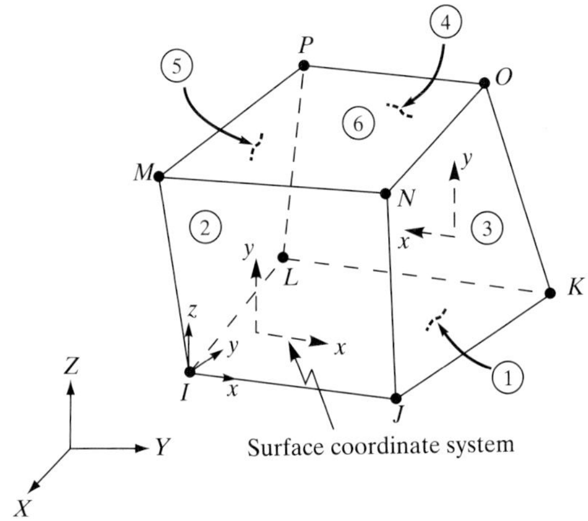

FIGURE 13.7 The SOLID45 element used by ANSYS.

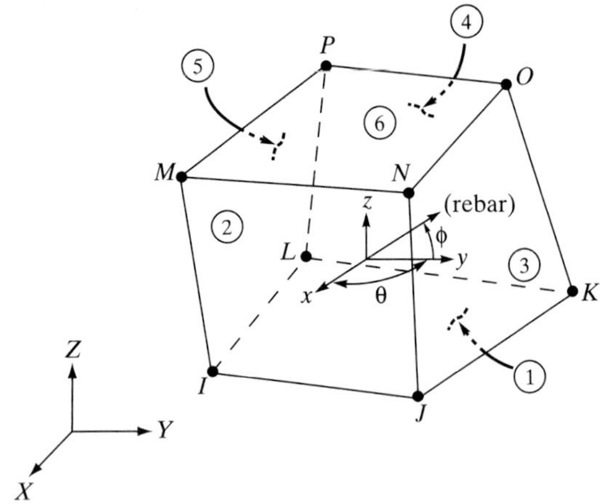

FIGURE 13.8 The SOLID65 element used by ANSYS.

analyze problems with or without reinforced bars. Up to three rebar specifications may be defined. The rebars are capable of plastic deformation and creep. The element has one solid material and up to three rebar materials. Rebar specifications include the material number; the volume ratio, which is defined as the ratio of the rebar volume to the total element volume; and the orientation angles. The rebar orientation is defined by two angles measured with respect to the element's coordinate system. The rebar capability is removed by assigning a zero value to the rebar material number.

The solution output consists of nodal displacements. Examples of additional elemental output include the normal components of the stresses in x-, y-, and z-directions, shear stresses, and principal stresses. The element's stress directions are parallel to the element's coordinate system.

SOLID72 is a four-node tetrahedral element, with each node having three translational degrees of freedom in the nodal x-, y-, and z-directions, as well as rotations about the nodal x-, y-, and z-directions, as shown in Figure 13.9. As in previous examples,

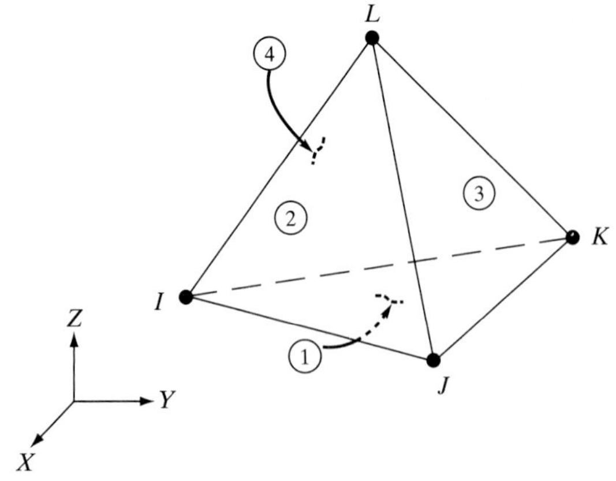

FIGURE 13.9 The SOLID72 element used by ANSYS.

the element's faces are shown by the circled numbers. Distributed surface loads (pressures) may be applied to the element's surfaces.

The solution output is similar to that of other structural-solid elements.

SOLID73 is an eight-node brick element that has three translational degrees of freedom in the nodal x-, y-, and z-directions, as well as rotations about the nodal x-, y-, and z-directions, as shown in Figure 13.10. The input data and the solution output are similar to those of elements discussed previously.

SOLID92 is a ten-node tetrahedral element that is more accurate than the SOLID72 element, but it requires more solution time. Each node has three translational degrees of freedom in the nodal x-, y-, and z-directions, as shown in Figure 13.11. This element may be used to analyze large-deflection, large-strain, plasticity, and creep problems.

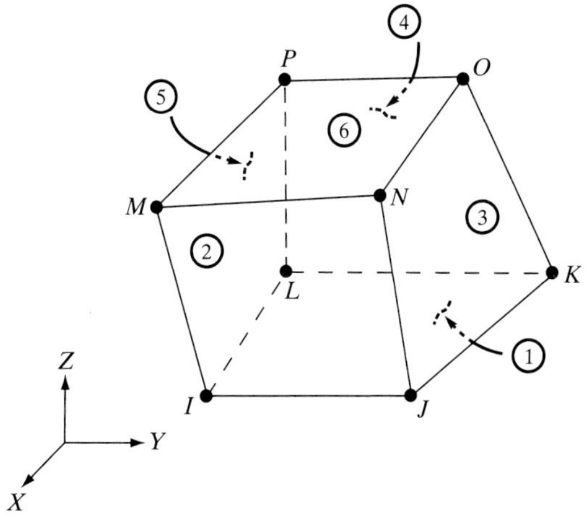

FIGURE 13.10 The SOLID73 element used by ANSYS.

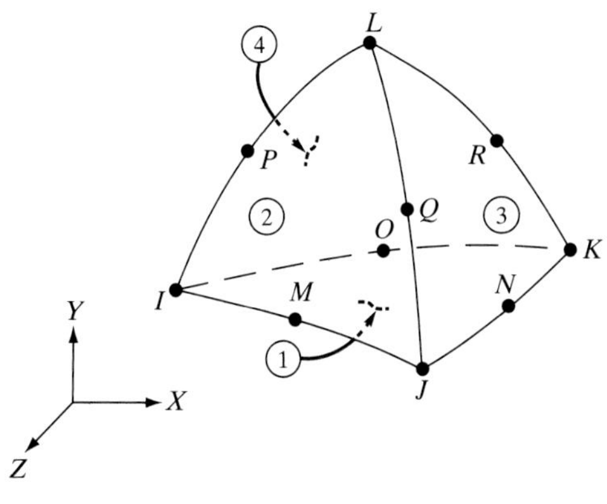

FIGURE 13.11 The SOLID92 element used by ANSYS.

13.7 BASIC SOLID-MODELING IDEAS*

There are two ways to create a solid model of an object under investigation: *bottom-up modeling* and *top-down modeling*. With *bottom-up modeling*, you start by defining keypoints first, then lines, areas, and volumes in terms of the defined keypoints. You can define keypoints on the working plane by the picking method or you can enter, in appropriate fields, the coordinates of the keypoints in terms of the active coordinate system. The keypoints menu is shown in Figure 13.12, and the command for creating keypoints is

main menu: **Preprocessor → Modeling → Create → Keypoints**

Lines, next in the hierarchy of bottom-up modeling, are used to represent the edges of an object. ANSYS provides four options for creating lines, as shown in Figure 13.13. With the splines options, you can create a line of arbitrary shape from a spline fit to a series of keypoints. You can then use the created line(s) to generate a surface with an arbitrary shape. The command for creating lines is:

main menu: **Preprocessor → Modeling → Create → Lines**

Using bottom-up modeling, you can define areas using the Area-Arbitrary submenu, as shown in Figure 13.14. The command for defining areas is:

main menu: **Preprocessor → Modeling → Create → Areas → Arbitrary**

There are five other ways by which you can create areas: (1) dragging a line along a path, (2) rotating a line about an axis, (3) creating an area fillet, (4) skinning a set of lines, and (5) offsetting areas. With the drag and rotate options, you can generate an area by dragging (sweeping) a line along another line (path) or by rotating a line about another line (axis of rotation). With the area-fillet command, you can create a constant-radius fillet tangent to two other areas. You can generate a smooth surface over a set of

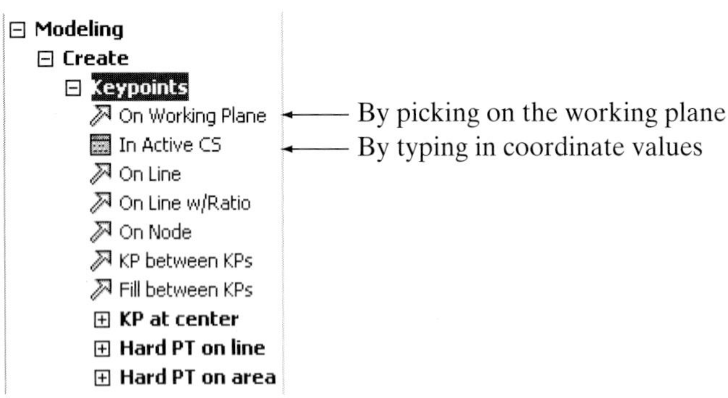

FIGURE 13.12 The Keypoints menu.

*Materials were adapted with permission from ANSYS documents.

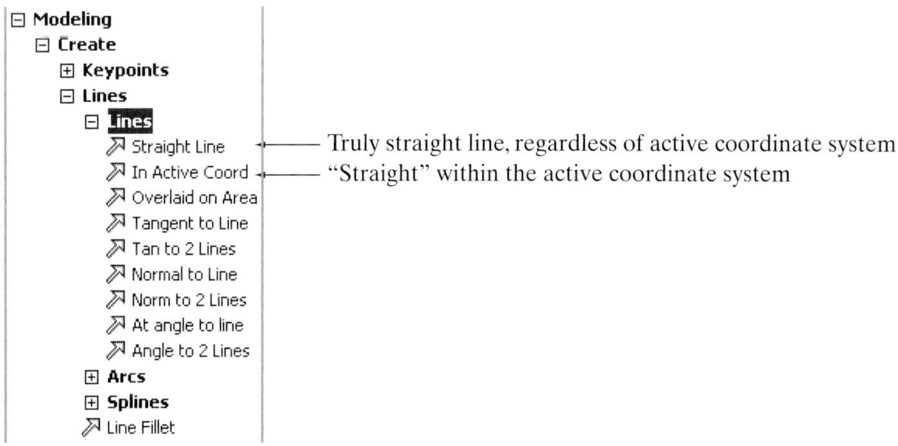

FIGURE 13.13 The Lines menu.

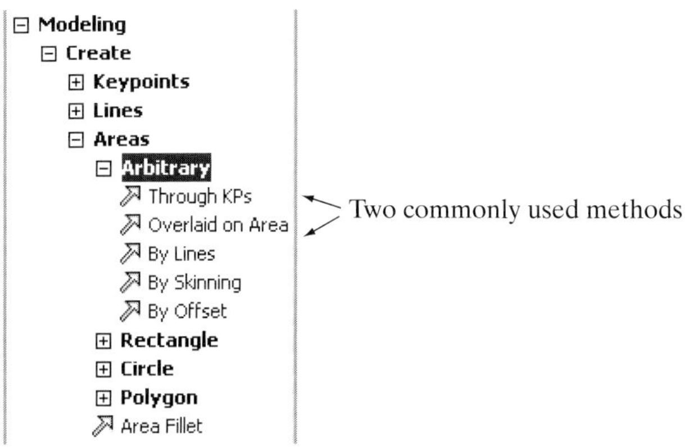

FIGURE 13.14 The Area-Arbitrary submenu.

lines by using the skinning command. Using the area-offset command, you can generate an area by offsetting an existing area. These operations are all shown in Figure 13.15.

You can define volumes using the bottom-up method by selecting the Volume submenu, as shown in Figure 13.16. The command for defining volume is:

main menu: **Preprocessor → Modeling → Create → Volumes → Arbitrary**

As with areas, you can also generate volumes by dragging (sweeping) an area along a line (path) or by rotating an area about a line (axis of rotation).

With top-down modeling, you can create three-dimensional solid objects using volume primitives. ANSYS provides the following three-dimensional primitives: block, prism, cylinder, cone, sphere, and torus, as shown in Figure 13.17.

Keep in mind that when you create a volume using primitives, ANSYS automatically generates and assigns numbers to areas, lines, and keypoints that bound the volume.

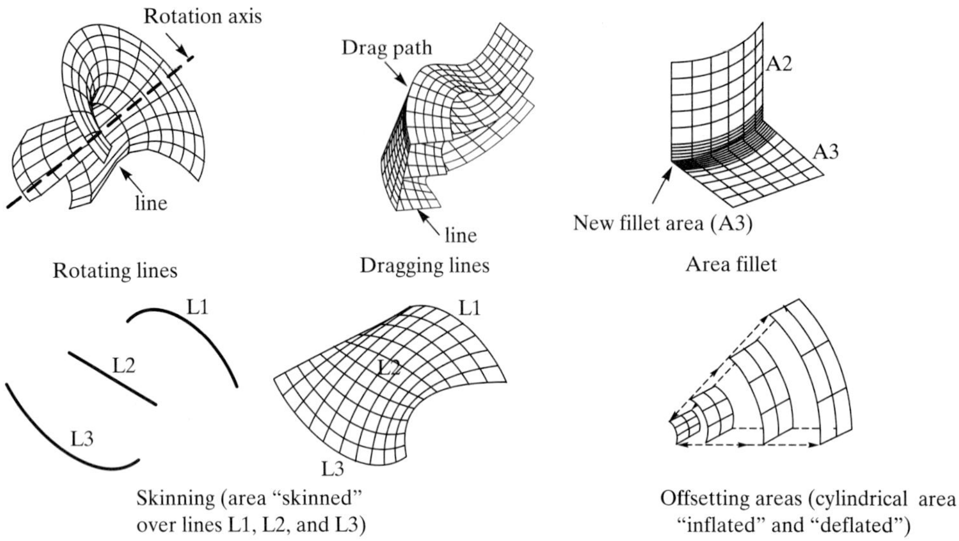

FIGURE 13.15 Additional area-generation methods.

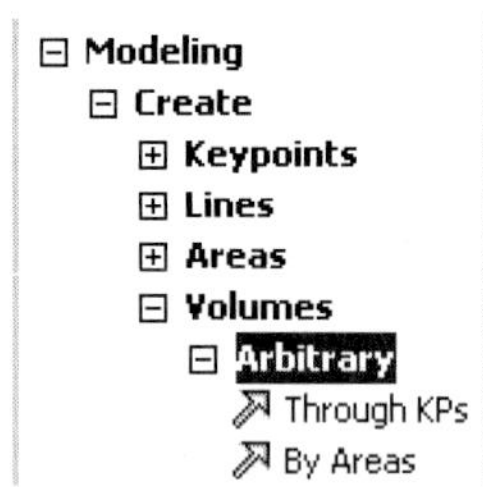

FIGURE 13.16 The Volume submenu.

Regardless of how you generate areas or volumes, you can use Boolean operations to add or subtract entities to create a solid model.

Meshing Control

So far, you have been using global element size to control the size of elements in your model. The GLOBAL-ELEMENT-SIZE dialog box allows you to specify the size of an element's edge length in the units of your model's dimensions. Let us consider other ways of controlling not only the size of elements, but also their shapes. Setting the element shape prior to meshing is important when using elements that can take on two shapes. For example, PLANE82 can take on triangular or quadrilateral shapes. Use the following command to see the dialog box for the meshing options (see Figure 13.18):

main menu: **Preprocessor → Meshing → Mesher Opts**

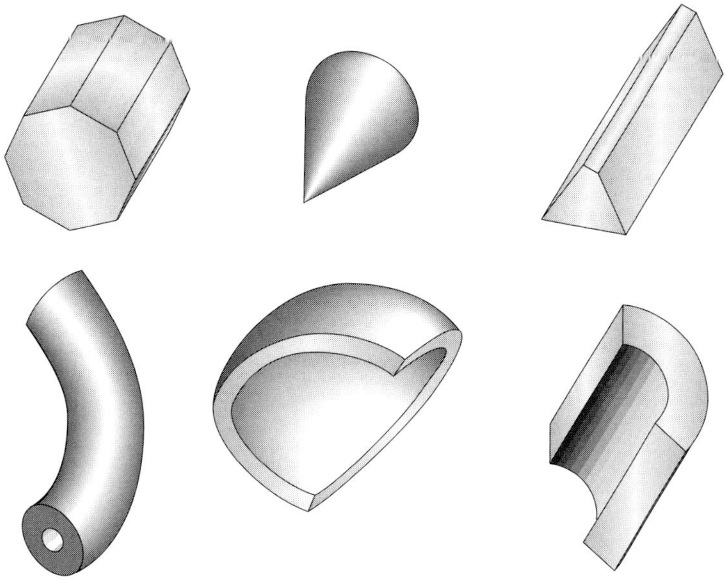

FIGURE 13.17 Examples of three-dimensional primitives.

Free Meshing Versus Mapped Meshing

Free meshing uses either mixed-area element shapes, all triangular-area elements, or all tetrahedral-volume elements. You may want to avoid using lower order triangular and tetrahedral elements (those without midside nodes) in analysis of structures, when possible. On the other hand, *mapped meshing* uses all quadrilateral-area elements and all hexahedral-volume elements. Figure 13.19 illustrates the difference between free and mapped meshing.

There are, however, some requirements that need to be met for mapped meshing. The mapped-area mesh requires that the area has three or four sides, equal numbers of elements on opposite sides, and an even number of elements for three-sided areas. If an area is bounded by more than four lines, then you need to use the *combine* command or the *concatenate* command to combine (reduce) the number of lines to four. The mapped-volume requirements are that the volume must be bound by four, five, or six sides, have an equal number of elements on the opposite side, and have an even numbers of elements if a five-sided prism or tetrahedron volume is involved. For volumes, you can *add* or *concatenate* areas to reduce the number of areas bounding a volume. Concatenation should be the last step before meshing. You cannot perform any other solid-modeling operations on concatenated entities. To concatenate, issue the following command (see Figure 13.20):

main menu: **Preprocessing → Meshing → Concatenate**

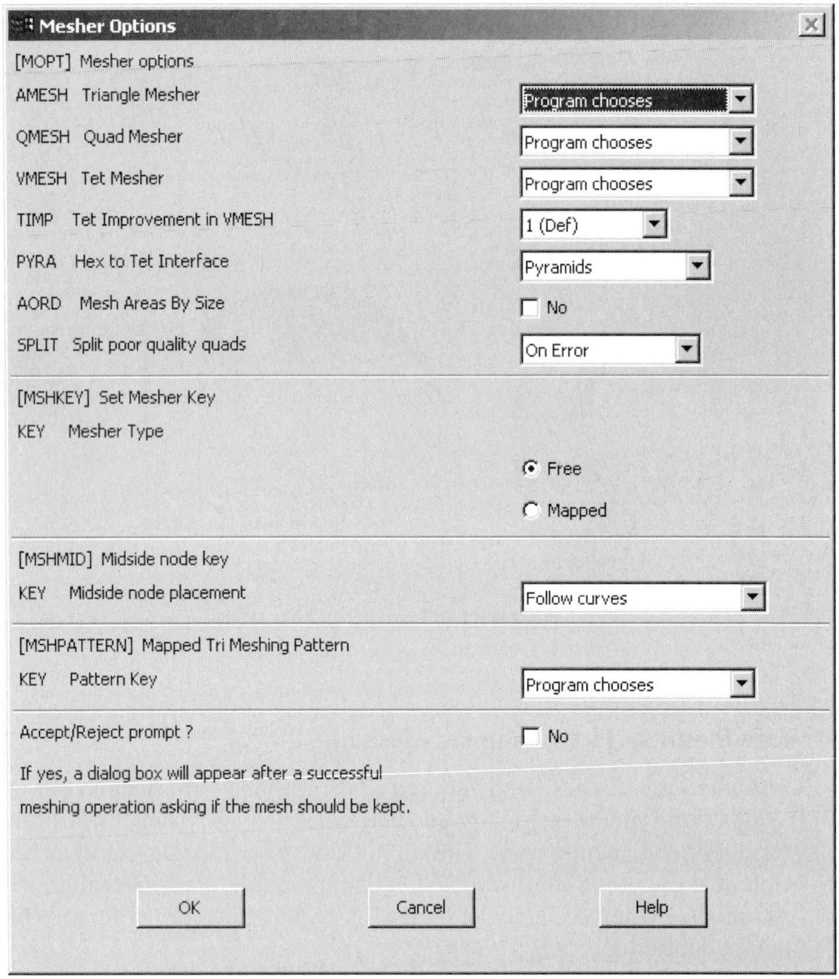

FIGURE 13.18 The dialog box for element shape.

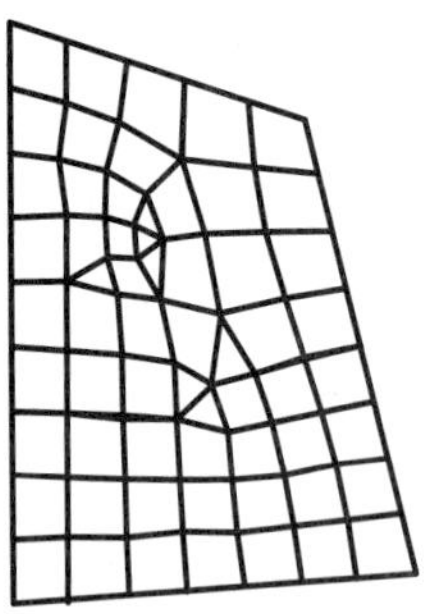

Free meshing

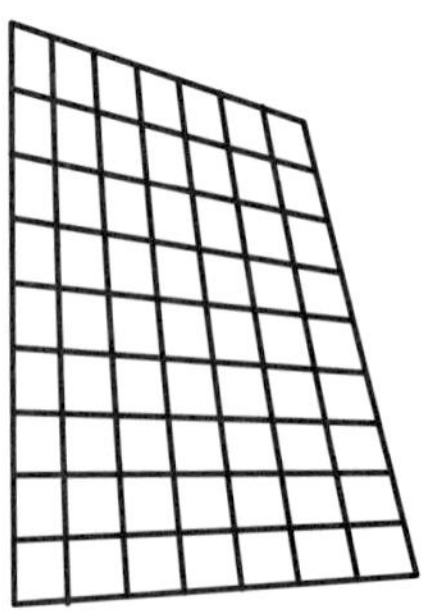

Mapped meshing

FIGURE 13.19 An illustration of the difference between free and mapped meshing.

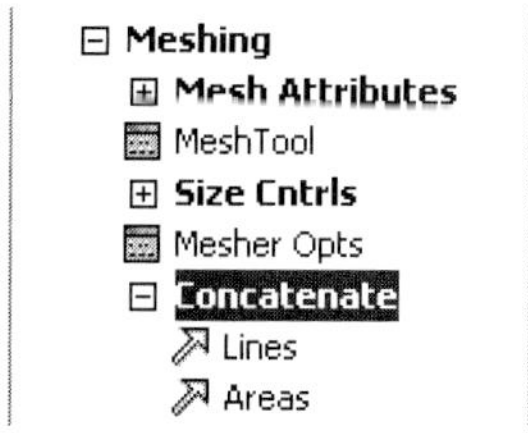

FIGURE 13.20 Concatenate dialog box.

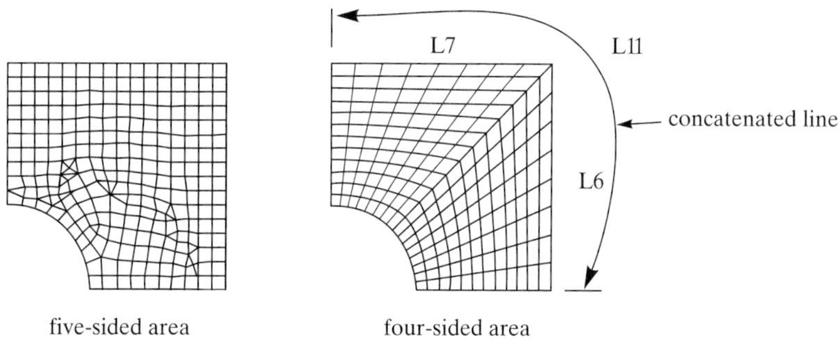

FIGURE 13.21 An example of free meshing and mapped meshing for an area.

Figure 13.21 shows an example of free and mapped meshing for an area. As a general rule, you want to avoid meshing poorly shaped regions and avoid creating extreme element-size transitions in your model. Examples of these situations are given in Figure 13.22.

If you are unhappy with the results of a meshed region, you can use the *clear* command to delete the nodes and elements associated with a corresponding solid-model entity. To issue the clear command, use the following sequence:

main menu: **Preprocessor → Meshing → Clear**

With the aid of an example, we will now demonstrate how to create a solid model of a heat sink, using area and extrusion commands.

EXAMPLE 13.1

Aluminum heat sinks are commonly used to dissipate heat from electronic devices. Consider an example of a heat sink used to cool a personal-computer microprocessor chip. The front view of the heat sink is shown in Figure 13.23. Using ANSYS, generate the solid model of the heat sink. Because of the symmetry, model only a quarter of the heat sink by extruding the shown frontal area by 20.5 mm.

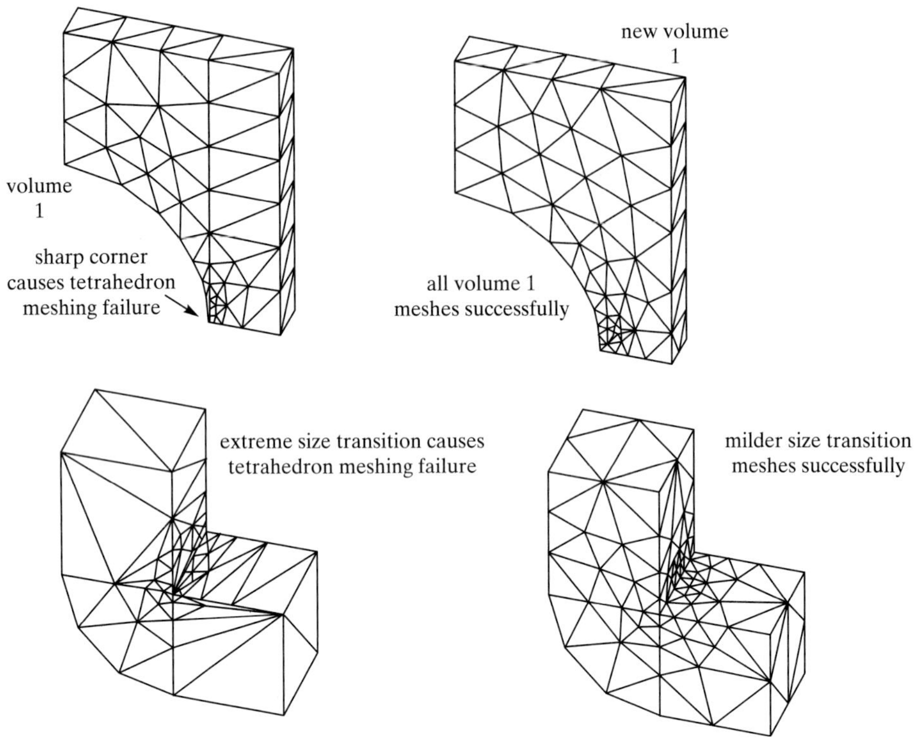

FIGURE 13.22 Examples of undesirable meshing situations.

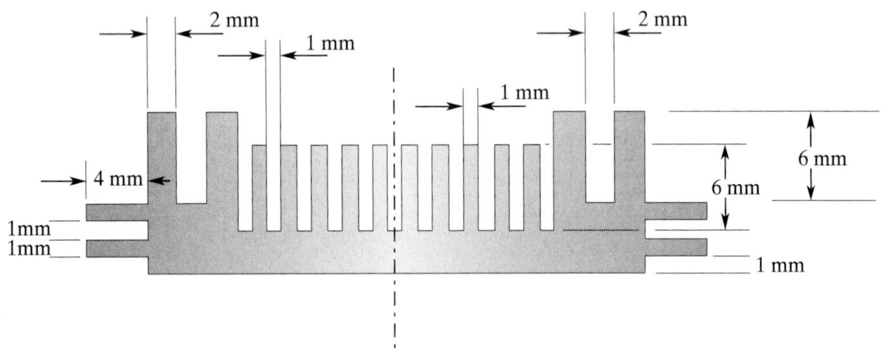

FIGURE 13.23 The front view of the heat sink in Example 13.1.

Enter the **ANSYS** program by using the Launcher. Type **Fin** (or a file name of your choice) in the **Jobname** entry field of the dialog box. Pick **Run** to start the Graphic User Interface (GUI).

Create a title for the problem. This title will appear on ANSYS display windows to provide a simple way of identifying the displays. To create a title, issue the following command:

utility menu: **File → Change Title** . . .

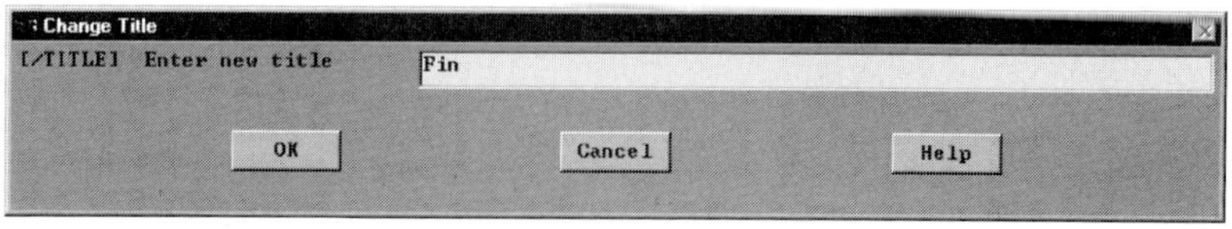

Set up the graphics area (i.e., workplane, zoom, etc.) with the following commands:

utility menu: **Workplane → Wp Settings** . . .

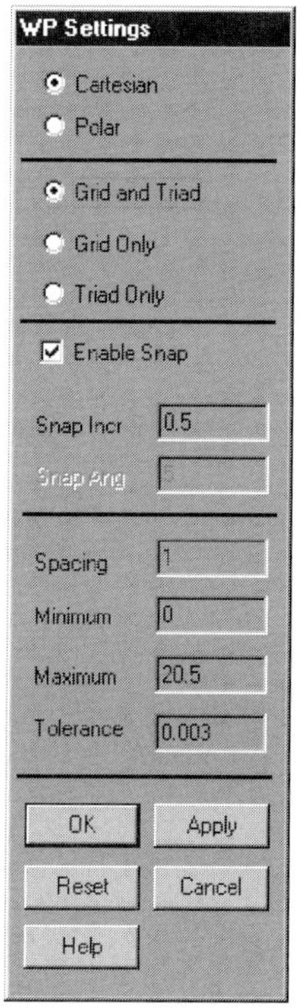

Toggle on the workplane by using the following command:

utility menu: **Workplane → Display Working Plane**

Bring the workplane to view by using the following command:

utility menu: **PlotCtrls → Pan, Zoom, Rotate** . . .

Click on the small circle until you bring the workplane to view. Then, create the geometry with the following command:

main menu: **Preprocessor → Modeling → Create → Areas → Rectangle → By 2 Corners**

On the work plane, pick the following locations or type the values in **WP X, WP Y, Width**, and **Height** fields:

[WP = 4, 0]

[Expand the rubber band up 2.5 and right 16.5]

[WP = 0, 1]

[Expand the rubber band up 1.0 and right 4.0]

[WP = 0, 3]

[Expand the rubber band up 1 and right 4.0]

[WP = 4, 2.5]

[Expand the rubber band up 1.5 and right 6.0]

[WP = 4, 4]

[Expand the rubber band up 6 and right 2]

[WP = 8, 4]

[Expand the rubber band up 6.0 and right 2.0]

[WP = 11, 2.5]

[Expand the rubber band up 6.0 and right 1.0]

[WP = 13, 2.5]

[Expand the rubber band up 6.0 and right 1.0]

[WP = 15, 2.5]

[Expand the rubber band up 6.0 and right 1.0]

[WP = 17, 2.5]

[Expand the rubber band up 6.0 and right 1.0]

[WP = 19, 2.5]

[Expand the rubber band up 6.0 and right 1.0]

OK

main menu: **Preprocessor → Modeling → Operate → Booleans → Add → Areas**

Pick All

main menu: **Preprocessor → Modeling → Operate → Extrude → Areas → Along Normal**

Pick or enter the area to be extruded, and then press the **Apply** button:

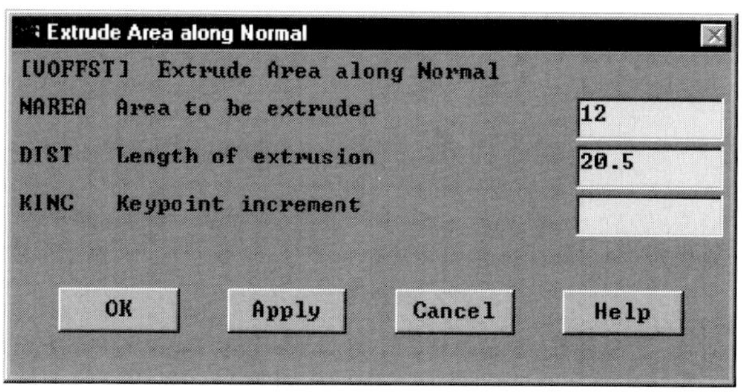

utility menu: **PlotCtrls → Pan, Zoom, Rotate** . . .

Press the **Iso** (Isometric view) button. You should then see the image in Figure 13.24.

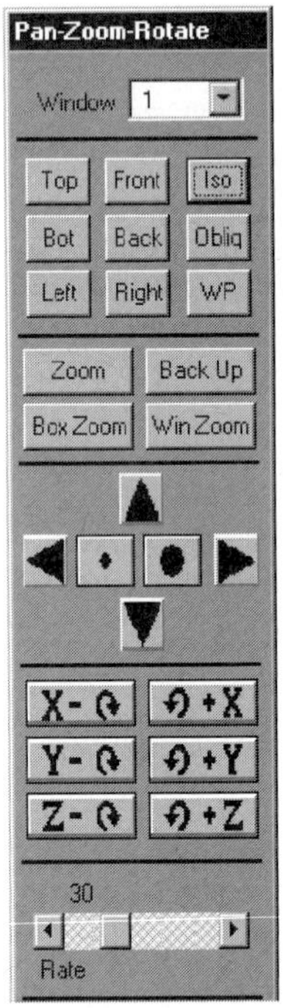

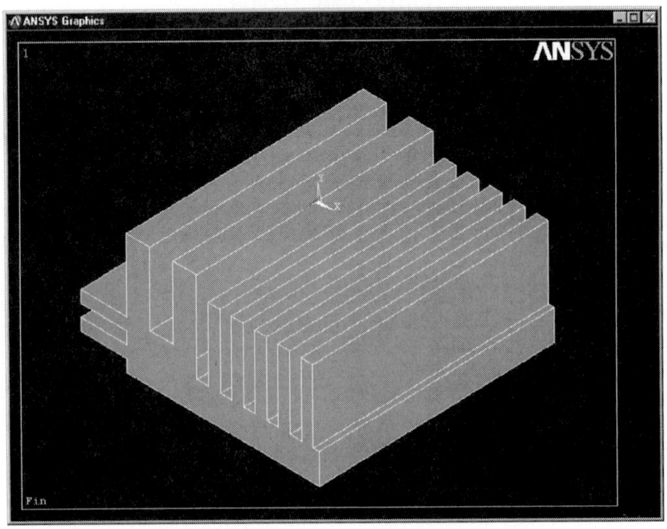

FIGURE 13.24 Isometric view of the heat sink.

Exit and save your results:

ANSYS Toolbar: **QUIT**

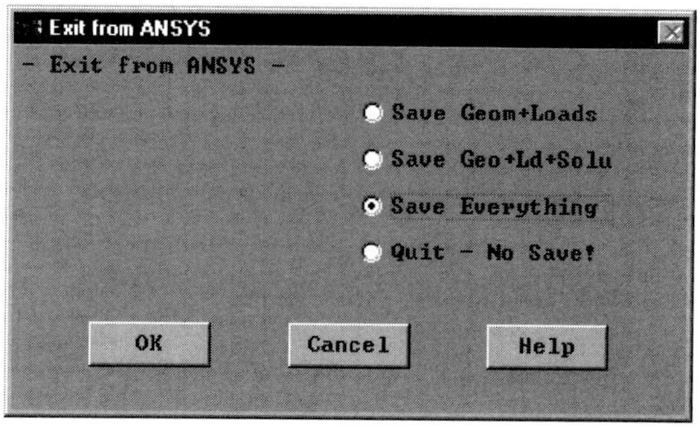

This example has demonstrated how to extrude an area along a normal direction to create a volume.

13.8 A THERMAL EXAMPLE USING ANSYS

EXAMPLE 13.2

A section of an aquarium wall with a viewing window has the dimensions shown in Figure 13.25. The wall is constructed from concrete and other insulating materials, with an average thermal conductivity of $k = 0.81\ \text{Btu/hr}\cdot\text{ft}\cdot{}^\circ\text{F}$. The section of the wall has a viewing window that is made of a six-inch-thick clear plastic with a thermal conductivity of $k = 0.195\ \text{Btu/hr}\cdot\text{ft}\cdot{}^\circ\text{F}$. The inside air temperature is kept at 70°F, with a corresponding heat transfer coefficient of $h = 1.46\ \text{Btu/hr}\cdot\text{ft}^2\cdot{}^\circ\text{F}$. Assuming a water-tank temperature of 50°F and a corresponding heat transfer coefficient of $h = 10.5\ \text{Btu/hr}\cdot\text{ft}^2\cdot{}^\circ\text{F}$, use ANSYS to plot the temperature distribution within the wall section. Note that the main purpose of this example is to show the selection capabilities of ANSYS and to show how to move the working plane when constructing three-dimensional models. Recall that the heat loss through such a wall may be obtained with reasonable accuracy from the equation $q = U_{\text{overall}}(T_{\text{inside}} - T_{\text{water}})$ and by calculating the overall U-factor for the wall.

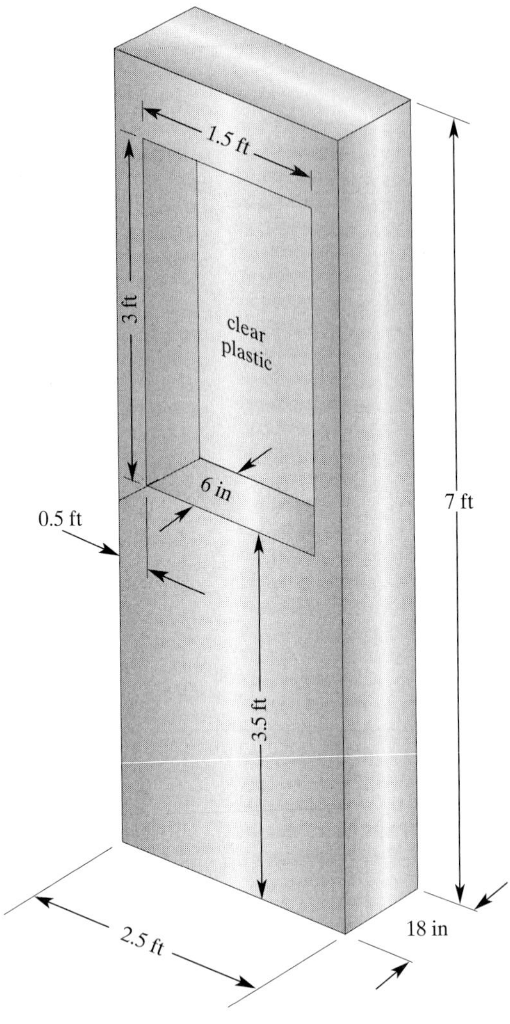

FIGURE 13.25 Dimensions of the wall and the clear-plastic viewing window of Example 10.2.

Enter the **ANSYS** program by using the Launcher. Type **Wall** (or a file name of your choice) in the **Jobname** entry field of the dialog box. Pick **Run** to start the GUI.

To create a title, issue the following command:

utility menu: **File → Change Title** . . .

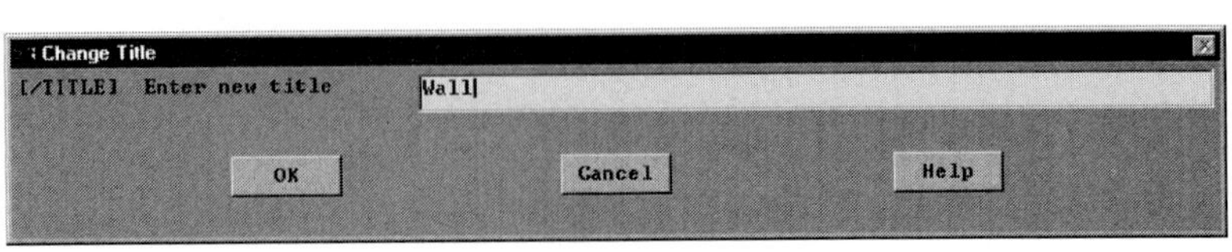

main menu: **Preprocessor → Element Types → Add/Edit/Delete**

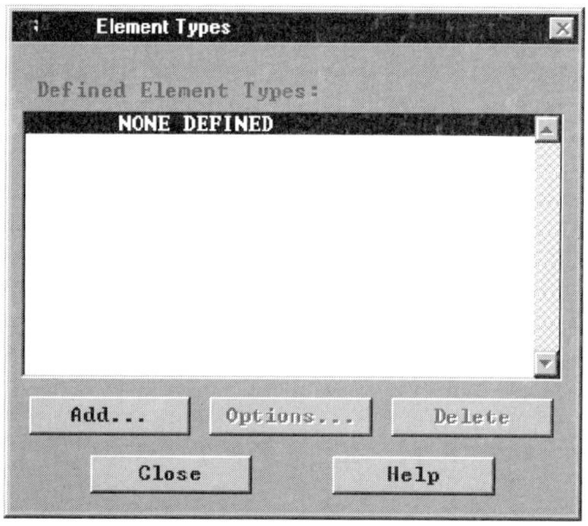

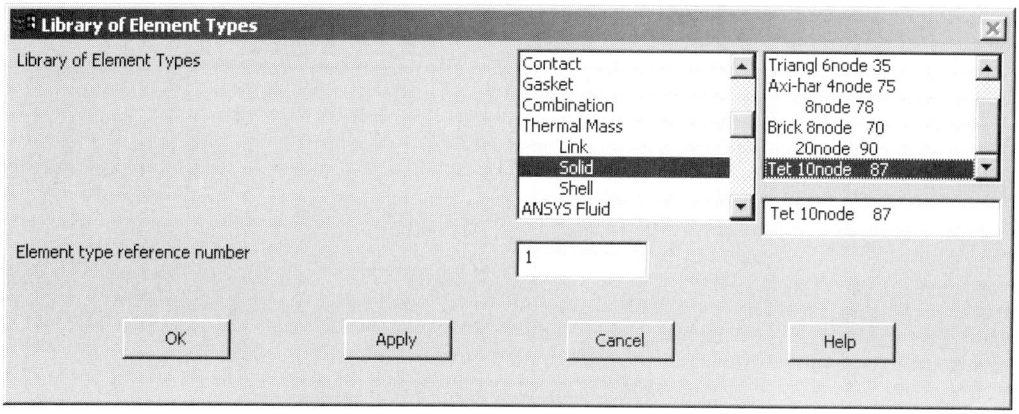

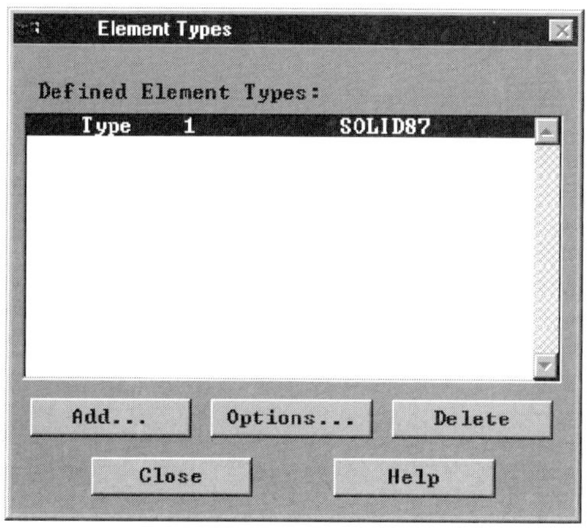

Assign thermal conductivity values for concrete and plastic with the following commands:

main menu: **Preprocessor** → **Material Props** → **Material Models** → **Thermal** → **Conductivity** → **Isotropic**

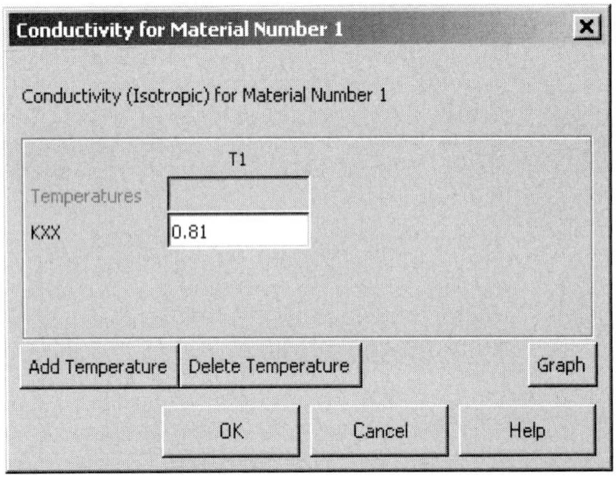

From the Define Material Model behavior window:

Material → **New Model** . . .

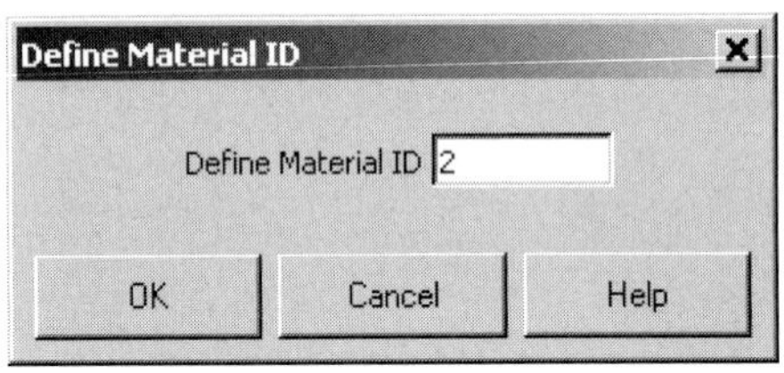

Double click on **Isotropic** and assign thermal Conductivity of Plastic

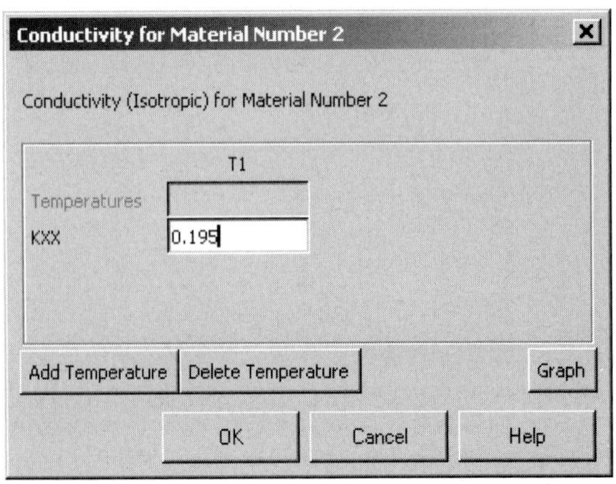

ANSYS Toolbar: **SAVE_DB**

Set up the graphics area (i.e., workplane, zoom, etc.) with the following commands:

utility menu: **Workplane → WP Settings** . . .

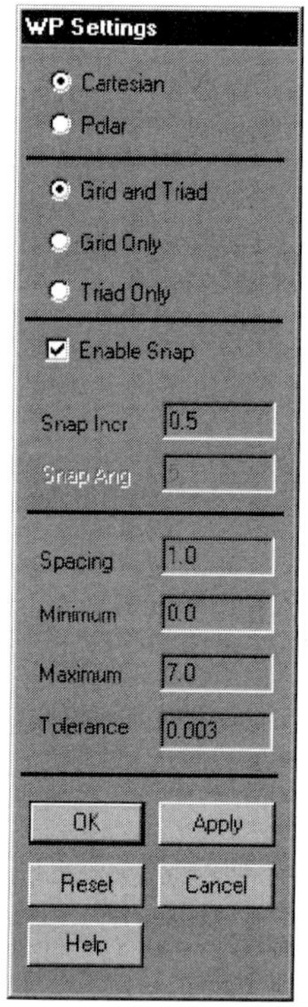

Toggle on the workplane by using the following command:

utility menu: **Workplane → Display Working Plane**

Bring the workplane to view by using the following command:

utility menu: **PlotCtrls → Pan, Zoom, Rotate** . . .

Click on the small circle until you bring the workplane to view. Then, press the **Iso** (Isometric view) button. Next, create the geometry with the following commands:

main menu: **Preprocessor → Modeling → Create → Volumes → Block → By 2 Corners &Z**

[WP = 0, 0]

[Expand the rubber up 7 and right 2.5]

[Expand the rubber band in the negative Z-direction to −1.5]

Create a volume, to be removed later, for the plastic volume:

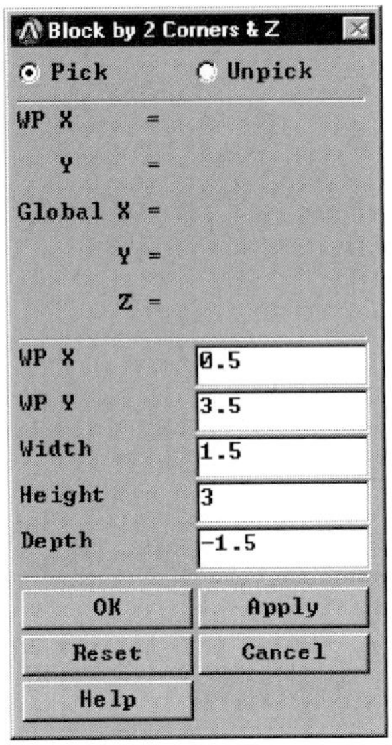

OK

main menu: **Preprocessor → Modeling → Operate → Booleans → Subtract → Volumes**

Pick Volume1 and **Apply**; then pick Volume2 and **Apply.**

OK

utility menu: **Plot → Volumes**

Create the plastic volume with the following command:

utility menu: **WorkPlane → Offset WP by Increments** . . .

In the X, Y, Z Offsets box, type in **[0, 0, −0.5]**.

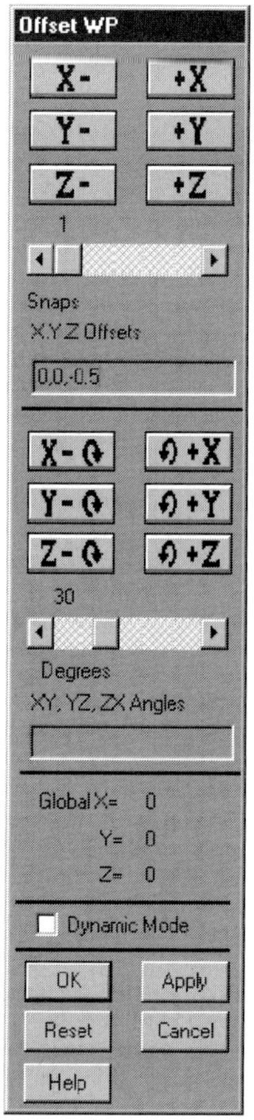

OK

Now, issue the following commands:

main menu: **Preprocessor** → **Modeling** → **Create** → **Volumes** → **Block** → **By 2 Corners** & **Z**

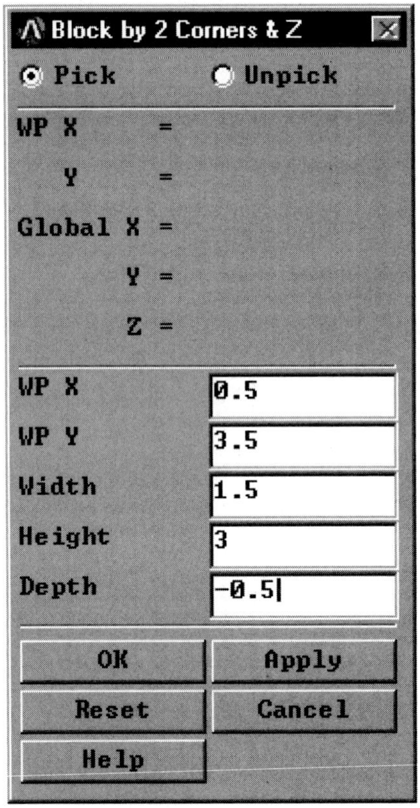

OK

main menu: **Preprocessor → Modeling → Operate → Booleans → Glue → Volumes**

Pick All

We now want to mesh the volumes to create elements and nodes, but first, we need to specify the element sizes. So, issue the following commands:

main menu: **Preprocessor → Meshing → Size Cntrls → Manual Size → Global → Size**

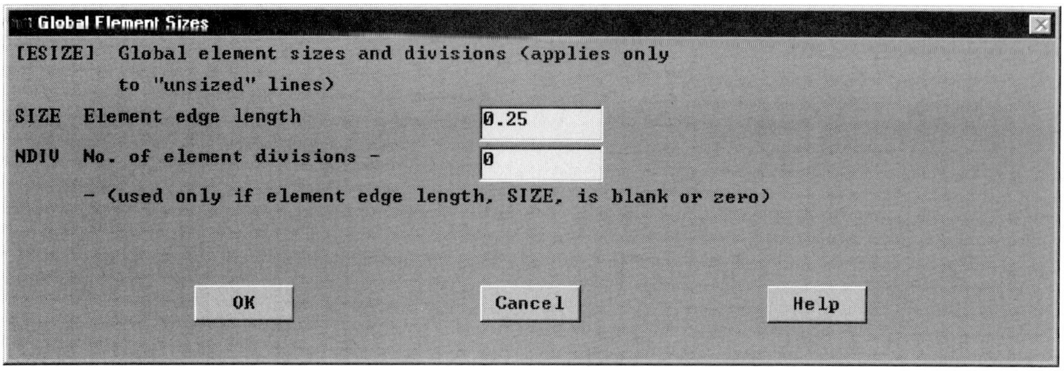

We also need to specify material attributes for the concrete and the plastic volumes before we proceed with meshing. To do so, we issue the following commands:

main menu: **Preprocessor** → **Meshing** → **Mesh Attributes** → **Picked Volumes**

[Pick the concrete part of the wall volume]

[Apply anywhere in the ANSYS graphics window]

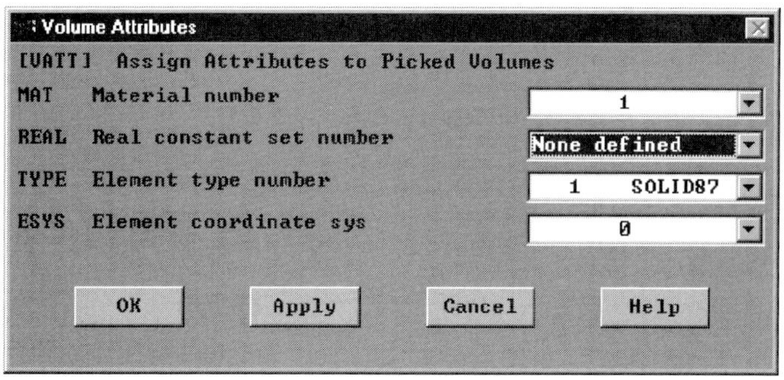

main menu: **Preprocessor** → **Meshing** → **Mesh Attributes** → **PickedVolumes**

[Pick the plastic volume]

[Apply anywhere in the ANSYS graphics window]

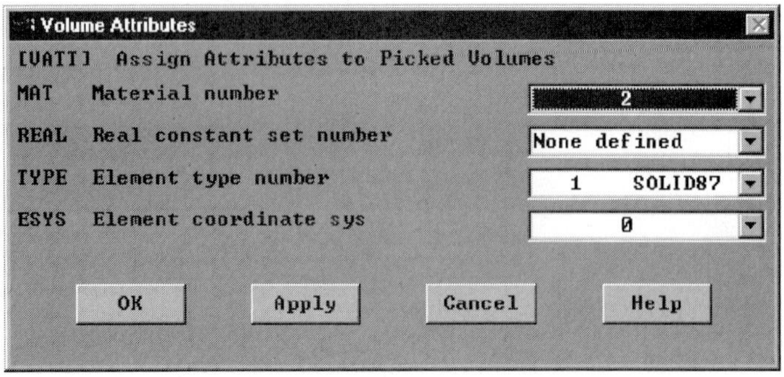

ANSYS Toolbar: **SAVE_DB**

We can proceed to mesh by issuing the following commands:

main menu: **Preprocessor** → **Meshing** → **Mesh** → **Volumes** → **Free**

Pick All

If you exceed the maximum number of elements allowed in the educational version of ANSYS try the following:

main menu: **Preprocessor** → **Meshing** → **Size Cntrls** → **Smart Size** → **Basic**

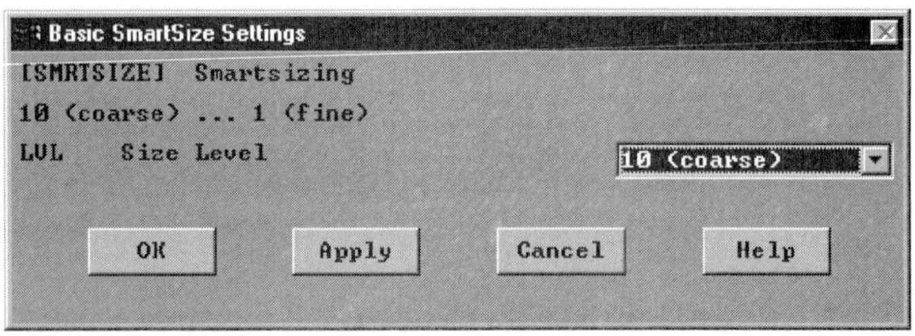

Now try to mesh the volumes again. To apply the boundary conditions, we first select the interior surfaces of the wall, including the clear plastic:

utility menu: **Select** → **Entities** . . .

In the **Min, Max** field, type **[0, −0.5]**, see Select Entities window. Choose all appropriate fields as shown.

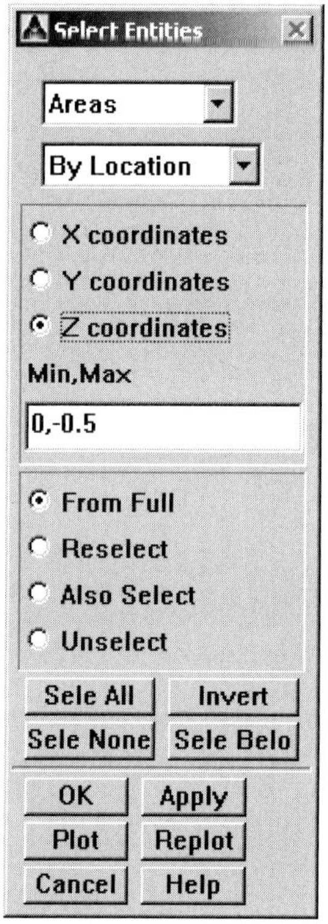

OK

utility menu: **Plot → Areas**

main menu: **Solution → Define Loads → Apply → Thermal → Convection → On Areas**

Pick All

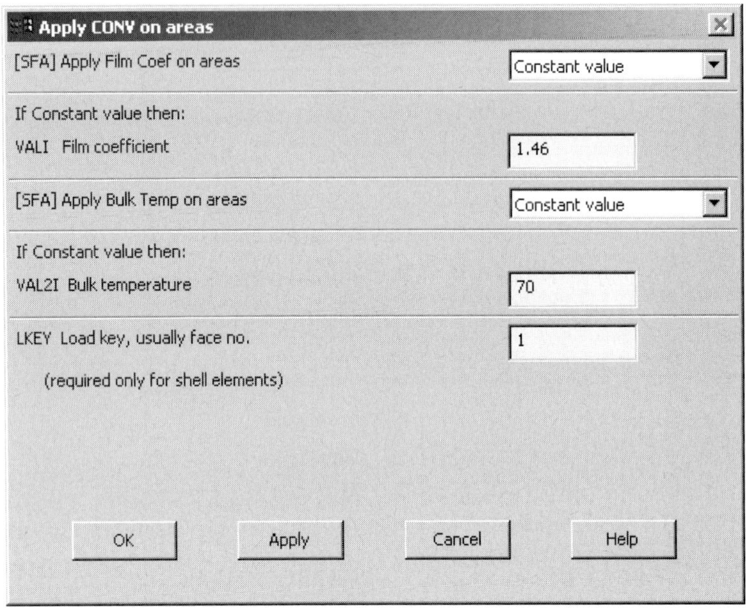

utility menu: **Select → Everything** . . .

utility menu: **Select → Entities** . . .

In the **Min, Max** field, type: **[− 1.0, − 1.5]**

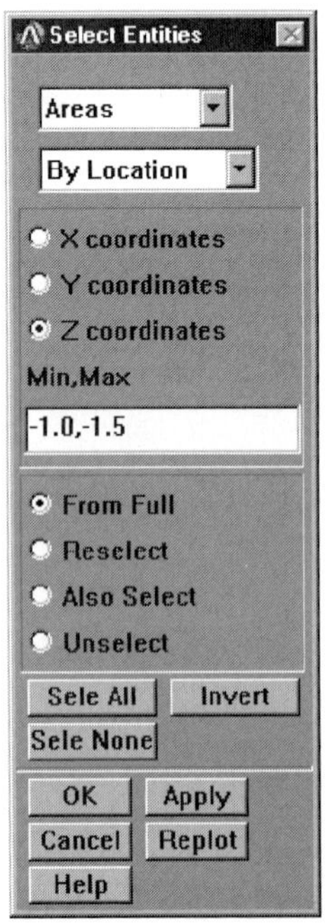

OK

utility menu: **Plot → Areas**

main menu: **Solution → Define Loads → Apply → Thermal → Convection → On Areas**

Pick All to specify the convection coefficient and temperature:

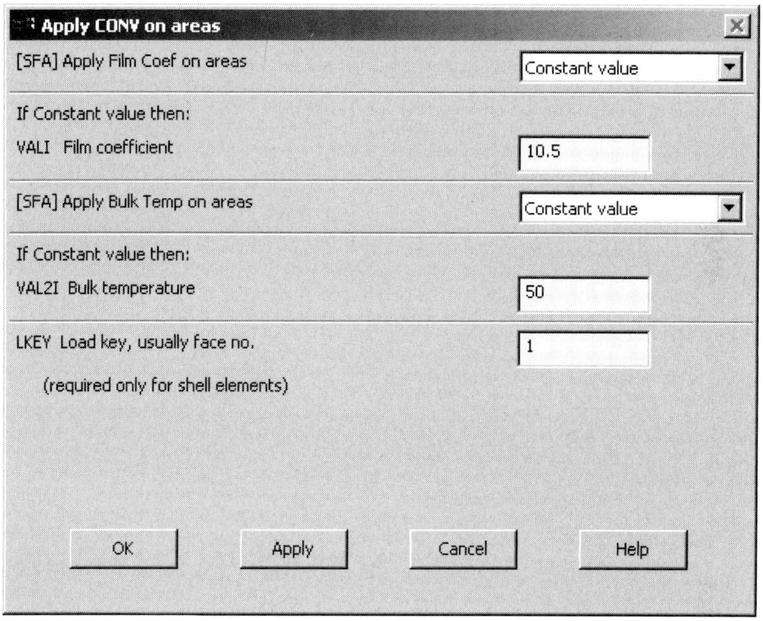

To see the applied boundary conditions, use the following commands:

utility menu: **PlotCntrls → Symbols** …

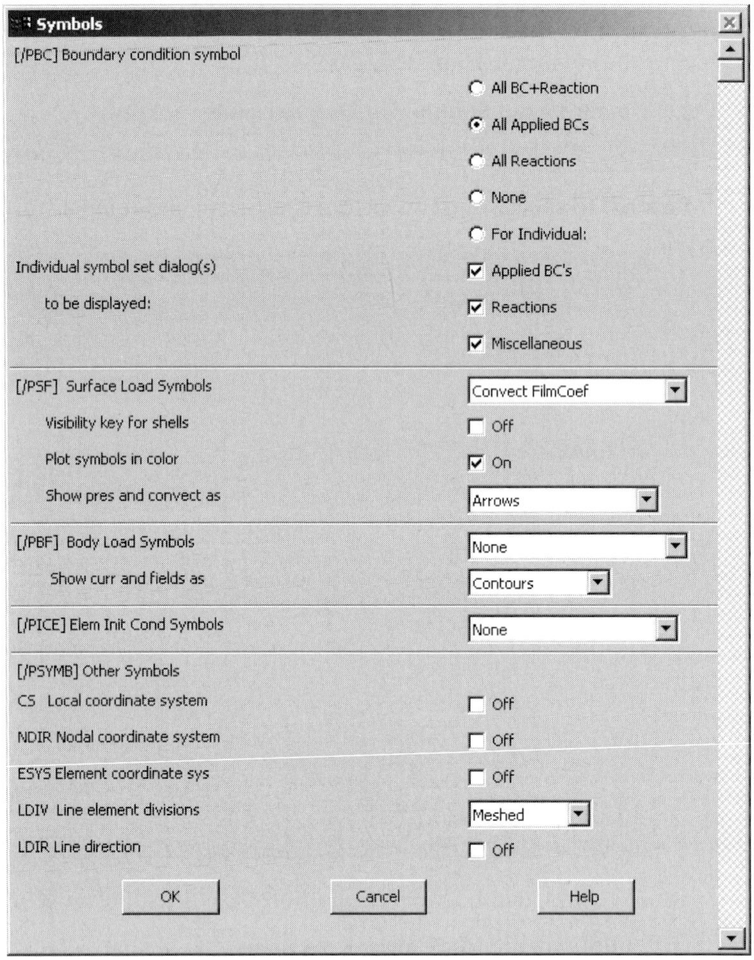

utility menu: **Select → Everything** . . .

utility menu: **Plot → Areas**

ANSYS Toolbar: **SAVE_DB**

Solve the problem:

main menu: **Solution → Solve → Current LS**

OK

Close (the solution is done!) window.

Close (the /STAT Command) window.

For the postprocessing phase, obtain information such as nodal temperatures and heat fluxes with the following commands (see Figure 13.26 and Figure 13.27):

main menu: **General Postproc → Plot Results → Contour Plot → Nodal Solu**

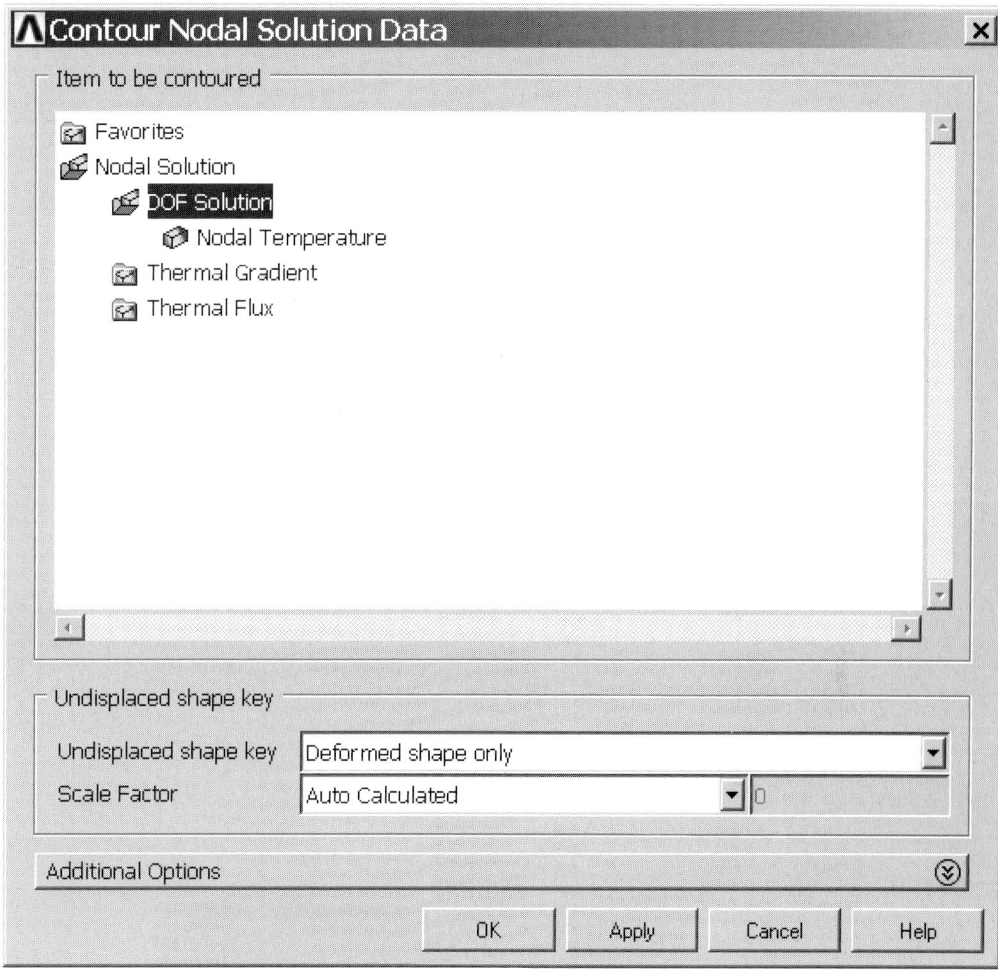

main menu: **General Postproc → Plot Results → Vector Plot → Predefined**

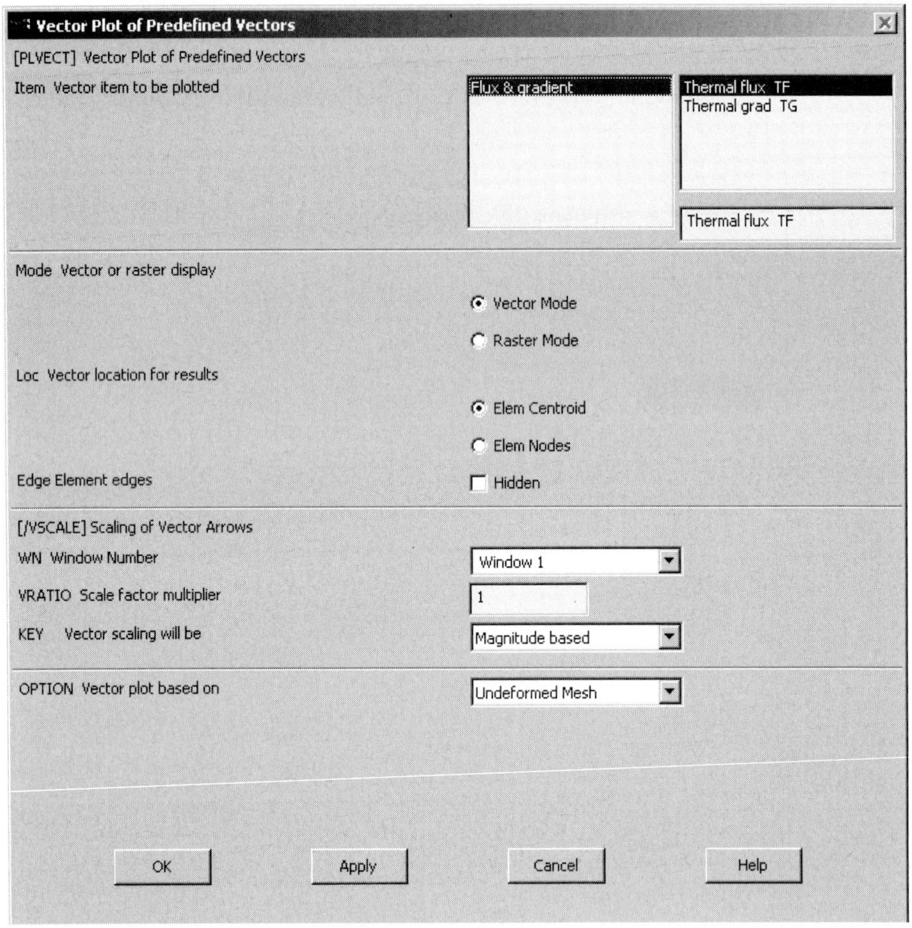

Exit and save your results:

ANSYS Toolbar: **QUIT**

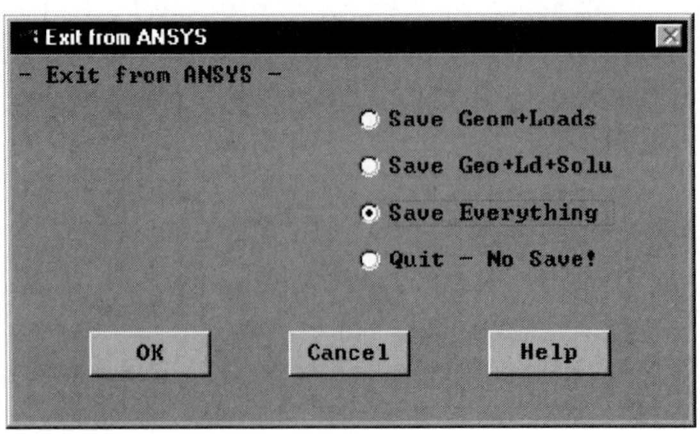

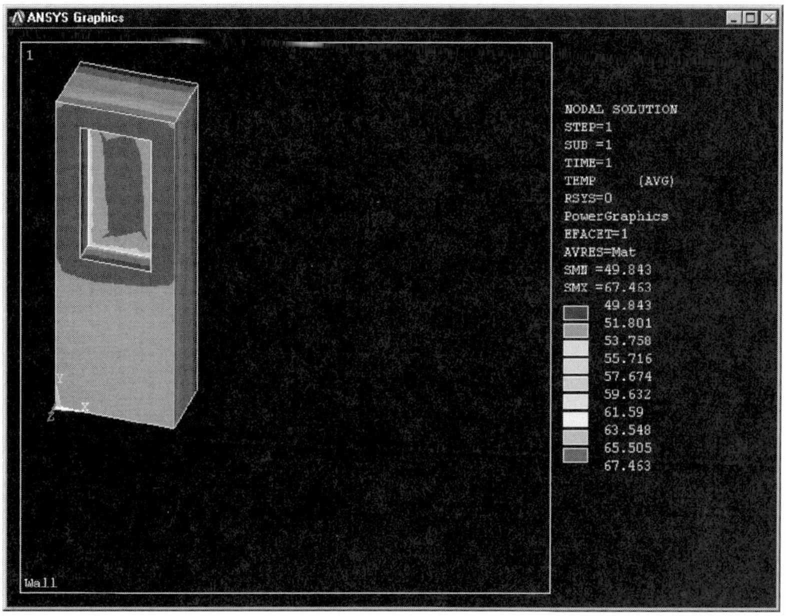

FIGURE 13.26 Temperature contour plot.

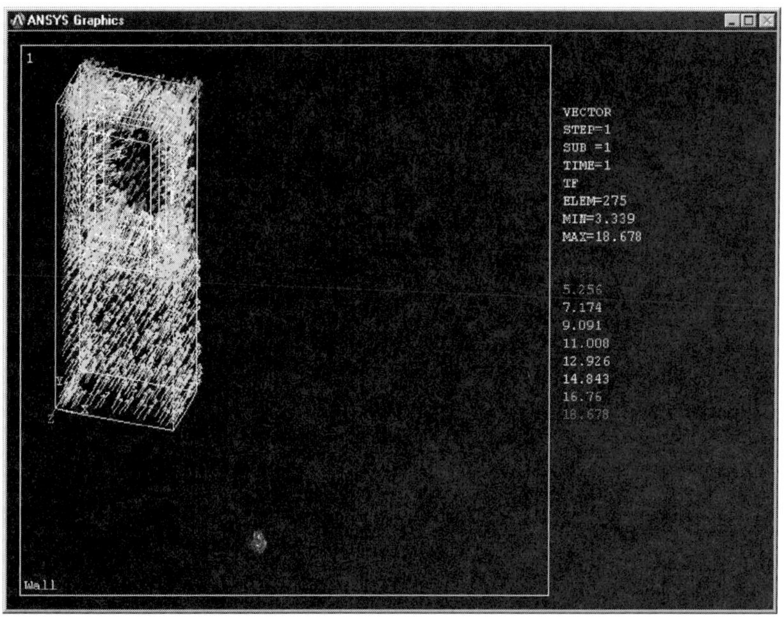

FIGURE 13.27 The heat flow vectors.

13.9 A STRUCTURAL EXAMPLE USING ANSYS

EXAMPLE 13.3

The bracket shown in Figure 13.28 is subjected to a distributed load of 50 lb/in^2 on the top surface. It is fixed around the hole surfaces. The bracket is made of steel, with a modulus elasticity of 29×10^6lb/in^2 and $\nu = 0.3$. Plot the deformed shape. Also, plot the von Mises stress distribution in the bracket.

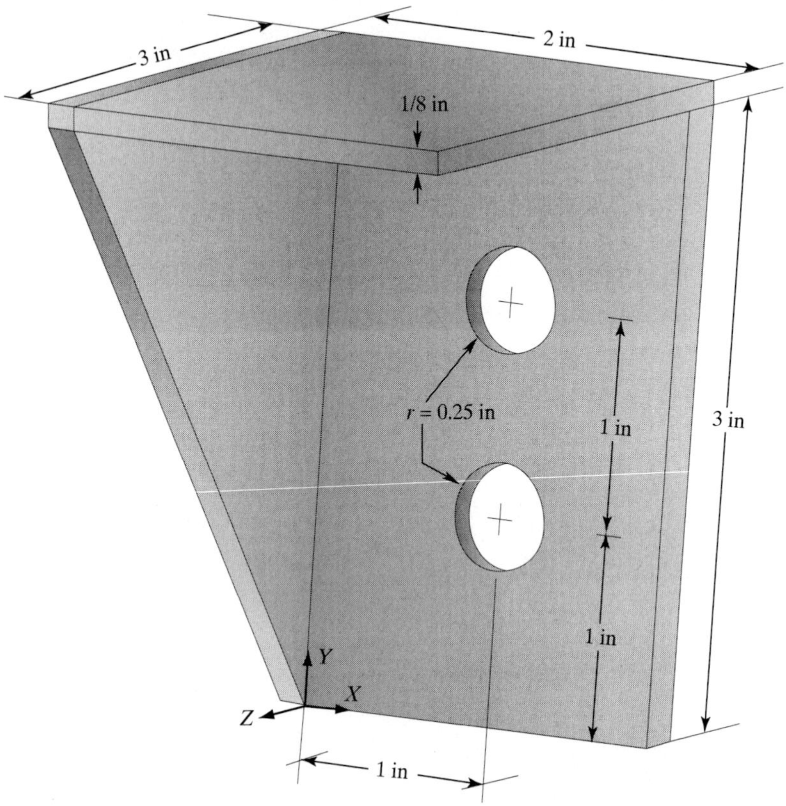

FIGURE 13.28 Dimensions of the bracket in Example 13.3.

The following steps demonstrate how to create the geometry of the problem: choose the appropriate element type, apply boundary conditions, and obtain nodal results:

Enter the **ANSYS** program by using the Launcher. Type **Brack3D** (or a file name of your choice) in the **Jobname** entry field of the dialog box. Pick **Run** to start the GUI.

Create a title for the problem.

utility menu: **File** → **Change Title** . . .

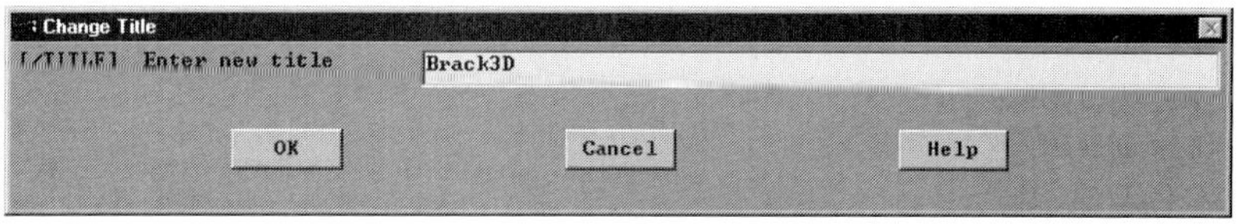

main menu: **Preprocessor → Element Type → Add/Edit/Delete**

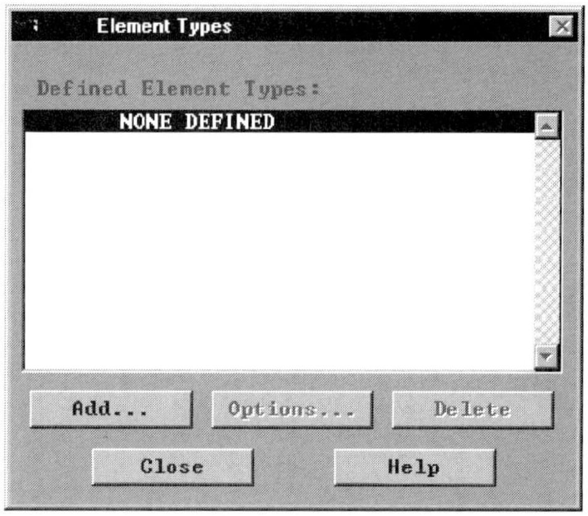

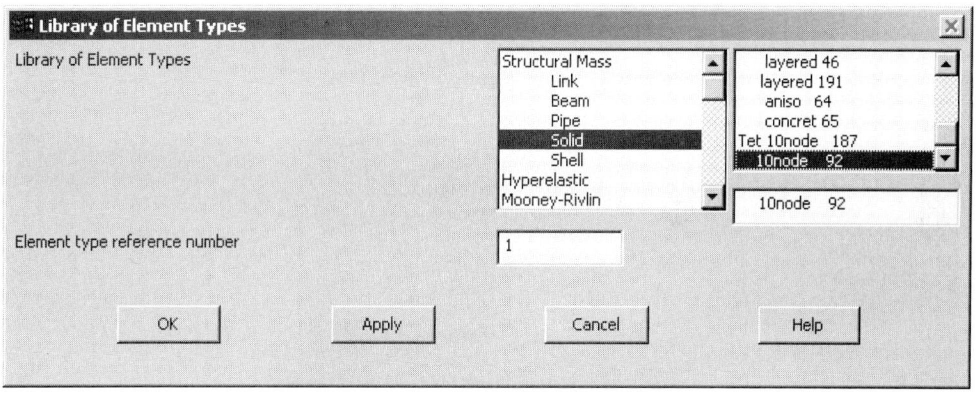

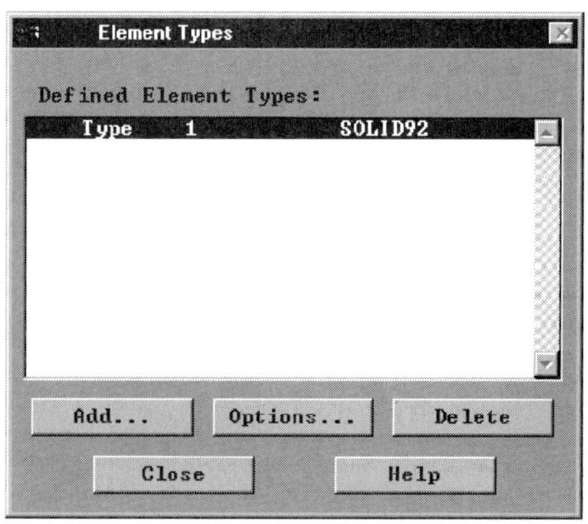

Assign the modulus of elasticity and Poisson's ratio with the following commands:

main menu: **Preprocessor → Material Props → Material Models → Structural → Linear → Elastic → Isotropic**

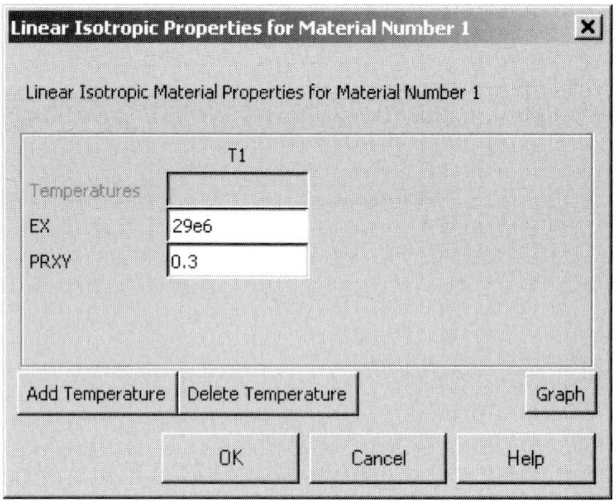

ANSYS Toolbar: **SAVE_DB**

Set up the graphics area (i.e. workplane, zoom, etc.) with the following commands:

utility menu: **Workplane → WP Settings** . . .

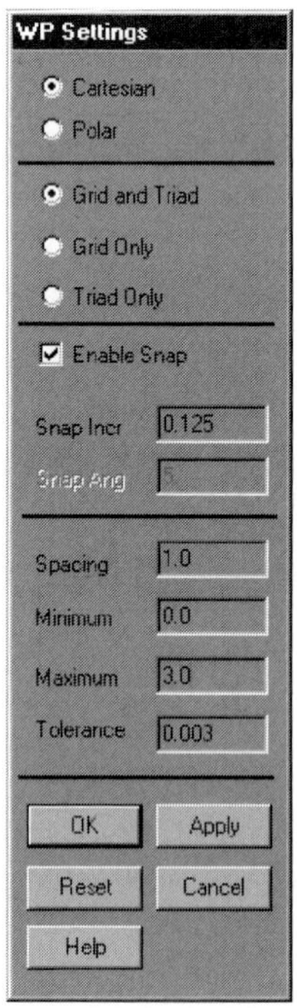

Toggle on the workplane by using the following command:

utility menu: **Workplane → Display Working Plane**

Bring the workplane to view by using the following command:

utility menu: **PlotCtrls → Pan, Zoom, Rotate** . . .

Click on the small circle until you bring the workplane to view. Then, press the **Iso** (Isometric view) button. Next, create the vertical plate by issuing the following commands:

main menu: **Preprocessor → Modeling → Create → Volumes → Block → By 2 Corners & Z**

[WP = 0,0]

[Expand the rubber band up 3.0 and to the right 2.0]

[Expand the rubber band in the negative Z-direction by 0.125]

OK

To create the holes, first we must create two cylinders, with the following commands:

main menu: **Preprocessor → Modeling → Create → Volumes → Cylinder → Solid Cylinder**

On the workplane, pick the following locations or type the values in **WP X, WP Y, Radius**, and **Depth** fields:

[WP = 1,1]

[Expand the circle to rad = 0.25]

[Expand the cylinder to a length of 0.125 in the negative Z-direction]

[WP = 1,2]

[Expand the circle to rad = 0.25]

[Expand the cylinder to a length of 0.125 in the negative Z-direction]

OK

Now, create the holes by subtracting from the vertical plate the volume of cylinders, with the following commands:

main menu: **Preprocessor → Modeling → Operate → Booleans → Subtract → Volumes**

Pick Volume-1 (the vertical plate) and **Apply**; then, pick Volume-2 and Volume-3 (the cylinders) and **Apply.**

OK

utility menu: **Plot → Volumes**

ANSYS Toolbar: **SAVE_DB**

Move and rotate the workplane and create the top plate with the following command:

utility menu: **Workplane → Offset WP by Increments** . . .

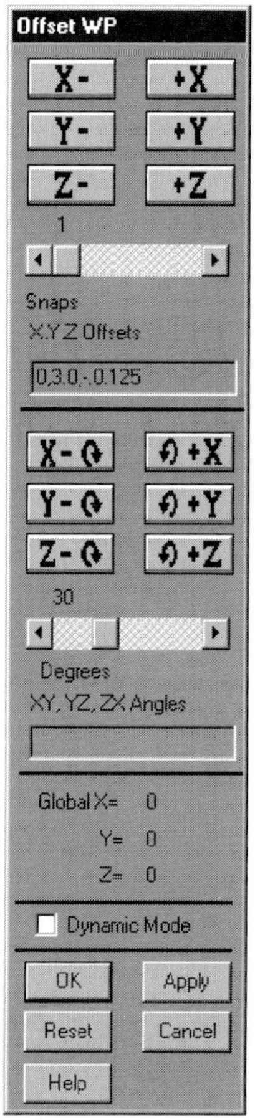

In the X, Y, Z Offsets box, type in **[0, 3.0, −0.125]**, and then **Apply.** To rotate the WP, move the Degrees Slider bar to **90** and then press the **+X rotation** button.

OK

utility menu: **PlotCtrls → Pan, Zoom, Rotate** . . .

Press the **Bot** (bottom view) button and issue the following commands:

main menu: **Preprocessor → Modeling → Create → Volumes → Block → By 2 Corners & Z**

[WP = 0,0]

[In the active workplane, expand the rubber band to 3.0 and 2.0]

[Expand the rubber band in the negative Z-direction by 0.125]

OK

utility menu: **WorkPlane → Align WP With → Global Cartesian**

utility menu: **Plot → Volumes**

utility menu: **WorkPlane → Offset WP by Increments . . .**

In the X, Y, Z Offsets box, type in **[0, 0, −0.125]**, then **Apply.** Rotate the workplane about the Y-axis. Move the Degrees Slider bar to **90** and then press the **−Y rotation** button.

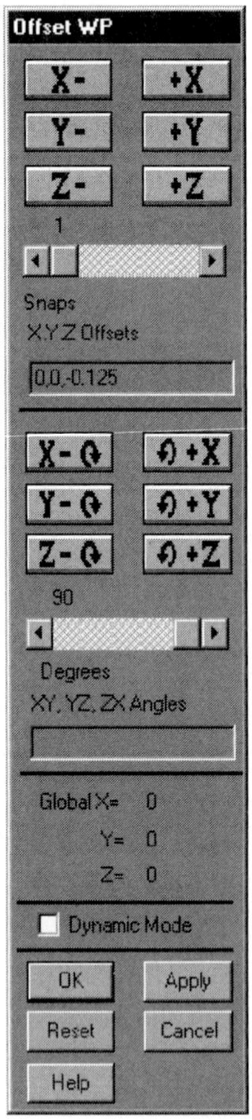

OK

utility menu: **PlotCtrls → Pan, Zoom, Rotate . . .**

Change the view to **Left** and issue the following commands:

main menu: **Preprocessor → Modeling → Create → Volumes → Prism → By Vertices**

[WP = 0,0]

[WP = 0, 3.125]

[WP = 3, 3.125]

[WP = 3.0, 3.0]

[WP = 0.125, 0]

[WP = 0,0]

Change the view to the isometric view by pressing the **Iso** button:

[Stretch the rubber band 0.125 in the Z-direction]

OK

utility menu: **Plot → Volumes**

utility menu: **PlotCtrls → Pan, Zoom, Rotate** . . .

Toggle on the dynamic mode; hold down the right button on the mouse and rotate the object as desired. Then, issue the following commands:

main menu: **Preprocessor → Modeling → Operate → Booleans → Add → Volumes**

Pick All

We now want to mesh the volumes to create elements and nodes, but first, we need to specify the element sizes. So, issue the following commands:

main menu: **Preprocessor → Meshing → Size Cntrls → Smart Size → Basic**

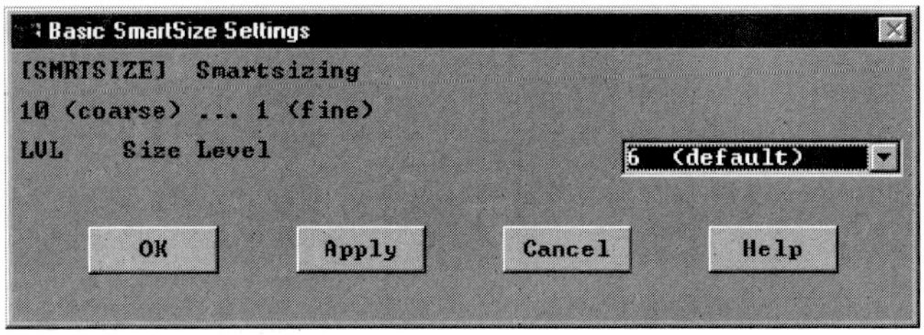

ANSYS Toolbar: **SAVE_DB**

main menu: **Preprocessor → Meshing → Mesh → Volumes → Free**

Pick All

Close

ANSYS Toolbar: **SAVE_DB**

Now, we need to apply boundary conditions. First, we will fix the periphery of the holes by using the following command:

utility menu: **PlotCtrls → Pan, Zoom, Rotate** . . .

Choose the **Front** view and issue the following commands:

main menu: **Solution → Define Loads → Apply → Structural → Displacement → On Keypoints**

Change the picking mode to "circle" by toggling on the **•Circle** feature. Now, starting at the center of the holes, stretch the rubber band until you are just outside the holes and Apply:

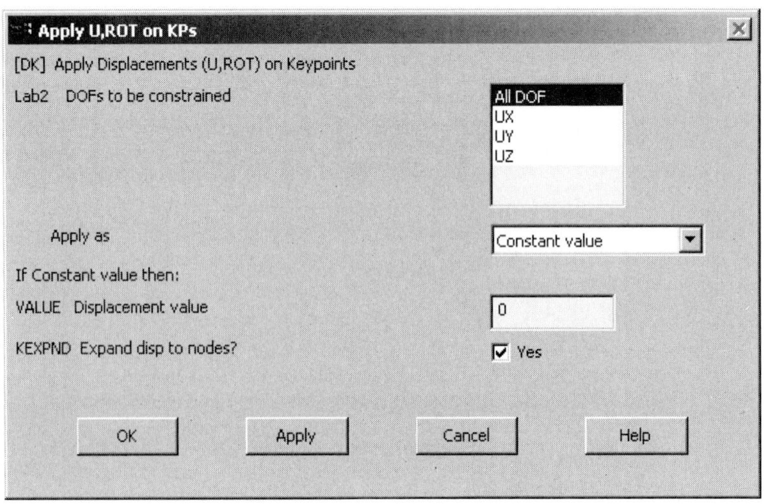

Choose the isometric view and issue the following commands:

utility menu: **Select → Entities . . .**

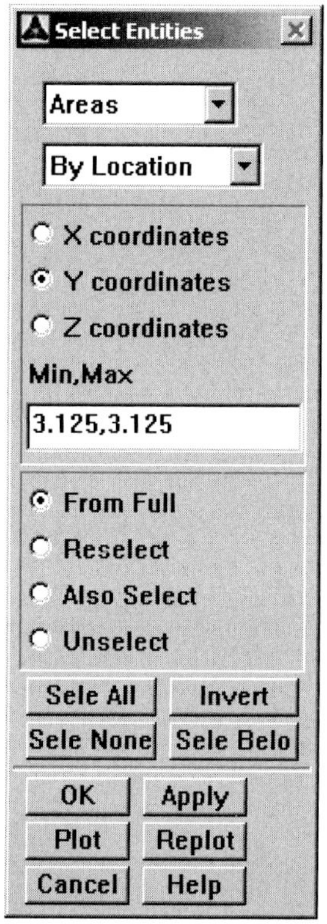

In the **Min, Max** field, type **[3.125, 3.125]**.

OK

utility menu: **Plot → Areas**

main menu: **Solution → Define Loads → Apply → Structural → Pressure → On Areas**

Pick All to specify the distributed load (pressure) value:

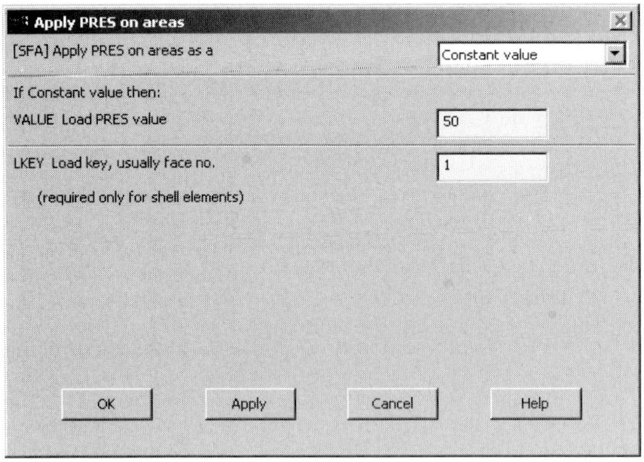

To see the applied boundary conditions, use the following commands:

utility menu: **PlotCtrls → Symbols . . .**

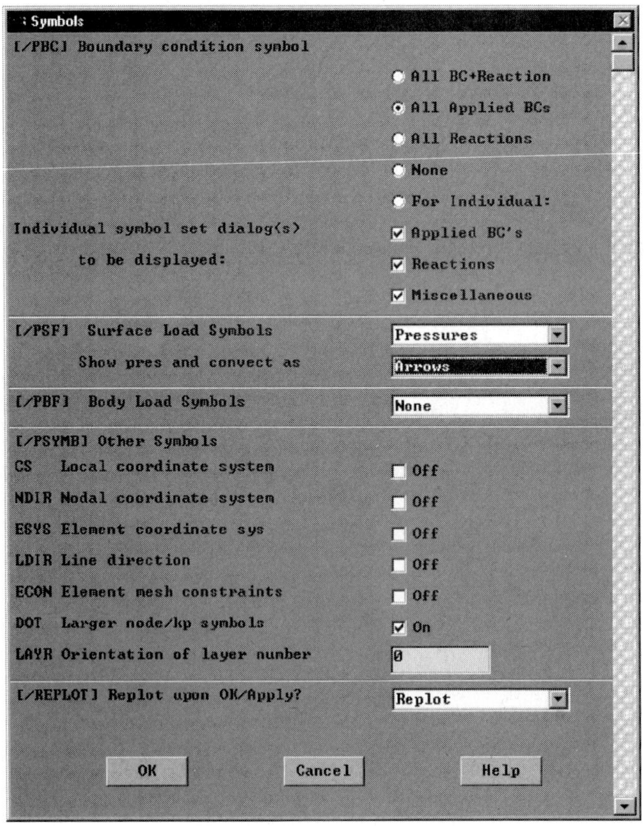

utility menu: **Select → Everything . . .**

utility menu: **Plot → Areas**

ANSYS Toolbar: **SAVE_DB**

Solve the problem:

main menu: **Solution → Solve → Current LS**

OK

Close (the solution is done!) window.

Close (the /STAT Command) window.

In the postprocessing phase, first plot the deformed shape by using the following commands (see Figure 13.29):

main menu: **General Postproc → Plot Results → Deformed Shape**

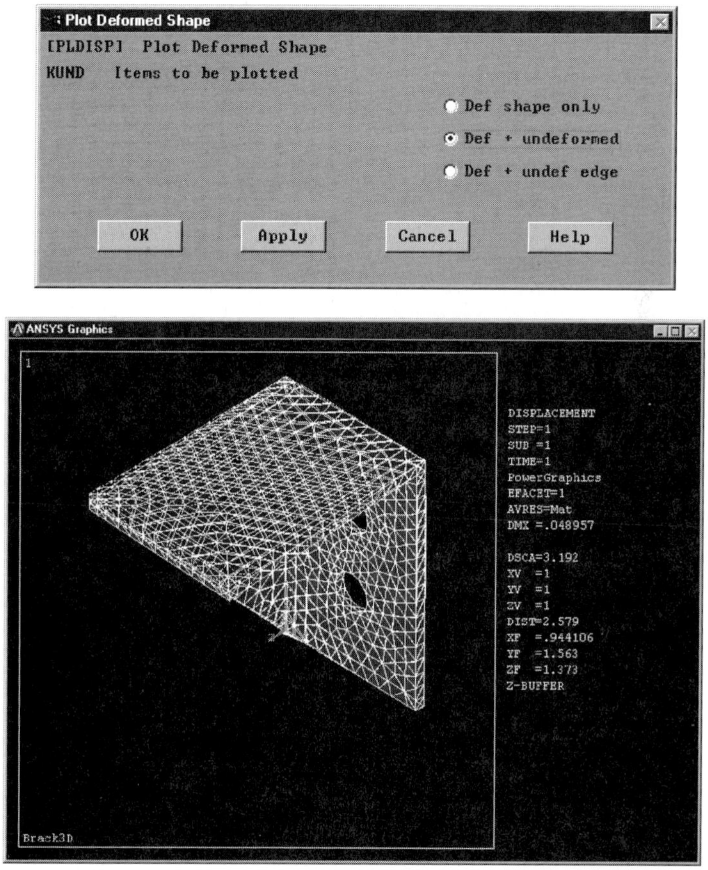

FIGURE 13.29 The deformed shape of the bracket.

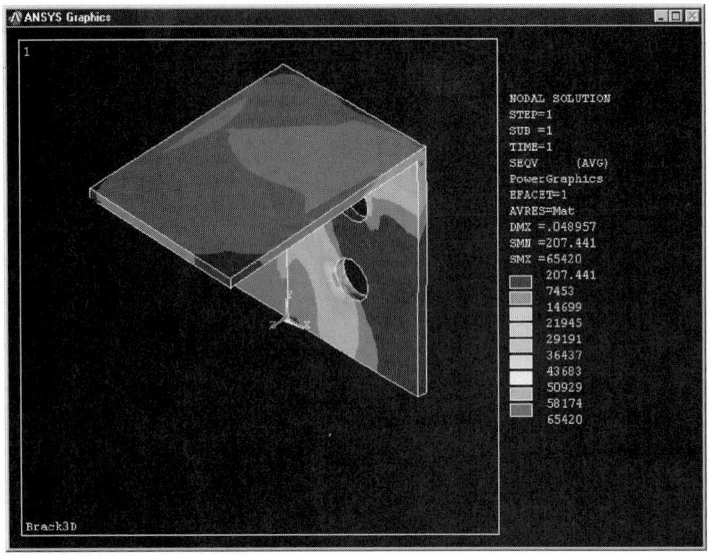

FIGURE 13.30 The von Mises stress distribution within the bracket.

Plot the von Mises stresses by using the following commands (see Figure 13.30):

main menu: **General Postproc → Plot Results → ContourPlot → Nodal Solu**

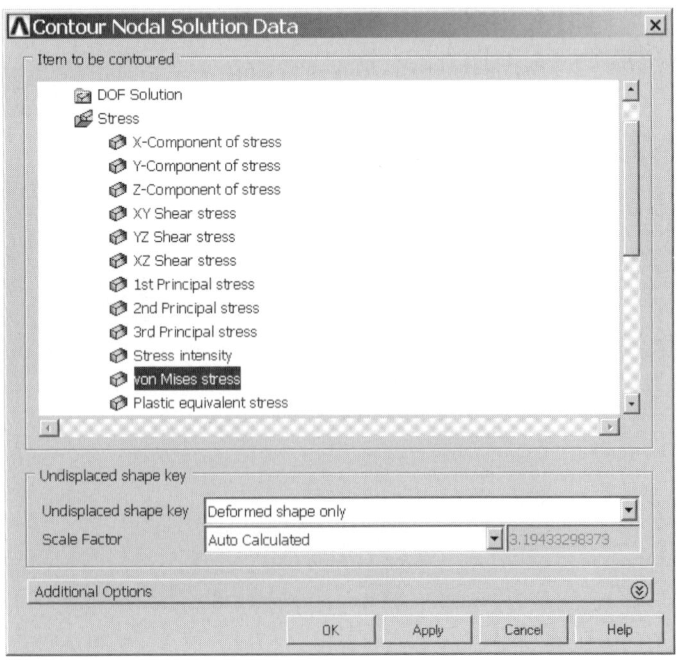

Exit and save your results:

ANSYS Toolbar: **QUIT**

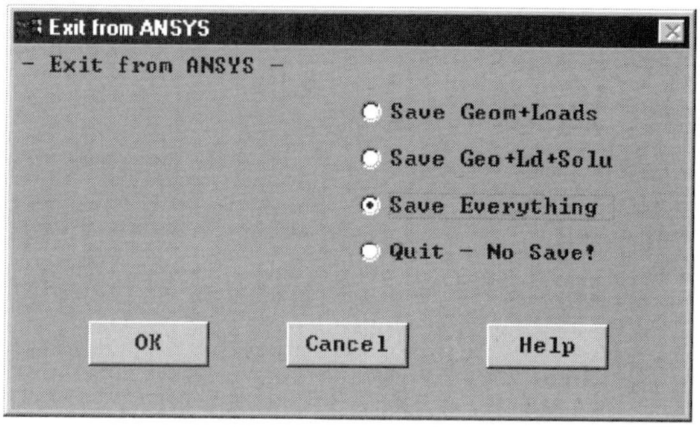

SUMMARY

At this point you should

1. know how the shape functions for a tetrahedral element are derived.
2. know how the stiffness matrix and load matrix for a tetrahedral element are derived.
3. be familiar with the eight-node brick element and its higher order counterpart, the twenty-node brick element.
4. be familiar with some of the structural-solid and thermal elements available through ANSYS.
5. understand the difference between the top-down and bottom-up solid-modeling methods.
6. be able to find ways to verify your FEA results.

REFERENCES

ANSYS User's Manual: Procedures, Vol. I, Swanson Analysis Systems, Inc.

ANSYS User's Manual: Commands, Vol. II, Swanson Analysis Systems, Inc.

ANSYS User's Manual: Elements, Vol. III, Swanson Analysis Systems, Inc.

Chandrupatla, T., and Belegundu, A., *Introduction to Finite Elements in Engineering*, Englewood Cliffs, NJ, Prentice Hall, 1991.

Zienkiewicz, O. C., *The Finite Element Method*, 3d. ed., New York, McGraw-Hill, 1977.

PROBLEMS

1. For a tetrahedral element, derive an expression for the stress components in terms of the nodal displacement solutions. How are the three principal stresses computed from the calculated stress component values?

2. Use ANSYS to create the solid model of the object shown in the accompanying figure. Use the dynamic-mode option to view the object from various directions. Plot the solid object in its isometric view.

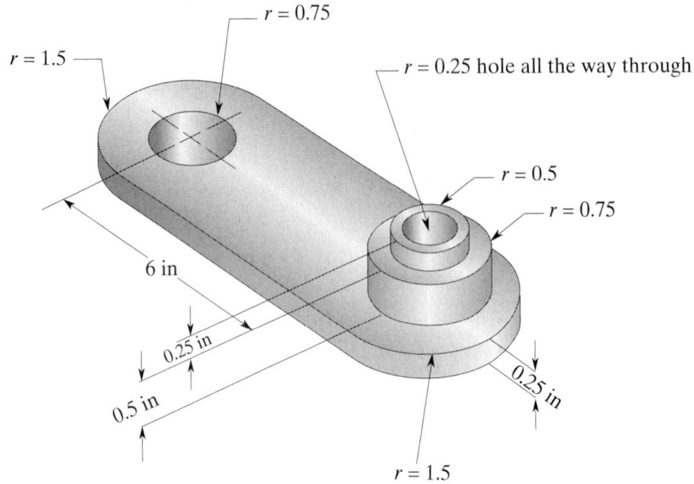

3. Use ANSYS to create a solid model of a foot-long section of a pipe with the internal longitudinal fins shown in the accompanying figure. Use the dynamic-mode option to view the object from various directions. Plot the object in its isometric view.

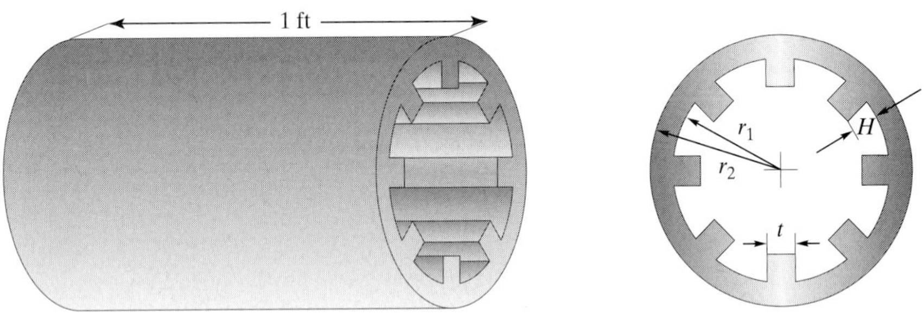

$r_1 = 2$ in

$r_2 = 2\frac{1}{4}$ in

$t = \frac{3}{4}$ in

$H = \frac{3}{4}$ in

4. Use ANSYS to create the solid model of the wall-mount piping support bracket shown in the accompanying figure. Use the dynamic-mode option to view the object from various directions. Plot the solid object in its isometric view.

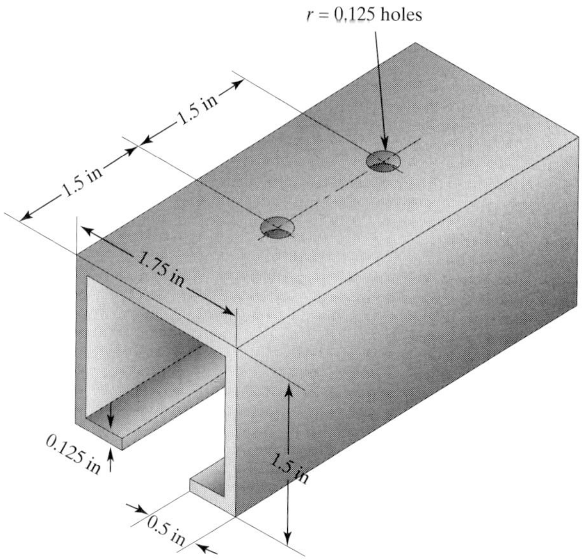

5. Use ANSYS to create the solid model of the heat exchanger shown in the accompanying figure. Use the dynamic-mode option to view the object from various directions. Plot the model of the heat exchanger in its isometric view.

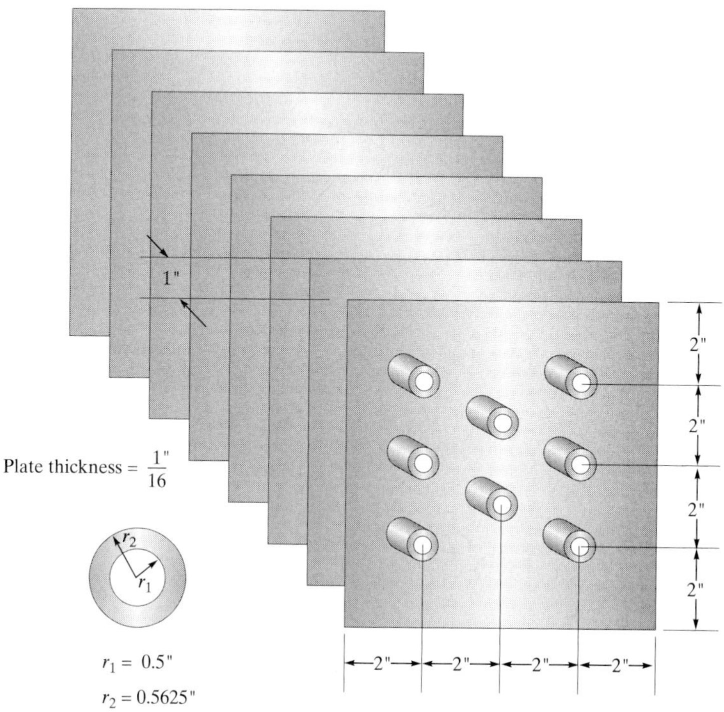

6. Use ANSYS to create the solid model of the wheel shown in the accompanying figure. Use the dynamic-mode option to view the object from various directions. Plot the object in its isometric view.

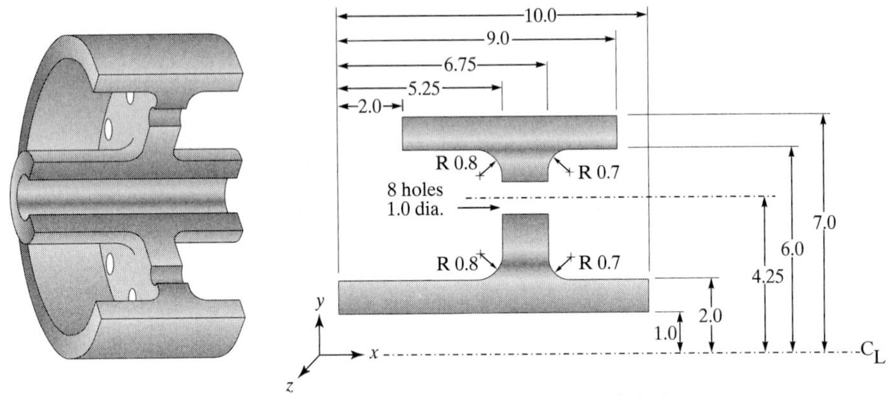

Dimensions are in inches.

7. Use ANSYS to create a solid model of a 100-mm-long section of a pipe with the internal longitudinal fins shown in the accompanying figure. Use the dynamic-mode option to view the object from various directions. Plot the object in its isometric view.

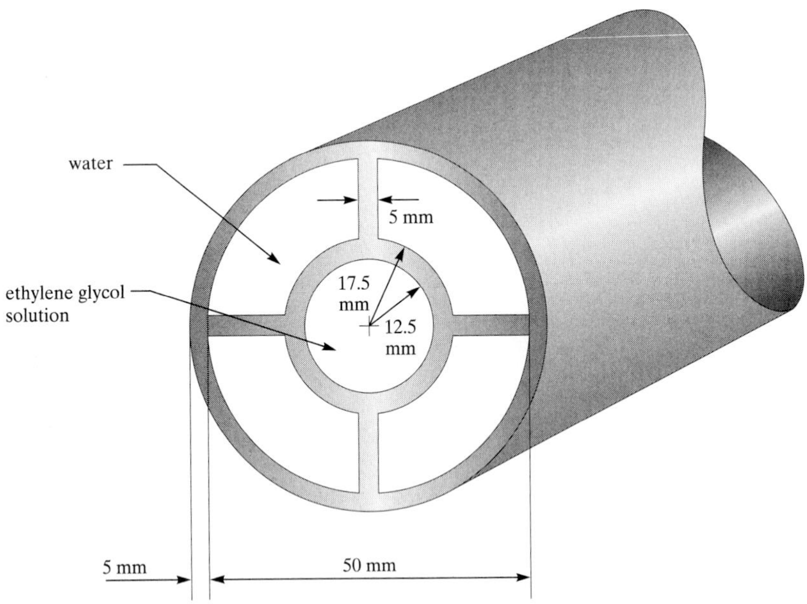

8. Using ANSYS, calculate and plot the principal stress distributions in the support component shown in the accompanying figure. The bracket is made of steel. It is fixed around the hole surfaces.

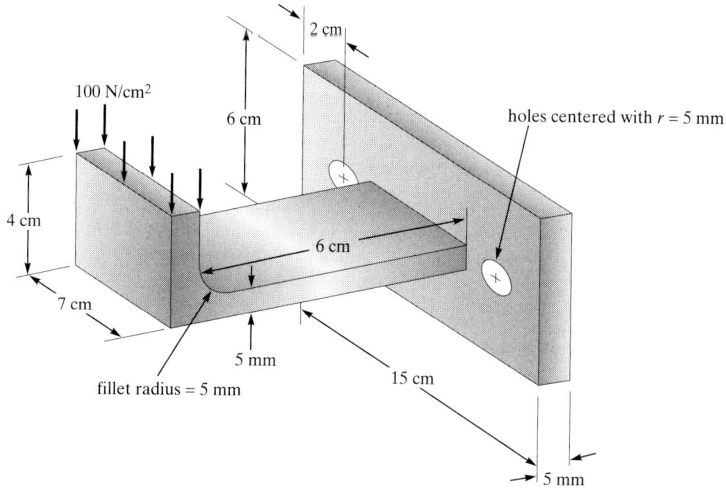

9. Using ANSYS, calculate and plot the von Mises stress distribution in the traffic signpost shown in the accompanying figure. The post is made of steel, and the sign is subjected to a wind gust of 60 miles/hr. Use the drag force relation $F_D = C_D A \frac{1}{2} \rho U^2$ to calculate the load caused by the wind, where F_D is the load, $C_D = 1.18$, ρ represents the density of air, U is the wind speed, and A gives the frontal area of the sign. Distribute the load on the section of the post covered by the sign. Could you model this problem as a simple cantilever beam and thus avoid creating an elaborate finite element model? Explain.

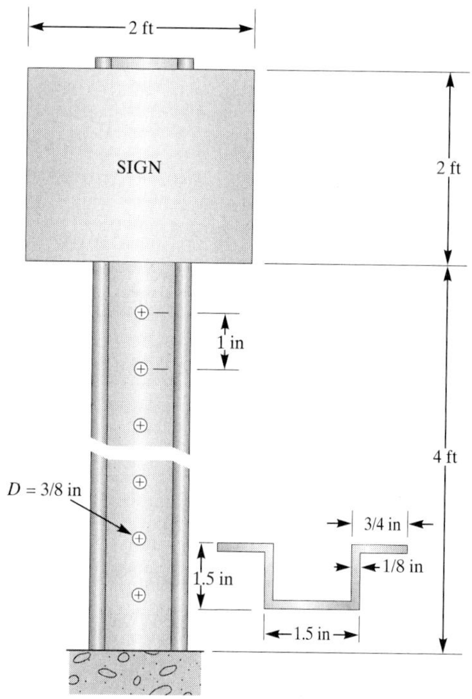

10. Determine the temperature distribution inside the aluminum heat sink in Example 13.1 if the surrounding air is at 25°C, with a corresponding heat transfer coefficient $h = 20\ \text{W/m}^2 \cdot \text{K}$. The heat sink sits on a chip that dissipates approximately $2000\ \text{W/m}^2$. Extrude the frontal area shown in the accompanying figure 20.5 mm to create a quarter model of the heat sink.

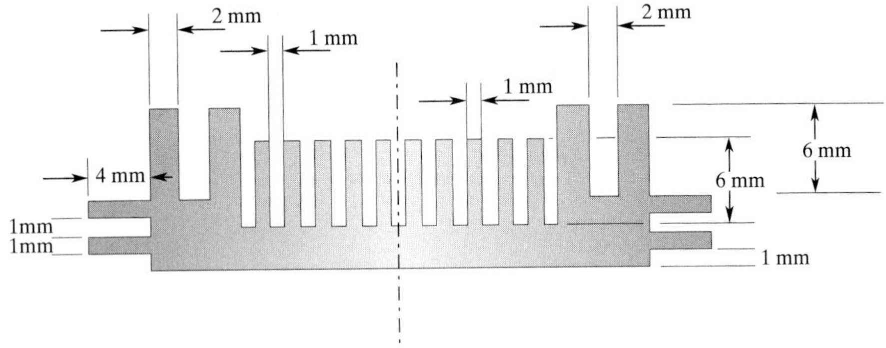

The front view of the heat sink in problem 10.

11. Imagine that by mistake, an empty coffee pot has been left on a heating plate. Assuming that the heater puts approximately 20 Watts into the bottom of the pot, determine the temperature distribution within the glass if the surrounding air is at 25°C, with a corresponding heat transfer coefficient $h = 15\ \text{W/m}^2 \cdot \text{K}$. The pot is cylindrical in shape, with a diameter of 14 cm and height of 14 cm, and the glass is 3 mm thick. Could you solve this problem using a one-dimensional conduction and thus avoid creating an elaborate three-dimensional model?

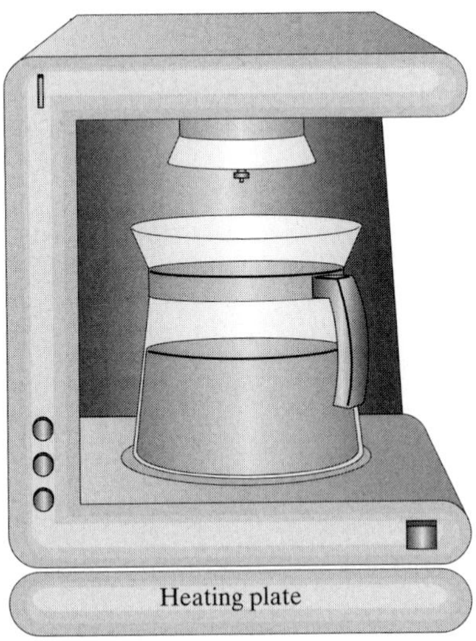

12. Using ANSYS, generate a three-dimensional model of a socket wrench. Take measurements from an actual socket. Use solid-cylinder, hexagonal-prism, and block primitives to construct the model. Make reasonable assumptions regarding loading and boundary conditions, and perform stress analysis. Plot the von Mises stresses. Discuss the type and magnitude of loading that could cause failure.

13. During the winter months, the inside air temperature of a room is to be kept at 70°F. However, because of the location of a heat register under the window, the temperature distribution of the warm air along the window base is nonuniform. Assume a linear temperature variation from 80°F to 90°F (over a foot long section) with a corresponding heat transfer coefficient $h = 1.46$ Btu/hr · ft^2 · °F. Also, assume an outside air temperature of 10°F and a corresponding $h = 6$ Btu/hr · ft^2 · °F. Using ANSYS, determine the temperature distribution in the window assembly, as shown in the accompanying figure. What is the overall heat loss through the window assembly?

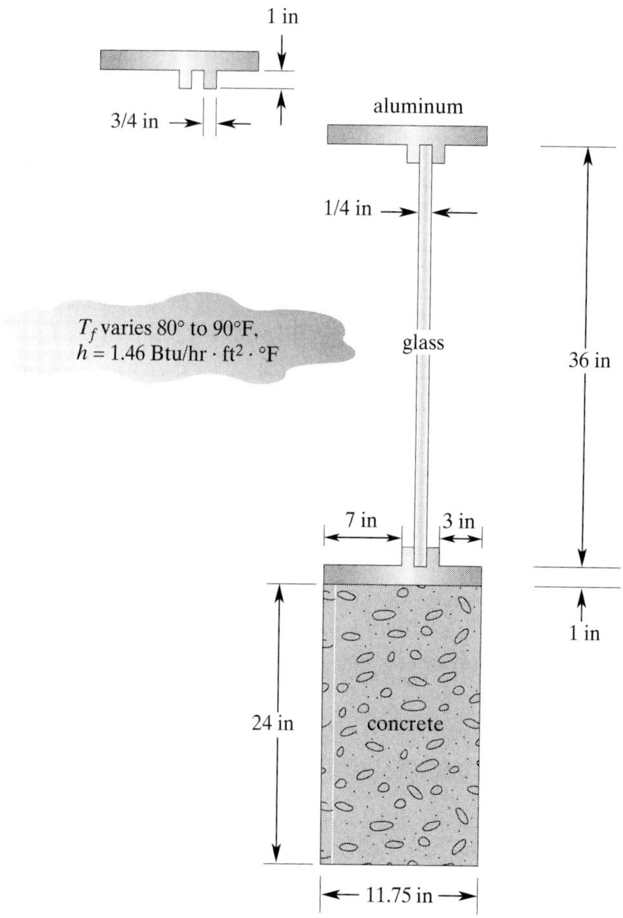

14. Using ANSYS, calculate and plot the principal stress distributions in the link component shown in the accompanying figure. The link is made of steel.

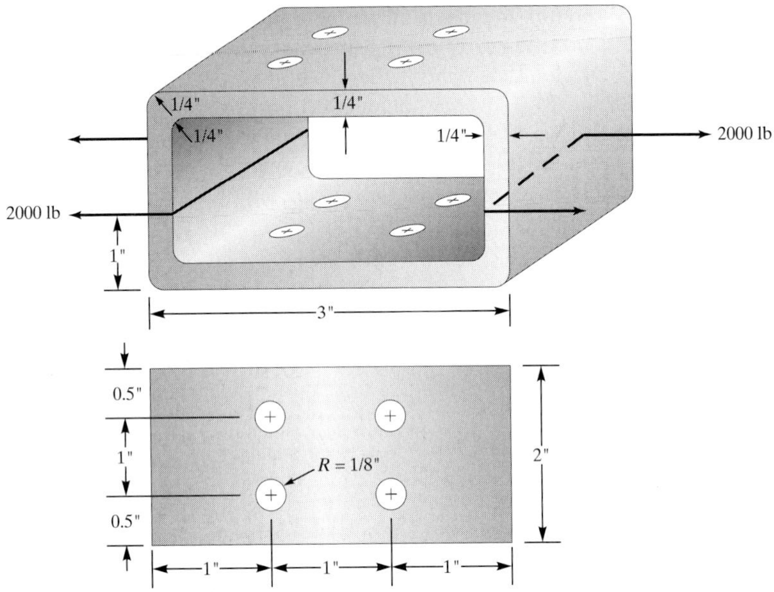

15. Design Problem Referring to one of the design problems in Chapter 10 (problem 15), each student is to design and construct a structural model from a $\frac{3}{8}'' \times 6'' \times 6''$ sheet of plexiglas material that adheres to the specifications and rules given in problem 15. Additionally, for this project, the model may have any cross-sectional shape. Examples of some common sections are shown in the accompanying figure.

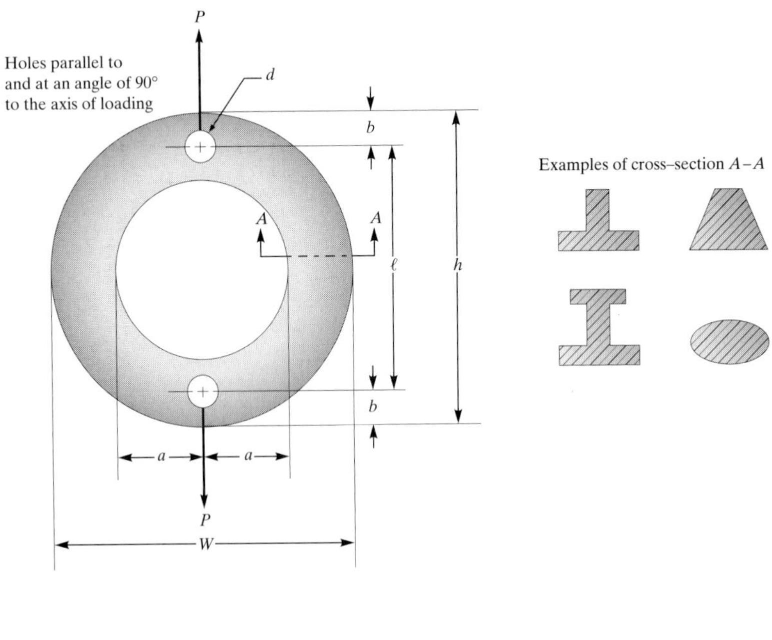

16. **Design Problem** Using a three-dimensional beam element in ANSYS, you are to size the cross sections of members of the frame shown in the accompanying figure. Use hollow tubes. The frame is to support the weight of a traffic light and withstand a wind gust of 80 miles/hr. Write a brief report discussing your final design.

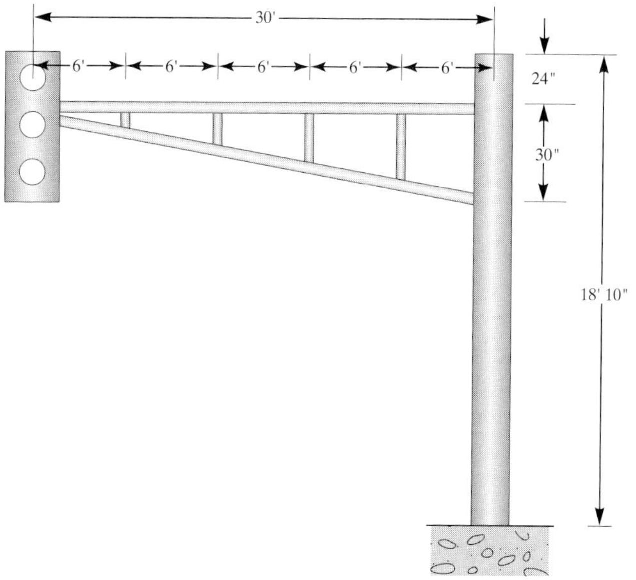

C H A P T E R 1 4

Design and Material Selection*

Engineers are problem solvers. In this chapter, we introduce you to the engineering design process. Engineers apply physical and chemical laws and principles along with mathematics to *design* millions of products and services that we use in our everyday lives: products such as cars, computers, aircraft, clothing, toys, home appliances, surgical equipment, heating and cooling equipment, health care devices, tools and machines that make various products, and so on. In recent years, the use of finite element analysis as a design tool has grown rapidly. As you have seen so far, finite element methods can be used to obtain solutions to a large class of engineering problems.

In this chapter, we look more closely at what the term *design* means and learn more about how engineers go about designing products. We discuss the basic steps that most engineers follow when designing a component or system.

As a design engineer, whether you are designing a machine part, a toy, a frame for a car, or a structure, the selection of material is an important design decision. There are a number of factors that engineers consider when selecting a material for a specific application. For example, they consider properties of material such as density, ultimate strength, flexibility, machinability, durability, thermal expansion, electrical and thermal conductivity, and resistance to corrosion. They also consider the cost of the material and how easily it can be repaired. Engineers are always searching for ways to use advanced materials to make products lighter and stronger for different applications.

In this chapter, we also look more closely at materials that are commonly used in various engineering applications. We discuss some of the basic physical characteristics of materials that are considered in design. We examine solid materials, such as metals and their alloys, plastics, glass, and wood, and those that solidify over time, such as concrete. We also investigate in more detail basic fluids, such as air and water, that not only are needed to sustain life, but also play important roles in engineering. Did you ever stop to think about the important role that air plays in food processing, driving power tools, or in your car's tire to provide a cushiony ride? You may not think of water as an engineering material either, but we not only need water to live, among other applications, we also need water to generate electricity in steam and hydroelectric power plants, and we use high pressurized water, which functions like a saw, to cut materials. The main topics discussed in Chapter 14 are

---

*Materials were adapted with permission from *Engineering Fundamentals by Moaveni (2002).*

14.1 ENGINEERING DESIGN PROCESS

Let us begin by emphasizing what we said earlier about what engineers do. Engineers apply physical laws, chemical laws and principles, and mathematics to *design* millions of products and services that we use in our everyday lives. These products include cars, computers, aircraft, clothing, toys, home appliances, surgical equipment, heating and cooling equipment, health care devices, tools and machines that make various products, and so on. Engineers consider important factors such as cost, efficiency, reliability, and safety when designing the products. Engineers perform tests to make certain that the products they design withstand various loads and conditions. Engineers are continuously searching for ways to improve already existing products as well. Engineers continuously develop new advanced materials to make products lighter and stronger for different applications. Let us now look more closely at what constitutes the *design process*. These are the basic steps that engineers, regardless of their background, follow to arrive at solutions to problems. The steps include

1. recognizing the need for a product or a service
2. defining and understanding the problem (the need) completely
3. doing preliminary research and preparation
4. conceptualizing ideas for possible solutions
5. synthesizing the findings
6. evaluating good ideas in more detail
7. optimizing solutions to arrive at the best possible solution
8. presenting the final solution

Keep in mind that these steps, which we will discuss soon, are not independent of one another and do not necessarily follow one another in the order they are presented. In fact, engineers often need to return to steps 1 and 2 when clients decide to change design parameters. Quite often, engineers are also required to give oral and written progress reports on a regular basis. Therefore, be aware that even though we listed presenting the final solution as step 8, it could well be an integral part of many other design steps. Let us now take a closer look at each step, starting with the need for a product or a service.

Step 1. *Recognizing the need for a product or a service.* All you have to do is look around to realize the large number of products and services—designed by engineers—that you use every day. Most often we take these products and services for granted until, for some reason, there is an interruption in the services they provide. Some of these existing products are constantly being modified to take advantage of new technologies. For example, cars and home appliances are constantly being redesigned to incorporate new technologies. In addition to the products and services already in use, new products are being developed every day for the purpose of making our lives more comfortable, more pleasurable, and less laborious. There is also that old saying that every time someone complains about a situation, or about a task, or curses a product, right there is an opportunity for a product or a service. As you can tell, the need for products and services exists; what one needs to do is to identify them. The need may be identified by you, the company that you may eventually (or already) work for, or by a third-party client that needs a solution to a problem or a new product to make what they do easier and more efficient.

Step 2. *Defining and understanding the problem.* One of the first things you need to do as a design engineer is to fully understand the problem. **This is the most important step in any design process.** If you do not have a good grasp of what the problem is or of what the client wants, you will not come up with a solution that is relevant to the need of the client. The best way to fully understand a problem is by asking many questions. You may ask the client questions such as

How much money are you willing to spend on this project?

Are there restrictions on the size or the type of materials that can be used?

When do you need the product or the service?

How many of these products do you need?

Questions often lead to more questions that will better define the problem. Moreover, keep in mind that engineers generally work in a team environment where they consult each other to solve complex problems. They divide up the task into smaller, manageable problems among themselves; consequently, productive engineers must be good team players. Good interpersonal and communication skills are increasingly important now because of the global market. You need to make sure you clearly understand your portion of the problem and how it fits with the other problems. For example, various parts of a product could be made by different companies located in different states or countries. In order to assure that all components fit and work well together, cooperation and coordination are essential, which demands good teamwork and strong communication skills. Make sure you understand the problem, and make sure that the problem is well defined before you move on to the next step. *This point can not be emphasized enough.* Good problem solvers are those who first fully understand what the problem is.

Step 3. *Doing preliminary research and preparation.* Once you fully understand the problem, as a next step you need to collect useful information. Generally speaking, a good place to start is by searching to determine if a product already exists that closely meets the need of your client. Perhaps a product, or components of a product, already have been developed by your company that you could modify to meet the need. You do not want to reinvent the wheel! As mentioned earlier, depending on the scope of project, some projects require collaboration with other companies, so you need to find out what is available through these other companies as well. Try to collect as much information as you can. This is where you spend lots of time not only with the client, but also with other engineers and technicians. Internet search engines are becoming increasingly important tools to search for such information. Once you have collected all pertinent information, you must then review it and organize the information in a suitable manner.

Step 4. *Conceptualizing ideas for possible solutions.* During this phase of design you need to generate some ideas or concepts that could offer reasonable solutions to your problem. In other words, without performing any detailed analysis, you need to come up with some possible ways of solving the problem. You need to be creative and perhaps develop several alternative solutions. At this stage of design, you do not need to rule out any reasonable working concept. If the problem consists of a complex system, you need to identify the components of the system. You do not need to look at details of each possible solution yet, but you need to perform enough analysis to see whether the concepts that you are proposing have merits. Simply stated, you need to ask yourself the following question: Would the concepts be likely to work if they were pursued further? Throughout the design process, you must also learn to budget your time. Good engineers have time management skills that enable them to work productively and efficiently. You must learn to create a milestone chart detailing your time plan for completing the project. You need to show the time periods and the corresponding tasks that are to be performed during these time periods.

Step 5. *Synthesizing the findings.* Good engineers have a firm grasp of the fundamental principles of engineering, which they can use to solve many different problems. Good engineers are analytical, detail-oriented, and creative. During this stage of design, you begin to consider details. You need to perform calculations, run computer models, narrow down the materials to be used, size the components of the system, and answer questions about how the product is going to be fabricated. You will consult pertinent codes and standards, and make sure that your design will be in compliance with them.

Step 6. *Evaluating good ideas.* Analyze the problem in more detail. You may have to identify critical design parameters and consider their influence in your final design. At this stage, you need to make sure that all calculations are performed correctly. If there are some uncertainties in your analysis, you must perform experimental investigation. When possible, working models must be created and tested. At this stage of the design procedure, the best solution must be identified from alternatives. Details of how the product is to be fabricated must be fully worked out.

Step 7. *Optimizing solutions.* Optimization means minimization or maximization. There are two broad types of design: a functional design and an optimized design. A functional design is one that meets all of the preestablished design requirements, but allows for improvement to be made in certain areas of design. We discuss design optimization in Chapter 15.

Step 8. *Presenting the final solution.* Now that you have a final solution, you need to communicate your solution to the client, who may be your boss, another group within your company, or an outside customer. You may have to prepare not only an oral presentation but also a written report. A reminder again—although we have listed the presentation as step 8 of the design process, quite often engineers are required to give oral and written progress reports on a regular basis to various groups. Consequently, presentation could well be an integral part of many other design steps.

14.2 MATERIAL SELECTION

Design engineers, when faced with selecting materials for their products, often ask questions such as

How strong will the material be when subjected to an expected load?

Would it fail, and if not, how safely would the material carry the load?

How would the material behave if its temperature were changed?

Would the material remain as strong as it would under normal conditions if its temperature is increased?

How much would it expand when its temperature is increased?

How heavy and flexible is the material?

What are its energy absorbing properties?

Would the material corrode?

How would it react in the presence of some chemicals?

How expensive is the material?

Would it dissipate heat effectively?

Would the material act as a conductor or as an insulator to the flow of electricity?

It is important to note that we have posed only a few generic questions; we could have asked additional questions had we considered the specifics of the application. For example, when selecting materials for implants in bioengineering applications, one must consider many additional factors, including

Is the material toxic to the body?

Can the material be sterilized?

When the material comes into contact with body fluid, will it corrode or deteriorate?

Because the human body is a dynamic system, we should also ask how the material would react to mechanical shock and fatigue?

Are the mechanical properties of the implant material compatible with those of bone to ensure appropriate stress distributions at contact surfaces?

These are examples of additional specific questions that one could ask to find suitable material for a specific application.

By now it should be clear that material properties and material cost are important design factors. In general, mechanical and thermophysical properties of a material depend on its phase. For example, as you know from your everyday experience, the density of ice is different from liquid water (the ice cubes float in liquid water), and the density of liquid water is different from that of steam. Moreover, the properties of a material in a single phase could depend on its temperature and the surrounding pressure. For example, if you were to look up the density of liquid water in the temperature range of, say, 4° C to 100° C, under standard atmospheric pressure, you would find that its density decreases with increasing temperature in that range. Therefore, properties of materials depend not only on their phase but also on their temperature and pressure. This is another important fact to keep in mind when selecting materials.

14.3 ELECTRICAL, MECHANICAL, AND THERMOPHYSICAL PROPERTIES OF MATERIALS

As we have been explaining up to this point, when selecting a material for an application, as an engineer you need to consider a number of material properties. In general, the properties of a material may be divided into three groups: electrical, mechanical, and thermal properties. In electrical and electronic applications, the electrical resistivity of materials is important. How much resistance to flow of electricity does the material offer? In many mechanical, civil, and aerospace engineering applications, the mechanical properties of materials are important. These properties include modulus of elasticity, modulus of rigidity, tensile strength, compression strength, the strength-to-weight ratio, modulus of resilience, and modulus of toughness. In applications dealing with fluids (liquids and gases), thermophysical properties such as thermal conductivity, heat capacity, viscosity, vapor pressure, and compressibility are important properties. Thermal expansion of a material, whether solid or fluid, is also an important design factor. Resistance to corrosion is another important factor that must be considered when selecting materials.

Material properties depend on many factors, including how the material was processed, its age, its exact chemical composition, and any nonhomogeneity or defect within the material. Material properties also change with temperature and time as the material ages. Most companies that sell materials will provide upon request information on the important properties of their manufactured materials. Keep in mind that when practicing as an engineer, you should use the manufacturer's material property values in your design calculations. The property values given in this and other textbooks should be used as typical values—not as exact values.

Electrical resistivity—The value of electrical resistivity is a measure of resistance of material to flow of electricity. For example, plastics and ceramics typically have high resistivity, whereas metals typically have low resistivity, and among the best conductors of electricity are silver and copper.

Density—Density is defined as mass per unit volume; it is a measure of how compact the material is for a given volume. For example, the average density of aluminum alloys is 2700 kg/m^3, and compared to steel density of 7850 kg/m^3, aluminum has a density which is approximately 1/3 of the density of steel.

Modulus of elasticity (Young's modulus)—Modulus of elasticity is a measure of how easily a material will stretch when pulled (subject to a tensile force) or how well the material will shorten when pushed (subject to a compressive force). The larger the value of the modulus of elasticity is, the larger the required force would be to stretch or shorten the material. For example, the modulus of elasticity of aluminum alloy is in the range of 70 to 79 GPa, whereas steel has a modulus of elasticity in the range of 190 to 210 GPa; therefore, steel is approximately three times stiffer than aluminum alloys.

Modulus of rigidity (shear modulus)—Modulus of rigidity is a measure of how easily a material can be twisted or sheared. The value of modulus of rigidly, also called shear modulus, shows the resistance of a given material to shear deformation. Engineers consider the value of shear modulus when selecting materials for shafts or rods that are subjected to twisting torques. For example, the modulus of rigidity or shear modulus for aluminum alloys is in the range of 26 to 36 GPa, whereas the shear modulus for steel is in the range of 75 to 80 GPa. Therefore, steel is approximately three times more rigid in shear than aluminum.

Tensile strength—The tensile strength of a piece of material is determined by measuring the maximum tensile load a material specimen in the shape of a rectangular bar or cylinder can carry without failure. The tensile strength or ultimate strength of a material is expressed as the maximum tensile force per unit cross-sectional area of the specimen. When a material specimen is tested for its strength, the applied tensile load is increased slowly. In the very beginning of the test, the material will deform elastically, meaning that if the load is removed, the material will return to its original size and shape without any permanent deformation. The point to which the material exhibits this elastic behavior is called the yield point. The yield strength represents the maximum load that the material can carry without any permanent deformation. In certain engineering design applications (especially involving brittle materials), the yield strength is used as the tensile strength.

Compression strength—Some materials are stronger in compression than they are in tension; concrete is a good example. The compression strength of a piece of material is determined by measuring the maximum compressive load a material specimen in the shape cube or cylinder can carry without failure. The ultimate compressive strength of a material is expressed as the maximum compressive force per unit cross-sectional area of the specimen. Concrete has a compressive strength in the range of 10 to 70 MPa.

Modulus of resilience—Modulus of resilience is a mechanical property of a material that shows how effective the material is in absorbing mechanical energy without going through any permanent damage.

Modulus of toughness—Modulus of toughness is a mechanical property of a material that indicates the ability of the material to handle overloading before it fractures.

Strength-to-weight ratio—As the term implies, it is the ratio of strength of the material to its specific weight (weight of the material per unit volume). Based on the application, engineers use either the yield or the ultimate strength of the material when determining the strength-to-weight ratio of a material.

Thermal expansion—The coefficient of linear expansion can be used to determine the change in the length (per original length) of a material that would occur if the temperature of the material were changed. This is an important material property to consider when designing products and structures that are expected to experience a relatively large temperature swing during their service lives.

Thermal conductivity—Thermal conductivity is a property of material that shows how good the material is in transferring thermal energy (heat) from a high temperature region to a low temperature region within the material.

Heat capacity—Some materials are better than others in storing thermal energy. The value of heat capacity represents the amount of thermal energy required to raise the temperature 1 kilogram mass of a material by 1°C or using U.S. Customary units, the amount of thermal energy required to raise one pound mass of a material by 1°F. Materials with large heat capacity values are good at storing thermal energy.

Viscosity, vapor pressure, and bulk modulus of compressibility are additional fluid properties that engineers consider in design.

Viscosity—The value of viscosity of a fluid represents a measure of how easily the given fluid can flow. The higher the viscosity value is, the more resistance the fluid offers to flow. For example, it requires less energy to transport water in a pipe than it does to transport motor oil or glycerin.

Vapor pressure—Under the same conditions, fluids with low vapor pressure values will not evaporate as quickly as those with high values of vapor pressure. For example, if you were to leave a pan of water and a pan of glycerin side by side in a room, the water will evaporate and leave the pan long before you would notice any changes in the level of glycerin.

Bulk modulus of compressibility—A fluid bulk modulus represents how compressible the fluid is. How easily can one reduce the volume of the fluid when the fluid pressure is increased? For example, it would take a pressure of 2.24×10^7 N/m^2 to reduce 1 m^3 volume of water by 1%, or said another way, to final volume of 0.99 m^3.

In this section, we explained the meaning and significance of some of the physical properties of materials. Mechanical and thermophysical properties of some materials are given in Appendices A and B. In the following sections, we examine the application and chemical composition of some common engineering materials.

14.4 COMMON SOLID ENGINEERING MATERIALS

In this section we will briefly examine the chemical composition and common application of some solid materials. We will discuss light metals, copper and its alloys, iron and steel, concrete, wood, plastics, silicon, glass, and composite materials. Most of you may have already taken a class in materials science and have an in-depth knowledge of atomic structure of various materials. Here our intent is to provide a quick review to materials and their applications.

Lightweight Metals

Aluminum, titanium, and magnesium, because of their small densities (relative to steel), are commonly referred to as lightweight metals. Because of their relatively high strength-to-weight ratios, lightweight metals are used in many structural and aerospace applications.

Aluminum and its alloys have densities that are approximately 1/3 the density of steel. Pure aluminum is very soft; thus it is generally used in electronics applications and in making reflectors and foils. Because pure aluminum is soft and has a relatively small tensile strength, it is alloyed with other metals to make it stronger, easier to weld, and to increase its resistance to corrosive environments. Aluminum is commonly alloyed with copper (Cu), Zinc (Zn), magnesium (Mg), manganese (Mn), Silicon (Si), and lithium (Li). Generally speaking, aluminum and its alloys resist corrosion; they are easy to mill and cut, and can be brazed or welded. Aluminum parts can also be joined using adhesives. They are good conductors of electricity and heat, and thus have relatively high thermal conductivity and low electrical resistance values. American National Standards Institute (ANSI) assigns designation numbers to specify aluminum alloys.

Aluminum is fabricated in sheets, plates, foil, rod, and wire, and is extruded into window frames or automotive parts. You are already familiar with everyday examples of common aluminum products, including beverage cans, household aluminum foil, non-rust staples in tea bags, building insulation, and so on.

Titanium has an excellent strength-to-weight ratio. Titanium is used in applications where relatively high temperatures exceeding 400°C up to 600°C are expected. Titanium alloys are used in the fan blades and the compressor blades of the gas turbine engines of commercial and military airplanes. In fact, without the use of titanium alloys, the engines on the commercial airplanes would have not been possible. Like aluminum, titanium is alloyed with other metals to improve its properties. Titanium alloys show excellent resistance to corrosion. Titanium is quite expensive compared to aluminum. It is heavier than aluminum, having a density which is roughly $\frac{1}{2}$ that of steel. Because of their relatively high strength-to-weight ratios, titanium alloys are used in both commercial and military airplane airframes (fuselage and wings) and landing gear components. Titanium alloys are becoming a metal of choice in many products; you can find them in golf clubs, bicycle frames, tennis racquets, and spectacle frames. Because of their excellent corrosion resistance, titanium alloys have been used in the tubing used in desalination plants as well. Titanium hips and other joints are examples of other applications where titanium is currently being used.

Magnesium is another lightweight metal that looks like aluminum. It has a silvery white appearance, but it is lighter than aluminum, having a density of approximately of 1700 kg/m^3. Pure magnesium does not provide good strength for structural applications and because of this, it is alloyed with other elements such as aluminum, manganese, and zinc to improve its mechanical characteristics. Magnesium and its alloys are used in nuclear applications, in drycell batteries, and in aerospace applications and some automobile parts as sacrificial anodes to protect other metals from corrosion.

Copper and Its Alloys

Copper is a good conductor of electricity and because of this property is commonly used in many electrical applications, including home wiring. Copper and many of its alloys are also good conductors of heat, and this thermal property makes copper a good choice for heat exchanger applications in air conditioning and refrigeration systems. Copper alloys are also used as tubes, pipes, and fittings in plumbing and heating applications. Copper is alloyed with zinc, tin, aluminum, nickel, and other elements to modify its properties. When copper is alloyed with zinc, it is commonly called *brass*. The mechanical properties of brass depend on the exact composition of percent copper and percent zinc. *Bronze* is an alloy of copper and tin. Copper is also alloyed with aluminum, referred to as *aluminum bronze*. Copper and its alloys are also used in water tubes, heat exchangers, hydraulic brake lines, pumps, and screws.

Iron and Steel

Steel is a common material that is used in the framework of buildings, bridges, the body of appliances such as refrigerators, ovens, dishwashers, washers and dryers, and cooking utensils. Steel is an alloy of iron with approximately 2% or less carbon. Pure iron is soft and thus not good for structural applications, but the addition of even a small amount of carbon to iron hardens it and gives steel better mechanical properties, such as greater strength. The properties of steel can be modified by adding other elements such as chromium, nickel, manganese, silicon, and tungsten. For example, chromium is used to increase the resistance of steel to corrosion. In general, steel can be classified into three broad groups: (1) the carbon steels containing approximately 0.015% to 2% carbon, (2) low-alloy steels having a maximum of 8% alloying elements, and (3) high-alloy steels containing more than 8% of alloying elements. Carbon steels constitute most of the world's steel consumption; thus, you will commonly find them in the body of appliances and cars. The low-alloy steels have good strength and are commonly used as machine or tool parts and as structural members. The high-alloy steels such as stainless steels could contain approximately 10% to 30% chromium and could contain up to 35% nickel. The 18/8 stainless steels, which contain 18% chromium and 8% nickel, are commonly used for tableware and kitchenware products. Finally, cast iron is also an alloy of iron that has 2% to 4% carbon. Note that the addition of extra carbon to the iron changes its properties completely. In fact, cast iron is a brittle material, whereas most iron alloys containing less than 2% carbon are ductile.

Concrete

Today **concrete** is commonly used in construction of roads, bridges, buildings, tunnels, and dams. What is normally called concrete consists of three main ingredients: *aggregate,* cement, and water. Aggregate refers to materials such as gravel and sand, and cement refers to the bonding material that holds the aggregates together. The type and size (fine to coarse) of aggregate used in making concrete varies depending on application.

The amount of water used in making concrete could also influence the strength of concrete. Of course, the mixture must have enough water so that the concrete can be poured and have a consistent cement paste that completely wraps around all aggregates. The ratio of amount of cement to aggregate used in making concrete also affects the strength and durability of concrete. Another factor that could influence the cured strength of concrete is the temperature of its surrounding when concrete is poured. Calcium chloride is added to cement when the concrete is poured in cold climates. The addition of calcium chloride will accelerate the curing process to counteract the effect of low temperature of the surrounding. You may have also noticed as you walk by newly poured concrete for driveways or sidewalks that water is sprayed onto the concrete for some time after it is poured. This is to control the rate of contraction of concrete as it sets.

Concrete is a brittle material, which can support compressive loads much better than it does tensile loads. Because of this, concrete is commonly *reinforced* with steel bars or steel mesh that consist of thin metal rods to increase its load bearing capacity, especially in the sections where tensile stress are expected. Concrete is poured into forms that contain the metal mesh or steel bars. Reinforced concrete is used in foundations, floors, walls, and columns. Another common construction practice is the use of *precast concrete.* Precast concrete slabs, blocks, and structural members are fabricated in less time with less cost in factory settings where surrounding conditions are controlled. The precast concrete parts are then moved to the construction site and are erected at site. This practice saves time and money. As we mentioned above, concrete has a higher compressive strength than tensile strength. Because of this, concrete is also *prestressed* in the following manner. Before concrete is poured into forms that have the steel rods or wires, the steel rods or wires are stretched; after the concrete has been poured and after enough time has elapsed, the tension in the rods or wires is released. This process, in turn, compresses the concrete. The prestressed concrete then acts as a compressed spring, which will become uncompressed under the action of tensile loading. Therefore, the prestressed concrete section will not experience any tensile stress until the section has been completely uncompressed. It is important to note once again the reason for this practice is that concrete is weak under tension.

Wood

Throughout history, **wood,** because of its abundance in many parts of the world, has been a material of choice for many applications. Wood is a renewable source, and because of its ease of workability and its strength, it has been used to make many products. Today, wood is used in a variety of products ranging from telephone poles to toothpicks. Common examples of wood products include hardwood flooring, roof trusses, furniture frames, wall supports, doors, decorative items, window frames, trimming in luxury cars, tongue depressors, clothespins, baseball bats, bowling pins, fishing rods, and wine barrels. Wood is also the main ingredient that is used to make various paper products. Whereas a steel structural member is susceptible to rust, wood, on the other hand, is prone to fire, termites, and rotting. Wood is anisotropic material, meaning that its properties are direction-dependent. For example, as you may already know, under axial loading (when pulled), wood is stronger in a direction parallel to a grain than it is in a direction across the grain. However, wood is stronger in a direction normal to the grain when it is bent. The properties of wood also depend on its moisture content; the lower the moisture content, the stronger the wood is. Density of wood is generally a good indication of how strong the wood is. As a rule of thumb, the higher the density of wood, the higher its strength. Moreover, any defects, such as knots, would affect the load carrying capacity of wood. Of course, the location of the knot and the extent of defect will directly affect its strength.

Timber is commonly classified as *softwood* and *hardwood*. Softwood timber is made from trees that have cones (coniferous), such as pine, spruce, and Douglas fir. Hardwood timber is made from trees that have broad leaves or have flowers. Examples of hardwoods include walnut, maple, oak, and beech. This classification of wood into softwood and hardwood should be used with caution, because there are some hardwood timbers that are softer than softwoods.

Plastics

In the later part of the 20th century, **plastics** have become increasingly the material of choice for many applications. They are very lightweight, strong, inexpensive, and easily made into various shapes. Over 100 million metric tons of plastic are produced annually worldwide. Of course this number increases as the demand for inexpensive, durable, disposable material grows. Most of you are already familiar with examples of plastic products, including grocery and trash bags, plastic soft drink containers, home cleaning containers, vinyl sidings, *polyvinyl chloride* (PVC) piping, valves, and fittings that are readily available in home improvement centers. Styrofoam™ plates and cups and plastic forks, knives, spoons, and sandwich bags are other examples of plastic products that are consumed every day.

Polymers are the backbones of what we call plastics. They are chemical compounds that have very large molecular chainlike structures. Plastics are often classified into two categories: *thermoplastics* and *thermosetting*. When heated to certain temperatures, the thermoplastics can be molded and remolded. For example, when you recycle Styrofoam dishes, they can be heated and reshaped into cups or bowls or other shaped dishes. By contrast, thermosets can not be remolded into other shapes by heating. The application of heat to thermosets does not soften the material for remolding; instead, the material will simply break down. There are many other ways of classifying plastics; for instance, they may be classified on the basis of their chemical composition, molecular structure, or the way molecules are arranged or their densities. For example, based on their chemical composition, polyethylene, polypropylene, polyvinyl chloride, and polystyrene are the most commonly produced plastics. A grocery bag is an example of a product made from high-density polyethylene (HDPE). However, note that in a broader sense, for example, polyethylene and polystyrene are thermoplastics. In general, the way molecules of a plastic are arranged will influence its mechanical and thermal properties. Plastics have relatively small thermal and electrical conductivity values. Some plastic materials, such as Styrofoam cups, are designed to have air trapped in them to reduce the heat conduction even more. Plastics are easily colored by using various metal oxides. For example, titanium oxide and zinc oxide are used to give a plastic sheet its white color. Carbon is used to give plastic sheets their black color, as is the case in black trash bags. Depending on an application, other additives are also added to the polymers to obtain specific characteristics such as rigidity, flexibility, enhanced strength, or a longer life span that excludes any change in the appearance or mechanical properties of the plastic over time. As with other materials, research is being performed every day to make plastics stronger and more durable and to control its aging process, to make plastics less susceptible to sun damage, and to control water and gas diffusion through them. The latter is especially important when the goal is to add shelf life to food that is wrapped in plastics.

Silicon

Silicon is a nonmetallic chemical element that is used quite extensively in the manufacturing of transistors, and various electronic and computer chips. Pure silicon is not found in nature; it is found in the form of *silicon dioxide* in sands and rocks or found combined with other elements such as aluminum or calcium or sodium or magnesium in the form commonly referred to as *silicates*. Silicon, because of its atomic structure, is an excellent *semiconductor*. A semiconductor is a material whose electrical conductivity properties can be changed to act either as a conductor of electricity or as an insulator (preventing electricity flow). Silicon is also used as an alloying element with other elements, such as iron and copper, to give steel and brass certain desired characteristics. Make sure not to confuse silicon with *silicones*, which are synthetic compounds consisting of silicon, oxygen, carbon, and hydrogen. You find silicones in lubricants, varnishes, and waterproofing products.

Glass

Glass is commonly used in products such as windows, light bulbs, TV CRT tubes, housewares such as drinking glasses, chemical containers, beverage and beer containers, and decorative items. The composition of the glass depends on its application. The most widely used form of glass is *soda-lime-silica* glass. The materials used in making soda-lime-silica glass include sand (silicon dioxide), limestone (calcium carbonate), and soda ash (sodium carbonate). Other materials are added to create desired characteristics for specific applications. For example, bottle glass contains approximately 2% aluminum oxide, and glass sheets contain about 4% magnesium oxide. Metallic oxides are also added to give glass various colors. For example, silver oxide gives glass a yellowish stain, and copper oxide gives glass its bluish, greenish color, depending on the amount added to the composition of glass. Optical glasses have very specific chemical compositions and are quite expensive. The composition of optical glass will influence its refractive index and its light dispersion properties. Glass that is made completely from silica (silicon dioxide) has properties that are sought after by many industries, such as fiber optics, but it is quite expensive to manufacture because the sand has to be heated to temperatures exceeding 1700°C. Silica glass has a low coefficient of thermal expansion, high electrical resistivity, and high transparency to ultraviolet light. Because silica glass has a low coefficient of thermal expansion, it can be used in high-temperature applications.

Ordinary glass has a relatively high coefficient of thermal expansion; therefore, when its temperature is changed suddenly, it could break easily due to thermal stresses developed by the temperature rise. Cookware glass contains boric oxide and aluminum oxide to reduce its coefficient of thermal expansion.

Fiberglass. Silica glass fibers are commonly used in fiber optics, the branch of science that deals with transmitting data, voice, and images through thin glass or plastic fibers. Every day, copper wires are being replaced by transparent glass fibers in telecommunications to connect computers together in networks. The glass fibers typically have an outer diameter of 0.125 mm (12 micron) with an inner transmitting core diameter of 0.01 mm (10 micron). Infrared light signals in the wavelength ranges of 0.8 to 0.9 m or 1.3 to 1.6 m; wavelengths are generated by light-emitting diodes or semiconductor lasers and travel through the inner core of glass fiber. The optical signals generated in this manner can travel to distances as far as 100 km without any need to amplify the optical signals again. Plastic fibers made of polymethylmethacrylate, polystyrene, or polycarbonate are also used in fiber optics. These plastic fibers are in general cheaper and more flexible than glass fibers. But when compared to glass fibers, plastic fibers require more amplification of signals due to their greater optical losses. They are generally used in networking computers in a building.

Composites

Because of their light weight and good strength, composite materials are becoming increasingly the materials of choice for a number of products and aerospace applications. Today you will find composite materials in military planes, helicopters, satellites, commercial planes, fast-food restaurant tables and chairs, and many sporting goods. They are also commonly used to repair bodies of automobiles. In comparison to conventional materials, such as metals, composite materials can be lighter and stronger. For this reason, composite materials are used extensively in aerospace applications. Composites are created by combining two or more solid materials to make a new material that has properties that are superior to those of individual components. *Composite materials* consist of two main ingredients: *matrix material* and *fibers*. Fibers are embedded in matrix materials, such as aluminum or other metals, plastics, or ceramics. Glass, graphite, and silicon carbide fibers are examples of fibers used in construction of composite materials. The strength of fibers is increased when embedded in a matrix material, and the composite material created in this manner is lighter and stronger. Moreover, in a single material, once a crack starts due to either excessive loading or imperfections in the material, the crack will propagate to the point of failure. On the other hand, in a composite material, if one or a few fibers fail, it does not necessarily lead to failure of other fibers or the material as a whole. Furthermore, the fibers in a composite material can be oriented in a certain direction or many directions to offer more strength in the direction of expected loads. Therefore, composite materials are designed for specific load applications. For instance, if the expected load is uniaxial, meaning that it is applied in a single direction, then all the fibers are aligned in the direction of the expected load. For applications expecting multidirection loads, the fibers are aligned in different directions to make the material equally strong in various directions.

Depending upon what type of host matrix material is used in creating the composite material, the composites may be classified into three classes: (1) polymer-matrix composites, (2) metal-matrix composites, and (3) ceramic-matrix composites. We discussed the characteristics of matrix materials earlier when we covered metals and plastics.

14.5 SOME COMMON FLUID MATERIALS

Fluid refers to both liquids and gases. Air and water are among the most abundant fluids on earth. They are important in sustaining life and are used in many engineering applications. We briefly discuss them next.

Air

We all need air and water to sustain life. Because air is readily available to us, it is also used in engineering as a cooling and heating medium in food processing, in controlling thermal comfort in buildings, as a controlling medium to turn equipment on and off, and to drive power tools. Understanding the properties of air and how it behaves is important in many engineering applications, including understanding the lift and the drag forces. Better understanding of how air behaves under certain conditions leads to design of better planes and automobiles. The earth's atmosphere, which we refer to as air, is a mixture of approximately 78% nitrogen, 21% oxygen, and less than 1% argon. Small amounts of other gases are present in earth's atmosphere, as shown in Table 14.1.

There are other gases present in the atmosphere, including carbon dioxide, sulfur dioxide, and nitrogen oxide. The atmosphere also contains water vapor. The concentration level of these gases depends on the altitude and geographical location. At higher altitudes (10 km to 50 km), the earth's atmosphere also contains ozone. Even though these gases make up a small percentage of earth's atmosphere, they play a significant role in maintaining a thermally comfortable environment for us and other living species.

TABLE 14.1 The composition of dry air

Gases	Volume by Percent
Nitrogen (N_2)	78.084
Oxygen (O_2)	20. 946
Argon (Ar)	0. 934
Small amounts of other gases are present in atmosphere, including:	
Neon (Ne)	0.0018
Helium (He)	0.000524
Methane (CH_4)	0.0002
Krypton (Kr)	0.000114
Hydrogen (H_2)	0.00005
Nitrous oxide (N_2O)	0.00005
Xenon (Xe)	0.0000087

Humidity. There are two common ways of expressing the amount of water vapor in air: *absolute humidity* (or *humidity ratio*) and *relative humidity*. The *absolute humidity* is defined as the ratio of mass of water vapor in a unit mass of dry air, according to

$$\text{absolute humidity} = \frac{\text{mass of water vapor (kg)}}{\text{mass of dry air (kg)}} \qquad (14.1)$$

For humans, the level of a comfortable environment is better expressed by *relative humidity,* which is defined as the ratio of the amount of water vapor or moisture in the air to the maximum amount of moisture that air can hold at a given temperature. Therefore, relative humidity is defined as

$$\text{relative humidity} = \frac{\text{amount of moisture in the air (kg)}}{\text{maximum amount of moisture that air can hold (kg)}} \qquad (14.2)$$

Most people feel comfortable when the relative humidity is around 30% to 50%. The higher the temperature of air the more water vapor the air can hold before it is fully saturated. Because of its abundance, air is commonly used in food processing, especially in food drying processes to make dried fruits, spaghetti, cereals, and soup mixes. Hot air is transported over the food to absorb water vapors and thus remove them from the source.

Understanding how air behaves at given pressures and temperatures is also important when designing cars to overcome air resistance or designing buildings to withstand wind loading.

Water

You already know that every living thing needs water to sustain life. In addition to drinking water, we also need water for washing, laundry, grooming, cooking, and fire protection. You may also know that 2/3 of the earth's surface is covered with water, but most of this water cannot be consumed directly; it contains salt and other minerals that must be removed first. Radiation from the sun evaporates water; water vapors form into clouds and eventually, under favorable conditions, water vapors turn into liquid water or snow and fall back on land and the ocean. On land, depending on the amount of precipitation, part of water infiltrates the soil, part of it may be absorbed by vegetation, and part runs as streams or rivers and collects into natural reservoirs called lakes. Surface water refers to water in reservoirs, lakes, rivers, and streams. Groundwater, on the other hand, refers to the water that has infiltrated the ground; surface and ground waters eventually return to the ocean, and the water cycle is completed. As we said earlier, everyone knows that we need water to sustain life, but what you may not realize is that water could be thought of as a common engineering material! Water is used in all steam-power generating plants to produce electricity. In a power plant, fuel is burned in a boiler to generate heat, which in turn is added to liquid water to change its phase to steam; steam passes through turbine blades, turning the blades, which in effect runs the generator connected to the turbine, creating electricity. The low-pressure steam liquefies in a condenser and is pumped through the boiler again, closing a cycle.

Water is also used as a cutting tool. High-pressure water containing abrasive particles is used to cut marble or metals. Water is commonly used as a cooling or cleaning agent in a number of food processing plants and industrial applications. Thus, water is not only transported to our homes for our domestic use, but it is also used in many engineering applications. So you see, understanding the properties of water and how it can be used to transport thermal energy, or what it takes to transport water from one location to the next, is important to mechanical engineers, civil engineers, manufacturing engineers, agricultural engineers, and so on.

SUMMARY

At this point you should

1. know the basic design steps that all engineers follow, regardless of their background, to design products and services. These steps are (1) recognizing the need for a product or a service, (2) defining and understanding the problem (the need) completely, (3) doing the preliminary research and preparation, (4) conceptualizing ideas for possible solutions, (5) synthesizing the results, (6) evaluating good ideas in more detail, (7) optimizing the solutions to arrive at the best possible solution, (8) and presenting the solution.
2. realize that economics plays an important role in engineering decision making.
3. understand that engineers select materials for an application based on characteristics of materials, such as strength, density, corrosion resistance, durability, toughness, the ease of machining, and manufacturability. Moreover, you need to understand that material cost is also an important selection criterion.
4. be familiar with common applications of basic materials, such as light metals and their alloys, steel and its alloys, composite materials, and building materials such as concrete, wood, and plastics.
5. be familiar with the application of fluids, such as air and water, in engineering. You should also be familiar with the composition of air and what the term humidity means.

REFERENCE

Moaveni, Saeed, *Engineering Fundamentals, an Introduction to Engineering,* Pacific Grove, CA, Brooks-Cole, 2002.

PROBLEMS

1. Investigate the use of various materials in construction of bicycle frames. Choose a bike frame and create a finite element model of it. Assume boundary conditions and loading, and perform stress analysis of the frame using various materials. Write a report discussing your findings.

2. Investigate the use of various materials in construction of tennis racquets. Choose a tennis racquet and create a finite element model of it. Assume boundary conditions and loading, and perform stress analysis of the racquet using various materials. Write a report discussing your findings.
3. Investigate the design of hip implants, including material characteristics. Create a finite element model of a hip and perform stress analysis of it. Write a report discussing your findings.
4. Investigate the design of wooden roof and floor trusses. Visit a lumberyard and obtain drawings of a roof, a floor truss, or both. Make a finite element model, including appropriate loading and boundary conditions for the truss, and perform stress analysis. Write a report discussing your findings, including appropriate materials that are commonly used.
5. Investigate the design of aluminum fins in applications dealing with dissipating heat from electronic devices. Obtain a heat sink used to cool a PC microprocessor, and generate a finite element model of it. Make appropriate boundary and heat load assumptions, and look at the thermal performance of the heat sink. Write a report discussing your findings, including appropriate materials for this application.
6. Investigate the design of window frames with and without thermal breaks. Generate a finite element model of a window frame with and without thermal breaks. Make assumptions for appropriate boundary conditions and thermal loads. Write a report discussing your findings.
7. Visit a hardware store and look at the design of shelf brackets. Create a finite element model of a bracket, and perform stress analysis. Make appropriate assumptions for the boundary conditions and loading. Write a report discussing your findings.
8. Investigate the design of ski lifts. Create a finite element model of a frame used in a ski lift after you have made appropriate assumptions for boundary conditions and loading. Perform stress analysis and write a report discussing your findings.
9. Generate a model of a tool such as a wrench. Take actual measurements and create a finite element model after you have made appropriate assumptions for the boundary conditions and loading. Write a report discussing your findings.
10. Investigate the design of an exercise machine, such as a universal weight machine. Create a finite element model of the frame or a component. Perform stress analysis after you have made appropriate boundary conditions and loading. Write a report discussing material options and your other findings.

CHAPTER 15

Design Optimization

The objectives of this chapter are to introduce the basic design optimization ideas and the parametric design language of ANSYS. The main topics discussed in Chapter 15 include the following:

15.1 Introduction to Design Optimization

15.2 The Parametric Design Language of ANSYS

15.3 Examples of Batch Files

15.4 An Example of an Optimization Batch File

15.1 INTRODUCTION TO DESIGN OPTIMIZATION

Optimization means minimization or maximization. There are two broad types of design: a functional design and an optimized design. A functional design is one that meets all of the preestablish design requirements, but allows for improvements to be made in certain areas of the design. To better understand the concept of a functional design, we will consider an example. Let us assume that we are to design a 10-foot-tall ladder to support a person who weighs 300 pounds with a certain factor of safety. We will come up with a design that consists of a steel ladder that is ten feet tall and can safely support the load of 300 lb at each step. The ladder would cost a certain amount of money. This design would satisfy all of the requirements, including those of the strength and the size and, thus, constitutes a functional design. Before we can consider improving our design, we need to ask ourselves what criterion should we use to optimize the design? Design optimization is always based on some criterion such as cost, strength, size, weight, reliability, noise, or performance. If we use the weight as an optimization criterion, then the problem becomes one of minimizing the weight of the ladder. For example, we may consider making the ladder from aluminum. We could also perform stress analysis on the new ladder to see if we could remove material from certain sections of the ladder without compromising the loading and factor of safety requirements.

Another important fact to keep in mind is that while an engineering system consists of various components, optimizing individual components that make up a system

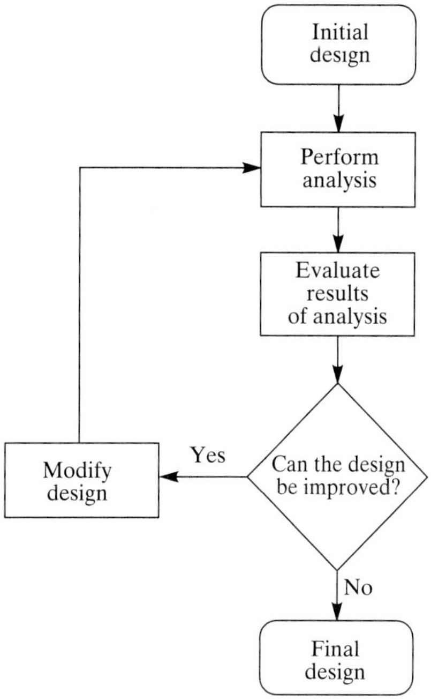

FIGURE 15.1 An optimization procedure.

does not necessarily lead to an optimized system. For example, consider a thermal-fluid system such as a refrigerator. Optimizing the individual components independently—such as the compressor, the evaporator, or the condenser—with respect to some criterion does not lead to an optimized overall system.

This chapter presents some basic ideas in design optimization of a component. We will focus only on weight as an optimization criterion. Traditionally, improvements in a design come from the process of starting with an initial design, performing an analysis, looking at results, and deciding whether or not we can improve the initial design. This procedure is shown in Figure 15.1.

In the past few decades, the optimization process has grown into a discipline that ranges from linear to nonlinear programming techniques. As is the case with any discipline, the optimization field has its own terminology. We will use the next two examples to introduce the fundamental concepts of optimization and its terminology.

EXAMPLE 15.1

Assume that you have been asked to look into purchasing some storage tanks for your company, and for the purchase of these tanks, you are given a budget of \$1,680. After some research, you find two tank manufacturers that meet your requirements. From Manufacturer A, you can purchase 16-ft^3-capacity tanks that cost \$120 each. Moreover, this type of tank requires a floor space of 7.5 ft^2. Manufacturer B makes 24-ft^3-capacity tanks that cost \$240 each and that require a floor space of 10 ft^2. The tanks will be placed in a section of a lab that has 90 ft^2 of floor space available for storage. You are looking for the greatest storage capability within the budgetary and floor-space limitations. How many of each tank (x_1, x_2) must you purchase?

First, we need to define the *objective function,* which is the function that we will attempt to minimize or maximize. In this example, we want to maximize storage capacity. We can represent this requirement mathematically as

$$\text{Maximize } Z = 16x_1 + 24x_2 \tag{15.1}$$

subject to the following constraints:

$$120x_1 + 240x_2 \leq 1680 \tag{15.2}$$

$$7.5x_1 + 10x_2 \leq 90 \tag{15.3}$$

$$x_1 \geq 0 \tag{15.4}$$

$$x_2 \geq 0 \tag{15.5}$$

In Eq. (15.1), Z is the objective function, while the variables x_1 and x_2 are called *design variables*. The limitations imposed by the inequalities in (15.2)–(15.5) are referred to as a set of constraints. Although there are specific techniques that deal with solving linear programming problems (the objective function and constraints are linear), we will solve this problem graphically to illustrate some additional concepts. The inequalities in (15.2)–(15.5) are plotted in Figure 15.2.

The shaded region shown in Figure 15.2 is called a *feasible solution region*. Every point within this region satisfies the constraints. However, our goal is to maximize the

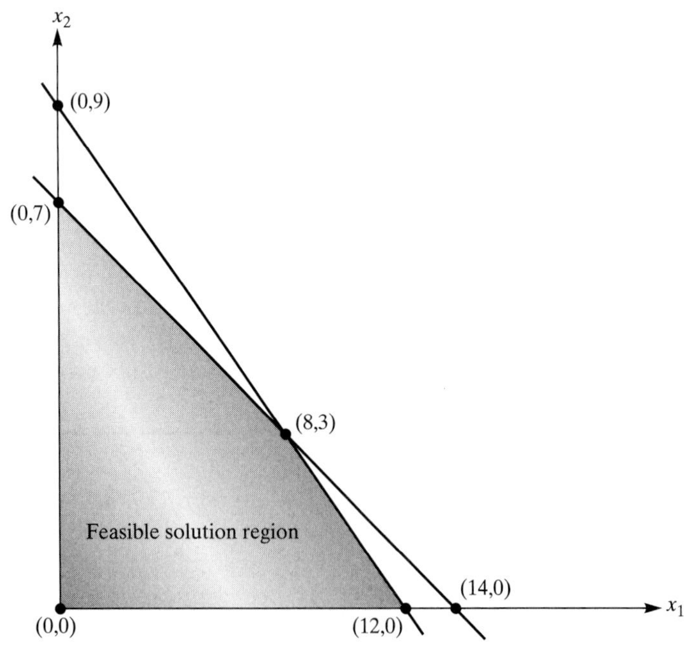

FIGURE 15.2 The feasible solution region for Example 15.1.

TABLE 15.1 Value of the objective function at the cornerpoints of the feasible region

Cornerpoints (x_1, x_2)	Value of $Z = 16x_1 + 24x_2$
0,0	0
0,7	168
12,0	192
8,3	200 (max.)

objective function given by Eq. (15.1). Therefore, we need to move the objective function over the feasible region and determine where its value is maximized. It can be shown that the maximum value of the objective function will occur at one of the cornerpoints of the feasible region. By evaluating the objective function at the cornerpoints of the feasible region, we see that the maximum value occurs at $x_1 = 8$ and $x_2 = 3$. This evaluation is shown in Table 15.1.

Thus, we should purchase eight of the 16-ft^3 tanks from Manufacturer A and three of the 24-ft^3 tanks from Manufacturer B to maximize the storage capacity within the given constraints.

Let us now consider a nonlinear example to demonstrate some additional terms.

EXAMPLE 15.2

Consider a wooden cantilever beam with rectangular cross section subject to the point loads shown in Figure 15.3. To satisfy safety requirements, the average stress in the beam is not to exceed a value of 30 MPa. Furthermore, the maximum deflection of the beam must be kept under 1 cm. Additional spatial restrictions limit the size of the cross section according to the limits 5 cm $\leq x_1 \leq$ 15 cm and 20 cm $\leq x_2 \leq$ 40 cm. We are interested in sizing the cross section so that it results in a minimum weight of the beam.

This problem is modeled by the objective function:

$$\text{Minimize } W = \rho g x_1 x_2 L \tag{15.6}$$

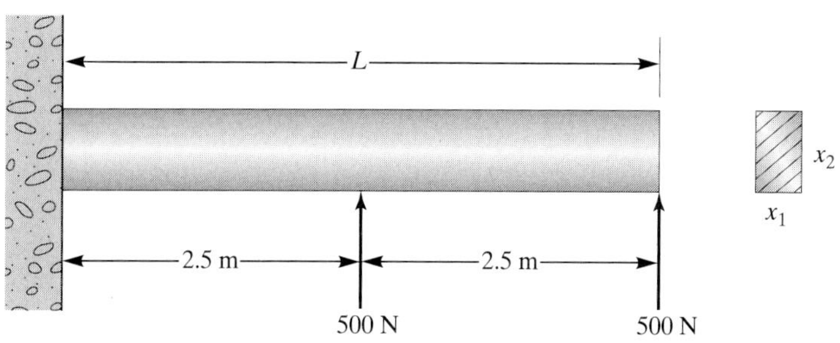

FIGURE 15.3 A schematic of the beam in Example 15.2.

Assuming constant material density, the problem then becomes one of minimizing the volume:

$$\text{Minimize } V = x_1 x_2 L \tag{15.7}$$

The constraints for this problem are

$$\sigma_{max} \leq 30 \text{ MPa} \tag{15.8}$$

$$\delta_{max} \leq 1 \text{ cm} \tag{15.9}$$

$$5 \text{ cm} \leq x_1 \leq 15 \text{ cm} \tag{15.10}$$

$$20 \text{ cm} \leq x_2 \leq 40 \text{ cm} \tag{15.11}$$

For this example, the variables x_1 and x_2 are called *design variables*; the σ-variable for stress and the δ-variable for deflection are called *state variables*. We will solve this problem using ANSYS, but first, let us look at the parametric design language, how to create a batch file, and optimization routines of ANSYS.

15.2 THE PARAMETRIC DESIGN LANGUAGE OF ANSYS*

You can define your own variables or choose one of the ANSYS-supplied parameters. User-named parameters, however, must adhere to the following rules: (1) User-named parameters must consist of one to eight characters and must begin with a letter; (2) a parameter may be assigned a numeric value, a character value, or another parameter, as long as the value of the other parameter is currently known to ANSYS; and (3) parameters can be of a scalar type or represent an array of values. Scalar parameters may be defined by using the following command:

utility menu: **Parameters → Scalar Parameters**

To use a parameter, input the parameter's name in the field where ANSYS expects a value. For example, to assign a modulus of elasticity value of 29×10^6 lb/in^2 to a machine part made of steel, you can define a parameter with the name STEEL and assign a value of 29e6 to it.

ANSYS allows the user to define up to 400 parameters. You can define character parameters by placing the characters in single quotes. For example, if you want to define a parameter by the name of Element and assign the characters PLANE42 to it, you can do so by typing: **Element = 'PLANE 42'.** You can obtain predefined parameters by using the following commands:

utility menu: **Parameters → Get Scalar Data → Parameters**

*Materials were adapted with permission from ANSYS documents.

You can also use thousands of ANSYS-supplied values as parameters. For example, you can retrieve nodal coordinates, node numbers, nodal displacements, nodal stresses, an element volume, and so on, and assign them to parameters. You can access the ANSYS-supplied parameters by using the command

utility menu: **Parameters → Get Scalar Data**

or by using the command

utility menu: **Parameters → Get Vector Data**

You can list the parameters that have been defined by using the following command:

utility menu: **List → Status → Parameters → Named Parameters**

You can use already-defined parameters to form an expression—for example: Area = Length* Width. When using parametric expressions in a command field, use parentheses to force operations to occur in the desired order. ANSYS also offers built-in functions that are a set of mathematical operations that return a single value. Examples include SIN, COS, LOG, EXP, SQRT, ABS. To make use of these functions, use the following command:

utility menu: **Parameters → Array Operations → Vector Functions**

Once you define the model in terms of design parameters, then you can run ANSYS's design-optimization routines interactively with the Graphical User Interface or by using a batch file. The batch mode is generally preferable because it offers a much quicker way to perform analyses. Up to this point, we have been running ANSYS interactively using the GUI. Using a text editor, you can also create an ANSYS batch file with all of the necessary commands to generate a model. The batch file is then submitted to ANSYS as a batch job. The usual procedure for design optimization consists of the following eight main steps:

1. *Create an analysis file to be used during looping.* You begin by initializing the design variables, building the model parametrically, and obtaining a solution. You then need to retrieve and assign to parameters the values that will be used as state variables and objective functions.
2. *Enter OPT and specify the analysis file.* At this point, you are ready to enter the OPT processor to begin optimization.
3. *Declare optimization variables.* Here, you define the objective function and specify which variables are design variables and which are state variables. ANSYS allows you to define only one objective function. You can use up to 60 design variables and up to 100 state variables in your model.
4. *Choose an optimization procedure.* The ANSYS program offers several different optimization procedures. The procedures are divided into *methods* and *tools*. The optimization methods of ANSYS deal with minimizing a single objective function. On the other hand, the optimization tools are techniques to measure and understand the design space of a problem. For a complete list of procedures available with ANSYS, along with the relevant theory behind each procedure, see ANSYS

online documents. You can also supply your own external procedure to ANSYS to be used during the optimization phase.

5. *Specify optimization looping controls.* Here, you specify the maximum number of iterations to be used with an optimization procedure.
6. *Initiate optimization analysis.*
7. *Review the resulting design sets and postprocess results.*

Throughout the book, up to this point, we have explained how to use ANSYS interactively. We now introduce the required steps to create a batch file. We then create a batch file for Example 15.2, and using this problem, we demonstrate the optimization steps.

Batch Files

You may recall from studying Chapter 8 that when you first enter ANSYS, you are at the Begin Level. From the Begin level, you can enter one of the ANSYS processors. Commands that give you entry to a processor always start with a slash (/). For example, the **/PREP7** command gives general access to the ANSYS preprocessor. You gain access to the general postprocessor by issuing the command **/POST1.** To move from one processor to another, you must first return to the Begin Level by exiting the processor you are currently in. Only then can you access another processor. To leave a processor and return to the Begin Level, you must issue the **FINISH** command.

The fundamental tool used to enter data and control the ANSYS program is the *command*. Some commands can be used only at certain places in your batch file, while others may be used in other processors. For example, you cannot use the **/PREP7** model-generating commands in other processors. The command format consists of one or more fields separated by commas. The first field always contains the command name. A command argument may be skipped by not specifying any data between the commas. In such cases, ANSYS substitutes the default data for the argument.

For long programs and to keep track of the flow of the batch file, you can document the batch file by placing comments within the file. A comment is indicated by exclamation mark (!), and thus, information beyond the exclamation point is interpreted as comments by ANSYS. Now, we will use the following examples to show how to create batch files. To better understand batch files, you are encouraged to duplicate the following batch files on your own.

15.3 EXAMPLES OF BATCH FILES

Use a plain text editor (without formats) to create your batch files. Note that ANSYS commands are case insensitive. After you have created a batch file, you can run it using ANSYS's **Batch** option or by using the **Interactive** option. If you decide to use the Interactive option then issue the following command:

utility menu: **File** → **Read Input From** . . .

then choose the text file (batch file) which contains the ANSYS commands. After running a batch file, see filename.**out** for output information. If you need more information about how to utilize a command, use ANSYS's Help menu.

Chapter 3, Example 3.1

```
/Title, Chapter 3, Example 3.1
/Prep7                  !To begin preprocessing (define the model)
Et, 1, link1            !Element type, 2-d truss element
R, 1, 8                 !Real constant, the value of area for this problem
Mp, ex, 1, 1.9e6        !Material property, Modulus of Elasticity
Mp, nuxy, 1, .3         !Material property, Poisson's ratio
N, 1, 0, 0              !Node 1 at 0, 0
N, 2, 36, 0             !Node 2 at 36, 0
N, 3, 0, 36             !Node 3 at 0, 36
N, 4, 36, 36            !Node 4 at 36, 36
N, 5, 72, 36            !Node 5 at 72, 36
/Pnum, node, 1          !Show numbers of the nodes for displays
Nplot                   !Plot the nodes
E, 1, 2                 !Element 1 defined by nodes 1 and 2
E, 2, 3                 !Element 2 defined by nodes 2 and 3
E, 3, 4                 !Element 3 defined by nodes 3 and 4
E, 2, 4                 !Element 4 defined by nodes 2 and 4
E, 2, 5                 !Element 5 defined by nodes 2 and 5
E, 4, 5                 !Element 6 defined by nodes 4 and 5
/Pnum, elem, 1          !Show numbers of the elements for displays
Eplot                   !Plot the elements
Finish                  !Exit Prep7 processor
/Solu                   !To begin the solution phase
D,1, all, 0             !Displacement of node 1 in all directions is zero
D, 3, all, 0            !Displacement of node 3 in all directions is zero
F, 4, fy, -500          !A force of 500 lb at node 4 in the negative y-dir.
F, 5, fy, -500          !A force of 500 lb at node 5 in the negative y-dir.
/Pbc, all, 1            !Show boundary conditions for displays
Eplot                   !Plot the elements (model) with the boundary conditions
Solve                   !Solve
Finish                  !Exit the Solution processor
```

```
/Post1                        !To begin postprocessing
Etable, axforce, smisc, 1     !Make an element table containing axial forces
Etable, axstress, ls, 1       !Make an element table containing axial stresses
/Pbc, all, 1
pldisp, 1                     !Plot the deformed shape
Pletab, axstress              !Plot the axial stresses
Prnsol, u, comp               !Print nodal displacements
Prrsol                        !Print reaction forces (solution)
Pretab                        !Print element table results
Finish                        !Exit the general postprocessor
/EOF                          !End of File
```

An edited version of the output is shown next.

```
***** POST1 NODAL DEGREE OF FREEDOM LISTING *****

LOAD STEP= 1 SUBSTEP= 1
TIME= 1.0000 LOAD CASE= 0

THE FOLLOWING DEGREE OF FREEDOM RESULTS ARE IN GLOBAL
COORDINATES

NODE    UX             UY             UZ
  1     0.0000         0.0000         0.0000
  2     -0.35526E-02   -0.10252E-01   0.0000
  3     0.0000         0.0000         0.0000
  4      0.11842E-02   -0.11436E-01   0.0000
  5      0.23684E-02   -0.19522E-01   0.0000

MAXIMUM ABSOLUTE VALUES
NODE       2           5          0        5
VALUE -0.35526E-02-0.19522E-01  0.0000   0.19665E-01

***** POST1 TOTAL REACTION SOLUTION LISTING *****

LOAD STEP=   1 SUBSTEP= 1
TIME= 1.0000   LOAD CASE= 0
```

THE FOLLOWING X,Y,Z SOLUTIONS ARE IN GLOBAL COORDINATES

NODE	FX	FY
1	1500.0	0.0000
3	1500.0	1000.0

TOTAL VALUES
VALUE 0.22737E-12 1000.0

***** POST1 ELEMENT TABLE LISTING *****

STAT ELEM	CURRENT AXFORCE	CURRENT AXSTRESS
1	−1500.0	−187.50
2	1414.2	176.78
3	500.00	62.500
4	−500.00	−62.500
5	−707.11	−88.388
6	500.00	62.500

MINIMUM VALUES

ELEM	1	1
VALUE	−1500.0	−187.50

MAXIMUM VALUES

ELEM	2	2
VALUE	1414.2	176.78

□

Chapter 4, Example 4.5

```
/Title, Chapter 4, Example 4.5
/Prep7
Et, 1, 3
R, 1, 7.65, 204, 12.22          !Real constant, area, area moment of inertia, beam height
Mp, ex, 1, 30e6
MP, Nuxy, 1, .3
N, 1, 0, 108
N, 2, 120, 108
N, 3, 120, 0
/pnum, node, 1
Nplot
E, 1, 2
```

```
E, 2, 3
/Pnum, elem, 1
Eplot
Finish
/Solu
D, 1, all, 0
D, 3, all, 0
P, 1, 2, 66.67,,                  !Distributed load of 66.67 lb/in
/Pbc, all, 1
Solve
Finish
/Post1
Etable, maxstress,nmisc, 1,3  !Make an element table containing maximum stresses
Nlist
Elist
Rlist
Mplist
Prnsol, u, comp
Prrsol
Pretab
/Pbc, all, 1
Pldisp, 1
Finish
/EOF
```

An edited version of the output is shown next.

```
***** POST1 NODAL DEGREE OF FREEDOM LISTING *****

LOAD STEP= 1   SUBSTEP= 1
TIME= 1.0000   LOAD CASE= 0

THE FOLLOWING DEGREE OF FREEDOM RESULTS ARE IN GLOBAL
COORDINATES

NODE               UX              UY          UZ
1 0.0000           0.0000          0.0000      0.0000
2 -0.28459E-03     -0.16359E-02    0.0000      0.16605E-02
3 0.0000           0.0000          0.0000      0.0000

***** POST1 TOTAL REACTION SOLUTION LISTING*****
```

```
LOAD STEP= 1    SUBSTEP= 1
TIME= 1.0000    LOAD CASE= 0

THE FOLLOWING X, Y, Z SOLUTIONS ARE IN GLOBAL COORDINATES

NODE     FX         FY        MZ
1        544.29     4524.0    0.10235E+06
3        -544.29    3476.4    19296.

TOTAL VALUES
VALUE    0.0000     8000.4    0.12164E+06

***** POST1 ELEMENT TABLE LISTING *****

STAT     CURRENT
ELEM     MAXSTRES
  1      2994.3
  2      728.26
```

□

Chapter 6, Example 6.4

```
/TITLE, Chapter 6, Example 6.4
/Prep7
ET, 1, 32              !1-D conduction element
ET, 2, 34              !1-D convection element
R, 1, 1                !Real constant, Unit area
R, 2, 1
MP, KXX, 1, 0.08       !Material Property, the value of thermal conductivity in the
                        first layer
MP, KXX, 2, 0.074      !Material Property, the value of thermal conductivity in the
                        second layer
MP, KXX, 3, 0.72       !Material Property, the value of thermal conductivity in the
                        third layer
MP, HF, 1, 40          !The value of heat transfer coefficient
N, 1,0,0
N, 2,0.05,0
N, 3,0.2,0
N, 4,0.3,0
N, 5,0.3,0
/pnum, node, 1
nplot
```

```
Type,1                   !You need to define the element type before defining elements
Mat,1                    !Material type 1
Real,1
E, 1, 2
Mat,2
E,2,3
Type,1
Mat,3
E,3,4
Type,2
Real,2
Mat,1
E,4,5
/pnum, elem, 1
Eplot
Finish
/solu
antype,0,new
solcontrol,0
NT, 1, TEMP, 200         !Temperature at node 1 is set at 200 C
NT, 5, TEMP, 30          !Convective temperature at node 5 is set at 30 C
/Pbc, all
Eplot
solve
Finish
/POST1
Nlist
Elist
Mplist
Prnsol
FINISH
/Eof
```

An edited version of the output is shown next.

NODE	X	Y	Z
1	0.0000	0.0000	0.0000
2	0.50000E-01	0.0000	0.0000
3	0.20000	0.0000	0.0000
4	0.30000	0.0000	0.0000
5	0.30000	0.0000	0.0000

LIST ALL SELECTED ELEMENTS. (LIST NODES)

ELEM	MAT	TYP	REL	NODES	
1	1	1	1	1	2
2	2	1	1	2	3
3	3	1	1	3	4
4	1	2	2	4	5

***** POST1 NODAL DEGREE OF FREEDOM LISTING *****

LOAD STEP= 1 SUBSTEP= 1
TIME= 1.0000 LOAD CASE= 0

NODE	TEMP
1	200.00
2	162.27
3	39.894
4	31.509
5	30.000

□

Chapter 9, Problem 10

```
/PREP7
/TITLE, Chapter 9, Problem 10
ET, 1, plane55          !Element Type, PLANE 55
MP, KXX, 1, 168         !Material Property, Thermal Conductivity
k, 1, 0, 0
k, 2, 0, 0.05
k, 3, 0.1,0.05
k, 4, 0.1, 0
```

```
L, 1, 2                  !Defining lines by connecting keypoints
L, 2, 3
L, 3, 4
L, 4, 1
AL, 1, 2, 3, 4           !Defining an area by connecting lines 1, 2, 3, 4, 5
ESIZE,,10                !This command defines the number of element divisions
AMESH, ALL               !This command generates nodes and area elements
FINISH
/solu
NSEL,S,LOC, X,0.1        !Selects a subset of nodes at X=0.1
D, All,Temp,80
ALLSEL                   !Selects everything before applying the other boundary
                         !condition and solving
SFL,2,Conv,50,,20        !This command applies the convective boundary condition on
                          line 2
solve
FINISH
/POST1
SET, 1
plnsol, temp             !Plot nodal solution
prnsol, temp             !Prints nodal solution
FINISH
/eof
```

An edited version of the output (partial solution) is shown next.

```
***** POST1 NODAL DEGREE OF FREEDOM LISTING *****

LOAD STEP= 1   SUBSTEP= 1
TIME= 1.0000   LOAD CASE= 0

NODE      TEMP
1         78.410
2         77.979
3         78.406
4         78.393
5         78.371
6         78.341
7         78.302
8         78.255
9         78.199
```

10	78.134
11	78.061
12	80.000
13	77.996
14	78.048
15	78.135
16	78.257
17	78.416
18	78.612
19	78.849
20	79.135
21	79.478
22	80.000
23	80.000
24	80.000
25	80.000
26	80.000
27	80.000
28	80.000
29	80.000
30	80.000
31	80.000
32	79.732
33	79.474
34	79.234
35	79.021
36	78.836
37	78.684
38	78.564
39	78.479

□

Chapter 10, Problem 1

```
/Title, Chapter 10, Problem 1
length=6.                !defining variable length
height=6.                !defining variable height
radius=.5                !defining variable radius
load=-1000.              !load in psi
/PREP7
ET, 1, plane82           !2-D plane element
MP, EX, 1, 30E6
MP, NUXY, 1, 0.3
```

```
K, 1, length, 0              !defining keypoints
k, 2, length, height
K, 3, 0, height
K, 4, 0, radius
K, 5, radius, 0
K, 6, 0, 0
L, 1, 2                      !defining the lines connecting the keypoints
L, 2, 3
L, 3, 4
LARC, 4, 5, 6, radius        !defining a circular arc
        !Larc, P1, P2, Pc, Rad
        !P1: Keypoint at one end of circular arc line
        !P2: Keypoint at other end of circular arc line
        !Pc: Keypoint defining plane of arc and center of curvature side
        !Rad: Radius of curvature of the arc
L, 5, 1
AL, 1, 2, 3, 4, 5            !defining an area by connecting lines 1, 2, 3, 4, 5
ESIZE,,6                     !This command defines the number of element divisions
AMESH, ALL                   !This command meshes the area and generates nodes and
                              area elements
FINISH
/solu
dl, 3, 1, symm               !This command specifies symmetry surfaces on a line seg-
                              ment
          !dl, line, area, lab
          !line: line number
          !area: area (number) containing the line
          !lab: symmetry label
dl, 5, 1, symm
nsel, x, length, length      !This command selects subset of nodes
sf, all, pres, load          !This command is used to define a pressure load
nall
solve
FINISH
/POST1
SET, 1
```

```
/pnum, kpoi, 1
/pnum, line, 1
/pnum, element, 1
/WIND, 1, LTOP              !Window 1 at left top corner
kplot
/NOERASE                    !Overlays displays
/WIND, 1, OFF               !TURN WINDOW 1 OFF
/wind, 2, rtop              !Window 2 at right top corner
lplot
/wind, 2, off
/WIND, 3, lbot              !Window 3 at left bottom
eplot
/wind, 3, off
/wind, 4, rbot
/pnum, element, 0
plnsol, s, x                !Plots the nodal solution
nsel, x, 0, 0
Prnsol, s, comp             !Prints the nodal solution result
FINISH
/Eof
```

□

Chapter 11, Example 11.7

```
/Prep7
/Title, Chapter 11, Example 11.7
ET,1,3
Mp, Ex, 1, 30e6
Mp, Prxy, 1, 0.3
Mp, Dens, 1, 0.000732    !Density=weight/(area*length*g)=26/(12)(32.2)(7.65)(12)
R, 1, 7.65, 204, 12.22
R, 2, 7.68, 301, 15.69
N, 1, 0, 0
N, 2, 0, 180
N, 3, 240, 180
N, 4, 240, 0
E, 1,2
```

```
Real, 2
E, 2, 3
Real, 1
E, 3, 4
Finish
/Solu
antype,modal,new     !Analysis type, modal
Modopt, lanb,6       !Modal (solution) options, Block Lanczos method, Number of
                      modes to extract
D, 1, all, 0         !Node 1 is fixed (zero displacement)
D, 3, all, 0         !Node 3 is fixed
Solve
Finish
/Post1
Set,list             !Defines the data set to be read and list from the results file
finish
/eof
```

□

Chapter 13, Problem 6 (partial solution, without the eight holes)

```
/Prep7
/Title, Chapter 13, Problem 6
K,1,0,1                  !Keypoint 1 at x=0 and y=1
K,2,10,1                 !Keypoint 2 at x=10 and y=1
Kgen, 2, all,,,0,1,0     !The Kgen command generates additional keypoints from
                          !an existing pattern
      !kgen, Itime, Np1, Np2, Ninc, DX, DY, DZ,Kinc, Noelem, Imove
      !Itime: number of sets to be generated including the original set
      !Np1, Np2, Ninc: set of keypoints defining pattern to be copied
      !If Np1=all, Np2 and Ninc are ignored and the pattern is all selected keypoints
      !DX, DY, DZ: geometric increments in active coordinate system between sets
      !Kinc: keypoint increment between generated sets.
      !Noelem: specifies if elements and nodes are also to be generated;
      !0=generate them; 1=do not generate
      !Imove: specifies whether keypoints will be moved or newly defined.
K,,5.25,2                      !When keypoint number is left blank, the lowest avail-
                                able number is used
```

```
K,, 6.75, 2
Kgen, 2, -2,,,0, 4, 0
K,, 2, 6
K,, 9, 6
Kgen, 2, -2,,, 0, 1, 0
/Pnum, kpoi, 1
kplot                    !Plot keypoints
L, 1, 2                  !Defines a line between two keypoints
L, 2, 4
L, 4, 6
L, 6, 8
L, 8, 10
L, 10, 12
L, 12, 11
L, 11, 9
L, 9, 7
L, 7, 5
L, 5, 3
L, 3, 1
/pnum, line, 1
Lplot                    !Plot lines
Lfillt, 9, 10, .8        !Generates a fillet line between two intersecting lines

        !Lfillt,Nl1,Nl2,Rad,Pcent
        !Nl1: number of the first intersecting line
        !Nl2: number of the second intersecting line
        !Rad: radius of the fillet
        !Pcent: number to be assigned to generated keypoint at fillet arc center
Lfillt, 11, 10, .8
Lfillt, 3, 4, .7
Lfillt, 5, 4, .7
Lplot
kgen, 2, 1, 2,, 0, -1, 0
kplot
```

arotat, 1, 2, 3, 15, 4, 16, 22, 21, 360 !Arotat generates areas by rotating a line pattern about an axis

!Arotat,Nl1,Nl2,Nl3,Nl4,Nl5,Nl6,Pax1,Pax2,Arc
!Nl1,Nl2,...,Nl6: list of lines (maximum of 6 lines) to be rotated
!Pax1,Pax2: keypoints defining the axis about which the line pattern is to be rotated
!Arc: Arc length in degrees

arotat, 5, 6, 7, 8, 9, 13, 22, 21, 360
arotat, 10, 14, 11, 12,,, 22, 21, 360
nummrg, kpoi !Merges keypoints
/view, 1, 5, 2, 5 !Defines the viewing direction

!View,Wn,Xv,Yv,Zv
!Wn: Window number
!Xv,Yv,Zv: The object is viewed along the line from point Xv,Yv,Zv in the global coords to global origin

aplot
finish
/eof □

15.4 AN EXAMPLE OF AN OPTIMIZATION BATCH FILES

We will now solve Example 15.2 using ANSYS. The batch file is as follows:

```
/PREP7          ! Gain access to the preprocessor
!
!     Initialize design variable parameters:
!
X1=2.0          ! Initialize variable X1, the width of the cross section
X2=3.0          ! Initialize variable X2, the height of the cross section
!
```

```
!      Define element type, area, area moment of inertia:
!
ET,1,BEAM3          ! Define element type; two-dimensional beam element selected
                    ! (ET=Element Type, element reference number=1)
AREA=X1*X2          ! Define the beam's cross-sectional area
IZZ=(X1*(X2**3))/12       ! Second moment of the area about the Z-axis
!
!      Assign real constants, modulus of elasticity:
!
R,1,AREA,IZZ,X2           ! Assign area, area moment of inertia, cross-
                            sectional height
                          ! (R=Real Constant—designating geometry properties
                            such as area, area moment of inertia, real constant
                            reference number=1)
MP,EX,1,30E6              ! Assign value of the modulus of elasticity
                          ! (MP=Material Property, EX=modulus of elasticity,
                            material reference number=1, value of modulus of
                            elasticity)
!
!      Create the geometry of the problem:
!
N,1,0,0                 ! Define node 1 at location X=0, Y=0
N,2,2.5,0               ! Define node 2 at location X=2.5, Y=0
N,3,5.0,0               ! Define node 3 at location X=5.0, Y=0
E,1,2           ! Define element 1 as having node 1 and node 2
E,2,3           ! Define element 2 as having node 2 and node 3
FINISH          ! Return to the Begin Level to access other processors
!
!      Apply boundary conditions, apply loading, and obtain solution:
!
/SOLU           ! Enter the Solver processor


ANTYPE,STATIC   ! Analysis type is static
D,1,ALL,0       ! Apply boundary conditions; all displacements at node 1
                  are zero
F,2,FY,500      ! Apply 500 lb at node 2 in the positive Y-direction
F,3,FY,500      ! Apply 500 lb at node 3 in the positive Y-direction
SOLVE           ! Solve the problem
FINISH          ! Return to the Begin Level to access other processors
!
!      Retrieve results parametrically:
!
/POST1                          ! Enter postprocessing
NSORT,U,Y                       ! Sort nodes based on UY deflection
*GET,DELTAMAX,SORT,,MAX         ! Assign DELTAMAX = maximum deflection
ETABLE,VOLU,VOLU                ! VOLU = volume of each element
ETABLE,SMAX_I,NMISC,1       ! SMAX_I = maximum stress at end I of each
                                       element
ETABLE,SMAX_J,NMISC,3       ! SMAX_J = maximum stress at end J of each
                                       element
SSUM
*GET,VOLUME,SSUM,ITEM,VOLU ! Parameter VOLUME = total volume
```

```
ESORT,ETAB,SMAX_I,,1        ! Sorts elements based on absolute value of
                              SMAX_I
*GET,SMAXI,SORT,,MAX        ! Parameter SMAX_I = maximum value of SMAX_I
ESORT,ETAB,SMAX_J,,1        ! Sorts elements based on absolute value of
                              SMAX_J
*GET,SMAXJ,SORT,,MAX        ! Parameter SMAX_J = maximum value of SMAX_J
SMAX=SMAXI>SMAXJ            ! Parameter SMAX = the greater of SMAXI and
                              SMAXJ
FINISH
!
!     Establish parameters for optimization:
!
/OPT
OPVAR,X1,DV,0.05,0.15              ! Parameter X1 is a design variable
                                   ! (See Eq. (15.10))
OPVAR,X2,DV,0.2,0.4                ! Parameter X2 is a design variable
                                   ! (See Eq. (15.11))
OPVAR,DELTAMAX,SV,0,0.01           ! Parameter DELTAMAX is a state variable
                                   ! (See Eq. (15.9))
OPVAR,SMAX,SV,0,30000000           ! Parameter SMAX is a state variable
                                   ! (See Eq. (15.8))
OPVAR,VOLUME,OBJ                   ! VOLUME is the objective function
OPTYPE,SUBP                        ! Use subproblem approximation method
OPSUBP,100                         ! Maximum number of iterations
OPEXE                              ! Initiate optimization
OPTYPE,SWEEP                       ! Sweep evaluation tool
OPSWEEP,BEST,5                     ! 5 evaluations per DV at best design set
OPEXE                              ! Initiate optimization looping
FINISH
```

The optimization procedure used in the problem we just solved included the subproblem approximation method, which is an advanced zero-order method, and the sweep-generation technique, which varies one design variable at a time over its full range using uniform design variable increments.

An edited version of the output is shown next.

```
***** ANSYS ANALYSIS DEFINITION (PREP7) *****

     PARAMETER X1 = 0.1000000

     PARAMETER X2 = 0.3000000

     ELEMENT TYPE 1 IS BEAM3 2-D ELASTIC BEAM
         KEYOPT(1-12) = 0 0 0 0 0 0 0 0 0 0 0 0

     CURRENT NODAL DOF SET IS UX UY ROTZ
     TWO-DIMENSIONAL MODEL

     PARAMETER AREA = 0.3000000E-01

     PARAMETER IZZ = 0.2250000E-03

     REAL CONSTANT SET 1 ITEMS 1 TO 6
0.30000E-01 0.22500E-03 0.30000 0. 0. 0.

     MATERIAL 1 EX = 0.1300000E+11

     NODE 1 KCS = 0 X,Y,Z= 0. 0. 0.

     NODE 2 KCS = 0 X,Y,Z= 2.5000 0. 0.

     NODE 3 KCS = 0 X,Y,Z= 5.0000 0. 0.

     ELEMENT 1 1 2

     ELEMENT 2 2 3

***** ROUTINE COMPLETED ***** CP = 0.551
```

```
***** ANSYS SOLUTION ROUTINE *****

 PERFORM A STATIC ANALYSIS
 THIS WILL BE A NEW ANALYSIS

 SPECIFIED CONSTRAINT UX FOR SELECTED NODES 1 TO 1
 BY 1
 REAL= 0. IMAG= 0.
 ADDITIONAL DOFS= UY ROTZ

 SPECIFIED NODAL LOAD FY FOR SELECTED NODES 2 TO 2
 BY 1

 REAL= 500.000000 IMAG= 0.

 SPECIFIED NODAL LOAD FY FOR SELECTED NODES 3 TO 3
 BY 1

 REAL= 500.000000 IMAG= 0.

****** ANSYS SOLVE COMMAND *****

     S O L U T I O N O P T I O N S

 PROBLEM DIMENSIONALITY.........2-D
 DEGREES OF FREEDOM.... UX UY ROTZ
 ANALYSIS TYPE..............STATIC (STEADY-STATE)

    L O A D S T E P O P T I O N S

 LOAD STEP NUMBER............. 1
 TIME AT END OF THE LOAD STEP..... 1.0000
 NUMBER OF SUBSTEPS............ 1
 STEP CHANGE BOUNDARY CONDITIONS....... NO
  PRINT OUTPUT CONTROLS..........NO PRINTOUT
 DATABASE OUTPUT CONTROLS..........ALL DATA WRITTEN
                  FOR THE LAST SUBSTEP
```

```
***** ANSYS RESULTS INTERPRETATION (POST1) *****

 SORT ON ITEM=U COMPONENT=Y ORDER= 0 KABS= 0 NMAX= 3

 SORT COMPLETED FOR 3 VALUES.

 *GET DELTAMAX FROM SORT ITEM=MAX  VALUE= 0.934829060E-02

 STORE VOLU FROM ITEM=VOLU FOR ALL SELECTED ELEMENTS

 STORE SMAX_I FROM ITEM=NMIS COMP= 1 FOR ALL SELECTED ELEMENTS

 STORE SMAX_J FROM ITEM=NMIS COMP= 3 FOR ALL SELECTED ELEMENTS

 SUM ALL THE ACTIVE ENTRIES IN THE ELEMENT TABLE

 TABLE LABEL TOTAL
 VOLU 0.150000
 SMAX_I 0.333333E+07
 SMAX_J 833333.

 *GET VOLUME FROM SSUM ITEM=ITEM VOLU         VALUE= 0.150000000

 SORT ON ITEM=ETAB COMPONENT=SMAX ORDER= 0 KABS= 1 NMAX= 2

 SORT COMPLETED FOR 2 VALUES.

 *GET SMAXI FROM SORT ITEM=MAX      VALUE= 2500000.00

 SORT ON ITEM=ETAB COMPONENT=SMAX ORDER= 0 KABS= 1 NMAX= 2

 SORT COMPLETED FOR 2 VALUES.

 *GET SMAXJ FROM SORT ITEM=MAX      VALUE= 833333.333

 PARAMETER SMAX = 2500000.

 EXIT THE ANSYS POST1 DATABASE PROCESSOR
```

```
***** ANSYS OPTIMIZATION ANALYSIS (OPT) *****

9 Parameters exist and design set No. 1 is established.

DV NAME= X1       MIN= 0.50000E-01  MAX= 0.15000     TOLER= 0.10000E-02

DV NAME= X2       MIN= 0.20000     MAX= 0.40000      TOLER= 0.20000E-02

SV NAME= DELTAMAX MIN=      0. MAX= 0.10000E-01       TOLER= 0.10000E-03

SV NAME= SMAX MIN= 0.        MAX= 0.30000E+08 TOLER= 0.30000E+06

Default OBJ tolerance set to 0.01*(CURRENT PARAMETER VALUE) = 1.5E-03.

OBJ NAME= VOLUME    TOLER= 0.15000E-02

ACTIVE OPTIMIZATION IS THE SUBPROBLEM APPROXIMATION METHOD

SUBPROBLEM APPROXIMATION OPTIMIZATION WILL PERFORM A MAXIMUM OF
100 ITERATIONS
UPON EXECUTION WITH A MAXIMUM OF 7 SEQUENTIAL INFEASIBLE SOLUTIONS

RUN OPTIMIZATION (SUBPROBLEM APPROXIMATION) WITH A MAXIMUM OF 100
ITERATIONS
AND 7 ALLOWED SEQUENTIAL INFEASIBLE SOLUTIONS.

>> BEGIN SUBPROBLEM APPROXIMATION ITERATION 1 OF 100 (MAX) <<<

>>>>>> SOLUTION HAS CONVERGED TO POSSIBLE OPTIMUM <<<<<<
 (BASED ON OBJ TOLERANCE BETWEEN FINAL TWO DESIGNS)

FINAL VARIABLES ARE

        SET 6
       (FEASIBLE)
DELTAMAX(SV) 0.93351E-02
SMAX (SV) 0.29249E+07
X1 (DV) 0.62267E-01
X2 (DV) 0.35148
VOLUME (OBJ) 0.10943
```

```
***** DESIGN SENSITIVITY SUMMARY TABLE ******

    VOLUME DELTAMAX SMAX
X1  1.690  -0.1304  -0.4032E+08
X2 0.4513 -0.7969E-01  -0.1533E+08

ACTIVE OPTIMIZATION TOOL IS SWEEP EVALUATION

SWEEP OPTIMIZATION TOOL WILL PERFORM 5 ITERATIONS PER DESIGN
VARIABLE UPON
EXECUTION ABOUT BEST DESIGN SET BASED ON 2 CURRENT DESIGN VARIABLES

RUN OPTIMIZATION (SWEEP DESIGNS) WITH A MAXIMUM OF 10 ITERATIONS
ABOUT DESIGN SET 6.

>>BEGIN SWEEP ITERATION 1 OF 10 <<<

>>>SWEEP SELECTION OPTIMIZATION COMPLETED AFTER 10 ITERATIONS<<<<<<
```

```
BEST VARIABLES ARE:

       SET 15
         (FEASIBLE)

DELTAMAX(SV)       0.94544E-02
SMAX   (SV)        0.29498E+07
X1    (DV)         0.62267E-01
X2    (DV)         0.35000
VOLUME (OBJ)       0.10897
```

```
*** EXIT FROM ANSYS DESIGN OPTIMIZATION (/OPT) ***

***** ROUTINE COMPLETED ***** CP = 4.035

***** END OF INPUT ENCOUNTERED *****

NUMBER OF WARNING MESSAGES ENCOUNTERED= 0
NUMBER OF ERROR MESSAGES ENCOUNTERED= 0
```

The above ANSYS output should give you a good idea of the steps that the program follows to move toward an optimized solution.

SUMMARY

At this point you should

1. have a good understanding of the fundamental concepts in design optimization, including the definitions of objective function, constraints, state variables, and design variables. You should also know what is meant by a feasible solution region.
2. know how to define and retrieve user-defined and ANSYS-supplied parameters.
3. know the basic steps involved in the optimization process of ANSYS.
4. be familiar with the creation of batch files.

REFERENCES

ANSYS User's Manual: Introduction to ANSYS, Vol. I, Swanson Analysis Systems, Inc.

ANSYS User's Manual: Procedures, Vol. I, Swanson Analysis Systems, Inc.

ANSYS User's Manual: Commands, Vol. II, Swanson Analysis Systems, Inc.

ANSYS User's Manual: Elements, Vol. III, Swanson Analysis Systems, Inc.

Hillier, F. S., and Lieberman, G. J., *Introduction to Operations Research*, 6th ed., New York, McGraw-Hill, 1995.

Rekaitis, G. V., Ravindran A., and Ragsdell, K. M., *Engineering Optimization—Methods and Applications*, New York, John Wiley and Sons, 1983.

PROBLEMS

1. Solve Problem 9 in Chapter 3 using a batch file.
2. Solve Problem 15 in Chapter 3 using a batch file.
3. Solve Problem 7 in Chapter 4 using a batch file.
4. Solve Problem 15 in Chapter 4 using a batch file.
5. Solve Problem 22 in Chapter 4 using a batch file.
6. Solve Problem 7 in Chapter 6 using a batch file.
7. Solve Problem 12 in Chapter 9 using a batch file.
8. Solve Problem 17 in Chapter 9 using a batch file.
9. Solve Problem 5 in Chapter 10 using a batch file.
10. Solve Problem 17 in Chapter 11 using a batch file.
11. Solve Problem 18 in Chapter 11 using a batch file.
12. Solve Problem 4 in Chapter 13 using a batch file.

APPENDIX A

Mechanical Properties of Some Materials

Average Mechanical Properties of Typical Engineering Materials[a] (SI Units)

Materials		Density ρ (Mg/m³)	Modulus of Elasticity E (GPa)	Modulus of Rigidity G (GPa)	Yield Strength (MPa) σ_Y Tens.	Comp.[b]	Shear	Ultimate Strength (MPa) σ_u Tens.	Comp.[b]	Shear	% Elongation in 50 mm specimen	Poisson's Ratio ν	Coef. of Therm. Expansion α $(10^6)/°C$
Metallic													
Aluminum Wrought Alloys	2014-T6	2.79	73.1	27	414	414	172	469	469	290	10	0.35	23
	6061-T6	2.71	68.9	26	255	255	131	290	290	186	12	0.35	24
Cast Iron Alloys	Gray ASTM 20	7.19	67.0	27	—	—	—	179	669	—	0.6	0.28	12
	Malleable ASTM A-197	7.28	172	68	—	—	—	276	572	—	5	0.28	12
Copper Alloys	Red Brass C83400	8.74	101	37	70.0	70.0	—	241	241	—	35	0.35	18
	Bronze C86100	8.83	103	38	345	345	—	655	655	—	20	0.34	17
Magnesium Alloy	[Am 1004-T61]	1.83	44.7	18	152	152	—	276	276	152	1	0.30	26
Steel Alloys	Structural A36	7.85	200	75	250	250	—	400	400	—	30	0.32	12
	Stainless 304	7.86	193	75	207	207	—	517	517	—	40	0.27	17
	Tool L2	8.16	200	78	703	703	—	800	800	—	22	0.32	12
Titanium Alloy	[Ti-6Al-4V]	4.43	120	44	924	924	—	1,000	1,000	—	16	0.36	9.4
Nonmetallic													
Concrete	Low Strength	2.38	22.1	—	—	—	12	—	—	—	—	0.15	11
	High Strength	2.38	29.0	—	—	—	38	—	—	—	—	0.15	11
Plastic Reinforced	Kevlar 49	1.45	131	—	—	—	—	717	483	20.3	2.8	0.34	—
	30% Glass	1.45	72.4	—	—	—	—	90	131	—	—	0.34	—
Wood Select Structural Grade	Douglas Fir	0.47	13.1	—	—	—	—	2.1[c]	26[d]	6.2[d]	—	0.29[e]	—
	White Spruce	3.60	9.65	—	—	—	—	2.5[c]	36[d]	6.7[d]		0.31[e]	—

[a] Specific values may vary for a particular material due to alloy or mineral composition, mechanical working of the specimen, or heat treatment. For a more exact value, reference books for the material should be consulted.

[b] The yield and ultimate strengths for ductile materials can be assumed equal for both tension and compression.

[c] Measured perpendicular to the grain.

[d] Measured parallel to the grain.

[e] Deformation measured perpendicular to the grain when the load is applied along the grain.

Source: *Mechanics of Materials*, 2nd ed., R. C. Hibbler, Macmillan, New York.

Average Mechanical Properties of Typical Engineering Materials[a] (U.S. Customary Units)

Materials		Specific Weight γ (lb/in³)	Modulus of Elasticity E (10^3) ksi	Modulus of Rigidity G (10^3) ksi	Yield Strength (10^3) ksi σ_u Tens.	Yield Strength Comp.[b]	Yield Strength Shear	Ultimate Strength (10^3) ksi σ_u Tens.	Ultimate Strength Comp.[b]	Ultimate Strength Shear	% Elongation in 50 mm specimen	Poisson's Ratio ν	Coef. of Therm. Expansion α (10^6)/°F
Metallic													
Aluminum Wrought Alloys	2014-T6	0.101	10.6	3.9	60	60	25	68	68	42	10	0.35	12.8
	6061-T6	0.098	10.0	3.7	37	37	19	42	42	27	12	0.35	13.1
Cast Iron Alloys	Gray ASTM 20	0.260	10.0	3.9	—	—	—	26	97	—	0.6	0.28	6.70
	Malleable ASTM A-197	0.263	25.0	9.8	—	—	—	40	83	—	5	0.28	6.60
Copper Alloys	Red Brass C83400	0.316	14.6	5.4	11.4	11.4	—	35	35	—	35	0.35	9.80
	Bronze C86100	0.319	15.0	5.6	50	50	—	95	95	—	20	0.34	9.60
Magnesium Alloy	[Am 1004-T61]	0.066	6.48	2.5	22	22	—	40	40	22	1	0.30	14.3
Steel Alloys	Structural A36	0.284	29.0	11.0	36	36	—	58	58	—	30	0.32	6.60
	Stainless 304	0.284	28.0	11.0	30	30	—	75	75	—	40	0.27	9.60
	Tool L2	0.295	29.0	11.0	102	102	—	116	116	—	22	0.32	6.50
Titanium Alloy	[Ti-6Al-4V]	0.160	17.4	6.4	134	134	—	145	145	—	16	0.36	5.20
Nonmetallic													
Concrete	Low Strength	0.086	3.20	—	—	—	1.8	—	—	—	—	0.15	6.0
	High Strength	0.086	4.20	—	—	—	5.5	—	—	—	—	0.15	6.0
Plastic Reinforced	Kevlar 49	0.0524	19.0	—	—	—	—	104	70	10.2	2.8	0.34	—
	30% Glass	0.0524	10.5	—	—	—	—	13	19	0.34	—	0.34	—
Wood Select Structural Grade	Douglas Fir	0.017	1.90	—	—	—	—	0.30[c]	3.78[d]	0.90[d]	—	0.29[e]	—
	White Spruce	0.130	1.40	—	—	—	—	0.36[c]	5.18[d]	0.97[d]	—	0.31[e]	—

[a] Specific values may vary for a particular material due to alloy or mineral composition, mechanical working of the specimen, or heat treatment. For a more exact value, reference books for the material should be consulted.

[b] The yield and ultimate strengths for ductile materials can be assumed equal for both tension and compression.

[c] Measured perpendicular to the grain.

[d] Measured parallel to the grain.

[e] Deformation measured perpendicular to the grain when the load is applied along the grain.

APPENDIX B

Thermophysical Properties of Some Materials

Thermophysical Properties of Some Materials (at room temperature or at the specified temperature) (SI units)

Material	Density (kg/m^3)	Specific Heat ($J/kg \cdot K$)	Thermal Conductivity ($W/m \cdot K$)
Aluminum (alloy 1100)	2740	896	221
Asphalt	2110	920	0.74
Cement	1920	670	0.029
Clay	1000	920	
Concrete (stone)	2300	653	1.0
Fireclay Brick	1790 @ 373 K	829	1.0 @ 473 K
Glass (soda lime)	2470	750	1.0 @ 366 K
Glass (lead)	4280	490	1.4
Glass (pyrex)	2230	840	1.0 @ 366 K
Iron (cast)	7210	500	47.7 @ 327 K
Iron (wrought)	7700 @ 373 K		60.4
Paper	930	1300	0.13
Soil†	2050	1840	0.5
Steel (mild)	7830	500	45.3
Wood (ash)	690		0.172 @ 323 K
Wood (mahogany)	550		0.13
Wood (oak)	750	2390	0.176
Wood (pine)	430		0.11

† Reference: Incropera, F., and Dewitt D., *Fundamentals of Heat and Mass Transfer,* 4th ed., New York, John Wiley and Sons, 1996.
Reference: ASHRAE Handbook: Fundamental Volume, American Society of Heating, Refrigerating, and Air-Conditioning Engineers, Atlanta, 1993.

APPENDIX C

Properties of Common Line and Area Shapes

TABLE C.1 Centroids of line segments.

Arc Segment 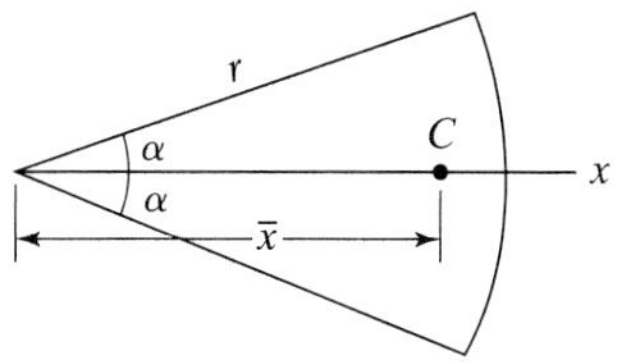	$\bar{x} = \frac{r\sin\alpha}{\alpha}$
Quarter and semicircular Arcs 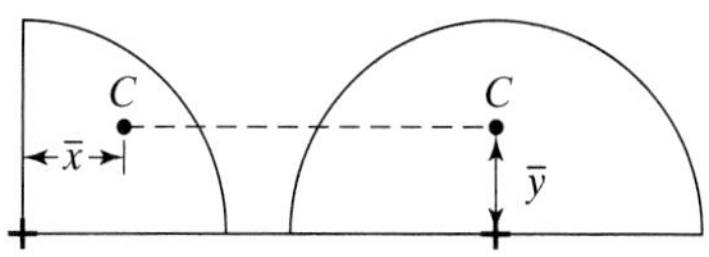	$\bar{x} = \frac{2r}{\pi}$ $\bar{y} = \frac{2r}{\pi}$

TABLE C.2 Centroids and second moments of area of common shapes.

Shape	Centroid	Area Moments of Inertia
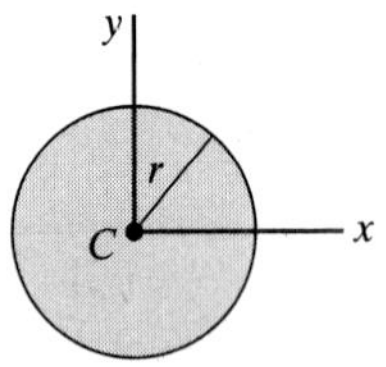	----	$I_x = I_y = \frac{\pi r^4}{4}$ $J_c = \frac{\pi r^4}{2}$
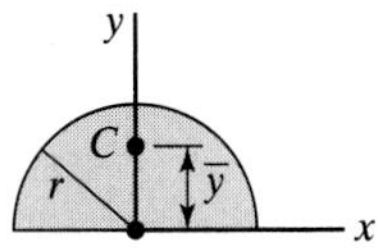	$\bar{y} = \frac{4r}{3\pi}$	$I_x = I_y = \frac{\pi r^4}{8}$ $J_c = \frac{\pi r^4}{4}$
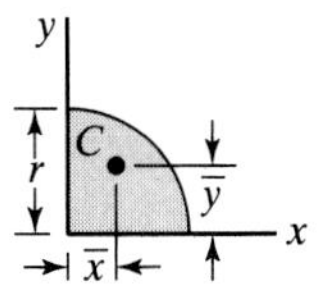	$\bar{x} = \bar{y} = \frac{4r}{3\pi}$	$I_x = I_y = \frac{\pi r^4}{16}$ $J_c = \frac{\pi r^4}{8}$
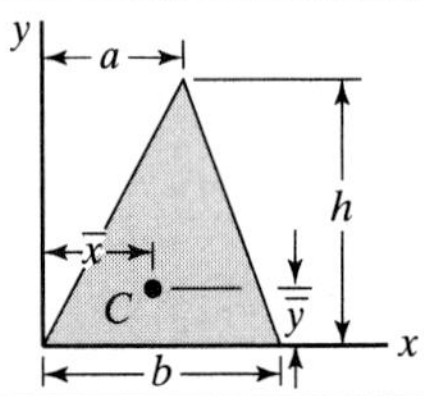	$\bar{x} = \frac{a+b}{3}$ $\bar{y} = \frac{h}{3}$	$I_x = \frac{bh^3}{12}$ $I_{\bar{x}} = \frac{bh^3}{36}$
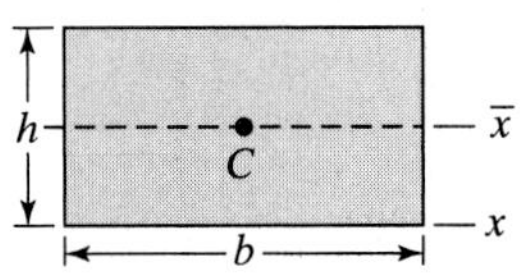	----	$I_x = \frac{bh^3}{3}$ $I_{\bar{x}} = \frac{bh^3}{12}$ $J_c = \frac{bh}{12}(b^2 + h^2)$

TABLE C.3 Mass moments of inertia of common shapes.

Slender rod 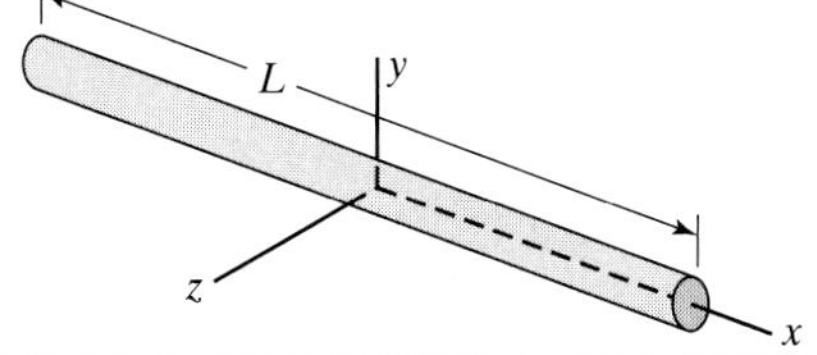 	$I_y = I_z = \frac{1}{12} mL^2$
Thin disk 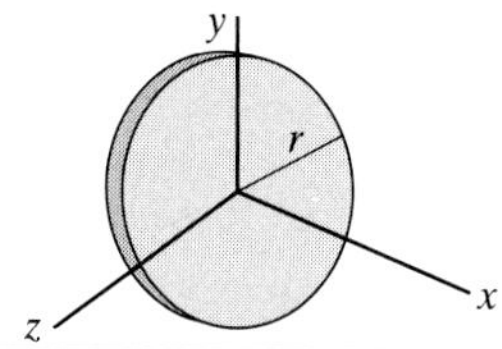 	$I_x = \frac{1}{2} mr^2$ $I_y = I_z = \frac{1}{4} mr^2$
Cylinder 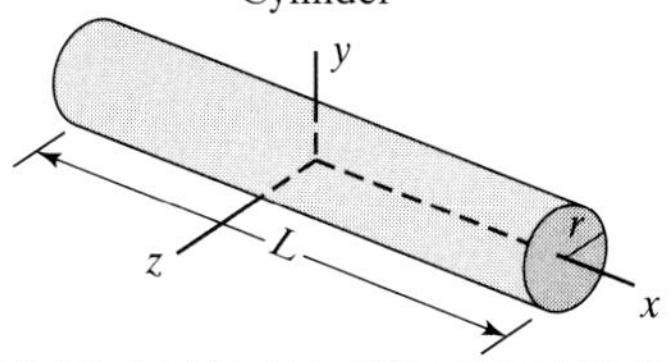 	$I_x = \frac{1}{2} mr^2$ $I_y = I_z = \frac{1}{12} m(3r^2 + L^2)$
Thin Plate 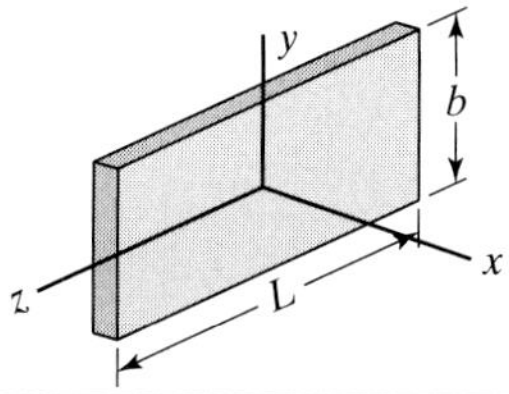 	$I_x = \frac{1}{12} m(b^2 + L^2)$ $I_y = \frac{1}{12} mL^2$ $I_z = \frac{1}{12} mb^2$
A rectangular bar 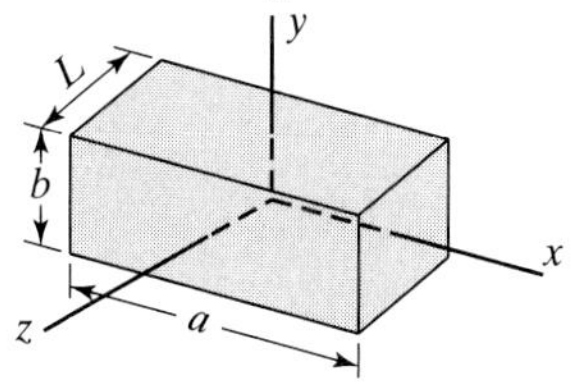 	$I_x = \frac{1}{12} m(b^2 + L^2)$ $I_y = \frac{1}{12} m(L^2 + a^2)$ $I_z = \frac{1}{12} m(a^2 + b^2)$
Sphere 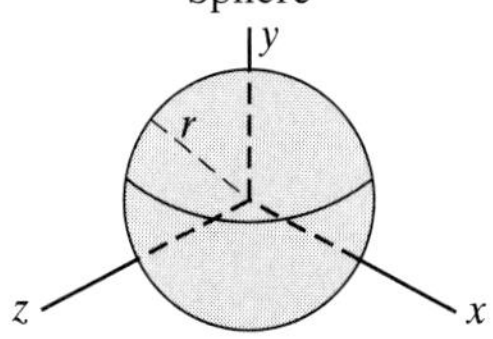 	$I_x = I_y = I_z = \frac{2}{5} mr^2$

APPENDIX D

Geometrical Properties of Structural Steel Shapes

Wide-flange sections or W shapes

				Flange		x–x axis			y–y axis		
	Area A	Depth d	Web thickness t_w	width b	thickness t_f	I	S	r	I	S	r
Designation*	in^2	in.	in.	in.	in.	in^4	in^3	in.	in^4	in^3	in.
W24 × 104	30.6	24.06	0.500	12.750	0.750	3100	258	10.1	259	40.7	2.91
W24 × 94	27.7	24.31	0.515	9.065	0.875	2700	222	9.87	109	24.0	1.98
W24 × 84	24.7	24.10	0.470	9.020	0.770	2370	196	9.79	94.4	20.9	1.95
W24 × 76	22.4	23.92	0.440	8.990	0.680	2100	176	9.69	82.5	18.4	1.92
W24 × 68	20.1	23.73	0.415	8.965	0.585	1830	154	9.55	70.4	15.7	1.87
W24 × 62	18.2	23.74	0.430	7.040	0.590	1550	131	9.23	34.5	9.80	1.38
W24 × 55	16.2	23.57	0.395	7.005	0.505	1350	114	9.11	29.1	8.30	1.34
W18 × 65	19.1	18.35	0.450	7.590	0.750	1070	117	7.49	54.8	14.4	1.69
W18 × 60	17.6	18.24	0.415	7.555	0.695	984	108	7.47	50.1	13.3	1.69
W18 × 65	16.2	18.11	0.390	7.530	0.630	890	98.3	7.41	44.9	11.9	1.67
W18 × 50	14.7	17.99	0.355	7.495	0.570	800	88.9	7.38	40.1	10.7	1.65
W18 × 46	13.5	18.06	0.360	6.060	0.605	712	78.8	7.25	22.5	7.43	1.29
W18 × 40	11.8	17.90	0.315	6.015	0.525	612	68.4	7.21	19.1	6.35	1.27
W18 × 35	10.3	17.70	0.300	6.000	0.425	510	57.6	7.04	15.3	5.12	1.22
W16 × 57	16.8	16.43	0.430	7.120	0.715	758	92.2	6.72	43.1	12.1	1.60
W16 × 50	14.7	16.26	0.380	7.070	0.630	659	81.0	6.68	37.2	10.5	1.59
W16 × 45	13.3	16.13	0.345	7.035	0.565	586	72.7	6.65	32.8	9.34	1.57
W16 × 36	10.6	15.86	0.295	6.985	0.430	448	56.5	6.51	24.5	7.00	1.52
W16 × 31	9.12	15.88	0.275	5.525	0.440	375	47.2	6.41	12.4	4.49	1.17
W16 × 26	7.68	15.69	0.250	5.500	0.345	301	38.4	6.26	9.59	3.49	1.12
W14 × 53	15.6	13.92	0.370	8.060	0.660	541	77.8	5.89	57.7	14.3	1.92
W14 × 43	12.6	13.66	0.305	7.995	0.530	428	62.7	5.82	45.2	11.3	1.89
W14 × 38	11.2	14.10	0.310	6.770	0.515	385	54.6	5.87	26.7	7.88	1.55
W14 × 34	10.0	13.98	0.285	6.745	0.455	340	48.6	5.83	23.3	6.91	1.53
W14 × 30	8.85	13.84	0.270	6.730	0.385	291	42.0	5.73	19.6	5.82	1.49
W14 × 26	7.69	13.91	0.255	5.025	0.420	245	35.3	5.65	8.91	3.54	1.08
W14 × 22	6.49	13.74	0.230	5.000	0.335	199	29.0	5.54	7.00	2.80	1.04

* Reported with a W, then the nominal depth in inches and the weight per foot.

Source: *Mechanics of Materials*, 2nd ed., R. C. Hibbler, Macmillan, New York.

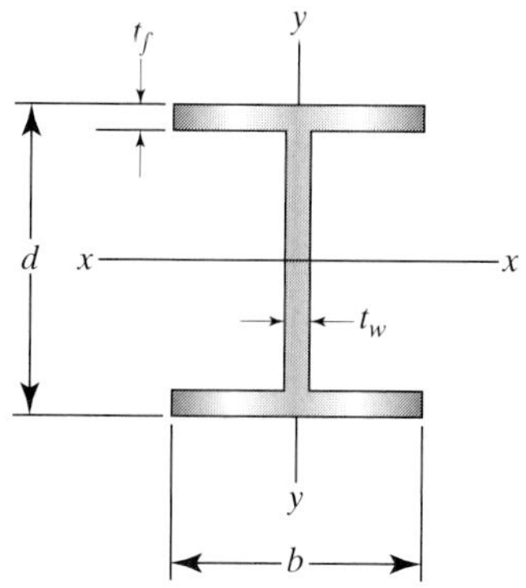

Wide-flange sections or W shapes

Designation*	Area A	Depth d	Web thickness t_w	Flange width b	Flange thickness t_f	x–x axis I	x–x axis S	x–x axis r	y–y axis I	y–y axis S	y–y axis r
	in^2	in.	in.	in.	in.	in^4	in^3	in.	in^4	in^3	in.
W12 × 87	25.6	12.53	0.515	12.125	0.810	740	118	5.38	241	39.7	3.07
W12 × 50	14.7	12.19	0.370	8.080	0.640	394	64.7	5.18	56.3	13.9	1.96
W12 × 45	13.2	12.06	0.335	8.045	0.575	350	58.1	5.15	50.0	12.4	1.94
W12 × 26	7.65	12.22	0.230	6.490	0.380	204	33.4	5.17	17.3	5.34	1.51
W12 × 22	6.48	12.31	0.260	4.030	0.425	156	25.4	4.91	4.66	2.31	0.847
W12 × 16	4.71	11.99	0.220	3.990	0.265	103	17.1	4.67	2.82	1.41	0.773
W12 × 14	4.16	11.91	0.200	3.970	0.225	88.6	14.9	4.62	2.36	1.19	0.753
W10 × 100	29.4	11.10	0.680	10.340	1.120	623	112	4.60	207	40.0	2.65
W10 × 54	15.8	10.09	0.370	10.030	0.615	303	60.0	4.37	103	20.6	2.56
W10 × 45	13.3	10.10	0.350	8.020	0.620	248	49.1	4.32	53.4	13.3	2.01
W10 × 30	8.84	10.47	0.300	5.810	0.510	170	32.4	4.38	16.7	5.75	1.37
W10 × 39	11.5	9.92	0.315	7.985	0.530	209	42.1	4.27	45.0	11.3	1.98
W10 × 19	5.62	10.24	0.250	4.020	0.395	96.3	18.8	4.14	4.29	2.14	0.874
W10 × 15	4.41	9.99	0.230	4.000	0.270	68.9	13.8	3.95	2.89	1.45	0.810
W10 × 12	3.54	9.87	0.190	3.960	0.210	53.8	10.9	3.90	2.18	1.10	0.785
W8 × 67	19.7	9.00	0.570	8.280	0.935	272	60.4	3.72	88.6	21.4	2.12
W8 × 58	17.1	8.75	0.510	8.220	0.810	228	52.0	3.65	75.1	18.3	2.10
W8 × 48	14.1	8.50	0.400	8.110	0.685	184	43.3	3.61	60.9	15.0	2.08
W8 × 40	11.7	8.25	0.360	8.070	0.560	146	35.5	3.53	49.1	12.2	2.04
W8 × 31	9.13	8.00	0.285	7.995	0.435	110	27.5	3.47	37.1	9.27	2.02
W8 × 24	7.08	7.93	0.245	6.495	0.400	82.8	20.9	3.42	18.3	5.63	1.61
W8 × 15	4.44	8.11	0.245	4.015	0.315	48.0	11.8	3.29	3.41	1.70	0.876
W6 × 25	7.34	6.38	0.320	6.080	0.455	53.4	16.7	2.70	17.1	5.61	1.52
W6 × 20	5.87	6.20	0.260	6.020	0.365	41.4	13.4	2.66	13.3	4.41	1.50
W6 × 15	4.43	5.99	0.230	5.990	0.260	29.1	9.72	2.56	9.32	3.11	1.46
W6 × 16	4.74	6.28	0.260	4.030	0.405	32.1	10.2	2.60	4.43	2.20	0.966
W6 × 12	3.55	6.03	0.230	4.000	0.280	22.1	7.31	2.49	2.99	1.50	0.918
W6 × 9	2.68	5.90	0.170	3.940	0.215	16.4	5.56	2.47	2.19	1.11	0.905

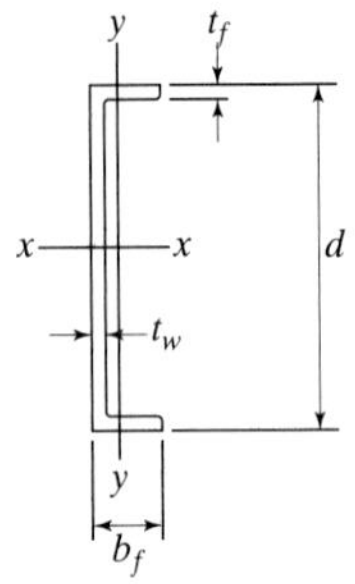

American standard channels or C shapes

Designation*	Area A	Depth d	Web thickness t_w		Flange width b_f		Flange thickness t_f		x-x axis I	x-x axis S	x-x axis r	y-y axis I	y-y axis S	y-y axis r
	in²	in.	in.		in.		in.		in⁴	in³	in.	in⁴	in³	in.
C15 × 50	14.7	15.00	0.716	11/16	3.716	3 3/4	0.650	5/8	404	53.8	5.24	11.0	3.78	0.867
C15 × 40	11.8	15.00	0.520	1/2	3.520	3 1/2	0.650	5/8	349	46.5	5.44	9.23	3.37	0.886
C15 × 33.9	9.96	15.00	0.400	3/8	3.400	3 3/8	0.650	5/8	315	42.0	5.62	8.13	3.11	0.904
C12 × 30	8.82	12.00	0.510	1/2	3.170	3 1/8	0.501	1/2	162	27.0	4.29	5.14	2.06	0.763
C12 × 25	7.35	12.00	0.387	3/8	3.047	3	0.501	1/2	144	24.1	4.43	4.47	1.88	0.780
C12 × 20.7	6.09	12.00	0.282	5/16	2.942	3	0.501	1/2	129	21.5	4.61	3.88	1.73	0.799
C10 × 30	8.82	10.00	0.673	11/16	3.033	3	0.436	7/16	103	20.7	3.42	3.94	1.65	0.669
C10 × 25	7.35	10.00	0.526	1/2	2.886	2 7/8	0.436	7/16	91.2	18.2	3.52	3.36	1.48	0.676
C10 × 20	5.88	10.00	0.379	3/8	2.739	2 3/4	0.436	7/16	78.9	15.8	3.66	2.81	1.32	0.692
C10 × 15.3	4.49	10.00	0.240	1/4	2.600	2 5/8	0.436	7/16	67.4	13.5	3.87	2.28	1.16	0.713
C9 × 20	5.88	9.00	0.448	7/16	2.648	2 5/8	0.413	7/16	60.9	13.5	3.22	2.42	1.17	0.642
C9 × 15	4.41	9.00	0.285	5/16	2.485	2 1/2	0.413	7/16	51.0	11.3	3.40	1.93	1.01	0.661
C9 × 13.4	3.94	9.00	0.233	1/4	2.433	2 3/8	0.413	7/16	47.9	10.6	3.48	1.76	0.962	0.669
C8 × 18.75	5.51	8.00	0.487	1/2	2.527	2 1/2	0.390	3/8	44.0	11.0	2.82	1.98	1.01	0.599
C8 × 13.75	4.04	8.00	0.303	5/16	2.343	2 3/8	0.390	3/8	36.1	9.03	2.99	1.53	0.854	0.615
C8 × 11.5	3.38	8.00	0.220	1/4	2.260	2 1/4	0.390	3/8	32.6	8.14	3.11	1.32	0.781	0.625
C7 × 14.75	4.33	7.00	0.419	7/16	2.299	2 1/4	0.366	3/8	27.2	7.78	2.51	1.38	0.779	0.564
C7 × 12.25	3.60	7.00	0.314	5/16	2.194	2 1/4	0.366	3/8	24.2	6.93	2.60	1.17	0.703	0.571
C7 × 9.8	2.87	7.00	0.210	3/16	2.090	2 1/8	0.366	3/8	21.3	6.08	2.72	0.968	0.625	0.581
C6 × 13	3.83	6.00	0.437	7/16	2.157	2 1/8	0.343	5/16	17.4	5.80	2.13	1.05	0.642	0.525
C6 × 10.5	3.09	6.00	0.314	5/16	2.034	2	0.343	5/16	15.2	5.06	2.22	0.866	0.564	0.529
C6 × 8.2	2.40	6.00	0.200	3/16	1.920	1 7/8	0.343	5/16	13.1	4.38	2.34	0.693	0.492	0.537
C5 × 9	2.64	5.00	0.325	5/16	1.885	1 7/8	0.320	5/16	8.90	3.56	1.83	0.632	0.450	0.489
C5 × 6.7	1.97	5.00	0.190	3/16	1.750	1 3/4	0.320	5/16	7.49	3.00	1.95	0.479	0.378	0.493
C4 × 7.25	2.13	4.00	0.321	5/16	1.721	1 3/4	0.296	5/16	4.59	2.29	1.47	0.433	0.343	0.450
C4 × 5.4	1.59	4.00	0.184	3/16	1.584	1 5/8	0.296	5/16	3.85	1.93	1.56	0.319	0.283	0.449
C3 × 6	1.76	3.00	0.356	3/8	1.596	1 5/8	0.273	1/4	2.07	1.38	1.08	0.305	0.268	0.416
C3 × 5	1.47	3.00	0.258	1/4	1.498	1 1/2	0.273	1/4	1.85	1.24	1.12	0.247	0.233	0.410
C3 × 4.1	1.21	3.00	0.170	3/16	1.410	1 3/8	0.273	1/4	1.66	1.10	1.17	0.197	0.202	0.404

* Reported with a C, then the nominal depth in inches and the weight per foot.

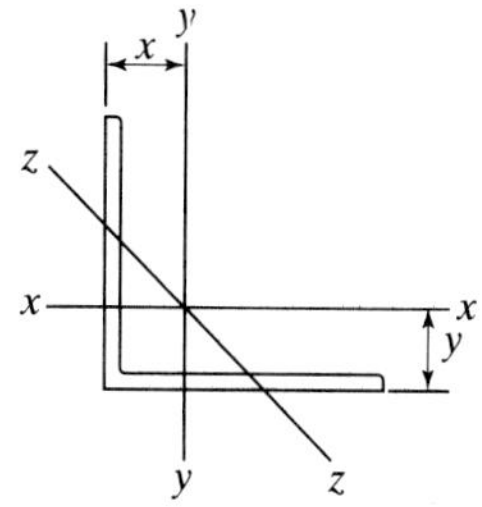

Angles having equal legs

Size and thickness	Weight per Foot	Area A	x–x axis				y–y axis				z–z axis
			I	S	r	y	I	S	r	x	r
in.	lb	in²	in⁴	in³	in.	in.	in⁴	in³	in.	in.	in.
L 8 × 8 × 1	51.0	15.0	89.0	15.8	2.44	2.37	89.0	15.8	2.44	2.37	1.56
L 8 × 8 × 3/4	38.9	11.4	69.7	12.2	2.47	2.28	69.7	12.2	2.47	2.28	1.58
L 8 × 8 × 1/2	26.4	7.75	48.6	8.36	2.50	2.19	48.6	8.36	2.50	2.19	1.59
L 6 × 6 × 1	37.4	11.0	35.5	8.57	1.80	1.86	35.5	8.57	1.80	1.86	1.17
L 6 × 6 × 3/4	28.7	8.44	28.2	6.66	1.83	1.78	28.2	6.66	1.83	1.78	1.17
L 6 × 6 × 1/2	19.6	5.75	19.9	4.61	1.86	1.68	19.9	4.61	1.86	1.68	1.18
L 6 × 6 × 3/8	14.9	4.36	15.4	3.53	1.88	1.64	15.4	3.53	1.88	1.64	1.19
L 5 × 5 × 3/4	23.6	6.94	15.7	4.53	1.51	1.52	15.7	4.53	1.51	1.52	0.975
L 5 × 5 × 1/2	16.2	4.75	11.3	3.16	1.54	1.43	11.3	3.16	1.54	1.43	0.983
L 5 × 5 × 3/8	12.3	3.61	8.74	2.42	1.56	1.39	8.74	2.42	1.56	1.39	0.990
L 4 × 4 × 3/4	18.5	5.44	7.67	2.81	1.19	1.27	7.67	2.81	1.19	1.27	0.778
L 4 × 4 × 1/2	12.8	3.75	5.56	1.97	1.22	1.18	5.56	1.97	1.22	1.18	0.782
L 4 × 4 × 3/8	9.8	2.86	4.36	1.52	1.23	1.14	4.36	1.52	1.23	1.14	0.788
L 4 × 4 × 1/4	6.6	1.94	3.04	1.05	1.25	1.09	3.04	1.05	1.25	1.09	0.795
L 3 1/2 × 3 1/2 × 1/2	11.1	3.25	3.64	1.49	1.06	1.06	3.64	1.49	1.06	1.06	0.683
L 3 1/2 × 3 1/2 × 3/8	8.5	2.48	2.87	1.15	1.07	1.01	2.87	1.15	1.07	1.01	0.687
L 3 1/2 × 3 1/2 × 1/4	5.8	1.69	2.01	0.794	1.09	0.968	2.01	0.794	1.09	0.968	0.694
L 3 × 3 × 1/2	9.4	2.75	2.22	1.07	0.898	0.932	2.22	1.07	0.898	0.932	0.584
L 3 × 3 × 3/8	7.2	2.11	1.76	0.833	0.913	0.888	1.76	0.833	0.913	0.888	0.587
L 3 × 3 × 1/4	4.9	1.44	1.24	0.577	0.930	0.842	1.24	0.577	0.930	0.842	0.592
L 2 1/2 × 2 1/2 × 1/2	7.7	2.25	1.23	0.724	0.739	0.806	1.23	0.724	0.739	0.806	0.487
L 2 1/2 × 2 1/2 × 3/8	5.9	1.73	0.984	0.566	0.753	0.762	0.984	0.566	0.753	0.762	0.487
L 2 1/2 × 2 1/2 × 1/4	4.1	1.19	0.703	0.394	0.769	0.717	0.703	0.394	0.769	0.717	0.491
L 2 × 2 × 3/8	4.7	1.36	0.479	0.351	0.594	0.636	0.479	0.351	0.594	0.636	0.389
L 2 × 2 × 1/4	3.19	0.938	0.348	0.247	0.609	0.592	0.348	0.247	0.609	0.592	0.391
L 2 × 2 × 1/8	1.65	0.484	0.190	0.131	0.626	0.546	0.190	0.131	0.626	0.546	0.398

APPENDIX E

Conversion Factors

Conversion Factors

Quantity	SI → US Customary	US Customary → SI
Length	1 mm = 0.03937 in	1 in = 25.4 mm
	1 mm = 0.00328 ft	1 ft = 304.8 mm
	1 cm = 0.39370 in	1 in = 2.54 cm
	1 cm = 0.0328 ft	1 ft = 30.48 cm
	1 m = 39.3700 in	1 in = 0.0254 m
	1 m = 3.28 ft	1 ft = 0.3048 m
Area	1 mm^2 = 1.55E–3 in^2	1 in^2 = 645.16 mm^2
	1 mm^2 = 1.0764E–5 ft^2	1 ft^2 = 92903 mm^2
	1 cm^2 = 0.155 in^2	1 in^2 = 6.4516 cm^2
	1 cm^2 = 1.07E–3 ft^2	1 ft^2 = 929.03 cm^2
	1 m^2 = 1550 in^2	1 in^2 = 6.4516E–4 m^2
	1 m^2 = 10.76 ft^2	1 ft^2 = 0.0929 m^2
Volume	1 mm^3 = 6.1024E–5 in^3	1 in^3 = 16387 mm^3
	1 mm^3 = 3.5315E–8 ft^3	1 ft^3 = 28.317E6 mm^3
	1 cm^3 = 0.061024 in^3	1 in^3 = 16.387 cm^3
	1 cm^3 = 3.5315E–5 ft^3	1 ft^3 = 28317 cm^3
	1 m^3 = 61024 in^3	1 in^3 = 1.6387E–5 m^3
	1 m^3 = 35.315 ft^3	1 ft^3 = 0.028317 m^3
Second Moment of Area $(length)^4$	1 mm^4 = 2.402E–6 in^4	1 in^4 = 416.231E3 mm^4
	1 mm^4 = 115.861E–12 ft^4	1 ft^4 = 8.63097E9 mm^4
	1 cm^4 = 24.025E–3 in^4	1 in^4 = 41.623 cm^4
	1 cm^4 = 1.1586E–6 ft^4	1 ft^4 = 863110 cm^4
	1 m^4 = 2.40251E6 in^4	1 in^4 = 416.231E–9 m^4
	1 m^4 = 115.86 ft^4	1 ft^4 = 8.631E–3 m^4

Conversion Factors (*continued*)

Quantity	SI → US Customary	US Customary → SI
Mass	1 kg = 68.521E–3 slug 1 kg = 2.2046 lbm	1 slug = 14.593 kg 1 lbm = 0.4536 kg
Density	1 kg/m^3 = 0.001938 $slug/ft^3$ 1 kg/m^3 = 0.06248 lbm/ft^3	1 $slug/ft^3$ = 515.7 kg/m^3 1 lbm/ft^3 = 16.018 kg/m^3
Force	1 N = 224.809E–3 lbf	1 lbf = 4.448 N
Moment	1 N · m = 8.851 in · lb 1 N · m = 0.7376 ft · lb	1 in · lb = 0.113 N · m 1 ft · lb = 1.356 N · m
Pressure, Stress, Modulus of Elasticity, Modulus of Rigidity	1 Pa = 145.0377E–6 lb/in^2 1 Pa = 20.885E–3 lb/ft^2 1 KPa = 145.0377E–6 Ksi	1 lb/in^2 = 6.8947E3 Pa 1 lb/ft^2 = 47.880 Pa 1 Ksi = 6.8947E3 KPa
Work, Energy	1 J = 0.7375 ft · lb 1 KW · hr = 3.41214E3 Btu	1 ft · lb = 1.3558 J 1 Btu = 293.071E–6
Power	1 W = 0.7375 ft · lb/sec 1 KW = 3.41214E3 Btu/hr 1 KW = 1.341 hp	1 ft · lb/sec = 1.3558 W 1 Btu/hr = 293.07E–6 KW 1 hp = 0.7457 KW
Temperature	$°C = \frac{5}{9}(°F - 32)$	$°F = \frac{9}{5}°C + 32$

APPENDIX F

An Introduction to MATLAB

In this Appendix we will introduce MATLAB, which is mathematical software available in most university computational labs today. MATLAB is a very powerful tool, especially for manipulating matrices; in fact, it was originally designed for that purpose. Many good textbooks discuss, MATLAB's capabilities for solving a full range of problems. Here our intent is to introduce only some basic ideas so that you can perform some essential operations. Before the introduction of electronic spreadsheets and mathematical software such as MATLAB, engineers wrote their own computer programs to solve engineering problems. Even though engineers still write computer codes to solve complex problems, they take advantage of built-in functions of solvers that are readily available with computational tools such as MATLAB. MATLAB is also versatile enough that you can use it to write your own program.

We begin by discussing MATLAB's basic makeup. We will explain how to input data or a formula in MATLAB and how to carry out some typical engineering computations. We will also explain the use of MATLAB's mathematical, statistical, and logical functions. Next we will discuss MATLAB's conditional statements and repetitive loops, and plotting the results of an engineering analysis using MATLAB. Finally, we will briefly discuss MATLAB's curve fitting and symbolic capabilities.

MATLAB—BASIC IDEAS

We begin by explaining some basic ideas; then once you have a good understanding of these concepts, we will use MATLAB to solve some engineering problems. As is the case with any new software you explore, MATLAB has its own syntax and terminology. A typical MATLAB Window is shown in Figure F.1. The main components of the MATLAB window in the default mode are marked by arrows and numbered as shown in Figure F.1.

1. **Menu bar:** contains the commands you can use to perform certain tasks, for example to save your Workspace, or to change the View settings.
2. **Current Directory:** shows the active directory, but you can also use it to change the directory.

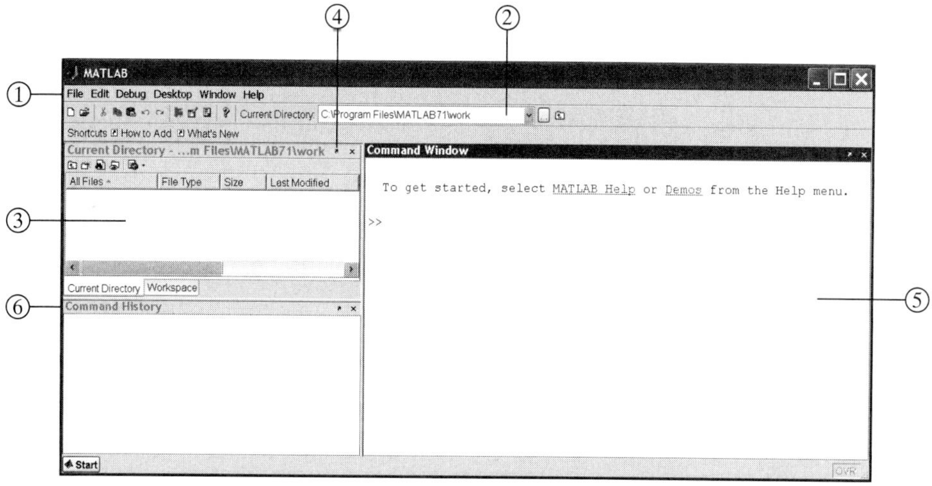

FIGURE F.1 The desktop layout for MATLAB.

3. **Current Directory Window:** shows all files, their types, sizes, and description in the Current Directory.
4. **Moving window outside the desktop:** clicking on this icon will move the window outside the desktop.
5. **Command Window:** This is where you enter variables and issue MATLAB commands.
6. **Command History:** shows the time and the date at which commands were issued during the previous MATLAB sessions. It also shows the history of commands in the current (active) session.

As shown in Figure F.1, MATLAB's desktop layout, in default mode, is divided into three windows: the Current Directory, the Command Window, and the Command History. You type (enter) commands in the Command Window. For example, you can assign values to variables or plot a set of variables. The Command History window shows the time and the date of the commands you issued during the previous MATLAB sessions. It also shows the history of commands in the current (active) session. You can also transfer old commands, which you issued during previous sessions, from the Command History window to the Command Window. To do this, move your mouse pointer over the command you want to move and then click the left button, and while holding down the button drag the old command line into the Command Window. Alternatively, when you strike the up-arrow key, the previously executed command will appear again. You can also copy and paste commands from the current Command Window, edit them, and use them again. To clear the contents of the Command Window, type **clc.**

Once in the MATLAB environment you can assign values to a variable or define elements of a matrix. For example, as shown in Figure F.2, to assign a value 5 to the

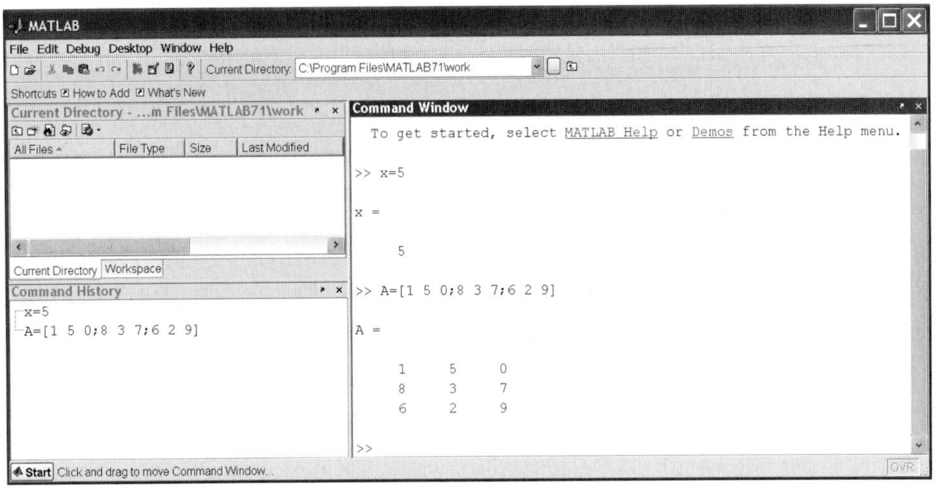

FIGURE F.2 Examples of assigning values or defining elements of a matrix in MATLAB.

variable x, in the Command Window, after the prompt sign >>, you simply type **x = 5.** The basic MATLAB's scalar (arithmetic) operations are shown in Table F.1.

TABLE F.1 MATLAB's basic scalar (arithmetic) operators

Operation	Symbol	Example: x=5 and y=3	Result
Addition	+	$x+y$	8
Subtraction	−	$x-y$	3
Multiplication	*	$x*y$	15
Division	/	$(x+y)/2$	4
Raised to a power	^	x^2	25

To define the elements of a matrix, for example $[A] = \begin{bmatrix} 1 & 5 & 0 \\ 8 & 3 & 7 \\ 6 & 2 & 9 \end{bmatrix}$ you type

A = [1 5 0;8 3 7;6 2 9]

Note that in MATLAB the elements of the matrix are enclosed in brackets **[ ]** and are separated by blank spaces, and the elements of each row are separated by a semicolon (**;**).

The format, disp, and fprintf Commands

MATLAB offers several commands to display the results of your calculation. The MATLAB **format** command allows you to display values in certain ways. For example, if you define $x = 2/3$, then MATLAB will display $x = 0.6667$. By default, MATLAB will display 4 decimal digits. If you want more decimal digits displayed, then you type *format long*. Now if you type x again, the value of x is displayed with 14 decimal digits, that is $x = 0.66666666666667$. You can control the way the value of x is displayed in several other ways as shown in Table F.2. Note that the format command does not affect the number of digits maintained when calculations are carried out by MATLAB. It only affects the way the values are displayed.

TABLE F.2 The format command options

MATLAB Command	How the Result of $x = 2/3$ is Displayed	Explanation
format short	0.6667	Shows four decimal digits—default format
format long	0.66666666666667	Shows 14 decimal digits
format rat	2/3	Shows as fractions of whole numbers
format bank	0.67	Shows two decimal digits
format short e	6.6667e−001	Shows scientific notation with four decimal digits
format long e	6.666666666666666e−001	Shows scientific notation with 14 decimal digits
format hex	3fe5555555555555	Shows hexadecimal
format +	+	Shows +, − or blank based on whether the number is positive, negative, or zero.
format compact		It suppresses the blank lines in the output.

The **disp** command is used to display text or values. For example, given **x = [1 2 3 4 5]**, then the command **disp(x)** will display 1 2 3 4 5. Or the command disp('Result = ') will display Result =. Note that the text that you want displayed must be enclosed within a set of single quotation marks.

The **fprintf** command offers a great deal of flexibility. You can use it to print text and/or values with a desired number of digits. You can also use special formatting characters such as **\n** and **\t** to produce linefeed and tabs. The following example will demonstrate the use of the fprintf command.

EXAMPLE F.1

In the MATLAB command window, type the following commands as shown.

```
x = 10
fprintf('The value of x is %g \n', x)
```

The MATLAB will display

```
The value of x is 10
```

A screen capture of the command window for Example F.1 is shown in the accompanying figure.

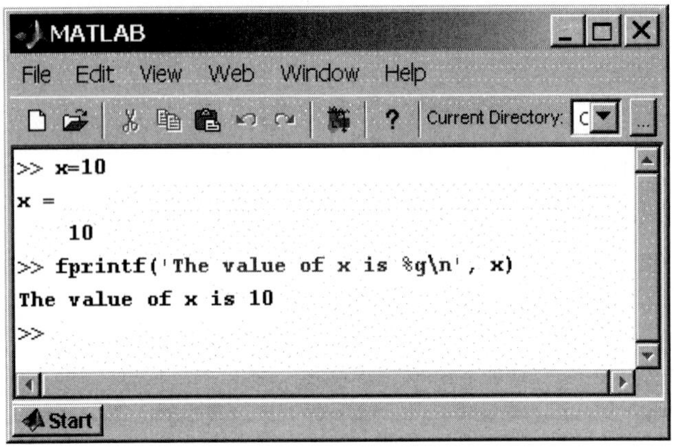

Note that the text and formatting code are enclosed within a set of single quotation marks. Also note that the %g is the number format character and is replaced by *x* value, 10. It is also important to note that MATLAB will not produce an output until it encounters the *\n*. Additional capabilities of disp and fprintf commands will be demonstrated using other examples later.

SAVING YOUR MATLAB WORKSPACE

You can save the workspace to a file by issuing the command: **save your_filename.** The your_filename is the name that you would like to assign to the workspace. Later you can load the file from the disk to memory by issuing the command: **load your_filename.** Over time, you may create many files; then you can use the **dir** command to list the contents of the directory. For simple operations you can use the Command Window to enter variables and issue MATLAB commands. However, when you write a program that is longer than a few lines, you use an M-file. Later in this appendix, we will explain how to create, edit, run, and debug an M-file.

GENERATING A RANGE OF VALUES

When creating, analyzing, or plotting data, it is often convenient to create a range of numbers. To create a *range* of data or a row matrix, you need to specify only the starting number, the increment, and the end number. For example, to generate a set of *x* values in the range of 0 to 100 in increments of 25 (i.e. 0 25 50 75 100), in the Command Window, type

x = 0:25:100

Note that in MATLAB language, the range is defined by a starting value, followed by a colon (**:**), the increment followed by another colon, and the end value. As another example, if you were to type

Countdown = 5:-1:0

then the Countdown row matrix would consist of values: 5 4 3 2 1 0.

CREATING FORMULAS IN MATLAB

You can use MATLAB to input engineering formulas and compute the results. When typing your formula, use parentheses to dictate the order of operation. For example, in MATLAB's command window, if you were to type **count=100+5*2,** MATLAB would perform the multiplication first, which results in a value of 10, and then this result would be added to 100, which would yield an overall value of 110 for the variable count. If, however, you want MATLAB to add the 100 to 5 first and then multiply the resulting 105 by 2, you should place parentheses around the 100 and 5 in the following manner **count=(100+5)*2,** which results in a value of 210. The basic MATLAB arithmetic operations are shown in Table F.3.

TABLE F.3 The basic MATLAB arithmetic operations.

Operation	Symbol	Example: $x = 10$ and $y = 2$	z, the result of the formula given in the example
addition	+	$z=x+y+20$	32
subtraction	−	$z=x-y$	8
multiplication	*	$z=(x*y)+9$	29
division	/	$z=(x/2.5)+y$	6
raised to a power	^	$z=(x\text{^}y)\text{^}0.5$	10

ELEMENT BY ELEMENT OPERATION

In addition to basic scalar (arithmetic) operations, MATLAB provides element-by-element operations and matrix operations. MATLAB's symbols for element-by-element operations are shown in Table F.4. To better understand their use, suppose you have measured and recorded the mass (kg) and speed (m/s) of 5 runners who are running along a straight path: **m=[60 55 70 68 72]** and **s=[4 4.5 3.8 3.6 3.1].** Note that the mass **m** array and speed **s** array, each have 5 elements. Now suppose you are interested in determining the magnitude of each runner's momentum. We can calculate the momentum for each runner using MATLAB's element-by-element multiplication operation in the following manner: **momentum=m.*s,** resulting in **momentum=[240 247.5 266 244.8 223.2].** Note the multiplication symbol (.*) for element-by-element operation.

TABLE F.4 MATLAB's element by element operations.

Operation	Arithmetic Operations	Equivalent Element by Element Symbol for the Operation
addition	+	+
subtraction	−	−
multiplication	*	.*
division	/	./
raised to a power	^	.^

To improve your understanding of the element-by-element operations, in MATLAB's Command Window type **a=[7 4 3 −1]** and **b=[1, 3, 5, 7]** and then try the following operations.

```
>> a+b
ans = 8 7 8 6
>> a−b
ans = 6 1 −2 −8
>> 3*a
ans = 21 12 9 −3
>> 3.*a
ans = 21 12 9 −3
>> a.*b
ans = 7 12 15 −7
>> b.*a
ans = 7 12 15 −7
>> 3.^a
ans = 1.0e+003 *
2.1870 0.0810 0.0270 0.0003
>> a.^b
ans = 7 64 243 −1
>> b.^a
ans = 1.0000 81.0000 125.0000 0.1429
```

Try the following example on your own.

EXAMPLE F.2

The following example will show how the density of standard air changes with temperature. It also makes use of MATLAB's element-by-element operations. The density of standard air is a function of temperature and may be approximated using the ideal gas law according to

$$\rho = \frac{P}{RT}$$

where

P = standard atmospheric pressure (101.3 kPa)

$R \equiv$ gas constant and its value for air is $\left(286.9\left(\frac{\text{J}}{\text{kg} \cdot \text{K}}\right)\right)$

$T \equiv$ air temperature in Kelvin

Using MATLAB, we want to create a table that shows the density of air as a function of temperature in the range of 0 °C (273.15 K) to 50 °C (323.15 K) in increments of 5 °C.

In MATLAB's command window we type the following commands.

```
>> Temperature = 0:5:50;
>> Density = 101300./((286.9)*(Temperature+273));
>> fprintf('\n\n');disp('Temperature(C)  Density(kg/m^3)');
   disp([Temperature', Density'])
```

In the commands shown above, the semicolon(;) suppresses MATLAB's automatic display action. If you type **Temperature = 0:5:50** without the semicolon at the end, MATLAB will display the values of Temperature in a row. It will show

Temperature = 0 5 10 15 20 25 30 35 40 45 50

The ./ is a special element by element division operation that tells MATLAB to carry the division operation for each of the temperature values.

In the **disp** command, the prime or the single quotation mark over the variables **Temperature′** and **Density′** will change the values of Temperature and Density, which are stored in rows, to column format before they are displayed. As explained in Chapter 2, in matrix operation the process of changing the rows into columns is called Transpose of matrix. The final results for Example F.2 are shown in Figure F.3. Note that the values of Temperature and Density are shown in columns. Also note the use of the **fprintf** and **disp** commands.

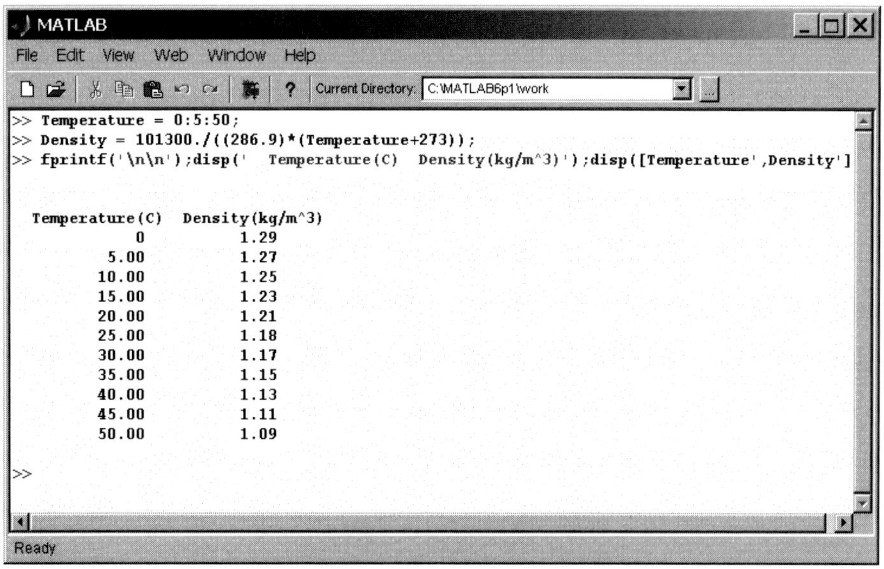

FIGURE F.3 The result of Example F.2.

In the above example, we used the element-by-element division operation. Other MATLAB element-by-element operation symbols are given in Table F.4.

MATRIX OPERATIONS

MATLAB offers many tools for matrix operations and manipulations. Table F.6. shows examples of these capabilities. Later in this appendix we will demonstrate a few MATLAB matrix operations with the aid of Examples F.9 and F.10.

TABLE F.6 Examples of MATLAB's matrix operations

Operation	Symbols or Commands	Example: A and B are matrices that you have defined
Addition	+	A+B
Subtraction	−	A−B
Multiplication	*	A*B
Transpose	*matrix name'*	A'
Inverse	inv(*matrix name*)	inv(A)
Determinant	det(*matrix name*)	det(A)
eigenvalues	eig(*matrix name*)	eig(A)
Matrix left division (uses Gauss elimination to solve a set of linear equations)	\	see Example F.10

EXAMPLE F.3

Using MATLAB, create a table that shows the relationship between interest earned and the amount deposited, as shown in Table F.5.

TABLE F.5 The relationship between interest earned and the amount deposited

Dollar Amount	Interest Rate			
	0.06	**0.07**	**0.075**	**0.08**
1000	60	70	75	80
1250	75	87.5	93.75	100
1500	90	105	112.5	120
1750	105	122.5	131.25	140
2000	120	140	150	160
2250	135	157.5	168.75	180
2500	150	175	187.5	200
2750	165	192.5	206.25	220
3000	180	210	225	240

To create a table that is similar to Table F.5, we type the following commands.

```
>> format bank
>> Amount = 1000:250:3000;
>> Interest_Rate = 0.06:0.01:0.08;
>> Interest_Earned = (Amount')*(Interest_Rate);
>> fprintf('\n\n\t\t\t\t\t\t\t Interest Rate');fprintf('\n\t Amount\t\t');...
fprintf(' \t\t %g ',Interest_Rate);fprintf('\n');disp([Amount',Interest_Earned])
```

On the last command line, note the three periods . . . (an ellipsis) represent a continuation marker in MATLAB. The ellipsis means there is more to follow on this command line. Note the use of **fprintf** and **disp** commands. The final result for Example F.3 is shown in Figure F.4.

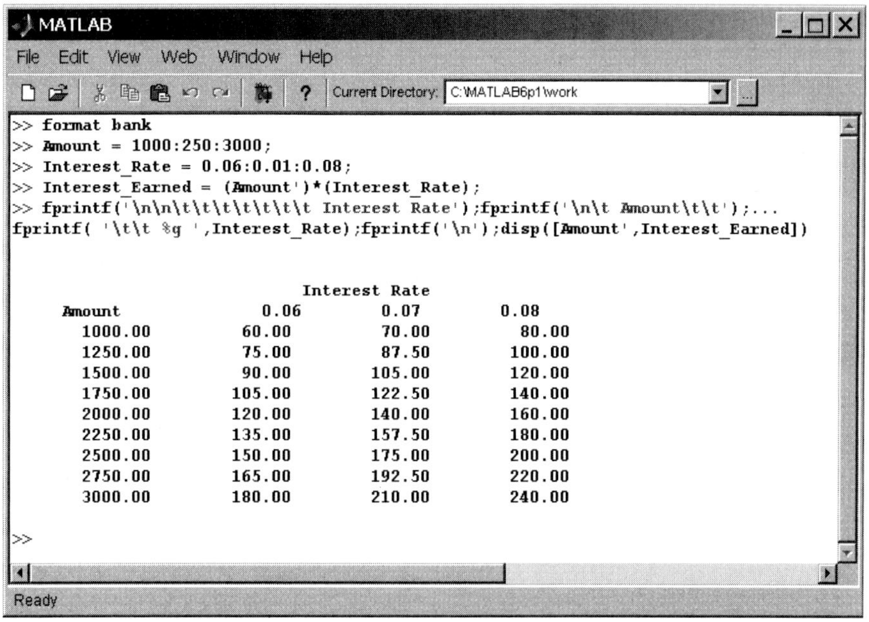

FIGURE F.4 The commands and result for Example F.3.

USING MATLAB FUNCTIONS

MATLAB offers a large selection of built-in functions that you can use to analyze data. The MATLAB functions are available in various categories including mathematical and trigonometric, statistical, and logical functions. In this appendix we will discuss some of the common functions. MATLAB offers a help menu that you can use to obtain information on various commands and functions. You can also type **help** followed by a command name to learn how to use the command.

Some examples of commonly used MATLAB functions, along with their proper use and descriptions, are shown in Table F.7. Refer to Example F.4 when studying Table F.7.

EXAMPLE F.4

The following set of values will be used to introduce some of MATLAB's built-in functions. Mass = [102 115 99 106 103 95 97 102 98 96].

When studying Table F.7, the results of the executed functions are shown under the "result of the example" column.

TABLE F.7 Some MATLAB functions that you may use in engineering analyses.

Function	Description of the Function	Example	Result of the Example
sum	It sums the values in a given array	sum(Mass)	1013
mean	It calculates the average value of the data in a given array.	mean(Mass)	101.3
max	It determines the largest value in the given array	max(Mass)	115
min	It determines the smallest value in the given array	min(Mass)	95
std	It calculates the standard deviation for the values in the given array	std(Mass)	5.93
sort	It sorts the values in the given array in ascending order	sort(Mass)	95 96 97 98 99 102 102 103 106 115
pi	It returns the value of π, 3.14151926535897...	pi	3.14151926535897...
tan	It returns tangent value of the argument. The argument must be in radians.	tan(pi/4)	1
cos	It returns cosine value of the argument. The argument must be in radians.	cos(pi/2)	0
sin	It returns sine value of the argument. The argument must be in radians.	sin(pi/2)	1

More examples of MATLAB's Functions are shown in Table F.8.

TABLE F.8 More examples of MATLAB functions.

sqrt(x)	Returns the square root of value x.
factorial(x)	Returns the value of factorial of x. For example, factorial(5) will return: (5)(4)(3)(2)(1) = 120.
Trigonometric Functions	
acos(x)	This is the inverse cosine function of x. It is used to determine the value of an angle when its cosine value is known.
asin(x)	This is the inverse sine function of x. It is used to determine the value of an angle when its sine value is known.
atan(x)	This is the inverse tangent of x function. It is used to determine the value of an angle when its tangent value is known.
Exponential and Logarithmic Functions	
exp(x)	Returns the value of e^x.
log(x)	Returns the value of natural logarithm of x. Note that x must be greater than 0.
log10(x)	Returns the value of common (base 10) logarithm of x.
log2(x)	Returns the value of base 2 logarithm of x.

EXAMPLE F.5

Using MATLAB, compute the average (arithmetic mean) and the standard deviation of the density of water data given in Table F.9.

TABLE F.9 Data for Example F.5

Group-A Findings $\rho(kg/m^3)$	Group-B Findings $\rho(kg/m^3)$
1020	950
1015	940
990	890
1060	1080
1030	1120
950	900
975	1040
1020	1150
980	910
960	1020

The final results for Example F.5 are shown in Figure F.5.
The MATLAB commands leading to results follow.

>> Density_A = [1020 1015 990 1060 1030 950 975 1020 980 960];
>> Density_B = [950 940 890 1080 1120 900 1040 1150 910 1020];
>> Density_A_Average = mean(Density_A)

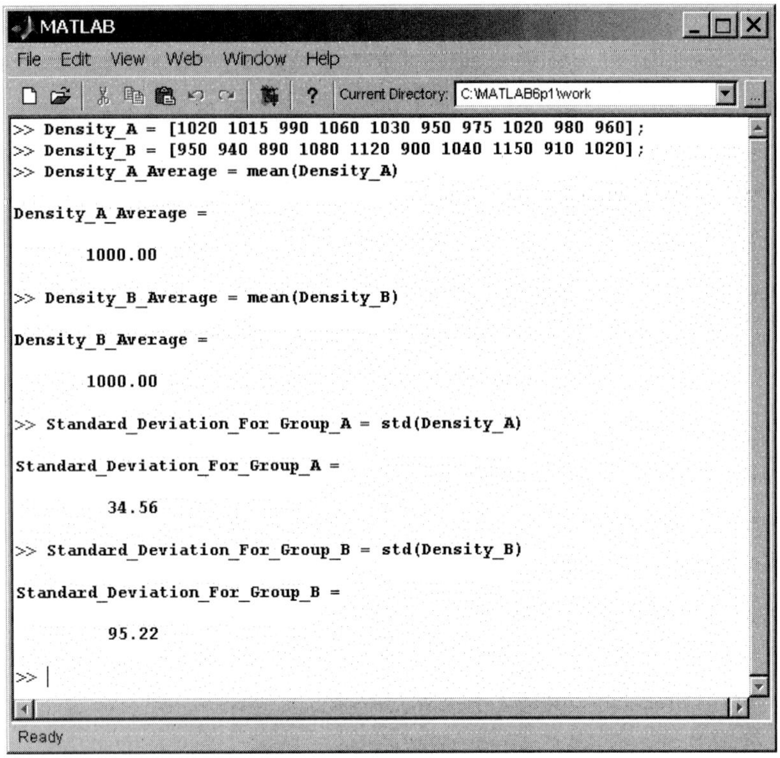

FIGURE F.5 MATLAB's Command Window for Example F.5.

```
Density_A_Average =
    1000.00
>> Density_B_Average = mean(Density_B)
Density_B_Average =
    1000.00
>> Standard_Deviation_For_Group_A = std(Density_A)
Standard_Deviation_For_Group_A =
    34.56
>> Standard_Deviation_For_Group_B = std(Density_B)
Standard_Deviation_For_Group_B =
    95.22
>>
```

THE LOOP CONTROL—*for* AND *while* COMMANDS

When writing a computer program, often it becomes necessary to execute a line or a block of your computer code many times. MATLAB provides ***for*** and ***while*** commands for such situations.

The *for* loop

Using the *for* loop, you can execute a line or a block of code a specified (defined) number of times. The syntax of a *for* loop is

> *for* index = start-value : increment : end-value
> a line or a block of your computer code
> *end*

For example, suppose you want to evaluate the function $y = x^2 + 10$ for x values of 22.00, 22.50, 23.00, 23.50, and 24.00. This operation will result in corresponding y values of 494.00, 516.25, 539.00, 562.25, and 586.00. The MATLAB code for this example then could have the following form:

```
x = 22.0;
for i = 1:1:5
     y=x^2+10;
     disp([x',y'])
     x = x + 0.5;
end
```

Note that in the preceding example, the index is the integer i and its start-value is 1, it is incremented by a value of 1, and its end-value is 5.

The *while* Loop

Using the *while* loop, you can execute a line or a block of code until a specified condition is met. The syntax of a *while* loop is

> *while controlling-expression*
> a line or a block of your computer code
> *end*

With the *while* command, as long as the controlling expression is true, the line or a block of code will be executed. For the preceding example, the MATLAB code using the *while* command becomes:

```
x = 22.0;
while x <= 24.00
        y=x^2+10;
        disp([x',y'])
        x = x + 0.5;
end
```

In the preceding example the <= symbol denotes less than or equal to. It is called a relational or comparison operator. We will explain MATLAB's logical and relational operators next.

USING MATLAB RELATIONAL OPERATORS AND CONDITIONAL STATEMENTS

In this section we will look at some of MATLAB's relational operators and conditional statements. The relational or comparison operators allow for testing of relative magnitude of various arguments. These relational operators are shown in Table F.10. We will use Example F.6 to demonstrate the use of MATLAB's conditional statements and relational operators.

TABLE F.10 MATLAB's Relational operators and their descriptions.

Relational Operator	Its Meaning
<	less than
<=	less than or equal to
= =	equal to
>	greater than
>=	greater than or equal to
~=	not equal to

The Conditional Statements—*if, else*

When writing a computer program, sometimes it becomes necessary to execute a line or a block of code based on whether a condition or a set of conditions is met (true). MATLAB provides *if* and *else* commands for such situations.

The *if* Statement

The *if* statement is the simplest form of a conditional control. Using the *if* statement, you can execute a line or a block of your program as long as the expression following the *if* statement is true. The syntax for the *if* statement is

if expression
  a line or a block of your computer code
end

For example, suppose we have a set of ten scores for an exam: 85, 92, 50, 77, 80, 59, 65, 97, 72, 40. We are interested in writing a code that shows that scores below 60 indicate failing. The MATLAB code for this example then could have the following form:

```
scores=[85 92 50 77 80 59 65 97 72 40];
for i=1:1:10
  if scores (i) <60
     fprintf('\t %g \t\t\t\t\t FAILING\n', scores (i))
  end
end
```

The *if, else* Statement

The **else** statement allows us to execute other line(s) of computer code(s) if the expression following the **if** statement is not true. For example, suppose, we are interested in showing not only the scores that indicate failing, but also the scores that show passing. We can then modify our code in the following manner.

```
scores=[85 92 50 77 80 59 65 97 72 40];
for i=1:1:10
    if scores (i) >=60
      fprintf('\t %g \t\t\t\t\t PASSING\n', scores (i));
    else
      fprintf('\t %g \t\t\t\t\t FAILING\n', scores (i))
    end
end
```

In MATLAB's Command Window, try the following examples on your own.

MATLAB also provides the **elseif** command that could be used with the **if** and **else** statements. Try MATLAB's help menu: type **help elseif** to learn about the **elseif** statement.

EXAMPLE F.6

The pipeline shown in Figure F.6 is connected to a control (check) valve that opens when the pressure in the line reaches 20 psi. Various readings were taken at different times and recorded. Using MATLAB's relational operators and conditional statement, create a list that shows the corresponding open and closed position of the check valve.

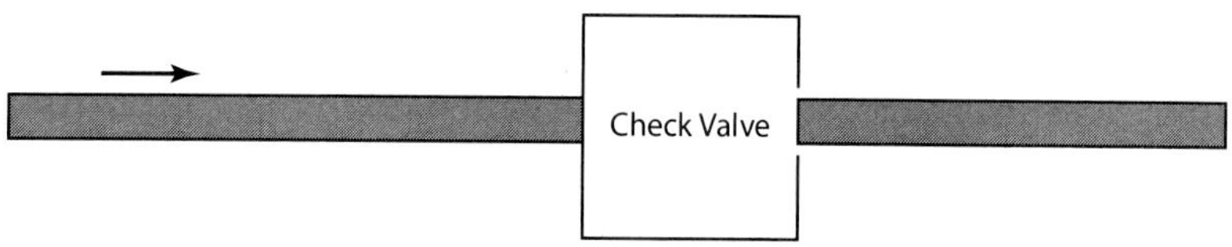

FIGURE F.6 A schematic diagram for Example F.6.

The solution to Example F.6 is shown in Figure F.7. The commands leading to the solution follow:

```
>> pressure=[20 18 22 26 19 19 21 12];
>> fprintf('\t Line Pressure (psi) \t Valve Position\n\n');for i=1:8
if pressure(i)>=20
fprintf('\t %g \t\t\t\t\t OPEN\n',pressure(i))
else
fprintf('\t %g \t\t\t\t\t CLOSED\n',pressure(i))
end
end
```

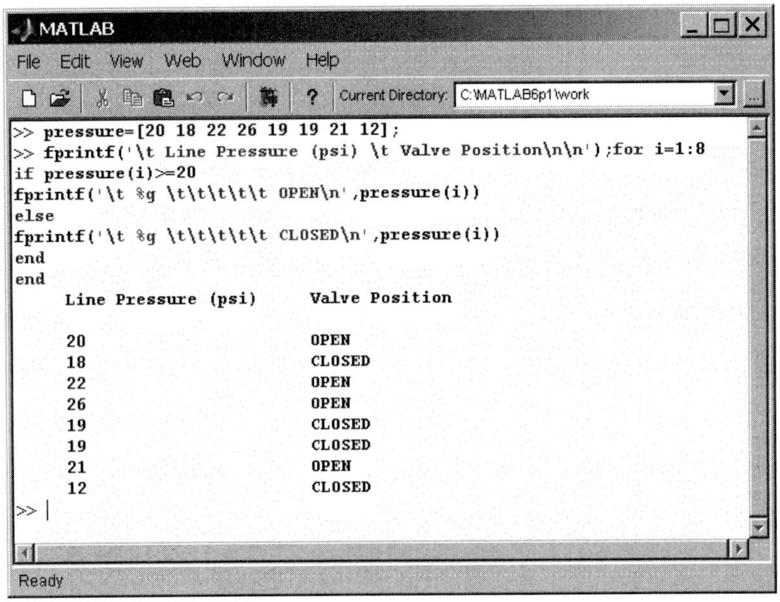

FIGURE F.7 The solution of Example F.6.

THE M-FILE

As explained previously, for simple operations you can use MATLAB's Command Window to enter variables and issue commands. However, when you write a program that is more than few lines long, use an M-file. It is called an M-file because of its *.m* extension. You can create an M-file using any text editor or using MATLAB's Editor/Debugger. To create an M-file, open the M-file Editor and MATLAB opens a new window in which to type your program. As you type your program, you will notice that MATLAB assigns line numbers in the left column of the window. The line numbers are quite useful for debugging your program. To save the file, simply click **File → Save** and type in the filename. The name of your file must begin with a letter and may include other characters such as underscore and digits. Be careful not to name your file the same as a MATLAB command. To see if a filename is used by a MATLAB command, type **exist ('file-name')** in the MATLAB's Command Window. To run your program, click on **Debug → Run** (or use the function key **F5**). Don't be discouraged to find mistakes in your program the first time you attempt to run it. This is quite normal! You can use the Debugger to find your mistakes. To learn more about debugging options, type *help debug* in the MATLAB's Command window.

EXAMPLE F.7

It has been said that when Pascal was 7 years old, he came up with the formula $\frac{n(n+1)}{2}$ to determine the sum of 1, 2, 3, . . . , through n. The story suggests that one day he was asked by his teacher to add up numbers 1 through 100, and Pascal came up with the answer in few minutes. It is believed that Pascal solved the problem in the following manner:

First, on one line he wrote the numbers 1 through 100, similar to

1 2 3 4. . . . . .99 100

Then on the second line he wrote the numbers backward

100 99 98 97. . . . . .2 1

Then he added up the numbers in the two lines, resulting in one hundred identical values of 101

101 101 101 101. . . . . .101 101

Pascal also realized that the result should be divided by 2—since he wrote down the numbers 1 through 100 twice—leading to the answer: $\frac{100(101)}{2} = 5050$. Later, he generalized his approach and came up with the formula $\frac{n(n+1)}{2}$.

Next, we will write a computer program using an M-file that asks a user to input a value for n and computes the sum of 1 through n. To make the program interesting, we will not make use of Pascal's formula; instead, we will use a *for* loop to solve the problem. We have used MATLAB's Editor to create the program and have named it **For_Loop_Example.m,** as shown in Figure F.8. In the shown program, the % symbol denotes comments. MATLAB will treat any text following the % symbol as comments. Also, note that you can find the Line (Ln) and Column (Col) numbers corresponding

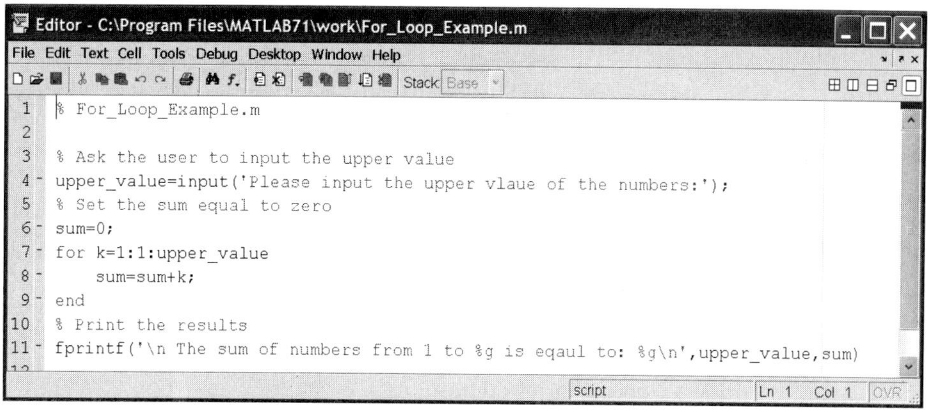

FIGURE F.8 The M-file for Example F.7.

to a specific location in your program by moving the cursor. The line and the column numbers are shown in the right-side bottom corner of the Editor window. As you will see, the knowledge of line and column numbers are useful for debugging your program. We run the program by clicking on **Debug → Run,** and the result is shown in Figure F.9.

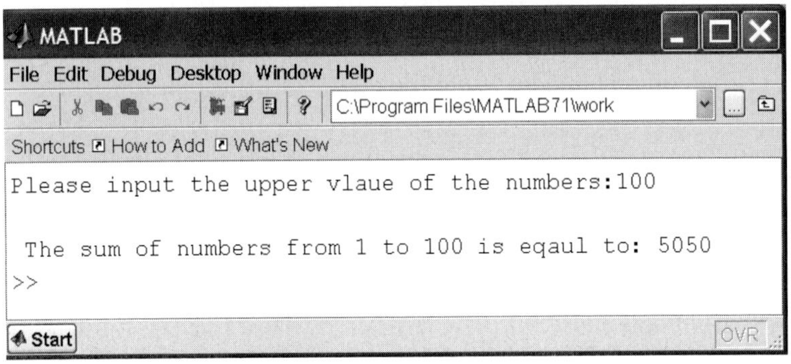

FIGURE F.9 The results of Example F.7.

PLOTTING WITH MATLAB

MATLAB offers many choices for creating charts. For example, you can create *x-y* charts, column charts (or histograms), contour, or surface plots. However, as an engineering student, and later as a practicing engineer, most of the charts that you will create will be *x-y* type charts. Therefore, we will explain in detail how to create an *x-y* chart.

EXAMPLE F.8

Starting with a 10 cm by 10 cm sheet of paper, what is the largest volume you can create by cutting out x cm by x cm from each corner of the sheet and then folding up the sides? This problem has a simple analytical solution, however, to demonstrate MATLAB's plotting features, we will use MATLAB to obtain the solution.

The volume created by cutting out x cm by x cm from each corner of the 10 cm by 10 cm sheet of paper is given by volume $= (10 - 2x)(10 - 2x)x$. Moreover, we know that for $x = 0$ and $x = 5$ the volume will be zero. Therefore, we need to create a range of x values from 0 to 5 using some small increments such as 0.1. We then plot the volume versus x and look for the maximum value of volume. The MATLAB commands that lead to the solution follow.

```
>> x = 0:0.1:5;
>> volume = (10-2*x).*(10-2*x).*x;
>> plot (x,volume)
>> title ('Volume as a function of x')
>> xlabel ('x (cm)')
>> ylabel ('Volume (cm^3)')
>> grid minor
>>
```

The MATLAB command window for Example F.8 is shown in Figure F.10. The plot of volume versus x is shown in Figure F.11.

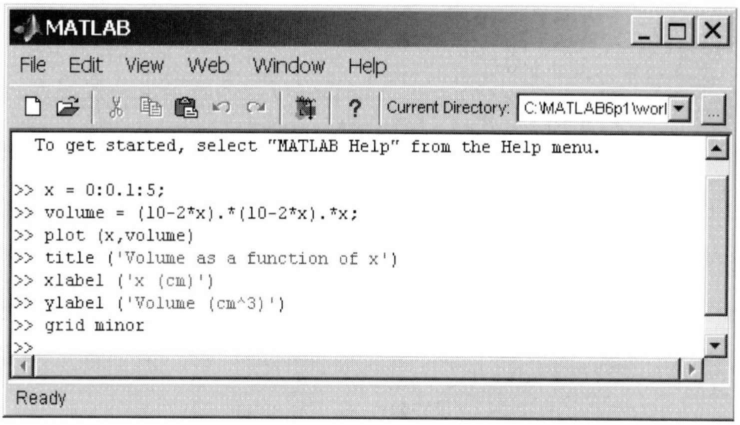

FIGURE F.10 The MATLAB command window for Example F.8.

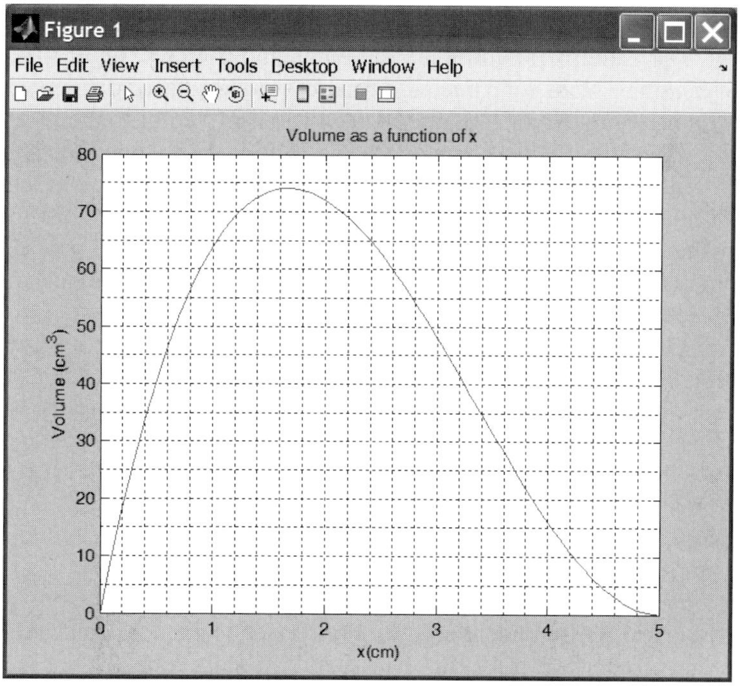

FIGURE F.11 The plot of volume versus x for Example F.8.

Let us now discuss the MATLAB commands that are commonly used when plotting data. The **plot(x,y)** command plots y values versus x values. You can use various line types, plot symbols or colors with the command **plot(x,y,s)** where s is a character string that defines a particular line type, plot symbol, or line color. The s can take on one of the properties shown in Table F.11.

TABLE F.11 MATLAB line and symbol properties

S	Color	S	Data Symbol	S	Line Type
b	blue	.	point	-	solid
g	green	O	circle	:	dotted
r	red	x	x-mark	-.	dash-dot
c	cyan	+	plus	--	dashed
m	magenta	*	star		
y	yellow	s	square		
k	black	d	diamond		
		v	triangle (down)		
		^	triangle (up)		
		<	triangle (left)		
		>	Triangle (right)		

For example, if you issue the command **plot(x,y, 'k*-')**, MATLAB will plot the curve using a black solid line with *marker shown at each data point. If you do not specify a line color, MATLAB automatically assigns a color to the plot.

Using **title('text')** command, you can add text on top of the plot. The **xlabel('text')** creates the title for the X-axis. The text that you enclose between single quotation marks will be shown below the X-axis. Similarly, the **ylabel('text')** command creates the title for the Y-axis. To turn on the grid lines, type the command **grid on** (or just **grid**). The command **grid off** removes the grid lines. To turn on the minor grid lines as shown in Figure F.11, type the command **grid minor.**

Generally, it is easier to use the graph property editor. For example, to make the curve line thicker, change the line color, and to add markers to the data points, with the mouse pointer on the curve, double-click the left mouse button. Make sure you are in the picking mode first. You might need to click on the arrow next to the print icon to activate the picking mode. After double-clicking on the line, you should see the line and the marker editor window. As shown in Figure F.12, we increased the line thickness from 0.5 to 2, changed the line color to black, and set the data point marker style to Diamond. These new settings are reflected in Figure F.13.

Next, we will add an arrow pointing to the maximum value of volume by selecting the **Text Arrow** under the **Insert** option (see Figure F.13), and add the text "Maximum volume occurs at $x = 1.7$ cm". These additions are reflected in Figure F.14.

We can also change the font size and style and make the title or the axes labels boldface. To do so we pick the object that we want to modify and then from the menu bar

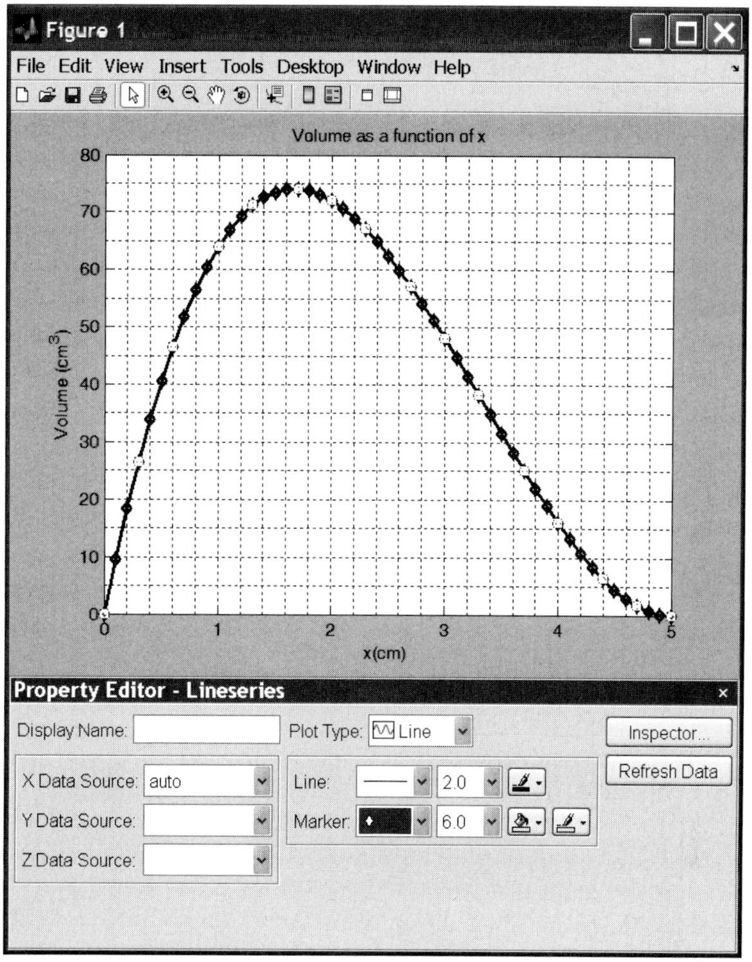

FIGURE F.12 The plot of Example F.8 with modified properties.

select **Edit** and then **Current Object Properties. . . .** Then using the property editor shown in Figure F.15 we can modify the properties of the selected object. We have changed the font size and the font weight of the title and the labels for Example F.8 and shown the changes in Figure F.16.

With MATLAB you can generate other types of plots including contour and surface plots. You can also control the x- and y-axis scales. For example, the MATLAB's **loglog(x,y)** uses the base 10 logarithmic scales for x- and y-axes. Note that x and y are the variables that you want to plot. The command **loglog(x,y)** is identical to the **plot(x,y),** except it uses logarithmic axes. The command **semilogx(x,y)** or **semilogy(x,y)** creates a plot with base 10 logarithmic scales for either only x-axis or y-axis. Finally, note that you can use the **hold** command to plot more than one set of data on the same chart.

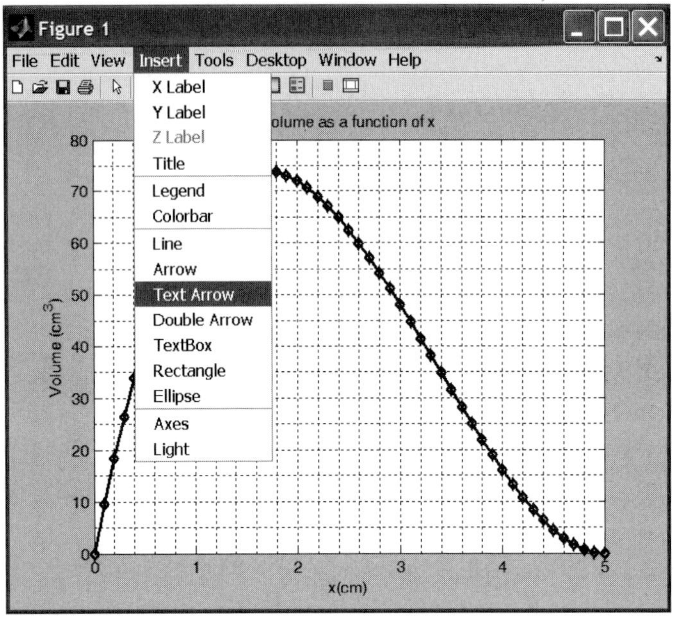

FIGURE F.13 Using the Insert Text Arrow or Insert Text options, you can add arrows or text to the plot.

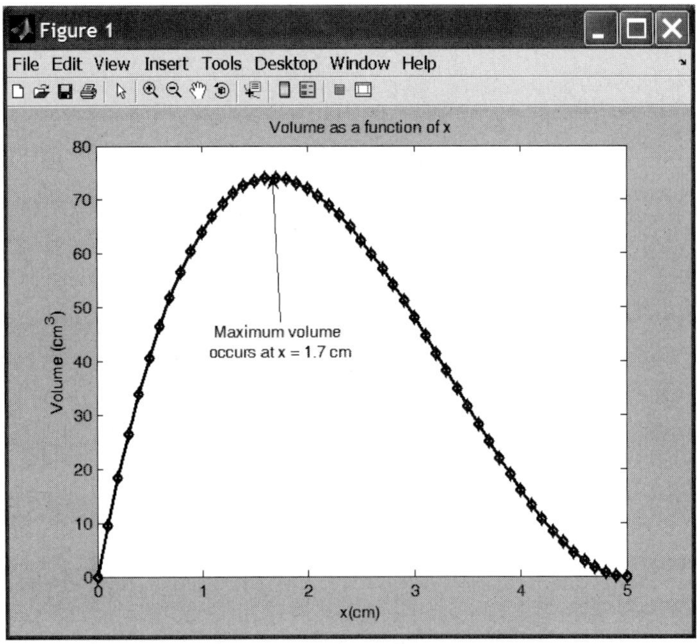

FIGURE F.14 The solution of Example F.8.

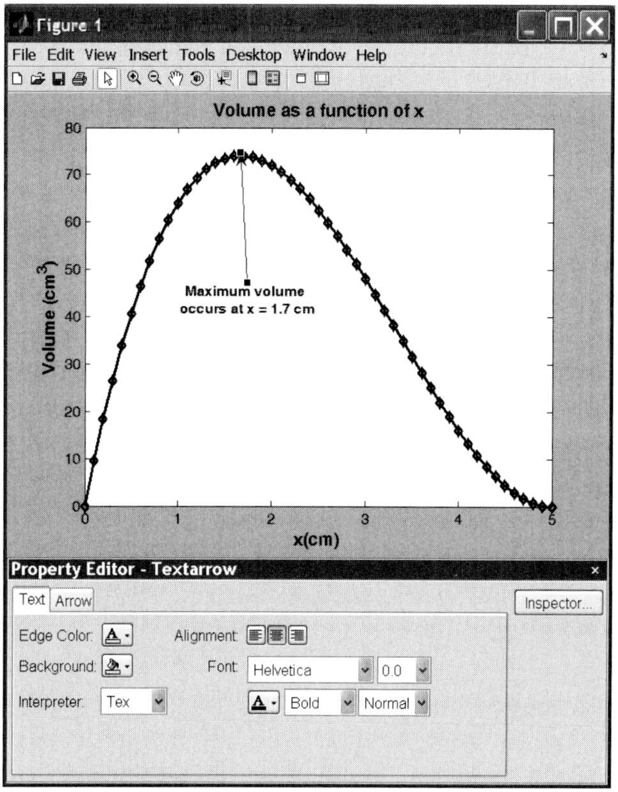

FIGURE F.15 MATLAB's Property Editor.

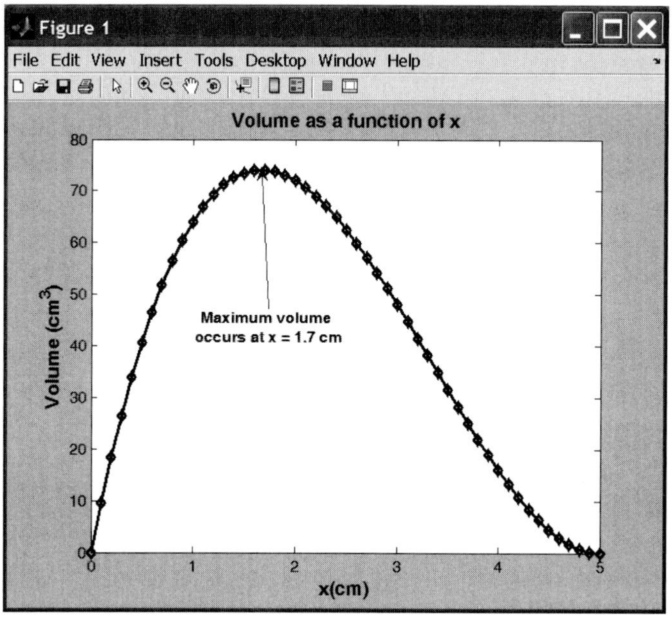

FIGURE F.16 The result of Example F.8.

A reminder, when creating an engineering chart, whether you are using MATLAB, Excel, other drawing software, or freehand, an engineering chart must contain proper labels with proper units for each axis. The chart must also contain a figure number with a title that explains what the chart represents. If more than one set of data is plotted on the same chart, the chart must also contain a legend or list that shows symbols used for different data sets.

EXAMPLE F.2 (revisited)

Using the results of Example F.2 create a graph showing the value of air density as a function of temperature.

The command window and the plot of density of air as a function of temperature are shown in the accompanying figures.

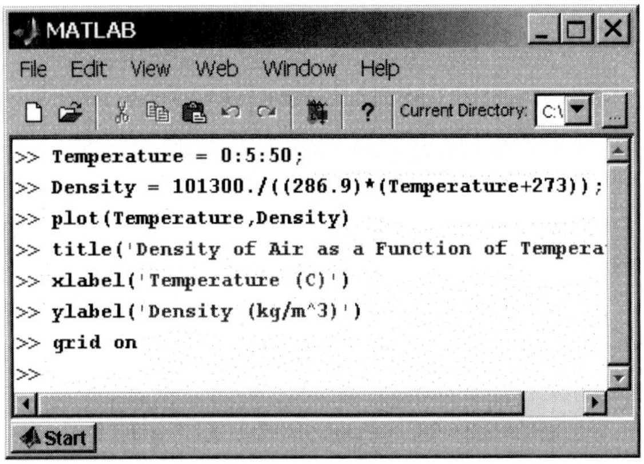

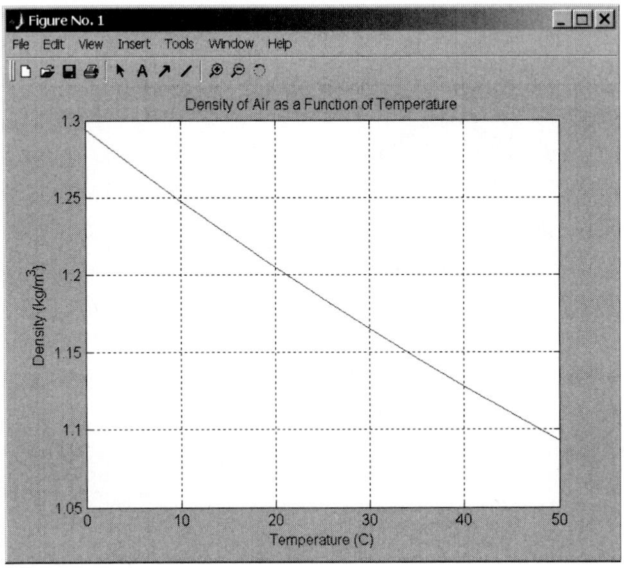

IMPORTING EXCEL AND OTHER DATA FILES INTO MATLAB

At times, it might be convenient to import data files that were generated by other programs, such as Excel, into MATLAB for additional analysis. To demonstrate how to import a data file into MATLAB, consider the Excel file shown in Figure F.17. The Excel file was created for Example F.8, with two columns, the *x* values and the corresponding volume. To import this file into MATLAB, from menu bar we select **File** then **Import Data . . .** and then go to the appropriate directory and open the file we want. The import Wizard window, as shown in Figure F.18, will appear next. Select the "Create vectors from each column using column names" and the Wizard will import the data and will save them in **x** and **volume** variables.

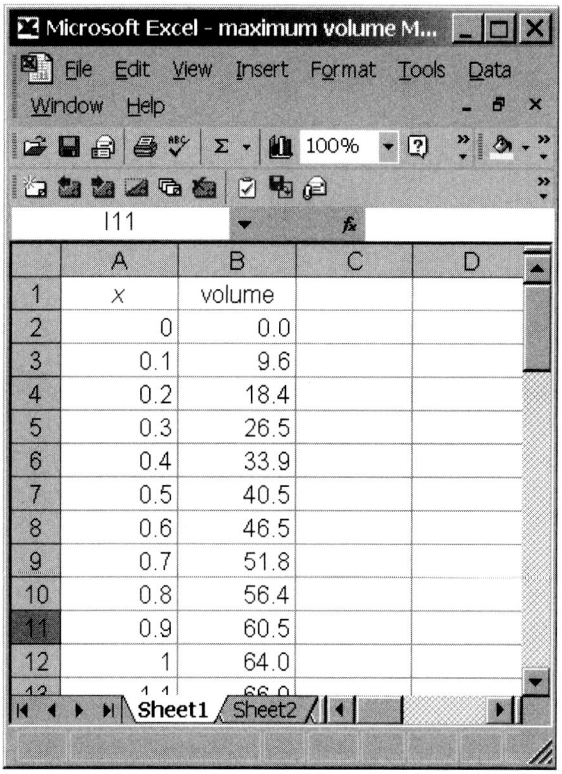

FIGURE F.17 The Excel data file used in the example.

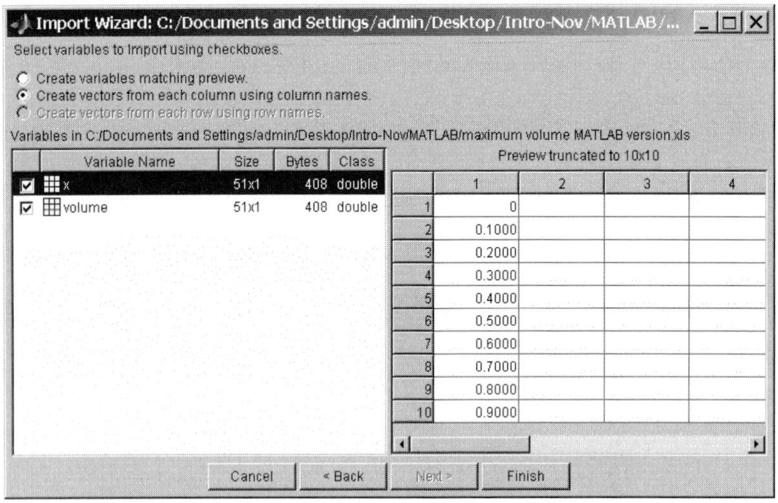

FIGURE F.18 MATLAB's Import Wizard.

Now let's say that we want to plot volume as a function of x. We then simply type the MATLAB commands that are shown in Figure F.19. The resulting plot is shown in Figure F.20.

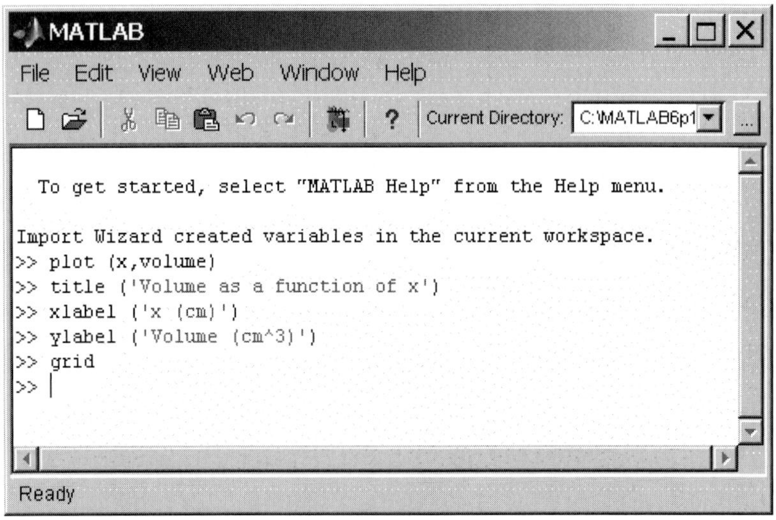

FIGURE F.19 The commands leading to the plot shown in Figure F.20.

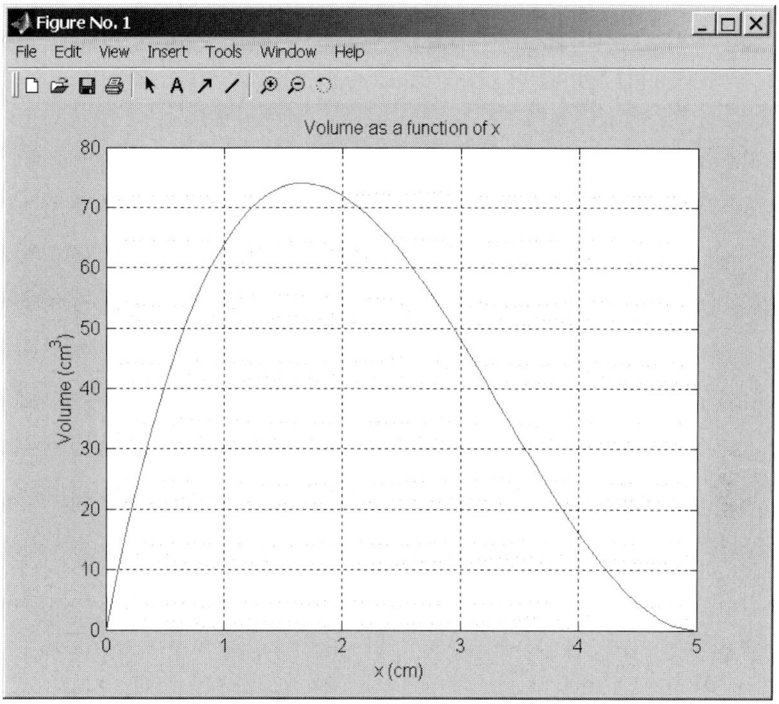

FIGURE F.20 Plot of volume versus x using data imported from an Excel file.

MATRIX COMPUTATIONS WITH MATLAB

As explained earlier, MATLAB offers many tools for matrix operations and manipulations. Table F.6 shows examples of these capabilities. We will demonstrate a few of MATLAB's matrix commands with the aid of the following examples.

EXAMPLE F.9

Given the following matrices: $[\mathbf{A}] = \begin{bmatrix} 0 & 5 & 0 \\ 8 & 3 & 7 \\ 9 & -2 & 9 \end{bmatrix}$, $[\mathbf{B}] = \begin{bmatrix} 4 & 6 & -2 \\ 7 & 2 & 3 \\ 1 & 3 & -4 \end{bmatrix}$, and $\{\mathbf{C}\} = \begin{Bmatrix} -1 \\ 2 \\ 5 \end{Bmatrix}$, using MATLAB, perform the following operations.

(a) $[\mathbf{A}] + [\mathbf{B}] = ?$, (b) $[\mathbf{A}] - [\mathbf{B}] = ?$, (c) $3[\mathbf{A}] = ?$, (d) $[\mathbf{A}][\mathbf{B}] = ?$, (e) $[\mathbf{A}]\{\mathbf{C}\} = ?$, (f) determinate of $[\mathbf{A}]$.

The solution to this problem follows. When studying these examples, note that the response given by MATLAB is shown in regular typeface. Information that the user need to type is shown in **boldface.**

To get started, select "MATLAB Help" from the Help menu.

```
>> A=[0 5 0;8 3 7;9 -2 9]

A =

     0     5     0
     8     3     7
     9    -2     9

>> B=[4 6 -2;7 2 3;1 3 -4]

B =

     4     6    -2
     7     2     3
     1     3    -4

>> C=[-1;2;5]

C =

    -1
     2
     5

>> A+B

ans =

     4    11    -2
    15     5    10
    10     1     5

>> A-B

ans =

    -4    -1     2
     1     1     4
     8    -5    13

>> 3*A

ans =

     0    15     0
    24     9    21
    27    -6    27

>> A*B

ans =
    35    10    15
    60    75   -35
    31    77   -60
```

```
>> A*C
ans =
    10
    33
    32
>> det(A)
ans =
    -45
    >>
```

EXAMPLE F.10

Solve the following set of equations using the Gauss elimination and by inverting the [**A**] matrix (the coefficients of unknowns) and multiplying it by {**b**} matrix (the values on the right hand side of equations).

$$2x_1 + x_2 + x_3 = 13$$
$$3x_1 + 2x_2 + 4x_3 = 32$$
$$5x_1 - x_2 + 3x_3 = 17$$

For the above problem, the coefficient matrix [**A**] and the right hand side matrix {**b**} are

$$[A] = \begin{bmatrix} 2 & 1 & 1 \\ 3 & 2 & 4 \\ 5 & -1 & 3 \end{bmatrix} \text{ and } \{b\} = \begin{Bmatrix} 13 \\ 32 \\ 17 \end{Bmatrix}$$

We will first use the MATLAB matrix left division operator \ to solve this problem. The \ operator solves the problem using Gauss elimination. We then solve the problem using the **inv** command.

To get started, select "MATLAB Help" from the Help menu.

```
>> A=[2 1 1;3 2 4;5 -1 3]
A =
    2    1    1
    3    2    4
    5   -1    3
>> b=[13;32;17]
b =
    13
    32
    17
```

```
>> x=A\b

x =

      2.0000
      5.0000
      4.0000

>> x=inv(A)*b

x =

      2.0000
      5.0000
      4.0000
```

Note if you substitute the solution ($x_1 = 2, x_2 = 5$, and $x_3 = 4$ into each equation, you find that they satisfy them. That is, $2(2) + 5 + 4 = 13$, $3(2) + 2(5) + 4(4) = 32$, and $5(2) - 5 + 3(4) = 17$.

CURVE FITTING WITH MATLAB

MATLAB offers a variety of curve fitting options. We will use Example F.11 to show how you can also use MATLAB to obtain an equation that closely fits a set of data points. For Example F.11, we will use the command **polyfit(x, y, n),** which determines the coefficients ($c_0, c_1, c_2, \ldots, c_n$) of a polynomial of order n that best fits the data according to:

$$y = c_0x^n + c_1x^{n-1} + c_2x^{n-2} + c_3x^{n-3} + \cdots + c_n$$

EXAMPLE F.11

Find the equation that best fits the following set of data points.

X	Y
0.00	2.00
0.50	0.75
1.00	0.00
1.50	−0.25
2.00	0.00
2.50	0.75
3.00	2.00

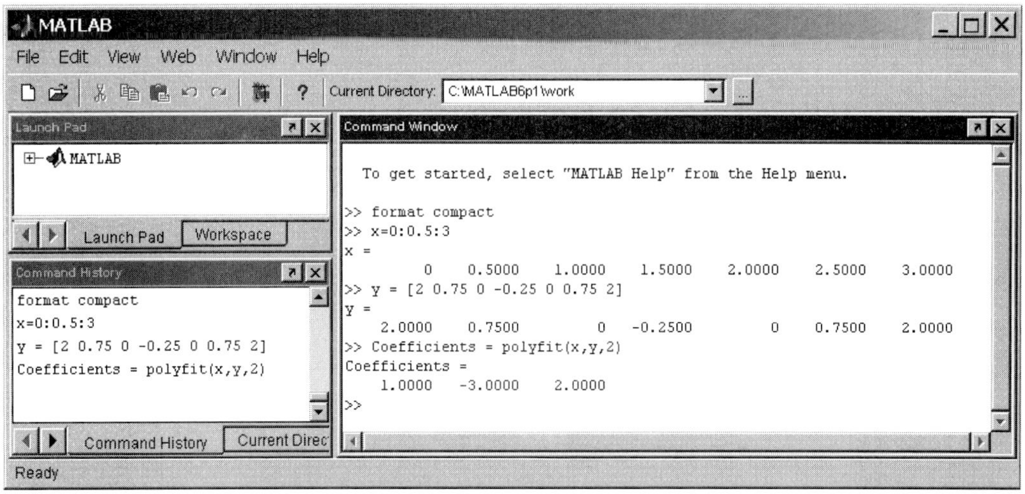

FIGURE F.21 The command window for Example F.11.

The plot of data points reveals that the relationship between y and x is quadratic (second order polynomial). To obtain the coefficients of the second order polynomial that best fits the given data, we will type the following sequence of commands. The MATLAB command window for Example F.11 is shown in Figure F.21.

```
>>format compact
>> x=0:0.5:3
>> y = [2 0.75 0 −0.25 0 0.75 2]
>> Coefficients = polyfit(x,y,2)
```

Upon execution of polyfit command, MATLAB will return the following coefficients, $c_0 = 1$, $c_1 = -3$ and $c_2 = 2$, which leads to the equation $y = x^2 - 3x + 2$.

SYMBOLIC MATHEMATICS WITH MATLAB

In the previous sections we discussed how to use MATLAB to solve engineering problems with numerical values. In this section we briefly explain the symbolic capabilities of MATLAB. In symbolic mathematics, as the name implies, the problem and the solution are presented using symbols such as x instead of numerical values. We will demonstrate MATLAB's symbolic capabilities using the following examples.

EXAMPLE F.12

We will use the following functions to perform the MATLAB's symbolic operations shown in Table F.12.

$$f_1(x) = x^2 - 5x + 6$$
$$f_2(x) = x - 3$$
$$f_3(x) = (x + 5)^2$$
$$f_4(x) = 5x - y + 2x - y$$

TABLE F.12 Examples of MATLAB's symbolic operations

Function	Description of the Function	Example	Result of the Example
sym	It creates a symbolic function	F1x = sym('x^2−5*x+6') F2x = sym('x−3') F3x = sym('(x+5)^2') F4x = sym('5*x−y+2*x−y')	F1x = x^2−5*x+6 F2x = x−3 F3x = (x+5)^2 F4x = 5*x−y+2*x−y
factor	When possible, it factorizes the function into simpler terms.	factor(Fx1)	(x−2)*(x−3)
simplify	It simplifies the function	simplify(F1x/F2x)	x−2
expand	It expands the function	expand(F3x)	x^2+10*x+25
collect	It simplifies a symbolic expression by collecting like coefficients	collect(F4x)	7*x−2*y
solve	It solves the expression for its values	solve(F1x)	x = 2 and x = 3
ezplot(f, min, max)	It plots the function f in the range of min and max	ezplot(F1x,0,2)	See Figure F.22.

SOLUTIONS OF SIMULTANEOUS LINEAR EQUATIONS

In this section we will show how you can use MATLAB's symbolic solvers to obtain solutions to a set of linear equations. Consider the following 3 linear equations with three unknowns, x, y, and z.

$$2x + y + z = 13$$
$$3x + 2y + 4z = 32$$
$$5x - y + 3z = 17$$

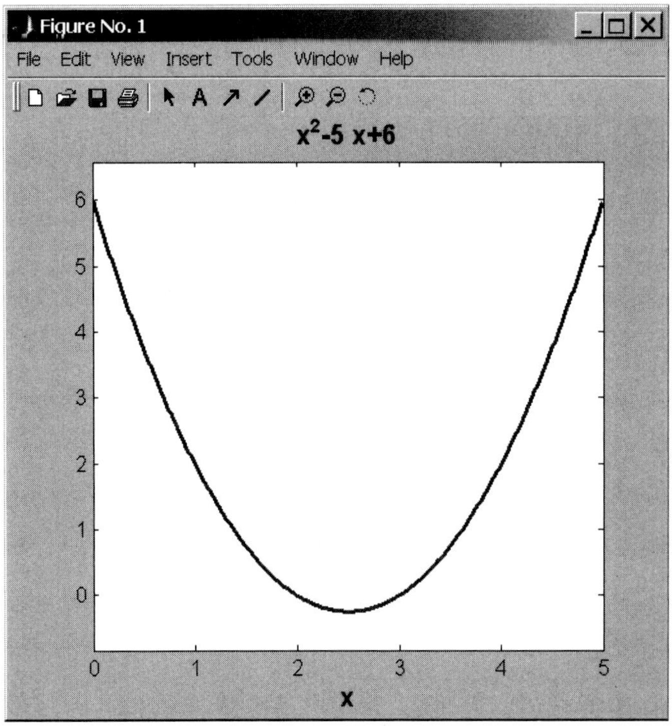

FIGURE F.22 The ezplot for Example F.12; see the last row in Table F.12.

In MATLAB, the solve command is used to obtain solutions to symbolic algebraic equations. The basic form of the **solve** command is **solve('eqn1','eq2', . . ,'eqn').** As shown below, we define each equation first and then use the **solve** command to obtain the solution.

```
>> equation_1 = '2*x+y+z=13';
>> equation_2 = '3*x+2*y+4*z=32';
>> equation_3 = '5*x-y+3*z=17';
>> [x,y,z]=solve(equation_1,equation_2,equation_3)
```

The solution is given by $x = 2$, $y = 5$, and $z = 4$. The MATLAB Command Window for this example is shown in Figure F.23.

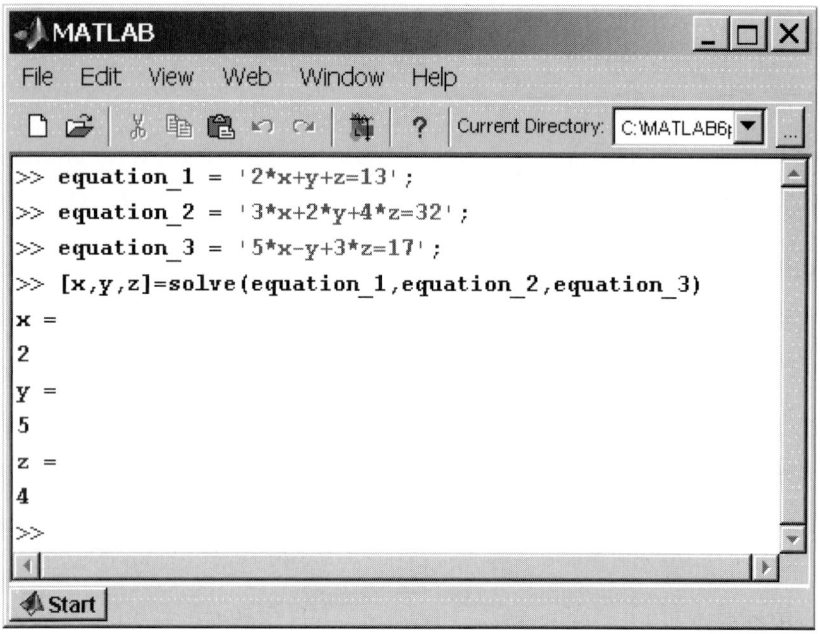

FIGURE F.23 The solution of the set of linear equations discussed in the example.

As we said at the beginning of this appendix, there are many good textbooks that discuss the capabilities of MATLAB to solve a full range of problems. Here our intent was to introduce only some basic ideas so that you can perform some essential operations or write a simple program to solve for the solution of your finite element model.

Index

A

B

C

D

E

F

I

K

L

M

N

O

P

Q

R

S

T

U

V

W

Y